Essentials of Stem Cell Biology

Essentials of Stem Cell Biology

DISCARD

EDITORS

Robert Lanza

John Gearhart
Brigid Hogan
Douglas Melton
Roger Pedersen
James Thomson
E. Donnall Thomas
Michael West

ELSEVIER

ACADEMIC
PRESS

AMSTERDAM • BOSTON • HEIDELBERG • LONDON
NEW YORK • OXFORD • PARIS • SAN DIEGO
SAN FRANCISCO • SINGAPORE • SYDNEY • TOKYO

Elsevier Academic Press
30 Corporate Drive, Suite 400, Burlington, MA 01803, USA
525 B Street, Suite 1900, San Diego, California 92101-4495, USA
84 Theobald's Road, London WC1X 8RR, UK

This book is printed on acid-free paper.

Copyright © 2006, Elsevier Inc. All rights reserved.

Library of Congress Cataloging-in-Publication Data
Application Submitted

British Library Cataloguing in Publication Data
A catalogue record for this book is available from the British Library

ISBN-13: 978-0-12-088442-1
ISBN-10: 0-12-088442-9

For all information on all Elsevier Academic Press publications
visit our Web site at *www.books.elsevier.com*

Printed in China

06 07 08 09 10 9 8 7 6 5 4 3 2

Contents

PART ONE
Introduction to stem cells

PART TWO
Basic biology/mechanisms

PART FIVE
Applications

PART SIX
Regulation and ethics

Contents

Contributors

Numbers in parentheses indicate the chapter number of the authors' contribution.

Alaa Adassi (26)
Stem Cell Institute University of Minnesota Medical School Mayo Mail Code 716 420 Delaware Street, SE Minneapolis, MN 55455

Russell C. Addis (40)
John Hopkins University, School of Medicine, Baltimore, MD

Michal Amit (37)
Stem Cell Center, Bruce Rappaport Faculty of Medicine, Technion — Israel Institute of Technology, Haifa, Israel

Peter W. Andrews (9, 45)
Dept. Biomedical Science, University of Sheffield, Western Bank, Sheffield, UK

Piero Anversa (57)
Cardiovascular Research Institute, New York Medical College, Valhalla, NY

Anthony Atala (15)
Wake Forest Institute for Regenerative Medicine, Wake University School of Medicine, Winston-Salem, NC

Joyce Axelman (40)
John Hopkins University, School of Medicine, Baltimore, MD

Yann Barrandon (59)
Laboratory of Stem Cell Dynamics, Department of Experimental Surgery, Lausanne, Switzerland

Steven R. Bauer (67)
Center for Biologics Evaluation and Research, FDA, Laboratory of Stem Cell Biology, Rockville, MD

Daniel Becker (53)
Department of Neurology, Vanderbilt University Medical Center, Nashville, TN

Nissim Benvenisty (43)
Department of Genetics, Institute of Life Sciences, The Hebrew University of Jerusalem, Givat-Ram, Jerusalem, Israel

Paolo Bianco (62)
National Institute of Dental and Craniofacial Research, NIH, Bethesda, MD

Helen M. Blau (28)
Baxter Laboratory in Genetic Pharmacology, Dept of Microbiology and Immunology, Stanford University School of Medicine, Stanford, CA

Susan Bonner-Weir (55)
Section on Islet Transplantation and Cell Biology, Joslin Diabetes Center, Boston, MA

Mairi Brittan (34)
Barts and the London, Queen Mary's School of Medicine and Dentistry, Whitechapel, London UK

Hal E. Broxmeyer (16)
Walther Oncology Center and Department of Microbiology and Immunology, Indiana University School of Medicine, Indianapolis, IN

Contributors

S. Bultman (8)
Department of Genetics, University of North Carolina, Chapel Hill, NC

Arnold I. Caplan (27)
Skeletal Research Center, Case Western Reserve University, Cleveland, OH

Melissa K. Carpenter (35, 42)
Robarts Research Institute, London, Ontario, Canada, and CyThera, Inc., San Diego, CA

Fatima Cavaleri (4)
Max Planck Institute for Molecular Biomedicine, Muenster, GERMANY

Connie Cepko (20)
Department of Genetics, Howard Hughes Medical Institute, Harvard Medical School, Boston, MA

Howard Y. Chang (50)
Dept of Biopharmaceutical Sciences, UCSF, San Francisco, CA

Xin Chen (50)
Dept of Dermatology and Genetics, Stanford University, Stanford, CA

Tao Cheng (7)
Massachusetts General Hospital, Harvard Medical School, Boston, MA

Gregory O. Clark (40)
Division of Endocrinology, John Hopkins University, School of Medicine, Baltimore, MD

Giulio Cossu (58)
Stem Cell Research Institute, Dibit, H.S. Raffaele, Milan, Italy

M. Gabriella Cusella De Angelis (58)
Stem Cell Research Institute, Dibit, H.S. Raffaele, Milan, Italy

Susana M. Chuva de Sousa Lopes (12)
HUBRECHT LABORATORIUM, Netherlands Institute for Developmental Biology, UTRECHT, The Netherlands

George Q. Daley (23)
Children's Hospital Boston, MA

Brian R. Davis (63)
Fund for Inherited Disease Research, Newtown, PA

Natalie C. Direkze (34)
Barts and the London, Queen Mary's School of Medicine and Dentistry, Whitechapel, London UK

Yuval Dor (33)
Department of Cellular Biochemistry and Human Genetics, The Hebrew University-Hadassah Medical School, Jerusalem, Israel

Craig Dorrell (32)
Oregon Health & Science University, Dept. of Molecular & Medical Genetics, Portland, Oregon

Joshua D. Dowell (54)
Herman B Wells Center for Pediatric Research, Indiana University School of Medicine, Indianapolis, IN

Jonathan S. Draper (45)
Dept. Biomedical Science, University of Sheffield, Western Bank, Sheffield, UK

Gregory R. Dressler (31)
Dept. of Pathology, University of Michigan, Ann Arbor, MI

Gabriela Durcova-Hills (39)
C/o Dilly Bradford The Wellcome Trust/Cancer Research, UK Institute of Cancer and Developmental Biology, University of Cambridge, Cambridge, UK

Martin Evans (36)
Cardiff School of Biosciences, Cardiff University, Cardiff UK

Margaret A. Farley (66)
Yale University Divinity School, New Haven, CT

Donna M. Fekete (20)
Department of Biological Sciences, Purdue University, West Lafayette, IN

Loren J. Field (54)
Herman B Wells Center for Pediatric Research and Krannert Institute of Cardiology, Indiana University School of Medicine, Indianapolis, IN

Donald W. Fink, Jr. (67)
Center for Biologics Evaluation and Research, FDA, Cell Therapy Branch, Rockville, MD

K. Rose Finley (51)
Howard Hughes Medical Institute, Children's Hospital, Boston, MA

Elaine Fuchs (21)
Howard Hughes Medical Institute, Laboratory of Mammalian Cell Biology and Development, The Rockefeller University, NY, NY

Margaret T. Fuller (5)
Stanford University School of Medicine, Dept. of Developmental Biology, Stanford, CA

Richard Gardner (1)
University of Oxford, Dept of Zoology, OXFORD, UK

John D. Gearhart (40)
Institute for Cell Engineering, John Hopkins University, School of Medicine, Baltimore, MD

Pamela Gehron Robey (62)
National Institute of Dental and Craniofacial Research, NIH, Bethesda, MD

Sharon Gerecht-Nir (25)
Stem Cell Center, Bruce Rappaport Faculty of Medicine, Technion — Israel Institute of Technology, Haifa, Israel

Victor Goldberg (60)
Dept. of Orthopaedics, Case Western Reserve University/University Hospitals of Cleveland, Cleveland, OH

Rodolfo Gonzalez (52)
Program in Stem Cell Biology (Developmental & Regeneration Cell Biology) The Burnham Institue, La Jolla, CA

Elizabeth Gould (19)
Department of Psychology, Princeton University, Princeton, NJ

Ronald M. Green (65)
Ethics Institute, Dartmouth College, Hanover, NH

Markus Grompe (32)
Oregon Health & Science University, Dept. of Molecular & Medical Genetics, Portland, Oregon

Alexandra Haagensen (55)
Section on Islet Transplantation and Cell Biology, Joslin Diabetes Center, Boston, MA

Susan Hawes (48)
Monash Institute of Medical Research, Monash University and the Australian Stem Cell Centre, Clayton, Victoria, Australia

Hiroshi Hisatsune (30)
Riken Center for Developmental Biology, Stem Cell Research Group, Center for Developmental Biology, Chuo-ku, Kobe, Japan

James Huettner (53)
Kennedy Krieger Institute, Baltimore, MD

Jerry I. Huang (60)
University Hospitals Research Institute, Cleveland, OH

Marko E. Horb (10)
Centre for Regenerative Medicine, Department of Biology & Biochemistry, University of Bath, Bath, UK

Jaimie Imitola (52)
Program in Stem Cell Biology (Developmental & Regeneration Cell Biology) The Burnham Institue, La Jolla, CA

Joseph Itskovitz-Eldor (25, 37)
Dept. of Ob/Gyn, Rambam Medical Center & Stem Cell Center, Bruce Rappaport Faculty of Medicine, Technion — Israel Institute of Technology, Haifa, Israel

Rudolf Jaenisch (11)
Whitehead Institute for Biomedical Research, Mass Institute of Technology, Cambridge, MA

Penny Johnson (9)
Dept. Biomedical Science, University of Sheffield, Western Bank, Sheffield S10 2TN, Great Britain

D. Leanne Jones (5)
Stanford University School of Medicine, Dept. of Developmental Biology, Stanford, CA

Jan Kajstura (57)
Cardiovascular Research Institute, New York Medical College, Valhalla, NY

Gerard Karsenty (24)
Baylor College of Medicine, Houston, TX

Pritinder Kaur (56)
The University of Melbourne, Epithelial Stem Cell Biology Laboratory, Peter MacCallum Cancer Institute, East Melbourne, Victoria, Australia

Candace L. Kerr (40)
Dept. of Gynecology and Obstetrics, John Hopkins University, School of Medicine, Baltimore, MD

Kathleen C. Kent (40)
John Hopkins University, School of Medicine, Baltimore, MD

Ali Khademhosseini (61)
Langer Lab, Chemical Engineering, Massachusetts Institute of Technology, Cambridge, MA

Chris Kintner (17)
The Salk Institute for Biological Studies, La Jolla, CA

Irina Klimanskaya (38)
Advanced Cell Technology, Inc., Worcester, MA

Nobuyuki Kondoh (30)
Riken Center for Developmental Biology, Stem Cell Research Group, Center for Developmental Biology, Chuo-ku, Kobe, Japan

Naoko Koyano-Nakagawa (17)
Department of Neuroscience, University of Minnesota, Minneapolis, MN

Jennifer N. Kraszewski (40)
John Hopkins University, School of Medicine, Baltimore, MD

Tilo Kunath (14)
Mount Sinai Hospital, Toronto, Ontario, CANADA

Mark A. LaBarge (28)
Lawrence Berkeley Nation Laboratory, Department of Cancer Biology, Berkeley, CA

Robert Langer (61)
Langer Lab, Chemical Engineering, Massachusetts Institute of Technology, Cambridge, MA

Annarosa Leri (57)
Cardiovascular Research Institute, New York Medical College, Valhalla, NY

Shulamit Levenberg (61)
Faculty of Biomedical Engineering Techion, Haifa, Israel, MA

S. Robert Levine (68)
Research Portfolio Committee, Juvenile Diabetes Research Foundation

John W. Littlefield (40)
John Hopkins University, School of Medicine, Baltimore, MD

Terry Magnuson (8)
Department of Genetics, University of North Carolina, Chapel Hill, NC

Yoay Mayshar (43)
Department of Genetics, Institute of Life Sciences, The Hebrew University of Jerusalem, Givat-Ram, Jerusalem, Israel

John W. McDonald (53)
Kennedy Krieger Institute, Baltimore, MD

Anne McLaren (13, 39)
Anne McLaren, c/o Dilly Bradford, The Wellcome Trust/Cancer Research UK Gurdon Institute, University of Cambridge, Cambridge, UK

Jill McMahon (38)
Harvard University, Cambridge, MA

Alexander Meissner (11)
Whitehead Institute for Biomedical Research, Mass Institute of Technology, Cambridge, MA

Douglas A. Melton (33)
Department of Molecular and Cellular Biology and Howard Hughes Medical Institute, Harvard University, Cambridge, MA

Christian Mirescu (19)
Department of Psychology, Princeton University, Princeton, NJ

N.D. Montgomery (8)
Department of Genetics, University of North Carolina, Chapel Hill, NC

Malcolm A.S. Moore (22)
Developmental Hematology, Memorial Sloan-Kettering Cancer Center, New York, NY

Mary Tyler Moore (68)
International Chairwoman, Juvenile Diabetes Research Foundation

Franz-Josef Mueller (52)
Program in Stem Cell Biology (Developmental & Regeneration Cell Biology) The Burnham Institue, La Jolla, CA

Christine Mummery (12)
HUBRECHT LABORATORIUM, Netherlands Institute for Developmental Biology, UTRECHT, The Netherlands

Andras Nagy (46)
Mount Sinai Hospital, Samuel Lunenfeld Research Institute, Toronto, Canada

Shin-Ichi Nishikawa (30)
Riken Center for Developmental Biology, Stem Cell Research Group, Center for Developmental Biology, Chuo-ku, Kobe, Japan

Satomi Nishikawa (30)
Riken Center for Developmental Biology, Stem Cell Research Group, Center for Developmental Biology, Chuo-ku, Kobe, Japan

Hitoshi Niwa (6)
Lab for Pluripotent Cell Studies, RIKEN Ctr for Developmental Biology, Chu-o-ku, Kobe C, Japan

Jitka Ourednik (52)
Program in Stem Cell Biology (Developmental & Regeneration Cell Biology) The Burnham Institue, La Jolla, CA

Vaclav Ourednik (52)
Program in Stem Cell Biology (Developmental & Regeneration Cell Biology) The Burnham Institue, La Jolla, CA

Kook I. Park (52)
Program in Stem Cell Biology (Developmental & Regeneration Cell Biology) The Burnham Institue, La Jolla, CA

Ethan S. Patterson (40)
John Hopkins University, School of Medicine, Baltimore, MD

Alice Pébay (41)
Monash Institute of Medical Research, Monash University and the Australian Stem Cell Centre, Clayton, Victoria, Australia

Martin F. Pera (41, 48)
Monash Institute of Medical Research, Monash University and the Australian Stem Cell Centre, Clayton, Victoria, Australia

Christopher S. Potten (2)
EpiStem Limited, Incubator Building, Manchester, UK

Sean L. Preston (34)
Barts and the London, Queen Mary's School of Medicine and Dentistry, Whitechapel, London UK

Nicole L. Prokopishyn (63)
Fund for Inherited Disease Research, Newtown, PA

Jean Pyo Lee (52)
Program in Stem Cell Biology (Developmental & Regeneration Cell Biology) The Burnham Institue, La Jolla, CA

Christopher Reeve (69)
The Estate of Mr. Christopher Reeve, C/O Maggie Goldberg, VP of Communications, Christopher Reeve Paralysis Foundation, Springfield, NJ

Ariane Rochat (59)
Laboratory of Stem Cell Dynamics, Department of Experimental Surgery, Lausanne, Switzerland

Nadia Rosenthal (29)
EMBL Mouse Biology Programme, European Molecular Biology Laboratory, Monterotondo (Rome), ITALY

Janet Rossant (14)
Mount Sinai Hospital, Toronto, Ontario, CANADA

Michael Rubart (54)
Krannert Institute of Cardiology, Indiana University School of Medicine, Indianapolis, IN

Maurilio Sampaolesi (58)
Stem Cell Research Institute, Dibit, H.S. Raffaele, Milan, Italy

Maria Paola Santini (29)
EMBL Mouse Biology Programme, European Molecular Biology Laboratory, Monterotondo (Rome), ITALY

David T. Scadden (7)
Massachusetts General Hospital, Harvard Medical School, Boston, MA

Hans Schöler (4)
Max Planck Institute for Molecular Biomedicine, Muenster, GERMANY

Michael J. Shamblott (40)
Institute for Cell Engineering, John Hopkins University, School of Medicine, Baltimore, MD

M. Minhaj Siddiqui (15)
Wake Forest Institute for Regenerative Medicine, Wake University School of Medicine, Winston-Salem, NC

Richard L. Sidman (52)
Program in Stem Cell Biology (Developmental & Regeneration Cell Biology) The Burnham Institue, La Jolla, CA

William B. Slayton (49)
University of Florida College of Medicine, Department of Pediatrics, Gainesville, FL

Evan Y. Snyder (52)
Program in Stem Cell Biology (Developmental & Regeneration Cell Biology) The Burnham Institue, La Jolla, CA

Gerald J. Spangrude (49)
University of Utah, Department of Medicine, Division of Hematology, Salt Lake City, UT

Lorenz Studer (18)
Stem Cell and Tumor Biology, Memorial Sloan-Kettering Cancer Center, New York, NY

M.A. Surani (47)
Wellcome Trust Cancer Research UK Gurdon Institute, University of Cambridge, Cambridge, UK

Yang D. Teng (52)
Program in Stem Cell Biology (Developmental & Regeneration Cell Biology) The Burnham Institue, La Jolla, CA

James A. Thomson (44, 50)
Dept of Biopharmaceutical Sciences, UCSF, San Francisco, CA

David Tosh (10)
Centre for Regenerative Medicine, Department of Biology
& Biochemistry, University of Bath, Bath, UK

Tudorita Tumbar (21)
Howard Hughes Medical Institute, Laboratory of
Mammalian Cell Biology and Development, The
Rockefeller University, NY, NY

Edward Upjohn (56)
The University of Melbourne, Epithelial Stem Cell Biology
Laboratory, Peter MacCallum Cancer Institute, East
Melbourne, Victoria, Australia

George Varigos (56)
The University of Melbourne, Epithelial Stem Cell Biology
Laboratory, Peter MacCallum Cancer Institute, East
Melbourne, Victoria, Australia

Catherine M. Verfaillie (3, 26)
Stem Cell Institute, University of Minnesota Medical
School, Minneapolis, MN

Zhongde Wang (11)
Whitehead Institute for Biomedical Research, Mass Institute
of Technology, Cambridge, MA

Gordon C. Weir (55)
Section on Islet Transplantation and Cell Biology, Joslin
Diabetes Center, Boston, MA

J.W. Wilson (2)
EpiStem Limited, Incubator Building, Manchester, UK

Nicholas A. Wright (34)
Barts and the London, Queen Mary's School of Medicine
and Dentistry, Whitechapel, London UK

Chunhui Xu (42)
Robart's Research Institute, London, Ontario, CANADA

Jun Yamashita (30)
Riken Center for Developmental Biology, Stem Cell
Research Group, Center for Developmental Biology, Chuo-
ku, Kobe, Japan

Holly Young (35, 42)
Robarts Research Institute, London, Ontario, CANADA

Jung U. Yoo (60)
University Hospitals Research Institute, Cleveland, OH

Laurie Zoloth (64)
Center of Bioethics, Northwestern University, Chicago, IL

Leonard I. Zon (51)
Howard Hughes Medical Institute, Children's Hospital,
Boston, MA

Thomas P. Zwaka (44)
Department of Molecular and Cellular Biology, Baylor
College of Medicine, Houston, TX

Robert Zweigerdt (54)
Institute for Transplantation Diagnostics and Cell
Therapeutics, Duesseldorf Medical School, Duesseldorf,
Germany

Preface

This abridgment of the *Handbook of Stem Cells* attempts to incorporate all the essential subject matter of the original two-volume edition in a single volume. The material has been reworked in an accessible format suitable for students and general readers interested in following the latest advances in stem cells. Although some extra language and chapters have been deleted, rigorous effort has been made to retain from the original two-volume set that which is pertinent to the understanding of this exciting area of biology.

The organization of the book remains largely unchanged, combining the prerequisites for a general understanding of adult and embryonic stem cells; the tools, methods, and experimental protocols needed to study and characterize stem cells and progenitor populations; as well as a presentation by the world's experts of what is currently known about each specific organ system. No topic in the field of stem cells is left uncovered, including basic biology/mechanisms, early development, ectoderm, mesoderm, endoderm, methods (such as detailed descriptions of how to derive and maintain animal and human embryonic stem cells), application of stem cells to specific human diseases, regulation and ethics, and patient perspectives.

To help those unfamiliar with the material and scientific terminology, *Essentials of Stem Cell Biology* includes a glossary of terms and suggested reading list with each chapter. This new volume has also been enhanced with over 150 full-color illustrations. The result is a comprehensive reference in an easily accessible and affordable format. It represents the combined effort of eight editors and more than 200 scholars and scientists whose pioneering work has defined our understanding of stem cells. We can only hope that the new knowledge and research outlined in this volume will help contribute to new therapies for cancer, heart disease, diabetes, and a wide variety of other diseases that presently afflict humanity.

Robert Lanza, M. D.
Boston, Massachusetts

Foreword

What can usefully be said about stem cells in a foreword to a collection of definitive articles by the world's experts, inasmuch as this book already covers every conceivable aspect of the subject? In response to this question, I shall attempt to place the work described here in the broader context of science — and of modern cell and developmental biology specifically.

My view of the science in this book comes from the perspective of someone who spent 30 years in universities probing the molecular details of basic cell processes. Over this time, our understanding of cells increased at a rate that startled even those most closely involved in the wave upon wave of new discoveries. This increase in knowledge was catalytic: as our understanding of cells advanced, it allowed new research tools to be developed that directly sped further advances in understanding, which in turn led to new tools, and so on. Consider, for example, the DNA chip technology described in Chapters 4 and 51, which allows an investigator to examine the expression of tens of thousands of genes simultaneously. Because hundreds of small steps were needed to move from the striking initial discovery of DNA hybridization in 1961 (Marmur and Doty, 1961) to this new technology, its development was unpredictable in advance.

It is the same for the advance of science itself, as emphasized repeatedly in the "Beyond Discovery" series of brief articles from the National Academy of Sciences. Designed to explain science to the general public, each of these eight-page documents traces the path leading to a breakthrough of great human benefit — such as the global positioning system (GPS) or the cure for childhood leukemia. In every case, the final discovery depended on knowledge developed over decades through the efforts of a large number of independent scientists and engineers. Each piece of knowledge, often seemingly useless on its own, was combined in unexpected ways with other knowledge to produce a final result whose power seems almost magical.

The great enterprise of science, sparked simply by human curiosity about how the world works — for example, an attempt to account for the motions of the planets in the night sky — has transformed our world. And because new knowledge builds on old knowledge, the pace continually accelerates as the amount of old knowledge increases. Thus, we should expect the inventions that benefit humans in this new century to be even more dramatic than those of the last century. But we can be equally sure of the futility of attempting to predict what they will be in advance.

What does all this have to do with stem cells? Personally, I become uncomfortable whenever I hear claims that describe the precise benefits to be derived from this research — especially when they are associated with a timeline. Nevertheless, the history of science makes it certain that the knowledge derived from research on stem cells will eventually lead to enormous benefits for human health, even if they are unpredictable. Eventually, we will be able to use our profound understanding of biology to grow new organs that can be safely transplanted into human patients, and the work with stem cells will no doubt make important contributions to this breakthrough. But the research outlined in this book is equally certain to contribute to cures for cancer and for a large number of other less famous diseases — many of mysterious origin — that are terrible afflictions for humanity.

Bert Vogelstein, the chairman of the National Academies committee that produced the report "Stem Cells and the Future of Regenerative Medicine," reminds us that "The stem cell debate has led scientists and nonscientists alike to contemplate profound issues, such as who we are and what makes us human," (National Research Council, 2002). The debate has also made it clear that the success of modern societies will depend on education systems that place a much higher value on conveying an understanding of the nature of science to everyone.

Accomplishing such a goal will require new recognition by scientists of their teaching responsibilities at the college level as well as of their different but critical roles in the support of inquiry-based science education for students from 5 to 18 years old (National Research Council, 1996). We must all face the realization that, even under the best of circumstances, the transitions to the science-centered education systems that our nations need will require decades. In the end, the life of a scientist must change — incorporating a much broader view of what it means to be a scientist and changing the way that most of us apportion our time. In a sense, therefore, we scientists should view the controversy over stem cell research as a healthy wake-up call — a call to action in a new century in which our startling discoveries will increasingly dominate the news.

Bruce Alberts, PhD

NOTES

Marmur, J., and Doty, P. (1961). Thermal renaturation of DNA. *J. Mol. Biol.* **3**, 585–594.

National Academy of Sciences. "Beyond Discovery" series. *Available at:* http://www.BeyondDiscovery.org.

National Research Council. (2002). "Stem Cells and the Future of Regenerative Medicine." National Academies Press, Washington, DC. *Available for purchase at:* http://www.nap.edu/catalog/10195.html.

National Research Council. (1996). "National Science Education Standards." National Academies Press, Washington, DC. *Available for purchase or download at:* http://www.nap.edu/catalog/4962.html.

Embryonic Stem Cells in Perspective

Biologists have explored the development of embryos of all sorts, from worms to humans, in search of the answer to the question of how a complex organism derives from a single cell, the fertilized egg. We now know many of the genes involved in regulating development in different species and find remarkable conservation of genetic pathways across evolution. We also have a good understanding of the logic of development — how the embryo repeatedly uses the same kinds of strategies to achieve cellular specialization, tissue patterning, and organogenesis. One common strategy of development is the use of the stem cell to help generate and maintain a given tissue or organ. A stem cell is a cell that, when it divides, can produce a copy of itself as well as a differentiated cell progeny. This self-renewal capacity underlies the ability of adult stem cells, such as hematopoietic stem cells and spermatogonial stem cells, to constantly renew tissues that turn over rapidly in the adult. The concept of the stem cell arose from the pioneering studies of Till and McCullogh on the hematopoietic stem cell and those of Leblond on spermatogenesis and the intestinal crypt. Even in tissues like the brain, where cells do not turn over so rapidly in the adult, there are long-lived quiescent stem cells that may be reactivated to repair damage.

Much current research is focused on the identification, characterization, and isolation of stem cells from the adult, with the hope that such cells may be useful for therapeutic repair of adult tissues either by exogenous cell therapy or by reactivation of endogenous stem cells. However, to date, most adult stem cells have restricted potential, and achieving indefinite proliferation and expansion of the stem cells in culture

is still not routine. During embryogenesis, cells are initially proliferative and pluripotent; they only gradually become restricted to different cell fates. The question of whether pluripotent stem cells exist in the embryo has been of interest for years. In mammals, it was known in the 1960s and 1970s that early mouse embryos, up to late gastrulation stages, could produce tumors known as teratocarcinomas when transplanted to ectopic sites, such as the kidney capsule. These tumors contain a variety of differentiated cells types, including muscle, nerve, and skin, as well as an undifferentiated cell type, the embryonal carcinoma (EC) cell. EC cells could be propagated in the undifferentiated state *in vitro*. Importantly, Pierce in 1964 showed that a single EC cell could regenerate a tumor containing both EC cells and differentiated progeny (Kleinsmith and Pierce, 1964), demonstrating that EC cells are the stem cells of the tumor.

What was the relevance of these tumor cells to normal development? Many studies were carried out in the 1970s showing that EC cells could reveal their pluripotency when injected back into early embryos. The best, most karyotypically normal EC cells could contribute to many different cell types in the resulting chimeras, including, in rare instances, the germ line. This led to excitement that these cells might be used to introduce new genetic alterations into the mouse and that normalization of tumorigenicity could be achieved by promoting differentiation of tumor cells. However, chimerism was often weak, and EC-derived tumors were a common feature of the chimeras (Papaioannouo and Rossant, 1984). Thus, although EC cells had remarkable properties of differentiation, they were still clearly tumor cells. In 1981, Martin

(1981) and Evans and Kaufman (1981) discovered that permanent pluripotent cell lines, known as embryonic stem (ES) cells, could be derived directly from the blastocyst in culture. This changed the whole perspective of the field. The differentiation of these cells, although they make teratomatous tumors in ectopic sites, is easier to control than that of EC cells. Dramatically, ES cells grown for many passages in culture can make an entire mouse when supported by tetraploid extraembryonic tissues (Nagy *et al.*, 1993). When such mice are made from robust hybrid cell lines, they show no enhanced tumor susceptibility and appear normal in all respects (Eggan *et al.*, 2001). All of these properties have made mouse ES cells an incredibly powerful tool for introducing alterations into the mouse genome and analyzing their effects (Rossant and Nagy, 1995).

Are ES cells true stem cells? The *in vivo* equivalent of the ES cell is unclear. ES cells resemble the cells of the primitive ectoderm or epiblast in their gene expression patterns and their pattern of tissue contribution in chimeras. Transcription factors, such as Oct4 and Nanog, that are required for formation and survival of the pluripotent cells in the embryo are also needed for ES survival (Chambers *et al.*, 2003; Mitsui *et al.*, 2003; Nichols *et al.*, 1998). However, *in vivo*, the epiblast only has a limited period of possible stem cell expansion before all cells differentiate into the tissues of the three germ layers at gastrulation. Germ cells, which are set aside at gastrulation, go on to provide the gametes that will impart pluripotency to the zygotes of the next generation. However, there is no evidence that germ cells are a special stem cell pool in the epiblast, which could be the ES equivalent. Rather, it appears that all epiblast cells have the capacity to form germ cells in the right environment (Tam and Zhou, 1996). Thus, the germ cell is just one of the differentiation options of epiblast.

In vitro, it is clear that ES cells can be expanded indefinitely in the undifferentiated state and still retain the capacity for differentiation. In this regard, they certainly have stem cell properties. However, there has not been a clear demonstration that a single cell can both self-renew and differentiate, as has been shown for EC cells. It is known that single cells are fully pluripotent, since chimeras made by injecting single ES cells into blastocysts show ES contributions to all fetal cell types analyzed (Beddington and Robertson, 1989). However, in some ways ES cells seem more like progenitor cells, where the population can be expanded by the right growth factor environment but all cells will differentiate when the supportive environment for self-renewal is removed. Practically, the difference is probably not important, but if true, it may be misleading to extrapolate from what we know about how ES cells maintain the proliferative state to other stem cells. The search for "stemness" genes and proteins may not be a useful undertaking until we agree on how to define different stem cell populations.

So much of the excitement about mouse ES cells has focused on their use as a tool for germ line transmission of genetic alterations, that the remarkable differentiation properties of these cells in culture have been underexplored. The

derivation of cell lines from early human embryos that seem to share many of the properties of mouse ES cells (Shamblott *et al.*; Thomson *et al.*, 1998) has refocused attention on the *in vitro* properties of ES cells. Many questions remain before ES cells can be transformed from an interesting biological system to a robust therapeutic modality for degenerative diseases. How similar are mouse and human ES cells, and how valid is it to use data from one to drive research in the other? How can ES cells be maintained through many passages in a truly stable state, where all cells are stem (or progenitor) cells, epigenetic programming is stable, and genetic abnormalities are minimal? How can ES cells be directed to differentiate reproducibly into given cell types, and how can differentiated progenitors be isolated and maintained? How can we ensure that ES cells will not be tumorigenic *in vivo*?

All this research on ES cells, both mouse and human, will provide new insights into embryonic development and new clues as to how to isolate and characterize new stem cells from different embryonic or adult tissues. Conversely, research on how normal embryonic development is regulated will provide new clues as to how to maintain and differentiate stem–progenitor cells in culture. The interplay between developmental biologists and stem cell biologists will be key to a fundamental understanding of stem cell development and its translation into therapeutic outcomes.

Janet Rossant, PhD

REFERENCES

Beddington, R. S. P., and Robertson, E. J. (1989). An assessment of the developmental potential of embryonic stem cells in the midgestation mouse embryo. *Development* **105**, 733–737.

Chambers, I., Colby, D., Robertson, M., Nichols, J., Lee, S., Tweedie, S., and Smith, A. (2003). Functional expression cloning of Nanog, a pluripotency sustaining factor in embryonic stem cells. *Cell* **113**, 643–655.

Eggan, K., Akutsu, H., Loring, J., Jackson-Grusby, L., Klemm, M., Rideout, W. M., III, Yanagimachi, R., and Jaenisch, R. (2001). Hybrid vigor, fetal overgrowth, and viability of mice derived by nuclear cloning and tetraploid embryo complementation. *Proc. Nat. Acad. Sci. USA* **98**, 6209–6214.

Evans, M., and Kaufman, M. H. (1981). Establishment in culture of pluripotential cells from mouse embryos. *Nature* **292**, 154–155.

Kleinsmith, L. J., and Pierce, G. B. (1964). Multipotentiality of single embryonal carcinoma cells. *Cancer Res.* **24**, 1544–1551.

Martin, G. R. (1981). Isolation of a pluripotent cell line from early mouse embryos cultured in medium conditioned by teratocarcinoma stem cells. *Proc. Nat. Acad. Sci. USA* **78**, 7634–7638.

Mitsui, K., Tokuzawa, Y., Itoh, H., Segawa, K., Murakami, M., Takahashi, K., Maruyama, M., Maeda, M., and Yamanaka, S. (2003). The homeoprotein Nanog is required for maintenance of pluripotency in mouse epiblast and ES cells. *Cell* **113**, 631–642.

Nagy, A., Rossant, J., Nagy, R., Abramow-Newerly, W., and Roder, J. C. (1993). Derivation of completely cell culture-derived mice from early-passage embryonic stem cells. *Proc. Nat. Acad. Sci. USA* **90**, 8424–8428.

Nichols, J., Zevnik, B., Anastassiadis, K., Niwa, H., Klewe-Nebenius, D., Chambers, I., Scholer, H., and Smith, A. (1998). Formation of pluripotent stem cells in the mammalian embryo depends on the POU transcription factor Oct4. *Cell* **95**, 379–391.

Papaioannou, V. E., and Rossant, J. (1984). Effects of the embryonic environment on proliferation and differentiation of embryonal carcinoma cells. *Cancer Surv.* **2**, 165–183.

Rossant, J., and Nagy, A. (1995). Genome engineering: the new mouse genetics. *Nat. Med.* **1**, 592–594.

Shamblott, M. J., Axelman, J., Wang, S., Bugg, E. M., Littlefield, J. W., Donovan, P. J., Blumenthal, P. D., Huggins, G. R., and cultured human primordial germ cells. *Proc. Nat. Acad. Sci. USA* **95**, 13726–13731.

Tam, P. P., and Zhou, S. X. (1996). The allocation of epiblast cells to ectodermal and germ line lineages is influenced by the position of the cells in the gastrulating mouse embryo. *Dev. Biol.* **178**, 124–132.

Thomson, J. A., Itskovitz-Eldor, J., Shapiro, S. S., Waknitz, M. A., Swiergiel, J. J., Marshall, V. S., and Jones, J. M. (1998). Embryonic stem cell lines derived from human blastocysts. *Science* **282**, 1145–1147.

Embryonic Stem Cells Versus Adults Stem Cells: Some Seemingly Simple Questions

As reflected by the contributions to this volume, we have been making tremendous strides in research on stem cells. As the same time, it has been surprisingly difficult to answer several seemingly simple questions.

How Do You Define a Stem Cell?

The textbook definition is that a stem cell is a cell that divides to generate one daughter cell that is a stem cell and another daughter cell that produces differentiated descendants. The definition readily fits a newly fertilized egg but begins to unravel as we move along the pathway of development. Totipotent embryonic stem (ES) cells can readily be recovered from the inner cell mass or the germinal ridge of embryos. But the window of time for recovering ES cells from the embryo is narrow — about day 4 to 6 for the inner cell mass and slightly later for the germinal ridge in mouse embryos. Where do the daughter cells that are immortal stem cells go after the window closes? One possible answer is that we may have been misled by the observation that ES cells are immortal if cultured under the appropriate conditions. *In vivo* they may have a limited life span, and they may gradually disappear as the embryo develops. A simpler and more appealing answer is that they probably become both the stem cells of the hematopoietic system and the more recently identified stem-like cells found in essentially every nonhematopoietic tissue of adult vertebrates.

What Are the Differences Between Stem-like Cells in Adult Tissues and ES Cells?

One answer is that ES cells can readily be shown to differentiate into essentially all cell phenotypes, whereas most isolates of adult stem cells from sources such as bone marrow stroma, fat, muscle, and nervous tissue have a more limited potential for differentiation. Also, most but not all isolates of adult stem cells have a more limited life span in culture than ES cells have. These distinctions, however, are valid only if we assume that the scientists who have worked with adult stem cells have been clever enough to devise all the experimental conditions for testing their potentials for differentiation and expansion. But are we that clever? What about the nuclear transfer experiments in which the nucleus of any completely differentiated cell can be reprogrammed to generate an ES cell if it is inserted into an enucleated embryo? Many nuclear transfer experiments fail, but the successful experiments say that any cell can become a stem cell if we are clever enough to send the correct signals from the cytoplasm to the nucleus. Therefore, the progression from a fertilized egg to an ES cell to an adult stem cell to a differentiated cell may be a continuum in which few if any steps are irreversible. If this concept is correct, the differences among ES cells, adult stem cells, and fully differentiated cells come down to questions of how many steps need to be reversed and how difficult they are to reverse to re-create a totipotent and immortal stem cell.

Are ES Cells or Adult Stem Cells Better Suited to Medical Therapies or Tissue Engineering?

Several distinguished scientists have offered simple answers to this question, some favoring ES cells and others adult stem cells. As time passes, it seems clear that we need far more research to answer it. Use of ES cells is hindered by the tumorigenicity of the cells and the danger of immune responses if they are used heterologously. Adult stem cells have not shown any tendency to becoming malignant, and several kinds of adult stem cells can be obtained in adequate amounts for autologous therapy. However, it is unlikely that one kind of stem cell will be ideal for all practical applications envisioned. ES cells may prove to be ideal for creating new organs through new protocols that will circumvent the current ethical and technical minefields. Adult stem cells may be more useful for repairing damage to tissues by trauma, disease, or perhaps uncomplicated aging. Recent observations are providing increasing evidence for the concept that adult stem cells are part of a natural system for tissue repair. The initial response to tissue injury appears to be proliferation and differentiation of stem-like cells endogenous to the tissue. After the endogenous stem-like cells are exhausted, non-hematopoietic stem cells from the bone marrow are recruited to the site of injury. Moreover, the data indicate that the adult stem cells that home to injured tissues repair the damage by two or three mechanisms: by differentiating into the appropriate cell phenotype, by providing cytokines and other factors to enhance recovery of endogenous cells, and perhaps by cell fusion, a process that may provide a rapid mechanism for differentiation of the stem cells.

Summary

We are at a remarkable stage in research with both ES and adult stem cells. One report after another destroys the dogmas of biology that still fill textbooks. There are limits on the potentials of the cells and their practical applications. But we are far from knowing the limits, particularly since we still cannot precisely define their critical features and we still depend on complex biological systems for testing them.

Darwin J. Prockop, MD, PhD

"Stemness": Definitions, Criteria, and Standards

Introduction

Stem cells have recently generated more public and professional interest than almost any other topic in biology. One reason stem cells capture the imagination of so many is the promise that understanding their unique properties may provide deep insights into the biology of cells as well as a path toward treatments for a variety of degenerative illnesses. And although the field of stem cell biology has grown rapidly, there exists considerable confusion and disagreement as to the nature of stem cells. This confusion can be partly attributed to the sometimes idiosyncratic terms and definitions used to describe stem cells. Although definitions can be restrictive, they are useful when they provide a basis for mutual understanding and experimental standardization. With this intention, I present explanations of definitions, criteria, and standards for stem cells. Moreover, I highlight a central question in stem cell biology, namely the origin of these cells. I also suggest criteria or standards for identifying, isolating, and characterizing stem cells. Finally, I summarize the notion of "stemness" and describe its possible application in understanding stem cells and their biology.

What Is a Stem Cell?

Stem cells are defined functionally as cells that have the capacity to self-renew as well as the ability to generate differentiated cells (Weissman *et al.*, 2001; Smith, 2001). More explicitly, stem cells can generate daughter cells identical to their mother (self-renewal) as well as produce progeny with more restricted potential (differentiated cells). This simple and broad definition may be satisfactory for embryonic or fetal stem cells that do not perdure for the lifetime of an organism. But this definition breaks down in trying to discriminate between transient adult progenitor cells that have a reduced capacity for self-renewal and adult stem cells. It is therefore important when describing adult stem cells to further restrict this definition to cells that self-renew throughout the life span of the animal (van der Kooy and Weiss, 2000). Another parameter that should be considered is potency: Does the stem cell generate to multiple differentiated cell types (multipotent), or is it only capable of producing one type of differentiated cell (unipotent)? Thus, a more complete description of a stem cell includes a consideration of replication capacity, clonality, and potency. Some theoretical as well as practical considerations surrounding these concepts are considered in this chapter.

SELF-RENEWAL

Stem cell literature is replete with terms such as "immortal," "unlimited," "continuous," and "capable of extensive proliferation," all used to describe the cell's replicative capacity. These rather extreme and vague terms are not very helpful, as it can be noted that experiments designed to test the "immortality" of a stem cell would by necessity outlast authors and readers alike. Most somatic cells cultured *in vitro* display a finite number of (less than 80) population doublings prior to replicative arrest or senescence, and this can be contrasted with the seemingly unlimited proliferative capacity of stem cells in culture (Houck *et al.*, 1971; Hayflick, 1973; Hayflick, 1974;

Sherr and DePinho, 2000; Shay and Wright, 2000). Therefore, it is reasonable to say that a cell that can undergo more than twice this number of population doublings (160) without oncogenic transformation can be termed "capable of extensive proliferation." In a few cases, this criteria has been met, most notably with embryonic stem (ES) cells derived from either humans or mice as well as with adult neural stem cells (NSCs) (Smith, 2001; Morrison *et al.*, 1997). An incomplete understanding of the factors required for self-renewal *ex vivo* for many adult stem cells precludes establishing similar proliferative limits *in vitro*. In some cases, a rigorous assessment of the capacity for self-renewal of certain adult stem cells can be obtained by single-cell or serial transfer into acceptable hosts, an excellent example of which is adult hematopoietic stem cells (HSCs) (Allsopp and Weissman, 2002; Iscove and Nawa, 1997). Adult stem cells are probably still best defined *in vivo*, where they must display sufficient proliferative capacity to last the lifetime of the animal. Terms such as "immortal" and "unlimited" are probably best used sparingly if at all.

CLONALITY

A second parameter, perhaps the most important, is the idea that stem cells are clonogenic entities: single cells with the capacity to create more stem cells. This issue has been exhaustively dealt with elsewhere and is essential for any definitive characterization of self-renewal, potential, and lineage.[1] Methods for tracing the lineage of stem cells are described in subsequent chapters. Although the clonal "gold standard" is well understood, there remain several confusing practical issues. For instance, what constitutes a cell line? The lowest standard would include any population of cells that can be grown in culture, frozen, thawed, and subsequently repassaged *in vitro*. A higher standard would be a clonal or apparently homogenous population of cells with these characteristics, but it must be recognized that cellular preparations that do not derive from a single cell may be a mixed population containing stem cells and a separate population of "supportive" cells required for the propagation of the purported stem cells. Hence, any reference to a stem cell line should be made with an explanation of their derivation. For example, it can be misleading to report on stem cells or "stem cell lines" from a tissue if they are cellular preparations containing of a mixed population, possibly contaminated by stem cells from another tissue.

POTENCY

The issue of potency maybe the most contentious part of a widely accepted definition for stem cells. A multipotent stem cell sits atop a lineage hierarchy and can generate multiple types of differentiated cells, the latter being cells with distinct morphologies and gene expression patterns. At the same time, many would argue that a self-renewing cell that can only produce one type of differentiated descendant is nonetheless a stem cell (Slack, 2000). A case can be made, for clarity, that a unipotent cell is probably best described as a progenitor. Progenitors are typically the descendants of stem cells, only they more constrained in their differentiation potential or

capacity for self-renewal and are often more limited in both senses.

DEFINITION

In conclusion, a working definition of a stem cell is a clonal, self-renewing entity that is multipotent and thus can generate several differentiated cell types. Admittedly, this definition is not applicable in all instances and is best used as a guide to help describe cellular attributes.

Where Do Stem Cells Come From?

The origin or lineage of stem cells is well understood for ES cells; their origin in adults is less clear and in some cases controversial. It may be significant that ES cells originate before germ layer commitment, raising the intriguing possibility that this may be a mechanism for the development of multipotent stem cells, including some adult stem cells. The paucity of information on the developmental origins of adult stems cells leaves open the possibility that they too escape lineage restriction in the early embryo and subsequently colonize specialized niches, which function to both maintain their potency as well as restrict their lineage potential. Alternatively, the more widely believed, though still unsubstantiated, model for the origin of adult stem cells assumes that they are derived after somatic lineage specification, whereupon multipotent stem cells–progenitors arise and colonize their respective cellular niches. In this section, I briefly summarize the origin of stem cells from the early embryo and explain what is known about the ontogeny of adult stem cells focusing attention on HSCs and NSCs.

STEM CELLS OF THE EARLY EMBRYO

Mouse and human ES cells are derived directly from the inner cell mass of preimplantation embryos after the formation of a cystic blastocyst (Papaioannou, 2001). This population of cells would normally produce the epiblast and eventually all adult tissues, which may help to explain the developmental plasticity exhibited by ES cells. In fact, ES cells appear to be the *in vitro* equivalent of the epiblast, as they have the capacity to contribute to all somatic lineages and in mice to produce germ line chimeras. By the time the zygote has reached the blastocyst stage, the developmental potential of certain cells has been restricted. The outer cells of the embryo have begun to differentiate to form trophectoderm, from which a population of embryonic trophoblast stem cells has also been derived in mice (Tanaka *et al.*, 1998). These specialized cells can generate all cell types of the trophectoderm lineage, including differentiated giant trophoblast cells. At the egg cylinder stage of embryonic development (embryonic day (E) 6.5 in mice), a population of cells near the epiblast (Figure 1) can be identified as primordial germ cells (PGCs), which are subsequently excluded from somatic specification or restriction (Saitou *et al.*, 2002). PGCs migrate to and colonize the genital ridges, where they produce mature germ cells and generate functional adult gametes. PGCs can be isolated either prior or subsequent to their arrival in the genital ridges and, when

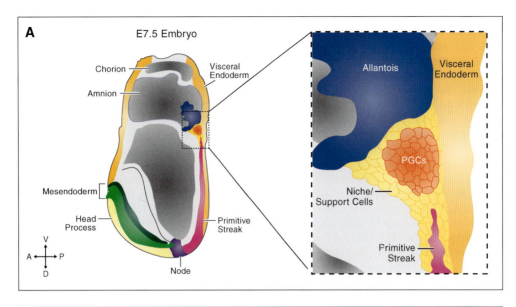

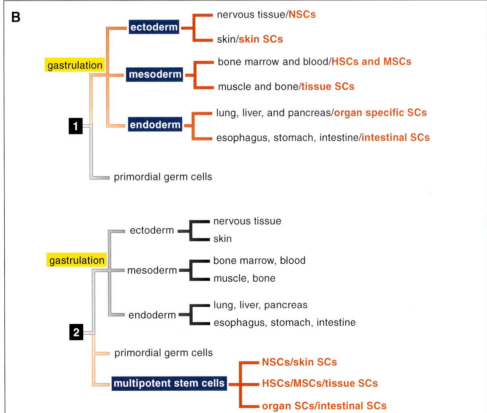

Figure 1. (A) Development of primordial germ cells. A schematic of an embryonic day 7.5 mouse embryo highlights the position of the developing primordial germ cells (PGCs) proximal to the epiblast. The expanded view on the right serves to illustrate the point that PGCs escape lineage commitment/restriction by avoiding the morphogenetic effects of migrating through the primitive streak during gastrulation. (B) Putative developmental ontogeny of stem cells. In lineage tree 1, the development of stem cells occurs after the formation of germ layers. These stem cells are thus restricted by germ layer commitment to their respective lineage (e.g., mesoderm is formed, giving rise to hematopoietic progenitors that become hematopoietic stem cells). Lineage tree 2 illustrates the idea that stem cells might develop similarly to PGCs, in that they avoid the lineage commitments during gastrulation and subsequently migrate to specific tissue and organ niches.

cultured with appropriate factors *in vitro*, can generate embryonic germ (EG) cells (Matsui *et al.*, 1992; Resnick *et al.*, 1992). EG cells have many of the characteristics of ES cells with respect to their differentiation potential and their contribution to the germ line of chimeric mice (Labosky *et al.*, 1994; Stewart *et al.*, 1994). The most notable difference between ES and EG cells is that the latter may display (depending upon the developmental stage of their derivation) considerable imprinting of specific genes (Surani, 1998; Sorani, 2001; Howell *et al.*, 2001). Consequently, certain EG cell lines are incapable of producing normal chimeric mice.

Importantly, no totipotent stem cell has been isolated from the early embryo. ES and EG cells generate all somatic lineages as well as germ cells but rarely if ever contribute to the trophectoderm, extraembryonic endoderm, or extraembryonic mesoderm. Trophectoderm stem (TS) cells have been isolated, and these only generate cells of the trophectoderm lineage. It remains to be seen whether cells can be derived and maintained from totipotent embryonic stages. Although our understanding of cell fates in the early embryo is incomplete, it appears that the only pluripotent stem cells found after gastrulation are PGCs (with the possible exceptions of multipotential adult progenitor cells [Jiang *et al.*, 2002] and teratocarcinomas). It may be that PGCs escape germ layer commitment during gastrulation by developing near the epiblast and subsequently migrate to positions inside the embryo proper. This developmental strategy may not be unique to PGCs, and it raises the interesting possibility that other stem cells might have similar developmental origins. Alternatively, it may be the case that adult stem cells are derived from PGCs. Although intriguing, it is important to stress that this idea lacks experimental evidence.

ONTOGENY OF ADULT STEM CELLS

The origin of most adult stem cells is poorly understood. With the issue of adult stem cell plasticity at the forefront, as described in this section, studies designed to elucidate the ontogeny of adult stem cells may help to reveal their specific lineage relationships and shed light on their plasticity and potential. Information on the origins of adult stem cells would also help to define the molecular programs involved in lineage determination, which may in turn provide insights into methods for manipulating their differentiation. To this end, I summarize what is known about the development of adult stem cells within the context of the hematopoietic and neural systems.

The development of hematopoietic cells in mice occurs soon after gastrulation (E7.5), although HSCs with the same activities as those in the adult have only been observed and isolated at midgestational stages (E10.5) (Orkin, 1996; Dzierzak, 2002; Weissman, 2000). These observations suggest that the embryo has a unique hematopoietic lineage hierarchy, which may not be founded by an adulttype HSC. Thus, hematopoiesis appears to occur at multiple times or in successive waves within the embryo, and the emergence of an HSC may not precede or be concomitant with the appearance of differentiated hematopoietic cells.

The first site of hematopoiesis in the mouse is the extraembryonic yolk sac, soon followed by the intraembryonic aorta–gonad–mesonephros (AGM) region. Which of these sites leads to the generation of the adult hematopoietic system and, importantly, HSCs is still unclear. Results from nonmammalian embryo-grafting experiments, with various findings in the mouse, suggest that the mammalian embryo, specifically the AGM, generates the adult hematopoietic system and HSCs (Kau and Turpen, 1983; Medvinsky *et al.*, 1993; Medvinsky and Dzierzak, 1996). Interestingly, the midgestational AGM is also the region that harbors migrating PGCs and is thought to produce populations of mesenchymal stem cells, vascular progenitors, and perhaps hemangioblasts (Molyneaux *et al.*, 2001; Minasi *et al.*, 2002; Alessandri *et al.*, 2001; Hara *et al.*, 1999; Munoz-Chapuli *et al.*, 1999). In the absence of studies designed to clonally evaluate the lineage potential of cells from the AGM, and without similarly accurate fate mapping of this region, it remains possible that all of the adult stem cell types thought to emerge within the AGM arise from a common unrestricted precursor. This hypothetical precursor could help to explain reports of nonfusion-based adult stem cell plasticity. The observed lineage specificity of most adult stem cells could likewise be attributed to the high-fidelity lineage restriction imposed on them by the specific niche they colonize or are derived from. Simple ideas such as these have not been ruled out by experimental evidence, underscoring both the opportunity and the necessity for further study of the developmental origins of adult stem cells.

A key lesson from studies of the developing hematopoietic system is that the appearance of differentiated cells does not tell us where or when the corresponding adult stem cells originate. Definitive lineage tracing, with assays of clonogenic potential, remains the method of choice for identifying the origin of stem cells. Another potential pitfall revealed by these studies is that the definition of the stem cell can make all the difference in its identification.

The development of NSCs begins with the formation of nervous tissue from embryonic ectoderm following gastrulation. Induction of the neural plate is thought to coincide with the appearance of NSCs as well as restricted progenitor types (Temple, 2001). The exact frequency and location of stem cells within the developing neuroepithelium remains unknown; specific markers must be discovered to fully unravel this question. An emerging view in the field is that embryonic neuroepithelia generate radial glial that subsequently develop into periventricular astrocytes and that these cells are the embryonic and adult NSCs within the central nervous system (Alvarez-Buylla *et al.*, 2001; Tramontin, 2003; Doetsch *et al.*, 1999; Gaiano and Fishell, 2002). Developing and adult NSCs also appear to acquire positional and temporal information. For example, stem cells isolated from different neural regions generate region-appropriate progeny (Kalyani *et al.*, 1998; He *et al.*, 2001; Anderson *et al.*, 1997). In addition, several studies suggest that temporal information is encoded within NSCs, that earlier stem cells give rise more frequently to neurons, and that more mature stem cells preferentially differentiate into glia (Temple, 2001; Qian *et al.*,

2000; White *et al.*, 2001). Moreover, more mature NSCs appear incapable of making cells appropriate for younger stages when transplanted into the early cerebral cortex (Desai and McConnell, 2000). Thus, the nervous system appears to follow a classical lineage hierarchy, with a common progenitor cell generating most if not all differentiated cell types in a regional- and temporal-specific manner. There may also be rare stem cells in the nervous system, perhaps not of neural origin, that have greater plasticity in terms of producing diverse somatic cell types and lacking temporal and spatial constraints (Weissman, 2000; Temple, 2001). There are several caveats that must be considered when describing the developmental origins of NSCs. First, disrupting the neuroepithelia to purify NSCs may have the undesirable effect of dysregulating spatial patterning acquired by these cells. Second, growth of purified NSCs in culture may reprogram the stem cells through exposure to nonphysiological *in vitro* culture conditions. Both of these problems can be addressed either by *in vivo* lineage tracing or by prospectively isolating NSCs and transplanting them into acceptable hosts without intervening culture. Carefully designed experiments promise to answer questions important not only for stem cell biology but also for neuroembryology and development. These include which features of the developmental program are intrinsic to individual cells, which differentiation or patterning signals act exclusively to instruct specific cell fates, and how developmental changes in cell-intrinsic programs restrict the responses of progenitors to cell-extrinsic signals.

How Are Stem Cells Identified, Isolated, and Characterized?

How stem cells are identified, isolated, and characterized are the key methodological questions in stem cell biology, so much so that subsequent chapters are devoted to addressing these problems in detail. Here, I briefly outline standards and criteria that may be employed when approaching the challenge of identifying, isolating, and characterizing a stem cell.

EMBRYONIC STEM CELLS

The basic characteristics of an ES cell include self-renewal, multilineage differentiation *in vitro* and *in vivo*, clonogenicity, a normal karyotype, extensive proliferation *in vitro* under welldefined culture conditions, and the ability to be frozen and thawed. In animal species, *in vivo* differentiation can be assessed rigorously by the ability of ES cells to contribute to all somatic lineages and produce germ line chimerism. These criteria are not appropriate for human ES cells; consequently, these cells must generate embryoid bodies and teratomas containing differentiated cells of all three germ layers. Moreover, as a stringent *in vivo* assessment of pluripotency is impossible, human ES cells must be shown to be positive for wellknown molecular markers of pluripotent cells. These markers are defined as factors expressed consistently, and enriched, in human ES cells (Brivanlou *et al.*, 2003). As a substitute for whole-animal chimerism, human ES cells could be tested for

their contributions to specific tissues when transplanted in discrete regions of nonhuman adults or embryos. A complementary analysis might include transplanting human ES cells into nonhuman blastocysts and evaluating their contribution to various organs and tissues, though this experiment has raised ethical concerns in some quarters. Finally, a practical consideration is the passage number of ES cells. Although it is important to establish the capacity of ES cells to proliferate extensively, it is equally important that lowpassage cells are evaluated experimentally to guard against any artifacts introduced through *in vitro* manipulation.

ADULT STEM CELLS

The basic characteristics of an adult stem cell are a single cell (clonal) that self-renews and generates differentiated cells. The most rigorous assessment of these characteristics is to prospectively purify a population of cells (usually by cell surface markers), transplant a single cell into an acceptable host without any intervening *in vitro* culture, and observe selfrenewal and tissue, organ, or lineage reconstitution. Admittedly, this type of *in vivo* reconstitution assay is not well defined for many types of adult stem cells. Thus, it is important to arrive at an accurate functional definition for cells whose developmental potential is assessed *in vitro* only. Above all, clonal assays should be the standard by which fetal and adult stem cells are evaluated because this assay removes doubts about contamination with other cell types.

Two concepts about the fate or potential of stem cells have moved to the forefront of adult stem cell research. The first is plasticity, the idea that restrictions in cell fates are not permanent but are flexible and reversible. The most obvious and extreme example of reversing a committed cell fate comes from experiments in which a terminally differentiated somatic cell generates to another animal following nuclear transfer or cloning (Solter, 2000; Rideout *et al.*, 2001). Nuclear transfer experiments show that differentiated cells, given the appropriate conditions, can be returned to their most primal state. Thus, it may not be surprising if conditions are found for more committed or specified cells to dedifferentiate and gain a broader potential. A related concept is that of transdifferentiation. Transdifferentiation is the generation of functional cells of a tissue, organ, or lineage that is distinct from that of the founding stem cell (Liu and Rao, 2003; Blau *et al.*, 2001). Important issues here are whether the cells proposed to transdifferentiate are clonal and whether the mechanism by which they form the functional cell requires fusion (Medvinsky and Smith, 2003; Terada *et al.*, 2002; Wang *et al.*, 2003; Ying *et al.*, 2002). Experiments designed to carefully evaluate these possibilities will yield insight into the nature of stem cells.

Stemness: Progress Toward a Molecular Definition of Stem Cells

Stemness refers to the common molecular processes underlying the core stem cell properties of self-renewal and the generation of differentiated progeny. Although stems cells in different cellular microenvironments or niches will by necessity have different physiological demands and therefore distinct molecular programs, there are likely certain genetic characteristics specific to and shared by all stem cells. Through transcriptional profiling, many of the genes enriched in ES cell, TS cell, HSC, and NSC populations have been identified (Ivanova *et al.*, 2002; Ramalho-Santos *et al.*, 2002; Tanaka *et al.*, 2002; Anisimov *et al.*, 2002; Luo *et al.*, 2002; Park *et al.*, 2002). By extending this approach to other stem cells and more organisms, it may be possible to develop a molecular fingerprint for stem cells. This fingerprint could be used as the basis for a molecular definition of stem cells that, when combined with their functional definition, would provide a more comprehensive set of criteria for understanding their unique biology. Perhaps more importantly, these types of studies could be used to help identify and isolate new stem cells. This goal is far from being accomplished, but the preliminary findings for specific stem cells have been described. The transcriptional profiling of stem cells has suggested that they share several distinct molecular characteristics. Stem cells appear to have the capacity to sense a broad range of growth factors and signaling molecules and to express many of the downstream signaling components involved in the transduction of these signals. Signal transduction pathways present and perhaps active in stem cells include TGF®, Notch, Wnt, and Jak/Stat family members. Stem cells also express many components involved in establishing their specialized cell cycles, either related to maintaining cell cycle arrest in G1 (for most quiescent adult stem cells) or connected to progression through cell cycle checkpoints promoting rapid cycling (as is the case for ES cells and mobilized adult stem cells) (Burdon *et al.*, 1999; Savatier *et al.*, 2002). Most stem cells also express molecules involved in telomere maintenance and display elevated levels of telomerase activity. There is also considerable evidence that stem cells have significantly remodeled chromatin acted upon by DNA methylases or transcriptional repressors of histone deacetylase and Groucho family members. Another common molecular feature is the expression of specialized posttranscriptional regulatory machinery regulated by RNA helicases of the Vasa type. Finally, a shared molecular and functional characteristic of stem cells appears to be their resistance to stress, mediated by multidrug resistance transporters, proteinfolding machinery, ubiquitin, and detoxifier systems.

Although in its infancy, the search for a molecular signature to define stem cells continues. We have begun to understand in general terms what molecular components are most often associated with stem cells. In the future, it may be possible to precisely define stem cells as a whole and individually by their telltale molecular identities. Until that time, stemness remains a concept of limited utility with tremendous potential.

ACKNOWLEDGMENTS

I would like to thank Jayaraj Rajagopal and Kevin Eggan for helpful discussion and suggestions. I apologize to those authors whose work was inadvertently overlooked or omitted because of space limitations.

Douglas A. Melton, PhD
Chad Cowen, PhD

REFERENCES

Alessandri, G., *et al.* (2001). Human vasculogenesis *ex vivo*: embryonal aorta as a tool for isolation of endothelial cell progenitors. *Lab. Invest.* **81**, 875–885.

Allsopp, R. C., and Weissman, I. L. (2002). Replicative senescence of hematopoietic stem cells during serial transplantation: does telomere shortening play a role? *Oncogene* **21**, 3270–3273.

Alvarez-Buylla, A., Garcia-Verdugo, J. M., and Tramontin, A. D. (2001). A unified hypothesis on the lineage of neural stem cells. *Nat. Rev. Neurosci.* **2**, 287–293.

Anderson, D. J., *et al.* (1997). Cell lineage determination and the control of neuronal identity in the neural crest. *Cold Spring Harb. Symp. Quant. Biol.* **62**, 493–504.

Anisimov, S. V., *et al.* (2002). SAGE identification of gene transcripts with profiles unique to pluripotent mouse R1 embryonic stem cells. *Genomics* **79**, 169–176.

Blau, H. M., Brazelton, T. R., and Weimann, J. M. (2001). The evolving concept of a stem cell: entity or function? *Cell* **105**, 829–841.

Brivanlou, A. H., *et al.* (2003). Stem cells: setting standards for human embryonic stem cells. *Science* **300**, 913–916.

Burdon, T., *et al.* (1999). Signaling mechanisms regulating selfrenewal and differentiation of pluripotent embryonic stem cells. *Cells Tiss. Organs* **165**, 131–143.

Desai, A. R., and McConnell, S. K. (2000). Progressive restriction in fate potential by neural progenitors during cerebral cortical development. *Development* **127**, 2863–2872.

Doetsch, F., *et al.* (1999). Subventricular zone astrocytes are neural stem cells in the adult mammalian brain. *Cell* **97**, 703–716.

Dzierzak, E. (2002). Hematopoietic stem cells and their precursors: Developmental diversity and lineage relationships. *Immunol. Rev.* **187**, 126–138.

Gaiano, N., and Fishell, G. (2002). The role of notch in promoting glial and neural stem cell fates. *Annu. Rev. Neurosci.* **25**, 471–490.

Hara, T., *et al.* (1999). Identification of podocalyxin-like protein 1 as a novel cell surface marker for hemangioblasts in the murine aorta–gonad–mesonephros region. *Immunity* **11**, 567–578.

Hayflick, L. (1973). The biology of human aging. *Am. J. Med. Sci.* **265**, 432–445.

Hayflick, L. (1974). The longevity of cultured human cells. *J. Am. Geriatr. Soc.* **22**, 1–12.

He, W., *et al.* (2001). Multipotent stem cells from the mouse basal forebrain contribute GABAergic neurons and oligodendrocytes to the cerebral cortex during embryogenesis. *J. Neurosci.* **21**, 8854–8862.

Houck, J. C., Sharma, V. K., and Hayflick, L. (1971). Functional failures of cultured human diploid fibroblasts after continued population doublings. *Proc. Soc. Exp. Biol. Med.* **137**, 331–333.

Howell, C. Y., *et al.* (2001). Genomic imprinting disrupted by a maternal effect mutation in the *Dnmt1* gene. *Cell* **104**, 829–838.

Iscove, N. N., and Nawa, K. (1997). Hematopoietic stem cells expand during serial transplantation *in vivo* without apparent exhaustion. *Curr. Biol.* **7**, 805–808.

Ivanova, N. B., *et al.* (2002). A stem cell molecular signature. *Science* **298**, 601–604.

Jiang, Y., *et al.* (2002). Pluripotency of mesenchymal stem cells derived from adult marrow. *Nature* **418**, 41–49.

Kalyani, A. J., *et al.* (1998). Spinal cord neuronal precursors generate multiple neuronal phenotypes in culture. *J. Neurosci.* **18**, 7856–7868.

Kau, C. L., and Turpen, J. B. (1983). Dual contribution of embryonic ventral blood island and dorsal lateral plate mesoderm during ontogeny of hemopoietic cells in *Xenopus laevis*. *J. Immunol.* **131**, 2262–2266.

Labosky, P. A., Barlow, D. P., and Hogan, B. L. (1994). Mouse embryonic germ (EG) cell lines: transmission through the germ line, and differences in the methylation imprint of insulin-like growth factor 2 receptor (*Igf2r*) gene compared with embryonic stem (ES) cell lines. *Development* **120**, 3197–3204.

Liu, Y., and Rao, M. S. (2003). Transdifferentiation: Fact or artifact. *J. Cell Biochem.* **88**, 29–40.

Luo, Y., *et al.* (2002). Microarray analysis of selected genes in neural stem and progenitor cells. *J. Neurochem.* **83**, 1481–1497.

Matsui, Y., Zsebo, K., and Hogan, B. L. (1992). Derivation of pluripotential embryonic stem cells from murine primordial germ cells in culture. *Cell* **70**, 841–847.

Medvinsky, A. L., *et al.* (1993). An early preliver intraembryonic source of CFU-S in the developing mouse. *Nature* **364**, 64–67.

Medvinsky, A., and Dzierzak, E. (1996). Definitive hematopoiesis is autonomously initiated by the AGM region. *Cell* **86**, 897–906.

Medvinsky, A., and Smith, A. (2003). Stem cells: fusion brings down barriers. *Nature* **422**, 823–835.

Minasi, M. G., *et al.* (2002). The mesoangioblast: A multipotent, selfrenewing cell that originates from the dorsal aorta and differentiates into most mesodermal tissues. *Development* **129**, 2773–2783.

Molyneaux, K. A., *et al.* (2001). Time-lapse analysis of living mouse germ cell migration. *Dev. Biol.* **240**, 488–498.

Morrison, S. J., Shah, N. M., and Anderson, D. J. (1997). Regulatory mechanisms in stem cell biology. *Cell* **88**, 287–298.

Munoz-Chapuli, R., *et al.* (1999). Differentiation of hemangioblasts from embryonic mesothelial cells? A model on the origin of the vertebrate cardiovascular system. *Differentiation* **64**, 133–141.

Orkin, S. H. (1996). Development of the hematopoietic system. *Curr. Opin. Genet. Dev.* **6**, 597–602.

Papaioannou, V. (2001). Stem cells and differentiation. *Differentiation* **68**, 153–154.

Park, I. K., *et al.* (2002). Differential gene expression profiling of adult murine hematopoietic stem cells. *Blood* **99**, 488–498.

Qian, X., *et al.* (2000). Timing of CNS cell generation: a programmed sequence of neuron and glial cell production from isolated murine cortical stem cells. *Neuron* **28**, 69–80.

Ramalho-Santos, M., *et al.* (2002). "Stemness": Transcriptional profiling of embryonic and adult stem cells. *Science* **298**, 597–600.

Resnick, J. L., *et al.* (1992). Long-term proliferation of mouse primordial germ cells in culture. *Nature* **359**, 550–551.

Rideout, W. M., 3rd, Eggan, K., and Jaenisch, R. (2001). Nuclear cloning and epigenetic reprogramming of the genome. *Science* **293**, 1093–1098.

Saitou, M., Barton, S. C., and Surani, M. A. (2002). A molecular program for the specification of germ cell fate in mice. *Nature* **418**, 293–300.

Savatier, P., *et al.* (2002). Analysis of the cell cycle in mouse embryonic stem cells. *Methods Mol. Biol.* **185**, 27–33.

Shay, J. W., and Wright, W. E. (2000). Hayflick, his limit, and cellular ageing. *Nat. Rev. Mol. Cell Biol.* **1**, 72–76.

Sherr, C. J., and DePinho, R. A. (2000). Cellular senescence: mitotic clock or culture shock? *Cell* **102**, 407–410.

Slack, J. M. (2000). Stem cells in epithelial tissues. *Science* **287**, 1431–1433.

Smith, A. G. (2001). Embryo-derived stem cells: of mice and men. *Annu. Rev. Cell Dev. Biol.* **17**, 435–462.

Solter, D. (2000). Mammalian cloning: advances and limitations. *Nat. Rev. Genet.* **1**, 199–207.

Stewart, C. L., Gadi, I., and Bhatt, H. (1994). Stem cells from primordial germ cells can reenter the germ line. *Dev. Biol.* **161**, 626–628.

Surani, M. A. (1998). Imprinting and the initiation of gene silencing in the germ line. *Cell* **93**, 309–312.

Surani, M. A. (2001). Reprogramming of genome function through epigenetic inheritance. *Nature* **414**, 122–128.

Tanaka, S., *et al.* (1998). Promotion of trophoblast stem cell proliferation by FGF4. *Science* **282**, 2072–2075.

Tanaka, T. S., *et al.* (2002). Gene expression profiling of embryoderived stem cells reveals candidate genes associated with pluripotency and lineage specificity. *Genome Res.* **12**, 1921–1928.

Temple, S. (2001). The development of neural stem cells. *Nature* **414**, 112–117.

Terada, N., *et al.* (2002). Bone marrow cells adopt the phenotype of other cells by spontaneous cell fusion. *Nature* **416**, 542–545.

Tramontin, A. D., *et al.* (2003). Postnatal development of radial glia and the ventricular zone (VZ): a continuum of the neural stem cell compartment. *Cereb. Cortex* **13**, 580–587.

van der Kooy, D., and Weiss, S. (2000). Why stem cells? *Science* **287**, 1439–1441.

Wang, X., *et al.* (2003). Cell fusion is the principal source of bone marrow-derived hepatocytes. *Nature* **422**, 897–901.

Weissman, I. L. (2000). Stem cells: units of development, units of regeneration, and units in evolution. *Cell* **100**, 157–168.

Weissman, I. L., Anderson, D. J., and Gage, F. (2001). Stem and progenitor cells: origins, phenotypes, lineage commitments, and transdifferentiations. *Annu. Rev. Cell Dev. Biol.* **17**, 387–403.

White, P. M., *et al.* (2001). Neural crest stem cells undergo cell-intrinsic developmental changes in sensitivity to instructive differentiation signals. *Neuron* **29**, 57–71.

Ying, Q. L., *et al.* (2002). Changing potency by spontaneous fusion. *Nature* **416**, 545–548.

1

Present Perspective and Future Challenges

R. L. Gardner

Introduction

Many researchers have worked to bring recognition to the enormous potential that cells of early embryonic origin possess for genetic modification of organisms, regenerative medicine, and investigation of facets of development that are difficult to explore *in vivo*. Historically, however, this field is firmly rooted in the pioneering work of Roy Stevens and Barry Pierce on mouse teratomas and teratocarcinomas, tumors that continued to be regarded with disdain by many mainstream pathologists and oncologists well after these workers had embarked on their studies. Stevens developed and exploited mouse strains with high incidences of such tumors in order to determine their cellular origins. Pierce focused his attention on the nature of the cell that endowed teratocarcinomas with the potential for indefinite growth, which the more common teratomas lacked. Conversion of solid teratocarcinomas to an ascites form proved to be a significant advance in enabling dramatic enrichment of the morphologically undifferentiated cells in such tumors, among which their stem cells were expected to be included.

Later, in 1964, an impressive experiment by Pierce and a colleague showed unequivocally that, upon transplantation to histocompatible adult hosts, individual morphologically undifferentiated cells could form self-sustaining teratocarcinomas that contained as rich a variety of differentiated tissues as their parent tumor. Hence, the embryonal carcinoma (EC) cell, as the stem cell of teratocarcinomas has come to be known, was the first self-perpetuating pluripotential cell to be characterized. Although teratocarcinomas were obtained initially as a result of genetically determined aberrations in the differentiation of male or female germ cells, it was found that they could also be established in certain genotypes of mice by grafting early embryos ectopically in adults. Adaptation of culture conditions soon followed to enable EC cells to be perpetuated in an undifferentiated state or induced to differentiate *in vitro*. Although the range of differentiation detected in these circumstances was more limited than *in vivo*, it could nevertheless be impressive. Research on murine EC cells, in turn, provided the impetus for obtaining and harnessing the human counterpart of these cells from testicular tumors for *in vitro* study.

One outstanding question regarding the use of murine EC cells as a model system for studying aspects of development remained, namely, the basis of their malignancy. Was this a consequence of genetic change, or was it simply because such "embryonic" cells failed to relate to the ectopic sites into which they were transplanted? The obvious way of addressing this question was to ask how EC cells behave when placed in an embryonic rather than an adult environment. This study was done independently in three laboratories by injecting the cells into blastocysts. The results from each laboratory led to the same rather striking conclusion: EC cells — which, if injected into an adult, would grow progressively and kill it — were able to participate in normal development following their introduction into the blastocyst. Using genetic differences between donor and host as cell markers, researchers found that EC cells were able to contribute to most, if not all, organs and tissues of the resulting offspring. Most intriguingly, according to reports from one laboratory, this could very exceptionally include the germ line. The potential significance of this finding was considerable in its implications for possible controlled genetic manipulation of the mammalian genome. It raised the prospect of being able to select for extremely rare events, thus bringing the scope for genetic manipulation in mammals closer to that in microorganisms.

There were problems, however. One was that the EC contribution in chimeric offspring was typically both more modest and more patchy than that of cells transplanted directly between blastocysts. Also, the chimeras frequently formed tumors; those that proved to be teratocarcinomas were often evident already at birth, suggesting that growth regulation of at least some transplanted EC cells failed altogether. Other chimeras developed more specific tumors such as rhabdomyosarcomas as they aged, which were also clearly of donor origin, thereby revealing that the transplanted EC cells had progressed further along various lineages before their differentiation went awry. In extreme cases, the transplanted EC cells disrupted development altogether, so that fetuses did not survive to birth. Although the best EC lines could contribute to all or most tissues of the body of chimeras, they did so exceptionally. Finally, the frequency with which colonization of the germ line could be obtained with EC cells was too low to enable them to be harnessed for genetic modification. It seemed likely, therefore, that the protracted process of generating teratocarcinomas *in vivo* and then adapting them to culture militated against the retention of a normal genetic constitution by their stem cells. If this was indeed the case, the obvious way forward was to see if such stem cells could be obtained in a less circuitous manner.

This issue prompted investigation of what happens when murine blastocysts are explanted directly on growth-inactivated feeder cells in an enriched culture medium. The result was the derivation of lines of cells indistinguishable from EC cells in both their morphology and the expression of various antigenic and other markers, as well as in the appearance of the colonies they formed during growth. Moreover, like EC cells, these self-perpetuating blastocyst-derived stem cells could form aggressive teratocarcinomas in both syngeneic and immunologically compromised nonsyngeneic adult hosts. They differed from EC cells principally by giving much more frequent and widespread somatic chimerism following reintroduction into the preimplantation conceptus and, if tended carefully, by also routinely colonizing the germ line. Moreover, when combined with host conceptuses whose development was compromised by tetraploidy, they could sometimes form offspring in which no host-derived cells were discernible. Thus, these cells, which exhibited all the desirable characteristics of EC cells and few of their shortcomings, came to be called embryonic stem (ES) cells. Once it had been shown that ES cells could retain their ability to colonize the germ line after *in vitro* transfection and selection, their future was assured. Surprisingly, however, despite the wealth of studies demonstrating their capacity for differentiation *in vitro*, particularly in the mouse, it was a long time before the idea of harnessing ES cells for therapeutic purposes took root. Thus, although Robert Edwards explicitly argued more than 20 years ago that human ES cells might be used thus, only within the past few years has this notion gained momentum, encouraged particularly by derivation of the first cell lines from human blastocysts.

Terminology

There is some confusion in the literature about terminology in discussing the range of different types of cells that ES cells are able to form, an attribute that, in embryological parlance, is termed their potency. Some refer to these cells as being totipotent because, at least in the mouse, they have been shown to be able to generate all types of fetal cells and, under certain conditions, entire offspring. This is inappropriate on two counts. First, totipotency is reserved by embryologists for cells that retain the capacity to form an entire conceptus and thus produce a new individual unaided. The only cells that have so far been shown to be able to do this are blastomeres from early cleavage stages. Second, murine ES cells seem unable to form all the different types of cell of which the conceptus is composed. Following injection into blastocysts, they normally generate only cell types that are products of the epiblast lineage. Although they can also form derivatives of the primitive endoderm lineage — which, for some obscure reason, they do much more readily *in vitro* than *in vivo* — they have never been shown to contribute to the trophectodermal lineage. Hence, a widely adopted convention is to describe ES cells as pluripotent stem cells to distinguish them from

stem cells like those of the hematopoietic system, which have a narrower but nevertheless impressive range of differentiative potential. Another source of confusion is the surprisingly common practice of referring to cells, particularly putative ES cells from mammals other than the mouse, as totipotent because their nuclei have been shown to be able to support development to term when used for reproductive cloning.

Another facet of terminology relates to the definition of an ES cell, which again is not employed in a consistent manner. One view, to which the author subscribes, is that use of this term should be restricted to pluripotent cells derived from pre- or peri-implantation conceptuses that can form functional gametes as well as the full range of somatic cells of offspring. Although there are considerable differences among strains of mice in the facility with which morphologically undifferentiated cell lines can be obtained from their early conceptuses, competence to colonize the germ line as well as somatic tissues seems nevertheless to be common to lines from all strains that have yielded them. This is true, for example, even for the nonobese diabetic (NOD) strain, whose lines have so far been found to grow too poorly to enable their genetic modification.

ES-like Cells in Other Species

As shown in Table 1-1, cell lines that can be maintained for variable periods *in vitro* in a morphologically undifferentiated state have been obtained from morulae or blastocysts in a variety of species of mammals in addition to the mouse. They have also been obtained from the stage X blastoderm in the chick and from blastulae in three species of teleost fish. The criteria employed to support claims that such lines are counterparts of murine ES cells are quite varied and often far from unequivocal. They range from maintenance of an undifferentiated morphology during propagation or expression of at least some ES cell markers, through differentiation into a variety of cell types *in vitro*, to production of histologically diverse teratomas or chimaerism *in vivo*.

What such ES-like (ESL) cells lines have in common with murine ES cells, in addition to a morphologically undifferentiated appearance, is a high nuclear–cytoplasmic ratio. Among the complications in assessing cell lines in different species is the variability in the morphology of the growing colonies. Although colonies of ESL cells in the hamster and rabbit are very similar to colonies of murine ES cells, those of most other mammals are not. This is particularly true in the human, whose undifferentiated ESL cell colonies closely resemble those formed by human EC cells of testicular origin, as do ESL cell colonies from other primates. In the marmoset, rhesus monkey, and human, ESL cells not only form relatively flattened colonies but also exhibit several differences from mouse ES cells in the markers they express. Because they closely resemble human EC cells in all these respects, the differences seem to relate to species rather than to cell type.

2

TABLE 1-1
Vertebrates from Which ES-like Cells Have Been Obtained

Species	Basis of validation[a]
Rat	CP but mouse ES contamination
	M&M
	CP
	M&M
	M&M
Golden hamster	IVD
Rabbit	M&M, IVD
	CP
Mink	T (but limited range of cell types)
	T (wide range of cell types)
	IVD
	M&M
Pig	IVD
	M&M
	CP
	CP
Sheep	M&M
	?[b]
	?[b]
Cow	M&M
	IVD
	IVD
	?
	CP
	CP
	IVD
Horse	IVD
Marmoset	IVD
Rhesus monkey	T
Human	T
Chicken	IVD (& CP including germ line but only with passage 1–3 cells)
Medaka	IVD
	CP
Zebra fish	IVD (limited) & CP (with short-term cultured cells)
Gilthead sea bream	IVD & (CP with short-term cultured cells)

[a]M&M: morphology and ES cell markers, IVD: differentiation *in vitro*, T: teratoma production *in vivo*, CP: chimera production by morula aggregation or blastocyst injection.
[b]Exhibited an ES-like morphology initially but rapidly acquired a more epithelial one thereafter.

In two studies in the sheep, colonies are reported to look like those formed by murine ES cells initially but to adopt a more epithelial-like appearance rapidly thereafter. This change in morphology bears an intriguing similarity to the transition in conditioned medium of murine ES cells to so-called epiblast-like cells, which is accompanied by loss of their ability to colonize the blastocyst. Given that this transition is reversible, whether a comparable one is occurring spontaneously in sheep clearly warrants further investigation.

In no species has the production of chimeras with ESL cells rivaled that obtained with murine ES cells. Where it has been attempted, both the rates and the levels of chimerism are typically much lower than are found with murine ES cells. An apparent exception is one report for the pig, in which 72% of offspring were judged to be chimeric. However, this figure is presented in an overview of work that remains unpublished, and no details are provided regarding the number of times the donor cells were passaged before they were injected into blastocysts. In a subsequent study in this species using ESL cells that had been through 11 passages, one chimera was recorded among 34 offspring. However, as the authors of this latter study point out, rates of chimerism of only 10 to 12% have been obtained following direct transfer of inner-cell mass cells to blastocysts in the pig. Hence, technical limitations may have contributed to the low success with ESL cells in this species.

The only species listed in Table 1-1 in which colonization of the germ line has been demonstrated is the chicken, but this was with cells that had been passaged only one to three times before being injected into host embryos. Hence, they do not really qualify as stem cells that can be propagated indefinitely *in vitro*. Consequently, in conformity with the terminology discussed earlier, morphologically undifferentiated cell lines in all species listed in Table 1-1 should be assigned the status of ESL cells rather than ES cells.

Generally, the initial strategy for attempting to derive ES cell lines in other species has been to follow the conditions that proved successful in the mouse, namely, the use of enriched medium in conjunction with growth-inactivated feeder cells and either leukemia inhibitory factor (LIF) or a related cytokine. Various modifications introduced subsequently include same-species rather than murine feeder cells and, in several species including the human, dispensing with LIF. Optimal conditions for deriving cell lines may differ from those for maintaining them. Thus, in one study in the pig, the use of same-species feeder cells was found to be necessary to obtain cell lines, though murine STO cells were adequate for securing their propagation thereafter. Feeder-free conditions were found to work best in both the medaka and the gilthead sea bream.[43–45] Moreover, the cloning efficiency of human ESL lines was improved in serum-free culture conditions.

Unexpectedly, despite being closely related to the mouse, the rat has proved particularly refractory to derivation of ES cell lines (see Table 1-1). So far, the only cell lines that have proved to be sustainable long term in this species seem to lack all properties of mouse ES cells, including differentiation potential, apart from colony morphology. Indeed, except for the 129 strain of mouse, establishing cell lines that can be propagated *in vitro* in a morphologically undifferentiated state seems almost more difficult in rodents than in most of the other vertebrates in which it has been attempted.

Overall, one is struck by species variability in the growth factors, status of conceptus or embryo, and other requirements for obtaining pluripotential cell lines in species other than the mouse. So far, one can discern no clear recipe for success.

Of course, obtaining cells that retain the capacity to colonize the germ line following long-term culture is essential only for genetically modifying animals in a controlled manner. Having cells that fall short of this but are nevertheless able to differentiate into a range of distinct types of cells *in vitro* may suffice for many other purposes.

Embryonic Germ Cells

The preimplantation conceptus is not the only source of pluripotential stem cells in the mouse. Sustainable cultures of undifferentiated cells that strikingly resemble ES cells in their colony morphology have also been obtained from primordial germ cells and very early gonocytes in this species. These cells, termed embryonic germ (EG) cells, have also been shown to be capable of yielding high rates of both somatic and germ-line chimerism following injection into blastocysts.

TABLE 1-2

Vertebrates from Which Embryonic Germ Cells Have Been Obtained

Species	Basis of validation[a]
Mouse	M&M
	CP
	CP (including germ line)
	CP (including germ line)
	IVD
Pig	CP
	CP (with transfected cells)
Cow	IVD (& short-term CP)
Human	IVD
Chicken	CP (including germ line, but cells Cultured for only five days)
	CP

[a]Abbreviations as listed in the footnote to Table 1-1.

These findings have prompted those struggling to derive ES cell lines in other species to explore primordial germ cells as an alternative for achieving controlled genetic modification of the germ line. As shown in Table 1-2, EG-like (EGL) cells have been obtained in several mammals as well as the chick, but as with ESL cells, their ability to participate in chimera formation has, with one exception, only been demonstrated at low passage. Moreover, although donor cells have been detected in the gonad of a chimera obtained from low-passage EGL cells in the pig, no case of germ-line colonization has been reported except with cells from chick genital ridges that were cultured for only five days. Even here, the proportion of offspring of the donor type was very low.

It is noteworthy, however, that even in the mouse rates of malformation and perinatal mortality appear to be higher in EG than in ES cell chimeras. This may relate to erasure of imprinting in the germ line, which seems to have begun by the time primordial germ cells have colonized the genital ridges or, for certain genes, even earlier. It is perhaps because of such concerns that the potential of EG cells for transgenesis in strains of mice that have failed to yield ES cells has not been explored. Interestingly, unlike in the mouse, EGL cell lines derived from genital ridges and the associated mesentery of 5- to 11-week human fetuses seem not to have embarked on erasure of imprinting. Obviously, it is important to confirm that this is the case before contemplating the use of such cells as grafts for repairing tissue damage in humans.

Future Challenges

The value of ES and ESL cells as resources for both basic and applied research is now acknowledged almost universally. Present barriers to exploitation of their full potential in both areas are considered in the next sections of this chapter, together with possible ways of addressing these barriers. Fundamental to progress is gaining a better understanding of both the nature and the basic biology of these cells.

BIOLOGY OF ES AND ESL CELLS

Germ-Line Competence

Although murine ES cells have been used extensively for modifying the genome, several problems remain that limit their usefulness in this respect. Among these problems is the loss of competence to colonize the germ line, a common and frustrating problem whose basis remains elusive. It is not attributable simply to the occurrence of sufficient chromosomal change to disrupt gametogenesis because it can occur in lines and clones found to be karyotypically normal. At present, it is not known whether it is because of the failure of the cells to be included in the pool of primordial germ cells or their inability to undergo appropriate differentiation thereafter, possibly as a consequence of perturbation of the establishment of genomic imprinting or its erasure. Even within cloned ES lines, cells have been found to be heterogeneous in expression of imprinted genes. Given that many ES cell lines are likely to have originated polyclonally from several epiblast founder cells, there is the further possibility that they might, *ab initio*, consist of a mixture of germ-line-competent and -noncompetent subpopulations.

Recent studies on the involvement of bone morphogenetic protein signaling in the induction of primordial germ cells have been interpreted as evidence against a specific germ cell lineage in mammals. Particular significance has been attached to experiments in which distal epiblast, which does not usually produce primordial germ cells, was found to do so when grafted to the proximal site whence these cells normally originate. However, because of the extraordinary degree of cell mixing that occurs in the epiblast before gastrulation, descendants of all epiblast founder cells are likely to be present throughout the tissue by the time of primordial germ cell induction. Hence, the possibility remains that competence for induction is lineage dependent, and thereby segregates to only some epiblast founder cells. Because ES cell lines are typically produced by pooling all colonies derived from a single blastocyst, they might originate from of a mixture of germ line-competent and -noncompetent founder cells.

Male ES cell lines have almost invariably been used in gene-targeting studies, even though this complicates work on X-linked genes whose inactivation may lead to cell-autonomous early lethality or compromise viability in the hemizygous state. Here, female (XX) lines would, in principle, offer a simpler alternative except that they are generally held to suffer partial deletion or complete loss of one X-chromosome after relatively few passages. However, the security of this conclusion is not clear because few references to their use have appeared in the literature since the early reports, in which consistent loss of all or part of one X was first documented. More recently, one of only two female lines tested was found to be germ-line-competent, but the entire donor-derived litters were unusually small, raising the possibility, not entertained by the authors, that the line in question was XO. Interestingly, female human ESL cell lines do not seem to show a similar propensity for X-chromosome loss.

Origin and Properties of ES and ESL Cells

It is evident from the earlier overview that considerable diversity exists even among eutherian mammals in the characteristics of cells from early conceptuses that can be perpetuated *in vitro* in a morphologically undifferentiated state. The reason for this is far from clear, particularly because most such cell lines have been derived at a corresponding stage — namely, the preimplantation blastocyst — often using inner cell mass tissue isolated therefrom. In the mouse, in contrast to their EC counterparts, ES cells have not been obtained from postimplantation stages, which argues that there is a rather narrow window during which their derivation is possible. What this relates to in developmental terms remains obscure, although the finding that ES cells can shift reversibly to a condition that shows altered colony morphology and gene expression, together with loss of ability to generate chimeras following blastocyst injection, offers a possible approach for addressing this problem. Whether the late-blastocyst stage sets the limit for obtaining ESL cell lines in other mammals has not yet been addressed critically.

Just as ES cell lines have been obtained from preblastocyst stages in the mouse, ESL cell lines have been obtained from such stages in other mammals. However, neither in the mouse nor in other species have the properties of cell lines derived from morulae been compared with those from blastocysts to see if they show consistent differences. Indeed, it remains to be ascertained whether the lines from morulae originate at an earlier stage in development rather than progressing to blastocyst or, more specifically, epiblast formation before doing so. Although it has been claimed that lines isolated from morulae have an advantage over those from blastocysts in being able to produce trophoblast, this has not actually been shown to be the case. However, species-, as opposed to stage-related differences, in the ability of cell lines to produce trophoblast tissue have been encountered. Early claims that mouse ES cells can form trophoblastic giant cells are almost certainly attributable to the short-term persistence of contaminating polar trophectoderm tissue. Thus, the production of such cells seems to be limited to the early passage of ES lines derived from entire blastocysts. It has never been observed with lines established from microsurgically isolated epiblasts. Although the situation is not clear in many species, in primates, differentiation of trophoblast has been observed routinely in ESL cell lines established from immunosurgically isolated inner cell masses. Moreover, differentiation of human cell lines to the stage of syncytiotrophoblast formation has been induced efficiently by exposing them to BMP4.

Pluripotency

A seminal characteristic of ES or ESL cells is their pluripotency. The most critical test of this — not practicable in some species, particularly the human — is the ability to form the entire complement of cells of normal offspring. This assay, originally developed in the mouse, entails introducing clusters of ES cells into conceptuses whose development has been compromised by making them tetraploid, either by suppressing cytokinesis or by fusing sister blastomeres electrically at the two-cell stage. ES cells are then either aggregated with the tetraploid cleavage stages or injected into tetraploid blastocysts. Some resulting offspring contain no discernible host cells. It seems likely that host epiblast cells are present initially and play an essential role in "entraining" the donor ES cells before being outcompeted, because groups of ES cells on their own cannot substitute for the epiblast or inner cell mass. Selection against tetraploid cells is already evident by the late-blastocyst stage in chimeras made between diploid and tetraploid morulae. Aggregating ESL cells between pairs of tetraploid morulae has been tried in cattle, but resulted in their contributing only very modestly to fetuses and neonates.

The second most critical test is whether the cells yield widespread, if not ubiquitous, chimerism in offspring following introduction into the early conceptus, either by injection into blastocysts or by aggregation with morulae. The third is the formation teratomas in ectopic grafts to histocompatible or immunosuppressed adult hosts, since it is clear from earlier experience with murine and human EC cells that a wider range of differentiation can be obtained in these circumstances than *in vitro*. For such an assay to be incisive, it is necessary to use clonal cell lines and thus ensure that the diversity of differentiation obtained originates from one type of stem cell rather than from a medley of cells with more limited developmental potential. Although teratoma formation has been demonstrated with clonal ESL cells in the human, this is not true for corresponding cell lines in other species. A note of caution regarding the use of teratomas for assessing pluripotency comes from work which found that hepatocyte differentiation depended not only on the site of inoculation of mouse ES cells but also on the status of the host. Thus, positive results were obtained with spleen rather than hind-limb grafts and only when using nude rather than syngeneic mice as hosts.

Conditions of Culture

ES and ESL cells are usually propagated in complex culture conditions that are poorly defined because they include both growth-inactivated feeder cells and serum. This complicates the task of determining the growth factor and other requirements necessary for their maintenance as well as for inducing them to form specific types of differentiated cells. Although differentiation of murine ES cells in a chemically defined medium has been achieved, their maintenance under such conditions has not. Murine ES cells can be both derived and maintained independently of feeder cells, provided that a cytokine that signals via the gp 130 receptor is present in the medium. However, whether the relatively high incidence of early aneuploidy recorded in the two studies in which LIF was used throughout in place of feeders is significant or coincidental is not clear. It is important to resolve this question in order to learn whether feeder cells serve any function other than acting as a source of LIF or a related cytokine. Production of extracellular matrix is one possibility. However,

species variability is also a factor here since LIF is not required for maintaining human ESL lines, whose cloning efficiency is actually improved by omission of serum, though feeder cells are required. The norm has been to use murine feeder cells both for obtaining and for perpetuating ESL cell lines in other mammals, including the human. Recently, however, there has been a move to use feeders of human origin for human ESL cells. This is a notable development because it would not be acceptable to employ xenogeneic cells for growing human ESL cell lines destined for therapeutic rather than laboratory use. The situation is somewhat confusing in the case of the pig; in one study, but not in others, porcine feeders were found to be necessary for deriving ESL cell lines that could then be perpetuated on murine STO cells. Moreover, among teleost fish, feeder-free conditions seem to be optimal for maintaining ESL cells in both the medaka and the sea bream but possibly not in the zebra fish.

Susceptibility Versus Resistance to Derivation

An area whose further investigation could be informative in facilitating the establishment of pluripotent stem cell lines in other species is the basis of susceptibility versus resistance to ES cell derivation in the mouse. Thus, although ES cell lines can be obtained easily in 129 mice and relatively so in C57BL/6 and a few additional strains (Table 1-3), other genotypes have proved more resistant. Notable among the latter is the NOD strain from which, despite considerable effort, genetically manipulable lines have not yet been obtained. This is not simply related to the susceptibility of this strain to insulin-dependent diabetes, because the ICR strain from which NOD was developed has proved to be equally refractory. However, refractoriness seems to be a recessive trait because excellent lines with high competence to colonize the germ line have been obtained from [NOD × 129]F1 epiblasts. Moreover, this is not the only example in which refractoriness has been overcome by intercrossing. Interestingly, marked differences in the permissiveness for ESL cell derivation have also been found among inbred strains of the medaka.

Human ESL Cell

Mouse EC and ES cells have been used extensively to study aspects of development that, for various reasons, are difficult to investigate in the intact conceptus. Exploiting corresponding cells for this purpose is even more pressing for gaining a better understanding of early development in our own species, given the relative scarcity of material, ethical concerns about experimenting on conceptuses, and statutory or technical limitations on the period for which they can be maintained *in vitro*. Obviously, in view of their provenance, human ESL cells are likely to provide a more apposite model system than human EC cells, which have mainly been used until recently. One concern here is that so-called spare conceptuses (i.e., those surplus to the needs of infertility treatment) are the sole source of material for producing human ESL cell lines. Because the conceptuses produced *in vitro* by in vitro

TABLE 1-3

Genotypes of ES Cells Other Than 129 for Which Germ-Line Transmission Has Been Demonstrated

Genotype
C57BL/6
C57BL/6N
C57BL/6JOla
[C57BL/6x CBA]F1
CBA/CaOla
BALB/c
DBA/1lacJ
DBA/1Ola
DBA/2N
C3H/He
C3H/Hen
FVB/N
CD1[a]
NOD
[NOD × 129/Ola]F1
129 × [129 × DDK]F1
PO[a]

[a]Outbred strains.

fertilization (IVF) or related techniques that are judged to be of the highest quality are selected for infertility treatment, those used for deriving ESL lines tend to be of lower quality. Does this matter so far as the properties of the resulting cell lines are concerned, particularly if their use therapeutically is contemplated? Is the ability to form a blastocyst that looks satisfactory morphologically adequate, or will it prove acceptable to produce conceptuses specifically for generating ESL cell lines, so that quality is less of a concern?

ES Cell Transgenesis

One important use of ES cell transgenesis is to obtain animal models of human genetic diseases. Because few would claim that the mouse is the ideal species for this purpose, the incentive for being able to undertake such studies in more appropriate or experimentally tractable mammals must remain a high priority. For example, given its widespread use for studying respiratory physiology, the sheep would be a more relevant species than the mouse as a model system for studying cystic fibrosis. However, unless pluripotential cells able to colonize the germ line can be obtained in other species, this approach to transgenesis will continue to be limited to the mouse. Although the feasibility of an alternative strategy — namely, genetically modifying and selecting cells that are not germ line-competent, such as fetal fibroblasts, then exploiting transfer of their nuclei to oocytes — has been demonstrated, it is extremely demanding technically and entails considerable fetal attrition.[93]

STEM CELL THERAPY

Potential Hurdles

One major interest in ESL cells is the prospect of exploiting them therapeutically to repair damage to tissues or organs resulting from disease or injury. This poses a host of new challenges, not all of which have received the attention they deserve. Perhaps the most obvious challenge is whether it will be possible to obtain efficient directed differentiation of stem cells to yield pure cultures of the desired type of more differentiated cells as opposed to a mixed population. If the latter proves to be the case, the rigorous purging of cultures of residual undifferentiated or inappropriately differentiated cells will be necessary. How this is approached will depend on whether any contamination of grafts is acceptable and, if so, how much. One way in which this particular problem has been circumvented in murine model systems for in vitro differentiation of ES cells is to transfect them with the coding region of a gene for an antibiotic resistance or fluorescent protein coupled to a promoter that is expressed only in the desired type of differentiated cell. Recent advances have made it possible to carry out similar genetic modification of human ESL cells. Although effective selection of the desired type of differentiated cell may be achieved with this approach, it remains to be seen whether use of genetically modified cells will be acceptable in a clinical as opposed to a laboratory context.

Another important issue in contemplating stem cell therapy is the cycle status of the desired type of cell. In certain cases, including cells that are not postmitotic in grafts may be highly undesirable or even hazardous. In others, the presence of such cells may be essential to meet the demands of tissue growth or turnover. The latter would depend on obtaining the differentiation of ESL cells to stem cells rather than to fully differentiated cells of the desired type. Given the growing evidence that tissue stem cells require a specific niche for their maintenance, this could prove difficult to achieve. Establishing and maintaining an appropriate niche in vitro to be able to enrich for tissue-specific stem cells is likely to pose a considerable challenge and will unquestionably depend on obtaining better knowledge of the normal biology of individual tissues.

Yet another important issue is whether engrafted cells will survive and function properly when placed in a damaged tissue or organ. When the donor cells are to provide a hormone, neural transmitter, or soluble growth factor, it may be possible to place them at some distance from the site of damage. However, when this is not practicable, there remains the question of whether transplanted cells will fare better than native ones in a tissue or organ seriously damaged by disease or injury. If they do not, how can one circumvent this difficulty, bearing in mind that achieving organogenesis in vitro is still a rather remote prospect? Regarding neurodegenerative disease, some progress has been made in "cleaning up" sites of tissue damage. For example, antibody-mediated clearance of plaques from the brain in transgenic mice overexpressing amyloid precursor protein has been demonstrated. However, such intervention may not be necessary in all cases. Transplanting differentiated murine ES cells enriched for putative cardiomyocytes to a damaged region of the left ventricle in rats led concomitantly to a reduction in size of this region and an improvement in the performance of the heart.

Therapeutic Cloning

Establishing ESL cell lines from blastocysts derived by nuclear replacement, so-called therapeutic cloning, has been widely advocated as a way of tailoring grafts to individual patients, thereby circumventing the problem of graft rejection. Although the feasibility of producing ES cells in this way has been demonstrated in the mouse, there is sharp division of opinion within the biomedical research community about whether such cells would be safe to use therapeutically. Particular concern centers on the normality of the donor genome regarding the epigenetic status of imprinted genes. Moreover, recent observations on chromosome segregation during mitosis in early cloned primate embryos has raised doubts about whether cloning by nuclear replacement will work in the human.

Embryonic Versus Adult Stem Cells

Concern about the use of early human conceptuses as a source of stem cells focused much attention on recent studies that suggest so-called adult stem cells are more versatile in their range of differentiation than has generally been supposed. There is a continuing lively debate about the interpretation of many findings, which do not at present justify the common assertion that adult cells render the use of ESL cells for therapeutic purposes unnecessary. Of particular concern is a growing body of evidence that adult cells may be changing their differentiated state not as independent entities but through fusing with cells of the type to which they are claimed to have converted.

There is a more general point that, with few exceptions — among which the hematopoietic system is the clearest example — evidence is lacking that cells from adult organs and tissue that can be propagated in culture actually functioned as stem cells *in situ*. Hence, the adoption of the term *stem cell* for cells from many adult sources is questionable. It is possible, if not likely, that cells that are strictly postmitotic in their normal environment can be induced to resume cycling when removed from it and placed in an enriched culture medium, which may contain growth factors to which they would not otherwise be exposed. Such cells might lack features of true stem cells, such as accurate proofing of DNA replication, conservation of turnover through transit amplification of differentiating progeny, and maintenance of telomere length. They might therefore be severely compromised in their ability to function in grafts.

Summary

Since the pioneering studies of Stevens and Pierce which pointed the way in the 1950s and 1960s, impressive progress has been made in harnessing stem cells of embryonic as opposed to fetal or adult origin for basic research and in exploring new approaches to regenerative medicine. There is, however, still a great deal to be learned about the origin and properties of such cells, as well as the control of their self-renewal versus differentiation, if we are to take full advantage of what they have to offer. The effort of acquiring the necessary knowledge will undoubtedly provide us with the further reward of gaining deeper insight into the biology of stem cells in general.

KEY WORDS

Blastocyst The late preimplantation stage of development when the embryo cavitates to form a hollow sphere bounded by a monolayer of trophectoderm to part of the inner surface of which is attached a disk of cells called the inner mass cell. The former contributes exclusively to placental tissues and the latter to both the fetus and placenta, plus additional extraembryonic membranes.

Chimera An animal composed of cells originating from two or more embryos produced, for example, by injecting embryonic stem cells into a blastocyst of different genotype.

Embryonic stem cell-like cells Stem cells that resemble murine embryonic stem cells, but whose potential to form all types of adults cells including germ cells has yet to be established.

Embryonic stem cells Stem cells derived from the preimplantation embryo with an unrestricted capacity for self-renewal that have the potential to form all types of adult cells including germ cells. Thus far, such cells have only been obtained from certain strains of mice.

Pluripotency The potential to form most or all types of specialized cells.

FURTHER READING

Amit, M., Carpenter, M. K., Inokuma, M. S., Chiu, C. P., Harris, C. P., Waknitz, M. A., Itskovitz-Eldor, J., and Thomson, J. A. (2000). Clonally derived human embryonic stem cell lines maintain pluripotency and proliferative potential for prolonged periods. *Dev. Biol.* **227**, 271–278.

Andrews, P. W., Przyborski, S. A., and Thomson, J. A. (2001). Embryonal carcinoma cells as embryonic stem cells. *In Stem cell biology* (D. R. Marshak, R. L. Gardner, and D. Gottlieb, eds.), pp. 231–265. Cold Spring Harbor Laboratory Press, New York.

Bradley, A., Evans, M. J., Kaufman, M. H., and Robertson, E. J. (1984). Formation of germline chimaeras from embryo-derived teratocarcinoma cells. *Nature* **309**, 255–256.

Brook, F. A., and Gardner, R. L. (1997). The origin and efficient derivation of embryonic stem cells in the mouse. *Proc. Natl Acad. Sci. USA* **94**, 5709–5712.

Edwards, R. G. (1982). The case for studying human embryos and their constituent tissues in vitro. *In Human conception in vitro* (R. G. Edwards and J. M. Purdy, eds.), pp. 371–388. Academic Press, London.

Evans, M. J., and Kaufman, M. H. (1981). Establishment in culture of pluripotential cells from mouse embryos. *Nature* **292**, 154–156.

Kiessling, A. A., and Anderson, Scott. (2003). *Human embryonic stem cells.* Jones and Bartlett, Publishers, Sudbury MA.

Marshak, D. R., Gardner, R. L., and Gottlieb, D. (2001). *Stem cell biology.* Cold Spring Harbor Laboratory Press, Cold Spring Harbor, NY.

Pierce, G. B. (1975). Teratocarcinoma: introduction and perspectives. *In Teratomas and differentiation* (M. I. Sherman and D. Solter, eds.), pp. 3–12. Academic Press, New York.

Robertson, E. J., Kaufman, M. H. Bradley, A., and Evans, M. J. (1983). Isolation, properties and karyotype analysis of pluripotential (EK) cell lines from normal and parthenogenetic embryos. *In Teratocarcinoma stem cells* (L. M. Silver, G. R. Martin, and S. Strickland, eds.), pp. 647–663. Cold Spring Harbor Laboratory Press, Cold Spring Harbor, NY.

2

The Development of Epithelial Stem Cell Concepts

C. S. Potten and J. W. Wilson

Introduction

In the 1950s and 1960s, all proliferating cells in the renewing tissues of the body were regarded as having an equal potential to self-maintain, one daughter cell on average from each division of a proliferative cell being retained within the proliferative compartment. Thus, all proliferating cells were regarded as stem cells. It proved somewhat difficult to displace this concept. Groundbreaking work by Till and McCulloch in 1961 provided the first clear evidence that for one of the replacing tissues of the body, the bone marrow, not all proliferative cells are identical. Their approach was to study the cells that were capable of repopulating hemopoietic tissues, following cellular depletion of the tissue by exposure to a cytotoxic agent, that is, radiation. Specifically, mice were irradiated to deplete their bone marrow of endogenous, functional hematopoietic precursors; then they were injected with bone marrow-derived precursors obtained from another animal. The exogenous cells were subject to a variety of treatments, prior to transplant. It was found that the hemopoietic precursors circulated in the host and seeded cells into various hemopoietic tissues, including the spleen. Those cells that seeded into the spleen and possessed extensive regenerative and differentiative potential grew by a process of clonal expansion to form macroscopically visible nodules of hemopoietic tissue, 10 to 14 days after transplant. By appropriate genetic or chromosome tracking (marking), it could be shown that these nodules were derived from single cells (i.e., they were clones) and that further clonogenic cells were produced within the clones. The colonies were referred to as spleen colonies, and the cells that form the colonies were called colony-forming units (spleen) (CFUs).

These experiments provided the theoretical basis for subsequent human bone marrow transplant studies. Through a variety of pre-irradiation manipulations and pre- and post-transplantation variables, this technique led to our current understanding of the bone marrow hierarchies or cell lineages and their stem cells. These studies showed that this tissue contained undifferentiated self-maintaining precursor cells that generated dependent lineages that were able to differentiate

down a range of different pathways, generating a variety of cell types. Recent studies have suggested that these CFUs are not the ultimate hemopoietic stem cells but are part of a stem cell hierarchy in the bone marrow.

Such clonal regeneration approaches have been subsequently developed for a variety of other tissues, notably by the imaginative approaches adopted by Rod Withers for epidermis, intestine, kidney and testis. These clonal regeneration approaches were summarized and collected in a book produced in 1985, but the field was initiated by work done by Withers. These approaches implicated hierarchical organizations within the proliferative compartments of many tissues. The stringency of the criteria defining a clone varied enormously depending as it did on the number of cell divisions required to produce the detectable clones. For epidermis and intestine, the stringency was high since the clones could be large and macroscopic, containing many cells resulting from many cell divisions. In fact, they were very similar in appearance to the spleen colony nodules.

One difficulty with the interpretation and generality of application to stem cell populations based on these clonal regeneration studies is the fact that, in order to see the regenerating clones, the tissue has to be disturbed, generally by exposure to a dose of radiation. This disturbance may alter the cellular hierarchies that one wishes to study and will certainly alter the nature (e.g., cell cycle status, responsiveness to signals, susceptibility to subsequent treatment) of the stem cell compartment. This has been referred to as the biological equivalent of the Heisenberg uncertainty principle in quantum physics. However, these clonal regeneration assays still provide a valuable and, in some places, unique opportunity to study some aspects of stem cell biology *in vivo*, that is, by using this approach to look at stem cell survival and functional competence under a variety of conditions.

A Definition of Stem Cells

Relatively few attempts have been made to define what is meant by the term *stem cells*, which has resulted in some confusion in the literature and the use of a variety of terms, the relationship between which sometimes remains obscure. These terms include *precursors*, *progenitors*, *founder cells*, and so on. The concept is further complicated by the use of terms such as *committed precursors* or *progenitors* and the sometimes confusing use or implication of the term *differen-*

tiation. One difficulty in defining stem cells is the fact that the definitions are often very context-dependent and, hence, different criteria are brought into the definition by embryologists, hematologists, dermatologists, gastroenterologists, and other specialists.

In 1990 in a paper in *Development*, we attempted to define a stem cell. This definition was, admittedly, formulated within the context of the gastrointestinal epithelium, but we felt it had a broader application. The definition still largely holds and can be summarized as follows. Within adult replacing tissues of the body, the stem cells can be defined as a small subpopulation of the proliferating compartment, consisting of relatively undifferentiated proliferative cells that maintain their population size when they divide, while at the same time producing progeny that enter a dividing transit population within which further rounds of cell division occur, together with differentiation events, resulting in the production of the various differentiated functional cells required of the tissue. The stem cells persist throughout the animal's lifetime in the tissue, dividing a large number of times; as a probable consequence of this large division potential, these cells are the most efficient repopulators of the tissue following injury. If this repopulation requires a reestablishment of the full stem cell compartment, the self-maintenance probability of the stem cells at division will be raised from the steady-state value of 0.5 to a value between 0.5 and 1, which enables the stem cell population to be reestablished while at the same time maintaining the production of differentiated cells to ensure the functional integrity of the tissue.

The consequences of this definition are obvious, namely, that stem cells are:

- Rare cells in the tissue, vastly outnumbered by the dividing transit population. and are the cells upon which the entire lineage and ultimately the tissue are dependent.
- The only permanent long-term residents of the tissue.
- Cells at the origin of any cell lineages or migratory pathways that can be identified in the tissue.

The concept of differentiation enters into the definition of stem cells, and this, too, often leads to confusion. In our view, differentiation is a qualitative and relative phenomenon. Cells tend to be differentiated relative to other cells, and hence adult tissue stem cells may, or may not, be differentiated relative to embryonic stem cells (a point of current debate, bearing in mind the controversy in the literature concerning bone marrow stem cell *plasticity*). Stem cells produce progeny that may differentiate down a variety of pathways leading to the concept of *totipotency* and *pluripotency* of stem cells in terms of their differentiation. This is actually a strange concept to apply to a stem cell since it is their progeny that differentiate and not the stem cells themselves. The fact that the progeny can differentiate down more than one differentiated lineage as is very obviously the case in the bone marrow, and results in bone marrow stem cells being referred to as pluripotent and the initial dividing transit cells that initiate a lineage that ultimately leads to specific differentiated cells, can be thought of as committed precursors for that lineage.

Some of the instructive signals for differentiation in the hemopoietic cell lineage are now well understood, but such signals for other tissues organized on a cell lineage basis have yet to be determined. There is much debate in the literature concerning the extent to which stem cells may be instructed to produce progeny of specific differentiated types and whether this is limited or unlimited. This topic is referred to as the degree of plasticity for stem cells. There are two very distinct issues here:

- The first is whether a stem cell like a bone marrow stem cell is ever instructed by its environment in nature, or in laboratory or clinical situations, to make an apparently unrelated tissue cell type such as a liver, intestinal, or skin cell and whether it can regenerate these tissues if they are injured. A subsidiary question is not whether this ever happens normally in nature, and whether we, as experimentalists or clinicians, can provide the necessary instructions or environment for this to happen in a controlled situation.
- The second issue relates not only to the stem cells but also to the early progeny of stem cells from, for example, the bone marrow, and whether these cells that circulate around the body and may end up in a distant tissue can ultimately express differentiation markers unrelated to the bone marrow cell lineages but specific to the tissue in which the cell then resides.

The former issue is one of plasticity of the bone marrow stem cells, and the latter may be more an issue of the plasticity of the bone marrow-derived cell lineages. If a bone marrow stem cell can ever be instructed to be a gastrointestinal stem cell, it should be capable of undertaking all the functional duties of a gastrointestinal stem cell, including the regeneration of the gastrointestinal epithelium if it is subsequently injured. The cloning of animals by nuclear transfer technology into egg cytoplasm clearly demonstrates that all nuclei of the body contain a full complement of DNA and that under the right environmental conditions this can be reprogrammed (or unmasked) by environmental signals to make all the tissues of the body. It should be remembered, however, that such cloning experiments, as Dolly the sheep, are rare and inefficiently produced events. They do, however, clearly indicate the enormous potential that can be achieved if we can provide the necessary instructive reprogramming signals. It should enable us in the future to reproducibly instruct any adult tissue stem cell to make any tissue of the body. If and when this becomes the case, the distinction between embryonic stem cells and adult tissue stem cells may disappear.

Hierarchically Organized Stem Cell Populations

The issue here is what determines the difference between a dividing transit cell and a stem cell, and whether that transition is an abrupt one or a gradual one. One can think of this transition as being a differentiation event that distinguishes a dividing transit cell from a stem cell. This is an old argument. Do differentiation signals act on preexisting stem cells, removing on average half the cells produced by previous symmetric divisions or, do the stem cells divide asymmetrically to produce a differentiated progeny at division and a stem cell? One possibility is that this distinction is made at the time that a stem cell divides. Indeed, do they need to divide to differentiate? In this case, such divisions must be regarded as asymmetric, with the dividing stem cell producing one stem cell (i.e., for self-maintenance) and one dividing transit cell. This type of asymmetric division may occur in some tissues such as the epidermis. If this is the case, however, the stem cell must also retain the potential to alter its self-maintenance probability, which for an asymmetric division is 0.5 in steady state, and adopt a value somewhat higher than this if stem cells are killed and require to be repopulated.

The current view regarding the bone marrow stem cells is that the transition between a stem cell and a dividing transit cell is a gradual one that occurs over a series of divisions within a cell lineage, which inevitably implies that one has a population of stem cells with a varying degree of stemness or, conversely, a varying degree of differentiation. For the bone marrow, one issue is whether experimentalists have ever identified the presence of the truly ancestral ultimate bone marrow stem cell. The difficulty here may be one of identifying and extracting such cells, the location of which is probably in the bone where they will be present in increasingly diminishing numbers, as one looks for the increasingly primitive cells.

Our current model for the gastrointestinal cellular organization, which is based on an attempt to accommodate as much experimental data as possible, is that the commitment to differentiation producing dividing transit cells does not occur at the level of the ultimate stem cell in the lineage but at a position two or three generations along the cell lineage. If such a concept is drawn as a cell lineage diagram, the proliferative units in the intestine, the crypts, each contains four to six cell lineages and, hence, four to six lineage ancestor stem cells but up to 30-second and third-tier stem cells, which under steady-state circumstances are inevitably displaced and moved toward the dividing transit compartment. But if damage occurs in one or more of the ultimate stem cells, they can assume the mantle of the ultimate stem cell and repopulate the lineage. This gives rise to the concept of actual and potential stem cells (see Figure 2-1), which is discussed later in this chapter.

An analogy can be drawn here with the hierarchical organization within an organization such as the army, a concept that was discussed at the time we were formulating the text for the development paper in which we defined stem cells. In a military battlefield environment, the hierarchically organized army is under the control and ultimately dependent upon the highly trained (or so one hopes) general. In the event that the general is killed in the battlefield, there may be a reasonably well-trained captain who can take over command and assume the insignia and uniform as well as the function of the general. In the event that the captain, too, should be killed, there may be less well-trained officers who will attempt to assume the mantle of command. Ultimately, the vast majority of the troops, the privates, would be insufficiently trained or experienced to be able to adopt the functional role of the com-

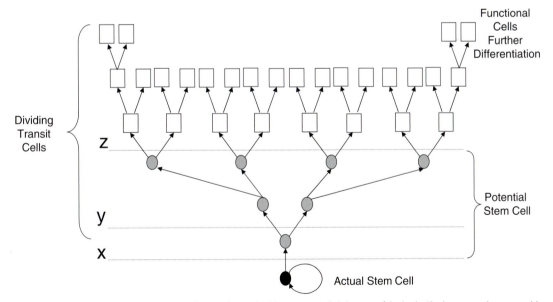

Figure 2-1. A typical stem cell-derived cell lineage that may be applicable to most epithelial tissues of the body. The lineage is characterized by a self-maintaining lineage ancestor actual stem cell (black) which divides and produces a progeny that enters a dividing transit population. The number of cell generations in the dividing transit population varies from tissue to tissue. The commitment to differentiation that separates the stem cell from the dividing transit population can occur at the point of the actual stem cell division (X), in which case the stem cells are dividing asymmetrical on average. This commitment may be delayed to point Y or Z, generating a population of potential stem cells that can replace the actual stem cell if it is killed. Under normal steady-state circumstances, the potential stem cells form part of the dividing transit population and are gradually displaced down the lineage, undergoing further differentiation events if required to produce the functional mature cells of the tissue.

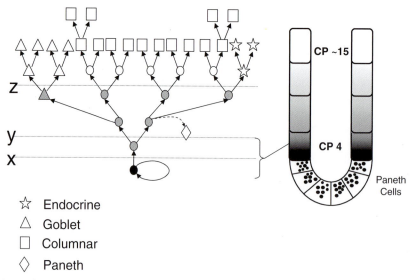

☆ Endocrine
△ Goblet
☐ Columnar
◇ Paneth

Figure 2-2. The cell lineage for the small intestinal crypts. It is postulated that each crypt contains four to six such lineages and, hence, four to six lineage ancestor actual stem cells and there are about six cell generations in each lineage with at least four distinct differentiated cell types being produced. The attractive feature of this cell biological model system is that the position of a cell in a lineage can be related to its topographical position in a longitudinal section through the crypt as shown on the right.

mander. However, the Dolly the sheep scenario suggests that occasionally a private, given a crash course in military strategy, might function as the officer in command. The analogy could be taken even further to relate to the apoptosis sensitivity that is seen in the gastrointestinal ultimate stem cells. These cells appear to adopt a strategy with complete intolerance to any genetic damage and a reluctance to undertake repair, since this may be associated with inherent genetic risk that they commit an altruistic suicide: the general who undergoes a nervous breakdown or serious injury and has to be removed from command.

In the small intestinal crypts, there have been no useful markers that permit the stem cells to be identified and, hence, studied. However, such markers are now being identified. In the absence of markers, the small intestine proved to be an invaluable biological model system to study stem cells because the cells of the intestinal cell lineage are arranged spatially along the long axis of the crypt. This can be demonstrated by cell migration tracking and mutational marker studies. As a consequence, the stem cells are known to be located at very specific positions in the tissue (crypts): the fourth–fifth cell position from the crypt base in the small intestine and at the very base of the crypt in the mid-colon of the large intestine (see Figure 2-2).

Skin Stem Cells

The first suggestion that the proliferative compartment of the epidermis, the basal layer, was heterogeneous and contained only a small subpopulation of stem cells came with the development of the skin macrocolony clonal regeneration assay developed by Withers. This was soon combined with other cell kinetic and tissue organization data to formulate the concept of the epidermal proliferative unit (EPU) (see Figure 2-3). This suggested that the basal layer consisted of a series of small, functionally, and cell lineage-related cells, with a spatial organization that related directly to the superficial functional cells of the epidermis, the stratum corneum. The concept indicated that the epidermis should be regarded as being made up of a series of functional proliferative units. Each unit had a centrally placed self-maintaining stem cell and a short stem cell-derived cell lineage (with three generations). The differentiated cells produced at the end of the lineage migrated out of the basal layer into the suprabasal layers in an ordered fashion, where further maturation events occurred, eventually producing the thin, flattened, cornified cells at the skin surface that were stacked into columns (like a pile of plates), with cell loss occurring at a constant rate from the surface of the column (Figure 2-3).

Such an organization is clearly evident in the body skin epidermis of the mouse, its ears, and a modified version of the proliferative unit can be clearly identified in the dorsal surface of the tongue. There has been, and continues to be, some debate as to whether this concept applies to human epidermis. In many sites of the human body a similar columnar organization can be seen in the superficial corneal layers of the epidermis. What is more difficult in humans is to relate this superficial structure to a spatial organization in the basal layer. However, the spatial organization seen in the superficial layers must have an organizing system at a level lower in the epidermis, and it does not seem unreasonable to assume that this is in the basal layer as is the case for the mouse epidermis.

Withers developed a macroscopic, clonal regeneration assay for mouse epidermis, which generates nodules very similar in appearance to spleen colonies. Subsequently, Al-Barwari developed a microscopic clonal assay that required a

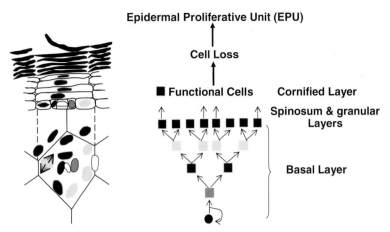

Figure 2-3. Diagrammatic representation of the cell lineage seen in the interfollicular epidermis and the relationship between the cell lineage and the spatial organization characterized as the epidermal proliferative unit (EPU), as seen in section view (upper portion of the figure on the left) and in surface view in epidermal sheets (lower portion of the figure on the left).

shorter time interval between irradiation and tissue sampling. Both techniques are fairly labor intensive and have not been used extensively. Together, these clonal regeneration assays were interpreted to indicate that only about 10% (or less) of the basal cells have a regenerative capacity (i.e., are stem cells).

The EPU stem cells must have an asymmetric division mode under steady-state cell kinetics because there is only one such cell per EPU. The epidermal microcolony assay developed by Al-Barwari suggests that following injury such as irradiation, surviving EPU stem cells can change their division mode from asymmetric to symmetric for a period of time to repopulate the epidermis (i.e., change their self-maintenance probability from 0.5 to a value higher than 0.5). Al-Barwari's observations also indicated that a significant contribution to re-epithelialization could come from the upper regions of the hair follicles. Studies on the structural organization of the epidermis following injury also made it clear that in order to reestablish the spatial distribution of stem cells, the epidermis undergoes a reorganization involving hyperplasia during which stem cells are redistributed and eventually establish their EPU spatial configurations.

The skin contains another important stem cell population, namely, that associated with the growing hair follicles. Hair is produced over a protracted period of time by rapid divisions in the germinal region of the growing hair follicle (termed an *anagen follicle*). This hair growth may be maintained for long periods of time — three weeks in a mouse (where the average cell cycle time may be 12 hours), months to years in humans, and more indefinite periods for some animal species such as Angora rabbits and Moreno sheep. This high level of cell division in the germinal matrix of the follicle, which has a considerable spatial polarity like the intestinal crypt, must have a fixed stem cell population residing in the lowest regions of the germinal matrix that can maintain the cell production for the required period of time. Very little is known about these stem cells. The complication with hair follicles is that in mouse and

human, the growing follicles eventually contain a mature hair, and cell proliferation activity ceases. The follicle shrinks and becomes quiescent (a telogen follicle). The simplest explanation here is that the telogen follicle, which consists of far fewer cells in total than in a growing follicle, contains a few quiescent hair follicle stem cells that can be triggered back into proliferation at the onset of a new hair growth cycle. However, as discussed below, there is some controversy concerning this concept.

It is now very clear that the skin contains a third stem cell compartment, which is located in the upper outer sheath of the hair follicle below the sebaceous glands. This is sometimes identifiable by virtue of a small bulge in the outer root sheath, and so this population of cells has been referred to as the bulge cells. A whole series of extremely elegant, but complicated, experiments has shown that these bulge cells possess the ability, under specialized conditions, to reform the hair follicle if it is damaged and also to contribute to the re-epithelialization of the epidermis. It is cells from this region of the follicle that were probably responsible for the epidermal re-epithelialization from follicles seen by Al-Barwari. Cells from the bulge can make follicles during development of the skin and also reestablish the follicles if they are injured.

The controversy concerns the issue of whether bulge stem cells, which are predominantly quiescent cells, ever contribute to the reestablishment of an anagen follicle under normal undamaged situations. The simplest interpretation is that these cells are not required for this process, since in order for this to happen some very complex cell division and cell migratory pathways have to be inferred. This goes somewhat against the concept of stem cells being fixed or anchored and also against the concept of keratinizing epithelia being a tightly bound strong and impervious barrier. What seems likely for the skin is that the EPU stem cell and the hair follicle stem cell have a common origin during the development of the skin from the bulge stem cells, which then become quiescent and are present as a versatile reserve stem cell population that

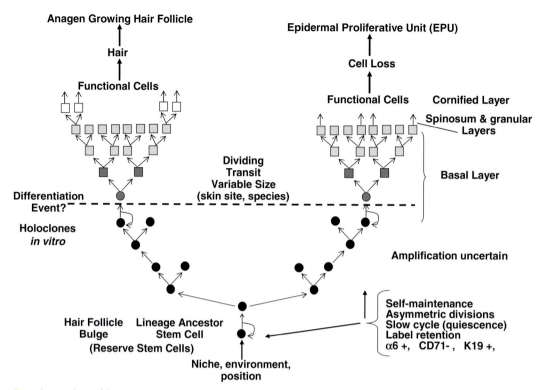

Figure 2-4. The complexity of the stem cell populations in mammalian skin as characterized in the mouse. A distinct cell lineage is proposed (a) for the interfollicular epidermis (EPU), (b) another for the matrix region of the growing hair follicle (anagen follicle), and (c) a potent reserve regenerative stem cell compartment which resides in the upper/outer root sheath or bulge region of the hair follicle. The stem cells in the bulge region can regenerate the epidermis, the hair follicle, and probably other structures such as the sebaceous glands.

can be called into action if the skin is injured and requires re-epithelialization (see Figures 2-4 and 2-5).

The Intestinal Stem Cell System

The intestinal epithelium, like all epithelia, is highly polarized and divided into discrete units of proliferation and differentiation. In the small intestine, the differentiated units are the finger-like villi protruding into the lumen of the intestine. These structures are covered by a simple columnar epithelium consisting of several thousand cells, which perform their specific function, become worn out, and are shed predominantly from the tip of the villus. There is no proliferation anywhere on the villus. The cell loss from the villus tip is precisely balanced in steady state by cell proliferation in units of proliferation at the base of the villi called crypts.

Each villus is served by about six crypts, and each crypt can produce cells that migrate onto more than one villus. The crypts in the mouse contain about 250 cells in total, 150 of which are proliferating rapidly and have an average cell cycle time of 12 hours. The cells move from the mouth of the crypt at a velocity of about 1 cell diameter per hour, and all this movement can be traced, in the small intestine, back to a cell position about 4 cell diameters from the base of the crypt. The very base of the crypt, in mice and humans, is occupied by a small population of functional differentiated cells called

Paneth cells. Cell migration tracking and innumerable cell kinetic experiments all suggest that the stem cells that represent the origin of all this cell movement are located at the fourth position from the base of the crypt in the small intestine, and right at the base of the crypt in some regions of the large bowel.

The crypt is a flask-shaped structure with about 16 cells in the circumferential dimensions. Mathematical modeling suggests that each crypt contains about five cell lineages and, hence, five cell lineage ancestor stem cells. Under steady-state kinetics, these cells are responsible for all the cell production, producing daughters that enter a dividing transit lineage of between six and eight generations in the small and large bowel, respectively (see Figures 2-1 and 2-2). The stem cells in the small intestine divide with a cycle time of approximately 24 hours and, hence, in the lifetime of a laboratory mouse divide about 1000 times. It is assumed that these cells are anchored or fixed in a microenvironmental niche that helps determine their function and behavior. The uniquely attractive feature of this model system from a cell biological point of view is that in the absence of stem cell specific markers, the behavior and characteristics and response to treatment of these crucial lineage ancestor cells can be studied by studying the behavior of cells at the fourth position from the bottom of the crypt in the small intestine. When this is done, one of the features that seems to characterize a small population of cells at

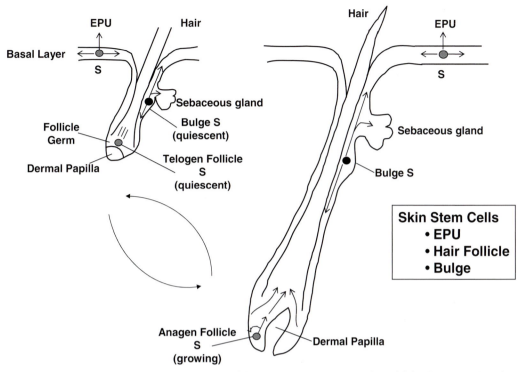

Figure 2-5. Diagrammatic representation of a growing anagen hair follicle and a resting or quiescent telogen follicle. The diagram shows the spatial distribution for the stem cell compartments shown in Figure 2–4.

this position (about five cells) is that they express an exquisite sensitivity to genotoxic damage such as are delivered by small doses of radiation. They appear to tolerate no DNA damage and activate a p53-dependent altruistic suicide (apoptosis). It is believed that this is part of the genome protection mechanisms that operate in the small intestine and account for the very low incidence of cancer in this large mass of rapidly proliferating tissue.

Clonal regeneration techniques also developed by Withers have been used extensively. These techniques suggest the presence of a second compartment of clonogenic or potential stem cells (about 30 per crypt) that possess a higher radioresistance and a good ability to repair DNA damage. These observations, together with others, suggest a stem cell hierarchy of the sort illustrated in Figures 2-1 and 2-2, with the commitment to differentiation that distinguishes dividing transit cells from stem cells occurring about three generations along the lineage. Virtually identical lineage structures can be inferred for the colonic crypts.

There has been an absence of stem cell-specific markers in the past, but some may now be available. Antibodies to Musashi-1, an RNA binding protein identified as playing a role in asymmetric division control in neural stem cells, appears to be expressed in very early lineage cells in the small intestine (see Figure 2-6).

Very recent studies have indicated that the ultimate stem cells in the crypt possess the ability to selectively segregate old and new strands of DNA at division and retain the old tem-

plate strands in the daughter cell destined to remain a stem cell. The newly synthesised strands which may contain any replication-induced errors are passed to the daughter cell destined to enter the dividing transit population and to be shed from the tip of the villus five to seven days after birth from division. Cairns developed this a concept in 1975. This selective DNA segregation process provides a second level of genome protection for the stem cells in the small intestine, protecting them totally from the risk of replication-induced errors, thus providing further protection against carcinogenic risk and an explanation for the very low cancer incidence in this tissue (see Table 2-1). This mechanism of selective DNA segregation allows the template strands to be labeled with DNA synthesis markers at times of stem cell expansion (i.e., during late tissue development and during tissue regeneration after injury). The incorporation of label into the template strands persists (label-retaining cells), thus providing a truly specific marker for the lineage ancestor cells (see Figure 2-6). Figure 2-6 also illustrates some other ways in which intestinal stem cells may be distinguished from their rapidly dividing progeny.

Stem Cell Organization on the Tongue

Oral mucosae are keratinizing, stratified epithelia, similar to epidermis in their structural organization. The dorsal surface of the tongue is composed of many small, filiform papillae that have a very uniform shape and size. Detailed histologi-

Stem Cell Identification/Responses

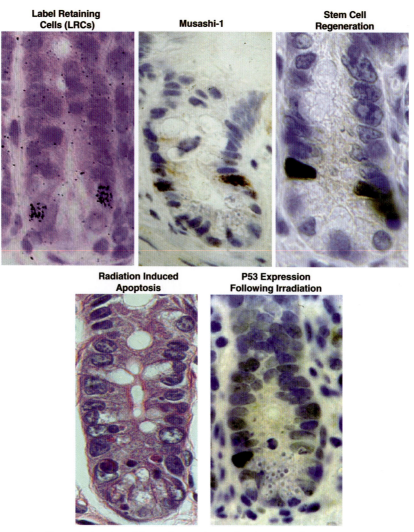

Label Retaining Cells (LRCs) **Musashi-1** **Stem Cell Regeneration**

Radiation Induced Apoptosis **P53 Expression Following Irradiation**

Figure 2-6. Photomicrographs of longitudinal sections of the small intestinal crypts from the mouse illustrating a range of possible ways of identifying the stem cell compartment. Making use of the selective strand, segregation hypothesis template strands of DNA can be labeled, generating label-retaining cells at the fourth position from the bottom of crypts. Musashi-1, an RNA binding protein, is expressed in early lineage cells and under some labeling conditions can show specificity for individual cells at around cell position 4. Part of the regenerative or potential stem cell compartment can be seen by S-phase labeling (Bromodeoyuridine labeling) at critical phases following cytotoxic injury when these cells are called into regenerative mode. The example shown here is a labeling pattern at 24 hours after two doses of 5 Fluorouracil when the only cells in S phase are a few cells scattered around the fourth position from the base of the crypt. As part of the genome protective mechanism, it is postulated that the ultimate lineage ancestor stem cells have an exquisite sensitivity to radiation and the induction of genome damage. When this happens, the cells commit suicide via apoptosis, which can be easily recognized and occurs at about the fourth position from the base of the crypt. These cells do not express p53 protein, at least at the times studied and as detectable by immunohistochemistry. However, some cells do express p53 protein at high levels following radiation exposure, and it is postulated that these are the surviving potential stem cells in cell cycle arrest to allow for repair prior to entering rapid regenerative cell cycles. Under appropriate immunohistochemical preparative procedures, individual wild-type P53 protein expressing cells can be seen at around cell position 4.

cal investigations, together with cell kinetic studies performed by Hume, showed that each papilla is composed of four columns of cells, two dominant and two buttressing columns. The dominant anterior and posterior columns represent modified versions of the epidermal proliferative units and are called tongue proliferative units. The cell migratory pathways were mapped (like the studies in the intestinal crypts), which enabled the position in the tissue from which all migration originated to be identified, this being the presumed location of the stem cell compartment. The lineage characterizing this epithelium is similar to that seen in the dorsal epidermis of the mouse — that is, self-replacing asymmetrically dividing stem cells, occurring at a specific position in the tissue, and producing a cell lineage that has approximately three generations (Figure 2-7). The stem cells here have a particularly pronounced circadian rhythm.

Generalized Scheme

For the major replacing tissues of the body, hierarchical or cell lineage schemes appear to explain the cell replacement processes. These schemes may involve isolated, single stem cells that under steady-state circumstances must be presumed to divide asymmetrically, producing a dividing transit population. The size of the dividing transit population differs dramatically from tissue to tissue, the number of generations defining the degree of amplification that the transit population provides for each stem cell division. This is related inversely to the frequency that stem cells will be found within the proliferating compartment (see Figure 2-8).

For some systems such as the bone marrow and the intestine, the commitment to differentiation that separates the

dividing transit compartment from the stem cell compartment appears to be delayed until a few generations along the lineage. This generates a stem cell hierarchy with cells of changing (decreasing) stemness or, conversely, increasing commitment, leading to the concept of committed precursor cells. In the small intestine, this delay in the commitment to differentiation to a dividing transit population provides the tissue with a reserve population of potential stem cells that can repopulate the tissue if the lineage ancestor cells are destroyed; this gives an added level of tissue protection in this extremely well-protected tissue.

With regard to the bone marrow, committed precursors, or even earlier cells, appear to circulate in the blood and may lodge in various tissues. Given appropriate microenvironments and local signals, some of these lodged cells may be instructed to differentiate down unusual pathways. This has prompted research into using such cells to repopulate the liver of patients with specific gene defects that result in life-threatening, hepatic metabolic deficiencies.

Although the transdifferentiation theory is attractive, recent research indicates that the apparent plasticity of stem cells may be less clear-cut. Transplantation experiments in mice with specific gene disorders suggest that transplanted bone marrow cells may "fuse" with liver cells and hence, complement any gene deficiency in the hepatocytes. These hybrid cells will be viable and undergo clonal expansion. Experimental findings do, indeed, show that cells forming functional liver tissue in the gene-deficient animals have specific genetic markers for both the donor and the host animal. Our concepts of stem cells clearly require further development and refinement.

TABLE 2-1

Why Do Small Intestinal Stem Cells Not Develop More Cancers?

When one considers that the tissue is:

- 3–4 times greater in mass (length)
- 1.5 times more rapidly proliferation
- 2–3 times more total stem cells
- 3–4 times more stem cell divisions in a lifetime

Compared with the large intestine:
The small intestine has 70 times fewer cancers.

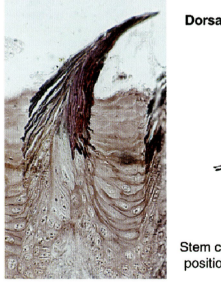
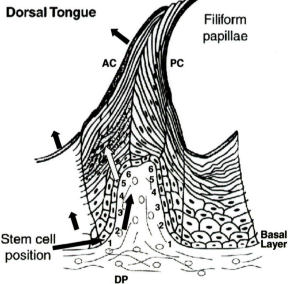

Figure 2-7. A histological section through the dorsal surface of the tongue (left panel) and a diagrammatic representation of this tissue showing the tongue proliferative units (the dominant anterior column AC, and posterior column PC). Cell migratory pathways have been identified based on cell positional analyses and cell marking and the location of the stem cells identified in the basal layer. The stem cells in this tissue express one of the strongest circadian rhythms in proliferation seen anywhere in the body.

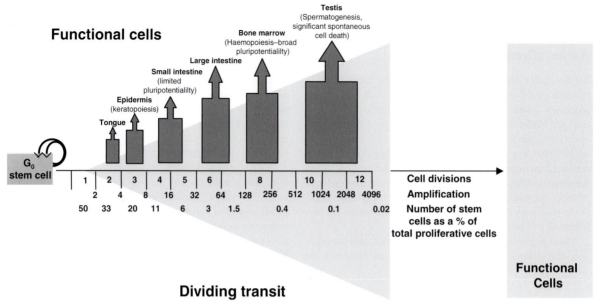

Figure 2-8. A diagrammatic representation of a stem cell-derived cell lineage showing the approximate positions for the number of cell generations in the dividing transit population for a range of murine tissues. Stratified keratinizing epithelia such as the tongue and epidermis tend to have the shortest lineages, and the bone marrow and the testis tend to have the longest lineages. Also shown is the degree of theoretical amplification that the dividing transit lineage provides for each stem cell division and the inverse relationship between the degree of amplification and the proportion of the proliferative compartment that the stem cells occupy.

Summary

Stem cell concepts have evolved dramatically over the last few years from the simple ideas in the literature in the mid-twentieth century. This has culminated in a rapid expansion of interest in both embryonic and adult tissue stem cells in the last five years with the development of interest in gene therapy and tissue engineering. This chapter explores the evolution of stem cell concepts as applied to adult epithelial tissues. These tissues are characterized by a high degree of polarization and very distinct cell maturation and migration pathways, which permit the identification of specific locations in the tissues that represent the origins of all this cell movement. Cells at the origin of the migratory pathways must represent the cells upon which the tissue is ultimately dependent and the cells that have a long-term (permanent) residence in the tissue, that is, the stem cells. A variety of cell kinetic studies, together with lineage tracking experiments, have indicated that in the intestine, the dorsal surface of the tongue, and interfollicular epidermis, the proliferative compartment of the tissue is divided into discrete units of proliferation each with its own stem cell compartment. In the skin, the evolving stem cell studies suggest at least three distinct stem cell populations providing a source of cells for the epidermis, for the growing hair follicle, and a reserve regenerative highly potent population in the upper follicle region. In the small intestine there are indications that the stem cell compartment itself is hierarchical, with a commitment to differentiation occurring two to three generations down the lineage, resulting in a population of actual stem cells that perform their function in steady state, and a population of potential stem cells that can be called into action if the actual stem cells are killed. Until recently, there have been no reliable markers for adult intestinal stem cells; however, new developments have indicated ways in which these cells may be identified. Cancer is rare in the small intestinal epithelium, which is surprising since this tissue represents a large mass with many stem cells dividing many times. This suggests that effective genome protective mechanisms have evolved, and some aspects of these mechanisms have now been identified.

KEY WORDS

Actual functional stem cells   The cells on which the tissue is ultimately dependent for day to day cell replacement.

Dividing transit cells   The amplifying cells derived from stem cells that continue to divide several times before undergoing terminal differentiation (maturation) into the functional cells of the tissue.

Epidermal proliferative unit   The functional group of proliferative basal cells derived from a single stem cell, together with the distally arranged functional differentiated cells.

Potential stem cell   Cell that retains the capacity to function fully as a stem cell if needed. Normally, these cells are displaced with time into the dividing transit population, but they retain the undifferentiated status of the ultimate stem cell until such time that they are displaced into the dividing transit populations.

Self-maintenance probability   The probability that stem cells make other stem cells on division. It applies to populations of stem cells rather than individual cells. In steady state it is 0.5, but during situations where stem cell populations expand it can be between 0.5 and 1.0.

Tongue proliferative unit A modified version of the epidermal proliferative unit identified in the filiform papillae on the dorsal surface of the tongue.

FURTHER READING

Hume, W. J., and Potten, C. S. (1976). The ordered columnar structure of mouse filiform papillae. *J. Cell Sci.* **22**, 149–160.

Marshman, E., Booth, C., and Potten, C. S. (2002). The intestinal epithelial stem cell. *Bioessays* **24**, 91–98.

Potten, C. S. (1974). The epidermal proliferative unit: the possible role of the central basal cell. *Cell Tissue Kinet.* **7**, 77–88.

Potten, C. S. (1998). Stem cells in gastrointestinal epithelium: numbers, characteristics and death. *Philos. Trans. R. Soc. Lond. B. Biol. Sci.* **353**, 821–830.

Potten, C. S. (2004). Radiation, the ideal cytotoxic agent for studying the cell biology of tissues such as the small intestine. *Radiation Research* **161**. (In press).

Potten, C. S., and Booth, C. (2002). Keratinocyte stem cells: a commentary. *J. Invest. Dermatol.* **119**, 888–899.

Potten, C. S., Booth, C., and Pritchard, D. M. (1997). The intestinal epithelial stem cell: the mucosal governor. *Int. J. Exp. Pathol.* **78**, 219–243.

Potten, C. S., and Hendry, J. H. (Eds.) (1985). *Cell clones: manual of mammalian cell techniques*, p. 235. Churchill Livingstone, Edinburgh.

Potten, C. S., and Loeffler, M. (1990). Stem cells: attributes, cycles, spirals, pitfalls and uncertainties. Lessons for and from the crypt. *Development* **110**, 1001–1020.

"Adult" Stem Cells: Tissue Specific or Not?

Catherine M. Verfaillie

Stem Cells: Definition

Stem cells are defined by the following three criteria. First, stem cells undergo self-renewing cell divisions; that is, they can give rise to at least one daughter cell that is identical to the initial cell, a characteristic required to maintain the stem cell pool. Second, stem cells undergo lineage commitment and differentiation, giving rise to more differentiated progenitors, precursor cells, and ultimately terminally differentiated cells. Differentiation is defined by acquisition of cell type specific morphological, phenotypic, and functional features. When differentiation is not restricted to a given tissue, stem cells are termed pluripotent, whereas most adult stem cells are considered multipotent; that is, they differentiate into multiple cell types that are, however, restricted to a given tissue. Third, stem cells repopulate in a robust fashion a given tissue *in vivo*. This requires that stem cells home to a given tissue, where they differentiate in response to specific cues to differentiate into cell types of that tissue that can take over the function of that tissue.

That stem cells exist in postnatal tissues has been recognized since the 1960s, with the first conceptual proof that blood or bone marrow (BM) contains cells that can rescue humans and animals from BM failure. Full characterization of hematopoietic stem cells (HSCs) was not accomplished, however, until the last decade, during which the phenotype of murine, and to a lesser extent human HSCs, was defined. Proof has been obtained that even a single murine HSC can fully reconstitute all blood cell types following transplantation in lethally irradiated animals and that progeny of such cells can reconstitute the hematopoietic system in secondary lethally irradiated recipients. HSCs therefore fulfill all characteristics of stem cells. Although HSCs are commonly obtained from postnatal tissues, such as BM and those circulating in the blood, they can also be obtained from prenatal tissues, including umbilical cord blood, BM, liver, the aorta-gonad mesonephros region, and the yolk sac. Although the degree of self-renewal may differ for cells from ontogenically earlier or later HSCs, all HSCs, regardless of ontogeny, have the same functional characteristics.

Since then, several other tissue-specific stem cells have been defined, as is described elsewhere in this book. For instance, neural stem cells (NSCs) can be found in postnatal as well as prenatal brain in a number of different neurogenic areas of the brain, including the subventricular zone and the rostromigratory pathway. Human NSCs, like HSCs, can be prospectively identified by selecting cells based on cell surface determinants, including AC133 and CD24. Culture of single AC133/CD24+ cells leads to the formation of neurospheres that differentiate into neurons, astrocytes, and oligodendrocytes. Transplantation *in vivo* in immunodeficient animals leads to differentiation to the same cell types. No data exists currently as to whether these cells can functionally reconstitute areas of the brain. Therefore, human NSCs do not totally fulfill the criteria for stem cells. However, murine or rat NSCs can functionally reconstitute at least some compartments of the brain, such as, for instance, the dopamine-producing cells in the substantia nigra, or result in remyelination in shiverer mice, caused by deletion of the animal model for demyelinating diseases.

Mesenchymal stem cells (MSCs), also termed marrow stromal cells (MSCs), have been isolated from multiple tissues, foremost from BM aspirates, but also subcutaneous adipose tissue and fetal lung. MSCs were first described by Fridensthein and colleagues as cells capable of giving rise to fibroblast-like colonies that could differentiate into osteocytes and adipocytes. Since then, the phenotype of MSCs present in bone tissue and BM and of cultured human MSCs has been elucidated, and antigenic determinants have been defined that allow selection of MSCs from rodent and human BM to almost homogeneity. Using single-cell sorting and ring cloning, investigators have shown that MSCs differentiate not only into osteocytes and adipocytes, but also chondrocytes, skeletal myocytes, and smooth muscle myocytes. In addition, grafting of MSC in animals with cartilaginous or bone defects results in engraftment and tissue-specific differentiation *in vivo*.

Usually, stem cells give rise to at least two differentiated cell types. However, some stem cells, such as endothelial and corneal stem cells, give rise to only one differentiated cell type. Recent studies have suggested that endothelial "stem cells" may persist into adult life, where they contribute to the formation of new blood vessels, in a process known during embryonal development as vasculogenesis, and not solely on angiogenesis, known as the formation of new blood vessels by sprouting from preexisting vessels. As for HSCs and NSCs, the phenotype and the *in vitro* and *in vivo* differentiation potential of angioblasts, which can be isolated from BM and peripheral blood, have been characterized.

Adult Stem Cells: Plasticity

Most tissue-specific stem cells are thus thought to be multipotent, but no longer pluripotent, like embryonic stem (ES) cells. Indeed, "adult" stem cells are generated during development beyond the stage of gastrulation. During gastrulation, pluripotent cells are fated to become mesoderm, endoderm, and ectoderm, and subsequently tissue-specific fate decisions are made. Therefore, adult stem cells have lost pluripotency and have acquired tissue-specific, restricted differentiation abilities. However, in the past few years more than 200 reports have suggested that presumed tissue-restricted stem cells may possess developmental capabilities resembling those of more immature, pluripotent cells such as ES cells. Although the recent reports of greater potency of adult stem cells have been received with great enthusiasm by the lay and scientific community, they have also met with a mostly healthy dose of skepticism. If the concept were true, this would suggest that our previous understanding of lineage commitment and restriction of differentiation potential of stem cells acquired during development may not be correct, and thus challenge the established dogmas developed in biology over the past century.

In this book, individual chapters describe in detail the findings of apparent plasticity of adult tissue-specific stem cells. It is therefore unnecessary to describe the individual studies in detail. Rather, we will here try to put the studies, pitfalls, and possible explanations for these observations in perspective.

Adult Stem Cell Plasticity: Criticisms and Pitfalls

The major criticism regarding the claim that stem cells are more potent than previously thought comes from the fact that the majority of studies describing plasticity do not fulfill the criteria commonly used to describe stem cells: (1) self-renewing single cells that (2) differentiate into functional progeny and (3) reconstitute a damaged organ *in vivo*. Furthermore, a number of potential technical difficulties in determining the donor origin of the presumed lineage-switched cells plague the interpretation of some of the findings.

CRITICISMS

To demonstrate clonality *in vitro*, several methods can be used, including limitation of dilution analysis, isolation of cells using cloning rings, single-cell deposition using either fluorescence-activated cell sorting or via micromanipulators, or retroviral marking and characterization of the viral insertion site in the host cell genome. As the chance for integration of a retrovirus in the same location in the host cell genome is less than 1/10,000–1/100,000, this method represents the most stringent assessment of clonal origin of differentiated progeny. Single-cell deposition by Flourescence activated cell sorting (FACS) or using micromanipulators is generally also considered fool-proof evidence of single-cell derivation of cell progeny, even though it is theoretically possible that two cells closely attached to one another may be co-

deposited. In contrast, isolation of single cells by limiting dilution or via cloning rings always includes the risk that two or more cells were isolated. For *in vivo* studies, single-cell transplantation or viral marking studies represent the only means to determine single-cell derivation of differentiated progeny. The majority of studies published to date have not definitively proven that the greater potency of adult stem cells can be ascribed to a single cell capable of differentiation into the tissue of origin but one or more additional studies. Therefore, they do not prove true stem cell plasticity. Notable exceptions include the study by Krause *et al.* (2001) in which a single homed HSC was shown to yield not only hematopoietic chimerism, but also cells present in epithelium of lung, liver, gastrointestinal tract, and skin. The level of contribution seen in the nonhematopoietic system by progeny of a single cell in the Krause study far exceeded that seen in a study by Wagers *et al.* (2002) in which a single HSC obtained by prospective phenotypic isolation from fresh murine BM reconstituted the hematopoietic system, but far fewer cells were detected in epithelial tissues.

Differences in the HSC used for grafting may explain these differences, in view of the notion that will be discussed later, that even highly purified stem cell populations may still be heterogeneous and represent a spectrum of different potentialities. Grant *et al.* showed that transplantation of a single HSC into irradiated recipients results not only in the reconstitution of the hematopoietic system, but also endothelial cells provided that injury was induced in the retinal capillary bed. Another example is the study by Jiang *et al.* (2002), showing that a single BM-derived multipotent cell injected in the blastocyst contributed to most, if not all, tissues of the ensuing mouse.

The nature of a differentiated cell is characterized by morphological, phenotypic, as well as functional features. The majority of studies published, however, have only shown that a cell acquires morphological and phenotypic characteristics of a novel cell type. Because cell-surface or intracellular antigenic determinants are not necessarily associated with a single-cell type, these criteria alone do not suffice to identify differentiated cells. For instance, although CD34 represents an antigen on hematopoietic cells, it has become clear that CD34 can also be found on angioblasts, or on progenitor cells from liver. Expression, moreover, is not stable, and cells commonly thought to express certain antigens may lose this expression following cell activation or cell proliferation. Spurious expression of antigens, or absorption of proteins from culture medium and, therefore possibly *in vivo* from serum or the microenvironment, in or on cells, may interfere with interpretation of lineage switch.

Better proof that differentiation in a tissue-specific manner has occurred is obtained if donor cells express a marker transgene such as Green Flourescent Protein (GFP) or β-Gal expressed from a tissue-specific promoter, as has been used to demonstrate muscle differentiation from BM cells. However, unlike the saying "it looks like a duck, walks like a duck and quacks like a duck," proof of differentiated cell function is required to claim lineage switch. Such proof has only been achieved in a few studies claiming stem cell plasticity. The first study showed that transplantation of highly enriched HSC leads to restoration of liver function in hereditary tyrosinemia type I (HT-I) mice, which have the fumary-lacetoacetate hydroylase gene deleted, which leads to liver failure unless the animals are maintained on 2-(2-nitro-4-trifluoromethylbenzoyl)-1,3-cyclohexanedione (NTBC). They demonstrated that grafting of highly enriched HSCs leads to independence of NTBC because of the generation of fumary-lacetoacetate hydrolase (FAH) expressing hepatic parenchymal cells. As mentioned earlier, Grant *et al.* showed that a single HSC can give rise to functioning endothelial cells. Likewise, there is evidence that a single BM cell differentiates into functioning endothelial cells, as well as into cells with morphological, phenotypic, and functional characteristics of hepatocytes.

Final proof for greater potency of adult stem cells is that the second cell type can robustly and functionally restore an organ *in vivo*. Some would say that this needs to be achieved in the absence of tissue damage. HSC can indeed engraft in the absence of BM damage, provided that very large doses of HSC are transplanted. Poor engraftment in the absence of irradiation is in part related to an issue of "space," that is, lack of stem cell niches available for engraftment, and may in part be due to lack of cytokine and other signals needed for robust clonal expansion of HSC. Because space and tissue-specific cues are more abundantly present in the presence of tissue damage, engraftment may occur better under circumstances of tissue damage. For instance, engraftment of HSC-derived cells in the endothelial capillary bed of the retina was only seen following vascular damage. Aside from space and proliferative signals, engraftment requires that stem cells are specifically attracted to the tissue, in a process that has been described as stem cell homing in hematopoietic transplantation. This is the result of chemokines and cytokines generated by the tissue and expression of the correct complement of adhesive ligands to which the engrafting cell can adhere, which increase when the tissue is inflamed.

Nevertheless, even in the setting of tissue damage such as lethal irradiation, or inflammatory and degenerative damage to muscle or other tissues, reported levels of engraftment of BM cells, enriched HSC, or other stem cells in tissues other than the tissue of origin have in general been low. One possibility is that the low numbers of lineage switched cells are a reflection not of plasticity at the stem cell level, but transdifferentiation of mature cells. If that is the case, plasticity will not have clinical implications. A second possibility is that the cues for clonal expansion are insufficient. Better understanding of those signals emanated by the microenvironment might then ultimately lead to clinically useful plasticity. Although significant progress has been made in understanding factors responsible for inducing self-renewal and differentiation of stem cells in their niche in lower species, such as in spermatogenesis in drosophila, the nature of the niche, and the nature of factors that govern self-renewal, lineage commitment, and differentiation of mammalian stem cells, are still unknown, even for the best characterized HSC.

TECHNICAL DIFFICULTIES

The majority of studies suggesting that plasticity exists have been done in rodent models, or using retrospective analyses of human tissues following BM or organ transplantation. Demonstration of donor origin of cells has depended on a number of rodent studies, and most human studies, on presence of the Y-chromosome in presumed lineage switched cells. Although hybridization with Y-chromosome specific probes has been widely accepted to detect donor cells in grafted animal and human tissues, when used to evaluate engraftment in tissue slices, care needs to be taken that specific hybridization is accomplished and that a signal ascribed to one cell nucleus is not the result of detection of a second nucleus in a plane underneath or above the one of that specific cell nucleus.

For rodent studies, presence of marker genes in donor cells, including β-galactosidase (β-Gal) and eGFP, has been used to demonstrate the donor origin of presumed lineage switched cells. While β-Gal as a transgene is theoretically easy to use, endogenous galactosidase present in lysosomes of mammalian cells may interfere with the specificity of donor cell detection, despite the fact that endogenous mammalian β-Gal has activity detectable at pH 6, whereas bacterial enzyme works well at higher pH, and most staining protocols readily distinguish between endogenous galactosidase activity. Although detection of GFP fluorescence is simple, because of autofluorescence of certain cell types, specific detection of GFP fluorescence may be difficult. One novel development in the use of transgenes has been that the β-Gal or the GFP gene can be expressed from tissue-specific promoters, allowing detection not only of donor origin of the cell, but also tissue-specific differentiation.

In analogy to HSC transplantation studies in which extensive use is made of the polymorphism in the Ly5 gene or the glucosephosphate isomerase gene to detect donor cells, studies describing plasticity have used donor cells expressing a cell surface marker that is lacking on recipient cells, such as CD26 on liver cells. Alternatively, one could exploit differences in major histocompatibility antigens between donor and recipient cells, although this is only applicable in the setting of allogeneic transplantation.

Adult Stem Cell Plasticity: Possible Explanations

To many, *stem cell plasticity* may be a new concept, but the idea is actually almost a century old. For instance, already in the late 1800s, it was recognized that epithelial changes occur in tissues in response to different stresses, which was termed *metaplasia*. For instance, a change from squamous epithelium to columnar epithelium due to gastric reflux occurs in Barrett's esophagus. Furthermore, numerous examples exist of lineage switch in lower species. In *Drosophila*, undifferentiated cells in imaginal discs are the precursors for legs and wings. When cells from one imaginal disc are transferred to another imaginal disc, positional identity is sometimes lost and the cells acquire the identity of the new location, a phenomenon known as transdetermination. Young Urodeles can regenerate whole limbs, thought to be the result of reactivation of specific homeobox genes in the regenerating blastema such as, for instance, MSX1. When MSX1 is expressed in murine myotubes, they undergo de-differentiation and can redifferentiate not only into myoblasts, but also into osteoblasts, chondrocytes, and adipocytes And then there is "Dolly."

What, then, are the mechanisms underlying the observed plasticity? Four plausible explanations may exist. First, there is evidence that stem cells/progenitor cells for a given organ may exist in a distant organ. It is well known that HSCs not only exist in the BM, but also circulate in the blood and can be harvested from distant organs such as muscle. Obviously, hematopoietic reconstitution following transplantation of non-purified or partially purified cells from muscle or other tissues that are contaminated with HSCs cannot be defined as plasticity.

Second, apparent plasticity can be the result of fusion between donor cells and recipient cells. That this is possible has long been established in the laboratory, since the creation of heterokaryons. This technique is commonly used for antibody production. During the last one to two years, this phenomenon has come to the forefront in the field of stem cell plasticity as fusion between cells *in vitro* and cells *in vivo* has been shown. This may explain at least some of the observations. In 2002, two independent groups showed that co-culture of ES cells with either BM cells or NSCs can lead to fused, tetraploid, and aneuploid cells that function like ES cells while maintaining expression of some of the genes from the BM cells or NSCs. The frequency of the fusion event was low ($1/10^4$–10^5), and this required considerable selectable pressure to occur. The resultant ES cells could form embryoid bodies *in vitro* and could contribute to some tissues when injected in the blastocyst. More recently, at least two groups demonstrated that the remarkable tissue replacement seen following HSC or BM transplantation into HT-I mice is in large part the result of fusion between HSC-derived monocytes and parenchymal hepatocytes, providing the hepatocytes with the missing FAH gene and allowing them to survive in the absence of NTBC. The frequency of fusion events was low ($1/10^5$ cells), and the cells only expanded *in vivo* in the setting of selectable pressure (i.e., removal of NTBC leading to the death of nonfused cells but survival of fused cells). Interestingly, the genetic program of the HSC-derived cell was silenced, and the donor nucleus had started to transcribe hepatic genes. Therefore, this demonstration of nuclear reprogramming from one cell fate to another could be considered as plasticity. However, what was not shown is that this happens at the stem cell level, and that a hepatic stem cell was generated from an HSC. Although one might consider cell fusion as one method to genetically correct tissues, the fact that the fused cells were, as was described for the fused cells *in vitro*, tetraploid and aneuploid, gives one pause.

How widespread is the fusion phenomenon in the field of stem cell plasticity? No studies have definitively shown that a donor stem cell fuses with a recipient stem cell, resulting in reprogramming at the stem cell level, with subsequent clonal growth of the fused stem cell. However, most investigators who have demonstrated stem cell plasticity are reevaluating the phenomenon to address this question specifically. As fused cells do contain more than 2N DNA, plasticity would be expected more often in tissues where multinucleated cells are common. These include multinucleated skeletal myotubes generated through cell fusion.

Although BM to skeletal muscle transdifferentiation may be caused by this phenomenon, some studies have shown that BM-derived cells can give rise to mononucleated muscle satellite cells, the progenitor for myoblasts, that subsequently fuse with resident muscle fibers. As the etiology of the cell contributing to the satellite compartment was not identified, it is not clear whether this constitutes stem cell plasticity or transfer of satellite cells present in the BM graft. Other tissues include cardiac muscle that become multinucleated by endoduplication, hepatic cells, and within the hematopoietic tissue, macrophages. It is thus possible that macrophages generated by hematopoietic cells can fuse with cells of other tissues. Then no new stem cells would be generated. Rather, one would find individual cells located outside of presumed stem cell niches in different organs, as has been described in a number of studies demonstrating plasticity.

Whether the single-donor-derived cells detected in numerous epithelia tissues where they apparently acquired characteristics of the novel tissue are the result of transdifferentiation or from cell fusion is not known. Some recent studies have elegantly shown that cells can be more definitively shown or eliminated by exploiting the ability of the recombinase gene, *Cre*, to excise DNA marked by flanking Lox-P sites. In these studies *Cre* expressed in the grafted cells from a universal promoter or a tissue-specific promoter can only activate, for instance, GFP, placed between Lox-P sites in recipient cells, if cell fusion occurs. However, again there are studies that demonstrated using the Cre-Lox system, that fusion may not explain all events of apparent lineage switch.

A third possible explanation for the observed plasticity is that different stem cell populations may be even more heterogeneous than previously thought. Over the last decade it has become clear that even the best characterized stem cell population, namely, HSCs, is more heterogeneous than previously thought. For instance, investigators have identified cell subpopulations in mouse and humans HSCs that are responsible for long-term hematopoietic repopulation, short-term hematopoietic repopulation, or cells that can repopulate the lymphoid or the myeloid lineage. It is therefore not inconceivable that aside from cells committed to long-term hematopoietic repopulation, even earlier stem cells, committed to a hemangioblast fate, a mesodermal fate, or even earlier truly pluripotent cells, may persist within the so-called HSC population, and co-purified using similar cell surface antigenic determinants. Then, differentiation would be to the hematopoietic lineage when the cell persists in its BM niche, but cells could differentiate into endothelium, other mesodermal cell lineages, or even ectodermal or endodermal cell types, when transferred into appropriate microenvironments. If that were the case, then this would constitute not plasticity, but rather persistence of more pluripotent stem cells.

The final possibility is that stem cells can truly be reprogrammed, in a manner similar to that observed during metaplasia, as is seen in lower species, or in nuclear transplantation. To prove this and to eliminate the possibility that plasticity is the reflection of heterogeneity at the stem cell level, lineage tracing of individual stem cells will be required.

Potential Use of Adult Stem Cells

What speaks most to the imagination is that stem cells can one day be used to replace defective body parts, by *in vivo* infusion, or by creation of bioartificial tissues. Although this may be possible in the future, the immediate potential benefit of stem cells lies in the fact that they constitute powerful tools to study cell self-renewal and differentiation. Investigators have used HSCs and, more recently, other defined stem cell populations and their intermediate committed progeny cells to identify the growth factor requirement of their development. This has yielded clinically used cytokines, including erythropoietin, granulocyte colony stimulating factor, keratinocyte growth factor, and many more. With the completion of the human and mouse genome projects, stem cells and their differentiated progeny can now be used to define genetic programs that need to be activated and inactivated for cell differentiation to occur, leading to further insight in their developmental programs and to the creation of both protein growth factors and small molecules that can activate such programs.

If stem cell plasticity is due to de- and re-differentiation, understanding the genetic mechanisms underlying these processes will be invaluable. "Cell reprogramming" is already underway in the clinical setting, for instance, by using demethylation agents and histone deacetlyases to reactivate fetal hemoglobin in patients with hemoglobinopathies. Insights in processes that reprogram one tissue-specific stem cell into another may then enhance the level of plasticity and make it perhaps clinically relevant.

Hematopoietic disorders and patients with other malignancies undergoing intensive chemo-/radiation-therapy have been treated for the last two to three decades with HSCs. With progress being made in defining the nature and differentiation potential of stem cell populations for other tissues, such as NSC, keratinocyte stem cells, and corneal stem cells, stem cell therapy may become a mainstay for treatment of inherited or acquired defects in these tissues. If studies indicating that adult stem cells may have greater differentiation potential can be confirmed, adult stem cells may be used to treat degenerative or genetic disorders of many more organs. Adult stem cells might then be used without prior differentiation *in vitro*, as there is so far no evidence that undifferentiated adult stem cells will cause tumor formation. Adult stem cells might also be used as autologous grafts, even though such personalized therapies may be prohibitively expensive. Furthermore, for acute illnesses, such as myocardial infarctions, or immune-based diseases, such as diabetes, allogeneic therapy may be needed, which will then require that strategies be developed to overcome immune rejection.

Summary

Until recently the dogma was that embryonic stem (ES) cells were the only pluripotent cells and that tissue specific stem cells found in postnatal life have a differentiation potential that is limited to a single organ-system. However, several recent reports have challenged this concept and have suggested that the stem cells residing in one postnatal organ can differentiate to cells of an entirely different organ-system and may thus be able to cross lineage and even germ layer boundaries. Although such greater potential has caused significant excitement in the scientific and non-scientific community, the reports have also been viewed with skepticism. This arises from several factors, including the low levels of putative trans-differentiation observed, the fact that some studies could not be repeated, and most of all the fact that this phenomenon would contradict the dogma that somatic stem cells have been lineage primed and committed early during development. Several theories have been proposed to explain the apparent plasticity of postnatal stem cells, most of which can be supported with scientific observations. In this chapter we reviewed criteria that characterize stem cells and that thus should be used to characterize plasticity of stem cells, discuss the evidence for stem cell plasticity, and then review potential mechanisms that may explain stem cell plasticity.

KEY WORDS

Cell fusion   Fusion between cells giving rise to "heterokaryons" in which genetic information of the donor cell can efface part or all of the genetic information and hence the fate of the acceptor cell.

De-differentiation and transdifferentiation   The loss of genetic information that provides cell identity with acquisition of a more primitive (de-differentiation, as in "Dolly") or different (as in apparent stem cell plasticity) cell identity.

Stem cell plasticity   Apparent ability of a stem cell/progenitor cell fated to a given tissue to acquire a differentiated phenotype of a different tissue.

FURTHER READING

Clarke, D. L., Johansson, C. B., Wilbertz, J., Veress, B., Nilsson, E., Karlstrom, H., Lendahl, U., and Frisen, J. (2000). Generalized potential of adult neural stem cells. *Science* **288**, 1660–1663.

Grant, M. B., May, W. S., Caballero, S., *et al.* (2002). Adult hematopoietic stem cells provide functional hemangioblast activity during retinal neovascularization. *Nat. Med.* **8**, 607–612.

Gussoni, E., Soneoka, Y., Strickland, C., Buzney, E., Khan, M., Flint, A., Kunkel, L., and Mulligan, R. (1999). Dystrophin expression in the mdx mouse restored by stem cell transplantation. *Nature* **401**, 390–394.

Jiang, Y., Jahagirdar, B., Reyes, M., Reinhardt, R. L., Schwartz, R. E., Chang, H.-C., Lenvik, T., Lund, T., Blackstad, M., Du, J., *et al.* (2002). Pluripotent nature of adult marrow derived mesenchymal stem cells. *Nature* **418**, 41–49.

Johansson, C. B., Momma, S., Clarke, D. L., Risling, M., Lendahl, U., and Frisen, J. (1999). Identification of a neural stem cell in the adult mammalian central nervous system. *Cell* **96**, 25–34.

Krause, D. S., Theise, N. D., Collector, M. I., Henegariu, O., Hwang, S., Gardner, R., Neutzel, S., and Sharkis, S. I. (2001). Multi-organ, multilineage engraftment by a single bone marrow-derived stem cell. *Cell* **105**, 369–377.

Lagasse, E., Connors, H., Al-Dhalimy, M., Reitsma, M., Dohse, M., Osborne, L., Wang, X., Finegold, M., Weissman, I. L., and Grompe, M. (2000). Purified hematopoietic stem cells can differentiate into hepatocytes *in vivo*. *Nat. Med.* **6**, 1229–1234.

Pittenger, M. F., Mackay, A. M., Beck, S. C., Jaiswal, R. K., Douglas, R., Mosca, J. D., Moorman, M. A., Simonetti, D. W., Craig, S., and Marshak, D. R. (1999). Multilineage potential of adult human mesenchymal stem cells. *Science* **284**, 143–147.

Spangrude, G., Heimfeld, S., and Weissman, I. (1988). Purification and characterization of mouse hematopoietic stem cells. *Science* **241**, 58.

Wagers, A. J., Sherwood, R. I., Christensen, J. L., and Weissman, I. L. (2002). Little evidence for developmental plasticity of adult hematopoietic stem cells. *Science*.

Wang, X., Al-Dhalimy, M., Lagasse, E., Finegold, M., and Grompe, M. (2001). Liver repopulation and correction of metabolic liver disease by transplanted adult mouse pancreatic cells. *Am. J. Pathol.* **158**, 1–9.

Ying, Q. Y., Nichols, J., Evans, E. P., and Smith, A. G. (2002). Changing potency by spontaneous fusion. *Nature* **416**, 545–548.

4

Molecular Bases of Pluripotency

Fatima Cavaleri and Hans Schöler

Introduction

Early mammalian embryogenesis is characterized by a gradual restriction in the developmental potential of the cells that constitute the embryo. The zygote and single blastomeres from a 2–4 cell morula are totipotent. As the embryo continues to cleave, the blastomeres lose the potential to differentiate into all lineages. The blastocyst is the first landmark of the embryo in which lineage restriction is apparent. At this stage, the outer cells of the embryo compact into the trophectoderm, from which the placenta will derive. The inner cells, termed inner cell mass (ICM), will give rise to all cell lineages of the embryo *proper* but cannot contribute to the trophoblast and thus are considered pluripotent. Once isolated and cultured *in vitro* under permissive conditions, the ICM may be propagated as an embryonic stem (ES) cell line. As a matter of fact, these cells are the *in vitro* substitutes for embryos in the search for the genetic switches and molecular mechanisms required to ensure pluripotency. Mutations affecting the ability of ES cells to self-renew or differentiate and contribute to distinct cell lineages provide the necessary tools to unravel the molecular network underlying pluripotency.

In an attempt to define the molecular basis underlying pluripotency, we will focus on four main areas:

1. Influence of extracellular factors on pluripotency and self-renewal (ligands, cytokines, receptors).
2. Signaling pathways activated in pluripotent cells (Jak-STAT and ERK cascades).
3. Gene transcriptional programs operating in pluripotent cells (mainly Oct4 and its target genes).
4. Gene function during development of the early mammalian embryo.

Cellular Models to Study Pluripotency

Three different types of pluripotent cell lines are currently available as cellular models for the study of mechanisms of pluripotency: EC, ES and EG.

EC: Embryonal carcinoma (EC) cells are historically the first pluripotent stem cells derived from mouse embryonic or fetal tissues. EC cells are self-renewing undifferentiated cells derived from teratocarcinomas —

gonadal malignant tumors containing undifferentiated cells mixed with various differentiated tissues of the three primary germ layers. Teratocarcinomas, first observed in some strain of mice as naturally developing tumors, were derived from preimplantation/pregastrulating embryos or primordial germ cells (PGCs) that had been ectopically grafted *in vivo*. EC cells can re-form teratocarcinomas *in vivo* and differentiate *in vitro* when cultured in suspension. They provide a suitable model system to study cellular commitment and differentiation, but have a major disadvantage: EC cells are tumor cells, and consequently they are typically aneuploid. Although they can integrate into a developing embryo and contribute to adult tissues, their contribution is extremely poor and lacks consistency and reproducibility. Most importantly, EC cell contribution to the germ line was shown to be a very rare event.

ES: EC cells set the technical stage for derivation and handling of ES cells a few years later. In 1981, two different scientists, Evans and Martin, reported the establishment of pluripotent cell lines from blastocyst stage mouse embryos. The pluripotentiality of these cells was considered proven by their capability to form teratocarcinoma upon subcutaneous injection into syngenic mice *in vivo* and to differentiate from embryoid bodies into tissues of all germ layers *in vitro*. Although prone to discard the Y-chromosome, ES cells are in general euploid and constitute an ideal, direct *in vitro* link to and from the embryo. In 1984, Bradley showed that pluripotent ES cells could be efficiently used to alter the mouse body composition and germ line through generations. A few years later the basis of modern mouse developmental genetics was laid down with the generation of genetically modified mice derived from ES cells that had been manipulated by retroviral or homologous recombination methods.

EG cells: In addition to EC and ES cells, a third type of pluripotent stem cells, called embryonic germ or EG cells, has been isolated from mouse. Migratory PGCs form colonies that are morphologically indistinguishable from ES colonies when grown on feeder cells in the presence of serum and a cocktail of growth factors, namely, LIF (leukemia inhibitory factor), bFGF (basic fibroblast growth factor) and SCF (stem cell factor). Like ES cells, EG cells show full developmental capacity, being able to differentiate into derivatives/lineages of all germ layers *in vitro*, form teratocarcinoma *in vivo*, and contribute to all tissues of chimeric mice, including germ line, upon injection into host blastocysts.

In summary, pluripotent murine embryo-derived ES and EG cells exhibit unique properties that make them an extremely powerful model system to unveil the molecular basis of plurypotency both *in vivo* and in *vitro*:

- Unlimited self renewal: they can be grown in large numbers and indefinitely passaged *in vitro*.
- Stable karyoype.
- Refractoriness to senescence, that is, they are virtually immortal.
- Highly efficient and reproducible differentiation potential: they can give rise to derivatives of all three embryonic germ layers *in vitro* and *in vivo*.
- Germ line colonization.
- Clonogenicity: they grow as separate colonies that can be expanded as independent subclones following genetic manipulations.
- High versatility to genetic manipulation without loss of pluripotency, as introduction of foreign DNA does not affect their ability to be fully integrated into the founder tissue of a host embryo or to colonize its germ line.

The Stem Cell Environment: Cytokines and Pluripotency

The stem cell niche: In many cases, the culture of stem cells has been complicated by the lack of knowledge of their cellular environment or niche. Hematopoietic stem cells, for example, were identified more than 40 years ago, in 1961, but conditions still have to be established to ensure their maintenance *in vitro*. In addition, they might require complicated three-dimensional arrangements of specific stromal cells in order to proliferate.

The ES cell niche: In contrast, ES cells are relatively easy to be technically handled. The establishment of the first murine embryonic stem lines was achieved by culturing early embryos on a layer of mitotically inactivated mouse fibroblasts. Without such a "feeder" layer, cultured embryonic cells would not remain pluripotent, *suggesting that fibroblasts either promote self-renewal or suppress differentiation, or both*.

LIF and other cytokines: Fibroblasts maintain pluripotency of ES cells by secreting a factor, which was identified as leukemia inhibiting factor (LIF), also known as differentiation inhibiting activity (DIA). LIF is a member of the interleukin-6 family of cytokines, including IL-6, oncostatin M (OSM), ciliary neurotrophic factor (CNTF), and cardiotrophin-1 (CT-1). The IL-6 family cytokines are structurally and functionally related. They act on a variety of cells (i.e., they are pleiotropic) and can mediate proliferation or differentiation or both according to the target cell types. For example LIF, OSM, and IL-6 are competent to induce myeloma growth and to inhibit macrophage differentiation of M1 cells. The redundancy in biological function is due mainly to the structural similarity of the receptor complex involved in signal transduction (see below).

Redundant cytokine functions and development: The absence of a developmental phenotype in IL-6-, LIF-, and CTNF-null mice confirms that IL-6-related cytokines are indeed functionally redundant. However, LIF mutant females are infertile, as the interaction between embryo and uterine wall (decidual reaction) strictly depends on a surge of estrogen on the fourth day of gestation, which coincides also with a surge in LIF production by the uterus. As a consequence, LIF −/− females fail to support embryo implantation, LIF −/− embryos can implant and develop to term in a normal uterus.

ES self-renewal dependence on cytokine supply can be attributed to several factors. LIF may influence the rate of cell proliferation or cell cycle progression and act on the stem cell phenotype by activating a signaling cascade that operates on the up- or down-regulation of genes that are exclusively expressed in "pluripotent" or differentiated cells, respectively. Analysis of the ES expression profile does not favor either one of the two above-mentioned explanations, since LIF withdrawal triggers disappearance and appearance of pluripotent and differentiated markers, respectively, within 24 h. A complete and systematic analysis of the target genes lying downstream of the LIF-induced signaling pathways is necessary to clarify the cytokine *modus operandi* on the ES cell phenotype.

Cytokine-Receptor Binding on ES Cells: Multiple Relay Stations

Structure: The receptors involved in the IL-6 family cytokine signaling cascade belong to the cytokine receptor class I family. The extracellular domain of all members of cytokine receptor class I family is composed of a variable number of fibronectin type III modules. Two of the fibronectin modules are conserved among all members of the family and constitute the cytokine-binding module. The cytoplasmatic domain of the receptor contains three conserved motifs, called box 1, box 2, and box 3 in a membrane proximal to distal order, and lacks intrinsic kinase activity. These three subdomains are responsible for transmitting the extracellular signal into the cytoplasm.

Dimerization: Binding of the IL-6 family cytokine to their cognate receptors leads to homodimerization of gp130 or heterodimerization of gp30 with the cytokine cognate receptor. For example, IL-6/IL6R and IL-11/IL-11-R complexes induce gp130 homodimerization. Both LIF and CT-1 bind to LIFR and induce LIFR/gp130 heterodimerization. CTNF engages LIFR/gp130 as a signaling competent complex through an association with CTNF/CTNFR, whereas OSM engages the LIFR/gp130 or OSMR/gp130 by binding to the gp130 portion of the heterodimers (see Figure 4-1). All the described receptor complexes share gp130 as the common component critical for signal transduction, which explains the observed redundancy in cytokine functions.

LIF and LIFR: LIF and LIFR have a reciprocal pattern of expression in the mouse blastocysts: the cytokine is expressed in the trophectoderm and the receptor in the ICM. This pattern of distribution was suggestive of a paracrine interaction

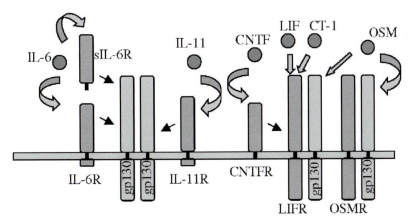

Figure 4-1. Schematic structure of the cytokine receptor complexes sharing gp130 as a common subunit. IL-6/IL6R and IL-11/IL-11-R complexes induce gp130 homodimerization. Both LIF and CT-1 bind to LIFR and induce LIFR/gp130 heterodimerization. CTNF engages LIFR/gp130 as a signaling competent complex through an association with CTNF/CTNFR, whereas OSM engages the LIFR/gp130 or OSMR/gp130 by binding to the gp130 portion of the heterodimers.

between trophoblast and ICM whereby production of LIF by trophoblast could sustain the pluripotent ICM.

Expression of cytokine receptors and ligands: Although gp130 is widely expressed in various tissues, ligand-specific receptor components display a more restricted expression. LIFR, OSMR, and CTNFR are expressed in ES cells; consequently, CT-1, OSM, and CNTF and LIF are interchangeable in preventing ES cell differentiation and supporting ES cell derivation and maintenance of ES cells in culture. IL-6 and IL-11 cannot substitute for CT-1; OSM and CNTF as the IL-6 and IL-11 receptors are not expressed in ES cells. However, IL-6 can prevent ES differentiation if delivered in conjunction with a soluble form of the IL-6 receptor, which retains ligand binding activity and capability to induce gp130 homodimerization (see Figure 4-1).

Genetic studies: LIFR-null embryos die shortly after birth, and exhibit reduced bone mass and profound loss of motoneurons. Embryos homozygous for the gp130 mutation die between 12 and 18 days postcoitum (dpc) because of placental, myocardial, hematological, and neurological disorders. CTFR-deficient mice exhibit perinatal death and display profound motor neuron deficits.

Receptor gene function in ES cells and diapause: The late embryonic lethality of the gp130 −/− fetuses is in conflict with the gp130 requirement for ES cell self-renewal *in vitro*. Delayed or quiescent blastocysts were used for the initial experiments of ES cell derivation. Lactating females can conceive while still nursing their pups but cannot support blastocyst implantation because they do not produce estrogen at the fourth day of gestation. Consequently, embryonic development is arrested and resumes under favorable conditions for optimal development. This phenomenon, termed *diapause*, can be artificially induced by ovariectomy after fertilization. The embryos reach the blastocyst stage, hatch from the zona pellucida, and float in uterus in a quiescent status for up to 4 weeks. In this scenario, the epiblast, which normally preserves its pluripotent status for about three days (from 3.5 dpc, when

it forms, up to 6.5 dpc, when gastrulation starts), can be maintained for longer time periods and resumes development when an estrogen-rich environment is established.

The possibility that cytokine receptors may then have an embryonic function in the quiescence embryo state was investigated. LIFR−/− and gp130 −/− delayed embryos are unable to resume embryogenesis after 12 and 6 days of diapause, respectively. The number of cells that constitute the ICM of delayed gp130 −/− blastocysts is gradually reduced by apoptosis during the 6-day period of diapause. Moreover, ICMs isolated from delayed gp130 null blastocysts cannot form a pluripotent outgrowth *in vitro*, as they differentiate exclusively into parietal endoderm.

Thus, it appears that maintenance of the epiblast during diapause is temporally dependent on different cytokines and that Gp130 plays a more critical role than LIFR in this process. Two models may explain why epiblast cells enter apoptosis in the absence of gp130 signaling:

1. gp130 relays a cell survival signal from the extracellular compartment to the nucleus. This model is supported by the anti-apoptotic activity of the transcription factor STAT3 in a variety of cells.
2. gp130 suppresses epiblast differentiation. In the absence of gp130 signal, epiblast cells may differentiate inappropriately (as shown by the endoderm formation solely) and consequently die.

Signal Transduction: Cascades to the Stem Cell Nucleus

Homo- or heterodimerization of gp130 results in the activation of receptor-associated kinases of the Janus family Jak1, Jak2 and Tyk2.

These tyrosine kinases constitutively interact with the conserved regions box 1 and box 2 of gp130. The receptor complex is inactive until ligand-induced receptor dimerization

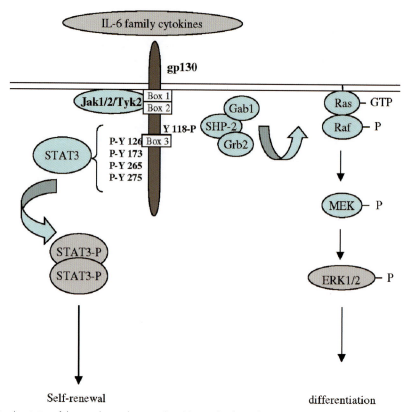

Figure 4-2. Schematic description of the signaling pathways induced by IL-6 family cytokine. Following gp130 hetero- or homodimerization, activated JAKs phosphorylate the intracellular domain of gp130 on Y126,173, 265, 275, and 118. STAT3 association with phosphorylated Y126–275 leads to STAT3 phosphorylation, dimerization, and translocation to the stem cell nucleus. Association of SHP-2 with phosphorylated Y118 leads via adaptor proteins to activation of the Ras pathway and translocation of ERK1/2 to nucleus. STAT3 activation induces ES self-renewal, whereas ERK activation causes cell differentiation.

brings the associated Jak kinases within sufficient proximity to allow *trans*phosphorylation and activation of the kinase catalytic domain. Activated Jaks phosphorylate specific tyrosines on the intracellular domain of gp130, creating docking sites for the recruitment of SH2 proteins to the activated receptor complex. When gp130 is phoshorylated, several signaling pathways are activated, involving STAT1 and STAT3, the SH2-domain containing tyrosine phosphatase (SHP2), ERK1 and ERK2 (extracellular signal receptor kinases or mitogen-activated kinases (MAPK), growth-factor receptor-bound protein (Grb) 2, Grb2-associated binder protein (Gab) 1, and phosphatidylinositol-3 kinase (PI3K) (see Figure 4-2).

STAT: Latent Transcription Factors Transmitting Signals

The STAT family: STAT proteins belong to a group of latent cytoplasmic transcription factors that play a central role in transmitting signals from the membrane to the nucleus, hence their name (signal transducers and activator of transcription). Seven major STAT proteins have thus far been identified in mouse (STAT1 to STAT6, including STAT5a and STAT5b).

With the exception of STAT4, which is restricted to myeloid cells and testis, STAT factors are ubiquitously expressed. They are activated in many cell types by a broad range of cytokines, growth factors, and interferons (IFNs), and they are substrates for tyrosine kinases of the Src and Jak families.

Structure: STAT proteins share several conserved structural and functional domains. A tetramerization and a leucine zipper-like domain are located at the amino terminus, followed by a DNA binding domain, a Src homology domain 3-like region (SH3, proline rich motif binding domain), a Src homology domain 2 (SH2), a critical site of tyrosine phosphorylation (Y705 in STAT3), and a carboxy-terminal transactivation domain. No evidence has emerged so far to suggest an SH3 function.

The SH2 domain plays three important roles:

- Recruitment to activated receptor complexes.
- Interaction with Jaks.
- STAT dimerization and DNA binding.

Regulation: The regulation of STAT signaling is mostly post-translational and involves both tyrosine and serine phoshorylation.

Phosphorylation of the conserved tyrosine (Y 701 for STAT1, Y705 for STAT3) results in dimerization of STAT1/STAT3 through the intermolecular interaction of the SH2 domains and the domain containing the phosphorylated tyrosine. STAT1 and STAT3 homo- and heterodimers translocate to the nucleus, where they activate gene transcription by binding to specific DNA sequences. Consistent with the requirement for dimerization-induced activation, the consensus binding sites are symmetrical dyad sequences. In order to achieve maximal transcriptional activity, the C-terminal transactivation domain of both STAT1 and STAT3 requires phoshorylation at serine 727 by MAPK family members, suggesting a cross talk between MAPK and JAK/STAT pathways.

STAT3: STAT3 was originally identified as an acute response factor which, upon IL-6-family cytokine stimulation, induces the expression of a variety of genes, referred to as *acute response genes*, whose expression dramatically increases with tissue injury and inflammation.

Targeted disruption of STAT3 gene *in vivo* leads to embryonic lethality. STAT3 null embryos develop into elongated egg cylinder but degenerate at around E7.0. At this stage wild-type embryos start to express STAT3 in the visceral endoderm. Embryonic lethality is then explained as a consequence of the failure in establishing metabolic exchanges between embryo and maternal blood. STAT3 thus plays a unique crucial role during embryonic development that cannot be compensated by other members of the STAT family.

LIF and STAT3: The initial studies conducted on LIF-dependent transcriptional activation in embryonic stem cells showed induction of a DNA binding activity that correlated with LIF treatment (Hocke *et al.*, 1995) (Boeuf *et al.*, 1997). Steady-state levels of STAT1, 3, 5, and 6 were unaffected by LIF treatment. In contrast, co-immunoprecipitation experiments indicated STAT3 as being a component of the tyrosine-phosphorylated complex formed upon LIF induction. In addition, tyrosine kinase inhibitors were shown to impair formation of the activated STAT3 complex and to alter undifferentiated ES cell morphology versus a differentiated phenotype.

STAT3: ES cell renewal and the undifferentiated state: Identification of STAT3 as a key determinant of ES renewal came from the elegant studies conducted in A. Smith's lab and T. Yokota's lab. Granulocyte colony-stimulating factor receptor (G-CSF-R) belongs to class I cytokine receptor family and is not expressed in ES cells. In order to characterize the functional role of the receptor intracellular domains or residues in signal transduction, chimeric receptors constituted of the extracellular domain of G-CSF-R fused to the transmembrane and cytoplasmic region of gp130 or LIFR were engineered. G-CSFR/gp130 and G-CSF-R receptors can support ES self-renewal at higher and lower efficiency, respectively, whereas the G-CSF-R/LIF-R chimera cannot support formation of stem cell colonies. This result suggests that gp130 is an essential component in signaling self-renewal in ES cells and that G-CSF-R and gp130 intracellular domains activate common signaling pathways.

The intracellular domain of gp130 harbors four consensus motifs **YXXQ** whose phosphorylated tyrosines were shown to function as docking sites for STAT3. Single or double mutation of these tyrosine residues (Y 126, 173, 265, and 275, enumeration starting from the transmembrane domain) did not affect self-renewal appreciably (see Figure 4-2). Rather, mutations of Y265/275 completely abolished STAT3-induced binding activity and formation of undifferentiated colonies. These data demonstrate that STAT3 docking sites are essential in mediating transmission of the signal from gp130 to STAT3 in self-renewing ES cells, although the tyrosine residues are not functionally equivalent. Moreover, conditional expression of STAT3F, containing a phenylanine substituting for Tyrosine 705, caused complete differentiation of ES cells even in the presence of LIF.

In the work from Matsuda, STAT3 activation was shown to be sufficient to maintain the undifferentiated status of mouse embryonic cells. In order to uncouple STAT activation from any signaling pathway induced at a membrane receptor by extracellular factors, the authors constructed a fusion protein composed of the STAT3 entire coding region and the ligand-binding domain of the estrogen receptor. The STAT3ER chimeric protein was specifically tyrosine 705 phosphorylated in the presence of the synthetic steroid ligand, 4-hydroxytamoxifen (4HT). Parental ES and ES expressing STAT3ER were grown in the presence of LIF or 4HT. The compact colonies formed by STAT3ER ES in the presence of either LIF or 4HT demonstrated that STAT3 activation is sufficient to maintain the undifferentiated phenotype of ES cells. Most importantly, besides being morphologically undifferentiated, cells grown for one month in 4HT widely contributed to all tissues of chimeric mice.

STAT3 independent ES cell renewal: The striking paradox created by STAT3- or gp130-null phenotypes and the strict dependence on gp130-STAT3 signaling pathways for maintenance of pluripotent ES cells is suggestive of the existence of alternative pathways governing pluripotency *in vivo*. Dani *et al.* presented some evidence of STAT3 independent signaling pathways operating in self-renewing ES cells. When ES cells are grown at high density in the absence of LIF, the newly differentiated cells start synthesizing LIF, which in turn allows the expansion of undifferentiated cells. In order to eliminate the LIF-dependent pathways of self-renewal, a LIF-deficient ES cell line was generated. Strikingly, LIF−/− cells can still produce undifferentiated colonies, though at lower efficiency than wt or heterozygous lines when induced to differentiate. ES cell renewal factor, or ESFR, can support the pluripotential character of ES cells upon blastocyst injection. Interestingly, ESFR does not operate via LIFR or gp130 because it is effective on LIFR-deficient ES cells and is not blocked by anti-gp130 antibodies. STAT3 is not induced in the presence of ESRF.

SHP-2/ERK Signaling

STAT 1 and 3 are only two of the downstream effector molecules induced by cytokine signaling via gp130. LIF treatment of ES cells increases MAP kinase activity and induces phosphorylation of ERK1 and ERK2. The bridging factor between cytokine receptor and MAP kinase is the widely expressed tyrosine phosphatase SHP-2.

SHP-2 structure: SHP-2 contains two N-terminal SH2 domains and a C-terminal catalytic domain. SHP-2 interacts with the intracellular domain of Gp130 through phosphorylated Tyrosine 118, located inside the consensus YSTV sequence. Recruitment to the activated receptor induces SHP-2 phosphorylation, which leads to an increase in phosphatase activity by preventing an intramolecular interaction between the SH2 and the catalytic domain of SHP2. In BAF-B03 pro-B cells, mutation of gp130 Tyrosine 118 into phenylalanine was shown to block SHP-2 phosphorylation and induction of ERK2 activation.

SHP-2-ERK signaling and ES cells: G-CSFR responsive ES cells provided an excellent tool for establishing the role played by the SHP-2-ERK signaling in ES cell propagation.

A mutant G-CSFR-gp130 chimeric receptor that cannot be phoshorylated on Y118 (and that consequently cannot induce SHP-2 phosphorylation) does not impair the self-renewal of ES cells (see Figure 4-2). In contrast, the mutated chimera makes ES more sensitive to G-CSF, as they can be maintained at a lower concentration of G-CSF (1000-fold less) than the one required by ES cells expressing the unmodified receptor. Moreover, the signal started at the phosphorylated Y118 mediates attenuation of activated STAT3, as shown by slower decay of phosphorylated STAT3.

Enhanced ES self-renewal is observed when either a catalytically inactive SHP-2 is overexpressed or ERK phosphorylation is chemically blocked. These results indicate that SHP-2 and ERK activation is not required for the maintenance of self-renewal signaling, but rather they inhibit it. This conclusion is confirmed by enhanced LIF sensitivity and increased proliferation rates observed in ES cells, and embryoid bodies derived thereof, expressing a ΔSH2-SHP-2 protein. However, cardiac/epithelial differentiation of SHP-2 mutant cells is inhibited and delayed, indicating that SHP-2 plays a positive role in ES differentiation. In conclusion, ES cell self-renewal is a consequence of the precise balance of antagonistic signaling pathways.

SHP2 communicates with ERKs: Activated ERKs undergo nuclear translocation, which enables them to modulate the activities of transcription factors that govern proliferation, differentiation, and cell survival. SHP-2 interaction with two adapter molecules, Grb2 and Gab1, provide two alternative routes that couple cytokines to the activation of ERK pathways.

SHP-2/Grb2 association leads to Ras activation through the GTP-GDP exchange protein SOS. SHP-2 has also been shown to associate with the scaffold Grb2-associated binder protein Gab1 and PI3-kinase and activate ERK MAP kinase via a Ras dependent pathway.

Grb2 and Gab1 adapter proteins: Like most adapter proteins, Grb-2 contains SH2 and SH3 domains. The SH2 domain mediates Grb2 binding to SHP-2 while the SH3 domain mediates interaction with SOS. Tyrosine phosphorylation of Gab1

following stimulation with cytokines depends on the gp130 site (Y118) docking SHP-2, but is independent from the gp130 C-terminal domain interacting with STAT3. Gp130-mediated ERK2 activation is enhanced by Gab1 expression and inhibited by a dominant negative Ras.

Functional ablation of Grb2 in ES cells: Null mutation of the Grb2 gene leads to embryonic lethality at around 7.5 dpc. The differentiation potential of Grb2-null ICM cells is compromised as cultured blastocysts lack either visceral or parietal endodermal cells. The ability of Grb2 to support endoderm differentiation is abrogated by mutation in the SH2 or SH3 domain. Interestingly, transformation of Grb2-deficient ES cells with an activated variant of Ras restores endoderm differentiation, indicating that the Grb2-Ras pathway is essential for early specification of the endoderm tissues.

The Consequences of LIFR/gp130 Interaction at the Gene Expression Level

The phenotype of a particular cell type is the culmination of a well-defined and unique pattern of gene expression characterized by the activation of some genes and repression of others. Therefore, elucidation of the molecular basis of a pluripotency is based primarily on the identification of the key transcription factors involved in regulating gene expression at the pluripotent "state." In a very simplistic model, genes that promote and maintain an undifferentiated cellular state would be expressed in pluripotent cells, while those activated during stem cell differentiation would be repressed. Conversely, in terminally differentiated cells, pluripotency-related genes would be silenced, whereas those required for the differentiated cell state would be expressed. Elements that influence the transcriptional regulation of gene expression include epigenetic modifications of the genome, such as methylation and histone deacetylation, which determine what types of interactions are allowed between DNA and transcription factors and modulating cofactors.

Oct4 Is a Key Transcription in Pluripotency

Oct4 is a transcription factor belonging to the class V of POU factors. The POU family of transcription factors binds to the octamer motif ATGC(A/T)AAT found in the regulatory domains of cell type-specific as well as ubiquitous genes. POU factors have a common conserved DNA binding domain, called the POU domain, which was originally identified in the transcription factors Pit-1, Oct-1, Oct-2, and Unc86. The POU domain is comprised of two structurally independent subdomains — the POU specific domain (POU$_S$) and the homeodomain (POU$_H$) — connected by a flexible linker of variable length.

The POU domain of Oct4 is characterized by several properties that confer the Oct4 protein an impressive versatility in the operational mode of transcriptional regulation:

- Flexible amino acid-base interactions allowing recognition of moderately variable cognate DNA elements.
- Variable orientation, spacing, and positioning of DNA-tethered POU subdomains relative to each other on the DNA, as evidenced by the different arrangement on the palindromic Oct factor recognition sequences.
- Cooperative binding of the two subdomains to the DNA.
- Interaction with other transcription factors and regulatory modulators.
- Post-translational modifications modulating Oct4 transactivation in various cell types.

The POU domain binds to DNA via interaction of the third "recognition" helix of the POU_H with bases in the DNA major groove at the 3′A/TTTA rich portion of the octamer site. The POU_H domain is structurally similar to other homeodomains. The POU_S domain exhibits site-specific, high-affinity DNA binding and bending capability. Both the POU_S and POU_H subdomains function as structurally independent units with cooperative high-affinity DNA-binding specificity. Functional cooperation between the two subdomains may occur indirectly via the DNA by overlapping base contacts from the two subdomains.

In addition to the DNA binding function, both the POU_H and POU_S subdomains can participate in protein–protein interactions. In ES cells, Oct4 activates gene transcription regardless of the octamer motif distance from the transcriptional initiation site. However, Oct4 can transactivate only from a proximal location in differentiated cells. In this scenario, interaction between the adenovirus protein E1A or human papillomavirus E7 oncoprotein and Oct4 POU domain is sufficient for Oct4 to elicit transcriptional activation from remote binding sites. E1A and E7 proteins would therefore mimic unidentified embryonic stem cell specific coactivators that serve a similar function in pluripotent cells.

Two domains spanning the N- and C-terminal portion of Oct4 protein define the transactivation capacity of the POU transcription factor. The N-terminal region (N domain) is a proline and acid residue-rich region, whereas the C-terminal domain (C domain) is a region rich in proline, serine, and threonine residues. The N-terminal domain can function as an activation domain in heterologous cell systems. However, the C-terminal domain of Oct4 exhibits a POU domain mediated cell type-specific function. Intramolecular interactions between the POU domain and C-terminal domain may lead to cell type-specific interactions with different cofactors or kinases. An interesting regulation model of the C-domain predicts that association between the Oct4 POU domain and other factors can alter the phosphorylation status of the protein and ultimately modulate the activity of the C-domain.

OCT4 Expression

IN THE MOUSE

Oct4, otherwise designated *Oct3*, or *POU5F1*, is a maternally inherited transcript that is developmentally regulated in mice. It is expressed at low levels in all blastomeres until the 4-cell stage at which time the gene undergoes zygotic activation resulting in high Oct4 protein levels in the nuclei of all blastomeres until compaction. After cavitation, Oct4 is maintained only in the inner cells (ICM) of the blastocyst and is down-regulated in the differentiated trophectoderm (TE). Following implantation, Oct4 expression is restricted to the primitive ectoderm (epiblast), although it is transiently expressed at high levels in cells of the forming hypoblast (primitive endoderm). During gastrulation, starting at 6.0–6.5 dpc, Oct4 expression becomes down-regulated in the epiblast in an anterior-posterior manner. From 8.5 dpc on, Oct4 becomes restricted to precursors of the gametes or primordial germ cells (PGCs). Oct4 is also expressed in undifferentiated mouse embryonic stem (ES), embryonic germ (EG), and embryonic carcinoma (EC) cell lines. ES and EC cell treatment with the differentiation-inducing agent retinoic acid (RA) induce rapid Oct4 down-regulation in both cell types. Oct4 is not expressed in differentiated tissues. Recently, a very low amount of Oct4 messenger was detected in multipotent adult precursor cells (MAPCs) and human breast cancer cells.

IN HUMANS

Oct4 is highly expressed in ICM cells relative to TE cells in discarded human embryos. It is also expressed in pluripotent human EC and ES/EG lines.

OCT4 Activity in Assays

Gene inactivation experiments indicate that Oct4 plays a determinant role in the specification of mouse pluripotent cells. However, Oct4 function is not confined to pluripotent cells. Extrapolation of the data obtained by manipulating Pou5f1 expression in ES cells indicates that Oct4 is a master regulator of cell fate in all three tissues of a preimplantation embryo.

EMBRYOS WITH NO OCT4

Oct4-deficient embryos die at the peri-implantation stage and form empty decidua (or implantation sites) that contain trophoblastic cells but are devoid of yolk sac or embryonic structures. The embryos develop up to the 3.5 dpc blastocyst stage, with an unaltered number of cells in the ICM or trophoblastic compartment compared with wild-type embryos, suggesting that in the absence of Oct4 protein, during early stages of development, cell proliferation is not affected. *In vitro* cultures of cells immunosurgically isolated from the inner region of Oct4 −/− 4 dpc blastocysts contain trophoblastic giant cells but do not contain a pluripotent outgrowth or extra-embryonic endoderm. These results indicate that the ICM of Oct4-deficient blastocyst is defined only by the stereo arrangement of not-pluripotent cells, which are diverted to a trophoblastic phenotype. These findings are consistent with the concept that Oct4 is essential in the establishment of the ICM pluripotency.

ES CELLS WITH LOW AND HIGH LEVELS OF OCT4

Based on Oct4 expression levels during the first two events of lineage commitment in the early embryo, it is not entirely surprising that a critical amount of Oct4 has recently been reported to be crucial for the maintenance of ES cell self-renewal.

ES cells containing a bacterial genetic system for the conditional expression and repression of Oct4 transgene allowed researchers to determine the cellular phenotype linked to variable levels of Oct4 protein. Briefly, ES cells in which one or both endogenous *Oct4* alleles had been inactivated were transfected with constructs for a tetracycline-regulated transactivator and a responsive *Oct4* transgene, whose precise level of expression was modulated by variable amounts of the antibiotic.

A 50% decrease in the endogenous Oct4 levels relative to that of undifferentiated ES cells resulted in the commitment of ES cells to trophoectoderm lineages, containing both proliferating and endoduplicating giant cells based on the culture conditions. This result is consistent with the phenotype of Oct-4–/– embryos and the observation that trophoblastic differentiation of outer cells of the morula is accompanied by Oct4 down-regulation.

However, an increase beyond the 50% threshold level of Oct4 leads to the concomitant differentiation of ES cells into extra-embryonic endoderm and mesoderm. Interestingly, LIF withdrawal leads to the specification of the same lineages. Less subtle changes in Oct4 level (both increase or decrease) do not affect ES cell self-renewal.

In conclusion, the precise level of Oct4 protein governs commitment of embryonic cells along three distinct cell fates (self-renewal, trophectoderm, or extra-embryonic endoderm and mesoderm).

The conditional Oct4-null ES line was used in a complementation essay to establish which domains of the Oct4 protein are sufficient to maintain ES cells. The complementation assay was based on the ability of a proper Oct4 molecule to rescue the self-renewal capability of cells that would otherwise differentiate due to Oct4 down-regulation. Oct4 is the only POU protein that has the ability to rescue the self-renewing phenotype, as Oct2 or Oct6 has no effect on cell fate in this system.

A truncated Oct4 protein containing the Oct4 POU domain and either the N- or C-terminal domain can support ES cell self-renewal as the wild-type counterpart. Furthermore, gene expression analysis revealed that Oct4 transactivation domains, though equivalent in sustaining the undifferentiated stem cell phenotype, elicit activation of different target genes. It would be interesting to determine the consequences of either the N or C domain deletion on pluripotency and embryo development *in vivo*.

Regulation of Oct4 Expression

Use of *Oct-4/LacZ* transgenes has allowed the identification of two distinct enhancer elements reciprocally driving the cell type-specific expression of *Oct4*. The distal enhancer or DE, located approximately 5 Kb upstream the promoter, regulates Oct4 expression in preimplantation embryos (morula and ICM), PGCs, and in ES, F9 EC, and EG cells. However, the DE is inactive in cells of the epiblast. The proximal enhancer or PE, located approximately 1.2 Kb upstream, directs Oct4 expression in the epiblast, including *Oct4* down-regulation in the anterior to posterior direction after gastrulation, and in P19 EC cells, including RA-dependent *Oct4* down-regulation (see Figure 4-3). Two similar inverted elements, the 2A site of DE and the 1A site of the PE, are bound by transcription factors

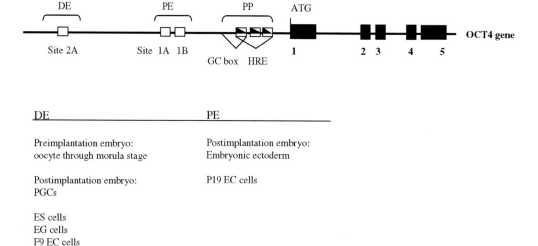

Figure 4-3. Genomic structure of the *Oct4* gene. Open boxes indicate the conserved regions included in the proximal and distal enhancer. Black boxes indicate the five exons of the gene. The cell specificity conferred by the two enhancers is also depicted.

in vivo in undifferentiated ES and EC cells. Upon RA-induced differentiation, protection of these two areas is lost. These data suggest that these elements are involved in regulating *Oct4* transcription probably through local signals and/or cell type-specific factors.

The Oct4 gene comprises a TATA-less promoter containing a GC-rich Sp1-like sequence and three hormone response element half sites — (HRE) R1, R2, and R3.

Oct4 and SF1 (steroidogenic factor-1) expression patterns are inversely correlated to the germ cell nuclear factor (GCNF) expression pattern in P19 cells. Both SF1 and GCNF are orphan nuclear receptor. During RA-induced differentiation of P19 cells, SF1/Oct4 down-regulation is followed temporally by GCNF up-regulation first and then by induction of the orphan receptors ARP-1/COUP-TFII and EAR-3/COUP-TF-I, which act as negative regulators of Oct4 promoter-driven transcription. SF1 binding to a site overlapping the R3 and part of the flanking R2 repeat in undifferentiated P19 cells can activate the Oct4 promoter in synergism with retinoic acid receptor (RAR). GCNF binds to the R2 repeat *in vivo* and represses transcription driven by the proximal enhancer-promoter (PEP) in P19 cells (see Figure 4-3).

Analysis of GCNF-deficient embryos has emphasized the impact of GCNF-induced repression of the Oct4 gene. Low GCNF expression is detected in the whole mouse embryo at 6.5 dpc. At 7.5 dpc, increasing GCNF and decreasing Oct4 mRNA levels are observed in neural folds and at the posterior of the embryo. As mentioned above, at the end of gastrulation Oct4 is maintained only in PGCs at the base of the allantois. In 8.0–8.5 dpc GCNF-deficient embryos, Oct4 mRNA is detected in the putative hindbrain region and posterior of the embryo, which indicates that loss of GCNF leads to loss of Oct4 repression in somatic cells and loss of GCNF-induced restriction of Oct4 in the germ line. The same phenotype is observed in GCNF-deficient mice containing a targeted deletion of the DNA binding domain of GCNF. These findings thus implicate GCNF in the restriction of the mammalian germ line and embryonic stem cell potency.

DNA methylation constitutes an important mechanism of gene expression regulation during embryogenesis. A global loss of DNA methylation occurs during cleavage prior to the 16-cell stage. Following implantation, a wave of *de novo* DNA methylation occurs in all genes, except those containing CpG islands. *In vivo* analysis of the PEP region of the *Oct4* gene reveals that *Oct4* is not methylated from the blastula stage up to 6.25 dpc, a time during which other genes undergo *de novo* methylation. Interestingly, a cis-specific demethylation element exists within the *Oct4* PE. Mutational analysis shows that site 1A within PE is involved in preventing methylation. The presence of this cis-specific demodification element causes demethylation of sequences that had been methylated prior to their introduction into EC cells and protects from *de novo* methylation *in vivo*. Interestingly, methylated PE sequences induce a decrease in transcriptional activation both of a reporter gene and the endogenous *Oct4* in P19 cells. According to the model proposed by Gidekel and Bergman, binding of trans-acting factors to the PE element offers protection against *de novo* methylation, which can begin soon after gastrulation, when these factors are down-regulated. At the end of gastrulation, these factors would be present only in the germ cell lineage, the sole lineage to express Oct4 by 8.5 dpc.

Target Genes

A few putative Oct4 target genes have been identified to date and are briefly discussed below (see Figure 4-4). Human chorionic gonadropin (hCG) is required for implantation and maintenance of pregnancy. HCG is secreted by the trophectoderm of peri-implantation blastocyst. Oct4 has been shown to silence the expression of α and β subunit genes of hCG in human choriocarcinoma cells. Oct4 binding to a unique octamer motif (ACAATAATCA) in the hCGβ-305/-249 promoter considerably reduces both hCG β mRNA and protein levels in JAr choryocarcinoma cells. Although Oct4 is a potent inhibitor of hCGα expression, no octamer binding site has been identified in the promoter region of hCGα.

Like CG in humans, tau interferon (IFN-τ) is needed to prevent regression of the maternal corpus luteum during the early stages of pregnancy in ruminant species. In cattle,

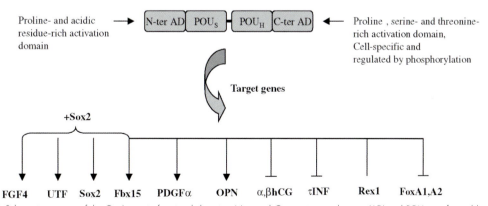

Figure 4-4. Schematic structure of the Oct4 protein functional domains: N-ter and C-ter activation domains (AD) and POU specific and homeodomains (POUs and POUH). Also shown is the Oct4 effect on the transcription of its target genes.

members of the multigene IFN-τ family are expressed in the TE from the blastocyst stage up to the beginning of the placentation process. Expression of IFN-τ gene is activated from an Ets-2 enhancer located in the promoter region at −79/−70. IFN-τ repression by Oct4 is specific, as neither Oct1 nor Oct2 interferes with Ets-2-induced activation of IFN-τ promoter in JAr cells. The mechanism of repression is based on protein–protein interactions between the Oct4 N-POU domains and the region localized between the activation and the DNA binding domain of Ets-2. These data, taken together with the observation that reduced Oct4 protein level triggers differentiation of murine ES cells into trophoectoderm, suggest that silencing of *Oct4* in the TE is a prerequisite for up-regulation of hCGs and IFN-τ.

The *Rex1 (Zfp-42)* gene, which encodes a zinc finger protein, is expressed at high levels in ES and EC cells, and is down-regulated upon RA treatment. Both processes are mediated by the octamer motif located in the promoter region of Rex1. Low levels of Oct4 protein are sufficient to activate Rex1 promoter in P19 RA-differentiated cells, whereas high Oct4 or Oct6 level inhibit transcription in F9 cells. Distinct Oct4 protein domains elicit the observed effects, suggesting different molecular mechanisms of Oct4 mediated transcriptional activation and repression.

Expression of the platelet-derived growth factor α receptor (PDGFαR) in undifferentiated human EC Tera2 cells depends on a canonical octamer motif within the gene promoter. Mutation of the octamer site has been shown to decrease the promoter activity and PDGFαR expression.

Oct4 Does Not Play *Solo*

Regulation of the FGF4 and UTF genes has provided insight into an Oct4-induced activation model based on parternship with the transcription factor Sox2.

Sox2 belongs to the *Sox* (Sry-related HMG box-containing) family of proteins that bind to DNA through the 79-amino acid HMG domain (High Mobility Group). In contrast to most DNA binding proteins, which access to DNA through the major groove, the HMG box interacts with the minor groove of the DNA helix and, as a consequence, induces a dramatic bend in the DNA molecule. As a result of this energetically high-cost interaction, Sox proteins bind to DNA with a high dissociation constant. Sox2 is co-expressed with Oct4 in the ICM of preimplantation embryos, ES cells, EC cells, and germ cells.

Distinct regulatory elements govern FGF4 gene expression in the mouse ICM, myotomes, and developing limbs. FGF4 expression in the ICM and in ES/EC cells is conferred from a distal enhancer localized in the 3′ UTR of the gene. Oct4 (or Oct1) and Sox2 bind to the respective cognate sites in the embryonic enhancer and form a unique ternary complex that elicits strong synergistic transcriptional activation. Formation of an active ternary complex is highly dependent on the spatial arrangement of the adjacent Sox and Oct binding sites on the DNA, since insertion of 3 base pairs between the Sox and Oct recognition sequences of the FGF4 gene severely impairs the enhancer function. The C-terminal domains of both Sox2 and Oct4 contribute to the functional activity of Sox2/Oct4 complex. Activity of the Oct4 C domain requires a Sox2/Oct4 complex. In fact, the synergistic action of Sox2 and Oct4 results from two distinct, yet concerted, events. The first event involves cooperative binding of Sox2 and Oct4 to the DNA via their respective DNA binding domains. The tethering of each factor to the enhancer region on FGF4 ameliorates the intrinsic activity of the activation domains of each protein. Upon formation of the ternary Sox2/Oct4 complex, novel DNA–protein and protein–protein interactions induce conformational changes that may lead to activation of latent domains and constitute a new, distinct platform for the recruitment of other co-activators. In line with this model, synergistic activation of FGF4 transcription by Sox2 and Oct4 in HeLa fibroblast cells is mediated by p300, a potential bridging factor, which should promote enhancer–promoter interactions. The embryonic FGF4 enhancer sequences are conserved in the mouse and human genes.

Embryonic stem cell-specific expression of UTF1 is regulated by the synergistic action of Sox2 and Oct4 as well. The binding sites for these two factors are arranged with no intervening spacing in the second intron of the UTF1 gene. Interestingly, one base difference in the canonical octamer binding sequence enables the recruitment of an active Oct4/Sox2 complex and prevents binding of a transcriptionally inactive complex containing Oct1 or Oct6. The sequence ACTAGCAT (canonical sequence: ATTA/TGCAT) is recognized specifically by Oct4, while Oct1 and Oct6 can bind to it only by exploiting half of the 3′ adjacent Sox site (AACAATG). The POU$_H$ domain of Oct4 is essential for Oct4 to exhibit unique DNA binding ability on the UTF1 consensus sequence.

A noncanonical binding site for Oct4 and Sox2 has been found in the 3′ regulatory region of the *Sox2* gene, called SRR2, and is involved in Sox2 expression in ES/EC cells. Sox2 and Oct4 or Oct6 can bind simultaneously, but not cooperatively, to SRR2. Oct6 or Oct4 (but not Oct1) have been shown to slightly increase Sox2-dependent transcription of the Sox2 gene.

The F-box containing protein 15 is expressed predominantly in ES cells. Embryonic expression of Fbx15 results from the cooperative binding of Oct4 and Sox2 to cognate sites juxtaposed in the ES specific enhancer. Mutation of either binding site abolishes the activity of the enhancer.

Analysis of Sox2-deficient embryos supports the hypothesis that functional association between Oct4 and Sox2 constitutes a new paradigm of gene activation in the early embryo. Sox2-deficient embryos fail to survive shortly after implantation, at 6.0 dpc, with abnormal implants showing disorganized extra-embryonic tissues and lacking Oct4-positive epiblast cells. At the egg cylinder stage, Sox2 is expressed in the epiblast and in the adjacent extra-embryonic ectoderm (ExE) of wild-type embryos. Postimplantation lethality cannot be attributed to lack of Sox2 expression in ExE, as wild-type ES cell can rescue development in Sox2-deficient mice.

Immunosurgically isolated Sox2-null ICMs are diverted to a trophoectodermal and endodermal phenotype. In Sox2-deficient blastocysts, Oct4 expression is detected by RT-PCR, indicating that a pluripotent ICM can be specified but not maintained. However, complete knockout of the Sox2 gene is not feasible during early development, as maternal Sox2 protein persists throughout preimplantation development. A Sox2-null ES cell line cannot be established, confirming the role of Sox2 in the maintenance of pluripotent stem cells.

The structural flexibility of the linker sequence between the two independent POU subdomains endows Oct4 with the ability to form homodimers on specific DNA sequences. A novel *Palindromic Oct factor Recognition Element* (PORE), composed of an inverted pair of homeodomain-binding sites separated by 5 bp (ATTTGaaatgCAAAT), has been identified in first intron of the EC/ES cell-specific osteopontin (OPN) gene enhancer.

OPN is an extracellular phosphoprotein which, binding to specific integrins, modulates cell migration and adhesion of the primitive endoderm cells. In the mouse embryo, OPN is co-expressed with Oct4 in ICM and forming primitive endoderm at 4.0 dpc, and is downregulated in the epithelial hypoblast of E 4.5 embryos. In EC and ES cells, OPN expression is inversely correlated with Sox2 protein levels.

The POU transcription factors Oct1, Oct6, and Oct4 bind to the DNA PORE element as monomers, homodimers and heterodimers. Oct4 monomer does not elicit transcriptional activation via the PORE in EC cells. Rather, activation of gene transcription is highly dependent on Oct4 dimerization on the PORE. In fact, mutation in the palindromic element of either homedomain-binding sites drastically impairs dimerization and transcriptional activation. Interestingly, Sox-2 represses Oct4 mediated OPN gene transactivation by binding to the cognate site adjacent to the PORE in the *OPN* intron. Sox-2 may interfere with Oct4 dimer formation.

Dimerization on the PORE creates a specific conformational surface that may be suitable for interaction with unidentified co-activators. In fact, Oct1/ or Oct-2/PORE dimers can interact and synergize in transcriptional activation with the lymphoid specific co-activator OBF-1. According to a model of the PORE Oct4 dimer structure, Isoleucine 21 (I21) of the POU_S and Serine 107 (S107) of the POU_H of two distinct Oct4 molecules make specific contact in the PORE dimer interface. Mutation of either residue impairs dimerization on the PORE element. S107 corresponds to a conserved phosphorylation site identified in Oct1 and Pit1. Phosphorylation of this serine has been shown to influence Oct1 binding to DNA and may provide an additional level of regulation of Oct4 dimer activity. Mutation of S107 into glutamate might structurally mimic a phosphorylated serine.

To date, OPN is the only known Oct4 target gene to be transcriptionally regulated by Oct4 dimers. Mutant mice with impaired dimerization of the Oct4 protein might be valuable in identifying new Oct4 dimer target genes and the function of these genes during early development.

Nanog

Two recent reports have highlighted the importance of a new player on the stage of pluripotency. Nanog, a divergent homeobox factor, is expressed *in vivo* in the interior cells of compacted morulae, in the ICM or epiblast of a preimplantation blastocyst, and in postmigratory germ cells. *In vitro* Nanog is a marker of all pluripotent cell lines, from ES (both murine and human) to EG and EC. Hence the name Nanog, after the Celtic land of the ever young Tir nan Og.

Nanog-deficient embryos die soon after implantation due to a failure in the specification of the pluripotent epiblast, which is diverted to endodermal fate. Similarly, *Nanog* deletion in ES cells causes differentiation into parietal/visceral endoderm lineages. These data demonstrate that Nanog is essential for maintenance of a pluripotent phenotype both *in vivo* and *in vitro* and that endoderm specification depends on Nanog down-regulation. Nanog overexpression renders ES cells independent from LIF/STAT3 stimulation for self-renewal. LIF and Nanog have an additive effect on the propagation of undifferentiated versus differentiated colonies. The differentiation potential of ES cells overexpressing Nanog is both reduced and retarded, but removal of Nanog transgene reverses the cells' status to that of the parental stem cell. *Nanog* expression does not seem to be regulated by Oct4, but the two homeo-factors work in concert in order to maintain a pluripotent phenotype.

FoxD3, FGF4, FGFR2, B-integrin

Genes whose disruption causes pre- and postimplantation embryonic lethality are the best candidates as gatekeepers of early development.

FoxD3, a transcription factor belonging to the forkhead family, is expressed in mouse and human ES cells, and during mouse embryogenesis in the epiblast and neural crest cells. FoxD3-deficient blastocysts present a regular pattern of *Oct4*, *Sox2*, and *FGF4* expression at 3.5 dpc. Nevertheless, mutant mouse embryos die at 6.5 dpc, showing a reduced epiblast, extended proximal extra-embryonic endodermic and ectodermic tissues, and lack of a primitive streak. These results indicate that FoxD3 is required for maintenance of the epiblast but not for differentiation of extra-embryonic tissues. This has been confirmed by the observation that FoxD3-deficient ICMs fail to expand after prolonged culture *in vitro*. Oct4 and FoxD3 proteins interact in solution, and Oct4 inhibits FoxD3 activation of the FoxA1 and FoxA2 endodermal promoters in a heterologous cell system.

Fibroblast growth factor 4 (FGF4) is produced by ICM cells and was first postulated to function in an autocrine fashion to promote the proliferation and expansion of the ICM. Later on it was discovered that it was involved in the patterning of the extra-embryonic ectoderm by stimulating the proliferation of trophoblast stem cells. FGF4- and fibroblast growth-factor receptor (FGFR) 2(FGFR2)-null embryos cannot form an egg cylinder and die soon after implantation. When cultured *in vivo*, their ICMs fail to expand and eventually degenerate. Although FGFR2 ES cells have not been derived to date, FGF4 –/– ES cells proliferate in the absence of the growth factor, suggesting that at least *in vitro* FGF4 is not required for ES cell proliferation.

B-integrin-deficient embryos lack a proper TE epithelium and blastocyst cavity and consequently fail to implant at the preimplantation stage.

Genetic deletion of *B-myb*, *Taube Nuss*, and *Chk1* lead to early embryonic lethality and severe defects in the outgrowth of the ICM. This phenotype can be attributed to inactivation of housekeeping genes, involved either in proliferation or cell cycle checkpoints or apoptosis. These processes affect the ICM more than TE cells due to the difference in the proliferation rate of the two lineages.

A Genetic Model for Molecular Control of Pluripotency

Oct4 cannot be considered to be a master gene for pluripotency, since it cannot prevent differentiation of ES cells upon LIF withdrawal. This finding implies that Oct4 and LIF probably activate two different pathways of gene activation, with the second relying on STAT3. The fact that both Oct4 up-regulation and LIF withdrawal lead to the same pattern of ES cell differentiation can be explained by assuming a cross-talk between the two pathways.

Hitoshi Niwa has suggested a model of the known molecular mechanisms controlling ES cell phenotype, which is outlined below:

Oct4 target genes can be subdivided into three categories:

A Those activated by Oct4 and Sox2 (FGF4, UTF).
B Those repressed by Oct4 (hCG α, β).
C Those activated by Oct4, but also repressed by a squelching mechanism when Oct4 is overexpressed (Rex1). This last group of genes is considered to comprise the cross-talk

junction between the Oct4 and the LIF-STAT3 signaling pathway, as they should be co-activated by Oct4 and unidentified X factors which lie downstream the STAT3 activation cascade.

STAT3 is hypothesized to activate ES "state" genes or to suppress endodermal/mesodermal genes or both.

As described above, activated STAT3 and subtle changes in the *Oct4* expression are compatible with maintenance of a pluripotent ES cell fate. In order to achieve a pluripotent status, group A and C genes need to be activated and group B genes need to be silenced by Oct4. Group B genes are activated only when Oct4 expression falls below the 50% threshold and are specific for the trophoectoderm lineages. A 50% increase in the expression of Oct4, or LIF withdrawal, induces down-regulation of group C genes, either by squelching of the X-co-activators, lying downstream the STAT3 pathway, or by down-regulation of STAT3-induced transcriptional program, leading to a differentiation into mesoderm/endoderm.

The validity of this model is supported by the existence of the E1A-like activities postulated to exist in ES cells, which may likely represent the mentioned co-activators X. Nanog would be incorporated within this model as an essential determinant of pluripotency, which induces ES state gene activation and/or repression of visceral/parietal state genes (see Figure 4-5A). Identification of Oct4 co-activators and Nanog/STAT3 target genes is required to enrich and validate the described transcriptional network.

In vivo Oct4 is essential for the specification of a pluripotent ICM. Nanog and Oct4 would be critical for maintenance of the epiblast during formation of the hypoblast. Postim-

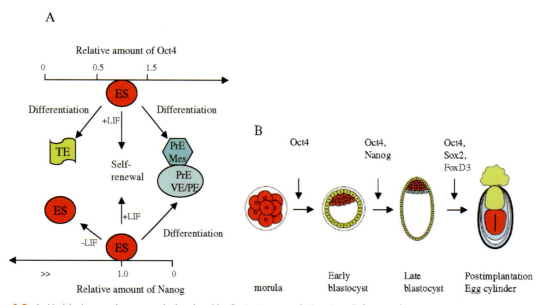

Figure 4-5. A. Model relative to the integrated roles played by Oct4, Nanog, and LIF on ES cells fate specification according to variable concentration of Oct4 and Nanog. B. Model relative to the role played by Oct4, Nanog, Sox2, and FoxD3 during early mouse development.

plantation maintenance of the epiblast would be dependent on Oct4, Sox2, and FoxD3 (see Figure 4-5B).

Plenty of questions regarding the mechanisms of pluripotency remain unanswered: how are the Oct4, STAT3, Nanog transcriptional pathways regulated, and how do they crosstalk? Are any other genes regulating pluripotency? And, in particular, is there any one master gene controlling pluripotency?

Discovery of such a master gene(s) would constitute the panacea of modern human regenerative medicine, as it would obviate the need for human cloning with all its genetic implications and ethical considerations. However, it appears that pluripotency is most likely achieved through the combination of properly sequenced processes that control chromatin accessibility, chromatin modifications, activation, and repression of specific genes. This is further complicated by potential sensitivity to subtle changes in gene expression levels.

Future Directions

The advent of DNA microchip technology has granted scientists an easy and rapid method of comparing the gene expression profiles of pluripotent and differentiated cells. In the last three years, a number of reports have obliterated the old concept that adult stem cells are restricted in their potential to only give rise to cell lineages of their tissue of origin. However, it appears that adult stem cell plasticity is mostly linked to environmental cues of the early embryonic blastocyst milieu. The transdifferentiation potential of adult stem cells *in vitro* is very low and inefficient. In addition, genetic manipulation of adult stem cells via homologous recombination has not been reported.

Euploid pluripotent ES cell lines have been derived from human blastocysts (Thomson *et al.*, 1998), which may eventually help to create a renewable source of donor cells with reduced immunogenicity for use in transplantation therapy. We are very far from using human ES cells in clinical trials. However, investigating the molecular mechanisms underlying pluripotency using the mouse model is the simplest present-day tool available for scientists that will eventually lead to human ES cell-based therapy.

KEY TERMS

Blastocyst Preimplantation mammalian embryo, consisting of an outer layer of trophectoderm cells, surrounding an inner cell mass.

Pluripotent Able to give rise to differentiated cells of all three germ layers. Cells of the inner cell mass and ES cells are pluripotent.

Stem cell A cell that is capable of both self-renewal and differentiation.

Totipotent Able to give rise to all cell types, including cells of the trophectoderm lineage. In mammals only the fertilized egg and early cleavage stage blastomeres are totipotent.

FURTHER READING

Boef, H., Hauss, C., Graeve, F. D., Baran N., and Kedinger, C. (1997). Leukemia inhibitory factor-dependent transcriptional activation in embryonic stem cells. *J. Cell Biol.* **138**, 1207–1217.

Bradley, A., Evans, M., Kaufman, M. H., and Robertson, E. (1984). Formation of germ line chimeras from embryo-derived teratocarcinoma cell lines. *Nature* **309**, 255–256.

Burdon, T., Chambers, I., Stracey, C., Niwa, H., and Smith, A. (1999). Signaling mechanisms regulating self-renewal and differentiation of pluripotent embryonic stem cells. *Cells Tissues Organs* **165**, 131–143.

Cavaleri, F., and Schöler, H. R. (2003). Nanog: a new recruit to the embryonic stem cell orchestra. *Cell* **113**, 551–552.

Dani, C., Chambers, I., Johnstone, S., Robertson, M., Ebrahimi, B., Saito, M., Taga, T., Li, M., Burdon, T., Nichols, J., and Smith, A. G. (1998). Paracrine induction of stem cell renewal by LIF-deficient cells: a new ES regulatory pathway. *Dev. Biol.* **203**, 149–162.

Evans, M. J., and Kaufman, M. H. (1981). Establishment in culture of pluripotential cells from mouse embryos. *Nature* **292**, 154–156.

Gidekel, S., and Bergman, Y. (2002). A unique developmental pattern of *Oct-3/4* DNA methylation is controlled by a *cis*-demodification element. *J. Biol. Chem.* **277**, 34,521–34,530.

Heinrich, P. C., Behrmann, I., Muller-Newen, G., Schaper, F., and Graeve, L. (1998). Interleukin-6-type cytokine signalling through the gp130/Jak/STAT pathway. *Biochemical Journal* **334**, 297–314.

Hirano, T., Ishihara, K., and Hibi, M. (2000). Roles of STAT3 in mediating the cell growth, differentiation and survival signals relayed through the IL-6 family of cytokine receptors. *Oncogene* **19**, 2548–2556.

Hocke, G. M., Cui, M. Z., and Fey, G. H. (1995). The LIF response element of the α macroglobulin gene confers LIF-induced transcriptional activation in embryonal stem cells. *Cytokine* **7**, 491–502.

Kolch, W. (2000). Meaningful relationships: the regulation of the Ras/Raf/MEK/ERK pathway by protein interactions. *Biochemical Journal* **351** Pt. 2, 289–305.

Martin, G. R. (1981). Isolation of a pluripotent cell line from early mouse embryos cultured in medium conditioned by teratocarcinoma stem cells. *Proc. Natl. Acad. Sci. USA* **78**, 7634–7638.

Niwa, H. (2001). Molecular mechanism to maintain stem cell renewal of ES cells. *Cell Structure & Function* **26**, 137–148.

Pesce, M., and Schöler, H. R. (2001). Oct-4: gatekeeper in the beginnings of mammalian development. *Stem Cells* **19**, 271–278.

Smith, A. G. (2001). Embryo-derived stem cells: of mice and men. *Annu Rev. Cell. Dev. Biol.* **17**, 435–462.

Taga, T., and Kishimoto, T. (1997). Gp130 and the interleukin-6 family of cytokines. *Annual Review of Immunology* **15**, 797–819.

Thomson, J. A., Itskovitz-Eldor, J., Shapiro, S. S., Waknitz, M. A., Swiergiel, J. J., Marshall, V. S., and Jones, J. M. (1998). Embryonic stem lines derived from human blastocysts. *Science* **282**, 1145–1147.

5

Stem Cell Niches

D. Leanne Jones and Margaret T. Fuller

Introduction

The ability of adult stem cells to both self-renew and produce daughter cells that initiate differentiation is the key to tissue homeostasis, providing a continuous supply of new cells to replace short-lived but highly differentiated cell types, such as blood, skin, and sperm. The critical decision between stem cell self-renewal and differentiation must be tightly controlled. If too many daughter cells initiate differentiation, the stem cell population may become depleted. Alternatively, unchecked stem cell self-renewal could abnormally expand the number of proliferative, partially differentiated cells in which secondary mutations could arise, leading to tumorigenesis. A detailed understanding of how the choice between stem cell self-renewal and the onset of differentiation is determined may facilitate the expansion of adult stem cell populations in culture while maintaining essential stem cell characteristics. This is a critical first step toward harnessing the potential of adult stem cells for tissue replacement and gene therapy.

Stem Cell Niche Hypothesis

Two classic developmental mechanisms can give rise to daughter cells that follow different fates. First, asymmetric partitioning of cell fate determinants in the mother cell can produce daughter cells that follow different cell fates, even though the two daughter cells reside in the same microenvironment. Second, the orientation of the plane of division can place two daughter cells in different microenvironments, which may then specify different cell fate choices through intercellular signaling. Ultimately, stem cell number, division, self-renewal, and differentiation are likely to be regulated by the integration of intrinsic factors and extrinsic cues provided by the surrounding microenvironment, now known as the *stem cell niche*.

The concept of the stem cell niche arose from observations that many adult stem cells, such as hematopoietic stem cells, lose the potential for continued self-renewal when removed from their normal cellular environment and the idea from developmental biology that different signaling microenvironments can direct daughter cells to adopt different fates. In the stem cell niche hypothesis, signals from the local microenvironment, or niche, specify stem cell self-renewal. If space

within the niche were limited such that only one daughter cell could remain in the niche, the other daughter cell would be placed outside of the niche, where it might initiate differentiation due to lack of self-renewal factors. However, if space within the niche is available or an adjacent is empty, both daughters of a stem cell division can retain stem cell identity. Therefore, the stem cell niche hypothesis predicts that the number of stem cells can be limited by the availability of niches with the necessary signals for self-renewal and survival. As a consequence, the niche provides a mechanism to control and limit stem cell numbers.

The existence of a stem cell niche has been proposed for several adult stem cell systems. The precise spatial organization of the stem cells with respect to surrounding support cells plays an important role in the ability of the niche to adequately provide proliferative and antiapoptotic signals and to exclude factors that promote differentiation. In each case, stem cells are in intimate contact with surrounding support cells that serve as a source of critical signals controlling stem cell behavior. Adhesion between stem cells and either an underlying basement membrane or the support cells themselves appears to play an important role in holding the stem cells within the niche and close to self-renewal signals. In addition, the niche could provide polarity cues to orient stem cells within the niche so that, upon division, one cell is displaced outside of the niche into an alternate environment that encourages differentiation.

In this chapter, we review current knowledge of the role of the niche and the control of stem cell self-renewal using the *Drosophila* male and female germ lines as model systems. We then extrapolate from the molecular mechanisms revealed in the analysis of *Drosophila* germ-line stem cell niches to suggest paradigms for controlling stem cell behavior within other adult stem cell systems.

Stem Cell Niches in the *Drosophila* Germ Line

The *Drosophila* germ line has emerged as a valuable model system, providing significant insight into the regulation of stem cell behavior and the importance of the stem cell niche. Although in many systems stem cells are rare and can be difficult to locate, the precise identity and location of the germline stem cells (GSCs) in the *Drosophila* ovary and testis are known. In addition, the availability of many mutants, a sequenced genome, and powerful genetic tools for cell type-specific ectopic expression has provided the opportunity to

43

address essential questions regarding how stem cells interact with their surrounding microenvironment.

In *Drosophila*, the ability to generate clones of cells that are genetically distinct from neighboring cells allows both lineage tracing and analysis of the effects of lethal mutations during late stages of the life cycle, when lethality would already have occurred in a entirely mutant animal. Lineage tracing by clonal marking analysis has led to the identification of GSCs in both the male and the female germ lines *in vivo*, within their normal environment. These genetically marked GSCs can be observed to continually produce a series of differentiating germ cells. Clonal analysis also allows the generation of mutant GSCs in an otherwise wild-type animal, allowing the analysis of a specific gene's function on stem cell maintenance, self-renewal, and survival.

In *Drosophila*, both male and female GSCs normally divide with invariant asymmetry, producing precisely one daughter stem cell and one daughter cell that will initiate differentiation. In both the ovary and the testis, GSCs are in intimate contact with surrounding support cells that provide critical self-renewal signals, maintenance signals, or both, thereby constituting a stem cell niche. Oriented division of stem cells is important for placing one daughter cell within the niche while displacing the other daughter cell destined to initiate differentiation outside of the germ-line stem cell niche.

GERM-LINE STEM CELL NICHE IN THE *DROSOPHILA* OVARY

The adult *Drosophila* ovary consists of approximately 15 ovarioles, each with a specialized structure, the germarium, at the most anterior tip (Figure 5-1A). Two to three GSCs lie at the anterior tip of the germarium, close to several groups of differentiated somatic cell types, including the terminal filament, cap cells, and inner germarial sheath cells (Figure 5-1A, Figure 5-1B). When a female GSC divides, the daughter cell that lies closer to the terminal filament and cap cells retains stem cell identity; the daughter cell that is displaced away from the cap cells initiates differentiation as a cystoblast. The cystoblast and its progeny undergo four rounds of cell divi-

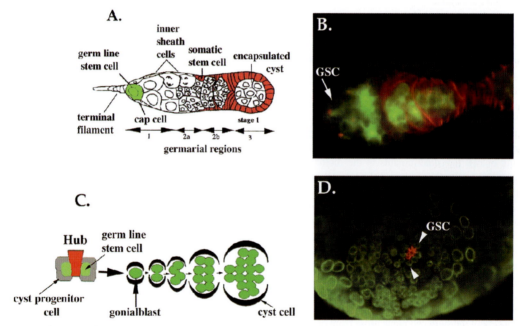

Figure 5-1. Germ-line stem cell niches in the *Drosophila* ovary and testis. (A) Schematic of a *Drosophila* germarium, which houses the germ-line stem cells (GSCs), anterior to the left and posterior to the right. The terminal filament, cap, and inner sheath cells express molecules important for the maintenance and self-renewal of female GSCs and comprise the stem cell niche. GSCs undergo asymmetric cell division, producing one daughter cell that will retain stem cell identity and one daughter cell, a cystoblast, that will initiate differentiation. As these divisions take place, the more mature cysts are displaced toward the posterior of the germarium. Cyst encapsulation by the somatic stem cell (SSC) derivatives occurs in region 2A–2B. Mature encapsulated cysts budding from the germarium make up region 3. (B) In the immunofluorescence image of a *Drosophila* germarium, germ cells are labeled with an antibody to the germ cell-specific protein, Vasa. Antibodies to the membrane protein α-spectrin label the somatic cells within the germarium, as well as a vesiculated, cytoplasmic ball-shaped structure known as the spectrosome in GSCs (arrow) and cystoblasts. (C) Schematic of the early steps in *Drosophila* spermatogenesis. GSCs surround and are in contact with a cluster of postmitotic, somatic cells known as the apical hub. The hub cells are a primary component of the male GSC niche. Each GSC is surrounded by two somatic stem cells, the cyst progenitor cells. The GSC undergoes asymmetric cell division, generating one daughter cell that will retain stem cell identity and one daughter cell, a gonialblast, which then undergoes four rounds of cell division with incomplete cytokinesis to produce 16 spermatogonia. The gonialblast is surrounded by cyst cells, which ensure spermatogonial differentiation. (D) In the immunofluorescence image of the apical tip of a *Drosophila* testis, the germ cells are labeled with an antibody to Vasa, and the somatic hub is labeled with an antibody to the membrane-associated protein, Fasciclin III. Eight GSCs (arrowheads) surround the apical hub.

sion with incomplete cytokinesis. Of the 16 germ cells, only one will become the oocyte; the other 15 cells become nurse cells, which support the growth of the oocyte. The terminal filament, cap cells, and inner germarial sheath cells express molecules that regulate critical aspects of GSC behavior, constituting the germ-line stem cell niche in the ovary.

Xie and Spradling (2000) directly demonstrated the existence of a functional stem cell niche that can program cells to assume stem cell identity in the ovary. Taking advantage of a mutation that increases the rate at which stem cells are lost, Xie and Spradling (2000) showed that an empty stem cell niche is quickly filled by the division of a neighboring stem cell. In this case, the mitotic spindle of the GSC became reoriented parallel to the terminal filament and cap cells, and the stem cell divided symmetrically so that both daughters of the stem cell division became stem cells. This kind of GSC replacement assay has not been performed for the *Drosophila* testis due to the much larger number of stem cells in that organ.

In the ovary, the vertebrate bone morphogenetic protein (BMP) 2/4 homolog Decapentaplegic (Dpp) is required for maintenance of GSCs. The cap cells and inner germarial sheath cells express *dpp*, which activates the Dpp signaling pathway in adjacent GSCs. Dpp binds and facilitates the association of type I and type II serine/threonine kinase receptors, allowing the type II receptor to phosphorylate and activate the type I receptor, which in turn phosphorylates the downstream mediator mothers against Dpp (Mad). Mad facilitates nuclear translocation of Medea (Med), a transcriptional activator that stimulates Dpp target gene expression.

Excessive Dpp signaling can block germ cell differentiation in the *Drosophila* ovary. Overexpression of *dpp* results in enlarged germaria filled with cells resembling GSCs. TGF-β pathway signaling is also required for long-term maintenance of GSCs. Loss of function mutations in the type I receptor *saxophone (sax)* shortens the half-life of GSCs from one month to one week and slows the rate of GSC divisions. Clonal analysis revealed that the downstream signaling components *mad* and *Med* are required cell autonomously in the germ line for maintaining the normal half-life of GSCs. In the current working model, Dpp, secreted from cap cells and inner germarial sheath cells, signals to regulate the maintenance and rate of division of female GSCs. As differentiating germ cells are also in contact with inner germarial sheath cells, which also express *dpp* mRNA, the model may require an additional mechanism to ensure germ cell differentiation even in the presence of Dpp. This raises the possibility that Dpp signaling from the niche may play a permissive rather than an instructive role in specifying stem cell maintenance.

The Piwi protein is also expressed in the terminal filament and cap cells and has been shown to act nonautonomously to support GSC maintenance in the *Drosophila* ovary. The *piwi* gene family, implicated in RNA silencing and translational regulation, plays crucial roles in stem cell maintenance in many organisms. In *Drosophila*, *piwi* mutant ovaries contain a normal number of primordial germ cells (PGCs) at the third

instar larval stage, but adult ovaries contained only a few differentiating germ cells. Overexpression of *piwi* in the soma leads to an increase in GSCs in the germarium, suggesting a role in GSC self-renewal. Piwi is also expressed in the germ line, where it appears to play a cell-autonomous role in controlling the rate of GSC division in the ovary. Null mutations in *piwi* also result in a failure to maintain *Drosophila* male GSCs, although the mechanism by which Piwi acts on GSCs in the male has not been established.

GERM-LINE STEM CELL NICHE IN THE *DROSOPHILA* TESTIS

The *Drosophila* adult testis is a long, coiled tube filled with cells at all stages of spermatogenesis. In adult *Drosophila melanogaster*, approximately nine GSCs lie at the apical tip of the testis, forming a ring that closely surrounds a cluster of postmitotic somatic cells called the hub (Figure 5-1C, 5-1D). When a male GSC divides, it normally produces one cell that will retain stem cell identity and one cell, called a gonialblast, that is displaced away from the hub and will initiate differentiation (Figure 5-1C). The gonialblast and its progeny undergo four rounds of transit-amplifying mitotic divisions with incomplete cytokinesis, creating a cluster of 16 interconnected spermatogonia.

In the *Drosophila* testis, signaling through the Janus Kinase–Signal Transducer and Activator of Transcription (JAK-STAT) pathway has been shown to specify stem cell self-renewal of male GSCs.[8,9] The somatic apical hub cells are a major component of the GSC niche in the testis. Hub cells express the ligand Unpaired (Upd), which activates the JAK-STAT pathway in the adjacent stem cells and specifies stem cell self-renewal. *Drosophila melanogaster* has one known JAK, encoded by the *hopscotch (hop)* gene and one known STAT, Stat92E. In males carrying a viable, male sterile *hop* allele, the initial round of germ cell differentiation occurs, but GSCs are lost soon after the first rounds of definitive stem cell divisions at the onset of spermatogenesis.[8,9,18] Mosaic analysis of homozygous mutant germ cells demonstrated that *Stat92E* activity is required cell autonomously in the germ line for stem cell self-renewal. Upd is normally expressed exclusively in the hub cells, and ectopic expression of *upd* in early germ cells resulted in an enlarged testis tip filled with thousands of small cells resembling GSCs and gonialblasts. Together, these data suggest that the hub cells contribute to the germ-line stem cell niche by secreting the ligand Upd, which specifies stem cell self-renewal by activating the JAK-STAT pathway in GSCs. Experiments in tissue culture suggest that Upd protein associates with the extracellular matrix upon secretion. If binding to the extracellular matrix restricts Upd diffusion *in vivo*, then only cells that maintain direct contact with the hub may receive sufficient levels of Upd to retain stem cell identity. Consistent with this hypothesis, activation of Stat92E, the sole *Drosophila* STAT homolog, is observed only in hub cells and the adjacent GSCs. *In situ* analysis showed that the gene encoding the receptor for Upd, *domeless*, is broadly expressed in the testis, excluded only from postmeiotic spermatocytes and spermatids.

COORDINATE CONTROL OF GERM-LINE STEM CELL AND SOMATIC STEM CELL MAINTENANCE AND PROLIFERATION

Multiple stem cell populations can reside within a common anatomical location — for example, hematopoietic and mesenchymal stem cells in the bone marrow. Coordinated control of the proliferation of different stem cell types may be especially important when the two stem cell types generate differentiated cell populations that must work together to maintain a tissue. The *Drosophila* female and male gonads provide excellent systems in which to study how the behavior of two such stem cell populations, somatic and germ line, can be coordinately controlled.

In the *Drosophila* ovary, the germarium houses a second population of stem cells in addition to GSCs. These somatic stem cells (SSCs) (Figure 5-1A, 5-1B) produce the many specialized follicle cells that cover each developing egg chamber. Lineage tracing, achieved through clonal analysis, demonstrated that SSCs are located several cell diameters from the female GSCs in the ovariole (Figure 5-1A). Each cyst of interconnected germ cells is encapsulated by somatic follicle cells in region 2A–2B before budding from the posterior end of the germarium (Figure 5-1A, 5-1B).

The Hedgehog (Hh) signal transduction pathway has been implicated in controlling the proliferation and differentiation of the SSCs and their progeny. Hh is strongly expressed in terminal filament and cap cells at the tip of the germarium (Figure 5-1A). Loss of *hh* activity reduced the number of somatic cells in the germarium. Consequently, fewer follicle cells are available to intercalate between adjacent germ-line cysts, resulting in an accumulation of unencapsulated cysts in the germarium. Overexpression of *hh* in the ovary leads to hyperproliferation of somatic cells, resulting in increased numbers of cells that separate adjacent egg chambers and increased numbers of specialized follicle cells at the poles of developing egg chambers. At this time, it is not clear how *hh*, expressed in the terminal filament and cap cells, might regulate SSC proliferation, as the SSCs lie several cell diameters away. It is possible that the SSCs receive the Hh signal directly or that Hh is also signaling through some other somatic cell type, the inner sheath cells, for example, to control the proliferation of SSCs indirectly (Figure 5-1A).

In the ovary, the *fs(1)Yb* gene may serve as an upstream regulator of both GSC and SSC proliferation in the *Drosophila* ovary. Mutations in *fs(1)Yb* lead to precocious differentiation of GSCs without apparent self-renewal. Consequently, ovarioles consist of several differentiating germ-line cysts and a germarium devoid of germ cells. There is a concomitant reduction in the number of somatic cells. Conversely, overexpression of *Yb* leads to increased numbers of both GSCs and somatic cells in the germarium.

Yb protein is expressed in the terminal filament and cap cells, and *Yb* mutants exhibit reduced expression of Hh and Piwi protein in cap cells and somewhat reduced expression in terminal filament cells. Loss of function of *Yb* results in loss of GSCs, similar to that observed in *piwi* mutants. Also, the phenotypes resulting from overexpression of *Yb* are similar to those seen upon ectopic expression of *piwi* and *Hh*. Based on these observations, Yb may regulate the expression of *piwi* and *Hh* within the GSC niche and, in doing so, coordinately control the behavior of GSCs and SSCs in parallel pathways.

In the *Drosophila* testis, a population of somatic stem cells, called cyst progenitor cells (CPCs), self-renew and generate the somatic cyst cells. The CPCs flank the male GSCs and directly contact the apical hub cells using thin cytoplasmic extensions (Figure 5-1C). As for GSCs, the daughter cell that remains adjacent to the hub retains stem cell identity, and the daughter cell displaced from the hub becomes a cyst cell and does not divide again. Two somatic cyst cells, which may be the functional equivalent of mammalian Sertoli cells, enclose each gonialblast and its progeny and play a major role in ensuring spermatogonial differentiation (Figure 5-1C).

In the testis, self-renewal of both GSCs and CPCs may be regulated by the same signal, the Upd ligand. In the testis, both the GSCs and somatic CPCs reside adjacent to the apical hub cells. The number of early somatic cells at the testis tip decreases dramatically in *hop* mutant testes. Reciprocally, the number of early somatic cells increases in response to ectopic expression of Upd in early germ cells. Upd secreted from the apical hub could signal directly to the somatic as well as to the GSC populations to specify stem cell self-renewal. Alternatively, Upd signaling to the germ line could control somatic stem cell proliferation indirectly by causing germ cells to send a second signal to neighboring CPCs to specify stem cell identity. CPCs and cyst cells are present in agametic testes, which supports the first model over the second. If Upd signals directly to both GSCs and CPCs to specify stem cell identity, then a requirement for a signal from the apical hub to direct stem cell self-renewal may serve to spatially coordinate an asymmetric outcome to stem cell divisions in both the somatic and germ-line lineages.

STRUCTURAL COMPONENTS OF THE NICHE

In *Drosophila* gonads, adhesion between GSCs and the surrounding support cells is important for holding stem cells within the niche, close to self-renewal signals and away from differentiation cues. GSCs in the ovary and testis appear to make direct cell–cell contact with surrounding support cells. Clusters of adherens junctions are observed between female GSCs and cap cells, as well as between male GSCs and the adjacent hub cells. Immunofluorescence analysis revealed that the *Drosophila* E-cadherin homolog Shotgun (Shg) and β-catenin homolog Armadillo (Arm) are highly concentrated at the interface between the GSCs and the cap cells in females and between the GSCs and the hub cells in the male. Song *et al.* recently demonstrated that recruitment into and maintenance of female GSCs within the niche requires the activity of both *shg* and *arm*. Removal of *shg* activity from the germ line using clonal analysis resulted in failure of female GSCs to be efficiently recruited into their niches in the developing ovary. Furthermore, *shg* mutant GSCs that were recruited to the niche were not maintained, suggesting that *D*E-cadherin-mediated cell adhesion is required for holding GSCs in their niche in the germarium, which in turn is required for efficient stem cell self-renewal.

Gap junctional intercellular communication via transfer of small molecules may also be involved in the survival and differentiation of early germ cells in the *Drosophila* ovary. Mutations in the *zero population growth (zpg)* gene, which encodes a germ-line-specific gap junction protein, result in loss of early germ cells at the beginning of differentiation in both males and females. Zpg protein is concentrated on the germ cell–soma interface in males and females and between adjacent germ cells in developing egg chambers. Transfer of small molecules, nutrients, or both from surrounding support cells to germ cells via gap junctions may be essential for the survival of early germ cells undergoing differentiation. The presence of gap junctions between female GSCs and adjacent support cells, coupled with the eventual loss of GSCs in *zpg* mutants, also suggests that signaling via gap junctions may play a role in stem cell maintenance or may help physically maintain GSCs in their niche.

In summary, the male and female germ lines of *Drosophila* have provided a genetic system in which to test the principles and investigate the basic underlying mechanisms of the stem cell niche theory. Clonal marking experiments conclusively identified GSCs *in situ* in both the testis and ovary, allowing the study of the relationship between these stem cells and their surrounding microenvironment. Several themes arising from the analysis of *Drosophila* male and female GSCs offer potential paradigms for analysis of stem cell niches in mammalian systems. First, stem cells are usually located adjacent to support cells that secrete factors required for maintaining stem cell identity: the hub cells at the apical tip of the testis and the cap cells in the germarium at the tip of the ovary. The signal transduction pathways involved in stem cell maintenance may

not be conserved between male and female GSC systems, nor are they necessarily conserved between GSC and SSC populations within the gonads. Second, cell–cell adhesion between GSCs and niche cells is required for stem cell maintenance, physically maintaining stem cells within the niche and ensuring that GSCs are held close to self-renewal signals emanating from the microenvironment.

Stem Cell Niches Within Mammalian Tissues

Specialized niches have been proposed to regulate the behavior of stem cells in several mammalian tissues maintained by stem cell populations, including the male germ line, the hematopoietic system, the epidermis, and the intestinal epithelium. Many of these niches share several characteristics with stem cell niches in the *Drosophila* germ line, specifically signaling molecules secreted from the surrounding microenvironment and cell adhesion molecules required for anchoring stem cells within the niche.

MAMMALIAN TESTIS

The seminiferous tubules of the mammalian testis are the site of ongoing spermatogenesis in the adult. In the embryo, PGCs divide and migrate to the genital ridges. In males, the PGCs, gonocytes, home to the basement membrane of the seminiferous tubules, differentiate into spermatogonial stem cells. The A_s (single) spermatogonia, the presumptive stem cells, are found close to several groups of supporting somatic cells, including the peritubular myoid and Sertoli cells, which may contribute to the stem cell niche (Figure 5-2).

Spermatogonial stem cells taken from a fertile mouse or rat can be transplanted into the seminiferous tubules of an

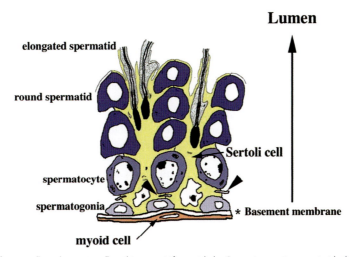

Lumen

elongated spermatid

round spermatid

Sertoli cell

spermatocyte

spermatogonia

* **Basement membrane**

myoid cell

Figure 5-2. Organization of germ cells and somatic cells within a seminiferous tubule. Spermatogenesis occurs inside the seminiferous tubules that make up the testis. The mammalian male germ-line stem cells (A_s spermatogonia) lie at the periphery of the seminiferous tubules adjacent to the basement membrane (asterisk), and differentiation proceeds through multiple stages, creating spermatogonia, spermatocytes, spermatids, and spermatozoa, which are released into the lumen of the tubule. The spermatogonia are in close association with several somatic cell types, including the peritubular myoid cells and Sertoli cells. Sertoli cells flank germ cells of all stages and are joined continuously around the tubule by tight junctions (arrowheads). Sertoli cells and myoid cells are strong candidates for cellular components of the stem cell niche within the testis.

immunodeficient mouse recipient. These exogenous stem cells are able to migrate through layers of differentiating germ cells and Sertoli cell tight junctions to find the stem cell niche along the basement membrane and establish colonies of donor-derived spermatogenesis. The availability of a spermatogonial stem cell transplantation assay has allowed characterization of the stem cell niche in the mammalian testis. For example, both the number of stem cells and the available niches increase with age and testis growth; the microenvironments within testes from immature pups were better at allowing colonization events, whether the donor stem cell was from an adult or from a pup.

Secreted signaling molecules that specifically direct self-renewal of mammalian male GSCs have not yet been identified. However, Sertoli cells produce a growth factor, glial cell line-derived neurotrophic factor (GDNF), that affects the proliferation of premeiotic germ cells, including stem cells. Depletion of stem cell reserves is observed in mice lacking one copy of GDNF. Conversely, mice overexpressing GDNF under the control of a promoter that drives preferential expression in the germ line show accumulation of undifferentiated spermatogonia that neither differentiate nor undergo apoptosis. Older GDNF-overexpressing mice regularly form non-metastatic testicular tumors, thus suggesting that GDNF contributes to paracrine regulation of spermatogonial proliferation and differentiation.

The development of a transplantation assay to test the function of mammalian male GSCs has provided a framework on which to begin the molecular characterization of spermatogonial stem cells and of the stem cell niche in the adult mammalian testis. Enrichment of stem cells using FACS sorting and monoclonal antibodies for specific surface markers, followed by transplantation of sorted populations, has led to identification of α6 integrin as a candidate surface marker for spermatogonial stem cells, raising the possibility that attachment of stem cells to the ECM may be important for stem cell maintenance. The association of spermatogonia and differentiating spermatocytes with Sertoli cells is likely to be mediated in part by adherens junctions, although the precise cadherin and cadherin-like molecules involved in this cell–cell interaction have not been conclusively identified. Meanwhile, Sertoli cells lining the tubules are continuously joined by tight junctions that regulate the movement of cells and large molecules between the basal compartment and the lumen of the seminiferous tubules (Figure 5-2).

HEMATOPOIETIC SYSTEM

The major anatomical sites of hematopoiesis change during ontogeny. Hematopoietic stem cells (HSCs) are first present in the embryonic yolk sac and the aorta–gonad–mesonephros (AGM) region, followed by the fetal liver and spleen. Just before birth, HSCs migrate to the bone marrow, where blood formation is maintained throughout the lifetime of the animal.

Characterization of the HSC niche and of the signaling molecules that influence HSC maintenance and self-renewal is in the initial stages. HSCs reside along the inner surface of the bone, and differentiating cells migrate toward the center of the bone marrow cavity. Recently, the bone-forming osteoblasts have been proposed to be a major component of the HSC niche, as increases in the number of osteoblasts led to a concomitant increase in the number of long-term HSCs. These osteoblasts secreted an elevated level of the Notch ligand Jagged-1, suggesting that activation of the Notch signaling transduction pathway in HSCs may support HSC proliferation. Furthermore, the cell adhesion molecule N-cadherin, which is expressed by the spindle-shaped N-cadherin+ CD45⁻ osteoblasts, may be responsible for holding HSCs within the niche and close to self-renewal and survival signals.

A stromal cell line that can maintain highly purified murine and human HSCs *in vitro* has been isolated and molecularly characterized. The AFT024 cell line was derived from murine fetal liver and can support HSC growth from four to seven weeks. These cultured stem cells retain the ability to reconstitute hematopoiesis *in vivo* after transplantation comparable to freshly purified HSCs. Studies of the growth factors secreted by this cell line, combined with the cell–cell and cell–ECM adhesion molecules present on the surface of these cells, may also serve as a source for candidate molecules that will be important components of the HSC niche in the bone marrow.

Recent studies have shown that signaling through the canonical Wnt pathway can direct HSC self-renewal *in vitro* and *in vivo*. Wnt is a secreted growth factor that binds to members of the Frizzled (Fz) family of cell surface receptors. The β-catenin molecule serves as a positive regulator of the pathway by mediating transcription in cooperation with members of the Lef–TCF transcription factor family. In the absence of a Wnt signal, cytoplasmic β-catenin is quickly degraded through the ubiquitin–proteasome pathway.

Transduction of HSCs with a retrovirus encoding a constitutively active β-catenin molecule resulted in self-renewal and expansion of HSCs in culture for at least four weeks and in some cases as long as one to two months under conditions in which control HSCs did not survive in culture beyond 48 hours. The cultured cells resembled HSCs morphologically and phenotypically and were capable of reconstituting the entire hematopoietic system of lethally irradiated mice when transplanted in limiting numbers. Proliferation of wild-type HSCs cultured in the presence of growth factors was blocked by a soluble form of the ligand-binding domain of the Fz receptor, suggesting that Wnt signaling is required for the proliferation response of HSCs to cytokines within their niche. Because no other cell types were present in these cultures, this result raises the possibility that a Wnt secreted from HSCs may act as an autocrine signal to promote HSC proliferation.

The cell–cell and cell–ECM adhesion molecules involved in anchoring HSCs within the bone marrow have not yet been identified. Interestingly, HSCs are mobile and detectable in the peripheral blood, spleen, and liver, suggesting that HSCs can migrate out of the niche. Although circulating HSCs and progenitor cells are quickly cleared from the peripheral blood, the number of bloodborne HSCs is fairly stable, suggesting that the flux of HSCs into and out of the blood is roughly equivalent. The mechanisms that recruit HSCs back into the niche after migration or homing of HSCs to the bone marrow after transplantation have not been clearly elucidated, although cellular adhesion molecules and chemokine receptors are likely involved. However, the mobility of HSCs suggests that adhesion between HSCs and niche cells may be highly regulated.

MAMMALIAN EPIDERMIS

The mammalian epidermis is comprised primarily of keratinocytes, a subpopulation of which are stem cells. Epidermal stem cells are multipotential; they produce progeny that differentiate into interfollicular epidermis and sebocytes and contribute to all the differentiated cell types involved in forming the hair follicle, including the outer root sheath, inner root sheath, and hair shaft.

It is not yet understood whether one "primordial" epidermal stem cell creates the stem–progenitor cell populations that maintain the interfollicular epidermis, the hair follicle, and sebaceous gland, or whether the stem cells that maintain each of these specific cell types are equivalent, with their fate determined by the local environment. However, accumulating evidence supports a model whereby the microenvironment, or niche, affects differentiation toward particular lineages. For example, cultured rat dermal papillae cells can induce hair follicle formation by rat footpad epidermis, in which follicles are not normally found. These data suggest that stem cells that normally maintain the interfollicular epidermis can be reprogrammed to act as hair follicle stem cells by signals emanating from the surrounding microenvironment. For this chapter, we consider the stem cells that generate the hair follicle and the interfollicular epidermis separately.

Hair Follicle

After placement and formation of the hair placode during mammalian embryonic development, the lower portion of the hair follicle cycles through periods of growth (anagen), regression (catagen), and quiescence (telogen). The proliferative cells that generate the inner root sheath and hair shaft are called matrix cells, a transiently dividing population of epithelial cells at the base of the hair follicle that engulfs a pocket of specialized mesenchymal cells, called the dermal papilla (Figure 5-3A).

By using multiple strategies, a stem cell niche for the mammalian epidermis has been located along the upper portion of the hair follicle in a region called the *bulge*. Specifically, the bulge is located along the outer root sheath, which is contiguous with the interfollicular epidermis (Figure 5-3A). As the hair follicle regresses during catagen, the dermal papilla comes into close proximity with the follicular bulge. It has been suggested that one or more signals from the dermal papilla may cause stem cells, transit-amplifying cells, or both in the bulge to migrate out and begin proliferating to regenerate the hair follicle.

Both in human and in mouse epidermis, β1 integrin expression is enriched in cells within the bulge region of the outer root sheath. Targeted disruption of the β1 integrin gene

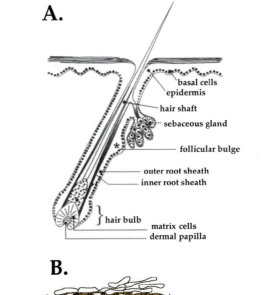

A.

basal cells
epidermis
hair shaft
sebaceous gland
follicular bulge
outer root sheath
inner root sheath
hair bulb
matrix cells
dermal papilla

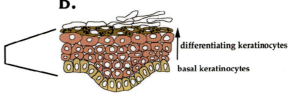

B.

differentiating keratinocytes
basal keratinocytes

Figure 5-3. (A) Schematic of the components of a hair follicle (modified from Watt, 2001). The follicular bulge (asterisk) has been proposed to act as a stem cell niche, which houses cells that can contribute to all the differentiated cell types involved in the formation of the hair follicle, including the outer root sheath, inner root sheath, and hair shaft. Stem cells within the bulge can also generate sebocytes and the cells that maintain the interfollicular epidermis. The bulge is located along the outer root sheath, which is contiguous with the interfollicular epidermis (dotted line). (B) Drawing of a cross-section of the interfollicular epidermis. The stem cells that maintain the interfollicular epidermis are within the basal layer of the epidermis and divide to produce the transit-amplifying cells, which undergo a process of terminal differentiation as they migrate toward the surface of the skin. Dead squamae are shed from the surface of the skin. Interfollicular epidermal stem cells are found in patches, surrounded by transit-amplifying cells that form an interconnecting network between the stem cell clusters.

in the outer root sheath cells did not disrupt the first hair cycle. However, proliferation of matrix cells was severely impaired, resulting in progressive hair loss and dramatic hair follicle abnormalities. Proliferation of interfollicular keratinocytes was also significantly reduced, and by seven weeks, these mice completely lacked hair follicles and sebaceous glands, suggesting that β1 integrin function is required for normal epidermal proliferation. It is possible that one role of β1 integrin is to anchor stem cells within the bulge, close to self-renewal signals.

To date, no candidate growth factors that might be secreted by cells in and around the bulge to control stem cell self-renewal have been definitively identified. However, both the Sonic hedgehog (Shh) and the Wnt/β-catenin signaling pathways have been shown to affect some aspects of cell proliferation and differentiation in the epidermis and epidermal appendages. Shh signaling appears to specify hair follicle placement and growth during embryogenesis, as well as postnatal follicle regeneration. Shh is expressed at the distal portion of the growing hair follicle, along one side of the matrix closest to the skin surface (Figure 5-3A). Interestingly, fibroblast growth factors (FGFs) and BMPs, expressed in the dermal papilla, affect hair follicle growth by regulating *Shh* expression in matrix cells (Figure 5-3A).

Loss of *Shh* function leads to disruption in hair follicle growth, while ectopic expression of Shh target genes induces follicular tumors. In addition, basal cell carcinomas, caused by mutations in downstream components of the Shh pathway, are composed of cells similar to hair follicle precursor cells. Shh expression in the epithelium results in the expression of target genes such as *Patched (Ptc)* in both the proliferating matrix cells and the adjacent dermal papilla. Because Shh targets are expressed in both the epidermis and the underlying dermal tissue, it is unclear whether the effect of Shh on epithelial cell proliferation is direct or indirect.

Wnt signaling also plays an important role in the formation of hair follicles during embryogenesis and postnatal specification of matrix-derived cells into follicular keratinocytes. Overexpression of a stabilized form of β-catenin in murine skin leads to the formation of ectopic hair follicles and hair follicle-derived tumors. Alternatively, a skin-specific knockout of β-catenin attenuated hair germ formation during embryogenesis and dramatically restricted specification of cell fates by the multipotent bulge stem cells after completion of the initial hair cycle. At the initiation of the second growth phase, β-catenin-deficient stem cells were incapable of differentiating into follicular epithelial cells and were restricted to producing interfollicular keratinocytes.

Several members of the Lef1–TcF family, as well as numerous Wnts, are expressed in the skin. Cells within the bulge and the lower outer root sheath express *Tcf3; Lef1* is highly expressed in the proliferative matrix cells and differentiating hair-shaft precursor cells. Experiments suggest that Tcf3 may act as a repressor to maintain characteristics of the bulge and the lower outer root sheath cells and may do so independently of binding to β-catenin. This implies that Tcf3 likely acts in a Wnt independent manner to direct the differentiation of cells within the bulge and lower outer root sheath. Lef1, on the other hand, requires binding to β-catenin and presumably activation by one or more Wnts to mediate its effects on hair follicle differentiation. Although these data reemphasize a role for Wnt signaling in hair follicle generation, there is still no evidence that Wnts directly control the proliferation, maintenance, or self-renewal of epidermal stem cells.

Stabilization of β-catenin through Wnt signaling, in concert with the activation of *Lef1* transcription by repression of BMP signaling by noggin, acted to repress E-cadherin expression and to drive follicle morphogenesis. Conditional removal of α-catenin also resulted in the arrest of hair follicle formation and the subsequent failure of sebaceous gland formation. Together, these results highlight the importance of Wnt signaling and of the regulation of adherens junction formation in the development and maintenance of hair follicles.

Interfollicular Epidermis

Stem cells in the bulge can migrate superficially to maintain the interfollicular epidermis. The stem cells that maintain the interfollicular epidermis are within pockets in the basal layer of the epidermis and divide to generate the transit-amplifying cells, which undergo a process of terminal differentiation as they migrate toward the surface of the skin (Figure 5-3B). Interfollicular epidermal stem cells are found in patches, surrounded by transit-amplifying cells that form an interconnecting network between the stem cell clusters.

Adhesion to the extracellular matrix promotes stem cell identity and prevents differentiation of keratinocytes. Human or mouse basal keratinocytes can be grown and cultured *in vitro*, and cultured adult human keratinocytes have been used as autografts in the treatment of burn victims for the past 20 years. The ability to culture keratinocytes *in vitro* has also allowed the development of potential strategies to use cutaneous gene therapy to correct various skin disorders and chronic wounds.

When cultured keratinocytes are placed in suspension, they immediately cease cell division and initiate differentiation. Keratinocytes express a variety of integrins, and although some are generally expressed, others are induced only during development, wounding, and disease. All cells within the basal layer of the epidermis express β1 integrin. However, the cells most likely to be stem cells within the interfollicular epidermis express surface levels of β1 integrin two- to threefold higher than those of transit-amplifying cells. High levels of β1 integrin with signaling through the mitogen-activated protein kinase cascade were demonstrated to promote stem cell identity in basal keratinocytes. As a result, a high level of β1 integrin is commonly used as a visual marker for stem cells within the interfollicular epidermis.

Activated β-catenin and signaling through TcF has been demonstrated to increase the proportion of stem cells in keratinocyte cultures. Cultured mouse keratinocytes exposed to Wnt3a- and noggin-conditioned media showed a significant increase in β-catenin and Lef1 levels and in localization of Lef1 to the nucleus. These two signaling pathways could act to reduce *E-cadherin* expression in interfollicular basal keratinocytes, similar to the manner in which they act to regulate *E-cadherin* expression during hair follicle morphogenesis. Interestingly, stem cells in the interfollicular epidermis have surface levels of E-cadherin lower than those of transit-amplifying cells.

Notch signaling has been suggested to promote differentiation of interfollicular keratinocytes. High levels of the Notch ligand, Delta1, in cultured human epidermal keratinocytes signal to adjacent cells to differentiate while the stem cells are protected from this signal. Notch signaling has also been demonstrated to stimulate differentiation in mouse epidermal cells. Accordingly, conditional removal of *Notch* from basal keratinocytes results in epidermal hyperplasia, suggesting that Notch negatively regulates epithelial stem cell proliferation and may act as a tumor suppressor in mouse epidermis. Interestingly, *Notch1* deficiency also results in increased levels of β-catenin and Lef1 and in the formation of basal cell carcinoma-like tumors.

GUT EPITHELIUM

The inner lining of the colon and small intestine is a simple columnar epithelium constantly renewed by the proliferation of stem cells residing within pockets, or crypts, along the intestinal wall. Intestinal cells leave the crypt at a rate of 200–300 cells/day and migrate onto ciliated villi that protrude into the gut lumen (reviewed by Winton, 2001; Figure 5-4A). No cell-specific marker has been characterized that allows conclusive identification and characterization of intestinal stem cells. However, lineage-tracing experiments have located the presumptive stem cells of both the small intestine and the colon near the base of each crypt. Within a crypt, approximately four to five stem cells generate transit-amplifying cells, which are capable of up to six transit divisions. Migration of these transit-amplifying cells out of the proliferative zone is required for the onset of differentiation (Figure 5-4B). The stem cells may be maintained at the base of crypts, embedded in the intestinal wall, for protection from toxins passing through the gut lumen.

Wnt signaling is involved in controlling the proliferation and differentiation of intestinal epithelial cells. In humans, mutations in the adenomatous polyposis coli *(APC)* gene, a negative regulator of Wnt signaling, are etiologically linked to the development of colorectal cancers. Also, constitutively active nuclear complexes of Tcf4–β-catenin are found in *APC$^{-/-}$* colon carcinoma cell lines or in cell lines that have a stable form of β-catenin, suggesting that hyperactivation of Tcf4 may contribute to cellular transformation. Loss of the Tcf4 transcription factor, which is expressed in the intestinal epithelium, leads to the depletion of stem cells and the failure to maintain the proliferative compartments in the intervillus pockets of the neonatal small intestine. This phenotype was not evident in the colon, suggesting that another TcF family member may act redundantly with or instead of Tcf4 in the large intestine.

If crypts serve as a niche to support the self-renewal of intestinal stem cells, this implies that cells near the intestinal stem cells may be the source of a secreted self-renewal signal. Nuclear β-catenin is found only in the cells at the base of the crypts within the adult mouse small intestine, suggesting activation of the Wnt pathway in these cells (Figure 5-4B). Therefore, mesenchymal cells underlying the crypt epithelium could be a source for a secreted Wnt ligand that could act as a paracrine signal to direct the proliferation of stem cells, progenitor cells, or both in the intestinal epithelium.

NEURAL STEM CELLS

Neurogenesis persists in particular regions of the adult brain, occurring in both the subventricular zone (SVZ) of the lateral ventricle and the hippocampus. Neural stem cells that have the capacity to self-renew and produce precursors that will

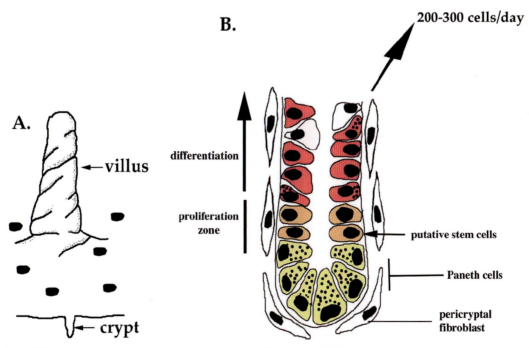

Figure 5-4. (A) Drawing showing the relationship between small intestinal villi and crypts. (B) Schematic representation of a small intestinal crypt. Lineage-tracing experiments in both the small intestine and the colon have led to the conclusion that stem cells reside near the crypt base. The approximate location of stem cells in the small intestine is at position 4 from the crypt base above the differentiated Paneth cells. Colonic crypts do not contain Paneth cells, and the stem cells in the colon have been localized to the base of the crypt. Other differentiated cells migrate out toward the lumen of the gut. Factors secreted by the pericryptal fibroblasts could contribute to stem cell maintenance and proliferation; therefore, these cells are likely candidates for cellular components of the stem cell niche within the gut.

differentiate into both neurons and glia can be cultured from the SVZ and hippocampus. When cultured *in vitro*, cells from these tissues can generate free-floating, spherical clusters called neurospheres that contain mixed populations of stem cells and precursor cells. Although growth factors such as FGF-2 and EGF can support the growth of neurospheres in culture, the physiologically relevant signaling molecules that support stem cell self-renewal in the adult brain have not yet been identified.

Cells isolated from many regions in the adult brain, including nonneurogenic regions, can generate neurons both *in vitro* and, after grafting back to neurogenic regions, *in vivo*. These data suggest that neural stem cells may be distributed throughout the adult central nervous system and that the local environment, or niche, may determine their developmental fate.

Astrocytes from both the SVZ and the hippocampus can provide neurogenic signals to progenitor cells, suggesting that astrocytes may be a critical component of the neural stem cell niche. Astrocytes from the adult spinal cord have no effect on the growth of neural stem cells in culture, indicating that astrocytes from different regions of the central nervous system exhibit different capabilities for regulating the fate choice of adult stem cells.

Summary

Stem cell niches have been proposed to play a critical role in the maintenance of stem cells in the male germ line, the hematopoietic system, the epidermis, the intestinal epithelium, and the adult nervous system. Characterization of these stem cell niches depends on the ability to identify stem cells *in vivo* in their normal environment. Through comparison of different stem cell systems, some themes emerge that indicate possible general characteristics of the relationship between stem cells and their supporting niche.

First, secreted factors elaborated by or induced by cells composing the stem cell niche can function to direct stem cell fate decisions. However, the precise signaling pathway or pathways may be different for each stem cell type and within each stem cell niche. Studies in *Drosophila* indicate that support cells adjacent to stem cells secrete factors required for maintaining stem cell identity and for specifying stem cell self-renewal. Both JAK-STAT signaling and TGF-β signaling have been implicated in the regulation of stem cell behavior by surrounding support cells in *Drosophila*. In mammals, the Wnt signal transduction pathway has been demonstrated to play a role in specifying stem cell self-renewal in HSCs, although the Wnt signal may be secreted from the stem cells themselves and may act in an autocrine loop to control stem cell proliferation. Wnt signaling may also be involved in directing the proliferation of stem cells, transit-amplifying cells, or both in the intestinal epithelium. However, the same signaling pathway may be exploited for distinct purposes in different stem cell systems. In the mammalian epidermis, Wnt signaling is likely involved in specifying the fate of hair follicle precursors rather than in specifying self-renewal of the multipotent stem cells in the bulge.

Second, cell adhesion is also emerging as an important characteristic of the interactions of stem cells with the niche. Adhesion between stem cells and niche cells is required for stem cell maintenance in the *Drosophila* male and female germ line, ensuring that GSCs are held close to self-renewal signals emanating from the niche. Attachment to niche cells or to a basal lamina may also be important for stem cell maintenance within adult mammalian tissues — hence the high levels of the β1 integrin characteristic of stem cells in the interfollicular epidermis and in the multipotent stem cells within the bulge region of the outer root sheath. Interestingly, targeted disruption of β1 integrin in cells within the bulge region of the outer root sheath severely impaired the proliferation of precursor cells that contributed to the interfollicular epidermis, hair follicle, and sebaceous glands. Thus, similar to the role of adherens junctions in maintaining *Drosophila* GSCs in the niche, β1 integrin-mediated adhesion may be required to hold multipotent epidermal stem cells within the niche and close to self-renewal signals. In the mammalian testis, α6 integrin has been identified as a cell surface marker for the enrichment of spermatogonial stem cells, although a specific role for α6 integrin in spermatogonial stem cell maintenance has not yet been directly demonstrated. Similarly, α6 integrin is expressed by basal keratinocytes in the epidermis; however, there is no strong correlation between α6 expression and proliferative potential. Therefore, although cell adhesion is frequently a conserved feature of stem cell maintenance in supportive niches, the specific types of junctions and cell adhesion molecules that play roles may differ among different stem cell niche systems.

Third, the precise cellular organization of stem cells with respect to surrounding support cells may play an important role in the regulation of appropriate stem cell numbers. In the *Drosophila* ovary and testis, where the stem cells normally divide with invariant asymmetry, the mitotic spindle is oriented to place the daughter cell that will retain stem cell identity within the stem cell niche; the daughter cell destined to differentiate is placed outside of the niche and away from self-renewal signals. Either attachment to niche cells or the extracellular matrix via junctional complexes or localized signals within the niche may provide polarity cues toward which stem cells can orient during division. This stereotyped division plane can in turn specify an asymmetric outcome to stem cell divisions, in which one daughter cell retains attachment to niche cells and the other is displaced out of the stem cell niche. As stem cells are definitively identified *in vivo*, in the context of their normal support cell microenvironment, it will be interesting to determine if stem cell divisions are likewise oriented in the seminiferous tubules, bone marrow, follicular bulge, and intestinal crypts and within neurogenic regions of the adult brain.

ACKNOWLEDGMENTS

The authors would like to thank Amy Wagers, Tony Oro, Alan Zhu, and Erin Davies for critical reading of the manuscript. D. Leanne Jones was a Lilly Fellow of the Life Sciences Research Foundation.

KEY WORDS

Adult stem cell Relatively undifferentiatied precursor cells that maintain the ability throughout adult life to proliferate, producing some progeny cells that maintain stem cell identity, renewing the stem cell population, and other progeny cells that initiate differentiation along one or more defined lineages.

APC The human *adenomatous polyposis coli* gene, a negative regulator of Wnt signaling. Also, a structural protein associated with the cytoplasmic face of adherens junctions. Mutations in the *APC* gene are linked to the development of colorectal cancer.

CPC Cyst progenitor cell — stem cell for the somatic cyst cells in the *Drosophila* testis.

GSC Germ-line stem cell.

HSC Hematopoietc stem cell.

Niche The local microenvironment that supports stem cell maintenance and self-renewal.

Wnt A secreted growth factor that binds to members of the Frizzled (Fz) family of cell surface receptors.

FURTHER READING

Alonso, L., and Fuchs, E. (2003). Stem cells in the skin: Waste not, Wnt not. *Genes Dev.* **17**, 1189–1200.

Gage, F. H. (2000). Mammalian neural stem cells. *Science* **287**, 1433–1438.

Gonzalez-Reyes, A. (2003). Stem cells, niches, and cadherins: A view from *Drosophila. J. Cell Sci.* **116 (Pt. 6)**, 949–954.

Kiger, A. A., Jones, D. L., Schulz, C., Rogers, M. B., and Fuller, M. T. (2001). Stem cell self-renewal specified by JAK-STAT activation in response to a support cell cue. *Science* **294**, 2542–2545.

Oro, A. E., and Scott, M. P. (1998). Splitting hairs: Dissecting roles of signaling systems in epidermal development. *Cell* **95**, 575–578.

Spradling, A., Drummond-Barbosa, D., and Kai, T. (2001). Stem cells find their niche. *Nature* **414**, 98–104.

Tulina, N., and Matunis, E. (2001). Control of stem cell self-renewal in *Drosophila* spermatogenesis by JAK-STAT signaling. *Science* **294**, 2546–2549.

Watt, F. M. (2001). Epidermal stem cells. *In Stem cell biology* (D. R. Marshak *et al.*, eds.), pp. 439–453. Cold Spring Harbor Press, Cold Spring Harbor, NY.

Watt, F. M., and Hogan, B. L. (2000). Out of Eden: stem cells and their niches. *Science* **287**, 1427–1430.

Winton, D. J. (2001). Stem cells in the epithelium of the small intestine and colon. *In Stem cell biology* (D. R. Marshak *et al.*, eds.). pp. 515–536. Cold Spring Harbor Press, Cold Spring Harbor, NY.

Xie, T., and Spradling, A. C. (2000). A niche maintaining germ line stem cells in the *Drosophila* ovary. *Science* **290**, 328–330.

Yamashita, Y., Jones, D. L., and Fuller, M. T. (2003). Orientation of asymmetric stem cell division by the APC tumor suppressor and centrosome. *Science* **301**, 1547–1550.

6

Mechanisms of Stem Cell Self-Renewal

Hitoshi Niwa

Self-renewal of embryonic stem (ES) cells is achieved by symmetrical cell division while maintaining pluripotency. This can be modulated by extrinsic factors such as a cytokine leukemia inhibitory factor (LIF) for mouse ES cells. External signals control gene expression by regulating transcription factors. *Oct-3/4* acts as a pivotal player in determining self-renewal or differentiation. However, it is still a mystery as to how self-renewal is achieved, since cell cycle regulation, apoptosis, and telomerase activity have not been analyzed well in ES self-renewal. More information is needed to reach a better understanding of self-renewal in ES cells.

If a question can be put at all, then it can also be answered.
Ludwig Wittgenstein

Stem cells possess mystical qualities. As a result, we feel that even if all possible scientific questions are answered, the main problem of stem cells has not been touched at all. However, what we can only do is ask proper questions and seek answers for them. We believe that if it can be said at all, it can be said clearly.

Self-Renewal of Pluripotent Stem Cells

The capacities for self-renewal and differentiation are the two characteristic potentials of stem cells. Self-renewal can be defined as making a complete phenocopy of stem cells through mitosis, which means that at least one daughter cell generated by mitosis possesses the same capacity of self-renewal and differentiation. In stem cell self-renewal, symmetric cell division generates two stem cells; asymmetric cell division results in one stem cell and either one differentiated progeny or a stem cell with a restricted capacity for differentiation. Self-renewal by symmetric cell division is often observed in transient stem cells appearing in early embryonic development to increase body size. In contrast, self-renewal by asymmetric cell division can be found in permanent stem cells in embryos in later developmental stages and in adults to maintain the homeostasis of the established body plan.

ES cells are pluripotent stem cells derived from pre- or peri-implantation embryos (Figure 6-1). The first ES cell lines were established from the mouse inner cell mass (ICM) of blastocyst-stage embryos in 1981 by Gail Martin as well as Martin Evans and Matthew Kaufman. A pluripotent stem cell population appears only transiently during early embryogen-

esis, so competency for deviation of ES cells exists in a very narrow range of developmental stages. In the case of mouse development, the pluripotent stem cell population is first seen as ICM by segregation of trophectoderm during the formation of blastocyst at 3.5 days postcoitum (dpc). At 4.5 dpc, a primitive endoderm layer can be seen at the surface of ICM, and the remaining pluripotent cell population covered by primitive endoderm is designated as epiblast. After implantation, epiblast cells start to proliferate rapidly and increase in size. At 6.0 dpc, apoptotic cell death eliminates the central part of the epiblast, resulting in the formation of an epithelialized monolayer of pluripotent stem cells designated as primitive ectoderm (Figure 6-2). Primitive ectoderm undergoes differentiation to embryonic germ layers through gastrulation, and it is there that cells lose pluripotency. After 7.0 dpc, only primordial germ cells retain latent pluripotency, which can be shown by the establishment of embryonic germ (EG) cells *in vitro* as reported by Yasuhisa Matsui *et al.*, 1991.

ES cells are not equivalent to the pluripotent stem cells in the ICM, although they are directly derived from the ICM. The ICM pluripotent stem cells divide slowly. In the delayed blastocyst generated by ovariectomy after fertilization, the doubling time of ICM–epiblast cells is estimated at 96 hours or longer. However, mouse ES cells grow more rapidly than these cells, and they display a doubling time of 12 to 14 hours. Such rapid growth of pluripotent stem cells is observed only in the epiblast after implantation, and it may be triggered by the signals from the primitive endoderm and extraembryonic ectoderm. Their doubling time at 5.0 dpc is 11 to 12 hours, almost the same as ES cells, and it reaches 4 to 5 hours at 6.0 dpc. Because the epithelial characteristics and altered pluripotency are evident in primitive ectoderm that cannot be found in ES cells, ES cells are most similar to the pluripotent stem cells in the epiblast at 5.0 dpc. Expression patterns of stage-specific marker genes also suggest that the pattern in ES cells is most similar to the epiblast between 4.75 and 5.0 dpc.

Molecular Mechanism to Retain ES Cell Self-Renewal

The ability of continuous self-renewal *in vitro* is one of the characteristic phenotypes of ES cells. As found in the case of other cellular phenotypes, it should be regulated by transcriptional control in the nucleus by extracellular signals (Figure 6-3). At the molecular level, self-renewal can be defined as the combinatorial phenomenon of keeping pluripotency and stimulating cellular proliferation.

EXTRACELLULAR SIGNALS FOR ES CELL SELF-RENEWAL

Prior to the establishment of ES cells in routine culture, researchers manipulated embryonal carcinoma (EC) cells. EC cells are pluripotent stem cells derived from a particular type of tumor, teratocarcinoma. This tumor consists of tissues derived from multiple, and often all three, germ layers. Teratocarcinomas are derived from ectopically migrated primordial germ cells, and they continuously grow by self-renewal and differentiation of the remaining pluripotent stem cells. Because many different types of cell lines had been established from various tumors, such an interesting characteristic of teratocarcinomas intrigued researchers and spurred them to establish pluripotent cell lines. In their efforts to do so, several important strategies for the culture of pluripotent stem cells were developed. The first EC cell line from a mouse teratocarcinoma was established in 1970. Thereafter, several techniques, such as co-culture with feeder cells, were developed to improve the proper self-renewal of EC cells *in vitro*.

One of the first mouse ES cell lines was established by the culture of ICM on feeder layers in the presence of an EC-conditioned medium, and another was derived from delayed blastocysts cultured on feeder cells. The role of feeder cells as a source of soluble growth factors was suggested by the efficient replacement of such cells with a medium conditioned by buffalo rat liver cells. It was then that a cytokine LIF was identified as a responsible factor mediating this phenomenon by Austin Smith *et al.* (1988). LIF had been originally identified as a cytokine-inducing differentiation and as preventing self-renewal of the particular leukemia cell line M1, but against ES cells it exhibited the opposite effect — inhibition of differentiation while retaining the capacity for self-renewal. Using recombinant LIF, researchers can maintain ES cells with pluripotency on gelatinized dishes during long-term culture *in vitro*. Moreover, new ES cell lines can be established under such conditions from blastocysts of the genetic background named 129, indicating that the presence of LIF is sufficient to maintain ES cell self-renewal in this case.

What is the role of LIF in maintaining ES cell self-renewal? Since the removal of LIF results in differentiation mainly toward primitive endoderm, one of its effects on ES cells is to inhibit differentiation. Reports have suggested that the action of LIF is limited to inhibiting differentiation without stimulating proliferation, but it is still hard to state this clearly because all of these experiments were done under crude experimental conditions in the presence of fetal calf serum (FCS) in the culture medium.

LIF belongs to the *IL-6* cytokine family, whose members share the transmembrane glycoprotein gp130 as a common component for signal transduction of their receptors. The high-affinity LIF receptor consists of a heterodimer of gp130 and LIF receptor-β (LIFRβ). LIFRβ possesses its own cytoplasmic domain homologous to that of gp130, but our experiment using the chimeric molecules consisting of the extracellular domain of granulocyte colony-stimulating factor receptor and the intracellular domain of either gp130 or

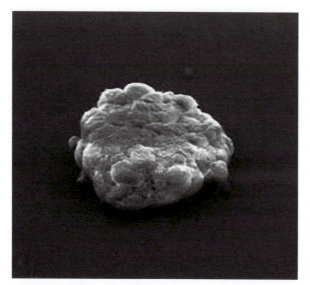

Figure 6-1. Scanning electromicroscope view of mouse ES cells. Mouse ES cells form compact colonies in which cells are in tight contact.

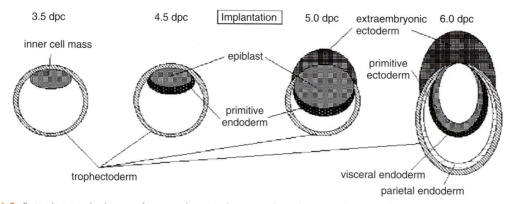

Figure 6-2. Peri-implantation development of mouse embryos. A pluripotent cell population is in the inner cell mass (ICM), epiblast, and primitive ectoderm. ES cells are most similar to the epiblast in their characteristics, although they are normally derived from the ICM.

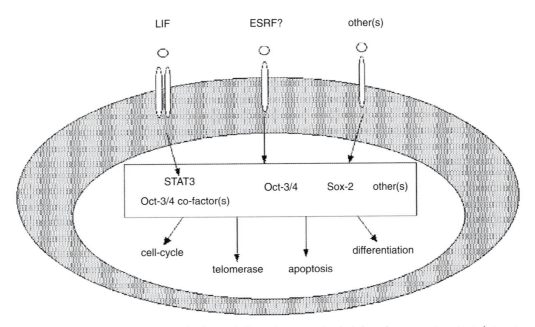

Figure 6-3. Molecular mechanism governing ES cell self-renewal. The mechanism can be divided into three categories: extrinsic factors, transcriptional regulators, and effectors.

LIFRβ revealed that only gp130 is responsible for signal integration to retain ES cell self-renewal. One of the major pathways of signal transduction using gp130 is the JAK–STAT pathway, and its importance in ES cells has been shown by a series of genetic manipulations. We demonstrated that the blockage of activation of the signal transduction molecule STAT3 by overexpression of its dominant-negative mutant in the presence of LIF induces differentiation similar to that induced by the withdrawal of LIF, indicating that STAT3 activation is essential for LIF action. On the other hand, it was shown that activation of STAT3 is sufficient to maintain ES cell self-renewal in the absence of LIF. They transduced ES cells with the chimeric molecule consisting of STAT3 and the ligand-binding domain of the mutant estrogen receptor, which can be dimerized by the artificial estrogen derivative tamoxifen. Their results showed that self-renewal of these ES cells can be maintained by tamoxifen without LIF as efficiently as with LIF.

Although the LIF action on mouse ES cells is drastic, its physiological action during development appears to be restricted. Elimination of the function of LIF, gp130, LIFRβ, or STAT3 by gene targeting did not interfere with the self-renewal of pluripotent stem cells during early embryogenesis. The role of gp130 in the pluripotent cell phenotype was evident only when the delayed blastocysts were carefully analyzed. The ICM of delayed blastocysts normally maintains pluripotency, but gp130$^{-/-}$ blastocysts could not maintain pluripotency during the delayed period. Since the maintenance of blastocysts in the uterus without implantation is a characteristic feature in rodents, the responsiveness of ES cells to gp130 signaling has its origin in this adaptive physiological function. Moreover, it may be the reason why LIFs

do not show obvious effects on ES cells of other species, especially primates. However, the function of the gp130–STAT3 pathway in germ cell development is evolutionarily conserved, since it can be found in invertebrates. This may suggest that the role of this system in rodent ES cells is a small evolutionary cooption derived from the maintenance of germ cells.

Although LIF is the only extrinsic factor to date for promoting ES cell self-renewal, its action is not unique. It was reported that the activity of a medium conditioned by parietal endoderm cells could replace LIF for short-term culture without activating STAT3, indicating that a different molecular mechanism can support ES cell self-renewal in the absence of LIF. Unfortunately, it remains a mystery as to what kind of signals mediate this phenomenon, since the responsible substance named ES cell renewal factor (ESRF) has not been identified. It is clear that neither LIF nor ESRF is sufficient to maintain ES cell self-renewal because the culture to detect their activity always contains FCS, a cocktail of several soluble factors. FCS can be replaced by artificial chemical components, but such a simple replacement is limited in the presence of feeder cells. In the feeder-free condition, ES cells can be maintained in high-density culture conditions but not in clonal-density conditions, indicating that a community effect is evident under such conditions. Since this community effect is masked in the presence of FCS or feeder cells, it can be conducted by soluble factors and may be replaced by cell–cell interaction using adherent molecules.

Recently, Qi-Long Ying *et al.* reported that a combination of LIF and bone morphogenic factor (BMP)-4 allows propagation of mouse ES cells without feeder and FCS, and we also established feeder- and serum-free culture of mouse ES cells by providing an excess amount of adrenocorticotropic hormone (ACTH) into the culture with LIF, indicating that ES cells require only a few factors for maintaining self-renewal. The contribution of the Wnt pathway was also reported, suggesting that there is functional overlap and cross-talk between different types of signals.

TRANSCRIPTIONAL REGULATION FOR ES CELL SELF-RENEWAL

STAT3

STAT3, activated by LIF, acts as a transcriptional regulator in nuclei. However, its function is not commonly found in various pluripotent stem cells. Many mouse and human EC cell lines propagate in an LIF-independent manner, as found in primate ES cells, and overexpression of dominant-negative STAT3 in LIF-independent EC cells does not affect their self-renewal. These data clearly rule out the possibility that independence against exogenous LIF is not simply caused by the presence of autocrine or intracellular activation of its signal transduction pathway. In mouse ES cells, we previously proposed that STAT3 might activate the expression of a partner of *Oct-3/4*, which has yet to be identified, to maintain ES cell self-renewal (see later sections of this chapter).

Oct-3/4

The molecular mechanism governing self-renewal and differentiation of pluripotent stem cells was first characterized using EC cells because of their longer history and convenience of culture *in vitro*. Many EC cell lines have been adapted to *in vitro* culture in the presence of FCS without particular supplemental factors such as feeders. These EC cells undergo differentiation synchronously with chemical inducers such as all-trans retinoic acid. Thus, many trials have been conducted using these EC cells to identify genes involving the transition from self-renewal to differentiation. In 1990, three different groups identified the same gene, which encodes a transcription factor of the POU family, by its specific expression in undifferentiated stem cells followed by down-regulation during differentiation. This gene is *Oct-3/4*, which was initially reported as *Oct-3* or *Oct-4*, encoded by *Pou5f1*.

Oct-3/4 expression is tightly restricted in stem cell populations during development. Its expression is detectable in totipotent and pluripotent cells — such as fertilized eggs, all blastomeres of morula, the ICM of blastocysts, the epiblast, and the primitive ectoderm — and then restricted in latent pluripotent cells in germ cell lineage, although retention of expression in differentiating cells is observed at 8.5 dpc, especially in neural cell lineage. The function of *Oct-3/4* in pluripotent stem cells was initially analyzed by a conventional gene-targeting strategy. According to the study by Jennifer Nichols *et al.* (1998) heterozygous *Oct-3/4*-deficient animals developed normally, but homozygous embryos obtained by their intercross exhibited developmental defects at the peri-implantation stage. The homozygous embryo was never recovered at the egg cylinder stage after implantation; however, implantation was not affected since homozygous embryos could be recovered from swelling deciduas at 5.0 dpc, and it was observed at the blastocyst stage at close-

to-expected Mendelian frequency. When the ICM was isolated by immunosurgery and cultured *in vitro*, cells underwent differentiation to trophectoderm, whereas the ICM derived from wild-type or heterozygous blastocysts formed a stem cell clump surrounded by parietal endoderm. These data indicated that the function of *Oct-3/4* was essential in establishing the proper pluripotency in the ICM of blastocysts, but it was still unclear whether its function was necessary for the self-renewal of established pluripotent stem cells.

Our functional analyses of *Oct-3/4* in ES cells revealed their unique characteristics. Loss-of-function phenotype of *Oct-3/4* evaluated by the combination of gene-targeting and tetracycline-inducible transgene expression revealed that its function is essential for the continuous propagation of stem cell populations. Moreover, loss of *Oct-3/4* function strictly determines the differentiated fate of ES cells toward trophectoderm, which is merely observed in normal culture conditions, as found in the ICM of *Oct-3/4*-null embryos. Once the essential nature of the *Oct-3/4* function in ES cell self-renewal was established, the next question was its sufficiency; however, this question was hard to address. Unexpectedly, overexpression of *Oct-3/4* induces differentiation toward primitive endoderm lineage.

Careful estimation of the threshold level to induce differentiation using tetracycline-regulatable transgenesis revealed that only a 50% increase over the normal expression level is sufficient. It was reported that overexpression of *Oct-3/4* represses its transcriptional activation using a squelching mechanism, and high-level expression of *Oct-3/4* is detected in primitive endoderm imaged on the surface of the ICM at 4.5 dpc, suggesting that the phenomenon observed in ES cells might have some physiological significance. Therefore, *Oct-3/4* can be regarded as a three-way switch to determine three different cell fates — pluripotent stem cells, primitive endoderm, and trophectoderm — in a dose-dependent manner. The original question about its sufficiency in ES cell self-renewal was finally addressed in *Oct-3/4*-null ES cells maintained by a tetracycline-regulatable *Oct-3/4* transgene. In these ES cells, *Oct-3/4* expression is maintained after the withdrawal of LIF, but they undergo normal differentiation events toward primitive endoderm in the absence of LIF, indicating that *Oct-3/4* expression is not sufficient to continue self-renewal without LIF. This result also revealed that the LIF–STAT3 axis does not form linear cascade with *Oct-3/4* to maintain ES cell self-renewal, since loss of *Oct-3/4* induces differentiation to trophectoderm in the presence of LIF.

In contrast to STAT3, such *Oct-3/4* function appears to be common in various pluripotent stem cells. Stem cell-specific expression of *Oct-3/4* is reported in primate ES cell lines, and overexpression of *Oct-3/4* induces differentiation in various mouse EC cell lines. However, *Oct-3/4* may be an evolutionarily new component of the genome because it can be found only in mammals. In zebra fish, the diverse POU family member POU2 was identified as an ortholog of *Oct-3/4*, indicating its rapid evolution in vertebrates because of the absence of *Oct-3/4*- and POU2-like sequences in invertebrate genomes.[51]

Sox-2

Sox-2 is a member of the Sry-related transcription factor family.[52] Its function in ES cells was first identified in relation to *Oct-3/4*. When the fibroblast growth factor-4 *(Fgf-4)* gene was identified as a possible target of *Oct-3/4*, its enhancer element specifically active in pluripotent stem cells was analyzed. It possesses binding elements for the *Sox* family members as well as the POU family members. Subsequently, the *Sox* family members expressed in ES cells were surveyed, and *Sox-2* was identified. *Sox-2* can be regarded as one of the cofactors of *Oct-3/4*, since it activates the transcription of target genes, such as *Fgf-4*, *Utf-1*, *Fbx-15*, and *Lefty-1* in cooperation with *Oct-3/4*. Moreover, *Sox-2* expression is regulated by *Oct-3/4* and *Sox-2*, indicating that a positive feedback mechanism may be involved in the maintenance of ES cell self-renewal.

Recently, *Sox-2* function in pluripotent stem cells was analyzed *in vivo* by gene targeting. According to the study by Ariel Avilion *et al.* (2003), heterozygous *Sox-2*-deficient embryos develop normally, but homozygous embryos stop development in the peri-implantation stage, as found in *Oct-3/4* mutants. However, the precise point at which they exhibit abnormalities is slightly later than the point in *Oct-3/4* mutants (abnormal embryos without an epiblast can be recovered at 6.0 dpc). A homozygous blastocyst looks healthier than that of *Oct-3/4* mutants, and the isolated ICM generates primitive endoderm as well as trophectoderm. Such a delay of abnormal phenotype may be caused by the persistence of maternal transcripts as discussed in the report, but it will be necessary to confirm the precise role of *Sox-2* in ES cells, as we did for *Oct-3/4*.

Nanog

Nanog (also reported as ENK) is a new member of the transcription factors whose functions are essential for keeping self-renewal identified by Kaoru Mitsui *et al.* (2003) and Ian Chambers *et al.* (2003). It encodes an NK2-family homeobox transcription factor and is named for Tir Na Nog, the name of the land of the ever-young in Celtic myth, because its forced expression in mouse ES cells allow self-renewal in the absence of LIF. However, Nanog expression is not regulated by STAT3 directly, and Nanog cannot replace the function of *Oct-3/4*, so its function is still mysterious. Loss of function in the embryo resulted in the peri-implantation lethality, and that in ES cells induced differentiation to parietal endoderm-like cells with up-regulation of Gata-6. Since overexpression of Gata-6 triggers a similar differentiation event and its expression is up-regulated after withdrawal of LIF or STAT3 activity, one of the possible functions of Nanog might be repression of the Gata-6 expression.

Transcription Factors Involving ICM Outgrowth

Many possible transcriptional factors have been reported in the involvement in self-renewal of pluripotent stem cells. It was reported that overexpression of the homeobox transcription factor Pem replaces the LIF dependency of mouse ES cells, but strong expression of Pem is observed in differenti-

ated cell types such as extraembryonic tissues, and Pem-null animals develop normally without abnormality in pluripotent stem cells during early embryogenesis. In contrast, inhibition of Ehox activity results in the maintenance of a stem cell phenotype in limiting concentrations of LIF, but it will be necessary to confirm its function *in vivo*.

A defect of ICM outgrowth found in *Oct-3/4* or *Sox-2* mutant embryos might be regarded as a landmark of gene function in pluripotent cell populations. However, such a phenotype may not always reflect the abnormality of pluripotent stem cells themselves. For example, a mutation of the signal transduction adapter protein Disabled-2 (Dab-2) resulted in a defect of ICM outgrowth in which only primitive endoderm cells were maintained after one week. However, Dab-2 expression is detectable in primitive endoderm but not in the ICM, and the homozygous embryos that die at 5.5 dpc exhibit abnormal migration of primitive endoderm cells. Why do Dab-2-null embryos show a growth defect of pluripotent cell clumps? We think that it might be because of the functional deficiency of primitive endoderm cells essential for maintaining pluripotent cells as epiblast. Similar dissociation between abnormal phenotype in ICM outgrowth and function in pluripotent stem cells was evident more clearly in *Fgf-4*. Homozygous *Fgf-4*-null embryos showed a defect in ICM outgrowth, but ES cells lacking *Fgf-4* function were established by serial gene targeting. Therefore, not only a cell-autonomous defect but also a non-cell-autonomous defect can produce the defect of ICM outgrowth. We think that the case of the forkhead family transcription factor, Foxd3, may be the latter case because its function on differentiation of primitive endoderm has been pointed out in ES cells.

EFFECTOR MOLECULES TO RETAIN ES CELL SELF-RENEWAL

Prevention of Differentiation

To maintain ES cell self-renewal, entering the differentiation process should be strictly prevented. During differentiation, expression of various genes is up-regulated in a lineage-specific manner. The GATA family transcription factors Gata-4 and Gata-6 are specifically up-regulated during differentiation to primitive endoderm induced by overexpression of *Oct-3/4* or withdrawal of LIF. Their function in primitive endoderm differentiation *in vivo* was confirmed by gene targeting. Loss of Gata-4 resulted in functional deficiency in visceral endoderm, whereas loss of Gata-6 affected differentiation of primitive endoderm. Interestingly, Junji Fujikura and co-workers in our group found that ectopic expression of either Gata-4 or Gata-6 in ES cells activates expression of endogenous Gata-4 and Gata-6 and induces differentiation toward parietal endoderm. These data strongly suggest that one of the functions of *Oct-3/4* in maintaining ES cell self-renewal is the prevention of differentiation by repressing genes inducing differentiation. In the case of GATA factors, repression may occur indirectly by activation of a repressor. In contrast, genes involving trophectoderm differentiation, such as the homeobox transcription factor Cdx-2, may be directly inhibited by *Oct-3/4* because their expression is rapidly up-regulated as *Oct-3/4* is repressed. We observed that ectopic expression of Cdx-2 induced differentiation toward trophectoderm (unpublished data), indicating the significance of a gatekeeper function of *Oct-3/4*.

Maintenance of Stem Cell Proliferation

Proliferation of ES cells is achieved through their self-renewal. As in other cell types, it should be regulated by controlling the cell cycle. However, as described in the next section, cell cycle regulation in ES cells is unusual. For example, it may lack a major break of cell cycle, since the retinoblastoma gene is kept in an inactivated state. Unfortunately, no *Oct-3/4* target gene involving cell cycle regulation has been identified. STAT3 can activate several genes involving cell cycle regulation, such as *c-myc* in M1 leukemia cells and MCF7 cells, but none of them has been identified as the targets in ES cells.

Recently, a novel Ras family member was identified by its specific expression in ES cells by Kazutoshi Takahashi *et al.* (2003). This gene, *E-Ras*, encodes a constitutively active form and stimulates phosphatidylinositol-3-OH kinase. Loss of ERas activity in ES cells results in slow proliferation and poor tumorigenicity after transplantation, suggesting that its function is not essential but important for rapid proliferation of mouse ES cells.

Unlimited propagation requires the maintenance of telomeres. ES cells possess a constitutive telomerase activity, and loss of this activity results in limited growth. ES cells lacking detectable telomerase activity by deletion of telomerase RNA showed a reduced growth ratio after more than 300 divisions and almost zero after 450 cell divisions. Since the induction of differentiation results in the reduction of telomerase activity, maintenance of this activity should be coupled with maintenance of ES cell self-renewal.

Regulation of apoptosis is also important for continuous growth of stem cells. Since stem cells have the proliferative ability for self-renewal *in vivo*, they must strictly control it to reduce the incidence of tumorigenesis. Undifferentiated ES cells show a higher incidence of apoptosis against various stresses, such as ultraviolet irradiation, blockage of cell cycle, and oxidative stress, than differentiated cells. Such high susceptibility to apoptosis may contribute to keeping the mutation ratio in undifferentiated cells as low as in the differentiated cells by eliminating the cells that might be damaged. Function of the tumor suppressor gene *p53* in ES cells contributes to hypersensitivity to UV irradiation, but the contribution of the *p53*-independent pathway was also suggested.

Self-Renewal as a Marker of "Stemness"

ADULT PLURIPOTENT STEM CELLS

Since self-renewal is a common feature of stem cells, the molecular mechanism governing it might be shared among different types of stem cells, and it could be regarded as a marker of "stemness." However, at present, stocked information for each stem cell line is still too little to highlight their overlap. Although *Oct-3/4* expression is observed in pluripotent stem cells, such as ES, EC, and EG cells, a faint level of expression, 1/1000 of ES cells, was reported in multipotent adult progenitor cells. It may not have functional significance, since a 50% reduction induces differentiation, or it may be significant if 1 in 1000 cells expresses the level of *Oct-3/4* found in ES cells. No other stem cells express detectable levels of *Oct-3/4*, although neural stem cells express *Sox-2* at the level found in ES cells. Even if the molecule is not shared, the principle such as a gatekeeper function might be shared by different transcription factors in different stem cells.

ADULT MULTIPOTENT STEM CELLS: CLASSICAL ADULT STEM CELLS

Hematopoietic stem cells (HSCs) were adult stem cells identified and purified first. Analyses of knockout mice identified several essential genes such as *Scl*, *Gata-2*, and *Bmi-1* for self-renewal of HSCs *in vivo*, and functional screening *in vitro* revealed the stimulatory effect of Hoxb4 on self-renewal of HSCs. Moreover, several soluble factors have been identified on their ability to accelerate the proliferation of HSCs, which include LIF and Wnt, but it is still impossible to maintain self-renewal of HSCs continuously *in vitro*, indicating that there is yet unidentified factor(s) essential for it. Long-term self-renewal has been succeeded for neural stem cells (NSCs), but the typical culture condition, the neurosphere culture, cannot provide pure population of NSCs since they form aggregate with differentiated progenies. There is no functionally essential gene reported previously, although several marker genes for NSCs are available such as Nestin and Musashi. In contrast, the essential role of the paired-class homeobox gene *Pax-7* was reported in the satellite cells of skeletal muscle, a candidate of muscle stem cells. Mesenchymal stem cells and endotherial stem cells are an intriguing source of variety of cell types because of their wide range of multipotency, but a molecular mechanism supporting their self-renewal has not been revealed, which might be due to their heterogeneity.

Summary

ES cell self-renewal has been analyzed in the past two decades using mouse ES cells, and many principal molecules have been identified. The isolation of human ES cells has accelerated studies on ES cells of various organisms except rodents, and they have revealed both the common and different characteristics of these ES cell lines with different origins. Since we now know that there is some difference in the mechanism to maintain self-renewal between human and mouse ES cells, we will need to characterize human ES cells carefully in comparison to mouse ES cells for their future application on regenerative medicine. However, although our knowledge is still far from complete in understanding the molecular mechanisms governing self-renewal, ES cells can be regarded as one of the best characterized among stem cells. Therefore, results from ES cell studies can provide good models for other stem cell systems.

KEY WORDS

Pluripotency A differentiation ability to differentiate varieties of cells that belong to all three germ layers, at least one cell type for each. In contrast, totipotency is defined as an ability to generate a whole animal autonomously, whereas multipotency is defined as an ability to give multiple cell types that belong to particular germ layers, not all three. Unipotency means an ability to give a single differentiation cell type that is most limited ability of stem cells such as germ-line stem cells.

Self-renewal A style of cell division characteristic for stem cells that give at least one daughter cell with the same differentiation ability as the parental stem cells. Symmetric self-renewal generates two stem cells, whereas the asymmetric one gives one stem cells and one differentiated progeny. On the molecular level, this event can be divided into two mechanisms for preventing differentiation and promoting cell division.

Stem cell A cell that has abilities to self-renew and differentiation. Stem cells can be found as transient populations during development and stable ones in adult tissue. Embryonic stem cells are specified as stem cells derived from the pluripotent stem cells in early-stage embryos.

FURTHER READING

Avilion, A. A., Nicolis, S. K., Pevny, L. H., Perez, L., Vivian, N., and Lovell-Badge, R. (2003). *Genes Dev.*, **17**, 126–140.

Chambers, I., Colby, D., Robertson, M., Nichols, J., Lee, S., Tweedie, S., and Smith, A. (2003). *Cell* **113**, 643–655.

Evans, M. J., and Kaufman, M. H. (1981). *Nature*, **292**, 154–156.

Fujikura, J., Yamato, E., Yonemura, S., Hosoda, K., Masui, S., Nakao, K., Miyazaki , J. I., and Niwa, H. (2002). *Genes Dev.* **16**, 784–789.

Martin, G. R. (1981). *Proc. Natl. Acad. Sci. USA* **78**, 7634–7638.

Matsui, Y., Toksoz, D., Nishikawa, S., Williams, D., Zsebo, K., and Hogan, B. L. (1991). *Nature*, **353**, 750–752.

Mitsui, K., Tokuzawa, Y., Itoh, H., Segawa, K., Murakami, M., Takahashi, K., Maruyama, M., Maeda, M., and Yamanaka, S. (2003). *Cell* **113**, 631–642.

Nichols, J., Zevnik, B., Anastassiadis, K., Niwa, H., Klewe-Nebenius, D., Chambers, I., Scholer, H., and Smith, A. (1998). *Cell*, **95**, 379–391.

Niwa, H. (2001). *Cell Struct. Funct.*, **26**, 137–148.

Niwa, H., Miyazaki, J., and Smith, A. G. (2000). *Nat. Genet.*, **24**, 372–376.

Smith, A. G., Heath, J. K., Donaldson, D. D., Wong, G. G., Moreau, J., Stahl, M., and Rogers, D. (1988). *Nature*, **336**, 688–690.

Takahashi, K., Mitsui, K., and Yamanaka, S. (2003). *Nature* **423** 541–545.

Ying, Q. L., Nichols, J., Chambers, I., and Smith, A. (2003). *Cell*, **115**, 281–292.

Cell Cycle Regulators in Stem Cells

Tao Cheng and David T. Scadden

Introduction

Adult stem cells have defined therapeutic roles evident in clinical bone marrow transplantation. The promise of broader therapeutic use for adult stem cells has been fueled by the recent controversial finding that cells derived from one tissue type may display phenotypic characteristics of other tissue types under appropriate environmental cues. Therapeutic efficacy of stem cells in part depends on their proliferation; therefore, strategies to manipulate them require understanding of their cell cycle control. A significant hurdle restricting broader use of adult stem cells is their limited number and differentiation in response to proliferative stimuli, thus compromising *ex vivo* expansion efforts. Cell cycle regulators play key roles in this process. In this chapter, we do not intend to detail the biochemical pathways of general cell cycle regulation because they were largely obtained from other model systems and have been extensively reviewed elsewhere. Instead, we focus on the distinct cell cycle kinetics in stem cell populations and its molecular base exemplified by the defining roles of the CKIs in murine HSCs. Admittedly, those studies do not give the whole picture with regard to how the cell cycle in stem cells is controlled. Nevertheless, they underscore the importance of further investigation of other cell cycle regulators in stem cell biology and offer new paradigms for therapeutic manipulations of stem cells.

Cell Cycle Kinetics of Stem Cells *in Vivo*

As largely modeled in the hematopoietic system, maintenance of mature cell production requires a cytokine-responsive progenitor cell pool with prodigious proliferative capacity and a much smaller population of stem cells intermittently giving rise to daughter cells, some of which constitute the proliferative progenitor compartment. Under activating conditions such as after transplantation, an increase in stem cell divisions takes place as evidenced by depletion of cycling cells using the S-phase toxin (5-fluorouracil [5-FU] or hydroxyurea). However, relative quiescence or slow cycling in the stem cell pool appears to be essential to prevent premature depletion under conditions of physiologic stress over the lifetime of the organism. Therefore, the highly regulated proliferation of HSCs occurs at a very limited rate under homeostatic conditions.

Stem cell proliferation has been directly measured by bromodeoxyuridine (BrdU)-labeling experiments, and cell cycle lengths have been estimated at approximately 30 days in small rodents, or only about 8% of the cells cycling daily. Similar analyses using population kinetics have estimated that stem cells replicate once per 10 weeks in cats. In higher order primates, the frequency of cell division in the stem cell pool has been estimated to occur once per year. However, it is still not clear whether the relative quiescence reflects a complete cell cycle arrest of most cells in the stem cell compartment, termed the *clonal succession model*, or a very prolonged G_1 or G_2 phase of cycling stem cells. Although the retrovirus-based clonal marking studies indicated a dormancy of most stem cells at a given time, which supports the clonal succession model, this view has been challenged by the competitive repopulation model and by BrdU incorporation in defined stem cell pools.

In contrast, the essential feature of the progenitor population is that it irreversibly develops into maturing cells and in the process undergoes multiple, rapid cell divisions. The progenitor cell pool essentially operates as a cellular amplification machine generating many differentiated cells from the few cells entering the system. Therefore, it is directly responsible for the number of terminally differentiated cells. It is also termed the transit, amplifying cell pool. The differences between the stem and progenitor cell populations in different stages has been regarded as a phentotype distinction marking the stage of a cell within the hematopoietic cascade. However, an alternative model recently proposed is that the specific position in a cell cycle determines whether a primitive cell functions as a stem or a progenitor cell. In this model, stimuli received at distinct positions in the cell cycle provoke proliferation–differentiation and yield either stem or progenitor cell outcomes, thereby challenging the traditional view of "hierarchy" within the hematopoietic differentiation. In either model, "stemness" associates with the limited rate of cell proliferation.

Relative arrest of the cell cycle distinguishes stem from progenitor cells in other tissue systems as well. In the central nervous system (CNS), for example, evidence suggests that the proliferative pools of adult neural progenitors are derived from a quiescent multipotent precursor or neuronal stem cell (NSC). Ablation of the proliferative zone containing the lineage-committed neuronal progenitor cell (NPC) can be repopulated from a small number of quiescent NSCs. Perhaps largely because of this quiescence, endogenous NSCs do not produce complete recovery in cases of severe injury, although they do participate in self-repair after brain damage. Another example is found in the mouse dermal stem cell population. There are about fourfold fewer cells in S-G₂/M in the stem cell population compared with the progenitor pool, although both cell populations constantly proceed through the cell cycle. In summary, the dichotomy of relative resistance to proliferative signals by stem cells and the brisk responsiveness by progenitor cells is generally believed to be a central feature of tissue maintenance, although the distinctions between stem and progenitor cells in many nonhematopoietic organs are yet to be fully defined.

Stem Cell Expansion *Ex Vivo*

The relative quiescence of stem cells may prevent their premature exhaustion, but it is problematic in the context of the *in vitro* expansion necessary for transplantation and gene therapy. Methods for inducing stem cell proliferation have long been sought as a means to expand the population of cells capable of repopulating the marrow of ablated hosts and to render stem cells transducable with virus-based gene transfer vectors. Although great effort has been made to directly expand stem cells using different combinations of hematopoietic growth factors (cytokine cocktails), few culture systems have been applied in the clinical setting at least in part because of the lack of proof that any of the culture conditions support expansion of long-term repopulating HSCs in humans. Gene-marking studies in large animals, including primate and human, indicate poor transduction in the stem cell compartment during long-term engraftment. These cytokine-based efforts to expand stem cells have often resulted in increased cell numbers but at the expense of multipotentiality and homing ability. Although data suggest that under some specific conditions murine HSCs may divide *in vitro*, net expansion is achieved in limited fashion and is always associated with, and often dominated by, cellular differentiation. Recent studies on the potent effect of Notch ligands and wnt proteins on stem cell expansion *in vitro* are promising. However, whether such "successful" protocols can be adapted to clinically useful human HSC expansion remains to be determined.

Alternative approaches aimed at targeting the negative regulatory cytokines have also been sought to activate or expand stem cells. Factors such as TGFβ-1 and macrophage inhibitory protein-1α (MIP-1α) have been noted to play a role in dampening hematopoietic cell growth kinetics. In particular, TGFβ-1 has been shown to be able to selectively inhibit the growth of HSCs and progenitor cells. Antisense oligonucleotides, or specific neutralizing antibodies, have been shown to permit primitive hematopoietic cell entry into the cell cycle and to enhance the efficiency of retroviral transduction into those cells. However, long-term engraftment after *in vitro* manipulation by these methods remains to be defined in an *in vivo* model. Furthermore, more efficient and specific methods, such as RNA interference technology, aimed to knock down the essential elements in the inhibitory circuits of stem cell expansion have yet to be developed.

Although HSCs may be induced to divide *in vitro*, it remains unclear which combination of factors is specific for stem cell proliferation without differentiation. The complex microenvironments ("niches") in which the stem cells reside and the intrinsic properties of HSCs in relation to the environmental cues are largely unknown. The ultimate success of stem cell expansion *in vitro* will require greater understanding of the interplay between stem cell and microenvironment and the signaling circuitry involved in achieving self-renewal.

Mammalian Cell Cycle Regulation and Cyclin-dependent Kinase Inhibitors

The molecular principles of cell cycle regulation have been largely defined in yeast, and with orthologous systems it is applicable to the mammalian cell cycle. A number of surveillance checkpoints monitor the cell cycle and halt its progression, mainly via the p53 pathway, when DNA damage occurs and cannot be repaired. In eukaryotic cells, factors that determine whether cells will continue proliferating or cease dividing and differentiate appear to operate mainly in the G₁ phase of the cell cycle (Figure 7-1). Cell cycle progression is regulated by the sequential activation and inactivation of CDKs. In somatic cells, movement through G₁ and into S phase is driven by the active form of the Cyclin D1, 2,3/CDK4, 6 complex and the subsequent phosphorylation retinoblastoma (Rb) protein. Once Rb is phosphorylated, the critical transcription factor, E2F-1, is partially released from an inhibited state and turns on a series of genes, including cyclin A and cyclin E, which form a complex with CDK2 and cdc25A phosphatase. The cdc25A is able to remove the inhibitory phosphates from CDK2, and the resultant cyclin E/CDK2 complex then further phosphorylates Rb, leading to a complete release of E2F and the transcription of multiple other genes essential for entry into S-phase and DNA synthesis. In parallel, the c-myc pathway also directly contributes to the G₁–S transition by elevating the transcription of genes for cyclin E and cdc25A (Figure 7-1). CDK activity is strictly dependent on cyclin levels that are regulated by ubiquitina-

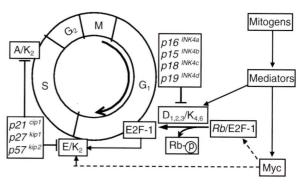

Figure 7-1. Cell cycle regulators in G₁ and S phases (description detailed in the text).

tion and subsequent proteolysis. On mitogenic stimulation, cyclin D serves as an essential sensor in the cell cycle machinery and interacts with the CDK4/6-Rb-E2F pathway.

In addition to regulation by cyclins and phosphorylation/dephosphorylation of the catalytic subunit, CDKs are largely controlled by CKIs. Two families of low molecular weight CKIs, Cip/Kip and INK4, are capable of interacting with CDKs to impair progression through the G_1 phase. The Cip/Kip family, which includes p21$^{Cip1/Waf1}$, p27^{kip1}, and p57^{Kip2} (p21, p27, and p57 hereafter), may interact with a broad range of cyclin–CDK complexes, whereas the INK4 family, p16^{INK4A}, p15^{INK4B}, p18^{INK4C}, and p19^{INK4D} (p16, p15, p18, and p19 hereafter), specifically inhibit CDK4 and CDK6 kinases. The detailed biochemical roles of CKIs have been reviewed by Sherr and Roberts (1999). Both families have been shown to have essential roles in arresting cell cycle progression in a number of model systems. Studies using antisense strategies have been able to release the cells in G_0 stage into the cell cycle. Knockout analysis in rodent models has provided a strong basis for further exploring those molecules in stem cell biology. Interestingly, the CKIs p27 and p18 have a profound effect on overall cellularity and organ size, resulting in a larger whole animal when either gene is knocked out.

In addition, there appears to be a distinct cell cycle control operating in stem cells to maintain their stemness. This has been shown in mouse embryonic stem (ES) cells with a "defective" Rb pathway and a nonresponsive p53 pathway. Finally, because stem, progenitor, and more differentiated cells share many common cytokine receptors, it is likely that the distinct cell cycle profile in stem cells must be mediated by either distinct upstream intracellular mediators or unique combinatorial relationships of common biochemical mediators limiting the intensity of signals to enter into cell cycle. Defining the basis for the participating mechanisms in the stem cell response requires stepwise analysis of individual cell cycle regulators and ultimately a systems approach to define how these cell cycle regulators interact with one another and intersecting signaling pathways.

Roles of Cyclin-dependent Kinase Inhibitors in Stem Cell Regulation

Although significant progress has been made in our understanding of how the cell cycle is regulated in a variety of other model systems, little is known about how the cell cycle is molecularly controlled in stem cells. Given the relative quiescence of stem cells *in vivo*, a reasonable starting point is the analysis of cell cycle inhibitors and whether reduction of their cell cycle blockade may be a mechanism for enabling stem cell entry into the cell cycle (Figure 7-2).

CKIs have been demonstrated to be involved in a number of stem and progenitor cell systems. *Dipio*, an analog of p21/p27 in *Drosophila*, has been reported to control embryonic progenitor proliferation. Recent studies of knockout mice and cells that lack a specific CKI have begun to clarify their unique activity in stem cell populations. An increased stem or progenitor cell potential has been found in p21–/– or p27–/–

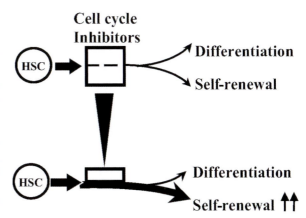

Figure 7-2. Model of targeting cell cycle inhibitors for enhancement of stem cell self-renewal.
Top: Under the homeostatic condition. *Bottom:* Potential novel approach for stem cell expansion by targeting the cell cycle inhibitors.

mice when dermal, neural, or otic tissues were assessed. In the hematopoietic system, most CKI family members have been found to be expressed in CD34$^+$ cells, although the expression patterns differ. One CKI with abundant mRNA in quiescent HSCs with reduced levels in progenitor populations is p21. Functional assessment of this CKI in stem cell biology has been carried out using p21–/– mice.

ROLES OF p21 IN STEM CELL REGULATION

In the absence of p21, HSC proliferation and absolute number were increased under normal homeostatic conditions. Exposing the animals to cell cycle-specific myelotoxic injury resulted in premature death because of hematopoietic cell depletion. Furthermore, self-renewal of primitive cells was impaired in serially transplanted bone marrow from p21–/– mice leading to hematopoietic failure. Therefore, p21 governs cell cycle entry of stem cells, and, in its absence, increased cell cycling leads to stem cell exhaustion. Under conditions of stress, restricted cell cycling is crucial to prevent premature stem cell depletion and hematopoietic death. These findings have been recently extended to human cells and to a nondevelopmental context. Using postnatal CD34$^+$ CD38$^-$ human cells, it was shown that interrupting p21 expression with lentivectors *ex vivo* resulted in expanded stem cell number, which was validated by increased function in the transplantation assay using irradiated NOD/SCID mice recipients. Such a study further supports an alternative paradigm for increasing HSC numbers by releasing the brake on cell cycle entry rather than focusing on combination of proproliferative cytokines. Importantly, these data further supported the notion that postnatal human stem cell proliferation can be uncoupled from differentiation in *ex vivo* settings.

Interestingly, the role of p21 has been paradoxically noted to positively affect proliferation capability following cytokine stimulation in progenitor cell pools. This may be due to the requirement for p21 to promote the association of CDK4 with

D-type cyclins. LaBaer and colleagues demonstrated that low concentrations of p21 promote assembly of active kinase complexes and thereby entry into cycle, whereas higher concentrations are inhibitory. The stoichiometry of p21 and cyclin–CDK complexes appears to be crucial in determining the relative effect on movement of the cell through late G_1 into S phase. This was further confirmed in a study showing that p21 and p27 are essential activators of cyclin D-dependent kinases in murine fibroblasts. Mantel *et al.* noted that bone marrow progenitor cells from mice proliferated poorly and formed few colonies with thymidine treatment except when transduced with a p21-encoding retroviral vector. Similarly, we noted a transient rise in p21 immediately following release of cell cycle arrest in 32D cells. Therefore, as observed in other systems, p21 has a dual function in the hematopoietic system depending on the differentiation stage and CDK complex type and status. In addition, complex roles of p21 in apoptosis or differentiation may participate in stem cell regulation, although these functions are yet to be thoroughly investigated.

Why p21 expression is elevated in HSCs is unclear, but two upstream regulators have been assessed. WTI is known to induce p21 transcription, and overexpression of WTI results in altered stem cell cycling and differentiation in primary hematopoietic cells. However, null mutant mice for WTI do not appear to have a stem cell defect. It is well known that p21 is also transcriptionally regulated by p53 and serves as a downstream mediator of cell cycle arrest induced by the p53 pathway. It would be logical to expect the HSC phenotype of the p53−/− animal to be similar to that of p21−/−. Interestingly, in the absence of p53, HSC function has been reported to be significantly enhanced under stress conditions in a manner opposite to that in the absence of p21. Because p53 mediates apoptosis in many cell types, the enhanced function of HSCs in the absence of p53 suggests that in some settings increased survival may dominate over accelerated proliferation of HSCs.

ROLES OF p27 IN STEM CELL REGULATION

Of particular interest to tissue regeneration is p27 because of its direct involvement in cell cycle-mediated hyperplasia. Direct flow cytometric analysis shows p27 expression in primitive cells and in more mature progenitors, supporting the hypothesis that p27 has a role in hematopoiesis. This role is also supported indirectly by improved retroviral transduction following knockdown of p27 with an antisense oligonucleotide. The p27 appears to accumulate at points in which signals for mitosis affect cell cycle regulators and has been shown to serve as an important regulator at a restriction point for mitogenic signals. Because progenitor cells have a robust response to growth factors, p27 likely plays a specific role in progenitor cell pools. Its role may be quite distinct from that of p21, which is shown to be a molecular switch for stem cell cycle control. Disruption of the p27 gene has resulted in a mouse with hyperplasia of a number of organs (including hematopoietic tissues) and spontaneously generated tumors of specific type.

Using p27−/− mice, researchers have reported that p27 does not affect stem cell number, cell cycling, or self-renewal but markedly alters progenitor proliferation and pool size. When competitively transplanted, p27-deficient stem cells generated progenitors that eventually dominated blood cell production. Thus, modulating p27 expression in a small number of stem cells may translate into effects on most mature cells, thereby providing a strategy for potentiating the impact of transduced cells in stem cell gene therapy. Such a dramatic effect of p27 absence on hematopoietic reconstitution was also observed in liver regeneration, and a specific role of p27 in the committed progenitor cells not at the stem cell level was also reported in the mouse CNS. Therefore, distinct roles for p27 and p21 have been defined in hematopoiesis, and indirect evidence suggests that these distinctions may be preserved across stem and progenitor pools from multiple different adult tissues.

OTHER CYCLIN-DEPENDENT KINASE INHIBITORS AND THE RETINOBLASTOMA PATHWAY IN STEM CELL REGULATION

One of the best studied pathways in cell cycle regulation is that of Rb, which directly interacts with Cyclin D and the INK4 proteins in early G_1 phase and serves as a critical and initial interface between mitogenic stimuli and cell fate commitment following division. A role for Rb in stem cell regulation is indirectly supported by the finding that ES cells do not have intact G_1 machinery but that acquisition of Rb pathway products induces the transition from symmetric to asymmetric cell division, which is a critical feature of mature stem cell function. Mice deficient in Rb are not viable and show defects in multiple tissue types, including the hematopoietic lineage.

Although deficient hematopoiesis in Rb−/− mice indicated that this protein might be critical to stem cell function, more definitive studies in the stem cell compartment have not been reported, likely because of the early lethality of the Rb null embryo. Instead, INK4 proteins closely associated with Rb have been studied in the context of stem cell biology. These studies include a recent report indicating that Bmi-1, an upstream inhibitor of p16^{INK4A}(p16) and p19^{INK4D} (p19), expression, is critical for HSC self-renewal. In the absence of Bmi-1, self-renewal of HSCs and neural stem cells is diminished, an effect dependent on the expression of p16. In mice engineered to be devoid of p16, a complex HSC phenotype has been observed, with a decreased number of stem cells in younger mice but enhanced self-renewal in serially transplanted animals.

The INK4 family member, p18^{INK4C} (p18 hereafter), is expressed in multiple tissue types including hematopoietic cells, the loss of which in mice results in organomegaly with higher cellularity and an increase in the incidence of tumors with advanced age or in the presence of carcinogens. Furthermore, it has been suggested that p18 is involved in the symmetric division of precursor cells in mouse developing brain. Recently, it has been reported that the absence of p18 increases HSCs. Similar to the p21 null setting, but unlike p21, an increase in stem cell self-renewal is observed.

Systematic evaluation of proximate molecular regulators of cell cycling is, therefore, yielding a complex picture of how each influences primitive cell function. Different members of the CKI subfamilies appear to play distinct roles in stem or progenitor cell populations. The function of these different CKIs appears to be highly differentiation-stage specific and confers an important level of regulation in stem or progenitor cells to maintain homeostasis. Cooperative effects between members of the two CKI subfamilies are likely, with evidence of such interplay between p15 and p27 now documented. How the CKIs exert distinct effects and the pathways converging on these regulators are the subject of ongoing study and will potentially provide further insight for manipulation of stem and progenitor populations. Whether these pathways are shared among primitive populations of cells in all tissues is not yet clear, but preliminary data suggest that such is the case.

RELATION BETWEEN CYCLIN-DEPENDENT KINASE INHIBITORS AND TRANSFORMING GROWTH FACTOR β-1

TGFβ-1 has been documented to have varied effects on hematopoietic cells, including enhancement of granulocytes proliferation in response to granulocyte-macrophage colony-stimulating factor or inhibition of progenitor cell responsiveness to other growth-promoting cytokines. The detailed roles of TGFβ-1 in signaling pathways and in hematopoiesis have been extensively reviewed elsewhere. TGFβ-1 has been extensively characterized as a dominant negative regulator of hematopoietic cell proliferation, including inhibiting primitive progenitor cells. Antisense TGFβ-1 or neutralizing antibodies of TGFβ-1 have been used to induce quiescent stem cells into the cell cycle and to augment retroviral gene transduction in conjunction with down-regulation of p27 in human CD34+ subsets. Based on the roles of CKIs in hematopoietic cells as described previously, the link between TGFβ-1 and CKIs in stem cell regulation has been recently addressed. TGFβ-1-induced cell cycle arrest has been shown to be mediated through p15, p21, or p27 in multiple cell lines or cell types, including human epithelial cell lines, fibroblast cells, and colon and ovary cancer cell lines. Recently, it was proposed that p21 and p27 are key downstream mediators for TGFβ-1 in hematopoietic cells, and a study examined whether p21 or p27 was a proximal mediator for TGFβ-1-induced cell cycle exit in primary hematopoietic cells. Using fine mapping of gene expression in individual cells, researchers documented TGFβ-1 and p21 to be up-regulated in quiescent, cytokine-resistant HSCs and also in terminally differentiated mature blood cells as compared with immature, proliferating progenitor cell populations. Type II TGFβ-1 receptors were expressed ubiquitously in these subsets of cells without apparent modulation.

To provide further biochemical analysis of whether the coordinate regulation of TGFβ-1 and p21 or p27 represented a dependent link between them, the cytokine-responsive 32D cell line was analyzed for p21 or p27 up-regulation following cell cycle synchronization and release in the presence or absence of TGFβ-1. Despite marked antiproliferative effects of TGFβ-1, neither the transcription of p21 mRNA nor the expression of p21 or p27 was altered. To corroborate these observations in primary cells, bone marrow mononuclear cells derived from mice engineered to be deficient in p21 or p27 were assessed. Both progenitor and primitive cell function was inhibited by TGFβ-1 equivalently in knockout and wild-type littermate controls. This data indicated that TGFβ-1 exerts its inhibition on cell cycling independent of p21 and p27 in primitive hematopoietic cells.

Other data have recently reported examining the Cip/Kip CKI family member p57 in hematopoietic progenitors. The absence of p57 was associated with a lack of responsiveness to TGFβ, failing to arrest cell cycling. Furthermore TGFβ was noted to induce p57 expression, arguing for a direct link between TGFβ and the cell cycle regulatory function of p57. In addition, Dao et al. reported that blocking TGFβ-1 could down-regulate p15 expression in human CD34+ cells and that TGFβ-1 may act through the INK4 family and the Cip/Kip family in hematopoietic cells. However, extensive biochemical analysis in primary hematopoietic cell subsets is needed to further address this question.

CKIS AND NOTCH

Notch1 has been well defined as a mediator of decisions at multiple steps in the hematopoietic cascade, including stem cell self-renewal versus differentiation. Variable effects on cell cycling have been reported, including inhibition of proliferation and, in contrast, maintenance of proliferation but with a decreased interval in the G_1 phase of the cell cycle. The latter observation has been followed up by more extensive analysis of an interaction between Notch1 and CKI regulation. It has been reported that the basis for Notch influencing G_1 may be through alteration in G_1–S check point regulator stability, specifically affecting the proteasome degradation of CKI, p27. The links of receptor mediated effectors of stem cell function and cell cycle regulators are, therefore, beginning to emerge and provide an essential component of the larger regulatory network.

Summary

Further therapeutic potential of stem cells is envisioned to be broadened if the biology of the stem cells can be exploited to permit efficient *ex vivo* manipulation and enhance repopulation *in vivo*. Given the relative quiescence of HSCs that has not been satisfactorily overcome by cytokine manipulation *in vitro*, direct intervention in the control of the cell cycle has been sought as an approach to disassociate cell proliferation from cell differentiation, thereby potentially bypassing a major hurdle in current stem cell expansion strategies. CKIs appear to be compelling candidates for the latter approach. In particular, we have learned that p21 and p27 govern the pool size of hematopoietic stem or progenitor cells, respectively, and their inhibitory roles in hematopoietic cells are not dependent on the action of TGFβ-1. Therefore, targeting specific CKIs together with TGFβ-1 may provide complementary strategies for enhancing hematopoietic stem or progenitor cell expansion and gene transduction. Controlled manipulation of specific CKIs directly or through their upstream mediators may also be relevant for the expansion or possible regeneration of other non-HSC pools.

As the roles of cell cycle regulators in the molecular control of stem cells are explored, many issues remain to be addressed. Much work is needed to delineate the roles of individual members of the cell cycle machinery within the context of tissue-specific stem cell types. Furthermore, how these relate to one another and to signal transduction pathways that operate uniquely in specific stem cell populations has yet to be elucidated. Coupling extrinsic signals to cycle control will be essential first steps before understanding how a stem responds to the complicated setting of its microenvironment. Piecing together the components and their interactions by either a reductionist or systems approach will offer targets for a more rational manipulation of the stem cell.

ACKNOWLEDGMENTS

This work was supported by National Institutes of Health (NIH) grants DK02761, HL70561 (T.C.), HL65909, and DK50234; the Burroughs Wellcome Foundation; and the Doris Duke Charitable Foundation (D.T.S.). Because of the limited space and the targeted topic, many related publications cannot be cited in this chapter. We very much appreciate the important contributions from many other investigators in the field. We thank Mathew Boyer for his assistance in preparing this manuscript.

KEY WORDS

CDK inhibitors (CKIs)   Intracellular molecules with low molecular weight specifically inhibiting the activities of cyclin-dependent kinases during cell cycle progression. There are two subfamilies of CKIs, including Cip/Kip ($p21^{Cip1/Waf1}$, $p27^{kip1}$, and $p57^{Kip2}$) and INK4 ($p16^{INK4A}$, $p15^{INK4B}$, $p18^{INK4C}$, and $p19^{INK4D}$).

Stem cell quiescence   Mitotic quiescence or slow cycling of adult stem cells in comparison with the lineage-committed progenitor cells. It is generally associated with G_0 or prolonged G_1 phase in cell cycle.

Stem cell self-renewal   The process by which a stem cell replicates itself. It is not a synonym of stem cell proliferation since the proliferation may be also accompanied by cell differentiation.

FURTHER READING

Bradford, G. B., Williams, B., Rossi, R., and Bertoncello, I. (1997). Quiescence, cycling, and turnover in the primitive hematopoietic stem cell compartment. *Exp. Hematol.* **25**, 445–453.

Cheng, T., Rodrigues, N., Dombkowski, D., Stier, S., and Scadden, D. T. (2000). Stem cell repopulation efficiency but not pool size is governed by $p27^{kip1}$. *Nat. Med.* **6**, 1235–1240.

Cheng, T., Rodrigues, N., Shen, H., Yang, Y., Dombkowski, D., Sykes, M., and Scadden, D. T. (2000). Hematopoietic stem cell quiescence maintained by p21cip1/waf1. *Science* **287**, 1804–1808.

Cheng, T., Shen, H., Rodrigues, N., Stier, S., and Scadden, D. T. (2001). Transforming growth factor beta 1 mediates cell-cycle arrest of primitive hematopoietic cells independent of p21(Cip1/Waf1) or p27(Kip1). *Blood* **98**, 3643–3649.

Harrison, D. E., Astle, C. M., and Lerner, C. (1988). Number and continuous proliferative pattern of transplanted primitive immunohematopoietic stem cells. *Proc. Natl. Acad. Sci. U. S. A.* **85**, 822–826.

Lemischka, I. R., Raulet, D. H., and Mulligan, R. C. (1986). Developmental potential and dynamic behavior of hematopoietic stem cells. *Cell* **45**, 917–927.

Pardee, A. B. (1989). G1 events and regulation of cell proliferation. *Science* **246**, 603–608.

Qiu, J., Takagi, Y., Harada, J., Teramoto, T., Rodrigues, N., Moskowitz, M., Scadden, D. T., and Cheng, T. (2004). Regenerative response in ischemic brain restricted by p21cip1/waf1. *J Exp Med* **199**, 937–45.114.

Sherr, C. J., and Roberts, J. M. (1999). CDK inhibitors: positive and negative regulators of G1-phase progression. *Genes Dev.* **13**, 1501–1512.

Yuan, Y., Shen, H., Franklin, D. S., Scadden, D.T., and Cheng, T. (2004, May). In vivo self-renewing divisions of haematopoietic stem cells are increased in the absence of the early G1-phase inhibitor, p18(INK4C). *Nat. Cell Biol.* **6**(5), 436–42.115.

8

Epigenetic Mechanisms of Cellular Memory During Development

N. D. Montgomery, T. Magnuson, and S. Bultman

Introduction

CHROMATIN AND TRANSCRIPTIONAL REGULATION DURING DEVELOPMENT

Regulation of gene expression is inherently more complicated in eukaryotes than in prokaryotes because the transcriptional machinery must recognize a chromatin template instead of naked DNA (Li, 2002; Jenuwein and Allis, 2001) (Figure 8-1). Consequently, chromatin-modifying factors must play crucial roles in transcriptional regulation and be involved in the ability of stem cells to proliferate and differentiate into genetically identical but functionally diverse cell types. This supposition is supported by the fact that mutations in chromatin-modifying factors generated by gene targeting in the mouse confer mutant phenotypes ranging from early embryonic lethality to neoplasia (Li, 2002). As described below, the significance of chromatin-modifying factors is also demonstrated by normal and abnormal epigenetic events that have been documented in cloning experiments.

THE IMPORTANCE OF CHROMATIN IS UNDERSCORED BY UNEXPECTED DEVELOPMENTAL DEFECTS IN CLONED ANIMALS

The first cloned cat, called Cc for either copy cat or carbon copy, is of particular interest because she has different coat-color markings than her genetic donor, despite being genetically identical. This observation indicates that DNA sequence alone is not sufficient to confer certain genetic traits. Instead, a heritable influence must exist that is not based on DNA sequence. This influence, referred to as *epigenetics*, is not as well understood as the genetic code, but it is becoming increasingly clear that it involves the ability of chromatin structure to affect transcription at the level of individual genes, clusters of genes, or even whole chromosomes. In the case of Cc, the apparent paradox of her unexpected coat-color markings can be reconciled by an epigenetic event called X-chromosome inactivation. Female mammals have two X-chromosomes, whereas males have only one. Consequently, one of the two X-chromosomes is inactivated stochastically

on a cell-by-cell basis in females during early embryogenesis and maintained in a clonal manner throughout the numerous cell divisions that occur during development.

The primary result of this process is to achieve the same level of X-linked gene activity in females as males (i.e., dosage compensation), but the stochastic nature of X-chromosome inactivation also results in individual females having unique distributions of cells where one or the other X-chromosome is inactivated. This mosaicism is usually not visibly evident, but it can be readily observed in Cc because her two X-chromosomes carry different alleles of a pigmentation gene partly responsible for calico coat color. The allele conferring orange color was inactivated in a different subset of pigment producing cells (i.e., melanocytes) than the cat from which she was derived, so Cc had a different pattern of orange, black, and white coloration.

Not surprisingly, cloning is a very inefficient process compared to *in vitro* fertilization or more conventional embryological manipulations. In fact, Cc was the only clone to survive out of over 200 nuclear transfer attempts, which is consistent with success rates of less than 1% up to 3% for cattle, sheep, goats, pigs, and mice. Many clones fail to reach the blastocyst stage and successfully implant into the uterus. Clones that progress beyond this stage often die *in utero* because of placental defects or die shortly after birth with increased birth weight, respiratory distress, and/or cardiovascular defects (Li, 2002). Furthermore, although it is too soon to assess the long-term health of cloned cattle, sheep, goats, and pigs that have survived, cloned mice are known to have impaired immune systems, become obese, and die prematurely.

Unlike X-chromosome inactivation, which occurred normally in Cc, the embryonic and postnatal lethality and the adult obesity in many other clones is probably due to abnormal epigenetic events stemming from inappropriate chromatin structure, leading to deregulated expression of *Oct4* and many other genes (Li, 2002). This notion is supported by evidence that cloned embryos sometimes exhibit aberrant patterns of DNA methylation at CpG dinucleotides and abnormal expression of both imprinted and nonimprinted genes. Many more chromatin-based abnormalities will probably be detected utilizing chromatin immunoprecipitation (ChIP) assays.

A

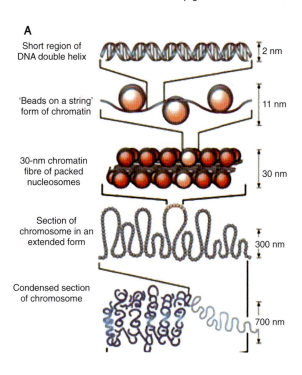

Short region of
DNA double helix — 2 nm

'Beads on a string'
form of chromatin — 11 nm

30-nm chromatin
fibre of packed
nucleosomes — 30 nm

Section of
chromosome in an
extended form — 300 nm

Condensed section
of chromosome — 700 nm

B

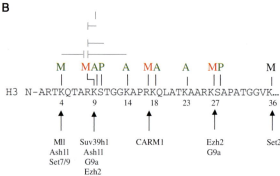

Figure 8-1. Chromatin and covalent histone modifications. **(A)** Schematic of chromatin structure. 147-bp segments of DNA (top level) wrap 1.65 times around histone octamers to form nucleosomes (second level). Arrays of nucleosomes assemble into 30-nm solenoid structures (third level), considered to be the fundamental unit of higher-order chromatin. Extensive looping (fourth level) and condensation (fifth level) brought about by various proteins result in the final configuration of the interphase nucleus, which undergoes further condensation to yield metaphase chromosome during mitosis (bottom panel). After returning to interphase, some regions of the genome remain highly condensed and are referred to as heterochromatin. Reproduced with permission from *Nature*. **(B)** Methylation (M), acetylation (A), and phosphorylation (P) of specific amino acids on H3 N-terminal tail. Position of lysines (K) are indicated. Except for R19 (black), which functions in a context-dependent manner, all of the modifications either promote (green) or inhibit (red) transcription. Moreover, K9 methylation is mutually exclusive with certain neighboring modifications (indicated above). Histone methyltranferases are shown below the residues they are able to modify.

COORDINATE ACTION OF CHROMATIN-MODIFYING FACTORS

Over the last few years, chromatin research has made tremendous progress by utilizing genetic, biochemical, and molecular approaches in a broad range of model organisms. An important theme that has emerged from these multidisciplinary efforts is that DNA methylation, histone modifications, and ATPase chromatin-remodeling complexes are functionally interdependent. It is becoming evident that these interactions also apply to DNA replication and repair during interphase of the cell cycle, chromosome segregation in mitosis, and recombination in meiosis. A second important theme is that chromatin-modifying factors from each of these three general categories underlie some of the most well-studied epigenetic processes: position-effect variegation (PEV), X-chromosome inactivation, Polycomb group (Pc-G) silencing and trithorax group (trx-G) activation, and monoallelic expression in both genomic imprinting and allelic exclusion. Recent work has provided considerable insight into the molecular mechanisms and has revealed striking similarities.

In the sections that follow, we review what is currently known about DNA methylation, the histone code, and ATPase chromatin-remodeling complexes. Subsequently, we discuss the role of these factors in PEV, X-chromosome inactivation, Pc-G silencing and trx-G activation, and imprinting.

Chromatin-Modifying Factors

DNA METHYLATION

Mammalian genomes are extensively methylated, particularly at cytosines of CpG dinucleotides, which are symmetrically methylated on complementary strands of DNA (Bird, 2002). Programmed changes in patterns of this cytosine methylation during embryogenesis suggest that methylation may play an important role in cell fate specifications. Shortly after fertilization, the male pronucleus is rapidly and actively demethylated. Interestingly, the female pronucleus is resistant to this active demethylation, perhaps because the paternal genome only becomes accessible to putative demethylases during the protamine-to-histone conversion. From the two-cell stage to the blastocyst stage, DNA methyltransferases are largely excluded from nuclei, and both the paternal and maternal genomes undergo passive demethylation, such that the genome of implantation stage embryos is globally demethylated. Important exceptions are imprinted genes and certain retroviral-like elements, which maintain their methylation status during cleavage divisions. After implantation, cells of the mouse embryo proper are remethylated. In contrast, mouse extraembryonic cells experience only modest remethylation, suggesting that DNA methylation plays a less prominent role in these tissues.

Independent waves of demethylation and remethylation also occur in the germ line. Primordial germ cells are actively demethylated in both sexes beginning around mouse embryonic day (E) 10.5 and finishing by E12.5. This demethylation erases the parental legacy of alleles and is presumably critical for imprinting. Subsequently, male germ cells are remethylated late in embryonic development during the leptotene stage of meiosis; female germ cells are remethylated postnatally, at around P6, during oocyte maturation.

Several DNA methyltransferase enzymes have been identified and characterized. These enzymes are separated into two functional classes — the *de novo* and maintenance methyltransferases. DNMT3A and DNMT3B are *de novo* methyltransferases responsible for remethylation in postimplantation embryos and in germ cells. *De novo* methyltransferase activity is stimulated by DNMT3L, which resembles DNMT3A and DNMT3B but lacks the methyltransferase catalytic domain and thus is thought to lack methyltransferase activity of its own. DNMT1 represents the second functional class of DNA methyltransferases. DNMT1 is a maintenance methyltransferase that prefers hemimethylated templates and is recruited to actively replicating DNA through an association with PCNA, the replication fork clamp. Once recruited, DNMT1 propagates the methylated state by copying methylation onto the nascent strand.

DNA methylation is not inherently repressive, and the transcriptional consequences of DNA methylation appear to be mediated by interplay with other chromatin-modifying factors. Cytosine methylation alters the substrate presented to DNA binding proteins, imposing allosteric constraints that favor some interactions but block others. Methylated CpG dinucleotides in and around promoters often block transcriptional activator binding and recruit chromatin-modifying enzymes that promote higher-order, repressive chromatin conformations. Methyl-CpG binding proteins, such as MBD2 and MeCP2, can recruit histone deacetylase, histone methyltransferase, and ATP-dependent nucleosome remodeling complexes, which all work in concert to restrict access of the general transcription machinery to the DNA.

The importance of DNA methylation to cell physiology is perhaps best illustrated by the pathologies that result when the process of establishing and reading these marks go awry. Much as the exquisite fidelity of DNA replication is not absolute, somatic cells accumulate epimutations during aging and carcinogenesis. A variety of tumor suppressors, including *RB* and *p16*, are silenced by aberrant methylation of promoter CpG islands in human tumors, and mutations in *DNMT3B* cause a disease called ICF (I̲mmunodeficiency, C̲entromeric region instability, and F̲acial abnormalities). In other diseases, methylation patterns are maintained more or less appropriately, but the cellular machinery responsible for recognizing those marks is disrupted. For instance, mutation of *MECP2* causes a neurological disorder known as Rett syndrome, and mice mutant for *Mbd2* exhibit behavioral defects, including compromised maternal nurturing of pups.

THE HISTONE CODE

DNA methylation is important for the proper expression of many genes and is perturbed in some diseases, but it cannot account for many other aspects of transcriptional regulation. First, most CpG islands are unmethylated, regardless of whether or not the associated gene is expressed. Second, many cell types are properly specified despite having globally hypomethylated genomes in *Dnmt1⁻/⁻* embryos. Lethality does not occur until closer to midgestation at E9.5–10.5. Third, and perhaps most importantly, the nematode *Caenorhabitis elegans* lacks detectable cytosine methylation, and the fruit fly *Drosophila melanogaster* has only trace levels restricted to the earliest stages of embryogenesis. Thus, evolutionary conserved mechanisms other than DNA methylation must be crucial for transcriptional regulation.

In this regard, histones are intimately associated with DNA and are among the most abundant and highly conserved proteins from yeast to human. Not only have the amino-acid sequences and three-dimensional structures been conserved to a remarkable extent but so have a variety of post-translational modifications. These modifications are covalent and consist of acetylation, methylation, phosphorylation, ubiquitination, and polyribosylation of specific residues of H2A, H2B, H3, and H4, usually in the N-terminal tails that protrude away from nucleosomes. For example, Figure 8-1B shows the position and nature of modifications that occur on H3 N-terminal tails. Each of these modifications either promotes or inhibits transcription. The situation is complicated by the fact that some of the opposing modifications inhibit each other and are therefore mutually exclusive. All of these modifications and the interplay that occurs between them are referred to as the histone code (Jenuwein and Allis, 2001). Briefly, enzymes such as histone acetyltransferases (HATs) and histone methyltransferases (HMTs) write the code by introducing specific modifications. These modifications then serve as docking sites for proteins that read and execute the code. For example, acetylated lysines can be recognized by the bromodomains of subunits of ATPase remodeling complexes to promote transcription. In contrast, certain methylated lysines can be bound by the chromodomain of HP1, or Pc-G proteins, to facilitate higher-order chromatin structures that inhibit transcription. Considerable interplay and a two-way flow of epigenetic information also occurs between histone modifications and DNA methylation. For example, just as DNA methylation can influence histone deacetylation, genetic screens in *Arabadopsis* and *Neurospora* have uncovered mutations in HMTs (*DIM5* and *KRYPTONITE*) that influence DNA methylation. It seems likely that this will also be the case in mammals, especially since mutations in other putative chromatin-modifying factors perturb DNA methylation (see the following section, ATPase Chromatin-Remodeling Complexes).

Covalent modifications of histones must be labile so that the transcriptional status of a particular gene can be reversed. HATs are counteracted by histone deacetylases (HDACs), and kinases are counteracted by phosphatases. However, an exception may be histone methylation. Even though lysines can be mono-, di-, or tri-methylated and gene activity increases when histone 3 lysine 4 (H3-K4) is converted from a di- to tri-methyl state, no histone demethylases have been identified which can remove these methyl groups or those from other lysine or arginine residues. Instead, there has been speculation that the most N-terminal amino acids, including methylated residues, might sometimes be proteolytically cleaved. Such a mechanism would not be very precise or dynamic but would reverse the effect of methylation until new histones are synthesized and assembled into nucleosomes during S phase. However, it is also possible that clipped tails might serve as a signal to be replaced by histone variants. Unlike core histones, some histone variants are expressed during G1 and G2 of the cell cycle and can be incorporated into nucleosomes of nondividing cells where they confer unique properties. For example, H3.3 differs from H3 at only four amino acids but has been associated with rDNA arrays and other transcriptionally active loci. In addition, CENP-A replaces H3 at centromeres, and H2AX is localized to double-strand breaks during DNA repair.

ATPASE CHROMATIN-REMODELING COMPLEXES

In addition to enzymes that covalently modify histones, there is a second evolutionarily conserved mechanism that modulates chromatin structure, which is carried out by SWI/SNF and a variety of other ATPase chromatin-remodeling complexes (Narlikar et al., 2002). SWI/SNF does not possess significant DNA binding ability of its own but is recruited to promoters by sequence-specific transcription factors. The energy derived from ATP hydrolysis allows the complex to alter the conformation and position of nucleosomes. DNA-histone contacts are broken, and histone octamers can be slid several hundred base pairs upstream or downstream. As a result, a core promoter can be made nucleosome-free and accessible to the RNA Polymerase II holoenzyme so that transcription can be initiated. However, SWI/SNF can be recruited to other loci by transcriptional repressors and have the opposite effect by inhibiting transcription. Gene expression profiling with whole-genome oligonucleotide arrays demonstrated that approximately 300 genes, or about 5% of the total number in the yeast genome, are regulated by SWI/SNF complexes in a positive or negative manner.

SWI/SNF complexes have been shown to work in concert with HAT complexes (Narlikar et al., 2002). Sequence-specific transcription factors can recruit HAT complexes to promoters to acetylate histone tails. Acetylated lysine residues can then serve as docking sites for bromodomains of SWI2/SNF2-related catalytic subunits, resulting in an increased affinity of SWI/SNF complexes for their chromatin targets. At other loci or at different stages of the cell cycle (late mitosis), the order is reversed, with SWI/SNF-related complexes recruited first and HAT complexes second. Genetic interactions between SWI/SNF (swi2/snf2) and HAT (gcn5) mutations in yeast indicate that the coordinated recruitment of their gene products must be important in vivo. Compared to single mutants, double mutants exhibit synergistic effects in the deregulated expression of downstream target genes and grow very slowly with mitotic defects on some backgrounds and are not viable on others. It is possible that these phenotypes are not due entirely to transcriptional deregulation because, much like histone modifying enzymes, ATPase chromatin-remodeling complexes have been implicated in DNA repair, replication, recombination, and mitosis.

Additional ATPase chromatin-remodeling complexes have been characterized that are distinct from SWI/SNF, although each one has a catalytic subunit with sequence similarity to the DNA-dependent ATPase domain of SWI2/SNF2 (Narlikar et al., 2002). At this point, more similarities than differences have been identified between the various complexes in vitro, and it is not clear whether the differences are relevant in vivo. Therefore, it is not clear why such diversity has been selected for during evolution, but it is tempting to speculate. Some complexes may modulate higher-order chromatin structure, whereas other complexes might act on nucleosome arrays. Different complexes could act at distinct promoters or overlap at a large subset of loci. It is also possible that two complexes perform fundamentally different tasks at the same promoter to influence transcriptional initiation, or one could act downstream to influence transcriptional elongation. A precedent for this sort of division of labor comes from a recent report that the FACT and SWR1 complexes disassemble nucleosomes in coding regions during transcriptional elongation.

The diversity of chromatin-remodeling complexes is likely to be even greater than is currently realized. Numerous SWI2/SNF2-related genes have been identified by reduced stringency hybridization or genome sequencing projects, and some of these genes probably encode "orphan" catalytic subunits of complexes that await identification and purification. Some of these putative catalytic subunits have interesting properties, such as the ability to regulate DNA methylation. A targeted mutation of Lsh (lymphoid specific helicase) results in a 50 to 70% reduction in cytosine methylation throughout the genome. Homozygotes die shortly after birth, possibly because of renal failure. (The gene is lymphoid-specific in adults but widely expressed in embryos.) Interestingly, expression and activity of de novo and maintenance DNA methyltransferases are unaffected. Instead, Lsh1 is expressed during S phase and may facilitate localization of Dnmt1 to hemimethylated DNA following replication or protect against demethylase activity. Mutations in ATRX, another SWI2/SNF2-like gene, reduce DNA methylation in rDNA arrays and other repeats and result in mental retardation, α-thalassemia, and fertility defects in humans.

Epigenetic Processes

POSITION-EFFECT VARIEGATION

Shortly after discovering the mutagenic properties of X-irradiation, Muller described the isolation of several radiation-induced mutations of the D. melanogaster white (w) eye-color gene in 1930. Interestingly, the mutant eyes contained patches of wild-type cells with red pigmentation intermingled with patches of mutant cells lacking pigmentation and appearing white. Six years later, it was demonstrated that the X-rays induced a chromosomal rearrangement in each mutant line, such as an inversion in In(1)w^{m4} (abbreviated w^{m4} for white mottled 4), that resulted in heterochromatin being juxtaposed with the w locus. The regulatory elements and coding sequence of the w gene were not perturbed in any of these mutants, but heterochromatin would spread across the breakpoint and silence transcription in a subset of eye progenitor cells during development. In contrast, the heterochromatin would not spread far enough to reach the *w* gene in other eye progenitor cells, and it would be transcribed at wild-type levels. Clonal expansion of these cell populations subsequently produced patches of white and red eye color, respectively. This silencing process must be stochastic because even genetically identical flies exhibit unique patterns of white and red coloration. This epigenetic process, which has been documented at other loci in *D. melanogaster* and in other organisms, is appropriately called position-effect variegation (PEV).

Because *w^m4* is such a visual example of PEV, it served as the foundation of genetic screens to identify genes that modify the extent of heterochromatin spreading in PEV. *E(var)* mutations were recovered that enhance PEV, resulting in increased heterochromatin spreading, and cause the *w* gene to be silenced in a greater percentage of eye cells. Therefore, *E(var)* mutants have eyes that are more white than typical *w^m4* flies. In contrast, *Su(var)* mutations were recovered that suppress PEV, inhibiting the spread of heterochromatin, and cause the *w* gene to be transcribed in a greater percentage of eye cells. These mutants have eyes that are more red than *w^m4* and therefore more closely resemble wild type. In these screens, most of the *E(var)s* and *Su(var)s* were isolated as dominant mutations. When homozygosed, more than half of these mutations are lethal, indicating that the majority of the corresponding gene products are essential.

Chromatin and Molecular Basis or PEV

Although *E(var)* and *Su(var)* gene products are not strictly defined by conserved features, most of those that have been characterized at the molecular level contain domains or motifs present in Pc-G, trx-G, or other chromatin-modifying factors. The connection between PEV and chromatin has also been strengthened by observations that histone, HAT, HDAC, Pc-G, and Trx-G mutations can behave as *E(var)s* or *Su(var)s*. On a mechanistic level, the most significant progress has come from analysis of Suv39h1 and HP1α which are mammalian orthologs of *D. melanogaster Su(var)3-9* and *Su(var)2-5* (which encodes heterochromatin protein 1 or HP1), respectively (Richards and Elgin, 2002). Suv39h1 methylates the lysine 9 residue of histone H3 (H3-K9) in pericentric regions of the genome, and this covalent modification serves as a docking site for the chromodomain of HP1α (Figure 8-2B). HP1α is a structural component of heterochromatin, and, once situated at a particular H3-K9 residue, it is thought to utilize a second domain, called the chromo shadow domain, to directly or indirectly bind another Suv39h1 molecule. This recruitment step results in H3-K9 methylation at an adjacent nucleosome and is followed by more HP1α binding, thereby creating a feedback loop such that heterochromatin can propagate or spread over Mb intervals throughout pericentric regions. This feedback loop is quite dynamic and is also able to maintain pericentric regions as heterochromatin throughout the numerous cell divisions that occur during embryogenesis and in adults. This form of cellular memory is accomplished

by the HP1α chromo shadow domain directly interacting with the molecular chaperone chromatin assembly factor (CAF-1), which is localized to replication forks by proliferating cell nuclear antigen (PCNA) during S phase and which incorporates newly synthesized histones into nascent nucleosomes.

The importance of *Suv39h1* and *Suv39h2* (a closely related paralog) has been confirmed by gene targeting. Single mutants do not exhibit a detectable phenotype, but double homozygotes often die during fetal development. As expected from the model presented above, double mutants are devoid of H3-K9 methylation in pericentric regions, but not elsewhere, which results in chromosomal instability in somatic tissues and the germline. As a result, surviving double mutants are runted, prone to B-cell lymphomas, and infertile. Neither *HP1α* nor two other *HP1*-related genes have been knocked out yet in the mouse, but mutations in *Drosophila Su(var)2-5* are lethal.

RNAi and Sequence Specificity of PEV

Suv39h1 does not exhibit any specificity other than the fact that it methylates H3-K9, so what directs its activity to histone tails located in pericentric chromosomal regions, known to be particularly rich in heterochromatin, but not other regions of the genome? Work from the *Schizosaccharomyces pombe* counterparts of *Su(var)3-9 (Clr4)* and *HP1 (Swi6)* suggest that an RNAi-like mechanism is crucial (Richards and Elgin, 2002; Jenuwein, 2002) (Figure 8-2A). In wild-type fission yeast, DNA repeats that are highly enriched in pericentric regions are transcribed in a bidirectional manner to produce 1.4- and 2.4-kbp double-stranded transcripts that are processed by dicer into approximately 22 nucleotide sense and antisense RNAs. Similar to short interfering RNAs (siRNAs) or microRNAs, these RNA oligonucleotides physically associate with argonaute and other proteins to comprise RNA-induced initiation of transcriptional gene silencing complexes (RITS). However, unlike conventional siRNAs, which move to the cytoplasm to bind and degrade cognate mRNAs, the heterochromatin-derived siRNAs remain in the nucleus. It is thought that they direct the RITSs to the corresponding DNA repeats and indirectly recruit Clr4, which methylates H3-K9 and enables Swi6 binding (Figure 8-2A).

The studies described above could be performed in a straightforward manner in *S. pombe* because the genes encoding the RNAi machinery exist as singletons. There is mounting evidence, however, that RNAi-like directed changes in chromatin structure exist in other organisms and might not be restricted to pericentric heterochromatin. dsRNAs with homology to transgenes can mediate DNA methylation and transcriptional silencing of transgenes in plants. In addition, mutations in Pc-G genes perturb RNAi and co-suppression in both *D. melanogaster* and *C. elegans*. Pc-G gene products have been implicated in chromatin structure and transcriptional silencing but not RNA degradation or post-transcriptional regulation. These results therefore suggest that RNA-directed silencing may not be carried out entirely at the post-transcriptional level but might also influence chromatin structure and act at the transcriptional level. This notion is

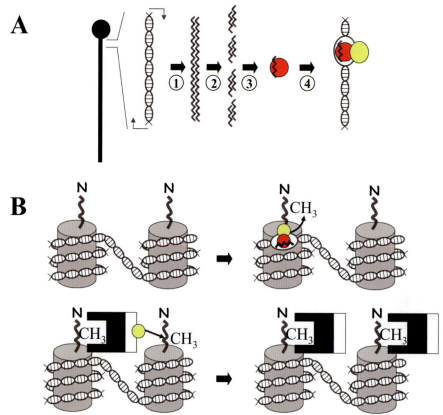

Figure 8-2. Specificity and propagation of epigenetic marks in PEV. **(A) Specificity:** DNA repeats in pericentric region are transcribed in a bidirectional manner (1). Double-stranded RNAs are cleaved by dicer (2) and processed by RNAi-like machinery (3) to yield RNA oligonucleotides that associate with RITS (red circle). RNA oligonucleotides are thought to interact with the segment of pericentric repeat that served as template (4) and directly or indirectly recruit Clr4 histone methyltransferase (yellow circle). **(B) Propagation:** Clr4 (yellow circle) methylates H3-K9 (CH3) (upper right panel), which serves as a docking site for the chromodomain of Swi6 (black part of box) (lower left panel). The chromo shadow domain of Swi6 (white part of box) is thought to directly or indirectly recruit more Clr4, which methylates H3-K9 on adjacent nucleosome (barrel) (lower left). Recruitment of more Swi6 to nascent H3-K9 methyl group enables next nucleosome to be modified (lower right panel).

supported by the finding that RNAi decreases hnRNA levels in addition to mRNA.

Although RNAi is thought to have arisen as a defense against RNA viruses and the mobilization of transposable elements, it is tempting to speculate that siRNAs may play a more general role in regulation of gene expression. It is conceivable that siRNAs provide sequence specificity to other enzymes involved in chromatin modifications. In this regard, only rarely have DNA methyltransferases and HMTs been shown to interact with sequence-specific activators or repressors. Ironically, Jacob and Monod proposed that the *lac* repressor might encode an RNA molecule and that RNAs might have an important role regulating operons, but this notion was never embraced and was subsequently dismissed as a possible mechanism for eukaryotic transcriptional regulation.

X-CHROMOSOME INACTIVATION

Similar to *Drosophila* PEV, certain chromosome rearrangements result in mosaic silencing of coat-color genes in female

mice. However, whereas *w* is juxtaposed to pericentric regions in *w^{m4}*, the pink-eye dilution (*p*) and albino (*c*) coat-color loci are juxtaposed to the X-chromosome in t(X;7) translocations (Figure 8-3). Because a stochastic choice is made to silence one of the two X-chromosomes to achieve dosage compensation in a process called X-chromosome inactivation, some cells choose to inactivate the translocated X and therefore silence *p* and *c*. Other cells choose to inactivate the intact X and express *p* and *c*. As with PEV, this choice is made early in development (around implantation) and then propagated clonally.

The similarities between PEV and X-chromosome inactivation extend to molecular and mechanistic levels. Both employ noncoding RNAs that direct similar chromatin modifications. In the case of X-chromosome inactivation, repressive epigenetic marks are targeted by *Xist*, a 17-kb, untranslated RNA that is transcribed from the inactive but not the active X-chromosome (Plath *et al.*, 2002). After being transcribed, *Xist* does not diffuse away from the inactive X (Xi) but coats the chromosome in *cis*, which allows it to

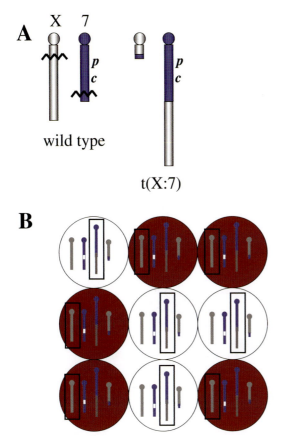

Figure 8-3. Mosaic coat color in t(X:7) mice. **(A)** Schematic of mouse chromosomes X (gray) and 7 (blue). The coat-color genes *pink-eyed dilution* (*p*) and *albino* (*c*) are located on chromosome 7. A reciprocal t(X:7) translocation fuses the two chromosomes (right). Translocation breakpoints are indicated with squiggles. **(B)** Schematic of t(X:7) coat-color mosaicism. Brown and white circles represent pigmented and unpigmented melanocytes, respectively. Intact X and chromosome 7 are shown to the left of X:7 derivatives. The white segment of the intact chromosome 7 represents *p* and *c* mutations (i.e., the cells are heterozygous for both genes). The black box indicates the inactivated X-chromosome. Pigment cells that inactivate the translocated X silence the only functional *p* and *c* alleles and do not produce pigment. Cells that inactivate the intact X maintain expression of the *p* and *c* genes on the translocated chromosome and produce pigment.

directly or indirectly localize chromatin-modifying complexes to the Xi. Complementary genetic approaches underscore *Xist*'s role in X-inactivation; a targeted mutation in *Xist* inhibits inactivation of the mutated chromosome; and *Xist* expressed from an autosomal transgene can lead to ectopic inactivation of the transgenic autosome, similar to t(X;7) translocations. Thus, *Xist* is both necessary and sufficient to initiate long-range silencing in *cis*. Because of this potent silencing capability, cells must count their X-chromosomes and repress *Xist* on one chromosome, the active X (Xa), per diploid cell. In the mouse, *Xist* repression on the Xa is accomplished by expression of *Tsix*, another large untranslated RNA that is partially antisense to *Xist* (Plath *et al.*, 2002). However, it is not clear whether *Tsix* is utilized in all mammals, and even

in the mouse, the exact mechanism by which *Tsix* blocks *Xist* is not known.

The Xi is cytologically distinguishable as late-replicating, condensed heterochromatin at the nuclear periphery. In recent years, antibodies have been developed that detect histone marks and chromatin-associated proteins, some of which are enriched on the heterochromatic Xi and underrepresented on the Xa. Staining of differentiating ES cells with these antibodies has established a sequence of epigenetic modifications that apparently initiate and maintain X-inactivation (Figure 8-4). Histones associated with the Xi are hypoacetylated, hypomethylated at H3-K4, and hypermethylated at H3-K9 and H3-K27. All of these marks are consistent with repressive higher-order chromatin conformations such as those found in pericentric heterochromatin. In addition, some reports have suggested that the histone variant macroH2A is enriched on the Xi and that a separate variant H2A.Z is excluded from the Xi. Finally, in mouse extraembryonic tissues, the Xi is DNA hypermethylated relative to the Xa. However, the Xi is not hypermethylated in mouse extraembryonic tissues, which undergo imprinted rather than stochastic X-inactivation. This discrepancy is consistent with a more prominent role for DNA methylation in the embryo proper.

A number of genes on the Xi escape X-chromosome inactivation. These loci contain epigenetic marks enriched on the Xa (e.g., histone hyperacetylation), are biallelically expressed, and replicate synchronously with the corresponding Xa alleles, indicating that they escape all aspects of X-inactivation. Some of these genes reside in the pseudoautosomal region that carries functional homologs on the Y-chromosome. Thus, these genes are essentially diploid in both males and females, precluding any need for dosage compensation. Other genes escape X-inactivation despite lacking homologs on the Y. Organisms appear to be less sensitive to dosage of these gene products. Presumably, escaping genes are protected from the spread of repressive chromatin marks by flanking insulator elements. However, the mechanisms of escape are poorly understood.

POLYCOMB AND TRITHORAX GROUPS

Polycomb Group (Pc-G) Silencing

Genetic screens in *Drosophila* have identified many Polycomb group (Pc-G) genes required for the proper expression of homeotic genes (Simon and Tamkun, 2002). Whereas gap and pair-rule gene products are required to establish homeotic gene expression, Pc-G factors are required for maintenance. Not unexpectedly, these distinct classes of regulatory factors act in concert. Gap and pair-rule genes encode DNA binding factors that bind to enhancers and promoters of homeotic genes in the *Antennapedia* (ANT-C) and *bithorax* (BX-C) complexes, and, in so doing, directly activate or repress transcription. However, these genes are transiently expressed and the corresponding gene products decay by mid-embryogenesis, yet homeotic gene expression is properly maintained throughout the remainder of development. Some gap and pair-rule proteins that repress transcription are thought to directly or indirectly recruit Pc-G factors before decaying. These Pc-G

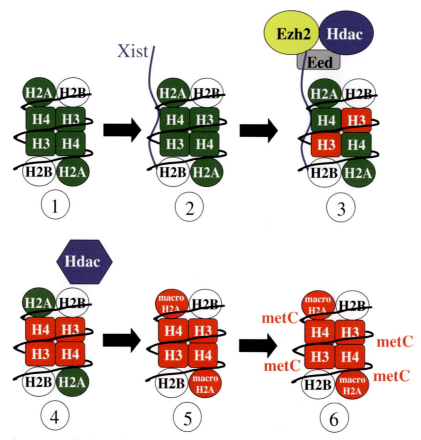

Figure 8-4. Sequential epigenetic marks during X-chromosome inactivation. For all histones (H2A, H2B, H3, and H4), green represents covalent modifications or variants associated with transcriptional activity, whereas red represents transcriptional repression. H2B is white because no covalent modifications or variants are known to be enriched on the inactive X-chromosome (Xi). (1) Initially, both X chromosomes are H3 and H4 hyperacetylated, H3-K4 hypermethylated, and DNA hypomethylated. (2) Xist coats one of the two X-chromosomes, which is destined to be the Xi, as the first step of the inactivation process. (3) Next, the Eed-Ezh2-Hdac mplex is recruited and mediates H3 deacetylation and H3-K9 and H3-K27 methylation. Concomitant with these Pc-G-mediated modifications, H3-K4 is demethylated. Subsequently, H4 is deacetylated (4), and macroH2A replaces H2A in many nucleosomes on the Xi (5). (6) Finally, CpGs on the Xi become methylated.

factors assemble into at least two complexes and alter chromatin structure. In so doing, the Pc-G complexes act as a form of cellular memory to keep inactive homeotic genes in a permanent "off" state over the course of many cell divisions (Simon and Tamkun, 2002). This regulation is crucial for proper anterior-posterior (A-P) patterning because homeotic genes, which are expressed in stem cell populations in the imaginal discs, control cell fate decisions in the head, the three thorax segments, and the eight abdominal segments.

Accordingly, Pc-G mutants exhibit ectopic homeotic gene expression and A-P patterning defects. In fact, going back to the original genetic screens, many Pc-G mutants were isolated as heterozygotes that have two or three thorax number one segments (T1-T1-T3 or T1-T1-T1) instead of the normal T1-T2-T3 pattern. These T2-to-T1 and T3-to T1 homeotic transformations are most evident in male flies because mating structures, called sex combs, which are normally found only on the first pair of legs emanating from T1, are found on the first two or all three pairs of legs. This observation accounts

for the names of quite a few Pc-G genes: Polycomb (*Pc*), extra sex combs (*esc*), additional sex combs (*Asx*), sex combs on the midleg (*Scm*), posterior sex combs (*Psc*), and so on. Moreover, when many of these same mutations are made homozygous, homeotic gene expression is altered to a greater extent, homeotic transformations are more severe and widespread, expression of other types of genes is perturbed, and embryonic or larval lethality often results. In addition, robust genetic interactions among mutations in distinct Pc-G genes exacerbate these mutant phenotypes.

The first suggestion that Pc-G factors might modulate chromatin structure came from the molecular analysis of PC and the realization that it shares a chromodomain with HP1. More recently, E(Z) has been shown to share a SET domain with SU(VAR)3-9 and have HMT activity. Coupled with conventional chromatography and protein–protein interaction data, these observations have led to a model analogous to Suv39h/Clr4 and HP1/Swi6 in PEV (Figure 8-5). Like PEV, one Pc-G member interacts with PCNA and may facilitate the

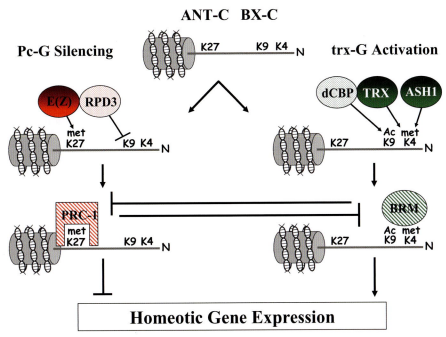

Figure 8-5. Homeotic gene regulation by Polycomb- and trithorax-groups. Schematic illustrating antagonistic roles of Pc-G silencing and trx-G activation in homeotic gene regulation. A single nucleosome in the ANT-C or BX-C is shown with an H3 N-terminal tail projecting to the right (top panel). Pc-G silencing is initiated by the addition of repressive covalent modifications: E(Z) methylates (met) K27, and RPD3 deacetylates K9 (upper left). PRC1 complex binds met K27 (lower left). trx-G activation is initiated by the addition of other covalent modifications: TRX and ASH1 methylate (met) K4, and dCBP acetylates (Ac) K9 (upper right). A SWI/SNF-related BRM complex is recruited to K4-methylated, K9-acetylated nucleosomes (lower right). PRC1 and BRM recruitment are mutually exclusive (lower panels). Whereas PRC1 recruitment represses homeotic gene transcription, BRM promotes expression (bottom panel).

reestablishment of repressive complexes following DNA replication. It is also quite possible that the spreading of Pc-G silencing is blocked by insulator elements. However, there are some differences. Pc-G complexes are distributed more widely throughout the genome (antibodies detect about 100 sites on polytene chromosomes) than Suv39h/Clr4 and HP1/Swi6, which are primarily restricted to pericentric regions. Pc-G complexes also act more locally. They apparently spread over tens of kb to around 100 kb, whereas Suv39h/Clr4 and HP1/Swi6 spread over Mbp intervals. Both of these differences could be due, in part, to *cis*-acting elements, called Polycomb Response Elements (PREs), which are enriched in the ANT-C and BX-C and can be bound by a Pc-G member (PHO) that associates with E(Z)-ESC complex. It also is not clear how other Pc-G factors contribute to the silencing process.

Mammalian Pc-G gene products are more numerous, due to duplication and divergence events during vertebrate evolution, but comprise similar complexes and perform similar roles in development. Null mutations of *Ezh2*, *Eed*, and *YY1*, all of which encode components of the first complex, result in early embryonic lethality. However, it is clear that the function of this complex is not restricted to early embryogenesis but is instead involved in a variety of biological processes in many different tissues. An *Ezh2* conditional mutation in B cells demonstrates that the wild-type gene product is required for VDJ recombination. *Ezh2* is also overexpressed in human

prostate cancer tumors, and, conversely, RNAi down-regulation of *Ezh2* inhibits the proliferation of prostate cancer cells *in vitro*. Although *Eed* null mutants exhibit severe A-P patterning defects during gastrulation, they die before many *Hox* genes are expressed. However, an ENU-induced hypomorph mutant line misexpresses *Hox* genes and exhibits homeotic transformations. It also has placental defects and has been implicated in T-cell lymphomas. It has also become clear that *Eed* plays a role in X-chromosome inactivation and genomic imprinting.

In contrast, null mutations of Pc-G genes that encode components of the mammalian PRC1 complex generally do not confer embryonic-lethal phenotypes. *MPc2/M33/Cbx2*, *Bmi1*, *Mel18*, and *Rae28/Mph1* homozygotes exhibit altered *Hox* gene expression and homeotic transformations, which are manifested by changes in vertebral identities, as well as cell proliferation defects in the hematopoietic system. One exception is *Ring1B/Rnf2*, but, although null mutants die during gastrulation, hypomorphs are viable and do exhibit altered *Hox* gene expression and homeotic transformations. Of all these genes, *Bmi1* is particularly noteworthy because it inhibits the *Ink4a* tumor suppressor locus and is required for continued proliferation of hematopoietic stem cells. It also regulates proliferation in more differentiated cells in several hematopoietic lineages. Gain-of-function mutations down-regulate both *Ink4a* gene products (the p16 and p19 cyclin-dependent kinase inhibitors) and cooperate with *c-Myc* in B- and T-cell lymphomas. Conversely, loss-of-function mutations lead to increased expression of p16 and p19 and decreased proliferation of erythrocytes, resulting in anemia. Moreover, this phenotype is suppressed by an *Ink4a* targeted mutation.

Trithrax Group (trx-G) Activation

Genetic screens in *Drosophila* identified loss-of-function mutations in *trithorax* (*trx*) and a number of other genes that suppress dominant *Pc* mutant phenotypes (Simon and Tamkun, 2002). This *trx* group (*trx*-G) of genes also encode proteins that have a variety of interesting domains and motifs, some of which are shared with *Pc*-G members but have the opposite effect on transcription. For example, the TRX protein has a SET domain, just like E(Z), but methylates H3-K4 instead of H3-K9/H3-K27 and is involved in transcriptional activation of homeotic genes instead of transcriptional repression (Figure 8-5). Consequently, mutations in *trx* or other trx-G genes result in decreased homeotic gene expression instead of ectopic expression. Of course, the fact that trx-G and Pc-G counteract each other at the molecular level is not surprising, but is expected, considering that the two groups have an antagonistic relationship at the genetic level (Simon and Tamkun, 2002).

Genetic and molecular interactions among other trx-G genes, which encode various chromatin-modifying factors, are common and important at the mechanistic level. TRX physically binds to another HMT called ASH1, and both TRX and ASH1 bind to the dCBP histone acetyltransferase. TRX also binds to a subunit of a SWI/SNF-related complex, called SNR1, which includes OSA, BRAHMA (BRM), and MOIRA (MOR). Like Suv39 and HP1 or E(Z) and PC, TRX and/or ASH1 may provide a covalently modified histone interface for the BRM complex (Figure 8-5). Mammalian counterparts of *trx*-G genes are important for many aspects of development, ranging from implantation to *Hox* gene expression, erythropoiesis, and neural tube closure at midgestation to postnatal fitness and cancer prevention. With respect to *Hox* gene regulation, *trx*-G and Pc-G factors counteract each other just as they do in *Drosophila*.

IMPRINTING

Mammals are diploid for all autosomal loci, inheriting one copy of each gene from each parent. Presumably, both alleles of most genes are utilized. For a growing number of characterized genes, however, only one allele is used per cell, a process termed *monoallelic expression*. For genes subject to monoallelic expression, each cell is functionally hemizygous, expressing either the maternally or paternally inherited allele but not both.

The best studied type of monoallelic expression is imprinting, whereby genes are expressed according to their parental origin (Reik and Walter, 2001). For example, only maternally inherited copies of *H19* are expressed (paternal alleles are silenced), and only paternally inherited copies of *Igf2* are expressed. Such parent-of-origin expression explains the inability of mammalian partheongenotes, gynegenotes, and androgenotes, as well as certain uniparental disomies, to develop to term. Imprinted genes tend to be tightly linked. To date, more than 10 clusters encompassing roughly 60 imprinted genes have been identified in the mouse (*www.mgu.har.mrc.ac.uk/imprinting/all_impmaps.html*). These clusters are generally associated with regions of DNA that are differentially methylated in sperm and oocytes. These regions, termed imprinting control regions (ICRs), appear to regulate entire imprinting clusters. Targeted deletion of ICRs leads to loss of imprinting (biallelic expression or silencing) of all of the genes in the cluster. In addition, *Dnmt1*$^{-/-}$ embryos exhibit loss of imprinting of almost all imprinted genes, and the progeny of both *Dnmt3L* and *Dnmt3a*$^{-/-}$, *Dnmt3b*$^{-/-}$ females exhibit loss of imprinting of maternally imprinted genes. These findings have led to a general consensus that DNA methylation is a critical part of the biochemical imprint that distinguishes maternal and paternal alleles at imprinted loci.

Some ICRs act as methylation-sensitive insulator elements. At the *H19/Igf2* imprinting cluster on mouse chromosome 7, the ICR includes clustered binding sites for the insulator CTCF (Figure 8-6). Here, allele-specific methylation drives allele-specific, methylation-sensitive insulator binding,

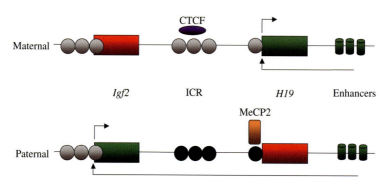

Figure 8-6. Coordinated transcription of two imprinted genes. Top: Maternal allele includes an unmethylated ICR (three white circles) between *Igf2* and *H19* and binds CTCF (blue oval), which blocks the association of *Igf2* with enhancers (three barrels) downstream of *H19*, leading to *Igf2* silencing (red). Because the CTCF binding sites do not separate *H19* and these enhancers, *H19* is transcribed (arrow, green). Bottom: Methylated paternal allele ICR is methylated (three black circles) and cannot bind CTCF, which facilitates interaction between *Igf2* and its downstream enhancers, leading to *Igf2* transcription (arrow, green). Conversely, *H19* promoter is methylated (1 black circle), recruits MeCP2 (orange box) and is silenced (red).

which reciprocally regulates *H19* and *Igf2* by blocking maternal *Igf2* with its downstream enhancer (Figure 8-6). Other ICRs are associated with noncoding RNAs. At the imprinted cluster on mouse chromosome 17, the noncoding, paternally transcribed *Air* RNA acts in *cis* to silence reciprocally imprinted genes, including *Igf2r*. Such a phenomenon is reminiscent of silencing of the X-chromosome by *Xist*, but the mechanism of RNA-mediated silencing at imprinted loci is poorly understood.

Recently, additional epigenetic marks that distinguish parental alleles at imprinted loci have begun to emerge. Nucleosomes associated with active alleles tend to be hyperacetylated on H3 and H4 tails and methylated at H3-K4. Conversely, histones associated with silenced alleles tend to be hypoacetylated and methylated at H3-K9. The enzymes responsible for these modifications at imprinted loci remain largely undefined. Loss-of-function mutations in the Pc-G gene *Eed* lead to biallelic expression of several imprinted genes, including *Mash2* and *Cdkn1c* in the Beckwith-Wiedeman syndrome (BWS) imprinting cluster on mouse chromosome 7. EED is associated with EZH2, a HMT with activity directed against H3-K9 and H3-K27. Future work will need to address allele-specific association of the latter mark at imprinted loci. Interestingly, allele-specific DNA methylation remains intact at genes affected in *Eed*[−/−] embryos. This observation suggests that covalent histone modifications at imprinted loci are either parallel to or downstream of DNA methylation and that allele-specific DNA methylation may drive allele-specific transcriptionally repressive chromatin marks.

Summary and Future Directions

Drawing upon several models of epigenetic inheritance, much has been learned about chromatin-modifying factors and how they act and interact to regulate transcription. It is becoming increasingly clear that these factors and their corresponding epigenetic marks dictate developmental potentials. The present challenge is to understand better chromatin-remodeling mechanisms in cell fate decisions, cell proliferation and differentiation, and plasticity. In addition to providing considerable insight into many aspects of embryogenesis and postnatal development, this knowledge should prove valuable for improving the prospects of stem cell technologies in the clinical setting.

KEY TERMS

Imprinting Differential expression of an allele depending upon maternal versus paternal inheritance.

Polycomb group silencing Chromatin-based gene silencing mechanism, originally identified for its role in *Hox* gene repression during development.

Position effect variegation Mosaic silencing of a gene due to a chromosomal rearrangement.

Trithorax group activation Chromatin-based mechanism required to counteract Polycomb group silencing and originally identified for its role in *Hox* gene expression during development.

X-chromosome inactivation Mammalian dosage compensation mechanism involving inactivation of one X-chromosome in females.

FURTHER READING

Bird, A. (2002). DNA methylation patterns and epigenetic memory. *Genes Dev.* **16**, 6–21.

Jenuwein, T., and Allis, C. D. (2001). Translating the histone code. *Science* **293**, 1074–1080.

Jenuwein, T. (2002). An RNA-guided pathway for the epigenome. *Science* **297**, 2215–2218.

Li, E. (2002). Chromatin modification and epigenetic reprogramming in mammalian development. *Nat. Rev. Genet* **3**, 662–673.

Narlikar, G. J., Fan, H. Y., and Kingston, R. E. (2002). Cooperation between complexes that regulate chromatin structure and transcription. *Cell* **108**, 475–487.

Plath, K., Mlynarczyk-Evans, S., Nusinow, D. A., and Panning, B. (2002). Xist RNA and the mechanism of X chromosome inactivation. *Annu. Rev. Genet* **36**, 233–278.

Reik, W., and Walter, J. (2001). Genomic imprinting: parental influence on the genome. *Nat. Rev. Genet* **2**, 21–32.

Richards, E. J., and Elgin, S. C. (2002). Epigenetic codes for heterochromatin formation and silencing: rounding up the usual suspects. *Cell* **108**, 489–500.

Simon, J. A., and Tamkun, J. W. (2002). Programming off and on states in chromatin: mechanisms of Polycomb and trithorax groups. *Curr. Opin. Genet. Dev.* **12**, 210–218.

Cell Fusion and the Differentiated State

Penny Johnson and Peter W. Andrews

Introduction

Cell fusion recently surfaced as an issue in analyzing data from experiments designed to demonstrate the pluripotency of various adult stem cells. However, the phenomenon of spontaneous cell fusion has been known for a considerable time, and experimentally induced cell fusion has been widely used for years, for both genetic analysis and studies of the differentiated and determined states.

In studies made over 40 years ago, Sorieul and Ephrussi (1961) found that the co-culture of two mouse cell lines with differing marker chromosomes resulted in the appearance of a significant proportion (up to 10%) of spontaneously derived hybrid cells during about three months of continuous culture. Subsequently, it was shown that specific drug selection techniques could be used to isolate such hybrid cells and the hypoxanthine-aminopterin-thymidine (HAT) selection system to eliminate parental cells separately deficient in hypoxanthine phosphoribosyl transferase (HPRT), and thymidine kinase (TK) was introduced. Genetic complementation resulting in expression of both enzymes allowed for survival of the hybrids. It was then found that a substantial loss of human chromosomes occurs in spontaneous hybrids between diploid human fibroblasts and mouse L-cells, while the complete set of mouse chromosomes is retained. This phenomenon of specific chromosome loss coupled with the subsequent discovery that cell fusion could be induced by inactivated Sendai virus, or by incubating cells in polyethylene glycol (PEG), provided a powerful new method for human gene mapping, which was widely used during the 1970s. Thus, by correlating the expression of human traits in such hybrids with the retention of particular human chromosomes, individual human genes could be located on specific chromosomes or even parts of chromosomes.

Hybrid Cells and Differentiated Phenotypes

Simultaneous with the development of techniques for producing hybrid cells, John Gurdon and his colleagues demonstrated that a nucleus obtained from a differentiated cell of a tadpole, when transferred to an enucleated frog oocyte, supported the development of an adult frog. The general conclusion from these studies was that all differentiated cells retain

full genetic potential for development and that the differentiated states of functionally distinct cell types must therefore arise from differential regulation of gene activity. What, then, would be the result of combining the genomes from different cell types in a single hybrid cell?

In 1965, Henry Harris had described the behavior of different nuclei in heterokaryons formed between various cell types. The most striking of these combinations was the fusion of HeLa cells with mature chicken erythrocytes in which, unlike mammalian erythrocytes, the chromatin is condensed and the nucleus is inactive. In the resulting heterokaryons, the chicken erythrocyte nuclei became active for both DNA and RNA synthesis and reexpressed chicken-specific genes.

This result was confirmed and shown to be the result of cytoplasmic factors. For example, chicken erythrocyte nuclei were found to be reactivated when introduced into enucleated cytoplasms. Furthermore, the reactivated chicken nucleus in cells reconstituted with fibroblast cytoplasm supported synthesis of chicken globin.

Attempts to discern general rules from the fusion of cells of different phenotypes, however, proved difficult. Despite the results with erythrocyte heterokaryons, an early conclusion was that the fusion of cells expressing two distinct states of differentiation frequently resulted in the loss of those differentiated functions in the hybrid cells. For example, hybrids formed between fibroblasts and melanomas did not produce melanin, and globin synthesis was not inducible in hybrids between fibroblasts and Friend erythroleukemia cells. On the other hand, hybrids between human leukocytes and mouse liver tumor cell lines sometimes expressed liver-specific proteins from the human genome contributed by the fibroblasts. Similarly, rat hepatoma–mouse lymphocyte or fibroblast hybrids often expressed mouse albumin. In other studies, hybrids between neuroblastoma cells, which exhibit a variety of neural features, and mouse L-cells, which are long-established, immortalized fibroblastoid cells, retained at least some of the neural features of the neuroblastoma parental cells, notably their electrical activity.

Apart from attempting to investigate the control of specific differentiated states, Harris and his colleagues, in particular, sought to use cell hybridization to establish the mode of genetic control of the transformed state of cancer cells, in particular whether the genetic changes that underlie the transformed state of tumor cells were the result of the loss or gain of gene function. In a conclusion that presaged the identification of tumor suppressor genes, they reported that malignancy acted as a recessive trait at the cellular level, as fusion of malignant cells with nonmalignant partners resulted in the formation of nonmalignant hybrid cells in which malignancy reappeared with subsequent chromosome loss. Clearly, such a result is not always the case, as might be inferred from our current knowledge of oncogenes as well as tumor suppressor genes. Perhaps the most notable exception to this rule was the formation of immortal and tumorigenic "hybridomas" that produced monoclonal antibodies following the fusion of terminally differentiated plasma cells and a lymphoid cell line.

Generally, it seemed that hybrids of cells with distinct phenotypes did not express a hybrid phenotype. Rather, they tended to express genes associated with one or other of the parent cells but not both. For example, mouse hepatoma–Friend erythroleukemia hybrids were described, which continued expressing liver functions but in which globin expression was extinguished. Furthermore, in some cases, the gene expression typical of one parental phenotype was activated from the genome of the other parental cell. However, no clear rules emerged as to which phenotype would predominate.

The mechanisms by which one phenotype predominates over another in such hybrids largely remains poorly understood. In some cases, at least, the genome of one contributing nucleus retained the capacity for reactivation of its tissue-specific genes even when extinguished in the initial hybrid cells. For example, in Chinese hamster fibroblast–rat hepatoma hybrids, rat liver functions were extinguished, only to reappear in some subclones on subsequent passage, possibly because of the loss of particular chromosomes. This result implies a stable modification to the genome responsible for the maintenance of its epigenetic state and not erased in hybrid cells. In other situations, experiments with phenomena such as imprinting and X-inactivation indicate that DNA methylation and histone acetylation can play a role in the heritable regulation of gene activity. Similar mechanisms are likely to play a role in the maintenance of a stable differentiated phenotype and might underlie results such as these.

On the other hand, dynamic regulatory factors must also be important because the early heterokaryon experiments clearly suggest that any repression of gene activity can be overcome by diffusible factors. Perhaps the earliest and clearest identification of a factor that can play a role in the dynamic regulation of gene activity was the discovery of the helix–loop–helix transcription factor, MyoD. The presence of MyoD alone, introduced into a cell by transfection with appropriate expression vectors, is sufficient to activate muscle-specific genes from several distinct cell types. MyoD is also subject to positive autoregulation so that, once expressed in a cell, it tends to maintain its own expression, thus establishing a dynamic system for maintenance of the muscle-differentiated state.

Nevertheless, and not surprisingly, even this story is not so simple. In some cells, MyoD is not sufficient to activate muscle gene expression; other somatic cell hybrid experiments show that MyoD itself is subject to negative regulators specified by loci elsewhere in the genome. If these patterns of regulation also apply to other key regulatory genes, it would not be unexpected that the outcome of fusion experiments between distinct types of differentiated cells would depend on the parental cells, the interactions of structural modifications to chromatin and DNA, and the dynamic, diffusible regulatory factors pertinent to those cells.

Hybrids of Pluripotent Cells

The behavior of hybrids of pluripotent cells has attracted considerable interest ever since lines of embryonal carcinoma (EC) cells were established *in vitro*. Indeed, the first description of established cultures of mouse EC cells included a report of the outcome of fusing EC cells and fibroblasts; in those experiments, the fibroblast phenotype predominated.

EC cells are the malignant, pluripotent stem cells of teratocarcinomas, which occur spontaneously in male mice of the 129 strain and in young human males, as testicular tumors of germ cell origin. Teratocarcinomas may contain an array of differentiated cell types corresponding to somatic cells of any of the three germ layers of the developing embryo or to extraembryonic cell types of the yolk sac or the trophoblast (in man but not mice). The stem cell status of EC cells was first clearly demonstrated experimentally by the landmark study by Kleithsmith and Pierce, and the relation of these cells to the inner cell mass or primitive ectoderm of the early stages of embryonic development, at least in the mouse, became evident through a series of studies during the 1970s. Those studies laid the foundations for the derivation of embryonic stem (ES) cells directly from the blastocyst of both mouse and primate, including human, embryos. ES cells are evidently the "normal" counterparts of tumor-derived EC cells in both mice and humans. A closely related pluripotent cell type is the embryonic germ (EG) cell, lines of which have been derived from cultures of primordial germ cells from the genital ridges of both mouse and human embryos.

In the late 1970s, several groups reported experiments to investigate the consequences of fusing EC cells with somatic, differentiated cells. In those experiments, hybrids of EC cells with thymocytes retained an EC phenotype. Others found similar results in fusions of EC cells with various lymphoid cells. Typically, the hybrids extinguished differentiated markers of the lymphoid parents and retained pluripotency. In several cases in which the parental EC cells had lost their ability to differentiate (nullipotent EC cells), the hybrids regained that ability. The most likely explanation for this result is that the nullipotent EC cells had accumulated mutations that limited their ability to differentiate. Such mutations would be expected to provide EC cells with a strong selective advantage because differentiation is typically accompanied by the loss of extended growth potential. Fusion with a wild-type cell could then allow genetic complementation and restore pluripotency, provided that the overall phenotype of the hybrid was that of an EC cell. Recently, a pluripotent human EC cell line, NTERA2, has been fused with a nullipotent line, 2102Ep. In that case, the hybrid cells did retain an ability to differentiate, suggesting that "loss-of-function" recessive mutations in 2102Ep were principally responsible for its inability to differentiate. However, although the hybrids differentiated, they did not give rise to neural differentiation, as the parental NTERA2 cells are well documented to do. Thus, the 2102Ep cells had evidently acquired dominant mutations that interfered with expression of that particular lineage in NTERA2 cells.

One circumstance in which the extinction of the somatic cell genes was variable was expression of the class 1, major histocompatibility complex (MHC) antigens. In mouse EC cells, the MHC antigens of the H2 complex are typically not expressed, although they are expressed in human EC cells. In some hybrids with lymphoid cells, expression of the somatic parent's H2 genes was suppressed; in others, it was not. In fusions of mouse EC cells with human lymphoid cells, most human chromosomes are lost, as generally occurs in mouse–human hybrids, and the hybrids retained an EC phenotype. However, in those retaining human chromosome 6, the human MHC antigens of the human lymphocyte antigen complex remained active.

Not all combinations of EC cells with somatic cells resulted in hybrids expressing an EC phenotype. Generally, hybrids with fibroblasts or fibroblastoid cells yielded fibroblastoid hybrids, as originally reported. In some hybrid combinations, for example, EC cell–hepatomas, both phenotypes were extinguished. However, intriguing results were obtained in EC cell–Friend erythroleukemia cell hybrids. Such hybrids generally expressed the phenotype of the erythroleukemia cells and were inducible for globin expression, which was specified by the EC-derived genome as well as the erythroleukemia genome. Nevertheless, when the EC cell parent was tetraploid, some hybrids retained an EC phenotype. Apparently, an increased dosage of the EC genome was sufficient to overcome the dominant effect of the erythroleukemia cell.

Reprogramming Somatic Cell Nuclei with EC, ES, or EG Cell Cytoplasm

Not only in frogs but also in mammals, nuclear transfer into enucleated eggs has resulted in the birth of live animals and illustrates the plasticity of nuclei derived from fully differentiated somatic cells. The ability of the egg or oocyte to "reprogram" the somatic cell nucleus to such a profound extent is remarkable. These experiments raised the prospect of deriving embryonic stem cells genetically identical to a prospective patient, who might receive transplants of differentiated derivatives of those stem cells to replace diseased or damaged tissues. However, the technique is demanding and inaccessible for many researchers, and the availability of human donor eggs is limited and ethically problematic. An attractive alternative for both study and practical application would be to "reprogram" somatic cells using pluripotent ES, EG, or EC cells. However, despite the extensive earlier studies of EC–somatic cell hybrids, none clearly demonstrated reprogramming of the genome derived from the somatic cell parent. It is formally possible that the somatic genome was silenced and that maintenance of the pluripotent state was actively

dependent on continued gene expression from the EC cell-derived genome alone.

In more recent experiments, EG cells have been fused with thymocytes from transgenic mice carrying a neo^R/lacZ transgene. As in the EC x thymocyte hybrids, EG–thymocyte hybrids expressing lacZ and drug resistance were isolated and retained a pluripotent EG-like phenotype. When the thymocytes came from transgenic mouse thymocytes engineered to express green fluorescent protein (GFP) under the transcriptional control of the Oct-4 promoter, the hybrids but not the thymocytes expressed GFP. Because Oct-4 is transcribed exclusively in germ cells, early embryos, and ES and EC cells, GFP should only be expressed when reprogramming of the thymocyte nucleus occurred. In addition, in both ES and EG hybrids with female thymocytes, and in similar EC hybrids, reactivation of the inactive thymocyte X-chromosome was observed. Thus, these experiments have now provided formal evidence that the somatic genome is indeed reprogrammed in these hybrids.

That the reprogramming ability is conserved between species was demonstrated in reciprocal fusions using mouse or human EC cells and cells of lymphocytic origin. Hybrids formed from fusion of the human T-cell line, CEM C7A, to mouse p19 EC cells expressed the endogenous human Oct-4 and Sox2 genes and adopted the morphology of the murine EC cells, with a high nuclear:cytoplasm ratio, prominent nucleoli, and growth in tight, well-defined adherent colonies. In the murine EC–human T-lymphocytes, there was considerable heterogeneity of expression between colonies and within a single colony over time. Thus, expression of human markers of cells of endoderm (collagen IV, laminin B1) and ectoderm (nestin) origin was noted in colonies that also expressed Oct-4 and Sox2. These data suggest that cross-species reprogramming had occurred in the hybrids and that subsequent spontaneous differentiation resulted in the human partner expressing the potential to adopt a fate not previously available to it. The reciprocal cross, using human 2102Ep EC cells and murine thymocytes, resulted in expression of the endogenous murine Oct-4 gene. Collectively, these data demonstrate that the ability to reprogram somatic nuclei is not the exclusive domain of the egg and oocyte but can also be achieved by pluripotent ES, EG, and EC cells.

Interestingly, in a fusion of thymocytes with ES or EG cells, both EG and ES cells are capable of reprogramming thymocytes but display different capacities for erasure of imprints. Imprinted genes are those for which the maternal and paternal alleles exhibit differential expression in somatic cells. Imprinting is established during gametogenesis and most likely involves methylation at specific loci. After fertilization, the imprinted pattern of those specific genes is retained in all cells of the developing embryo except that, necessarily, such imprints are erased in the primordial germ cells (PGCs) and are reinstated prior to the completion of gametogenesis in preparation for the next generation. Thus, in ES cells derived from the blastocyst, monoallelic expression of imprinted genes is evident, but imprinting is absent from EG cells that have been derived from PGC after imprints have been erased. In EG–thymocyte hybrids, the H19 and p57Kip2 loci, normally methylated on the paternal allele and preferentially expressed from the maternal allele in thymocytes, were denuded of the paternal imprints. Similarly, the maternally methylated Peg1/Mest allele was demethylated in the EG–thymocyte hybrids, and both maternal and paternal alleles were expressed. Furthermore, ordinarily methylated but not imprinted genes, such as Aprt, Pgk2, and globin, were demethylated in these hybrids just as they are in PGCs. Thus, the EG cells appeared to possess the same capacity as PGC to erase imprints. By contrast, methylation at the imprinted loci, H19 and Igf2r, was maintained in ES–thymocyte hybrids. Similarly, it was also found that imprinting was not erased in EC–thymocyte hybrids, although the nonimprinted allele, normally silent in thymocytes, was expressed in the hybrids. These data strongly suggested that although extensive remodeling of the chromatin sufficient to establish pluripotency occurred, reprogramming does not extend to imprinting if the pluripotent partner itself retains imprinting. The erasure of imprints was also found to be dominant in ES–EG hybrids.

These observations on epigenetic imprinting in stem cell hybrids have relevance to the development of embryos derived after somatic nuclear transfer to enucleated oocytes, and fusions using ES, EG, and EC cells may serve as useful models of the process. The efficiency with which somatic nuclei generate living clones is notoriously low. Dolly the sheep, the first mammal cloned from a fully differentiated adult cell nucleus, was the single success in more than 300 nuclear transfer experiments. Epigenetic abnormalities that are the result of inappropriate imprinting are probably one important reason for the poor success rate and questionable health of clones living to term. A constellation of 70 to 80 genes was shown to be inappropriately expressed in mouse embryos cloned from adult nuclei. In addition, many cloned embryos are subject to developmental abnormalities indicative of faulty imprinting at several loci.

During normal embryogenesis, PGCs, which eventually differentiate into male or female gametes, are erased of imprinting. During the development of the gametes in embryos resulting from normal fertilisation, correct imprinting is reestablished. However, because a cloned individual does not result from fertilization by mature sperm of an egg but rather from a donated diploid nucleus injected into an enucleate egg, the process of erasure and reinstatement of correct imprinting never occurs. The cell fusion experiments using ES, EG, and EC cells previously described may help to elucidate the process of imprinting and reprogramming and to provide further information about why embryo cloning is a risky and inefficient process.

Cell Fusion and the Demonstration of Stem Cell Plasticity

Recently, there have been myriad discussions and reports regarding adult stem cells and their apparent capacity not only to regenerate their own tissue of origin but also to regenerate lineages other than those from which they derive. For example, hematopoietic stem cells and mesenchymal stem cells have been reported to generate neurons, muscle, hepatocytes, and a host of other tissues. Similarly, fetal neural cells have been reported to create hematopoietic and other tissues. A degree of controversy has surrounded these reports, and scientists have been at pains to design experiments that unequivocally demonstrate that a single such adult stem cell can generate cells of more than one lineage.

Cell fusion occurs spontaneously under appropriate conditions *in vivo,* as well as *in vitro,* and *in vivo* can be a normal physiological process during the development of several organs and tissues. For example, myoblast–myoblast and myoblast–myotube fusions occur as part of normal skeletal muscle development, and fusion of trophoblast cells to form the giant cells occurs in the development of the placenta. Recent genetic studies in *Drosophila* and *Caenorhabditis elegans* have determined some of the genes required for these developmental fusion events.

The existence of genetically regulated, physiologically relevant cell fusion *in vivo,* as well as evidence of reprogramming of somatic cells following cell fusion, may confound many of the observations of adult stem cell plasticity. Two separate examples have recently been reported of spontaneous fusion of cells, in the absence of an external fusogenic agent, between populations of pluripotent stem cells and other cell types kept in co-cultures. For example, in mixed cultures using murine ES and genetically tagged bone marrow cells grown under conditions favoring the outgrowth of hybrids it was revealed that ES–bone marrow hybrids could be isolated at a frequency of 10^{-5}–10^{-6}. Spontaneous fusion was dependent on the presence of IL-3 and the leukemia inhibitory factor, which obligates cytokines for hematopoietic lineages and the maintenance of pluripotency and self-renewal in ES cells, respectively. In similar experiments using mixed cultures of murine ES and genetically marked cells taken from fetal and adult mouse brain, it was also found that spontaneous hybrid formation occurred at ten-fold frequency. In both reports, resulting hybrids displayed the morphology and pluripotency of the ES partner. Where once the somatic partner in these hybrids was restricted in its developmental fate to producing the repertoire of hematopoietic lineages or central nervous system lineages, respectively, fusion with the ES cell allowed expression of new potential fates. These spontaneously formed hybrid cells were capable of contributing to all three germ layers after injection into blastocysts. Thus, in any demonstration of lineage plasticity of adult stem cells, it becomes important to rule out the possibility of fusion with another endogenous pluripotent stem cell that reprograms the test cell.

A further development in cell fusion has recently been documented in two reports examining the nature of hepatocytes derived from hematopoietic stem cells, previously one of the most robust arguments for adult stem cell plasticity. Using mice engineered to be susceptible to liver degeneration through tyrosinemia because of ablation of the fumarylacetoacetate hydrolase gene, the regeneration of hepatocytes from transplanted, lineage-depleted bone marrow was unequivocally shown. However, ordinarily, the liver contains a subset of polyploid cells, suggesting that cell fusion may be a characteristic feature of this organ. The two followup studies report that the hepatocyte outgrowth as a result of hematopoietic stem cell transplantation almost certainly occurred because of the fusion of hematopoietic cells with preexisting hepatocytes. In another study, some of the regenerated liver cells were diploid rather than tetraploid, as would be expected when two diploid cells fuse. This suggests that it is possible to generate a diploid cell from a tetraploid hybrid and still maintain the reprogrammed phenotype.

The identity of the fusion partner derived from bone marrow was not determined in these reports. Nevertheless, the authors speculate that it is probably not the hematopoietic stem cell itself but a later progenitor. In the liver regeneration model, the authors speculate that the hematopoietic partner cells are phagocytic Kupffer cells, macrophages, B- or T-cells; this view is supported by the finding that liver regeneration occurred after the hematopoietic lineages had been repopulated. However, because liver injury did not mobilize a similar response, it is unlikely that this is a generalized response to tissue damage; rather, it may be the result of the artificial constraints of the experimental design.

Others have observed indications of engraftment in distant tissues by cells introduced during the transplantation of donated organs or tissues. For example, Y-chromosome-positive cells with apparent neuronal or cardiomyocyte function have been noted in the brains and hearts, respectively, of female bone marrow transplant recipients. These observations have been interpreted as evidence of transdifferentiation of the incoming donor cells and subsequent population of host organs. However, given the findings described previously, it now seems plausible that these may have been the result of spontaneous fusion of host and donor cells.

Summary

The study and use of cell fusion has an extensive history. It has been an invaluable tool in somatic cell genetics, and it has highlighted the complexity of the mechanisms that regulate the maintenance of the determined and differentiated states. Although it is difficult to make strong claims that the study of cell hybrids has contributed substantially to understanding those mechanisms, the ability of pluripotent stem cells to reprogram somatic nuclei to a primitive, pluripotent state increases the potential of achieving somatic cell reprogramming in an efficient manner for therapeutic purposes without resorting to nuclear transfer to oocytes — so-called therapeutic cloning. The phenomenon of spontaneous fusion is also an issue that now has to be addressed in the analysis of any claims for plasticity of otherwise lineage-restricted, adult stem cells.

ACKNOWLEDGMENTS

This work was supported partly by grants from the Wellcome Trust, Yorkshire Cancer Research, and the BBSRC.

KEY WORDS

Cell fusion The process whereby the cell membrane between two juxtaposed cells breaks down and reforms to incorporate the cytoplasm and nuclei of both cells into a single, viable cell.

Heterokaryon The immediate result of fusing two different cells which yields a single cell with two or more identifiable intact nuclei, one from each parental cell.

Hybrid cell The cell that arises when a heterokaryon resulting from cell fusion enters mitosis, with breakdown of the nuclear membranes of the individual nuclei and formation of a single nucleus in each daughter cell containing the genetic material from the original parental cells. Note, however, that there is often significant genome loss during this and subsequent divisions, so the hybrid cell may not contain all of the genetic material from each parental cell.

FURTHER READING

Davidson, R. L. (1973). Control of the differentiated state in somatic cell hybrids. *Symp. Soc. Dev. Biol.* **31**, 295–328.

Gurdon, J. B. (1962). The developmental capacity of nuclei taken from intestinal epithelial cells of feeding tadpoles. *J. Embryol. Exp. Morph.* **10**, 622–640.

Harris, Henry. (1995). *The cells of the body: A history of somatic cell genetics.* Cold Spring Harbor Press, Cold Spring Harbor, NY.

Kennett, R. H. (1979). Cell fusion. *Methods Enzymol.* **58**, 345–359.

Kleinsmith, L. J., and Pierce, G. B. (1964). Multipotentiality of single embryonal carcinoma cells. *Cancer Res.* **24**, 1544–1552.

Sorieul, S., and Ephrussi, B. (1961). Karyological demonstration of hybridisation of mammalian cells *in vitro*. *Nature* **190**, 653–654.

Terada, N., Hamazaki, T., Oka, M., Hoki, M., Mastalerz, D. M., Nakano, Y., Meyer, E. M., Morel, L., Petersen, B. E., and Scott, E. W. (2002). Bone marrow cells adopt the phenotype of other cells by spontaneous cell fusion. *Nature* **416**, 542–545.

Weiss, M. C., and Green, H. (1967). Human–mouse hybrid cell lines containing partial complements of human chromosomes and functioning human genes. *Proc. Nat. Acad. Sci. USA* **58**, 1104–1111.

Weiss, M. C., Sparkes, R. S., and Bertolotti, R. (1975). Expression of differentiated functions in hepatoma cell hybrids: IX. Extinction and reexpression of liver-specific enzymes in rat hepatoma–Chinese hamster fibroblast hybrids. *Somatic Cell Genet.* **1**, 27–40.

Ying, Q. L., Nichols, J., Evans, E. P., and Smith, A. G. (2002). Changing potency by spontaneous fusion. *Nature* **416**, 545–548.

10

How Cells Change Their Phenotype

David Tosh and Marko E. Horb

Introduction

Until recently, it was thought that once a cell had acquired a stable differentiated state, it could not change its phenotype. We now know this is not the case, and over the past few years a plethora of well-documented examples have been presented whereby already differentiated cells or tissue-specific stem cells have been shown to alter their phenotype to express functional characteristics of a different tissue. In this chapter, we examine evidence for these examples, comment on the underlying cellular and molecular mechanisms, and speculate about possible directions of research.

Metaplasia and Transdifferentiation: Definitions and Theoretical Implications

Metaplasia is defined as the conversion of one cell type to another, and it can include conversions between tissue-specific stem cells. *Transdifferentiation*, on the other hand, refers to the conversion of one differentiated cell type to another and should therefore be considered a subset of metaplasia. Historically, metaplasia has been the term used by pathologists, but in recent years transdifferentiation has become the favored term, even when discussing the conversion of tissue-specific stem cells to unexpected lineages. Within the medical community, the idea of metaplasia is not controversial, but in the scientific community, some skepticism still surrounds the phenomenon of transdifferentiation — it being attributed to tissue culture artifacts or cell fusion. Nevertheless, it is important to study metaplasia and transdifferentiation to gain a better understanding about the regulation of cellular differentiation, which may lead to new therapies for a variety of diseases, including cancer.

Why Study Transdifferentiation?

Regardless of which definition is applied, we consider the study of transdifferentiation and metaplasia to be important for four reasons. First, it allows us to understand the normal developmental biology of the tissues that interconvert. Most transdifferentiations occur in tissues that arise from neighboring regions in the developing embryo and are therefore likely to differ in the expression of only one or two tran-

Essentials of Stem Cell Biology
Copyright © 2006, Elsevier, Inc.
All rights reserved.

scription factors. If the genes involved in transdifferentiation can be identified, then this might shed some light on the developmental differences that exist among adjacent regions of the embryo.

Second, it is important to study metaplasia because it predisposes to certain pathological conditions, such as Barrett's metaplasia (see later sections for further details). In this condition, the lower end of the esophagus contains cells characteristic of the intestine, and there is a strong predisposition to adenocarcinoma. Therefore, understanding the molecular signals in the development of Barrett's metaplasia will help to identify the key steps in neoplasia and may provide us with potential therapeutic targets as well as diagnostic tools.

Third, understanding transdifferentiation will help to identify the master switch genes and thus allow us to reprogram stem cells or differentiated cells for therapeutic purposes.

The fourth reason is that we will be able to identify the molecular signals for inducing regeneration and therefore to promote regeneration in tissues that otherwise do not regenerate (e.g., limb regeneration).

Examples of the Phenomenon

Despite the controversy surrounding transdifferentiation, there are numerous examples that exist in both humans and animals; here we will focus on a select few. The examples we have chosen to examine in detail include the conversions of pancreas to liver, liver to pancreas, esophagus to intestine, iris to lens, and bone marrow to other cell types. Other chapters in this book describe some of these as well as additional examples in more detail.

PANCREAS TO LIVER

The conversion of pancreas to liver is one well-documented example of transdifferentiation. This type of conversion is not surprising since both organs arise from the same region of the endoderm and both are thought to arise from bipotential cells in the foregut endoderm. In addition, the organs share many transcription factors, displaying their close developmental relationship. The appearance of hepatocytes in the pancreas can be induced by different protocols, including feeding rats a copper-deficient diet with the copper chelator Trien, overexpressing keratinocyte growth factor in the islets of the pancreas, or feeding animals a methionine-deficient diet and exposing them to a carcinogen. It has also been observed naturally in a primate, the vervet monkey. Although the functional nature of the hepatocytes has been examined in detail, until recently, the molecular and cellular basis of the switch from pancreas to liver was poorly understood.

Two *in vitro* models have been produced for the transdifferentiation of pancreas to liver. The first model uses the pancreatic cell line AR42J, and the second uses mouse embryonic pancreas tissue in culture; both rely on the addition of glucocorticoid to induce transdifferentiation. AR42J cells are amphicrine cells derived from azaserine-treated rats, and they express both exocrine and neuroendocrine properties. That is, they are able to synthesize digestive enzymes and express neurofilament. The dual nature of this cell line is evident in that when exposed to glucocorticoid, they initially enhance the exocrine phenotype by producing more amylase, but when cultured with hepatocyte growth factor and activin A, the cells convert to insulin-secreting β-cells. These properties of the AR42J cells suggest that they are an endodermal progenitor cell type with the potential to become exocrine or endocrine cell types.

The transdifferentiated hepatocytes formed from pancreatic AR42J cells express many of the proteins normally found in adult liver — for example, albumin, transferrin, and transthyretin. They also function as normal hepatocytes; in particular, they are able to respond to xenobiotics (e.g., they increase their catalase content after treatment with the peroxisomal proliferator, ciprofibrate). Although the mouse embryonic pancreas also expresses liver proteins after culture with dexamethasone, it is not clear whether the same cellular and molecular mechanisms are in operation as in the AR42J cells. It is possible that, rather than the hepatocytes arising from already differentiated cell types, the liver-like cells are derived from a subpopulation of pancreatic stem cells.

To determine the cell lineage of hepatocyte formation from pancreatic AR42J cells, we performed a lineage experiment based on the perdurance of green fluorescent protein (GFP) and used the pancreatic elastase promoter. After the transdifferentiation, some cells that expressed GFP also contained liver proteins (e.g., glucose-6-phosphatase). This result suggests that the nascent hepatocytes must have once had an active elastase promoter; therefore, they were differentiated exocrine cells. To elucidate the molecular basis of the switch in cell phenotype, we determined the expression of several liver-enriched transcription factors. Following treatment with dexamethasone, C/EBPβ became induced, the expression of the exocrine enzyme amylase was lost, and liver genes (e.g., glucose-6-phosphatase) were induced. These properties of C/EBPβ make it a good candidate to be an essential factor involved in the transdifferentiation of pancreas to liver. Indeed, C/EBPβ is sufficient to transdifferentiate AR42J cells to hepatocytes. Therefore, C/EBPβ appears to be a good candidate for the master switch gene distinguishing liver and pancreas.

LIVER TO PANCREAS

The numerous examples of pancreas-to-liver transdifferentiation suggest that the reverse switch should also occur readily; nevertheless, examples of this type of conversion are infrequent. The presence of pancreatic tissue in an abnormal location is known as heterotopic, accessory, or aberrant pancreas, and the frequency has been reported to range from 0.6 to 5.6%. In most cases (70–90%), the heterotopic pancreas is found in the stomach or intestine and is considered to be an embryological anomaly. In contrast, intrahepatic pancreatic heterotopia has only been reported in six individuals, comprising less than 0.5% of all cases of heterotopic pancreas. In general, heterotopic pancreatic tissue can be composed of exocrine, endocrine, or both types of cells. In almost every case of pancreatic heterotopia in the liver, however, only exocrine cells are present; only one case describes the presence of endocrine cells. Unlike the other cases of accessory pancreas, these rare incidents of intrahepatic pancreatic tissue cannot be explained as the result of a developmental error. In fact, in most of these cases, the patients were diagnosed with cirrhosis, suggesting that the heterotopic pancreas arose as a metaplastic process. Results with the animal models concur.

In other animals, pancreatic exocrine tissue can be induced in the liver by feeding rats polychlorinated biphenyls or by exposing trout to various carcinogens, such as diethylnitrosamine, aflatoxin B_1, or cyclopropenoid fatty acid. In these examples, the hepatic exocrine tissue is most often associated with tumors or injury, such as hepatocellular carcinomas (which arise from hepatocytes) or cholangiolar neoplasms (which arise from the bile duct), or adenofibrosis. Much like the human cases, these results suggest that during carcinogenesis, a metaplastic event occurs that generates pancreatic tissue. Indeed, pancreatic metaplasia in trout can be inhibited by the addition of the glucosinolate indole-3-carbinol, a known anticancer agent. Whether inhibiting metaplasia prevents neoplasia remains to be determined. The ability of one cell (liver) to transdifferentiate into another (pancreas), no matter how rare, suggests that it should be possible to identify the molecular signals involved in switching a cell's phenotype and thus to learn how to control and direct this conversion for therapeutic purposes.

Two recent reports have shown that it is possible to experimentally convert liver cells into pancreatic cells. Each has used a different approach to bring about transdifferentiation — by changing either the extracellular or the intracellular environment. In the first example, hepatic oval cells were isolated and maintained in tissue culture media supplemented with leukemia inhibitory factor (LIF). Upon the removal of LIF and the addition of high concentrations of glucose (23 mM) in the medium, the oval cells transdifferentiated to pancreatic cells. The oval cells were converted into a variety of pancreatic cell types, including glucagon, insulin, and pancreatic polypeptide-expressing cells. Functionally, these oval cell-derived endocrine cells were able to reverse hyperglycemia in streptozotocin-induced diabetes. The mechanism whereby glucose induces the transdifferentiation is not known, though previously it was shown that glucose can promote the growth and differentiation of β-cells in the normal pancreas; perhaps a similar mechanism operates here.

In the second example, hepatic cells (either *in vivo* or *in vitro*) were induced to transdifferentiate by overexpression of a superactive form of a known pancreatic transcription factor, Pdx1. Pdx1 is expressed early in the endoderm prior to overt morphological development of the pancreas and has been shown to play a fundamental role in the development of the entire pancreas. Although a previous study showed that continuous overexpression of the unmodified Pdx1 in the liver increased hepatic insulin production, it is not known if this represents a true transdifferentiation or simply the activation of the insulin gene.

In the study using modified Pdx1, the transdifferentiation appears to be relatively complete in that both exocrine and endocrine cells were produced, including insulin, glucagon, and amylase-expressing cells. On its own, Pdx1 is unable to convert liver to pancreas and requires an extra activation domain, VP16. This might be because of the lack of appropriate protein partners or the presence of some inhibitory proteins in the liver cells, since sequence-specific transcription factors are known to require other tissue-specific coactivators to stimulate transcription. To overcome this problem, the VP16 activation domain was fused to Pdx1 — the VP16 activation being able to activate transcription directly by binding to various coactivators as well as the basal transcription machinery, eliminating the need for other tissue-specific proteins. When Pdx1–VP16 was overexpressed in the liver, using the liver-specific promoter transthyretin, it was able to induce the transdifferentiation of liver to pancreas. This is an example whereby a known transcription factor essential for pancreas development was engineered to act as a master switch gene.

In conclusion, the preceding results demonstrate that it is possible to transdifferentiate liver cells, whether fully differentiated or not, into pancreatic cells. Since the liver has the ability to regenerate, this tissue can provide an abundant resource for the production of pancreatic cells with the aim of curing diabetes. The first study shows that an extracellular factor, glucose, can be used; the second reveals that a modified intracellular tissue-specific transcription factor, Pdx1–VP16, is sufficient. These studies suggest that if we can identify the key factors involved in the physiological regulation of the adult pancreas as well as in embryonic development, this will play an important role in understanding and promoting the transdifferentiation of various cell types into pancreas. However, overexpression of a single transcription factor may not be sufficient; the ability to change a cell's phenotype may therefore require the use of modified or engineered transcription factors (e.g., Pdx1–VP16) that can artificially recruit the transcriptional machinery.

BARRETT'S METAPLASIA

Barrett's metaplasia (or Barrett's esophagus, as it is sometimes called) is the clinical situation in which intestinal cells are found in the tissue of the lower end of the esophagus. In the strictest terms, it is the conversion of stratified squamous epithelium to columnar epithelium and is characterized by the presence in biopsy material of acid mucin-containing goblet cells. The importance of Barrett's metaplasia stems from the rise in apparent incidence of the disease and its risk associated with the development of adenocarcinoma of the esophagus. It is not known why metaplastic cells have a greater disposition toward neoplastic progression.

One of the proposed mechanisms for development of Barrett's metaplasia is gastrooesophageal reflux. It is assumed that prolonged acid reflux from the stomach (generally with bile acids) promotes damage to the epithelia at the end of the esophagus. Presumably, in the early stages of the disease, the normal stratified squamous epithelium is replaced. Eventually, reepithelialization results in the formation of columnar as opposed to stratified squamous epithelium, most likely because of repeated exposure to an acid environment. It is not clear whether the different intestinal cell types formed in the esophagus arise from the same stem cell in the basal layer or whether there is a transdifferentiation of columnar cells to goblet cells. It is also not understood why some patients with efflux disease do not go on to develop Barrett's metaplasia.

The molecular events underlying Barrett's metaplasia are not well understood. However, good candidate genes include the *caudal*-related homeobox genes *cdx1* and *cdx2*. Several lines of evidence support this statement. First, both genes are expressed in the intestine (but not in the stomach or in the esophagus) and are known to be important in regulating intestine-specific gene expression. Second, colonic epithelium is transformed to squamous epithelium (similar to the esophagus) in mice haploinsufficient for *cdx2*. Third and more recently, it was found that ectopic expression of *cdx2* in the stomach can induce intestinal metaplasia, and there is some evidence for early expression of *cdx2* in patients with Barrett's metaplasia. These results suggest that *cdx* genes may provide a target for therapeutic intervention in Barrett's metaplasia.

REGENERATION

The idea of regenerative medicine or tissue engineering mainly implies using tissue-specific stem cells to replace damaged or "lost" organs. However, as more examples of the plasticity of differentiated cells become known, it may be more feasible to use them than to use stem cells. Historically, the classical example of transdifferentiation occurs during lens regeneration in newts, where the dorsal iris-pigmented epithelium (IPE) is converted to lens. In other species, it may be a different source of cells that undergoes transdifferentiation to lens, such as the outer cornea in *Xenopus laevis*. In a similar fashion, regeneration of the neural retina occurs by transdifferentiation of the retinal-pigmented epithelium (RPE) in various vertebrates. In both cases, the regeneration has been shown to occur in adult newts as well as in other vertebrate embryos, including chick, fish, and rat.

Upon removal of the lens (lentectomy), only the dorsal IPE transdifferentiates into lens in three phases — dedifferentiation, proliferation, and transdifferentiation — even though both the dorsal and ventral IPE have the potential to become lens. It has been shown that the dorsal and not the ventral IPE expresses Pax6, Prox1, and FGFR-1, suggesting that these factors may play an important role in inducing the transdifferentiation. Indeed, inhibition of FGFR-1 will block the transdifferentiation of IPE to lens, and in *Xenopus,* it has been shown that FGF-1 can induce the transdifferentiation of outer cornea to lens. In agreement with this, retinal regeneration in chick embryos can be promoted by the addition of either FGF-1 or FGF-2 but not other growth factors such as TGFβ. These results show that the microenvironment in which a cell resides plays a critical role in regulating transdifferentiation.

Why one cell changes its phenotype (dorsal IPE) but another does not (ventral IPE) in response to exogenous growth factors may depend on the competence of each cell (e.g., the presence of the appropriate receptors). It is possible that FGFs may be responsible for inducing de-differentiation of the tissue and that the expression of particular transcription factors induces both proliferation and transdifferentiation. This classical model of lens regeneration illustrates that the molecular signals involved in transdifferentiation can be identified and used to promote regeneration of cells normally considered unable to alter their phenotype.

BONE MARROW TO OTHER CELL TYPES

Some cell-type conversions using bone marrow-derived stem cells have been shown to occur across what was previously considered to be germ-line boundaries (i.e., mesoderm to endoderm). In this situation, it is not evident whether the cell must first become a different stem cell and then differentiate along a different pathway or if it directly transdifferentiates to another phenotype. However, some doubt has been cast on these observations, and it has been suggested that the result is caused by an artifact from fusion of the circulating hematopoietic stem cells with resident cells. (See chapters 3, 9, and 32 in this volume for a description of these types of transdifferentiations.)

De-differentiation as a Prerequisite for Transdifferentiation

A question arises: If transdifferentiation is to occur, must the parent cell lose its phenotype before acquiring a new identity? In some examples (IPE to lens), there is an intermediate phenotype in which the cells do not express markers for either cell type (Figure 10-1B). However, examples of direct transdifferentiation do occur. Perhaps the best example is the transdifferentiation of pancreas to liver (pancreatic exocrine to hepatocyte) (Figure 10-1A). Whether a cell undergoes transdifferentiation directly, through a de-differentiated state, or through a stem cell may vary depending on which cell types are being studied (Figure 10-1). In other words, does the parent cell contain the necessary information to change its phenotype directly, or does it require the synthesis of new proteins? In direct transdifferentiation, the cell's competency is already established, and it is the removal of an inhibitor or the addition of an activator that pushes the fate of the cell over the final hurdle. For de-differentiation and stem cell intermediates, it may be necessary to establish the competency of the parent cell before it can undergo transdifferentiation. Further studies examining the transdifferentiation potential

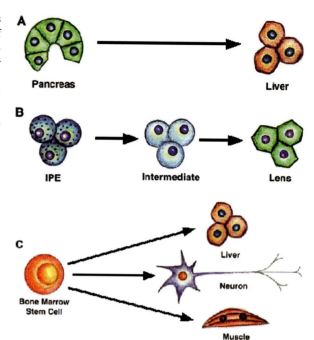

Figure 10-1. Examples of transdifferentiation. Transdifferentiation can occur in different stages: (A) Transdifferentiation of pancreas to liver can occur without cell division or an intermediate phenotype. (B) Transdifferentiation of pigment epithelium to lens requires an intermediate stage in which the cell does not possess the characteristics of either phenotype. (C) The pluripotency of stem cells is shown by their ability to convert to different cell lineages. In this example, the switch is direct and there is no conversion to another tissue-specific stem cell.

of individual transcription factors and various cell types will help to bring about an understanding of the rules of transdifferentiation.

How to Change a Cell's Phenotype Experimentally

The ability to change a cell's phenotype will greatly facilitate the design of therapies for diseases such as diabetes, liver failure, and neurodegenerative disorders (e.g., Parkinson's disease). We suggest six steps to follow to try and change a cell's phenotype experimentally.

1. IDENTIFY POTENTIAL FACTORS TO INDUCE TRANSDIFFERENTIATION

Transdifferentiation may be achieved in several ways using extracellular growth factors, individual transcription factors, or combinations of the two. An understanding of how individual organs or cell types form will help to identify those molecular factors that can be used to direct the transdifferentiation of other cell types. We believe that those factors essential for the initial development of an organ will work best, since they sit at the top of the hierarchy of the signaling cascade. Functional screens, such as those used previously to identify novel mesoderm-inducing factors, should aid the identification of new factors with the potential to direct transdifferentiation.

2. CHOOSE A CELL TYPE TO CONVERT

In many cases of transdifferentiation, only certain cells can undergo a specific transdifferentiation, suggesting that there are restrictions on whether a cell has the competence to undergo transdifferentiation. Therefore, choosing the cell type initially is important. We suggest that using closely related cell types will greatly improve the chances for transdifferentiation. Examples include the use of pancreatic AR42J cells in the conversion of pancreas to liver. Ultimately, either primary cultures of well-defined cell types or *in vivo* experiments will be necessary for this to be of therapeutic use.

3. CHOOSE THE METHOD OF OVEREXPRESSION

It is important to determine whether continuous or limited overexpression of a particular factor is required. On the one hand, tissue-specific promoters will allow the factor to be expressed for only a relatively short time; upon transdifferentiation, the promoter will no longer be active. Ubiquitous promoters, on the other hand, will express the chosen factor continuously and may produce undesired results. Many transcriptional regulators are expressed only transiently and require a strict temporal regulation for proper development to occur. For example, continuous overexpression of *Hlxb9* in the normal pancreas interferes with the differentiation of both exocrine and endocrine cells. Thus, the use of the constitutive promoters, such as cytomegalovirus, may not be suitable. We believe that the use of tissue-specific promoters is best suited to the type of experiments described here to prevent the chosen factor from interfering with the proper differentiation of the new phenotype.

4. IDENTIFY WHETHER A MODIFICATION OF THE FACTOR IS REQUIRED

It is possible that after identifying the factor, no transdifferentiation occurs. There may be several reasons for this, but we suggest testing a superactive version of the same factor before dismissing it. The easiest way to do this is to make use of a strong activation domain that has already been characterized, such as VP16. Whether it is fused to the N- or C-terminus should not matter, but we suggest that the VP16 be fused to the whole open reading frame of the transcription factor and not just to the DNA-binding domain.

5. CHARACTERIZE THE NEW PHENOTYPE

Initially, use of a reporter construct will greatly help the identification of a successful conversion — for example, the use of the elastase promoter driving GFP in the conversion of liver to pancreas. It is best to initially use a promoter expressed in all cell types in a particular organ rather than one specific to an individual cell type within that organ. Three questions can be addressed under this heading. First, is the transdifferentiation specific to producing a single cell type or, if the organ contains numerous cell types, are several types produced? In the case of the transdifferentiation of liver to pancreas, ectopic

expression of Pdx1–VP16 produces more than one type of pancreatic cell. Second, what is the identity of the new cell type, and can the loss of the other phenotype be demonstrated? Third, is the transdifferentiation stable? For a cell-type conversion to be a true transdifferentiation, the phenotype of the new cell must be stable.

6. TEST THE TRANSDIFFERENTIATION ACTIVITY IN OTHER CELL TYPES

As mentioned previously, in many instances, only a subset of cells can undergo a particular transdifferentiation. Therefore, it is essential to identify which cells can be transdifferentiated and why one cell is able to respond but another is not. An understanding of the competence of individual cells will lead to a greater understanding of what is necessary for transdifferentiation to occur — for example, is de-differentiation a necessary prerequisite? The question of whether it will be possible to direct the transdifferentiation of cells into one particular cell type, such as an insulin-secreting β-cell, remains to be seen.

Summary

The recent demonstration that adult stem cells and even differentiated cells are more versatile than previously thought means that abundant tissue for therapeutic purposes can be obtained from these cell types. Since some scientists and members of the public are averse to the use of embryonic stem cells for any form of clinical treatment, using either adult stem cells or differentiated cells induced to transdifferentiate may obviate the use of embryonic tissue. In the end, the ethical issues raised by some may not apply to transdifferentiated cells.

ACKNOWLEDGMENTS

We would like to thank the Medical Research Council, the Wellcome Trust, and the BBSRC for funding. We would like to thank Lori Dawn Horb for the artwork.

KEY WORDS

Master Switch Gene   A transcription factor that is sufficient to induce the transdifferentiation of one cell type to another; is associated with the loss of one set of differentiated properties (e.g., pancreatic) and the gain of another phenotype (e.g., hepatic). Master Switch Genes are normally involved in coordinating the differentiation of a particular cell fate through the activation of various downstream targets.

Metaplasia   The conversion of one cell type to another; can include conversions between tissue-specific stem cells.

Transdetermination   The all-or-none conversion of one regenerating body structure into another.

Transdifferentiation   The conversion of one differentiated cell type to another; a subset of metaplasia.

FURTHER READING

Eguchi, G., and Kodama, R. (1993). Transdifferentiation. *Curr. Opin. Cell. Biol.* **5**, 1023–1028.

Okada, T. S. (1991). *Transdifferentiation: flexibility in cell differentiation*. Clarendon Press, Oxford.

Shen, C. N., Slack, J. M. W., and Tosh, D. (2000). Molecular basis of trans-differentiation of pancreas to liver. *Nat. Cell. Biol.* **2**, 879–887.

Slack, J. M. W. (1985). Homoeotic transformations in man: implications for the mechanism of embryonic development and for the organization of epithelia. *J. Theor. Biol.* **114**, 463–490.

Tosh, D., and Slack, J. M. W. (2002). How cells change their phenotype. *Nat. Rev. Mol. Cell. Biol.* **3**, 187–194.

11

Nuclear Cloning and Epigenetic Reprogramming

Zhongde Wang, Alexander Meissner, and Rudolf Jaenisch

Introduction

Successful cloning by nuclear transfer (NT) requires the reprogramming of a differentiated genome into a totipotent state that can reinitiate normal embryogenesis (Figure 11-1). Embryonic genes silenced in the donor nucleus must be reactivated, and donor nucleus specific genes detrimental to the totipotent state need to be silenced. A chromatin structure that ensures such gene expression patterns has to be established in the donor genome. This process, broadly defined as epigenetic reprogramming, must occur within hours to a few days following NT to allow the development of a reconstructed embryo. Faulty epigenetic reprogramming in cloned embryos leads to widespread irregularities in gene expression that might result in the developmental abnormalities and embryonic lethality frequently observed in cloning. Indeed, cloning by NT is characterized by extremely low efficiency in all species to which this technique has been applied. Most clones die before birth, and the rare clones that survive to term or adulthood display a range of developmental abnormalities. Among these abnormalities are circulatory problems, respiratory distress, obesity, immune dysfunction, kidney or brain malformation, and early death. Noticeably, an increase in placental and birth weight, a cross-species phenotype referred to as large offspring syndrome (LOS), is often observed. At present, little is known about the events that take place during reprogramming and about the molecules in the egg cytoplasm responsible for this process.

A better understanding of the molecular mechanisms governing epigenetic reprogramming would provide great insights into the developmental abnormalities associated with cloning. In this chapter, we discuss recent advances in the understanding of the molecular and cellular aspects of epigenetic reprogramming following NT. We begin by reviewing the role of DNA methylation, a key epigenetic modification known to control normal development and gene expression. Then, we discuss the aberrant DNA methylation patterns observed in clones and how they might lead to abnormalities in the animals. Finally, we review what is known about factors that may affect epigenetic reprogramming.

Epigenetics and Epigenetic Reprogramming in Cloning

After fertilization, the genetic content of a zygote is inherited by all somatic cells of the developing organism. However, only a subset of the genes is active in a given cell type. For normal development to proceed, it is essential to turn on the appropriate genes and turn off genes not required in a particular cell. This process generally involves DNA methylation and chromatin modifications that impose stable but reversible marks on the genome. Such stable alterations resulting in differential gene expression are often referred to as *epigenetic*.

Because each somatic nucleus within an organism has acquired a certain tissue-specific epigenetic state during development, cloning from somatic cells requires the resetting of a differentiated nucleus to a totipotent, embryonic ground state. One likely explanation for cloning-associated abnormalities is inadequate epigenetic reprogramming of the donor genome. Microarray experiments have shown that hundreds of genes are abnormally expressed in newborn cloned animals. Furthermore, DNA methylation patterns in cloned preimplantation embryos and in tissues from cloned animals have been shown to be aberrant compared with controls. The most direct evidence for the notion that cloning phenotypes are epigenetic rather than genetic comes from the observation that the abnormal phenotypes of cloned animals are not transmitted to their offspring.

DNA METHYLATION DURING NORMAL DEVELOPMENT

DNA methylation provides heritable information to the DNA that is not encoded in the nucleotide sequence. In higher eukaryotes, DNA methylation is the only covalent modification of the DNA. It occurs at position 5 of the pyrimidine ring of cytosines and is almost exclusively restricted to CpG dinucleotides in somatic cells. In contrast, embryonic stem (ES) cells and early embryos seem to contain significant amounts of non-CpG methylation (mostly CpA). Currently, the functional role of this non-CpG methylation is not clear.

DNA methylation has been implicated to participate in a diverse range of cellular functions and pathologies, including tissue-specific gene expression, cell differentiation, genomic imprinting, X-chromosome inactivation, regulation of chromatin structure, carcinogenesis, and aging. In general, methylation is found in CpG-poor regions; CpG-rich areas (CpG

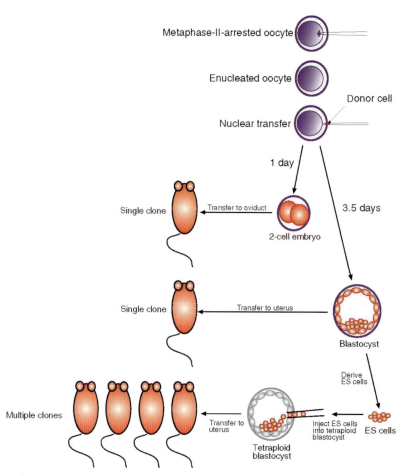

Metaphase-II-arrested oocyte

Enucleated oocyte

Donor cell

Nuclear transfer

1 day

Single clone ← Transfer to oviduct — 2-cell embryo

3.5 days

Single clone ← Transfer to uterus — Blastocyst

Derive ES cells

Multiple clones ← Transfer to uterus — Tetraploid blastocyst ← Inject ES cells into tetraploid blastocyst — ES cells

Figure 11-1. Generation of cloned mice. The metaphase spindle is removed from a metaphase-II-arrested oocyte using micromanipulators and the donor nucleus is injected directly into the cytoplasm. Three different approaches to subsequently generate cloned mice are depicted. First, to minimize the time in culture 2-cell embryos can be transferred to the oviduct of a recipient female. Second, to further assess the developmental potential of the reconstructed embryos in vitro, clones can be cultured until the blastocyst stage (day 3.5) and then be transferred to the uterus of a recipient female. Finally, ES cells can be derived from the Inner Cell Mass (ICM) of the cloned blastocyst and by using tetraploid embryo complementation multiple identical clones that are solely derived from the donor ES cells can be generated.

islands) seem to be protected from this modification and are generally associated with active genes. This is consistent with the fact that methylated CpG islands are found on the inactive X-chromosome and on the silenced allele of imprinted genes. The methyl group is positioned in the major groove of the DNA, where it can easily be detected by proteins interacting with the DNA. The effects of DNA methylation on chromatin structure and gene expression are likely mediated by a family of proteins that share a highly conserved methyl CpG-binding domain (MBD). Two of these, Mecp2 and Mbd1, have been suggested to be involved in transcriptional repression based on biochemical observations that they form complexes with histone deacetylases and other proteins important for chromatin structure.

DNA methylation patterns are extremely dynamic in early mammalian development. Within 1 to 2 cell divisions after fertilization, a wave of global demethylation takes place. It has been suggested that the paternal genome is actively demethylated during the period of protamine–histone exchange and that the maternal genome subsequently becomes demethylated, presumably through a passive DNA replication mechanism. By the morula stage, methylation is found only in some repetitive elements and imprinted genes. After implantation, genomewide methylation levels increase dramatically, establishing a differential pattern between the cells of the inner cell mass and those of the trophectoderm and resulting in the formation of methylation patterns found in the adult. Primordial germ cells (PGC) also undergo global demethylation. In contrast to demethylation during preimplantation, all parental-specific epigenetic marks are erased in the PGC by embryonic day 13–14. As a result, PGC and diploid germ cells are the only cell types in which the paternal and maternal genomes are equivalent. Upon initiation of gametogenesis, PGC remethylation begins, and the parental-specific methylation patterns that will code for monoallelic expression of imprinted genes are established.

Maintenance and establishment of DNA methylation are accomplished by at least three independent catalytically active DNA methyltransferases: Dnmt1, Dnmt3a, and Dnmt3b. There are two isoforms of Dnmt1: an oocyte-specific isoform (Dnmt1o) and a somatic isoform. Somatic Dnmt1 is often referred to as the "maintenance" methyltransferase because it is believed to be the enzyme responsible for copying methylation patterns after DNA replication. The oocyte-specific isoform of Dnmt1 is believed to be responsible for maintaining but not for establishing maternal imprints. The Dnmt3 family (Dnmt3a, 3b, 3l, and several isoforms) is required for the *de novo* methylation that occurs after implantation, for the *de novo* methylation of newly integrated retroviral sequences in mouse ES cells, and for the establishment of imprints (Dnmt3l). It was recently shown that Dnmt3a has a strong preference for unmethylated DNA.

The essential role of DNA methylation in mammalian development is highlighted by the fact that mutant mice lacking each of the enzymes (generated by gene targeting) are not viable and die either during early embryonic development (Dnmt1 and Dnmt3b) or shortly after birth (Dnmt3a). The knockout of Dnmt3l leads to male infertility and the failure to establish imprints in female eggs. Disruption of Dnmt2 did not reveal any obvious effects on genomic DNA methylation. The biological role of Dnmt2 is still elusive; however, a possible role in centromere function has been suggested.

ABNORMAL DNA METHYLATION PATTERNS IN CLONES

Considering the fundamental role of DNA methylation in development, it seems likely that any cloned embryo will need to recapitulate a functional pattern of epigenetic modifications to proceed through normal embryogenesis. Several research groups have investigated DNA methylation patterns in cloned embryos and reported finding abnormalities in DNA methylation. In cloned bovine embryos, satellite sequence methylation levels are closer to the donor cells than to control embryos. However, methylation patterns of single-copy gene promoters in cloned bovine blastocysts appeared to be normally demethylated. In addition, the satellite sequences, not the single-copy genes, showed more methylation in the trophectoderm than in the inner cell mass of cloned bovine blastocysts. Using antibodies against 5-methyl cytosine, two independent studies showed that the cloned bovine embryos did not undergo global demethylation in early embryogenesis and even showed precocious *de novo* methylation, with abnormally hypomethylated euchromatin and abnormally hypermethylated centromeric heterochromatin. These findings suggest that different chromosomal regions might respond differently to demethylation in the egg cytoplasm. Interestingly, when the same satellite sequences examined in bovine were analyzed in a different species (porcine), methylation levels at the blastocyst stage of cloned embryos were more comparable to those of fertilized control embryos, suggesting species-specific differences. A recent study that analyzed several imprinted genes in cloned murine blastocysts showed

that most of the examined genes displayed aberrant methylation and expression patterns.

It would be interesting to know if aberrant methylation patterns during preimplantation development contribute to the low efficiency of generating clones and to what extent the clones can tolerate such variation. Unfortunately, analyzing the methylation status in preimplantation embryos provides only indirect correlations, preventing satisfactory resolution of this question.

To establish a potential correlation between global DNA methylation levels and the developmental potential of cloned embryos, one study compared the genomewide methylation status among spontaneously aborted cloned fetuses, live cloned fetuses, and adult clones in bovine. When genomewide cytosine methylation levels were measured by reverse-phase high performance liquid chromatograph (HPLC), they found that a significant number of aborted fetuses lacked detectable levels of 5-methylcytosine. In contrast, when seemingly healthy adult, lactating clones were compared to similarly aged lactating cows produced by artificial insemination, comparable DNA methylation levels were observed. The authors suggested that the survivability of cloned cattle is related to the global DNA methylation status. All evidence suggests that a correct global methylation status is required for development. However, subtle changes might be compatible with normal development and result only in minor or no phenotypes. For example, by applying restriction landmark genome scanning (RLGS) in two seemingly healthy cloned mice, it was shown that methylation patterns at several sites in each clone differed from those in the controls.

The reason for the frequent abnormal DNA methylation patterns in cloned embryos is still unclear. It is likely that, because of the epigenetic difference between the somatic donor cell and the gametes, the somatic nucleus responds differently to the egg cytoplasm, affecting subsequent events during embryogenesis. For example, the highly coordinated demethylation process in the pronuclei of the maternal and paternal genome upon fertilization might not happen appropriately in the somatic donor genome following NT. It is not clear whether all of the somatic epigenetic marks imposed by DNA methylation during differentiation can be removed from the donor nucleus. Any failure to demethylate the DNA sequences normally demethylated during early cleavage stages of development might be stably passed on to progeny cells. Another possible explanation for the aberrant methylation patterns in clones may result from the ectopic expression of the somatic form of Dnmt1 in the egg- and cleavage-stage cloned embryos. In the mouse oocyte and preimplantation embryo, the oocyte-specific form (Dnmt1o), but not the longer somatic form, is expressed. It has been shown that a translocation of Dnmt1o between nucleus and cytoplasm is tightly regulated during murine preimplantation development. In contrast, cloned preimplantation mouse embryos were reported to aberrantly express the somatic form of the Dnmt1 gene, and the translocation of Dnmt1o was absent. As mentioned previously, DNA methyltransferases (Dnmt1, 3a, 3b, and 3l) play important roles in setting up and maintaining

DNA methylation patterns. It is reasonable to speculate that dysregulation of any of these enzymes in clones may alter DNA methylation patterns.

These abnormal DNA methylation patterns could result in embryo lethality or phenotypic abnormality. Little is known about the developmental role of dynamic changes in DNA methylation during preimplantation, although recently, the importance of early embryonic methylation patterns in setting up the structural profile of the genome was shown. The failure to establish correct methylation patterns early in development might therefore have far-reaching effects on the chromatin structure. Interestingly, mouse embryos deficient for Dnmt1 and Dnmt3b die around embryonic day 9.5, but Lsh mutant mice die only after birth, despite showing a substantial loss of methylation throughout the genome. This would suggest that, at least for embryogenesis, high levels of DNA methylation are not essential.

Factors That May Affect Epigenetic Reprogramming in Cloning

Studies have shown that epigenetic reprogramming seems to be incomplete in most, if not all, of the clones. In this section, we discuss some of the aspects of the donor genome shown to directly affect cloning efficiency. In addition, we discuss the role of the recipient egg cytoplasm in reprogramming a donor genome.

EPIGENETIC STATE OF THE DONOR GENOME

Mature gametes have the full potential to initiate embryogenesis as a result of epigenetic modifications acquired during gametogenesis. In contrast, cloning shortcuts this process by omitting all the epigenetic modifications acquired during germ cell development. A somatic donor nucleus has an epigenetic state radically different from that of the zygote.

Cloning Efficiency and Developmental Stage of the Donor Cell

Clones have been successfully produced using cells derived from different developmental stages in several species. An active research topic has been whether cloning efficiency depends on the developmental stages of the animals from which the donor cells are used. In amphibians, the developmental stage of a donor cell directly correlates with cloning

efficiency. Cells from undifferentiated blastula were found to be more efficient to clone than cells from differentiated gastrula, neurula, or tail bud stage cells. As a result, no frog has been cloned from an adult donor cell so far. In mammals, however, the comparison of somatic cells from different stages of mammalian development has generated controversial results. For example, in bovine, the development of clones both at preimplantation (blastocyst formation rate) and at postimplantation stages are similar when fetal, newborn, and adult cells were compared, regardless of the donor cell types used. Yet others have shown that blastocysts generated from cultured bovine fetal cells have higher success rates for both pregnancy and calving compared to those derived from cultured adult cells. An unexplained exception found in these experiments was that among all the fetal and adult cells compared, adult cumulus cells produced the highest pregnancy and calving rates.

In mice, a consistent difference in the cloning efficiency between embryonic cells (2.4%), fetal cells (1.0% for female and 2.2% for male), and adult cells (0.5% for female and 1.7% for male) was found. In rabbits, it was found that nuclei from morula cells and fetal fibroblasts are more efficient than the nuclei from fibroblasts derived from young or aged animals when used to produce cloned blastocysts. Because cells from different tissues and even from the same tissue at a particular stage of development are not necessarily in the same epigenetic state, we believe that the comparison is informative only if the exact differentiation state of the donor cells is known.

Cloning Efficiency and the Differentiation State of the Donor Cell

Our laboratory recently compared clones derived from mouse ES and cumulus cells for the extent of epigenetic reprogramming by monitoring the activation of Oct-4 and 10 other embryonic genes. Significantly, all ES clones expressed these genes normally, but 38% of the cumulus cell-derived clones failed to do so. In addition, it was found that blastocysts derived from ES cells develop to term between a ten-fold and twenty-fold higher efficiency than those derived from somatic cells (Table 11-1). Because ES cells express Oct-4 and other pluripotency genes, they likely require little or no epigenetic reprogramming following NT to support embryogenesis.

This raised the question of whether fully differentiated cells have the potential to be reprogrammed. Somatic tissues

TABLE 11-1

Differentiation States of Donor Cells and Their Cloning Efficiencies

Donor Cell Types	Percentage of NT Embryos Developed to Morula or Blastocyst Stage	Percentage of Clones Developed to Term per Transferred Blastocysts	References
ES cell	10–20%	30–50%	Rideout et al., 2000; Eggan et al., 2001 and 2002
Cumulus cell or fibroblast	60–70%	1–3%	Wakayama et al., 1998 and 1999
B- or T-cell	4%	ND	Hochedlinger et al., 2002; Jaenisch et al., 2003

harbor cell types with different epigenetic states, such as somatic stem cells, progenitor cells, and fully differentiated cells. Because no definitive marker for the differentiation state of a donor cells was used in previous somatic cell cloning experiments, the possibility that most somatic clones produced thus far were derived from rare adult stem cells could not be excluded. To clarify these issues, our laboratory recently cloned mice from fully differentiated B and T cells in which the genetic rearrangements of the immunoglobulin and TCR genes were used as stable markers for the identity and differentiation state of the donor cells. This result is the first unequivocal demonstration that fully differentiated cells have the potential to be reprogrammed to a totipotent state. However, the cloning efficiency was very low by using these donor cells, and mice could not be produced by directly transferring the cloned blastocysts into uteri of recipient mice. Rather, in a first step, ES cells were derived from the cloned blastocysts; monoclonal mice were generated in a second step by injecting the cloned ES cells into tetraploid blastocysts (see Figure 11-1).

It was also found that B and T cells are much less efficient than other types of somatic cells (cumulus and fibroblast cells) in the production of cloned blastocysts (see Table 11-1). This observation, along with the results obtained from the comparison of cloning efficiencies between somatic cells and ES cells, suggests that the differentiation state of the donor genome has a direct effect on epigenetic reprogramming. It will be interesting to determine whether fully differentiated cells from other tissues are different in their cloning efficiency. These experiments might help to explain the effect of tissue-specific epigenetic modifications on the reprogramming process.

Donor Cell Type-Specific Abnormalities in Clones. Several groups have compared cells derived from different tissues to investigate whether cloning phenotypes are donor cell-type specific. In bovine, for example, it was found that calves cloned from cumulus and oviduct cells were not overweight, as frequently observed in clones from other cell types. It was found that mice cloned from Sertoli cells tend to die prematurely from hepatic failure, tumors, or both, whereas mice cloned from cumulus cells often become obese. These findings in mammals have extended the discoveries reported in early amphibian cloning experiments, in which cloning phenotypes were found to be correlated with the donor cell types used for NT. Consistent with these studies, our laboratory has recently shown, through microarray experiments, that a subset of genes, which were abnormally expressed in cloned animals, were donor cell-type specific. This is the first molecular evidence showing that the tissue origins of donor cells could directly influence their epigenetic reprogramming.

Environmental Effects on the Donor Genome

It is widely established that epigenetic states can be influenced by environmental cues, both *in vitro* and *in vivo*. For example, the DNA methylation status in mouse tissues can be affected by diet and aging. Moreover, it has been shown that imprinted gene expression in ES cells can be altered during cell culture.

Thus, cloning from cells in which imprinting marks have been altered would result in abnormal imprinted gene expression in the cloned animals. This is because parental-specific imprinting patterns can be established only during germ cell development, not by any of the postzygotic developmental stages. Indeed, mice cloned from mouse ES cells showed abnormal expression patterns for some of the examined imprinted genes. Cell culture effects were also found in somatic cell cloning experiments, in which different porcine fibroblast subclones derived from the same primary cell line resulted in different developmental potential when used as nuclear donors. Because methylation is progressively lost during *in vitro* culture of fibroblasts, such variation among cell lines in producing clones could be the result of changes in DNA methylation.

Genetic Background of the Donor Genome

Our laboratory investigated the influence of genetic background on cloning efficiency by comparing mouse ES cells with either inbred or hybrid backgrounds as donor cells. Results showed that F1 hybrid ES cells were more efficient than inbred ES cells to clone. Although all of the inbred clones that survived to term died at birth because of respiratory stress, a certain percentage of F1 clones survived. However, because similar differences were also found between mice entirely derived from inbred ES and those derived from F1 ES cells by ES–tetraploid complementation, the beneficial effect from the more heterogeneous genetic background of F1 ES cells is not limited to cloning. Therefore, it is possible that the respiratory stress experienced by most inbred clones at their births was caused by their delayed development rather than by a failure of epigenetic reprogramming.

Cell Cycle Stage of the Donor Genome and Reprogramming Efficiency

It has been suggested that donor cells arrested at G0 of the cell cycle are crucial for cloning somatic cells by NT. However, mammals have been cloned from G1 and M phase donor cells. To investigate which cell cycle stage is more advantageous in reprogramming a somatic nucleus, a recent study compared the development of bovine clones produced from fibroblasts either at the G0 (high-confluence treatment) or G1 ("shake-off" treatment) stage of the cell cycle. There was no difference in the blastocyst formation rate of these two groups. However, when postimplantation development of these clones was examined in 50 recipients, five calves were obtained from clones derived from G1 cells, but none of the G0 clones survived beyond 180 days of gestation. The authors suggested that the donor cell cycle stage is important for the development of clones — in particular, that G1 cells are more amenable to supporting late-gestation stage development. The underlying mechanisms for the correlation between donor cell cycle and cloning efficiency remain elusive.

RECIPIENT CYTOPLASM

Matured eggs in most mammalian species are arrested at the metaphase of the second meiosis (MII), caused by high levels of maturation promoting factor (MPF) activity. Upon fertilization or by some artificial stimuli, MPF activity in the eggs starts to decline, releasing eggs from the MII arrest to finish the cell cycle. This process is commonly referred to as egg activation. Both unactivated (MII, high MPF activity) eggs and activated eggs (low MPF activity) have been used as recipients for cloning experiments in mammals. The transfer of an interphase nucleus into an enucleated MII egg results in premature chromosome condensation (PCC) and nuclear envelope breakdown (NEBD), which is induced by high levels of MPF activity in the MII eggs. PCC and NEBD can cause chromosome damage if a nucleus with an incompatible cell cycle stage is transferred into the MII egg. Elongated chromosomes with single- and double-stranded chromatids will form by PCC when G0/G1 and G2 phase nuclei, respectively, are transferred into MII eggs. In these situations, no DNA damage seems to occur. When mitotic chromosomes were transferred into MII cytoplasm, the chromosomes remained condensed. However, extensive chromosome fragmentation, termed *pulverization,* occurred when an S-phase nucleus was used. Thus, to maintain a normal diploid genome in the cloned embryos, coordination between the cell cycle stage of the donor nucleus and the recipient egg needs to be considered. These observations suggested that a donor nucleus at any cell cycle stage, with the exception of S phase, is compatible with an egg arrested at the MII stage of oogenesis.

Yet, some reports suggest that activated eggs could be considered "universal recipients." This is because activated eggs have lost MPF activity and do not induce PCC and NEBD, regardless of the cell cycle stage of the donor cells. Consequently, chromatin damage is avoided. Indeed, cloned goats, sheep, and cows have been produced using activated eggs as recipients. However, no mouse was cloned when nuclei from cleavage mouse embryos were transferred into enucleated zygotes. Furthermore, when somatic cell nuclei (cumulus cells) were transferred to enucleated zygotes, severe chromosome damage occurred in all of the cloned embryos, and the authors suggested that endonuclease activity of the recipient cytoplasm was responsible for the fragmentation of the donor chromatin. At present, there is no evidence that activated eggs can be used as "universal recipients," particularly not in mice. However, these different observations of mouse and farm animals remain unsolved.

In addition to the cell cycle compatibility between donor nucleus and recipient cytoplasm, which is important for maintaining intact diploid genomes of cloned embryos, the potential difference in reprogramming activities in the unactivated and activated egg cytoplasm needs to be considered. Several studies have been performed to address this issue. In bovine, one study found that, although a cloned calf could be produced from cumulus cells using MII eggs all clones arrested before or at the eight-cell stage when activated eggs were used. Similarly, another group found that bovine somatic clones (from skin fibroblasts) could be produced only by MII eggs but not by activated eggs. Interestingly, the latter study also found that embryonic nuclei (from blastomeres) can be reprogrammed by both MII and activated eggs. This differential requirement for the egg cytoplasmic environment by somatic and embryonic nuclei is consistent with the idea that different epigenetic states of donor nuclei require different degrees of reprogramming.

It is believed that the MPF activity, present in unactivated eggs but not in activated eggs, is important for the reprogramming of somatic genomes. It has also been proposed that an MII egg allows more time than a zygote for the donor nucleus to be remodeled before the first cell cycle starts. Others have speculated that the enucleated zygotes fail to reprogram somatic genomes because reprogramming activities associated with pronuclei are removed during the zygote enucleation step.

In summary, it remains unclear how the egg cytoplasm reprograms a differentiated genome. Identifying the molecular nature of the reprogramming activity might help to improve cloning efficiency.

Summary

Accumulated evidence suggests that incomplete epigenetic reprogramming probably occurs in all clones. In this chapter, we discussed faulty epigenetic reprogramming in clones, as well as the factors that may affect the reprogramming. We only focused on the abnormalities of DNA methylation patterns in clones because the potential roles of other chromatin modifications such as histone acetylation and methylation in NT have yet to be addressed in more detail.

Despite our limited understanding of epigenetic reprogramming, we know that even fully differentiated cells can be reprogrammed by the egg cytoplasm to a totipotent or less differentiated state. Thus, a somatic cell nucleus from a patient could be transferred into an enucleated egg to produce ES cells possessing the potential to differentiate into different somatic cells useful for cell therapy. We believe that the abnormalities found in clones will unlikely interfere with the therapeutic applications of NT because problems inherent to the NT technology do not impede the generation of functional cells for tissue repair. In our laboratory, we successfully derived ES cells from a tissue isolated from an immune-deficient adult mouse by NT and subsequently repaired the gene responsible for such a disease. This result established a paradigm for the treatment of a genetic disorder by combining NT with gene therapy.

ACKNOWLEDGMENTS

We would like to thank Konrad Hochedlinger, Kevin Eggan, Caroline Beard, Robert Blelloch, and Teresa Holm for critical reading of the manuscript. Zhongde Wang is supported by a postdoctoral fellowship from the Lalor Foundation; Alexander Meissner is supported by a Boehringer Ingelheim Ph.D. fellowship; and Rudolf Jaenisch is supported by the National Cancer Institute (Grant #5 R37 CA84198).

KEY WORDS

DNA methylation A biological process of modifying DNA molecules by covalently adding a methyl group to the position 5 of the pyrimidine ring of cytosines in DNA. DNA methylation plays important roles in controlling genome stability, regulating gene expression, and silencing viral genomes.

Epigenetic reprogramming A process by which the epigenetic state of the genome of a cell is partially or completely converted to that of another type of cell. In nuclear transfer, it generally refers to the process of reversing the epigenetic state of a differentiated donor genome back to a totipotent embryonic state.

Epigenetics A term describing the phenomena that certain inheritable, yet reversible, traits of a cell are not determined by the primary DNA sequence but by specific higher-order chromatin organizations of its genome.

FURTHER READING

Bird, A. (2002). DNA methylation patterns and epigenetic memory. *Genes. Dev.* **16**, 6–21.

Campbell, K.H., Alberio, R., Lee, J.H., and Ritchie, W.A. (2001). Nuclear transfer in practice. *Cloning Stem. Cells* **3**, 201–208.

Hochedlinger, K., and Jaenisch, R. (2002). Nuclear transplantation: lessons from frogs and mice. *Curr. Opin. Cell Biol.* **14**, 741–748.

Jaenisch, R., and Bird, A. (2003). Epigenetic regulation of gene expression: how the genome integrates intrinsic and environmental signals. *Nat. Genet.* **33** Suppl, 245–254.

Jaenisch, R. (2004). Human cloning—the science and ethics of nuclear transplantation. *N. Engl. J. Med.* **351**, 2787–2791.

12

Differentiation in Early Development

Susana M. Chuva de Sousa Lopes and Christine L. Mummery

Introduction

During the first cleavage divisions, totipotent blastomeres of the mammalian embryo segregate and eventually become committed to the extraembryonic, somatic, and germ-line lineages, losing developmental potency. In mice and humans, pluripotent embryonic cells can be isolated from early embryos and maintained in culture as embryonic stem (ES) cells. In this chapter, we review the development of the mammalian embryo during preimplantation and the earliest postimplantation stages as a basis for understanding developmental potency of ES cells, although it is important to realize that, despite nearly 40 years of research, a number of basic questions on early mammalian development still remain unanswered. On the one hand, this is due to the small size of early mammalian embryos and the effort that is required to obtain them in the large numbers necessary, until recently, for analysis of gene and protein expression. On the other hand, it is because implantation (and later placentation), through which the embryo establishes (and maintains) a physical connection with the mother, makes embryos relatively inaccessible; it has not been possible to mimic implantation *in vitro* or monitor these developmental stages *in vivo*. However, advances in techniques for analysis of gene expression in small sample sizes, *in vitro* fertilization, clonal analysis of cultured embryos, and the use of genetic markers are providing new clues on key developmental events and allowing important questions to be addressed.

Among the most important recent conclusions is the realization that early mammalian development in terms of timing and mechanisms of axis formation is not as entirely distinct from that in lower vertebrates as was previously thought.

Preimplantation Development

In mammals, fertilization occurs in the oviduct where sperm encounters and fuses with the oocyte. As a result, the oocyte nucleus, which had been arrested in metaphase II, completes meiosis, and the two parental pronuclei fuse to form the diploid zygotic nucleus (Figure 12-1A). Progressive demethylation of both paternal and maternal genomes begins after fertilization, leading later to epigenetic reprogramming.

Essentials of Stem Cell Biology
Copyright © 2006, Elsevier, Inc.
All rights reserved.

Transcription of the embryonic genome starts at the two-cell stage in mice and at the four- to eight-cell stage in humans. Until then the embryo relies solely on maternal mRNA, but after activation of the embryonic genome, maternal transcripts are rapidly degraded, although maternally encoded proteins may still be present and functionally important. The embryo continues cleaving without visible growth (Figure 12-1).

CELL POLARIZATION OCCURS DURING COMPACTION

At the 8-cell stage in mice and the 8- to 16-cell stage in humans, the embryo undergoes a process known as *compaction* to become a morula, a compact smooth spherical structure (Figure 12-1D). All blastomeres flatten, maximize their contacts, and become polarized. Their cytoplasm forms two distinct zones: the *apical domain* accumulates endosomes, microtubules, and microfilaments, whereas the *nucleus* moves to the basal domain. Furthermore, gap junctions form basally ensuring communication between blastomeres, and numerous microvilli and tight junctions are formed apically.

The next cleavage plane of some blastomeres is perpendicular to their axis of polarity, resulting in two cells with different phenotypes. One daughter cell is located inside the embryo, is small and apolar, and contains only basolateral elements. The other daughter cell is located at the surface of the embryo, is larger and polar, and contains the entire apical domain of the progenitor cell and some basolateral elements. These polar cells inherit the region containing the tight junctions, thereby creating a physical barrier between the inner apolar cells and the maternal environment.

BLASTOCYST FORMATION (CAVITATION)

After compaction, the presumptive trophectoderm (TE) cells form the outer layer of the embryo. Intercellular contacts strengthen between these cells, and a true epithelium is formed. This thin single-cell layer develops a continuum of junctional complexes, including gap junctions, desmosomes, and tight junctions. Furthermore, the composition of the basal and apical membranes becomes more distinct, with Na^+/K^+-ATPases accumulating in the basal membrane. These ion pumps actively transport sodium ions into the embryo, which leads to accumulation of water molecules. A fluid-filled cavity, the blastocoelic cavity, is thus created on one side of the embryo in a process known as cavitation (Figure 12-1E). The presumptive inner cell mass (ICM) cells stay closely associated during this process, not only because of gap junctions, tight junctions, and interdigitating microvilli between

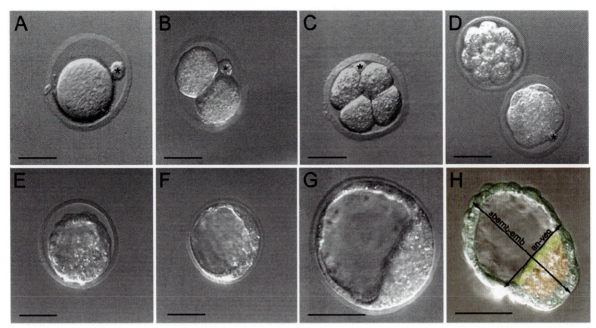

Figure 12-1. Mouse preimplantation development. After fertilization, the two parental pronuclei fuse to form the zygote (A). The embryo cleaves, forming a two-cell (B), four-cell (C), and eight-cell embryo. The embryo then undergoes compaction to become a smooth spherical structure, the morula (D). Note that the second polar body remains attached to the embryo (*). The blastocoelic cavity then develops on one side of the embryo to form an early blastocyst (E). The cavity enlarges, occupying most of the expanded blastocyst (F, G). Around embryonic day (E) 4.5, the late blastocyst reaches the uterus, "hatches" from the zona pellucida, and is ready to implant (H). The late blastocyst consists of three cell subpopulations: the trophectoderm (green), the inner cell mass (orange), and the primitive endoderm (yellow). In the blastocyst three axes can be defined: the embryonic-abembryonic (abemb-emb), the animal-vegetal (an-veg), and a third axis on the same plane but perpendicular to the an-veg axis. (Photomicrographs courtesy of B. Roelen.)

the cells but also because processes from TE cells fix the ICM to one pole of the embryo and partially isolate it from the blastocoel. The intercellular permeability seal of the TE cells prevents fluid loss, and as a consequence the blastocoelic cavity gradually expands to occupy most of the blastocyst between the 64- and 128-cell stage (Figure 12-1E to G). At this stage, the embryo is not radially symmetric around the embryonic-abembryonic axis but is bilaterally symmetric (slightly oval).

The outer TE layer and the ICM are composed of descendants of the outer and inner cell population of the morula, respectively. The TE in turn consists of two subpopulations: the polar TE contacts the ICM and the mural TE surrounds the blastocoelic cavity. The TE descendants give rise to extraembryonic structures such as the placenta but do not contribute to the embryo proper. The cells of the ICM that contact the blastocoelic cavity differentiate to primitive endoderm, which is also an extraembryonic tissue. Furthermore, the ICM gives rise not only to the embryo proper but also to the visceral yolk sac, amnion, and the allantois, a structure that will form the umbilical cord. An overview of cell lineage relationships in the early mouse is shown in Figure 12-2. During preimplantation development (3 to 4 days in mice, 5 to 7 days in humans), the embryo has traveled through the oviduct inside the zona pellucida, a protective glycoprotein coat. Reaching the uterus, the blastocyst "hatches" from the zona pellucida and is ready to implant (Figure 12-1H). Stages of

mouse and human preimplantation development are summarized in Table 12-1.

AXIS SPECIFICATION DURING PREIMPLANTATION IN THE MOUSE

In lower vertebrates, the body axes are already specified in the undivided egg or very soon thereafter, whereas in mammalian embryos, axis specification was thought to be completed only during gastrulation. This view was supported by the observation that the mammalian embryo is extremely plastic, ignoring disturbances such as the removal or reaggregation of blastomeres. The prevailing concept, therefore, became one of no embryonic prepatterning before gastrulation. Recent studies, however, have suggested that the mammalian zygote may in fact be polarized and that the body axes specified at the time of fertilization are similar to lower vertebrates. In the mouse zygote, the position of the animal pole, marked by the second polar body, or the sperm entry point, which triggers Ca^{2+} waves, have been discussed as defining the plane of first cleavage. However, it is still unclear whether the positions of these two cues are directly responsible for the zygote polarity and subsequent position of the first cleavage plane. Alternatively, zygote polarity and the positions of the second polar body and the sperm entry point might be determined by an intrinsic asymmetry already present in the oocyte.

The first cleavage plane coincides with the embryonic-abembryonic boundary of the blastocyst, and, interestingly,

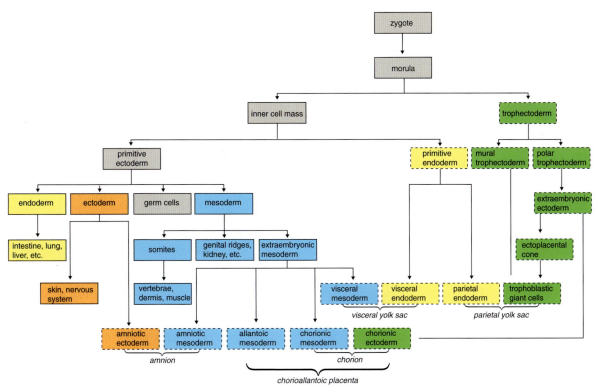

Figure 12-2. Cell lineages in mouse development. Trophectoderm-derived tissues are depicted in green, endoderm-derived tissues in yellow, ectoderm-derived tissues in orange, and mesoderm-derived in blue. The cell/tissues represented in grey are regarded as pluripotent. All extraembryonic tissues are enclosed by hatched lines, whereas embryonic tissues are enclosed by solid lines. (Adapted from Hogan *et al.*, 1994)

TABLE 12-1

Summary of Mouse and Human Preimplantation Development

Stage (M)	Time	Stage (H)	Time	Developmental Processes
Zygote	0–20 h	zygote-2 cell	0–60 h	Axis determination
2-cell	20–38 h	4–8 cell	60–72 h	Activation of embryo genome
4-cell	38–50 h			
8-cell	50–62 h	8–16 cell	~3.5 d	Compaction
16-cell	62–74 h		~4.0 d	Two phenotypically different cells emerge
32-cell	~3.0 d	32 cell	~4.5 d	Blastocoelic cavity forms (cavitation)
64-cell	~3.5 d		~5.5 d	Blastocyst consists of two cell populations (ICM and TE)
128–256-cell	~4.5 d	166–286 cell	~6.0 d	Part of ICM differentiates to PrE; hatching, followed by implantation

During mouse and human development, the timing of each cleavage division is dependent on environmental factors (*in vitro* versus *in vivo*), individual variation, and mouse strain. The cleavage times presented here are ranges from several published sources. Adapted from Hogan *et al.* (1994) and Larsen (1997). d, days; h, hours; H, human; ICM, inner cell mass; M, mouse; PrE, primitive endoderm; TE, trophectoderm.

the fates of the two blastomeres are distinguishable and can be anticipated. The blastomere containing the sperm entry site generally divides first and contributes preferentially to the embryonic region of the blastocyst, whereas its sister cell preferentially forms the abembryonic region. Parthenogenic eggs that do not contain a sperm entry point are able to divide and develop to blastocysts, although the two blastomeres do not tend to follow different fates. This indicates that, although during normal development the site of sperm penetration

correlates with the later spatial arrangement of the blastocyst, it is not essential for patterning the embryo.

There is a clear topographic relationship between the blastocyst and the zygote. The blastocyst has three defined axes, which correlate with the position of the second polar body and the plane of the first cleavage. Although the blastocyst axes also correlate with the three axes of the uterine horn, a relationship with the body axes of the future fetus is less clear.

DEVELOPMENTAL POTENCY OF THE EARLY MOUSE EMBRYO

In the mouse, both blastomeres of a two-cell stage embryo transplanted separately into foster mothers develop into identical mice. To assess the developmental potential of each blastomere of four-cell and eight-cell mouse embryos, Susan Kelly in 1977 combined isolated blastomeres with genetically distinguishable blastomeres of the same age, creating chimeric composites. Each blastomere was shown to contribute extensively to both embryonic and extraembryonic tissues (TE and visceral yolk sac) and to generate viable and fertile mice. This indicated that at these developmental stages all blastomeres are still totipotent. However, Andrzej Tarkowski and Joanna Wróbleswska in 1967 showed that isolated four-cell and eight-cell stage blastomeres were able to develop to blastocysts and implant, but were incapable of generating viable fetuses. This may be explained by the fact that a defined number of cell divisions (five) occurs before blastocyst formation. Thus, in contrast to the normal 32-cell blastocyst, isolated blastomeres from four-cell and eight-cell embryos resulted in 16-cell and 8-cell blastocysts, respectively. Tarkowski and Wróbleswska postulated that it is the position of a cell in the blastocyst that determines its fate: cells at the surface of the embryo become TE, whereas cells enclosed in the embryo become ICM. Blastocysts generated from isolated four- and eight-cell blastomeres contain progressively fewer cells in the ICM, making it likely that a minimum number of ICM cells is necessary for survival beyond the blastocyst stage.

Although different phenotypically, the two-cell subpopulations in the 16-cell morula are still plastic and able to produce cells of the other lineage provided they are at the correct position in the embryo, that is, inside or at its surface. Cells of the ICM of 32- and 64-cell embryos are also still capable of contributing to all tissues of the conceptus (embryonic and extraembryonic) and are thus totipotent. The potency of TE cells has been difficult to determine because TE cells are not easy to isolate (tightly connected with each other) and because they are not readily integrated inside the embryo (low adhesiveness). After the 64-cell stage, the ICM loses totipotency.

Once the embryo has implanted (up to embryonic day [E]7.0), embryonic cells (including the primordial germ cells formed slightly later in development) lose their ability to contribute to the embryo when introduced directly into a host blastocyst. Remarkably, when introduced into genetically identical adult mice, epiblast cells are able to generate teratocarcinomas, tumors that contain tissues derived from the three germ layers (endoderm, mesoderm, and ectoderm) and an embryonal carcinoma (EC) stem cell population. EC cells are then able to form mouse chimeras when introduced into blastocysts, suggesting that, although pluripotency is lost in the epiblast, it can be regained to a certain extent. Similarly, primordial germ cells isolated from E8.5 mouse embryos cultured to become embryonic germ (EG) cells and adult hematopoietic and neural stem cells are able to regain pluripotency and can contribute to the embryo when introduced into blastocysts.

GENES IMPORTANT DURING PREIMPLANTATION MOUSE DEVELOPMENT

Before implantation, the embryo is relatively self-sufficient and can, for example, develop *in vitro* in simple culture media without growth factor supplements. Only relatively few mutations (specific gene deletions, insertions, and more extensive genetic abnormalities) have been reported to result in preimplantation lethality (Table 12-2). The reasons for this are not clear, but one may be that the initial presence of maternal transcripts in the zygote effectively results in maternal rescue. Ablation of specific maternal transcripts in the zygote is not always feasible using conventional knockout techniques because deficiency in candidate genes often results in lethality before adulthood. To date, only a limited number of maternal-effect genes involved in preimplantation development have been identified (see Table 12-2).

Interestingly, most genes transcribed during preimplantation development are detected immediately after genome activation and continue to be transcribed, resulting in mRNA accumulation. Therefore, to trigger the different specific developmental events during preimplantation, post-transcriptional regulation may play an important role.

ES cells are derived from the ICM; therefore, it is not surprising that ES and ICM cells express common genes. Some of these genes have been described as being necessary for maintaining the undifferentiated phenotype of ES cells and could be expected to play important roles in the segregation of the pluripotent ICM from the differentiated TE cell population. However, when deleted in the mouse, most of those genes appear to be crucial during implantation or gastrulation but not during the preimplantation period when both ICM and TE are formed. Most pertinent in this respect are the genes for leukemia inhibitory factor (LIF) and LIF receptors. Although mouse ES cells are highly dependent on LIF for maintenance of pluripotency in culture, deletion of neither receptor nor ligand genes appears to affect the pluripotency of the ICM at the blastocyst stage. Interestingly, *in vivo* LIF appears important for regulation of implantation (see the section on implantation).

The transcription factor Oct4 has the best-characterized involvement in regulating potency in mammals. Oct4 is initially expressed by all blastomeres, but expression becomes restricted to the ICM as the blastocyst forms (Figure 12-3). Thereafter, a transient up-regulation of Oct4 occurs in the ICM cells that differentiate to primitive endoderm. Interestingly, expression levels of Oct4 in mouse ES cells also regulate early differentiation choices, mimicking events in the blastocyst: mouse ES cells lacking *Oct4* differentiate to TE, whereas a twofold increase in Oct4 expression leads to endoderm and mesoderm formation. Mouse embryos deficient in *Oct4* are unable to form mature ICM and die around implantation time.

Other genes described as being involved in cell fate determination during preimplantation development include *Taube*

TABLE 12-2
Lethal Mouse Mutations Affecting Differentiation During Early Development

Gene/Locus	Mutant Phenotype	References
Zar1	Zygote arrest	Wu et al. (2003) Nat Genet **33**, 187
Hsf1	Zygote to 2-cell stage arrest	Christians et al. (2000) Nature **407**, 693
NPM2	Zygote to 2-cell stage arrest	Burns et al. (2003) Science **300**, 633
Mater	2-cell stage arrest	Tong et al. (2000) Nat Genet **26**, 267
Tcl1	4- to 8-cell stage arrest	Narducci et al. (2002) PNAS **99**, 11712
Stella	Failure to form blastocysts	Payer et al. (2003) Curr Biol **13**, 2110
Wt1 (Wilms' tumor 1)	Zygotes fail to undergo mitosis (background dependent)	Kreidberg et al. (1999) Mol Reprod Dev **52**, 366
C^{25H} (pid)	2- to 6-cell stage embryos fail to undergo mitosis	Lewis (1978) Dev Biol **65**, 553
Tgfb1	2- to 4-cell arrest (background dependent)	Kallapur et al. (1999) Mol Reprod Dev **52**, 341
E-cadherin (uvomorulin)	Defects in compaction	Larue et al. (1994) PNAS **91**, 8263; Riethmacher et al. (1995) PNAS **92**, 855
Trb (Traube)	Defects in compaction	Thomas et al. (2000) Dev Biol **227**, 324
Mdn (morula decompaction)	Defects in compaction	Cheng and Costantini (1993) Dev Biol **156**, 265
Om (ovum mutant)	Failure to form blastocysts (background dependent)	Wakasugi et al. (1967) J Reprod Fertil **13**, 41; Baldacci et al. (1992) Mamm Genome **2**, 100
SRp20	Failure to form blastocysts	Jumaa et al. (1999) Curr Biol **9**, 899
t^{12}, t^{w32}	Failure to form blastocysts	Bennett (1975) Cell **6**, 441; Smith (1956) J Exp Zool **132**, 51
T^{hp} (hairpin)	Failure to form blastocysts	Babiarz (1983) Dev Biol **95**, 342
Ts (Tail short)	Failure to form blastocysts	Paterson (1980) J Exp Zool **211**, 247
α-E-catenin	Failure to form blastocysts (TE defect)	Torres et al. (1997) PNAS **94**, 901
Vav	Blastocysts fail to hatch	Zmuidzinas et al. (1995) EMBO J. **14**, 1
Os (oligosyndactyly)	Metaphase arrest at the early blastocyst stage	van Valen (1966) J Embryol Exp Morphol **15**, 119; Magnuson and Epstein (1984) Cell **38**, 823
Brg1	Abnormal blastocyst development (fail to hatch)	Bultman et al. (2000) Mol Cell **6**, 1287
A^x (lethal nonagouti)	Abnormal blastocyst development	Papaioannou and Mardon (1983) Dev Genet **4**, 21
l(5)-1	Abnormal blastocyst development	Papaioannou (1987) Dev Genet **8**, 27
t^{wPa-1}	Abnormal blastocyst development	Guenet et al. (1980) Genet Res **36**, 211
PL16	Abnormal blastocyst development	Sun-Wada et al. (2000) Dev Biol **228**, 315
CpG binding protein (CGBP)	Abnormal blastocyst development	Carlone and Skalnik (2001) Mol Cell Biol **21**, 7601
Thioredoxin (Txn)	Abnormal blastocyst development	Matsui et al. (1996) Dev Biol **178**, 179
Gpt	Abnormal blastocyst development	Marek et al. (1999) Glycobiology **9**, 1263
Ltbp2	Abnormal blastocyst development	Shipley et al. (2000) Mol Cell Biol **20**, 4879
Hba^{thJ}	Decreased TE cell number	Hendrey et al. (1995) Dev Biol **172**, 253
A^y (lethal yellow)	Defects in TE formation	Papaioannou and Gardner (1979) J Embryol Exp Morphol **52**, 153
Evx1	Defects in TE formation	Spyropoulos and Capecchi (1994) Genes Dev **8**, 1949
Eomes (Eomesodermin)	Defects in TE formation	Russ et al. (2000) Nature **404**, 95
Cdx2	Defects in TE formation	Chawengsaksophak et al. (1997) Nature **386**, 84
Egfr	Defects in ICM formation (background dependent)	Threadgill et al. (1995) Science **269**, 230
β1 integrin	Defects in ICM formation	Stephens et al. (1995) Genes Dev **9**, 1883; Fässler and Meyer (1995) Genes Dev **9**, 1896
Lamc1	Defects in ICM formation	Smyth et al. (1999) J Cell Biol **144**, 151
B-myb	Defects in ICM formation	Tanaka et al. (1999) J Biol Chem **274**, 28067
Fgf4	Defects in ICM formation	Feldman et al. (1995) Science **267**, 246
Fgfr2	Defects in ICM formation	Arman et al. (1998) PNAS **95**, 5082
Taube Nuss (Tbn)	Defects in ICM formation	Voss et al. (2000) Development **127**, 5449
Oct4 (Oct3, Pou5f1)	Defects in ICM formation	Nichols et al. (1998) Cell **95**, 379
Nanog	Defects in ICM formation	Mitsui et al. (2003) Cell **113**, 631; Chambers et al. (2003) Cell **113**, 643

The table is divided into three sections. The top lists maternal-effect genes. Embryos lacking these genes develop normally in heterozygous but not homozygous mothers. The middle section includes genes and loci that, when deleted, cause embryonic lethality before implantation. The lower section includes genes and loci that, when deleted, cause embryonic lethality during implantation but before the formation of the egg cylinder. Embryos deficient in most of these genes develop to normal blastocysts and are able to hatch and implant, but the whole embryo or selectively the TE or ICM (ICM- or primitive endoderm-derived cells) degenerates soon thereafter, leading to resorption. Extensive genetic abnormalities are not included in the table. ICM, inner cell mass; TE, trophectoderm.

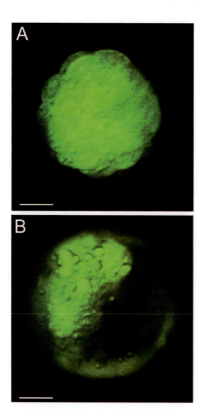

Figure 12-3. Oct4 expression at morula and blastocyst stages. GFP expression driven by distal elements of the Oct4 promoter (kindly supplied by H. Schöler) was used here to mimic endogenous Oct4 expression. In the morula, all blastomeres express high levels of Oct4 (A). In the early blastocyst, the inner cell mass expresses high levels of Oct4, whereas weaker expression is observed in trophectoderm cells (B).

an intimate and highly regulated cross-talk between mother and mammalian embryo makes implantation a complex process.

Reaching the uterus, the blastocyst hatches from the zona pellucida and the TE cells become adhesive, expressing integrins that enable the embryo to bind the extracellular matrix (ECM) of the uterine wall. The mouse embryo adheres to the uterine wall via the mural TE cells of the abembryonic region and is slightly tilted. In contrast, human embryos bind through the embryonic region. Once attached to the uterus, trophoblast cells secrete enzymes that digest the ECM, allowing them to infiltrate and start uterine invasion. At the same time, the uterine tissues surrounding the embryo undergo a series of changes collectively known as the *decidual response*. These changes include formation of a spongy structure known as decidua, vascular changes leading to the recruitment of inflammatory and endothelial cells to the implantation site, and apoptosis of the uterine epithelium.

THE MURINE TROPHECTODERM AND PRIMITIVE ENDODERM CELLS

Apoptosis occurring in the uterine wall gives TE cells the opportunity to invade the decidua by phagocytosing dead epithelial cells. At about E5.0, the mural TE cells cease division but continue endoreduplicating their DNA to become primary trophoblastic giant cells. This cell population is joined by polar TE cells that migrate around the embryo and similarly become polytene (secondary trophoblastic giant cells). However, other polar TE cells continue dividing and remain diploid, giving rise to the ectoplacental cone and the extraembryonic ectoderm that pushes the ICM into the blastocoelic cavity (Figure 12-4).

During implantation, the primitive endoderm layer forms two subpopulations: the visceral endoderm (VE) and the parietal endoderm (PE), both of which are extraembryonic tissues. The VE is a polarized epithelium closely associated with the extra-embryonic ectoderm and the ICM (Figure 12-4A); later in development, it contributes to the visceral yolk sac. PE cells migrate largely as individual cells over the TE (Figure 12-4A) and secrete large amounts of ECM to form a thick basement membrane known as Reichert's membrane. The PE cells together with the trophoblastic giant cells and Reichert's membrane form the parietal yolk sac.

DEVELOPMENT OF THE MURINE INNER CELL MASS TO THE EPIBLAST

The ICM located between the recently formed extraembryonic ectoderm and the VE gives rise to all cells of the embryo proper. During implantation, the ICM organizes into a pseudostratified columnar epithelium (also referred to as primitive or embryonic ectoderm, epiblast, or egg cylinder) surrounding a central cavity, the proamniotic cavity (Figure 12-4). Signals from the VE, including BMPs, are responsible for apoptosis in the core of the epiblast leading to its cavitation. Between E5.5 and E6.0, the proamniotic cavity expands to the extraembryonic ectoderm, forming the proamniotic canal (Figure 12-5).

nuss, *B-myb*, *Nanog*, *Cdx2*, and *Eomes* (see Table 12-2). Both *Taube nuss* and *B-myb* homozygous-deficient mice develop to normal blastocysts. At the time of implantation, however, *Taube nuss*$^{-/-}$ ICM cells undergo massive apoptosis and the embryo becomes a ball of trophoblast cells; in *B-myb* knockout mice, the ICM also degenerates, but the reason for this is unclear. *Taube nuss* and *B-myb* seem to be necessary for ICM survival, whereas *Oct4* is required for establishment and maintenance of the ICM identity but not cell survival. *Nanog* is expressed exclusively in the ICM and whereas *Oct4* prevents TE differentiation, *Nanog* prevents differentiation of ICM to primitive endoderm. In agreement, *Nanog*$^{-/-}$ blastocysts are formed, but the ICM in culture differentiates into endoderm. In contrast, *Cdx2* and *Eomes* appear to be involved in trophoblast development, and embryos lacking these genes die soon after implantation probably because of defects in the trophoblast lineage.

From Implantation to Gastrulation

The mechanisms used by the mammalian embryo to implant are species dependent, contrasting with the general developmental steps during the preimplantation period. In addition,

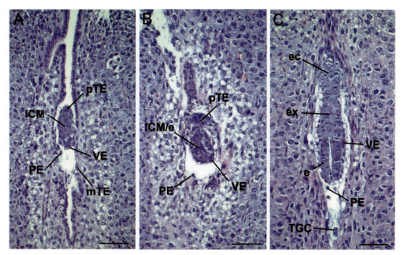

Figure 12-4. Tissue formation and movements during and shortly after implantation of the mouse embryo (E5.0–E5.5). During implantation, the cell division rate in the embryo increases, leading to rapid growth (A–C). The primitive endoderm cells segregate into visceral endoderm (VE) and parietal endoderm (PE). The polar trophectoderm cells (pTE) form the ectoplacental cone (ec) and the extraembryonic ectoderm (ex). pTE cells together with mural trophectoderm cells (mTE) contribute to form the trophoblastic giant cells (TGC). The inner cell mass (ICM) cavitates and organizes into an epithelium known as the epiblast (e).

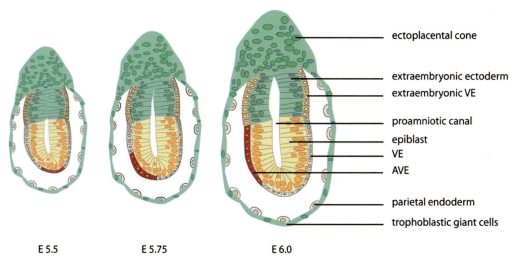

Figure 12-5. Tissue formation and movements in the pregastrulation mouse embryo (E5.5–E6.0). During this period, the extraembryonic ectoderm organizes into an epithelium. The proamniotic cavity initially restricted to the epiblast now expands into the extraembryonic ectoderm, forming the proamniotic canal. At E5.5, the most distal visceral endoderm cells (red) express a different set of markers than the surrounding visceral endoderm (VE). These distal VE cells move from the distal tip to surround the prospective anterior part of the epiblast and form the anterior visceral endoderm (AVE). The VE surrounding the extraembryonic ectoderm consists of a columnar epithelium, whereas the VE cells surrounding the epiblast are more flattened. (Adapted from Lu *et al.*, 2001)

After implantation, a wave of *de novo* methylation occurs, leading to epigenetic reprogramming (finished by E6.5). This affects the entire genome to a different extent in embryonic and extraembryonic lineages. After implantation, the rate of cell division increases, followed by rapid growth. At E4.5, the ICM consists of approximately 20 to 25 cells, at E5.5 the epiblast has about 120 cells, and at E6.5 it consists of 660 cells.

At E6.5, gastrulation starts with the formation of a morphologically visible structure in the future posterior side of the embryo, the primitive streak. During this complex process, the three definitive germ layers are formed; the germ line is set aside; and extraembryonic mesoderm that contributes to the visceral yolk sac, placenta, and umbilical cord is generated. An overview of tissue formation and movement during mouse gastrulation is shown in Figure 12-6.

THE HUMAN EMBRYO

Human development during implantation and gastrulation is significantly different from that of the mouse. Briefly, the

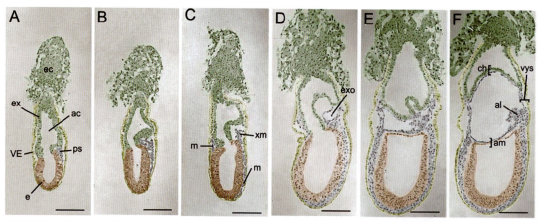

Figure 12-6. Tissue formation and movements during the gastrulation of the mouse embryo (E6.5–E7.5). Gastrulation begins with the formation of the primitive streak (ps) in the posterior side of the E6.5 embryo at the junction of the extraembryonic ectoderm (ex) and epiblast (e) (A). As more cells ingress through the streak, it elongates toward the distal tip of the embryo, between epiblast and visceral endoderm (VE) (B). While the newly formed embryonic mesoderm (m) moves distally and laterally to surround the whole epiblast, the extraembryonic mesoderm (xm) pushes the extraembryonic ectoderm upwards and to the center (C, D). The extraembryonic mesoderm develops lacunae, creating a mesoderm-lined cavity known as exocoelom (exo). The exocoelom enlarges, and as a consequence, the tissue at the border of extraembryonic and embryonic ectoderm fuses, dividing the proamniotic cavity (ac) in two and forming the amnion (am) and the chorion (ch) (E). The layer of extraembryonic mesoderm and the visceral endoderm together form the visceral yolk sac (vys). At the posterior side of the embryo the allantois (al) and the primordial germ cells are formed (E, F). The tissues colored green (the extraembryonic ectoderm and ectoplacental cone (ec)) are derived from the trophectoderm. The tissues in yellow are derived from the primitive endoderm and epiblast cells that passed through the streak, generating the definitive endoderm. These definitive endoderm cells intercalate with the visceral endoderm in the region of the streak but also form a larger patch of cells at the distal part of the embryo. This patch of cells moves anteriorly, displacing the anterior visceral endoderm, which moves toward the extraembryonic region of the embryo. The tissues colored orange are derived from the inner cell mass and remain ectoderm. The tissues colored blue are formed during gastrulation and represent primitive streak and mesoderm-derived tissues (excluding the primordial germ cells present at the basis of the allantois). For the lineages of early mouse development see Figure 12-2.

human trophoblast cells invade the uterine tissue and form the syncytiotrophoblast, a tissue similar to the mouse giant trophoblast cells. However, the trophoblast cells that contact the ICM and the blastocoelic cavity remain single cells and diploid, and are known as cytotrophoblasts. These cells proliferate and fuse with the syncytiotrophoblast. In humans, no structure equivalent to the murine extraembryonic ectoderm is formed.

Human primitive endoderm cells, also known as hypoblast cells, form on the surface of the ICM and proliferate. Some of these cells migrate to line the blastocoelic cavity leading to the formation of the primary yolk sac and Heuser's membrane. The human primary yolk sac is not equivalent to the murine parietal yolk sac, although both are transient structures. Moreover, it is still unclear whether human embryos develop a PE-like cell type. Paralleling the formation of the murine Reichert's membrane, a spongy layer of acellular material known as the extraembryonic reticulum is formed between cytotrophoblast and Heuser's membrane. Thereafter, the extraembryonic reticulum is invaded by extraembryonic mesoderm. The origin of this tissue in humans is still unknown. The extraembryonic mesoderm proliferates to line both Heuser's membrane and cytotrophoblast. The extraembryonic reticulum then breaks down and is replaced by a fluid-filled cavity, the chorionic cavity.

A new wave of hypoblast proliferation generates cells that contribute to the formation of the definitive yolk sac. This new structure displaces the primary yolk sac, which buds off and

breaks up into small vesicles that remain present in the abembryonic pole. The definitive yolk sac in humans is equivalent to the visceral yolk sac in the mouse.

The human ICM organizes into a pseudostratified columnar epithelium and cavitates, producing the amniotic cavity. The ICM cells that lie on the hypoblast are known as the epiblast and give rise to the embryo proper. The ICM cells that contact the trophoblast form the amnion. The human embryo forms a bilaminar embryonic disc, similar to chick embryos, and patterns of cell movement during gastrulation are conserved between chick and humans.

With such diversity in extraembryonic structures supporting the development of the ICM in mice and humans, it is not surprising that ES cells derived from mice and humans are not similar. They differ in developmental potency, for example, in their ability to differentiate to TE. Human ES cells can form TE in culture, but under normal circumstances mouse ES cells do not. Furthermore, mouse ES cells in culture have recently been shown to develop into cells with primordial germ cell properties. In turn, these cells can develop into sperm- or oocyte-like cells. The potential of these cells to fertilize or to be fertilized and generate viable mice is still unknown. It is not yet known whether human ES cells have this potential to form primordial germ cells in culture. Mouse and human ES cells also express different cell surface markers; have different requirements in culture for self-renewal; and respond differently to growth and differentiation cues, the most striking being the response to LIF.

IMPLANTATION: MATERNAL VERSUS EMBRYONIC FACTORS

In mice, the presence of the blastocyst in the uterus is sufficient to trigger ovarian production of progesterone and estrogen. These two hormones are absolutely required for embryo survival because they prime the uterus for implantation and decidualization. The uterus starts producing LIF and members of the epidermal growth factor (EGF) family, including EGF, heparin-binding EGF, transforming growth factor-alpha (TGFα), and amphiregulin. Those molecules, together with Hoxa10, induce the production of cyclo-oxygenase (COX) enzymes, the rate-limiting enzymes in the production of prostaglandins. The embryo, on the other hand, also produces important molecules that act in autocrine and paracrine ways. With a few exceptions, all of these factors and corresponding receptors play crucial roles during this period and when deleted in the mouse lead to lethality during or soon after implantation.

The suppression of the maternal immune response is also essential during implantation but is still incompletely understood. TE cells, the only cell population of the conceptus that physically contacts maternal cells, have developed several mechanisms to avoid rejection. Examples are the production of numerous factors and enzymes, including indoleamine 2,3-dioxygenase (IDO) by the TE cells that suppress the maternal immune system and the lack of polymorphic class I and II major histocompatability complex (MHC) antigens in TE cells.

THE ROLE OF EXTRAEMBRYONIC TISSUES IN PATTERNING THE MOUSE EMBRYO

Extraembryonic tissues not only are necessary for nutrition and regulating implantation during development but also play crucial roles in patterning the embryo before and during gastrulation. Unequivocal evidence for this role comes from the analysis of chimeric embryos generated from blastocysts colonized with ES cells. In chimeras, ES cells preferentially colonize epiblast-derived tissues. It is, therefore, possible to generate embryos with extraembryonic tissues of one genotype and epiblast-derived tissues of another genotype. For example, nodal is expressed embryonically and extraembryonically (depending on the developmental stage). Furthermore, *nodal*-deficient embryos fail to gastrulate. It was thus initially difficult to distinguish embryonic from extraembryonic functions. However, when *nodal*$^{-/-}$ ES cells were introduced into wild-type blastocysts, the extraembryonic tissues were wild-type, whereas epiblast-derived tissue lacked *nodal*. The developing chimera was essentially normal until midgestation, suggesting that the presence of *nodal* (exclusively) in the extraembryonic tissues is sufficient to rescue embryonic patterning.

In contrast to the extensive mixing of epiblast cells, labeled primitive endoderm cells develop as more coherent clones, consistent with the function of the VE in embryo patterning. The primitive endoderm cells in the vicinity of the second polar body preferentially form VE cells surrounding the epiblast, whereas cells away from the second polar body preferentially form VE cells surrounding the extraembryonic ectoderm.

At E5.5, the most distal VE cells are characterized by the expression of the homeobox gene *Hex*. This cell population migrates toward the prospective anterior side of the embryo during the next day of development, producing an endodermal stripe known as the anterior visceral endoderm (AVE) (Figure 12-5). The AVE is therefore the first clear landmark of an anterior–posterior axis in the embryo, preceding the formation of the primitive streak, at the opposite side of the embryo during gastrulation.

Before gastrulation, the extraembryonic ectoderm signals to the proximal epiblast, inducing expression of several genes important for posterior proximal identity. In contrast, signals from the distal VE (and later the anteriorly migrating AVE) seem to inhibit the expression of that same set of genes, restricting in this way the posterior fate.

Both VE and VE-like cell lines secrete signals that are able to induce differentiation of mouse and human ES cells at least toward cardiomyocytes. Making use of the tissues or sequences of signal transduction pathways used by the embryo for its own patterning and differentiation might be an efficient way to study and even direct ES cell differentiation.

ACKNOWLEDGMENTS

We are grateful to J. Korving, M. Reijnen, L. Tertoolen, J. Heinen, and F. Vervoordeldonk for technical assistance and to B. Roelen and L. Defize for careful reading and useful comments on the manuscript. S.M.C.S.L. was supported by Fundação para a Ciência e Tecnologia, Portugal (SFRH/BD/827/2000).

KEY WORDS

Blastocyst   Very early animal embryo consisting of a spherical outer epithelial layer of cells known as the trophectoderm (that forms the placenta), a clump of cells attached to the trophectoderm known as the inner cell mass (from which stem cells are derived), and a fluid-filled cavity, the blastocoel. When fully expanded, the blastocyst "hatches" from the zona pellucida in which it has developed and implants in the uterus of the mother.

Chimera   Organism made up of cells from two or more different genetic donors. In general, two genetically different very early embryos (at the morula stage) are aggregated to form one single embryo or, alternatively, embryonic stem cells are allowed to attach to a morula or are injected inside a blastocyst. Embryonic stem cells integrate preferentially in embryonic tissues, whereas the extraembryonic tissues are derived from the recipient embryo.

Differentiation   Process during embryogenesis or in pluri/multipotent cells in culture in which an increase in the complexity or organization of a cell or tissue results in a more specialized function.

Implantation   Process by which the mammalian blastocyst physically connects with the uterus of the mother. It involves the displacement of the uterine epithelium cells and extracellular matrix degradation by proteolytic enzymes secreted by the embryo. Implantation occurs only in mammalian development and is necessary to provide the growing embryo with sufficient protection and metabolic needs.

FURTHER READING

Beddington, R. S., and Robertson, E. J. (1999). Axis development and early asymmetry in mammals. *Cell* **96**, 195–209.

Hogan, B., Beddington, R., Costantini, F., and Lacy, E. (1994). *Manipulating the mouse embryo: a laboratory manual.* 2nd ed. Cold Spring Harbor Laboratory Press, New York.

Kelly, S. J. (1977). Studies of the developmental potential of 4- and 8-cell stage mouse blastomeres. *J. Exp. Zool.* **200**, 365–376.

Larsen, W. J. (1997). *Human Embryology.* 2nd ed. Churchill Livingstone, New York.

Lu, C. C., Brennan, J., and Robertson, E. J. (2001). From fertilization to gastrulation: Axis formation in the mouse embryo. *Curr. Opin. Genet. Dev.* **11**, 384–392.

Pedersen, R. A. (1986). Potency, lineage, and allocation in preimplantation mouse embryos. *In Experimental approaches to mammalian embryonic development* (J. Rossant and R. A. Pedersen, eds.), pp. 3–33. Cambridge University Press, Cambridge, UK.

Tarkowski, A. K., and Wróblewska, J. (1967). Development of blastomeres of mouse eggs isolated at the 4- and 8-cell stage. *J. Embryol. Exp. Morphol.* **18**, 155–180.

Zernicka-Goetz, M. (2002). Patterning of the embryo: The first spatial decisions in the life of a mouse. *Development* **129**, 815–829.

Primordial Germ Cells in Mouse and Human

Anne McLaren

Introduction

WHAT ARE PRIMORDIAL GERM CELLS?

The germ cell lineage terminates in the differentiation of the gametes (eggs and spermatozoa). In mammals the lineage arises in the extraembryonic mesoderm at the posterior end of the primitive streak. From here the germ cells migrate to the two genital ridges, which later form the gonads. During this period, they proliferate at a steady rate and are known as primordial germ cells (PGCs). Once in the genital ridges, they may be termed gonocytes. Proliferation ceases in the male genital ridge when the germ cells transiently arrest in G_0/G_1 as prospermatogonia and in the female genital ridge when they enter prophase of the first meiotic division as oocytes. These events, which occur before birth in both mouse and human, mark the initiation of the lengthy processes of spermatogenesis and oogenesis, respectively, and the end of the primordial phase of germ cell development.

PRIMORDIAL GERM CELLS ARE NOT STEM CELLS

Stem cells are commonly defined as cells with a choice: they can divide to form either two cells like themselves (self-renewal) or one cell like themselves and one that is embarking on a pathway of differentiation (asymmetric division). PGCs do not at any stage constitute a stem cell population: each of the cell divisions that they undergo (9 or 10 in the mouse, more in the human) moves them further along their developmental trajectory. At a later stage in the male germ cell lineage, a true stem cell population forms: spermatogenic stem cells in the testis divide slowly to form a self-renewing population throughout the lifetime of the male, giving rise to waves of proliferating spermatogonia that enter meiosis as spermatocytes, then differentiate into spermatids, and finally become mature spermatozoa. PGCs can give rise *in vitro* to pluripotent stem cell populations that will proliferate indefinitely, given appropriate culture conditions, but PGCs *in vivo* are not stem cells.

Origin of the Germ Cell Lineage

IDENTIFICATION OF PRIMORDIAL GERM CELLS

In many mammals, including mouse and human, PGCs can be readily identified by staining for alkaline phosphatase

activity. Tissue nonspecific alkaline phosphatase (TNAP) is expressed in many tissues, but activity is markedly higher in germ cells than in the surrounding somatic cells throughout the primordial period. The function of the enzyme is not clear, because it has been shown that disruption of the *Tnap* gene does not appear to affect germ cell development. PGCs have also been identified histologically and by electron microscopy and, more recently, also by a variety of genetic markers (see the section on gene expression below). The allantois, once it has started to grow, forms a convenient landmark for spotting the cluster of PGCs at its base (about 8 days postcoitum [dpc] in the mouse). According to the study by Ginsburg *et al*. (1990), the cluster could first be identified, in the same location but before an allantois was apparent, at 7.25 dpc, in mid-gastrulation.

EARLY STUDIES

The failure to identify germ cells in the mouse embryo by morphology, alkaline phosphatase activity, or any other feature at any stage of development earlier than mid-gastrulation was frustrating because in many invertebrates and lower vertebrates the germ line could be traced back to very early development, even to the egg. In *Drosophila* the germ line derives from the pole plasm, incorporated in the pole cells, the first cells to form; in *Caenorhabditis elegans* the germ cell determinants (P granules) present in the egg are asymmetrically segregated during the first four cell divisions, to form the germ line; in *Xenopus* and other anuran amphibia, the germ plasm apparent in the egg can be followed into the germ cells as they form. Expression of homologs of the germ-line-specific gene *Vasa* is localized in "germ plasm" from early cleavage stages onward in both zebra fish and chick. No such lineage could be detected in the mouse or any other mammal. Chimera studies were unable to identify any cells in the preimplantation embryo that were uniquely associated with the germ line. Circumstantial evidence had led some early workers to assert that the germ cell lineage in mammals had an extraembryonic origin. Subsequently, however, various lines of evidence, in particular the transplantation studies of Gardner and Rossant (1979), established beyond doubt that the ancestors of the germ cell lineage were to be found in the epiblast, which is derived from the inner cell mass of the blastocyst, not from the outer trophectoderm.

TIME AND PLACE OF LINEAGE DETERMINATION

Fate-mapping of the mouse epiblast has been achieved by injecting a long-lasting fluorochrome into single epiblast cells of embryos removed from the uterus in early gastrulation. Subsequent culture of the embryo for 48 hours reveals the tissue fate of the clonal descendants of the injected cell. The study of Lawson and Hage (1994) established that only the most proximal cells of the epiblast, those close to the extraembryonic ectoderm, included PGCs among their descendants. No clone consisted only of PGCs, proving that at the time of injection (6.0 or 6.5 dpc), the injected cells were not lineage-restricted. Indeed, PGCs never constituted more than a small proportion of each marked clone, and marked PGCs only made up a small proportion of the total number of TNAP-positive PGCs. Clonal analysis revealed that the most likely time of germ cell specification was 7.2 dpc, approximately the time at which a TNAP-positive cluster of putative germ cells had first been identified in the extraembryonic mesoderm.

Clones without any PGCs contained more cells than those with PGCs, and the fewer the PGCs the larger was the total number of cells. This suggested that germ cell determination was associated with an increase in cell doubling time, to about 16 hours for PGCs, in contrast to 6 to 7 hours for the surrounding somatic extraembryonic mesoderm cells. Counts of PGCs on successive days had previously given a similar figure of 16 hours for PGC doubling time in the period 8.5–13.5 dpc.

In *Amphibia*, the germ cell lineage can be traced back to the egg in frogs and toads (*Anura*), but in Urodeles it is induced in mid-gastrulation, in the mesoderm, similar to Mammals. A. D. Johnson has proposed that the Urodele/Mammal mode of development is the more primitive. The early-determination mode, characterized by germ plasm, must then have been derived several different times in evolution. This view is supported by the expression pattern of germ line-specific genes such as *Vasa* and *Dazl*, which are similar in Urodeles and Mammals and also in primitive fishes but very different in *Anura*, chick, zebra fish, *Drosophila*, and *Caenorhabditis*.

SIGNALING FACTORS

Bmp4 homozygous mutants die around the time of gastrulation. Using a strain combination that survived until late gastrulation, Hogan's group showed that the homozygotes had no allantois and no PGCs. BMP4 protein, known to act as a signaling molecule, is normally expressed in the extraembryonic ectoderm (Lawson *et al.*, 1999). Chimeras made between normal embryonic stem (ES) cells and *Bmp4*-negative embryos, in which the epiblast contained both cell types but the extraembryonic tissues (which do not incorporate ES cell derivatives) were all BMP4-negative, also lacked both allantois and PGCs. Hence, *Bmp4* expression in the extraembryonic ectoderm is required for the establishment of the germ cell lineage, implicating a *Bmp4*-dependent signal to the immediately adjacent proximal epiblast cells. *Bmp8b* mutants have a reduced number of PGCs, suggesting that BMP8b is also involved in signaling, perhaps in interaction with BMP4. The molecular details of the signaling pathway have still to be established.

Further evidence that signaling from the extraembryonic ectoderm is required for germ cell determination comes from

experiments in which distal epiblast cells, which would normally give rise to neurectoderm, were transplanted at 6.5 dpc to a proximal location. Some of the PGCs that formed expressed the genetic marker carried by the donor embryo. This finding suggests that all epiblast cells in early gastrulation have the potential to develop into PGCs, if they receive the appropriate signals. Signals emanating from the extraembryonic ectoderm predispose the neighboring epiblast cells to a germ cell fate, but they do not determine that fate, because the clonal analysis referred to in the previous section on time and place of lineage restriction showed that, for some epiblast cells, only a small proportion of their descendants became PGCs. Once the cells have passed through the primitive streak and reached the cluster region, a further signal or signals may be required to complete the specification process and halt further movement.

GENE EXPRESSION

For decades, the high level of expression of *Tnap* was the only useful genetic marker for the initial stages of the mammalian germ cell lineage. *Oct4* is expressed diffusely during gastrulation, so PGCs are not distinguished from the surrounding tissues until about 8.0 dpc. Other useful PGC markers (*SSEA1*, *mouse vasa homolog*, *Dazl*) are expressed somewhat later.

According to the study of Saitou, Barton, and Surani (2002), in which single-cell cDNA libraries were made from the cluster region of 7.0–7.5 dpc mouse embryos, PGC libraries could be distinguished from somatic cell libraries by the relatively high level of expression of *Tnap* and the absence of expression of *Hoxb4*. Two new germ cell–specific genes were isolated from the PGC libraries. These were termed *fragilis* and *stella*. *fragilis* is expressed from about 6.0 dpc on, in the proximal region of the epiblast. Its expression pattern shifts as the epiblast cells move toward the primitive streak, and by 7.2 dpc it is concentrated around the cluster region. Once the PGCs begin to move away, *fragilis* is downregulated, although it comes on again later. *In vitro*, *fragilis* expression can be induced in any epiblast tissue, if placed in proximity to extraembryonic ectoderm. It belongs to a widely distributed family of interferon-inducible genes, of which other members code for proteins showing homotypic adhesion and changes in cell cycle regulation. In contrast, *stella*, is a novel gene, not part of a gene family. It carries a nuclear-localization signal. It is up-regulated in the cluster region, where *fragilis* expression is strongest, at the time when PGCs can first be visualized. *stella* maintains its PGC-specific expression during germ cell migration and into the genital ridges. Thus, it provides a valuable new germ cell–specific genetic marker.

Migration

ROUTE

As the germ cell cluster breaks up, at about 8.0 dpc, the PGCs appear to migrate actively into the visceral endoderm, which carries them along as it invaginates to form the hind gut.

Although initially ventral, they become distributed around the hind gut and pass dorsally into the genital ridges (gonad primordia), through the body wall, or up the dorsal mesentery and round the coelomic angle at 10.0–11.0 dpc. The entire period of migration has been visualized by Wylie's group on a video of PGCs carrying a transgenic green fluorescent protein marker. The PGCs show locomotory behavior throughout, which is nondirectional in the wall of the hind gut where they are carried along passively, but strongly directional as they leave the hind gut and enter the genital ridges. Once in the ridges, the locomotory behavior ceases.

GUIDANCE MECHANISMS

Little is known of the mechanism that encourages the PGCs to enter the visceral endoderm (rather than the allantois, or some other extraembryonic mesodermal region) as the cluster breaks up. Contact guidance may play a part in facilitating exit from the endodermal hind gut wall because the adhesion properties (particularly to laminin) change at this time. As the PGCs move toward the genital ridges, they have been reported to contact one another through long cellular processes, forming a loose network. The molecular nature of the directional signals that attract them toward the genital ridges has yet to be definitively established, but in mice as well as in zebra fish there is recent evidence that the chemokine–receptor pair stromal cell-derived factor 1 (SDF-1) and its G-protein-coupled receptor (CXCR4) act to guide migrating PGCs.

GENE EXPRESSION

Certain gene products need to be present if PGC migration is to take place normally. Abnormal migration patterns are readily detected because they result in subfertility or sterility. Two classic mouse mutants, *White-spotting* and *Steel*, both show sterility in the homozygous condition and defects in hematopoiesis and pigment cell migration. *White-spotting* (W) codes for c-kit, a cell surface receptor; *Steel* (SL) codes for its ligand, stem cell factor. If either element of this signal transduction pathway is defective, PGC migration is disturbed, proliferation is affected, programmed cell death (apoptosis) occurs, and few germ cells reach the genital ridges. Other genes known to be required for normal PGC migration include *gcd* (*germ cell deficient*), *β1 integrin*, and *Fgf8*.

Germ Cells in the Genital Ridge

PHENOTYPE

Mouse PGCs migrate actively into the genital ridges between 10.5 and 11.5 dpc. Soon after entry into the ridges, their motile phenotype changes to a rounded shape, they up-regulate E-cadherin, movement ceases, and they form loose groups. Gene expression changes; *mvh* (*mouse vasa homolog*) and *gcna* (*germ cell nuclear antigen*) act as useful germ cell markers at this stage. Proliferation continues for another couple of days, still with a doubling time of about 16 hours. However, by 12.5 dpc, germ cells in both female and male embryos are entering the premeiotic cell cycle, mitosis ceases, and meiotic genes such as *Scp3* (*Synaptonemal complex protein 3*) are up-regulated.

A detailed quantitative study of germ cells in the human fetal ovary, before and during entry into meiosis, was made by Baker (1963).

SEX DETERMINATION

In a female genital ridge, the germ cells proceed at about 13.5 dpc into prophase of the first meiotic division, passing through leptotene, zygotene, and pachytene stages before arresting in diplotene, in primordial follicles, shortly after birth. Germ cells enter into first meiotic prophase at the same time in ectopic locations outside the genital ridge, such as in the adrenal primordium, in male and in female embryos. Germ cells isolated from both male and female embryos at 11.5 dpc or earlier will enter meiosis at this time also in cultured aggregates of lung tissue or even on a feeder layer *in vitro*. These observations indicate that germ cell entry into meiosis occurs cell autonomously, at a time that appears to be intrinsic rather than determined by an extrinsic signal.

In contrast, in the male genital ridge the germ cells at about 13.5 dpc down-regulate meiotic genes such as *Scp3* and enter mitotic arrest in G_1/G_0 as prospermatogonia. This change in cell fate determination is not cell autonomous but is induced by the somatic cells, probably the Sertoli cells, of the male genital ridge. The somatic signal that blocks entry into meiosis has not yet been identified, but possible candidates are prostaglandin D and LDL. In XY embryos, the testis-determining gene on the Y-chromosome (*Sry*) is expressed in the supporting cell lineage for about 36 hours, from 10.5 to 12.0 dpc. *Sox9*, the presumed target of *Sry*, is expressed from 11.5 dpc on and is required for the differentiation of Sertoli cells from the supporting cell lineage and for their subsequent maintenance. By 12.5 dpc, germ cells in the male genital ridge are enclosed in cords, lined with Sertoli cells, and surrounded by peritubular myoid cells that have migrated in from the mesonephric region in response to an *Sry*-dependent signal. According to Adams and McLaren (2002), germ cells removed from a female genital ridge at 12.5 dpc or earlier and aggregated with somatic tissue from a 12.5 dpc male genital ridge will enter mitotic arrest rather than meiotic prophase, but by 13.5 dpc they are already committed to the female pathway of development. The male germ cells, however, are already committed to the male pathway by 12.5 dpc and develop as prospermatogonia even when aggregated with female genital ridge cells.

X-CHROMOSOME REACTIVATION

In female mammals, one or other of the two X-chromosomes, at random, is inactivated during gastrulation in all cells of the somatic lineages and also in the germ cell lineage. During germ cell migration, PGCs in both male and female embryos have just a single X-chromosome active (dosage compensation). Once in the genital ridges, however, the silent X in XX germ cells is reactivated, in both human and mouse oocytes, and remains active throughout oogenesis. Reactivation occurs not only in the female genital ridge but also in XX germ cells experimentally introduced into a male genital ridge. *Xist* is involved in the initiation of X-chromosome inactivation: the only gene to be expressed on the inactive X, it produces a stable transcript that coats the whole chromosome and can be visualized by RNA fluorescent *in situ* hybridization (FISH). Once the germ cells are in the genital ridge, *Xist* is down-regulated and the transcript disappears, consistent with reactivation of the silent X chromosome.

EPIGENETIC CHANGES

Patterns of DNA methylation are imposed during gastrulation on the nascent somatic cell lineages but not on PGCs, perhaps because of their extraembryonic birthplace. Global methylation in PGCs is still further decreased once the germ cells are in the genital ridges. Imprinted genes (i.e., genes in which only the paternal or only the maternal allele is expressed, but not both) are characterized by differential DNA methylation at specific CpG sites. Genes shown to be imprinted in the mouse are not necessarily imprinted in the human (e.g., *Igf2r*). In the germ cell lineage, the previous genomic imprint has to be erased, and a new imprint is established according to the sex of the embryo. The differential site-specific methylation in some imprinted genes has been examined in mouse germ cells by bisulfite sequencing. It decreases during PGC migration or shortly after entry into the genital ridges. At 10.5 dpc, some site-specific methylation is present, much less at 11.5 dpc, and very little at 12.5 dpc. New imprints are established before or after birth: in the female germ line, for example, different imprinted genes acquire their new methylation pattern at different stages of oogenesis. According to Obata *et al.* (2002), 12.5 dpc female germ cells are able to undergo genomic imprinting in organ culture. Their nuclei are then capable of supporting development to full term after transfer to an enucleated mature oocyte, followed by *in vitro* fertilization.

CELL AUTONOMOUS OR INDUCED?

As we have seen, many features of PGCs change once they have entered the genital ridges. To what extent are these changes caused by signals that emanate from the somatic tissues? We know that this is so for the block to entry into meiosis in the male genital ridge (see the section on sex determination) and also for the expression of *mvh*, which depends on contact with genital ridge somatic tissue. Conversely, we know that entry of germ cells into the first meiotic prophase occurs cell autonomously unless they are exposed to 12.5 dpc or later male genital ridge tissue (see the section on sex determination). Also, Resnick's group finds that expression of *gcna* occurs cell autonomously, at a preprogrammed time, apparently requiring neither entry into the genital ridges nor even migration. As yet, there is no evidence bearing on this question, in relation either to X-chromosome reactivation or to epigenetic changes.

Embryonic Germ Cells

EMBRYONIC GERM CELL DERIVATION

Early attempts to establish long-term cultures of mouse PGCs met with little success, even when feeder cells were used.

Chromosomally stable stem cells were eventually obtained by combining three growth factors in the cultures: stem cell factor (SCF), fibroblast growth factor 2 (bFGF), and leukemia inhibitory factor (LIF). These stem cells were termed embryonic germ cells (EGCs) to distinguish them from embryonic stem (ES) cells. EGCs, like ES cells, proved capable of indefinite proliferation in culture and were pluripotent both *in vitro*, and *in vivo* in chimeras in which they colonized all cell lineages including the germ line. EGC lines have been derived from PGCs before and during migration (8.5, 9.5 dpc) and also from PGCs in the genital ridge but only up to 12.5 dpc. EGC lines have also been derived in some other mammalian species, including from human PGCs. Like mouse EGCs, human EGCs show high levels of alkaline phosphatase activity and express SSEA1. According to Shamblott *et al.* (1998), they also express SSEA4, TRA-1-60, and TRA-1-81. (See Chapter 41.)

XIST EXPRESSION

In PGCs, *Xist* may code for a stable transcript (migrating XX PGCs) or expression may be entirely absent (XY PGCs; XX PGCs after X-chromosome reactivation in the genital ridge). Undifferentiated EGC lines derived from the PGCs are always characterized by an unstable *Xist* transcript, visualized by RNA FISH as a small dot overlying the locus. A similar situation is seen in ES cells. Once the EGCs or ES cells start to differentiate, the unstable transcript disappears and is replaced by a stable transcript or absence of expression, according to whether the cell is XX or XY.

EPIGENETIC CHANGES

The epigenetic status of imprinted genes in EGCs might be expected to reflect that in the PGCs from which they were derived. However, this is not so. Site-specific differential methylation proved to be absent from most imprinted genes, in EGC lines derived not only from 11.5 but also from 9.5 dpc germ cells. This apparent conflict with the PGC data can be resolved if one assumes that the epigenetic changes involved in erasure of imprints continue after the cells are put into culture, reaching an imprint-free state once the EGC line is established. When these imprint-free EGCs have been used for making chimeras, some of the chimeras have shown growth retardation and skeletal defects (ribs and spine). Germ-line transmission, however, has been reported with EGCs derived from 11.5 dpc as well as 8.5 dpc PGCs. In human EGC lines, the epigenetic status of undifferentiated cells has not been examined, but after differentiation to fibroblast-like cells, three imprinted genes showed monoallelic expression (as in normal somatic tissues), and the fourth, *Igf2*, showed some relaxation of imprinting. These results suggest that there may be a considerable difference in the timing of imprint erasure between humans and mice. According to Baker (1963), there is also a difference between humans and mice in the timing of entry into meiosis: relative to colonization of the genital ridge, entry into meiosis is later and less well synchronized in human than in mouse germ cells.

Summary

The germ cell lineage in the mouse is not predetermined but is established during gastrulation, in response to signalling molecules acting on a subset of epiblast cells that move through the primitive streak together with extra-embryonic mesoderm precursors. After migration to the site of the future gonads, germ cell sex determination is achieved, with germ cell phenotype in male and female embryos diverging. Evidence suggests that all germ cells spontaneously take the female pathway, entering prophase of the first meiotic division five or six days after the birth of the germ cell lineage, with the exception of those located in the embryonic testis, which exit the cell cycle in response to some inhibitory signal and remain in G_1/G_0 until after birth, when spermatogenesis begins. Site-specific DNA methylation of imprinted genes is erased in germ cells at about the time of entry into the future gonads, and new imprints are established later. In culture, germ cells respond to certain growth factors by proliferating indefinitely. These immortalized embryonic germ (EG) cell lines are chromosomally stable and pluripotent, closely resembling the embryonic stem (ES) cell lines derived from blastocyst-stage embryos. Human EG cell lines have also been made.

KEY WORDS

Bisulfite sequencing A technique whereby the methylation of each CpG site in a defined stretch of DNA is determined.

Epigenetic Factors involved in influencing gene expression during development, without affecting DNA base sequence.

Gastrulation The period of embryonic development during which the definitive body plan is laid down. Cells from the primitive ectoderm pass through the "primitive streak" region to form a third layer, the mesoderm, between ectoderm and endoderm.

Meiosis The process by which germ cells (i.e., those in the ovaries or testes) divide to produce gametes. In meiosis I, homologous chromosomes exchange genetic material. In meiosis II, the two resulting diploid cells (i.e., which contain two sets of chromosomes) with their recombined chromosomes divide further to form two haploid gametes (i.e., which contain only one set of chromosomes).

Pluripotent Capable of giving rise to all the cell types in the fetus but not able on its own to form a fetus.

FURTHER READING

Adams, I. R., and McLaren, A. (2002). Sexually dimorphic development of mouse primordial germ cells: switching from oogenesis to spermatogenesis. *Development* **129**, 1155–1164.

Baker, T. G. (1963). A quantitative cytological study of germ cells in human ovaries. *Proc. Roy. Soc. B* **158**, 417–433.

Gardner, R. L., and Rossant, J. (1979). Investigation of the fate of 4.5 day post-coitum mouse inner cell mass cells by blastocyst injection. *J. Embryol. Exp. Morphol.* **52**, 141–152.

Ginsburg, M., Snow, M. H. L., and McLaren, A. (1990). Primordial germ cells in the mouse embryo during Gastrulation. *Development* **110**, 521–528.

Lawson, K. A., Dunn, R. R., Roelen, B. A., Zeinstra, L. M., Davis, A. M., Wright, C. V., Korving, J. P., and Hogan, B. L. (1999). Bmp4 is required for the generation of primordial germ cells in the mouse embryo. *Genes Dev.* **13**, 424–436.

Lawson, K. A., and Hage, W. J. (1994) Clonal analysis of the origin of primordial germ cells in the mouse. *In Ciba Foundation Symposium 182, Germ Line Development* (J. Marsh and J. Goode, eds.), pp. 68–91. John Wiley & Sons, Chichester, England.

McLaren, A. (1988). The developmental history of female germ cells in mammals. *In Oxford reviews of reproductive biology*, Chapt. 4, Vol. 10, pp. 162–179. Oxford University Press.

McLaren, A. (2003). Primordial germ cells. *Dev. Biol.* **262**, 1–15.

Obata, Y., Kono, T., and Hatada, I. (2002). Maturation of mouse fetal germ cells *in vitro. Nature* **418**, 497–498.

Saitou, M., Barton, S., and Surani, M. A. (2002). A molecular programme for the specification of germ cell fate in mice. *Nature* **418**, 293–300.

Shamblott, M. J., Axelman, J., Wang, S., Bugg, E. M., Littlefield, J. W., Donovan, P. J., Blumenthal, P. D., Huggins, G. R., and Gearhart, J. D. (1998). Derivation of pluripotent stem cells from cultured human primordial germ cells. *Proc. Natl. Acad. Sci.* **95**, 13726–13731.

Witschi, E. (1948). Migration of the germ cells of human embryo from the yolk sac to the primitive gonadal folds. *Contrib. Embryol. Carnegie Inst.* **32**, 67–80.

Wylie, C. (1999). Germ cells. *Cell* **96**, 165–174.

14

Stem Cells in Extraembryonic Lineages

Tilo Kunath and Janet Rossant

Introduction

The establishment of mouse embryonic stem (ES) cells provided an excellent model to study embryonic lineages in culture and *in vivo*. These pluripotent cells can contribute to all embryonic tissues, including the germ line, in chimeras. Furthermore, tissues from all three germ layers have been successfully formed from ES cells in culture. These abilities have made ES cells an attractive culture system for studying the regulators of lineage-specific determination and differentiation. However, the inability of ES cells to differentiate into cells of the trophoblast lineage has precluded them from being a cell culture model for this essential extraembryonic lineage. The establishment of trophoblast stem (TS) cell lines from mouse blastocysts and early trophoblast tissue has provided such a model. TS cells can be grown indefinitely in culture, can differentiate into trophoblast subtypes, and can contribute exclusively to the trophoblast lineage in chimeras. This chapter reviews development of the extraembryonic trophoblast lineage in the mouse, considers evidence for the existence and the location of trophoblast progenitor populations *in vivo*, and describes the work that led to the establishment of TS cell lines. Some applications of this cell culture system are reviewed here; detailed protocols for the establishment and maintenance of TS cell lines can be found in Chapter 45 of this volume. Finally, the extraembryonic lineage and a representative cell culture model are described briefly.

Trophoblast Lineage

TROPHOBLAST DEVELOPMENT

An early priority of all mammalian embryos is to establish the extraembryonic lineages. The first such lineage to form is the trophoblast. In the mouse, this occurs by embryonic day (E) 3.5 when the preimplantation morula embryo cavitates to form the blastocyst. The outer sphere of 40–50 trophectoderm (TE) cells of the early blastocyst contains the sole precursors of the entire trophoblast lineage. An eccentrically located clump of cells within the trophectodermal sphere is the inner cell mass (ICM). One day later, a second extraembryonic lineage, primitive endoderm (PrE) forms at the exposed surface of the ICM. The remainder of the ICM is the primitive ectoderm or epiblast. These three lineages are the foun-

dations for development of the entire conceptus, and two of them (TE and PrE) are restricted to the extraembryonic compartment. The primitive ectoderm will also produce several extraembryonic tissues, such as the amnion, the allantois, and the mesodermal compartment of the definitive yolk sac (Figure 14-1).

The derivatives of the trophoblast lineage are essential for the survival of the embryo in the maternal uterine environment. It mediates implantation into the uterus and establishes a barrier for nutrient and waste exchange. It comprises a major portion of the placenta, where trophoblast cells take on endocrine and immunological roles in addition to the primary task of supplying the embryo with nutrient-rich blood and removing wastes. A role in embryo patterning is also beginning to emerge for this versatile lineage.

As mentioned previously, the trophoblast lineage is unambiguously present at blastocyst formation as the TE. There is a distinction between TE in contact with the ICM, or polar TE, and TE surrounding the blastocoel, or mural TE. The most distal or abembryonic cells in the mural TE are the first to differentiate into primary trophoblast giant cells followed by cells laterally to border of the ICM. The giant cells are aptly named, since they reach extraordinary sizes and possess very high DNA content. These cells undergo rounds of DNA synthesis (S phase) without intervening mitoses in a process known as endoreduplication. The giant cells express several genes of the prolactin family, including placental lactogen I (*PL-I*) early in giant cell differentiation. Later, they express *PL-II*.

Proliferating TE cells at the proximal region of the conceptus generate several trophoblast structures. At E6.5, trophoblast tissue in direct contact with the embryonic ectoderm forms the extraembryonic ectoderm (ExE). These cells are diploid and highly proliferative. Continuous with the ExE and immediately above it is the ectoplacental cone (EPC). The rate of proliferation is less in this tissue than in ExE, but the cells are still diploid. The EPC loses expression of several genes associated with ExE, such as *Fgfr2*, *Cdx2*, and Eomesodermin (*Eomes*), and it initiates expression of other genes, such as *Tpbp* (formerly *4311*). The outer periphery of the EPC has a lower mitotic index than the core, and these outer cells contribute secondary giant cells to the growing parietal yolk sac as well as to the placenta itself (Figure 14-2). During gastrulation, posterior mesoderm migrates into the extraembryonic region and becomes associated with trophoblast tissue of the ExE to form the chorion. This trophoblast tissue is now referred to as chorionic ectoderm (ChE) (Figure 14-2). At E8.5–9.0, the mesoderm-derived allantois fuses with the

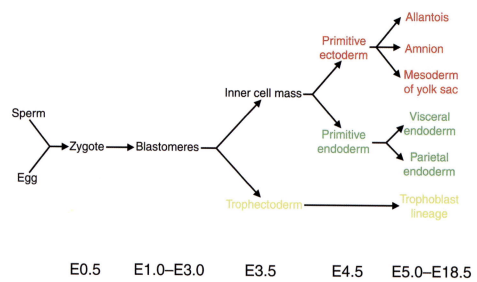

E0.5   E1.0–E3.0   E3.5   E4.5   E5.0–E18.5

Figure 14-1. Extraembryonic lineages during mouse development. The primitive ectoderm, primitive endoderm, and trophoblast lineages are shown in red, green, and yellow, respectively. After fertilization, the zygote divides to produce blastomeres that eventually segregate into two lineages at blastocyst formation (E3.5): inner cell mass and trophectoderm. About one day later, the inner cell mass is further subdivided into the primitive ectoderm (or epiblast) and primitive endoderm. The primitive ectoderm will form all three germ layers of the embryo proper (not shown) and will contribute to several extraembryonic tissues: the allantois, the amnion, and the mesodermal part of the definitive (or visceral) yolk sac. The primitive endoderm differentiates into two major types of extraembryonic endoderm: visceral and parietal endoderm. The trophectoderm is the precursor to all trophoblast lineages in the placenta as well as to the giant cell layer of the parietal yolk sac.

chorion, which further fuses with the base of the EPC, resulting in occlusion of the EPC cavity. These are the initial steps for forming a mature chorioallantoic placenta (Figure 14-2). In this mature tissue, the extraembryonic mesoderm and ChE combine to form the labyrinth, the site of nutrient and waste exchange with the maternal blood supply. The EPC differentiates into the spongiotrophoblast, a supporting tissue that is in intimate contact with the labyrinthine trophoblast. Peripheral giant cells are in direct contact with the maternal decidua and are the invasive cells of the trophoblast. The maternal vasculature invades the spongiotrophoblast layer, and the endothelium is replaced with endovascular trophoblast as it enters the labyrinth.

Numerous genetic mutants have been characterized that affect various aspects of placental development, and they have been summarized in two recent reviews. A few mutants of note are described here. The T-box gene, *Eomes*, is expressed in the TE of the blastocyst and the ExE of the early postimplantation embryo. A homozygous null mutation in this gene resulted in an embryonic lethal phenotype at the periimplantation stage. Blastocysts implanted into the uterus but developed no further. The phenotype was caused by a cell-autonomous role for *Eomes* in trophoblast tissue as determined by chimeric analysis. The *Cdx2* mutation may have a similar defect to *Eomes*$^{-/-}$, since the gene is also expressed in early trophoblast derivatives, and the knockout results in lethality at the time of implantation. Detailed analysis of *Cdx2*$^{-/-}$ embryos revealed a lack of mature TE resulting in a failure of implantation (Strumpf *et al.*, 2005), a more severe

phentotype than *Eomes*$^{-/-}$ embryos. However, the precise phenotype has not been reported yet. The orphan nuclear receptor, *Errβ* (also known as *Esrrb*), is expressed in the ExE and the ChE. Targeted deletion of this gene resulted in embryonic lethality at E9.5 with a failure of the chorion and an overproduction of giant cells. This gene is not required for the initial period of trophoblast proliferation, but it is required for its maintenance. The bHLH gene, *Hand1*, has an essential role in trophoblast giant cells. This gene is expressed in giant cells and functionally promotes their differentiation. *Hand1*-deficient embryos die of placental failure because of a block in giant cell differentiation. The lessons from these and other mutations point to the importance of lineage-specific transcription factors during trophoblast development.

FGF SIGNALING AND TROPHOBLAST PROGENITOR CELLS *IN VIVO*

The polar TE at the embryonic pole of the blastocyst remains diploid and proliferative, and the mural TE differentiates into postmitotic giant cells. Sustained proliferation of polar TE cells is dependent on interactions with the ICM and later with the epiblast. Transplanting an ectopic ICM to the mural TE region inhibits giant cell differentiation and induces a zone of proliferation. The dividing cells were determined by genetic markers to be of TE origin and not from the transplanted ICM. A model emerged whereby trophoblast proliferation was dependent on signals from the ICM, and in the absence of this stimulus, the default pathway of giant cell differentiation would ensue. In agreement with this, the trophoblast-derived

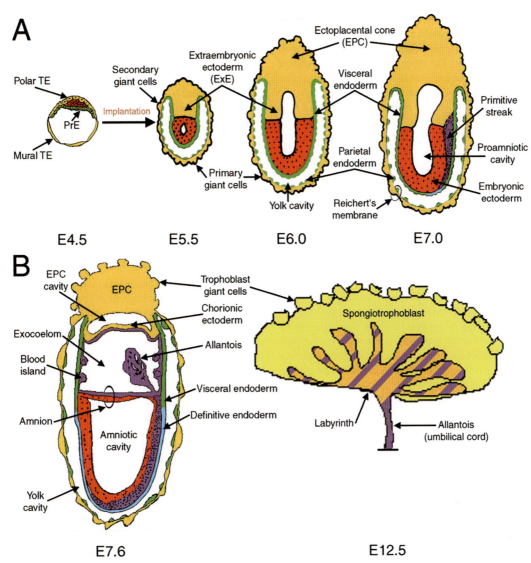

Figure 14-2. A. Postimplantation development. (A) Implantation to gastrulation. Based on their relative position to the inner cell mass, the trophectoderm (TE) cells of the E4.5 blastocyst have segregated into mural and polar TE. The polar TE grows into the blastocoel to produce the extraembryonic ectoderm and outward to form the ectoplacental cone (EPC). The mural TE differentiates into primary giant cells. Secondary giant cells line the outer surface of the EPC and contribute cells peripherally to the growing parietal yolk sac. At E6.5, the primitive streak forms at the posterior region of embryonic ectoderm adjacent to the extraembryonic ectoderm. By E7.0, definitive endoderm is beginning to emerge from the leading edge of the primitive streak. The primitive endoderm expands to line the entire surface of the former blastocoel, now known as the yolk cavity. The cells adjacent to the giant cell layer differentiate into parietal endoderm, and the cells in contact with the extraembryonic ectoderm and embryonic ectoderm become visceral endoderm. (B) Later development of extraembryonic lineages. By E7.5, posterior mesoderm has moved into the extraembryonic region and contributed to the chorion, amnion, and visceral yolk sac tissues. Three cavities are now present: the EPC cavity, the exocoelom, and the amniotic cavity. The interaction of visceral endoderm and extraembryonic mesoderm induces blood islands in the definitive yolk sac region. The allantois, emerging from the posterior end of the embryo into the exocoelom, fuses with the chorion that also combines with the EPC to form the chorioallantoic placenta (E12.5). Chorionic ectoderm and mesoderm combined with allantoic mesoderm generate the labyrinth. The spongiotrophoblast is mostly derived from the EPC. The definitive endoderm has replaced most of the visceral endoderm in the embryonic region after E7.5. B.

ExE and EPC differentiate into giant cells when explanted into culture. Contact with embryonic ectoderm, by placing the ExE or EPC into the amniotic cavities of E7.5 embryos ("embryonic pocket" culture), inhibited giant cell differentiation of the ExE only. Thus, the trophoblast tissue nearest the epiblast, the ExE, can respond to proliferation signals from the embryo, but the more distal EPC is refractory to them. This is the first indication of where the trophoblast progenitors may reside *in vivo*.

Clues to the identity of the embryo-derived signal came from expression and genetic studies of the fibroblast growth factor (FGF) signaling pathway. Embryos mutant for the *Fgf4* gene die shortly after implantation. Since this ligand is expressed in the ICM, it could have an autocrine function in maintaining ICM cells, a paracrine role in the proliferation of polar TE cells, or both. The reciprocal expression pattern of *Fgfr2* in the TE and the ExE lent support for the latter model. Two different targeted deletions of the *Fgfr2* gene resulted in different phenotypes on the same genetic background. The deletion of exon 9, the alternatively spliced IIIc exon, resulted in a peri-implantation lethal phenotype similar to the *Fgf4* mutation. Surprisingly, deletion of the complete Ig III loop (exons 7, 8, and 9) resulted in a less severe phenotype, with lethality occurring at E10.5 because of a failure of the placental labyrinth. The IIIc deletion could be a dominant–negative mutation, since there is a possibility that a soluble FGFR2 IIIb variant is produced. However, heterozygous mice were normal and fertile.

On the other hand, the complete Ig III deletion might be a hypomorphic allele. They show by Western analysis that a truncated protein with the remaining two Ig loops and a complete intracellular domain is produced. Although it cannot bind FGF ligands, it may participate in some ligand-independent signaling. Regardless of which *Fgfr2* allele is the complete loss-of-function mutation, an essential role for FGF signaling in trophoblast development is unmistakable. An *Fgfr2 IIIb* isoform-specific knockout resulted in a phenotype that did not have trophoblast defects as reported for the two *Fgfr2*-null alleles. The mice died at birth with severe lung and limb abnormalities. In addition, the placental rescue of an *Fgfr2*-null allele by tetraploid aggregation resulted in a phenocopy of the *IIIb*-specific mutation. This implicates the IIIc isoform of FGFR2 as the trophoblast-specific receptor and predicts that a *IIIc*-specific mutation should show trophoblast defects similar to one of the null mutations. The *Grb2* and *FRS2α* mutations may also have defects in the initial period of proliferation of the polar TE by disabling the FGF signal transduction pathway. Grb2 is an adaptor for several signal pathways, but FRS2α is specific to FGF signaling.

Based on gene expression studies, FGF-dependent trophoblast progenitors are likely present from E3.5 to E8.5 (for five days). *Fgfr2* expression is down-regulated in the ChE by E8.5, and *Eomes*, an essential gene for the early trophoblast, is also repressed. A more precise prediction of when and where early trophoblast progenitors reside came from studies on the distribution of activated diphosphorylated mitogen-activated protein kinase (dpMAPK) during early development. These studies revealed regions of the embryo where MAPK was activated, a common target in several signal transduction pathways. The early ExE (E5.5) was positive for dpMAPK, but most of the epiblast and visceral endoderm (VE) were not. The region of ExE positivity become more restricted to the trophoblast cells closest to the epiblast as development continued. This ring of dpMAPK staining persisted in the ChE after gastrulation occurred and contact with the epiblast was lost. The MAPK activation in ExE and ChE was attributed to FGF signaling, since a specific inhibitor (SU5402) abolished staining in this region and not others. The dpMAPK+ trophoblast cells in the ExE–ChE may be pro-

genitors for the trophoblast lineage, but they only exist for a fraction (<25%, or 5 of 19 days) of embryogenesis. In this situation, it may be more appropriate to refer to this subset of trophoblast cells as multipotent progenitors. Nevertheless, we refer to the FGF-dependent cultured cell lines derived from this tissue as stem cells, just as ES cell lines are considered stem cells. In their established culture conditions, ES and TS cells can be maintained in their respective primitive states from which they can differentiate into several cell types.

As will be detailed later in this chapter, FGF-dependent TS cell lines can be derived from blastocysts and early postimplantation trophoblast tissue. Studies by Uy *et al.* (2002) used TS cell line derivations as an assay to determine which tissues could generate the characteristic TS cell cultures. Although this may not precisely determine where the multipotent progenitors exist *in vivo*, it should definitively exclude regions and embryological times in which they do *not* exist. The ability to derive TS cell lines was demonstrated for embryos from the early blastocyst stage to as late as the 10-somite pair stage (~E8.0). From postimplantation embryos, TS cell lines could only be derived from ExE or ChE and not from the EPC or embryonic tissues; they could not be derived from trophoblast tissue at later stages. It was also noted that TS cell lines could be derived from all regions of the ExE, including the cells adjacent to the EPC (farthest from the epiblast). Later in development, the ChE, a trophoblast tissue not in direct contact with the epiblast, efficiently produced TS cell lines. This suggested that the inductive FGF signal supplied by the embryo at this stage is diffusible over a long range or that it is produced from other tissues, such as the extraembryonic mesoderm or endoderm. The areas of dpMAPK positivity and the regions where TS cell lines could be derived did not perfectly coincide. TS cell cultures could be obtained from ExE cells distant from the epiblast and regions of the chorion that were not positive for dpMAPK. However, *Fgfr2* expression did persist in these dpMAPK– regions, suggesting that cells could respond to an FGF signal but were not doing so *in vivo*. For this reason, trophoblast cells in ExE–ChE that are positive for dpMAPK may more accurately represent the trophoblast progenitor population *in vivo*.

The current model proposes that FGF4 produced by the ICM is necessary for the proliferative maintenance of the polar TE and that the lack of FGF signaling in the mural region results in giant cell differentiation. After implantation, the epiblast continues to produce FGF4 that signals to the overlying ExE, resulting in MAPK activation and maintenance of the trophoblast progenitor population. The FGFR2 IIIc isoform is predicted to be the trophoblast-specific FGF receptor *in vivo*. Essential transcription factors for the maintenance of this population include *Cdx2*, *Eomes*, and *Errβ*. After gastrulation and the formation of the chorion, portions of the ChE remain positive for dpMAPK, but the source and identity of the FGF signal is unclear at this point (Figure 14-3).

The FGF-dependent progenitors are not present after chorioallantoic fusion, but the placenta is still required to

A

B

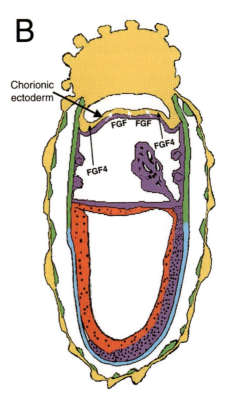

Figure 14-3. A. Model of early trophoblast development. (A) During early postimplantation development, FGF4 produced by the epiblast signals to the overlying extraembryonic ectoderm (white arrows) using the FGFR2 IIIc isoform. This leads to activation of the MAPK pathway, as indicated by diphosphorylated MAPK (dpMAPK). Downstream transcription factors essential for the maintenance of this lineage include Eomes, Cdx2, and Errβ. (B) After gastrulation, the chorionic ectoderm is far from the embryo proper but maintains dpMAPK activity. FGF4 may be diffusing over a long range, or a different FGF may be supplied by the underlying extraembryonic mesoderm.

undergo a substantial increase in size and is likely under the control of different signaling pathways. The TGFβ signaling pathway activated by Nodal is a possible candidate for this signal. An insertional mutation at the *nodal* locus resulted in a placental defect with excessive giant cell formation and a lack of spongiotrophoblast and labyrinth development. Unlike *Fgfr2* or the transcription factors *Cdx2*, *Eomes*, and *Errβ*, *nodal* is not expressed in the early ExE or ChE. Its expression in the placenta is specific to the spongiotrophoblast layer (an EPC descendant), begins at E9.5, and continues until term. Interestingly, a compound null–hypomorph *nodal* mutation resulted in a less severe placental defect in which the giant cell and spongiotrophoblast layers were expanded at the expense of the labyrinth. This suggests that Nodal-dependent progenitors may be residing in the labyrinth and that the source of the proliferative signal has shifted from the embryo to the spongiotrophoblast.

TS Cell Lines

DERIVATION OF TS CELL LINES

The culture of ExE or EPC tissue in most conditions resulted in giant cell differentiation. This led to the proposal that the default state of early trophoblast tissue is giant cells. The accumulating evidence that FGF signaling was important for trophoblast development and could prevent giant cell transformation led Dr. Satoshi Tanaka to revisit ExE explant cultures. FGF4 reduced not only giant cell formation in ExE explants, but also the amount of EPC-like differentiation. Encouraged by these results, ExE from E6.5 embryos was disaggregated into single cells and plated on embryonic fibroblasts (EMFIs) in the presence of FGF4 and its essential cofactor heparin. This resulted in the formation of flat epithelial colonies that could be passaged indefinitely. Identical cell lines could also be derived from E3.5 blastocyst outgrowths (Tanaka *et al.*, 1998). Chapter 45 of this volume provides detailed protocols on the derivation and maintenance of these TS cell lines.

TS cell lines expressed *Fgfr2* and the transcription factors characteristic of trophoblast progenitors *in vivo*: *Cdx2*, *Eomes*, and *Errβ*. Removal of FGF4, heparin, or the EMFIs resulted in giant cell differentiation (Figure 14-4A and 14-4B) and complete down-regulation of these four markers. During the differentiation process, markers of intermediate trophoblast tissue, such as *Mash2* and *Tpbp*, were expressed. The cell culture-derived giant cells exhibited increased ploidy and expressed the giant cell marker, *Pl-I*, like their *in vivo* counterparts. TS cells did not express the ES cell or early epiblast marker *Oct4*, the mesoderm marker *Brachyury*, or the endoderm marker *Hnf-4*. FACS analysis and visual inspection of the cultures revealed that TS cells cultures are heterogeneous even when grown in stem cell conditions (Figure 14-4C and 14-4D). TS cell cultures could also be maintained in EMFI-conditioned medium, suggesting that EMFIs produce a soluble factor or factors important for TS cell self-renewal.

The developmental potential of TS cells was tested in chimeras generated by blastocyst injections. Green fluorescent protein (GFP)-transgenic TS cells were analyzed in chimeras from E6.5 to term (E18.5). All chimeras had TS cell contributions exclusively in trophoblast tissues (Figure 14-4E and 14-4F). There were no examples of contributions to the embryo proper or to the visceral yolk sac and amnion, two extraembryonic membranes that do not contain trophoblast cells. In addition, all subtypes of the trophoblast lineage could be colonized. In early embryos, GFP-TS cells were in the ExE, EPC, and giant cells. In older conceptuses, the labyrinth and spongiotrophoblast had GFP-TS cell contributions. The exclusive, tissue-specific restriction observed in chimeras indicates that TS cells are committed to the trophoblast lineage and retain the potential to differentiate into its many cell types.

Although TS cells can produce some remarkable chimeras with high contributions to the trophoblast lineage, improvements in its efficiency are required before this can be used as a routine assay to test the potential of manipulated TS cell lines. The percentage of observing chimeras from the most efficient method (blastocyst injection) is 45% of embryos recovered. However, the frequency of obtaining *high-contribution* chimeras is between 5 and 10%. There may be several reasons for this low efficiency. First, the heterogeneity of the TS cell cultures suggests that only a fraction of the cells may be true stem cells. Any method to identify cells with greater stem cell potential and separate them from differentiated cells should improve the efficiency and extent of contribution during chimera production. Diploid cells can be separated from higher ploidy cells by fluorescence-activated cell sorting (FACS) with Hoechst 33342 dye, and a reduction in heterogeneity is observed by morphology and FACS analysis. However, this procedure only excludes one of the trophoblast subtypes: polyploid giant cells. A more precise method of isolation — for example, FACS for a specific cell-surface marker such as FGFR2 or high-efflux properties such as the side population — should further narrow the stem cell population. To date, attempts to rescue a genetic trophoblast defect by TS cell chimerism have not been reported, but identifying and isolating TS cells with maximum developmental potential should help in this area.

TS CELLS AS MODELS OF TROPHOBLASTS

The TS cell culture system was used to investigate the role of the orphan nuclear receptor, ERRβ, through the use of a small molecule inhibitor. Diethylstilbestrol (DES) is a synthetic estrogen that is a potent agonist for the classical estrogen receptors, ERα and ERβ. It was demonstrated that DES has the opposite effect on the three ER-related receptors: ERRα, ERRβ, and ERRγ. ERRs are constitutively active transcription factors and do not require a ligand for activation. Since TS cells and ExE express high levels of *Errβ*, the effects of DES treatment were investigated. DES specifically induced TS cell differentiation toward a trophoblast giant cell phenotype, and estradiol had no effect. To investigate the effect of DES treatment on trophoblast tissue *in vivo*, pregnant mice were fed DES during early placentation (E4.5 to 8.5), and their placentas were examined at E9.5. Strikingly, the DES-treated placentas did not have the usual trilayered placenta consisting of spongiotrophoblast, labyrinth, and giant cells. Instead, they consisted of multiple layers of giant cells similar to *Errβ*-mutant placentas. This strongly suggested that the effects of DES in the placenta and in TS cells are mediated through only one of the ERRs: ERRβ. This illustrates that ERRβ is essential for the maintenance of TS cells in culture and the trophoblast progenitor population *in vivo*.

In an effort to identify novel trophoblast-specific genes, a cDNA microarray analysis was performed by comparing TS cell, ES cell, and EMFI RNA. Total RNA from these samples was used to prepare cDNA hybridized to the NIA mouse 15K cDNA clone set in the laboratory of Minoru Ko *et al.* Lists of genes particular to each cell type were generated, and several novel trophoblast lineage genes were investigated further by Northern blotting. Two novel genes (*Mm. 320575* and *Mm. 46582*) and a *Greul1*-homolog were specifically expressed in TS cells and giant cells but not in ES cells or EMFIs. These genes were considered trophoblast-lineage genes and not TS cell-specific genes, since they were all expressed in giant cells.

Since the first report of TS cell lines, several laboratories have used this stem cell culture system for a diverse array of studies. A very interesting result was obtained when *Oct4* was conditionally repressed in ES cells. The *Oct4*-repressed cells transformed into trophoblast giant cells at the expense of embryonic lineages that ES cells normally produce under differentiative conditions. Even more striking was the establishment of TS cell cultures when FGF4 was added to ES cells during the down-regulation of *Oct4*. This implicated Oct4 as a major repressor of the trophoblast lineage. Avilion *et al.* used the derivation of TS cell lines as an assay to determine the necessity of the high-mobility group box transcription factor, *Sox2*, for this lineage. They determined that *Sox2* is an essential gene for the trophoblast lineage, since null TS cells could not be derived. Adelmann *et al.* derived TS cells mutant for the hypoxia-responsive transcription factor *Arnt*. They went on to show that *Arnt*$^{-/-}$ TS cells were deficient in forming the intermediate EPC-like trophoblast cells during differentiation in culture. Yan *et al.* (2001) used TS cells to study the effects of retinoic acid (RA) treatment in culture. They also found that TS cells skipped the intermediate, *Mash2+* trophoblast subtype and differentiated directly into giant cells when treated with RA. Ma *et al.* found that transfection of TS cells with *nodal* decreases giant cell formation and that JunB activ-

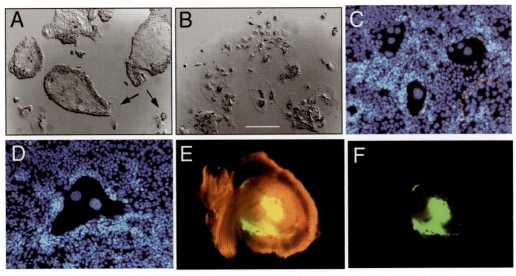

Figure 14-4. A. TS cells and a TS cell chimera. TS cells grown in (A) stem cell and (B) differentiative conditions. (A) Differential interference contrast (DIC) micrograph of several TS cell colonies grown in stem cell conditions. They form tight epithelial sheets with occasional differentiated giant cells (arrows) at their periphery. (B) DIC micrograph of six-day differentiated TS cells showing a drastic change in cell morphology to the characteristic giant cells. The bar is 50 μm. (C and D) Two examples of confluent stem cell cultures of TS cells stained with Hoechst 33342. The huge nuclei and vast cytoplasm of the giant cells are evident. There are also cells with intermediate sized nuclei (dots in panel C, for example). (E and F) An E8.5 TS cell chimera generated by blastocyst injection of green fluorescent protein — TS cells. The embryo with placenta was observed under UV fluorescence with (E) partial bright-field and (F) dark-field optics.

ity increases it. Female TS cells were used by Mak *et al.* (2000) to show that paternally imprinted X-chromosome inactivation observed in the female trophoblast lineage *in vivo* is maintained in TS cells. Furthermore, the imprinting is maintained by stable association of the Polycomb proteins, Eed and Enx1, with the inactive X-chromosome. The role of α7 integrin in trophoblast attachment to different extracellular substrates was investigated with the TS cell culture system. Suppressor of cytokine signaling 3 (*SOCS3*)-mutant TS cells were derived and shown to have an increased tendency to form giant cells at the expense of EPC-like cells, similar to the phenotype observed for the *SOC3*$^{-/-}$ placenta *in vivo*. A targeted mutation against the mitochondrial transmembrane guanosine triphosphatase, mitofusin 2 (*Mfn2*), resulted in a giant cell defect and lethality by E11.5. *Mfn2*$^{-/-}$ TS cell lines were derived and shown to have severe morphological defects in their mitochondria. They were spherical and small, instead of the long mitochondrial tubules observed in wild-type TS cells. Finally, Shiota *et al.* compared the DNA methylation status of CpG islands in TS cells and differentiated TS cells (giant cells) using the restriction-landmark genomic-scanning method. These various reports illustrate that TS cell lines provide a useful model to study the trophoblast lineage in the mouse.

Extraembryonic Endoderm Lineage

EXTRAEMBRYONIC ENDODERM DEVELOPMENT

The PrE lineage first makes its appearance on the blastocoelic surface of the ICM at E4.5. The developmental path taken by the PrE depends on the tissue with which it interacts. The PrE in contact with the extraembryonic and embryonic ectoderm differentiates into VE, and the PrE in contact with the trophoblast giant cell layer differentiates into parietal endoderm (PE) (Figure 14-2). The PE collaborates with the giant cell layer to form the intervening Reichert's membrane. The combination of the PE, giant cells, and a thick basement membrane is the parietal yolk sac. The VE is a complex lineage with roles in nutrient delivery, embryonic cavitation, anterior induction, and hemangioblast induction.

Once the initial PrE cells are established, the GATA factors, GATA4 and GATA6, are major players in the maintenance and elaboration of this lineage. *GATA4*-mutant embryos form VE but express elevated levels of *GATA6*, suggestive of some functional redundancy. The embryos die around E9.5 with heart defects. The *GATA6* mutation is more severe, with serious VE defects and lethality occurring between E5.5 and 7.5. An instructive role for these proteins in extraembryonic endoderm specification was shown by ectopic expression in ES cells.

The transition of VE to PE first occurs at the periphery of the ICM as the cells migrate onto the inner surface of the mural TE at the blastocyst stage. The VE grows as an epithelial layer with coherent growth, and the PE cells grow individually and scatter on the giant layer as they lay down matrix proteins. The differentiation of VE to PE is considered by some to be the first epithelial-to-mesenchymal transition in development. Studies in embryonal carcinoma cells identified cyclic adenosine monophosphate as an inducer of PE, if the cells were first differentiated into VE by RA treatment. The ligand, parathyroid hormone-related protein (PTHrP), was later suggested to be an endogenous inducer of PE. PTHrP is produced by the giant cell layer, and PE cells express its receptor, the G-protein-coupled receptor, type I PTHrP-R, at high levels. The zinc-finger transcription factor, Snail, was identified as an immediate early target of PTHrP signaling in ES cells and was shown to be expressed in PE *in vivo*. *Snail* and the highly related gene, *Slug*, are important for epithelial–mesenchymal transitions in several species. In *Snail*^{-/-} embryos, the initial mesoderm is formed, but it remains epithelial-like in agreement with the defined role of *Snail* in epithelial-mesenchymal transitions. *Snail*^{-/-} embryos are smaller than wild-type littermates at E7.5 and die before E8.5. A phenotype in the VE-to-PE transition was not reported on, but such a defect would not be inconsistent with the time of lethality.

EXTRAEMBRYONIC ENDODERM PROGENITORS?

PrE is considered to be the progenitor of both VE and PE, but this primitive cell type only persists one day after implantation (E5.5). After this point, both extraembryonic endoderm layers (VE and PE) continue to grow extensively. Chimera studies have shown that PrE cells have the potential to contribute to both VE and PE in a single chimera. Observations such as the sequential appearance of VE cells followed by PE cells during the differentiation of ES and EC cells have biased opinions on how these cell types arise *in vivo*. These observations led to a model in which VE cells continually produce more PE cells in the marginal zone (the proximal region of yolk cavity). However, these experiments do not conclusively prove such a relationship *in vivo*. It is possible that the VE and PE layers have separate and self-sufficient cell populations after an early point in development (e.g., by E6.5). One possibility is that PE progenitors exist at the distal tip of the parietal yolk sac and that they contribute descendants that migrate up toward the marginal zone. The VE layer may have a similar progenitor zone to maintain its population. Mitotic indices in both tissues measured at E7.5 were similar. An alternative theory puts extraembryonic endoderm stem cells in the marginal zone between the VE and the PE. Although this region has been proposed to be the site of VE-to-PE transition, direct lineage analysis of these cells has not been performed. Real-time imaging of embryo–embryo chimeras (harboring different fluorescent transgenes) should answer some of these questions. The recent derivation of extraembryonic endoderm (XEN) cell lines from mouse blastocysts has provided a cell culture model to address some of the issues discussed above (Kunath *et al.*, 2005). XEN cells can be cultured indefinitely, express markers of the extraembryonic endoderm lineage, and exclusively contribute to this lineage in chimeras.

Summary

Gene expression studies (e.g., *Eomes*), TS cell line derivation potential, and MAPK activity in the embryo have led to a precise prediction of when and where trophoblast multipotent

progenitors exist *in vivo*. The ExE cells close to the epiblast and a discrete population of ChE are the most likely locations for these early progenitors. However, by E8.5, the FGF-dependent progenitors do not exist and may be replaced by a different class of multipotent cells. The extraembryonic VE and PE layers exhibit substantial growth during development, but a defined progenitor or stem cell population has yet to be identified in this cell lineage.

KEY WORDS

Blastocyst A mammalian embryo, just prior to implantation, consisting of a hollow ball of cells surrounding a smaller clump of cells known as the inner cell mass. The blastocyst is the source material to derive embryonic stem (ES) trophoblast stem (TS), and extraembryonic endoderm (XEN) cell lines.

Primitive endoderm An extraembryonic lineage that forms part of the yolk sac. The primitive endoderm does not contribute to endodermal tissues, such as the gut, liver, or pancreas. These are derived from definitive endoderm, a distinct lineage from primitive endoderm.

Trophectoderm The outer layer of cells of the blastocyst. Trophectoderm cells are the sole precursors to the trophoblast lineage of the placenta, and they do not contribute cells to the embryo proper.

Trophoblast A general term used to describe all the cell types of the developing and mature placenta derived from the trophectoderm. In the mouse, the trophoblast lineage would include extraembryonic ectoderm, ectoplacental cone, trophoblast giant cells, chorionic ectoderm, labyrinthine trophoblast, and spongiotrophoblast.

FURTHER READING

Bielinska, M., Narita, N., and Wilson, D. B. (1999). Distinct roles for visceral endoderm during embryonic mouse development. *Int. J. Dev. Biol.* **43**, 183–205.

Cross, J. C. (2000). Genetic insights into trophoblast differentiation and placental morphogenesis. *Semin. Cell Dev. Biol.* **11**, 105–113.

Fujikura, J., Yamato, E., Yonemura, S., Hosoda, K., Masui, S., Nakao, K., Miyazaki, J., and Niwa, H. (2002). Differentiation of embryonic stem cells is induced by GATA factors. *Genes Dev.* **16**, 784–789.

Kunath, T., Arnaud, D., Uy, G. D., Okamoto, I., Chureau, C., Yamanaka, Y., Heard, E., Gardner, R. L., Avner, P., and Rossant, J. (2005). Imprinted X-inactivation in extra-embryonic endoderm cell lines from mouse blastocysts. *Development* **132**, 1649–1661.

Kunath, T., Strumpf, D., and Rossant, J. (2004). Early trophoblast determination and stem cell maintenance in the mouse-a review. *Placenta* **25 Suppl**, S32–38.

Mak, W., Baxter, J., Silva, J., Newall, A. E., Otte, A. P., and Brockdorff, N. (2002). Mitotically stable association of polycomb group proteins eed and enx1 with the inactive X chromosome in trophoblast stem cells. *Curr. Biol.* **12**, 1016–1020.

Rossant, J., and Cross, J. C. (2001). Placental development: Lessons from mouse mutants. *Nat. Rev. Genet.* **2**, 538–548.

Rossant, J., and Ofer, L. (1977). Properties of extra-embryonic ectoderm isolated from postimplantation mouse embryos. *J. Embryol. Exp. Morphol.* **39**, 183–194.

Strumpf, D., Mao, C.-A., Yamanaka, Y., Ralston, A., Chawengsaksophak, K., Beck, F., and Rossant, J. (2005). Cdx2 is required for correct cell fate specification and differentiation of trophectoderm in the mouse blastocyst. *Development* **132**, 2093–2102.

Tanaka, S., Kunath, T., Hadjantonakis, A. K., Nagy, A., and Rossant, J. (1998). Promotion of trophoblast stem cell proliferation by FGF4. *Science* **282**, 2072–2075.

Uy, G. D., Downs, K. M., and Gardner, R. L. (2002). Inhibition of trophoblast stem cell potential in chorionic ectoderm coincides with occlusion of the ectoplacental cavity in the mouse. *Development* **129**, 3913–3924.

Yan, J., Tanaka, S., Oda, M., Makino, T., Ohgane, J., and Shiota, K. (2001). Retinoic acid promotes differentiation of trophoblast stem cells to a giant cell fate. *Dev. Biol.* **235**, 422–432.

Amniotic Fluid-Derived Pluripotential Cells

Paolo deCoppi, M. Minhaj Siddiqui and Anthony Atala

Introduction

Human amniotic fluid has been used in prenatal diagnosis for more than 70 years. It has proved to be a safe, reliable, and simple screening tool for a variety of developmental and genetic diseases. However, there is now evidence that amniotic fluid may have utility beyond its use as a diagnostic tool and may be a source of a powerful therapy for a multitude of congenital and adult disorders. A subset of cells in amniotic fluid has been isolated and found capable of maintaining prolonged undifferentiated proliferation as well as of differentiating into multiple-tissue types encompassing the three germ layers. It is possible that we will soon see the development of therapies using progenitor cells isolated from amniotic fluid for the treatment of newborns with congenital malformations as well as therapies for adults using cryopreserved amniotic fluid.

In this chapter, we describe several experiments that have isolated and characterized pluripotent progenitor cells from amniotic fluid. We also provide various lineages that these cells have been differentiated into and directions in this area of research.

Amniotic Fluid and Amniocentesis

The first reported amniocentesis took place in 1930 when attempts were being made to correlate the cytologic examination of cell concentration, count, and phenotypes in the amniotic fluid to the sex and health of the baby. Since then, the development of techniques of karyotype and the discovery of reliable diagnostic markers such as α-fetoprotein, as well as the development of ultrasound-guided amniocentesis, have greatly increased the reliability of the procedure as a valid diagnostic tool as well as the safety of the procedure.

Amniocentesis provides a safe method of isolating cells from the fetus, which can then be karyotyped and examined for chromosomal abnormalities. In general, the protocol consists of acquiring 10 to 20 ml of fluid using a transabdominal approach. Amniotic fluid samples are then centrifuged, and the cell supernatant is resuspended in culture medium. Approximately 10^4 cells are seeded on 22×22 mm cover slips. Cultures are grown to confluence for three to four weeks in 5% CO_2 at 37°C, and the chromosomes are characterized from mitotic phase cells.

Amniocentesis is performed typically around 16 weeks of gestation, although in some cases it may be performed as early as 14 weeks when the amnion fuses with the chorion and the risk of bursting the amniotic sac by needle puncture is minimized. Amniocentesis can be performed as late as term. The amniotic sac is usually noticed first by ultrasound around the 10-week gestational time point.

Amniotic fluid cell culture consists of a heterogeneous cell population displaying a range of morphologies and behaviors. Studies on these cells have characterized them into many shapes and sizes varying from 6 to 50 μm in diameter and from round to squamous in shape. Most cells in the fluid are terminally differentiated along epithelial lineages and have limited proliferative and differentiation capabilities. Previous studies by Cremer *et al.* (1981) have noted an interesting composition of the fluid consisting of a heterogeneous cell population expressing markers from all three germ layers.

Much research has been conducted on the source of these cells and on the fluid itself. Current theories suggest that the fluid is largely derived from urine and peritoneal fluid from the fetus as well as from some ultrafiltrate from the plasma of the mother entering though the placenta. The cells in the fluid have been shown to be overwhelmingly from the fetus and are thought to be mostly cells sloughed off the epithelium, digestive, and urinary tract of the fetus as well as off the amnion.

Our laboratory investigated the possibility of isolating a progenitor cell population from amniotic fluid. The amniotic fluid was from normal fetuses obtained using a transabdominal approach from 14 to 21 weeks of gestation. Initially, male fetuses were used to preclude the possibility of maternal-derived cells.

Isolation and Characterization of Progenitor Cells

A pluripotential subpopulation of progenitor cells in the amniotic fluid can be isolated through positive selection for cells expressing the membrane receptor c-kit, which binds to the ligand stem cell factor. Roughly 0.8 to 1.4% of cells in amniotic fluid have been shown to be c-kit[pos] in analysis by fluorescence-activated cell sorting.

The progenitor cells maintain a round shape for one week after isolation when cultured in nontreated culture dishes. In this state, they demonstrate a very low proliferative capability. After the first week, the cells begin to adhere to the plate and change their morphology, becoming more elongated and proliferating more rapidly to reach 80% confluence with a need for passage every 48 to 72 hours. No feeder layers are required for either maintenance or expansion. The progenitor cells derived from amniotic fluid show a high self-renewal capacity with >300 population doublings, far exceeding Hayflick's limit. The doubling time of the undifferentiated cells is noted to be 36 hours, with little variation with passages.

These cells have been shown to maintain a normal karyotype at late passages and have normal G1 and G2 cell cycle checkpoints. They demonstrate telomere length conservation in the undifferentiated state as well as telomerase activity even in late passages as shown by Bryan *et al.* (1998). Analysis of surface markers shows that progenitor cells from amniotic fluid expressed human embryonic stage-specific marker SSEA4 and the stem cell marker OCT4, but they did not express SSEA1, SSEA3, CD4, CD8, CD34, CD133, C-MET, ABCG2, NCAM, BMP4, TRA1-60, or TRA1-81, to name a few. This expression profile is of interest as it demonstrates expression by the amniotic fluid-derived progenitor cells of some key markers of embryonic stem cell phenotype but not the full complement of markers expressed by embryonic stem cells. This hints that the amniotic cells are not as primitive as embryonic cells and yet maintain greater potential than most adult stem cells. Work by Thomson *et al.* (1998) shows that, although the amniotic fluid progenitor cells form embryoid bodies *in vitro* that stain positive for markers of all three germ layers, these cells do not form teratomas *in vivo* when implanted in immunodeficient mice. Lastly, cells expanded from a single cell maintain similar properties in growth and potential as the original mixed population of the progenitor cells.

Differentiation Potential of Amniotic Progenitor Cells

The progenitor cells derived from human amniotic fluid are pluripotent and have been shown to differentiate into osteogenic, adipogenic, myogenic, neurogenic, endothelial, and hepatic phenotypes *in vitro*. Each differentiation has been performed through proof of phenotypic (Figure 15-1) and biochemical (Figure 15-2) changes consistent with the differentiated tissue type. We will describe each set of differentiations separately.

ADIPOCYTES

To promote adipogenic differentiation, the progenitor cells can be induced in dexamethasone, 3-isobutyl-1-methylxanthine, insulin, and indomethacin. The progenitor cells cultured with adipogenic supplements change their morphology from elongated to round within 8 days. This coincides with the accumulation of intracellular droplets. After 16 days in culture, more than 95% of the cells have their cytoplasm filled with lipid-rich vacuoles.

Adipogenic differentiation also demonstrates the expression of an adipogenic-specific transcription factor and of lipoprotein lipase, as noted by Kim *et al.* (1998) and Rosen *et al.* (1999). Expression of these genes is noted in the progen-

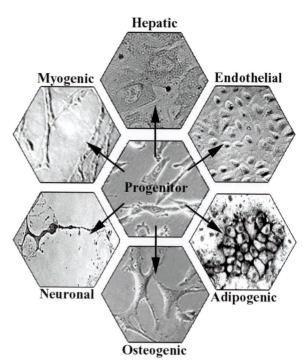

Figure 15-1. The isolated progenitor cells were capable of differentiation into multiple cell types, including muscle, liver, endothelial cells, adipocytes, osteoblasts, and neurons.

itor cells under adipogenic conditions but not in undifferentiated cells (Figure 15-2B).

ENDOTHELIAL CELLS

The amniotic fluid progenitor cells can be induced to form endothelial cells by culture in endothelial basal medium on gelatin-coated dishes. Full differentiation is affected with one month in culture; however, phenotypic changes are noticed within one week of initiation of the protocol. Human-specific endothelial cell surface marker (P1H12), factor VIII (FVIII), and kinase insert domain-containing receptor are specific for differentiated endothelial cells. The differentiated cells stain positively for FVIII, KDR, and P1H12. The amniotic fluid-derived progenitor cells do not stain for endothelial-specific markers and, once differentiated, are able to grow in culture and form capillary-like structures *in vitro*. These cells also express molecules that are not detected in the progenitor cells on RT-PCR analysis (Figure 15-2F).

HEPATOCYTES

For hepatic differentiation, the progenitor cells are seeded on Matrigel- or collagen-coated dishes at different stages and cultured in the presence of hepatocyte growth factor, insulin, oncostatin M, dexamethasone, fibroblast growth factor 4, and monothioglycerol for 45 days. After 7 days of the differentiation process, cells exhibit morphological changes from an elongated to a cobblestone appearance. The cells show positive staining for albumin at day 45 after differentiation and also express the transcription factor HNF4α, the c-met recep-

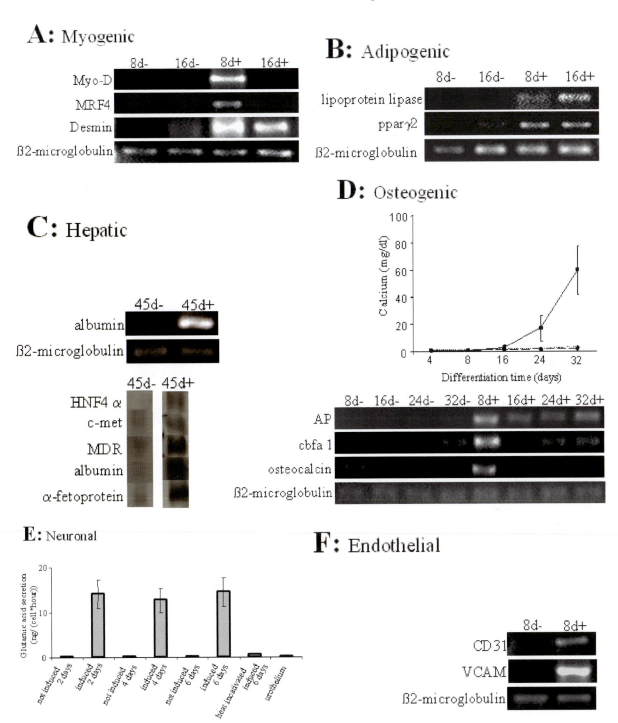

Figure 15-2. The differentiated cell types expressed functional and biochemical characteristics of the target tissue. (A) Myogenic-induced cells showed a strong expression of *desmin* expression at day 16 (lane 4). *MyoD* and *MRF4* were induced with myogenic treatment at day 8 (lane 3). Specific PCR-amplified DNA fragments of *MyoD*, *MRF4*, and *desmin* could not be detected in the control cells at days 8 and 16 (lanes 1 and 2). (B) Gene expression of *pparγ2* and *lipoprotein lipase* in cells grown in adipogenic-inducing medium was noted at days 8 and 16 (lanes 3 and 4). (C) RT-PCR revealed an up-regulation of *albumin* gene expression. Western blot analyses of cell lysate showed the presence of the hepatic lineage-related proteins HNF-4α, c-met, MDR, albumin, and α-fetoprotein. Undifferentiated cells were used as negative control. (D) Osteogenic-induced progenitor cells showed a significant increase of calcium deposition starting at day 16 (solid line). No calcium deposition was detected in progenitor cells grown in control medium or the negative control cells grown in osteogenic conditions (dashed line). RT-PCR showed presence of *cbfa1* and *osteocalcin* at day 8 and confirmed the expression of *AP* in the osteogenic-induced cells. (E) Only the progenitor cells cultured under neurogenic conditions showed the secretion of glutamic acid in the collected medium. The secretion of glutamic acid could be induced (20 minutes in 50-mM KCl buffer). (F) RT-PCR of progenitor cells induced in endothelial medium (lane 2) showed the expression of *CD31* and *VCAM*.

tor, the MDR membrane transporter, albumin, and α-fetoprotein (Figure 15-2C).

MYOCYTES

Myogenic differentiation is induced in the amniotic fluid-derived progenitor cells by culture in media containing horse serum and chick embryo extract on a thin gel coat of Matrigel. To initiate differentiation, the presence of 5-azacytidine in the media for 24 hours is necessary. Phenotypically, the cells can be noted to organize themselves into bundles that fuse to form multinucleated cells. These cells express sarcomeric tropomyosin and desmin, both of which are not expressed in the original progenitor population.

Interestingly, the development profile of cells differentiating into myogenic lineages mirrors a characteristic pattern of gene expression seen with embryonic muscle development, as shown in work by Rohwedel *et al.* (1994) and Bailey *et al.* (2001) (Figure 15-2A).

NEURONAL CELLS

For neurogenic induction, the amniotic progenitor cells are induced in dimethyl sulfoxide, butylated hydroxyanisole, and neuronal growth factor. Progenitor cells cultured in neurogenic conditions change their morphology within the first 24 hours. Two cell populations are apparent: morphologically large, flat cells and small, bipolar cells. The bipolar cell cytoplasm retracts toward the nucleus, forming contracted multipolar structures. Over the subsequent hours, the cells display primary and secondary branches and cone-like terminal expansions. The induced progenitor cells show a characteristic sequence of expression of neural-specific proteins, including those expressing neuroepithelial, neuron, and glial differentiation (Figure 15-2E).

OSTEOCYTES

Osteogenic differentiation was induced in the progenitor cells with use of dexamethasone, β-glycerophosphate, and ascorbic acid-2-phosphate. The progenitor cells maintained in this medium demonstrated phenotypic changes within 4 days with a loss of spindle-shape phenotype and a development of an osteoblast-like appearance with finger-like excavations into the cytoplasm. At 16 days, the cells aggregated, showing typical lamellar bone-like structures. In terms of functionality, these differentiated cells demonstrate a major feature of osteoblasts, which is to precipitate calcium. Differentiated osteoblasts from the progenitor cells are able to produce alkaline phosphatase (AP) and to deposit calcium consistent with bone differentiation. The undifferentiated progenitor cells lacked this ability.

The progenitor cells in osteogenic medium express specific genes implicated in mammalian bone development in a pattern consistent with the physiological analog (Figure 15-2D).

Future Directions

Much interesting work remains to be done with this cell population. *In vitro*, there remain a few cell types that have not

been investigated but are of great interest scientifically and therapeutically. *In vivo* work to complement the *in vitro* differentiations demonstrating the capacity of these cells to functionally supplement normal tissue will highlight the true clinical potential of these cells. Of great interest may be cell types traditionally senescent in differentiated form, as these cells could be expanded in an undifferentiated form and differentiated into the cell type of interest in large numbers. These progenitor cells, if seeded on a scaffold, could potentially be differentiated into the desired cell types. Such a mixture of correctly differentiated cells could process local cues to structure themselves into highly complex formations in the scaffold much as they do during development. For such reasons, the amniotic progenitor cells may also have practical use in tissue engineering of organs.

The ease of maintenance, proliferation, and differentiation of the amniotic progenitor cells also provides great promise as potential cells that could be used for other purposes such as investigation into development pathways or drug screening. Many experiments to probe the exact potential of these cells and fully characterize their source will be beneficial, as they will help to define realistic goals and applications for use of these cells.

Summary

The pluripotent progenitor cells isolated from amniotic fluid present an exciting possible contribution to the field of stem cell biology and regenerative medicine and may be an excellent source for research and therapeutic applications. The embryonic and fetal progenitor cells have better potential for expansion than adult stem cells. For this reason, they could represent a better source for therapeutic applications in which large numbers of cells are needed. The ability to isolate the progenitor cells during gestation may also be advantageous for babies born with congenital malformations. Furthermore, the progenitor cells can be cryopreserved for future self-use. When compared with embryonic stem cells, the progenitor cells isolated from amniotic fluid have many similarities: They can differentiate into all three germ layers, they express common markers, and they preserve their telomere length. However, the progenitor cells isolated from amniotic fluid have, in our opinion, considerable advantages. They easily differentiate into specific cell lineages, they do not need feeder layers to grow, and they do not require the sacrifice of human embryos for their isolation, thus avoiding the current controversies associated with the use of human embryonic stem cells. The discovery of these cells has been recent, and a great deal of work remains to be done on the characterization and use of these cells. Initial results have been promising and are sure to lead to interesting developments.

KEY WORDS

Karyotype The chromosome characteristics of an individual cell or of a cell line, usually presented as a systematized array of metaphase chromosomes from a photomicrograph of a single cell nucleus.

Pluripotent cells   Primordial cells that may still differentiate into various specialized types of tissue elements (e.g., mesenchymal cells).

Progenitor cells   In development, a parent cell that gives rise to a distinct cell lineage by a series of cell divisions.

Telomere   A specialized nucleic acid structure found at the ends of linear eukaryotic chromosomes.

Teratoma   Malignant tumor thought to originate from primordial germ cells or misplaced blastomeres that contains tissues derived from all three embryonic layers, such as bone, muscle, cartilage, nerve, tooth buds, and various glands.

FURTHER READING

Bailey, P., Holowacz, T., and Lassar, A. B. (2001). The origin of skeletal muscle stem cells in the embryo and the adult. *Curr. Opin. Cell Biol.* **13**, 679–689.

Black, I. B., and Woodbury, D. (2001). Adult rat and human bone marrow stromal stem cells differentiate into neurons. *Blood Cells Mol. Dis.* **27**, 632–636.

Brace, R. A., Ross, M. G., and Robillard, J. E. (1989). *Fetal and neonatal body fluids: the scientific basis for clinical practice.* Perintology Press, New York.

Bryan, T. M., Englezou, A., Dunham, M. A., and Reddel, R. R. (1998). Telomere length dynamics in telomerase-positive immortal human cell populations. *Exp. Cell Res.* **239**, 370–378.

Cremer, M., Schachner, M., Cremer, T., Schmidt, W., and Voigtlander, T. (1981). Demonstration of astrocytes in cultured amniotic fluid cells of three cases with neural-tube defect. *Hum. Genet.* **56**, 365–370.

Dunn, J. C., Yarmush, M. L., Koebe, H. G., and Tompkins, R. G. (1989). Hepatocyte function and extracellular matrix geometry: long-term culture in a sandwich configuration. *FASEB J.* **3**, 174–177.

Hinterberger, T. J., Sassoon, D. A., Rhodes, S. J., and Konieczny, S. F. (1991). Expression of the muscle regulatory factor MRF4 during somite and skeletal myofiber development. *Dev. Biol.* **147**, 144–156.

Jaiswal, N., Haynesworth, S. E., Caplan, A. I., and Bruder, S. P. (1997). Osteogenic differentiation of purified, culture-expanded human mesenchymal stem cells *in vitro. J. Cell Biochem.* **64**, 295–312.

Kim, J. B., Wright, H. M., Wright, M., and Spiegelman, B. M. (1998). ADD1/SREBP1 activates PPAR-γ through the production of endogenous ligand. *Proc. Nat. Acad. Sci. USA* **95**, 4333–4337.

Patapoutian, A., Yoon, J. K., Miner, J. H., Wang, S., Stark, K., and Wold, B. (1995). Disruption of the mouse MRF4 gene identifies multiple waves of myogenesis in the myotome. *Development* **121**, 3347–3358.

Rohwedel, J., Maltsev, V., Bober, E., Arnold, H. H., Hescheler, J., and Wobus, A. M. (1994). Muscle cell differentiation of embryonic stem cells reflects myogenesis *in vivo*: developmentally regulated expression of myogenic determination genes and functional expression of ionic currents. *Dev. Biol.* **164**, 87–101.

Rosen, E. D., Sarraf, P., Troy, A. E., Bradwin, G., Moore, K., Milstone, D. S., Spiegelman, B. M., and Mortensen, R. M. (1999). PPAR-γ is required for the differentiation of adipose tissue *in vivo* and *in vitro. Mol. Cell* **4**, 611–617.

Schwartz, R. E., Reyes, M., Koodie, L., Jiang, Y., Blackstad, M., Lund, T., Lenvik, T., Johnson, S., Hu, W.S., and Verfaillie, C. M. (2002). Multipotent adult progenitor cells from bone marrow differentiate into functional hepatocyte-like cells. *J. Clin. Invest.* **109**, 1291–1302.

Thomson, J. A., Itskovitz-Eldor, J., Shapiro, S. S., Waknitz, M. A., Swiergiel, J. J., Marshall, V. S., and Jones, J. M. (1998). Embryonic stem cell lines derived from human blastocysts. *Science* **282**, 1145–1147.

Woodbury, D., Schwarz, E. J., Prockop, D. J., and Black, I. B. (2000). Adult rat and human bone marrow stromal cells differentiate into neurons. *J. Neurosci. Res.* **61**, 364–370.

16

Cord Blood Hematopoietic Stem and Progenitor Cells

Hal E. Broxmeyer

Introduction

CORD BLOOD TRANSPLANTATION AND BANKING

Hematopoietic stem and progenitor cells from umbilical cord blood have been used to transplant more than 5000 recipients with various malignant or genetic disorders since the first transplant, performed in October 1988. This cord blood transplant successfully cured the disordered and fatal hematological manifestations of Fanconi anemia; the male recipient of human lymphocyte antigen (HLA)-matched donor cord blood cells from a female sibling is alive and well more than 17 years after the transplant. This and subsequent cord blood transplants using sibling cells were the result of extensive laboratory-based studies and the first proof-of-principle cord blood bank, established in the author's laboratory, that suggested the feasibility of such transplants with cells previously considered waste material except for some routine clinical testing needs. Since those initial clinical studies and banking efforts, numerous cord blood banks have been developed worldwide, allowing the extension of cord blood transplantation to situations using HLA-matched and partially HLA-matched cord blood cells from unrelated and related allogeneic donors.

ADVANTAGES AND DISADVANTAGES OF CORD BLOOD FOR TRANSPLANTATION

Cord blood transplantation can be used, especially in children, to treat a multiplicity of malignant and nonmalignant disorders currently treatable by bone marrow transplantation. One obvious advantage of cord blood as a source of transplantable stem cells is the lower incidence of graft-vs.-host disease compared to that of bone marrow, allowing the use of cord blood with a greater HLA-disparity than is usually acceptable for bone marrow transplantation. However, engraftment of neutrophils and, even more so, of platelets is delayed after cord blood, compared with bone marrow, transplantation. The limiting numbers of stem–progenitor cells in single collections of cord blood, the immature nature of these rare repopulating cells, the difficulty of cord blood progenitors to program themselves toward differentiation, or all of these factors may be responsible for this relatively delayed blood cell engraft-

ment. Although cord blood has been used successfully to transplant adults, the limited number of collected cord blood cells has limited the number of cord blood transplants performed in adults. The use of multiple, unrelated cord blood units is among a number of procedures being considered to alleviate the problem of limiting donor cord blood cell numbers, but not enough information is yet available to validate this concept. Several attempts to increase cord blood stem cell numbers through *ex vivo* expansion efforts and transplantation of "expanded cells" have not yet resulted in encouraging clinical results.

This chapter focuses on the functional characteristics of cord blood stem and progenitor cells for proliferation, self-renewal, and homing, three important functions for clinical transplantation. Current information on these functional activities of cord blood stem and progenitor cells helps to explain successes with cord blood transplantation. However, efforts to manipulate these cells for enhanced functional activity may prove efficacious in extending the clinical usefulness of cord blood transplantation, and new information in the field of hematopoiesis will be described that addresses this possibility.

Characteristics and Cryopreservation of Cord Blood Stem and Progenitor Cells

CYCLING STATUS AND RESPONSES TO GROWTH FACTORS

Initial laboratory studies showed that the frequency and proliferative capacity of hematopoietic progenitor cells in cord blood was enhanced compared to that found in bone marrow. Although cord blood progenitors were in a slow (G_0/G_1) cell cycle state, they responded rapidly to the proliferation-inducing signals from growth factors. These growth factors include the granulocyte–macrophage colony-stimulating factor (GM-CSF), macrophage (M)-CSF, granulocyte (G)-CSF, erythropoietin (Epo), thrombopoietin (TPO), the potent costimulating cytokines steel factor (SLF, also called stem cell factor), and Flt3-ligand (FL). SLF and FL activate their respective tyrosine kinase receptors c-kit and Flt3 and synergize with many other growth factors and themselves to enhance proliferation of cord blood progenitors. SLF, with various CSFs, helped determine the enhanced frequency of immature subsets of progenitor cells in cord blood. The combination of SLF and FL, alone and with other cytokines, helped elucidate the efficient recovery of high proliferative potential progenitor cells from cord blood stored frozen for 15 years. These proliferative characteristics, as well as *in vivo* studies in mice with the nonobese-diabetic severe-combined immunodeficiency (NOD–SCID) genotype, demonstrated the superiority of cord blood to bone marrow stem cells for engraftment.

NUMBER OF CORD BLOOD CELLS REQUIRED FOR DURABLE ENGRAFTMENT

Several parameters, including nucleated cellularity and progenitor cell content of cord blood collections, have been used to predict the engrafting capability of these cells. One group provided evidence that the content of progenitor cells was a better predictor of speed of engraftment than the nucleated cellularity for cord blood transplantation, but most clinicians still rely on nucleated cellularity as one of several parameters, including HLA-typing, when determining whether or not to use a specific collection of cord blood for transplantation. A collection in the range of $\geq 2 \times 10^7$ nucleated cord blood cells per kilogram of body weight is the cutoff many transplanters feel most comfortable using, although successful transplants have been reported with collections containing as few as 5×10^6 nucleated cells/kg.

CRYOPRESERVATION OF CORD BLOOD

Almost all cord blood transplants have used cord blood that was first frozen for cryopreservation and storage in a public or private cord blood bank. Thus, the success of cord blood transplantation has relied and will continue to rely heavily on the belief that cord blood stem and progenitor cells can be cryopreserved and recovered efficiently in terms of the quantity and quality of the stem and progenitor cells. That this is so was first reported by the author's group for short-term freezes and then for longer term stored cells. The ultimate test is to use the frozen cord blood collection in a clinical transplant setting. A cord blood has been stored and used successfully in a clinical setting probably no longer than in the 5 to 7 year range. The longest surviving recipient of a transplant with cord blood from a frozen and stored collection has lived an additional 17 plus years. That was the recipient of the first cord blood transplant, performed in October 1988.

Three groups have reported the recovery of progenitors from cord blood stored frozen 12 to 15 years. The most extensive study demonstrated an average recovery (+/– 1SD) after 15 years storage of defrosted nucleated cells, granulocyte–macrophage (CFU-GM), erythroid (BFU-E), and multipotential (CFU-GEMM) progenitors, respectively, of 83 ± 12, 95 ± 16, 84 ± 25, and 85 ± 25. This was based on analysis of the same samples pre- and postfreeze, using the same culture conditions for progenitor cell analysis, and was comparable to the efficiency of recovery of these cells after 10 years of storage. The intact functional capabilities of these defrosted progenitors were highlighted by the extensive proliferative capacities of CFU-GM, BFU-E, and CFU-GEMM, which, respectively, generated colonies of up to 22,500, 182,500, and 292,500 cells after stimulation in a semisolid methylcellulose culture medium with Epo, GM-CSF, IL-3, and SLF. CFU-

GEMM colonies could be replated in secondary dishes, with resultant CFU-GEMM colonies as large as those formed in the primary culture dishes. $CD34^+CD38^-$ cord blood cells isolated from the defrosts of the 15-year frozen cells demonstrated more than a 250-fold *ex vivo* expansion of progenitor cells.

Perhaps of greater relevance to clinical transplantation, $CD34^+$ cells isolated from these defrosts were able to engraft NOD–SCID mice with a frequency equal to that of freshly isolated cord blood $CD34^+$ cells. Thus, it appears that cord blood can be stored frozen at least 15 years with the high likelihood that they will be able to engraft human recipients. A recent study suggested that measurement of progenitor cell recovery, as assessed by colony assays, is a more valid indicator than numbers of viable nucleated cells for cryopreserved cells. This interpretation was based on the finding that research cord blood collections intentionally subjected to an overnight thaw and refreeze did not form colonies of progenitors, yet they demonstrated the viability of nucleated cells in the 68 to 98% range, as determined by Trypan Blue exclusion. Although cryopreservation efforts have been relatively successful, studies continue to optimize methods for the cryopreservation of cord blood stem cells.

Cord Blood Transplantation Problems and Possible Countermeasures

BACKGROUND TO PROBLEMS

Although cord blood has been successfully used for stem cell transplantation, most transplants have been done in children. Many clinicians feel that the limiting numbers found in single collections of cord blood preclude their routine use in adults. Moreover, even in children, cord blood transplantation is associated with delayed engraftment of neutrophils, and especially of platelets, increasing the hospitalization time of the transplanted recipients. Although one means of dealing with the issue of limiting numbers of stem cells in cord blood collection may be to use multiple cord blood units for transplantation into single recipients, efforts to enhance the clinical utility of cord blood for both children and adults will likely require a greater understanding than is currently available of the self-renewal, proliferation, and homing characteristics of hematopoietic stem cells in general and of cord blood stem cells in particular. Information in these areas will clearly enhance prospects for successful *ex vivo* expansion of stem cells and for engrafting capabilities of these cells as well as of nonexpanded stem cells.

GENOMICS AND PROTEOMICS OF CORD BLOOD STEM–PROGENITOR CELLS

A comprehensive genomic and proteomic profile of cord blood, compared to that of adult bone marrow and mobilized peripheral blood, could shed light on the proliferative, self-renewal, and homing potentials of stem cells. Genomic profiling of stem cells has begun for cells from several sources, including cord blood. In addition, first attempts at proteomic profiling are being reported. However, these reports are limited to CD34⁺ cell populations, which contain hematopoietic progenitors as well as stem cells.

A problem inherent in analysis of CD34$^+$ cells is that they are not a pure population of stem cells; they contain progenitor cells, the stem cell content is likely a minor proportion of the total population, and this fraction is composed of more than stem and progenitor cells. Even the more highly purified population of human CD34$^+$CD38$^-$ cells is not nearly as pure in stem cell content. Unfortunately, the field of phenotypic characterization of human stem and progenitor cells is not at the level of that obtained for murine stem and progenitor cells. Genomic and proteomic information on human stem cells will therefore have to be interpreted cautiously, with the understanding that what is detected may not be specific for or even present in the stem cell population. Until human stem cell populations are better characterized in terms of a phenotype that can consistently recapitulate detection of stem cell function, and until more in-depth analysis of cord blood stem cell genomics and proteomics is elucidated, we can rely on functional analysis of the responses of cord blood stem cells for self-renewal, proliferation, and homing.

EX VIVO EXPANSION

Current Knowledge

Although cord blood progenitor cell populations have been extensively expanded *ex vivo* by many different investigators, it is not yet clear that human stem cells have been expanded much, if at all. In fact, loss of stem cell function has been reported after *ex vivo* expansion procedures. It may be relevant in certain circumstances to use *ex vivo*-expanded hematopoietic progenitors in clinical settings. However, use of *ex vivo*-expanded stem cells will likely require a greater understanding of the self-renewal process of stem cells. Recent studies suggest we may be closer to understanding cytokines and intracellular signaling events that influence the self-renewal process of stem cells.

Cytokines and Intracellular Molecules Implicated in Self-Renewal of Stem Cells

Numerous cytokines and chemokines, and their receptors, have been identified that act alone and together to modulate proliferation of myeloid progenitor cells. Information on cell cycle regulation and cell cycle checkpoints is also accumulating for progenitor cells.

A limited number of cytokines and intracellular signaling molecules have been implicated in proliferation, self-renewal, or both, of hematopoietic stem cells. These include, but are not necessarily limited to, the following ligands and their receptors: SLF/c-kit, FL/FLT3, Notch ligands–Notch, and Wnt3a–Frizzled. Intracellular molecules implicated include the following: p21cip1/waf1, Hoxb4–Pbx1, Bmi1, and a stromal cell-derived membrane protein, mKirre. Nanog, Stat3, and Hex have been implicated in the growth and differentiation of embryonic stem cells. It is possible that intracellular molecules involved in the regulation of embryonic stem cells

also play a role in the proliferation, self-renewal, or both, of hematopoietic stem cells. Stat3 has been linked to *in vivo* regulation of hematopoiesis, and serine phosphorylation of Stat3 has been linked to proliferation of progenitor cells in response to combined stimulation by SLF and GM-CSF or IL-3. Overexpression of Hex is associated with enhanced proliferation of myeloid progenitor cells. Stat5 is crucial for FL synergistic stimulation of myeloid progenitor.

Other cytokines that may influence stem cell proliferation, self-renewal, or both, are TPO, Oncostatin M, and IL-20. TPO, an early-acting cytokine, has been implicated in hemangioblast development and is one of the ingredients, with SLF and FL, used by investigators to *ex vivo* expand hematopoietic stem and progenitor cells. Oncostatin M is a T helper cell 1 produced cytokine that regulates progenitor cell homeostasis. IL-20 is a new member of the interleukin family. IL-20 is a candidate stem cell effector molecule that has selectivity for CFU-GEMM among myeloid progenitor cells. IL-20 enhances numbers of CFU-GEMM from human and mouse bone marrow and human cord blood in the presence of SLF and Epo *in vitro*. It has no effect *in vitro* on erythroid, granulocyte–macrophage, or megakaryocyte progenitors. IL-20 transgenic mice have increased numbers and cell cycling of CFU-GEMM but not of other myeloid progenitors. The administration of IL-20 to normal mice significantly increases only CFU-GEMM numbers and cell cycling. This is the first cytokine reported with such specificity. Because CFU-GEMM can be replated *in vitro* under appropriate cytokine conditions (SLF, –/+ cord blood plasma), suggesting limited self-renewal capacity for CFU-GEMM, it is possible that IL-20 may also have proliferative- and/or self-renewal-enhancing effects on stem cells. IL-20 binds to both IL-20 receptor (R) types I and II. IL-20R type 1 is composed of IL-20Rα and IL-20Rβ subunits; IL-20R type II is composed of the IL-20Rβ subunit and one subunit of the IL-22R.

Interestingly, although both IL-19 and IL-24 bind the murine Baf3 cell lines engineered to express Type I or Type II IL-20Rs and stimulate proliferation of the appropriate receptor-containing cells, they did not demonstrate an effect on murine or human myeloid progenitor cells. The IL-20R through which IL-20 is acting is not yet clear, nor is it clear how its mechanisms of action are mediated. Both IL-20RI and IL-20RII elicit intracellular signals using Stat3. Stat3 plays an essential role in maintaining innate immunity. It also is a signaling pathway involved in self-renewal signals induced by leukemia inhibitory factor in embryonic stem cells and, with the JAK2 pathway, promotes self-renewal in *Drosophila* germ-line stem cell spermatogonia divisions. When complemented by a second signal from either SLF/c-kit or FL/Flt3, Stat3 promotes self-renewal of primary multipotential hematopoietic cells.

Cell Cycle Checkpoints, Asymmetry of Division, and Self-Renewal

Self-renewal of hematopoietic stem cells is a poorly understood event. Little is known of the mechanisms mediating stem cell self-renewal. Self-renewal requires cell division without loss of stemness and pluripotentiality in at least one of the daughter cells. This is sometimes referred to as asymmetric cell division. Maintenance, expansion, or loss of stem cells will depend on the populations of stem cells produced or lost through symmetric or asymmetric divisions. The process by which cell fate determinants are segregated at cell division depends on polarization and segregation across the mitotic spindle; daughters of cell division can inherit similar or very different cellular contents. Proper functioning of the mitotic spindle and its relative positioning is extremely important for regulation of self-renewal. Cell cycle checkpoints are necessary to maintain the progression of cell division events in a linear and ordered fashion. Several cell cycle checkpoints have been described. The mitotic spindle assembly checkpoint (MSAC) ensures that the cell cycle does not progress from metaphase to anaphase until all paired sister chromatids are arranged properly across the metaphase plate. This alignment establishes the plane of division. It also establishes the plane of polarity necessary for cell-fate determinant segregation (e.g., self-renewal or differentiation). The MSAC is probably critical for proper regulation of self-renewal and differentiation, and mitotic checkpoint proteins may be involved.

The cyclin-dependent kinase modulator, p21cip1/waf1, has been implicated in the SLF (Steel Factor-stem cell factor (SCF)) synergistic stimulation of the proliferation of progenitor cells and in the functioning of stem cells. Synergistic stimulation of cells by combinations of cytokines is likely important to the *in vivo* functioning of both progenitor and stem cells. Of the cytokines that influence stem cell proliferation, self-renewal, or both, they have all been shown to work only or more efficiently when used with other cytokines, such as SLF. p21cip1/waf1, because of its role in SLF synergy, could be important for stem cell function. Because p21cip1/waf1 has been linked to proper functioning of the MSAC, it seems reasonable that the MSAC is involved in the proliferation, self-renewal, or both, of stem cells. Moreover, because p21cip1/waf1 is linked to cytokine synergy and cytokine synergy likely influences stem cell function, cytokines may act on stem cells through p21cip1/waf1 and the MSAC.

In addition to effects on growth, cytokines have been implicated in the survival–antiapoptosis of hematopoietic stem and progenitor cells. One such molecule is a cysteine, another amino acid, cysteine (CXC) motif chemokine, stromal cell-derived factor-1 (Sdf1)/Cxcl12, which signals and induces its activities through the receptor Cxcr4. It is possible that (Sdf1)/Cxcl12 can augment the *ex vivo* expansion of cord blood stem cells when used with one or more of the stem cell active growth factors.

Implications of Self-Renewal

Should any of the described growth factors, survival enhancing factors, or intracellular signaling molecules recently implicated in the regulation of hematopoietic, embryonic, or both types of stem cells prove useful in *ex vivo* expansion of hematopoietic stem cells, such activities may allow enhanced usefulness and broadness of applicability of cord blood transplantation.

Homing of Stem and Progenitor Cells

IMPORTANCE OF HOMING

A potential problem associated with *ex vivo*-expanded hematopoietic stem cells may be changes in the characteristics of the homing receptors, adhesion molecules, or both, on these cells that could occur during their time in cell culture. Assessment of the engrafting capability of *ex vivo*-cultured stem cells may underestimate stem cell numbers and activity in a Non Obese Diabetic–Severe Combined Immunodeficiency (NOD–SCID) or similar mouse model for human stem cells. Newer methods that use direct injection of the human cells into the marrow, rather than intravenous injection, may allow the detection of stem cells that cannot accurately home to the marrow. Such intrafemoral transplantation of NOD–SCID mice revealed a short-term repopulating human cord blood cell with a unique phenotype. Although this modification of the NOD–SCID human stem cell repopulating assay may allow a more accurate quantitation of the numbers of stem cells in the test inoculation, the lack of appropriate homing capacity in the NOD–SCID mice might indicate a potential problem with the homing capacity of these cells in human clinical transplantation. Thus, greater insight into the homing capacities and mechanisms involved would undoubtedly allow enhanced transplantation of donor cells in human recipients.

SDF1/CXCL12–CXCR4 AND CD26 IN HOMING OF STEM CELLS

It has been recently reported that stem cells home with absolute efficiency. This, however, appears highly unlikely and is not consistent with the work of several other groups or of actual clinical transplantation results. Perhaps only a subset of stem cells home with absolute efficiency. It is possible to greatly increase the homing and engrafting capacity of long-term marrow competitive engrafting and self-renewing mouse bone marrow stem cells by decreasing the dipeptidylpeptidase IV (DPPIV) activity of CD26, or by eliminating CD26, on hematopoietic stem and progenitor cells. DPPIV truncates Sdf1/Cxcl12 into an inactive form that does not have chemotactic activity but that can block the chemotactic activity of full-length Sdf1/Cxcl12. The Sdf1/Cxcl12–Cxcr4 axis has been implicated in chemotaxis, homing, and mobilization. Either inhibiting CD26 activity with a peptidase such as Diprotin A or Val-Pyr, or functionally deleting CD26 results in greatly reduced G-CSF-induced mobilization of progenitor cells. It is believed that the mechanism of this latter effect is caused by the inactivation or elimination of CD26, which blocks the truncating effect of CD26 on Sdf1/Cxcl12, thus allowing stromal cell-produced Sdf1/Cxcl12 to maintain a greater "holding" action on Cxcr4-expressing stem and progenitor cells.

Based on this belief, it was hypothesized that inactivation or elimination of CD26 on stem cells might enhance the homing capacity of exogenously infused stem cells. Experimental evidence supported this hypothesis as Diprotin A-treated wild-type mouse marrow cells or marrow cells from

CD26–/– mice demonstrated short-term homing and long-term marrow competitive repopulating capacities of these stem cells greater than the capacities of wild-type marrow cells treated with control medium or the capacities of untreated CD26+/+ marrow cells. It is possible that the enhanced marrow repopulating capability of stem cells in which CD26 is inactivated or eliminated may result partly from the enhanced stem cell survival activity of nontruncated–inactivated Sdf1/Cxcl12. Treatments such as inactivation or loss of CD26 activity that have the capacity to enhance the homing–engrafting capability of hematopoietic stem cells may allow greater engrafting capability with limiting numbers of stem cells, such as found in cord blood, and may also enhance the homing–engrafting capability of *ex vivo*-expanded stem cells. Either possibility would likely enhance the effectiveness of cord blood transplantation, potentially resulting in more routine cord blood transplantation for adults and possibly in the use of single cord blood collections for multiple transplant recipients.

Summary

The enhanced frequency and quality of hematopoietic stem cells in cord blood at the birth of a baby have endowed cord blood with the capacity to cure a variety of malignant and genetic disorders. Cord blood transplantation works in both children and adults, but it has been used mainly in children because of the apparent limiting numbers of stem–progenitor cells in single cord blood collections. Being able to *ex vivo* expand, increase, or both, the homing efficiency of stem cells from cord blood would increase the usefulness and applicability of cord blood for transplantation. Although attempts at *ex vivo* expansion of stem cells for clinical cord blood transplantation have been disappointing thus far, new information regarding factors and intracellular signaling molecules involved in the regulation of hematopoietic stem cell activity and in the growth of embryonic stem cells (which may be translatable to hematopoietic stem cells), as well as advancements in understanding and manipulating the homing capacities of stem cells, offers hope that we may soon be ready to enhance stem cell transplantation in general and cord blood stem cell transplantation in particular.

KEY WORDS

Chemokine A family of cytokines originally defined by their chemoattractant properties, but having additional functional activity on cells.

Cord blood transplantation Infusion of umbilical cord blood containing hematopoietic stem and progenitor cells into recipients to replace the blood cell system.

Cytokine Biologically active molecule that occurs naturally and influences the survival, proliferation, migration, or other aspects of immature or mature cells.

Hematopoietic progenitor cell The progeny of a stem cell that has little or no self-renewal capacity but is committed to produce mature blood cell types of one or more lineages (e.g., granulocytes, macrophages, erythrocytes, megakaryocytes, lymphocytes, etc.)

Hematopoietic stem cell A cell that can make more of itself (= self-renew) and give rise to all the blood-forming tissue.

FURTHER READING

Bagby, G. C., Jr., and Henrich, M. (1999). Growth factors, cytokines and the control of hematopoiesis. *In Hematology: basic principles and practice*, 3rd edition (R. Hoffman, S. Shattil, B. Furie, H. Cohen, L. Silberstein, and P. McGlave, eds.), pp. 154–201. Churchill Livingstone, Philadelphia.

Ballen, K., Broxmeyer, H. E., McCullough, J., Piaciabello, W., Rebulla, P., Verfaillie, C. M., and Wagner, J. E. Ballen. (2001). Current status of cord blood banking and transplantation in the United States and Europe. *Biol. Blood Marrow Transplant.* **7**, 635–645.

Broxmeyer, H. E. (2004). Proliferation, self-Renewal, and survival characteristics of cord blood hematopoietic stem and progenitor cells. *In Cord blood: biology, immunology, and clinical transplantation* (H. E. Broxmeyer, ed.), Chapter 1, pp. 1–21. Amer. Association of Blood Banking, Bethesda, MD.

Broxmeyer, H. E., and Smith, F. O. (2004). Cord blood hematopoietic cell transplantation. *In Thomas' hematopoietic cell transplantation*, 3rd ed. (Blume, K. G., Forman, S. J., Appelbaum, F. R., eds.), pp. 550–564. Blackwell Sciences, Malden, MA.

Broxmeyer, H. E., Srour, E. F., Hangoc, G., Cooper, S., Anderson, J. A., and Bodine, D. (2002). High efficiency recovery of hematopoietic progenitor cells with extensive proliferative and ex-vivo expansion activity and of hematopoietic stem cells with NOD/SCID mouse repopulation ability from human cord blood stored frozen for 15 years. *Proc. Natl. Acad. Sci. USA* **100**, 645–650.

Burns, C. E., and Zon, L. I. (2002). Portrait of a stem cell. *Developmental Cell* **3**, 612–613.

Christopherson, K. W., II, and Broxmeyer, H. E. (2004). Hematopoietic Stem & Progenitor Cell Homing, Engraftment, & Mobilization in the context of the CXCL12/SDF-1 — CXCR4 Axis. *In Cord blood: biology, Immunology, Banking, and clinical transplantation* (H. E. Broxmeyer, ed.), Chapter 4, pp. 65–86. America Association of Blood Banking, Bethesda, MD.

Morrison, S. J., Shah, N. M., and Anderson, D. J. (1997). Regulatory mechanisms in stem cell biology. *Cell* **88**, 287–298.

Smith, F. O., Srour, E. F., and Broxmeyer, H. E. (2004). Ex-Vivo Expansion and Gene Transduction of Cord Blood Stem Cells. *In Cord blood: biology, immunology, banking, and clinical transplantation* (H. E. Broxmeyer, ed.), Chapter 6, pp. 125–150. America Association of Blood Banking, Bethesda, MD.

Tao, W., and Broxmeyer, H. E. (2004). Towards a molecular understanding of hematopoietic stem and progenitor cells. *In Cord blood: biology, immunology, banking, and clinical transplantation* (H. E. Broxmeyer, ed.), Chapter 5, pp. 87–123. America Association of Blood Banking, Bethesda, MD.

Neurogenesis in the Vertebrate Embryo

Chris Kintner and Naoko Koyano-Nakagawa

Introduction

A long-standing objective in the field of developmental biology is to determine how the diverse cell types that comprise the central nervous system (CNS) are generated during embryonic development. This issue has been difficult to address because the CNS is comprised of different cell types such as neurons and glia and because the cellular composition of neural tissue varies enormously depending on its position along the body axis. Nonetheless, recent studies, mainly on the developing spinal cord, have revealed a rudimentary picture of the mechanisms that govern cell-type diversity in the vertebrate CNS. These studies reveal that many of these mechanisms act in the early embryo when neural precursors first arise.

In this chapter, we describe the early events that govern the formation of neural precursors and their differentiation into neurons in the developing vertebrate CNS. We first explain how the precursor cells for the CNS arise in the vertebrate embryo and how they differ from those for other developmental lineages. We next describe the role of the proneural genes as critical regulatory factors that promote the differentiation of neural precursors into neurons. Finally, we explain how the process of neural patterning may control the fate of neural precursors by regulating the activity of the proneural genes. According to the general model emphasized in this chapter, (1) neural precursors in the CNS are already restricted in their fate when they form in the embryo as a consequence of the patterning processes that specify their position along the neuraxes and (2) patterning genes trigger the proneural gene cascade at the proper time and place, thus determining patterns of neuronal differentiation. This model is likely to influence future studies of cell-type diversification in CNS development, with the goal of manipulating embryonic and adult stem cells to restore damaged neural tissue in a therapeutic setting.

Embryonic Induction and the Establishment of Neural Tissue

The progenitor cells for the vertebrate CNS first appear in development with the formation of the neural plate from a portion of the ectoderm (also known as the epiblast) during gastrulation (Figure 17-1). The neural plate subsequently forms a tube consisting of neuroepithelial cells (NECs) arranged around a central lumen that extends along the anterior-posterior axis (Figure 17-2A, B). In addition to the neural plate, the ectoderm also forms the neural crest, a migrating population of precursors that move throughout the embryo, generating both neural and nonneural cell types. In addition, ectodermal cells produce placodal structures that contain neural precursors for sensory ganglia as well as neurons in the ear and nose. Finally, other regions of the ectoderm contain precursors for nonneural tissue, most prominently in ventral regions where they generate the skin. Thus, neural precursors arise in the vertebrate embryo at gastrulae stages when the ectoderm is subdivided into regions with different developmental fates, a process governed by inductive tissue interactions between the ectoderm and another region called the organizer.

In their classic experiment, Mangold and Spemann showed that the ventral ectoderm of a host embryo could be induced to form a complete nervous system when exposed to a transplanted piece of tissue called the *organizer*. Subsequently, this process, called *neural induction*, has been described in chick and mouse embryos, suggesting that it is a key feature of neural tissue formation in all vertebrates, including, by extension, human embryos. Additional embryological experiments were subsequently instrumental in showing that neural tissue is specified during neural induction by two sets of signals. One set neuralizes the ectodermal cells, thus causing them to form nerve cells rather than skin cells. A second set patterns the neuralized ectoderm, thus determining subregions that will form nerve cells of a brain type or spinal cord type, for example. As these two sets of signals have been identified and studied, it has become clear that neuralization and patterning are intimately linked. Indeed, the idea that a generic neural lineage exists is likely to be misleading because position plays such a prominent and early role in specifying cell fate in the CNS. To explain this, one needs to consider how neuralizing and patterning signals act during neural induction.

Neuralization of the Ectoderm

In amphibians, the ectoderm can be easily isolated from blastula-stage embryos and placed in culture, where it differentiates into skin but not into neural tissue. However, dissociating the isolated ectoderm into individual cells has been known since Holtfreter's experiments to "induce" neural differentiation, suggesting that ectodermal cells can generate neural precursors even in the absence of inductive signals from

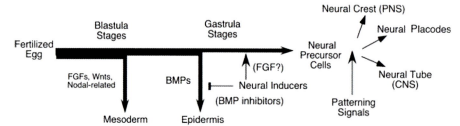

Figure 17-1. *Default model of neural induction.* Following fertilization, a region of the early embryo generates the ectoderm or epiblast, which responds to patterning signals as development progresses from left to right. At blastula stages, these signals include those that induce mesodermal derivatives in posterior regions of the embryo using growth factors such as FGF, Wnts, and nodal-related families. At gastrulae stages, the ectoderm on the ventral side is induced to become epidermis by BMP signaling. However, a region of ectoderm avoids BMP signaling through inhibitors produced by the organizer, producing neural tissue. This neural tissue responds to a variety of patterning signals that divide it into different neural fates.

the organizer. A molecular understanding of this phenomenon was uncovered in *Xenopus* embryos during the study of signaling molecules that play prominent roles in axis determination. One of these molecules is the bone morphogenetic proteins (BMPs): members of the TGF-β superfamily of growth factors that play a key role in patterning the embryo along the dorsal-ventral axis. Surprisingly, the inhibition of BMP signaling has been found to be the critical event required for converting ectoderm into neural tissue (Figure 17-1). When reagents that block the BMP signaling pathway are introduced into isolated ectoderm, they effectively convert it into neural tissue. Conversely, adding back BMPs as soluble ligands to dissociated ectodermal cells effectively blocks neural differentiation and promotes the formation of epidermal tissue.

Finally, BMP inhibitors such as noggin, chordin, and follistatin have been identified that bind and antagonize BMP signaling extracellularly. These inhibitors are potent neural inducers, are expressed at quite high levels in organizer tissues, and underlie the molecular basis of organizer activity revealed in Mangold and Spemann's experiment. These observations have led to the so-called default model for neural induction in which ectoderm is neuralized when inhibitors produced by the organizer block BMP signaling before and during gastrulation (Figure 17-1).

The default model also takes into account that growth factor signaling is required for the production of other embryonic cell lineages, such as those that generate mesodermal derivatives. Induction of mesodermal derivatives occurs before gastrulation, mediated by different families of growth factors such as the Wnts, the fibroblast growth factors (FGFs), and the nodal-related members of the TGF-β superfamily (Figure 17-1). Significantly, ectoderm can be induced to produce mesodermal tissue if exposed to these factors at the appropriate stage. Thus, embryonic cells may only generate neural tissue if they avoid a series of signaling events that promote their differentiation along nonneural lineages (Figure 17-1).

A major caveat to the default model is that other signaling pathways may act during neural induction to neuralize the ectoderm. For example, genetic experiments in mice show that some neural tissue forms even when neural induction has been disabled by mutations in the BMP inhibitors or by

removal of the organizer. This remaining neural tissue is an indication that additional pathways operate in embryos to specify neural tissue. Other results in chick experiments suggest that FGF is more effective than BMP inhibitors at inducing neural tissue in epiblast cells adjacent to the neural plate. In *Xenopus* embryos, the ectoderm also forms neural tissue when exposed to FGF at the appropriate stage, although whether or not FGF is normally required for neural induction remains controversial. FGF action in this case may be mediated through Smad10, whose function is critical for the formation of neural precursors in *Xenopus* embryos. In addition, FGF signaling is required for maintaining neural precursor cells in culture and for regulating their differentiation *in vivo*. Together, these results point to a role for FGF signaling in the formation of neural precursors as a means of regulating their differentiation.

The Wnt signaling pathway has also been implicated in the formation of neural precursors in early embryos. Reagents that block Wnt signaling in frog embryos antagonize neural tissue formation, and blocking Wnt signaling in chick embryos allows the epiblast to respond to FGF and form neural tissue. Inhibiting Wnt activity in ES cells also potentiates neural cell formation. In summary, these results may indicate that the inhibition of BMP signaling is not sufficient for embryonic cells to form neural tissue. Nonetheless, they do not necessarily mean that the default model is incorrect. With further study, for example, these other pathways may contribute to neural induction in the same way as the BMP inhibitors: by preventing BMP activity or the activity of other signals that promote a nonneural state.

Neural Patterning

The neuroectoderm of the neural plate produces the NECs of the neural tube, thus forming the neural progenitors that will generate the various neurons and glia that comprise the CNS. As the neural tube forms, the NECs are morphologically homogeneous, perhaps giving the mistaken impression that they are generic neural precursors at this stage. To the contrary, their homogeneous appearance belies NECs' status as an already diverse population of progenitor cells as a result of neural patterning. To illustrate this point, the following

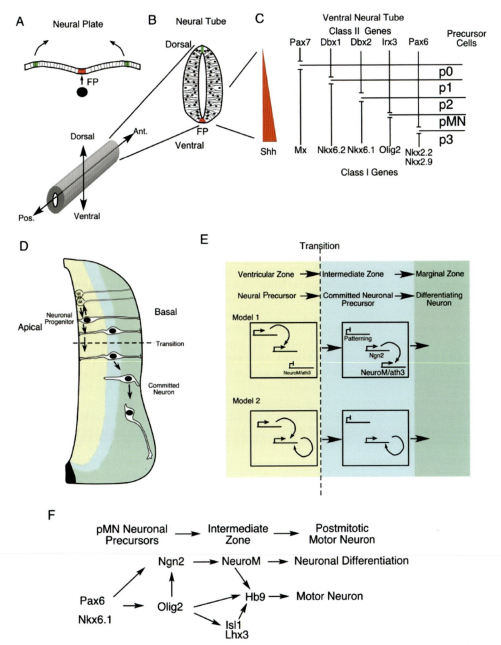

Figure 17-2. *Patterning and neurogenesis in the ventral spinal cord.* (A) The neural plate forms during gastrulation as a thickening of the ectoderm into a neuroepithelium, at which time patterning along the dorsal-ventral axis begins with the establishment of the floor plate (FP) and the roof plate (RP). (B) As the neural plate forms a neural tube, the NECs are patterned by signals emanating in part from the floor and roof plates. (C) In ventral spinal cord, this patterning consists of a gradient of Shh secreted by the floor plate, which activates or represses the expression of the class I and II homeodomain proteins (one exception is Olig2, a bHLH repressor) in a concentration-dependent fashion. MX is a hypothetical protein that has been proposed to contribute to ventral patterning by analogy with other class I proteins. (D) Neurogenesis within the neural tube is organized along a third developmental axis that corresponds to the apico-basal orientation of the neuroepithelium. Progenitor cells consist of NECs localized along the apical surface of the ventricular zone (yellow) within which their nucleus transverses during the cell cycle. During cell division, the two daughter cells separate at the apical surface, with recent studies suggesting that they maintain contact with the basal surface. A precursor undergoes neuronal differentiation when its nucleus migrates laterally (blue), exiting the cell cycle. Terminal neuronal differentiation is likely to be completed when a precursor delaminates from the neuroepithelium by detaching from ventricular surface, and migrating laterally into the marginal zone (green). (E) Expression of proneural genes during the different phases of neurogenesis as shown by shading used in part D. Proneural genes such as the neurogenins are expressed in dividing precursors within the ventricular zone (yellow). Neuronal commitment occurs in the intermediate zone (blue) when the levels of neurogenin are sufficiently high (in model 2) or when a downstream bHLH gene is activated (model 1). In either case, commitment causes the precursor to make the transition to a postmitotic neuron (blue) that undergoes terminal neuronal differentiation (green). (F) Integration of neuronal differentiation with neuronal subtype specification. In the ventral spinal cord, patterning leads to expression of both the proneural protein Ngn2 and HD transcription factors, which cooperate to activate the expression of the motor neuron determinate HB9. This cooperation integrates a program of neuronal differentiation promoted by Ngn2, and perhaps NeuroM, along with a program of motor neuron differentiation. (Part C is adapted from Shirasaki *et al.*).

141

description will focus on the spinal cord, where perhaps the most is known about how patterning influences the formation and fate of neural precursors.

As NECs form the neural plate, they are exposed to a variety of signals that specify their position within the nervous system along two major cardinal axes. In the spinal cord, one of these axes, dorsal-ventral (D-V), depends largely on signals produced by two specialized midline structures, one at the ventral pole of neural tube (the floor plate) and the other at the dorsal pole (the roof plate) (Figure 17-2B). By acting as morphogens, the signals produced by these so-called organizing centers subdivide the NECs of the neural tube into domains with different developmental fates. Specifically, the floor plate cells secrete a protein, called Sonic hedgehog (Shh, Figure 17-2B), which induces or suppresses at different concentration thresholds the expression of genes, usually those encoding homeodomain (HD) transcription factors, in NECs lying at different positions in the ventral spinal cord (Figure 17-2C). In this manner, the gradient of Shh activity subdivides the ventral NECs into at least five distinct areas by activating or suppressing the expression of different transcription factors, which then sharpen into nonoverlapping zones by cross-repression (Figure 17-2C). NECs in one of these zones (pMN, Figure 17-2C) produce somatic motor neurons, and those in the other zones produce various classes of interneurons. These patterning events along the D-V axis take place in the context of a similarly complex patterning of the neural tube along a second orthogonal anterior-posterior (A-P) axis that also begins at neural plate stages.

Although the signals and target genes mediating this patterning are less well understood, it is clear that they intersect with D-V patterning to generate a Cartesian coordinate system in which NECs express a unique code of transcription factors that determine cell fate at each point along the neuraxes. This code, for example, ensures that motor neurons form in response to Shh in a ventral domain along the entire spinal cord but that different motor neuron subtypes form at each A-P axial level. In summary, neural patterning of the ventral spinal cord as well as other regions of the neural tube sets up a diverse pattern of gene expression within NECs that is already apparent when they form at neural plate stages. This pattern of gene expression is thought to be a major determinant of NEC fate, dictating when and where neurons and glia form.

Proneural Gene Cascade: A Downstream Target of Neural Patterning

How then does patterning of the NECs described previously dictate precise patterns of neuronal differentiation? The key finding addressing this question has come from the discovery of a class of basic helix-loop-helix (bHLH) proteins encoded by the so-called proneural genes. As transcriptional activators, the proneural proteins are thought to activate gene expression necessary both for the differentiation of precursors into neurons and for neuronal cell-type specification, thus acting as a molecular switch of differentiation (Table 17-1).

The vertebrate proneural bHLH genes fall into two families based on the homology to bHLH genes originally identified in *Drosophila* as mutations that block neural differentiation (Table 17-1). One smaller family consists of those related to the *Drosophila* achaete-scute genes, such as *Mash1*. The second, larger family encodes proteins related to *Drosophila* atonal and can be subdivided structurally into three subfamilies: the neurogenin (Ngn)-like, the NeuroD-like, and the atonal-like. Expression of these different subfamilies occurs in precise spatial and temporal patterns both within the dividing NECs and within cells that have initiated neuronal differentiation. When eliminated by targeted mutation, loss of specific proneural bHLH genes results in deletion of specific populations of neurons. However, the loss of neurons is likely to be much more severe when multiple members are simultaneously eliminated, indicating that the proneural genes have overlapping function as found in *Drosophila*. Because of this genetic redundancy, it is difficult to test experimentally whether all neuronal differentiation is driven by proneural gene action. However, in gain-of-function experiments, proneural proteins are potent inducers of neuronal differentiation when ectopically expressed not only in neural precursors but also in some nonneural tissues.

The proneural proteins function to initiate many of the physiological changes that occur when NECs undergo terminal neuronal differentiation (Table 17-1). One such function is to promote cell cycle exit, an irreversible set of events incurred by all NECs as they form neurons. NECs initiate cell cycle exit when their nuclei move to the lateral edge of ventricular zone, where they enter the G$_0$ phase and eventually delaminate out of the neuroepithelium (Figure 17-2D). Ectopic expression of the proneural proteins in NECs or in tissue culture models of NECs causes rapid cell cycle arrest, although some subtypes of proneural proteins promote this transition better than others.

The mechanism by which the proneural proteins initiate irreversible cell cycle exit appears to be quite complex and is an area of active research. This mechanism may involve direct protein–protein interactions with cell cycle machinery or alternatively transcriptional changes in expression of genes that encode cell cycle regulators, such as cyclin-dependent kinase inhibitors p21, p57, and p27. Proneural proteins also function to activate the expression of genes associated with all subtypes of neurons, such as those that encode neuronal isoforms of the cytoskeletal proteins, channels involved in membrane excitability, and proteins involved in axon guidance. Significantly, genes encoding the proneural proteins are expressed in neural precursors and in committed neurons but often are transient in expression and lost as neurons mature. Thus, proneural proteins may initiate expression of panneuronal genes directly and then maintain expression indirectly by activating a downstream transcriptional network. This network may include not only transcriptional activators of the neuronal genes but also transcriptional repressers that relieve the repression of neuronal genes. For example, transcriptional enhancers for many of the panneuronal differentiation genes contain binding sites for a repressor, called REST/NRSF,

TABLE 17-1
Known Functions or Expression Patterns of bHLH Proteins in Neuronal Differentiation

Group	Subgroup	bHLH name	Comments
Acheate-scute	Acheate-scute-like	Mash1	Required for autonomic neuronal differentiation
			Autonomic and olfactory neuron differentiation
			Coordination of differentiation in ventral forebrain
			Determination gene in olfactory sensory neurons
			Promotes neuronal fate and inhibits astrocytic fate in cortical progenitors
			Negative indirect autoregulation of the promoter
			Retinal development
		Xash1	*Xenopus* homolog of Mash1, expressed in anterior regions of the CNS
		Cash1	Chick homolog of Mash1 and Xash1; has similar expression pattern
		Xash3	Only in *Xenopus*, early neural plate expression
		Cash4	Only in chick, early proneural gene in posterior CNS
Atonal	Nscl-like	Nscl1, Nscl2	Expressed during "late" phases of neuronal commitment
	Neurogenin-like	Xngn1	Promotes neurogenesis in both neuroectoderm and ectoderm; overexpression in developing embryos induces various downstream targets such as Ath-3, Xcoe2, Delta, MyT1, NeuroD, Tubulin, Neurofilament, Hes6, XETOR, and NKL
		Ngn1/NeuroD3/ Math4C	Sensory lineage
			Required for proximal cranial sensory ganglia
			Neurogenesis in developing dorsal root ganglia
			Specification of dorsal interneurons by cross-inhibition with Math1
			Differentiation in olfactory sensory neurons
			Inhibits gliogenesis
		Ngn2/Math4A	Required for epibranchial placode-derived cranial sensory ganglia
			Promotes neuronal fate and inhibits astrocytic fate in cortical progenitors
			Induced by and cross-regulates and Pax6
		Ngn3	Promotes gliogenesis in the spinal cord
	NeuroD-like	NeuroD/β2	Converts *Xenopus* ectoderm into neurons
			Cell fate, determination, differentiation, and survival in neural retina
			Required for differentiation of the granule cells in cerebellum and hippocampus
			Survival of inner ear sensory neurons
			Neurite outgrowth
		NeuroD2/NDRF	Required for development and survival of CNS neurons
		Math2/Nex1	Expressed in postmitotic cells of the brain
			Induces differentiation of PC12 cells and expression of *GAP-43* gene
		Xath2	Expressed in postmitotic cells of stage 32+ *Xenopus* dorsal telencephalon
		Math3/NeuroD4/ NeuroM	Expressed in transition stage in neurogenesis
			Amacrine cell specification in the retina
			Cooperates with Lim-HD proteins to specify motor neurons
		Xath3	Converts ectoderm into a neural fate
			Promotes sensory neuron marker expression
	Atonal-like	Math1	Required for cerebellar granule neuron development
			Required for generation of inner ear hair cells
			Required for proprioceptor pathway development
			Specification of dorsal interneuron subpopulation
		Xath1	Expressed in hindbrain; induces neuronal differentiation in ectoderm
		Math5	Promotes retinal ganglion cell fate through brn-3b
			Retinogenesis, regulated by Pax6
		Xath5	Retinal ganglion cell fate
			Regulates neurogenesis in olfactory placode
		Math6	Promotes neuronal fate at the expense of glial fate
Olig	Olig-like	Olig1, Olig2	Specification of motor neurons and oligodendrocytes
			Motor neurons specification in combination with ngn2
		Olig3	Transiently expressed in different types of progenitors of embryonic CNS
E12	E12	E12/E47	Dimerization partner of various bHLH proteins

which acts to extinguish the expression of these genes in nonneuronal cells as well as in neuronal precursors. This repression is presumably blocked in neurons by a mechanism involving the proneural proteins.

Another function associated with the proneural proteins during neuronal differentiation is the inhibition of gene expression required for astroglia or oligodendrocyte differentiation. Neural precursors first generate neurons and then switch to produce both types of glia at later stages, suggesting that glial differentiation genes need to be repressed in neural precursors during neurogenesis. Studies using cultured neural stem cells indicate that the proneural proteins inhibit astroglia differentiation in neural precursors not by binding DNA but by competing for critical coactivators required to induce the expression of glial genes such as glial fibrillary acidic protein. In addition, proneural proteins can interfere with growth factor induction of glial differentiation by binding to and inhibiting components of the CNTF signaling pathway. Ectopic expression of proneural proteins also suppresses the formation of oligodendrocyte precursors that normally arise within discrete regions of the neural tube after neurogenesis is largely complete. Thus, one function of the proneural proteins is to prevent cells from expressing genes necessary for glial differentiation while activating those required for neuronal differentiation.

Finally, the proneural proteins are involved not only in promoting changes associated with generic neuronal differentiation but also in activating gene expression required for neuronal subtype specification. Since proneural proteins fall into several subfamilies with distinct sequence differences, one possibility is that a given subfamily is specialized to promote the differentiation of a particular type of neuron. Indeed, in *Drosophila* there is strong evidence that the achaete-scute class of proneural proteins induces one type of external sense organ and the atonal class induces another. Similar differences have been described for vertebrate proneural proteins, suggesting that they are designed partly to activate different downstream targets associated with neuronal cell-type specification. The best-understood example of this occurs during the specification of motor neurons using the expression of an HD transcription factor called HB9 (Figure 17-2F). Expression of HB9 is activated only where neural precursors in the ventral neural tube exit the cell cycle and produce motor neurons (intermediate zone in Figure 17-2D).

Analysis of the enhancer required for this activation reveals an element with closely aligned binding sites for proneural proteins as well as for two HD proteins, Islet1 and Lhx2, known to be required for motor neuron differentiation. Binding of these factors cooperatively activates expression of HB9, thus driving motor neuron differentiation. Significantly, although some proneural proteins can cooperate to activate HB9 expression, others cannot. Similar links have been made between the patterning of NECs and the expression of proneural proteins in the dorsal spinal cord. In this case, neighboring domains of NECs produce different classes of interneurons by expressing distinct members of the proneural bHLH family. Thus, these observations strongly suggest that

proneural proteins function to execute generic neuronal differentiation as well as to activate the downstream targets genes needed for neuronal cell-type specification.

Potential Links Between Neural Patterning and Neurogenesis Control

Because the proneural proteins behave as a molecular "switch" that promotes neuronal differentiation, the way their activity is regulated has important consequences for determining the fate of NECs. In some cases, the key element in this switch is a bHLH cascade in which the expression of one class of proneural protein in NECs can trigger neuronal differentiation by activating the expression of a downstream proneural gene (Figure 17-2E, model 1).

Alternatively, the key element may be in the form of a threshold in which only high levels of proneural gene expression in NECs are sufficient to trigger neuronal differentiation (Figure 17-2E, model 2). In either case, sufficiently high activity of proneural proteins in NECs promotes exit from the cell cycle and terminal neuronal differentiation. Conversely, if the activity or expression of proneural proteins is inhibited, NECs seem to revert to a ground state in which they have the option to divide and either become a neuron at a later time or serve as the source of progenitor cells for various glia at even later stages (Figure 17-2D). Thus, proneural protein activity is not only a key factor in determining the onset and duration of neurogenesis but is also key in maintaining proper balance between the number of NECs that undergo terminal neuronal differentiation and the number that are retained in a progenitor mode, thus maintaining a progenitor cell pool for later-born neurons or for glia. Not surprisingly, many of the factors that control the fate of NECs seem to converge on the expression or activity of the bHLH proteins, including the patterning genes described previously.

Analysis of the enhancer that drives expression of the proneural gene, *Ngn2*, in the spinal cord has revealed several discrete elements responsible for different spatial and temporal expression patterns in NECs. These elements are likely to be driven by transcription factors whose expression is spatially restricted in NECs during neural patterning. For example, in the ventral spinal cord, the first neurons to be generated are motor neurons, and their generation in chick spinal cord is correlated with the early expression of Ngn2 within a narrow ventral domain of NECs. As already explained, this region of NECs is patterned by Shh signaling, which induces the expression of a key transcription factor, called Olig2, within the motor neuron-producing area of ventral NECs (Figure 17-2C and F). When ectopically expressed in the embryonic spinal cord, Olig2 induces ectopic motor neuron differentiation and does so partly by inducing ectopic and precocious expression of Ngn2. Significantly, motor neurons arise in response to ectopic Olig2, with kinetics similar to those they normally do in the ventral neural tube. Expressing high levels of Ngn2 along with Olig2 short-circuits this time course, resulting in rapid motor neuron differentiation. Thus, the interactions among the patterning gene, *Olig2*, and the proneural protein, *Ngn2*, seem to be key in promoting motor neuron differentiation. Since *Olig2* is a bHLH repressor, its regulation of *Ngn2* expression seems to be indirect, perhaps through the regulation of inhibitors of proneural gene expression such as those described in later sections of this chapter.

Thus, the general emerging principle is that the fate of NECs during neurogenesis is established using interactions between patterning genes and proneural proteins. The remaining challenge is to determine how the bHLH cascade is engaged in a myriad of ways to produce the appropriate number and types of neurons that comprise each region of the CNS along the neuraxes. This challenge, though daunting in its complexity, is likely to revolve around the large number of factors that seem to regulate the expression or activity of the proneural proteins.

Regulation of Proneural Protein Expression and Activity

One striking feature of the bHLH proteins is their ability to feedback and autoactivate expression of themselves or to activate a downstream bHLH gene (Figure 17-2E). As a result, direct or indirect changes in the strength of this positive feedback loop is one avenue that can be exploited during the process of patterning to control neuronal differentiation (Figure 17-3). The following section reviews some of the prominent regulators of the bHLH cascade that have been described and are likely to be the focus of future research in this area.

Members of the Id family of bHLH proteins contain a dimerization domain but are unable to bind DNA. Since the

proneural proteins bind DNA as heterodimers with the ubiquitously expressed E proteins, they are inactivated when they instead form nonfunctional dimers with the Id proteins. Targeted mutations in Idl and Id3 causes premature neuronal differentiation in mice, demonstrating that these proteins negatively regulate the differentiation of neural precursors, most likely by inhibiting the activity of the proneural proteins. The factors critical in regulating the expression of Id proteins are not known, but one potentially significant input is repression of these genes by the patterning genes that promote neurogenesis. In addition, expression of these genes is likely to be a target of the Notch signaling pathway, which plays a prominent role in regulating neurogenesis, as described later in this chapter.

A well-established family of proteins that negatively regulate neurogenesis are the bHLH transcriptional repressors called Esr, Hes, Her, Hrt, and Hey, depending on their species of origin and the subfamily classification of their structure. Many of these genes were isolated and named based on homology to genes in *Drosophila*, called Hairy or Enhancer of Split, which play important roles in regulating proneural genes during fly neural development. Functional analysis of these bHLH repressors shows that they potentially antagonize the activity of the proneural proteins by several mechanisms: interacting directly protein to protein, competing with the proneural proteins for their binding sites (the E-box) in DNA, or binding distinct DNA elements (the N-box or high-affinity repressor sites) in enhancers targeted by the proneural proteins.

An extremely large body of literature highlights the importance of bHLH repressors as potent regulators of the proneural proteins during neurogenesis. For example, targeted muta-

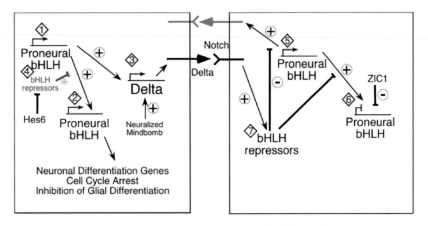

Differentiating Neuron   **Inhibited Precursor**

Figure 17-3. *Possible points of regulation of the bHLH cascade by patterning genes.* Differentiation of NECs depends on the bHLH cascade whose activity can be regulated in ways that promote neuronal differentiation, as shown on the left, or which keep the cell in a precursor state, as shown on the right. The bHLH proteins activate the expression of Delta (left panel), thus inhibiting neuronal differentiation non-cell autonomously by activating the Notch receptor in neighboring cells (right panel). Products encoded by the patterning genes can potentially influence the activity of the proneural proteins in a variety of ways, as indicated by the numbers enclosed in diamonds. They can promote neuronal differentiation by promoting the expression of the neurogenins (1), the downstream bHLH (2), or the expression or activity of Delta (3), thereby enhancing lateral inhibition. Alternatively, neuronal differentiation can be enhanced by inhibiting the activity of proteins such as the bHLH repressors (4) that inhibit the activity of the bHLH proteins and thus neuronal differentiation. Patterning genes may prevent neuronal differentiation by inhibiting the activity of the neurogenins (5) or the downstream bHLHs (6), as shown for Zic1. Finally, patterning may inhibit neurogenesis by promoting the expression of the bHLH repressors (7).

tions of Hes1 or Hes5 in the mouse result in precocious and increased numbers of precursors undergoing neuronal differentiation, and ectopic expression of these factors in *Xenopus* or zebra fish strongly inhibits neurogenesis in gain-of-function experiments. In some cases, the bHLH repressors seem to regulate neurogenesis within relatively uniform domains of NECs. In other cases, the expression of the repressor bHLH proteins is controlled by the Notch signaling pathway during a local patterning process, called *lateral inhibition*, which influences the ability of NECs to undergo differentiation.

During lateral inhibition, the expression of the bHLH repressors is likely to be directly regulated by the Notch signal transduction pathway through binding sites for a DNA binding protein referred to here as Suppressor of Hairless, or Su(H). In the absence of Notch signaling, Su(H) acts as a transcriptional repressor thought to actively inhibit the expression of the bHLH repressors. However, upon Notch receptor activation by ligand binding, the Notch intracellular domain (ICD) is released from the membrane, moves to the nucleus, and converts Su(H) from a repressor into an activator, thus rapidly inducing gene expression. As a consequence, activating the Notch pathway induces the expression of repressor bHLH genes and thereby inhibits neurogenesis, and inhibiting the Notch pathway enhances the levels of neuronal differentiation within a pool of neural precursors. Significantly, the proneural proteins are potent activators of a least one Notch ligand related to *Drosophila* Delta. Thus, proneural proteins not only promote neuronal differentiation cell autonomously but also, by activating Delta, inhibit neuronal differentiation in their neighbors non-cell autonomously (Figure 17-3).

The interaction of the proneural proteins with the Notch pathway appears to be a critical factor in determining the number of neurons generated from NECs within a given region of the neural tube. As a result, one can imagine a scenario where patterning genes act by targeting the activity of the Notch pathway, perhaps by targeting several proteins known to be Notch modulators. Activity of the Notch receptor, for example, is modulated by post-translational modification mediated by glycosyltransferases encoded by the vertebrate homologs of the *Drosophila* Fringe gene. The Fringe homologs are dynamically expressed within neural precursor populations, where they may influence the activity of the Notch pathway. Another Notch modulator expressed in neural precursors is a small ankyrin repeat protein, called NRARP, which promotes the turnover of Notch ICD. Indeed, numerous mechanisms have been proposed to change the half-life of Notch ICD, thus altering the efficacy of Notch activity. Finally, another potential mechanism of modulating Notch activity is by changing the activity, expression, or both of the ligands. In this respect, an important factor in controlling ligand activity appears to be their removal from the cell surface following ubiquitination by specific E3 ligases. In all, modulation of Notch activity is likely to be one way in which the output of the patterning genes could target the activity of the proneural proteins during neurogenesis.

In addition to negative regulators of proneural protein activity, the patterning genes could influence neurogenesis by regulating the expression of genes whose products promote the activity of proneural proteins. For example, proneural proteins induce the expression of a bHLH protein, called Hes6, which is distantly related to the repressor bHLH proteins described previously. As a target of the proneural proteins, Hes6 is expressed ubiquitously in neural precursors in regions where neuronal differentiation occurs within neurogenic epithelium. However, in contrast to the other repressor bHLH proteins, Hes6 promotes neurogenesis in ectopic expression experiments, and it seems to do so by antagonizing the activity of the repressor bHLH proteins. Thus, targeting Hes6, a repressor of repressors, could conceivably be a means of regulating the efficacy of the bHLH cascade. A similar scenario applies to the HLH proteins called EBF/Olf-1/Coe, whose expression is activated in neural precursors by the proneural bHLH proteins and which can promote neurogenesis in some assays. How these transcription factors modulate the activity of the proneural proteins is not known, but their expression is a potential target of regulation by patterning genes.

Another significant class of transcription factors that may link patterning and the bHLH cascade falls into a family of related Krüppel-like C2H2 zinc-finger proteins, including Glil-3, Zicl-5, and Nkl. The *Gli* genes are the vertebrate homologs of *Drosophila Cubiutis Interruptus*, the downstream transcriptional mediators of Shh. Given the importance of Shh signaling in regulating neurogenesis in the spinal cord as well as in other regions of the CNS, the Gli proteins are likely to have a role in regulating the bHLH cascade. Indeed, both the Gli proteins and the closely related Zic and NKL proteins have been shown to have positive and negative effects on neurogenesis when overexpressed in *Xenopus* embryos. The mechanism by which these transcription factors regulate neurogenesis is mostly unknown. A major exception was revealed by analysis of a proneural gene in chick and mouse called *Mathl* and *Cathl*, respectively. *Cathl/Mathl* is expressed in NECs in the dorsal neural tube where it drives the differentiation of dorsal interneurons. The neuronal enhancer of *Mathl* contains a site for autoregulation as well as a binding site for Zicl, which inhibits the activity of the autoregulatory site. In this manner, Zicl prevents *Mathl* from activating its own promoter and inducing neuronal differentiation. How Zicl is regulated in this case is not fully understood, but interestingly, the expression of other proneural genes does not appear to respond to Zicl in the spinal cord. Thus, the Gli/Zic/Nkl family of proteins could contribute to the temporal and regional regulation of the proneural proteins as a downstream consequence of patterning.

The patterning of NECs may also result in changes in proneural protein activity by mechanisms involving post-translational modifications. For example, bHLH proteins can be regulated by phosphorylation. A specific example of this regulation has been demonstrated for the proneural protein, NeuroD, in *Xenopus* embryos. *Xenopus* NeuroD contains a consensus phosphorylation site for the regulatory kinase, GSK3β, which, when mutated, dramatically changes NeuroD's ability to promote neuronal differentiation. One possibility is that GSK3β regulates this by using Wnt signaling, thereby changing the efficacy of proneural activity. Proneural activity might also be regulated by targeted protein turnover using degradation by the ubiquitin–proteasome pathway. Although this form of regulation has not been examined thoroughly in the context of neurogenesis, it is likely that proneural proteins, like other bHLH transcription factors, will be targeted by ubiquitin ligases for degradation in a regulated manner. Finally, a relatively new and exciting level of regulation is likely to occur at the level of RNA. The recent identification of small, interfering microRNAs provides a compelling means of coordinated regulation of gene expression during differentiation. Finally, regulation of RNA activity during neurogenesis may occur through RNA-binding proteins, many of which are expressed in neural precursors in response to proneural gene activity. Future work will undoubtedly uncover additional links with the patterning of neural precursors and the regulation of proneural activity at both the RNA and protein levels.

Summary

The development of the CNS can be represented as a series of fate choices progressively made by embryonic cells in response to both intrinsic and extrinsic cues. One of the first fate decisions is made in the ectoderm, where cells form the NECs of the neural plate rather than differentiating into non-neural tissues. This choice apparently can occur by default, suggesting that embryonic cells can form NECs in the absence of extrinsic instructions. However, a key process during neural induction is neural patterning during which a complex network of gene expression is established along the neuraxis, thereby specifying the position and subsequent fate of NECs. These complex genetic networks, many of which involve HD transcription factors, dictate patterns of neurogenesis by controlling when and where NECs undergo neuronal differentiation. Significantly, the patterning genes appear to regulate neurogenesis by converging on the activity of the proneural bHLH proteins, which function as molecular switches to initiate neuronal differentiation by promoting cell cycle arrest, expression of neuronal differentiation genes, suppression of glial differentiation genes, and activation of neuronal subtype genes. Thus, the neural precursors for the CNS initially choose their fate by default, but neural patterning is instrumental in instructing their subsequent neuronal fate by establishing a complex code of gene expression that drives the bHLH cascade at the proper time and place.

KEY WORDS

Neural Induction   Early events in the vertebrate embryo that lead to the formation of the neuroepithelium of neural tube, thus creating the progenitors cells that give rise to the neurons and glia comprising the central nervous system.

Neural patterning   Developmental processes in which neural precursors are endowed with a positional identity, thus enabling them to give rise to subtypes of neurons that comprise distinct regions of neural tissue along the anterior-posterior and dorsal-ventral axis of the nervous system.

Neurogenesis   The events that occur when neural precursors leave the cell cycle and activate a program of terminal neuronal differentiation.

Proneural proteins   A class of basic–helix–loop–helix transcription factors that are known to play critical roles in promoting neurogenesis within progenitor cells.

FURTHER READING

Anderson, D. J. (2001). Stem cells and pattern formation in the nervous system; the possible versus the actual. *Neuron* **30**, 19–35.

Barolo, S., and Posakony, J. W. (2002). Three habits of highly effective signaling pathways: principles of transcriptional control by developmental cell signaling. *Genes Dev.* **16**, 1167–1181.

Bertrand, N., Castro, D. S., and Guillemot, F. (2002). Proneural genes and the specification of neural cell types. *Nat. Rev. Neurosci.* **3**, 517–530.

Hamburger, V. (1988). *The heritage of experimental embryology: Hans Spemann and the organizer.* Oxford University Press, Oxford.

Harland, R. (2000). Neural induction. *Curr. Opin. Genet. Dev.* **10**, 357–362.

Helms, A. W., and Johnson, J. E. (2003). Specification of dorsal spinal cord interneurons. *Curr. Opin. Neurobiol.* **13**, 42–49.

Jessell, T. M. (2000). Neuronal specification in the spinal cord: inductive signals and transcriptional codes. *Nat. Rev. Genet.* **1**, 20–29.

Mumm, J. S., and Kopan, R. (2000). Notch signaling: from the outside in. *Dev. Biol.* **228**, 151–165.

Schweisguth, F. (2004). Regulation of Notch signaling activity. *Curr. Biol.* **19**, R129–138.

Streit, A., and Stern, C. D. (1999). Neural induction. A bird's eye view. *Trends Genet.* **15**, 20–24.

18

The Nervous System

Lorenz Studer

Introduction

Over the last few years embryonic stem (ES) cells have emerged as a powerful tool to study brain development and function. The recent isolation of human ES (hES) cells has stimulated ES cell research directed toward cell therapeutic applications. The central nervous system (CNS) has been proposed as one of the prime targets for ES cell therapies, due to early successes in directing ES cell fate toward neural lineages, the experience with fetal tissue transplantation, and the devastating nature of many CNS diseases with very limited treatment options. Some of the most striking advantages of ES cells compared with any other cell type are extensive self-renewal capacity and differentiation potential, access to the earliest stages of neural development, and ease of inducing stable genetic manipulations.

Neural Development

The neural plate is derived from the dorsal ectoderm and is induced by "organizer" signals derived from the underlying notochord. The dominant model of neural induction is the default hypothesis. This hypothesis states that in the absence of BMP signaling during early gastrulation, neural tissue is formed spontaneously, while exposure to BMP signals causes epidermal differentiation. Accordingly, signals emanating from the organizer essential for neural induction are BMP inhibitors such as Chordin, Noggin, Follistatin, and cerberus. However, FGF signals emanating from precursors of the organizer prior to gastrulation have been implicated in providing a "pre-pattern" for neural induction via activation of Sox3 and ERNI. Other players during neural induction are IGF and Wnt signals.

After the formation of the neural plate, cells undergo a well-defined set of morphological and molecular changes leading to the formation of neural folds and neural tube closure. This is followed by orchestrated waves of neural proliferation and differentiation. Of particular importance in determining specific neural fates are signals that provide regional identity both in the antero-posterior (AP) and dorso-ventral (DV) axis and that define domains of distinct expression of homeodomain proteins and bHLH transcription factors. The leading hypothesis of AP axis specification states that anterior fates are default during early neural induction, while FGF, Wnt, and retinoid signals actively posteriorize cell fates. DV identity is determined by the antagonistic action of Sonic hedgehog (Shh) secreted ventrally form the notochord and floor plate, and BMPs and Wnt signals dorsally. Ample evidence from explant and ES cell differentiation studies confirms a concentration-dependent role of Shh to define specific progenitor domains within the neural tube. A role for BMPs in dorsal neural patterning has been suggested from explant and ES cell differentiation studies. *In vivo* genetic ablation studies and work in transgenic mice overexpressing the BMP receptor type 1a under control of the regulatory elements of the nestin gene are pointing to a role for BMPs in dorsal patterning. However, loss-of-function studies of the BMP receptors suggest a much more limited role for BMPs in dorsal patterning, affecting choroids plexus development only. Wnt signals also contribute to dorsal patterning, particularly the establishment of the neural crest.

Subsequent differentiation of patterned neural precursor cells occurs in a stereotypic fashion, with neurons being born first followed by astroglial and oligodendroglial differentiation. Onset of neuronal differentiation is controlled via inhibition of the Notch pathway that represses proneural bHLH genes (for review). Astrocytic fate is established via activation of Jak/STAT signals, which exert an instructive effect on the multipotent neural progenitor to drive astrocytic differentiation. However, recent findings on the neurogenic properties of radial glial and the identification of adult neural stem cells as a cell expressing astrocytic markers suggest a more complex and dynamic interaction between neural stem cell and astrocytic fates. Oligodendrocytes were believed to derive from bi-potent glial precursors termed O2A progenitors or from other glially committed precursors. However, more recent data suggest a lineage relationship between motoneuron and oligodendrocytes in spinal cord by their shared requirement for Olig2 expression. A review of the developmental signals that specify various neuronal subtypes is beyond the scope of this text, but some of the signals involved will be discussed. For a more in-depth discussion of this topic, a number of comprehensive reviews are available.

Neural Stem Cells

Neural stem cells have been isolated from both the developing and adult brain. Over 10 years of intensive research have demonstrated self-renewal of neural stem cells and multilineage differentiation into neurons, astrocytes, and oligodendrocytes. However, neural stem cells do not efficiently give rise to all the various neuron types present in the adult brain and are largely limited to the production of GABA and gluatamate neurons. The isolation and propagation of neural stem cells can be achieved by selective growth and proliferation conditions. The most commonly used method is the neurosphere assay. Under these conditions, neural precursors are grown as free-floating spheres in the presence of EGF and FGF-2. Human neurosphere cultures are often also supplemented with leukemia inhibiting factor (LIF). Neurospheres in rodents can be formed from single cells, and the capacity for neurospheres formation is often used as an assay to demonstrate stem cell properties. For example, the prospective isolation of neural stem cells via surface marker expression (AC133, Lex1, and combination of surface markers) was based on the ability of these cells to form neurospheres *in vitro*. These data need to be interpreted cautiously because neurosphere formation is not necessarily a true test of the stem cell identity. Neurospheres contain many differentiated cells in addition to the progenitor/stem cell population. Recent studies in the adult SVZ have demonstrated that neurospheres are formed more efficiently from transient amplifying precursors than from true stem cells in the adult SVZ.

An alternative to neurospheres is the monolayer culture technique whereby neural precursor/stem cells are grown on an attachment matrix such as fibronectin or laminin in the presence of FGF2. These conditions are more amenable to study the precise lineage relationship of individual cells, and complete cell lineage trees have been worked out under such conditions. One of the most important limitations of current neural stem cell technology is the limited control over neural patterning and neuronal subtype specification. The derivation of midbrain dopamine neurons has served as a model for these difficulties. Functional midbrain dopamine neurons can be derived from short-term expanded precursor cells isolated from the early rodent and human midbrain. However, long-term expansion causes a dramatic loss in the efficiency of midbrain dopamine neuron generation. Several strategies have been developed in an attempt to overcome these problems ranging from exposure to complex growth-factor cocktails changing oxygen levels to the transgenic expression of Nurr1, a key transcription factor during midbrain dopamine neuron development. However, none of the approaches has succeeded in deriving fully functional midbrain dopamine neurons from long-term expanded neural stem cells.

Neural Differentiation of Mouse ES Cells

NEURAL INDUCTION

There are at least three main strategies for the neural induction of ES cell *in vitro*: systems based on Embryoid-body (EB), stromal feeder mediated neural induction, and protocols based on default differentiation into neural fates (Figure 18-1).

Protocols Based on Embryoid Body

EBs are formed upon aggregation of ES cells in suspension culture. The interactions of cells within the EB causes cell differentiation mimicking gastrulation. Accordingly, derivatives of all three germ layers can be found in EBs. Various modifications of the basic protocol have been developed to enhance neural induction and to select and expand EB-derived neural precursors.

The first EB-mediated neural differentiation protocol was based on exposure to retinoic acid (RA) for 4 days following 4 days of EB formation in the absence of RA (the so-called 4–/4+ protocol). In addition to neural induction, RA treatment also exhibits a strong caudalizing effect on AP patterning mediated through activating the *Hox* gene cascade.

An alternative EB-based strategy is the exposure to conditioned medium derived from a hepatocarcinoma cell line (HepG2), which appears to induce neuroectodermal fate directly. Accordingly, HepG2-treated aggregates do not express endodermal or mesodermal markers but apparently give rise directly to neural progeny. The active component within HepG2-conditioned medium is not known, although data suggest that at least two separable components are responsible for this activity.

A third EB-based strategy makes use of neural-selective growth conditions. EB progeny is kept under minimal growth conditions in serum-free medium containing insulin, transferrin, and selenite (ITS medium). Under these conditions, most EB-derived cells die, and a distinct population of immature cells emerges that expresses increasing levels of the intermediate filament nestin. These nestin+ precursors can be replated and directed toward various neuronal and glial fates using a combination of patterning, survival, and lineage-promoting factors (see the sections Derivation of ES-Derived Neurons and ES-Derived Glia).

Stromal Feeder Mediated Neural Induction

Bone marrow-derived stromal cell lines have been used for many years to support the growth of undifferentiated hematopoietic stem cells. More recently, it has been reported that several stromal cell lines exhibit neural inducing properties in co-culture with mouse ES cells. Stromal cell lines with the highest efficiencies of neural induction are typically at the preadipocytic stage of differentiation and are isolated from the bone marrow (e.g., PA-6, MS5, S17) or the aorta-gonad-mesonephros (AGM) region. The molecular nature of this stromal-derived inducing activity remains unknown. However, the efficiency and robustness of neural induction using stromal feeders are high compared to alternative protocols, and differentiation occurs without apparent bias in regional specification.

Neural Differentiation by Default

Co-culture free, direct neural differentiation protocols are based on the default hypothesis that absence of signals in primitive ectodermal cells will lead to neural differentiation. Two independent studies with mouse ES cells confirmed that under minimal conditions, in the absence of BMP but in the

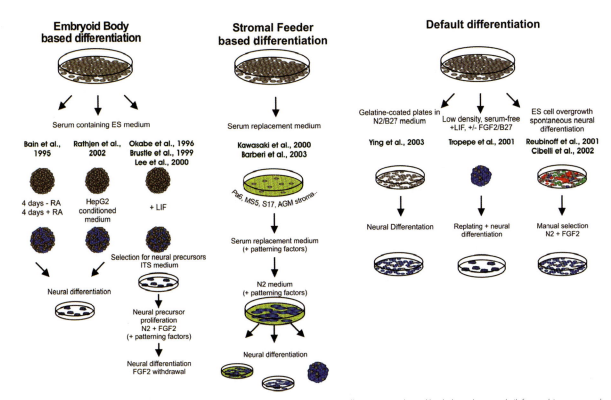

Figure 18-1. Basic techniques for inducing neural differentiation in embryonic stem cells *in vitro. Embryoid-body based protocols (left panels)* are initiated by the aggregation of undifferentiated ES cells. Neural differentiation (represented in blue) is promoted either by exposure to RA, HepG2 (hepatocarcinoma cell line) conditioned medium, or via neural selection in defined medial. Classic studies are cited (in green) for each of the three main EB-based strategies. *Stomal feeder mediated differentiation* is obtained upon plating undifferentiated ES cells at low density on stromal feeder cell lines derived from the bone marrow of the AGM region of the embryo. Serum-free conditions are required throughout the protocol. Conditions can be readily adapted to achieve neural subtype specific differentiation for a large number of CNS cell types. Classic studies are cited (in green) for PA6 and MS5 mediated differentiation. *Neural differentiation* is achieved by reducing endogenous BMP signals via plating cells at low density under minimal medium conditions or in the presence of the BMP antagonist noggin. Neurally committed cells can also be mechanically isolated and propagated from plates exhibiting spontaneous neural differentiation after overgrowth of ES cells.

presence of endogenous FGF signals, neural induction does occur in nonadherent or adherent monocultures.

DERIVATION OF ES-DERIVED NEURONS

Spontaneous differentiation into neurons occurs rapidly upon neural induction of mouse ES cells. Neuronal subtype specification can be influenced by the mode of neuronal induction. This is particularly the case for RA induction protocols. The basic strategy for achieving neuronal subtype specification is based on mimicking patterning events that define AP and DV patterning events *in vivo.*

Midbrain Dopaminergic Neurons

Derivation of midbrain dopamine neurons from ES cells has been of particular interest because of the clinical potential for dopamine neuron transplants in Parkinson's disease. Protocols for the dopaminergic differentiation of mouse ES cells are based on studies in explants that identified FGF8 and SHH as critical factors in midbrain dopamine neurons specification. The effect of SHH/FGF8 on ES-derived neural precursors was first described using an EB-based five-step differentiation pro-

tocol. Under these conditions, up to 34% of all neurons expressed tyrosine hydroxylase (TH), the rate-limiting enzyme in the synthesis of dopamine. A further increase in dopamine neuron yield (nearly 80% of all neurons expressing TH) was obtained in Nurr1 overexpressing ES cells (17). Midbrain dopaminergic differentiation was also obtained using co-culture of ES cells on the stromal feeder cell line (PA6)(81), with 16% of all neurons expressing TH in the absence of Shh and FGF8. These results were initially interpreted as PA6 exhibiting a specific patterning action that promotes dopamine neuron fate. However, later studies demonstrated that neural precursors induced on stromal feeders can be shifted in AP and DV identity and reach a yield of up to 50% of all neurons expressing TH without a need for transgenic Nurr1 expression.

Numbers of TH neurons need to be interpreted carefully in all *in vitro* differentiation studies as TH is an unreliable marker for identifying dopamine neuron. TH is expressed in other catecholaminergic neuron types such as noradrenergic and adrenergic cells and can be induced in many cell types under various unspecific external stimuli such as stress or hypoxia. It is therefore essential to use additional markers to confirm dopamine neuron identity and to perform functional studies *in vitro* and *in vivo*.

Serotonergic Neurons

The developmental origin of serotonergic neurons is closely related to that of midbrain dopamine neurons. Both neuronal subtypes are dependent on signals emanating from the isthmic organizer. Accordingly, serotonergic neurons are a major "contaminating" neuronal subtype in protocols aimed at the derivation of midbrain dopaminergic cells. Application of FGF4 preceding FGF8 and Shh application ectopically induces serotonergic neurons in explant culture, and FGF4 exposure has been shown to enhance serotonergic differentiation in the five-step protocol (17) and in stromal feeder-based protocols. Novel strategies to refine serotonergic differentiation may come from developmental studies on the role of the transcription factor Lmx1b in serotonergic differentiation and from findings in zebra fish identifying the elongation factor foggy and the zinc finger protein too few as important determinants of serotonergic fate.

Motor Neurons

Development of spinal motor neurons has been studied in great detail using a variety of loss- and gain-of-function approaches. Early studies have demonstrated that cells expressing markers of motoneurons can be generated using an EB induction protocol in combination with RA exposure (2–/7+). More systematic approaches using RA exposure in combination with Shh treatment have yielded ES-derived motoneurons at high efficiency and provided an example of how developmental pathways can be harnessed to direct ES cell fate *in vitro*. The use of an ES cell line expressing GFP under the control of the HB9 promoter allowed simple identification and purification of ES-derived motoneurons. The *in vivo* properties of these cells were demonstrated by contribution to the motoneuron pool after transplantation into the developing chick spinal cord.

Efficient derivation of motoneurons has also been achieved using stromal feeder protocols in combination with SHH and RA treatment. The next challenges for *in vitro* motoneuron differentiation protocols will be the selective generation of motoneurons of distinct AP and columnar identity. Other important factors for clinical translation in the treatment of spinal cord injury or amyotrophic lateral sclerosis (ALS) include strategies to modulate axonal outgrowth, target selection, and specificity of muscle innervation. One particularly useful approach might be the genetic or pharmacological manipulation of ES-derived motor neurons in an effort to overcome growth inhibitors present in an adult environment.

GABA Neurons

GABA cells are the main inhibitory neuron type within the brain and are the dominant neuron type in the basal forebrain, particularly in the striatum. The presence of GABAergic neurons during ES cell differentiation *in vitro* has been reported under various conditions, including the classic 4–/4+ EB-based differentiation protocols, which yield approximately 25% GABA neurons (100). The presence of GABA neurons has also been reported under default neural induction conditions.

Directed differentiation to GABA neurons has been achieved using a stromal feeder-based approach. Neural induction on MS5 is followed by neural precursor proliferation in FGF2 and subsequent exposure to Shh, FGF8. The delayed application of FGF8 and Shh promotes ventral forebrain identities as determined by the expression of the forebrain specific marker FOXG1B (BF-1(101)) and the increase in GABAergic differentiation. GABA neurons are implicated in a wide variety of neurological disorders, including Huntington's disease, epilepsy, and stroke.

Glutamate Neurons

Glutamate neurons can be readily obtained at high efficiencies from mouse ES cells. For example, in the Bain protocol, approximately 70% of all neurons are glutamatergic, and ES-derived neurons with NMDA and non-NMDA receptor subtypes have been described. Very similar neuron subtype compositions have been obtained with various related protocols. More detailed phsyiological data on glutamatergic neurons have been reported after co-culture of ES derived neurons on hippocampal brain slices. Interestingly, this study suggested a possible bias toward establishing AMPA-versus NMDA-type synaptic contacts.

Other Neuronal and Neural Subtypes

Although the presence of about 5% glycinergic neurons has been reported using the classic 4–/4+ EB protocol, no directed differentiation protocols have been developed for the derivation of this neuron type. Other interesting neural types generated from ES cells are precursors of the otic anlage. These precursors were obtained by culturing EBs for 10 days in EGF and IGF followed by bFGF expansion. After transplantation of these precursors *in vivo*, differentiation was observed into cells expressing markers of mature hair cells (106). Derivation of radial glial cells from mouse ES cells provides another interesting assay system to probe neuronal and glial lineage relationships in early neural development, and a protocol for deriving highly purified radial glia-like cells has been reported recently.

Neural Crest Differentiation

The neural crest is a transient structure formed from the most dorsal aspects of the neural tube of the vertebrate embryo. It contains migratory cells that form the peripheral nervous system, including sensory, sympathetic, and enteric ganglia, large parts of the facial skeleton, as well as various other cell types, including Schwann cells, melanocytes, and adrenomedullary cells. ES cells provide a powerful assay to study neural crest development *in vitro*. The main strategy for deriving neural crest-like structures from ES cells is based on BMP exposure following neural induction. This can be achieved in mouse and partly nonhuman primate ES cells using the PA6 stromal feeder cell system. This study showed the development of both sensory and sympathetic neurons in a BMP dose-dependent manner. The derivation of smooth muscle cells required growth in chicken extract in combination with BMP withdrawal. No melanocytes or Schwann cells could be obtained under these conditions.

Another recent study suggests efficient neural crest induction, including Schwann cell differentiation using an EB-based multistep differentiation protocol in combination with BMP2 treatment. Neural crest formation has also been reported in HepG2-mediated neural differentiation protocols upon exposure to staurosporine previously reported to induce avian neural crest development. However, characterization of neural crest progeny was limited to morphological observations and expression of Sox10. Future studies are required to provide more data on the ability of ES progeny to differentiate in all neural crest derivatives and the role of Wnt signals for neural crest specification *in vitro*.

ES-DERIVED GLIA

Neural progenitors derived from mouse ES cells can be readily differentiated into astrocytic and oligodendroglial progeny under conditions similar to those described from primary neural precursors. The first reports on the glial differentiation of mouse ES cells were based on the 4–/4+ EB protocols or multistep EB differentiation protocols. Most of the glial progeny under these conditions are astrocytes with only a few immature oligodendrocytes present. However, subsequent studies have defined conditions for the selective generation of both astrocytes and oligodendrocytes.

Oliogodendrocytes

Highly efficient differentiation into oligodendrocytes was reported first using a modified multistep EB-based protocol. ES-derived neural precursors were expanded with FGF2, followed by FGF2 + EGF and FGF2 + PDGF. These conditions yielded a population of A2B5+ glial precursors that are capable of differentiation into both astrocytic (~36% GFAP+) and oligodendrocytic (~38% O4+) progeny upon mitogen withdrawal. The 4–/4+ EB RA induction protocol has recently been optimized for the production of oligodendrocytic progeny. This study demonstrated efficient selection of neural progeny by both positive (Sox1-eGFP) and negative (Oct4-HSV-thymidine kinase) selection. Oligodendrocytic differentiation is achieved in RA-induced EBs after expansion in FGF2 and followed by dissociation and replating in serum-free medium containing FGF2 and Shh. The final step involved Shh and FGF2 withdrawal and the addition of PDGF and thryoidhormone (T3). Under these conditions, ~50% of all cells express oligodendroglial markers. Stromal feeder mediated induction, initially thought to bias toward neuronal progeny, can also be readily adapted to derive oligodendrocytes at very high efficiencies.

Astrocytes

Highly efficient differentiation of ES cells into astrocytes has been reported using stromal feeder mediated neural induction, followed by sequential exposure to FGF2, bFGF/EGF, EGF/CNTF, and CNTF. Over 90% of all cells expressed the astrocytic marker GFAP under these conditions. Significant numbers of GFAP cells were also obtained using HepG2-mediated neural differentiation or multistep EB protocols. Glial progenitors obtained with a multistep EB protocol were recently "transplanted" *in vitro* into hippocampal slices and revealed that full physiological maturation of ES-derived astrocytes can be achieved upon interaction with an appropriate host environment.

LINEAGE SELECTION

Lineage selection based on surface markers or the cell-type-specific expression of promoter-driven selectable markers provides an alternative approach to directed *in vitro* differentiation protocols. The use of genetic markers in ES cells is particularly attractive owing to the ease of inducing stable genetic modifications and the availability of large libraries of transgenic and gene-targeted mice and ES cells. Efficient purification of neural progeny from mouse ES cells *in vitro* was demonstrated via positive selection using a Sox1-EGFP knock-in cell line and combined with negative selectable marker controlled by endogenous Oct4 locus. Other ES lines used successfully for the genetic identification and purification of neural precursors and neuronal and glial progeny *in vitro* include tau-eGFP knockin cells, a GFAP transgenic cells, GAD-lacZ knock-cells, BF1-lacZ knockin cells, and ES cell lines driven under the regulatory elements of the Nestin gene. Promoter-driven lineage selection for motor neuron fate *in vitro* includes ES cells with eGFP knocked into the Olig2 locus or expressed as a transgene under the HB9 promoter.

Neural Differentiation of Human and Nonhuman Primate ES Cells

Neural differentiation potential was readily observed when primate ES, or ES-like cells were first established in both monkey and hES cells, as well as in human EG cells and monkey parthenogenetic stem cells. However, the derivation of purified populations of neural progeny from hES cells required more systematic studies. A highly efficient protocol for the neural differentiation of hES cells was based on a modified multistep EB approach. ES cells are aggregated short-time (~4 days) and subsequently replated under serum-free conditions in the presence of FGF-2. Under these conditions, neural precursors can be readily identified based on the formation of multilayered epithelia termed rosettes. These neural precursor cells can be enzymatically separated from the surrounding cell types and grown to purity under neurosphere-like conditions in the presence of FGF-2 that yield neurons and astrocytes upon differentiation. The main neuronal subtypes are GABAergic and glutamatergic phenotypes. A similar spectrum of differentiated progeny was obtained when hES-derived neural precursors were obtained after spontaneous neural differentiation by overgrowth of undifferentiated cells.

Rosette-like structures were manually isolated under microscopic view and subsequently grown and passaged under neurosphere-like conditions. A third strategy that yielded differentiated cell populations enriched for neural precursors was based on a modified EB RA induction protocol, followed by lineage selection in attached cultures using serum-free conditions and supplementation with bFGF.

Protocols that yield better control over neuronal subtype specification are still in development, but first examples have been provided for the derivation of midbrain dopamine neurons from nonhuman primate ES. Interestingly, PA6-mediated neural induction caused differentiation into retinal pigment epithelial cells in addition to neural differentiation. More recently, human ES cells have been coaxed into neuronal populations with 70% TH expressing cells. This protocol was based on the neural induction of human ES cells by MS5 stroma, followed by sequential exposure to Shh/FGF8 and differentiation in the presence of AA, BDNF, and GDNF. One interesting difference between human and mouse ES cell differentiation is the presence of large numbers of rosette-like structures during neural differentiation of human ES cells that are rarely observed in mouse ES progeny.

Developmental Perspective

Highly reproducible *in vitro* differentiation protocols and an increasing number of available ES cell reporter lines provide powerful tools for establishing cell-based developmental screens. In addition, large-scale gene-trapping approaches have yielded individual ES cells lines mutant for a large proportion of the mouse genome (http://baygenomics.ucsf.edu/; http://www.lexicon-genetics.com). Several ES cell-based *in vitro* differentiation screens have been carried out to assay for genes involved in neural/neuronal induction. One such screen using kinase-directed combinatorial libraries has identified a molecule with potent inhibitory activity for GSK3β during neural differentiation of mouse ES and P19 cells. The result suggested that activation of Wnt signaling via increased β-catenin levels and activation of downstream targets, including LEF1/TCF1, increases neuronal differentiation.

In agreement with these findings, Wnt1 has been identified as a downstream target during RA-mediated neural differentiation of P19 cells. In contrast, a recent report identified the Wnt antagonist Sfrp2 during neural differentiation of EBs exposed to RA, and Wnts have been described as an essential component in maintaining undifferentiated ES cells. The availability of more comprehensive genomic tools will facilitate the establishment of detailed gene expression profiles in ES cell progeny and functional maps to complement existing data. Early gene-trapping studies that utilized *in vitro* differentiation screens prior to or instead of *in vivo* studies may also see a revival using much more refined neural differentiation protocols. The availability of high-throughput functional genomic approaches such as RNAi-based gene knockdown screens in *Drosophila* and *C. elegans* could be readily adapted to the analysis of ES cell differentiation *in vitro* and may provide unprecedented opportunities for studying develop-

ment, particularly in human ES-based systems where no acceptable *in vivo* assays exist.

Therapeutic Perspectives

One driving force behind deciphering the developmental program that controls cell fate specification is the hope that such insights can be harnessed for generating specialized cells for therapy. However, despite the excitement over the potential of ES cells in neural repair, there are only few examples where such approaches have been tested in animal models of disease.

PARKINSON'S DISEASE

One of the most widely discussed applications is the derivation of unlimited numbers of dopamine neurons from hES cells for the treatment of Parkinson's disease (PD). PD is particularly attractive for cell transplantation due to the relatively defined pathology affecting primarily midbrain dopamine neurons. At the onset of clinical symptoms, the majority of midbrain dopamine neurons have already died, providing further rationale for a cell replacement approach. The first ES cell-based study that showed functional improvement in 6OHDA lesioned rats, an animal model of Parkinson's disease, was based on the transplantation of low numbers of largely undifferentiated mouse ES cells isolated after short-term differentiation in EB cultures. Spontaneous differentiation into large numbers of neurons with midbrain dopamine characteristics was observed. However, the clinical relevance of this approach is limited due to the high rate of tumor formation (>50% of the animals with surviving grafts developed teratomas).

Remarkable functional improvement was obtained after transplantation of dopamine neurons derived from Nurr1 overexpressing mouse ES cells. This study reported behavioral restoration in 6OHDA lesioned rats and demonstrated *in vivo* function via electrophysiological recordings from grafted dopamine neurons in slices obtained from the grafted animals. However, transgenic expression of Nurr1 raises safety concerns that may preclude clinical translation. Another study showed behavioral recovery with dopamine neurons derived from naïve mouse ES cells. Both regular mouse ES cells and ES cells derived after nuclear transfer (ntES cells) from a cloned blastocyst were used in this study. Successful grafting of ntES-derived dopamine neurons provided a first example of therapeutic cloning in neural disease. Differentiation of hES cells into midbrain dopamine neurons has been reported very recently. However, there have been no reports on the *in vivo* functionality of these cells.

Several issues remain to be addressed before the first clinical trials with hES cell derived dopamine in Parkinson's disease can take place. Over 20 years of fetal tissue research have demonstrated that fetal midbrain dopamine neurons can survive and function long term (>10 years) in the brain of Parkinson's patients. However, these studies have also shown limited efficacy in placebo-controlled clinical trials and have demonstrated potential for side effects. The stem cell field will have to learn from the fetal tissue transplantation trials and to better define and address the critical parameters that will take cell therapy in PD to the next level. The derivation of highly purified populations of substantia nigra type dopamine neurons from hES cells is an important first step on this road.

HUNTINGTON'S DISEASE

Similar to PD, the pathology in Huntington's disease (HD) affects a selective neuronal cell population, the GABAergic medium spiny neurons in the striatum. Fetal tissue transplantation trials provide proof of concept for cell-based therapies in treatment of HD. However, unlike in PD, grafted cells are required to reconnect not to local targets within the striatum but to project from the striatum to the targets in the globus pallidus and the substantia nigra pars reticulata. Although the identification of the molecular defect in HD as a unstable expansion of CAG repeats in the IT15 gene suggest many possible alternative therapeutic approaches, it also provides the research community with genetic models of the disease that allow careful evaluation of all therapeutic options. ES-derived GABAergic neurons have not yet been tested in any animal model of HD, but the *in vitro* derivation of GABA neurons with forebrain characteristics has been achieved at high efficiencies. Another interesting avenue for stem cells in HD is the derivation of hES cells from embryos with HD mutations identified during preimplantation diagnostics. Such HD ES cell lines could provide invaluable insights into the selective vulnerability of the striatal GABAergic cell population.

SPINAL CORD INJURY AND OTHER MOTONEURON DISORDERS

Traumatic or degenerative injuries to the spinal cord are often devastating and irreversible. Cell replacement using stem cells has been touted as a prime application of stem cell research. However, the complexity of cell therapy in spinal cord injury is enormous and far from resolved. Motor neurons are one of the main cell types affected in spinal cord injury and in various degenerative diseases such as amyotrophic lateral sclerosis (ALS, or Lou Gehrig's disease). The efficient derivation of motoneurons from mouse ES cells has been demonstrated with both EB-based and stromal feeder protocols. However, the functionality of ES-derived neurons *in vivo* has only been addressed via xenografts into the developing chick spinal cord. The behavior of ES-derived motor neurons in the adult CNS and in animal models of spinal cord injury or disease remains to be tested.

Prior to the availability of directed differentiation protocols, there were reports on functional improvement in animal models of spinal cord injury after grafting ES progeny. In one such study, dissociated 4−/4+ mouse EBs derived from D3 or Rosa26 ES cells were transplanted into the spinal cord 9 days after a crush injury. The grafted cells differentiated *in vivo* into oligodendrocytes, neurons, and astrocytes and induced significant functional improvement in BBB scores as compared to sham-injected animals. However, the mechanism by which functional improvement was obtained remains controversial. Based on the ability for efficient differentiation into

oligodendrocytes *in vivo*, it was suggested that remyelination of denuded axons might be a key factor. Functional improvement was also reported after grafting EB-derived cells obtained from human EG cells. EB-derived cell populations derived from human EG cells were implanted into the CSF of rats after virus-induced neuronopathy and motor neuron degeneration, a model of ALS. While a very small number of transplanted cells started to express markers compatible with motor neuron fate, most cells differentiated into neural progenitor or glial cells. It was concluded that the functional improvement was due to enhancing host function rather than to reestablishing functional connections with graft-derived motoneurons.

STROKE

Very little information is available on the behavior of ES cell progeny in animal models of stroke. A recent study showed that grafted ES cells can survive in a rat stroke model induced by transient ischemia via occlusion of the middle cerebral artery. The goal of this study was noninvasive imaging of the grafted cells using high-resolution MRI after transfection with superparamagnetic iron-oxide particles. The authors provided evidence for extensive migration of the grafted cells along the corpus callosum toward the ischemic lesion. However, no phenotypic analyses of the differentiated cell types were performed, and no functional effects were measured. Cell transplantation efforts for the treatment of stroke are complicated by the variability of the affected cell populations depending on stroke location.

DEMYELINATION

The capacity of mouse ES-derived progeny to remyelinate *in vivo* has been demonstrated after transplantation of highly purified ES-derived glial progenitors into the spinal cord of *md* rats, an animal model of Pelizaeus-Merzbacher syndrome. This study showed impressive *in vivo* differentiation results and yielded large grafts comprised of myelinating oligodendrocytes. However, the grafted cells were not able to extend the short lifespan of these animals, precluding detailed functional analyses. A second study demonstrated remyelination after grafting purified oligosphere cultures derived from 4−/4+ EBs into the spinal cord of shiverer mice or into the chemically demyelinated spinal cord.

Important challenges for the future are the derivation of functional oligodendrocytes from hES cells and the demonstration of functional benefits *in vivo*, including strategies to obtain remyelination over more extended CNS regions. In clinical terms, the transplantation of ES-derived oligodendrocytes into models of multiple sclerosis would be of particular interest. However, a successful strategy will require sophisticated strategies to overcome host-mediated factors that prevent oligodendrocytes maturation as well as strategies that address the autoimmune nature of the disease.

OTHER DISEASES

Several disease models that have been approached with fetal neural progenitors have not yet been tested using ES cell-based approaches. These include epilepsy and enzymatic deficiencies such as lyosomal storage diseases. Other CNS disorders such as Alzheimer's disease have been touted as future applications for ES cell therapy. However, at the current stage of research, the challenges for cell therapy in Alzheimer's disease seem to be overwhelming, and it appears more likely that the role for ES cells might be in providing cellular models of disease rather than for cell replacement. Some early attempts toward achieving this goal have studied neural differentiation in ES cells that exhibit the disease-causing mutant of APP knocked into the endogenous APP locus.

Conclusion

The development of protocols that allow the directed differentiation from ES cells to specific neural fates provides an essential basis for all cell-based approaches in neural repair. Although these protocols have become routine for mouse ES cells, work with hES cells lags behind. However, it is likely that these difficulties will be overcome within the next few years and that some of the very first clinical ES cell applications will be within the CNS. Beyond the role in regenerative medicine, ES cell *in vitro* differentiation protocols will become an essential tool for gene discovery and will serve as a routine assay of neural development. The availability of libraries of ES cells with specific mutations or expressing specific transgenes will be a great asset for such studies. *In vitro* ES cell differentiation will also provide unlimited sources of defined neural subtypes for pharmacological assays in drug screening and toxicology. One of the most important contributions might be the availability of ES cells as a basic research tool to unravel the complex signals that govern the development from a single pluripotent ES cells to the amazing diversity of cell types that comprise the mammalian CNS.

KEY WORDS

Embryoid body   Clusters derived from aggregated embryonic stem cells. Embryoid-bodies formation initiates a process that has similarities to gastrulation and leads to differentiation of ES cells into cells of the three germ layers (ectoderm, endoderm, mesoderm).

Neural patterning   Developmental mechanism to specify regional identity of neural precursor cells (e.g., forebrain versus midbrain versus spinal cord identity; or dorsal versus ventral identity). Neural patterning is often controlled by soluble proteins that act in a concentration-dependent manner to specify naïve progenitor cells.

Neural rosettes   Term used to describe the tube-like columnar epithelial structures appearing during early stages of neural differentiation from ES cells. Particularly prominent are these structures during neural differentiation of primate ES cells. Rosettes correspond developmentally to the neural plate and early neural tube stage.

Neurospheres   Nonadherent clusters of neural progenitor cells grown *in vitro*. Culture system for the propagation of neural stem cells.

Prospective isolation   A strategy to isolate defined cell types based on a set of predetermined cell specific marker using fluorescence activated cell sorting (FACS) or other techniques that allow physical separation of cells.

FURTHER READING

Barberi, T., Klivenyi, P., Calingasan, N. Y., Lee, H., Kawamata, H., Loonam, K., Perrier, A. L., Bruses, J., Rubio, M. E., Topf, N., Tabar, V., Harrison, N. L., Beal, M. F., Moore, M. A., and Studer, L. (2003). Neural subtype specification of fertilization and nuclear transfer embryonic stem cells and application in parkinsonian mice. *Nat. Biotechnol.* **21**, 1200–1207.

Brustle, O., Jones, K. N., Learish, R. D., Karram, K., Choudhary, K., Wiestler, O. D., Duncan, I. D., and McKay, R. G. (1999). Embryonic stem cell-derived glial precursors: A source of myelinating transplants. *Science* **285**, 754–756.

Ding, S., Wu, T. Y., Brinker, A., Peters, E. C., Hur, W., Gray, N. S., and Schultz, P. G. (2003). Synthetic small molecules that control stem cell fate. *Proc. Natl. Acad. Sci. USA* **100**, 7632–7637.

Kawasaki, H., Mizuseki, Nishikawa, S., Kaneko, S., Kuwana, Y., Nakanishi, S., Nishikawa, S., and Sasai, Y. (2000). Induction of midbrain dopaminergic neurons from ES cells by stromal cell-derived inducing activity. *Neuron* **28**, 31–40.

Kim, J. H., Auerbach, J. M., Rodriguez-Gomez, J. A., Velasco, I., Gavin, D., Lumelsky, N., Lee, S. H., Nguyen, J., Sanchez-Pernaute, R., Bankiewicz, K., and McKay, R. (2002). Dopamine neurons derived from embryonic stem cells function in an animal model of Parkinson's disease. *Nature* **418**, 50–56.

Lee, S.-H., Lumelsky, N., Studer, L., Auerbach, J. M., and McKay, R. D. G. (2000). Efficient generation of midbrain and hindbrain neurons from mouse embryonic stem cells. *Nat. Biotechnol.* **18**, 675–679.

McDonald, J. W., Liu, X. Z., Qu, Y., Liu, S., Mickey, S. K., Turetsky, D., Gottlieb, D. I., and Choi, D. W. (1999). Transplanted embryonic stem cells survive, differentiate and promote recovery in injured rat spinal cord. *Nature Med.* **5**, 1410–1412.

Perrier, A. L., Tabar, V., Barberi, T., Rubio, M. E., Bruses, J., Topf, N., Harrison, N. L., and Studer, L. (2004). From the Cover: Derivation of midbrain dopamine neurons from human embryonic stem cells. *Proc. Natl. Acad. Sci. USA* **101**, 12543–12548.

Wichterle, H., Lieberam, I., Porter, J. A., and Jessell, T. M. (2002). Directed differentiation of embryonic stem cells into motor neurons. *Cell* **110**, 385–397.

Zhang, S. C., Wernig, M., Duncan, I. D., Brustle, O., and Thomson, J. A. (2001). In vitro differentiation of transplantable neural precursors from human embryonic stem cells. *Nat. Biotechnols* **19**, 1129–1133.

Neuronal Progenitors in the Adult Brain: From Development to Regulation

Christian Mirescu and Elizabeth Gould

Introduction

The adult mammalian brain has vastly reduced regenerative potential compared to the developing brain. Nevertheless, cell proliferation occurs in the adult brain, as does neurogenesis, and cells that give rise to neurons and glia *in vivo* and *in vitro* have been identified in the adult CNS. As an understanding of the factors that regulate stem cell proliferation and neurogenesis in the adult brain deepens, the development of methods directed at repairing the damaged brain through transplantation of cultured cells or through induction of endogenous neurogenesis in selected neuronal populations may lead to novel therapeutic strategies. This chapter will review the evidence that neural stem cells exist in the adult brain and discuss the factors that determine whether these cells divide and, if so, what is the fate of their progeny.

History of Stem Cells in the Adult CNS

Although studies dating back to the 1960s have documented neurogenesis in the adult mammalian brain, the isolation of cells with stem cell properties, that is, multipotent and self-renewing, from adult neural tissue did not occur until much later. In the early 1990s, reports demonstrated that cultured cells from the adult rodent striatum produced both neurons and glia, effectively documenting multipotentiality. These initial findings by Reynolds and Weiss (1992) from the University of Calgary were followed by more specific localization of stem cells within this brain region, this time identifying proliferating cells as residents of the subventricular zone (SVZ), which may have been included in some of the early cultures since the SZV borders the striatum. Since that time, numerous reports have identified cells from adult animals with the potential to produce neurons and glia when grown in culture or transplanted into other brain regions. In addition to the SVZ, such cells have been isolated from a variety of locations in the CNS, including the spinal cord, the hippocampus, and the neocortex, as well as the striatum. The self-renewing capability of these cells was reported in 1996 by McKay and colleagues from the Natural Institutes of Health (NIH) as well as Arturo Alvarez-Buylla and co-workers, demonstrating con-

Essentials of Stem Cell Biology
Copyright © 2006, Elsevier, Inc.
All rights reserved.

clusively that these cells had the characteristics of stem cells when grown *in vitro*. Studies since that time have sought to identify stem cells *in vivo* with some success.

Glial Characteristics of Neural Stem Cells

Beginning with studies of neurogenesis in adult birds and then extending to those in adult and developing mammals, several *in vivo* reports have suggested that stem cells have the characteristics of glia. These observations further expand the burgeoning list of functions attributed to glia and may lend insight into the production of neurons in the adult brain.

RADIAL GLIA

In adult birds, the germinal zone of the lateral ventricles resembles that of the embryonic ventricular zone (VZ) of mammals. Adult-generated neurons appear to use radial glia processes as migratory guides to their final destination in the rostral forebrain. The spatial and temporal relationships of radial glia to new neurons were observations that led to the first suggestion that radial glia may themselves serve as stem cells. That is, radial glia may give rise not only to new glia but also to neurons.

More direct evidence that glia serve as neural stem cells in the avian VZ comes from retroviral lineage studies, which have shown that radial glia exhibit multipotent differentiation. In the mammalian embryo, the VZ generates postmitotic cells that also take a rostral migratory route in forebrain development. Recent studies have shown that neuronal progenitors are radial glia in the embryonic mammalian forebrain. During subsequent neonatal development, the radial glia of the VZ disappear and are replaced by multiciliated ependymal cells that come to line the walls of the maturing lateral ventricles. Coinciding with this time, mitotically active cells begin to appear in the subventricular zone (SVZ). These dividing cells show ultrastructural characteristics of radial glia and furthermore express the radial glial marker RC-2. That the postnatal SVZ progenitors share properties similar to the radial glia of the embryonic VZ suggests that the VZ may serve to seed the mature SVZ with a self-replenishing population of adult neural stem cells. In adulthood, the fate of SVZ cells resembles the movement of cells born during development, as postmitotic cells appear to migrate toward anterior regions of the brain along the rostral migratory pathway. Similarities in the site of origin, migratory route, and ultimate position of adult-generated cells in avian species suggest that processes active in embryonic development are conserved in germinal regions of the adult brain, and that radial glia may be antecedents to adult neural stem cells in mammals as well.

Nevertheless, parallels drawn across species and observations of structural similarities to embryonic VZ cells are only indirect evidence that radial glia actually serve as primary neuronal precursors. More direct evidence supports this view; radial glia (isolated by fluorescent-activated cell sorting) have been observed to give rise to both neurons and astroglia. Furthermore, radial glia retrovirally tagged with green fluorescent protein (GFP) in the developing rodent brain have been found to have mitotic activity and to subsequently generate neurons. Such reports more directly demonstrate that radial glia progenitors are not an exclusive characteristic of the avian brain. In mammals as well, these cells also appear to be neural stem cells.

Despite the evidence that radial glia serve as progenitors in the adult avian and reptilian brain as well as in the developing mammalian brain, it is less likely that new neurons arise from radial glia in the adult mammalian brain. This is because radial glia are absent in the mature mammalian brain, as they are thought to differentiate into astroglia following development. Thus, in adult mammalian germinal regions, there must exist a distinct type of progenitor, perhaps descended from embryonic radial glia, which retains the self-renewing and multilineage potentials of neural stem cells *in vivo*.

ASTROGLIA

Numerous recent studies suggest that progenitor cells in the adult mammalian brain share characteristics with astroglia. Temporary destruction of neuroblasts and immature precursor cells in the adult brain by treatment with an antimitotic agent spares some astroglia in the SVZ. Within hours after halting cell division, astroglia begin to divide, and then several days later, migrating neuroblasts destined for the olfactory bulb, and possibly other regions, emerge. Finally, by 10 days later, the orientation and organization of chains of migrating neuroblasts resemble that of normal adult mice. Thus, astroglia appear to function as neural progenitors *in vitro* and *in vivo* and appear critical for recovery of germinal activity within the SVZ.

In addition to the SVZ as an identified region of adult germinal activity, the subgranular zone (SGZ) of the hippocampal dentate gyrus appears to contain cells with stem cell-like characteristics in adulthood. The SGZ lies between the granule cell layer (GCL) and the hilus of the dentate gyrus. Astroglia within the SGZ share ultrastructural features with those of SVZ astroglia. When dividing cells are labeled with [^{3}H]-thymidine or with the thymidine analog bromodeoxyuridine (BrdU) in adulthood, a large proportion of adult-generated cells in the dentate gyrus exhibit an astrocytic phenotype soon thereafter. The majority of BrdU-labeled GFAP-positive astroglia disappear rapidly, coinciding with an increase in the number of BrdU-labeled GFAP-negative cells, suggesting that SGZ astroglia may undergo a transition from a glial to a nonglial phenotype. In order to examine whether this transition actually gives rise to the production of new neurons in adulthood, antimitotic treatment was also used. Similar to the observed effects of such treatment in the SVZ, some astroglia survive and begin to divide soon after treatment. The administration of BrdU or [^{3}H]-thymidine two days after treatment, during SGZ astrocyte proliferation, yields labeled granule neurons, astroglia, and oligodendrocytes both one and five months later. Thus, it appears that SGZ astroglia share neural stem cell properties similar to SVZ astroglia, namely, the capacity for self-renewal and multilineage differentiation.

Adult Neurogenesis *in Vivo*

Numerous studies have conclusively demonstrated the production of new neurons in the brains of adult mammals. The case for SGZ and SVZ as germinal regions for new neurons in the adult brain is very strong. Collectively, this evidence supports the view that new neurons (i.e., cells with morphological, ultrastructural, biochemical, and electrophysiological characteristics of neurons) are added to the dentate gyrus and olfactory bulb of adult mammals. Evidence for adult neurogenesis in other brain regions, including the neocortex, striatum, olfactory tubercle, amygdala, and substantia nigra, exists, but these findings await further investigation. In the light of *in vitro* studies demonstrating cells with stem cell-like characteristics in the parenchyma of regions other than the dentate gyrus and SVZ, adult neurogenesis in other brain structures would not be unexpected.

During development, neurogenesis is regulated by the complex interplay of peptide signaling molecules and neurotransmitters. Since neurogenesis persists in some regions after development is complete, it is possible that regulatory mechanisms of this process are preserved as well. The following sections focus on the influence of various signaling molecules such as growth factors, neurotransmitters, and hormones on the production of new neurons in the adult brain.

NEUROTROPHINS

Brain-derived neurotrophic factor (BDNF) and neurotrophin-3 (NT-3) are members of a structurally related family of growth factors that function to prevent the death of embryonic neurons during development. Across the lifespan, neurotrophin expression varies dramatically. Embryonic expression of NT-3 appears highest in regions of the CNS where proliferation, migration and differentiation are ongoing. Levels then decrease with maturation. By contrast, BDNF expression is lowest in developing regions and increases with maturation. Both BDNF and NT-3 expression, and their preferred receptors TrkB and TrkC, have been demonstrated in cultures of cortical progenitor cells and *in vivo* at the onset of cortical neurogenesis, suggesting a possible direct action of these systems on neural progenitors. When regional BDNF expression is specifically enhanced by an adenoviral vector, a substantial increase of adult-generated neurons occurs in the mouse olfactory bulb, a known neurogenic region. In addition, BDNF overexpression results in adult neuron production in the caudate putamen, normally a nonneurogenic site in adulthood. By contrast, in hippocampal-derived stem cell clones, neurotrophins alone are only minimally effective in stimulating neurogenesis. Yet, in the presence of retinoic acid, both BDNF and NT-3 promote the acquisition of a neuronal fate by progenitors. These results suggest that neurotrophins may have regionally distinct and synergistic effects on adult neural stem cells.

INSULIN-LIKE GROWTH FACTOR

Neurogenic effects of insulin-like growth factor (IGF) have been identified in cultures of embryonic neural stem cells as well as *in vivo* using transgenic models overexpressing or lacking specific IGF encoding genes. In proliferative cultures of embryonic precursors, insulin and IGFs promote the proliferation of neural stem cells and the acquisition of a neuronal phenotype. In mice and humans, disruption of the IGF-I gene is associated with profound retardation of brain growth. In contrast, overexpression of IGF-I results in larger brains marked by greater numbers of neurons. Similar effects on proliferation and neuronal lineage appear conserved in adulthood, as peripheral IGF-I infusion (for 6 days) increases the number of proliferating cells and the fraction of new cells exhibiting neuronal characteristics in the adult hippocampus of rodents.

FIBROBLAST GROWTH FACTOR

Extensive evidence suggests that basic fibroblast growth factor (bFGF) elicits neurogenic effects in culture as well as in widespread regions of the prenatal brain, such as the cerebral cortex, the neonatal cerebellum, and the SVZ. When administered during adulthood, bFGF also exerts a mitogenic effect in both the SVZ and olfactory system. However, no effect of bFGF has been found in the adult hippocampus, suggesting that the germinal effect of bFGF might be ubiquitous during development but regionally specific in adulthood. When bFGF is delivered to neonatal rats on P1, increases in hippocampal DNA content and enhanced production of BrdU-positive cells are observed beyond the early postnatal period. Thus, bFGF appears to influence olfactory and SVZ neurogenesis throughout life, but in the dentate gyrus only during development.

GLUTAMATE

The bulk of excitatory inputs throughout the mammalian brain utilize glutamate as a neurotransmitter. The majority of inputs into the dentate gyrus GCL are glutamatergic afferents originating from the entorhinal cortex. Along with regions of the hippocampus other than the dentate gyrus, granule neurons of the dentate gyrus express *N*-methyl-D-aspartate receptors (NMDAr). In development, NMDA receptor blockade results in profound changes in the dentate gyrus and SVZ. Specifically, NMDA receptor antagonists increase cell proliferation in both germinal regions of rat pups. Similarly, in adulthood, both lesions of the entorhinal cortex and injections of either a competitive or noncompetitive NMDAr antagonist have been demonstrated to potentiate the birth and overall density of granule neurons. In contrast, injections of NMDA rapidly reduce the number of dividing cells in this region. However, a direct influence of glutamatergic neurotransmission on proliferating cells in the adult dentate gyrus is not likely because granule cell progenitors do not express the NR1 NMDAr subunit, which is obligate for receptor function.

SEROTONIN

Recent evidence suggests that serotonin (5-HT) exerts an influence on adult neurogenesis. Inhibition of 5-HT synthesis via parachlorophenylalanine (PCPA) administration and selective lesion of serotonergic neurons reduce numbers of adult-generated cells in the dentate gyrus and the SVZ. Similarly, the selective blockade of 5-HT1a receptors potently decreases cell proliferation in the dentate gyrus, suggesting that the serotonergic influence on adult hippocampal cell production is mediated (at least in part) by this high-affinity 5-HT receptor. Furthermore, treatment with a variety of antidepressants, which work (at least in part) through 5-HT, increases the production of new granule neurons in the dentate gyrus.

ADRENAL STEROIDS

Recent studies have also identified numerous hormones that have regulatory effects on adult neurogenesis, with adrenal steroids as the most widely researched neuroendocrine regulator of cell proliferation. In the dentate gyrus, reductions in circulating adrenal steroids produced by adrenalectomy are associated with increased numbers of proliferating cells within the adult dentate gyrus. In contrast, increases in exogenous corticosterone (CORT) reduce adult cell proliferation in the dentate gyrus of both neonatal and adult rats. In agreement with such results, various naturalistic stressors exert suppressive effects on cell proliferation in the prenatal, perinatal, and mature dentate gyrus. Such experiences reduce both proliferative activity and immature neuron numbers via an adrenal steroid dependent mechanism that appears to operate via NMDAr activation. Recent evidence also suggests that chronic stress and adverse early life experiences exert more lasting effects on adult neurogenesis. In addition to stress, aging has also been associated with diminished adult cell production in adult rodents. Adrenalectomy in aged rats appears capable of restoring proliferative activity to levels similar to those of young rats, suggesting that adrenal hormones also mediate aged-associated effects on the dividing dentate gyrus. Such effects, however, do not seem to underlie age-related cognitive decline in hippocampal-dependent tasks.

OVARIAN STEROIDS

Ovarian hormones have also been found to influence the rate of cell proliferation in the dentate gyrus. In adult female rats, natural periods of high estradiol (proestrus) are associated with rapid increases in cell production. In contrast, at longer survival time points in both wild-reared and laboratory-reared meadow voles, cell proliferation and survival have been reported to be inversely related to the levels of circulating estradiol. Thus, it appears to be that estradiol exerts a transient stimulatory effect and subsequent suppressive influence on dentate gyrus division. The estradiol-induced suppression appears to be mediated by adrenal steriods, as CORT levels increase in response to elevated estradiol.

Summary

Studies carried out over the past decade have identified cells with stem cell-like properties in the adult mammalian brain. The bulk of evidence suggests that these cells exhibit characteristics of glial cells, a finding that may elucidate basic mechanisms of neurogenesis and make isolating stem cells a more tractable problem. The production of new neurons in the adult brain has been shown to be regulated by growth factors, neurotransmitters and hormones. A more detailed understanding of the potential interactions among these modulators of adult neurogenesis may enable the controlled manipulation of neuron production in the damaged brain.

Numerous neurological conditions are associated with the loss of neural cells, perhaps most notably Alzheimer's disease and Parkinson's disease. The possibility that trauma and neurodegenerative disorders can be treated either by transplanting neural stem cells or by stimulating endogenous neurogenesis in existing populations of adult neural stem cells has theoretically powerful clinical potential. The key advantage of transplanting cell lines is that they can be genetically engineered, allowing for customization depending on the specific disease. On the other hand, strategies designed to enhance endogenous neurogenesis (perhaps with regional-specificity) would have the potential for clinical efficacy without requiring invasive techniques. Although it is presently unclear which manipulations will lead to specific types of clinical improvements, what is certain is that the presence of neural stem cells in the adult CNS has opened up a new frontier of scientific exploration.

KEY WORDS

Adult neurogenesis   The addition of new neurons to the adult brain, a process consisting of the proliferation, differentiation, and survival of neural stem cells.

Dentate gyrus   Subregion of the hippocampus in the medial temporal lobe, which contains the subgranular zone, the neurogenic site of the dentate gyrus.

Neural stem cell   Cell within the CNS which retains the capability of (1) self-renewal and (2) multilineage potential (i.e., ability to produce a variety of cell types; in the brain, predominantly neurons and glia).

Subventricular zone   Regions of the forebrain bordering the cerebral ventricles and identified as a neurogenic site.

Trophic factor   Growth substance produced by target cells, which promotes the proliferation of progenitor cells and/or the survival of their neuronal progeny by preventing programmed cell death (i.e., apoptosis).

FURTHER READING

Cameron, H. A., Hazel, T. G., and McKay, R. D. (1998). Regulation of neurogenesis by growth factors and neurotransmitters. *J. Neurobiol.* **36**, 287–306.

Doetsch, F., Caille, I., Lim, D. A., Garcia-Verdugo, J. M., and Alvarez-Buylla, A. (1999). Subventricular zone astrocytes are neuronal stem cells in the adult mammalian brain. *Cell* **97**, 703–716.

Johe, K. K., Hazel, T. G., Muller, T., Dugich-Djordjevic, M. M., and McKay, R. D. (1996). Single factors direct the differentiation of stem cells from the fetal and adult central nervous system. *Genes Dev.* **10**, 3129–3140.

Kozorovitskiy, Y., and Gould E. (2003). Adult neurogenesis: a mechanism for brain repair? *J. Clin. Exp. Neuropsychol.* **25**, 721–732.

Lie, D. C., Song, H., Colamarino, S. A., Ming, G. L., and Gage, F. H. (2004). Neurogenesis in the adult brain: new strategies for central nervous system diseases. *Ann. Rev. Pharmacol. Toxicol.* **44**, 399–421.

Lindvall, O., Kokaia, Z., and Martinez-Serrano, A. (2004). Stem cell therapy for human neurodegenerative disorders-how to make it work. *Nature Med.* **10 Suppl**, S42–50.

Picard-Riera, N., Nait-Oumesmar, B., and Baron-Van Evercooren, A. (2004). Endogenous adult neural stem cells: limits and potential to repair the injured central nervous system. *J. Neurosci. Res.* **76**, 223–231.

Reynolds, B. A., and Weiss, S. (1992). Generation of neurons and astrocytes from isolated cells of the adult mammalian central nervous system. *Science* **255**, 1707–1710.

Romanko, M. J., Rola, R., Fike, J. R., Szele, F. G., Dizon, M. L., Felling, R. J., Brazel, C. Y., and Levison, S. W. (2004). Roles of the mammalian subventricular zone in cell replacement after brain injury. *Prog. Neurobiol.* **74**, 77–99.

Seri, B., Garcia-Verdugo, J. M., McEwen, B. S., and Alvarez-Buylla, A. (2001). Astrocytes give rise to new neurons in the adult mammalian hippocampus. *J. Neurosci.* **21**, 7153–7160.

Tramontin, A. D., Garcia-Verdugo, J. M., Lim, D. A., and Alvarez-Buylla, A. (2003). Postnatal development of radial glia and the ventricular zone (VZ): a continuum of the neural stem cell compartment. *Cereb. Cortex* **13**, 580–587.

20

Sensory Epithelium of the Eye and Ear

Connie Cepko and Donna M. Fekete

Introduction

Humans rely heavily on both vision and hearing. Unfortunately, both deteriorate with age, partly because of the death of cells in the primary sensory organs, the eye and the ear. In addition, the frequency of disease genes that affect one or both of these modalities is relatively high. Stem cells able to replace some of the dying cells, either *in situ* or through engraftment, have been a hope for some time. In part, this is because there are few effective therapies for diseases of these tissues. Work aimed at identifying retinal and otic stem cells has been undertaken with more energy in the last several years because of the exciting findings for stem cells elsewhere in the body. Here, we review these recent findings in the context of normal development.

Introduction to Progenitor and Stem Cells in the Retina

The retina has served as a model of central nervous system (CNS) anatomy, physiology, and development (Rodieck 1998). Most studies aimed at understanding its development have concerned the production of the retinal neurons and glia from retinal progenitor cells. These cells were originally shown by lineage analysis to be multipotent throughout development, capable of generating both neurons and glia, even in a single, terminal cell division. Retinal progenitor cells do not appear to be totipotent except for the earliest progenitor cells, when clones can comprise all retinal cell types. Moreover, retinal progenitor cells do not appear to be able to proliferate extensively *in vivo* or following explantation and exposure to different culture conditions. Recent studies have been aimed at finding retinal stem cells (Tropepe *et al.* 2000 and Haruta *et al.* 2001). These studies have been conducted along the two lines established in the search for stem cells elsewhere in the CNS. One approach has been to search for mitotic cells capable of generating retinal neurons in the adult *in vivo*. The other approach has been to culture cells in growth factors. Both types of experiments have begun to yield promising answers, but much more needs to be done.

The Optic Vesicle Generates Diverse Cell Types That Can Undergo Transdifferentiation

To appreciate some of the intriguing observations concerning retinal stem cells, a review of the tissues that derive from the optic vesicle is needed (Barishak 2001). The optic vesicle is an evagination of the neural tube where the diencephalon and telencephalon meet. The vesicle at first protrudes as a simple evagination when the neural tube forms. Soon thereafter, the vesicle undergoes an invagination to form a two-layered optic cup. The outer cup will form a nonneural structure, the retinal pigmented epithelium (RPE), as well as with other support structures of the eye (Figure 20-1). The RPE is a single layer of epithelial cells heavily pigmented to capture stray light that passes through the retina. It performs several support functions, including such highly specialized functions as the isomerization of *trans* to *cis* retinal to allow the photopigments, the opsins, to continue to capture light. The RPE expresses many specific gene products. A recent transcriptome analysis conducted using serial analysis of gene expression (SAGE) showed that 40% of the RPE SAGE tags did not have a corresponding cDNA in GenBank. This is a much higher rate of unknowns than is seen in other tissues (e.g., the human retina).

The inner wall of the optic cup forms the neural retina (Figure 20-1). The primary sensory cells are the photoreceptors (PRs), which comprise two types, the rods and the cones. The rods are active under dim light, and the cones are active under daylight conditions. In addition, there are several types of interneurons, horizontal cells, amacrine cells, and bipolar cells and an output neuron, the retinal ganglion cell. The retina also has one glial cell that spans the retinal layers, the Müller glial cell. During the early phase of retinal neurogenesis, retinal progenitor cells produce the various retinal neurons in a conserved fashion, typically beginning with production of ganglion cells and finishing with production of rod PRs, bipolar interneurons, and Muller glia. The production of these cells begins in the center of the retina and proceeds to the retinal periphery, or margin. In amphibians and teleost fish, there is continual growth at the margin throughout the life of the animal in a region termed the ciliary marginal zone, or CMZ. In addition, in fish, there is a late wave of production of rod PRs as the retina expands.

The developmental sequence at the periphery of the retina is complex and not at all understood at a molecular level,

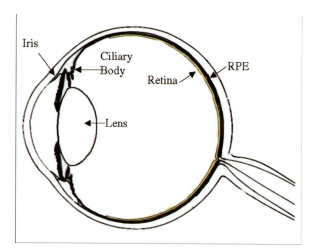

Figure 20-1. The eye is a complex tissue, developing from cells originating from the neural tube, the neural crest, the surface ectoderm, and the mesoderm. The retina is the neurosensory tissue that originates from the inner layer of the optic cup, and the RPE originates from the outer optic cup layer. Both the inner and outer optic cup layers contribute to the formation of the ciliary body and iris. The ciliary body also comprises cells from the neural crest, which form the ciliary muscle. The iris muscles derive from the outer layer of the optic cup. Stem cells have been isolated from the ciliary body and iris, and Muller glia in the retina have been found to divide and generate neurons in the early posthatch chick.

although it is an important region for stem cells. The margin is the fold that develops following invagination of the primary optic vesicle. Following the initial formation of the morphologically simple folds at the periphery, where the presumptive RPE and retina meet, some rather unusual morphogenetic events, including transdifferentiation, take place to form several anterior support structures for the eye. The ciliary body, with the associated pars plicata and pars plana, as well as the iris, form from this area (Figure 20-1). The pars plana and plicata each comprise two epithelial layers, one pigmented and one unpigmented. They are the site of attachment of the zonules, or suspensory ligaments of the lens. The unpigmented epithelial layer of the pars plicata and plana is continuous with the retina, and the pigmented layer is continuous with the RPE. The tight apposition of these two epithelial layers allows regulation of secretion from the ciliary body, as it is highly vascularized with a rather leaky type of blood vessel. Beyond secretion of aqueous (through the pars plicata) and vitreous humor (through the pars plana), the ciliary body also controls the shape of the lens. Neural crest-derived muscles form within the ciliary body and contract and relax the ligaments surrounding the lens during lens accommodation.

The iris is the shutter that opens and closes to allow more or less light to penetrate the eye. It includes a pigmented epithelial layer derived from the margin of the optic cup that is continuous with the RPE. It also has an initially unpigmented epithelial layer, the inner or posterior layer of epithelium, which is continuous with the retina. This layer gradually becomes pigmented, however, and additional pigmentation of

the iris is contributed by neural crest-derived melanocytes. Remarkably, the pupil is opened and closed by muscles that derive from the margin of the optic cup, the only ectodermally derived muscles in the body. This occurs because of transdifferentiation as initially pigmented cells separate from the epithelial sheet, proliferate, and form muscles. Thus, the retinal margin develops to serve several functions, and most of these diverse cell types derive from both the outer and the inner walls of the early optic cup.

Classical embryological experiments with birds, fish, amphibians, and mammals revealed a great deal of plasticity among the ocular tissues. For example, extirpation of most — but importantly, not all — of the retina leads to transdifferentiation of the RPE into the retina. This capacity exists only until embryonic day (E) 4 in chickens and E14 in mice. In urodeles, it can occur throughout life. In chicks, it was found to be induced by fibroblast growth factor (FGF) in RPE cultures. *In vivo*, it is not clear whether FGF is involved in the initial distinction between the retina and RPE, but delivery of FGF8 *in vivo* can trigger transdifferentiation. Wolffian regeneration in newts is a remarkable process by which the dorsal iris can regenerate a lens, originally not derived from the optic vesicle but from the surface ectoderm. This type of regeneration has not been seen in birds or mammals. All of these examples reveal that end-stage differentiated cells are not necessarily committed or irreversibly differentiated. Perhaps most remarkably, as explained later in this chapter, the pigmented cells derived from this area of the optic cup display broad developmental potential in adult mammals in that they are the source of retinal stem cells.

In Vivo Neurogenesis in the Posthatch Chicken

A search for stem cells in the developed retina was recently conducted by Reh and colleagues in the chicken (e.g. Fischer and Reh 2000 and 2001). Retinal neurogenesis in most of the chick retina is complete by E12. (Fischer and Reh 2000 and 2001) examined the posthatch chick (i.e., >E21) for the incorporation of bromodeoxyuridine (BrdU). They found that two areas could be labeled. In a normal retina without injury, the P7 retina was labeled in the ciliary margin, reminiscent of the aforementioned findings for amphibians and fish. These cells were followed using their BrdU label and were found to incorporate into the inner nuclear layer (INL), generating bipolar and amacrine neurons. No cells were found in the outer nuclear layer (ONL), the layer containing PRs. Antigens consistent with the INL fates were also observed. The newly generated cells appeared progressively more centrally as harvests were made later and later. These findings suggest that the CMZ cells were generating more retinal neurons to accommodate additional growth of the eye, previously thought to occur only through an expansion of the volume of the vitreous cavity and stretching of the retinal tissue. However, this does not occur throughout the life of the chicken as it does in amphibians and fish, since the growth stops a few weeks posthatch. The number of mitotic cells within the CMZ was not increased following injection of a toxin, N-methyl–D-aspartate (NMDA), unlike the response seen in the *Xenopus* eye. However, injection of 100-ng doses of epidermal growth factor (EGF), insulin, or insulin-like growth factor 1 (IGF-1), but not FGF, did increase the mitotic activity in this area. In addition, if insulin was applied with FGF2, cells with gene expression profiles and processes consistent with the ganglion cell fate were observed.

Another site of BrdU incorporation in the posthatch chick could be identified following application of the toxin NMDA, which primarily targets amacrine cells. If BrdU was applied two days after injection of NMDA at P7, incorporation of BrdU into Muller glial cells in the central retina was observed. Labeling with BrdU at one day or three days following toxin administration led to few BrdU-labeled cells. It thus appears that a process triggered by toxin administration requires approximately two days to stimulate one or two rounds of cell division. This response is also developmentally limited, as few BrdU-labeled cells were observed after P14, and the response was lost in a central-to-peripheral manner, similar to the initial wave of neurogenesis in the retina. Muller glial cells do possess the ability to undergo *reactive gliosis*, a phenomenon associated with various types of retinal damage in adult mammals and birds. Reactive gliosis occurs in astrocytes throughout the CNS and is characterized by limited cell division; expression of intermediate filament proteins, such as vimentin and glial fibrillary acidic protein (GFAP); and increased process outgrowth. It has not been established that this type of cell division leads to production of neurons elsewhere in the CNS or in the retina of adult birds.

The BrdU-labeled Muller glial cells induced by toxin treatment were found in the ONL and INL, whereas Muller glial nuclei are typically only found in the INL. Some of these BrdU-labeled cells co-expressed two markers of retinal progenitor cells, Chx10 and Pax6, and some expressed a bHLH gene, *CASH-1*, a marker of early retinal progenitor cells. A small percentage (<10%) of the BrdU$^+$ cells subsequently were found to have neuronal morphology and to express markers of amacrine cells and bipolar cells. Transiently, however, many cells expressed a neurofilament (NF) marker, normally expressed on horizontal and ganglion cells in the retina. This appeared to be transient as the number of NF$^+$ cells decreased, and markers and morphology consistent with mature ganglion and horizontal cells were not seen. Researchers did not find markers of PR cells. Many of the BrdU-labeled cells persisted in what appeared to be an arrested state for at least 12 days. They explored the possibility that the *in vivo* environment was limited for production of PR cells by culturing the toxin-treated retinas, but they were similarly unable to observe the genesis of PR cells, though they did find that toxin-treated retinas proliferated *in vitro* more than the untreated controls.

The division of Muller glia, following by genesis of neurons in the chick, can also be stimulated by growth factors in the absence of added toxins. Application of both insulin and FGF2 by intraocular injection, starting at P7 and continuing for three days, led to the production of many mitotic Muller

glial cells. Fourteen days after the last injection, some BrdU$^+$ cells showed markers of amacrine and perhaps ganglion cells, and others showed markers of Muller glia. Similar to the findings following NMDA injections, no markers of PR cells were observed. To explore whether the type of toxin, and thus the target cell killed by the toxin, combined with growth factor injection might give a more specific replacement of the targeted cells, researchers injected several types of toxins with insulin and FGF2. When ganglion cells were targeted by the toxins, kainic acid, or colchicine, more cells with ganglion cell markers were observed. These studies lead to the hope that replacement of the specific cells that die in various retinal diseases, such as ganglion cells in glaucoma, might be effectively replaced following stimulation of stem cells with the right cocktail of factors.

Several groups have searched for proliferation in the uninjured retina of mammals. Injection of BrdU intraperitoneally into 4 week-old rats for 5 days was done to determine if cells were proliferating. The only incorporation that was seen in the ciliary margin — that is, no labeling of Muller glial cells was observed. Opossum and mouse (as well as quail) were examined and no incorporation was found in centrally located Muller glia. Some incorporation of BrdU was seen in the ciliary margin of quail, though less than in the chicken, and a few labeled cells in the ciliary margin of the opossum, but no labeled cells in the ciliary margin were found in the mouse.

Radial glial cells, astrocytes, and Muller glia have been shown to share some antigens with progenitor cells. This resemblance is more extensive than previously appreciated. A systematic SAGE analysis of gene expression in the developing and mature murine retina was carried out. There were 85 genes preferentially expressed in Muller glia in the mature retina. Of these genes, the majority were also found in retinal progenitor cells. Some of these genes, such as cyclin D3, undoubtedly show that Muller cells retain the ability to divide but other genes are more enigmatic. Nonetheless, the studies from Reh and colleagues mentioned previously and the SAGE data argue that Muller glia should be explored further as a source of cells that might replace dying neurons. This notion is in keeping with the idea that radial glial cells elsewhere in the developing CNS, as well as astrocytes in the mature forebrain, can serve as neuronal progenitor cells and stem cells, respectively.

Growth of Retinal Neurospheres from the Ciliary Margin of Mammals

Reynolds and Weiss demonstrated in 1992 that cultures of CNS tissue in the presence of EGF, FGF, or both would lead to balls of cells with indefinite proliferation capacity, or *neurospheres*. Neurospheres were subsequently shown to produce neurons, astrocytes, and oligodendrocytes and were thus identified as originating from neural stem cells. Several groups have applied these protocols to the retina. Cultures from ocular tissues of mouse, rat, human, and cow have been made using FGF and EGF. Retinal neurospheres have been recovered and appear to have cells with indefinite proliferation potential, multipotency, and possibly totipotency.

Tropepe *et al.* (2000) reported that the ciliary body from adult mice had the most enriched source of retinal stem cells. They examined E14 RPE and retina and adult retina, RPE, iris, and ciliary body for production of neurospheres in the presence of FGF, EGF, or both. They were unable to recover any neurospheres with the proliferative capacity of stem cells from the embryonic or adult retina or from the adult iris, RPE, and ciliary muscle. A few neurospheres were recovered from the embryonic RPE, which included the peripheral margin, the precursor to the ciliary body. In the adult, neurospheres were only recovered from the pigmented cells of the ciliary body (Figure 20-1), termed the pigmented ciliary margin (PCM). Although there was some recovery from the E14 RPE, the number per eye increased tenfold in the adult PCM compared to the entire RPE of the E14 retina. This curious finding suggests that these cells are formed at the end of development and not the beginning, the opposite of what one might have predicted. Alternatively, during the maturation of the PCM, there is an expansion of a few early stem cells. Tropepe *et al.* found that the stem cells were either rare or hard to culture, as only 0.6% of the adult PCM-plated cells would produce a neurosphere in the presence of FGF2. They did not require exogenous FGF, as they could arise, albeit at a reduced frequency, in its absence. This was presumably because of endogenous FGF, as addition of anti-FGF antibody reduced the formation of neurospheres.

Because the PCM neurospheres originated with the pigmented cells of the ciliary body, it was of interest to determine whether pigment was necessary for their ability to be stem cells. This was examined in albinos, where they were isolated at a comparable frequency to that of pigmented animals. The RPE and ciliary body normally do not express Chx10. However, upon the genesis of neurospheres, cells began to express this retinal marker. If cultured under conditions that favor differentiation, the cells turned on markers of PRs, bipolar cells, and glia. However, no markers of amacrine and horizontal cell interneurons, or of ganglion cells, were seen. It is significant that rhodopsin, a definitive marker of a PR cell, can be expressed. Rhodopsin is not expressed by other CNS neurospheres or neural cell lines derived from other CNS locations. Retinal progenitor cells normally do not make oligodendrocytes, and the oligodendrocyte marker, O4, was not observed.

There is a naturally occurring null allele of the paired-type homeobox gene, *Chx10*, in mice (*orj*). This mutant has a retina and RPE approximately tenfold smaller than wild-type mice, with an expanded ciliary margin. When cultures were made from the ciliary margin of *orj*, approximately fivefold more neurospheres were recovered. These spheres were approximately one-third the size of the wild-type spheres, in keeping with the finding that *Chx10* is required for the full proliferation of retinal progenitor cells. Thus, *Chx10* must be involved with the allocation or regulation of the number of stem cells in the PCM.

Retinal neurospheres from the ciliary body of rats have also been isolated from postmortem human and bovine eyes. The rat neurospheres were dependent upon FGF2 and could not be recovered from the retina, the RPE, or the nonpigmented portion of the ciliary epithelium. Haruta *et al.* (2001) also isolated retinal stem cells from the rat eye, but they used iris tissue rather than ciliary margin tissue. As shown in Figure 20-1, the iris and ciliary body are adjacent to each other at the margin of the eye. The iris cells were cultured as an explant in the presence of FGF, rather than as dissociated cells. Tropepe *et al.* had reported previously that, unlike the ciliary body cells, dissociated cells from the iris did not generate neurospheres. Cells that migrated out of the iris explants were able to express some neural markers, such as NF200, but not PR-specific markers. However, if the cells were transduced with *crx*, a homeobox gene important in PR differentiation, approximately 10% of the cells expressed rhodopsin and recoverin, two markers of PR cells. The ciliary body-derived neurospheres can generate cells bearing the same retinal markers, including rhodopsin, without transduction of *crx*. The difference may be caused by culture conditions. The ciliary body-derived cells can be grown as spheres, and this environment may support PR development without a need for *crx* transduction. When Haruta *et al.* cultured ciliary body-derived cells as monolayers, as they cultured iris-derived cells, the ciliary body-derived cells did not express PR markers.

Haruta *et al.* point out that tissue from the iris can be readily obtained for autologous grafts. It is far more difficult to obtain tissue from the ciliary body, with an accompanying risk of damage to the ciliary body. However, because Tropepe *et al.* reported that postmortem human and bovine PCM could produce spheres at low frequency, it is possible that humans could be used as a source of donor cells. This would not be as immunologically compatible as an autologous graft, but it nonetheless might suffice, for it is still not clear how well tolerated retinal grafts might be.

That the pigmented ciliary body and iris cells are the source of retinal stem cells is rather surprising. It was expected that the periphery of the eye might be the area where stem cells would reside because this is the location of stem cells in amphibians and fish. In the CMZ of amphibians and fish, the nonpigmented cells contiguous with the retinal epithelium, located nearest to the retina, make more retinal cells. However, stem cells have not been recovered from this area in mammals. An alternative prediction might have been that retinal stem cells would be the RPE cells. As explained previously, the RPE is quite plastic early in development in mammals and chicks and throughout life in Urodeles, maintaining the ability to make retinal cells in response to certain conditions. However, the adult mammalian RPE has not been shown to generate neurospheres when cultured under the conditions described previously. Moreover, even the embryonic RPE does not supply many neurospheres in the neurosphere culture conditions. In contrast to the preceding predictions, the most robust source of retinal stem cells in the adult mammal is the ciliary body pigmented cells. These cells are immediately adjacent to, and contiguous with, the RPE, but they are not RPE cells, at least in terms of function, as described previously. We do not have markers that would help to further define them. They derive from the outer walls of the optic cup, and the inner walls normally produce the retina in normal development. It should be noted, however, that whether the iris stem cells are derived from the inner or outer walls of the optic cup is not clear at this time. Although the iris stem cells are pigmented, both the inner and outer walls of the optic cup normally develop pigmentation in the iris. Both iris- and ciliary body-derived stem cells are pigmented, which may provide a useful marker for their prospective isolation.

Prospects for Stem Cell Therapy in the Retina

A large number of the diseases of the retina are caused by degeneration of PR cells. Approximately 40% of the genes identified as human disease genes that lead to blindness are rod specific. Many of these diseases nonetheless lead to the loss of cone PR cells. This nonautonomous death of cones is the reason for loss of daylight, high-acuity vision. Thus, replacement of dying rods, or the retardation of the death of rods, might prevent or slow the death of cones. Replacement of dying cones themselves is another potential therapeutic approach. This is particularly appropriate when the etiology of the disease is not clear, as is the case in the most prevalent disease, age-related macular degeneration. The source of either rod or cone PR cells could be the endogenous stem cells themselves. The best scenario would be the stimulation of the division of Muller glia, which are distributed through the retina, followed by the induction of PR differentiation.

Unfortunately, as noted previously, Müller glia have not been observed to generate PR cells in the chick or any mammal that has been investigated. Nonetheless, future studies might lead to a manipulation that would stimulate PR production by Muller glia. A second source of PR cells might be the endogenous stem cells in the ciliary body or iris, left *in situ*. Although these cells can produce PR cells when cultured, they have not been shown to generate PR cells *in situ*. In addition, unless the cells could be made to migrate and cover the central retina, where most of our high-acuity vision occurs, they would not lead to retention or recovery of high-acuity vision. Still, if they could lead to retention of peripheral vision, some therapeutic benefit would be realized. Finally, engraftment of stem cells or of PR cells generated *in vitro* by stem cells can be attempted. The problems of graft rejection, if not an autologous graft, would have to be confronted, but they might not be as difficult to overcome as engraftment to sites in the periphery. Furthermore, iris-derived cells might be used as an autologous graft. Preliminary data from Tropepe *et al.* of engraftment of the neurospheres derived from the PCM are promising. Injection of such cells into the vitreous body of postnatal day 0 rats led to the formation of many PR cells, expressing rhodopsin, in the ONL. If such cells could be formed in a diseased retina, then two

possible benefits might be realized. One would be to simply prevent further degeneration of endogenous PR cells, which, as mentioned previously, can die by nonautonomous processes. The second benefit might be that the engrafted PR cells synapse with second-order neurons and provide vision themselves. To date, this has not been achieved. The concern here is that the site of engraftment might not support synaptogenesis — particularly at the advanced stage when much of the retina has degenerated, likely the stage when such therapies would be attempted. Nonetheless, such strategies are worth pursuing, particularly now that there are stem cells that can be manipulated to generate retinal cells; our understanding of the processes of the development of retinal cells has similarly been advancing.

Development and Regeneration of Tissues Derived from the Inner Ear

The entire vertebrate inner ear derives from the otic placode, a thickening of the dorsolateral surface ectoderm immediately lateral to the hindbrain. Like the lens and olfactory placodes, the otic placode invaginates and pinches off to form a single-layered ball of cells, now called the otic vesicle. From this simple epithelium, a large variety of tissues and cells types arise. The otic ectoderm is neurogenic for the first-order neurons of the eighth cranial ganglion, the statoacoustic ganglion. The ganglion neuroblasts are the earliest recognizable cell type; they delaminate from the otic ectoderm at the otic cup stage even before it completes vesicle formation. The otic vesicle is also sensorigenic, generating six to eight different sensory patches. Inner ear sensory organs subserve hearing and balance and are differentiated according to their function. There are three major classes of sensory organs: macula, crista, and acoustic. Acoustic organs vary substantially in structure and sensitivity across the vertebrates, reaching a pinacle in the mammalian organ of Corti. In all inner ear sensory organs, the mechanosensory hair cells are interspersed among a field of supporting cells essential for hair cell survival and function.

Finally, beyond the sensory patches, several types of nonsensory tissues derive from the otic epithelium. The most highly differentiated is the tissue that secretes the extracellular fluid, called endolymph, which bathes the apical surfaces of all otic epithelial cells. Endolymph contains an unusually high concentration of potassium ions. The tissue responsible for endolymph production is anatomically complex, highly vascularized, and endowed with many ion pumps and channels. Other nonsensory epithelia flank the sensory organs and may contribute, with the supporting cells, to the secretion of the specialized extracellular matrices perched above the hair cells to enhance their mechanosensitivity.

Lineage studies have yet to reveal all possible relationships among the constellation of inner ear cells, so we do not yet know if individual otic placode cells are truly pluripotent for all inner ear cell types. We are confident that mechanoreceptors and their supporting cells share a common progenitor in the bird and zebra fish inner ear and in the regenerating salamander lateral line. There is also evidence that sensory and nonsensory cells can be clonally related in the chicken and mouse ears. Furthermore, otic neurons and sensory cells can be related in the bird. To date, there is no direct evidence for the existence of a true otic stem cell — that is, one that divides asymmetrically to replicate itself while generating a daughter with an alternate cell fate or fates. Nonetheless, lineage studies indicate that multipotent progenitor cells constitute a normal feature of inner ear development, leaving open the possibility that similar cells may lurk in mature ears, where they might be poised to expand with appropriate signals or culture conditions.

A variety of growth factors and growth factor receptors have been associated with developing inner ears. However, it is important to distinguish between growth factors that may regulate cell proliferation and those that may influence cell fate specification in other ways, such as the role of FGFs in otic induction. Furthermore, many of the growth factor assays are performed *in vitro*, with the inherent risk that cells and tissues can change their growth factor responsiveness depending upon culture conditions This caveat notwithstanding, members of several growth-factor families can enhance cell proliferation of developing otocyst or neonatal inner ear tissues, either alone or in combinations, when presented in culture. These include bombesin, EGF, FGFs, GGF2, heregulin, insulin, IGFs, PDGF, and TGFα.

In Vivo Neurogenesis in Postembryonic Animals

PROLIFERATION IN NORMALS (OR AFTER GROWTH FACTOR TREATMENT)

The vestibular maculae are involved in sensing gravity, and in fishes and amphibians these organs continue to increase in size throughout life. Here, we focus on the sensory organs of warm-blooded vertebrates, where there is a marked contrast between birds and mammals in the timing of inner ear organogenesis. In both classes, sensory organs stop generating new cells approximately midway through embryogenesis. A notable exception is the vestibular maculae of birds, where there is ongoing addition and death of cells well beyond hatching. Cell counts or BrdU labeling have led to estimates that hair cells turn over with half-lives of 20–52 days in the chicken utricular macula. Within two weeks of hatching, nearly 500 cells in the saccular macula and 1400 cells in the utricular macula may be added per day with the steady-state addition of 850 hair cells per utricle per day reached 60 days after hatching. However, these numbers are apparently achieved in the absence of an amplifying progenitor. Rather, most progenitors divide once to produce a sibling pair consisting of one hair cell and one nonhair cell (presumed to be the supporting cell). BrdU$^+$ clusters exceeding three cells are rare 2 to 4 months after labeling. Thus, if there is a self-renewing stem cell pool in the mature avian macula, it consists of cells that divide on an extremely slow timescale. Interestingly, the neuronal colony-forming cells of the olfactory epithelium,

thought to be the true stem cell of this sensory organ, are also rare (1 in 3600 purified progenitors) and divide at a very slow rate. Ongoing receptor cell turnover (and regeneration) in the olfactory epithelium utilizes a transient amplifying progenitor pool that in turn generates a population of immediate neuronal precursors; the latter will divide symmetrically to make differentiated olfactory receptor cells.

In contrast to lower vertebrates, in mammals the vestibular macula is quiescent from birth. So, too, are the auditory organs of both birds and mammals. For example, BrdU injections failed to label cells in the adult mouse organ of Corti. However, in mutant mice that lack particular cyclin-dependent kinase inhibitors, cell turnover continues in postnatal animals. Dividing cells are observed several weeks after birth among hair cell (p19$^{Ink4d-/-}$) or supporting cell (p27$^{kip1-/-}$) layers. Prolonged mitosis in the p27^{kip1}-null is accompanied by the differentiation of supernumerary hair cells and supporting cells. In both mutants, many hair cells eventually undergo apoptosis, leading to hearing loss. Nonetheless, these data suggest that the differentiated organ of Corti, which has never been shown to regenerate naturally, has the potential to harbor cells that can divide and differentiate under appropriate circumstances. In this context, it is intriguing that a subset of cells in both auditory and vestibular organs of the neonatal mouse express the neural stem cell marker, nestin (Figure 20-2). Nestin is rapidly down-regulated about one week after birth, although it persists in nonsensory cochlear cells to day 15.

Several growth factors and cytokines can affect cell proliferation in undamaged, mature inner ear sensory organs or

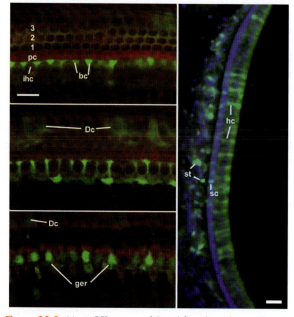

Figure 20-2. Nestin-GFP in organ of Corti (left) and utricular macula (right) of P5 mouse. In the cochlea, GFP is seen in border cells (bc) surrounding inner hair cells (ihc), Deiters cells (Dc) surrounding outer hair cells (outer hair cell rows 1, 2, and 3 are indicated), and greater epithelial ridge cells (ger). In the utricular macula, GFP is seen in stromal cells (st), supporting cells (sc) and hair cells (hc). Figure courtesy of Ivan Lopez (UCLA).

within sheets of sensory epithelial supporting cells. Insulin, IGF-1, or IL-1β enhanced proliferation of normal chicken utricular cells *in vitro*, and FGF-2 reduced proliferation. Rodent utricular cells proliferated in response to GGF2, EGF plus insulin, neu differentiation factor, TGFα, or TGFα plus insulin but lost response to heregulin by adulthood.

PROLIFERATION AFTER DESTRUCTION OF CELLS

The strongest evidence that progenitor cells reside in quiescent sensory organs of warm-blooded vertebrates comes from regeneration studies. Beginning with ground-breaking work in the late 1980s, many studies have shown that damaged sensory organs will regenerate new hair cells in chickens, largely through a proliferative mechanism. Thus, despite the absence of ongoing turnover, supporting cells of the auditory organ mount a vigorous mitotic response to damaging conditions: up to 15% of them enter the cell cycle to generate both hair cells and new supporting cells. Whether supporting cells coexist alongside self-renewing stem cells remains an open question. Data suggest that only 1 to 4% of cycling cells in the regenerating basilar papilla will divide more than once within a three-day window after ototoxic drug treatment. Even among this pool, ongoing proliferation appears to be extremely modest, although lineage analysis has not yet been performed to provide an unequivocal measure of clonal expansion. Like the basilar papilla, chicken macular cells also divide and differentiate in response to hair cell loss. The proliferation index of the drug-damaged macula rises in the presence of TGFα or TNFα.

Once again, we contrast mammals to lower vertebrates: The mammalian macula has only a weak proliferative response to hair cell loss, with scant evidence that hair cells can be regenerated through a cycling intermediate. Instead, the maculae primarily restore their hair cells through direct transdifferentiation of supporting cells or through self-repair of sublethally damaged hair cells. The limited proliferation that accompanies recovery may serve to replace transdifferentiated supporting cells rather than hair cells. Hair cell recovery is promoted by the addition of TGFα, IGF-1, retinoic acid, and brain-derived neurotrophic factor *in vivo* or *in vitro*. The weak proliferative response of the cultured, drug-damaged macula averaged 26 BrdU+ cells per sensory organ. This proliferation was enhanced tenfold by addition of heregulin (a member of the neuregulin family) and to a lesser extent by EGF or TGFα alone or with insulin. Neither hair cells nor supporting cells of the mammalian organ of Corti responded to heregulin, although more remote nonsensory epithelial cells of the cochlea did.

TRANSCRIPTION FACTOR REQUIREMENTS

Although many cell types are generated from otic epithelium, relatively few transcription factors have been definitively associated with cell fate specification in the ear. *NeuroD, Neurogenin-1*, and *Eya1* are essential for otic ganglion cell fate. *Brn3a/Brn3.0* is needed for ganglion cell survival and differentiation. *Math-1* is required for hair cell development and survival, and *Pou4f3/Brn3c/Brn3.1* is needed for subsequent

hair cell differentiation. Ectopic delivery of *Math1* leads to ectopic production of partially differentiated hair cells in the mammalian ear, both in the adult guinea pig and in cultures of postnatal rat sensory organs. Thus, some cells retain the capacity to switch to a hair cell fate even into adulthood.

In Vitro Expansion of Otic Progenitors

Several groups have used immortalizing oncogenes to isolate cell lines from the developing inner ear and explore their differentiation potential. Efforts to expand unadulterated otic progenitor pools are just beginning, using either immature or differentiated otic epithelium as starting material. The work is presented in order based on age of the starting tissue.

Work in the laboratories of Segil and Groves at the House Ear Institute in Los Angeles defined culture conditions (EGF plus periotic mesenchyme) that permit E13.5 cochlear progenitors to persist in culture long after they would normally become postmitotic. Mitotic progenitors generate islands of Math-1+ hair cells, with numbers of hair cells continuing to increase two weeks after plating.

Malgrange and collaborators dissociated cells from the newborn rat organ of Corti, approximately 5 to 6 days after sensory progenitors become postmitotic *in vivo*. Filtering through fine (15-μm) nylon mesh resulted in the isolation of a population of small cells, 98% of which expressed nestin. Beginning with cell suspensions, spherical colonies called *otospheres* developed when cultured in the presence of EGF, FGF2, or both factors. After only two days of culturing, BrdU+ progenitors expressing the hair cell marker myosin VIIA were observed, albeit in very small numbers (one per colony on average). About two cells per colony were immunopositive for markers of the supporting cells of the organ of Corti. The number of myosin VIIA+ cells increased to five cells per colony by 14 days, often appearing as a coherent islet. Some ultrastructural features of hair cells were evident in rare cells after both 2 and 14 days *in vitro*. Hair cell differentiation was not enhanced by switching the otospheres to "adherent" culture conditions that had been reported to induce neuronal differentiation in neurospheres. It is important to note that the method used in this study did not ensure that each otosphere originated from a single progenitor cell.

A study by Zhao reports culturing spheres from the isolated organ of Corti of adult guinea pig. Initially, the spheres did not express nestin or a supporting cell marker, cytokeratin. Serum or EGF supplements, or long-term culture, induced nestin expression and allowed differentiation of a small number of cells as hair cells (calretinin+, myosin VIIA+, prestin+, or Brn3.1+), supporting cells (cytokeratin+ or connexin+), neurons (NF+ or βIII-tubulin+), or astrocytes (GFAP+). The appearance of differentiated hair cells was extremely rare, at less than 1% of the cells in long-term cultures.

Most promising to date are the generation and differentiation of spherical cultures from single cells isolated from the utricular macula of the adult mouse. Taking their cue from the growth factor responsiveness of damaged vestibular sensory organs, Heller's laboratory used EGF, IGF-1, and bFGF to enhance sphere formation. The combination of EGF plus IGF-1 was most effective. Nestin[+], sphere-forming cells could be dissociated into single cells and then expanded into new spheres through several rounds. Approximately 2.5 spheres could be formed at each passage, suggesting that a small number of cells retain sphere-renewal capability. To induce differentiation, spheres were moved to adherent culture conditions in the presence of serum, but then grown for 14 days in serum-free conditions. The cells down-regulated nestin and other markers of the early otic vesicle and up-regulated markers of several differentiated cell types. Cells expressing hair cell markers were present in up to 15% of differentiated cells. Many showed features consistent with rudimentary stereociliary bundle formation and were surrounded by cells with an expression profile consistent with supporting cells. Math-1[+] cells co-labeled with BrdU, indicating that these hair cells arose from a proliferative progenitor. Significantly, neurospheres grown from the subventricular zone of the mouse forebrain, cultured under identical conditions, did not generate hair cells. This suggests that the utricular stem cells have a special capacity to form inner ear mechanoreceptors. Macular-derived spheres also produced a significant percentage of cells with neuronal (6%) or astrocytic (35%) phenotypes, cell types normally absent from the macular epithelium. Spheres could also generate an array of ectoderm, mesoderm, or endoderm derivatives when the cells were delivered into the amniotic cavity of stage 4 chicken embryos. The incidence of sphere-forming stem cells was rare even under optimal growth conditions: 0.07% of plated cells. This is consistent with the absence of BrdU-labeling of adult sensory organs in the mouse and suggests that stem cells may be both rare and quiescent *in vivo*.

Recently, Heller's group also defined culture conditions that induced ES cells to form spheres containing many BrdU[+], nestin[+] cells. Under growth conditions, the cells expressed nestin and markers of early otic vesicle. Under differentiating conditions, early otic markers plummeted, and markers of hair cells and supporting cells rose. This is extremely encouraging as it suggests that it may be unnecessary to start with endogenous ear tissue to generate otic progenitors for therapeutic purposes.

Prospects for Therapy

As methods for culturing otic stem cells become established, one can ask whether the addition of different transcription factors (such as Math1) can induce differentiation of one cell type over another. Stem cells or differentiated cells derived from various sources could then be implanted back into the animal to ask whether the cells will integrate and provide restoration of function in animal models of inner ear cell loss. The ear has some definite advantages for delivery of cells or gene transfer vectors, such as viruses. Surgical approaches to the fluid compartment of the inner ear provide access to the inner ear hair cells without requiring systemic delivery. For example, it is possible to inject through the round window delivering substances, such as neurotrophins, that can influ-

ence survival of sensory tissues or ganglion cells. Delivery of cells that release soluble molecules, such as growth factors, could potentially provide functional restoration without necessarily restoring structural integrity. On the other hand, structural integration of replacement mechanoreceptors will probably be essential, even if extremely difficult, in view of the precision with which hair cell stereocilia must interact with the nonsensory matrices overlying them. Replacement of functional ganglion neurons, rather than sensory receptor cells, may be less problematic. We anticipate considerable progress in these and related therapeutic approaches over the next decade, although substantial technical hurdles remain.

ACKNOWLEDGMENTS

D. M. Fekete thanks S. Heller, H. B. Zhao, and T. Nakagawa for sharing unpublished data and I. Lopez for Figure 20-2.

KEY WORDS

Ciliary margin   The peripheral region of the eye where retinal stem cells reside. In fish and amphibians, retinal cells are generated *in vivo* from these stem cells throughout life. It is also the area from which retinal stem cells have been isolated in mammals.

Otosphere   Spherical colonies generated by dissociation and culturing of inner ear sensory cells from the newborn rat organ of Corti.

FURTHER READING

Barishak, Y. R. (2001). *Embryology of the eye and its adnexa*, 2nd ed. Karger, Basel.

Corwin, J. T., and Warchol, M. E. (1991). Auditory hair cells: Structure, function, development, and regeneration. *Annu. Rev. Neurosci.* **14**, 301–333.

Fekete, D. M., and Wu, D. K. (2002). Revisiting cell fate specification in the inner ear. *Curr. Opin. Neurobiol.* **12**, 35–42.

Fischer, A. J., and Reh, T. A. (2000). Identification of a proliferating marginal zone of retinal progenitors in postnatal chickens. *Dev. Biology* **220**, 197–210.

Fischer, A. J., and Reh, T. A. (2001). Muller glia are a potential source of neural regeneration in the postnatal chicken retina. *Nat. Neurosci.* **4**, 247–252.

Haruta, M., Kosaka, M., Kanegae, Y., Saito, I., Inoue, T., Takahashi, M., Honda, Y., Kageyama, R., and Nishida, A. (2001). Induction of photoreceptor-specific phenotypes in adult mammalian iris tissue. *Nat. Neurosci.* **4**, 1163–1164.

Holley, M. (2002). Application of new biological approaches to stimulate sensory repair and protection. *Br. Med. Bull.* **63**, 157–169.

Li, H., Roblin, G., and Heller, S. (2003). Pleuripotent stem cells from the adult mouse inner ear. *Nature Medicine* **9**, 1293–1299.

Oesterle, E. C., and Hume, C. R. (1999). Growth factor regulation of the cell cycle in developing and mature inner ear sensory epithelia. *J. Neurocytol.* **28**, 877–887.

Rodieck, R. W. (1998). *The first steps in seeing*. Sinauer, Sunderland, MA.

Tropepe, V., Coles, B. L., Chiasson, B. J., Horsford, O. J., Elia, A. J., McInnes, R. R., and van der Kooy, D. (2000). Retinal stem cells in the adult mammalian eye. *Science* **287**, 2032–2036.

21

Epithelial Skin Stem Cells

Tudorita Tumbar and Elaine Fuchs

Introduction

Skin is a complex tissue made of a structured combination of cell types. It has an enormous regenerative capacity and contains several different kinds of stem cells (SCs). Fundamental differences likely exist between embryonic skin stem cells and adult skin stem cells. In this chapter we focus on epithelial skin stem cells (ESSCs) in postnatal mouse skin. Nonhuman studies allow for high flexibility and increased depth of study and are less restricted by ethical issues. ESSCs are thought to reside in a specialized area of the hair follicle called *the bulge*. Cells in this compartment possess the ability to differentiate, at least under stress conditions, into different cell lineages to regenerate not only the hair follicle but also the sebaceous gland (SG) and the epidermis. In adult skin, stem cells are thought to be the slow cycling cells that retain Bromo-deoxy-uridine (BrdU) DNA label over long periods of time. How these cells maintain their properties of self-renewal and differentiation remains largely unknown. Whether these cells are fundamentally different from their progeny, or whether it is simply their location, the so-called niche that instructs them to behave as stem cells, remains an open question. Recent work has begun to elucidate signaling mechanisms that control the fate of these cells in mouse skin. The challenge to isolate slowly cyling cells and to identify markers of ESSC, which may specify unique characteristics of these cells, is now overcome, opening new and exciting avenues for future insight into understanding what maintains these cells in a dormant but potent state.

A Brief Introduction to Mouse Skin Organization

The complex process of constructing the protective cover of the body starts around day 9 of mouse embryonic life (E9). Through a succession of signal exchanges between the ectoderm and mesoderm, a very structured tissue emerges designed to seal and protect the body of the animal against a diverse range of environmental assaults. The barrier function, or the sealing of the body from the external environment, is essential for the survival of the animal and is fully completed at E18, the day before the mouse is born. Hair follicle morphogenesis starts around E13 and takes place in waves until just

after birth. The hair follicles are specified embryologically, and, consequently, the maximum number of hair follicles that an animal will have for the rest of its life is determined before birth.

Mature skin is composed of two main tissues: the *epidermis* and its appendages, largely composed of specialized epithelial cells (keratinocytes) and the *dermis*, largely composed of mesenchymal cells. Epidermis consists of an innermost basal layer (BL) of mitotically active keratinocytes that express diagnostic keratins K5 and K14. As these cells withdraw from the cell cycle and commit to a program of terminal differentiation, they remain transcriptionally active and move upward toward the skin surface. As they enter the spinous layers, they first protect themselves by switching from expression of K5 and K14 to K1 and K10, which form keratin filaments that bundle and provide a robust inner strength to the cells. As they complete this process, they next go about producing the barrier by synthesizing and depositing proteins, such as involucrin, loricrin, SPRR, and others beneath the plasma membrane. In addition, as the cells enter the granular layer, they begin to produce and package lipids into lamellar granules, and they produce filaggrin, which bundles the keratin filaments even further to form cables. As the cells complete these tasks, an influx of calcium activates transglutaminases, which cross-link involucrin and its associates into a cornified envelope that serves as a scaffold for organization of the extruded lipids into external bilayers. Metabolically inert, these cells undergo an apoptotic-like loss of their nuclei and organelles, resulting in flattened, dead squames. These highly keratinized cells provide a seal to the body surface from which they are eventually sloughed, replaced by differentiating cells that are continuously moving outward. Epidermis regenerates itself every few weeks and has a remarkable ability to heal, expand, or retract in response to environmental cues.

Epidermal appendages, including the hair follicle and the sebaceous gland (SG) (Figure 21-1), are inserted deep into the dermis. The hair follicle is a very complex structure made of at least eight different cell types. The hair shaft is located in the middle of the follicle and grows upward, "breaking" the surface of the skin. Concentric layers of cells surrounding the shaft form the outer and inner root sheaths (ORS, IRS) (Figure 21-1). The basal layer of the epidermis is contiguous with the outer root sheath (ORS) of the epidermis, which the IRS degenerates in the upper portion of the follicle, liberating the hair shaft.

The basal epidermal layer and the ORS share a number of common biochemical markers, including K5, K14, and $\alpha 6 \beta 4$

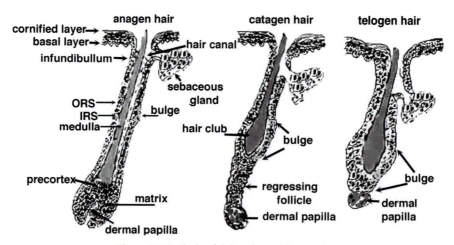

Figure 21-1. The hair follicle cycle in adult mouse skin.

integrins, and they represent a potent proliferative compartment of the skin. The highest proliferation, however, is seen in the matrix cells at the bulb of the follicle, which generate the IRS and the hair cells. Matrix cells surround a pocket of specialized mesenchymal cells, the dermal papilla (DP), with potential hair growth inductive properties. Matrix cells express a number of transcription factors (e.g., Lef1 and Msx-2) implicated in hair follicle differentiation. To some extent, IRS cells resemble epidermal granular cells, but their granules are uniquely composed of trichohyalin. In contrast, hair cells are more similar to stratum corneum because they are metabolically inert as they complete differentiation; however, they express a unique set of hair-specific keratins. Finally, the SG is located in the upper portion of the hair follicle, just above the arrector pili muscle and is made of fat-containing cells that will release their lipid content into the hair canal. The mitotically active cells of the SG express K5 and K14.

At birth, the morphogenesis of the hair follicle is almost complete. Certainly by day 4 of postnatal life all hair follicles in a mouse skin have reached maturity, and hair shafts start to appear at the skin surface. In postnatal life, the ORS of maturing follicles widens a bit on one side, creating a bulge. Located below the SG, and near or at the juncture of the arrector pili muscle, this specialized region is thought to be the compartment where epithelial stem cells may reside.

Stem cells in postnatal skin are required for self-renewal of the epithelial tissue. Like epidermis, cells of the postnatal hair follicles are in a state of flux, undergoing perpetual cycles of growth, regression, and rest. In relative synchrony at least for the early hair cycles, mouse hairs continue to grow and complete anagen, the growth phase of the hair cycle. Around 17 days of postnatal development, anagen ceases and massive cell death occurs in the bulb of the hair follicle, the catagen phase. Follicle cells below the bulge area are destroyed, except for the DP, which moves upward by virtue of its attachment to a shrinking basement membrane that separates the epithelial and mesenchymal compartments. The DP comes to rest just beneath the bulge. At this stage, hair follicles enter

their rest, or telogen phase. The new anagen is reinitiated around 21 days of life, and hair follicles again start performing this precise choreography of changes. Yet unidentified signals from the surrounding environment and from the DP are thought to be necessary to reactivate stem cells to restart the process of follicle morphogenesis.

Adult mouse skin contains epithelial stem cells located in at least two different compartments: epidermis and hair follicle. This review focuses primarily on the hair follicle stem cells located in the bulge area of the hair follicle. Despite their location in the hair follicle ORS, cells within this compartment have been shown to give rise to all the different hair cell lineages, the SG, and the epidermis.

The Bulge as a Residence of Epithelial Skin Stem Cells

It has been postulated that stem cells should have evolved special mechanisms of protecting their DNA against accumulation of replication errors, which may otherwise result in a high rate of tissue cancer. These mechanisms may involve either slow or rare cell cycling and/or asymmetrically segregating the newly synthesized DNA into the non-stem cell progeny (the immortal DNA strand hypothesis). To date, the validity of these hypotheses remains uncertain. In the hair follicle, the cell cycle times of ESSCs relative to transit amplifying (TA) (progeny) cells have been difficult to measure with precision. However, the concept of SCs as rarely dividing cells is in good agreement with the low mitotic activity detected in the bulge.

In a similar epithelial system, there are some suggestions that intestinal epithelial stem cells might be segregating their newly synthesized DNA strands asymmetrically so that adult stem cells always inherit the original strand of DNA. In the future, it will be important to assess whether asymmetric divisions might also occur in the bulge. However, the lack of markers for SC and the apparent scarcity of SC divisions in postnatal skin have made this issue technically difficult to address.

What is certain is that a population of slow-cycling and/or asymmetrically dividing cells, likely to be the stem cells, is detected in the skin epithelium. When mouse pups are administered tritiated thymidine or BrdU from 3 to 6 days of postnatal life, the label is incorporated into the newly replicated DNA and is found abundantly distributed in the BL and ORS. When labeling is subsequently followed by 4 to 8 weeks of chase, transient amplifying cells quickly divide and dilute the label, as they move up into the differentiated layers of hair follicle and epidermis, eventually being eliminated from the tissue. In contrast, the slow-cycling cells retain the label and are kept within the tissue. Through this approach, label-retaining cells (LRCs) have been localized predominantly to the bulge area of the murine hair follicle, with fewer scattered cells in the basal layer of the epidermis or elsewhere in the epidermis or hair follicle. If the LRCs in the skin epithelium are stem cells, then the bulge represents a major stem cell compartment of postnatal skin epithelium. Recently, a novel strategy was developed to fluorescently tag and isolate infrequently cycling stem cells in tissues. The method was based upon driving expression of a green fluorescent protein (GFP) fused to histone H2B in a cell type-specific and tetracycline-regulatable fashion. When expression is shut off, all dividing and differentiating cells dilute out the label and/or are sloughed from the tissue, leaving only the infrequently cycling cells brightly labeled. The system was successfully used in skin to monitor the label-retaining cells (LRCs) in the bulge and to isolate and characterize these cells.

With BrdU as a means of labeling the bulge cells, LRCs have been shown to contribute to the formation of both lower hair follicle and the upper region of the hair follicle ORS, called the *infundibulum*. This is based on the observation that LRCs can be found exclusively in the bulge area during the telogen phase of the hair cycle, but, on initiation of a new anagen, cells containing less BrdU signal, presumably derived from division of bulge cells, were also found in the lower and upper ORS, matrix, and even medulla (in the hair shaft). The correlation with the bulge was indirect with a single-label experiment. Using a double-label technique that marked the upper ORS cells based on distinct division time, Lavker and colleagues were able to show a flux of the infundibular (upper follicle above the bulge) cells into the basal layer of the epidermis in neonates and in wounded adult skin.

By using the tetracycline-regulatable H2B-GFP pulse and chase system, LRCs were both efficiently and sufficiently brightly labeled such that progeny could be tracked for ~eight divisions. At the start of the new hair cycle (early anagen), transiently amplifying progeny of bulge LRCs could be found in the matrix in the follicle bulb. Remarkably, even the terminally differentiating hair and inner root sheath cells could be assigned on the basis of fluorescence as being cellular derivatives of the bulge. Moreover, in response to a wounding stimulus, these H2B-GFP LRCs appeared to exit the bulge and migrate toward the wound site. In transit to replenish the damaged epidermis, these cells also appeared to proliferate, deposit a fresh basement membrane, and change some biochemical properties.

The bulge area of the rat vibrissae has been shown to contain multipotent cells that have tissue morphogenesis ability. Micro-dissected rat vibrissae bulge tissue and cells transplanted onto the back of athymic mice yielded entire hair follicles, SG, and epidermis. Bulge cells also exhibited the highest colony-forming ability when cells were placed in tissue culture, and this was true regardless of hair follicle cycle stage. In other follicle regions, efficient colony-forming and morphogenetic cells were not found except for the bulb of the rat vibrissae, where such cells were detected only in early anagen. In a recent study by Blanpain and colleagues (2004), bulge cells isolated from pelage follicles harboring the actin-GFP transgene were plated at limiting dilution and grown in a culture dish. Single colonies were isolated, and the descendants from these individual bulge-cell colonies were mixed with newborn dermal cells and grafted onto the back of nude mice. The newly forming skin contained epidermis, sebaceous glands, and entire hair follicles harboring the actin-GFP transgene, which demonstrated their derivation from an original bulge cell. These experiments not only suggest multipotency and the self-renewing properties of bulge cells, but also indicate their preservation during tissue culture expansion, an important requirement for future clinical applications.

There is a paradox between the slow-cycling features of bulge stem cells in skin and the ability of cultured cells isolated from the bulge region to form large colonies over a two-week period. A priori it is possible that stem cells in culture divide more rapidly either through the trauma of the isolation process or through their exposure to culture medium. Although the formation of hair follicles from grafted cells indicated that each colony *in vitro* must have some expansion of stem cells, the bulk of each colony may be rapidly dividing TA cells derived from these stem cells.

It is not yet clear precisely what relation exists between LRCs identified in the bulge and the cells dissected from the bulge that give rise to cloned colonies or grafted follicles. There is some indication, however, that LRCs are the cells that form colonies in culture. In a recent commentary, C. S. Potten cautioned about the use of *in vitro* approaches for stem cell studies. The study of stem cells outside of their tissue has been called by Potten "the biologic version of the Heisenberg principle." In other words, properties attributed to stem cells (i.e., colony forming or grafting) cannot be studied "without altering the tissue and in so doing altering the stem cell state of the tissue." Conversely, most LRCs could be innocent bystanders in the process and may never participate in the stem cell self-renewal or hair follicle morphogenesis. Finally, it is also possible that in the tissue there are stem cells that do not possess LRC properties. Thus, though intriguing as a potential stem cell marker, label retention must still be considered circumstantial and clearly dependent on the label-chase scheme used. It is possible that researchers may be overlooking the existence of stem cells that are not label retaining but that are nevertheless critical in tissue regeneration. Future studies will be necessary to evaluate this possibility.

Recent studies using retroviral transduction of keratinocytes have revisited the concept that the bulge contains a multipotent population of stem cells that is the normal source of tissue regeneration. In one study, Ghazizadeh and colleagues dermabraded mouse skin and transduced it with a retrovirus encoding the β-galactosidase reporter gene, followed by a 36-week chase with five cycles of depilation-induced hair follicle cycles. If the entire skin epithelium is generated from multipotent, long-lived stem cells within the bulge, then the distribution of β-Gal positive cells might be expected to be uniform across the different cell lineages in the hair follicle and epidermis. The result was puzzling because, even after the long chase and repeated stimulation of stem cells, only 30% of the hair follicles were uniformly blue. The rest of the follicles were positive in the ORS, IRS, or SG but not in all three locations. Moreover, there were defined units of blue in the epidermis far away from the hair follicle.[55] Similar results were obtained when the skin of mosaic mice obtained by aggregation of two stem cell types of different genetic background was analyzed.

These results leave the door open to the notion that there could be multiple classes of stem cells that are long lived, each with restricted potency that may not even reside in a single niche (the bulge). That said, it seems equally plausible that the follicles with mosaic patterns of β-Gal expression may arise from perhaps more easily infectible cells that have left the niche and are already committed to a particular lineage. Another contributing factor may be that the mosaicism arises from a chromatin inactivation mechanism, referred to as position effect variegation that often operates to spontaneously silence retroviral or transgene promoter activity. This mechanism may operate differently in stem cells and their committed progeny, leading to a nonrandom silencing pattern of gene expression of the reporter, in which the transduced retroviral DNA is silenced with higher efficiency in stem cells.

Despite these unresolved issues, the evidence is compelling that (1) the bulge contains a pool of LRCs that participate in the regeneration of the hair follicle and of the wounded skin; (2) bulge cells form large colonies in culture; and (3) bulge cell progeny can regenerate entire hair follicles, SG, and epidermis. These cells have many of the characteristics expected of epithelial skin stem cells. Keeping the "principle of uncertainty" in mind, we now consider the different models of stem cell activation and stem cell function in the skin epithelium.

Models of Epithelial Stem Cell Activation

Early data on hair follicle growth and stem cell function suggested that the bulb of the hair follicle, containing matrix cells, is the residence of stem cells. This hypothesis was hard to explain because of the cycling nature of the bulb, which undergoes extensive apoptosis in catagen, leaving only a small strand of epithelial cells connecting the DP and the bulge. In

addition, the hair follicle can completely regenerate even after the bulb is surgically removed.

Evidence that the bulge contains a large pool of LRCs, together with the appealing location of the bulge as a "niche" at the base of the permanent portion of the hair follicle, suggested a novel model for hair follicle growth, the *bulge activation hypothesis* (Figure 21-2C). In this model, stem cells are activated by yet unidentified signals transmitted through direct contact with the DP. Such contact occurs at the end of each hair cycle, as the surrounding dermal sheath shrinks during the apoptotic phase and drags the DP upward until it comes into contact with the bulge. Although this contact does not appear to be the sole source of stimulation, it seems to be a necessary stimulus to activate one or more bulge cells to divide, resulting in TA cells with decreasing levels of stemness.

At least in neonate and wounded adult skin, stem cell activation may be accompanied by a flux of upper ORS cells to the epidermis. When cells exit the bulge and migrate upward, they take on the fate of a TA epidermal or SG cell. When they migrate downward during early anagen, they give rise to a population of TA ORS and matrix cells that in turn further specialize to form the IRS, cortex, and medulla. When matrix TA cells cease dividing, perhaps through exhaustion of their proliferative capacity, the bulb of the hair follicle undergoes apoptosis (catagen phase). Mutations in a number of genes, including those encoding the transcription factors hairless and RXRα, result in a failure of the DP to be dragged upward at the end of the first postnatal hair cycle. The consequence is a block in bulge activation and a loss of all subsequent hair cycles.

The bulge activation hypothesis does not in itself explain why bulge cells and DP can sometimes sit adjacent to each other for extended periods of time, without stem cell activation, but further experiments may bring more evidence in support of this model.

A second model for stem cell activation is the *cell migration or the traffic light hypothesis*. This model is based on the observation that clonogenic and morphogenic cells in the rat vibrissae are found in the bulge at any hair follicle stage, but they are found at the base of the bulb in late catagen and early anagen. The telogen phase is short, if not absent, from vibrissae, and a new vibrissae cycle is initiated before the DP has moved upward completely to come into contact with the bulge. In investigating this process, Oshima *et al.* (1988) discovered that stem cells or their more or less committed progeny appeared to migrate along the ORS to come in contact with the DP, which then seemed to signal their pathway to differentiate. During catagen and early anagen, these bulge cell derivatives accumulated at the base of the hair follicle in a relative undifferentiated state. Later in anagen, this "traffic stop" appeared to be removed, and cells at the base of the bulb seemed to progress to become matrix and then differentiate into IRS and hair shaft cells. While in transit along the basement membrane surrounding the ORS, these cells were not able to form colonies or engraft successfully, perhaps because of their low concentration. In this model, stem cells

172

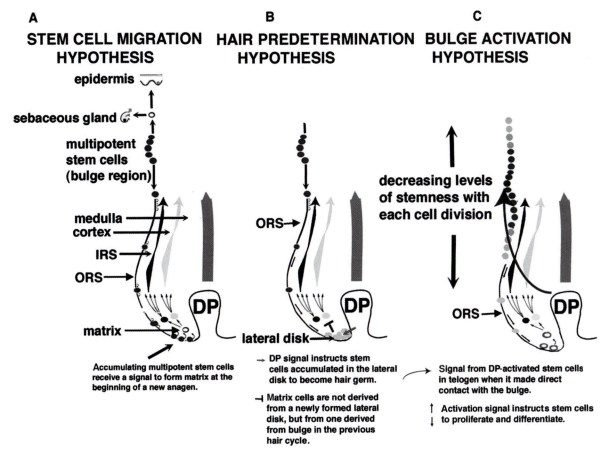

A

STEM CELL MIGRATION HYPOTHESIS

B

HAIR PREDETERMINATION HYPOTHESIS

C

BULGE ACTIVATION HYPOTHESIS

epidermis

sebaceous gland

multipotent stem cells (bulge region)

medulla
cortex

IRS

ORS

matrix

Accumulating multipotent stem cells receive a signal to form matrix at the beginning of a new anagen.

ORS

lateral disk

→ DP signal instructs stem cells accumulated in the lateral disk to become hair germ.

⊣ Matrix cells are not derived from a newly formed lateral disk, but from one derived from bulge in the previous hair cycle.

decreasing levels of stemness with each cell division

ORS

→ Signal from DP-activated stem cells in telogen when it made direct contact with the bulge.

↑ Activation signal instructs stem cells
↓ to proliferate and differentiate.

Figure 21-2. Models of stem cell activation in the hair follicle.

or their immediate progeny leave the niche and with an estimated speed of ~100 μm/day, they migrate with minimal division along 2 mm of ORS to reach the bulb (Figure 21-2A).

A third model derives from an integrative reassessment of the literature, with a novel interpretation of the accumulated data. Known as *hair follicle predetermination*, this model is based on multiple studies of hair follicle growth, and though speculative, it offers an alternative explanation for conflicting results. In this model, during mid- and late anagen, SCs are stimulated to leave the bulge niche and migrate away to form the ORS but not the matrix or the rest of the hair. It is postulated that these SC progeny then accumulate at the base of the follicle, adjacent to the DP, where they are modified to form a "lateral disc" in the bulb (Figure 21-2B). During telogen, the lateral disc maintains contact with the DP, and upon the next anagen, it proliferates (upward) and forms the new matrix and the inner layers of the hair follicle. During this phase, newly activated bulge cells give rise to the ORS and replenish the lateral disc. Therefore, the new hair follicle is formed from lateral disc cells that are predetermined for this role during the previous hair cycle. Based on the recent H2B-GFP LRC study, it would seem that a modification of the prede-

termination model is now warranted to explain how, possibly as a secondary step following the initial "lateral disk" activation, the H2B-GFP LRCs are stimulated to exit the bulge area, proliferate, and participate in the formation of the matrix, all within a single anagen. Finally, some models postulate the presence of multiple populations of nonbulge skin epithelial stem cells that are long lived and either unipotent or of interchangeable potential (hair cells can make epidermis and/or SG and vice versa).

Molecular Fingerprint of the Bulge-Putative Stem Cell Markers

One of the major problems in the advancement of the stem cell field has been the lack of specific biochemical markers characteristic for stem cells despite extensive searches. This has raised the possibility that stem cells may have few, if any, unique features, but instead it is their niche that instructs stem cells to behave differently than their offspring. Recently, infrequently cycling bulge cells have been isolated based on their H2B-GFP retention, and their transcriptional profile has been compared with progeny in the basal layer of the epidermis and

ORS of the hair follicle. Probing a third of the mouse genome, 154 mRNAs scored as being upregulated by greater than twofold in the bulge LRCs relative to the BL/ORS. Similar studies using either the K15-GFP transgenic mice or cell surface markers to purify bulge cells have recently corroborated many of these results.

Although a detailed analysis of all these newly identified factors will take time to complete, immunofluorescence microscopy has already indicated that a number of the proteins encoded by these bulge-upregulated mRNAs are expressed within the bulge but are not necessarily specific for the brightest LRCs within the bulge. Based upon the comparison made, these mRNAs cannot be considered as exclusive markers for the bulge, and indeed some are expressed in other cell types within the skin. In addition, recent isolation and characterization of bulge cells attached to the basement membrane versus those residing internally, have revealed heterogeneity in the molecular profiles of bulge cells. Whether these differences relate to a natural progression of bulge cells from attached to detached to exit/commitment is a fascinating possibility as yet unexplored.

Traditionally, some of the best characterized markers relating to stem cell function are integrins. $\beta1$ and $\alpha6$ integrins are increased in keratinocytes in culture with high proliferative capacity, and they also appear to be enhanced in the bulge area, relative to the lower and upper segments of the anagen phase follicle. Interestingly, both $\beta1$ and $\alpha6$ integrin mRNAs are among the 200 mRNAs identified in three different populations of stem cells when compared with their TA progeny, suggesting that the correlation between integrin levels and "stemness" may extend to stem cells beyond those of the skin.

The relation between integrin levels and stem cells is intriguing because it suggests the possibility that stem cells may be kept tight within the niche by adherence to each other and to the basement membrane. The need for stem cell progeny to migrate along the basement membrane surrounding the epidermis and its appendages may be fulfilled by a reduction in cell-substratum anchorage through down-regulation of integrins. Alternatively, however, it could be that, because cell proliferation is dependent on integrins, cells with high proliferative capacity by nature have elevated integrin levels. The extent to which regulation of integrins in the bulge is a reflection of hair growth mechanisms versus stem cell characteristics per se is an important issue, and one that has not yet been unequivocally resolved. Interestingly, up-regulation of cell migration associated integrin $\beta6$ in the ORS during anagen, when bulge and ORS cells migrate downward to make the new hair follicle, seems to suggest the second possibility.

What cellular mechanisms regulate differences in integrin levels? Are there specific stem cell cues that prompt these changes in integrin levels in TA cells to allow them to exit the niche and migrate? These questions remain to be addressed. In the meantime, integrins are thought to play a major role in the skin, in general, regardless of stem cells in regulating epidermal adhesion, growth, and differentiation.

Other studies reported that proteins expressed differentially in the bulge. They include keratins K15 and K19, both of which are also present in a large fraction of basal layer cells; CD71 low; S100 proteins; E-cad low; p63; and CD34. Of these, CD34, S100A6, and S100A4 were all up-regulated at the level of mRNA expression in the bulge.

Although it is found throughout the mitotically active cells of the skin, and hence is not restricted to the bulge, a p53 homolog protein called p63 may play a role in stem cell function. P63 null mice exhibit defects in epidermal proliferation, leading to the possibility that p63 might function in repressing epidermal growth factor (EGF) receptor and other cell cycle-regulated genes. Interestingly, a naturally occurring dominant negative isoform of p63 is induced in the stratified epithelial layers, where it appears to promote cell cycle withdrawal and commitment to terminal differentiation.

Another putative regulator of stem cell function is c-myc. Transgenic mice with elevated c-myc levels in the basal layer and ORS, which encompass both stem and TA skin cells, show epidermal hyperproliferation, severely impaired wound healing, and loss of hair. The recent microarray profiling of bulge LRCs adds new candidates that may be involved in regulating the transition between infrequently cycling skin stem cells and their transit-amplifying progeny. As these candidates are systematically tested, new inroads into our understanding of stem cell activation are likely to emerge.

Cell Signaling in Epithelial Skin Stem Cells

What coaxes stem cells to become hair follicles rather than epidermis is one of the most fascinating questions in the skin stem cell field. Many of the answers are still not in, but hints have begun to emerge in the past five years. It has long been known that in embryonic development, as in postnatal hair cycling, this decision to coax cells toward the hair pathway involves critical cross-talk between the epithelial and mesenchymal cells. Interestingly, similar signals seem to be used in specification of not only hair but also nail, mammary glands, and teeth.

In embryonic development of skin and its appendages, wnt and bone morphogenic pathway (bmp) signaling have surfaced as two major players that are critical for normal morphogenesis. Gene targeting of Lef1, β-catenin sonic hedgehog, and the noggin inhibitor of the BMP pathway result in reduction, loss, or developmental impairment of hair follicles.

The wnt signaling acts through a large family of soluble wnt morphogens that recognize a specific receptor family known as frizzles to stimulate β-catenin, a dual protein that acts at the crossroads between cell adhesion and cell signaling. A wnt signal results in an inhibition of β-catenin degradation, leading to an accumulation of the protein above and beyond what is required for adherens junctions. β-catenin can then complex with members of the Lef/Tcf family of HMG DNA binding proteins and affect expression of transcription of downstream target genes. A truncated, constitutively active form of β-catenin expressed at high levels in the ORS of basal layer results in excess skin and *de novo* initiation of hair follicle buds in postnatal skin, a characteristic normally specific to the embryonic skin. Taken together, these data suggest that wnt signaling may induce adult stem cells into a state more specific for their embryonic relatives or that the wnt signaling has a major role in specifying hair follicle fate to relatively undifferentiated cells.

In the adult hair follicle, two members of the Tcf/Lef1 family are expressed in strategic places. Lef1 is in the matrix, precortex, and DP, whereas Tcf3 is in the bulge and the ORS of skin epithelium. mRNA profiling revealed that Tcf3 mRNA is also up-regulated in bulge LRCs, suggesting that this regulation is at the level of gene expression. The localization of Tcf3 in the bulge is intriguing because a related protein Tcf4 is found in the intestinal stem cell compartment, or crypt, where it appears to play a role in intestine stem cell maintenance. *In vitro*, Tcf3 acts as a repressor in the absence of wnt signaling, and, in its presence, it can be converted to an activator. *In vivo*, transgenic expression of the repressor forms of Tcf3 both in stem cells and TA cells of the skin results in a lethal phenotype, in which epidermal basal cells are conferred ORS characteristics. Taken together, these findings suggest that Tcf3 in the bulge acts as a repressor. Although target genes for Tcf3 have not yet been identified in the skin, one candidate is c-Myc, an established Tcf/Lef1 target gene, which induces hyperproliferation and appears to deplete the stem cell compartment of the skin when c-Myc is overexpressed in transgenic mice.

In contrast to Tcf3, Lef1 is expressed in the matrix but accumulates more strongly in the nuclei of precortical cells. The progenitor cells of the hair shaft precortex expresses a bank of hair-specific keratin genes that possess Lef1 binding sites in their promoters. Precortex also shows nuclear β-catenin, and it expresses a wnt-responsive reporter gene TOPGAL. These cells are thus likely to receive a wnt signal, and wnt expression is seen in this region of the anagen hair follicle. Interestingly, when expressed in hair precursor cells in transgenic mice, dominant negative forms of Lef1 result in the production of SG cells, while loss of β-catenin results in production of epidermal cells at the expense of hair differentiation (Figure 21-3). In contrast, the transcription factor GATA3 was recently demonstrated to be indispensable for inner root sheath differentiation. Curiously, both Lef1 and GATA3 have been implicated in hematopoietic stem cell lineage determination.

Although wnts are clearly necessary for β-catenin stabilization, to obtain a wnt response, a cell must express Tcf/Lef factors. Recently, Jamora *et al.* showed that noggin, a bmp inhibitor, can induce Lef1 expression in keratinocytes *in vitro*, a feature corroborated in keratinocytes and skin epithelium null for the BMP receptor-1a. When these cells are also treated with Wnt3a, they affect the transcription of wnt responsive reporter genes. In most cases, this combination leads to transactivation, but curiously, the E-cadherin promoter seems to be down-regulated by these factors. *In vivo*, down-regulation of the E-cadherin promoter accompanies induction of hair

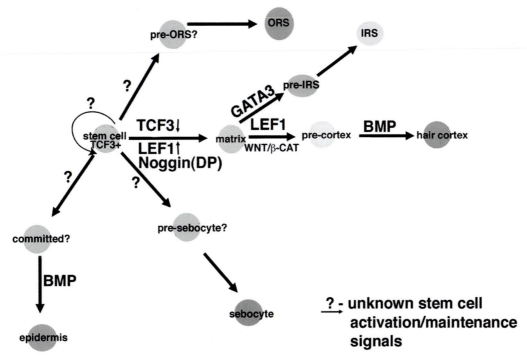

Figure 21-3. Schematics of signaling pathways in stem cell activation and fate choice.

placodes in embryogenesis and secondary hair germs in cycling hair follicles, and transgenic elevation of E-cadherin blocks follicle morphogenesis. Taken together, these findings suggest a model whereby activation of wnt signaling and inhibition of BMP signaling work together to maintain wnt responsiveness in skin epithelial cells and show that, through down-regulation of E-cadherin, the cells are able to undergo epithelial cell–cell remodeling, necessary for follicle morphogenesis. Moreover, the findings suggest that adult and embryonic stem cells may use similar signals to control their differentiation.

An intriguing feature of the wnt signaling pathway is that when constitutively activated, it leads to elevated levels of tissue progenitor cells in skin, in brain, and in intestine. In addition, purified Wnts stimulate isolated hematopoietic stem cells to proliferate in culture, and following skeletal muscle injury, Wnts appear to mobilize resident stem cells during the regeneration process. Taken together, these findings implicate the wnt signaling pathway in the self-renewal of stem cells and/or their transit-amplifying progeny. Consistent with these observations is the recent finding that the growth-restricted environment of the bulge is associated with an up-regulation of mRNAs more typically affiliated with wnt-inhibition. Future studies will be required to assess whether this tantalizing correlation is functionally significant.

Commentary and Future Directions

Probing more deeply into the molecular mechanisms of stem cell maintenance and lineage determination in the skin is now possible using newly identified factors that distinguish putative bulge stem cells from their progeny. Interestingly, a large fraction of these factors are either secreted or transmembrane factors that are likely involved in interactions with the environment. Some of these have been previously associated with stimulation of different cell types that coincidentally surround the bulge area, whereas others are likely involved in extracellular matrix and basement membrane organization. Still others are candidates for creating a local gradient of signaling molecules that could maintain stem cells within the bulge area in a growth and differentiation inhibited environment. These findings are in agreement with the bulge being an entity with tissue morphogenesis activity upon grafting, and begin to unravel the molecular mechanisms and the identity of the cells within the bulge that perform this activity. If future tests of these candidates support these predictions, then the concept of preformed niches, in which stem cells find their "home," may have to be reconsidered. In this regard, somatic stem cells may be even more potent than initially surmised and may be actively involved in building and maintaining their own niches through adult life.

For skin stem cells in particular, many of the most interesting questions still remain unaddressed. What specific genes are responsible for stem cell maintenance? How will their expression change during the hair cycle? How will their patterns change as stem cells either exit from their niche and migrate *in vivo* or are removed from their niche and cultured

in vitro? How does wounding influence the status of stem cells, and what particular signals penetrate the bulge to instruct cells to exit and help replenish the epidermis? How strong is the correlation between slow-cycling cells and stem cells? Are there long-term and short-term stem cells, and, if so, what makes them different? Do stem cells of the skin undergo asymmetric or symmetric divisions?

As we begin to learn more about the stem cells of the skin and the cues that these cells respond to, a clear picture should emerge as to how stem cells are coaxed along specific lineages, and how their niche impacts on this decision-making process. As we gain further insights into these issues, we will begin to explore the relation between adult and embryonic skin stem cells and the parallels between skin stem cells and other stem cells of the body, including pluripotent embryonic stem cells. It is the ultimate goal of the stem cell field to understand enough about the various unipotent, multipotent, and pluripotent cells of the body to make major inroads in the interface between stem cell biology and human medicine.

KEY WORDS

Hair cycle Succession of growth, regression, and rest that the hair follicle undergoes during the life of an animal.

Label retaining cells Cells that incorporate labeled nucleotides during the S phase and retain it preferentially over extended periods of time due to slow or infrequent divisions.

Stem cell activation Process that stimulates stem cells to exit their dormant, quiescent, undifferentiated state, and begin to produce more committed cells to participate in tissue growth or repair.

FURTHER READING

Blanpain, C., Lowry, W. E., Geoghegan, A., Polak, L., and Fuchs, E. (2004). Self-renewal, multipotency, and the existence of two cell populations within an epithelial stem cell niche. *Cell* 3; **118**(5), 635–648.

Cotsarelis, G., Sun, T. T., and Lavker, R. M. (1990). Label-retaining cells reside in the bulge area of pilosebaceous unit: implications for follicular stem cells, hair cycle, and skin carcinogenesis. *Cell* **61**, 1329–1337.

Fuchs, E. (1990). Epidermal differentiation. *Curr. Opin. Cell. Biol.* **2**, 1028–1035.

Hardy, M. H. (1992). The secret life of the hair follicle. *Trends Genet.* **8**(2), 55–61.

Oshima, H., Rochat, A., Kedzia, C., Kobayashi, K., and Barrandon, Y. (2001). Morphogenesis and renewal of hair follicles from adult multipotent stem cells. *Cell* **104**, 233–245.

Panteleyev, A. A., Jahoda, C. A., and Christiano, A. M. (2001). Hair follicle predetermination. *J. Cell Sci.* **114**, 3419–3431.

Potten, C. S., and Booth, C. (2002). Keratinocyte stem cells: A commentary. *J. Invest. Dermatol.* **119**, 888–899.

Potten, C. S., and Morris, R. J. (1988). Epithelial stem cells in vivo. *J. Cell Sci. Suppl.* **10**, 45–62.

Taylor, G., Lehrer, M. S., Jensen, P. J., Sun, T. T., and Lavker, R. M. (2000). Involvement of follicular stem cells in forming not only the follicle but also the epidermis. *Cell* **102**, 451–461.

Tumbar, T., Guasch, G., Greco, V., Blanpain, C., Lowry, W. E., Rendl, M., and Fuchs, E. (2004). Defining the epithelial stem cell niche in skin. *Science* **303**(5656), 359–363.

The Ontogeny of the Hematopoietic System

Malcolm A. S. Moore

Historic Perspective

The origin and potentiality of the hematopoietic stem cell has been debated for over a hundred years with two main concepts: the monophyletic hypothesis recognizing a common stem cell for all lympho-myeloid lineages, and the polyphyletic hypothesis recognizing a variety of distinct stem cells. The current consensus recognizes a pluripotential lympho-myeloid stem cell and a hierarchy of progressively lineage-restricted progenitor cells with a major bifurcation at the level of the common lymphoid and common myeloid progenitor. One early assumption had been that each developing hematopoietic or lymphoid organ produces its own complement of stem cells, with morphological studies variously proposing an *in situ* origin in hematopoietic tissues from undifferentiated mesenchyme or endothelium and in primary lymphoid organs such as thymus or avian bursa of Fabricius, a derivation from epithelium. Over 35 years ago, this view was challenged in studies in avian embryos that showed developing hematopoietic and lymphoid organs were colonized from the outset by an inflow of stem cells from the bloodstream. Since the yolk sac (YS) was identified at that time as the first site of hematopoiesis, it was proposed as the site of origin of stem cells that colonized all subsequent hematopoietic and lymphoid organs. This hypothesis was challenged by later studies that demonstrated an embryonic origin of hematopoietic or lymphoid cells that later appeared in spleen, marrow, thymus, and bursa of Fabricius.

In mammalian systems, hematopoiesis is clearly initiated in the YS, and subsequently in the aorta–gonad–mesonephros (AGM) region in association with the dorsal aorta and vitelline veins where the first stem cells arise with the capacity to engraft adult recipients. There has been extensive debate as to the role of the YS in mammalian hematopoietic ontogeny with, on the one hand, the site viewed as the initial source of stem cells that then migrated into the embryo, and, on the other hand, the other extreme that views the YS as a transient source of primitive erythroid progenitors, with definitive generation stem cells arising exclusively in the AGM region. There is increasing evidence that stem cells arise independently in both sites and that both contribute to the subsequent colonization of the fetal liver.

Sites of Initiation of Primitive and Definitive Hematopoiesis and Vasculogenesis

The sites of hematopoiesis, and the primary sites of lymphopoiesis, with times of initiation of activity in development, are shown in Table 22-1.

YOLK SAC DEVELOPMENT

The YS forms during gastrulation which begins in the mouse embryo at day 6.5 (E6.5). Mesodermal cells destined for extraembryonic sites exit the posterior primitive streak and subdivide the embryo into three separate cavities by the neural plate stage (E7.5). The central cavity, the exocoelom, becomes completely lined with mesoderm and, where this is adjacent to visceral endoderm, visceral YS forms. Between E7 and E7.5, mesodermal cells in the visceral YS proliferate and form mesodermal cell masses that are the precursor of the blood islands. Central cells accumulate hemoglobin, while the outer cells flatten and form endothelium. Lineage tracing experiments show that the hematopoietic mesoderm arises from posterior primitive streak mesoderm. Tissue recombination studies in the chick embryo also indicated that YS hematopoiesis requires diffusible signals from extraembryonic endoderm (hypoblast) that is analogous to visceral endoderm of the mouse. Indian hedgehog (Ihh) and smoothened (Smo), a receptor component essential for all Hedgehog signaling, are required for yolk sac development. Ihh appears to be the endodermal signal inducing hematopoietic and vascular specification of YS mesoderm, acting via induction of bone morphogenic protein-4 (BMP-4) in mesoderm that in turn induces hemato-vascular development (Figure 22-1).

Onset of Primitive and Definitive Hematopoiesis in the Yolk Sac

The YS is the site of both primitive erythropoiesis and macrophage production and also of definitive multilineage, and myeloid and erythroid lineage-restricted, progenitor cell production. The wave of primitive erythropoiesis begins in the YS at E7.5, with nucleated red cells producing embryonic globin predominating in the circulation through E14.5 and with eventual replacement by fetal liver-derived red blood cells (RBC) expressing adult globins by E15.5–16.6. In humans, β-like embryonic globin (hemoglobin ε) is expressed first in embryonic nucleated RBCs in YS blood islands. Subsequently, fetal globins (hemoglobin Aγ and Gγ) are expressed in definitive RBCs developing in the fetal liver. Finally, adult δ- and β-globins are expressed around the time of birth within bone marrow-derived RBCs. The zinc finger transcription factor, Erythroid Kruppel-like Factor (EKLF), plays a role in coordinating erythroid cell proliferation and hemoglobinization, participating in the switch from embryonic to fetal or fetal to adult β-globin expression.

A population of primitive high-proliferative potential progenitors high proliferative potential progenitor colony forming cells (HPP-CFC) were detected at E8 (1–8 somites) in the mouse embryo, exclusively in the YS. This remained the predominant site of expansion (> 100-fold) of these multipotent precursors until E10.5–11.5 when there was a dramatic increase of HPP-CFC in the circulation and liver, with a concomitant drop in the YS. Upon secondary replating, these HPP-CFC exhibited the stem cell features of self-renewal and potential to generate definitive erythroid and macrophage progeny. Separation of YS and embryo at the early E8 stage, before a common circulation is established, showed that HPPs were found exclusively in the YS and examination of the blood at E8.5 suggested that they entered the circulation from YS together with the first erythroblasts. Definitive erythroid (BFU-E) and myeloid (CFU-GM) progenitors and

TABLE 22-1
Comparative Chronology of Hematopoietic Development

Initiation of Hematopoiesis or Lymphopoiesis (days)	Mouse	Human	Chicken
Yolk sac	7.5	18	2
Aortic/AGM region	9.5–10	27	3
Liver	11	42	No
Spleen	13	48	8
Bone marrow	15	77	11
Thymus	11	40	7
Bursa of Fabricius	No	No	14
Onset of circulation	8.2	24	2.5

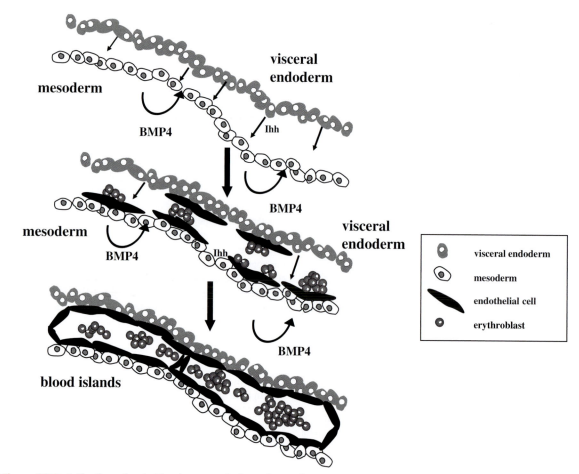

Figure 22-1. Activation of primitive hematopoiesis and vasculogenesis in the developing yolk sac. Indian hedgehog (*Ihh*) is secreted from visceral endoderm to the target extraembryonic (yolk sac) mesoderm, where it activates expression of *bmp4*. BMP4 protein in turn feeds back to extraembryonic mesoderm (in an autocrine or paracrine manner), activating genes such as *Flk1*, *CD34*, *SCL/tal-1*, and *AML1*. This is associated with formation of hematopoietic and vasculature stem and progenitor cells, possibly via a common precursor, the hemangioblast. Reproduced with permission from Baron, M. H. (2001). *J Hematotherapy & Stem Cell Research* **10**, 587–594.

mast cell precursors are detected in the YS at E8 but do not differentiate there; rather, they migrate to the fetal liver where BFU-E/CFU-E initiate definitive erythropoiesis.

The YS does not contain cells capable of long-term multilineage engraftment of adult irradiated mice until after such cells have appeared in the AGM region at E10.5. Although it has been argued that the AGM region is the sole site of origin of "definitive" HSC capable of engrafting adult recipients, and the presence of these cells in the YS at later stages represents an immigrant population, total body quantitation and tissue distribution analysis of competitive repopulating HSC detected by limiting dilution analysis argues for two independent sources of "definitive" HSC, both participating in seeding the fetal liver.

Support for this view was obtained in co-culture studies of either E8–8.5 YS or intraembryonic splanchnopleuric mesoderm (P-Sp) on a stromal cell line generated from the AGM region. The conventional definition of an HSC is that it must possess extensive self-renewal potential and multilineage differentiation potential. By this criteria, the YS contains HSC as early as E9 as demonstrated in a number of studies showing that CD34[+], c-Kit[+], and CD38[+] YS cells at this stage, when injected directly into the liver of busulphan-conditioned neonatal mice, gave long-term lympho-myeloid reconstitution. Secondary engraftment in sublethally irradiated mice demonstrated that the YS-derived HSC acquired normal marrow homing properties.

The argument that the "true" stem cells must have the ability to repopulate the adult mouse after intravenous injection is not relevant to normal ontogeny since the only requirement of the initial HSC population is that it be capable of colonizing the liver. Within that environment it may then acquire marrow homing features that permit it to become an "adult" stem cell. The molecular basis for the distinction between the embryonic and adult-type stem cell is unknown but may involve adhesion receptors, including CD34, CD44, selectins, and α- and β-integrins. In this context, β_1-integrin-deficient HSC fail to seed both fetal and adult hematopoietic organs.

The CXCR4 receptor and its ligand, the chemokine SDF-1, mediates HSC chemotaxis and is critical for HSC homing to marrow but not for fetal liver colonization. Co-culture of YS on AGM stroma confers adult engraftment potential, suggesting an inducing role for stromal cells in this region, particularly mesenchymal cells on the floor of the dorsal aorta. As currently envisaged, this inductive influence confers HSC potential in locally generated hematopoietic clusters. However, it is conceivable that early YS HSC traffic into the dorsal aorta and come under the influence of the AGM stroma, thus acquiring adult engraftment potential.

MACROPHAGE AND MICROGLIAL ONTOGENY

In murine ontogeny, primitive macrophage progenitors appear in the proximal region of the egg cylinder associated with expression of SCL/tal-1 and GATA-1 at E7. They increase until the early somite stage (E8.25), and then they decline sharply to undetectable levels by the 20 somite stage. Definitive lineage multipotent precursors with macrophage poten-

tial (HPP-CFC) as well as more restricted definitive CFU-GM and CFU-M appear at E8.25. Separation of YS and embryo at 8.25 (before the circulation) showed that HPP-CFC were found exclusively in YS. Once the circulation is established, HPP-CFC can be found in the circulation, indicating migration into the embryo.

In avian studies, chick-quail chimeras established as early as E2.5 show numerous YS-derived cells, including macrophages, within the embryo vasculature and within mesenchyme. At late E2, numerous scattered CD45[+] cells appear in YS and in the blood, exiting from the circulation through the endothelium and rapidly invading the whole embryo. These cells express high levels of CD45 and correspond exactly to cells identified as monocyte-macrophages. YS-derived macrophages showing marked acid phosphatase activity and phagocytic capacity appear within the neural tube, liver anlage, and nephric rudiments, beginning even before the circulation is established. In the mouse, the macrophage progenitors appearing in the YS at E7.5–8 migrate into the mesenchyme surrounding the brain rudiment. Initial migration occurs before a circulation is established and is interstitial, but after E8.2 seeding occurs via the vasculature. These YS-derived macrophages continue to proliferate and generate the brain microglia, which ultimately comprises 10% of the brain. The primitive generation YS-derived macrophages develop at E7–8 in the absence of a monocyte or promonocyte intermediary stage and differ from adult macrophages in the pattern of enzymes produced. Peroxidase-positive promonocytes of the definitive lineage appear in YS at E10.

ONTOGENY OF THE VASCULATURE

Morphogenesis of blood vessels is defined by a sequential pattern of gene expression in which SCL/Tal1 and Flk1 are expressed first at the angioblast stage, followed by PECAM, CD34, VE-cadherin, and later Tie2, while SCL expression is down-regulated in endothelium of mature vessels. As somatogenesis begins, vascular development is asymmetric, centered in YS blood islands, and in the embryo proper where angioblasts begin to coalesce into the aorta. By the murine 3 somite stage, vascular development has spread through the YS and extended aortic tubes are visible. Beginning at the 4 somite stage, erythroblasts disperse through the distal yolk sac and a few erythroblasts appear in the head and tail region of the embryo where the embryonic and YS vasculature meet. The circulation onset is defined as erythroblast movement within vessels and is initiated at the murine 4–6 somite stage during E8.25.

INITIATION OF HEMATOPOIESIS IN THE AORTA-GONAD MESONEPHROS REGION

Avian Intra-aortic and Para-aortic Hematopoiesis

In birds, the first intraembryonic hematopoiesis occurs at E3 in endothelium-associated intra-aortic clusters of markedly basophilic cells with prominent nucleoli, and subsequently (E6–8) in diffuse hematopoietic "para-aortic foci" located in the dorsal mesentery ventral to the aorta. Hematopoietic progenitors are detected in the aortic wall at E3–4. At E2, prior to the appearance of the intra-aortic clusters of CD45$^+$VEGFR-2^{-ve} hematopoietic cells, the aortic endothelium consisted entirely of CD45^{-ve}, VEGFR-2$^+$ flat endothelial cells. Chick-quail marker studies showed that the intra-aortic clusters developed in situ and not by migration from the YS. Acetylated low-density lipoprotein (Dil-Ac-LDL) labeling indicated that the hematopoietic clusters originated from aortic endothelium, while retroviral vector marking indicated that the para-aortic foci were derived from intra-aortic clusters. Mesodermal subdivision transplantation showed that the endothelium of the roof and sides of the dorsal aorta arose from somite mesoderm that generates "pure" angioblasts, while the floor develops from splanchnopleural mesoderm that generates progenitors with dual hematopoietic and angiogenic potential.

In the human embryo, hematopoietic clusters containing CD34+ cells appear in the aorta at 27 days gestation, 8 to 9 days after initation of YS hematopoiesis and 6 days after the onset of the circulation (Figure 22-2). The precursors of the hematopoietic clusters in the periaortic region were considered "endothelial" based on expression of CD34 and CD31 (PECAM-1), absence of CD45, and uptake of Dil-Ac-LDL. In mice, hematopoietic clusters appear at E10.5 in the ventral floor of the dorsal aorta. They appear to be in direct contact with underlying mesenchyme where the endothelium is interrupted. However, ultrastructural studies in humans show that the endothelial basal lamina is intact where it interacts with the cells in the hematopoietic clusters, and the hematopoietic and endothelial cells are interconnected by tight junctions (Figure 22-2). VE-cadherin-expressing cells in the AGM have

hematopoietic potential, and VE-cadherin is expressed on the luminal aspect of vascular endothelium but not in CD45$^+$ aorta-associated hematopoietic clusters. VE-cadherin is essential in stabilizing tight junctions between endothelial cells, suggesting the endothelial origin of hematopoiesis. Similar populations of CD34$^+$CD45^{-ve} cells in the human para-aortic region are labeled with the endothelial marker *Ulex Europus*, and generate both von Willebrand (vW)$^+$ endothelium and hematopoietic cells. This blood-forming "endothelium" is present in the human embryo AGM region at 28 days with a peak frequency at 31 days and absence by 44 days. Culture of the paraortic splanchnopleura (P-Sp) from 21 to 26-day human embryos on marrow stromal cells generated hematopoietic cells, demonstrating the hematogenic potential of P-Sp mesoderm that precedes the appearance of "hemogenic endothelium." The pre "endothelial" mesodermal precursor is CD34^{-ve}, CD45^{-ve}, and VEGFR-2/KDR$^+$.

Since cells present in the human embryo AGM prior to 28 days lacked direct hematopoietic potential, it was postulated that they must receive some inductive signal from surrounding tissue to induce this potential. A discrete region of densely packed mesenchymal cells lies beneath the ventral floor of the dorsal aorta and is 3 to 4 layers thick in mouse and 5 to 7 layers in humans (Figure 22-2). These mesenchymal cells are interconnected by tight junctions, express smooth muscle α-actin (SMα-A), and may be precursors of vascular smooth muscle. BMP-4 may play a crucial role in the induction of hematopoietic potential in the AGM region, much as it does in the YS. BMP-4 is polarized to the ventral wall of the dorsal aorta in the mesenchymal layer underlying the intra-aortic hematopoietic clusters where it could induce P-Sp mesoderm to become hemangioblasts and HSC (Figure 22-2).

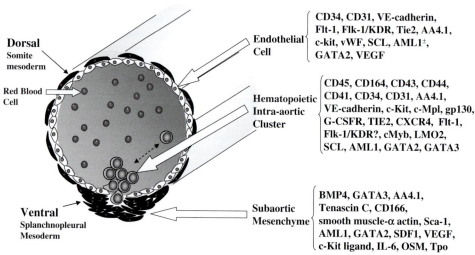

Figure 22-2. Intra-aortic hematopoietic clusters formation in the floor of the dorsal aorta based on data from E10.5 murine and E27 human studies. The dorsal endothelium of the aorta derived from angioblasts develop from somite mesoderm, whereas the ventral "hemogenic endothelium," or hemangioblast, develops from splanchnopleural mesoderm. The dorsal endothelium expresses markers associated with differentiated endothelium. The hematopoietic intra-aortic cluster expresses surface antigens and cytoadhesion molecules found on hematopoietic stem and progenitor cells, hematopoiesis-related transcription factors, and receptors for hematopoietic and angiogenic growth factors. The expression of Flk-1 on intra-aortic clusters is controversial. The subaortic mesenchyme expresses cytokines, chemokines, and cytoadhesion molecules and has molecular features of vascular smooth muscle.

The mesenchymal inductive influence extends beyond the induction of multilineage hematopoietic precursors and potentially includes induction of development of B and T lineage-restricted progenitors, and of HSC capable of long-term repopulation of adult mice. The mouse AGM region at E10.5 was identified as a source of CFU-S, preceding their appearance in YS. CFU-S precursors were present in both AGM and YS by E9 since cultures of either tissue at this stage generated CFU-S. *In vivo* engraftment was obtained with cells from the E10.5 dorsal aorta and vitelline/umbilical artery. Quantitation of long-term repopulating stem cell numbers by limiting dilution assay has shown that they first appear in the AGM at E10.5 and are present in equal numbers in AGM, YS, circulation, and early liver rudiment by E11 (Figure 22-3). The

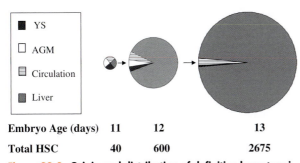

	YS
	AGM
	Circulation
	Liver

Embryo Age (days)	11	12	13
Total HSC	40	600	2675

Figure 22-3. Origin and distribution of definitive hematopoietic stem cells (numbers of competitive repopulating units (CRU) detected by limiting dilution competitive long-term repopulation assay in adult irradiated mice, adjusted for a calculated seeding efficiency of 10%). The figure shows absolute numbers in each tissue at daily intervals from E11 to E13 in the mouse embryo. Modified after Kumaravelu, P., *et al. Development* **129**, 4891–4899.

colonization of the fetal liver by stem cells is initated by a wave of migration from the AGM peaking at E11, followed by a second wave from the YS (Figures 22-4, 22-5). Cells capable of long-term lympho-myeloid engraftment have been reported to be generated in organ culture of P-Sp isolated prior to the onset of the circulation (E7.5–8), whereas cultured YS failed to generate such HSC. However, other investigators were able to generate repopulating HSC in co-culture of equally early P-Sp and YS, with a stromal cell line derived from the E10.5 AGM region.

The various inductive influences of the subaortic mesenchme could cause bipotential precursors located in the floor of the aorta to preferentially adopt a hematopoietic fate. Alternatively, endothelial cells may de-differentiate and switch to a hematopoietic fate in response to local or transient signals. A third possibility is that hemangioblasts or their precursors may migrate secondarily into the aortic wall, either from the underlying mesenchyme or from the circulation, in response to a chemokine gradient. Endothelial cells, angioblasts, and HSC express CXCR4, and mesenchyme adjacent to the dorsal aorta expresses high levels of its ligand SDF-1, indicating a role for this chemokine pathway in both vasculogenesis and migration of hemangioblastic precursors to the floor of the aorta.

Studies using embryonic stem (ES) cells differentiating for 2.5 to 3.5 days have demonstrated the transient development of cells capable of forming blast cell colonies (BL-CFC) in the presence of VEGF and the c-Kit ligand/stem cell factor (SCF) that precede the appearance of hematopoietic colony-forming cells. Cells in the blast colony express a number of genes common to hematopoietic and endothelial lineages (CD34, SCL, Flk-1) and on replating give rise to endothelial progenitors and both definitive and primitive hematopoietic

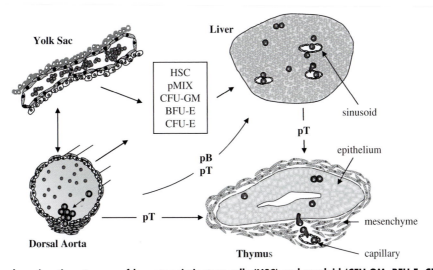

Figure 22-4. Vascular migration streams of hematopoietic stem cells (HSC) and myeloid (CFU-GM, BFU-E, CFU-E, pMIX) and lymphoid (pT, pB) progenitor cells between the dorsal aorta/AGM, the yolk sac blood islands, the embryonic liver, and the thymus, occurring between E10.5 and E11.5 in the mouse embryo. Immigrant cells (shown as basophilic blast cells) detach from the YS blood islands and from the aortic hematopoietic clusters, enter the circulation, and subsequently egress from the microvessels within the hepatic rudiment, or within the perithymic mesenchyme, and accumulate and proliferate within these developing organs.

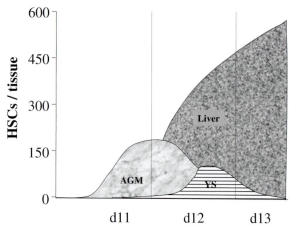

Figure 22-5. Schematic representation of colonization of the embryonic liver of the mouse embryo with HSC as determined by CRU assay in adult irradiated mice and adjusted for seeding efficiency. The numbers are based on the numbers that the AGM and YS are able to generate *in vitro*. *In vivo*, the high cumulative activity of the AGM region and the YS may provide the liver with a high proportion of definitive HSC. The data suggests consecutive colonization of the embryonic liver with HSC from the AGM region and the YS. Reproduced with permission from Kumaravelu, P., *et al. Development* **129**, 4891–4899.

progenitors. The thromobopoietin (Tpo) receptor c-Mpl was detected by day 3 of embryoid body formation when hemangioblasts first arise. Tpo alone supported Bl-CFC formation and nearly doubled the number of BL-CFC when combined with VEGF and SCF. Since hematopoietic and endothelial development are not extinguished by targeted inactivation of *c-Mpl* or *Tpo* genes, there must be some redundancy of cytokine pathways acting on hemangioblasts. Other overlapping pathways active on hemangioblasts include VEGF/Flk-1 c-Kit/SCF, BMP-4, and TGFβ. The ES results support the concept of a hemangioblast as the precursor of both endothelium and hematopoiesis. Tpo and c-Mpl transcripts are present in the early YS prior to appearance of the first blood islands, and c-Mpl is expressed on embryonic blood vessels and aorta. Other cytokines stimulating hemangioblasts are expressed in the AGM region, including VEGF, SCF, and BMP-4.

The respective roles of hemangioblast and "hemogenic endothelium" for initiating hematopoiesis in the AGM or YS is debated. A number of these studies identify endothelium as the source of hematopoiesis based on isolation of cells expressing markers associated with endothelium (VE-cadherin, CD31/PECAM, Flk-1, Flt-1, Tie-2, *Ulex Europus* binding, Dil-Ac-LDL uptake). None of these markers is unique to endothelium and possibly could be expressed on hemangioblasts or prehematopoietic mesoderm, whereas some are expressed on HSC. Direct derivation of hematopoietic cells from mature, already differentiated, endothelium has been reported in Ac-LDL-DiI labeling studies. Molecular analysis of single cKit+CD34+ cells from the E11 mouse AGM region, when populations of engraftable HSC as well as lymphoid and myeloid committed progenitors are present, showed

that the majority expressed hematopoietic-specific transcription factors AML-1, GATA-2, PU-1, and LMO2. The majority of cells also express the G-CSF receptor, while a minority express the erythropoietin receptor. The myeloid-specific gene MPO was expressed in 90% of cells, while the erythroid specific β-globin was expressed in 50% indicating that genes of mutually exclusive differentiation lineages may be expressed simultaneously in a single primitive cell, prior to lineage commitment.

AML1 plays a critical role in establishing definitive hematopoiesis in the AGM. AML1 expression is initiated in mesenchymal cells at the distal tip of the allantois, and endothelial cells in ventral portion of paired dorsal aorta at E8.5 and in both endothelium and mesenchyme of the ventral AGM and in the intra-aortic hematopoietic clusters between E9.5–11.5. Up to a third of the cells recovered from the AGM and the vitelline and umbilical arteries at E11.5 were AML1+, and this population contained all HSC capable of engrafting irradiated recipients. These HSC were initially CD45^{-ve} at E10.5, but by E11.5, HSC activity was present in both CD45+ and CD45^{-ve} fractions.

In AML1-deficient mice, HSC and intra-aortic clusters were absent, whereas in AML1 hemizygous mice, hematopoietic clusters were reduced in number and size and HSC numbers and distribution were altered, indicating a gene dosage effect. Haploinsufficiency of AML1 results in an earlier appearance of engraftable HSC in the YS with E10 YS cells reconstituting the hematopoietic system of irradiated adult mice. At the same time, there is premature termination of HSC activity in the AGM explants consistent with a change in the balance of HSC emergence, migration, and/or maintenance. With hemozygous levels of AML1, the YS may autonomously generate HSC simultaneous to their generation in the AGM, or AGM may autonomously generate abundant HSC that immediately and rapidly migrate to the YS where they are detected in abundance, or AML1 insufficiency blocks emigration of HSC from the YS resulting in accumulation there and a deficiency in the AGM.

Stem Cell Migration to Later Sites of Hematopoiesis

ONTOGENY OF LIVER HEMATOPOIESIS

Lineage tracing studies support the derivation of the liver from ventral foregut endoderm that is induced by signals from pericardium and septum transversum mesenchyme, to proliferate and adopt a hepatic state. The fetal liver stroma consists of cells that express features of epithelium (cytokeratin-8), mesenchyme (vimentin, osteospondin), and vascular smooth muscle (α-smooth muscle actin). These cells are supportive of long-term HSC proliferation, and their presence in the liver coincides with the duration of hepatic hematopoiesis. At late gestation when hematopoiesis declines, these cells are replaced by epithelial cells resembling mature hepatocytes and a minority of myofibroblasts. Oncostatin M, produced by hematopoietic cells within the liver, has been implicated in inducing hepatic maturation of the epithelial-mesenchymal stromal cells with loss of hematopoietic support capacity.

Fetal liver isolated before the 28 somite stage (E9.5) and grafted beneath the kidney capsule of adult mice resulted in survival of hepatic tissue but no hematopoietic elements were present. Administration of hematopoietic cells into the circulation of recipient mice resulted in multilineage hematopoietic engraftment of the implanted fetal liver. Grafts of liver isolated at >28 somite stage showed autonomous hematopoiesis, defining the time of initiation of HSC entry into the liver at E9.5–10. In humans the number of BFU-E drops abruptly in the YS at 35 days, and simultaneously they appear in the liver, reflecting oriented migration. The definitive BFU-E and CFU-E that are generated in the YS do not differentiate in that site and most likely seed the fetal liver and rapidly establish definitive erythropoiesis. This is supported by the observation that large numbers of CFU-E appear simultaneously with BFU-E at the onset of hepatic hematopoiesis, and embryonic blood contains significant numbers of definitive hematopoietic progenitors immediately prior to liver development.

There is a daily logarithmic increase in CFU-GM and BFU-E in the liver between 10 and 13 days paralleled by an extensive expansion of CFU-S. HSC detected in long-term competitive repopulation assays were reported at E12 in the liver, increasing 38-fold to E16 and decreasing thereafter. Limiting dilution long-term repopulation assays have demonstrated at least one stem cell, as detected by the competitive repopulating in the liver, circulation, AGM region, and YS at E11, with an increase in the liver to a 50-fold increased assay at E12 and a 250-fold by E13 (Figure 22-3). The 24-hour seeding efficiency of murine fetal liver HSC into adult marrow as measured by limiting dilution competitive repopulation assays is ~10%, which is essentially identical to the seeding of adult marrow HSC. Thus, ~10 HSC initiate hematopoisis by seeding the hepatic rudiment. These numbers could be explained by colonization from the AGM, but at E12 the YS makes a significant contribution, and these HSC may not require processing in the AGM but can mature *in situ*. Relative to adult bone marrow, Fetal liver HSC provide long-term lympho-myeloid repopulation of adult mice fivefold more efficiently than adult marrow, a fact explained by their sevenfold higher concentration and the observation that fetal HSC clones generated ~3-fold more cells than marrow HSC.

In conclusion, the liver appears to be colonized by an early wave of committed and multipotent progenitors and macrophages that may be predominantly of YS origin, and two waves of HSC, the initial one from the AGM arriving at E10, reaching a maximum at E11, and disappearing by E13, and on E12 a second wave arrives from the YS (Figure 22-5). It is possible that angioblasts also migrate from the yolk sac into the liver to initiate hepatic vasculogenesis. There is no evidence of hemangioblasts or hematogenic endothelium in fetal liver, and sorted hepatic endothelial cells isolated immediately prior to onset of human hepatic hematopoiesis (E27) were devoid of hematopoietic potential.

ONTOGENY OF BONE MARROW

The primordium of the bone marrow cavity develops following penetration by perichondrial mesenchymal cells and blood vessels into the zone of calcified cartilage in the central region of the long bones. Hypertrophic chondrocytes secrete VEGF that recruits vascular cells to penetrate the perichondrium and bring along osteoblast precursors and circulating hematopoietic cells, including primitive macrophages. Following resorption of the cartilaginous matrix, the developing marrow cavity appears as a network of connective tissue and a plexus of widely dilated veins. Marrow hematopoiesis is initiated by accumulation of large numbers of undifferentiated basophilic blast cells within the dilate marrow capillaries, beginning at E11–12 in chick, E17 in mouse, and E70–77 in humans. Subsequently, separate and distinct areas of erythropoiesis and granulopoiesis develop. Avian parabiosis studies demonstrated that circulating stem cells colonize the developing marrow beginning at E11–12. Following intravenous injection of tritiated thymidine-labeled embryonic YS or spleen cells into chick embryos, significant numbers of labeled blast cells localized in the marrow at E11. Cells from these tissues also repopulated the marrow when injected into irradiated embryos. In the mouse, CFU-GM/BFU-E and CFU-S and HSC appear in the femoral marrow at E17, and the populations double every 34 hours. The marrow progressively expands in the first two months of postnatal life, associated with a decline in hepatic hematopoiesis in the first week of life and a decline of splenic hematopoiesis after the third week.

Proof that fetal liver stem/progenitor cells were responsible for colonizing the marrow was provided in studies in which rats were injected *in utero* at E16 with a retroviral vector. Clonal identification of viral integration sites showed that fetal liver-derived clones appeared in the marrow and circulated throughout the life of the animals. The role of SDF-1 produced by marrow stromal cells in the chemoattraction of CXCR4[+] HSC is well established in the adult, and the failure of development of marrow hematopoiesis in mice with inactivation of the CXCR4/SDF-1 pathway strongly suggests that the initial wave of HSC migration is also SDF-1 dependent.

ONTOGENY OF THE SPLEEN

The splenic primordium appears as a dense syncytial-like mesenchymal thickening in the dorsal mesogastrium. The mesenchymal condensation is interspersed with vascular spaces where circulating blood comes into direct contact with mesenchymal reticulum cells. At the earliest stage, corresponding to E8 in chick embryos and E13 in mouse, large immature cells characterized by intense cytoplasmic basophila and prominent nucleoli are observed in both the vascular spaces and the perivascular mesenchyme of the spleen. Within 24 hours these basophilic cells appear scattered throughout the mesenchyme and frequently extend long tails of cytoplasm between the reticulum cells for some distance from the main body of the cells. By 72 hours granulopoiesis is extensive together with erythropoietic foci. Lymphopoiesis is initiated around birth, and myelopoietic activity progressively declines thereafter. Sex chromosome marker studies in parabiosed and twin embryos demonstrated extensive chimerism in the splenic rudiment of the chick embryo as early as E12. Cell-labeling studies confirmed that circulating cells colonized the avian splenic rudiment as early as E8, and together with embryonic hematopoietic cell reconstitution of irradiated embryos, indicated that splenic colonizing cells were present in the YS and circulation at the time of initiation of splenic hematopoiesis. Labeled thymic lymphocytes began to localize in the spleen with high efficiency by E17 coinciding with the initiation of lymphopoiesis.

The origin of cells initially colonizing the mammalian spleen rudiment is most likely the fetal liver since the AGM and yolk sac are no longer hematopoietic at this stage, and the marrow has not yet developed. CFU-GM/BFU-E, CFU-S, and HSC are detected by E15 in the splenic rudiment and increase in absolute numbers through to 3 weeks postnatally, then progressively decline as the spleen ceases to be a myelopoietic organ.

Cell Migration to Primary Lymphoid Organs

ONTOGENY OF THE THYMUS

The thymus is an example of interaction between mesenchymal (neural-crest derived) and epithelial (endodermal) tissues. Notch signaling may play a role in the induction of thymic epithelium. In the mouse embryo at E9.5, Notch receptors and the Notch ligand, Jagged1, are expressed in the third pharyngeal pouch, at the initiation of thymic organogenesis. The thymic anlage at E11 consists of stratified epithelium that over the following 24 hours changes to clustered epithelium and begins to express high levels of MHC class II antigen and of Delta1, the Notch ligand that plays a critical role in T cell development. Comparable development of the thymic rudiment is seen at ~35 days in the human embryo and 7–8 days in the chick embryo. Chromosome marker studies in parabiotic chick embryos demonstrated very high levels of thymic lymphoid chimerism following establishment of a YS vascular union at 6–7 days of incubation but not following a later established chorioallantoic vascular anastamosis, indicating that circulating stem cells colonized the rudiment at 6–7 days. Subsequent studies showed that the avian thymus is colonized in three waves during embryogenesis at E6, E12, and E18. Progenitors in the first wave came in part from para-aortic foci, with the second and subsequent waves coming from marrow and, to a minor extent, spleen.

The role of the YS as a source of thymic immigrant cells was suggested by studies in which 7-day YS cells were injected into irradiated chick embryos and colonized the thymus. However, sex-mismatched YS and embryo chimeras generated at 33 to 55 hours of incubation showed that the thymus as well as the bursa, spleen, and marrow were populated by cells of the sex of embryo and not by YS. Studies in chick-quail chimeras also showed that the primary lymphoid organs were colonized by cells that originate in the para-aortic region. It appears that in the avian system the YS is colonized at a very early stage by cells originating in the para-aortic region, and like the mammalian fetal liver, may serve as a site for expansion of lymphoid progenitors that secondarily migrate to the thymus and bursa of Fabricius. During initial thymic colonization (beginning E11 in the mouse, E7 in the chick), which precedes the onset of vascularization by 48 hours, blood-borne precursors must leave adjacent pharyngeal vessels, traverse perithymic mesenchyme, and basement membrane surrounding the epithelial rudiment to enter the thymus (Figure 22-4). Approximately 20 T cell precursors enter the thymus at E11–12, 300 between E12–13, and 3000 between E13–14.

Evidence for thymic chemotactic factors have been provided in functional transfilter cell migration studies that have shown that alymphoid thymic lobes attract cells from fetal liver fragments. MHC class II$^+$ epithelial cells are source of chemoattractant factors, and response is dependent on G-coupled receptors. A number of chemokines and their receptors are expressed either on immigrant cells or epithelium, and one example, TECK, is chemotactic for thymocytes. Chemokine signaling can induce expression of metalloproteinases (MMP9) on immigrant cells that serve to digest extracellular matrix and basement membrane material and facilitate penetration of the avascular epithelium. The failure of thymic development in the Nude mouse is due to a loss of function mutation in the *Foxn1* transcription factor that is essential for thymic epithelial development and results in failure of lymphoid precursors to enter the epithelium from the perithymic mesenchyme.

The lymphopoietic paths of intrathymic development occurs over two weeks with four phenotypically and genetically distinct phases. The first stage (lineage double negative stage 1) is CD4^{-ve}, CD8^{-ve}, CD25^{-ve}, CD44^{hi} and gives rise to T, B, dendritic cell (DC), and NK; the second stage is CD4^{-ve}, CD8^{-ve}, CD25$^+$, CD44^{hi} T, or DC only. Stage 3 is CD4^{-ve}, CD8^{-ve}, CD25$^+$, CD44^{lo} and absolutely committed to T. The final stage is pre-CD4$^+$, CD8$^+$, CD25$^-$. A 4000-fold expansion occurs in the first three phases, and a 250-fold expansion in the early pre-CD4$^+$, CD8$^+$double positive stage.

Origin and Commitment of Thymic Colonizing Cells

The YS and AGM region and the early stage of fetal liver are all sources of lympho-myeloid stem cells (HSC) at the time of thymic colonization. Yet extensive studies using *in vitro* thymic lobe cultures to identify thymic colonizing cells suggests that they are not HSC and are to some degree T committed prior to entry into the thymus. Lymphoid commitment may coincide with the first appearance of hematopoietic clusters in the vascular endothelium of the AGM region at E9.5–10. At this stage the CD45$^+$, cKit$^+$, CD34$^+$ population contains multilineage progenitors, as well as progenitors restricted to T (pT) or T/NK plus macrophage (pTm), B (pB), or B plus macrophage (pBm) but not pBT. A common lymphoid progenitor (CLP) identified in adult mouse marrow as a precursor of both B and T lineage can be distinguished from the HSC and common myeloid progenitor (CMP) by expression of the IL7Rα, in addition to being Sca-1$^+$, cKit$^+$, Lin^{-ve}. It appears that in fetal development a CLP stage comparable to the adult does not exist. The fetal pT cells have T cell receptor (TCR) rearrangements and appear in the fetal blood and liver at E11–12, where they greatly outnumber pB. This coincides with the period of initial thymic colonization when pT in the liver initially increase and then rapidly decline in number, while pB, initially rare, rapidly increase in number (Figure 22-4).

Cells isolated from the perithymic mesenchyme during initial colonization were IL7Rα$^+$, Lin^{-ve}. Consistent with the paucity of Notch ligand in the fetal liver environment, expression of Notch target genes, *Hes-1* and *pre-Tα*, were not detectable in either fetal liver or perithymic IL7Rα$^+$, Lin^{-ve} populations, indicating that Notch signaling was not activated in these cells prior to thymic entry. In contrast, Notch target genes were detected in precursors that had entered the epithelial environment where the Notch ligands Jagged and Delta are readily detected. This suggests that a Notch influence on the T lineage does not occur until cells enter the thymus. However, it is also possible that pre-thymic commitment to the T lineage is induced by factors other than Notch but is only revealed as a result of Notch signaling in the thymus.

To understand lymphoid lineage specification from the HSC, it is best to think of it as a progressive bias along the T, B, or myeloid pathway without abrupt transition steps. The first cell to enter the thymus is predominantly T directed but retains some myeloid and B potential. The latter has been revealed by the ability of these cells to undergo delayed B cell development in culture and by the appearance of B cell development in the thymus of mice with null mutations of Delta. The pB and the adult CLP are more biased to B than T differentiation, and the former retain some myeloid potential. The expression of Pax5 transcription factor has been correlated with B cell differentiation and progressive downregulation of myeloid and T specific genes.

ONTOGENY OF B LYMPHOCYTES

Avian Bursa of Fabricius

In the avian system, the bursa of Fabricius primordium appears by E5 as an epithelial thickening at the ventrocaudal contact of the cloaca with external ectodermal epithelium. As both bursal and thymic anlagen develop at sites of endodermal-ectodermal interaction, this origin may be of some fundamental importance in the early induction of primary lymphoid organs. By E10–11 longitudinal folds of lining epithelium project into the lumen, and by E12–13 epithelial buds project into the underlying mesenchyme. Bursal lymphopoiesis begins by E14, and at this stage numbers of large basophilic cells are observed in the blood vessels of the mesenchyme adjacent to the epithelial buds. These cells transit through the epithelial basement membrane and localize within the epithelium follicle where they proliferate extensively, differentiating into typical bursal B lymphocytes. Chromosome marker studies in parabiosed embryos and embryonic grafts of bursal

rudiments demonstrated that the bursa was colonized by circulating lymphoid precursors that entered at E13–14. Injection of embryos at E10–14 with tritiated thymidine-labeled embryonic YS, spleen marrow, and blood cells (but not thymocytes) showed rapid localization of labeled basophilic cells in the bursal epithelium, with subsequent proliferation and lymphoid differentiation. The "receptivity" of the bursal epithelium for labeled cells declined abruptly at E15. Parallel injection studies into E14 irradiated embryos using sex chromosome marking and confirmed that the above embryonic hematopoietic tissues, but not the thymus, contained bursal lymphoid precursors. By using specific markers expressed on restricted B and T precursors, it was shown that cells present on the E14 spleen or marrow that colonized the bursa upon injection were already B committed with VJ recombination, and distinct from T lineage precursors. Since such B-restricted precursors appear to be present in the circulation and in all hematopoietic organs at the time of bursal colonization, the precise site of B lineage commitment is unclear. As a result of sex-mismatched and chick-quail yolk sac-embryo chimeras, it is known that the B precursors ultimately derive from the intraembryonic para-aortic region, possibly as early as E2, although their subsequent expansion and B commitment may occur within the YS and then the spleen.

Mammalian B Cell Ontogeny

The fetal liver has long been considered the initial site of B cell commitment beginning at E14 in the mouse, and from then on, the B progenitors expand in a synchronous wave-like pattern reaching a peak in the perinatal stage. It was also generally accepted that early B cell progenitors expressed CD45R/B220 prior to CD19 expression. However, it has become recognized that B-lineage gene expression (DJH rearrangements and Ig germ line transcripts) could be detected in P-Sp/AGM and liver before E12.5. Furthermore, Rag-2 and Rag-1 and CD19 gene expression has been found at E9 in YS and P-Sp. By functional and phenotypic analysis AA4.1$^+$, FcγR$^+$ or AA4.1$^+$, Sca-1$^+$, B220^{-ve} cells with B/macrophage potential can be detected in E12.5–13 fetal liver. A novel population of cKit$^+$, AA4.1$^+$, CD34$^+$, Sca-1^{-ve}, CD45$^+$, CD19$^+$ cells (70% expressing IL7Rα but negative for CD45R/B220) were detected in E11 P-Sp/AGM and liver that differentiate exclusively into B cells. Numerically, at E11 these very early pro-B progenitors comprised ~100 in the AGM region and ~500 in the liver, with smaller numbers in the blood and yolk sac (Figure 22-4).

The use of co-culture on marrow stromal lines such as S17 or ST2, often with the addition of cytokines such as IL7 and cKit ligand, have been used to assay for B lymphoid potential. Such studies have identified the emergence of B- or B/macrophage-restricted progenitors in the AGM region at E10, coinciding with the development of the aortic hematopoietic clusters. The number of progenitors capable of generating B cells in the AGM at E10 (30–35 somite pairs) was 10 and increase 400-fold by E12 where they were located in the fetal liver within the CD45$^+$, Kit$^+$, Sca-1^{lo}, IL7Rα$^+$ population.

185

There has been a long, and as yet unresolved debate concerning the role of the YS in generating pB. Early studies with Ig allotype markers showed that YS cells transplanted into adult mice generated B cells, and *in utero* injection of E8–10 YS resulted in postnatal B and T chimerism. Chromosome marked E11 YS cells were shown to provide long-term lymphoid (T and B) and myeloid engraftment in mice. Cells from YS as early as E8–E9, when co-cultured on marrow stroma, have been reported to generate cells with B and T lymphoid potential as measured *in vitro* and following transplantation into immundeficient mice. The variable and often conflicting data relating to ontogenetic origin of B cells can be attributed, in part, to culture variables, the use of different stromal lines, and particularly the use of supplemental cytokines such as IL-7 and c-Kit ligand. Certain of these conditions may provide an environment capable of supporting lymphoid differentiation from hemogenic endothelium or hemangioblasts; others may support differentiation from lymphomyeloid stem cells and pMix; and yet others can expand only those cells that have undergone B cell commitment (pB or pBm). By restricting the criteria for a committed B cell progenitor to cells expressing CD45, and probably CD19, IL7Rα, and c-Kit, they first appear within the E10 AGM region where they may traffic to the YS as well as the early liver. It is unknown whether the more primitive cell populations in the YS that have B lymphoid potential can migrate into the embryo and undergo pB commitment in the AGM region.

Migration of B progenitors from the AGM to the liver begins at E10, with subsequent increase due to continued immigration and intrahepatic expansion (Figure 22-4). Since HSC are also colonizing the liver at this time, the possibility of intrahepatic generation of additional pB from these more primitive cells cannot be excluded.

KEY WORDS

AGM region   Aorto-gonad mesonephros region. The first site of intraembryonic hematopoiesis and the site of origin of hematopoietic stem cells that can function in the adult.

Hemangioblast   Common progenitor of both hematopoietic and endothelial lineage. Present in the yolk sac blood islands and in the floor of the dorsal aorta in early development. Presence in the adult marrow is controversial.

Hematopoietic stem cell   Self-renewing primitive cell capable of generating all blood cell lineages including lymphoid and myeloid populations.

Progenitor cells   Lineage-restricted primitive hematopoietic and lymphoid precursors, generated from the pluripotent stem cell but with no self-renewal potential.

Yolk sac blood islands   First site of both primitive and definitive hematopoietic development in mammals.

FURTHER READING

Baron, M. H. (2001). Molecular regulation of embryonic hematopoiesis and vascular development: A novel pathway. *J. Hematotherapy & Stem Cell Research* **10**, 587–594.

Cai, Z., de Bruijn, M., Ma, X., Dortland, B., Luteijn, T., Downing, J. R., and Dzierzak, E. (2000). Haploinsufficiency of AML1 affects the temporal and spatial generation of hematopoietic stem cells in the mouse embryo. *Immunity* **13**, 423–431.

Douagi, I., Vieira, P., and Cumano, A. (2002). Lymphocyte commitment during embryonic development in the mouse. *Seminars in Immunology* **14**, 361–369.

Kumaravelu, P., Hook, L., Morrison, A. M., Ure, J., Zhao, S., Zuyev, S., Ansell, J., and Medvinsky, A. (2002). Quantitative developmental anatomy of definitive haematopoietic stem cells/long-term repopulating units (HSC/RUs): role of the aorta-gonad-mesonephros (AGM) region and the yolk sac in colonization of the mouse embryonic liver. *Development* **129**, 4891–4899.

McGrath, K. E., Koniski, A. D., Malik, J., and Palis, J. (2003). Circulation is established in a step-wise pattern in the mammalian embryo. *Blood* **101**, 1669–1676.

Marshall, C. J., and Thrasher, A. J. (2001). The embryonic origins of human hematopoiesis. *Brit. J. Haematol.* **112**, 838–850.

Montecino-Rodriquez, E., and Dorshkind, K. (2003). To T or not to T: reassessing the common lymphoid progenitor. *Nature Immunology* **4**, 100–101.

Moore, M. A. S. (2003). *In vitro* and *in vivo* hematopoiesis. In *Atlas of Blood Cells: Function and Pathology* (D. Zucker-Franklin *et al.*, eds.). 3rd ed. Edi. Ermes, Milan, Italy.

Moore, M. A. S., and Owen, J. J. T. (1967). Experimental studies on the development of the thymus. *J. Exp. Med.* **126**, 715–726.

Tavian, M., Robin, C., Coulombel, L., and Peault, B. (2001). The human embryo, but not its yolk sac, generates lympho-myeloid stem cells: Mapping multipotential hematopoietic cell fate in intraembryonic mesoderm. *Immunity* **15**, 487–495.

23

Hematopoietic Stem Cells

George Q. Daley

Embryonic Stem Cells and Embryonic Hematopoiesis

Over two decades of research in mice has established that embryonic stem cells (ESCs) can give rise to all differentiated cell types in the adult organism and that ESCs are thus pluripotent. *In vitro*, ESCs undergo spontaneous aggregation and differentiate to form cystic embryoid bodies (EBs). These teratoma-like structures consist of semi-organized tissues, including contractile cardiac myocytes, striated skeletal muscle, neuronal rosettes, and hemoglobin-containing blood islands. In the last decade and a half, this *in vitro* system has been exploited to study differentiation events in a number of tissues and has begun to be used for discovery and characterization of small molecule pharmaceuticals. The availability of ES cells from the human might make it possible to produce specific differentiated cell types for replacement cell therapies to treat a host of degenerative diseases.

Stem cells from the embryo have fundamentally distinct properties when compared to stem cells from somatic tissues of the developed organism. The hematopoietic stem cell (HSC) is the best-characterized somatic stem cell in the adult. This rare cell residing in the bone marrow gives rise to all blood cell lineages and will reconstitute the lympho-hematopoietic system when transplanted into lethally irradiated animals. Bone marrow transplantation is widely employed for the treatment of congenital, malignant, and degenerative diseases. Although adult HSCs are an important target for genetic modification, success has been limited by difficulty expressing genes in HSCs and by the challenge of maintaining and expanding HSCs in culture. Moreover, recent concerns have been raised about the safety of gene therapy with retroviruses. Thus, harnessing ES cells as a source of HSCs would make it easier to genetically modify stem cell populations *ex vivo*, to discover small molecules that impact blood development, to study genetic and epigenetic influences on hematopoietic cell fate, and to empower preclinical models for gene and cellular therapy.

A critical question is whether current methods for *in vitro* differentiation of ESCs produce HSCs capable of long-term blood formation in adults. All protocols for *in vitro* differentiation of ES cells published to date appear to recapitulate the yolk sac stage of hematopoietic commitment, with question-

able developmental maturation into adult-type somatic HSCs. The first blood cells detected in the yolk sac of the embryo and in EBs *in vitro* are primitive nucleated erythrocytes. These cells express embryonic forms of hemoglobin with a left-shifted oxyhemoglobin dissociation curve, adapted to the low-oxygen environment of the embryo. Embryonic forms of hemoglobin serve as markers of primitive, embryonic erythropoiesis. Later in development, both yolk sac and EBs produce a variety of more differentiated myeloid cell types and enucleated red blood cells that express adult globins and are typical of circulating mature blood. This suggests that yolk sac precursors are capable of making the transition from the primitive to definitive hematopoietic programs, but the extent to which yolk sac-derived progenitors contribute long term to hematopoiesis in the adult remains controversial. Experimental manipulation of yolk sac blood progenitors can reveal a latent potential for hematopoietic engraftment in adults. When marked yolk sac cells from one embryo are transplanted into other embryos, yolk sac progenitors can contribute to blood formation in the adult. Yet yolk sac progenitors fail to engraft if injected directly into the adult. However, when directly injected into the liver of myeloablated newborn mice, highly purified CD34+/c–Kit+ progenitors isolated from mouse yolk sac will provide long-term blood production throughout adulthood. Apparently, the newborn liver retains the embryonic hematopoietic microenvironment and supports the developmental maturation of the yolk sac stem cells. Yolk sac cells can also engraft in irradiated mice if they are first cultured on a supportive stromal cell line taken from the aorta–gonad–mesonephros region of the embryo. Here once again, the yolk sac cells can be "educated" by the stromal cells to adopt an adult profile. So indeed, hematopoietic progenitors from yolk sac appear capable of sustaining hematopoiesis in adults, but based on recent evidence from a number of groups, it appears likely that yolk sac progenitors contribute to embryonic blood formation and thereafter yield to a distinct source of definitive HSCs that arise within the embryo proper. Compelling evidence suggests that the definitive HSCs responsible for lifelong hematopoiesis arise chiefly from a separate and distinct source in the aorta–gonad–mesonephros region of the developing embryo.

Blood Formation in Embryoid Bodies

As for yolk sac progenitors, it has proven exceedingly difficult to demonstrate that ES-derived hematopoietic progenitors can repopulate adult mice. This resistance of embryonic progenitors from EBs to engraft in the adult is reminiscent of the block to engraftment observed for yolk sac progenitors and likewise is believed to reflect the developmental immaturity of ES-derived HSCs, and the different microenvironments of the embryo and the adult. Whether *in vitro* differentiation of ESCs will promote formation of AGM-like HSCs remains unresolved. In order to model hematopoietic transplantation from ES cell sources, this challenge must be overcome.

Although EBs show a temporal wave of primitive followed by definitive hematopoiesis, the nature of the ES-derived HSCs has been a subject of great interest. The delineation of hematopoietic development within EBs has been pioneered by Keller and colleagues, whose seminal contributions have made the hematopoietic program among the most well-defined aspects of *in vitro* ES cell differentiation. Their work has defined the most primitive hematopoietic progenitor in EBs as the blast colony forming cell (BL-CFC), a transient cell with both primitive erythroid potential and the capacity to generate definitive erythrocytes and multilineage myeloid colonies upon replating. It also has the potential to form endothelial cells and as such has been defined as the hemangioblast. To date there is no data demonstrating lymphoid potential for this cell type. Only one group has reported repopulation of irradiated adult mice with cells taken from EBs fortuitously timed to harbor the maximal number of BL-CFCs, but in this report there were no markers to demonstrate that lymphoid and myeloid cells developed from a single cell committed to the hematopoietic lineage. Thus, to date the relationship of the BL-CFC to definitive HSCs remains ill-defined, and whether EBs support the development of AGM-like HSCs remains an open question.

Philadelphia chromosome translocation in both lymphoid and myeloid lineages, showing that the multilineage differentiation of stem cells is preserved despite BCR/ABL transformation.

We hypothesized that BCR/ABL would enable us to transform an ES-derived HSC, engraft mice, and determine the extent of lymphoid and myeloid differentiation *in vivo*. We introduced BCR/ABL into differentiated murine ES cells and cultured a primitive hematopoietic blast cell that generated nucleated erythroblasts *in vitro*, mimicking yolk sac blood formation. We picked and expanded single-cell clones of these cells, verified clonality by retroviral integration, injected irradiated mice, and observed successful lymphoid–myeloid engraftment in primary and secondary animals. The erythroid progenitors from engrafted mice expressed only adult globins, suggesting that the cells underwent developmental maturation to the definitive hematopoietic program *in vivo*. BCR/ABL expression enabled adult engraftment by altering the cell's homing properties, complementing a missing cytokine signal, or blocking apoptosis, allowing the ES-derivatives to acclimatize to the adult microenvironment and to differentiate into multiple hematopoietic lineages. These results provided the first definitive demonstration of the embryonic HSC (ε-HSC) that arises *in vitro* during ES differentiation. This cell is a common progenitor of both primitive embryonic erythropoiesis (yolk sac type) and the definitive adult lymphoid–myeloid hematopoietic stem cell. Although the BCR/ABL-transformed clones produce colonies with a comparable morphology to the BL-CFC and show primitive erythroid potential, their precise relationship to the hemangioblast is unclear, as the BCR/ABL-transformed clones do not appear to have endothelial potential.

Transformation of an EB-Derived HSC by BCR/ABL

We set out to address the question of whether a lymphoid–myeloid HSC developed within differentiating EBs by attempting to transform that putative cell. We borrowed from our experience with the disease chronic myeloid leukemia (CML), the classical pathologic condition of adult hematopoietic stem cells that is caused by the BCR/ABL oncoprotein. Several biological properties unique to the BCR/ABL oncoprotein made it particularly well suited for addressing the nature of primitive blood progenitors within EBs. Induction of CML requires expression in pluripotent hematopoietic stem cells. Multilineage hematopoiesis is monoclonal in most patients at the time of CML diagnosis, demonstrating that the fusion oncoprotein BCR/ABL endows leukemic stem cells with a competitive repopulation advantage over normal stem cells. Despite expression in a wide array of tissues, transgenic mice that carry BCR/ABL in their germ line develop only hematopoietic malignancies, reflecting the tropism of BCR/ABL for hematopoietic cell types and the sparing of nonhematopoietic tissues. Patients with CML harbor the

Promoting Hematopoietic Engraftment with STAT5 and HoxB4

Although BCR/ABL transformation targets a rare cell in ES cell cultures with lymphoid–myeloid developmental potential, the engrafted mice succumbed to leukemia, prompting us to explore means for isolating the ε-HSC without inducing the transformed phenotype. Our approach to generating normal, nontransformed blood progenitors from ES cells involved a combination of two strategies: (1) expression of single proteins in signaling pathways activated by BCR/ABL (e.g., STAT5), postulating that activation of downstream targets would be less disruptive to cell physiology than transformation by the complete oncoprotein; and (2) conditional expression of candidate genes using a novel ES cell line engineered to express the gene of interest from a tetracycline regulated promoter such that genetic effects could be induced and then reversed. The Ainv15 ESC line we created expresses the tet-dependent transcriptional transactivator protein from an active genomic locus (*ROSA26*).

Any gene of interest can be inserted with high targeting efficiency into an expression cassette located within the active

HPRT gene locus. Genes that are targeted correctly become resistant to neomycin (G418), and the inserted gene is expressed only in the presence of the potent tetracycline analog doxycycline and can be rapidly silenced following doxycycline removal. We chose to express the *STAT5* transcriptional regulator and the homeobox gene *HoxB4* in this system, owing to the central role of STAT5 in BCR/ABL and cytokine receptor signaling, and to extensive prior evidence from the Humphries group of the role of *Hox* genes in hematopoiesis and the unique properties of *HoxB4*, which was previously shown to enhance hematopoietic engraftment without inducing leukemia. We differentiated the modified ES cells into EBs and activated gene expression by adding doxycycline to the culture medium between differentiation days 4 and 6, timed to coincide with the maximal generation of primitive multipotential hematopoietic colonies. After 6 days, the EBs were dissociated, and the cells were cultured on the OP9 stromal cell line, previously shown to enhance production of hematopoietic progenitors from mES cells. Expansion of hematopoietic blast cells was observed upon *STAT5* and *HoxB4* gene induction, and vigorously growing colonies of hematopoietic cells were detected only in the presence of doxycycline. These cells were harvested, plated in semisolid media plus cytokines, and shown to produce a variety of blood cell colony types, with only the most primitive multipotential colonies significantly expanded. The cells, which also express the green fluorescent protein (GFP), were injected intravenously into irradiated syngeneic or immunodeficient mice.

Contributions of GFP+ cells to the peripheral blood were then monitored by flow cytometry, and specific lymphoid and myeloid cell populations were scored by antibody staining against specific cell surface differentiation antigens, by forward and side scatter properties, and by direct microscopic examination of cells cyto-centrifuged onto cover slips. In these experiments, both *STAT5* and *HoxB4*-expressing cells engrafted in mice and generated both lymphoid and myeloid populations in circulating blood. Interestingly, contributions of the *STAT5*-stimulated cells appeared to be transient, despite continued gene induction *in vivo* (through inclusion of doxycycline in the drinking water of the mice). Engraftment with *HoxB4*-expressing cells persisted in primary animals even in the absence of gene induction, and the cells of primary animals could be transplanted into secondary animals, suggesting self-renewal of a long-term reconstituting hematopoietic stem cell. Examination of peripheral blood smears from engrafted mice showed no evidence of abnormal hematopoiesis, although rare animals succumbed to hematologic malignancies from donor cells, suggesting some tendency for the genetically modified ESCs to undergo transformation *in vivo*. Retroviral delivery of *HoxB4* directly to populations of cells dissociated from EBs after 4 to 6 days of differentiation also succeeded in generating expanding cultures of hematopoietic cells that engrafted in irradiated mice. These data demonstrate that expression of the *STAT5* or *HoxB4* gene in differentiating cultures of ES cells yields hematopoietic engraftment in irradiated mice, with *HoxB4*

showing the most promise for stable engraftment in primary and secondary animals.

The mechanisms by which *STAT5* and particularly *HoxB4* drive hematopoietic engraftment from ESCs remain unclear. By driving cell proliferation, both genes might serve to increase the numbers of otherwise vanishingly rare HSCs above a threshold of detection. Alternatively, *HoxB4* might be altering cell fate, a known effect of homeobox genes, by promoting a transition from primitive to definitive HSC fate. RT-PCR analysis of *HoxB4*-expressing EB-derived cells in comparison to hematopoietic progenitors isolated from precirculation yolk sac confirmed the detection of markers of definitive hematopoiesis in *HoxB4* expressing cells, including adult-type globin. Moreover, *HoxB4*-transduced cells expressed *CXCR4*, the chemokine receptor implicated in hematopoietic stem cell homing to the bone marrow, and *Tel*, the transcription factor implicated in migration from the fetal liver to the adult bone marrow microenvironment. *HoxB4* does not appear to be expressed in the precirculation yolk sac, but is detected in primitive populations of CD34+ bone marrow cells from the adult. We therefore also tested whether expression of the *HoxB4* gene by retroviral infection in yolk sac progenitors would endow them with engraftment potential in adults. As for EB-derived cells, *HoxB4* expression in yolk sac progenitors induced dramatic expansion on OP9 stromal cultures and stable hematopoietic engraftment in adult mice. Yolk sac-derived progenitors could also be transplanted into secondary animals. These data support the hypothesis that activation of *HoxB4* endows embryonic hematopoietic progenitors with the potential to engraft in adult hematopoietic microenvironments, and therefore may be critical to the transition from the embryonic to adult hematopoietic program.

Lymphoid potential and transient lymphoid reconstitution of engrafted immunodeficient mice has been demonstrated, showing convincingly that EB-derived cells can show full hematopoietic differentiation potential. Native EB-cell derived hematopoietic progenitors function at best inefficiently to reconstitute the adult host, or may develop in only limited numbers under idiosyncratic culture conditions. If indeed distinct progenitors contribute to primitive yolk sac and definitive AGM-type hematopoiesis in the embryo, then distinct progenitors might arise at spatially and temporally distinct sites during EB formation from ES cells *in vitro*. Identifying the precise culture conditions that promote differentiation of ESCs into robust adult-type definitive HSCs *in vitro* remains a critical goal.

To date, gene modification has been required to demonstrate development of lymphoid–myeloid HSCs from differentiating ES cells *in vitro*, yet even gene modification results in inefficient engraftment. Given the technical challenges and risks inherent in genetic modification, it would be preferable to derive engraftable HSCs in as natural a process as possible by mimicking the developmental pathways of the embryo. Applying principles that specify blood formation to *in vitro* systems might enable enhanced blood formation in a safe and efficient manner.

Promoting Blood Formation *in Vitro* with Embryonic Morphogens

During gastrulation in the mouse, secreted signaling molecules induce distinct cell fates from developing mesoderm. The earliest stages of blood formation occur in the extraembryonic mesoderm of the yolk sac, where blood islands form surrounded by endothelial cells and closely opposed to the visceral (primitive) endoderm. Several of the players that regulate the transcriptional control of hematopoiesis have been identified (e.g., *SCL*, *AML1/CBFa2*, and several *Hox* genes), largely due to their involvement at translocation breakpoints in leukemia. Gene knockout studies of these factors have validated their role in hematopoiesis. More recently, a small number of secreted factors have been identified that act as early embryonic inducers of mesodermal fate in the hematopoietic lineage. The most interesting of these are hedgehog factors and bone morphogenetic protein 4 (BMP4).

Indian hedgehog (Ihh) is a member of the hedgehog family of signaling molecules that play diverse roles in patterning early embryonic events. Baron and colleagues have shown that early hematopoietic activity in the developing murine yolk sac is dependent upon signals from the adjacent primitive/visceral endoderm. Recently, they demonstrated that Ihh was produced by visceral endoderm sufficient to mediate this induction. Ihh could respecify neuroectoderm to hematopoietic fate, and blocking Ihh function by anti-hh antibodies abrogated hematopoietic development. Baron and colleagues showed that Ihh induction of hematopoiesis in murine embryo explants led to expression of BMP4. BMP4, a member of the TGF-β family, has been implicated as a potent ventralizing factor and inducer of hematopoietic mesoderm in *Xenopus* development. A related hedgehog factor, Sonic hedgehog (Shh), has been shown by Bhatia and colleagues to enhance the expansion of human hematopoietic repopulating cells assayed by transplantation in immunodeficient mice. BMP4 and other members of the BMP family have also been implicated in regulation of proliferation and survival of primitive human hematopoietic populations. Antibodies to hh and noggin, a specific antagonist of BMP4 signaling, abrogated the proliferative effects of Shh in human hematopoietic cell culture. It has been shown that Ihh is expressed by visceral endoderm in developing EBs.[41] There is a single report that addition of BMP4 to differentiating cultures of rhesus ES cells augments formation of hematopoietic clusters. Most recently, Bhatia and colleagues have shown that BMP4 can enhance hematopoietic potential from human ES cells.

These studies offer the first hints that directed differentiation of human ES cells into HSCs might be feasible. However, numerous questions persist: will HSCs derived from hES cells function normally? Will they reconstitute normal immune function and remain nontumorigenic *in vivo*? Can immunologic issues be circumvented through nuclear replacement or gene modification? The hurdles for therapeutic applications of ES-derived cells remain high, but no matter what, the *in vitro* differentiation system will remain an important model for investigations of blood formation and embryonic development.

KEY WORDS

Embryoid bodies Cystic teratoma-like structures consisting of semi-organized tissues representing all three embryonic germ layers. Formed when embryonic stem cells are removed from culture conditions that inhibit differentiation, and are allowed to aggregate and differentiate.

Hemangioblast (blast colony forming cell- BL-CFC) A long-theorized but only recently identified embryonic cell representing a common progenitor of both the endothelial and hematopoietic lineages. Can be readily detected in differentiating cultures of embryonic stem cells as the first mesodermal element committed to the hematopoietic lineage. Assayed by colony formation in methylcellulose supplemented with vascular endothelial growth factor and stem cell factor (the BL-CFC).

Primitive versus definitive hematopoiesis The first wave of embryonic blood development. It occurs in the yolk sac and consists principally of nucleated erythrocytes that express embryonic globins. The primitive wave is believed to supply the needs of the embryo and to be later supplanted by a definitive class of hematopoietic stem cells that arise in the para-aortic region of the developing embryo proper. Definitive hematopoiesis generates mature myeloid and lymphoid lineages for the life of the animal.

FURTHER READING

Choi, K., Kennedy, M., Kazarov, A., Papadimitriou, J. C., and Keller, G. (1998). A common precursor for hematopoietic and endothelial cells. *Development* **125**, 725–732.

Doetschman, T. C., Eistetter, H., Katz, M., Schmidt, W., and Kemler, R. (1985). The *in vitro* development of blastocyst-derived embryonic stem cell lines: formation of visceral yolk sac, blood islands, and myocardium. *J. Embryol. Exp. Morphol.* **87**, 27–45.

Kennedy, M., Firpo, M., Choi, K., Wall, C., Robertson, S., Kabrun, N., and Keller, G. (1997). A common precursor for primitive erythropoiesis and definitive haematopoiesis. *Nature* **386**, 488–493.

Kyba, M., Perlingeiro, R. C., and Daley, G. Q. (2002). Hoxb4 confers definitive lymphoid–myeloid engraftment potential on embryonic stem cell and yolk sac hematopoietic progenitors. *Cell* **109**, 29–37.

Medvinsky, A., and Dzierzak, E. (1996). Definitive hematopoiesis is autonomously initiated by the AGM region. *Cell* **86**, 897–906.

Nakano, T., Kodama, H., and Honjo, T. (1994). Generation of lymphohematopoietic cells from embryonic stem cells in culture. *Science* **265**, 1098–1101.

Perlingeiro, R. C., Kyba, M., and Daley, G. Q. (2001). Clonal analysis of differentiating embryonic stem cells reveals a hematopoietic progenitor with primitive erythroid and adult lymphoid–myeloid potential. *Development* **128**, 4597–4604.

Potocnik, A. J., Kohler, H., and Eichmann, K. (1997). Hemato-lymphoid *in vivo* reconstitution potential of subpopulations derived from *in vitro* differentiated embryonic stem cells. *Proc. Nat. Acad. Sci. USA* **94**, 10,295–10,300.

Yoder, M. C. (2001). Introduction: spatial origin of murine hematopoietic stem cells. *Blood* **98**, 3–5.

24

Cell Differentiation in the Skeleton

Gerard Karsenty

Introduction

The vertebrate skeleton is composed of two specialized tissues, cartilage and bone, surrounding and protecting a third one, the bone marrow. Each of these two skeletal tissues contains a specific cell type of mesenchymal origin, chondrocytes in cartilage and bone-forming osteoblasts in bone. Both chondrocytes and osteoblasts fulfill unique functions that are critical for the growth, maintenance, and integrity of the skeleton. Chondrocytes are required for longitudinal growth of the bones and synthesize the cartilaginous scaffolds on which osteoblasts can deposit bone matrix during both development and postnatal growth. Osteoblasts synthesize a collagen-rich matrix that has the unique property to become eventually mineralized. Defect of differentiation or function of any of these two cell types has tremendous consequences as reflected by the large number of genetic and acquired diseases of skeleton. Several transcription factors as well as a complex interplay of secreted molecules have been identified that control the progression from a pluripotent progenitor to a functional cell.

Skeletogenesis

Formation of the vertebrate skeleton begins with aggregation of undifferentiated mesenchymal cells in structures that prefigure each future skeletal element. These skeletal condensations, also called *anlage*, form around 9.5 days postcoitum (dpc) during mouse development and become visible histologically around 10.5 dpc. Assembly of these skeletal condensations is orchestrated by molecules that control migration and association of undifferentiated cells (Fig. 24-1). Although some particularities exist between craniofacial, axial, and limb patterning, the pathways controlling this patterning process show many similarities. Once formed, skeletal condensations can form bones, following either an intramembranous or an endochondral ossification process. Bones formed by intramembranous ossification include the frontal, parietal, and parts of the temporal and occipital bones, the majority of the facial bones, and the lateral part of the clavicles. In these flat bones cells of the condensations expand to form a membranous structure; then cells differentiate directly into osteoblasts. They maintain and then increase their secretion of type I collagen while initiating expression of noncollagenous proteins more specific of the osteoblast such as *Bone sialo protein* and *Osteocalcin.*

In the rest of the skeleton, skeletogenesis proceeds by endochondral ossification, a multistep process. By 12.5 dpc of mouse development, cells in the center of the skeletal condensations differentiate into chondrocytes, expressing type II collagen and aggrecan. Between 14.5 dpc and 16.5 dpc, as the condensation elongates the most inner chondrocytes differentiate into hypertrophic chondrocytes expressing type X collagen. In the meantime, cells at the periphery of the condensation flatten and begin to differentiate into osteoblasts that express type I collagen. This structure, termed the bone collar, is most apparent in the region closer from the hypertrophic zone of the condensation. The second step of the endochondral ossification process is marked by two coordinated events. While hypertrophic chondrocytes start to die through apoptosis, blood vessels invade the region they occupied, bringing in differentiating osteoblasts from the bone collar. The bone matrix synthesized by osteoblasts then replaces the cartilaginous matrix. This ossification spreads centripetally until much of the cartilage is replaced. Only a small region at either end of the growing bone will remain cartilaginous. This structure, called the growth plate cartilage, will control longitudinal growth of the bone until the end of puberty. As the newly deposited bone matrix mineralizes, osteoclast precursors coming from the bloodstream invade the center of the forming structure and begin to differentiate under the influence of factors secreted by osteoblast progenitors. They start to resorb the bone matrix, forming the internal space that will contain the bone marrow.

Chondrocyte Differentiation

The study of chondrogenesis has largely benefited from two particularities of this process. First, differentiation of chondrocytes induces significant morphological changes that facilitate identification of distinct maturational stages. Second, most of the differentiation process is constantly recapitulated in the growth plate cartilage in skeletal element developing by endochondral ossification (Fig. 24-2). Cells at the earliest stage of maturation are small, and they are located in the most distal area, termed the *reserve* or *resting zone*. They are considered to be chondrocyte progenitors. Immediately below this resting zone emerges a proliferative zone, where cells are slightly larger, flatter, and rapidly dividing. The length of these columns of rapidly dividing chondrocytes is a major determinant of bone longitudinal growth. Proliferating chondrocytes express type II collagen and aggrecan. However, the most proximal cells (also called prehypertrophic chondrocytes) also express Indian hedgehog (Ihh) and the receptor for both parathyroid hormone (PTH) and PTH-related peptide (PTHrP). The most proximal zone appearing on the growth plate cartilage contains hypertrophic chondrocytes. They express type X collagen but not any more type II collagen, and eventually they undergo apoptosis.

Sox9, a member of the high-mobility-group (HMG) family of transcription factors, is the earliest known molecule to be required for chondrocyte specification. Sox9 was first identified as the gene inactivated in campomelic dysplasia (CD), a dominant genetic disorder characterized by skeletal malformation and sex reversal. Subsequently, Sox9 was found to activate the expression of the type II, type IX, and type XI collagen genes in chondrocytes (61–64). Mouse genetic studies clarified the role of Sox9 during chondrogenesis. They showed that Sox9-/- cells, but not wild-type (wt) embryonic stem (ES) cells, are excluded from the skeletal condensations indicating that Sox9 is required for their formation. Moreover, teratomas derived from Sox9-/- but not from wt ES, failed to develop cartilage in mouse chimeras, and Sox9-/- cells are abnormally sensitive to apoptotic signals. Two other HMG family members, L-Sox5 and Sox6, are also playing a key role during chondrocyte maturation. Both genes were found to be co-expressed with Sox9 in precartilaginous condensations but continue to be expressed in hypertrophic chondrocytes. Sox5 and Sox6 have redundant function in vivo, and single null mutant mice are virtually normal. However, targeted deletion of both L-Sox5 and Sox6 leads to late embryonic lethality due to generalized chondrodysplasia demonstrating their essential role in control of chondrocyte differentiation.

In contrast to this group of activators, NFATl, a member of the family of nuclear factor of activated T cells, acts as a repressor of chondrocyte differentiation. Indeed, its overexpression in chondrocytic cell lines suppresses expression of chondrocyte molecular markers, and NFAT1-deficient mice develop ectopic cartilage in the joints. The newly formed cartilage contains ordered and columnar chondrocytes with distinct morphologies and is eventually replaced by bone, recapitulating the process of endochondral ossification. FGFs, a class of secreted factors, also inhibit chondrocyte differentiation. Because of the large number of FGF members expressed at every step of skeleton formation, it is via the functional analysis of their receptors (FGFRs) that the role of these factors has been best characterized. Fgfr3 inactivation in mice causes increased chondrocyte proliferation, while activating mutations in FGFRS are found in chondrodystrophic patients with shortened proliferating zone at the growth plate cartilage. These observations indicate that one or several FGF(s), acting through this receptor, inhibits chondrocyte proliferation.

Chondrocyte proliferation hypertrophy, the last stage of chondrocyte maturation, is controlled by a negative feedback loop involving two growth factors, PTHrP and Indian hedgehog (Ihh). PTHrP, secreted by the chondrocytes of the perichondrium, signals to the PTHrP receptor (PPR) expressed by proliferating chondrocytes to stimulate their proliferation. When escaping PTHrP spatial range of activity, cells become prehypertrophic; they begin synthesizing Ihh, stop proliferating, and start differentiating in hypertrophic chondrocytes. Accordingly, chondrocyte-specific constitutive expression of PPR in mice causes a delay in conversion of proliferative chondrocyte into hypertrophic chondrocyte, and an activating mutation in the human PPR causes a short-limb dwarfism inherited disorder. In turn, Ihh secretion by prehypertrophic chondrocytes directs cells of perichondrium to up-regulate their synthesis of PTHrP, thereby indirectly slowing down the pace of chondrocyte hypertrophy. The presence of an increased fraction of hypertrophic chondrocytes in Inn-deficient mice confirmed this role.

One transcription factor, Runx2, formerly Cbfal, also controls hypertrophic differentiaition. Runx2 was originally identified for its role as an osteoblast differentiation factor (see below). Besides an absence of osteoblasts, Runx2-deficient mice have proximal long bones populated mostly by resting and proliferative chondrocytes, but they show no hypertrophic chondrocytes and no type X collagen expression. In contrast, targeted overexpression of Runx2 into nonhypertrophic chondrocytes of wild-type mice or of Runx2-deficient mice leads to ectopic endochondral bone formation and specific rescue of the lack of hypertrophic chondrocytes, respectively. This latter rescue occurs without rescue of the arrest of osteoblast differentiation, indicating that Runx2 control of chondrocyte hypertrophy is independent of its osteoblast differentiation ability.

Lastly, it has been shown that hypertrophic chondrocytes control vascular invasion, a step required for continuation of the endochondral ossification process. Indeed, cartilage resorption by MMP9 releases VEGF, an angiogenic factor secreted by hypertrophic chondrocytes, which in turn regulates vascular invasion. Recently, the analysis of mice deficient for connective tissue growth factor (CTGF) showed that this factor would be regulating extracellular matrix remodeling and vasculogenesis via its control of VEGF and MMP9 expression.

Osteoblast Differentiation

Unlike what is the case for chondrocytes, osteoblast differentiation is not marked by phenotypic changes in vivo. The only mark of differentiation is slow to appear and lies outside the cell: fully differentiated osteoblasts can produce a matrix that becomes mineralized. This absence of morphological characteristics at the cellular level implies that one has to rely on gene expression studies to assess osteoblast differentiation. Here again the osteoblast has only a few specific markers. In fact, there is only one truly osteoblast-specific structural gene, Osteocalcin, which is only expressed by fully differentiated osteoblas.

Runx2 is the earliest regulator of osteoblast differentiation identified to date. It belongs to a family of transcription factors related to runt, a regulator of neurogenesis and sexual differentiation in Drosophila. Runx2 is expressed in every osteoblast progenitor cell, regardless of its embryonic origin. Runx2-deficient mice have a normally patterned skeleton that

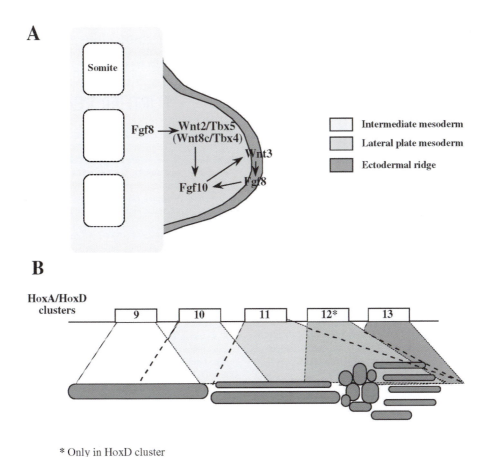

A

Somite

Fgf8 → Wnt2/Tbx5
(Wnt8c/Tbx4)

Wnt3

Fgf10 ← Fgf8

☐ Intermediate mesoderm
☐ Lateral plate mesoderm
☐ Ectodermal ridge

B

HoxA/HoxD
clusters

9 10 11 12* 13

* Only in HoxD cluster

Figure 24-1. Patterning of skeletal elements in the limbs. (A) An initial series of interactions between Fgf and Wnt signaling pathways induces limb bud formation and early outgrowth. **(B)** Limb skeletal elements are later specified by the proximo-distal expression of distinct genes of the HoxA and HoxD clusters. The area of influence of each particular set of genes (boxes) is marked with a distinct shade of grey.

is entirely cartilaginous. The only skeletal elements missing are those formed through intramembranous ossification. At these sites only a membranous structure exists; since there is no cartilaginous stage in the development of these structures, no cartilaginous scaffold can exist. Interestingly, Runx2 heterozygous mutant mice present hypoplastic clavicles and a severe delay of the closure of the fontanelles. These features recapitulate a human condition called Cleidocranial dysplasia (CCD), and mutations into the *Runx2* gene cause CCD. That Runx2 haploinsufficiency results in a significant phenotype further emphasizes its importance during osteoblast differentiation. Recently, mice harboring an inactivation of Cbfβ (a small protein without DNA binding and transcriptional abilities known to interact with Runxl, another Runx factor required for hematopoiesis, lethal only perinatally), were analyzed. In contrast with the identical phenotypical abnormalities observed in Runxl-deficient and classic Cbfβ-deficient mice, these mice did not reproduce the skeletal phenotype of the Runx2-deficient mice. In fact, their phenotype was rather similar to Runx2 haploinsufficiency. This observation and the absence of obvious heterodimerization between Runx2 and

Cbfβ *in vitro* suggest that Cbfβ may play an accessory role during osteoblast differentiation.

Osx, a zinc finger-containing nuclear factor, has recently been identified as another regulator of osteoblast differentiation. *Osx* is specifically expressed in osteoblasts, and Osx-deficient mice lack osteoblasts. *Osx* is not expressed in Runx2-null mice, yet Runx2 is expressed normally in Osx-deficient mice, indicating that Osx acts downstream of Runx2. Besides Cbfal and Osx, two broadly expressed members of the API family of transcription factors, ΔFosB and Fra-1, have been reported to act as positive regulators for osteoblast differentiation. ΔFosB is an alternative spliced transcript of FosB lacking FosB C-terminal domain. Generalized overexpression of ΔFosB in transgenic mice leads to an increase in osteoblast differentiation. Similarly, forced ubiquitous expression of Fra-1 causes an increase in osteoblast proliferation.

Secreted molecules have also been shown to control osteoblast differentiation. For instance, in the absence of Ihh, Runx2 expression is not induced in cells of the bone collar, and osteoblast diffentiation is blocked in skeletal elements undergoing endochondral ossification. These observations

193

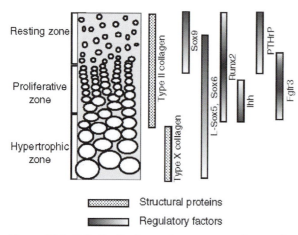

Resting zone

Proliferative zone

Hypertrophic zone

Type II collagen

Type X collagen

Sox9

L-Sox5, Sox6

Runx2

Ihh

PTHrP

Fgfr3

Structural proteins

Regulatory factors

Figure 24-2. Cell distribution and gene expression at the embryonic growth plate cartilage.

indicate that Ihh is a determinant of early osteoblast differentiation and that this secreted factor may act via its regulation of Runx2 expression. FGFs also regulate osteoblast differentiation. Indeed, inactivation of FGF18 in mice causes a general delay of bone formation affecting both skeletal elements forming by intramembranous and endochondral ossification.

KEY WORDS

Chondrocyte Cell type of mesenchymal origin present in cartilage; can be nonhypertrophic or hypertrophic.

Mesenchymal condensation Aggregation of undifferentiated mesenchymal cells prefiguring a future skeletal element.

Osteoblasts Cell type of mesenchym,al origin present in osteoblast and responsible for bone formation.

FURTHER READING

Bi, W., Deng, J. M., Zhang, Z., Behringer, R. R., and de Crombrugghe, B. (1999). Sox9 is required for cartilage formation. *Nat. Genet.* **22**, 85–89.

Bi, W., Huang, W.-H., Whitworth, D. J., Deng, J. M., Zhang, Z., Behringer, R., and de Crombrugghe, B. (2001). Haploinsufficiency of Sox9 results in defective cartilage primordia and premature skeletal mineralization. *Proc. Natl. Acad. Sci. USA* **98**, 6698–6703.

Ducy, P. (2000). Cbfa1: a molecular switch in osteoblast biology. *Dev. Dyn.* **19**, 461–471.

Hall, B. K., and Miyake, T. (2000). All for one and one for all: condensations and the initiation of skeletal development. *Bioesssays* **22**, 138–147.

Karsenty, G., and Wagner, E. F. (2002). Reaching a genetic and molecular understanding of skeletal development. *Dev. Cell.* **2**, 389–406.

Kronenberg, H. M. (2003). Developmental regulation of the growth plate. *Nature* **423**, 332–336.

Nakashima, K., Zhou, X., Kunkel, G., Zhang, Z., Deng, J. M., Behringer, R. R., and de Crombrugghe, B. (2002). The novel zinc finger-containing transcription factor osterix is required for osteoblast differentiation and bone formation. *Cell* **108**, 17–29.

Ornitz, D. M., and Marie, P. J. (2002). FGF signaling pathways in endochondral and intramembranous bone development and human genetic disease. *Genes Dev.* **16**, 1446–1465.

Vu, T. H., Shipley, J. M., Bergers, G., Berger, J. E., Helms, J. A., Hanahan, D., Shapiro, S. D., Senior, R. M., and Werb, Z. (1998). MMP-9/gelatinase B is a key regulator of growth plate angiogenesis and apoptosis of hypertrophic chondrocytes. *Cell* **93**, 411–422.

25

Human Vascular Progenitor Cells

Sharon Gerecht-Nir and Joseph Itskovitz-Eldor

Human Vascular Development, Maintenance, and Renewal

During the third week of human embryonic development, blood vessels are formed in conjunction with blood islands within the yolk sac mesoderm. During this process, blood islands develop alongside the endoderm, which segregate into individual hemangioblasts that are surrounded by flattened endothelial precursor cells. The hemangioblasts mature into the first blood cells, while the endothelial precursors develop into blood vessel endothelium. At the end of the third week, the entire yolk sac, the chorionic villi, and the connecting stalk are vascularized. New vascular formation, termed *vasculogenesis*, takes place within the embryo on day 18. During this process, the underlying endoderm secretes substances that cause some cells of the splanchnopleuric mesoderm to differentiate into angioblasts. These mesodermal angioblasts then flatten into endothelial cells and coalesce, resulting in small vesicular structures referred to as angiocysts. The latter fuse to form networks of angioblastic cords that later unite, grow, and invade embryonic tissues to create the arterial, venous, and lymphatic channels. The spread of vessel networks occurs by three processes:

1. Continuous fusion of the angiocysts — vasculogenesis.
2. Sprouting of new vessels from existing ones — angiogenesis.
3. Assimilation of new mesodermal cells into the existing vessel's wall.

Although the yolk sac is the first supplier of blood cells to the embryonic circulation, the role of blood cell production is taken over by a series of embryonic organs, such as the liver, spleen, thymus, and bone marrow. Therefore, two major concepts were suggested: the bipotential hemangioblast produces the primitive erythroid and endothelial progenitor cells, and the hemogenic endothelium gives rise to hematopoietic stem cells and endothelial progenitors. The ability of these progenitors of hematopoiesis (blood cell repopulation) or vasculogenesis or angiogenesis is yet to be determined.

VASCULOGENIC EMBRYONIC CELLS

Endothelial-Hematopoietic Cell Relationship

In human embryos, homogenic endothelium could be observed and isolated in both the extraembryonic and intra-

embryonic regions. CD34+ cells were detected on a human vascular system in the yolk sac and embryo at 23 days of gestation. At 35 days of gestation, CD34 was uniformly expressed at the luminal portion of the endothelial cells in developing intraembryonic blood vessels. Furthermore, in this stage, some CD34+ cells, which were round and packed into cell clusters, were found in close apposition to the endothelium in the preumbilical region of the embryo. These cells also expressed CD45 and CD31 (platelet endothelial cell adhesion molecule 1 [PECAM1]), introducing the idea that hematopoietic cells arise intrinsically in the ventral wall of human embryonic arteries. Culture examination of the hematopoietic potential of embryonic tissue rudiments revealed that the hematopoietic potential present inside the embryo included both lymphoid and myeloid lineages, whereas the yolk sac exhibited only myeloposieses potential, thus questioning its ability to contribute to definitive hematopoiesis in the adult human. In a more recent work, endothelial cells of 3- to 6-week-old human embryos were found to express CD31, CD34, and vascular endothelial cadherin (VE-Cad). Panleukocyte alloantigen CD45 expression was restricted to hematopoietic cells, which were very rarely present inside blood vessels that contained almost only erythrocytes. Furthermore, CD45+ cells disseminated in very low numbers within tissues, albeit somehow more densely in the subaortic mesenchyme. The isolation of CD31+CD34+CD45− cells from both yolk sac and aorta was shown to possess the potential of myelo-lymphoid differentiation in culture. Isolated CD34+ CD31− cells from 11- to 12-week-old human embryo differentiated into both CD34+ CD31− and CD34+ CD31+ cells. The latter were found to be capable of forming a network of capillary-like structures.

Endothelial-Smooth Muscle Cell Relationship

Evidence for the relationship between the endothelium and smooth muscle cells (SMCs) is mostly available from animal studies. Early periendothelial SMCs associated with embryonic endothelial tubes have been shown to trans-differentiate from the endothelium, up-regulating markers of the SMC phenotype (both surface markers and morphology). Yamashita *et al.* discovered common mouse embryonic vascular progenitors that differentiate into endothelial and SMCs by the expression of an entire set of SMC markers and the surrounding of endothelial channels when injected into chick embryos. The earliest evidence in humans was recently shown on umbilical vein endothelium-derived cells that differentiated into SMCs through *in vitro* culturing with a fibroblast growth factor.

BIPOTENCY OF PROGENITORS IN ADULT PERIPHERAL BLOOD

Various studies have been conducted in an attempt to identify an adult hemangioblast, the progenitor for both endothelial and hematopoietic cells. The methodology is to isolate a specific cell population from peripheral blood and to look for any endothelial outgrowth, manifested by circulating endothelial cells. Using this approach, researchers showed that CD34[+] cells isolated from postnatal life possessed endothelial features and angiogenic capabilities. A specific fraction of those cells also expressed KDR (human Flk-1) and AC133, which indicated the existence of a population of functional endothelial precursors. AC133[+] circulating cells were shown to differentiate into adherent endothelial cells *in vitro* and to form new blood vessels *in vivo*. A thorough examination revealed that most circulating endothelial cells isolated by their P1H12 expression (endothelial cell surface glycoprotein), originated from vessel walls and have a limited growth capability. Recently, CD34[+] KDR[+] cells isolated from adult bone marrow were shown to have bilineage differentiation potential: they generated hematopoietic cells, endothelial cells, and cells expressing both types of markers.

There is little evidence in humans of the origin and differentiation process of SMCs. Mature bovine vascular endothelium has been shown to give rise to SMCs via "transitional" cells, co-expressing both endothelial and SMC-specific markers. In humans, a recent report showed that mononuclear cells isolated from adult human peripheral buffy coat blood, show features or either SMCs or endothelial cells, depending on specific *in vitro* culture conditions.

Human Embryonic Stem Cells as a Source for Vascular Progenitors

SPONTANEOUS DIFFERENTIATION

In vivo Differentiation

Undifferentiated human embryonic stem cells (hESCs) form teratomas once injected into severe combined immunodeficient (SCID) mice. Within these teratomas various blood vessels are formed (Figure 25-1). The small-diameter vessels located at the center of the teratomas originate in humans. Therefore, during teratoma formation from hESCs, two parallel vascular processes occur: (1) angiogenesis of host vasculature into the forming human teratoma and (2) vasculogenesis of spontaneously differentiating hESCs.

In vitro Differentiation

Human embryoid bodies (hEBs) formed from spontaneous differentiation of hESCs comprise representatives of all three embryonic layers. CD34, which is considered to be an early human endothelial marker is clearly expressed by surrounding endothelial cells that form voids within 1-month-old hEBs (Figure 25-2). Exploring the early organization of CD34[+] cells within 10- to 15-day-old hEBs revealed two types of cell arrangements: the first is a typical three-dimensional (3D) vessel formation, and the second is a cryptic arrangement that

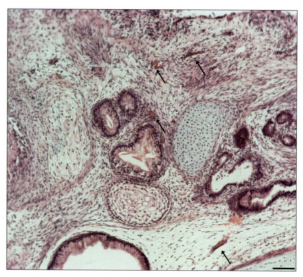

Figure 25-1. Teratoma vasculature. Human embryonic stem cells (hESCs) formed teratomas once injected into severe combined immunodeficient (SCID) mice. Various blood vessels (arrows) could be observed within the formed teratoma, including human and mouse-originating vessels. Bar = 100 μm

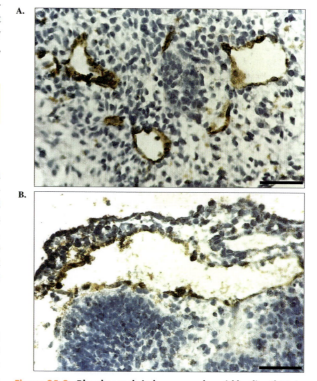

Figure 25-2. Blood vessels in human embryoid bodies (hEBs). Histological sections of 1-month-old hEBs stained with anti-CD34 revealed (a) the formation of relatively small vessels and (b) larger blood vessels. Bar = 100 μm

is occasionally somewhat difficult to associate with typical 3D-vessel formation (Figure 25-3A). However, both elongated and round cells stained for CD34 represent both endothelial and hematopoietic cells, respectively (Figure 25-3B).

CD31 has proven particularly useful because of its abundant early expression in mouse vascular development. It was shown that CD31 is expressed in developing hEBs. Exploring the early organization of CD31+ cells within 10- to 15-day-old hEBs revealed a typical 3D-vessel formation in different locations within the hEBs (Figure 25-4A–B). At this time point, a few round cells, probably early hematopoietic progenitors (Figure 25-4C), also expressed CD31. Furthermore,

unlike mouse ESCs, Flk1 was shown to be expressed in undifferentiated hESCs and increased very slightly during differentiation. Other endothelial markers, namely, VE-Cad and CD31 increase during the first week of hEB development. Thus, to study human vasculogenesis, various vascular markers should be examined and evaluated.

INDUCED DIFFERENTIATION

Both spontaneous and induced differentiation of hESCs can occur in two-dimensional (2D) and 3D cultures, that is, cell adherence and aggregation. Approaches for the promotion of specific vascular differentiation of hESCs include the following:

1. Genetic manipulation, also known as gene targeting — *knockin* and *knockout* specific angiogenic receptors and relevant transcription factors.
2. Exogenic factors — administrating the cell cultures with specific and known angiogenic and hematopoietic factors.
3. Matrix-based cultures — culturing the differentiating cells on matrices known to support vascular-cell cultures.
4. Co-culture — culturing the differentiating cells with specific stromal cell lines that may promote vascular differentiation.

Hematopoietic differentiation of hESCs was induced once hESCs were co-cultured with bone marrow cell line S17 or yolk sac endothelial cell line C166. Differentiation of hESCs cultured on S17 cells resulted in the formation of various types of hematopoietic colonies and even terminally differentiated hematopoietic cells, such as erythroid cells, myeloid cells, and megakaryocytes. Further enrichment of hematopoietic colony forming was achieved by sorting CD34+ cells from differentiation hESCs.

Nishikawa and colleagues (2001) showed that during mouse development both hematopoietic and SMC differentiation are linked to the development of endothelial precursors. Based on their approach, induction of vascular differentiation from hESCs was examined. Different lines of hESCs were seeded on type IV collagen matrices for 14 days with differentiation medium. These culture conditions induced the expression of all vascular endothelial growth factor (VEGF) isomers, VEGF receptors KDR, growth factor Ang2 and its receptor Tie2, and CD31-specific marker of endothelial cells. Some of the differentiated cells formed muscle-vascular arrangements, which were found to express smooth muscle α-actin (SMA). For the induction of a more defined population, single-cell suspensions of hESCs were used. Examination of the differentiated population revealed that the majority of the surviving cells expressed specific endothelial and endothelial-hematopoietic markers.

To induce lineage-specific differentiation, administration with human VEGF, platelet-derived growth factor BB (PDGF-BB), or specific hematopoietic cytokines was examined. VEGF treatment brought about maturation of the cells into endothelial cells producing von Willebrand factor (vWF) and with high lipoprotein metabolism. Up-regulation of the expression of SMC markers was achieved once supplemented with human PDGF-BB. In addition, colony formation unit

A.

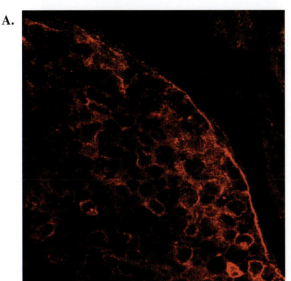

B.

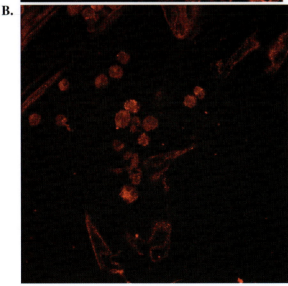

Figure 25-3. CD34+ cells in whole-mount human embryoid bodies (hEBs). 10- to 15-day-old EBs stained for CD34 revealed (a) occasionally cryptic, atypical vessel arrangement and (b) positive elongated and round cells. (×600).

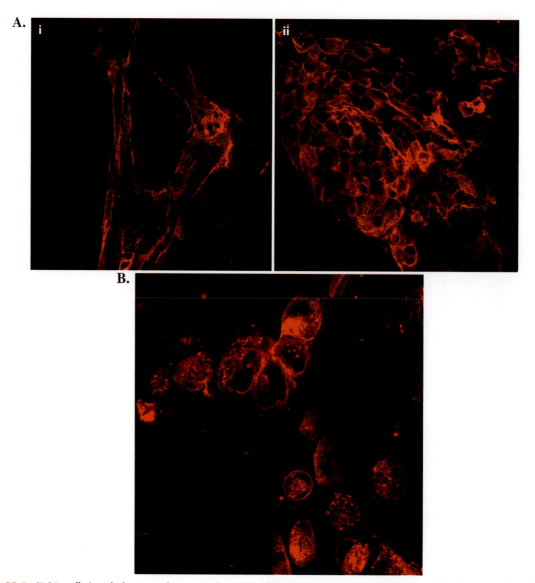

Figure 25-4. **CD31⁺ cells in whole-mount human embryoid bodies (hEBs).** 10- to 15-day-old-hEBs stained for CD31 revealed (a) different vascular organization within the EBs such as (1) elongated and (2) curved and (b) a few positive round cells. (x600)

assays revealed the progenitor populations ability to form different hematopoietic colonies (Figure 25-5). Formation of tube-like structures and sprouting was observed, once seeding the differentiated cells within 3D collagen and Matrigel gels (Figure 25-6). Histological sections showed the formation of a tube-like network and penetration of endothelial cells into the Matrigel. Electron microscopic examination further revealed typical arrangements of the endothelial cells within the matrigel, with the typical lipoprotein capsules and Weibel-Palade bodies in the cells' cytoplasm, and the presence of lumens in the cords. The presence of lumens in the cords points to the ability of the progenitor cells to differentiate into functional endothelial cells with lipoprotein metabolism and vWF production and also to form tube-like structures in suitable conditions.

ACKNOWLEDGMENTS

We thank Professor Raymond Coleman for histology instructions and assistance, Anna Ziskind for technical assistance, and Hadas O'Neill for editing. The hESC research was supported in part by NIH contract grant number 1RO1Hl73798-01.

KEY WORDS

Angiogenesis   Maturation and remodeling of the primitive vascular plexus into a complex network of large and small vessels.
Embryoid bodies   Aggregates of spontaneously differentiating embryonic stem cells that recapitulate aspects of early embryonic development.

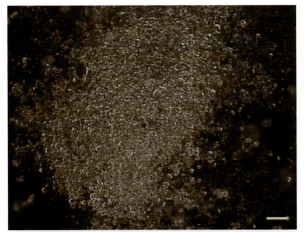

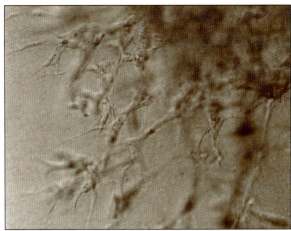

Figure 25-6. **Tube-like structures and sprouting.** An enriched progenitor population seeded within Matrigel sprouted and formed vessel-like structures. (×100)

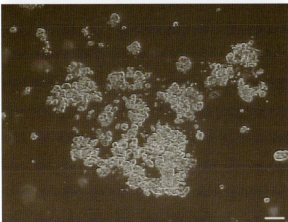

Vasculogenesis Generation of new blood vessels in which endothelial cell precursors undergo differentiation, expansion, and coalescence to form a network of primitive tubules.

FURTHER READING

Carmeliet, P. (2000). Mechanisms of angiogenesis and arteriogenesis. *Nat. Med.* **6**, 389–395.

Gerecht-Nir, S., Ziskind, A., Cohen, S., and Itskovitz-Eldor, J. (2003). Human embryonic stem cells as an in vitro model for human vascular development and the induction of vascular differentiation. *Lab Invest.* **83**, 1811–1820.

Itskovitz-Eldor, J., Schuldiner, M., Karsenti, D., Eden, A., Yanuka, O., Amit, M., Soreq, H., and Benvenisty, N. (2000). Differentiation of human embryonic stem cells into embryoid bodies comprising the three embryonic germ layers. *Mol. Med.* **6**, 88–95.

Kaufman, D. S., Hanson, E. T., Lewis, R. L., Auerbach, R., and Thomson, J. A. (2001). Hematopoietic colony-forming cells derived from human embryonic stem cells. *Proc. Natl. Acad. Sci. USA* **98**, 10716–10721.

Larsen, W. J. (1998). *Essentials of human embryology.* 2nd ed. Churchill Livingsone, New York.

Nishikawa, S. I. (2001). A complex linkage in the development pathway of endothelial and hematopoietic cells. *Curr. Opin. Cell Biol.* **13**, 673–678.

Peault, B., and Tavian, M. (2003). Hematopoietic stem cell emergence in the human embryo and fetus. *Ann. N. Y. Acad Sci.* **996**, 132–140.

Risau, W., and Flamme, I. (1995). Vasculogenesis. *Annu. Rev. Cell. Dev. Biol.* **11**, 73–91.

Yamashita, J., Itoh, H., Hirashima, M., Ogawa, M., Nishikawa, S., Yurugi, T., Naito, M., and Nakao, K. (2000). Flk1-positive cells derived from embryonic stem cells serve as vascular progenitors. *Nature* **408**: 92–96.

Yancopoulos, G., Davis, S., Gale, N., Rudge, J., Wiegand, S., and Holash, J. (2000). Vascular specific growth factors and blood vessel formation. *Nature* **407**, 242–248.

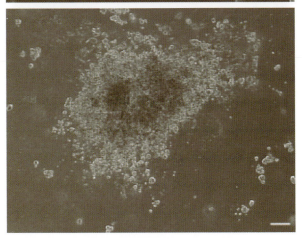

Figure 25-5. **Hematopoietic colonies.** Enriched progenitor population cultivated in semisolid media formed different types of hematopoietic colonies.

Hemangioblast/Hemogenic endothelium A precursor cell that gives rise to endothelial and hematopoietic cells.

Mesoderm The middle embryonic germ layer, between the ectoderm and endoderm, from which connective tissue, muscle, bone, cartilage, and blood vessels develop.

Multipotent Adult Progenitor Cells

Alaa Adassi and Catherine M. Verfaillie

Introduction

This chapter provides updated information regarding a rare cell population which we have named *multipotent adult progenitor cells*, or MAPCs. In 2001–2002, we published a series of papers demonstrating that while attempting to select and culture mesenchymal stem cells (MSCs) from human and subsequently mouse and rat bone marrow (BM), we accidentally identified a rare population of cells that has characteristics unlike most adult somatic stem cells in that they appear to proliferate without senescence and have pluripotent differentiation ability *in vitro* and *in vivo*.

Phenotype of Bone Marrow MAPC

MAPC can be cultured from human, mouse, and rat bone marrow (BM). Unlike MSC, MAPCs do not express major histocompatibility (MHC)-class I antigens, do not express or express only low levels of the CD44 antigen, and are CD105 (also endoglin, or SH2) negative. Unlike hematopoietic stem cells (HSC), MAPCs do not express CD45, CD34, and cKit antigens, but like HSC, MAPCs express Thy1, AC133 (human MAPC), and Sca1 (mouse), albeit at low levels. In the mouse, MAPCs express low levels of stage-specific embryonic antigen (SSEA)-1 and low levels of the transcription factors Oct4 and Rex1, known to be important for maintaining embryonic stem (ES) cells undifferentiated and known to be down-regulated when ES cells undergo somatic cell commitment and differentiation.

Isolation of MAPC from Other Tissues and Other Species

We have also shown that MAPC can be cultured from mouse brain and mouse muscle. The differentiation potential and expressed gene profile of MAPC derived from the different tissues appear to be highly similar. These studies used whole brain and muscle tissue as the initiating cell population, therefore containing more than neural cells and muscle cells, respectively. The implications of this will be discussed below. Studies are ongoing to determine whether cultivation of MAPC from other organs is possible and whether culture of MAPC, like ES cells, is mouse-strain dependent.

Nonsenescent Nature of MAPC

Unlike most adult somatic stem cells, MAPCs proliferate without obvious signs of senescence and have active telomerase. In humans, MAPC telomeres are 3 to 5 kB longer than in neutrophils and lymphocytes, and telomere length is not different when MAPCs are derived from young or old donors. This suggests that MAPCs are derived from a population of cells that either have active telomerase *in vivo* or that are highly quiescent *in vivo*, and therefore have not yet incurred telomere shortening *in vivo*. In human and rat MAPC cultures, we have not yet seen cytogenetic abnormalities. However, several subpopulations of mouse MAPCs have become aneuploid, even though additional subpopulations thawed subsequently were cytogenetically normal. This characteristic of mouse MAPCs is not dissimilar from that of other mouse cell populations, including mouse ES cells.

Stringent Culture Conditions Required for Maintenance of the Undifferentiated State of MAPC

The culture of MAPC is technically demanding, however. Major factors that play a role in successful maintenance of MAPC include cell density, CO_2 concentration and pH of the medium, lot of fetal calf serum that is used, and even the type of culture plastic that is used. Control of cell density appears to be species specific: mouse and rat MAPCs need to be maintained at densities between 500 and 1000 cells/cm^2, whereas human MAPCs need to be maintained between 1500 and 3000 cells/cm^2. It is not known why MAPCs tend to differentiate to the default MSC lineage when maintained at higher densities. However, for MAPC to have clinical relevance, this will need to be overcome. Gene array and proteomics studies are ongoing to identify the contact and/or soluble factors that may be responsible for causing differentiation when MAPCs are maintained at higher densities. These very demanding technical skills can, however, be "exported" from the University of Minnesota; after training at the University, several investigators have successfully isolated human and rat MAPC.

In Vitro Differentiation Potential of MAPC

Last year we published the fact that human, mouse, and rat MAPC can be successfully differentiated into typical mesenchymal lineage cells, including osteoblasts, chondroblasts, adiopcytes, and skeletal myoblasts. In addition, human, mouse, and rat MAPC can be induced to differentiate into cells with morphological, phenotypic, and functional characteristics of endothelial cells, and morphological, phenotypic, and functional characteristics of hepatocytes. Since then, we have also been able to induce differentiation of MAPC from mouse bone marrow into cells with morphological, phenotypic, and functional characteristics of neuroectodermal cells. Differentiation of MAPC to cells with neuroectodermal characteristics occurred by initial culture in the presence of basic fibroblast growth factor (bFGF) as the sole cytokine, followed by culture with FGF-8b and Sonic hedgehog (Shh), and then brain-derived neurotrophic factor (BDNF). Differentiation using these sequential cytokine stimuli was associated with activation of transcription factors known to be important in neural commitment *in vivo* and differentiation from NSC and mES cells *in vitro*. Cells staining positive for astrocyte, oligodendrocyte, and neuronal markers were detected. Neuron-like cells became polarized, and as has been described in most studies in which ES cells or NSC were differentiated *in vitro* to a midbrain neuroectodermal fate using FGF8 and Shh, approximately 25% of cells stained positive for dopaminergic markers, 25% for serotonergic markers, and 50% for GABA-ergic markers. Subsequent addition of astrocytes induced further maturation and prolonged survival of the MAPC-derived neuron-like cells, which now also acquired electrophysiological characteristics consistent with neurons, namely, voltage-gated sodium channels and synaptic potentials.

Interestingly, the lineage that continues to be elusive is cardiac myoblasts, despite the fact that mouse MAPC injected in the blastocyst contribute to the cardiac muscle. Although a number of *in vitro* differentiation conditions induce expression of Nkx2.5, GATA4, and myosin heavy chain mRNA and proteins, we have been unable to induce differentiation of MAPC to cells with the typical functional characteristic of cardiac myoblasts, that is, spontaneous rhythmic contractions or beating, a differentiation path that is almost a default differentiation pathway for mouse ES cells. The reason for the lack of functional cardiac myoblast properties is currently unknown.

A last comment regarding *in vitro* differentiation of MAPC is that, in contrast to differentiation of ES cells *in vitro*, the final differentiated cell product derived from MAPC is commonly >70–80% pure. This should allow using these *in vitro* differentiation models for gene and drug discovery. For instance, in a recently published study, we compared the expressed gene profile in human MAPC induced to differentiate to osteoblasts and chondroblasts, two closely related cell lineages. We could demonstrate that, although a large number of genes are co-regulated when MAPCs differentiate to these two lineages, specificity in differentiation can readily be detected. For instance, a number of known and yet to be fully characterized transcription factor mRNAs were differentially expressed during the initial phases of differentiation. Studies are ongoing to further define the role of these genes in lineage-specific differentiation. These studies do, however, exemplify the power of this model system to study lineage-specific differentiation *in vitro*.

Degree of Pluripotency

We have shown that transfer of 10–12 mouse MAPC into mouse blastocysts results in the generation of chimeric mice. When 10–12 MAPCs, expanded for 50–55 population doublings, were injected, approximately 80% of offspring was chimeric, with the degree of chimerisms varying between 1 and 40%. Cells found in different organs acquire phenotypic characteristics of the tissue. For instance, MAPC-derived cells detected in the brain of chimeric animals differentiate appropriately in region-specific neurons, as well as astrocytes and oligodendrocytes. Whether MAPC can do this without the help of other cells in the inner cell mass (i.e., can generate a mouse by tetraploid complementation) is not yet known. Also not yet known is whether MAPCs contribute to the germ line when injected in the blastocyst.

Postnatal Contribution to Tissues

Neither human nor mouse MAPC-injected IM in SCID mice have led to the development of teratomas (unpublished observations). Similarly, we have not yet detected donor-derived tumor formation following IV injection of human or mouse MAPC in NOD-SCID animals. However, when mouse undifferentiated MAPCs are administered IV to NOD-SCID mice, engraftment in the hematopoietic system as well as epithelia of gut, liver, and lung is seen. One of the animals developed a host-tumor, an occurrence seen frequently in aging NOD-SCID mice. We detected the contribution of MAPC-derived endothelium to tumor vessels.

Although PCR analysis for β-galactosidase in mouse–mouse transplants yielded positive signals in many tissues, we believe that this is mainly due to contaminating blood cells. When tissues were carefully examined for tissue-specific differentiated MAPC progeny, we could not detect MAPC progeny in brain, skeletal muscle, cardiac muscle, skin, or kidneys. Lack of engraftment in brain, skeletal, and cardiac muscle may be due to the fact that transplants were done in noninjured animals, where the blood brain barrier is intact and where little or no cell turnover is expected in muscle. More difficult to explain is the absence of MAPC-derived progeny in skin, possibly the organ with the greatest cell turnover. Studies are ongoing to trace the homing behavior of MAPC following infusion in noninjured animals and injured animals, which may shed light on these observations.

When endothelial cells generated from human MAPC by incubation *in vitro* with VEGF were infused in animals in which a tumor had been implanted underneath the skin, we detected enhanced tumor growth and found that up to 30% of the tumor vasculature was derived from the human endothelial cells. Similarly, wounds in the ears of these animals as a result of ear tagging contained human endothelial cells.

In yet another *in vivo* study, we evaluated the effect of human MAPC in a rat stroke model. Cortical brain ischemia was produced in male rats by permanently ligating the right middle cerebral artery distal to the striatal branch. Animals were placed on cyclosporine-A and 2 weeks later, 2×10^5 human MAPC were injected around the infarct zone. As controls, animals received normal saline or MAPC-conditioned medium. Limb placement test and tactile stimulation test were blindly assessed 1 week before brain ischemia, 1 day before transplantation, and 2 and 6 weeks after grafting. The limb placement test included eight subtests described by Johansson and co-workers. In a tactile stimulation test, a small piece of adhesive tape was rapidly applied to the radial aspect of each forepaw. The rats were then returned to their home cages, and the order of the tape removal (i.e., left versus right) was recorded. Three to five trials were conducted on each test day. Each trail was terminated when the tapes were removed from both forepaws or after 3 minutes. Animals were subsequently sacrificed to determine the fate of the human cells injected in the brain. After 2 and 6 weeks, animals that received human MAPC scored statistically significantly better in the limb placement test as well as tactile stimulation test compared with animals that received only CSA, or were injected with normal saline or MAPC-conditioned medium. The level of recuperation of motor and sensory function was 80% of animals without stroke. When the brain was examined for presence and differentiation of human MAPC to neuroectodermal cells, we found that human MAPCs were present but remained rather immature. Therefore, we cannot attribute the motor and sensory improvement to region-specific differentiation to neuronal cells and integration of neurons derived from MAPC in the host brain. Rather, the improvement must be caused by trophic effects emanated by the human MAPC to either improve vascularization of the ischemic area, to support survival of the remaining endogenous neurons, or to recruit neuronal progenitors from the host brain. These possibilities are currently being evaluated.

Possible Mechanisms Underlying Multipotent Adult Progenitor Cells

Currently, we do not fully understand the mechanism(s) underlying the culture selection of MAPC. We have definitive data to demonstrate that the pluripotency of MAPC is not due to co-culture of several stem cells. First, using retroviral marking studies, we have definitive proof that a single cell can differentiate *in vitro* to cells of mesoderm, both mesenchymal and nonmesenchymal, neuroectoderm and hepatocyte-like cells, and this for human, mouse and rat MAPC. Second, we have shown that a single mouse MAPC is sufficient for generation of chimeric animals. Indeed, we published a report that one third of the animals born from blastocysts in which a single MAPC was injected were chimeric, with chimerism degrees varying between 1 and 45%. This therefore rules out that the pluripotent nature of these cells is due to coexistence in culture of multiple somatic stem cells.

A second possibility for the greater degree of differentiation potential would be that cells undergo fusion and through this mechanism acquire greater pluripotency. Fusion has been shown to be responsible for apparent ES characteristics of marrow and neural stem cells that had been co-cultured with ES cells *in vitro* and more recently for the apparent lineage switch of bone marrow cells to hepatocytes when hematopoietic cells were infused in animals with hereditary tyrosinemia due to lack of the fumarylacetoacetate hydroxylase (FAH) gene. In the former two studies, the majority of genes expressed in the marrow or neural cell that fused with ES cells were silenced, and the majority of the genes expressed in ES cell were persistently expressed. Likewise for the bone marrow-hepatocyte fusion, the majority of genes expressed normally in hematopoioetic cells (except the FAH gene) were silenced, whereas genes-expressed hepatocytes predominated. Finally, the cells generated were in general tertaploid or anueploid.

We do not believe that this phenomenon underlies the observation that MAPCs are pluripotent. Cultivation and differentiation *in vitro* (in general, except the final differentiation step for neuroectoderm) does not require that MAPCs are co-cultured with other cells, making the likelihood that MAPCs are the result of fusion very low. Smith *et al.* suggested in a recent commentary that MAPCs could be caused by fusion of multiple cell types early on during culture leading to reprogramming of the genetic information and pluripotency. However, we have no evidence that MAPCs, even early during culture, are tetraploid and/or aneuploid, making this possibility less likely. Nevertheless, studies are ongoing to rule this out. The *in vivo* studies were not set up to be fully capable of ruling out this possibility. However, a number of findings suggest that fusion may not likely be the cause for the engraftment seen postnatally, nor the chimerism in the blastocyst injection experiment. The frequency of the fusion event described for the ES-BM, ES-NSC, and HSC-hepatocyte fusion was in general very low (i.e., 1/100,000 cells). Expansion of such fused cells could only be detected when drug selection was applied in the *in vitro* systems, and withdrawal of NTBC (2-(2-nitro-4-trifluoro-methylbenzoyl)-1,3-cyclohexanedione) in the FAH mouse model was used to select for cells expressing the FAH gene. The percent engraftment seen in our postnatal transplant models was in the range of 1 to 9%. The chimerism seen in blastocyst injection studies ranged between 33% and 80% when 1 and 1 and 10–12 MAPCs were injected, respectively. These frequencies are significantly higher than what has been described for the fusion events *in vitro* with ES cells and *in vivo* in the HSC-hepatocyte fusion studies. Furthermore, in contrast to what was described in the papers indicating that fusion may be responsible for apparent plasticity, all *in vivo* studies done with MAPC were done without selectable pressure, mainly in noninjured animals. Therefore, it is less likely that the pluripotent behavior of MAPC *in vivo* is due to fusion between the MAPC and the tissues where they engraft/contribute to. However, specific studies are currently being designed to formally rule this possibility out.

Currently, we do not have proof that MAPCs exist as such *in vivo*. Until we have positive selectable markers for MAPC, this question will be difficult to answer. If the cell exists *in vivo*, one might hypothesize that it is derived, for instance, from primordial germ cells that migrated aberrantly to tissues outside the gonads during development. It is also possible, however, that removal of certain (stem) cells from their *in vivo* environment results in "reprogramming" of the cell to acquire greater pluripotency. The studies on human MAPC suggest that such a cell that might undergo a degree of reprogramming is likely a protected (stem) cell *in vivo*, as telomere length of MAPC from younger and older donors is similar, and significantly longer than what is found in hematopoietic cells from the same donor. The fact that MAPC can be isolated from multiple tissues might signify that stem cells from each tissue might be able to be reprogrammed. However, as was indicated above, the studies in which different organs were used as the initiating cell population for generation of MAPC did not purify tissue-specific cells or stem cells. Therefore, an alternative explanation is that the same cells isolated from bone marrow that can give rise to MAPC in culture might circulate and be collected from other organs. However, until now we have been unsuccessful in isolating MAPC from blood or from umbilical cord blood, which argues against this phenomenon. Finally, cells selected from the different organs could be the same cells resident in multiple organs, such as MSC that are present in different locations, or cells associated with tissues present in all organs such as, for instance, blood vessels. Studies are ongoing to determine which of these many possibilities is correct.

We believe that MAPCs would have clinical relevance whether they exist *in vivo* or are created *in vitro*. However, understanding the nature of the cell will impact how one would approach their clinical use. If they exist *in vivo*, it will be important to learn where they are located and to determine whether their migration, expansion, and differentiation in a tissue-specific manner can be induced and controlled *in vivo*. If they are a culture creation, understanding the mechanism underlying the reprogramming event will be important, for that might allow this phenomenon to happen on a more routine and controlled basis.

KEY WORDS

Adult stem cell   Cell derived from postembryonic stage tissue with self-renewal ability, and ability to generate all cell types of the tissue where it was derived from but not other tissues.

Multipotent adult progenitor cell   Cell cultured from postembryonic stage bone marrow, and possibly other tissues, with extensive self-renewal ability and ability to generate most, if not all, cell types from the three germ layers of the embryo (i.e., mesoderm, endoderm, and ectoderm).

FURTHER READING

Jiang, Y., Jahagirdar, B., Reyes, M., *et al.* (2002). Pluripotent nature of adult marrow derived mesenchymal stem cells. *Nature* **418**, 41–49.

Nichols, J., Zevnik, B., Anastassiadis, K., *et al.* (1998). Formation of pluripotent stem cells in the mammalian embryo depends on the POU transcription factor Oct4. *Cell* **95**, 379–391.

Reyes, M., Lund, T., Lenvik, T., Aguiar, D., Koodie, L., and Verfaillie, C. M. (2001). Purification and ex vivo expansion of postnatal human marrow mesodermal progenitor cells. *Blood* **98**, 2615–2625.

Reyes, M., Dudek, A., Jahagirdar, B., Koodie, K., Marker, P. H., and Verfaillie, C. M. (2000). Origin of endothelial progenitors in human post-natal bone marrow. *J. Clin. Invest.* **109**, 337–346.

Schwartz, R. E., Reyes, M., Koodie, L., *et al.* (2002). Multipotent adult progenitor cells from bone marrow differentiate into functional hepatocyte-like cells. *J. Clin. Invest.* **96**, 1291–1302.

Terada, N., Hamazaki, T., Oka, M., *et al.* (2002). Bone marrow cells adopt the phenotype of other cells by spontaneous cell fusion. *Nature* **416**, 542–545.

Wang, X., Willenbring, H., Akkari, Y., *et al.* (2003). Cell fusion is the principal source of bone-marrow-derived hepatocytes. *Nature* **422**, 897–901.

Mesenchymal Stem Cells

Arnold I. Caplan

Introduction

In the mid-1980s, my laboratory started experiments to purify mesenchymal stem cells (MSCs) from adult human bone marrow. There were several reasons for initiating our approach and for our success. It is important to briefly review these events because they established the basis for the current state of the field and the basis for interesting refinements, new observations, and new directions.

From 1968 to 1972, my laboratory developed and refined the methods for liberating and culturing undifferentiated mesenchymal cells from the developing limb buds of 4.5-day-old chick embryos. These embryonic chick limb bud cells multiplied in culture, and depending on the culture conditions and the initial seeding densities, they differentiated into muscle, cartilage, bone, and various connective tissues. At intermediate plating densities, bone-forming cells called osteoblasts were observed to develop, and at a 2.5-fold increase in initial plating density of these same cells, the plates were filled with cartilage with little or no evidence of osteoblast differentiation. The chondrocytes, which formed the cartilage tissue, were shown to continuously modulate the synthesis of the major cartilage water structuring molecule, a proteoglycan called aggrecan; these same changes have also been observed for human articular cartilage. Thus, these cell cultures mimic the full range of cartilage changes from young to old from bouncy water-filled shock absorbers to the leathery, tough coverings of the joint surface.

In 1965, Professor Marshall R. Urist published his landmark paper on the purification of factors from demineralized bovine bone that induced cartilage and subsequent bone formation in various *in vivo* implantation systems and *in vitro*. The basis for Urist's assays was the observation that demineralized bone chips, when implanted into muscle or subcutaneous pouches in test animals, caused the formation of ectopic bone. Urist coined the term *bone morphogenetic proteins* (BMPs) to describe the presumptive inductive molecules that stimulated this tissue outcome.

In 1978, Glenn Syftestad joined my lab as a postdoc following a stay in Urist's lab. Syftestad was determined to purify BMPs from bovine bone, and the embryonic chick limb bud mesenchymal cell system was an obvious way to test for the chondrogenic stimulating activity (CSA) within the extract of demineralized bovine bone. The intermediate plating density where cartilage was never observed became the assay system for identifying and purifying chondrogenic inductive molecules from extracts of adult bone by observing chondrogenic differentiation under conditions in which no cartilage formed normally. Several papers and patents were published on these efforts and centered on a 31-kDa protein. We can now surmise that CSA was a heterodimer of BMPs; unfortunately, we, like Urist and his colleagues, were never able to sequence the proteins or to clone the genes. John Wozney and his collaborators at the Genetics Institute in Cambridge, Massachusetts (now part of Wyeth, Pharmaceutical Company) were the first to do this.

In the course of studying CSA, we did many *in vivo* implantations and observed ectopic cartilage and bone formation. During the course of these studies in the mid-1980s, it occurred to me that cells like the multipotent embryonic chick limb mesenchymal cells must reside or must come to the implantation site in the adult tissue in order for cartilage and bone to form. I call these adult cells MSCs. By this time, several investigators had proposed that osteogenic (bone-forming) progenitor cells reside in bone marrow. This seemed obvious since all bone is in contact with marrow elements. The experiments of Maureen Owen — and, in particular, her review article in 1988 that summarized A. J. Friedenstein's early work and included the possibility that fat cell progenitors were also in marrow — opened the door for speculation that all of the mesenchymal phenotypes could arise from this same class of progenitors in pathways analogous to those deduced for the blood-forming cells (hematopoiesis) (Figure 27-1).

In this context, several years earlier, my laboratory was deeply involved in the study of bone formation in the embryonic chick limb bud *in vivo* and in an attempt to compare it to *in vitro* events. A young and talented graduate student, Philip Osdoby, had wasted 18 months working with this limb bud mesenchymal cell culture system, because we followed the dogma of the day that cartilage was the progenitor of bone as evidenced by endochondral bone formation and growth plate dynamics. We studied this by plating embryonic chick limb cells at high density to encourage them to form mounds of cartilage. No matter how we varied the culture medium or its additives, we could not get these cultures to form bone. Osdoby made a key observation in his superb electron micrographs of these cartilage cultures that when rare, mineralized deposits (osteoid) were observed, they were never located in the cartilage nodules, but always in the fibrous tissue between these massive mounds of cartilage. We then realized that

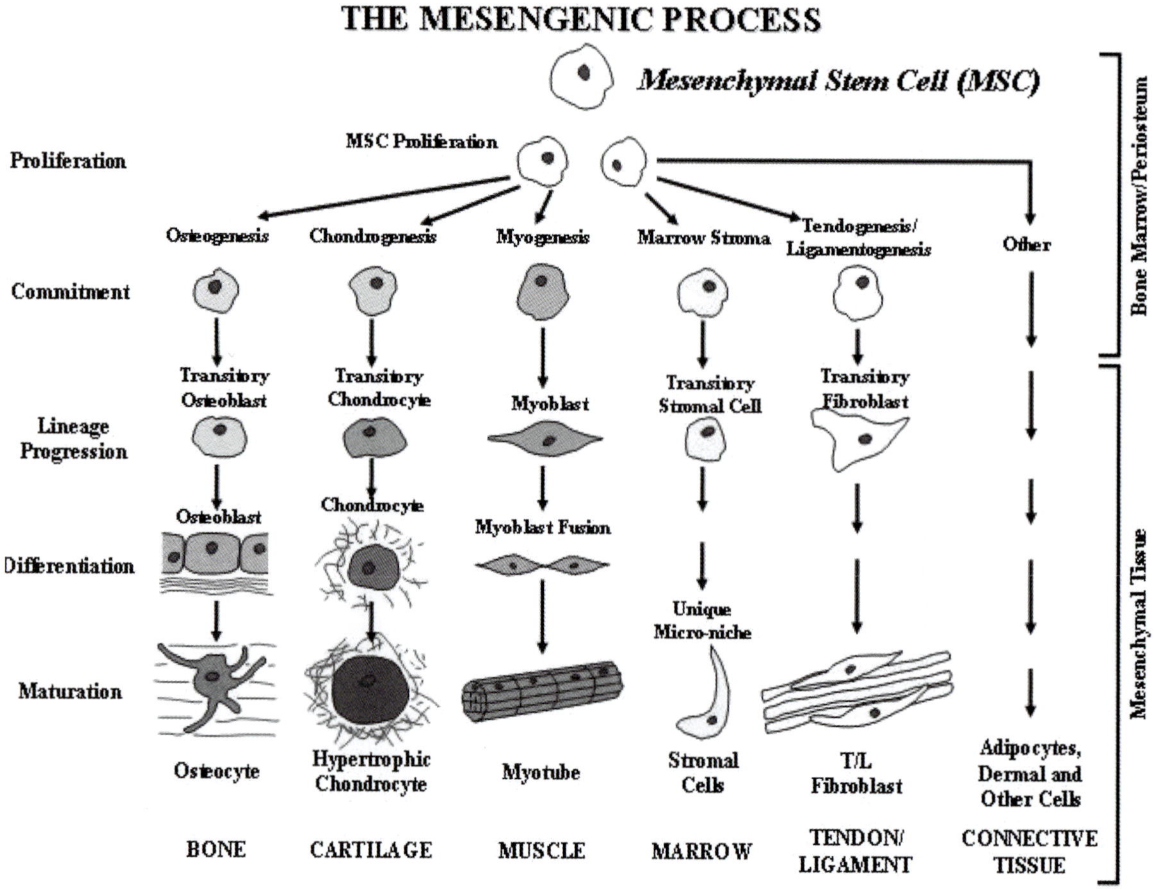

THE MESENGENIC PROCESS

Figure 27-1. The mesengenic process depicts mesenchymal progenitor cells entering distinct lineage pathways that contribute to mature tissues.

osteoblasts formed independently of cartilage and away from it. By plating the cells at density where cartilage never formed, we could observe many osteoblasts.

When we went back to the whole developing embryonic limb bud and looked at the location of first bone formation, it was outside and away from the already formed cartilaginous rods. Later, based on this early work of Osdoby and the subsequent morphological descriptions of first bone formation, Scott Bruder raised monoclonal antibodies to the tediously dissected first bone collars of the embryonic tibia and used these to establish the lineage transitions of mesenchymal progenitors to osteoblasts and osteocytes.

Thus, from this early work on embryonic mesenchymal cells differentiation, we had the osteogenic and chondrogenic lineages. Others had worked out many of the details of myogenic lineages, and quite interestingly, the suggestion had been made that hematopoietic support tissue (the marrow stroma) also contained cellular elements that provided differential support for separate arms of the hematopoietic pathway. The summary diagram in Figure 27-1 is therefore a composite of many individuals' efforts and was fashioned to mimic the fundamental aspects of the hematopoietic pathway; I called this multilineage pathway the *mesengenic process*. I

named the multipotent progenitor the *mesenchymal stem cell* because all of the lineage pathways depicted in Figure 27-1 were mesenchymal in origin. The reasons for calling this cell a stem cell were rather loose but were primarily inferred from the observation that the cells we isolated could be induced into different lineage pathways. Thus, these "stemmed" from a common source.

Adult mesenchymal tissues all have unique turnover dynamics — some relatively rapid, like bone, and others relatively slow, like cartilage. In all cases, turnover means that differentiated cells in the tissue die and are replaced by newly created cells. For bone, this death and birth can take place when large sheets of osteoblasts are fabricating sheets of osteoid that will eventually mineralize. Several full generations of cells may come and go while a sector of bone is being fabricated.

LOCATION OF MSCs

By the preceding logic, MSCs must be everywhere mesenchymal tissue turns over. To date, mesenchymal progenitors have been isolated from marrow, muscle, fat, skin, cartilage, and bone. Moreover, every blood vessel in the body has mesenchymal cells on the tissue side of the vessel (some of these are MSCs), although names like pericytes have been given to these multipotent cells.

TEST FOR MSCs

The tests for MSCs are straightforward: Populations of cells are placed in medium conditions that include exposure to strong phenotypic inducers. Following phenotypic induction, the expressional activity and morphology of the cells are monitored, and a judgment is made that a specific phenotype is present based on the analysis of indicator molecules, morphologies, or both. Of course, *in vivo*, we have no unique markers for MSCs, although Stephen Haynesworth and I long ago isolated three monoclonal antibodies (SH2, SH3, and SH4) that preferentially bind to MSCs compared to other hematopoietic cells.

PLASTICITY

One recent experimental complication is that we can take mesenchymal cells of a known phenotype, for example, an adipocyte with fat droplets inside, and expose these cells to inductive medium for another phenotype, for example, osteoblasts. These newly induced cells stop profucing fat-specific molecules and start producing bone-specific molecules in osteogenic inductive medium. The capacity to be induced from one lineage into a new lineage is referred to as "plasticity." Once this is recognized, the functional test for an MSC becomes less stringent because non-MSCs, like the adipocytes cited previously, and MSCs can be induced to become osteoblasts. For emphasis, this is an experimental situation and may not be relevant to *in vivo* events. However, a cell's capacity for plasticity clouds the issue of identifying MSCs by these functional assays for multiple mesenchymal phenotypes.

MSCs DECREASE WITH AGE

The body's need for MSCs changes dramatically depending on the age of the individual. In embryos, mesenchymal tissues form from a relatively high local density of progenitors within a very loose, watery extracellular matrix. The massive differentiation of these groupings of progenitors into differentiated cells, and thus into tissues, involves the substantial up-regulation of synthesis of complex and tissue-specific extracellular matrices. The cell-to-matrix ratios go from high to low during these differentiation events. Moreover, the matrix adds to the many-fold increase in the gross size of the tissues being formed. In this process, many of the progenitor cells are directly converted to differentiated cells, and the relative MSC numbers dramatically decrease in the embryo. Similarly, from birth to the teens, the increase in body-part sizes is many-fold; this also must involve the direct conversion of some of the MSCs into differentiated phenotypes. In this regard, it seems reasonable to expect a mesenchymal tissue injury of a child below the age of 5 years to regeneratively repair because of relatively high MSC levels; in a 20-year-old, the same tissue might only repair with fibrous scar tissue. To take this comparison even further, it would seem reasonable to expect that 50-year-olds would have fewer MSCs locally than 20-year-olds to repair mesenchymal tissue injures.

Since we have no unique probes for MSCs, we cannot determine the whole-body titers of MSCs. As an estimate, we can measure the colony-forming units — fibroblasts from marrow aspirates or marrow biopsies. Even in this assay, the differences in marrow aspiration techniques, medium additives, and growth-medium serum levels (and prescreening batches of serum) preclude establishing an experimental consensus on the absolute numbers of MSCs in marrow. What everyone seems to agree upon is the *decrease* of MSCs in marrow with age.

Based on this information, I propose six periods of differing MSC titers. Each period would be expected to have a different capacity for tissue expression and regenerative repair. The different periods are: (1) embryology, fertilization to birth, (2) neonatology, birth to 6 years, (3) teens, 7 to 20 years, (4) peak skeletal performance, 20 to 30 years, (5) Midlife, 35 to 55 years, and (6) Late life, 60 to 90 years.

The important distinction is that, although HSC titers and potency are relatively constant from birth to late life, MSC titers change by one or two orders of magnitude. This highlights the fundamental issue that both the definitions and the functional biology of MSCs and HSCs are expected to be different.

Tissue-Engineered Regeneration

With the preceding information as a base logic, it follows that the regenerative repair of certain mesenchymal tissues will require the augmentation of MSC numbers at the repair sites in post-teen adults. This can be accomplished in several ways depending on the age and health status of the individual. For years surgeons have used freshly harvested whole marrow to augment local skeletal tissue healing, and methods to concentrate and enrich the marrow are being introduced commercially. Key to this latter use, and to the use of culture-expanded MSCs, are the type and properties of the scaffolds used to isolate, concentrate, and/or deliver the cells to the repair site. Such composite grafts form the backbone of the new scientific discipline of tissue engineering.

Isolating and Culturing MSCs

MSCs have been isolated from several species and tissues, but the most well characterized and probably the purest preparation is from human bone marrow. The studies to purify MSCs from human marrow were initiated in the mid-1980s by Stephen Haynesworth in my laboratory. The key to our initial and rapid success was the lot of fetal bovine serum used to initiate these studies. The original batch of serum was selected from several lots by a selection assay using embryonic chick limb bud mesodermal cells in culture. At that time, we would prepare such cultures weekly from 200 dozen fertile eggs; batches of serum varied greatly from the same commercial supplier and from many suppliers. Based on color alone, many batches, by experience, were not suitable. (A dark brown or greenish tint indicated hemolyzed blood, and batches with these colors were not suitable.) The serum accounts for three key aspects of the MSC purification–expansion protocol: selective attachment to tissue culture plastic to provide discrete colonies of MSCs; optimal mitotic expansion of the cells; and maintenance of the undifferentiated or stem cell state of the MSCs.

Today, we use a laborious derivative of our initial, published screening protocol. In this regard, it is more difficult to screen batches of serum in that only 1 in 20 to 30 lots is suitable for human MSCs. This is because some suppliers blend batches to optimize their use in hybridoma cultures, making it more difficult to screen for a suitable lot for MSCs. Moreover, every animal preparation requires its own lot of serum with the batch for human being unsuitable for rat, rabbit, or mouse marrow MSCs. (In some cases, batches of serum optimal for human MSC are toxic to mouse MSCs.) Thus, if an unscreened batch of serum is used or if an agent such as dexamethasone is added to the basic growth medium, results for colony counts and for cellular response profiles to other bioactive agents will differ from those obtained in my laboratory. In evaluating other reports using MSCs, information pertaining to the serum and other medium components is crucial.

In the early 1990s, Haynesworth, our colleagues, and I organized the transfer of the MSC technology from Case Western Reserve University to a newly formed company called Osiris Therapeutics, Inc. (OTI, now in Baltimore, Maryland). The scientists at OTI have repeated and refined our basic technology and have added to the database of human MSCs. They showed that by second passage the MSC preparation was homogeneous by fluorescence-activated cell-sorting criteria using more than 100 cell-surface antibodies and observed that clones could be obtained that showed multipotency for different phenotypic lineages. Our efforts to use

mouse, rat, rabbit, dog, goat, and sheep marrow MSCs for tissue engineering were also transferred to OTI, where various preclinical animal models extended the results first obtained through my laboratory. OTI expanded the safety studies and initial human clinical trials started at Case Western Reserve University and University Hospitals of Cleveland.

Tissue-Engineering Uses of MSCs

In mature adults, marrow MSCs are used to supply progenitors for normal turnover and for repair–regeneration of tissue damage. The basis for this statement is the obvious involvement of MSCs in marrow stroma in support of normally active hematopoiesis and of MSCs in normal bone turnover to both directly support the formation of osteoblasts and indirectly support the formation of osteoclasts through two different lineage pathways (mesengenesis and hematogenesis, respectively). Thus, the use of culture-expanded MSCs to massively reform or repair certain tissues far exceeds their normal use and availability. In this case, we have adopted the logic that, to accomplish reconstruction of tissues that cannot normally provide suitable numbers of progenitors themselves, we must provide these progenitors from exogenous sources. Concomitantly, we must attempt to mimic selected principles of embryonic development and recapitulate these events in adults. The most important aspect of embryonic mesengenesis is that the progenitor cell to matrix ratio is very high just prior to lineage pathway entrance and that the exact limits or *edges* of the neotissue are defined. Moreover, we now recognize that each site and each tissue have a unique sequence of inductive events and boundary conditions. Next, I briefly describe the use of MSCs for specific tissue regeneration, taking this embryonic recapitulation logic into account as depicted in Figure 27-2.

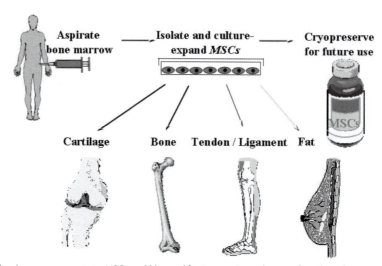

Figure 27-2. From an iliac bone marrow aspirate, MSCs could be used for tissue-engineered repair of cartilage, bone, or tendon or for plastic-surgical implant of fat from fresh or frozen-stored cells.

BONE REPAIR

The first and most obvious use of MSCs is in the area of bone regeneration in sites where the body cannot organize this activity, such as in nonunions. Critical size defects in nonunion models have shown that culture-expanded marrow MSCs in a porous, calcium phosphate, ceramic delivery vehicle are capable of regenerating structurally sound bone, where whole marrow or the vehicle alone cannot satisfactorily accomplish this repair. These preclinical models include the use of human MSCs in a femoral nonunion model in an immuno-compromised rat that will not reject the human cells.

CARTILAGE REPAIR

Cartilage is an avascular tissue incapable of regeneration–repair of even small defects in adults. Although chondrocytes have been used to attempt to repair large cartilage defects, it is difficult to integrate the neotissue with that of the host. In this case, we have pioneered the use of hyaluronan (HA) scaffolds because of the high content of HA in the embryonic mesenchyme of precartilaginous tissues. These HA-based scaffolds are being tested in human clinical trials in Europe using autologous chondrocytes. These scaffolds provide an inductive microenvironment for MSCs to enter the chondrogenic lineage, and HA-scaffold breakdown products (oligomers) appear to facilitate the integration of neotissue with that of the host.

MARROW REGENERATION

Injected MSCs make it back to the bone marrow to refabricate injured marrow stroma. This observation is the basis for a multilocation clinical trial to add back autologous MSCs to chemotherapy–radiation patients receiving bone marrow transplants.

MUSCLE REGENERATION

We were the first to document that congenic (i.e., normal) MSCs could be injected into a specific muscle of the muscular dystrophy (*mdx*) mouse to cure it by providing newly synthesized dystrophin to the affected myotubes. In this case, the donor MSCs differentiated into skeletal myoblasts, fused with the host myotubes, and caused the synthesis and distribution of dystrophin. Moreover, labeled MSCs injected into injured (infarct model) rat or pig heart appear to differentiate into cardiac myocytes.

FAT

MSCs have been induced into the adipocyte pathway and can massively accumulate fat droplets. Although not used in tissue engineering models, we propose that bags of autologous fat are more appropriate for some plastic surgeries than bags of saline or silicone.

TENDON REPAIR

Studies initiated in my laboratory and later completed at OTI and the University of Cincinnati clearly establish the use of autologous MSCs for repair of tendons. The innovative design of an MSC-contracted collagen gel around a restorable suture held in tension allows tendon defects to be spanned by these composite constructs under normal loads. In this case, the same MSC able to differentiate into cartilage or bone develops into appropriate tendon tissue at Achilles or patellar tendon sites.

GENE THERAPY

The studies with the *mdx* mouse cited previously establish the capacity of allo-MSCs to cure genetic defects. This principle has been translated to the clinical use of allogeneic bone marrow transplantation with or without additional culture-expanded MSCs to attempt to cure polysaccharide storage diseases or *Osteogenesis imperfecta* with substantial success. In addition, we long ago showed that human MSCs could be transfected with a retroviral construct without affecting their differentiation capacity. Molecules coded for by these gene inserts could be shown to be present many months after introduction into the *in vivo* model. Thus, MSCs hold the potential for a variety of gene therapy applications.

VASCULAR SUPPORT

Not well studied is the relationship between vascular endothelial cells and MSCs. It is clear that multipotent mesenchymal progenitor cells are present with blood vessels (pericytes). More important are the details of the interaction between these vascular support cells and angiogenesis–vasculogenesis. Certainly, in embryonic angiogenesis, MSCs play a strong role. In adults, for example, it is now clear that vascular endothelial cells in bone make skeletal inductive molecules, such as BMPs. In addition, MSCs make vascular endothelial growth factor. Based on these and other observations, we predict that, although nearly all tissues contain blood vessels, each tissue has specifically differentiated endothelial cells and perhaps even unique MSC-derived vascular support cells.

Summary

Adult bone marrow contains mesenchymal progenitor cells that may be called MSCs. Whether these preparations are composed of a pure population of stem cells or a spectrum of cells ranging from pluripotent stem cells to lineage-committed progenitors, it seems clear that these culture-expanded cells have the impressive potential to be used in tissue-engineered regeneration and repair of a variety of tissues. In some cases, as in bone nonunions, these cells are capable of differentiating directly into osteogenic cells. In the cases where they are used in cardiac (infarct) or brain (stroke) therapy, these cells appear to provide instructive and inductive microenvironments, inhibit scarring, and stimulate blood vessel regeneration for facilitated repair of critical lesions. That mesenchymal cells appear to be unusually plastic in their developmental capabilities opens the door for their use in several cell-based therapies.

ACKNOWLEDGMENTS

My thanks to my colleagues and collaborators at Case Western Reserve University and elsewhere for their assistance and encouragement. The National Institutes of Health generously supported much of the experimentation reviewed here.

KEY WORDS

Mesenchymal stem cells Cells that have the capacity to form different mesenchymal tissues given exposure to different inductive agents.

Mesenchyme The middle layer of an embryo, the mesoderm and the tissues that form (bone, cartilage, muscle, fat, etc.).

Osteoblast A cell formed by a multistep differentiation sequence that becomes capable of fabricating bone.

Tissue engineering A new science that encompasses aspects of the natural and physical sciences that focuses on the repair or regeneration of various tissues.

FURTHER READING

Caplan, A. I. (1984). Cartilage. *Sci. Am.* **251**, 84–94.

Caplan, A. I. (1991). Mesenchymal stem cells. *J. Ortho. Res.* **9**, 641–650.

Caplan, A. I. (2003). Embryonic development and the principles of tissue engineering. *In Novartis Foundation: tissue engineering of cartilage and bone*. John Wiley & Sons, London.

Goldberg, V. M., and Caplan, A. I. (2004). Principles of tissue engineering and regeneration of skeletal tissues. *In Orthopedic tissue engineering basic science and practice*. (V. M. Goldberg and A. I. Caplan, eds.). Marcel Dekker, New York.

Horwitz, E., Prockop, D., Fitzpatrick, L., Koo, W., Gordon, P., Neel, M., Sussman, M., Orchard, P., Marx, J., Pyeritz, R., and Brenner, M. (1999). Transplantability and therapeutic effects of bone marrow-derived mesenchymal cells in children with osteogenesis imperfecta. *Nature Medi.* **5**, 309–313.

Urist, M. R. (1994). The search for and discovery of bone morphogenetic protein (BMP). *In Bone grafts derivatives and substitutes* (M. R. Urist *et al.*, eds.), pp. 315–362. Butterworth Heinemann, London.

28

Skeletal Muscle Stem Cells

Mark A. LaBarge and Helen M. Blau

Introduction

In recent years, study of the muscle stem cell (MuSC) has been revitalized because of several provocative reports of potential plasticity of function within the heterogeneous MuSC population that can participate in processes ranging widely from hematopoiesis to osteogenesis, adipogenesis, and myogenesis. There are also reports suggesting that cells from the circulation and from the vasculature give rise to MuSCs and ultimately to skeletal muscle fibers. Here we discuss the identification and function of MuSCs in adult animals, their elusive and complex biochemical and functional heterogeneity, and their essential role in muscle regeneration throughout an organism's lifespan. We also review the plethora of recent unorthodox reports suggesting that cells responsible for muscle regeneration may originate from tissues other than skeletal muscle in adult animals, such as the bone marrow. Finally, we hypothesize, based on studies of muscular dystrophy and aging, how the MuSC niche could recruit and alter the fate of cells that enter that niche.

Tissue-specific stem cells in adult animals have been described most widely in tissues with high turnover, such as blood, or in tissues with a highly diverse cellular composition, such as the central nervous system (CNS). These tissues have a need for either frequent replenishment or for frequent remodeling; both are processes that lend themselves to the activity of a proliferative subpopulation of cells with the capacity to replace any cell type within that tissue. The classical characteristics used to define a cell as a stem cell derive from the best studied system, hematopoiesis. Accordingly, stem cells must be self-renewing, highly proliferative, and capable of differentiating into at least one other cell type. Unlike blood, skin, and CNS tissues, skeletal muscle fibers, the essence of muscle tissue, do not have an apparently rapid turnover. Because muscle tissue possesses a highly specialized cellular architecture and is prone to being damaged by physiologic use throughout life, its persistence and function necessitates the ability to regenerate.

Skeletal muscles are composed of bundles of muscle fibers (myofibers) that are large, terminally differentiated, multinucleate cells formed by the fusion of mononucleate MuSCs. Myofibers can generally be grouped into two differ-ent types based on function: fast or slow contracting. This distinction depends largely on the composition of the myosin heavy-chain (MyHC) isoforms that they express. As shown by interspecies grafting techniques or lineage tracing experiments during embryogenesis, MyHC-expressing mononucleate cells that produce muscle fibers in the limbs originate from the mesodermal somites. By contrast, the muscles of the craniofacial region originate from the sometomeres. To form the muscles of the limb, mononucleate myocytes migrate from the somites into the limb buds in two waves. In mouse, primary muscle fibers are formed at embryonic day 13 (E13) and are followed by the formation of secondary muscle fibers at E16, which surround and align themselves in parallel with the primary fibers. In the limb, these two phases of myogenesis are accompanied by changes in fiber type: primary fibers are relatively large in diameter and express slow MyHCs, whereas secondary fibers are smaller and express fast MyHCs when they are first formed. Eventually, the primary or secondary origin of the fibers cannot be distinguished because a mosaic of both fast and slow fibers of similar size make up the muscle at birth. Even the MyHC composition within an individual fiber can differ because myosins are encoded by distinct nuclei and maintained in nuclear domains. Myoblasts isolated from mice and cloned express all MyHCs regardless of their muscle of origin, suggesting that regulation is imposed *in vivo*.

The myogenic process is regulated by a well-known cascade of basic helix–loop–helix transcription factors, known as muscle regulatory factors (MRFs), expressed sequentially during myogenic development. Myf-5 and MyoD are expressed during the monocyte stage and then decline during differentiation and are followed by the expression of myogenin and MRF4. On their formation, myofibers express myosins, actins, and other proteins that make up the contractile apparatus and the complex cell surface array of dystroglycans, integrins, and dystrophin. The constant threat of damage in adult animals to these structurally complex, postmitotic cells resulting from exercise, chemical agents, or genetic deficiencies suggests that the need for a regenerative pool of cells within the muscle is profound. This need is particularly apparent when a muscle's pool of regenerative cells is exhausted or becomes nonfunctional, as happens in muscular dystrophies such as Duchenne's, the aging process, or following high doses of γ-irradiation. When MuSCs are inadequate in number or function, the result is progressive muscle degeneration and atrophy.

The Original Muscle Stem Cell — The Satellite Cell

The canonical MuSC in adult animals, designated *satellite cell*, was anatomically defined in 1961 by transmission electron microscopy (TEM) studies of the peripheral region of muscle fibers in the tibialis anticus muscle of the frog. The discovery of the satellite cell heralded the birth of the field of muscle regeneration. Satellite cells have a high ratio of nucleus to cytoplasm and are intimately juxtaposed to muscle fibers; they are resident in their own membrane-enclosed compartment, between the sarcolemma of the myofiber and the surrounding basal laminal membrane. So intimately are the satellite cells associated with the myofiber that they are impossible to discern from myonuclei within the fiber by conventional light microscopy. As a result, definitive identification requires extremely sensitive techniques such as TEM. By grafting quail somites into chicken embryos, organisms with similar developmental time courses but distinct nuclear morphologies, it was shown that these cells in the adult muscle are descended from the cells of the somites. Before the discovery of the satellite cell, it was unclear whether there existed a mononucleate cell with the sole purpose of repairing damaged muscle fibers or whether the nuclei of damaged fibers underwent a process whereby they replicated and ensheathed themselves in their own membrane, thereby proceeding to participate in their own repair. The former model is widely accepted, but it has yet to be demonstrated conclusively. Thus far, it has proven impossible to observe the same satellite cell in its characteristic anatomic position divide asymmetrically, renew itself, and then differentiate and give rise to a myonucleus in a myofiber. However, as described later in this chapter, several lines of evidence strongly suggest that the satellite cell is a MuSC that does just that.

First, electron microscopy (EM) studies showed that after a single injection of [³H] thymidine, only a small number of satellite cells, not myonuclei, had incorporated the radioactive label, suggesting that satellite cells were quiescent most of the time and were the only muscle associated cells that proliferated. A subsequent study showed that in the extensor digitorum longus (EDL) muscle of mice, only a small proportion of the satellite cell nuclei would incorporate [³H] thymidine but no myonuclei within fibers. However, after transplantation of the labeled EDL into the muscle bed of another animal, the myonuclei of the myofibers of the host animal had [³H] thymidine labeled myonuclei. The damage inflicted to the muscle bed during the transplantation procedure caused the satellite cells of the donor EDL to proliferate and differentiate, contributing donor myonuclei to the host myofibers. Taken together, these data suggest that satellite cells were stimulated to multiply and contribute to muscle in response to tissue damage.

The second line of evidence that satellite cells were MuSC derived from experiments that showed that, like other populations of proliferative cells, satellite cells are sensitive to γ-irradiation, which renders them unable to meet the demands imposed by increased weight or exercise. When part of the tibialis anterior (TA) muscle is surgically removed, the demands on the neighboring EDL muscle are increased, and because of the additional weight and exercise, the EDL increases in mass as a result of a hypertrophic adaptive response. The gram weight of the EDL increases, as does the average myofiber diameter compared with the EDL of the contralateral leg that did not have the TA resected. Interestingly, if one limb was treated with a high dose of γ-irradiation (25 Gy) before TA resection, the EDL could not adapt as well and was both reduced in mass and had significantly smaller myofiber diameters than nonirradiated controls. These results provided evidence that the activity of satellite cells is linked to the hypertrophic response, an observation confirmed by others.

A third line of investigation that implicated satellite cells as the primary regenerative cell of skeletal muscle derived from the characterization of animal models with muscular dystrophy. One of the most commonly used mouse models of Duchenne's muscular dystrophy (DMD) is the mdx mutant, which carries a point mutation in the dystrophin gene that creates a translational block leading to a truncated protein. As described previously, dystrophin is a key protein in the membrane-bound dystrophin glycoprotein complex (DGC), which is thought to fortify the plasma membrane of the muscle against the intense shearing forces that are generated during daily exercise. The muscles of the mdx mice have spontaneously revertant muscle fibers that express dystrophin resulting from a compensatory point mutation that corrects the translational block. As the mice age, revertant fibers are found in clusters, in which each constituent myofiber harbors the same compensatory mutation. The compensatory mutations found in different bundles are unique to each bundle. This finding can be explained by experiments in which limiting dilutions of β-galactosidase (β-gal) encoding retroviruses were used to demonstrate that a satellite cell infected with a single retrovirus has the ability to proliferate and migrate laterally, participating in regeneration of not only its own fiber but also of nearby fibers, leading to bundles of fibers regenerated by the progeny of the same satellite cell. Taken together, these data suggest that the bundles of revertant fibers in the mdx mouse are the result of fusion and differentiation of a clonally expanded satellite cell.

Fourth, a mouse model used to study pathogenesis in dystroglycan-based muscular dystrophies, another constituent of the DGC, has inadvertently provided a system that transiently labels satellite cells as they fuse with muscle fibers. A mouse was engineered such that a transgenic muscle creatine kinase promoter, which is only active in differentiated myofibers, would drive the expression of the Cre-recombinase protein, and the gene-encoding dystroglycan was flanked by LOX sites. In this model, the satellite cells expressed normal levels of dystroglycan until they fused with the existing muscle fibers, at which time the dystroglycan transgene in the newly contributed myonuclei was excised by the Cre enzyme already present in the myofiber cytoplasm. The Cre-induced LOX-recombination event did not occur immediately following fusion, and during the lag-time dystroglycan was transiently expressed in the muscle fiber. The result was that the usually strong onset of this type of dystrophy was delayed. Moreover, the transient presence of the satellite cell-derived dystroglycan strongly suggests that satellite cells were fusing with the myofibers.

Fifth, MuSCs are typically isolated on collagen-coated or gelatin-coated plates from crude preparations of skeletal muscle. When isolated by this method, they are usually termed *myoblasts*. These cells, if grown at clonal density, can differentiate into multinucleate myotubes when cultured in low-mitogen media, can fuse with existing myofibers when injected *in vivo*, and are considered to be the highly proliferative descendants of satellite cells. However, because of the method by which they were isolated, their exact origin and their relation to the anatomically defined satellite cell have been unclear. In addition to isolation from crude muscle preparations by protease digestion, myoblasts can also be isolated from individual physically dissociated myofibers plated in tissue culture, a method that has tried to demonstrate a direct link between satellite cells and myoblasts. The evidence that cells in the satellite cell position existed underneath the basal lamina in such single isolated myofiber preparations was obtained by scanning EM. Studies of isolated myofibers also showed that mononucleate cells migrated off the fibers, forming colonies that differentiated into myotubes in culture. Unfortunately, the methodology used and the absence of specific markers made it difficult to confirm that the cells that migrated off the muscle fibers were initially in the appropriate anatomic position, underneath the basal lamina juxtaposed to the myofiber sarcolemma. As a result, the evidence that satellite cells beget myoblasts is strong but not conclusive.

Certain physiologic and biochemical consequences that correlate with satellite cell activity, or lack of activity, suggest that these cells are essential to muscle regeneration. One of the first physiologic studies showing the dependence on MuSCs for an adaptive response in the EDL muscle is described earlier. In addition, in the fast fibers of the cat plantaris muscle, denervation and consequent immobilization cause muscle atrophy, leading both to a decrease in myonuclear number and fiber diameter compared with controls. Conversely, innervated plantaris muscle bearing increased weight exhibited augmentation of both myonuclear number and fiber diameter. These results are consistent with the interpretation that the satellite cells respond to exercise-induced muscle damage or stress. Finally, experiments were conducted in which myoblasts transduced with a nuclear localized β-gal expressing retrovirus, proliferated in tissue culture, and then injected into the TA and EDL muscles of an mdx host. When single muscle fibers were later isolated from the muscles, dystrophin expression was generally located in the myofiber in a domain surrounding a β-gal-expressing donor nucleus, demonstrating a potentially beneficial biochemical consequence of MuSC activity, the restoration of dystrophin expression.

Taken together, these experiments strongly support the hypothesis that satellite cells are MuSCs, because they meet the criteria of self-renewal, proliferation, and ability to give rise to more than one cell type (satellite cells, myoblasts, and muscle fibers). Like hematopoietic stem cells (HSCs), satellite cells are capable of robust proliferation as evidenced *in vitro* and *in vivo*; they self-renew following isolation in culture or serial transfer in mice; and they are capable of differentiating into myotubes in tissue culture and fusing with myofibers *in vivo*. These findings strongly suggest that the canonical MuSC, the satellite cell, is largely responsible for regeneration and maintenance of skeletal muscle.

Functional and Biochemical Heterogeneity among Muscle Stem Cells

During the study of MuSCs and their role in muscle regeneration, an apparent functional and biochemical heterogeneity within the population has emerged. Differences between MuSCs isolated at different ages from distinct myofiber types, fast and slow, and from muscles with distinctive embryonic origins, the limbs versus the masseter, are characterized by distinct physiologic and biochemical phenotypes. The purpose of this complexity remains unclear. Are the same cells observed at different time points, or are there distinct subpopulations of MuSCs (e.g., one reserve population that gives rise to another that is poised to participate in the regeneration of specific muscles at specific stages of development)?

Several lines of evidence suggest that satellite cells are functionally heterogeneous. For example, a population of satellite cells was reported that underwent mitosis at least once in 32 hours, was committed to a myogenic fate, and readily differentiated, whereas another much smaller population did not enter mitosis but was postulated to give rise to the first population following asymmetric divisions. Others have reported a radiation-resistant population of satellite cells that could be activated by an injection of the myotoxic snake venom, notexin. Interestingly, the notexin-responsive population is absent in the mdx mouse. Additional observations of the C2C12 mouse myoblast cell line and human primary myoblasts also suggested the existence of two distinct MuSC populations, one that gave rise to the other, which became quiescent instead of differentiating. In the case of C2C12 myoblasts, transcription factors were correlated with cells cultured in proliferative conditions (Myf-5$^+$ and MyoD$^+$) and in differentiation conditions (myogenin$^+$ and MRF4$^+$). However, even in differentiation media, a very small subpopulation of cells remained mononucleated but ceased to express any Myf-5 or MyoD. These mononucleate cells recapitulated the properties of their parental population when recultured in proliferative conditions and subsequently switched to conditions that favored differentiation. Taken together, these studies suggest functional heterogeneity among myogenic cells.

Biochemical heterogeneity among MuSCs has also been documented. Single-cell reverse transcriptase polymerase chain reaction (RT-PCR) was used to determine gene expression in quiescent and activated satellite cells by isolating single myofibers and injecting each with toxin to cause necrosis of the fiber, so that the remaining associated viable cells were thought to be quiescent satellite cells that could be readily collected for analysis. The results showed that all quiescent satellite cells expressed the tyrosine kinase receptor cMet, whereas only 20% expressed the adhesion protein m-cadherin, and few expressed the transcription factors Myf-5 or MyoD. On the other hand, "activated" satellite cells that crawled off the fiber and proliferated expressed all four of the MRF transcripts (Myf-5, MyoD, myogenin, and MRF4) and cMet and m-cadherin. In another study, the cell-surface antigen, CD34, was shown to be expressed on quiescent satellite cells, whereas following activation an alternatively spliced isoform was transiently expressed. In contrast to the first study, the second study found that Myf-5 and m-cadherin were detected on quiescent CD34$^+$ satellite cells. The authors speculated that CD34 plays a role in maintaining a quiescent state, and those CD34$^+$/Myf-5$^+$ satellite cells are the descendants of CD34$^-$/Myf-5$^-$ satellite cells.

Taken together, the cell-surface markers and transcription factors cMet$^+$/m-cadherin$^+$/CD34$^+$/Myf-5$^+$/MyoD$^+$ characterize most committed and activated satellite cells. (The transcription factor Pax7 is also implicated as a marker of these cells.) By contrast the quiescent parent cells are far less well defined: cMet$^+$/m-cadherin$^\pm$/CD34$^\pm$/Myf-5$^\pm$/MyoD$^-$. Moreover, not all of the markers are specific to satellite cells. For example, some cells of the blood are also known to express cMet receptor, and, therefore, it is only a good marker of satellite cells in the context of their anatomic position: membrane encircled and juxtaposed to a myofiber. The only direct evidence that the quiescent population gives rise to the activated and committed form of satellite cell is derived from the *in vitro* experiments conducted with the C2C12 cell line, not primary cells. Clearly, lineage studies and more markers are needed to define unambiguously the heterogeneity of MuSC and their etiology.

Unorthodox Origins of Skeletal Muscle

The observed heterogeneity within the MuSC population and the lack of conclusive evidence that satellite cells are the only MuSC raise the possibility that other cells associated with muscle tissue also give rise to MuSCs. During the past 30 years, several reports have suggested that alternative sources of MuSC may include the thymus, dermis, vasculature, synovial membrane, and the bone marrow. This section focuses on evidence that MuSCs can derive from cells of the bone marrow because the other tissue sources have already been extensively reviewed by Grounds *et al.* (2002).

A number of investigators have demonstrated that following a marrow transplant, bone marrow-derived cells (BMDCs) are present in diverse tissues in mice and humans, where they express characteristic tissue-specific proteins. These tissues include heart, epithelium, liver, skeletal muscle, and brain. These combined results suggest that repair of tissues, including skeletal muscle, may derive from cells other than tissue-specific stem cells that derive from bone marrow.

As early as 1967 and 1983, the idea that muscle could derive from circulating cells was tested, but the experiments

failed to detect such transitions because of technical limitations in the available methods of detection, such as detection of isozymes on starch gels. More recent studies have benefited from sensitive markers of individual bone marrow cells that can be detected by *in situ* hybridization, β-gal or green fluorescent protein (GFP) expression, as well as the use of sensitive instruments such as laser scanning confocal microscopes and flow cytometers. Using bone marrow from a transgenic mouse in which β-gal expression was under the control of the muscle-specific promoter for myosin light chain 3F (MLC3F-nLacZ), experimenters showed that marrow-derived cells contributed to muscle fibers damaged by cardiotoxin. Either following intramuscular injection of adherent or nonadherent bone marrow fractions into the tissue or bone marrow transplantation, donor-derived myonuclei were detected after 2 weeks. When bone marrow of wild-type mice was transplanted into lethally irradiated dystrophic mdx recipients, a somewhat greater frequency of dystrophin-expressing fibers was observed relative to nontransplanted mdx mice. These experiments suggested that cells of the bone marrow could respond to cues from damaged muscle and contribute to muscle regeneration. However, the frequency of the donor-derived fibers was low and generally did not exceed 0.3% even after 10 months.

Occasionally, mononuclear BMDCs were observed in regenerating muscle of transplant recipients that were seemingly nonhematopoietic and were not incorporated into regenerating myofibers. This finding suggests that bone marrow cell contribution to mature muscle fibers might occur through an intermediate, mononucleate MuSC-like stage. This hypothesis was confirmed when following whole bone marrow transplantation of GFP-labeled marrow into lethally irradiated wild-type hosts, GFP-labeled cells were observed on single muscle fibers isolated from transplant recipients in the appropriate anatomic position for satellite cells expressing the characteristic markers desmin, Myf-5, cMet receptor, and, a newly defined maker α7-integrin. In one study, these cells, when isolated, were heritably altered and were capable of self-renewal and proliferation as myoblast-like cells; some fused to form myotubes in culture or following direct injection into host muscle. In these studies, two-thirds of the endogenous satellite cells died at the radiation doses necessary for marrow transplant, resulting in a vacated niche to which the BMDC could contribute. BMDC contribution to host myofibers was very modest, approximately 0.3%, on a par with previous reports. To contribute substantially to mature myofibers in the host mouse, a second insult, voluntary exercise-induced stress, led to a 4% contribution of BMDC GFP$^+$ satellite cells to myofibers, a 20-fold increase over nonexercised mice. These data demonstrate that cells of the bone marrow can give rise to MuSCs that persist long term, self-renew, and regenerate muscle in response to the same cues as endogenous satellite cells.

Taken together, these data provide strong evidence that cells within bone marrow can act as precursors to MuSCs, but their precise nature and origin remain to be determined. Possibilities include (1) the existence of MuSCs within the marrow; (2) HSCs that give rise to both blood and other mesenchymal cell types such as muscle; and (3) a precursor that gives rise to HSCs, MuSCs, and other mesenchymal stem cells. Although myogenic cells had previously been shown to give rise to other mesodermal cell types such as adipocytes, cartilage, and osteocytes in culture, the possibility that they could give rise to circulating mesodermal cells such as HSCs was not thought to occur. However, in some studies a side population (SP) of cells was identified by flow cytometric analysis (FACS) in bone marrow that excludes Hoechst dye more than other marrow-derived cells and gave rise to skeletal muscle and blood, whereas reports also appeared suggesting that SP cells isolated from muscle tissue (muscle-SP) could give rise to blood and muscle.

Further studies have suggested that CD45$^+$ muscle-SP cells, a pan-hematopoietic marker, primarily give rise to blood, whereas CD45$^-$ muscle-SP cells give rise to muscle, indicating that the former derived from circulating HSCs and the latter from MuSCs or a cell population with similar "stem-like" properties within muscle. Analogous studies by others of CD45$^+$/Sca-1$^+$ muscle-SP showed that they did not express Myf-5 and gave rise only to blood colonies, whereas Sca-1$^+$/CD45$^-$ muscle-SP gave rise only to muscle *in vitro*. However, on injection into muscle *in vivo*, the Sca-1$^+$/CD45$^+$/Myf-5$^-$ muscle SP cells expressed Myf-5, and differentiation into muscle was observed. One could envision a model in which cells within the bone marrow give rise to vascular-associated Sca-1$^+$/CD45$^+$ cells, which in turn give rise directly to MuSCs or a muscle-SP intermediate that persists and participates in muscle and possibly other mesenchymal tissue regeneration. Taken together, these results are provocative in that they suggest a closer relationship among adult mesodermal tissues, blood, and muscle than was previously recognized. However, a direct lineage from any single cell of the bone marrow to a BMDC-derived MuSC has yet to be demonstrated.

The Muscle Stem Cell Niche

Tissue-specific stem cells occupy niches, microenvironments that instruct and support stem cell self-renewal, proliferation, and differentiation, providing specific cellular neighbors, signaling molecules, and extracellular matrix components. From studies of muscle aging and muscular dystrophy, it is suggested that MuSCs are wedded to their niches. Satellite cells become dysfunctional not only because of exhaustion of their replicative capacity but also because of changes in their microenvironment imposed on them because of disease and aging. Complex mechanisms are likely to dictate how the milieu of factors in a MuSC niche might affect the recruitment and contribution of circulating cells to the MuSC population.

A portrait of the MuSC niche can be envisioned based on knowledge gained from the collective set of mutations that cause muscular dystrophy and from the signals that are released on muscle damage, both of which are known to affect satellite cell behavior. A number of dystrophies result from defects in constituents of the dystrophin glycoprotein complex (DGC) (e.g., dystrophin, α7-integrin, dysferlin, caveolin-3, sarcoglycans, dystroglycans) and the extracellular proteins that associate with the DGC (e.g., laminin α2). It is generally thought that the DGC fortifies the myofiber membrane against repetitive shearing forces encountered during normal exercise and mediates communication between the myofiber and the extracellular environment. When the DGC is disrupted, the muscle undergoes rapid cycles of degeneration and regeneration. Before and immediately following the onset of most dystrophies, satellite cells are capable of successfully regenerating muscle. However, as the disease progresses, the satellite cells become senescent and cease to participate in regeneration. This is certainly true for DMD, one of the most severe forms of human muscular dystrophy. The cells usurp their proliferative potential in repair of muscle in a manner similar to that which accompanies aging, only on a much shorter time scale. Other factors that appear to characterize the MuSC niche are secreted factors, such as insulin-like growth factor (IGF)-1 and IGF-2, both of which help mediate muscle hypertrophy and are known to stimulate satellite cell proliferation and differentiation. By contrast, more transforming growth factor-β (TGF-β) is expressed in Duchenne's affected muscle and is correlated with an inhibition of satellite cell proliferation. The composition of molecules within the MuSC niche is likely to play a role in instructing MuSCs during muscle regeneration and during the aging process and possibly plays a role both in the recruitment and reprogramming of BMDCs en route to becoming MuSCs.

The process of bone marrow-derived myogenesis is typically associated with a response to damage: genetic, chemical, or physical. It is known that high doses of γ-irradiation do not injure myofibers. Thus, it is intriguing that BMDCs still contribute to myofibers (albeit at a low frequency) in otherwise genetically normal, uninjured mice. All the effects of γ-irradiation remain unknown but include (1) damage of the regenerative satellite cells with proliferative potential, not the postmitotic myofibers; (2) a vacated niche leading to a regenerative deficit compared with normal muscle; and (3) a release of factors that could modify the vacated MuSC niche, leading to occupation of that niche by circulating stem cells that are instructed by the local environment to assume the role of the lost cells. The end result is that BMDC are recruited into muscle tissue, assuming a myogenic function that allows them to participate in regeneration, thereby reducing the regenerative deficit in irradiated skeletal muscle.

Several lines of evidence suggest that γ-irradiation can also positively affect cell proliferation, in addition to the inhibitory effects described previously for satellite cells. After injection of a myogenic cell line into host muscles, tumors were found to form more rapidly in γ-irradiated muscle than in nonirradiated muscle and the effect was dose-dependent. Moreover, in contrast to damaging high-energy γ-irradiation, low-energy laser irradiation (LELI) was shown to promote survival of both MuSCs and myofibers. These findings fit well with those of others who argue that when the environment of precancerous cells is changed by transient insults, such as irradiation, conditions are created that favor proliferation and the cells divide uncontrollably. Thus, modification of the niche or microenvironment of a cell, including growth factors, cytokines, and adhesion molecules, can influence its ability to promote proliferation and plasticity, as in the case of the conversion of bone marrow to MuSC-like cells and fibers as described previously. Following injury or stress to the muscle or its stem cells, the proportion of satellite cells derived from bone marrow found within the total MuSC population could increase and participate to a greater extent in regeneration, serving as a backup reservoir to the tissue specific stem cells.

Summary

For more than 40 years the satellite cell has been considered the primary mediator of muscle regeneration in postnatal animals. This was first based on anatomic criteria using EM, then by use of antibodies to specific proteins, and ultimately by the ability of the cells to contribute to the repair of locally damaged muscle fibers. However, recent evidence suggests that this interpretation may be an oversimplification and that more than one source of cell may aid in muscle regeneration or that a heretofore unrecognized lineage(s) may contribute to myofibers, either via the satellite pool or directly. First, both functional and biochemical heterogeneity has now been shown. Second, other myogenic populations have been implicated in muscle repair, for example, muscle SP cells. Finally, cells that are not of myogenic origin appear to be capable of repairing muscle, such as meso-angioblasts, BMDCs, and possibly HSCs, or their more multipotent precursors.

It is unclear at present whether these cells originate from a similar cellular origin and represent different points in development or whether there exist multiple distinct cell types each of which has the capacity to participate in muscle regeneration. Because cells from widely disparate nonmuscle origins have been observed to have myogenic potential, it is particularly important to understand their role in normal muscle regeneration and their relationship to the anatomically described satellite cell. The interrelationships of these cells with their diverse markers must be investigated further and compared with one another and the classical satellite cell MuSC.

In conjunction with lineage tracing studies, a more global approach to defining a cell as a MuSC appears warranted, including consideration of the *functional* properties that are classically ascribed to entities designated as satellite cells that make them essential for muscle regeneration, whereas their origin and location are perhaps less essential: (1) they are highly proliferative and capable of fusing with existing muscle fibers or forming muscle fibers *de novo*, (2) they possess the necessary molecular constituents to respond appropriately to the regenerative needs of a muscle environment, (3) their absence or inability to function properly results in deleterious effects such as atrophy or myopathy, (4) they express key intracellular and cell surface regulatory proteins characteristic of muscle before their fusion with existing muscle fibers, and (5) in the absence of an experimentally manipulated environment their default fate is skeletal muscle. Thus, the challenge for the next few years will be to understand the role played during skeletal muscle regeneration by circulating cells, vascular associated cells, and cells from other nonmuscle tissues. Even if cells that originate from these sources do not play a significant role in normal muscle repair, their further study will illuminate the range of possible outcomes for those cells. Perhaps an understanding of the factors that govern their behavior will lead to information that is ultimately useful for developing new biomedical treatments of myopathies.

ACKNOWLEDGMENTS

We are very grateful for the support provided to H.B. by the following grants and foundations: Baxter Foundation; Ellison Medical Foundation AG-SS-0817-01; Aventis Pharma/Gencell S.A.; and NIH grants AG09521, AG20961, HL65572, HD018179. M.L. is supported by NIH grants AG20961, HD018179.

KEY WORDS

MuSC *Mu*scle *s*tem *c*ells; mononucleate cells that function to regenerate damaged myofibers through a process of proliferation, followed by subsequent fusion with existing or nascent myofibers.

Myofiber Large multinucleate cells that contain a contractile apparatus, composed primarily of actin and myosin family proteins, to facilitate movement. Muscle groups are composed bundles of myofibers.

Satellite cell An anatomical designation of a mononuclear cell that sits upon a myofiber juxtaposed to the plasma membrane of the myofiber, yet is ensheathed by the basal laminal membrane, which also surrounds the myofiber.

FURTHER READING

Blau, H. M., Brazelton, T. R., and Weimann, J. M. (2001). The evolving concept of a stem cell: entity or function? *Cell* **105**, 829–841.

Buckingham, M., Bajard, L., Chang, T., Daubas, P., Hadchouel, J., Meilhac, S., Montarras, D., Rocancourt, D., and Relaix, F. (2003). The formation of skeletal muscle: From somite to limb. *J. Anat.* **202**, 59–68.

Cohn, R. D., and Campbell, K. P. (2002). Molecular basis of muscular dystrophies. *Muscle Nerve* **23**, 1456–1471.

Grounds, M. D., White, J. D., Rosenthal, N., and Bogoyevitch, M. A. (2002). The role of stem cells in skeletal and cardiac muscle repair. *J. Histochem. Cytochem.* **50**, 589–610.

Mauro, A. (1961). Satellite cell of skeletal muscle fibers. *J. Biophys. Biochem.* **9**, 493–495.

Spradling, A., Drummond-Barbosa, D., and Kai, T. (2001). Stem cells find their niche. *Nature* **414**, 98–104.

Watt, F. M., and Hogan, B. L. (2000). Out of Eden: stem cells and their niches. *Science* **287**, 1427–1430.

Zammit, P., and Beauchamp, J. The skeletal muscle satellite cell: stem cell or son of stem cell? *Differentiation* **68** (4–5), 193–204.

Stem Cells and the Regenerating Heart

Nadia Rosenthal and Maria Paola Santini

Introduction

Against a backdrop of the seemingly limitless possibilities for endogenous or supplementary stem cells to restore aging or damaged tissues, the restricted regenerative capacity of the mammalian heart remains a perplexing exception. The regenerative response launched by other injured organs involves local populations of self-renewing precursor cells or recruitment of circulating stem cells to replace or repair the injured areas. In response to functional stress, the heart can increase its muscle mass through cellular hypertrophy, but the damaged heart needs a rapid response to repair damage to the muscle wall and maintain adequate blood flow to the rest of the body. Paradoxically, this most critical organ cannot restore the muscle loss that accompanies myocardial infarction and ischemia-reperfusion injury. Instead, interruption of the coronary blood supply results in apoptosis and fibrotic scar formation at the cost of functional muscle. As a result, the remaining cardiomyocytes undergo cellular hypertrophy, leading to decompensated function and congestive heart failure, an increasingly prevalent disease in the industrialized world.

What underlies the inability of the mammalian heart to rebuild itself in response to injury? Where are the stem cells of this vital organ, and, if there are none, how does it maintain its structural and functional integrity for decades? The impediment to adult mammalian cardiac regeneration has been attributed to distinct embryonic history. The heart is the first fully differentiated structure to form and function during vertebrate development. The primitive heart tube, composed of contracting cardiomyocytes lined by a layer of endocardial cells, ensures the establishment of a circulatory system that is critical to support rapid rates of embryonic growth. The progressive acquisition of the cardiac phenotype by precursor cells starts early in the primitive streak stage, so that by the time a cardiac crescent is fully formed, there are coordinated contractions in the primitive heart tube. Actively contracting fetal cardiomyocytes must continue to divide to provide for further growth of the embryonic heart. This phase of cardiomyocyte division ends soon after birth, when increases in myocardial mass are achieved largely through cellular hypertrophy, at which point the heart is considered fully developed.

In contrast to mammalian skeletal muscle, which regenerates injured tissue through activation of quiescent myogenic

precursor or multipotent adult stem cell populations, the heart does not appear to retain equivalent reserve cell populations to promote myofiber repair. The relative paucity of progenitor cells residing within the heart may impose severe limits on replacement of damaged myocardium. Thus, the prevailing assumption has been that the heart cannot regenerate as well as other organs because it does not maintain a sufficiently robust progenitor cell population.

Recruiting Circulating Stem Cell Reserves

The relative scarcity of progenitor cells residing in the adult myocardium has prompted a search for a renewable source of circulating somatic progenitor cells that might home to the heart in response to damage. The presence of such cell populations has gained credibility from observations of sex-mismatched cardiac human transplants in which a female heart is transplanted into a male host. In these patients, the presence of the Y-chromosome marks host-derived cells in the transplanted heart. Various numbers of Y-chromosome–positive myocytes and coronary vessels in the transplanted heart's male cells could be found. Cell fusion of host cells with donor cardiac cells, as has been proven for other regenerating tissues, was ruled out by the presence of a single X-chromosome. The presence of differentiated host cells in the transplanted tissues proves the existence of migratory precursor cells that are induced to differentiate by the cardiac milieu. Although this phenomenon could be a response to organ transplantation, it may also reflect a normal homeostatic process for the maintenance of cardiac muscle and coronary vasculature.

The lack of information regarding the precise origin of donor cells in these human transplants has prompted animal experiments in which stem cells isolated from bone marrow were enriched for various surface markers. A stem cell-enriched side population (SP) can be isolated by the relative efflux of Hoechst dye through MDR1, a P-glycoprotein capable of extruding dyes, toxic substances, and drugs. Movement of bone marrow SP cells can be traced if they are isolated from donor mice expressing a genetic marker such as lacZ and used to reconstitute lethally irradiated recipient mice. In such experiments, very few marked cells can normally be found in extrahematopoietic tissues of the reconstituted animals. However, in reconstituted mice that were subsequently subjected to coronary artery occlusion, lacZ marked cells could be found in vascular endothelium and cardiomyocytes of the border zone adjacent to the infarct. In other studies, cell populations expressing c-kit, the receptor for stem cell factor, were isolated from bone marrow and injected directly into the border zone of an experimentally induced infarct, where they migrated into the damaged region, differentiated into cardiomyocytes and vascular cells, and partially replaced necrotic myocardium. To what extent these different subsets of marked bone marrow cells represent the same cell population as migratory Y-chromosome-containing donor cells in the human transplant studies remains to be determined.

Insufficient revascularization represents another major impediment to the reconstitution of ischemic myocardial tissue and the prevention of further scar tissue formation. Although angiogenesis within the infarcted area is an integral component of the remodeling process, the capillary network is normally unable to support the greater demands of the hypertrophied myocardium. Fortunately, adult bone marrow contains endothelial precursors that resemble embryonic angioblasts that, if sufficiently mobilized, can participate in revascularization of the ischemic tissue. Recent human trials have confirmed that intracoronary infusion of endothelial progenitor cells in patients with acute myocardial infarction is associated with significant beneficial effects on postinfarction left ventricular remodeling processes, regional contractile function of the infarcted segment, and coronary blood flow reserve in the infarct artery.

Whatever the provenance and potential of cardiac progenitors, it is clear that, at least in mammals, the relatively poor recruitment of circulating stem cells to the site of myocardial injury limits the body's ability to aid in the repair process. Numerous chemotactic signals associated with inflammation, including cytokines and adhesion molecules, are preferentially expressed by the infarct border zone and may improve stem cell homing as well. Indeed, increasing evidence supports the notion that chemokines play a central role in directing cells from the bone marrow to ischemic myocardium, in which case enhancement of stem cell trafficking could be harnessed to amplify the natural homing process.

The Elusive Cardiac Stem Cell

The adult mammalian heart has long been considered a postmitotic organ without an endogenous population of stem cells, which instead contain a relatively constant number of myocytes that cease to divide shortly after birth and remain constant into senescence. A long-standing view is that the inability of differentiated cardiac myocytes to reenter the cell cycle may present the ultimate impediment to the heart's regenerative capacity and may be responsible for the drastic effects of acute and chronic myocyte death in the surviving myocardium after infarction.

The fact that cardiac myocytes are multinucleated and polyploid in many mammalian species has complicated interpretation of any observed DNA synthesis that might represent myocyte proliferation. Nevertheless, it has been argued that in the face of massive cardiomyocyte apoptosis and necrosis,

the diseased heart could not continue to function in the absence of new myocyte formation. Increases in myocyte number do not provide information about the origin of these new cells, however. Although most of the controversy surrounding mammalian heart regeneration has focused on the evidence for or against the replication of existing myocytes, cells capable of differentiating into a myocyte in the adult heart could originate through the commitment of precursor cells to the myocyte lineage, through replication of preexisting myocytes, or by a combination of these two mechanisms.

Decades of frustrated searching for a resident cardiac stem cell population have recently yielded more encouraging results through the application of methods used to study stem cells in the adult haematopoietic compartment. Side populations of cells resembling those isolated from the bone marrow have been found in adult rodent myocardium expressing the corresponding transport proteins. Cell-surface proteins that mark stem cell populations in other tissues are also found on a subpopulation of undifferentiated precursor cells in the adult heart. In certain instances, cells isolated from adult heart by virtue of their stem cell markers not only express appropriate markers but behave like cardiac progenitor cells *in vitro*, giving rise to clones that express biochemical markers of myocytes, smooth muscle, and endothelial cells.

It remains to be seen how these cells relate to a population of rare, small cycling cardiomyocytes that retain the capacity to proliferate in response to damage and are continuously renewed by the differentiation of stem-like cells as a normal function of cardiac homeostasis. The origins of cycling myocardial cells may be attributed to recently characterized Lin−/c-kit+ cells isolated from the adult rat heart that retain stem cell characteristics. These cells are self-renewing, clonogenic, multipotent *in vitro* and *in vivo*, and give rise to myocytes, smooth muscle, and endothelial vascular cells. When injected into an ischemic rat heart, a population of these cells or their clonal progeny reconstitutes large portions of the injured myocardial wall. The regenerated myocardium contains small myocytes that present the anatomical, biochemical, and functional properties of young myocytes. These data support the notion that myocyte renewal in the adult mammalian heart occurs constantly, albeit at a very low rate. The possibility that endogenous cardiac stem cells can be mobilized to migrate from their niche within the healthy heart to support regeneration of diseased myocardium has exciting implications for therapeutic intervention.

Evolving Concepts of Regeneration

The limited restorative capacity of the adult mammalian heart has been attributed to the loss of cardiomyocyte versatility soon after birth. Emerging concepts of regeneration as an evolutionary variable are dramatically illustrated by the relatively robust proliferative capacity of the injured heart in other vertebrate species. The dramatic regeneration of urodele amphibian limb and lens extends to their robust repair of injured myocardium. Unlike their mammalian counterparts, adult newt cardiomyocytes can readily proliferate after injury and contribute to the functional regeneration of the damaged heart. Newts repair their hearts efficiently in response to cardiac damage, leaving none of the dysfunctional scar tissue typical of the post-infarct mammalian myocardium. The recognition that cardiac tissue in other vertebrates can undergo extensive repair has prompted the proposal that regeneration may be a primordial attribute that has been lost during mammalian evolution.

Recent studies of newt regeneration point to a mechanism whereby the rebuilding of damaged tissue could couple acute response to injury with local activation of plasticity in surrounding tissues and/or activation of stem cell pools. A transient activity generated by thrombin, a critical component of the clotting cascade that ensures hemostasis and triggers other events of wound healing, has been linked to cell cycle reentry in multinucleate myotubes during urodele limb regeneration. Thus, selective activation of thrombin protease action in response to injury link response to damage and initiation of regenerative growth. It remains to be seen if such a linkage can be established in the regenerating myocardium.

Regeneration of the zebra fish heart offers a more genetically accessible model for dissecting the molecular basis of cardiac repair. After surgical removal of the ventricular apex and rapid clotting at the site of amputation, proliferating cardiac myofibers replace the clot and regenerate missing tissue, with minimal scarring. The requirement for cell cycle reentry in this model was supported by the decreased regeneration and increased fibrosis in a temperature-sensitive mutant of a mitotic checkpoint kinase, mps1.

It is still formally possible that the activation of cardiac progenitor cells underlies the extraordinary capacity of the adult zebra fish to restore extensive portions of the heart. Regeneration has been traditionally assumed to involve the recapitulation of genetic pathways employed during embryonic development. However, a novel profile of genes activated during zebra fish cardiac regeneration, distinct from that associated with cardiogenesis, is consistent with the emerging concept that differentiated myocytes can reenter the cell cycle and proliferate in response to heart injury. This would provide evidence for true epimorphic regeneration in the vertebrate heart, and argues for a clear distinction between mechanisms at work during regeneration versus development. If cardiac progenitor cells are indeed involved in this process, they may play a more instructive role in reforming damaged heart tissue.

The regenerative potential of the mammalian heart is a rapidly evolving concept. In the near future, cardiac repair is likely to be augmented through a number of avenues (Figure 29-1). The dramatic improvements that exogenously administered progenitor cells can affect in both animal and human myocardial repair underscore their therapeutic potential. Although resident cardiac progenitor cell populations have now been identified, the insufficiencies of endogenous stem cells to alleviate acute and chronic damage to mammalian cardiac tissue remain to be overcome. Recent advances in our understanding have uncovered an unexpected dynamism in

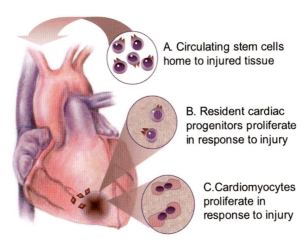

Figure 29-1. Several modes of cardiac regeneration supported by recent studies. (A) Circulating stem cells home to the infarcted area, guided by chemoattractive mechanisms, and participate in multiple functions, including neoangiogenesis and myocyte renewal. (B) Injury stimulates the proliferation of resident cardiac progenitors, which resemble tissue stem cells in their phenotype and ability to participate in cardiac renewal. (C) A subset of competent cardiomyocytes reenter the cell cycle to replace and rebuild missing tissue. Although these endogenous regenerative modes are not mutually exclusive, they provide different possibilities for therapeutic intervention.

A. Circulating stem cells home to injured tissue

B. Resident cardiac progenitors proliferate in response to injury

C. Cardiomyocytes proliferate in response to injury

cardiac homeostasis and highlight the heterogeneous proliferative potential of resident cardiomyocytes. Enhancing the functional regeneration of this most obdurate of organs raises the exciting prospect that regenerative processes in other tissues of the adult mammalian soma might be similarly harnessed to fend off the ravages of aging and disease in a new paradigm of self-renewal.

KEY WORDS

Cardiac hypertrophy Cellular response of cardiac cells to an increase in biomechanical stimuli, characterized by increased cell size, enhanced protein synthesis, and sarcomere organization. Prolonged maladaptive hypertrophy is associated with heart failure.

Fibrosis Formation of fibrous scar tissue after damage of an organ with complete substitution of the cells constituting the organ with collagenous material. Extended fibrosis leads to organ impairment and loss of functionality.

Regeneration Capacity of certain adult vertebrates to repair damaged organs in a well-defined spatial and temporal plan that reconstitutes the original organ.

Stem cell A cell capable of indefinite self-renewal and differentiation into diverse and contextually specific cell lineages.

FURTHER READING

Beltrami, A. P., Barlucchi, L., Torella, D., Baker, M., Limana, F., Chimenti, S., Kasahara, H., Rota, M., Musso, E., Urbanek, K., Leri, A., Kajstura, J., Nadal-Ginard, B., and Anversa, P. (2003, September 19) Adult cardiac stem cells are multipotent and support myocardial regeneration. *Cell* **114**(6), 763–776.

Bettencourt-Dias, M., Mittnacht, S., and Brockes, J. P. (2003, October 1). Heterogeneous proliferative potential in regenerative adult newt cardiomyocytes. *J. Cell Sci.* **116**(Pt. 19), 4001–4009.

Brockes, J. P., and Kumar, A. (2002). Plasticity and reprogramming of differentiated cells in amphibian regeneration. *Nat. Rev. Mol. Cell Biol.* **3**, 566–574.

Goodell, M. A., Rosenzweig, M., Kim, H., Marks, D. F., DeMaria, M., Paradis, G., Grupp, S. A., Sieff, C. A., Mulligan, R. C., and Johnson, R. P. (1997). Dye efflux studies suggest that hematopoietic stem cells expressing low or undetectable levels of CD34 antigen exist in multiple species. *Nat. Med.* **3**, 1337–1345.

Itescu, S., Kocher, A. A., and Schuster, M. D. (2003). Myocardial neovascularization by adult bone marrow-derived angioblasts: strategies for improvement of cardiomyocyte function. *Heart Fail. Rev.* **8**, 253–258.

Mathur, A., and Martin, J. F. (2004, July 10). Stem cells and repair of the heart. *Lancet* **364**(9429), 183–192.

Oh, H., Chi, X., Bradfute, S. B., Mishina, Y., Pocius, J., Michael, L. H., Behringer, R. R., Schwartz, R. J., Entman, M. L., and Schneider, M. D. (2004, May). Cardiac muscle plasticity in adult and embryo by heart-derived progenitor cells. *Ann. N.Y. Acad. Sci.* **1015**, 182–189.

Poss, K. D., Keating, M. T., and Nechiporuk, A. (2003, February). Tales of regeneration in zebrafish. *Dev. Dyn.* **226**(2), 202–210.

Quaini, F., Urbanek, K., Beltrami, A. P., Finato, N., Beltrami, C. A., Nadal-Ginard, B., Kajstura, J., Leri, A., and Anversa, P. (2002). Chimerism of the transplanted heart. *N. Engl. J. Med.* **346**, 5–15.

Urbich, C., and Dimmeler, S. (2004, August 20). Endothelial progenitor cells: characterization and role in vascular biology. *Circ. Res.* **95**(4), 343–353.

Potential of ES Cell Differentiation Culture for Vascular Biology

Hiroshi Hisatsune, Nobuyuki Kondoh, Jun Yamashita, Satomi Nishikawa, and Shin-Ichi Nishikawa

Introduction

Recent progress in vascular biology has been largely supported by two experimental systems that allow manipulation of *in vivo* angiogenesis: (1) the various systems to observe neoangiogenesis and (2) gene-targeting technology. The first has been useful to test both positive and negative molecules that affect the process of angiogenesis, and the second has been used to characterize the function of a particular gene under normal circumstances. Through *in vivo* studies investigators have identified a large number of molecules involved in angiogenesis and have characterized their functions in various situations, including embryonic vascular development. Although this list is still expanding, a new problem has been recognized: phenotypes induced by the *in vivo* manipulations are not necessarily sufficient to specify the function of objective molecules, particularly at the cellular level. Except for such molecules as vascular endothelial growth factor (VEGF) and its receptor, whose role in the development of the vascular system has been specified at both the organism and cellular levels, many of the listed molecules can be defined no further than as molecules required for vascular remodeling. An important reason for this problem is the deficit of our understanding of the function of each molecule at the cellular level. Indeed, it is difficult to describe how a given molecule is involved in the formation of a complex vascular system without knowing its function at the cellular level. Hence, development of experimental systems that enable investigators to evaluate the function of a given molecule at the cellular level is an urgent issue in this field.

Previously, most studies of the cell biology of vascular endothelial cells (EC) used endothelial cell lines that were generated and propagated *in vitro* by various methods. Although those cell lines have been useful for many purposes, it has also been widely recognized that, although many of them are not transformed, they have lost some features that are characteristic of the normal EC. In an attempt to ameliorate problems of established EC cell lines, we have prepared relatively normal EC from embryonic stem (ES) cells because ES cells have been proven to differentiate to all somatic cells in the human body. Thus, we expected that EC, particularly those cells present in the developing embryo, may be obtained

from ES cell differentiation cultures. In this chapter, we describe our data relating to EC and its progenitor that can be prepared in the ES cell differentiation culture.

Cultures for Embryonic Stem Cell Differentiation

ES cell lines are possible because of the discovery that leukemia inhibitory factor (LIF) can inhibit differentiation of ES cells, while maintaining their proliferation. Hence, induction of ES cell differentiation is achieved by transferring immature ES cells to conditions without LIF. The most popular method for inducing ES cell differentiation has been to use embryoid body (EB) as an environment for differentiation. By this method, ES cells are first rendered to form aggregates and are transferred to the LIF(−) differentiation condition. Two epithelial cell layers — visceral endoderm on the outside and ectoderm on the inside — are spontaneously formed from these aggregates. Various cell lineages have been demonstrated to be generated within EB without addition of exogenous cytokine, indicating that EB can provide an embryo-like microenvironment that contains the necessary molecules for ES cell differentiation. Indeed, the two-layered structure with visceral endoderm and embryonic ectoderm is reminiscent of the egg–cylinder stage embryo.

Although EB culture has advantages such as ease of performance and reproducibility, a clear disadvantage of this method is that it requires a complex three-dimensional (3D) structure that is difficult to manipulate and monitor. To overcome this problem, attempts have been made to induce EB differentiation in a two-dimensional (2D) plane. In an attempt to induce ES cell differentiation to the hematopoietic cell (HPC), Nakano *et al.* reported a culture of ES cells on OP9 stromal cell line. In this stromal cell-dependent culture, ES cells undergo proliferation and differentiation without forming 3D structures such as EB. In this culture, ES cells were cultured on OP9 that expressed the molecules required for proliferation and differentiation of hematopoietic stem cells. Of interest is the striking variation in the differentiation-supporting ability among stromal cell lines, particularly the repertoire of lineages generated in culture. Initially, Nakano *et al.* ascribed the activity of OP9 to support preferential differentiation to the HPC lineage to the deficit of macrophage colony-stimulating factor (M-CSF) expression in the OP9 stromal cell line. (OP9 is a stromal cell line that was established from the *op/op* mouse bearing a null mutation in the M-CSF gene). However, later study demonstrated that OP9 can induce preferential differentiation of lateral mesoderm as compared with other stromal cell lines. Indeed, PA6, which can also support proliferation and differentiation of hematopoietic stem cells as efficiently as OP9, was shown to be defective in inducing the lateral mesoderm. Instead, PA6 was shown to be an efficient stromal cell to support neuronal differentiation. Hence, variation among stromal cell lines will be useful for understanding the molecular mechanisms regulating each distinct cell lineage and will provide a powerful method to steer ES cell differentiation.

With regard to the differentiation to EC and HPC, we have shown that collagen IV matrix is sufficient for ES cell differentiation to these lineages through lateral mesoderm. In this culture, ES cells are spread on a 2D plane coated by collagen IV and undergo differentiation to lateral and paraxial mesoderm. This result indicates that neither 3D structure nor stromal cells are required for induction of early intermediate stages from ES cells. Our study showed that ES cells spontaneously proliferate and differentiate under a simple culture condition, although the cell density is the critical factor influencing the outcome.

Because the next step for ES cell differentiation culture is to develop chemically defined culture conditions, it is significant that ES cells can undergo differentiation under a simple culture condition. Although serum-containing medium can support the differentiation to lateral mesoderm without exogenous growth factors, the mesoderm induction does not occur in the serum-free medium unless appropriate cytokines such as bone morphogenetic protein (BMP) are added (our unpublished observation). Thus, it is expected that the simplicity of this culture system will be useful to identify molecular requirements for induction of each cell lineage.

Markers for Defining Intermediate Stages during Endothelial Cell Differentiation

To date, no ES cell differentiation cultures can steer the ES cell differentiation at will to select a particular pathway, although the PA6 stromal cell line can induce as high as 90% nestin[+] cells. Nonetheless, in every currently available culture method including ours, multiple cell lineages are generated simultaneously on induction of differentiation. The most significant problem this situation raises is the difficulty in monitoring the events in the culture before the appearance of such cells as erythrocytes that can be characterized easily. This problem can be ameliorated if intermediate stages in a particular differentiation pathway are distinguished from other cell lineages. Although a number of stage- and lineage-specific markers have been developed for defining early intermediate stages, most markers require fixation of cells and, thus, are difficult to apply to living cells.

Basically, two methods are available to purify living cells at various intermediate stages. The first method is to use drug-based selection. By this method, drug-resistant genes are expressed under the control of the gene regulatory unit that can direct the gene expression at a particular stage and in a particular cell lineage. After induction of ES cell differentiation, any cells of interest can be selected by adding the appropriate concentration of drug in the culture, if the drug-resistant gene is expressed specifically. The other method is to use cell sorting. To apply this technology, cells of interest have to be defined by the expression of cell surface markers that can be detected by labeled antibodies or fluorescent protein driven by a cell-specific promoter. Concerning the differentiation to EC, both methods have been shown to be useful. Because any depletion of unwanted cells by drug selections requires some incubation time, drug-based selection is not suitable for selection of intermediate stages that appear only transiently. For this purpose, cell sorting technology is more useful. Moreover, cell sorting technology can combine multiple markers for the definition of the cells, which is difficult to perform by drug-based selection, although it is theoretically possible.

Although cell sorting can use multiple markers simultaneously, few markers are available for sorting intermediate stages that appear during early embryogenesis. In this context, we believe that ES cell differentiation to EC might be the best characterized process with multiple markers. As summarized in Figure 30-1, we defined distinct intermediate stages during differentiation to EC by using multiple markers. In this scheme, mature EC is defined as the stage expressing all EC markers such as Flk1, VE-cadherin, platelet endothelial cell adhesion molecule (PECAM), and CD34. Analysis of the time course of the expression of these molecules demonstrated that they appear successively in an orderly manner. Moreover, immature stages before mesoderm differentiation are defined as cells negative in these markers but positive in E-cadherin expression that is down-regulated on mesoderm differentiation. Taken together, E-cadherin[+]Flk1[-] embryonic ectoderm cells differentiate first to E-cadherin[-]Flk1[+]VE-cadherin[-] mesoderm. This population then diverges to primitive erythrocytes, smooth muscle cells, and immature EC, but only the EC is Flk1[+]VE-cadherin[+]PECAM[-]CD34[-]. This ES population then expresses successively PECAM and eventually CD34. Each stage can be distinguished and sorted by the cell sorter, and the order of each stage during EC differentiation can be confirmed by sorting and short-term culture. As few as 2 to 3 days are required for completing all the process after the Flk1[+] mesoderm appears; dissection of this process can only be attained by cell sorting. Because of the short course, the process can be transferred to a chemically defined condition, thereby determining the molecular requirement for differentiation. Indeed, we have recently demonstrated that this process can be supported to some extent under a serum-free condition.

Utility of Embryonic Stem Cell Culture for Cell Biology of Endothelial Cells

Because a number of reliable methods for inducing EC from ES cells and markers to purify the cells in this differentiation

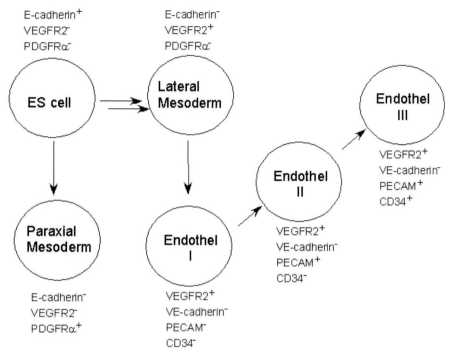

E-cadherin$^+$
VEGFR2$^-$
PDGFRα^-

E-cadherin$^-$
VEGFR2$^+$
PDGFRα^-

ES cell

Lateral Mesoderm

Endothel III

VEGFR2$^+$
VE-cadherin$^-$
PECAM$^+$
CD34$^+$

Endothel II

VEGFR2$^+$
VE-cadherin$^-$
PECAM$^+$
CD34$^-$

Paraxial Mesoderm

Endothel I

E-cadherin$^-$
VEGFR2$^-$
PDGFRα^+

VEGFR2$^+$
VE-cadherin$^-$
PECAM$^-$
CD34$^-$

Figure 30-1. Intermediate stages during differentiation of endothelial cells from embryonic stem (ES) cells. Five intermediate stages have been identified during differentiation of endothelial cells from ES cells. Each stage is defined in terms of the expression of the surface molecules indicated in the figure.

pathway are available, endothelial cells generated from ES are now ready for cell biology use.

CLONOGENIC ASSAY OF ENDOTHELIAL CELLS

An important system that emerged from investigation of ES cell differentiation culture is a clonogenic assay of the vascular progenitor using the OP9 stromal cell. A protocol of this method was described in detail in the report by Hirashima *et al.* Briefly, the cell population that contains vascular progenitors is seeded onto the OP9 stromal cell layer. After 4 to 7 days of incubation, vascular progenitors form a colony that can be detected by staining with EC specific markers (Figure 30-1). From the single-cell deposion analysis, it is likely that this assay can support clonogenic growth of even a single progenitor. Moreover, this assay is effective not only for the progenitors derived from ES cells but also for those present in actual embryos. In addition, a similar culture system using OP9 has been used for culturing the embryonic fragment to investigate the process of vascular development.

Many stromal cell lines could not be used for this assay. Indeed, as far as we have tested, no other cell lines were able to give better results than OP9 for this purpose. Because no exogenous molecules are required for this assay, it is likely that all essential molecules required for the clonogenic growth of EC such as VEGF are expressed by OP9. Our studies showed that many of the important angiogenic molecules are expressed by OP9. Moreover, we think that the expression level and balance of angiogenic molecules expressed in a given cell line are important factors in determining its apti-

tude for the clonogenic assay. Indeed, addition of exogenous VEGF to OP9 culture induces dispersion of the EC colony, rendering the quantification of the EC colony difficult. Nonetheless, the OP9 cell line provides a suitable condition for enumerating the number of clonogenic EC progenitors from both ES cell differentiation cultures and the actual embryos. Because the proliferative capacity of EC may decrease on maturation, it should be noted that the proportion of EC progenitors clonable by this assay would decrease during embryogenesis. However, further studies are needed.

ENDOTHELIAL CELL DIVERSIFICATION

It is established that EC diversify quickly into different subsets such as arterial, venous, or capillary EC during development of the vascular system. Recent studies have uncovered the molecular network involved in this process, in which Notch and Eph/Ephrin signaling pathways play key roles. Because the failure of EC diversification arising from defects of these signal pathways results in failure in the formation of an organized architecture of vascular network, generation of EC diversity is an essential step for vascular remodeling. Although the role of Notch and Eph/Ephrin in the vascular remodeling genes is demonstrated in the mice bearing null mutations in those genes, not much has been understood concerning the function of each molecule in EC diversification. ES cell differentiation culture, in its inherent potential to deal with differentiating EC, should be the most suitable experimental system to investigate the process of EC diversification. Indeed, we have demonstrated that some diversity is gener-

223

ated in our culture system. For instance, we are able to distinguish EC in terms of their potential to give rise to the hematopoietic cell lineage. The number of markers that can be used for defining the diversity, however, is not enough. For instance, Eph/Ephrin and Notch regulate the process of specification of venous and arterial EC, implicating the usefulness of surface expression of these molecules in monitoring the EC diversity. However, there is still a deficiency of reliable antibodies to these molecules, which can be used for surface staining. Hence, this culture system has not gained popularity in studying EC diversification, although it should have great potential.

CELLULAR BEHAVIOR OF ENDOTHELIAL CELLS

The method of preparing EC in ES cell differentiation culture should have enormous potential in the study of the cell biologic aspect of EC behavior. Although many EC cell lines such as Human Unbrical Vein Endothelial Cell (HUVEC) are available for the cell biology of EC, it has been recognized that those cell lines have lost properties that are characteristic of EC during *in vitro* propagation. Thus, before using the EC cell lines for cell biology, the extent to which the EC cell lines represent normal EC must be determined. Another way to obtain normal EC would be to purify distinct subsets of EC from developing embryos. However, although it is theoretically possible, it is not easy to sort enough cells from the developing vascular systems of embryos. Because accumulating evidence suggests that EC prepared in the ES cell cultures maintains many properties of normal EC, this is a good, presently available method for preparing normal EC for cell biology. Moreover, as discussed previously, it is possible to differentially prepare distinct EC subsets, although further attempt to produce various markers for EC diversity is yet to be required. In this sense, we are well aware that exploitation of ES cell-derived EC may be at the very initial stage. However, some pilot studies have already succeeded in indicating the usefulness of this method. In the final section of this chapter, we describe some examples of these studies.

As described in a previous section, OP9 provides a microenvironment that supports clonogenic proliferation of EC progenitors. A spontaneous balance of angiogenic factors established spontaneously in this culture allows EC progenitors to grow and form a round colony-like sheet in which cells adhere to each other by a VE-cadherin-mediated adherence junction. Thus, the colony-like EC sheets are useful in detecting various molecules that enhance the motility of EC because enhancement of EC motility induced the dispersion of each EC in the colony-like sheet, thereby disturbing their round shape. As shown in Figure 30-2, stable round EC sheets are transformed into clusters of dispersed EC by addition of exogenous VEGF. Moreover, we have recently shown that the VEGFR3 signal inhibits dispersion of the EC colony, probably negatively modulating the signal pathway of VEGFR2, which enhances EC motility.

The cell–cell junction is another factor affecting the integrity of the EC sheet. Immunostaining of the EC sheet formed on OP9 contains both adherence junction mediated by

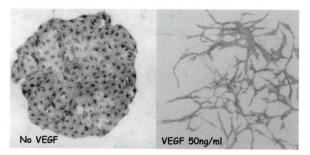

Figure 30-2. Clonogenic growth of endothelial progenitors on the OP9 stromal cell layer in the presence or absence of vascular endothelial growth factor (VEGF). VEGFR2⁺E-cadherin⁻ lateral mesoderm cells (see Figure 30-1) that were generated *in vitro* from embryonic stem (ES) cells were sorted and cultured on an OP9 stromal cell layer in the presence or absence of VEGF. In the absence of VEGF, the cells grew to form a round colony. Addition of VEGF induces dispersal of endothelial cells. Endothelial cells are identified by platelet endothelial cell adhesion molecule (PECAM) expression.

VE-cadherin and tight junction mediated by a set of claudin (our unpublished observations). A number of claudins are detected in the EC–EC junction, but Claudin 5 is the major molecule. Consistent with previous studies using EB, the addition of an antagonistic mAb to VE-cadherin disrupts the EC–EC junction, thereby disturbing formation of the round EC sheet. When this system is combined with various reagents that allow monitoring of various molecules associated with the cell–cell junction, more subtle change in the EC–EC junction can also be detected.

Finally, this system is useful for detecting signals that induce the change of cell shape. For instance, VEGF is an effective signal. As shown in Figure 30-2, the addition of exogenous VEGF in this culture enhances the motility of EC. In addition, through this treatment the shape of EC is changed from round to elongated. Currently, we do not know which signals induce shape changes of EC, but we expect that it would be valuable to use this culture system to screen the vasoactive substances.

This system can be useful to screen drugs that may inhibit or enhance those three activities of EC that are detectable in this culture. For instance, it would be possible to screen drugs that show inhibitory effect on VEGF-induced EC dispersion. Such drugs may have a potential for the maintenance of the integrity of the EC layer even in the presence of a strong VEGF signal that disturbs the EC integrity. Similarly, it would be possible to find drugs that specifically affect, for instance, the cell motility, while leaving proliferation and shape unchanged.

Concluding Remarks

In this chapter, we describe an *in vitro* system that can induce EC differentiation from ES cells and also the potential of these induced EC for cell biologic analyses of the behavior of EC. Although we have worked on this system for several years, we realize that it is still in the process of development. Nonetheless, it is important that the EC generated in this culture system behaves like the normal vascular progenitors that can be integrated in the vascular system that is induced in the *in vivo* neoangiogenesis setting. Thus, the EC supplied by this system is a rare source of normal EC. We encourage the reader to consider this system for investigating the cell biology of EC. We are ready to help introduce this method in the reader's work.

KEY WORDS

Embryonic stem (ES) cell   Cells that are derived from the inner cell mass of the blastocyst and propagated *in vitro* with their differentiation being arrested. ES cells have been established in mouse, primate, and human, but conditions for maintaining undifferentiated ES cells vary among species.

Fluorescent activated cell sorter   A technology extremely useful for purifying intermediates during ES cell differentiation. Gene expression profile is distinct among distinct cell types. As some of such genes characterizing a particular cell type are expressed on the cell surface, fluorescence dye-conjugated antibodies to those molecules can be used for distinguishing and purifying the living cells in terms of expression of a set of molecules on the cell surface.

Vascular endothelial growth factor (VEGF)   An essential molecule for proliferation and differentiation of endothelial cells. VEGF binds to both VEGF-receptor (VEGFR)1 and VEGFR2. For inducing endothelial cells in ES cell differentiation cultures, VEGFs have to be added when VEGFR2$^+$ cells appear, which is 3.5 to 4 days after induction of ES cell differentiation.

FURTHER READING

Carmeliet, P. (2000). Mechanisms of angiogenesis and ateriogenesis. *Nature Med.* **6**, 389–395.

Freshney, R. I. (2000). Culture of animal cells: A manual of basic technique, 4th edition. Willy-Liss, Inc.

Gilbert, Scott F. *Developmental biology*. 6th ed. *General understanding on embryogenesis.*

Turksen, K. Embryonic stem cells: methods and Protocols. *Methods in Molecular Biology* pp. **185**.

Cell Lineages and Stem Cells in the Embryonic Kidney

Gregory R. Dressler

Introduction

The goal of developmental biology is to understand the genetic and biochemical mechanisms that determine cell growth and differentiation and the three-dimensional patterning of a complex organism. This knowledge can reframe the pathogenesis of human diseases such as cancer within a developmental context and can also be applied to create new therapies for the regeneration of damaged tissues. Given the emphasis on stem cells throughout this book, it would seem prudent to discuss the origin of the embryonic kidney with respect to pluripotent renal stem cells and how they may differentiate into the multiplicity of cell types found in an adult kidney. Unfortunately, no such renal embryonic stem (ES) cells have been conclusively identified. Instead, the kidney develops from a region of the embryo, over a relatively long window of time, while undergoing a sequential anterior to posterior transition that reflects in part the evolutionary history of the organ. The questions that then arise are, at what stage are cells fated to become kidney cells, and when are they restricted in their developmental potential? To address these issues, it is necessary to review the early morphogenesis of the urogenital system. We can then appreciate which genes and molecular markers contribute to early renal development and what cell lineages arise from the region of the embryo devoted to creating the urogenital system.

The Anatomy of Kidney Development

A brief summary of renal patterning and the origin of the renal progenitor cells must begin with gastrulation. In vertebrates, the process of gastrulation converts a single pluripotent sheet of embryonic tissue, the epiblast or embryonic ectoderm, into the three primary germ layers, the endoderm, the mesoderm, and the ectoderm (Figure 31-1A). In mammals, gastrulation is marked by a furrow called the *primitive streak*, which extends from the posterior pole of the epiblast. Extension of the primitive streak occurs by proliferation and migration of more lateral epiblast cells to the furrow, followed by invagination through the furrow and migration back laterally under the epiblast sheet. At the most anterior end of the primitive streak is the node or organizer, called Hensen's node in the

chick, and the functional equivalent to the blastopore lip or Speeman's organizer in the amphibian embryo. The node is a signaling center that expresses a potent combination of secreted factors for establishing the body axes and left–right asymmetry. The node is positioned at the anterior pole of the primitive streak at approximately the midpoint of the epiblast. The more anterior epiblast generates much of the head and central nervous system and does not undergo gastrulation in the same manner. Once the streak reaches the node, it begins to regress back to the posterior pole. During this process of primitive streak regression, the notochord is formed along the midline of the embryo and just ventral to the neural plate. The notochord is a second critical signaling center for dorsal ventral patterning of both neural plate and paraxial mesoderm. The axial mesoderm refers to the most medial mesodermal cells, which in response to regression of the streak become segmented into somites, blocks of cells surrounded by a simple epithelium. At the stage of the first somite formation, going medial to lateral, the notochord marks the midline, the somites abut the notochord on either side, and the unsegmented mesoderm is termed intermediate near the somite and lateral plate more distally (Figure 31-1B). It is this region of intermediate mesoderm within which the kidney will form that is the primary focus of this chapter.

The earliest morphologic indication of unique derivatives arising from the intermediate mesoderm is the formation of the pronephric duct, or primary nephric duct. This single-cell thick epithelial tube runs bilaterally beginning at around the twelfth somite in birds and mammals. The nephric duct extends caudally until it reaches the cloaca. As it grows, it induces a linear array of epithelial tubules, which extend medioventrally and are thought to derive from periductal mesenchyme (Figure 31-2). The tubules are referred to as pronephric or mesonephric, depending on their position and degree of development, and represent an evolutionarily more primitive excretory system that forms transiently in mammals until it is replaced by the adult or metanephric kidney. Along the nephric duct, there is a graded evolution of renal tubule development, with the most anterior, or pronephric tubules, being very rudimentary, and the mesonephric tubules becoming well developed with glomeruli and convoluted proximal tubule-like structures. In contrast, the pronephros of the zebra fish larvae is a fully developed, functional filtration unit with a single midline glomerulus. Amphibian embryos such as *Xenopus laevis* have bilateral pronephric glomeruli and

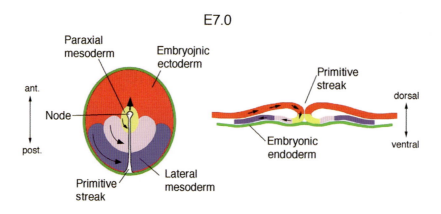

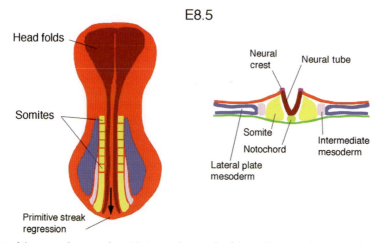

Figure 31-1. The origin of the intermediate mesoderm. (A) At gastrulation, cells of the epiblast, or embryonic ectoderm, migrate through the primitive streak. In the mouse, the single sheet of embryonic ectoderm lines a cup-shaped egg cylinder. The left panel is a planar representation of a mouse epiblast looking down into the cup from above. The streak begins at the posterior and moves toward the anterior. Fate mapping studies indicate that lateral plate mesoderm originates from the posterior epiblast, whereas axial mesoderm is derived from more anterior epiblast cells. The intermediate mesoderm most likely originates from cells between these two regions. At the mid-streak stage, the expression of lim1 can already be detected before gastrulation in cells of the posterior epiblast. The schematic on the right is a cross section through the primitive streak, showing the formation of mesoderm as cells of the epiblast migrate toward the streak, then invaginate, and reverse direction underneath the epiblast sheet. (B) At the time of primitive streak regression, the notochord is formed along the ventral midline. Paraxial mesoderm begins to form segments, or somites, in an anterior to posterior direction. The lateral plate mesoderm consists of two sheets called the somatopleure (dorsal) and the splanchnopleure (ventral). The region between the somite and the lateral plate is the intermediate mesoderm, where the first renal epithelial tube will form.

tubules that are functional until replaced by a mesonephric kidney in the tadpole. In fact, it is not altogether obvious in mammals where to draw the distinction between pronephric tubules and mesonephric tubules. Mature mesonephric tubules are characterized by a vascularized glomerulus at the proximal end of the tubule that empties into the nephric duct. The most anterior and posterior mesonephric tubules are more rudimentary, and the most posterior tubules are not connected to the duct at all.

The adult kidney, or metanephros, is formed at the caudal end of the nephric duct when an outgrowth, called the *ureteric bud* or *metanephric diverticulum*, extends into the surrounding metanephric mesenchyme. Outgrowth or budding of the epithelia requires signals emanating from the mesenchyme. Genetic and biochemical studies indicate that outgrowth of the

ureteric bud is mediated by the transmembrane tyrosine kinase RET, which is expressed in the nephric duct, and the secreted neurotrophin glial derived neurotrophic factor (GDNF), which is expressed in the metanephric mesenchyme. Once the ureteric bud has invaded the metanephric mesenchyme, inductive signals emanating from the bud initiate the conversion of the metanephric mesenchyme to epithelium (Figure 31-3). The induced, condensing mesenchymal cells aggregate around the tips of the bud and will form a primitive polarized epithelium, the renal vesicle. Through a series of cleft formations, the renal vesicle forms first a comma and then an S-shaped body, whose most distal end remains in contact with the ureteric bud epithelium and fuses to form a continuous epithelial tubule. This S-shaped tubule begins to express genes specific for glomerular podocyte cells at its most proximal

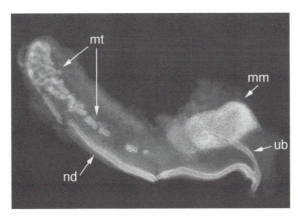

Figure 31-2. Expression of Pax2 at the time of metanephric induction. One side of the intermediate mesoderm-derived nephric chord was microdissected from an E11.5 mouse embryo and stained with anti-Pax2 antibodies. The micrograph shows Pax2 in the nephric duct (nd) and the ureteric bud (ub) branching from the posterior duct. Pax2 is in mesonephric tubules (mt), even those more posterior that are not connected to the nephric duct. At the posterior end, Pax2 is in the metanephric mesenchyme cells (mm), surrounding the ureteric bud, that has not yet begun to form epithelia.

end, markers for more distal tubules near the fusion with the ureteric bud epithelia, and proximal tubules markers in between. Endothelial cells begin to infiltrate the most proximal cleft of the s-shaped body as the vasculature of the glomerular tuft takes shape. At this stage, the glomerular epithelium consists of a visceral and parietal component, with the visceral cells becoming podocytes and the parietal cells the epithelia surrounding the urinary space. The capillary tuft consists of capillary endothelial cells and a specialized type of smooth muscle cell, termed the *mesangial cell*, whose origin remains unclear.

As, or while these renal vesicles are generating much of the epithelia of the nephron, the ureteric bud epithelia continues to undergo branching morphogenesis in response to signals derived from the mesenchyme. Branching follows a stereotypical pattern and results in new mesenchymal aggregates induced at the tips of the branches, as new nephrons are sequentially induced. This repeated branching and induction results in the formation of nephrons along the radial axis of the kidney, with the oldest nephrons being more medullary and the younger nephrons located toward the periphery. However, not all cells of the mesenchyme become induced and convert to epithelia; some cells remain mesenchymal and migrate to the interstitium. These interstitial mesenchymal cells, or stromal cells, are essential for providing signals that maintain branching morphogenesis of the ureteric bud and survival of the mesenchyme.

From a stem cell perspective, defining the population of cells that generate the kidney depends in part on which stage one considers. At the time of metanephric mesenchyme induction, there are at least two primary cell types, the mesenchyme and the ureteric bud epithelia. Although these cells are phenotypically distinguishable, they do express some common markers and share a common region of origin. As develop-

ment progresses, it was thought that most of the epithelium of the nephron was derived from the metanephric mesenchyme, whereas the branching ureteric bud epithelium generates the collecting ducts and the most distal tubules. This view has been challenged by cell lineage tracing methods *in vitro*, which indicate some plasticity at the tips of the ureteric bud epithelium such that the two populations may intermingle. Thus, at the time of induction, epithelial cells can convert to mesenchyme, just as the mesenchymal aggregates can convert to epithelia. Regardless of how the mesenchyme is induced, the cells are predetermined to make renal epithelia. Thus, their potential as renal stem cells has begun to be explored. To understand the origin of the metanephric mesenchyme, we begin with the patterning of the intermediate mesoderm.

Genes That Control Early Kidney Development

Genetic studies in the mouse have provided substantial new insights into the regulatory mechanisms underlying renal development. Although many genes can affect the growth and patterning of the kidney, of particular importance with respect to potential renal stem cells are the genes that control formation of the nephric duct epithelia and proliferation and differentiation of the metanephric mesenchyme (Table 31-1).

GENES THAT DETERMINE THE NEPHROGENIC FIELD

From a perspective of stem cells and potential therapy, the adult or metanephric kidney should be a primary focus. However, the early events controlling the specification of the renal cell lineages may be common among the pronephric and mesonephric regions. Indeed, many of the same genes expressed in the pronephric and mesonephric tubules are instrumental in early metanephric development. Furthermore, the earliest events that underlie regional specification have been studied in more amenable organisms, including fish and amphibians, in which pronephric development is less transient and of functional significance.

Although formation of the nephric duct is the earliest morphologic evidence of renal development, the expression of intermediate mesodermal specific markers precedes nephric duct formation temporally and marks the intermediate mesoderm along much of the anterior-posterior (A-P) body axis. The earliest markers specific of intermediate mesoderm are two transcription factors of the Pax family (Figure 31-4), *Pax2* and *Pax8*, which appear to function redundantly in nephric duct formation and extension. The homeobox gene *lim1* is also expressed in the intermediate mesoderm but is initially expressed in the lateral plate mesoderm before becoming more restricted. Genetics implicates all three genes in some aspects of regionalization along the intermediate mesoderm. In the mouse, *Pax2* mutants begin nephric duct formation and extension but lack mesonephric tubules and the metanephros. *Pax2/Pax8* double mutants have no evidence of nephric duct formation and do not express the *lim1* gene. Null mutants in *lim1* also lack the nephric duct and show reduced ability to

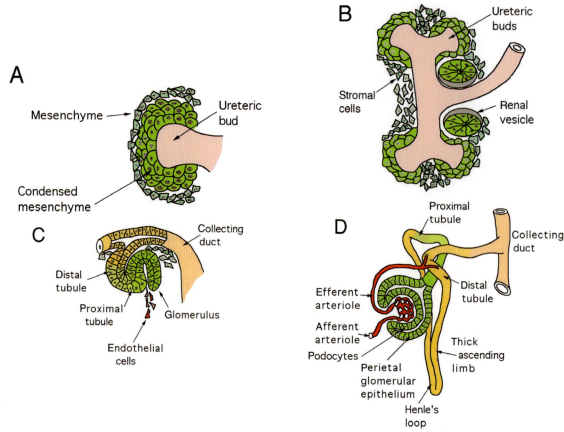

Figure 31-3. The sequential conversion of metanephric mesenchyme to renal epithelia. A schematic of the condensation and polarization of the metanephric mesenchyme at the tips of the ureteric bud epithelia is shown. (A) Epithelial precursors aggregate at the tips, whereas stromal cells remain peripheral. (B) The initial aggregates form a primitive sphere, the renal vesicle, as branching ureteric bud epithelia cells extend outward to induce a new aggregate. Stromal cells begin to migrate into the interstitium. (C) At the S-shaped body stage, the mesenchymal-derived structure is fused to the ureteric bud epithelial, which will make the collecting ducts and tubules. The proximal cleft of the S-shaped body is invaded by endothelial cells. Expression of glomerular, proximal tubule, and more distal tubule specific markers can be seen at this stage. (D) The architecture of the nephron is elaborated. Podocyte cells of the visceral glomerular epithelium contact the capillary tuft, and the glomerular basement membrane is laid down. The proximal tubules become more convoluted and grow into the medullary zone to form the descending and ascending limbs of Henle's loop. The distal tubules and collecting ducts begin to express markers for more differentiated, specialized epithelia.

differentiate into intermediate mesoderm specific derivatives. The reduced expression of *Pax2* in *lim1* mutants could explain the inability of these cells to differentiate into urogenital epithelia. In *Pax2/Pax8* double null embryos, however, the lack of lim1 expression may also be an integral part of the phenotype. Because lim1 expression precedes Pax2 and Pax8, is spread over a wider area in the pregastrulation and post-gastrulation embryo, and is Pax independent, it seems likely that maintenance and restriction of lim1 expression within the intermediate mesoderm requires activation of the *Pax2/8* genes at the 5–8 somite stage. In the chick embryo, ectopic nephric ducts can be generated within the general area of the intermediate mesoderm upon retrovirally driven Pax2b expression. These ectopic nephric ducts are not obtained with either lim1 or Pax8 alone, suggesting that Pax2 is sufficient to specify renal tissue. Strikingly, the ectopic nephric ducts paralleled the endogenous ducts and were not found in more paraxial or lateral plate mesoderm. This would suggest that

Pax2's ability to induce duct formation does require some regional competence, perhaps only in the lim1 expressing domain.

If Pax2/8 and lim1 restriction in the intermediate meso-derm are the earliest events that distinguish the nephrogenic zone from surrounding paraxial and lateral plate mesoderm, the question then remains as to how these genes are activated. In the axial mesoderm, signals derived from the ventral noto-chord pattern the somites along the dorsal-ventral axis. Similar notochord-derived signals could also pattern meso-derm along the medio-lateral axis. In the chick embryo, the notochord is dispensable for activation of the *Pax2* gene in the intermediate mesoderm. Instead, signals derived from the somites, or paraxial mesoderm, are required for activation of Pax2, although the nature of such signals is not clear.

In addition to somite-derived signals, additional signals may be required for the formation of epithelia within the pre-disposed intermediate mesoderm. Activation of Pax2/8 is

TABLE 31-1

Genes that Regulate Early Kidney Cell Lineages

Gene	Expression	Mutant Phenotype
lim1	Lateral plate Nephric duct	No nephric duct, no kidneys
Pax2	Intermediate mesoderm Nephric duct	No mesonephric tubules No metanephros
Pax2/Pax8	Intermediate mesoderm	No nephric duct, no kidneys
WT1	Intermediate mesoderm Mesenchyme	Fewer mesonephric tubules Apoptosis of mesenchyme
Eya1	Metanephric mesenchyme	No induction of mesenchyme
wnt4	Mesenchymal aggregates	No polarization of aggregates
bmp7	Ureter bud and metanephric mesenchyme	Developmental arrest postinduction Some branching, few nephrons
FoxD1	Metanephric mesenchyme Interstitial stroma	Developmental arrest, few nephrons Limited branching
pod-1	Stroma and podocytes	Poorly differentiated podocytes
pdgf(r)	S-shaped body	No vascularization of glomerular tuft

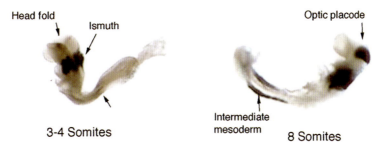

Figure 31-4. The activation of Pax2 expression in the intermediate mesoderm. One of the earliest markers for the nephrogenic region Pax2 expression is activated around the four to five somites in the cells between the axial and lateral plate mesoderm. The embryos shown carry a Pax2 promoter driving the *LacZ* gene, and expression is visualized by staining for beta-galactosidase activity. By the eight-somite stage, Pax2 marks the growing intermediate mesoderm even before the nephric duct is formed. Pax2 is also expressed in parts of the nervous system, especially the midbrain–hindbrain junction, known as the rhombencephalic ismuth, and in the optic placode and cup.

found more anterior to and precedes formation of the nephric duct. Thus, not all Pax2 positive cells will make the primary nephric duct. In the chick embryo, Bmp4 expressed in the overlying ectoderm is necessary for formation of the nephric duct, which initiates adjacent to somites 10–12, significantly more posterior to the initial Pax2 positive domain.

If Pax2/8 mark the entire nephric region, additional factors must specify the position of elements along the A-P axis in the intermediate mesoderm. Such patterning genes could determine whether a mesonephric or metanephric kidney is formed within the Pax2 positive domain. Among the known A-P patterning genes are members of the *HOX* gene family. Indeed, mice that have deleted all genes of the Hox11 paralogous group have no metanephric kidneys, although it is not clear whether this is truly a shift in A-P patterning or a lack of induction. A-P patterning of the intermediate mesoderm may also depend on the FoxC family of transcription factors. *Foxc1* and *Foxc2* have similar expression domains in the presomitic and intermediate mesoderm, as early as E8.5. As nephric duct extension progresses, Foxc1 is expressed in a

dorso-ventral gradient, with the highest levels near the neural tube and the lower levels in the BMP4 positive ventro-lateral regions. In *Foxc1* homozygous null mutants, the anterior boundary of the metanephric mesenchyme, as marked by GDNF expression, extends rostrally. This results in a broader ureteric bud forming along the A-P axis and eventual duplication of ureters. Similar defects are observed in compound heterozygotes of *Foxc1* and *Foxc2*, indicating some redundancy and gene dosage effects. Thus, *Foxc1* and *Foxc2* may set the anterior boundary of the metanephric mesenchyme, at the time of ureteric bud outgrowth, by suppressing genes at the transcriptional level.

GENES THAT FUNCTION AT THE TIME OF METANEPHRIC INDUCTION

Pax2 and Pax8 are co-expressed in the nephric duct, but the Pax2 expression domain is broader and encompasses the mesonephric tubules and the metanephric mesenchyme. Thus, Pax8 mutants have no obvious renal phenotype, Pax2 mutants have a nephric duct but no mesonephric tubules or metanephros, and Pax2/8 double mutants lack the nephric duct completely. Thus, either Pax2 or Pax8 is enough for duct formation, but Pax2 is essential for conversion of the metanephric mesenchyme into epithelia.

Pax2 mutants have no ureteric buds because they do not express GDNF in the mesenchyme and fail to maintain high levels of RET expression in the nephric duct. Despite the lack of bud, the metanephric mesenchyme is morphologically distinguishable in Pax2 mutants and expresses some specific markers, such as Six2. *In vitro* recombination experiments using Pax2 mutant mesenchyme, surgically isolated from E11 mouse embryos, and heterologous inducing tissues indicate that *Pax2* mutants are unable to respond to inductive signals. Thus, *Pax2* is necessary for specifying the region of intermediate mesoderm destined to undergo mesenchyme-to-epithelium conversion. In humans, the necessity of Pax2 function is further underscored because the loss of a single *Pax2* allele is associated with renal-coloboma syndrome, which is characterized by hypoplastic kidneys with vesicoureteral reflux.

A second essential gene for conversion of the metanephric mesenchyme to epithelia is *Eya1*, a vertebrate homolog of the *Drosophila eyes absent* gene. In humans, mutations in the *Eya1* gene are associated with branchio-oto-renal syndrome, a complex multifaceted phenotype. In mice homozygous for an *Eya1* mutation, kidney development is arrested at E11 because ureteric bud growth is inhibited and the mesenchyme remains uninduced, although Pax2 and WT1 expression appears normal. However, two other markers of the metanephric mesenchyme, Six2 and GDNF expression, are lost in the *Eya1* mutants. The loss of GDNF expression most probably underlies the failure of ureteric bud growth. However, it is not clear whether the mesenchyme is competent to respond to inductive signals. The *eyes absent* gene family is part of a conserved network that underlies cell specification in several other developing tissues. Eya proteins share a conserved domain but lack DNA binding activity. The Eya proteins can localize to the nucleus and affect target gene expression through direct interactions with the Six family of DNA binding proteins.

The Wilms tumor suppressor gene, *WT1,* is another early marker of the metanephric mesenchyme and is essential for its survival. Expression of WT1 is regulated spatially and temporally in a variety of tissues and is further complicated by the presence of at least four isoforms, generated by alternative splicing. In the developing kidney, WT1 can be found in the uninduced metanephric mesenchyme and in differentiating epithelium after induction. Initial expression levels are low in the metanephric mesenchyme, but become upregulated at the s-shaped body stage in the precursor cells of the glomerular epithelium, the podocytes. High WT1 levels persist in the adult podocytes. In the mouse, *WT1* null mutants have complete renal agenesis because the metanephric mesenchyme undergoes apoptosis and the ureteric bud fails to grow out of the nephric duct. The arrest of ureteric bud growth is most probably due to lack of signaling by the *WT1* mutant mesenchyme. As in *Pax2* mutants, the mesenchyme is unable to respond to inductive signals even if a heterologous inducer is used *in vitro*.

The Establishment of Additional Cell Lineages

At the time of ureteric bud invasion, at least two cell lineages appear to be established: the metanephric mesenchyme and the ureteric bud epithelia. As branching morphogenesis and induction of the mesenchyme progress, additional cell lineages are evident. The early E11.5 mouse metanephros contains precursors for most all cell types, including endothelial, stromal, epithelial, and mesangial cells. However, it is far from clear whether these cell types share a common precursor or whether the metanephric mesenchyme is a mixed population of precursors. The latter point may well be true for the endothelial lineage. Although transplantation studies with lineage markers indicate that the vasculature can be derived from E11.5 metanephric kidneys, the Flk1 positive endothelial precursors have been observed to be closely associated with the ureteric bud epithelium, shortly after invasion, and are probably not derived from the metanephric mesenchyme. The lineages most likely to share a common origin within the metanephric mesenchyme are the stromal and epithelial lineages. The maintenance of these two lineages is essential for renal development because the ratio of stroma to epithelia is a critical factor for the renewal of mesenchyme and the continued induction of new nephrons.

EPITHELIA VERSUS STROMA

What early events in the induced mesenchyme separate the stromal lineage from the epithelial lineage? In response to inductive signals, Pax2 positive cells aggregate at the tips of the ureteric bud. Activation of the *Wnt4* gene in these early aggregates appears critical to promote polarization. *Wnt* genes encode a family of secreted peptides that are known to function in the development of a many tissues. Mice homozygous for a *wnt-4* mutation exhibit renal agenesis resulting from growth arrest shortly after branching of the ureteric bud. Although some mesenchymal aggregation has occurred, there is no evidence of cell differentiation into a polarized epithelial vesicle. Expression of *Pax2* is maintained but reduced. Thus, *wnt-4* may be a secondary inductive signal in the mesenchyme that propagates or maintains the primary induction response in the epithelial lineage. The transcription factor FoxD1/BF-2 is expressed in uninduced mesenchyme and becomes restricted to those cells that do not undergo epithelial conversion after induction. FoxD1 expression is found along the periphery of the kidney and in the interstitial mesenchyme, or stroma. After induction, there is little overlap between FoxD1 and the Pax2 expression domain, prominent in the condensing pretubular aggregates. Clear lineage analysis is still lacking, although the expression patterns are consistent with the interpretation that mesenchyme cells may already have partitioned into a FoxD1 positive stromal precursor and a Pax2 positive epithelial precursor before or shortly after induction. Mouse mutants in *FoxD1* exhibit severe developmental defects in the kidney that point to an essential role for *FoxD1* in maintaining growth and structure.

Early ureteric bud growth and branching is unaffected, as is the formation of the first mesenchymal aggregates. However, at later stages (E13–14) these mesenchymal aggregates fail to differentiate into comma- and s-shaped bodies at a rate similar to wild-type. Branching of the ureteric bud is greatly reduced at this stage, resulting in the formation of fewer new mesenchymal aggregates. The fate of the initial aggregates is not fixed because some are able to form epithelium and most all express the appropriate early markers, such as Pax2, wnt4, and WT1. Nevertheless, it appears that the FoxD1 expressing stromal lineage is necessary to maintain growth of both ureteric bud epithelium and mesenchymal aggregates. Perhaps factors secreted from the stroma provide survival or proliferation cues for the epithelial precursors, in the absence of which the non–self-renewing population of mesenchyme is exhausted.

Some survival factors that act on the mesenchyme have already been identified. The secreted transforming growth factor beta (TGFβ) family member BMP7 and the fibroblast growth factor-2 (FGF2) in combination dramatically promote survival of uninduced metanephric mesenchyme *in vitro*. FGF2 is necessary to maintain the ability of the mesenchyme to respond to inductive signals *in vitro*. BMP7 alone inhibits apoptosis but is not sufficient to enable mesenchyme to undergo tubulogenesis at some later time. After induction, exogenously added FGF2 and BMP7 reduce the proportion of mesenchyme that undergoes tubulogenesis while increasing the population of FoxD1 positive stromal cells. At least after induction occurs, there is a delicate balance between a self-renewing population of stromal and epithelial progenitor cells, the proportion of which must be well regulated by both autocrine and paracrine factors. Whether this lineage decision has already been made in the uninduced mesenchyme remains to be determined.

The role of stroma in regulating renal development is further underscored by studies with retinoic acid receptors. Vitamin A deficiency results in severe renal defects. In organ culture, retinoic acid stimulates expression of RET to dramatically increase the number of ureteric bud branch points, thereby increasing the number of nephrons. However, it is the stromal cell population that expresses the retinoic acid receptors (RARs), specifically RARα and RARβ2. Genetic studies with *RARα* and *RARβ2* homozygous mutant mice indicate no significant renal defects when either gene is deleted. However, double homozygotes mutant for both *RARα* and *RARβ2* exhibit severe growth retardation in the kidney. These defects are due primarily to decreased expression of the RET protein in the ureteric bud epithelia and limited branching morphogenesis. Surprisingly, overexpression of RET with a HoxB7/RET transgene can completely rescue the double RAR mutants. Thus, stromal cells may provide paracrine signals for maintaining RET expression in the ureteric bud epithelia. Reduced expression of stromal cell marker FoxD1, particularly in the interstitium of RAR double mutants, supports this hypothesis.

CELLS OF THE GLOMERULAR TUFT

The unique structure of the glomerulus is intricately linked to its ability to retain large macromolecules within the circulating bloodstream while allowing for rapid diffusion of ions and small molecules into the urinary space. The glomerulus consists of four major cell types: the endothelial cells of the microvasculature, the mesangial cells, the podocyte cells of the visceral epithelium, and the parietal epithelium. The development of the glomerular architecture and the origin of the individual cell types are just beginning to be understood.

The podocyte is a highly specialized epithelial cell whose function is integral to maintaining the filtration barrier in the glomerulus. The glomerular basement membrane separates the endothelial cells of the capillary tufts from the urinary space. The outside of the glomerular basement membrane, which faces the urinary space, is covered with podocyte cells and their interdigitated foot processes. At the basement membrane, these interdigitations meet to form a highly specialized cell–cell junction, called the *slit diaphragm*. The slit diaphragm has a specific pore size to enable small molecules to cross the filtration barrier into the urinary space, while retaining larger proteins in the bloodstream. The podocytes are derived from condensing metanephric mesenchyme and can be visualized with specific markers at the s-shaped body stage. Although a number of genes are expressed in the podocytes, only a few factors are known to regulate podocyte differentiation. These include the *WT1* gene, which is required early for metanephric mesenchyme survival but whose levels increase in podocyte precursors at the s-shaped body stage. In the mouse, complete *WT1* null animals lack kidneys, but reduced gene dosage and expression of WT1 result in specific podocyte defects. Thus, the high levels of WT1 expression in podocytes appear to be required and make these precursor cells more sensitive to gene dosage. The basic helix–loop–helix protein Pod1 is expressed in epithelial precursor cells and in more mature interstitial mesenchyme. At later developmental stages, Pod1 is restricted to the podocytes. In mice homozygous for a *Pod1* null allele, podocyte development appears arrested. Normal podocytes flatten and wrap their foot processes around the glomerular basement membrane. *Pod1* mutant podocytes remain more columnar and fail to fully develop foot processes. Because Pod1 is expressed in epithelial precursors and in the interstitium, it is unclear whether these podocyte effects are due to a general developmental arrest because of the stromal environment or a cell autonomous defect within the *Pod1* mutant podocyte precursor cells.

Within the glomerular tuft, the origin of the endothelium and the mesangium is uncertain. At the S-shaped body stage, the glomerular cleft forms at the most proximal part of the S-shaped body, furthest from the ureteric bud epithelium. Vascularization of the developing kidney is first evident within this developing tuft. Kidneys excised at the time of induction and cultured *in vitro* do not exhibit signs of vascularization, leading to the presumption that endothelial cells migrate to the kidney some time after induction. However, hypoxygenation or treatment with vascular endothelial growth factor (VEGF) promotes survival or differentiation of endothelial precursors in these same cultures, suggesting that endothelial precursors are already present and require growth differentiation stimuli. *In vivo* transplantation experiments using lacZ expressing donors or hosts also demonstrate that the E11.5 kidney rudiment has the potential to generate endothelial cells, although recruitment of endothelium is also observed from exogenous tissue depending on the environment. The data are consistent with the idea that cells within the E11.5 kidney have the ability to differentiate along the endothelial lineage. Presumptive angioblasts are dispersed along the periphery of the E12 kidney mesenchyme, with some cells invading the mesenchyme along the aspect of the growing ureteric bud. At later stages, Flk-1 positive angioblasts localize to the nephrogenic zone, the developing glomerular cleft of the s-shaped bodies, and the more mature capillary loops, whereas VEGF localizes to the parietal and visceral glomerular epithelium. It appears that endothelial cells originate independent of the metanephric mesenchyme and invade the growing kidney from the periphery and along the ureteric bud.

The mesangial cells are located between the capillary loops of the glomerular tuft and have been referred to as *specialized pericytes*. The pericytes are found within the capillary basement membranes and have contractile abilities, much like a smooth muscle cell. Whether the mesangial cell is derived from the endothelial or epithelial lineage remains unclear. However, genetic and chimeric analyses in the mouse have revealed a clear role for the platelet-derived growth factor receptor (PDGFR) and its ligand platelet-derived growth factor (PDGF). In mice deficient for either PDGF or PDGFR, a complete absence of mesangial cells results in glomerular defects, including the lack of microvasculature in the tuft. PDGF is expressed in the developing endothelial cells of the glomerular tuft, whereas the receptor is found in the presumptive mesangial cell precursors. In ES cell chimeras of *Pdgfr* $^{-/-}$ and $^{+/+}$ genotypes, only the wild-type cells can contribute to the mesangial lineage. This cell autonomous effect indicates that signaling from the developing vasculature promotes proliferation and/or migration of the mesangial precursor cells. Also, expression of PDGFR and smooth muscle actin supports a model in which mesangial cells are derived from smooth muscle of the afferent and efferent arterioles during glomerular maturation.

What Constitutes a Renal Stem Cell?

The issue of renal stem cells is beginning to draw more attention as new information regarding development and lineage specification is unraveled. A prospective renal stem cell should be self-renewing and able to generate all of the cell types in the kidney. In simplest terms, all of the cells in the kidney could be generated by a single stem cell population as outlined in Figure 31-5. Despite the potential, several issues are outstanding. Even at the earliest stage of kidney development there are already two identifiable cell types. The presence of early endothelial precursors at the time of induction

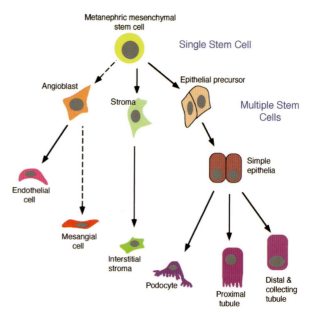

Figure 31-5. The major cell lineages of the kidney. Whether the kidney arises from a single renal stem cell or from multiple independent lineages remains to be determined. However, the basic differentiation scheme is becoming clearer. The cell lineage relationships are outlined schematically, with dotted lines reflecting ambiguity in terms of direct lineages. The metanephric mesenchyme contains angioblasts, stromal, and epithelial precursors. Whether angioblasts arise from mesenchymal cells or are a separate lineage that surround the mesenchyme is not entirely clear. Similarly, the origin of the mesangial cell is not well defined. Stromal and epithelial cells may share a common precursor, the metanephric mesenchyme, but segregate at the time of induction. The epithelial cell precursors, found in the aggregates at the ureteric bud tips, generate almost all of the epithelial cell types in the nephron.

would make a third distinct cell type. If the three earliest cell types can indeed by derived from the metanephric mesenchyme, a single stem cell may indeed exist and continue to proliferate as development progresses. At present, the data suggest that stroma and epithelia may share a common origin, whereas endothelial cells and their potential smooth muscle derivatives constitute a second lineage. However, even if there are three separate lineages already demarcated within the metanephric mesenchyme, the most relevant with respect to the repair of renal tissue is the epithelial lineage. Thus, if we consider the possibility of an epithelial stem cell the following points would be among the criteria for selection: (1) the cells would most likely be a derivative of the intermediate mesoderm, (2) the cells would express a combination of markers specific for the metanephric mesenchyme, and (3) the cells should be able to contribute to all epithelial components of the nephron, *in vitro* and *in vivo*.

Unlike ES cells, it seems improbable that cells from the intermediate mesoderm can be cultured indefinitely without additional transformation or immortalization taking place. Embryonic fibroblasts can be cultured from the mouse, but, in nearly every case, a limited number of cell divisions occur.

The problem is apparent even *in vivo* because the E11 metanephric mesenchyme is essentially quiescent and does not proliferate in the absence of induction. However, growth conditions that mimic induction might be able to allow for the mesenchyme cells to proliferate while suppressing their differentiation into epithelium. Alternatively, it may be possible to differentiate ES cells into intermediate mesodermal cells using combinations of growth and patterning factors.

At present, the complexity of the kidney still impedes progress in the area of tissue and cell-based therapies. Not only must the right cells be made but they must be able to organize into a specialized three-dimensional tubular structure capable of fulfilling all of the physiologic demands put on the nephrons. Developmental biology can provide a framework for understanding how these cells arise and what factors promote their differentiation and growth. Although we may not be able to make a kidney from scratch, it seems within the realm of possibility to provide the injured adult kidney with cells or factors to facilitate its own regeneration. Given the high incidence and severity of acute and chronic renal insufficiency, such therapies would be most welcome indeed.

ACKNOWLEDGMENTS

I thank the members of my laboratory for valuable discussion regarding this topic, particularly Pat Brophy, Yi Cai, and Sanj Patel. G.R.D. is supported by NIH grants DK54740 and DK39255 and a grant from the Polycystic Kidney Research Foundation.

KEY WORDS

Intermediate mesoderm A region of mesoderm extending anterior-posterior between the axial, or somitic, mesoderm and the lateral plate mesoderm, marked by expression of the *Pax2* and *Pax8* genes.

Kidney induction The activation of the epithelial specific program in the metanephric mesenchyme in response to signals emanating from the invading ureteric bud.

Metanephric mesenchyme The region of posterior intermediate mesoderm adjacent to the nephric ducts that is able to generate epithelial cells upon induction.

Renal stem cell A hypothetical metanephric mesenchymal cell that could potentially generate all the derivatives found in the nephron, including tubular epithelia, mesangia, stroma, and endothelia.

Renal vesicle The first polarized epithelial derivative of the induced metanephric mesenchyme that is found abutting the branching tips of the ureteric bud.

FURTHER READING

Al-Awqati, Q., and Oliver, J. A. (2002). Stem cells in the kidney. *Kidney Int.* **61**, 387–395.

Batourina, E., Gim, S., Bello, N., Shy, M., Clagett-Dame, M., Srinivas, S., Costantini, F., and Mendelsohn, C. (2001). Vitamin A controls epithelial/mesenchymal interactions through Ret expression. *Nat. Genet.* **27**, 74–78.

Bouchard, M., Souabni, A., Mandler, M., Neubuser, A., and Busslinger, M. (2002). Nephric lineage specification by Pax2 and Pax8. *Genes Dev.* **16**, 2958–2970.

Dressler, G. R. (2002). Development of the excretory system. *In Mouse development: patterning, morphogenesis, and organogenesis* (Ja.T.P.T. Rossant, ed.), pp. 395–420. Academic Press, San Diego, CA.

Dudley, A. T., Godin, R. E., and Robertson, E. J. (1999). Interaction between FGF and BMP signaling pathways regulates development of metanephric mesenchyme. *Genes Dev.* **13**, 1601–1613.

Kuure, S., Vuolteenaho, R., and Vainio, S. (2000). Kidney morphogenesis: cellular and molecular regulation. *Mech. Dev.* **92**, 31–45.

Oliver, J. A., Barasch, J., Yang, J., Herzlinger, D., and Al-Awqati, Q. (2002). Metanephric mesenchyme contains embryonic renal stem cells. *Am. J. Physiol. Renal Physiol.* **283**, F799–809.

Qiao, J., Cohen, D., and Herzlinger, D. (1995). The metanephric blastema differentiates into collecting system and nephron epithelia *in vitro*. *Development* **121**, 3207–3214.

Stark, K., Vainio, S., Vassileva, G., and McMahon, A. P. (1994). Epithelial transformation of metanephric mesenchyme in the developing kidney regulated by Wnt-4. *Nature* **372**, 679–683.

Torres, M., Gomez-Pardo, E., Dressler, G. R., and Gruss, P. (1995). Pax-2 controls multiple steps of urogenital development. *Development* **121**, 4057–4065.

Wellik, D. M., Hawkes, P. J., and Capecchi, M. R. (2002). Hox11 paralogous genes are essential for metanephric kidney induction. *Genes Dev.* **16**, 1423–1432.

Adult Liver Stem Cells

Craig Dorrell and Markus Grompe

Organization and Functions of Adult Mammalian Liver

The liver consists of several separate lobes and represents about 2% of human and 5% of murine body weight. It is the only organ with two separate afferent blood supplies. The portal vein brings in venous blood rich in nutrients and hormones from the splanchnic bed (intestines and pancreas), and the hepatic artery provides oxygenated blood. Venous drainage is into the vena cava. The bile secreted by hepatocytes is collected in a branched collecting system, the biliary tree, which drains into the duodenum. The hepatic artery, portal vein, and common bile duct enter the liver in the same location, the porta hepatis.

Knowledge of the cellular organization of the liver is essential for understanding hepatic stem cell biology. The main cell types resident in the liver are hepatocytes, bile duct epithelium, stellate cells (formerly called Ito cells), Kupffer cells, vascular endothelium, fibroblasts, and leukocytes. The hepatic lobule (illustrated in Figure 32-1) is the fundamental functional unit of the liver. The portal triad, consisting of a small portal vein, hepatic artery branch, and bile duct, is located on the perimeter. Arterial and portal venous blood enter here, mix, and then flow past the hepatocytes toward the central vein in the middle of the lobule. Liver sinusoids are the vasculature connecting the portal triad vessels and the central vein. In contrast to other capillary beds, sinusoidal vessels have a fenestrated endothelium, thus permitting direct contact between blood and the hepatocyte cell surface. In two-dimensional images, rows of hepatocytes oriented from portal to central form a hepatic plate. A channel formed by adjacent hepatocytes forms a bile canaliculus, which serves to drain secreted bile toward the bile duct in the portal triad. Bile secreted by the hepatocytes is collected in bile ducts, lined by duct epithelial cells. The canal of Hering represents the connection between the bile canaliculi and the bile ducts, at the interface between the lobule and the portal triad. Stellate cells represent about 5 to 10% of the total number of hepatic cells. In addition to storing vitamin A, they are essential for the synthesis of extracellular matrix proteins and produce many hepatic growth factors that play an essential role in the biology of liver regeneration. Kupffer cells, which are liver-resident tissue macrophages, also comprise about 5 % of the total liver

cell count. These cells are hematopoietic in origin (bone marrow derived) but are capable of replicating within the liver itself.

The liver is responsible for the intermediary metabolism of amino acids, lipids, and carbohydrates, the detoxification of xenobiotics, and the synthesis of serum proteins. In addition, the liver produces bile that is important for intestinal absorption of nutrients, as well as the elimination of cholesterol and copper. All of these functions are primarily executed by hepatocytes. The biochemical properties and pattern of gene expression are not uniform among all hepatocytes; "metabolic zonation" describes the different properties of periportal (adjacent to the portal triad) and pericentral (adjacent to the central vein) hepatocytes. For example, periportal hepatocytes express urea cycle enzymes and convert ammonia to urea, whereas pericentral hepatocytes express glutamine synthase and utilize ammonia to generate glutamine.

Liver Stem Cells

The liver is known to have a very high capacity for regeneration. In fact, mammals (including humans) can survive surgical removal of up to 75% of the total liver mass. The original number of cells is restored within 1 week and the original tissue mass within 2 to 3 weeks. Liver size is also controlled by prevention of organ overgrowth. Hepatic overgrowth can be induced by a variety of compounds such as hepatocyte growth factor (HGF) or peroxisome proliferators, but the liver size returns to normal very rapidly after removal of the growth stimulus. The role of liver stem cells in regeneration has been controversial, but it is now accepted that these cells are important for the repair of specific types of liver damage. In general, however, the cell types and mechanisms responsible for liver repair are determined by the type of liver injury. In addition, tissue replacement by endogenous cells (i.e., regeneration) must be distinguished from reconstitution by transplanted donor cells (i.e., repopulation). Thus, the definition of liver stem cells might include (A) cells responsible for normal tissue turnover, (B) cells that give rise to regeneration after partial hepatectomy, (C) cells responsible for progenitor-dependent regeneration, (D) transplantable liver repopulating cells, and (E) cells that produce in hepatocyte and bile duct epithelial phenotypes *in vitro*. Liver stem cells will be discussed according to each of these definitions.

CELLS RESPONSIBLE FOR NORMAL LIVER TISSUE TURNOVER

The mechanism by which the adult liver maintains its structure and function during normal cell turnover has been studied

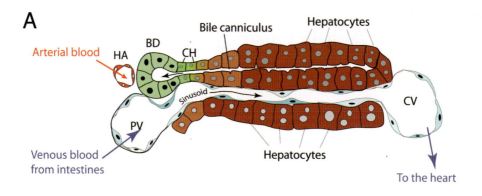

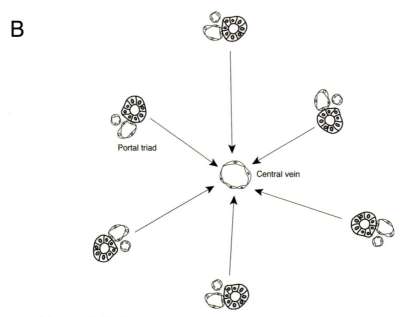

Figure 32-1. Structure of the hepatic lobule.
(A) The portal triad consists of bile ducts (BD), hepatic artery (HA), and portal vein (PV). Mixed blood from the hepatic artery and portal vein flows past hepatocytes through the sinusoids, covered with fenestrated endothelial cells to the central vein (CV). Bile produced by the hepatocytes is collected in the bile canniculus and flows toward the bile duct. The Canal of Hering (CH) is the junction between the hepatic plate and the bile ducts. This is the region where oval cell precursors may reside. (B) Each hepatic lobule consists of one central vein and six surrounding portal triads.

for many years. One of the main models, the *streaming liver model*, posits that normal liver turnover is similar to regeneration in the intestine, with young hepatocytes being born in the portal zone and migrating toward the central vein. Thus, different patterns of gene expression periportal and pericentral hepatocytes would reflect the maturation/differentiation process during this migration. However, recent work has provided strong evidence against the streaming liver hypothesis. First, the gene expression pattern in hepatocytes is now known to be a function of the direction of blood flow; when surgically reversed so that portal blood enters the lobule through the central vein and exits via the portal vein, the pattern also reverses. Therefore, lobular zonation is best explained by metabolite-induced gene regulation rather than lineage progression. Second, cell tracking via retroviral marking studies or X-chromosome inactivation patterns has shown no evidence of hepatocyte migration during normal turnover. A simpler model in which local proliferation of mature cells accounts for normal tissue turnover is now widely accepted. To date, no requirement for progenitor cells in normal liver homeostasis has been shown.

CELLS THAT GIVE RISE TO REGENERATION AFTER PARTIAL HEPATECTOMY

The process of liver regeneration after partial hepatectomy has been well studied in different experimental contexts. During this procedure, specific lobes are removed intact without damage to the lobes left behind. Within a week, the residual lobes grow to compensate for the mass of the resected lobes, although the removed lobes do not regrow. As with normal liver turnover, there is no evidence for the involvement or requirement for stem cells in this process. Classic thymidine-labeling studies have shown that virtually all hepatocytes in the remaining liver divide once or twice to restore the original cell number within 3 to 4 days. The earliest labeled hepatocytes are seen 24 hours after partial hepatectomy, and the peak of thymidine incorporation occurs at 24–48 hours depending on the species. Interestingly, there is zonal variation depending on how much tissue is removed. When only 15% of the liver is surgically removed, periportal (zone 1) hepatocytes divide preferentially, whereas cell division is seen equally in all three zones after 75% partial hepatectomy. Following the hepatocytes, other hepatic cell types also undergo a wave of mitosis, restoring the original number of all liver cells within one week.

Some of the stimulatory and inhibitory factors that control the initiation of regeneration after partial hepatectomy have been identified. Examples include hepatocyte growth factor (HGF), interleukin 6 (IL-6), tumor necrosis factor-alpha (TNF-α), transforming growth factor-alpha (TGF-α), and epidermal growth factor (EGF). Nonpeptide hormones such as triiodothyronine and norepinephrine also have significant roles in the regenerative response after liver injury. It is currently unknown whether any of these factors are also important for progenitor-dependent liver regeneration or engraftment and expansion of liver stem cells (see below).

Less knowledge exists about the mechanisms by which hepatocyte cell division and liver regeneration are halted after the appropriate liver mass has been restored. Some evidence suggests that transforming growth factor-beta 1 (TGF-β1) may be important in termination of liver regeneration. However, the exogenous signals (endocrine, paracrine, or autocrine) required for sensing the total liver cell mass and negatively regulating its size remain unknown.

CELLS RESPONSIBLE FOR PROGENITOR-DEPENDENT REGENERATION

After certain types of liver insult, small cells with oval-shaped nuclei and a high nuclear/cytoplasmic size ratio emerge in the portal zone, proliferate extensively, and migrate into the lobule. These cells, which usually become differentiated hepatocytes, are morphologically defined as oval cells. Oval cells appear to be the offspring of a more primitive cell in the Canal of Hering (Figure 32-1). Oval cell proliferation and differentiation therefore demonstrate progenitor-dependent liver regeneration, and the cell that gives rise to oval cell progeny can be considered a *facultative liver stem cell*. In the rat, chronic liver injury by chemicals such as DL-ethionine create an oval cell-stimulating environment (see Table 32-1). Because many of the compounds that induce oval cell proliferation are DNA-damaging agents and/or known carcinogens, oval cells have been regarded as precancerous. A common feature — probably a prerequisite — of progenitor-dependent liver regeneration is that existing hepatocytes cannot proliferate. Thus, progenitor-dependent regeneration is a backup repair mechanism. Oval cells express markers of both bile duct epithelium (CK19) and hepatocytes (albumin). In addi-

TABLE 32-1
Induction of Progenitor-Dependent Liver Regeneration

Chemical/Manipulation	Species
Dipin	Mouse
3,5-diethoxycarbonyl-1,4-dihydrocollidine (DDC)	Mouse
Phenobarbital + cocaine + p.H.	Mouse
Choline-deficient diet + DL-ethionine	Mouse
2-Acetylaminofluorene (AAF)	Rat
Diethylnitrosamine (DEN)	Rat
Solt-Farber Model: DEN + AAF + p.H.	Rat
Modified Solt-Farber Model: AAF + p.H.	Rat
Choline-deficient diet + DL-ethionine	Rat
D-galactosamine + p.H.	Rat
Lasiocarpine + p.H.	Rat
Retrorsine + p.H.	Rat

p.H. = partial hepatectomy.

tion, in the rat they express high levels of α-fetoprotein and thus resemble fetal hepatoblasts (see below). Oval cells are bipotential *in vitro* and retain the ability to differentiate into both the bile duct epithelial and hepatocyte lineages. Because of their similarity to hematopoietic blasts and their bipotential differentiation capacity, oval cells are regarded as a progenitor population. Therefore, oval cell precursors located in the Canal of Hering are likely candidates to be liver repopulating stem cells.

Until recently, it has been difficult to induce oval cell proliferation in the mouse and take advantage of the powerful genetics in this organism. Now, however, several protocols have been developed which result in progenitor-dependent hepatocyte regeneration in the mouse. One particularly useful regimen utilizes the chemical 3,5-diethoxycarbonyl-1,4-dihydrocollidine (DDC). Although the identification and classification of murine oval cells have been hampered by a paucity of useful cell markers, antibodies such as A6 have proven useful. An example is the recent discovery using transgenic mice that TGF-β1 inhibits A6$^+$ oval cell proliferation.

Table 32-1 shows a list of conditions that induce oval cell proliferation in the rat and mouse. However, oval cell proliferation has also been described in a variety of human liver diseases. Oval cells are found in disorders associated with chronic liver injury and are frequently identified at the edges of nodules in liver cirrhosis. The cell-surface marker OV6, which is expressed on both human and rat oval cells, has aided their identification. Rat oval cells induced by a classic carcinogen regimen express some genes associated with primitive hematopoietic cells, such as stem cell factor (SCF) and its receptor, the c-kit tyrosine kinase. SCF and c-kit are both expressed during the early stages of oval cell proliferation after partial hepatectomy in the rat 2-acetylaminofluorene(AAF)/partial hepatectomy model, but neither simple partial hepatectomy nor AAF administration alone has this effect. Thy-1, another marker differentially expressed in hematopoietic populations, is also expressed on rat oval cells.

The expression of hematopoietic stem cell (HSC) markers in oval cells is not unique to the rat. Human oval cells isolated from patients with chronic biliary diseases were found to express CD34 as well as the bile duct marker cytokeratin 19 (CK-19). In addition, cells that are c-kit+ but negative for hematopoietic markers have also been identified in human pediatric liver disease. In the mouse, the HSC marker Sca-1 has been observed on oval cells. Therefore, multiple independent studies support the concept that hepatic oval cells, but not regenerating hepatocytes, can express genes also found in HSC. This phenotypic connection inspired the hypothesis that oval cell precursors can be bone marrow-derived (see below).

Oval cells are defined primarily by their morphologic appearance. However, there can be variation in the marker genes expressed at different times following oval cell induction. In addition, different induction regimens can also produce phenotypically different cells. Retrorsine blocks the division of mature hepatocytes but does not result in the emergence of classic oval cells that are α-fetoprotein and OV6 positive. Instead, foci of small hepatocyte-like cells emerge and eventually result in organ reconstitution. Like oval cells, however, these express both hepatocyte and bile duct markers. Thus, oval cells may exist as a heterogeneous population.

TRANSPLANTABLE LIVER REPOPULATING CELLS

The most rigorous assessment of stem cell function is to test the ability of cells to repopulate the appropriate organ and restore its function. The HSC, for example, was defined by its ability to reconstitute all blood lineages in lethally irradiated hosts. In the 1990s, similar repopulation assays were developed for the liver in various animal models. During this liver repopulation, a small number of transplanted donor cells become engrafted in the liver and then proliferated to replace >50% of the liver mass. Thus, it is now possible to perform experiments analogous to those in the hematopoietic system, including cell isolation, competitive functional repopulation, serial transplantation, and retroviral marking. It should be emphasized that liver repopulation is usually limited to replacement of hepatocytes; repopulation of the biliary system is usually inefficient.

The widely used animal models for liver repopulation studies are summarized in Table 32-2. In all cases, liver repopulation by transplanted cells is based on a powerful selective advantage for the transplanted cells over the preexisting host hepatocytes. In many models, this selection is achieved by genetic differences (transgene/knockout), but host cell DNA damage has also been used successfully, particularly in the rat.

Hepatocytes as Liver Repopulating Cells

In the hematopoietic system, repopulation experiments with purified fractions of total bone marrow were used to identify subpopulations with high reconstitution activity. Similar experiments performed with liver cells have underlined a key difference between the mature cells of these two organs. Unlike the hematopoietic system, where differentiated cells

TABLE 32-2
Animal Models for Liver Repopulation

Model	Selective Pressure	Species
Albumin-urokinase transgenic	Urokinase mediated hepatocyte injury	Mouse
Fah knockout	Accumulation of toxic tyrosine metabolite	Mouse
Albumin-HSVTK transgenic	HSVTK mediated conversion of gancyclovir to toxin	Mouse
Mdr3 knockout	Bile acid accumulation	Mouse
Bcl2-transgenic donor cells	Fas ligand (Jo2) induces apoptosis in hepatocytes not expressing Bcl2	Mouse
Retrorsine conditioning	Host hepatocytes are inactivated by retrorsine (DNA damage)	Rat/Mouse
Conditioning by radiation	Host hepatocytes are inactivated by X rays (DNA damage)	Rat

have limited proliferative potential, differentiated hepatocytes are potent repopulating cells.

In the *Fah* mutant mouse model, size fractionation, retroviral marking, and competitive repopulation have shown that large, binucleated hepatocytes are the primary agents of liver repopulation. Furthermore, hepatocytes with 2n, 4n, and 8n DNA content all have repopulation capacity. These experiments suggest that the fully differentiated hepatocytes that comprise most of the organ are highly efficient at liver repopulation and have a stem cell-like capacity for proliferation.

Liver Repopulation by Non-hepatocytes

Although hepatocytes are potent, serially transplantable liver repopulating cells, other cell types are also capable of repopulating the liver. As with endogenous liver regeneration, multiple cell types can play a role. The following will describe liver repopulation by several nonhepatocyte cell types: fetal hepatoblasts, oval cells, pancreatic liver progenitors, and hematopoietic stem cells.

Liver Repopulation with Fetal Hepatoblasts. During mammalian development, hepatoblasts appear in the fetal liver bud. These cells, which express α-fetoprotein as well as hepatocyte (albumin) and biliary (CK19) markers, have been described as fetal liver stem cells. Retrorsine-stimulated progenitors in fetal rat liver exist as at least three distinct subpopulations of hepatoblasts (bipotential, unipotential hepatocytic, and unipotential ductal) at embryonic days (ED) 12–14. After transplantation, the bipotential cells proliferate in retrorsine-treated recipients, but unipotent cells grow even in untreated rats. These fetal cells proliferate more readily than their adult counterparts, indicating an enhanced proliferative and differentiative potential.

Liver Repopulation by Oval Cells. Oval cells share the bipotential differentiation capacity of fetal hepatoblasts and may therefore be useful repopulating agents. Upon transplantation of hepatic- or pancreatic-derived rat oval cells, modest proliferation and differentiation into mature hepatocytes is observed even under nonselective conditions. Similar experiments performed using DDC-treated liver and *Fah* mutant recipients have shown that oval cells have extensive liver repopulation capacity and can rescue the metabolic defect in these animals. Thus, oval cells may be attractive cell therapeutics.

Liver Repopulation by Pancreatic Progenitors. During embryogenesis, the main pancreatic cell types develop from a common endodermal precursor located in the ventral foregut, including ducts, ductules, acinar cells, and the endocrine α, β, and δ cells. The main epithelial cells of the liver, hepatocytes, and bile duct epithelium (BDE) are also thought to arise from the same region of the foregut endoderm. This tight relationship between liver and pancreas in embryonic development has raised the possibility that a common hepato-pancreatic precursor (*pancreatic liver stem cell*) may persist in adult life in both the liver and pancreas. The earliest description of pancreatic-derived hepatocytes was in hamsters treated with a pancreatic carcinogen, N-nitroso-bis(2-oxopropyl)amine (BOP) in 1981. Similarly, rare clusters of hepatocyte-like cells were found in rats treated with PPARα agonist Wy-14643. The best known example is the emergence of hepatocytes in copper-depleted rats after re-feeding of copper. In this system, weanling rats are fed a copper-free diet for 8 weeks, causing complete acinar atrophy, and then are re-fed copper. Within weeks, cells with hepatocellular characteristics emerge from the remaining pancreatic ducts. This notion is also supported by the expression of hepatocellular markers in human pancreatic cancers. More recently, a specific cytokine has been identified as a candidate to drive this process. Transgenic mice carrying the keratinocyte growth factor (KGF) gene driven by an insulin promoter consistently develop pancreatic hepatocytes. Thus, the existence of pancreatic-resident liver precursors has been shown in several different mammalian species and under multiple experimental conditions.

There is also evidence for the opposite relationship between the two organs: pancreatic precursors in the liver. In some liver tumors, particularly cholangiocarcinomas, expression of pancreatic markers such as amylase or lipase is observed. Similarly, cultured rat oval cell lines derived from liver can be differentiated *in vitro* into cells with a β-cell-like capacity for regulated insulin secretion. Such cells are also able to rescue diabetic rats in transplantation experiments. Others have demonstrated the emergence of insulin-secreting cells *in vivo* in the liver of animals treated with vectors expressing developmental pancreatic transcription factors.

Transplantation experiments have been conducted to test the pancreatic liver stem cell hypothesis. As mentioned above, pancreatic oval cells induced by copper depletion in the rat can yield morphologically normal hepatocytes *in vivo*. In addition, *Fah* knockout mice have been transplanted with wild-type pancreatic cells. Although only ~10% of transplanted *Fah* mutant mice survived long term and had extensive liver repopulation, more than half of the recipients had donor-derived (Fah⁺) hepatocyte nodules. These results likely reflect the rarity of these progenitors. Based on the number of

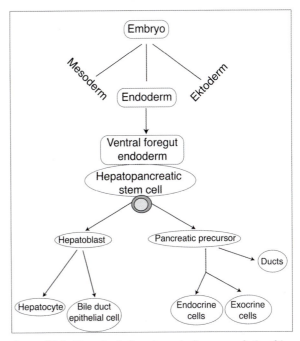

Figure 32-2. Hypothetical embryonic lineage relationships between liver and pancreas.
The differentiation sequence from the totipotent endoderm to the liver and pancreatic lineages is depicted. The ventral foregut endoderm cell which gives rise to both pancreas and liver during development may persist as a hepatopancreatic stem cell in adult life.

hepatocytes formed, it was calculated that ~1/5000 cells in the donor population are progenitors.

A model for the lineage relationships between hepatic and pancreatic cells is illustrated in Figure 32-2. This model predicts the existence of an adult hepato-pancreatic stem cell with capabilities similar to the embryonic ventral foregut endodermal precursor.

Liver Repopulation with Bone Marrow-derived Progenitors. The bone marrow of adult mammals contains cells with a variety of differentiation capacities, including the HSC and many distinct blood progenitor types. However, marrow also contains mesenchymal stem cells (MSC) capable of differentiating into chondrocytes, osteoblasts, and other connective tissue cell types. Epithelial precursors may also be present. Bone marrow and whole liver transplantation between genetically distinct animals have been used to demonstrate that a portion of rat liver oval cells are derived from the marrow donor, and murine hepatocytes have also been shown to express donor markers after blood or marrow transplantation. Recently, several reports have indicated that donor-derived epithelial cells are also present in human patients after a gender-mismatched bone marrow transplantation. In the absence of functional selection, however, these cells were identified by morphology and the expression of hepatocyte markers.

HSC as hepatocyte precursors. In order to define the marrow cell type responsible for the donor-derived hepatocytes in *Fah* knockout recipients, isolated cell populations were transplanted. Purified HSC with a KTLS phenotype (c-kit^{high}Thy^{lo}Lin^{neg}Sca-1$^+$) at limiting dilution were found to be sufficient for this activity. This study also demonstrated extensive liver repopulation and a complete correction of tyrosinemia, demonstrating conclusively that bone marrow-derived hepatocytes function normally in mice. The metabolic correction involved hepatocyte-specific parameters such as plasma amino acid levels, bilirubin conjugation and excretion, and serum transaminase levels.

A different approach to stem cell purification was used to show that a single murine HSC can not only reconstitute the hematopoietic system of recipient mice but also contribute to multiple other tissues, including liver epithelium, at a low frequency. It has also been shown that human HSC, particularly those isolated from umbilical cord blood, can give rise to hepatocyte-like cells in a xenotransplantation setting. Several groups have found that when NOD/SCID mice or pre-immune sheep are engrafted with human CD34$^+$ cells, human cells expressing albumin can be found in the recipient liver. It has been suggested that only the subpopulation of HSC that express the receptor for complement molecule C1q (C1qR(p)) is capable of giving rise to hepatocytes. However, interpretation of these xenotransplantation experiments is hampered by the lack of a functional assay.

MAPC as hepatocyte progenitors. Despite the evidence that KLTS HSC are the primary cell type responsible for the emergence of bone marrow-derived liver epithelial cells *in vivo*, they are not the only bone marrow-derived cells that can differentiate toward the hepatocytic lineage. Multipotent adult progenitor cells are a unique population of adult stem cells that can be isolated from the marrow of multiple mammalian species including human, rat, and mouse. MAPCs are telomerase positive and grow stably in culture for many passages if kept at low density. These cells have properties similar to embryonic stem cells in that they can be differentiated toward several different lineages — including the hepatocytic lineage — under the appropriate *in vitro* conditions. Over a two-week period, the majority of MAPCs can be driven to acquire multiple hepatocyte functions including urea synthesis, albumin secretion, and Phenobarbital-inducible cytochrome p450 induction.

When MAPCs are injected into blastocysts, the resulting embryos show chimerism in many tissues including liver. Furthermore, transplantation of MAPCs into adult mice yield hepatocyte-like donor cells in the liver. However, their ability to functionally correct liver disease has not yet been tested.

Physiologic significance of bone marrow-derived hepatocytes. Despite the unambiguous finding that fully functional bone marrow-derived hepatocytes exist, there has been considerable controversy regarding their functional importance in liver injury. In one model, illustrated in Figure 33-3A, replenishment of liver cells from the bone marrow is an important injury response pathway, particularly in progenitor-dependent regeneration (i.e., the oval cell response). Thus, hepatocytes would be generated either by direct differentiation of HSC or indirectly via an oval cell intermediate. The

opposing model, illustrated in Figure 34-3B, suggests that oval cell precursors are strictly tissue-resident stem cells and that the bone marrow contributes very few liver epithelial cells, even during injury. To determine whether liver damage enhances the transition of HSC to hepatocytes, the degree of cell replacement after BMT in mice has been measured not only in healthy animals, but also in the context of pre-existing hepatocyte injury or oval cell regeneration. The frequency of bone marrow-derived hepatocytes was not higher in the acute liver injury seen *Fah* knockout mice than in healthy control livers. In addition, no contribution from bone marrow precursors to the DDC-induced oval cell reaction was detected. Together these results strongly suggest that HSCs do not serve as epithelial cell precursors in all forms of hepatic injury. It remains to be determined whether HSCs play a significant role in any clinically relevant hepatic injury models to be examined in the future.

Mechanism for the formation of bone marrow-derived hepatocytes. Three basic mechanisms for the repopulation of liver by bone marrow-derived cells can be considered. First, bone marrow could theoretically harbor a specialized endodermal precursor capable of producing hepatocytes and other epithelial cells. This would be an endodermal counterpart to the mesenchymal stem cell. Second, hepatocytes and blood cells could be derived from a common multipotential stem cell by hierarchical differentiation (Figure 32-3A, trans-differentiation model). Third, bone marrow-derived hepatocytes might be derived not from differentiation at all, but rather by cell fusion (Figure 32-3B, fusion model). This possibility was raised by the observation that hematopoietic cells can spontaneously fuse with embryonic stem cell *in vitro* and produce multiple tissues in chimeric mouse embryos.

In order to address these hypotheses, transplantation experiments were designed in which genetic markers for both donor marrow and host hepatocytes were used in the *Fah* knockout model. The results from this study were clear: the vast majority of bone marrow-derived hepatocytes contained genetic information from both the donor and host, indicating cell fusion. Cytogenetic analysis of female to male gender-mismatched transplanted indicated a high frequency of tetraploid XXXY and hexaploid XXXXYY karyotypes, as predicted for fusion. These observations have been confirmed in other model systems and tissue types. Furthermore, the fusion donor for hepatocytes and muscle cells has been identified as myeloid. However, other experimenters have reported that hepatocytes can derive from bone marrow progenitors by a fusion-independent mechanism (i.e., true differentiation). Recent studies using xenotransplantation of human cord blood cells into NOD/SCID mice or pre-immune sheep, and transplantation of transgenic murine bone marrow have suggested that the observed cells producing human albumin did not derive by fusion. These models do not require functional engraftment and usually feature very low frequencies of bone marrow-derived hepatocyte-like cells, complicating a comparison of the "pro-fusion" and "anti-fusion" results. Nevertheless, fusion-independent mechanisms may prove to be important.

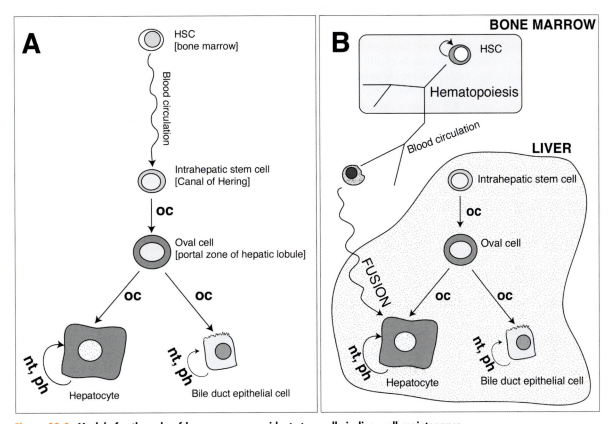

Figure 32-3. Models for the role of bone marrow resident stem cells in liver cell maintenance.
The various categories of liver regeneration (see text) are indicated: nt = normal tissue turnover, ph = regeneration after partial hepatectomy, oc = progenitor dependent regeneration (oval cell response).

(A) **Transdifferentiation model.** The anatomic location of the cells is given in square brackets. Hematopoietic stem cells (HSC) give rise to hepatocytes by trafficking to the liver via the blood circulation to then engraft and differentiate into epithelial cell precursors (intrahepatic stem cells). These cells can become activated during progenitor-dependent liver regeneration (oval cell response) and give rise to oval cells. Oval cells differentiate into hepatocytes or bile ducts. (B) **Fusion model.** There is no direct lineage relationship between HSC and liver epithelial cells. Cell fusion between circulating progeny of the HSC and hepatocytes are responsible for the emergence of bone marrow-derived liver epithelial cells. Progenitor-dependent liver regeneration (oc) occurs by activation of intrahepatic liver stem cells, which are not derived from the bone marrow.

CELLS THAT PRODUCE IN HEPATOCYTE AND BILE DUCT EPITHELIAL PHENOTYPES IN VITRO

Several *in vitro* models for hepatic stem cell growth and differentiation have been developed. Two general approaches to the *in vitro* study of liver stem cells can be distinguished. First, cell sorting can be applied to isolate putative liver stem cells on the basis of cell surface marker expression. These cells can then be cultured, and their growth and differentiation potential determined *in vitro*. Second, immortal cell lines can be derived from liver tissue by extensive *in vitro* manipulation and growth. To date, very little work has been done on the prospective isolation of hepatic progenitors by cell sorting. In contrast, putative liver progenitor cell lines from several mammalian species have been isolated and propagated in tissue culture. These systems aid the experimental understanding of factors controlling hepatocyte and duct cell differentiation, and may permit the generation of large numbers of hepatocytes *in vitro* for therapeutic transplantation.

In addition to the creation of hepatocyte progenitor cell lines, several investigators have devised strategies to conditionally immortalize differentiated hepatocytes. To date, however, the only documented therapeutic effects in animal models have been achieved with these cell lines or primary hepatocytes.

Prospective Isolation of Hepatocyte Progenitors by Cell Sorting

Recently, fetal and neonatal mouse liver cells have been fractionated by flow sorting and analyzed for their proliferative as well as differentiation capacity. This has revealed a multipotent population of cells in ED 13.5 mouse liver which are c-kit⁻, CD45⁻, c-met⁺, Ter119⁻, CD49f⁺/low. *In vitro*, they have extensive proliferative capacity and differentiate into either hepatocytic or biliary lineages. Interestingly, the same cells can also express multiple pancreatic markers when cultured long term or transplanted into a pancreatic environment *in vivo*. Further refinement and application of this approach should lead to a more precise definition of these progenitors.

Rat Cell Lines. The WB-344 cell line was clonogenically derived from nonparenchymal rat liver cells (probably from the Canal of Hering) and is probably the most intensely studied liver stem cell line. Although WB-344 cells can be cultured indefinitely *in vitro*, they retain the ability to differentiate into morphologically normal hepatocytes after transplantation without forming tumors. To date, WB-344 cells are the only cells to fulfill this stringent criterion and may represent a true liver stem cell line. Nonetheless, little is known about the molecular mechanisms regulating the stem cell to hepatocyte transition, and functional liver repopulation with this cell line has not yet been reported.

Multiple laboratories have isolated oval cell lines from carcinogen-treated rats. Consistent with the proposed role of oval cells in the formation of hepatocarcinoma, these cell lines form tumors upon transplantation into immunodeficient recipients.

Mouse Cell Lines. Transgenic mice that overexpress the HGF receptor c-met permit routine establishment of liver cell lines. Two morphologically distinct types of cells emerge from such cultures. Both grow extensively in culture under certain media conditions but can be induced to differentiate with the appropriate signals. Clonal cell lines with the "epithelial" morphology resemble hepatocytes and give rise to only hepatocyte-like offspring. In contrast, "palmate" clones can give rise to two distinct lineages depending on the differentiation conditions used. *In vivo* transplantation and differentiation of these cells have not yet been reported.

HBC-3 cells are a novel bipotential cell line derived from ED 9.5 mouse liver. This clonal cell line can be induced to differentiate into hepatocytes by DMSO treatment.

Another bipotential line, BMEL, was derived from a later developmental time point (ED 14). These cells have been shown to contribute both hepatocyte and bile duct engraftment upon transplantation to albumin urokinase plasminogen activator/SCID transgenic mice.

Pancreas-Derived Cell Lines. Several laboratories have produced cell lines from the adult pancreas and used them in transdifferentiation studies. Permanent cell lines with epithelial characteristics have been established from both liver and pancreas under similar culture conditions. These lack markers of mature pancreas or liver. However, their morphology and expression profiles for aldolase, lactate dehydrogenase, and cytokeratins were indistinguishable, suggesting that primitive epithelial cells from both organs are very similar.

Others established permanent "duct" epithelial cell lines also from the rat, which expressed cytokeratins 8 and 19 and carbonic anhydrase. Upon transplantation to a rat, hepato-cyte-specific markers were induced, and pancreatic duct markers were silenced. These data are consistent with the hypothesis of a common endodermal hepato-pancreatic stem cell.

KEY WORDS

Bone marrow-derived hepatocyte An engrafted hepatocyte with a chromosomal or expression marker that identifies it as possessing genetic material from a (usually transplanted) bone marrow cell; can be the product of a hematopoietic cell/hepatocyte fusion event, differentiation by a hematopoietic progenitor with hepatic lineage potential or differentiation by a hepatic-committed marrow-resident progenitor.

Liver stem cell A poorly characterized cell proposed as the precursor of oval cells, or to comprise the most primitive subset of oval cells; the term is also used to refer broadly to all liver progenitor cells.

Oval cell A small cell with an oval-shaped nucleus and a high nuclear/cytoplasmic size ratio that emerges in the portal zone under conditions of chronic liver damage; can differentiate into the hepatocyte or bile duct epithelial lineages and thus is a bipotential liver progenitor.

Pancreatic liver stem cell A cell located in the pancreas that can generate hepatocytes to support liver repair upon transplantation; may be a bipotential hepato-pancreatic precursor or an extrahepatic progenitor committed to hepatic differentiation.

FURTHER READING

Camargo, F. D., Chambers, S. M., and Goodell, M. A. (2004). Stem cell plasticity: from transdifferentiation to macrophage fusion. *Cell Prolif.* **37**, 55–65.

Fausto, N. (2004). Liver regeneration and repair: hepatocytes, progenitor cells, and stem cells. *Hepatology* **39**, 1477–1487.

Grompe, M. (2003). The role of bone marrow stem cells in liver regeneration. *Semin. Liver Dis.* **23**, 363–372.

Grompe, M. (2003). Pancreatic-hepatic switches in vivo. *Mech. Dev.* **120**, 99–106.

Gupta, S., and Chowdhury, J. R. (2002). Therapeutic potential of hepatocyte transplantation. *Semin. Cell Dev. Biol.* **13**, 439–446.

Lagasse, E., Connors, H., Al-Dhalimy, M., Reitsma, M., Dohse, M., Osborne, L., Wang, X., Finegold, M., Weissman, I. L., and Grompe, M. (2000). Purified hematopoietic stem cells can differentiate into hepatocytes in vivo. *Nat. Med.* **6**, 1229–1234.

Saxena, R., Theise, N. D., and Crawford, J. M. (1999). Microanatomy of the human liver-exploring the hidden interfaces. *Hepatology* **30**, 1339–1346.

Sell, S., and Ilic, Z. (1997). *Liver stem cells.* Landes Bioscience/Chapman & Hall, Austin, New York.

Thorgeirsson, S. S. (1996). Hepatic stem cells in liver regeneration. *Faseb J.* **10**, 1249–1256.

33

Pancreatic Stem Cells

Yuval Dor and Douglas A. Melton

Introduction

From a clinical perspective, the pancreas is an important focus of stem cell research because it is an attractive target for cell replacement therapy. In type I diabetes, the insulin-producing β-cells that reside in the pancreatic islets of Langerhans are destroyed by autoimmune attack, and it is thought that self-renewing stem cells could provide an unlimited source of β-cells for transplantation. Such therapeutic efforts require the prospective isolation of stem cells, with the potential to produce β-cells and the development of methods to direct their expansion and differentiation.

From a developmental biology perspective, the role of stem cells in the pancreas is a fascinating problem. New cells are produced during adulthood, but their origin is not clear. Much of the field is focused on the identification of β-cell progenitors and the characterization of their molecular requirements, but it is not known what role such cells play during pancreas maintenance and regeneration or whether the adult pancreas contains a population of true stem cells. Regardless, information about the specification of the β-cell fate from undifferentiated progenitors will be important in directing the differentiation of stem cells *in vitro*.

In this chapter, we review evidence for the existence and identity of pancreatic progenitor and stem cells and describe the criteria for experimental demonstration of such cells.

Definition of Stem Cells and of Progenitor Cells

The most rigorous definition of a stem cell is a cell that, upon proliferation, produces some progeny that have the same developmental potential (a process called self-renewal) as well as other progeny that have a more restricted developmental potential. Such cells may be present transiently during embryonic development or persistently during the entire life of the organism, but their defining property is self-renewal. A progenitor cell, on the other hand, is any cell that generates another differentiated cell type.

Another important notion in stem cell biology with potential relevance to the pancreas is that of the facultative stem cell. These cells are thought to be functional, differentiated cells in an adult organ that can de-differentiate in response to a specific signal (usually believed to be tissue damage) and

then differentiate into another cell type. For example, the liver is thought to contain facultative stem cells that reside in the bile ducts and are capable of producing hepatocytes. Thus, the bile duct cells provide an effective reservoir of liver progenitors in case of tissue damage, although they may not meet the criteria of being stem cells at the single-cell level.

How can we demonstrate that a particular cell is a stem cell? Formally, such a demonstration must be based on clonal analysis: A single cell has to be followed over time in order to show that it can produce more stem cells as well as differentiated cells. The gold standard for the identification of stem cells was set in the hematopoietic system, where clonal analysis has been carried out *in vitro* (by subcloning individual colonies) as well as *in vivo* (by serial transplantation of single stem cells into lethally irradiated mice). By contrast, the identification of progenitor cells is an easier task. It requires the demonstration using some sort of lineage-tracing experiment that an undifferentiated cell population generates differentiated cells.

Even when stem cells cannot be identified or isolated in a particular organ, their existence may be inferred from kinetic studies of 5′-bromo-2′-deoxyuridine (BrdU) incorporation. Because stem cells are believed to be slowly dividing, the presence of label-retaining cells can point to the anatomical location of a stem cell niche. Such an analysis was carried out in self-renewing organ systems such as skin and hair.

In light of these definitions, in the next sections we discuss the evidence for stem cells and progenitors during pancreas development and adult life.

Progenitor Cells during Embryonic Development of the Pancreas

The adult pancreas contains three major cell types: exocrine cells, organized in acini, that secrete digestive enzymes; duct epithelial cells that flush these enzymes to the duodenum; and endocrine cells, organized in the islets of Langerhans, that secrete hormones to the blood. The islets, accounting for ~1% of the cells in the adult pancreas, contain four major cell types that secrete different hormones: α, β, δ and pp-cells secreting glucagon, insulin, somatostatin, and pancreatic polypeptide, respectively. In recent years, a detailed cellular and molecular understanding of pancreas development has emerged (Figure 33-1). Although many questions remain, one recurring theme is the essential role played by progenitor cells in organogenesis. Early patterning of the gut tube generates a sheet of epithelial cells that express the homeobox

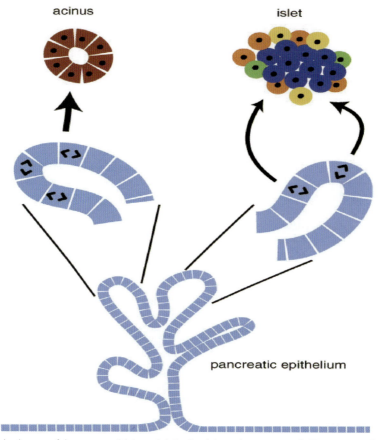

Figure 33-1. Embryonic development of the pancreas. Pdx1⁺ epithelial cells of the early pancreas (pale blue); acinar cells (red); and α-, β-, δ-, and pp-cells (indicated by orange, blue, yellow and green in the islet). The proposed plane of division of endocrine and exocrine progenitors is indicated.

gene *pdx1* around embryonic day 9 (E9) in the mouse. These epithelial cells bud from the gut tube, proliferate, and branch to form a tubular structure around E12.5 from which cells differentiate and organize into the exocrine and endocrine tissues of the mature pancreas. Numerous experiments demonstrate that this process is based upon the proliferation and stepwise differentiation of progenitor cells. As shown by lineage-tracing experiments using tamoxifen-induced cre recombinase, the early *pdx1⁺* cells (which by E10.5 also express the transcription factor p48/ptf1) produce all cell types in the adult pancreas. Later on, progenitors for duct, acinar, and endocrine lineages segregate, proliferate, and differentiate. Endocrine progenitors migrate from the tubular structure of the early pancreas and coalesce to form the islets of Langerhans just before birth. During that period, they undergo further restrictions in their potential, marked by the transient expression of several genes. Most notably, neurogenin3 expression marks progenitor cells for all endocrine lineages. Down-regulation of progenitor markers and expression of the hormone genes mark the terminal differentiation of endocrine cells. A similar mechanism of gradual commitment, though less well defined, is thought to act in the developing exocrine pancreas.

How do pancreatic progenitors choose their fates? A longstanding hypothesis is that the axis of mitosis in epithelial cells of the early pancreas correlates with, or even determines, the identity of daughter cells (Figure 33-1). When the plane of division is perpendicular to the lumen, both daughter cells remain epithelial (symmetric division) and may lobulate to form exocrine acini. When the plane of division is parallel to the lumen, one daughter cell detaches from the lumen and may become an endocrine progenitor cell (asymmetric division). This hypothesis, though untested, supports the popular view that adult pancreatic stem cells reside in the ducts. More recently, genetic evidence has suggested that cell–cell interactions mediated by the Notch pathway affect lateral specification of progenitor cells at multiple developmental junctions. Accordingly, mutations in Notch pathway genes can lead to accelerated, premature endocrine differentiation on the expense of the exocrine pancreas. These two models for specification may be compatible, as the Notch pathway is known to affect the plane of division in neural progenitors.

What is the relative contribution of progenitors and differentiated cells to proliferation of the embryonic pancreas? Although this question has not been addressed directly, the fate of hormone-expressing cells that appear early in pancreas development (E9.5–10.5) was recently examined. These cells, many of which coexpress multiple endocrine hormones, were initially thought to replicate and generate the mature islets. However, lineage-tracing and ablation experiments suggest that the multiple-hormone-expressing cells do not contribute to the adult endocrine pancreas. Rather, mature endocrine cells appear to be derived from midgestation progenitors.

Following this progenitor-based formative stage, rapid growth of the pancreas in late gestation and early postnatal life is thought to involve gradually less differentiation of progenitor cells and more replication of fully differentiated cells (see later sections of this chapter).

Although the identity and importance of progenitors during pancreas development is clear, there is no indication for true self-renewal of cells during this period. The recent demonstration of heterogeneity among the pdxl[+] cells of the early pancreas suggests that lineage segregation occurs very early during pancreas organogenesis. Furthermore, no clonal analysis of the embryonic pancreas has been carried out *in vivo*. *In vitro* clonal analysis demonstrated the common origin of the exocrine and endocrine pancreas but not self-renewal. Therefore, it is fair to conclude that progenitor cells play a major role in pancreas formation. However, there is no evidence for self-renewing pancreatic stem cells during embryonic development. We now turn to the postnatal growth and maintenance of the pancreas, with special attention to β-cells. Do progenitor cells of the embryonic type persist in adult life in an active or latent form? Is there evidence for stem cells capable of generating new pancreatic cells in the adult?

Progenitor Cells in the Adult Pancreas

When describing stem/progenitor cells in the adult pancreas, it is important to deal separately with two questions: First, what is the turnover rate of the different components of the pancreas during postnatal growth, throughout adult maintenance, and in response to injury? Second, are the new cells derived from stem/progenitor cells (neogenesis) or from replicating differentiated cells? It must be remembered that an impressive capacity for homeostatic maintenance or even regeneration does not indicate neogenesis. For example, dramatic regeneration of the injured liver can occur solely by proliferation of differentiated hepatocytes without a requirement for stem cells.

Evidence from Cell Dynamics. After birth, the growth of the pancreas continues but slows significantly around 1 month of age in the mouse and rat. However, even in old animals there is a measurable rate of cell birth in all pancreatic compartments. In the β-cell compartment, where most studies have been done, the replication rate falls from ~5% in 4-week-old animals to ~0.1% in mice older than 3 months. Embryonic-type progenitors (based on expression patterns) are not seen in the normal adult animal (with the possible exception of rare islet cells generated from neurogenin3[+] progenitors).

In spite of its low basal turnover rate, significant hyperplasia is seen in the adult pancreas under certain physiologic and pathologic conditions. For example, during pregnancy the β-cell mass doubles, a response attributed to a combination of

cell hypertrophy and cell proliferation. More dramatically, several reports have documented an ability of the β-cell compartment to recover from genetically programmed, autoimmune, surgical, or chemical damage.

Can the new cells, in normal homeostasis or in a regeneration setting, be fully accounted for by the replication of differentiated cells? Empirically, for the β-cell compartment, can the number of BrdU[+] pulse-labeled β-cells explain the accumulation of BrdU[+] β-cells following continuous administration of BrdU? If not, the presence of undifferentiated progenitors must be invoked. This type of analysis has, however, been proven difficult because it requires reliable values for several elusive parameters: What is the duration of the S phase and the total cell cycle in a specific compartment? What is the death rate, and how long does it take for a dying cell to be cleared? Furthermore, cell number is not easily inferred from total cell area as obtained by immunostaining because of cellular hypertrophy. Thus, a study of β-cell dynamics must include β-cell counting by fluorescence-activated cell sorting or by careful histological analysis across the whole pancreas, a criterion that is not always met.

The most comprehensive effort in this direction was carried out by Finegood *et al.*, who studied β-cell dynamics throughout the lifespan of the rat. Their results imply a significant contribution of progenitor cells to the β-cell mass in the first weeks after birth and then a shift to tissue maintenance by slow replication of β-cells. In addition, significant β-cell neogenesis was deduced in a similar study of chronic hyperglycemia in rats. These kinetic studies suggest that adult pancreatic progenitors exist. However, they do not help determine the molecular and anatomical origins of these cells.

Another argument for the existence of β-cell progenitors is the identification of single β-cells embedded in the adult exocrine tissue. These isolated cells are reportedly more frequent following insults, which led to the notion that they are newly generated from progenitors that reside in ducts or acini (see later sections of this chapter). The new β-cells are then believed to coalesce into islets in a mechanism resembling developmental islet morphogenesis. However, careful analysis is required to distinguish this interpretation from other potential explanations (e.g., disintegration of existing islets). In fact, recent genetic lineage tracing experiments suggest that the small clusters of β cells are derived from β cells rather than from stem/progenitor cells.

A combination of BrdU pulse-chase experiments and genetic labeling of differentiated cells (for example, using an inducible cre recombinase) may allow a direct comparison between the contribution of progenitors and the contribution of differentiated cells to pancreas growth and maintenance.

Many experiments neglect the kinetic aspects of neogenesis and focus on histological identification of progenitors, based on expression markers, in certain anatomical locations. However, without lineage tracing, these studies cannot demonstrate the fate of the putative progenitors or determine their importance. In the next sections, we describe the most notable proposals regarding the identity of adult pancreatic stem-progenitor cells.

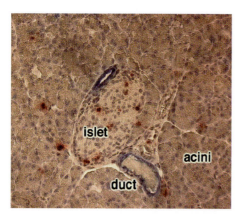

Figure 33–2. Histology of adult pancreas. Exocrine tissue (acini and ducts) and endocrine tissue (islet of Langerhans) are shown. Section from a 4-week-old mouse pancreas, pulse labeled with BrdU (red). Ducts are stained blue. Note the close association between the ducts and the islet. Original magnification = 200×.

DUCTS

It is widely believed that adult pancreatic stem/progenitor cells are located in the duct epithelium. Indeed, cells expressing β-cell markers (insulin, glut-2, pax6, isl1, and HNF3β) can often be found embedded in or adjacent to adult ducts following an insult to the pancreas.

Moreover, duct cell replication (as assessed by BrdU incorporation) is increased after such insults. In addition, adult duct preparations are claimed to be capable of endocrine differentiation *in vitro* (see later sections of this chapter). Conceptually, the appearance of endocrine cells in or near ducts is interpreted as recapitulation of embryonic pancreas morphogenesis, where endocrine progenitors bud from the epithelium. However, adult ducts are not necessarily identical in their genetic program to the embryonic tubes that should be referred to as "duct-like structures." Indeed, recent lineage experiments indicate that the definitive ductal and endocrine lineages are separated as early as E12.5. With regard to endocrine cells seen budding from ducts, caution is required when suggesting dynamic interpretations to static histological snapshots (Figures 33-2 and 33-3).

The preceding description shows the importance of specific duct markers for the assessment of the ductal origins of endocrine cells. Although several duct markers have been found (such as carbonic anhydrase, cystic fibrosis transmembrane-conductance regulator, and cytokeratin-19), so far none has been translated to a useful lineage marker in transgenic mice.

In summary, it seems that ducts can elicit a proliferative response to tissue damage and that endocrine markers are occasionally expressed in cells of the ductal epithelium. However, the fate of these cells and their relative contribution to the endocrine or exocrine pancreas has yet to be demonstrated.

ACINI

Starting from the observation of isolated β-cells embedded in exocrine tissue, other investigators have proposed that acinar

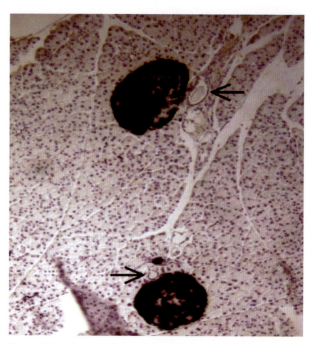

Figure 33–3. Lineage labeling of the endocrine pancreas. Section from the pancreas of a 4-week-old ngn3-cre; Z/AP double-transgenic mouse, stained for alkaline phosphatase activity (dark blue). In this mouse, cre-mediated recombination in *ngn3*⁺ embryonic endocrine progenitor cells leads to heritable expression of the human placental alkaline phosphatase gene. Only islet cells are labeled, suggesting that ducts (arrows) and acini never expressed the *ngn3* gene. Original magnification = 100×.

cells can transdifferentiate, with or without replication, into endocrine cells. This possibility is supported by *in vitro* studies showing that an acinar cell line can adopt endocrine features under certain conditions. However, as in the case of duct-embedded progenitors, the lineage of these cells was not followed *in vivo*; therefore, their origin and fate could not be determined. In this case, however, good lineage markers exist, so the possible acinar origin of endocrine cells can be directly tested. Preliminary experiments with an Elastase promoter-driven Cre recombinase have suggested that, under normal conditions, Elastase-expressing cells do not produce endocrine or duct cells. It remains to be tested, however, whether acinar cells contain "facultative" endocrine progenitors activated upon injury.

Intraislet Progenitors. Several groups have suggested the existence of intraislet progenitor cells capable of proliferation and differentiation to β-cells. These reports are based on the expression of putative stem-progenitor cell markers in islets. One such proposal is that the expression of Nestin, a marker of neuronal progenitors, labels intraislet endocrine progenitors with a potential for *in vitro* differentiation into several fates. However, a recent lineage analysis using Cre recombinase driven by the Nestin promoter showed that pancreatic endocrine cells do not form by the differentiation of Nestin⁺ cells. In addition, Nestin was recently shown to be expressed in the mesenchymal but not in epithelial cells of the embryonic pancreas, further questioning its relevance for pancreatic lineages.

Others have documented the coexpression of pancreatic hormones in islets following a diabetogenic insult. For example, It was proposed that following streptozotocin treatment, somatostatin[+] cells and glucagon[+] cells generate β-cells by proliferation and differentiation through somatostatin[+]pdxl[+], and glucagon[+]glut-2[+] intermediate cell types. These proposals have yet to be confirmed by lineage analysis.

Finally, clinical transplantation of islets may provide important clues about islet cell dynamics. Islet grafts into the portal vein (as done with diabetic patients) or under the kidney capsule (as done routinely with rodents) survive for many months and provide glycemic control. Although it is clear that cells in the grafts die and proliferate, it is not known if new endocrine cells in the graft are derived only from replicating differentiated cells or also from progenitors. Since purified islets with minimal exocrine tissue are used for transplantations, any indication for the existence of progenitors in the graft would point to an intraislet source.

Bone Marrow. A recent study found a significant contribution of bone marrow-derived cells to the β-cell compartment in the absence of tissue damage other than irradiation. To facilitate the detection of marrow-derived β-cells, marrow was taken from transgenic mice expressing Cre recombinase under the insulin promoter as well as enhanced green fluorescent protein (eGFP) preceded by a Cre-removable transcriptional stop signal. Thus, bone marrow cells of the donor mice should express eGFP only if they turn on the insulin promoter. By sorting eGFP[+] cells in pancreatic islets of recipient mice, up to 3% of all β-cells were identified from a donor origin two months after marrow transplantation. The elegant design of the experiment was such that the fusion of marrow cells with recipient β-cells could be ruled out. These surprising results still await independent confirmation.

OTHER EVIDENCE FOR ADULT PANCREATIC PROGENITOR CELLS

It has been reported that cells in the adult pancreas can reconstitute a degenerating liver deficient in the enzyme fumarylacetoacetate hydrolase. This observation suggested the existence of progenitors in the pancreas capable of liver differentiation. However, more recent results indicate that liver reconstitution by bone marrow cells represents complementation of defective hepatocytes by cell fusion rather than transdifferentiation of marrow cells to liver. Thus, it is likely that liver reconstitution by pancreatic cells represents the "fusibility" rather than the plasticity of these cells.

In summary, indirect evidence suggests that endocrine progenitor cells exist, or can appear, in the adult pancreas following injury. Careful lineage analysis has yet to demonstrate the existence, origin, fate, and importance of these cells to pancreas homeostasis. Similar to the situation in the embryo, there is no evidence for the existence of self-renewing stem cells in the adult pancreas.

Very recently, we have used a genetic lineage analysis system to study the dynamics of adult pancreatic β-cells. This analysis shows that adult β-cells are maintained largely by self-duplication rather than stem cell differentiation, during normal life in the mouse as well as during regeneration from pancreatectomy. The results cast doubt on the idea that stem cells contribute significantly to the formation of new β-cells during adult life. We cannot rule out the existence of adult pancreatic stem cells; however, it seems that the contribution of such putative cells to new β-cells is minimal. In addition, facultative pancreatic stem cells might exist and be recruited following other types of injury. The same system can now be used for testing claims about stem cell differentiation into β-cells, *in vitro* and *in vitro*.

Forcing Other Tissues to Adopt a Pancreatic Phenotype

From a therapeutic perspective, artificial induction of a β-cell fate in other cell types is as useful as finding and expanding endogenous pancreatic progenitors. Moreover, nonpancreatic cells, if indeed competent to adopt a β-cell program, may be more accessible for expansion and *in vitro* manipulations. Efforts in this direction have focused on tissues thought to be capable of pancreatic differentiation because of a common lineage origin or a similar genetic program. Here, we explain the pancreatic connection of the most popular starting tissues for such experiments.

LIVER TO PANCREAS

Several observations led to the notion that the adult liver may be a close relative of the pancreas in terms of the steps required for switching fates.

First, both organs are derived from the endoderm of the embryonic gut tube. Moreover, studies by Zaret and colleagues suggest that a bipotential liver–pancreas progenitor cell population exists by e8 in the ventral foregut endoderm. Using embryo tissue explantation experiments, they have shown that fibroblast growth factor signaling from the cardiac mesoderm can divert these progenitors from a default pancreatic fate to a liver fate. However, lineage or clonal analysis was not performed to formally demonstrate that bipotential cells exist.

Second, transdifferentiation of pancreas to liver, in which clusters of hepatocytes emerge in the pancreas, occurs spontaneously in humans and can be induced experimentally in rats. In cultured pancreatic cells, liver transdifferentiation was demonstrated by activation of a single transcription factor (C/EBP-β) without a requirement for cell division. These observations suggest that adult liver and pancreas cells may retain a degree of plasticity that allows for transdifferentiation to occur. Alternatively, a rare stem-progenitor cell population in these organs, yet to be identified, may be responsible for the metaplastic phenomenon.

Notably, no evidence has been found for deep similarities in the genetic program of the adult pancreas and liver. Based on these observations, several groups have tried to experimentally convert liver to pancreas by enforced expression of key pancreatic genes. One candidate for imposing a pancreatic fate is pdxl, the most upstream component of the pancreatic program. Indeed, adenoviral-mediated transfer of pdxl to adult mice was reported to induce low-level expression of insulin or exocrine pancreas markers in the liver.

A conceptual problem with simple overexpression of pdxl is the need for specific transcriptional cofactors to be present. For example, pdxl is known to require the Pbx and Meis proteins for proper function. With the hope of bypassing the need for pdxl cofactors absent from the liver, a pdxl-VP16 fusion gene was recently constructed. Transgenic expression of pdxl-VP16 (but not unmodified pdxl) in livers of *Xenopus* tadpoles led to significant expression of pancreatic endocrine and exocrine genes in the liver, concomitant with down-regulation of liver genes. The use of a liver-specific promoter (from the transthyretin gene) and the finding that the same construct induces a pancreatic phenotype in a hepatocyte cell line suggest that the responding cells are differentiated hepatocytes rather than uncommitted progenitors or stem cells.

In summary, it seems that ectopic expression of pancreatic transcription factors in the embryonic and perhaps the adult liver can lead to activation of a pancreatic program. The available data suggest that the mechanism is direct hepatocyte transdifferentiation, rather than activation of a common liver–pancreas progenitor. In addition to the therapeutic potential of the approach, analysis of the molecular details may provide significant insights into the problems of competence and potential.

NEURONS TO PANCREAS

Although arising from different lineages during embryogenesis, the endocrine pancreas and the nervous system share a significant portion of their developmental programs. For example, most transcription factors participating in the stepwise differentiation of the endocrine pancreas are also expressed in the developing brain. Some examples are isl1, Hb9, neurogenin3, NeuroD, Nkx2.2, Nkx6.1, pax6, and brn4. Notable exceptions are the absence of pdxl as well as insulin expression in the brain. However, the insulin promoter does contain neuronal regulatory elements that are repressed in the context of the endogenous insulin locus but give rise to neural expression when driving transgenes. The significance of this intriguing similarity is not fully understood. One possibility is that it reflects a common ancestral lineage for brain and endocrine pancreas, which diverged during evolution as the endocrine program came under pdxl control. This idea is supported by the finding that insulin and glucagon are expressed in specific neurons in the fly (which lacks a pancreas). Thus, neuronal progenitors are the closest known relatives of pancreatic progenitors in terms of their genetic program. This suggests that neuronal progenitors (easily derived from ES cells) may be only a few steps from adopting a pancreatic fate. However, attempts to induce ES-derived neural progenitors to a pancreatic fate have so far failed.

GUT TO PANCREAS

Several recent studies explored the potential of the nonpancreatic segments of the digestive tract to adopt a pancreatic fate. Ectopic expression of pdxl in the embryonic gut outside the normal expression domain of pdxl led to a partial adoption of a pancreatic fate, judged by gene expression and cell budding out of the epithelium. Similarly, ectopic expression of neurogenin3 in the embryonic chick or mouse led to the appearance of endocrine cells expressing glucagon and somatostatin (but not insulin). Inactivation of ptfla-p48, a transcription factor normally expressed in pancreatic progenitors, led to a cell-autonomous switch in the reverse direction, namely, the conversion of pancreas to duodenal epithelium. These results suggest that even after the onset of segment-specific genetic programs, the embryonic gut epithelium remains plastic. It is not known if such plasticity is retained in the adult intestine.

In Vitro Studies

Numerous attempts have been made to generate β-cells *in vitro* from dissociated pancreatic tissue using the available information on the identity of putative pancreatic stemprogenitors. Accordingly, pancreatic duct preparations were cultured, and the appearance of differentiated cell types was monitored. Some groups have reported on the generation of endocrine cells from duct cultures, and some even claimed to correct hyperglycemia in animals by grafting the *in vitro* generated islet-like clusters. Others have used islet preparations as a starting material. These studies, however, share several methodological problems that complicate their interpretation. First, the number of endocrine cells at the beginning and at the end of culture period was not always measured carefully. Second, all studies were carried out on enriched rather than pure cell populations, thus not ruling out the presence of contaminating cells from unintended compartments. Third, clonal analysis was not performed. Fourth, many of these studies used insulin immunostaining as a marker for β-cell differentiation and supplied large amounts of exogenous insulin in the culture medium. Therefore, they may have scored for insulin uptake rather than *de novo* synthesis.

Perhaps a more promising approach to the problem is the use of embryonic pancreas as a starting material, since it is clear that it contains progenitor cells that can proliferate and differentiate. This, however, has proved a difficult task as well. Much of the difficulty may result from the lack of proper culture conditions that mimic the embryonic environment.

Summary

The search for pancreatic stem or progenitor cells may provide insights into the basic mechanisms of organ homeostasis as well as a promising therapeutic approach for diabetes. It is clear that pancreatic progenitor cells exist and play a major role during embryonic development, although much has to be learned about the mechanisms and molecules controlling them. There is no evidence, however, for pancreatic stem cells capable of self-renewal.

As for the adult pancreas, no conclusive evidence shows the importance or mere existence of pancreatic progenitor or stem cells in the adult organism. On the contrary, it appears that at least for pancreatic beta cells, the major mechanism for tissue maintenance is self-duplication rather than stem cell differentiation. It is possible, however, that under certain stresses progenitor cells appear and may contribute to the maintenance of the pancreas. Most evidence points to the pancreatic ducts as the likely niche of such "facultative" progenitors. Development of culture conditions that support the growth and differentiation of dissociated pancreatic tissue, as well as *in vivo* lineage analysis systems, are the most needed tools for tackling this problem.

Other cell types may retain a degree of plasticity that allows for pancreatic differentiation under artificial conditions. Embryonic stem cells are the only cells available *in vitro* that can definitely differentiate into pancreatic cell types; however, this potential has yet to be demonstrated *in vitro*. Theoretical considerations and some experimental evidence suggest that neuronal progenitors, liver cells, and intestinal epithelial cells may also be forced to adopt a pancreatic fate.

KEY TERMS

Genetic lineage tracing The use of permanent labeling, usually DNA recombination, to mark a defined cellular population and trace its progeny.

Neurogenin3 A transcription factor expressed in endocrine progenitor cells of the embryonic pancreas.

pdx1 Pancreatic and duodenal homeobox gene-1; a transcription factor essential for pancreas development. During embryogenesis, pdx1 is expressed in stem/progenitor cells, giving rise to all cell types in the adult pancreas. During postnatal life, its expression becomes restricted to beta cells.

FURTHER READING

Bonner-Weir, S., and Sharma, A. (2002). Pancreatic stem cells. *J. Pathol.* **197**, 519–526.

Bouwens, L. (1998). Transdifferentiation versus stem cell hypothesis for the regeneration of islet beta-cells in the pancreas. *Microsc. Res. Tech.* **43**(4), 332–336.

Deutsch, G., Jung, J., Zheng, M., Lora, J., and Zaret K. S. (2001). A bipotential precursor population for pancreas and liver within the embryonic endoderm. *Development* **128**(6), 871–881.

Dor, Y., Brown, J., Martinez, O.I, and Melton, D.A. (2004). Adult pancreatic β cells are formed by self-duplication rather than stem cell differentiation. *Nature* **429**, 41–46.

Finegood, D.T., Scaglia, L., and Bonner-Weir, S. (1995). Dynamics of beta-cell mass in the growing rat pancreas. Estimation with a simple mathematical model. *Diabetes.* **44**(3), 249–256.

Gu, G., Dubauskaite, J., and Melton, D.A. (2002). Direct evidence for the pancreatic lineage: NGN3+ cells are islet progenitors and are distinct from duct progenitors. *Development.* **129**(10), 2447–2457.

Kodama, S., Kuhtreiber, W., Fujimura, S., Dale, E. A., and Faustman, D. L. (2003). Islet regeneration during the reversal of autoimmune diabetes in NOD mice. *Science* **302**, 1223–1227.

Murtaugh, L. C., and Melton, D. A. (2003). Genes, signals, and lineages in pancreas development. *Annu. Rev. Cell. Dev. Biol.* **19**, 71–89.

Pelengaris, S., Khan M., and Evan, G. I. (2002). Suppression of Myc-induced apoptosis in beta cells exposes multiple oncogenic properties of Myc and triggers carcinogenic progression. *Cell* **109**(3), 321–334.

Shapiro, A. M., Lakey, J. R., Ryan, E. A., Korbutt, G. S., Toth, E., Warnock, G. N., Kneteman N. M., and Rajotte, R. V. (2000). Islet transplantation in seven patients with type 1 diabetes mellitus using a glucocorticoid-free immunosuppressive regimen. *N. Engl. J. Med.* **343**(4), 230–238.

34

Stem Cells in the Gastrointestinal Tract

Sean L. Preston, Natalie C. Direkze, Nicholas A. Wright, and Mairi Brittan

Turnover of the epithelial cell lineages within the gastrointestinal tract is a constant process, occurring every two to seven days under normal homeostasis and increasing after damage. This process is regulated by multipotent stem cells, which generate all gastrointestinal epithelial cell lineages and can regenerate whole intestinal crypts and gastric glands. The stem cells of the gastrointestinal tract are as yet undefined, although it is generally agreed that they are located within a "niche" in the intestinal crypts and gastric glands. Studies of allophenic, tetraparental chimeric mice and targeted stem cell mutations suggest that a single stem cell undergoes an asymmetrical division to produce an identical daughter cell, thus replicating itself, and a committed progenitor cell, which further differentiates into an adult epithelial cell type. The discovery of stem cell plasticity in many tissues, including the ability of transplanted bone marrow to transdifferentiate into intestinal subepithelial myofibroblasts, provides a potential use of bone marrow cells to deliver therapeutic genes to damaged tissues, for example, in the treatment of mesenchymal diseases in the gastrointestinal tract, such as fibrosis and Crohn's disease. Studies are beginning to identify the molecular pathways that regulate stem cell proliferation and differentiation into adult gastrointestinal cell lineages, such as the Wnt and Notch–Delta signaling pathways, and to discover the importance of mesenchymal-epithelial interactions in normal gastrointestinal epithelium and in development and disease. Finally, despite some dispute, a strong case can be made that intestinal neoplasia arises as a result of a series of mutations in stem cells. The mechanism and direction of the spread of this mutated clone in the gastrointestinal mucosa are hotly disputed, and central to this argument is the position and nature of the gastrointestinal stem cell.

Introduction

There has been a tremendous increase in interest in stem cell biology and its potential applications in recent years. Although this interest has been galvanized by the exploitation of research in embryonic stem cells, it is interesting to note that what might be called the *intestinal stem cell community*, albeit small, has been working productively for some 40 to 50 years. With stem cells now claiming considerable attention, in retrospect, many of the basic tenets that govern

our understanding of organ-specific stem cells have come from studies of the gastrointestinal tract and the hematopoietic system.

With regard to the gastrointestinal tract, a large body of evidence shows that multipotential stem cells are found in specific zones, or niches, within gastric glands and intestinal crypts, composed of and maintained by myofibroblasts in the adjacent lamina propria. In this chapter, we review evidence that these multipotential stem cells generate all gastrointestinal epithelial cell lineages through committed precursor cells housed in the proliferative compartments of intestinal crypts and gastric glands, a concept that has had a long and difficult gestation. Notwithstanding their obvious significance, the gastrointestinal stem cells remain elusive and unidentified, mainly because of a lack of accepted morphological and functional markers at the single-cell level. We also explore concepts of stem cell number, location, and fate, and we touch on the ability of gastrointestinal stem cells to regenerate cell lineages of whole intestinal crypts and villi after damage. The luminal gut shows regional specializations of function — the stomach primarily for absorption and the intestinal mucosa for both absorption and secretion. This is reflected by variation in the adult cell lineages native to each tissue, and it is thereby consistent that stem cell fate within each tissue is also different.

We are beginning to understand the mechanisms that govern such variation. Controversial recent findings regarding stem cell plasticity in the gastrointestinal tract are examined in this chapter. Because of their longevity, putatively the same as that of the organism itself, stem cells are often viewed as the target cells for carcinogens and the cells of origin for spontaneous tumors. Recent thought on the location of stem cells in the colon has sparked debate concerning the possible pathways of morphological progression of transformed stem cells. This includes a *top-down* proliferation of mutated stem cells located within intercryptal zones on the mucosal surface, which spread downward into the adjacent crypts. Contrasting this is a *bottom-up* theory of the upward proliferation of mutated stem cells in the crypt base to produce dysplastic crypts that replicate and expand by crypt fission. This brings other facets of gut biology into sharp focus — the mechanisms of crypt reproduction; the clonal architecture of normal and dysplastic gastric glands, intestinal crypts, and their derivative tumors; and the role that stem cells take in these events.

In this chapter, we propose that the stem cells accumulate the multiple genetic events leading to tumorigenesis, and we explore the manner by which such mutated clones spread in gastrointestinal epithelia. Many of these concepts are being explored at the level of molecular regulatory pathways, including the signaling pathways of Wnt and transforming growth factor β (TGFβ). Our concepts of stem cell biology impinge considerably on our understanding of how these tumors arise.

Gastrointestinal Mucosa Contains Multiple Lineages

In the small intestine, the epithelial lining forms numerous crypts and larger, finger-shaped projections called *villi*. In the colon there are many crypts, which vary in size throughout the colon; the shortest is in the ascending colon. Overall, four main epithelial cell lineages exist in the intestinal epithelium: the columnar cells, the mucin-secreting cells, the endocrine cells, and Paneth cells in the small intestine. Other less common cell lineages are also present, such as the caveolated cells and membranous or microfold cells. Columnar cells, with apical microvilli, are the most abundant epithelial cells, termed *enterocytes* in the small intestine and *colonocytes* in the large intestine. "Goblet" cells containing mucin granules — and thus producing swollen, goblet-shaped cells — are found throughout the colonic epithelium, secreting mucus into the intestinal lumen. Endocrine, neuroendocrine, or enteroendocrine cells form an abundant cell population distributed throughout the intestinal epithelium; these cells secrete peptide hormones in an endocrine or paracrine manner from their contained dense core of neurosecretory granules. Paneth cells are located almost exclusively at the crypt base of the small intestine and ascending colon, contain large apical secretory granules, and express several proteins — including lysozyme, tumor necrosis factor, and the antibacterial cryptins (small molecular-weight peptides related to defensins).

In the stomach, the epithelial lining forms long, tubular glands divided into foveolus, isthmus, neck, and base regions. Gastric foveolar or surface mucus cells are located on the mucosal surface and in the foveola. They contain tightly packed mucous granules in the supranuclear cytoplasm and do not possess a theca. The mucus neck cells are situated within the neck and isthmus of the gastric glands and contain apical secretory mucin granules. The peptic-chief or zymogenic cells are located in the base of the glands in the fundic and body regions; they secrete pepsinogen from oval zymogenic granules. The parietal or oxyntic, acid-secreting cells are located in the body of the stomach in the base of the glands. These cells have many surface infoldings, or canaliculi, which form a network reaching almost to the base of the gland. Endocrine cell families include the enterochromaffin-like cells in the fundus or body that produce histamine; the gastrin-producing cells are a major component of the antral mucosa.

The intestinal crypts and gastric glands are enclosed within a fenestrated sheath of intestinal subepithelial myofibroblasts (ISEMFs). These cells exist as a syncytium that extends throughout the lamina propria and merges with the pericytes of the blood vessels. The ISEMFs are closely applied to the intestinal epithelium and play a vital role in epithelial-mesenchymal interactions. ISEMFs secrete hepatocyte growth factor (HGF), TGFβ, and keratinocyte growth factor (KGF), but the receptors for these growth factors are located on the epithelial cells. Thus, the ISEMFs are essential for the regulation of epithelial cell differentiation through the secretion of these and possibly other growth factors. Platelet-derived growth factor-α (PDGF-A), expressed in the intestinal epithelium, acts by paracrine signaling through its mesenchymal receptor, PDGFR-α, to regulate epithelial–mesenchymal interactions during development. Studies of mice with targeted deletions in the PDGF-A or PDGFR-α genes have shown defects in normal proliferation and differentiation of PDGFR-α positive mesenchymal cells. Typically, ISEMFs are α-smooth muscle actin-positive (αSMA+) and desmin-negative, but some myofibroblasts also express myosin-heavy chains. It has been proposed that these cells form a renewing population, migrating upward as they accompany the epithelial escalator. Although they appear to proliferate and migrate, they move relatively slowly and then move off into the lamina propria to become polyploid.

Epithelial Cell Lineages Originate from a Common Precursor Cell

Little is known of the location and fate of the stem cells within the gastrointestinal tract because of the lack of distinctive and accepted stem cells markers, although they are usually said to appear undifferentiated and can be identified operationally by their ability to repopulate crypts and glands after damage.

The Unitarian hypothesis, derived from work by Cheng and Leblond states that all the differentiated cell lineages within the gastrointestinal epithelium emanate from a common stem cell origin. Although widely propounded, until very recently little definitive evidence existed to underpin this hypothesis. Moreover, gastrointestinal endocrine cells have subsequently been proposed to derive from migrating neuroendocrine stem cells in the neural crest, a concept that still has its adherents. Although studies of quail neural crest cells transplanted into chick embryos, or experiments where the neural crest is eradicated, show gut endocrine cells to be of endodermal origin, it has subsequently been suggested that the endoderm is colonized by neuroendocrine-programmed stem cells from the primitive epiblast, which generate gut endocrine cells. This was not ruled out by chick–quail chimera experiments. Therefore, other models must be used to ascertain the gut endocrine cell origins, such as the chimeric mouse studies described later.

Several lines of evidence suggest that stem cells reside in the base of the crypts of Lieberkuhn in the small intestine, just superior to the Paneth cells (approximately the fourth or fifth cell position in mice). In the large intestine, they are presumed to be located in the midcrypt of the ascending colon and in the crypt base of the descending colon. However, within the gastric glands, migration of cells is bidirectional from the neck–isthmus region to form the simple mucous epithelium of the foveolus or pit, and cells migrate downward to form parietal cells and chief cells. Therefore, the stem cells are believed to be within the neck–isthmus region of the gastric gland. The Unitarian hypothesis is now supported by a considerable body of research.

Single Intestinal Stem Cells Regenerate Whole Crypts Containing All Epithelial Lineages

The ability of intestinal stem cells to regenerate epithelial cell populations of entire intestinal crypts and villi following cytotoxic treatment has been demonstrated using the crypt microcolony assay. Four days after irradiation, sterilized crypts undergo apoptosis and disappear, but they can be identified by remaining radio-resistant Paneth cells at the crypt base. At higher radiation dose levels, only single cells survive in each crypt, since a unit increase in radiation leads to unit reduction in crypt survival. Survival of one or more clonogenic cells in a crypt after radiation ensures crypt persistence, and there is regeneration of all epithelial cell populations of that crypt and, in the small intestine, of the overlying villi. Therefore, following cytotoxic damage, a single surviving stem cell can produce all cell types of the intestinal epithelium to reproduce a crypt.

Mouse Aggregation Chimeras Show That Intestinal Crypts Are Clonal Populations

Mouse embryo aggregation chimeras are readily made, wherein the two populations can be easily distinguished. The lectin Dolichos biflorus agglutinin (DBA) binds to sites on the B6-derived but not on the Swiss Webster (SWR)-derived cells in C57BL/6J Lac (B6) ↔ SWR mouse embryo aggregation chimeras, and it can be used to distinguish the two parental strains in gut epithelium. The intestinal crypts in each chimera studied were either positive or negative for DBA, and there were no mixed crypts in the tens of thousands studied. Therefore, each crypt forms a clonal population. This is the case for Paneth, mucous, and columnar cells, although it was not possible to detect the markers in endocrine cells because of their inability to bind the lectin on their surface. In neonatal C57BL/6J Lac (B6) ↔ SWR chimeras, there were mixed (i.e., polyclonal) crypts for the first two weeks after birth, suggesting that multiple stem cells exist during development. However, all crypts ultimately become derived from a single stem cell between birth and postnatal day 14, so-called *monoclonal conversion*. This apparent cleansing or "purification" of crypts could be caused by the stochastic loss of one stem cell lineage or by the segregation of lineages because of an extremely active replication of crypts by fission, which occurs at this developmental period. These findings were confirmed in mice bearing an X-linked defective gene for glucose-6-phosphate dehydrogenase (*G6PD*), to exclude the possibility that crypts from distinct strains segregate differentially during organogenesis.

In the stomach, the situation is similar, though more complex. Epithelial cell lineages in the antral gastric mucosa of the mouse stomach, including the endocrine cells, derive from a common stem cell. Identification of the Y-chromosome by *in situ* hybridization in XX–XY chimeric mice showed that gastric glands were also clonal populations. These findings were confirmed in CH3 ↔ BALB/c chimeric mice, where each gastric gland was composed of either CH3 or BALB/c cells. Thus, we might advance the general hypothesis that gastric glands in the mouse, in addition to the intestinal crypts, are clonally derived. In addition, by combining immunohistochemistry for gastrin, an endocrine cell marker, with *in situ* hybridization to detect the Y-chromosome, the male regions of the gastric glands were shown to be almost exclusively Y-chromosome positive with gastrin-positive endocrine cells, whereas the female areas in the chimeric stomach were gastrin positive and Y-chromosome negative. These results finally negate the concept that gut endocrine cells originate from a separate stem cell pool.

X-inactivation mosaic mice expressing a *lacZ* reporter gene were used to study clonality of gastric glands in the fundic and pyloric regions of the developing mouse stomach. As in the intestine, most glands are initially polyclonal, with three or four stem cells per gland, but they become monoclonal during the first six weeks of life — again either by purification of the glands, where division of one stem cell eventually overrides all other stem cells, or by gland fission. A population of approximately 5 to 10% of mixed, polyclonal glands persists into adulthood. The significance of these mixed glands with an increased stem cell number is not known, but it is possible that they do not undergo fission or have reduced fission rates. Perhaps they even have an increased stem cell number during development and therefore maintain a higher number of stem cells after crypt fission.

Somatic Mutations in Stem Cells Reveal Stem Cell Hierarchy and Clonal Succession

Somatic mutations at certain loci allow us to study stem cell hierarchy and clonal succession within the gastrointestinal tract. Mutations in the *DLb-1* on chromosome 11 are one good example of this; C57BL/6J ↔ SWR F1 chimeric mice show heterozygous expression of a binding site on intestinal epithelial cells for the DBA lectin. This binding site can be abolished when the *Dlb-1* locus becomes mutated either spontaneously or by the chemical mutagen ethyl nitrosourea (ENU). After ENU treatment, crypts emerge that initially are partially and then are entirely negative for DBA staining. Perhaps the simplest explanation for this phenomenon is that a mutation occurs at the *Dlb-1* locus in a stem cell within the small intestinal crypt. This mutated cell could expand stochastically to produce a clone of cells that cannot bind DBA and remain unstained. If this is the case, then a single stem cell can generate all the epithelial lineages within an intestinal crypt of the small intestine.

A "knockin" strategy at the *Dlb-1* locus can also be used to explain the preceding findings. If SWR mice do not express a DBA-binding site on their intestinal epithelial cells but can be induced to bind DBA by ENU treatment, wholly DBA+ or DBA− intestinal crypts would result. Visualization of the morphology, location, and longevity of mutant clones in crypts and villi of ENU-treated mice led to proposals that *committed epithelial progenitor* cells exist in mouse intestinal crypts. These transitory committed progenitor cells — the columnar cell progenitors and the mucus cell progenitors — evolve from pluripotential stem cells and then differentiate further into adult intestinal epithelial cell types.

Not much is known about the mechanisms that regulate the proliferation of these progenitor cells, but administration of proglucagon-derived, glucagon-like peptide 2 (GLP-2) to SWR mice was found to induce intestinal epithelial growth and repair specifically by stimulating the columnar cell progenitors, resulting in increased crypt and villus size in the normal small intestine. The receptor for GLP-2 (GLP2R) was recently shown to be located on enteric neurons and not on the gut enteroendocrine cells and in the brain as previously thought. GLP-2 activation of enteric neurons produces a rapid induction in *c-fos* expression, which signals growth of columnar epithelial cell progenitors and stem cells that generate adult columnar cell types. There is no stimulatory effect on the mucus cell lineage; instead, it is stimulated by KGF. Thus, the committed progenitor cells are involved in regeneration of damaged epithelia, possibly through a neural regulatory pathway.

In a further model, mice that heterozygously express a defective *G6PD* gene have a crypt-restricted pattern of G6PD expression, thus confirming that the intestinal crypts are derived from a single stem cell. Moreover, mice treated with the colon carcinogen dimethylhydrazine (DMH) or ENU also develop crypts that initially are partially and later are wholly negative for G6PD. The partially negative crypts could conceivably result from the mutation of a cell in the dividing transit population of the crypt that lacks stem cell properties. This is supported by the observation that these partially negative crypts are transient and decrease in frequency parallel to an increase in wholly negative crypts. Conversely, such partially negative crypts could become wholly negative by stochastic expansion of a mutant stem cell. Wholly negative crypts would then be a clonal population derived from this mutant stem cell.

After administration of a mutagen, in both the *Dlb-1* and the G6PD models, the time taken for the decrease in partially mutated crypts and the emergence of entirely negative crypts to reach a plateau is approximately 4 weeks in the small intestine and up to 12 weeks in the large intestine. This difference was initially thought to be due to cell cycle time differences between the colon and the small intestine. However, a favored explanation can be found in the *stem cell niche hypothesis*. This suggests that multiple stem cells occupy a crypt with random cell loss after stem cell division and was originally formulated as the *stem cell zone hypothesis*. The number of stem cells may be larger in the small intestine than in the large

intestine, causing the difference in time taken for phenotypic changes following mutagen treatment as the mutant stem cell expands stochastically. An alternative hypothesis might lie in crypt fission: the rate of fission at the time of mutagen administration was higher in the colon than in the small intestine. During crypt fission, when crypts divide longitudinally, selective segregation of the two cell populations could occur, "cleansing" the partially mutated crypts by segregating the mutated and nonmutated cells and by duplicating the wholly negative crypts to create monoclonal crypts.

Perhaps the best evidence for the clonality of human intestinal crypts, and the stem cell derivation of all contained epithelial cell lineages, comes from studies of the colon of a rare XO–XY patient who received a prophylactic colectomy for familial adenomatous polyposis (FAP). Nonisotopic *in situ* hybridization (NISH) using Y-chromosome-specific probes showed the patient's normal intestinal crypts to be composed almost entirely of either Y-chromosome positive or Y-chromosome negative cells, with about 20% of crypts being XO. Immunostaining for neuroendocrine-specific markers and Y-chromosome NISH used in combination showed that crypt neuroendocrine cells shared the genotype of other crypt cells. In the small intestine, the villus epithelium was a mixture of XO and XY cells, in keeping with the belief that the villi derive from stem cells of more than one crypt. Of the 12,614 crypts examined, only 4 crypts were composed of XO and XY cells, which could be explained by nondisjunction with a loss of the Y-chromosome in a crypt stem cell. There were no mixed crypts at patch boundaries. These observations agree with previous findings using chimeric mice that intestinal crypt epithelial cells, including neuroendocrine cells, are monoclonal and derive from a single multipotential stem cell. Consequently, the hypothesis that enteroendocrine cells and other differentiated cell types within the colorectal epithelium share a common cell of origin (the Unitarian hypothesis) appears to apply to both mice and humans. These observations have recently been confirmed in Sardinian women heterozygous for a defective *G6PD* gene.

It has been proposed that insight into stem cell organization can be gained from study of the methylation pattern of nonexpressed genes in the colon. In the normal human colon, methylation patterns are somatically inherited endogenous sequences that randomly change and increase in occurrence with aging. Examination of methylation tags of three neutral loci in cells from normal human colon showed variation in sequences between crypts and mosaic methylation patterns within single crypts. Multiple unique sites were present in morphologically identical crypts; for example, one patient had no identical methylation sequences of one gene within any of the crypts studied, even though all sequences were related. This indicates that some normal human colonic crypts are quasiclonal with multiple stem cells per crypt. Differences in methylation tags can highlight relationships among cells in a crypt where less closely related cells show greater sequence alterations and where closely related cells have similar methylation patterns. Sequence differences suggest that crypts are maintained by stem cells, which are randomly lost and replaced in a stochastic manner, eventually leading to a "bottleneck" effect in which all cells within a crypt are closely related to a single stem cell descendant. This reduction to the most recent common crypt progenitor is predicted to occur several times during life, superficially resembling the clonal succession of tumor progression.

In situ analyses of glandular clonality in the human stomach have been more problematic. Studies of polymorphisms on X-linked inactivation genes have revealed that in fundic and pyloric glands in the human female stomach, the pyloric glands appear homotypic and thus are monoclonal, although about half of the fundic glands studied were heterotypic for the HUMARA locus and were consequently polyclonal. This finding suggests that a more complex situation occurs in humans than the studies of gastric gland clonality in chimeric mice indicate. However, we have seen that some glands in the mouse remain polyclonal throughout life.

Bone Marrow Stem Cells Contribute to Gut Repopulation after Damage

The hematopoietic bone marrow stem cell is of mesodermal origin, and its functionality and cell surface markers have been well characterized. When transplanted into lethally irradiated animals and humans, as in clinical bone marrow transplantation, it has been long considered that the hematopoietic stem cell colonizes host tissues to form only new erythroid, granulocyte–macrophage, megakaryocyte, and lymphoid lineages. Although earlier studies suggest that vascular endothelium could derive from transplanted donor marrow, more recent studies not only have confirmed these earlier proposals concerning endothelial cells but also have indicated that adult bone marrow stem cells possess a considerable degree of plasticity and can differentiate different cell types, including hepatocytes, biliary epithelial cells, skeletal muscle fibers, cardiomyocytes, central nervous system cells, and renal tubular epithelial cells. These pathways can be bidirectional, as muscle and neuronal stem cells can also apparently form bone marrow.

It also appears that selection pressure induced by target organ damage can intensify the efficacy of this process as bone marrow stem cells differentiate into cardiomyocytes, endothelial cells, and smooth muscle cells in mice, with ischemic cell death following myocardial infarction and coronary artery occlusion. Bone marrow stem cells have also been shown to differentiate into pancreatic β-cells, and possibly more persuasively, fully differentiated cells can transdifferentiate into other adult cell types without undergoing cell division. For example, exocrine pancreatic cells can differentiate into hepatocytes *in vitro*. Furthermore, isolated potential hepatic stem cells from fetal mouse livers, which differentiate into hepatocytes and cholangiocytes when transplanted into recipient animals, can form pancreatic ductal acinar cells and intestinal epithelial cells when transplanted directly into the pancreas or duodenal wall. Thus, the conventional view that bone marrow stem cells generate cell types of a single lineage (i.e., all formed elements in the peripheral blood) has

been rectified in favor of the findings that adult bone marrow stem cells are highly plastic and can differentiate into many cell types within various organs.

These observations have raised the possibility of regeneration of a failing organ by transplanting an individual's own bone marrow stem cells to colonize and repopulate the diseased tissue, thus avoiding the allograft reaction. In apparent proof of principle, fumarylacetoacetate hydrolase (FAH)-deficient mice, which resemble type 1 tyrosinemia in humans, are rescued from liver failure by transplantation of purified hematopoietic stem cells that become morphologically normal hepatocytes, express the FAH enzyme, and therefore are functionally normal.

Several reports show that bone marrow cells can repopulate both epithelial and mesenchymal lineages in the gut. The colons and small intestines of female mice that received a bone marrow transplant from male mice donors were analyzed, as well as gastrointestinal biopsies from female patients with graft-versus-host disease following bone marrow transplant from male donors. Bone marrow cells frequently engraft into the mouse small intestine and colon and differentiate to form ISEMFs within the lamina propria. *In situ* hybridization confirmed the presence of Y-chromosomes in these cells; their positive immunostaining for α-SMA and negativity for desmin, the mouse macrophage marker F4/80, and the hematopoietic precursor marker CD34 determined their phenotype as pericryptal myofibroblasts in the lamina propria derived from transplanted bone marrow. This engraftment and transdifferentiation occurred as early as 1 week after bone marrow transplantation. Almost 60% of ISEMFs were bone marrow-derived 6 weeks after the transplantation, indicating that transplanted bone marrow cells are capable of a sustained turnover of the ISEMF cells in the lamina propria. Y-chromosome-positive ISEMFs were also seen in the human intestinal biopsy material. Lethally irradiated female mice given a male bone marrow transplant and a subsequent foreign-body peritoneal implant formed granulation tissue capsules containing myofibroblast cells derived from the hematopoietic stem cells of the transplanted bone marrow. This suggests that myofibroblasts may generally derive from bone marrow cells.

A growing number of reports have been issued showing that bone marrow cells can repopulate gastrointestinal epithelial cells in animals and humans. Bone marrow-derived epithelial cells were found in the lung, gastrointestinal tract, and skin 11 months after transplantation of a single hematopoietic bone marrow stem cell in the mouse. In the gastrointestinal tract, engrafted cells were present as columnar epithelial cells in the esophageal lining, a small intestinal villus, colonic crypts, and gastric foveola. No apparent engraftment into the pericryptal myofibroblast sheath was reported, and as only a single hematopoietic stem cell was transplanted, it is possible that the ISEMFs derive from mesenchymal stem cells within transplanted whole bone marrow. It is, however, generally believed that stromal cell populations do not survive following bone marrow transplantation — although if an empty niche exists, as after irradiation in the gut, engraftment might occur. Two very recent reports have claimed that local application of bone marrow stem cells, either directly injected into the stomach and duodenum or applied to the mucosa after the induction of experimental colitis, can also apparently lead to epithelial transdifferentiation.

In biopsies from female patients who had undergone sex-mismatched hematopoietic bone marrow transplantation, *in situ* hybridization for a Y-chromosome-specific probe with immunohistochemical staining for cytokeratins demonstrated mucosal cells of donor origin in the gastric cardia. Moreover, four long-term bone marrow transplant survivors showed multiple engraftment of esophageal, gastric, small intestinal, and colonic epithelial cells by donor bone marrow cells up to eight years after transplantation, emphasizing the long-term nature of this transdifferentiation.

It would be impossible to finish even such a short section on this topic without mentioning the mechanisms and significance of such phenomena. At first considered to be caused by transdifferentiation or lack of lineage fidelity — commonly seen in invertebrates, during gastrulation, or during organogenesis — it is becoming clear that such changes are neither simple nor readily reproducible. Some labs have been unable to reproduce earlier findings, and there are claims that adult tissues are contaminated with bone marrow precursors. Finally, the fusion of a transplanted bone marrow cell with an indigenous adult cell has been proposed as the mechanism by which bone marrow stem cells acquire the phenotype of target cell lineages. Initially, cell fusion was seen as a rare event, occurring only *in vitro* and under circumstances of extreme selection. In the FAH model described previously, cell fusion was recently shown to be common, and it cannot be ruled out as the main mechanism by which bone marrow stem cells transform to functional hepatocytes. However, the genetic mechanisms whereby gene expression is switched off in the recipient cells — followed by clonal expansion to repopulate large areas of the liver, for example — are as yet unclear. We should also recall that several tissues in the mouse are polyploid, such as the liver and the acinar cells of the exocrine pancreas. Other studies in which mixed-sex bone marrow transplants have been used to show plasticity have not reported evidence of cell fusion in animals or humans. For example, bone marrow-engrafted cells in the human stomach, intestine, buccal mucosa, and pancreatic islet cells showed a normal complement of X- and Y-chromosomes. Whatever the mechanism, it is clear that the most important criterion for altered lineage commitment, that of function, has only been fulfilled in a few models, such as the FAH model and possibly postinfarction cardiomyocyte engraftment. The future in this field will prove interesting.

Gastrointestinal Stem Cells Occupy a Niche Maintained by ISEMFs in the Lamina Propria

Stem cells within many tissues are thought to reside within a niche formed by a group of surrounding cells and their extra-cellular matrices, which provide an optimal microenvironment for the stem cells to function. The identification of a niche within any tissue involves knowledge of the location of the stem cells; as we have seen, this has proved problematical in the gastrointestinal tract. Some authors maintain that to prove that a niche is present, the stem cells must be removed and subsequently replaced while the niche persists, providing support to the remaining exogenous cells. Although this has been accomplished in *Drosophila*, such manipulations have not yet been possible in mammals. In this context, the survival of a single epithelial cell following cytotoxic damage to intestinal crypts in the microcolony assay is interesting, as many of the intestinal subepithelial myofibroblasts are also lost after irradiation, although sufficient numbers may remain or be replaced by local proliferation or migration from the bone marrow to provide a supportive niche for the surviving stem cell or cells. The ISEMFs surround the base of the crypt and the neck–isthmus of the gastric gland, a commonly proposed location for the intestinal and gastric stem cell niches, respectively. It is proposed that ISEMFs influence epithelial cell proliferation and regeneration through epithelial-mesenchymal cross-talk and that they ultimately determine epithelial cell fate.

There has been a long quest for markers of stem cells in the intestine. The neural RNA binding protein marker Musashi-1 (Msi-1) is a mammalian homolog of a *Drosophila* protein evidently required for asymmetrical division of sensory neural precursors. In the mouse, Msi-1 is expressed in neural stem cells and has recently been proposed as the first intestinal stem cell marker because of its expression in developing intestinal crypts and specifically within the stem cell region of adult small intestinal crypts. This is further substantiated by its expanded expression throughout the entire clonogenic region in the small intestine after irradiation.

The regulatory mechanisms of stem cell division within the niche to produce, on average, one stem cell and one cell committed to differentiation is as yet unknown, although there is no shortage of potential models. In the stem cell zone hypothesis, the bottom few cell positions of the small intestinal crypt are occupied by a mixture of cell types: Paneth, goblet, and endocrine. The migration vector is toward the bottom of the crypt. Above cell position 5, cells migrate upward, although only the cells that divide in the stem cell zone beneath are stem cells. Other models envisage the stem cells occupying a ring immediately above the Paneth cells, although there is little experimental basis for such an assertion since "undifferentiated" cells of similar appearance are seen among the Paneth cells in thin sections. Moreover, there is no difference in the expression of Msi-1 and Hes1 — a transcriptional factor regulated by the Notch signaling pathway also required for neural stem cell renewal and neuronal lineage commitment — in undifferentiated cells located in either the stem cell zone or immediately above the Paneth cells. This suggests that both populations may have the same potential as putative stem cells.

The number of stem cells in a crypt or a gland is presently unknown. Initially, all proliferating cells were believed to be stem cells. Although clonal regeneration experiments using the microcolony assay indicated that intestinal crypts contained a multiplicity of stem cells, it was clear that this was less than the proliferative cellularity. Proposed stem cell numbers have varied from a single stem cell to 16 or more. Others have proposed that the number of stem cells per crypt varies throughout the crypt cycle, with the attainment of a threshold number of stem cells per crypt being the signal for fission to occur. Although all cells in a crypt are initially derived from a single cell, as shown by the chimeric and X-inactivation experiments discussed previously, mutagenesis studies argue strongly for more than one stem cell per crypt, with stochastic clonal expansion of a mutant clone. A three-stem-cell colonic crypt has been suggested on this basis.

In organisms such as *Drosophila* and *Caenorhabditis elegans*, stem cell divisions are known to be asymmetric. We have no such firm concept in the mammalian gut, although there is some evidence to support the proposal. By labeling DNA template strands in intestinal stem cells with tritiated thymidine during development or tissue regeneration, and by labeling newly synthesized daughter strands with bromodeoxyuridine, segregation of the two markers can be studied. The template DNA strand labeled with tritiated thymidine is retained, but the newly synthesized strands labeled with bromodeoxyuridine become lost after the second division of the stem cell. This indicates not only that asymmetric stem cell divisions occur but also that by discarding the newly synthesized DNA, which is prone to mutation, into the daughter cell destined to differentiate, a mechanism of stem cell genome protection is afforded.

When a stem cell divides, the possible outcomes are that two stem cells (P) are produced, that two daughter cells destined to differentiate (Q) cells are produced, or that there could be an asymmetric division resulting in one P and one Q cell. These are sometimes called p, q, and r divisions or p and q divisions. If $p = 1$ and $q = 0$, then regardless of the number of stem cells per crypt, the cells are immortal and there will be no drift in the niche with time. Such a situation is called *deterministic*. However, if $p < 1$ and $Q > 0$ (i.e., a stochastic model), there will be eventual extinction of some stem cell lines and a drift toward a common stem cell from which all other cells derive. We previously described the variation in methylation patterns or "tags" that occur in human colonic crypts and explained that crypts apparently show several unique tags. The variance of these unique tags was compared with those expected using a variety of models, including no drift with aging (the deterministic model), drift with immortal stem cells with divergence (the numbers of unique tags are proportional to the stem cell number), drift with one stem cell per crypt, and a stem cell niche with more recent divergence (with loss of stem cells occurring proportional to the time since divergence). Multiple unique tags were found in some crypts, and the number of unique tags increased with the number of markers counted, which favors random tag drift and multiple stem cells per crypt. The variances were consistent with drift in immortal stem cells, where N (the number of stem cells) = 2, but favored a model where $0.75 < P < 0.95$ and $N < 512$.

Thus, the data supported a stochastic model with multiple stem cells per crypt. However, as in many such attempts, several major assumptions are necessary, such as a constant stem cell number. It is clear that variation in both P and N occur in this model. However, this analysis is consistent with another experiment where the time taken for monoclonal conversion, or the "clonal stabilization time," of OAT+/− individuals to convert to OAT−/− cells following irradiation was found to be about one year. Assuming 64 stem cells per niche and $P = 0.95$, the mean time for conversion should be some 220 days. The same assumptions suggest a bottleneck, where all stem cells are related to the most recent common ancestral cell, occurring every 8.2 years.

Multiple Molecules Regulate Gastrointestinal Development, Proliferation, and Differentiation

Although the molecular mechanisms by which pluripotent stem cells of the gastrointestinal tract produce differentiated cell types are not clearly understood, an increasing number of genes and growth factors have been identified that regulate development, proliferation, and differentiation as well as development of tumors. These are expressed by intestinal mesenchymal and epithelial cells and include members of the fibroblast growth factor family, epidermal growth factor family, TGFβ insulin-like growth factors 1 and 2, HGF–scatter factor, Sonic and Indian hedgehog, and PDGF-α, among others.

Wnt/β-catenin Signaling Pathway Controls Intestinal Stem Cell Function

The Wnt family of signaling proteins is critical during embryonic development and organogenesis in many species. There are 16 known mammalian *Wnt* genes, which bind to receptors of the frizzled (Fz) family, 8 of which have been identified in mammals. The multifunctional protein β-catenin normally interacts with a glycogen synthase kinase 3-β (GSK3-β), axin, and adenomatous polyposis coli (APC) tumor suppressor protein complex. Subsequent serine phosphorylation of cytosolic β-catenin by GSK3-β leads to its ubiquitination and to its proteasomal degradation, thereby maintaining low levels of cytosolic and nuclear β-catenin. Wnt ligand binding to its Fz receptor activates the cytoplasmic phosphoprotein disheveled, which in turn initiates a signaling cascade resulting in increased cytosolic levels of β-catenin. β-Catenin then translocates to the cell nucleus, where it forms a transcriptional activator by combining with members of the T-cell factor/lymphocyte enhancer factor (Tcf/LEF) DNA binding protein family. This activates specific genes, resulting in the proliferation of target cells, for example, in embryonic development (Figure 34-1). In addition to its role in normal embryonic development, the Wnt/β-catenin pathway plays a key role in malignant transformation. Mutations of the APC tumor suppression gene are present in up to 80% of human sporadic co-

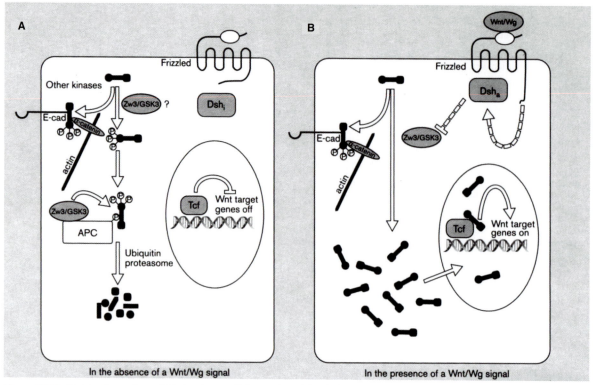

Figure 34-1. Wnt signaling pathway. (A) In the absence of Wnt signaling, disheveled is inactive (Dshi), and *Drosophila* zeste-white 3, or its mammalian homolog glycogen synthase kinase 3 (Zw3/GSK3), is active. β-Catenin (black dumbbell), through association with the APC–Zw3/GSK3 complex, undergoes phosphorylation and. degradation by the ubiquitin-proteasome pathway. Meanwhile, TCF is bound to its DNA binding site in the nucleus, where it represses the expression of genes such as Siamois in *Xenopus*. (B) In the presence of a Wnt signal, disheveled is activated (Dsha), leading to inactivation of Zw3/GSK3 by an unknown mechanism. β-catenin fails to be phosphorylated and is no longer targeted into the ubiquitin–proteasome pathway. Instead, it accumulates in the cytoplasm and enters the nucleus by an unknown pathway, where it interacts with TCF to alleviate repression of the downstream genes and provide a transcriptional activation domain. Willert, K., Nusse, R. *Curr Opin Genet Dev.* 1998 Feb;8(1):95–102

lorectal tumors. This mutation prevents normal β-catenin turnover by the GSK3β/axin/APC complex. This results in increased nuclear β-catenin/Tcf/LEF gene transcription and a subsequent increase in β-catenin-induced Tcf/LEF transcription. One of the main functions of APC appears to be the destabilization of β-catenin. Free β-catenin is one of the earliest events, or perhaps even the initiating event, in tumorigenesis in the murine small intestine and in the human colon. Many genes, including *c-myc, cyclin D1, CD44, c-Jun, Fra-1,* and *urokinase-type plasminogen receptor* have been identified as targets of the β-catenin/Tcf/LEF nuclear complex, although the precise mechanisms that lead to carcinogenesis are not entirely understood.

The Tcf/LEF family of transcription factors has four members; Tcf-1, LEF1, Tcf-3, and Tcf-4. Tcf-4 is expressed in high levels in the developing intestine from embryonic day (E) 13.5 and in the epithelium of adult small intestine, colon, and colon carcinomas. When there is loss of function of APC or mutations in β-catenin, increased β-catenin/Tcf-4 complexes are formed that lead to uncontrolled transcription of target genes. Mice with targeted disruption of the *Tcf-4* gene

have no proliferating cells within their small intestinal crypts and lack a functional stem cell compartment. This suggests that Tcf-4 is responsible for establishing stem cell populations within intestinal crypts; this in turn is thought to be activated by a Wnt signal from the underlying mesenchymal cells in the stem cell niche. Chimeric ROSA26 mice expressing a fusion protein containing the high-mobility group box domain of Lef-1 linked to the transactivation domain of β-catenin (B6Rosa26 <> 129/Sv(Lef-1/β-cat) display increased intestinal epithelial apoptosis. This occurs specifically in 129/Sv cells throughout crypt morphogenesis, unrelated to enhanced cell proliferation. On completion of crypt formation and in adult mice, there is complete loss of all 129/Sv cells. Stem cell selection appears to be biased toward the unmanipulated ROSA26 cells in these chimeras, suggesting that "adequate threshold" levels of β-catenin during development permit sustained proliferation and selection of cells, establishing a stem cell hierarchy. Increased β-catenin expression appears to induce an apoptotic response. Thus, the stem cell niche is unaffected by increased Lef-1/β-catenin during intestinal crypt development.

Transcription Factors Define Regional Gut Specification and Intestinal Stem Cell Fate

HOX GENES DEFINE REGIONAL GUT SPECIFICATION

Mammalian homeobox genes *Cdx-1* and *Cdx-2* display specific regional expression in developing and mature colon and small intestine. During embryogenesis, Cdx-1 localizes to the proliferating cells of the crypts and maintains this expression during adulthood. The Tcf-4 knockout mouse does not express Cdx-1 in the small intestinal epithelium; thus, the Wnt/β-catenin complex appears to induce Cdx-1 transcription in association with Tcf-4 during the development of intestinal crypts. Mice heterozygous for a Cdx-2 mutation develop colonic polyps composed of squamous, body, and antral gastric mucosa, with small intestinal tissue. Proliferation of Cdx-2-colonic cells with low Cdx-2 levels can produce clones of cells phenotypically similar to epithelial cells of the stomach or small intestine. This could indicate a possible homeotic shift in stem cell phenotype. Region-specific genes such as *Cdx-1*, *Cdx-2*, and *Tcf-4* appear to define the morphological features of differential regions of the intestinal epithelium and regulate the proliferation and differentiation of the stem cells.

In the future, it is clear that functional genomics will have an increasing role to play in the study and identification of intestinal stem cells. Studies of a consolidated population of stem cells isolated by laser-capture microdissection from germ-free transgenic mice lacking Paneth cells. There were no less than 163 transcripts enriched in these stem cells compared with normal crypt-base epithelium, which contains a predominance of Paneth cells. The profile showed prominent representation of genes involved in c-myc signaling, as well as in the processing, localization, and translation of mRNAs. Similar studies in the mouse stomach showed that growth-factor response pathways are prominent in gastric stem cells; examples include insulin-like growth factor. A considerable fraction of stem cell transcripts encode products required for mRNA processing and cytoplasmic localization. These include numerous homologs of *Drosophila* genes needed for axis formation during oogenesis.

Gastrointestinal Neoplasms Originate in Stem Cell Populations

We can use the development of colorectal carcinoma as a paradigm. The concept of the adenoma–carcinoma sequence, whereby adenomas develop into carcinomas, is now widely accepted, and most colorectal carcinomas are believed to originate in adenomas. The initial genetic change in the development of most colorectal adenomas is thought to be at the *APC* locus, and the molecular events associated with these stages are clear: a second hit in the *APC* gene is sufficient to give microadenoma development, at least in FAP. There are basically two models for adenoma morphogenesis, both of which closely involve basic concepts of stem cell biology in the

colon. In the first, recently formulated, mutant cells appear in the intracryptal zone between crypt orifices, and as the clone expands, the cells migrate laterally and downward, displacing the normal epithelium of adjacent crypts. A modification of this proposal is that a mutant cell in the crypt base, classically the site of the stem cell compartment, migrates to the crypt apex where it expands. These proposals are based on findings in some early non-FAP adenomas, where dysplastic cells were seen exclusively at the orifices and on the luminal surface of colonic crypts. Measurement of loss of heterozygosity (LOH) for APC and nucleotide sequence analysis of the mutation cluster region of the APC gene carried out on microdissected, well-orientated histological sections of these adenomas showed that half the sample had LOH in the upper portion of the crypts, most with truncating APC mutations. Only these superficial cells showed prominent proliferative activity, with nuclear localization of β-catenin indicating an APC mutation only in these apical cells. Earlier morphological studies have drawn attention to the same appearances. This top-down morphogenesis has wide implications for concepts of stem cell biology in the gut. It is clear that most evidence indicates that crypt stem cells are found at the origin of the cell flux, near the crypt base. These proposals, however, either reestablish the stem cell compartment in the intercryptal zone or make the intercryptal zone a favored locus where stem cells, having acquired a second hit, clonally expand.

An alternative hypothesis proposes that the earliest lesion is the unicryptal or monocryptal adenoma, where the dysplastic epithelium occupies an entire single crypt. These lesions are common in FAP, and although they are rare in non-FAP patients, they have been described. Here, a stem cell acquires the second hit, and then it expands, stochastically or more likely because of a selective advantage, to colonize the whole crypt. Thus, such monocryptal lesions should be clonal. Similar crypt-restricted expansion of mutated stem cells has been well documented in mice after ENU treatment and in humans heterozygous for the *OAT* gene, where after LOH, initially half and then the whole crypt is colonized by the progeny of the mutant stem cell. Interestingly, OAT+/OAT− individuals with FAP show increased rates of stem cell mutation with clustering of mutated crypts. Thus, in sharp contrast, the mutated clone expands not by lateral migration but by crypt fission in which the crypt divides, usually symmetrically at the base or by budding. Several studies have shown that fission of adenomatous crypts is the main mode of adenoma progression — predominantly in FAP, where such events are readily evaluated but also in sporadic adenomas. The non-adenomatous mucosa in FAP, with only one APC mutation, shows a large increase in the incidence of crypts in fission. Aberrant crypt foci, thought to be precursors of adenomas, grow by crypt fission, as do hyperplastic polyps. This concept does not exclude the possibility that the clone later expands by lateral migration and downward spread into adjacent crypts, but with the initial lesion the monocryptal adenoma, this model of morphogenesis is conceptually very different.

Recent work supporting the bottom-up spread of colorectal adenomas looked at a number of small (<3 mm) tubular adenomas. Here, nuclear accumulation of β-catenin was seen, indicating loss of function of one of the genes in the Wnt pathway, most likely APC, with subsequent translocation of β-catenin to the nucleus. Serial sections showed that the β-catenin nuclear staining extended to the bottom of the crypts and was present in crypts in the process of crypt fission. β-catenin expression was particularly marked in the nuclei of buds. At the surface, there was a sharp cutoff between the adenomatous cells in the crypt that showed nuclear β-catenin and those surface cells that did not. The adjacent crypts were filled with dysplastic cells containing nuclear β-catenin, which were not confined to the upper portions of the crypts. In larger adenomas, there was unequivocal evidence of surface cells growing down and replacing the epithelium of normal-looking crypts. Crypt fission was rare in normal and noninvolved mucosa and usually began with basal bifurcation at the base of the gland, whereas in adenomas, fission was commonly asymmetrical with budding from the superficial and mid-crypt. Multiple fission events were frequently observed in adenomas. The crypt fission index (the proportion of crypts in fission) in adenomas was significantly greater than that in noninvolved mucosa.

It is usually stated that adenomas do not display a stem cell architecture, but recent observations show that adenomatous crypts in early sporadic adenomas show superficial similarities to normal crypts in the distribution of their proliferative activity. Observations on possibly older adenomas have suggested that maximum proliferative activity is found toward the top of the crypts, indicating that that migration kinetics are reversed, with cells flowing towards the bottom of the crypt. These observations are corroborated by the report of increased apoptosis at the bottom of adenomatous crypts. Such a distribution could support a top-down mechanism. But there is evidence from examining the methylation histories of cells in adenomas for a discrete stem cell architecture. Moreover, although crypt mitotic scores are significantly greater in adenomas than in noninvolved mucosa and normal controls, the zonal distribution of mitoses in adenomatous crypts mitoses is evenly distributed throughout the crypt, which does not suggest a concentration of dysplastic cells in the tops of adenomatous crypts.

Examination of the adenomas of an XO–XY individual with FAP showed that none exceeded 2.5 mm in diameter. The monocryptal adenomas showed either the XO or the XY genotype and hence are clonal proliferations, as would be expected from the observation that crypts are clonal. Many microadenomas showed the XY genotype, and none of these early lesions showed mixture of XO and XY nuclei occupying the same crypt. It was shown that 76% of the microadenomas were polyclonal. There were also sharp boundaries at the surface between adjacent adenomatous crypt territories.

What are the implications of these considerations for our concepts of the single (stem) cell origin or clonality in colorectal adenomas? We have seen that crypts are clonal units; thus, these lesions would be polyclonal because of the mixture of clonal crypts and clonal adenoma — though in this instance, they have different clonal derivation. A study of

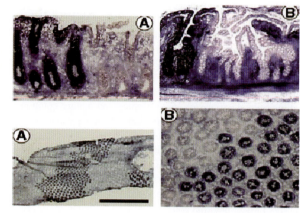

Figure 34-2. Visualization of X-inactivation patches directly by enzyme histochemistry in surgical resections from Sardinian females heterozygous for the G6PD Mediterranean mutation (563 C 3 T), previously shown to have reduced G6PD enzyme activity. Heterozygosity for the G6PD Mediterranean mutation was confirmed by polymerase chain reaction analysis of genomic DNA followed by MBOII restriction endonuclease. (A) G6PD staining in longitudinal sections of colonic crypts. In individual crypts, all epithelial cells show a similar staining pattern. (B) Longitudinal sections of crypts and villi in small intestinal mucosa stained for G6PD activity. Epithelial cells in individual crypts show a similar staining pattern, but the villous epithelium shows areas of positive and negative staining, confirming its polyclonal derivation. (C) G6PD staining in colonic patches in low power. (D) High-power view of large patches of crypts with irregular patch borders. Bar indicates 2 mm. Novelli, M., et al. Proc Natl Acad Sci USA. 2003 Mar 18;100(6):3311–4. 27

both sporadic and FAP adenomas, using X-linked restriction fragment-length polymorphisms, showed that such lesions were apparently monoclonal in origin. On the other hand, the X-linked patch in the colon is large — and can be in excess of 450 crypts in diameter (see Figure 34-2). So, unless an adenoma grows over a patch boundary and involves crypts on either side of that boundary, X-inactivation analysis will always show that such lesions are monoclonal.

On the "bottom-up" proposals, expansion of a clonal monocryptal adenoma by crypt fission would inevitably lead to a monoclonal microadenoma and thus adenoma. However, as mentioned before, studies on the XO–XY individual, who has a mean patch size of 1.48 crypts, indicated that some 76% of adenomas were polyclonal, supported by observations in Rosa26–Min chimeric mice. This could be explained by the transformation of noninvolved crypts by transformed stem cells.

So what is happening to the stem cells in the development of these early, monocryptal lesions? It has been shown that adenomatous crypts in FAP contained two stem cell lines, APC–/+ and APC–/–, and this was interpreted to mean that the APC–/– cells were expanding stochastically within the same crypt. There is good evidence that such monoclonal conversion occurs as a stochastic process. It has previously been shown that a quantitative analysis of age-related methylation suggests that crypts are maintained by niches containing multiple stem cells. Random stem cell loss with replacement suggests that all niche lineages except one will become extinct. It is fairly clear that clonal succession is related to tumor progression. Evidence from the methylation patterns of colonic crypts argues for clonal succession occurring in a stem cell niche without mutation, genetic alterations aiding clonal expansion of a mutant clone might confer a growth advantage. Several mouse genotypes exist in which crypt morphology — and therefore implicitly stem cell proliferative behavior — is normal, such as APC+/−, TP53+/−, Trp53−/−, Apc+/−, Trp53−/−, MLH1−/0, Apc+/−, Mlh1−/−, and Tgfbr2+/−. However, more subtle alterations, such as an increased incidence of crypt fission that may be dependent on stem cell behavior, might well be missed. These are special cases, so consider the situation in a normal individual, where stem cells are wild-type (APC+/+) and one APC allele is lost in a single stem cell. This APC+/− stem cell, unless it possesses a growth advantage, could be lost by the stochastic ejection from the niche. Indeed, it could be argued that this will happen in the overriding majority of instances, but the APC+/− cell could survive and the niche will be populated by progeny of this APC+/− cell, which will resemble an FAP crypt. A further hit in this crypt will lead to the formation of the monocryptal adenoma on the model proposed previously.

Others have also concluded that migrating crypt epithelial cells in the upper part of the crypt are the primary targets for transformation by APC mutation, and this has received some experimental support. At the bottom of the crypt, progenitor cells accumulate nuclear β-catenin and express Tcf target genes as a result of Wnt stimulation from surrounding basal pericryptal myofibroblasts. In normal crypts, cells that reach the mid-crypt region down-regulate β-catenin/Tcf, resulting in cell cycle arrest and differentiation. Cells that bear a β-catenin or an APC mutation do not respond to signals controlling β-catenin/Tcf activity, and these cells continue to behave as crypt progenitor cells in the surface epithelium, generating microadenomas. Computer modeling has been used to suggest that an expansion in the crypt stem cell population explained the putative proliferative abnormality in FAP, namely, an upward shift in the proliferative compartment toward the top of the crypt. Simulation of labeling index distribution curves from FAP crypts using a single mechanistic design was able to fit the data from both control and FAP crypts, indicating that the proliferative abnormality does not alter the rate of cell cycle proliferation, differentiation, or apoptosis of proliferating crypt cells. Instead, it suggests an expansion in the crypt stem cell population sufficient to explain the observed proliferative abnormality in FAP. Thus, β-catenin signalling in the colonic crypt controls the number of stem cells. The stem cell population is expanded in FAP crypts because of a germ-line APC mutation activating Tcf-4. Any increase in the size of the stem cell population might be expected to result in an increase in the rate of crypt fission.

Crypt fission is therefore an essential event in the expansion of mutated clones in adenomas. Although the morphology of this process is distinct, the molecular mechanisms that govern it are far from clear. We further conclude that the initial event in the genesis of colorectal adenomas, of both sporadic and FAP adenomas, is the monocryptal adenoma. Initial growth occurs through crypt fission, and spread into adjacent crypt territories is a later, secondary event.

Summary

The cells of the gastrointestinal tract undergo constant renewal and respond to damage by regeneration and repopulation. Each region of the gastrointestinal tract is morphologically distinct with its own repertoire of cell types. Although the stem cells are the most important cells of the gastrointestinal tract, responsible for the production of every other cell type in the gastrointestinal mucosa, they have not yet been closely characterized. There is, initially at any rate, a single stem cell in every intestinal crypt or gastric gland that indirectly generates a clone containing further stem cells, transit amplifying and differentiated cells, through the production of committed progenitor cells. This cell also produces new crypts by crypt fission, repairs entire crypts and villi when damaged, and generates gastrointestinal tumours. The stem cell or cells occupy a niche, formed by mesenchymal cells such as the ISEMFs, and extracellular matrix molecules, which regulate epithelial stem cells through mesenchymal-epithelial cross-talk.

The molecular events that regulate the development of the gastrointestinal tract and epithelial cell turnover in the normal tissue and in formation of carcinomas are beginning to be identified. It is clear that the Wnt/β-catenin signaling pathway and downstream molecules such as APC, Tcf-4, Fkh-6, Cdx-1, and Cdx-2 are vital for normal gastrointestinal stem cell function. We are beginning to identify molecular pathways that determine further proliferation of committed progenitor cells into specific epithelial cell lineages: The Notch–Delta signaling pathways involving Hes1 and Math1 transcription factors regulate differentiation of goblet, Paneth, and enteroendocrine cells in the small intestine. Factors secreted and expressed by the mesenchymal cells (KGF, HGF, etc.) that regulate gastrointestinal mucosal development and epithelial proliferation are rapidly becoming identified. Epithelial and intestinal subepithelial myofibroblast lineages apparently derived from bone marrow, presenting the possibility of delivering therapeutic genes to the damaged lamina propria — for example, in diseases that cause fibrosis, such as Crohn's disease — and even repopulating the damaged gut. Finally, we conclude that the colonic stem cell is pivotal in understanding mechanisms of tumorigenesis in the colon. The isolation and characterization of gastrointestinal stem cells is a priority in gut biology.

KEY WORDS

Crypt clonality Intestinal crypts, wherein all cell lineages are originally derived from a single cell during embryonic development.

Gastrointestinal stem cells Founder cells capable of limitless self-renewal throughout the lifetime of their host. Gastrointestinal stem cells are also responsible for the maintenance of tissue homeostasis, by producing daughter cells committed to multipotent differentiation to form all adult lineages within the gastrointestinal tract.

Gastrointestinal stem cell signaling Elucidation of the components involved in molecular pathways, such as Wnt/β-catenin, which regulate cell proliferation and differentiation in the gastrointestinal tract, will provide an insight into the mechanisms of normal epithelial cell turnover and neoplastic development.

Gastrointestinal tumor morphogenesis The mode of dysplastic growth within the gastrointestinal tract is a topic of heated debate, with two main conflicting theories. The first "bottom-up" theory proposes that neoplastic lesions arise in the base of the crypt, the "stem cell zone" and proliferate upwards. The alternative, "top-down" theory states that lesions arise within an "intercryptal" zone, and spread laterally and down into adjacent crypts.

Stem cell plasticity (in the gastrointestinal tract) Evidence suggests that a population of stem cells in the adult bone marrow can engraft within the gastrointestinal tract and form both epithelial and mesenchymal lineages, with an increased propensity in damaged tissue. However, the mechanisms by which bone marrow cells form gastrointestinal lineages are unresolved.

FURTHER READING

Brittan, M., and Wright, N. A. (2004). Stem cells in gastrointestinal structure and neoplastic development. *Gut* **53**, 899–910.

Garcia, S. B., Novelli, M., and Wright, N. A. (2000). The clonal origin and clonal evolution of epithelial tumours. *Int. J. Exp. Pathol.* **81**(2), 89–116.

Giles, R. H., vanEs, J. H., and Clevers, H. (2003). Caught up in a Wnt storm: Wnt signaling in cancer. *Biochemica et Biophysica Acta* **1653**, 1–24.

Preston, S. L., Wong, W. M., Chan, A. O., Poulsom, R., Jeffery, R., Goodlad, R. A., Mandir, N., Elia, G., Novelli, M., Bodmer, W. F., Tomlinson, I. P., and Wright, N. A. (2003). Bottom-up histogenesis of colorectal adenomas: origin in the monocryptal adenoma and initial expansion by crypt fission. *Cancer Res.* **63**(13), 3819–3825.

Wagers, A. J., and Weissman, I. L. (2004). Plasticity of adult stem cells. *Cell* **116**, 639–648.

Characterization of Human Embryonic Stem Cells

Holly Young and Melissa K. Carpenter

Pluripotent embryonic stem cells (ESCs) have indefinite replicative potential and the ability to differentiate into derivatives of all three germ layers that arise in mammalian development: ectoderm, endoderm, and mesoderm. Mouse ESCs were first isolated over 20 years ago by removing the inner cell mass (ICM) from mouse preimplantation embryos and serially passaging these cells on inactivated mouse feeder layers. The resulting cell lines showed unrestricted proliferative potential and were pluripotent. These ESCs differentiated into derivatives of all three germ layers, and upon injection into blastocysts, they contributed to all the tissue types in the resulting chimera. Because these cell lines can be genetically modified, they have been largely used for the generation of transgenic or knockout mice, but they also provide a unique tool for studying early development of the embryo.

The first human ESC (hESC) lines were derived by Thomson *et al.* (1998) from the ICM of surplus human embryos resulting from assisted reproductive procedures. Since the first report in 1998, hESC lines have been derived in multiple laboratories throughout the world (e.g., see NIH registry Web site at <wa>http://escr.nih.gov<t>). At present, there are many hESC lines. However, not all lines have been fully characterized. The defining features of ESCs include; derivation from preimplantation embryos, pluripotency, capacity for prolonged proliferation, and self-renewal, a normal euploid karyotype, and the expression of distinctive markers. Not only the derivation methods but also the culture conditions and handling of ESC lines determines their successful maintenance. Existing hESC lines were derived in different laboratories using different procedures, which may result in cell lines with different qualities. Although some of the lines have been characterized extensively, most cell lines have not been available long enough to allow full characterization. Therefore, further study will be required to definitively determine their differences and similarities.

ESCs grow as tightly compacted colonies of undifferentiated cells. The undifferentiated cells require a substrate such as feeder cells or matrix as well as specialized media. Although mouse ESCs were originally derived and maintained on feeder-layers, they can also be maintained in feeder-free culture conditions in which the media are supplemented

with leukemia inhibitory factor (LIF). Activation of the LIF pathway appears to be required for self-renewal of mouse ESCs. The human cells do not have the same response to LIF, suggesting that mouse ESCs and hESCs require different signals to maintain self-renewal and pluripotency. Typically, hESCs are instead maintained in media containing fetal bovine serum or serum replacement media supplemented with basic fibroblast growth factor and require the presence of feeder cells or conditioned medium from feeders. These culture systems maintain undifferentiated hESCs in distinct colonies, but the culture contains a heterogeneous population of cells that has been difficult to characterize consistently.

Identification of reliable markers for the characterization of hESCs is of great importance in order to exploit their potential. The comparison of expression patterns of mouse ESCs and hESCs presents similarities and dissimilarities between the species. Sato *et al.* (2003) compared the transcriptional profile of one hESC line (H1) to the published profiles for mouse ESCs. This study identified a set of 918 genes enriched in undifferentiated hESCs compared to hESCs that had undergone differentiation. This set of genes included ligand–receptor pairs for the FGF, TGF-β–BMP, and Wnt signaling pathways. Some of these components corresponded to genes previously identified as *stemness* genes, indicating that a *core molecular program* may exist. In contrast, Ginis *et al.* (2004) found a number of differences between expression profiles of mouse ES cells (D3) and three hES cell lines (H1, H9, and I6) using immunocytochemistry, reverse transcription-polymerase chain reaction (RT-PCR), and microarray analysis. As summarized in Table 35-1, ES cells from both species expressed transcription factors associated with pluripotent cells, such as Oct-4, Sox2, Tert, Utf1, and Rex-1. Using RT-PCR and immunocytochemistry, significant differences were found in expression of vimentin, β-tubulin, α-fetoprotein, eomesodermin, HEB, ARNT, Foxd3, and the LIF receptor. Focused microarrays revealed profound differences in cell cycle regulation, control of apoptosis, and cytokine expression. These findings confirm that there are many differences between mouse and hES cells and demonstrate the need to use multiple methodologies to characterize these differences.

To date, hESC lines have been characterized using markers previously used to characterize other pluripotent stem cell populations such as human embryonal carcinoma (EC) cells and mouse ESCs. Similar to mouse ESCs, the hESCs express alkaline phosphatase-related antigens. The undifferentiated hESCs express the globoseries glycolipid antigens designated stage-specific embryonic antigen (SSEA-3 and SSEA-4), which are expressed by human EC cells but not by mouse ESCs. In contrast to the mouse ESCs, the undifferentiated hESCs lack expression of a lactoseries oligosaccharide antigen SSEA-1. The keratin sulfate-related antigens (tumor-recognition antigens), TRA1-60 and TRA1-81, expressed by human EC cells are also expressed by hESCs. Remarkably, all of the hESC lines derived to date show similar expression of these markers (Table 35-2). Although the expression of these markers has been well characterized in pluripotent cells, the function of these markers has yet to be determined.

We compared the expression of these markers in four different cell lines derived in a single laboratory and maintained in feeder-free conditions using quantitative analyses. We found that an average of 80–95%, 91–94%, and 88–93% of the cells in cultures expressed SSEA-4, Tra1-60, and Tra1-81, respectively. The overall expression of surface markers is similar, although with extensive analysis, some small but statistically significant differences in the expression of markers among cell lines become apparent. Since hESCs have unlimited proliferative capacity, we also performed quantitative analyses of marker expression in several hESC lines over long-term culture in feeder-free conditions. hESC cultures from individual lines maintain high levels of expression (>80% of the cells) even after one year in continuous culture. Although cultures from the various lines appear to be more similar than different, it is important to recognize that hESC cultures contain undifferentiated hESCs alongside a small proportion of differentiated cells, making it extremely difficult to determine more subtle differences among cell lines. Further analysis will be required to determine whether variations in marker expression correlate with other characteristics of the cells, such as the capacity to differentiate.

hESCs express integrins that allow the cells to grip the matrix and each other. Extracellular matrix components are quite important in the early developing embryo. Laminin is the first extracellular matrix protein expressed in the two- and four-cell stage mouse embryo, and the laminin receptor is highly expressed in murine ES and EC cells. This information, and the finding that hESCs require a matrix for mainte-

TABLE 35-1
Comparison of Mouse and Human ES Cells

	Mouse ES	Human ES
Alkaline Phosphatase	+	+
SSEA-1	+	−
SSEA-3	−	+
SSEA-4	−	+
TRA-1-60	−	+
TRA-1-81	−	+
OCT 3/4	+	+
SOX 2	+	+
REX 1	+	+
TERT	+	+
FGF4	+	+
UTF-1	+	+
FOXD3	+	−
CX45	+	+
CX43	+	+
BCRP-1	+	+
LIFR	+	−
gp130	+	+
STAT3	+	+
Nanog	+	+

TABLE 35-2
Expression of Markers on Human ES Cell Lines

Cell Line		Karyotype	SSEA-1	SSEA-3	SSEA-4	TRA-1-60	TRA-1-81	Alkaline Phosphatase	CD90	AC133	OCT4	hTERT	Cripto	GCTM-2	TG343	Genesis	GDF3	Feeder-free culture	Cryopreservation	SOX2	Rex1	UTF-1	Nanog
H1	WA01	XY	−	+	+	+	+	+	+	+	+	+	+					+	+	+	+	+	
H7	WA07	XX	−	+	+	+	+	+	+	+	+	+	+					+	+				
H9	WA09	XX	−	+	+	+	+		+	+	+	+						+	+	+	+	+	
HES-1	ES01	XX	−	+	+	+	+	+			+		+	+	+	+	+		+				
HES-2	ES02	XX	−	+	+	+	+	+			+		+	+	+	+	+		+				
HES-3	ES03	XX	−	+	+	+	+	+			+		+	+	+	+	+		+				
HES-4	ES04	XY	−	+	+	+	+	+			+		+	+	+	+	+		+				
HES-6	ES06	XY	−		+	+		+			+		+	+	+	+	+		+				
HSF-6	UC06	XX	−	+	+																		
hESBGN.01	BG01	XY	−	+	+	+	+	+			+	+								+	+	+	
hESBGN.02	BG02	XY	−	+	+	+	+	+			+	+								+	+	+	
Miz-hES-1*	MI01	XY	−	+/−	+	+	+	+			+												

nance of the undifferentiated phenotype, led us to evaluate the expression of integrins in hESCs maintained on feeders or in feeder-free conditions. Cells in both conditions express high levels of the integrins called α6 and β1, moderate levels of integrin α2, and low levels of integrins α1, α2, α3, and β4, all noted for binding to laminin. This is consistent with the finding that the cells grow well on laminin and matrigel (the major component of matrigel is laminin). hESCs are also connected to each other by gap junctions as indicated by the expression of connexin 43. The gap junctions are pores in the cell membrane that physically connect the cells. This was confirmed using dye transfer studies demonstrating functional gap junctions between the hES cells.

As previously mentioned, expression of transcription factors is also used to characterize the hESCs. The POU transcription factor, Oct-4, is expressed in pluripotent cell populations such as early embryo cells, blastomeres, and germ cells. *In vitro*, Oct-4 is expressed in pluripotent cells such as EC, ES, and embryonic germ cells. Furthermore, it has been demonstrated that Oct-4 is necessary to retain pluripotency in murine embryonic cells and that the level of expression can influence the status of pluripotency and direct the cells toward specific phenotypes. In hESCs derived in various laboratories, Oct-4 is expressed in the undifferentiated cells and is downregulated upon differentiation. Other markers include hTert, the catalytic component of telomerase, cripto, a co-receptor essential in embryonic development, and the newly identified nanog, a factor involved in stem cell self-renewal. Quantitative comparison of the transcription factors Oct-4, hTert, and cripto shows similar expression among examined cell lines. The level of expression of these factors is maintained for more than one year in continuous culture.

Several studies have used microarrays to compare expression of markers among cell lines. In our lab, the H1, H7, and H9 lines analyzed by microarray show no major differences between lines. These cultures, all at relatively low passage (p28–p37), contain abundant undifferentiated colonies surrounded by stroma-like cells. When later passage cultures are analyzed, some small differences are detected. This most likely results directly or indirectly from a loss of stroma-like cells in later passage cultures. Further analysis using microarrays has been performed, comparing six hESC lines derived in different labs. All lines have significant homology in gene expression, including expression of Sox2, Nanog, GTCM-1, connexin 43, Oct-4, and cripto. These data indicate that hESC lines obtained from different sources and maintained in similar culture conditions share similar expression profiles. Again, it should be noted that hESC cultures represent a heterogeneous population of cells, containing undifferentiated cells as well as some spontaneously differentiated cells, and that differences in culture conditions may alter the signature profile of any cell line. As such, a detailed analysis must be considered to profile all existing lines.

Cytogenetic analysis of hESC lines show stable XX and XY karyotypes from multiple labs. Our lab has used G-banding to show that the cells retain a normal karyotype after one year in culture. We found that as many as 20% of the cultures contained some aneuploid cells. However, very few cultures showed high levels of aneuploidy. It is unclear whether the presence of aneuploid cells in the hESC cultures will alter the fundamental characteristics of the cells or affect growth rate, cell cycle regulation, and capacity to differentiate. The occasional identification of atypical cells, albeit at a low frequency, highlights the importance of careful monitoring.

Considerations for the Derivation of New Human Embryonic Stem Cell Lines

Our knowledge of appropriate phenotypic and molecular markers and of functional criteria to evaluate and compare various hESC lines is becoming more sophisticated. Still, although most well-characterized cell lines share fundamental expression patterns, these lines also demonstrate distinct differentiation and growth patterns that have yet to be quantified. This raises the possibility that the differing approaches used to derive these cell lines may account for some of these unique properties. Unlike mouse ESC lines, variations in stage and quality of embryos and in genetic background of the donors used to derive hESCs are not as tightly controlled and may contribute to differences in the lines. Establishing and expanding newly derived hESC cultures is a complex *in vitro* selection procedure involving numerous manipulations and processing. Therefore, the conditions used to derive hESC lines may determine properties such as cell cycle status, growth-factor responsiveness, epigenetic status, and differentiation capacity. Current information on the behavior of established hESC lines is important to consider when moving toward derivation of new cell lines and understanding the molecular basis that may account for different hESC properties. Based on the available information, we have listed some of the experimental parameters that may affect the nature of hESC lines in Table 35-3. As new cell lines are derived, it will be important to determine which (if any) of these parameters affects the status of the hESCs.

TABLE 35-3

Considerations for the Derivation of New Human ES Cell Lines

Factors for Consideration	Experimental Variations
I Removal and isolation of intracellular mass (ICM)	— Immunosurgery of trophoectoderm — Laser ablation of trophoectoderm — Direct "hatching" of ICM
II Stage of quality of embyro	— Fresh vs. frozen embryo — Clinical "grade" of embryo — Developmental stage e.g.) 8 cell vs. blastocyst
III Culture conditions for establishment and expansion	— Direct co-culture on MEF — Direct co-culture on human feeders — Use of MEF-CM and matrix — Serum-free conditions with defined factors

Summary

Overall, several useful markers are available for the assessment of hESCs. However, it is unclear whether these markers will be predictive of the differentiative capacity of the hESCs. Continued characterization of the existing and newly created hESCs is required to understand the mechanisms involved in retaining the undifferentiated state of the hESCs as well as the mechanisms involved in differentiation.

FURTHER READING

Carpenter, M. K., Rosler, E., Fisk, G., Brandenberger, R., Ares, X., Miura, T., Lucero, M., and Rao, M. S. (2004). Properties of four human ES cell lines maintained in a feeder free culture system. *Dev. Dyn.* **229**, 243–258.

Ginis, I., Luo, Y., Miura, T., Thies, R., Brandenburg, R., Gerecht-Nir, S., Amit, M., Hoke, A., Carpenter, M. K., Itskovitz-Eldor, J., and Rao, M. S. (2004). Differences between human and mouse embryonic stem cells. *Dev. Bio.* **269**, 360–380.

Rosler, E., Fisk, G., Ares, X., Irving, J., Miura, T., Rao, M. S., and Carpenter, M. K. (2004). Long term culture of human embryonic stem cells in feeder-free conditions. *Dev. Dyn.* **228**, 259–274.

Sato, N., Sanjuan, I. M., Heke, M., Uchida, M., Naef, F., and Brivanlou, A. H. (2003). Molecular signature of human embryonic stem cells and its comparison with the mouse. *Dev. Biol.* **260**, 404–413.

Thomson, J. A., Itskovitz-Eldor, J., Shapiro, S. S., Waknitz, M. A., Swiergiel, J. J., Marshall, V. S., and Jones, J. M. (1998). Embryonic stem cell lines derived from human blastocysts. *Science* **282**, 1145–1147.

Isolation and Maintenance of Murine Embryonic Stem Cells

Martin Evans

Mouse embryonic stem (ES) cells were first isolated more than 20 years ago. Their isolation and the conditions for their growth and maintenance are well described both in the original references and in numerous reviews and methods texts. This chapter is therefore written more as a commentary than a cookbook.

I first discuss the growth and maintenance of cultures of mouse ES cells because their isolation depends on the secure ability to maintain them. I would recommend that an established ES cell line be used to optimize growth conditions before attempts are made to isolate new ones.

ES cells grow well in culture and are not particularly fastidious about the media, but it must be remembered that ES cell cultures are essentially primary cultures. Therefore, it is necessary to use conditions of tissue culture that maintain their primary properties and do not select variants. Any growth in nonoptimal conditions will lead to selective pressure and the appearance of better-growing but worse-differentiating strains. In some cases, these may be recognized as chromosomal variants, but in others there is no gross karyotypic change identifying the altered cell strain. They will certainly, however, be less able to differentiate and will produce abnormal chimeras.

Maintenance of ES Cells

There are major requirements for ES cell culture in the context of maintaining their totipotency necessary for genetic engineering and gene-targeting projects:

- The cells are maintained in their undifferentiated state.
- The cells maintain their normal capability and range of differentiation.
- The cells retain a normal karyotype, a prerequisite for germ-line transmission.

To maintain the stem cell state (suppression of differentiation), the cells need to be cultured with mitotically inactivated feeder cells or in the presence of leukemia inhibitory factor (LIF). Each method has advantages and disadvantages. Feeders provide a more robust approach, for the supplementation of media by growth factors will lead to a more pulsatile

addition of the factors and will depend more on meticulous attention. Conversely, sufficient provision of feeders requires setting them up and inactivating at each passage. Growth in the presence of LIF may seem less troublesome and makes certain types of experiments much easier, but it can have its disadvantages. Some ES cell lines seemingly grow better on feeders than in the presence of LIF, whereas others can be switched between the two conditions with little trouble.

By far, the best practical method for maintaining totipotent cells in tissue culture remains the use of fibroblast feeders (STO or primary embryo fibroblast). For any serious long-term investment in ES cells, it is wise to expand and freeze ES cells over several passages, checking the karyotype (and germ-line transmission) at each passage. Use should be made only of ES cells with proven karyotype and germ-line transmission properties for critical experiments.

ES cell cultures are essentially primary cultures, and it is necessary therefore to use conditions of tissue culture that maintain their primary properties and do not select variants. It appears that extremes of growth conditions, either by plating the cells too sparsely or by allowing the cultures to become too dense and exhausting their medium, will encourage loss of the desired normal characteristics of ES cells. Either situation sets up conditions in which selection for abnormal (fast-growing, aneuploid) cells can occur. Healthy cultures of ES cells grow with a population doubling time of 15 to 20 hours. In practice, therefore, it is necessary to subculture ES cells about every three days and to renew the medium regularly.

Inasmuch as ES cells are not particularly fastidious about the medium in which they are grown, the optimum formulation I have used is Dulbecco's modified Eagles medium (DMEM) high-glucose, low-pyruvate formulation mixed 1:1 with Ham's F12M medium. This peculiar mix is a compromise between the high-yielding DMEM, originally designed for maximal growth of tissue-culture cell lines for virus production, and the finely balanced Ham's F12, originally developed for the clonal growth of cells. I originally supplemented the DMEM with nonessential amino acids (NEAA) and a mixture of nucleosides (adenosine, guanosine, cytidine, and uridine) to a final concentration of $30\,\mu M$ and thymidine to a final concentration of $10\,\mu M$. This supplement has not been tested properly when used in a DMEM/F12 mix, but it nevertheless may have an advantageous effect upon the primary cell isolation. ES cells grow happily in DMEM alone, but they grow better in DMEM + NEAA and better still in DMEM/F12. The only disadvantage of the latter is that it becomes acidified faster and the medium needs to be changed more frequently. These media may be prepared conveniently "in house" from premixed powder. Alternatively, 10× concentrates may be used, or a medium may be bought ready to use at 1× concentration. The water used must be of the highest purity, and glassware cleanliness is critical. These latter points have proved to be the most important in establishing new ES cell lines. I suspect that the most damaging contaminants are detergent residues and that their effects are not necessarily seen in bulk passage cultures. The most useful tool for optimization of media, sera, and other conditions is the cloning efficiency test (explained later in this chapter), which should also be used for any troubleshooting.

Media

The medium recommended for ES cells is as follows:

- DMEM/F12 (or DMEM)
- Glutamine (stored frozen as 100× concentrate of 200 mM)
- β-mercaptoethanol to a final concentration of 10^{-4} M either 4-μl neat in 500 ml or from a 10^{-2} M (100×) concentrate. (Prepared by adding 72-μl to 100-ml PBS)
- 10% calf serum and 10% foetal calf serum
- NEAA from 100× concentrate
- Nucleosides from 100× concentrate (if required)
- Antibiotics (if desired) — use only penicillin, streptomycin, kanamycin, or gentamicin. Do not use antimycotics.

Sera

Both fetal calf serum and newborn calf serum are used. It is a mistake to imagine that the higher price of fetal calf serum makes it better. Some newborn sera are excellent, and some fetal calf sera are extremely toxic to ES cells. Empirical selection is imperative. All serum should be carefully batch tested using a cloning efficiency test (explained later in this chapter) and should be ordered in a sufficient lot for the experiments planned. Look both for lack of toxicity (aim for a cloning efficiency of 20% or more) and growth promotion (the size of colonies). Aim to buy new serum batches that equal or exceed the quality of the presently used (control) sera. Be cautious with sera that show toxicity at high (e.g., 40%) concentrations; they can be excellent at lower levels, but other sera can be just as good without extra toxicity.

Colony-Forming Assay for Testing Culture Conditions

The colony-forming assay described should be used to test all the components of the media as well as some culture procedures (e.g., viability of cells after different regimens of dissociation):

- Set up experimental media in six-well cluster dishes (2 ml/well) at twice the intended final concentration of additives,

and equilibrate in the incubator at 37°C, 5% CO_2, in air before use. If you are using feeder cells, these need to be added either at this stage or with the sample cells. Use duplicate wells for each condition and appropriate controls (e.g., a known "good" serum batch). For FCS batch testing, use growth media supplemented with the FCS batch at 5%, 10%, and 20%. All other components (e.g., mercaptoethanol concentrations and batches of media) may be tested in the same way.

- Disaggregate ES cell stock culture on day 2 after plating (i.e., semiconfluent), ensuring as close to a single-cell suspension as possible. In a test of sera, remember, if using trypsin, to inactivate by a wash in serum containing medium before redispensing into serum-free medium. Count and resuspend in growth medium at a density of 10^3 cells/ml.
- Add 2-ml cell suspension to each well (making 4-ml total volume 1× concentration of additives), and return to the incubator.
- Incubate 6 to 8 days at 37°C, 5% CO_2 in air.
- Fix and then stain plates with Giemsa. ES cell colonies can be identified by characteristic morphology and dark staining properties. Differentiated colonies are paler in color. (It is useful to check sample colony appearance before fixing and staining by inverted phase contrast.) Count the total number of colonies (the plating index) and the proportion of ES cell colonies. In tests of LIF or feeders, the maintenance of undifferentiated colonies is important.

I still recommend the use of inactivated feeder layers for the maintenance of ES cells. I have found that the attraction and convenience of feeder-free methods are outweighed by the frequency with which stocks deteriorate under these conditions and would therefore prefer that an important seed stock be maintained with feeders. It is possible to use a belt-and-braces policy — to use feeders and added LIF. For feeders, use either STO cells or primary mouse embryo fibroblasts.

STO cells are routinely grown in DMEM supplemented with 10% newborn calf serum. Remember that STO cells are effectively 3T3 cells and should be passaged promptly when they reach confluence to prevent the accumulation of non-contact-inhibited cells in the population. Passage STOs at 5×10^3 cells per cm^2 and expect a harvest of up to 20 times this amount. If a confluent 10 cm Petri dish yields >10^7 cells, then the cells are losing their contact inhibition and will no longer produce a good feeder layer. Either replace them with a new batch or (and effectively) clone them out by seeding at ~100 cells per dish and select a flat clone to establish a new stock. Some workers prefer to use primary embryo fibroblasts to prepare feeder layers rather than STO cells, because they feel that better ES cell growth can be obtained. This may, however, reflect abused STO cells rather than an intrinsic superiority of primary embryo fibroblasts.

Feeders are prepared in the following manner:

- Remove the medium from a confluent 10-cm dish of STO cells, and replace it with DMEM/10% NCS plus 10 μg/ml of mitomycin C. (Stock mitomycin C is made at 2 mg/ml with PBS and can be stored at 4°C for two weeks.)

- Incubate plates for 2 to 3 hours. Avoid longer exposure to mitomycin C.
- Remove mitomycin C from the STO cells, and wash each plate three times with 10-ml PBS.
- Trypsinize, resuspend, and pellet the cells by centrifugation. This is an important washing step.
- Resuspend in growth medium and seed onto gelatinized tissue culture plates at 5×10^4 cells per square centimeter. These can be kept for later use (up to a week), or the inactivated cells can be stored for later use in suspension in a serum-containing medium at 4°C for a similar time.

ES Cell Passage Culture

Passage and maintenance of the ES cells is straightforward; bear in mind the following:

The cells should not be allowed to become so confluent that the medium becomes excessively acidified. Given the chance, they will continue dividing until they start to kill themselves by overcrowding and exhaustion of the medium; avoid this. Expect to harvest about 3×10^5 cells per square centimeter and plate at $2 \times 10^4/cm^2$. Feed regularly and passage every three days. It is important that you seed the cells as a suspension of single cells with few, if any, aggregates. Such aggregates will initiate differentiation and can severely complicate the ES culture for many purposes. Cells are best disaggregated by careful washing with PBS (remember, you have 20% serum in the medium, and this needs to be fully removed) followed by trypsin/EGTA (0.125% trypsin with 10^{-4} M EGTA in PBS — an improvement on trypsin/EDTA). Incubation at room temperature is sufficient. (It is also possible to disaggregate ES cells by a prolonged incubation in PBS/EGTA, but their viability is reduced.) Immediately after the trypsin incubation, add growth medium and pipette up and down a few times to generate single cells. Because it is important that the cells are well disaggregated, it is good to check by counting in a hemocytometer. You should see a mainly single-cell suspension with round, phase-bright cells. Dead cells or clumps are obvious; ragged and larger cells are feeders.

Isolation of New ES Cell Lines

Contrary to some well-expressed opinions, clearly it is possible to isolate ES cells from a variety of mouse strains and not just from strain 129. I append a prescriptive description of ES cell isolation (given later in this chapter), but first it is useful to consider the background and to point out that isolation is possible from the cleavage to early postimplantation stages and using different procedures.

Embryonic development starting from a single zygote proceeds through extensive cell proliferation and progressive cellular differentiation. At early stages some cells have wide prospective fates, but these are not necessarily self-renewing populations. Pluripotential stem cells were first identified experimentally in the mouse as the stem cell of teratocarcinomas by Kleinsmith and Pierce, who were able to passage

the tumor by transferring single cells isolated from embryoid bodies. These testicular teratocarcinomas in mice arise spontaneously in specific genetic strains and were extensively studied by Stevens (1970). The stem cells of these tumors arise from the primordial germ cells during gonad formation, and one question is whether this represents a parthenogenetic activation. This idea — that the teratocarcinoma stem cells might be embryo related rather than germ cell related — was tested directly by transplanting early embryos to adult testes or kidneys. Passageable, progressively growing teratocarcinomas were formed after transplantation of 1–3.5 day preimplantation embryos. From some of these tumors, clonal *in vitro* differentiating embryonal carcinoma cell lines were isolated and characterized. These were the direct forerunners of ES cells and their isolation. They demonstrated their pluripotentiality by forming well-differentiated teratocarcinomas upon reinjection into mice. These also differentiated extremely well *in vitro* via an embryo-like route. A series of observations of the properties of these and similar embryonal carcinoma (EC) cells started to provide convincing evidence of their homology with the pluripotential cells in the normal embryo, which were able to generate teratocarcinomas under conditions of ectopic transplantation. There seemed to be every reason to suppose that direct isolation into tissue culture should be possible.

Teratocarcinomas can also be formed from postimplantation mouse embryos. The embryonic part of the postimplantation mouse embryo undergoes gastrulation and forms an embryonic mesoderm by invagination from the ectoderm; this lies between the ectoderm and an embryonic endoderm. Skreb and his colleagues isolated these three layers of the rat embryo by microdissection and tested their developmental potency by ectopic transplantation. Only the embryonic ectoderm produced teratomas with multiple tissue types and was hence pluripotential, but these were not progressively growing teratocarcinomas. When these experiments were repeated with gastrulating mouse embryos, it was discovered that in contrast to the rat, transplantable teratocarcinomas were formed. At least, therefore, in the mouse, pluripotential stem cells could be recovered until about the end of gastrulation. Direct isolation of ES cells, however, from such a late stage has not been reported.

Taking these comparisons and molecular data into consideration, we have argued that mouse ES cells are — regardless of the route of isolation — homologous to the early postimplantational epiblast rather than the inner cell mass (ICM), as is often erroneously stated.

The culture conditions — feeder cells and media — had been refined by culturing both mouse and human teratocarcinoma EC cells. Evans and Kaufman, by using implantationally delayed blastocysts as the embryo source, succeeded in establishing cultures of pluripotential cells directly without an *in vivo* tumor step. Subsequent studies showed isolation of such cell lines from both inbred and outbred strains, from normal 3.5-day blastocysts, and those that had been implantationally delayed. Martin in an independent work showed that isolated 3.5-day ICMs may also be used. These cells, which have become known as ES cells, share the properties of indefinite proliferative capacity, embryonic phenotype, and differentiative capacity with their forerunners — the EC cells; in addition, being primarily derived, they may be kept entirely karyotypically normal.

Eistetter reported that ES cell cultures may be established from disaggregated 16–21 cell morulae with an apparent immediate growth of the colonies from some of the explanted single cells. Up to four separate ES cell colonies were founded from a single embryo.

The latest stage of isolation into cultures of ES cells from embryos has been reported by Brook and Gardner, who used 4.5-day-old hatched, peri-implantational embryos flushed from the uterus. They microdissected the epiblast cells from both the primary endoderm and the trophectoderm. Both whole epiblasts and those that had been disaggregated into single cells readily generated ES cultures — if they were not left in contact with the endoderm. Brook and Gardner also reconfirmed the benefit of using delayed blastocysts, showing that the isolated epiblasts of these were the most efficient source of ES cell cultures.

Method for Derivation of ES Cells

I would still recommend the original method of explantation of implantationally delayed blastocysts as the most effective method. It works well and involves no microdissection or immunosurgery.

First, it is essential to pay attention to the optimization of the culture conditions, as described previously. If things don't work, suspect the purity of the media and particularly any possible contamination by detergent residues. Seemingly stupid problems such as aliquoting the medium into small batches and using untested vessels (e.g., bijou bottles) can arise. Suspect specific items that have not been tested in the ES cell cultures — for instance, the cleanliness of the glass Pasteur pipettes from which micropipettes are pulled for manipulation of the embryos and growth of ES cell colonies.

Delayed blastocysts:

- Mate mice by caging together overnight and observing mating plugs in the morning. Separate the plugged females.
- On day 2 (counting the day of plug as day 0), it is necessary to remove estrogen activity but preserve progesterone. This was originally done by ovariectomy followed by injection of Depo-Provera, but we have introduced the less invasive, simpler, and considerably improved option of using the antiestrogenic effects of tamoxifen. The mice are treated with 10-μg tamoxifen and 1-mg Depo-Provera. Dissolve the tamoxifen in ethanol to make a 100× stock, and dilute this in sesame oil before injecting a dose of 10 μg intraperitoneally. At the same time, administer a dose of 1-mg Depo-Provera subcutaneously. Kill the animals, and recover the delayed blastocysts by flushing from the uterus between day 6 and day 8.
- Prepare feeder layers preincubated in an ES cell medium. (Although it is routine to pretreat plastic Petri dishes with

gelatine as an aid to ES cell culture, it is probably more important here than elsewhere.) I find that 1.6-cm four-well cluster dishes are convenient, but possibly better are 3-cm Petri dishes, each enclosed in a 10-cm plastic Petri dish (this aids handling and helps to minimize evaporation from the small volumes of medium).

- The embryos will enter diapause and may be recovered by flushing from the uterus 6 to 12 days later. They will be large, slightly ragged, hatched blastocysts, usually with a visible ICM.
- Using a drawn-out Pasteur pipette, recover the blastocysts and place them on preincubated feeder layers in full ES cell medium.
- Incubate for about four days, observing daily.
- When the blastocyst has attached and spread out with a visibly growing ICM derivative (but before this becomes encapsulated in a thick endodermal layer — if you see this happening, you have left it too long), carefully remove the medium and carefully wash twice in PBS with 10^{-4}-M EGTA.
- Wash with PBS/EGTA containing 0.125% trypsin and aspirate, leaving a thin wetting layer of the trypsin solution. Very briefly incubate at 37°C.
- Have a drawn-out Pasteur pipette with a tip aperture of 20 to 50 microns ready and filled with full ES cell medium. Observe under a dissecting microscope. When the cells are loosening but before everything swims off the dish, carefully flood the medium over the ICM area from the drawn-out pipette and then suck it up.
- This should neatly disaggregate it into single cells and small (2- to 4-cell) clumps — don't aim to make a complete single-cell suspension at this stage. Blow these cells out beneath the surface of the medium of a preincubated feeder dish or well in the ES cell medium.
- Incubate this dish for 7 to 10 days, observing carefully. Do not be tempted to feed daily, but if it seems necessary, replace about half the medium after 5 days. If many small ES cell colonies are becoming established, let them grow to a reasonable size before passaging 1 : 1 onto new feeders; thereafter, you should be able to grow it in the normal manner. (The size of the colonies depends upon the size of the cell aggregate that founded them — if they have come from a single cell, they can be left for 10 days before passage.) If few ES cell colonies appear, it is useful to repeat the trypsinization and passage by drawn-out Pasteur pipette on each one before trusting a bulk passage.

With 129 mice and similar strains, and if all is going well, you should be able to establish new ES cell lines from up to half of the explanted blastocysts. It is useful to make a freeze stock at the earliest possibility. There are usually plentiful cells by passage 4–6.

Summary

Although it is relatively easy to create new cell lines, their full validation — stability in culture, karyotype, chimera-forming ability, and germ line potential — takes a lot of time and work. It is, therefore, important to have sufficient characterized stocks of particular passage and to freeze batches to allow useful repeatable studies to be undertaken.

FURTHER READING

Evans, M. J. (1972). The isolation and properties of a clonal tissue culture strain of pluripotent mouse teratoma cells. *J. Embryol. Exp. Morphol.* **28**(1), 163–176.

Evans, M. J. (1981). Origin of mouse embryonal carcinoma cells and the possibility of their direct isolation into tissue culture. *J. Reprod. Fertil.* **62**(2), 625–631.

Evans, M. J. and Hunter, S. (2002). Source and nature of embryonic stem cells. *C. R. Biol.* **325**(10), 1003–1007.

Evans, M. J., and Kaufman, M. H. (1981). Establishment in culture of pluripotential cells from mouse embryos. *Nature* **292**(5819), 154–156.

MacLean, ••, Hunter, S., and Evans, M. (1999). Non-Surgical Method for the Induction of Delayed Implantation and Recovery of Viable Blastocysts in Rats and Mice by the Use of Tamoxifen and Depo-Provera. *Molecular Reproduction and Development* **52**(1), 29–32.

Martin, G. R. (1981). Isolation of a pluripotent cell line from early mouse embryos cultured in medium conditioned by teratocarcinoma stem cells. *Proc. Natl. Acad. Sci. USA* **78**(12), 7634–7638.

Martin, G. R., and Evans, M. J. (1975). Differentiation of clonal lines of teratocarcinoma cells: formation of embryoid bodies *in vitro. Proc. Natl. Acad. Sci. USA* **72**(4), 1441–1445.

Smith, A. G., Heath, J. K., Donaldson, D. D., Wong, G. G., Moreau, J., Stahl, M., and Rogers, D. (1988). Inhibition of pluripotential embryonic stem cell differentiation by purified polypeptides. *Nature* **336**(6200), 688–690.

Stevens, L. C. (1970). The development of transplantable teratocarcinomas from intratesticular grafts of pre- and postimplantation mouse embryos. *Dev. Biol.* **21**(3), 364–382.

Williams, R. L., Hilton, D. J., Pease, S. Willson, T. A., Stewart, C. L., Gearing, D. P., Wagner, E. F., Metcalf, D., Nicola, N. A., and Gough, N. M. (1988). Myeloid leukaemia inhibitory factor maintains the developmental potential of embryonic stem cells. *Nature* **336**(6200), 684–687.

Isolation, Characterization, and Maintenance of Primate ES Cells

Michal Amit and Joseph Itskovitz-Eldor

Introduction

During the 1970s, multipotent cell lines were isolated from the stem cells of mouse teratocarcinomas. Some of these embryonal carcinoma (EC) cell lines were shown to differentiate *in vitro* into a variety of cell types, including muscle and nerve cells. When grown in suspension, EC cells tend to aggregate into cell clusters known as embryoid bodies (EBs), in which part of the cells differentiate spontaneously. They also had been shown to differentiate *in vivo*, forming teratocarcinomas following their injection into recipient mice. EC cells served for years as models for research on development, establishing all the methodological know-how needed for the isolation and maintenance of embryonic stem (ES) cells (Andrews *et al.*, 2001).

The first mammalian ES cell lines were derived by M. J. Evans and M. H. Kaufman in 1981 from mouse blastocysts. Since then, mouse ES (mES) cells have played a key role in developmental studies. Unlike EC cells, ES cells are regarded as pluripotent, thus possessing the ability to differentiate into each cell type of the adult body. Following their injection into blastocysts, mES cells were reported to integrate into all fetal germ layers, including the germ line. In some cases, they were reported to develop into mature chimeric animals. Furthermore, several mES cell lines had been shown to form entire viable newborns when injected into tetraploid embryos or heat-treated blastocysts (Smith, 2001).

Because of the unique features of mES cells and their proven differentiative abilities, much effort has been invested in the derivation of ES cell lines from additional animal species. ES cell lines and ES cell-like lines have been successfully isolated from rodents such as golden hamsters, rats and rabbits, domestic animals, and three nonhuman primates (Amit and Itskovitz-Eldor, 2003). None of the ES cell lines cited here have been shown to demonstrate all features of mES cells; bovine ES cell lines were not shown to create teratomas following injection into severe-combined immunodeficiency (SCID) mice. The ability of nonhuman primate ES cell lines to integrate into all three germ layers during embryonic development using the blastocyst injection model has not been tested. Therefore, the mES cell model is still the most potent research model among all existing ES cell lines.

The first successful derivation of human ES (hES) cell lines was reported by Thomson and colleagues in 1998. The availability of hES cell lines provides a unique new research tool with widespread potential clinical applications (Figure 37-1). hES cells may be used for various study areas, such as early human development, differentiation processes, and lineage commitment. The availability of differentiation models of hES cells to specific lineages, similar to the existing ones of mES cells, may lead to future uses of these cells in cell-based therapy.

Since the first report on hES cell isolation to date, additional groups have reported the derivation of hES cell lines (summarized in Table 37-3). The increasing numbers of available hES cell lines indicate that the derivation of these lines is a reproducible procedure with reasonable success rates.

This chapter focuses on techniques for the derivation, characterization, maintenance and differentiation of primate ES cells.

What Are Primate ES Cells?

Based on the well-characterized mES cells, a list of ES cells features was created, which includes the following criteria: (1) can be isolated from the inner cell mass (ICM) of the blastocyst; (2) are capable of prolonged undifferentiated proliferation in culture; (3) exhibit and maintain normal diploid karyotypes; (4) are pluripotent, that is, are able to differentiate into derivatives of the three embryonic germ layers; (5) are able to integrate into all fetal tissues during embryonic development following injection into the blastocyst, including the germ layer; (6) are clonogenic, that is, single ES cells have the ability to form a homogeneous line harboring all parental ES cell line features; (7) express high levels of *OCT4* — a transcription factor known to be involved in their process of self-maintenance; (8) can be induced to differentiate after continuous culture at the undifferentiated state; (9) remain in the S phase of the cell cycle for most of their life span; and (10) do not show X-chromosome inactivation (NIH report on human ES cells).

According to accumulating knowledge, primate ES cells meet most of these criteria. Both human and nonhuman primate ES cell lines were derived from embryos at the blastocyst stage (summarized in Table 37-2). All primate ES cell lines were shown to be capable of continuous culture at an undifferentiated state, but expression of high levels of *OCT4*

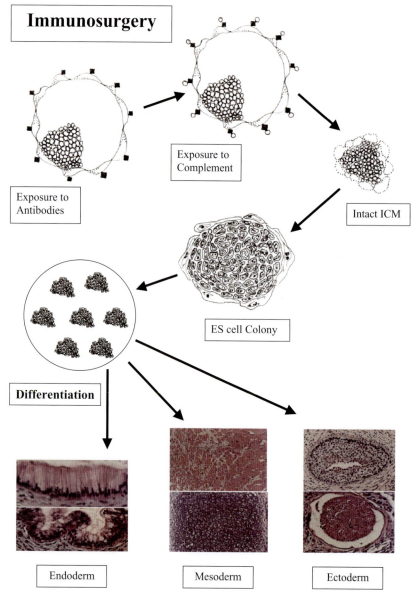

Figure 37-1. Derivation, differentiation, and possible applications of hES cells. For the isolation of hES cell lines, the ICM of the blastocyst is selectively removed and further cultured on mitotically inactivated MEF. When removed from the MEF feeder layer, the resulting hES cells may differentiate into representative tissues of the three embryonic germ layers both *in vitro* and *in vivo*.

had been demonstrated only in hES cells. The morphology of primate ES cells resembles that of mES cells: they create small and round colonies, although primate ES cell colonies seem somewhat less dense (Figure 37-2). On the single-cell level, there is no notable difference: like mES cells, primate ES cells are small and round, exhibiting a high nucleus-to-cytoplasm ratio, with one or more prominent nucleoli and typical spacing between the cells (Figure 37-2). Their pluripotency had been demonstrated both *in vitro* by the formation of EBs and *in vivo* in teratomas in which the cells differentiated into all three primary germ-layer derivatives. The karyotypes of primate ES cell lines were found to be normal diploid

karyotypes even after prolonged culture. Cases of karyotype instability are scarce, suggesting that they represent random changes, which often occur in cell culture. Primate ES cells have also been shown to be clonogenic with resultant single-cell clones demonstrating all ES cell features (Amit and Itskovitz-Eldor, 2003).

For ethical reasons, the ability of hES cells to integrate into fetal tissues during embryonic development cannot be tested. The specific stage of the cell cycle in which primate ES cells spend most of their time has not been reported so far, nor has the status of the X-chromosome inactivation, although the availability of normal 46 XX primate ES cell lines at the same

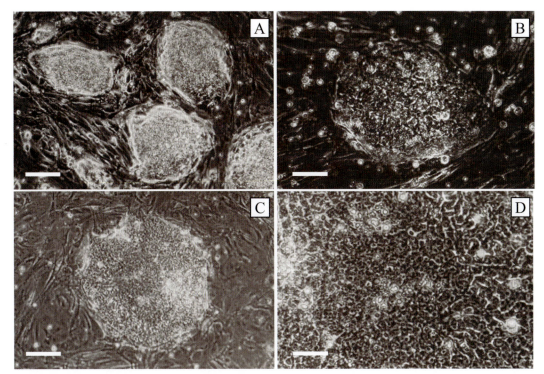

Figure 37-2. Colonies of undifferentiated cells from different ES cell lines grown on MEFs. (A) Mouse ES cell colony, (B) Rhesus ES cell colony, and (C) human ES cell colony. (D) High magnification photograph of hES cells; note the high nucleus-to-cytoplasm ratio, the presence of one or more nucleoli, and the typical spacing between the cells.

ratio as XY lines may indicate that these lines demonstrate normal X-chromosome inactivation. The features of primate ES cells are summarized in Table 37-1.

The potential of primate ES cells in the research on early primate embryonic development, and the development of research models for chronic diseases and for transplantation medicine, encourage scientists to invest great efforts in the isolation of nonhuman primate ES cells. The list and features of the reported nonhuman primate ES cell lines are summarized in Table 37-2. To date, there are more than two dozen established and well-characterized hES cell lines in several laboratories worldwide. The characteristics of the existing hES cell lines are summarized in Table 37-3.

Techniques for the Derivation of hES Cell Lines

The derivation of hES cell lines is a relatively simple procedure, with success rates of up to 30% and 50% for mice and humans, respectively. Embryonic cell lines are derived from embryos at the blastocyst stage. The isolation can be conducted using three methods: immunosurgical isolation, mechanical isolation, or by the use of an intact embryo.

IMMUNOSURGICAL ISOLATION

In the 1970s, Solter and Knowles developed a procedure known as immunosurgery for the derivation of some of the

TABLE 37-1
Characteristics of Primate ES Cells

1. Isolated from the ICM of the blastocyst.
2. Capable of prolonged undifferentiated proliferation in culture.
3. Exhibit and maintain normal diploid karyotype.
4. Pluripotent, that is, able to differentiate into derivatives of the three embryonic germ layers, even after being cultured continuously at the undifferentiated stage.
5. Clonogenic, that is, each single ES cell has the ability to form a homogeneous line harboring all parental ES cell line features.
6. Express high levels of *OCT4* — a transcription factor known to be involved in the self-maintenance process of ES cells.

existing EC cell lines and for the research on early embryonic development. This method of selectively isolating the ICM from the blastocyst laid the groundwork for the first ES cell line derivation in 1981 from mouse blastocysts.

The process of immunosurgery includes several stages. Initially, the glycoprotein outer layer of the zona pellucida (ZP) is dissolved by Tyrodes solution or pronase enzyme. The exposed embryo is then incubated for approximately 30 minutes in antihuman whole serum antibodies, which attach to any human cell. Penetration of the antibodies into the blastocyst is prevented because of cell–cell connections within the outer layer of the trophoblasts, leaving the ICM untouched.

TABLE 37-2

Characteristics of the Existing Nonhuman Primate ES Cell Lines

Line	Karyotype	EBs Formation	Formation of Teratomas	Continuous Culture	Staining with Undifferentiating Markers	Reference
R278 Rhesus ES cells	42, XY	Not available	+	>12 months	+	Thomson et al., 1995 and 1998
R366 Rhesus ES cells	42, XY	+	+	>3 months	+	Thomson et al., 1995 and 1998
R367 Rhesus ES cells	42, XY	Not available	+	>3 months	+	Thomson et al., 1995 and 1998
R394 Rhesus ES cells	42, XX	Not available	+	Not available	Not available	Thomson et al., 1998
R420 Rhesus ES cells	42, XX	Not available	+	Not available	Not available	Thomson et al., 1998
R456 Rhesus ES cells	42, XX	Not available	Not available	Not available	Not available	Thomson et al., 1998
R460 Rhesus ES cells	42, XY	Not available	Not available	Not available	Not available	Thomson et al., 1998
Cj11 Marmoset ES cells	46, XX	+	Not available	>12 months	+	Thomson et al., 1996
Cj25 Marmoset ES cells	46, XX	+	Not available	>3 months	Not available	Thomson et al., 1996
Cj28 Marmoset ES cells	46, XY	+	Not available	>3 months	Not available	Thomson et al., 1996
Cj33 Marmoset ES cells	46, XX	+	Not available	>3 months	Not available	Thomson et al., 1996
Cj35 Marmoset ES cells	46, XX	+	Not available	>3 months	Not available	Thomson et al., 1996
Cj36 Marmoset ES cells	46, XX	+	Not available	>3 months	Not available	Thomson et al., 1996
Cj39 Marmoset ES cells	46, XX	+	Not available	>3 months	Not available	Thomson et al., 1996
Cj62 Marmoset ES cells	46, XX	+	Not available	>12 months	+	Thomson et al., 1996
CMK5 Cynomolgus monkey	40, XY	+	+	>3 months	+	Suemori et al., 2001
CMK6 Cynomolgus monkey	40, XY	+	+	>6 months	+	Suemori et al., 2001
CMK7 Cynomolgus monkey	40, XX	+	Not available	> month	+	Suemori et al., 2001
CMK9 Cynomolgus monkey	40, XX	+	Not available	>3 months	+	Suemori et al., 2001

Sources: Thomson et al. (1995), *Proc. Natl. Acad. Sci. USA* **92**, 7844–7848; Thomson et al. (1996), *Biol. Reprod.* **55**, 254–259; Thomson and Marshall, (1998), *Curr. Top. Dev. Biol.* 38, 133–165; and Suemori et al. (2001), *Dev. Dyn.* 222, 273–279.

After rinsing off any antibody residue, the blastocyst is transferred into a guinea pig complement containing medium, and incubated once more until cell lysis is notable (see the schematic drawing of the procedure in Figure 37-1 and pictures in Figure 37-3A–C). Since the ZP allows the penetration of both antibodies and guinea pig complement, it may be alternatively removed postlysis by complement proteins. Following the selective removal of the trophectoderm, the intact ICM is further cultured on mitotically inactivated mouse embryonic fibroblasts (Figure 37-3).

MECHANICAL ISOLATION

ES cell lines can be derived directly from cultured blastocysts either by the mechanical dissection and partial removal of the trophoblast layer with 27G needles (Figure 37-3D) or by plating a zona-free whole embryo on mitotically inactivated mouse embryonic fibroblasts (MEFs). When trophoblasts are only partially removed or not removed at all, embryos attach to the feeder layer and flatten, permitting continuous growth of the ICM with the remaining surrounding trophoblasts as a monolayer. When the ICM reaches sufficient size, it is selectively removed and propagated (Figure 37-3E).

"LATE-STAGE" EMBRYOS

Early mammalian embryonic development *in vivo* requires the division of the blastocyst's ICM into two layers shortly after implantation: a layer of primitive endoderm, which generates the extraembryonic endoderm, and a layer of primitive ectoderm, which generates the embryo and some extraembryonic derivatives. The mammalian embryo continues to develop. At

TABLE 37-3

Main Features of the Existing Human ES Cell Lines

Line	Karyotype	EBs Formation	Formation of Teratomas	Continuous Culture	Staining with Undifferentiated Markers	Reference
H1	46, XY	+	+	>6 months	+	Thomson *et al.*, 1998
H7	46, XX	+	+	>6 months	+	Thomson *et al.*, 1998
H9	46, XX	+	+	>8 months	+	Thomson *et al.*, 1998
H13	46, XY	+	+	>6 months	+	Thomson *et al.*, 1998
H14	46, XY	+	+	>6 months	+	Thomson *et al.*, 1998
hES-1	46, XX	+	+	64 passages	+	Reubinoff *et al.*, 2000
hES-2	46, XX	+	+	44 passages	+	Reubinoff *et al.*, 2000
hES-3	46, XX	+	+	250 doubling	+	Internet site ES cell international
hES-4	46, XY	+	+	250 doubling	+	Internet site ES cell international
hES-5	46, XY	Not available	Not available	100 doubling	+	Internet site ES cell international
hES-6	46, XX	+	+	100 doubling	+	Internet site ES cell international
hES on human feeders	46, XY	Not available	+	>10 passages	+	Richards *et al.*, 2002
BG01	Not available	Not available	Not available	Not available	+	Internet site Bresagen
BG02	Not available	Not available	Not available	Not available	+	Internet site Bresagen
BG03	Not available	Not available	Not available	Not available	+	Internet site Bresagen
BG04	Not available	Not available	Not available	Not available	+	Internet site Bresagen
HSF-1	Not available	Not available	Not available	Not available	Not available	Internet site University of California
HSF-6	46, XX	Not available	Not available	Not available	+	Internet site University of California
Miz-hES-1	Not available	Not available	Not available	125 passages	+	NIH registry
ES-76	Not available	Not available	Not available	20 passages	+	Lanzendorf *et al.*, 2001
ES-78.1	Not available	Not available	Not available	6 passages	+/−	Lanzendorf *et al.*, 2001
ES-78.2	Not available	Not available	Not available	6 passages	+/−	Lanzendorf *et al.*, 2001
I3	46, XX	+	+	110 passages	+	Amit and Itskovitz-Eldor, 2002
I4	46, XX	+	To be examined	66 passages	+	Amit and Itskovitz-Eldor, 2002
I6	46, XY	+	+	120 passages	+	Amit and Itskovitz-Eldor, 2002
I8	46, XX	+	+	>6 months	+	Amit and Itskovitz-Eldor, Unpublished
I9	46, XX	+	+	>6 months	+	Suss-Toby *et al.* (2004), *Hum. Reprod.* **19**, 670–675.
I10	46, XX	+	+	>6 months	+	Amit and Itskovitz-Eldor, Unpublished
J3	46, XY	+	+	> a year	+/−	Amit and Itskovitz-Eldor, Unpublished

The following Internet sites were used:
http://stemcells.nih.gov/registry/index.asp
http://escells.ucsf.edu/
http://bresagen.com.au/
http://www.wicell.org
http://escellinternational.com/stem/celltable.htm
    And the following references: Thomson *et al.* (1998), *Science* **282**, 1145–1147; Reubinoff *et al.* (2000), *Nat. Biotechnol.* **18**, 399–404; Richards *et al.* (2002), *Nat. Biotechnol.* **20**, 933–936, Lanzendorf *et al.* (2001); *Fertility and Sterility* **76**, 132–137; Amit and Itskovitz-Eldor (2002), *J. Anat.* **200**, 225–232; Suss-Toby *et al.* (2004), *Hum. Reprod.* **19**, 670–675.

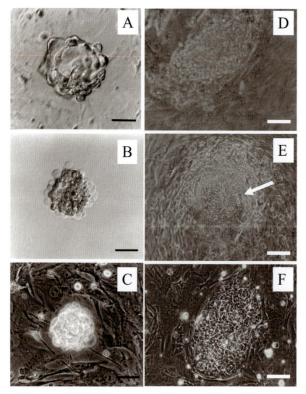

Figure 37-3. Methods for hES cell lines derivation. (A) Human blastocyst after ZP removal by Tyrode's solution, during exposure to rabbit antihuman whole antiserum. (B) Embryo after exposure to guinea pig complement; the ICM is surrounded by cells after lysis. (C) The intact ICM on mitotically inactivated MEFs. (D) Growing ICM after mechanical removal of the trophoblast. (E) Embryo placed in whole on MEFs after six days of culture. The growing ICM (white arrow) is surrounded by fibroblast-like cells. (F) Resulting hES cells colony cultured on MEFs. Bar = 50 μm. Pictures a–c are from Amit and Itskovitz (2002).

day 14 post fertilization, gastrulation occurs in which cells become progressively restricted to a specific lineage. Thus, the pluripotency of these cells is lost. In other words, pluripotent or totipotent cells exist and proliferate in the mammalian embryo for a short window of time, whose exact borders were never defined.

The pluripotency of human postimplantation embryonic cells, between the time of implantation and the gastrulation process, has never been examined previously. Following their plating in whole on MEFs, as mentioned earlier, embryos most often flatten, and a small ICM colony, surrounded by a monolayer of trophoblasts, can be observed (Figure 37-3E). In some cases, however, the ICM continues to grow and to develop small cysts. Small and flat structures of pluripotent-resembling cells can be recognized.

Once selectively removed, these cells are further cultured and propagated on fresh MEF plates. The resulting cell line exhibit the main characteristics of hES cell lines: (1) prolonged undifferentiated proliferation; (b) pluripotency, that is, the creation of EBs when grown in suspension or teratomas after injection into SCID beige mice; (3) maintenance of

normal karyotypes after being continuously cultured for several months; and (4) partial or full expression of typical surface markers. The key question that arises — whether pluripotent cells derived from "postimplantation" embryos are significantly different from hES cells isolated using the traditional methods — has no clear answer so far. Further research is needed to fully characterize these cell lines.

DERIVATION OF HES CELL LINES HARBORING SPECIFIC GENETIC DEFECTS

The ability of hES cells to differentiate into each cell type of the adult body may be used for research on the nature and course of specific diseases. Models established by the use of hES cell lines carrying specific genetic defects may be highly effective in the development of drug or gene therapy designed to treat these diseases.

Two methods of obtaining such lines are (1) genetic manipulation of existing hES cell lines and (2) derivation of hES cell lines from genetically compromised embryos. Since the chance that donated surplus embryos from the *in vitro* fertilization (IVF) program carry genetic diseases is relatively low, the preferable source of donated embryos would be non-retrieved embryos from the preimplantation genetic diagnosis (PGD) program. PGD is designed for couples who are carriers of genetic diseases to ensure the transfer of healthy embryos to the uterus by their examination prior to implantation. For this purpose, the embryos are grown *in vitro* to the 6–8 cell stage, at which point one or two cells are removed and analyzed either by polymerase chain reaction (PCR) or by fluorescence *in situ* hybridization (FISH).

In our experience, post-PGD embryos continue to develop *in vitro* to the blastocyst stage. Five hES cell lines that were isolated from donated post-PGD blastocysts were derived and found to possess hES cell features. One line was found to harbor the Van Waardenburg disease (deletion at the *PAX3* gene), and the other had myotonic dystrophy. The cell lines carrying genetic diseases could therefore be used for the development of *in vitro* models for these disorders.

Derivation of hES Cell Subclones

hES cell lines are derived from the ICM, which may not represent a homogeneous cell population. To eliminate the possibility that the pluripotency of the lines reflects a collection of several distinct committed multipotential cell types in the culture, parental lines have to be single-cell cloned. The main aim of the first derivation of hES single-cell clones was to establish the pluripotency of the parental lines, but as with mES cells, single-cell cloning has additional advantages.

METHOD FOR DERIVING SINGLE-CELL CLONES

To derive single-cell clones, hES cells are trypsinized to single cells, and each cell is plated in a separate well of 96-well plates. Alternatively, single cells may be plated in separate culture cylinders. After approximately two weeks of growth, the resulting colonies are passaged and propagated. When cloning is conducted on hES cells, the rates of success depend highly on the culture medium. Several culture media were tested to clone the first parental hES cell lines: medium supplemented with either fetal bovine serum (FBS) or serum replacement, and either with or without human recombinant basic fibroblast growth factor (bFGF). The highest cloning rates were obtained when serum-free growth conditions supplemented with bFGF were used. Although success rates were relatively low (up to 1%), they have been found to be the most suitable conditions for the clonal derivation of hES single-cell lines. Human ES single-cell lines, H-9.1 and H-9.2 derived by Amit *et al.* in 2000 under these conditions maintained ES cells features: they proliferated continuously as undifferentiated cells for prolonged periods (eight months), they maintained stable and normal karyotypes, they differentiated into advanced derivatives of all three embryonic germ layers *in vitro* in EBs and *in vivo* in teratomas, and they expressed high levels of telomerase activity. Thus, the pluripotency of hES single cells has been established.

To date, many single-cell clones from five parental ES cell lines — H-1, H-9, H-13, I-3, and I-6 — have been derived. Interestingly, in the same culture conditions, most parental lines had the same cloning efficiency of 0.5%, with two exceptions: line H-1 had a lower cloning efficiency of 0.16%, and the cloning of line I-4 proved unsuccessful. The variation among the existing parental cell lines in respect to their cloning efficiency remains to be determined.

ADVANTAGES OF SINGLE-CELL CLONES

In addition to proving the pluripotency of single hES cells, single-cell clones may have further features. First, they are easier to grow and manipulate than the parental lines. Second, they form homogeneous cell populations, which may be instrumental for the development of research models based on gene knockout or targeted recombination. The transfected or knockout cells could be cloned and analyzed individually. Clones that express the desired genotype could be further cultured and used for research. The main disadvantage of this strategy is the relatively low cloning efficiency (0.5–1%), which results in a rather difficult model to obtain when coupled with reduced successful recombination rates. The advantage of this strategy, however, lies in the prolonged culture abilities, which make possible extended periods of research use once a satisfactorily manipulated single-cell clone is created.

Any future application of hES cells for scientific or therapeutic purposes will depend on their karyotypic stability. The first reports on hES cell line derivation specifically state that, like mES cells, their karyotypes remain normal after continuous culture. Although in some culture conditions hES cell lines retain normal karyotypes even after prolonged culture of over 107 passages, random karyotypic instability may still occur. The karyotype of one culture of line I-6, for instance, became abnormal after 150 passages (30 months) of continuous culture on MEFs. Since the first reports on hES cell line derivations, reports on karyotypic instability have accumulated. Amit *et al.* (2000) examined the karyotype of parental line H-9 after 7, 8, 10, and 13 months of continuous culture and found that only

once after seven months of continuous culture, 4 out of the 20 cells examined demonstrated abnormal karyotypes. Eiges and colleagues reported in 2001 on two cells with 17-chromosome trisomy in a stably transfected clone. Based on the growing data on hES cells and the experience with mES cells, it is reasonable to assume that a subpopulation with an abnormal karyotype will acquire a selective growth advantage and take over the culture. Therefore, the periodical cloning of cultured hES cells for the purpose of maintaining a homogeneous euploid population may be needed but will be infrequent.

Methods for hES Cell Culture

The traditional methods used for culturing hES cells have not changed dramatically since their development for the derivation and culture of ECs and mES cells (for more information, see Robertson, 1987). Unlike mES cells which can be cultured on gelatin with the addition of the leukemia inhibitory factor (LIF) to the culture medium, until recently, hES cells could be cultured on mitotically inactivated MEFs only. The feeder layer has a dual role: firstly, as the term implies, it supports ES cell growth; secondly, it prevents spontaneous differentiation of ES cells during culture. Although hES cells require meticulous care, they can be cultured in large numbers and frozen and thawed with reasonable survival rates (for more information, see Amit and Itskovitz-Eldor, 2003). Since their first derivation in 1998, the culture systems of hES cells have enjoyed three major advantages in the basic culture conditions: (1) Amit *et al.* (2000) demonstrated the ability to grow these cells under serum-free conditions; (2) Richards and colleagues reported the use of human feeder layers as substitutes to MEFs in 2003; and (3) maintenance of the cells in an undifferentiated state in feeder-free conditions was reported in Xu *et al.*'s study in 2001.

HUMAN FEEDERS

In the future, hES cells may be directly applied in cell-based therapies. Any clinical use of these cells will require compliance with FDA guidelines. During hES cell culture on MEFs with medium supplemented with FBS, there is a risk of exposing the cells to retroviruses or other pathogens. One solution to this problem is the isolation and culture of hES cells in an entirely animal-free environment.

Recently, Richards and colleagues presented an animal-free system for the production and growth of hES cell lines. In the culture medium proposed, FBS was replaced with a supplement of 20% human serum, and a human feeder layer replaced the MEFs. Co-culture with both human embryo-derived feeder layers or Fallopian tube epithelial feeder layers was found to support hES cell growth and isolation of an hES cell line.

Another animal-free culture system for hES cells reported by Amit *et al.* (2003) consisted of foreskin feeder layers and a medium supplemented with serum replacement. After more than 105 passages (more than 300 doublings), the three hES cell lines grown in these conditions, I-3, I-6, and H-9, exhibited all hES cells features, including expression of typical surface markers and transcription factor *OCT4*, differentiation into representative tissues of the three embryonic germ layers both in EBs and in teratomas, high telomerase activity after 46 passages of culture, and maintenance of normal karyotypes. The morphology of the hES cell colony grown on foreskin feeder layer is slightly different from that grown on MEFs in terms of its long and elliptic organization (Figure 37-4A). No difference was found among the 12 foreskin fibroblast lines tested for their ability to support prolonged and undifferentiated proliferation of hES cells. A recent publication by Hovatta and colleagues demonstrated the derivation of new hES cell lines using foreskin fibroblasts as feeders and a medium supplemented with fetal calf serum.

There are several advantages to the use of foreskin fibroblasts as feeder layers. The serum replacement enables better-defined culture conditions. Furthermore, unlike embryo-derived or Fallopian tube epithelial human feeder layers which can grow to reach a certain limited passage, foreskin fibroblasts can grow to 42 passages. While exploring different foreskin fibroblast lines, no difference was found between the ability of high-passage human foreskin feeders and the ability of low-passage feeders to support the growth of hES cells, even after several cycles of freezing and thawing. These feeders may therefore have an advantage when large-scale growth of hES cells is required and pathogens screening is essential.

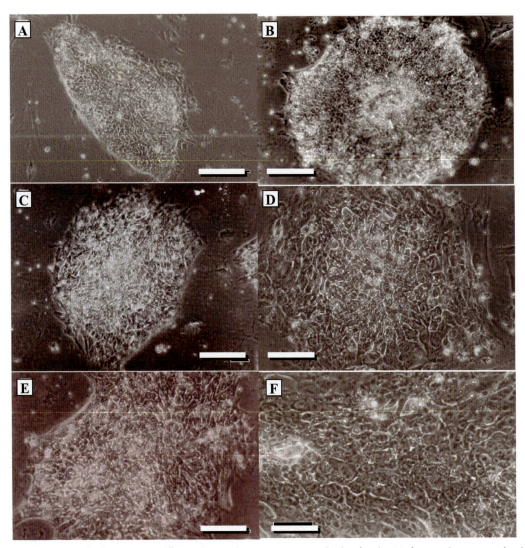

Figure 37-4. Human ES cell colonies grown in different culture conditions. (A) Human ES cell colony from line I-6 after several passages on foreskin matrix. Bar = 75 μM. (B) Example of the morphology of an undifferentiated ES cell colony from clone I-3.2, grown for 12 passages on Matrigel matrix using MEF conditioned medium. Bar 50 = μM. (C) Human ES cell colony from line I-6 after several passages on MEF matrix. Bar = 75 μM. (D) Colony I-3 grown in feeder-free conditions for 21 passages. Bar = 50 μM. (E) Human ES cell colony from line I-6 after several passages on Matrigel. Bar = 75 μM. No condition medium added. (F) Cells from cell line I-3 grown in feeder-free conditions for 20 passages. Bar = 38 μM.

FEEDER-FREE CULTURE OF hES CELLS

For their large-scale growth and therapeutic applications, the ideal culture method for hES cells would be to grow them on an animal-free matrix using a serum-free medium. The first step toward this solution was reported by Xu *et al.* (2001), who demonstrated a culture system in which hES cells were grown on matrigel, laminin, or fibronectin using 100% MEF-conditioned medium, supplemented with serum replacement. The major shortcomings of this system are the possible exposure of the cells to mouse pathogens through the condition medium, the absence of a well-defined condition medium, and the need for a simultaneous growth of MEFs. Richards *et al.* also tested the ability of human feeder-conditioned medium (embryo-derived, Fallopian tube epithelial, or foreskin fibroblast feeder layers) to sustain a continuous undifferentiated proliferation of hES cells grown on collagen I, human extracellular matrix, or laminin. These conditions were found inferior to the use of human feeder cells.

Recently, Amit *et al.* reported further advancement in the ability to culture hES cells for prolonged periods without the use of both feeder cells and conditioned media. With the combination of human recombinant transforming growth factors beta, bFGF, and serum replacement, and adhesion matrices like fibronectin, hES cells can be grown for more than 35 passages maintaining all ES cell features. Examples of undifferentiated hES cell colonies grown on the different matrices in these conditions are illustrated in Figure 37-4, and resultant cells types after differentiation in teratomas are shown in Figure 37-5.

The combination of serum replacement, human fibronectin, and human recombinant growth factors comprises a well-defined culture system for hES cells, which facilitates future uses of these cells in research practices and provides a safer alternative for future clinical applications.

Primate versus mES Cells

Although primate and mES cells share most of the features of ES cells, there are some differences among them. The most distinct difference is the colonies' morphology. Unlike mES cells which form compact and piled colonies, all primate ES cells were reported to form a "flat" morphology (Figure 37-2). On the single-cell level, however, there is no difference: both mouse and primate ES cells demonstrate typical spaces between the cells, a high nucleus-to-cytoplasm ratio, and the presence of one or more nucleoli (Figure 37-2).

Several surface markers were shown to be related to the undifferentiated stage of cells. Primate ES cells have been found to express different surface markers than mES cells. Although mES cells highly express surface marker stage-specific embryonic antigen-1 (SSEA-1), nonhuman primate ES cells and hES cells do not express this marker. In addition, hES cells and nonhuman primate ES cells strongly express SSEA-4, tumor recognition antigen (TRA)-1-60, and TRA-1-81, and they weakly express SSEA-3, with the exception of cynomolgus ES cells, which do not express SSEA-3; mES

cells never express these markers. Examples of the surface antigen expression are demonstrated in Figure 37-6.

In 1985, Rastan and Robertson reported that most mES cell lines are 40, XY. The reduced numbers of XX mES cell lines were related in this study to the lack of X-chromosome inactivation and X-chromosome deletions in mES cells. Interestingly, no difference was found in the XX-to-XY ratio among primate ES cell lines. This finding may be linked to different timing of the X-chromosome inactivation during the early embryonic development of the mouse and primate.

Williams *et al.* found that LIF maintained undifferentiated proliferation of mES cells with the absence of a MEF feeder layer. Both hES cells and nonhuman primate ES cell lines failed to maintain undifferentiated morphology without a MEF feeder layer and with LIF. This finding may indicate that different pathways govern the self-maintenance procedures of ES cells isolated from different species.

Another remarkable difference between mES cells and primate ES cells is the ES cells' ability to differentiate into trophoblasts *in vitro*, a characteristic that rarely appears in mES cells. Mouse trophoblasts do not secrete chorionic gonadotropin (CG), a hormone that was shown to be a crucial signal for maternal recognition of pregnancy secreted by the trophoectoderm. Both rhesus and marmoset ES cells were shown by Thomson and Marshall (1998) to secrete increasing levels of CG when allowed to differentiate. The cynomolgus ES cells do not secrete CG, as has been demonstrated by Suemori and co-workers. Human ES cells were reported by Thomson and Marshall to differentiate into hCG secreting trophoblasts *in vitro*.

The differences between mouse and primate ES cells may reflect the differences in the embryonic development of these different species. Therefore, mES cells may be found to be less suitable for research on early primate embryonic development. One advantage of using nonhuman primate ES cells is their potential contribution to the creation of transgenic monkeys, which may serve as models for chronic illnesses. (For additional reading on the differences between mouse and human ES cells, read Thomson and Marshall, 1998.)

Primate ES Cells Differentiation Systems

During their two decades of existence, many *in vitro* differentiation models were developed for mES cells. Similar models are gradually being created for primate ES cells. Most existing methods for directing mES cells differentiation include the formation of EBs as one of the initial steps, as is the case with the differentiation of mES cells into hematopoietic cells, neurons, and cardiomyocytes. Apparently, this step encourages ES cells to differentiate and consequently increases the rate and efficiency of differentiation. Itskovitz-Eldor and colleagues reported that human ES cells, like mES cells, spontaneously create EBs when cultured in suspension, including cystic EBs, which contain derivatives of the three embryonic germ layers. All nonhuman primate ES cell lines were also reported to create EBs *in vitro*,

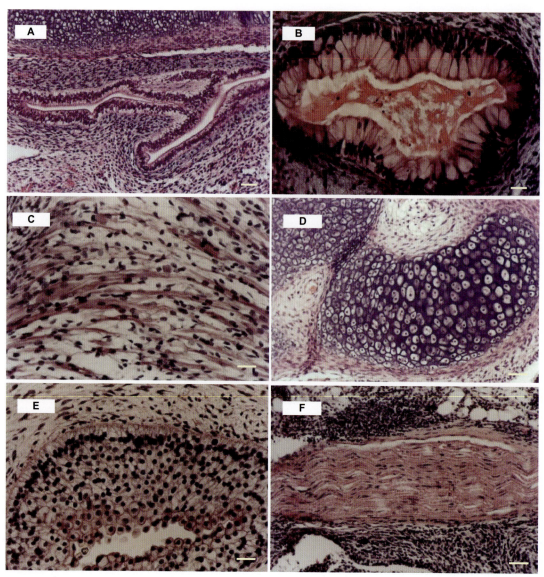

Figure 37-5. Histological sections from teratomas formed by hESCs grown on Matrigel, MEF matrix, or fibronectin in SCID-beige mice. (A) Columnar epithelium. (B) Mucus-secreting surface epithelium resembling the stratified epithelium found in the stomach. (C) Embryonal myotubes. (D) Cartilage. (E) Stratified epithelium. (F) Myelinated nerve. Hematoxylin and eosin staining. For A, D, E, and F Bar = 50 μM, for B and C Bar = 25 μM.

although some of them seemed less organized than mouse-derived EBs.

Primate ES cells can also differentiate *in vivo*. Following their injection into the hind muscle of SCID mice, they spontaneously create teratomas in which they differentiate into representative tissues of the three embryonic germ layers. They have been shown to differentiate into bone and cartilage tissue, striated muscles, gut-like structures, structures resembling fetal glomeruli, and neural rosettes. In EBs, ES cells differentiate as groups of cells or into simple structures; in teratomas, primate ES cells can create more complex and well-organized organ-like structures such as hair follicles, salivary gland, and teeth buds. The development of organ-like structures in teratomas requires cooperation between cells and tissues derived from different germ layers. (for additional information on mES cell differentiation, see Smith, 2000, and on human ES cells, see Chiu and Rao, Chs. 8–12.)

One of the milestones in creating directed-differentiation systems for mES cells was to genetically manipulate these cells. The efficient and stable transfection models can be used both for creating pure populations of cells and for inducing directed differentiation. Additional uses may be to "flag" the cells to permit their recognition in histological sections and for research models aimed at examining the role of specific genes during specific differentiation or embryonic development.

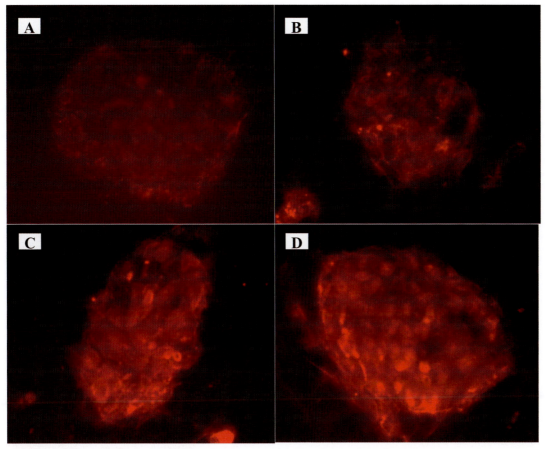

Figure 37-6. Immunostaining of ES cells with typical surface markers. (A) Mouse ES cell colony stained with SSEA-1. (B) Rhesus ES cells stained with SSEA4 and (C) TRA-1-60. (D). I-8 hES cell colony stained with TRA-81.

Stable transfection protocols were already developed for hES cells. In Eiges and colleague (2001), various transfection agents were used while Ma *et al.* and Gropp and co-workers used lentivirus vectors. Similar models were developed by the research groups of Asano and Hanazono for nonhuman primate ES cells.

To date, several studies have been published on lineage-specific differentiation of nonhuman primate ES cells in both spontaneous and directed-differentiation models. These include induced differentiation into hematopoietic cells using bone morphogenic protein 4, spontaneous differentiation into endoderm derivatives, and spontaneous and induced differentiation into neural cells.

Human ES cells had been shown to spontaneously differentiate into cardiomyocytes, endothelial cells, and insulin-secreting cells. Directed differentiation models for hES cells include neural precursors , hematopoietic cells, cardiomyocytes , trophoblasts, and hepatocyts. The data presented here demonstrate the feasibility of developing differentiation models for primate ES cells similar to those of mES cells.

ACKNOWLEDGMENTS

The authors thank Mrs. Hadas O'Neill for editing the manuscript. The research conducted was partly supported by NIH grant R24RR18405.

FURTHER READING

Amit, M., Carpenter, M. K., Inokuma, M. S., Chiu, C. P., Harris, C. P., Waknitz, M. A., Itskovitz-Eldor, J., and Thomson, J. A. (2000). Clonally derived human embryonic stem cell lines maintain pluripotency and proliferative potential for prolonged periods of culture. *Dev. Biol.* **227**(2), 271–278.

Amit, M., and Itskovitz-Eldor, J. (2002). Derivation and spontaneous differentiation of human embryonic stem cells. *J. Anat.* **200**(Pt 3), 225–232.

Amit, M., and Itskovitz-Eldor, J. (2003). Subcloning and alternative methods for the derivation and culture of human embryonic stem cells. *In Human embryonic stem cells* (A. Y. Chiu and M. Rao, eds.), pp. 127–144. Human Press, Totowa, NY.

Amit, M., Margulets, V., Segev, H., Shariki, K., Laevsky, I., Coleman, R., and Itskovitz-Eldor, J. (2003). Human feeder layers for human embryonic stem cells. *Biol. Reprod.* **68**, 2150–2156.

Amit, M., Shariki, C., Margulets, V., and Itskovitz-Eldor, J. (2004). Feeder and serum-free culture of human embryonic stem cells. *Biol. Reprod.* **70**, 837–845.

Andrews, W. P., Przyborski, S. A., and Thomson, J. A. (2001). Embryonal carcinoma cells as embryonic stem cells. *In Stem cell biology* (D. R. Marshak, R. L. Gardner, and D. Gottlieb, eds.), pp. 231–266. Cold Spring Harbor Laboratory Press, Cold Spring Harbor, NY.

Asano, T., Hanazono, Y., Ueda, Y., Muramatsu, S., Kume, A., Suemori, H., Suzuki, Y., Kondo, Y., Harii, K., Hasegawa, M., Nakatsuji, N., and Ozawa, K. (2002). Highly efficient gene transfer into primate embryonic stem cells with a simian lentivirus vector. *Mol. Ther.* **6**, 162–168.

Eiges, R., Schuldiner, M., Drukker, M., Yanuka, O., Itskovitz-Eldor, J., and Benvenisty, N. (2001). Establishment of human embryonic stem cell-transfected clones carrying a marker for undifferentiated cells. *Curr. Biol.* **11**, 514–518.

Hovatta, O., Mikkola, M., Gertow, K., Stromberg, A. M., Inzunza, J., Hreinsson, J., Rozell, B., Blennow, E., Andang, M., and Ahrlund-Richter, L. (2003). A culture system using human foreskin fibroblasts as feeder cells allows production of human embryonic stem cells. *Hum. Reprod.* **18**, 1404–1409.

Itskovitz-Eldor, J., Schuldiner, M., Karsenti, D., Eden, A., Yanuka, O., Amit, M., Soreq, H., and Benvenisty, N. (2000). Differentiation of human embryonic stem cells into embryoid bodies comprising the three embryonic germ layers. *Mol. Med.* **6**, 88–95.

Rastan, S., and Robertson, E. J. (1985). X-chromosome deletions in embryo-derived (EK) cell lines associated with lack of X-chromosome inactivation. *J. Embryol. Exp. Morphol.* **90**, 379–388.

Richards, M., Fong, C. Y., Chan, W. K., Wong, P. C., and Bongso, A. (2002). Human feeders support prolonged undifferentiated growth of human inner cell masses and embryonic stem cells. *Nat. Biotechnol.* **20**, 933–936.

Robertson, E. J. (1987). Embryo-derived stem cell lines. *In Teratocarcinomas and embryonic stem cells: a practical approach* (E. J. Robertson, ed.), pp. 71–112. IRL Press, Oxford.

Smith, A. G. (2000). Embryonic stem cells. *In Stem cell biology* (D. R. Marshak, R. L., Gardner, and D. Gottlieb, eds.), pp. 205–230. Cold Spring Harbor Laboratory Press, Cold Spring Harbor, NY.

Solter, D., and Knowles, B. B. (1975). Immunosurgery of mouse blastocyst. *Proc. Natl. Acad. Sci. USA* **72**, 5099–5102.

Suemori, H., Tada, T., Torii, R., Hosoi, Y., Kobayashi, K., Imahie, H., Kondo, Y., Intani, A., and Nakatsuji, N. (2001). Establishment of embryonic stem cell lines from cynomolgus monkey blastocysts produced by IVF or ICSI. *Dev. Dyn.* **222**, 273–279.

Thomson, J. A., Itskovitz-Eldor, J., Shapiro, S. S., Waknitz, M. A., Swiergiel, J. J., Marshall, V. S., and Jones, J. M. (1998) Embryonic stem cell lines derived from human blastocysts. *Science* **282**, 1145–1147 [erratum in *Science* (1998) 282, 1827].

Thomson, J. A., and Marshall, V. S. (1998). Primate embryonic stem cells. *Curr. Top. Dev. Biol.* **38**, 133–165.

Williams, R., Hilton, D., Pease, S., Wilson, T., Stewart, C., Gearing, D., Wagner, E., Metcalf, D., Nicola, N., and Gough, N. (1988). Myeloid leukemia inhibitory factor maintains the developmental potential of embryonic stem cells. *Nature* **336**, 684–687.

Xu, C., Inokuma, M. S., Denham, J., Golds, K., Kundu, P., Gold, J. D., and Carpenter, M. K. (2001). Feeder-free growth of undifferentiated human embryonic stem cells. *Nat. Biotechnol.* **19**, 971–974.

Approaches for Derivation and Maintenance of Human ES Cells: Detailed Procedures and Alternatives

Irina Klimanskaya and Jill McMahon

Introduction

Deriving human embryonic stem (hES) cells is a challenging endeavor. Although derivation of mouse ES cell lines has become a common procedure, the limited number of currently available hES cell lines is testament to the difficulties encountered at various stages of their derivation and maintenance. The common techniques for maintenance of hES cells involving mechanical passaging of the cells using collagenase or dispase are often complicated and introduce variability in the growth potential of the hES cells. Distributors of the available hES cell lines frequently recommend attending special training courses prior to working with the hES cell lines they provide.

In our lab, we recently established and characterized 17 hES cell lines, described in Cowan *et al.* (2004). These lines have been adapted to trypsinization, which simplifies the passaging of the cells and generates cells in sufficient numbers to permit experimentation. All of these lines can be successfully frozen and thawed using very simple procedures, with a recovery rate of 10% or higher.

Techniques for deriving and maintaining pluripotent human and mouse ES cells in culture have been described by a variety of labs, and there are notable similarities and differences. For the derivation and maintenance of hES cell lines in our lab, we adapted previously published methods and developed an approach that consistently produced new cell lines and that proved to be easily taught to other investigators.

This chapter describes the aspects of derivation and maintenance of hES cells that we found to be helpful in generating hES cell lines including the equipment used, preparation and quality control of media and reagents, cell passaging techniques, and other aspects of hES cell morphology and behavior.

Setting Up the Lab

EQUIPMENT

Initial steps in the derivation process are conducted under a dissecting microscope. We make an effort to keep embryos

and dishes containing the early mechanically passaged dispersions at 37°C. Dishes brought out of the incubators are set on 37°C slide warmers or viewed using microscopes fitted with heated stages. The mechanical dispersions necessitate having the dishes open for extended periods. Since the cultures are vulnerable to contamination during this time, we have the dissecting microscope within a bench-top laminar flow hood. A high-quality stereo microscope with a wide-range zoom is essential for the mechanical dispersion of colonies, including the inner cell mass (ICM) outgrowth; it permits an overall assessment of each plate and evaluation of the morphology of each colony when doing the mechanical passaging.

The equipment used in our lab is as follows:

Stereomicroscope for microdissection: A Nikon SMZ-1500 with the magnification range 10–100x works well with its easy zoom and the positioning mirror that regulates the depth and contrast of the image. A whole 35-mm dish can be scanned for colony morphology, and the zoom permits selection with precision of the parts of the colony that are the best for dissection.

Inverted microscope: A Nikon TE 300, or any regular inverted cell culture microscope, set up with phase and Hoffman modulation contrast (HMC) optics with phase objectives of 4×, 10×, 20×, and HMC 20× and 40×. HMC is recommended for viewing ES cells and is required for embryo evaluation.

Heated microscope stage for both stereo and inverted microscope: A (Nikon) slide warmer keeps extra PMEF plates at 37°C during mechanical dispersions.

Bench-top laminar flow hood with a HEPA filter: Vertical hoods by Terra Universal (Anaheim, California). Horizontal models sometimes produce too much vibration. We found that these vertical hoods were tall enough to accommodate a dissecting microscope and were very convenient and reliable.

Tissue culture incubator: All parameters (CO_2 concentration, humidity, and temperature) need to be checked daily with external monitoring equipment.

External monitoring equipment:
- Surface thermometer
- Mercury liquid-immersed thermometer

- Hygrometer
- CO_2 monitor and gas calibration kit (GD 444 series by CEA Instruments, Emerson, New Jersey), calibrated regularly

QUALITY ASSURANCE OF EQUIPMENT

Consistency in growth conditions is very important for the development of the embryos and the growth of the hES cells. The checklist of parameters monitored daily includes the percentage of CO_2 (5.0), temperature (37°C), and humidity of the incubators (>90%). A checklist with daily readings is very helpful for timely recalibration if an undesirable trend is noticed. Warming rings and platforms are constantly monitored with surface thermometers.

Two incubators are set aside for derivations and expansion of new lines. The incubators are checked prior to any new derivation round by growing mouse embryos from the two-cell stage to blastocyst; a passing score requires 90% to go to blastocyst. Cultures at early stages, prior to being frozen, are split between the two incubators as a protection against incubator failure. The incubators are not opened frequently, thereby maintaining steady growth parameters.

STERILITY

Some aspects of the derivation of hES cell lines put the associated work under more stringent sterility requirements than those of any typical cell culture lab. These include the limited availability of frozen human embryos, the labor-intensive nature of derivation and expansion, the team effort involved, and the long periods in culture until the newly established lines can be expanded and safely frozen. In addition, because hES cells are prone to spontaneous differentiation under unfavorable conditions, many labs prefer not to use antibiotics in cell culture media.

The reagents should either be purchased sterile from the manufacturer or be filter-sterilized in the lab. Most of the cell culture supplies can be bought sterile. However, everything that is sterilized in-house by autoclaving or by dry heat needs to be quality controlled with biological indicators (spore strips from Steris, Mentor, Ohio).

If it comes to the worst, the triple-action drug Normocin (see Media Components) appears to be tolerated by hES cells without significant changes in their pluripotency or growth rate and permits the rescue of contaminated cultures.

Preparing and Screening Reagents

The hES cell lines in our facility were derived and continue to be grown on primary mouse embryo fibroblast (PMEF) monolayers. We derived the lines in media containing Serum Replacement and Plasmanate, a component of the medium used in the IVF field for thawing human embryos and fetal bovine serum (FBS) at early stages of derivation.

MEDIA COMPONENTS

- KO-DMEM (Invitrogen, Cat. No. 10829)
- DMEM high glucose (Invitrogen, Cat. No. 11960-044)

- Serum Replacement (Invitrogen, Cat. No. 10828): Each lot needs to be tested, but as a guide, we found that the lots with osmolarity higher than 470 mOsm/kg and endotoxicity lower than 0.9 EU/ml were the best. Upon thawing, make single-use aliquots and freeze.
- Plasmanate (Bayer, Cat. No. 613-25): Each lot needs to be tested.
- FBS (Hyclone, Cat. No. SH30070.02): Each lot needs to be tested. Heat inactivate, if desired, and freeze in aliquots.
- β-mercaptoethanol, 55 mM (1000×) solution (Invitrogen, Cat. No. 21985-023)
- Non-essential amino acids (NEAA), 100x solution (Invitrogen, Cat. No. 11140050)
- Penicillin–streptomycin, 100× solution (Invitrogen, Cat. No. 15070-063)
- Glutamax-I, 100× solution, a stable dipeptide of L-glutamine and L-alanyl, a glutamine substitute (Invitrogen, Cat. No. 35050-061)

Penicillin–streptomycin and Glutamax-I are kept in frozen single-use aliquots.

- Basic fibroblast growth factor (bFGF) (Invitrogen, Cat. No. #13256-029): Add 1.25 ml of hES cell growth media without leukemia inhibitory factor (LIF) or bFGF to a vial containing 10 µg of bFGF. This makes an 8 µg/ml stock solution. Increasing final bFGF concentration to 8–20 ng/ml can be beneficial for the cells, especially at early stages of derivation, after thawing or when the cells are grown at low density. Make 120-µl aliquots and freeze.
- Human LIF (Chemicon International, Cat. No. LIF1010)
- 0.05% trypsin–0.53 mM EDTA (Invitrogen, Cat. No. 25300-054)
- Gelatin from porcine skin (Sigma, Cat. No. G1880)
- Phosphate-buffered saline (PBS), Ca2+, Mg2+-free (Invitrogen, Cat. No. 14190-144)
- Normocin, an antibiotic active against gram +/– bacteria that also has antimycoplasma and antifungi activity (Invivogen, San Diego, California; Cat. No. ant-nr-2; comes as a 500× solution)

MEDIA RECIPES

Bottles of media that are opened frequently become alkali rapidly; we suggest making smaller quantities that will last approximately a week.

PMEF Growth Medium

To a 500-ml bottle of high-glucose DMEM, add:

- 6-ml penicillin–streptomycin
- 6-ml Glutamax-I
- 50-ml FBS

hES Cell Basal Medium

To a 500-ml bottle of KO-DMEM, add:

- 6-ml penicillin–streptomycin
- 6-ml Glutamax-I
- 6-ml NEAA
- 0.6-ml β-mercaptoethanol

hES Cell Derivation Medium

Use this medium at early stages of ICM outgrowth. It has higher LIF and bFGF concentration and contains FBS. You can switch to hES cell growth medium when a steady growth of colonies has been reached (usually passage 2–4).

To 100 ml of basal medium, add:

- 5-ml Plasmanate
- 5-ml Serum Replacement
- 5-ml FBS
- 230 µl of human LIF (final concentration 20 ng/ml)
- 120 µl of bFGF stock solution (final concentration 8 ng/ml), or more (up to 20 ng/ml)
- Sterilize by 0.22-µm filtration

hES Cell Growth Medium

To 200 ml of basal medium, add:

- 20-ml Plasmanate
- 20-ml Serum Replacement
- 240 µl of human LIF for 10 ng/ml, or 480 µl for 20 ng/ml
- 120 µl of bFGF stock solution (final concentration 4 ng/ml) for 1× bFGF, or more if a higher concentration is desired
- Sterilize by 0.22-µm filtration

Gelatin

Dissolve 0.5 g of gelatin in 500 ml of warm (50–60°C) Milli-Q water. Cool to room temperature, and sterilize by 0.22-µm filtration. This makes 0.1% solution.

Mitomycin C

Add 2 ml of sterile Milli-Q water to a vial (2 mg) of lyophilized mitomycin C (Sigma, Cat. No. M 0503); this makes 1 mg/ml stock solution. The solution is light sensitive and is good for one week at 4°C.

SCREENING MEDIA COMPONENTS

It is important to be consistent with screening, aliquoting, and storage of the media components. Various lots of Serum Replacement, Plasmanate, and FBS should be screened, preferably on hES cells. The screening of Serum Replacement lots should be done prior to the first lot running out so that an evaluation of its qualities can be compared side by side to the previous lot. Some newly derived hES cells will die out with a change in lots at the initial stages.

Screening of FBS, Plasmanate, or Serum Replacement

This test is based on a published procedure for screening FBS lots for mouse ES cell work according to Robertson (1987). This approach can be used for screening any combination of reagents or for finding the best concentrations for media supplements. The quality of the reagents is assessed by counting the number of colonies, evaluating the morphology of the hES cells, and staining for alkaline phosphatase activity as detected with Vector Red Kit (Vector Laboratories, Burlingame, California).

1. Prepare 12-well plates with PMEFs. For each lot tested, you will need at least 12 wells to vary the concentration of the component being tested from the working concentration to high enough concentrations to evaluate toxicity: 8%, 10%, 20%, and 30%. Each concentration is done in triplicate and compared to media components at working concentrations known to support hES cell growth.

2. Split hES cells onto 12-well plates with a ratio of 1:6–1:10. The difference in reagents will be more noticeable when the cells are started at low density. However, some cell lines grow very slowly and differentiate when they are kept at a low density. Therefore, adjust the splitting ratio to the specific hES cell line. Resuspend the cells in a small volume of basal medium and add equal volumes of cell suspension to each well of the test plate; pipette up and down in each well or slowly move the plate in perpendicular directions for even distribution of the cells. Do not rotate because doing so will move most of the freshly added cells to center.

3. Change the medium daily and evaluate colony morphology under the microscope. Human ES cells grow in flat, tightly packed colonies with sharp refractory borders. The colonies appear deep red when stained for alkaline phosphatase activity. In differentiating colonies, the cells are more loosely packed, with diffuse borders, and stain pinker. Usually, the difference in the conditions being tested becomes more obvious as colonies grow bigger. However, as the colonies grow larger, they can begin to touch each other, and they tend to differentiate. Staining one of the triplicate wells for alkaline phosphatase activity prior to seeing signs of differentiation is advisable. Continue with the other wells in each set for another day or two before staining (see Figure 38-1 for a sample test).

Adequate recordkeeping for all commercial and in-house prepared lots of reagents are helpful for troubleshooting should the hES cells begin to exhibit differentiated morphology.

Preparing PMEF Feeders

We grow our hES cells on PMEF feeders that have been mitomycin C treated to generate stable monolayers. The PMEFs are made by standard procedures using 12.5 days postcoitus (dpc) (ICR) mouse embryos as described in Robertson (1987). The 12.5 dpc embryos are eviscerated, but the heads are left on during tissue disruption in trypsin; plating density is 1.5 embryos per 150-mm plate. PMEFs are expanded once after the initial plating (1:5 split) and then frozen (P1). The growth rate of PMEFs and their performance as feeders decreases as they go through multiple passages; therefore, thawed PMEFs are only passaged once (P2) for expansion purposes prior to mitomycin C treatment, at which point a new vial of PMEFs would be thawed.

MITOMYCIN C TREATMENT AND PLATING

Mitomycin C is added to the media of a confluent plate of PMEFs at a concentration of 10 µg/ml and incubated at 37°C for three hours. The cells are harvested by trypsinization and plated on gelatinized plates in PMEF growth medium. In

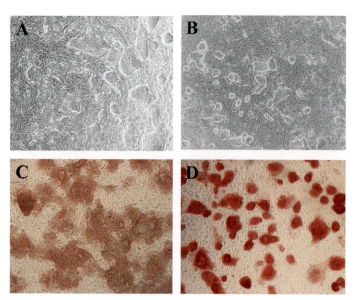

Figure 38-1. Media testing. (A and C) Comparison of 16% serum replacement with (B and D) 8% Serum Replacement and 8% plasmanate. The quality of the media supplements is assessed by (A and B) evaluating the morphology of the colonies under phase contrast, as described in the text, followed by (C and D) staining for alkaline phosphatase activity. Note that although the morphology of the colonies in panels A and B is comparable, the activity of alkaline phosphatase is higher in (D) the medium with both Serum Replacement and plasmanate. Magnification 40×.

serum-free hES cell growth media, the PMEFs may appear less confluent because of the spindle-like shape the cells take on. To ensure a confluent monolayer, we recommend a plating density of 50,000 to 60,000 cells/cm². We prefer to use plates of PMEFs no longer than 3–4 days after mitomycin C treatment.

Mechanical Passaging of hES Cell Colonies

Many established hES cell lines are passaged with collagenase or by dispase in conjunction with mechanical dispersion. Mechanical dispersion can provide colonies of "perfect" morphology, as it permits one to selectively pick undifferentiated colonies or even undifferentiated parts from differentiated and overgrown colonies, but it is time consuming and does not yield large numbers of cells, thus limiting expansion of the hES cell lines. Nevertheless, this procedure is invaluable at early stages of derivation or as a means of producing more homogeneously undifferentiated plates of cells for expansion or for adaptation to trypsin. It is also a tool for a "rescue operation" in critical situations when the success of salvaging a few colonies means saving an hES cell line.

MATERIALS NEEDED

Flame-Pulled Thin Capillaries

We use an alcohol or gas burner to pull presterilized glass Pasteur pipettes into finely drawn capillaries. The capillaries are broken by hand into angled tips, the shape of a hypodermic needle. The diameter of the capillaries may vary, but the best results are achieved when they are 10 to 100 ES cells in

diameter; this is how large the colony pieces are going to be. The choice of diameter depends on the operation. For instance, to do initial dispersion of an ICM outgrowth or to target undifferentiated parts of a colony, a diameter of 10 to 30 cells would be used.

Mouth-Controlled Suction Device

Similar to a mouth pipette used for embryo transfer, this device consists of a mouthpiece (Meditech International, Cat. No. 15601 P), rubber tubing, and a 0.22-μm syringe filter with a rubber tubing adapter for the Pasteur pipette. It provides precision in all manipulations for colony dispersions.

MECHANICAL DISPERSION

The procedure is similar to vacuuming. Gentle dispersion of the colony is achieved by simultaneously cutting off the pieces with the angled end of the capillary, very lightly moving them off, and sucking them in. With the opening of the capillary positioned nearly horizontal to the bottom of the dish, begin moving from the sides toward the colony center, chopping off and gently sucking in each piece. The light suction helps detach colony pieces and is applied at all times as you move from the periphery of the colony, collecting the colony parts. If the whole colony is coming from the monolayer in one piece, it is probably differentiated and should be discarded.

When the desired number of colonies is dispersed, blow out the pieces into the same plate for ICM dispersion or into a freshly prepared plate (see later sections of this chapter for details about preparing the receiving plate). To avoid having all the colonies stick to each other in the center of the plate, move the plate gently from side to side; do not swirl. (Figure 38-2 shows examples of mechanically passaged cultures,

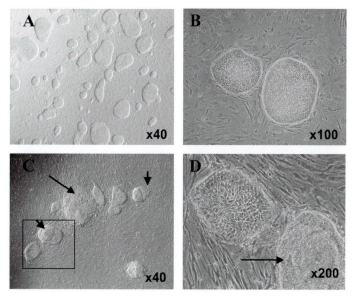

Figure 38-2. Morphology of mechanically dispersed hES cells. Colonies in panels A and B show no signs of differentiation. Note that the cells are small and tightly packed and that the colonies have sharp borders. In panel C, the long arrow points to a much differentiated colony that should not be dispersed; short arrows show partially differentiated colonies that can be dispersed and may produce undifferentiated colonies at the next passage. Panel D shows higher magnification of the framed colonies in panel C, shifted 90 degrees. The arrow points to a partially differentiated colony, which has become multi-layered in the middle. The other colony is undifferentiated and is similar to the colonies shown in panels A and B.

pointing out parts of colonies that have differentiated and should be avoided when passaging the culture.)

Derivation of hES Cells

Many factors that influence whether an isolated ICM will produce an hES cell line are not fully understood. Some of the factors to consider are what stage the embryo was frozen in and by what procedure, the length of time the embryo must be in culture to generate a blastocyst, the culture conditions, and the quality of both the ICM and the trophectoderm. When an embryo is ready for immunosurgery must be determined empirically, usually occurring between day 5 and day 7. Any embryo that has undergone cavitation and has a relatively intact trophectoderm is a candidate for immunosurgery.

IMMUNOSURGERY

The process of immunosurgery was performed essentially as described by Solter and Knowles (1975). It involves removing the zona pellucida with Acidic Tyrode's solution, incubating the embryo in an antibody that binds to the trophectoderm and preferably not to the ICM cells (especially important for the embryos with nonintact trophectoderm), and then lysing the trophectoderm cells with complement. The dead cells that surround the ICM are removed by sucking the ICM through a narrow capillary. The isolated ICM is put on a prepared PMEF for further growth and dispersion.

Materials needed:

- Acidic Tyrode's (Specialty Media, Cat. No. MR-004.D)
- Rabbit anti-human RBC antibody (purified IgG fraction, Inter-Cell Technologies, Hopewell, New Jersey, Cat. No.

AG 28840): Aliquoted and stored at $-80°C$; freshly diluted $1:10$ in hES cell derivation medium
- Complement (Sigma, Cat. No. S1639): Aliquoted and stored at $-80°C$; freshly diluted $1:10$ in derivation medium
- Capillaries for embryo transfer: Thinly drawn capillaries (approximately the diameter of the ICM) for the trophectoderm removal

Prepared Mitomycin C-Treated PMEF Plates

For the initial ICM outgrowth, change the medium on a four-well plate of mitomycin C-treated PMEFs to hES cell derivation medium the night before the immunosurgery to let it get conditioned by the PMEFs, final volume of $250\,\mu l$.

Instead of conditioning it overnight, the derivation medium can be supplemented with 30% of an hES cell-conditioned medium. To collect an hES cell-conditioned medium, add medium to a near-confluent culture of hES cells with good morphology (see Figure 38-2A for an example of colony density and morphology), leave for 24 hours, collect the medium, filter, and store for 2–3 days at $4°C$.

IMMUNOSURGERY PROCEDURE

1. Each embryo is processed separately. A dish is prepared with a series of 30-μl microdrops, three for each step: Acidic Tyrode's, anti-human RBC antibody, complement, and three drops of derivation medium for each wash. The drops are covered with embryo-tested mineral oil and are equilibrated in the CO_2 incubator for 60 minutes.
2. Under the dissecting microscope, transfer the embryo into the first Acidic Tyrode's drop for a quick (1–2 seconds) wash, then move into the second drop. Watch the embryo

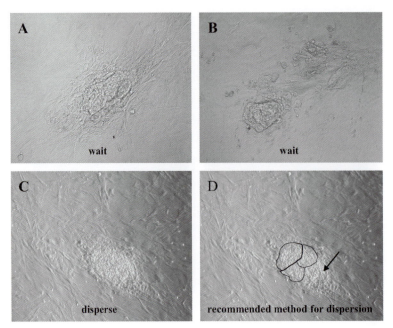

Figure 38-3. Initial ICM outgrowth. The initial outgrowth of the ICM rarely shows typical ES colony morphology and often includes many differentiated-looking cells. When no obvious ES cell-like colonies large enough for dispersion can be located, more time is required before the first colony dispersion can be done. (A and B) The initial ICM outgrowth of two future hES lines. At this stage, when the dispersion is attempted, the outgrowth and the PMEF monolayer come up together, so it is better to wait before dispersing. (C) The ICM is ready for dispersion when a colony of ES-like cells is large enough to be dispersed into several pieces, leaving 20 to 50% of the outgrowth on the original place for future regrowth. (D) Lines transcribe the number of pieces recommended for the dispersion of this colony. A narrow capillary is used, and a small part of the colony is left untouched (arrow). Magnification 100×.

closely; as soon as the zona pellucida thins and is nearly dissolved, move the embryo into the next series of hES cell medium drops. Move the embryos through the first antibody drop into the second and third drops; then put the dish into the incubator for 30 minutes.

3. Transfer the embryos through three drops of derivation medium and through three drops of the complement solution as described previously; incubate in the last drop of complement in the CO_2 incubator for 15 minutes and check for any "bubbling" trophoblast cells. If no cells show signs of lysis, or if only a few cells are bubbling, continue the incubation and recheck in five minutes. The embryo should be transferred to the drop of derivation medium as soon as all trophoblast cells are lysed or if no new bubbling cells appear after rechecking; the total incubation in complement should not exceed 30 minutes.

4. Gently pass the embryo through the opening of a thinly drawn capillary (about the diameter of the ICM); the lysed trophectoderm cells should detach after 1–2 passes.

5. Wash the ICM in the drops with derivation medium and place into the prepared well of a four-well plate. The ICM should attach within 24 hours.

ICM DISPERSION

At early stages of derivation, we recommend doing the first dispersion as soon as at least 2–3 colony pieces can be

obtained from the initial outgrowth (Figure 38-3). The dispersed colonies may be left in the same well or moved to a new well. If there are only a few colony pieces (1–5), they should be placed close to each other but with enough space to permit growth. It is better to disperse colonies before they grow into contact with each other and prior to signs of differentiation, such as becoming multilayered.

When the colony growth is slow, change two-thirds of the medium every 2–3 days to keep it conditioned at all times. As more colonies appear, change two-thirds of the medium daily and increase volume to 500 µl/well of a four-well plate.

Even if an original colony looks differentiated or comes off as a single piece, when replated, it usually gives an outgrowth of hES cells. When doing the initial dispersion, part of the original colony should be left untouched as a backup, especially if the picked pieces are transferred into a new well. Expect it to grow back in 1–2 days; the new outgrowth can be picked and recombined with previously picked cells. Multiple harvests can be obtained from the initial outgrowth. It is critical at this stage to expand the number of colonies slowly and steadily. See Figure 38-4 as an example of the length of time between dispersions and appearance of the cultures during the process of derivation. In this case, no immunosurgery was done because the trophectoderm was not sufficiently intact.

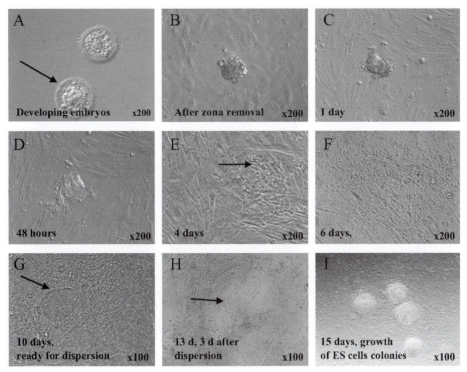

Figure 38-4. Early stages of hES cells derivation. (arrow in panel A) A blastocyst of poor quality, grade 3, underwent (B) the removal of the zona pellucida. (C) The next day, after being plated on the PMEF monolayer, it has attached. (D) Forty-eight hours after plating, the ICM appears smaller, possibly because of some cell death. (E) Four days after plating, an outgrowing group of cells is visible with some small cells in the middle (arrow), (F) which become less visible two days later. (G) Ten days after plating, a small colony of ES-like cells (arrow) has formed within the large group of differentiated-looking cells and is now large enough for dispersion. (H) Original ICM outgrowth two days after it was dispersed; note regrowth of small ES-like cells in the cleared area (arrow), which are ready for another dispersion. (I) Formation of ES cell colonies from the recombined first and second dispersion.

Maintenance of Established hES Cell Cultures

Usually, once steady growth of colonies is reached, use of the hES cell derivation medium is discontinued and the cultures are maintained in hES cell growth medium. For established cultures, remove one-third of the medium from growing culture, put it on a new PMEF dish, and add two-thirds volume of fresh medium. Change two-thirds of the medium daily; the medium should not turn yellow. Cultures should be expanded gradually by the progression from relatively sparsely populated four-well dishes to confluent four-well dishes to 35-mm dishes. Throughout the process, the cultures should be observed daily, differentiated colonies should be removed, and undifferentiated colonies should be dispersed as necessary (see Figure 38-5 for example approaches to be used for dispersion of colonies with different morphologies).

By the time the cells are growing on a 35-mm dish or a six-well plate, it is usually sufficient to disperse 50 to 100 average-sized colonies to populate a new well. In 1 to 2 days, it may be necessary to disperse some of the larger colonies, leaving the pieces in the same well. Usually, mechanical passaging needs to be done every 5 to 6 days, but several larger colonies may need to be dispersed daily.

ADAPTATION OF hES CELLS TO TRYPSIN

Our experience with trypsinization of 20 newly established by us hES cell lines as well as of H1, H7, and H9 hES cells lines described in Thomson *et al.* (1998), demonstrates that after the initial adaptation of the lines to trypsin, this procedure can be robust and yield large quantities of hES cells that exhibit all the properties of pluripotential cells. Trypsinized cells retain undifferentiated colony morphology, express characteristic molecular markers (i.e., *Oct-4*, alkaline phosphatase, SSEA-3, SSEA-4, TRA-1-60, and TRA-1-81), differentiate into three germ layers *in vitro* and in teratomas, and maintain normal karyotypes.

Newly derived hES cells may be successfully passaged with trypsin as early as passage 2–3 from a four-well plate. However, trypsinization is not always successful, and several attempts may be necessary before the cells are adapted to trypsin; always keep a backup well of mechanically passaged cells.

The safest approach is to begin with a subconfluent 35-mm well of a six-well plate of colonies with good morphology.

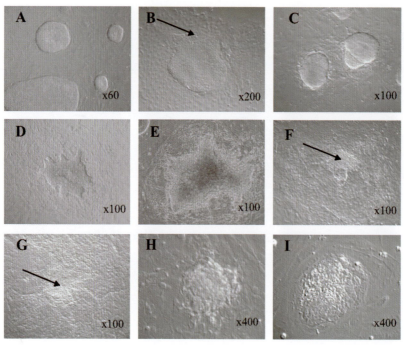

Figure 38-5. Approaches for dispersion of colonies with different morphologies. Various morphologies of hES cell colonies encountered at early stages of derivation when mechanical dispersion can be the tool of choice used to selectively pick undifferentiated colony parts. (A) All colonies are undifferentiated and can be mechanically passaged. (B) The colony that has few signs of differentiation (arrow) and is surrounded by differentiated cells; the undifferentiated part is easily separated from surrounding differentiated cells. (C) Partially differentiated, multilayered colonies (the centers are thickened and yellowish in color) can be mechanically dispersed into several small pieces but may result in both differentiated and undifferentiated colonies. (D–F) All these colonies are more extensively differentiated; a thin layer of differentiated cells covers them like a veil. They can be cut into pieces through the top cell layer and passaged, and they may yield undifferentiated colonies. (G–I) These colonies are badly differentiated; the arrow on panel G shows a group of undifferentiated cells within the large differentiated area. If this one must be saved, wait a few days for this group to increase in size.

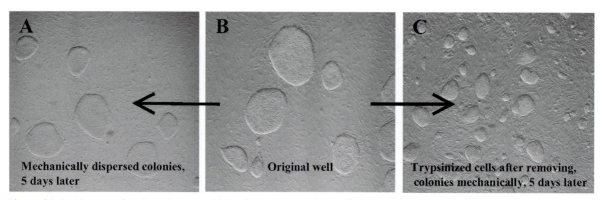

Figure 38-6. Adaptation of mechanically passaged hES cells to trypsin. (A) After five days of growth, colonies were mechanically dispersed and transferred to a fresh plate of PMEFs. (B) Morphologically, they are similar to the original plate. (C) The remaining colonies in the original plate, which were passaged with trypsin and plated onto a plate of the same growth area, show actively growing colonies and will probably be ready for passaging in 1–2 days.

Mechanically pick 50 to 100 colonies and transfer them into a new well as a backup. Differentiated colonies may be removed mechanically prior to trypsinization. Trypsinize the remaining colonies in the original well and plate into the same diameter well. The cells should be ready for the next split in 5 to 7 days (Figure 38-6). For the second trypsinization, split 1:3. After this step, the cells can usually be trypsinized routinely without problems, but a mechanical backup always should be maintained until the cells are frozen and the test vials are successfully thawed.

Trypsinization

Generally, hES cells recover from trypsinization better when they are not dispersed to single cells but remain as small

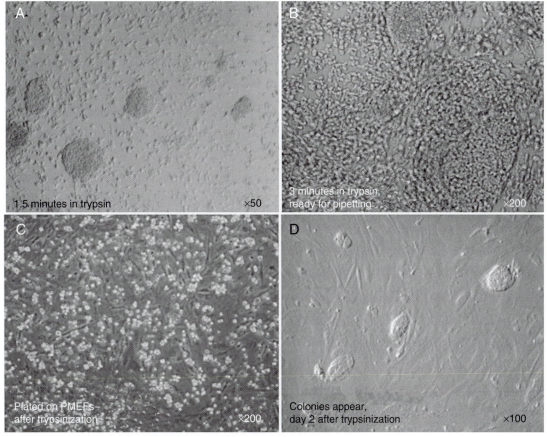

Figure 38-7. Passaging by trypsinization. (A) After 1.5 minutes in trypsin, the PMEFs look shrunken, and the hES colonies have loosened a little. (B) After 3 minutes in trypsin at higher magnification, the colonies are less compact; this is a good time to begin pipetting. Please note that depending on the density of the colonies, the days after passaging, and the degree of differentiation, the time required to reach this stage may vary and needs to be worked out empirically. (C) Suspension of hES cells after replating onto fresh PMEFs. Note that most cells are in small cell aggregates. (D) Small colonies begin to appear on day 2 after plating (at a 1 : 3 ratio). The time at which the first colonies are seen may vary depending on the splitting ratio and the size of the cell aggregates.

clumps of approximately 2 to 20 cells. The procedure works best when hES cell colonies are dispersed by a combination of enzymatic digestion and pipetting; we do the pipetting before the PMEF monolayer, and the colonies turn into a single-cell suspension. The time in trypsin required for the cells to detach varies depending on the hES cell density, degree of differentiation, age of the culture, temperature of trypsin, and so on. Therefore, instead of providing a fixed incubation time in trypsin, we recommend checking the appearance of the hES culture under the microscope and empirically working out the best incubation time for each plate (Figure 38-7).

1. Warm the trypsin in a 37°C water bath; keep it warm until ready for the procedure.
2. Rinse the cells with PBS two times (1–2 ml per 35-mm dish).
3. Add 1 ml of trypsin to each 35-mm dish. Incubate in the hood at room temperature for several minutes, usually 2 to 5 minutes, frequently checking the cells under the micro-

scope. The cells are ready for mechanical dispersion when the PMEFs begin to shrink; the colonies should round up but remain attached. Some cells may begin to detach and float (Figures 38–7A and 38–7B).
4. Prepare a centrifuge tube with 10 ml of warm PMEF medium.

Note: It is necessary to use PMEF medium to inactivate the trypsin because our hES cell medium is serum free.

Tilt the plate and begin to gently pipette the trypsin solution up and down with an automatic 1-ml pipetteman (Gilson type), pouring it over the cell monolayer at an angle. Properly digested cells should detach easily, leaving visible clear gaps in the monolayer where the trypsin solution was poured. If no such gaps appear, leave it for another 1 to 2 minutes and test again. Expect the monolayer to detach after several repetitions. On cell cultures less than 5 days old, you should be able to completely disperse the monolayer, but if the culture is older or very dense, there may be some undigested material that can be discarded. Usually, it takes 5 to 10 pipetting

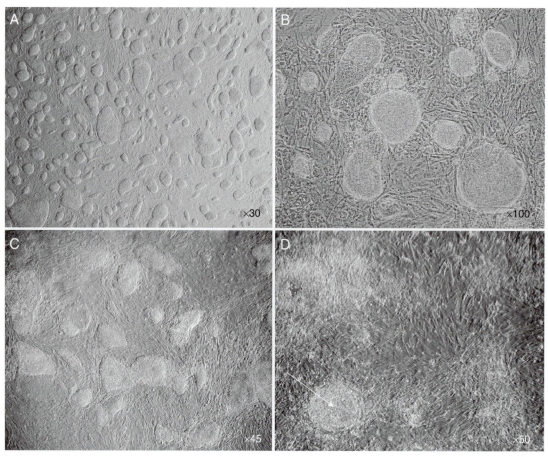

Figure 38-8. Evaluation of the culture prior to trypsinization. (A and B) Mostly undifferentiated hES cells are ready for trypsinization. (C) The colonies are a little overgrown and show signs of differentiation but still can be safely passaged with trypsin. (D) Badly differentiated hES cells (an arrow points to a colony) can still be rescued by mechanically picking colony pieces and passaging.

strokes to break the colonies into small clumps of cells (see Figure 38-7C for the approximate clump size). Extensive pipetting should be avoided.

5. Transfer the trypsinized cell suspension into the prepared centrifuge tube; centrifuge for 5 minutes at 160 g.
6. Aspirate the medium, and resuspend the pellet in hES cell medium, again avoiding extensive pipetting to preserve small cell aggregates and to replate at the desired ratio. The colonies should become visible in 1 to 2 days, depending on the splitting ratio and the clump size (Figure 38-7D).

Human ES cultures passaged with trypsin can be maintained in an undifferentiated state. However, if conditions are unfavorable because of changes in media quality, a splitting ratio that is too high or low, or problems with PMEF quality, the cultures can have a degree of differentiation that should be evaluated prior to the next trypsinization (Figure 38-8).

Freezing hES Cells

Many of the established hES cells have low recovery rates upon thawing, as low as 0.1 to 1%. This may be because of the method of passaging the cells. Mechanical picking or using collagenase dispersion usually results in large cell aggregates, which presumably do not get cryopreserved as efficiently as smaller clumps. Trypsinized cells in our lab have a recovery rate of about 10 to 20% or higher and do not require more complicated procedures such as vitrification (Figure 38-9).

FREEZING MEDIUM

The best recovery rate was observed in freezing medium consisting of 90% FBS–10% dimethyl sulfoxide (DMSO). However, *Oct-4* expression in the thawed cells was lower than in cells frozen in hES cell growth medium with 10% DMSO. Nevertheless, by the next passage, the expression and distribution of Oct-4 and other markers of undifferentiated cells were indistinguishable between these two freezing conditions. We routinely use the 90% FBS–10% DMSO medium.

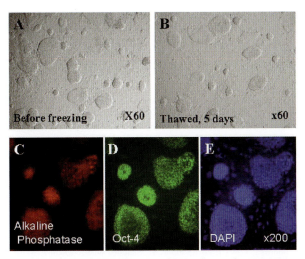

Figure 38-9. Freezing and thawing of hES cells. (A) The approximate density and morphology of the colonies of hES cells at freezing. (B) When thawed onto the same diameter plate, the colonies should be ready for the next split in 5–7 days. (C and D) Thawed hES cells show high expression of *Oct-4* and alkaline phosphatase. (E) same field as C and D, DAPI-stained.

FREEZING PROTOCOL

Select a high-quality confluent culture with good morphology for freezing. We also recommend taking a picture of a sample field and staining for molecular markers characteristic of undifferentiated hES cells for future reference.

Materials needed:

- Chilled freezing medium: 90% FBS, 10% DMSO
- Cryovials, labeled with the line, passage number, and date
- Cryovial rack (rack with ice reservoir by Corning)
- Styrofoam rack from packaging for 15-ml centrifuge tubes
- −80°C freezer

1. Trypsinize the cells; centrifuge in PMEF medium (see earlier explanation).
2. Resuspend the pellet in the cold freezing medium. We recommend freezing one confluent 35-mm plate per vial in 0.5 ml of freezing medium. Work quickly and keep the cells on ice after the addition of the freezing medium.
3. Aliquot cell suspension into prechilled freezing vials and sandwich the vials between two Styrofoam racks; tape to prevent the two racks from separating and transfer to a −80°C freezer overnight. Transfer the cryovials to liquid nitrogen for long-term storage.

Thawing hES Cells

Thawing hES cells is a relatively simple procedure. The main rule to follow is to do everything quickly.

PREPARATION

1. Prepare mitomycin C-treated PMEFs a day before thawing.

2. Make a thawing medium. We use 70% hES cell growth medium supplemented with 2× hLIF and 8 ng/ml bFGF with 30% hES cell- or PMEF-conditioned medium.
3. Change the medium on the PMEF plate to the thawing medium; equilibrate in the CO_2 incubator for one hour. For 35-mm plates, use 1.5 ml medium; for four-well plates, use 0.5-ml medium per well.
4. Prepare a 50-ml conical tube with 10 to 15 ml of warm hES cell growth medium.

THAWING

1. Thaw the vial in a 37°C water bath, constantly agitating while ensuring that the neck of the vial is above the water level. Check the content of the vial after about 40 seconds and at 10-second intervals until only a small piece of ice remains.
2. Quickly spray the vial with 70% isopropanol, then using a 1-ml pipetteman, add warm hES cell medium to the contents of the vial dropwise with gentle agitation. Do it quickly but very gently. Immediately transfer the contents into the prepared 50-ml tube with warm hES cell medium; centrifuge at 160 g for 5 minutes.
3. Remove the medium completely without touching the pellet.
4. Add 0.5 ml of hES cell thawing medium, gently resuspend the cells using a 1-ml pipetteman (2 to 4 repetitions), and transfer to prepared PMEF plates with equilibrated hES cell-thawing medium. Spread the cells evenly throughout the well by moving the plate several times in two directions at 90 degrees to each other; avoid swirling.
5. Check the cells the next day; if there are many dead cells or the medium has changed color, change two-thirds of the medium; otherwise, do not change it for another day.
6. The colonies usually begin to appear in 3 to 4 days and can be ready for splitting in 5 to 10 days (Figure 38-7).

hES Cell Quality Control

Although the morphology of hES cells is often used for evaluating the quality of the culture and its readiness for passaging or freezing, this criterion alone cannot be used for an assessment of ES cell pluripotency. Staining for the expression of *Oct-4* or alkaline phosphatase even in colonies of "perfect" morphology can result in one or both of these markers appearing in the cells only at the periphery of the colony. It is important, therefore, to regularly assess the cells by analyzing the expression of markers of pluripotent cells. We look at *Oct-4*, SSEA-3, SSEA-4, TRA 1-60, and TRA 1-81 by immunostaining, or we perform an enzyme assay for alkaline phosphatase. The procedures for such assays and available antibodies are described elsewhere.

KEY WORDS

Conditioned medium Medium left in contact with cultured cells, usually for a prolonged period of time.

Immunosurgery A method of removing the trophectoderm of a blastocyst using antibodies bound to the surface antigens of the trophectoderm and complement.

PMEF (primary mouse embryonic fibroblasts) A mixed population of cells derived from dispersed mouse embryos and cultured for a limited number of passages under conditions favoring the growth of fibroblasts.

ACKNOWLEDGMENTS

We gratefully acknowledge the financial support of the Howard Hughes Medical Institution and the guidance and encouragement of Dr. Douglas Melton, the recipient of this HHMI grant, without whom this work could not have been accomplished. We would also like to thank our collaborators at Boston IVF, especially Dr. Jeannine Witmyer, for contributing their invaluable experience in the culture and evaluation of early embryos. A special thank you to Jocelyn Atienza and Chad Cowan for their ongoing work in developing and characterizing the hES cell lines reported on in this chapter. We are grateful to Drs. Martin Pera and Susan Lanzendorf for sharing helpful tips on the derivation and maintenance of hES cells. Lastly, we would like to thank Dr. Andy McMahon for his generous support throughout this project.

FURTHER READING

Cowan, C. A., Klimanskaya, I., McMahon, J., Atienza, J., Witmyer, J., Zucker, J. P., Wang, S., Morton, C. C., McMahon, A. P., Powers, D., and Melton, D. A. (2004). Derivation of embryonic stem cell lines from human blastocysts. *New Engl. J. Med.* **350**, 1353–1356.

Reubinoff, B. E., Pera, M. F., Fong, C. Y., Trounson, A., and Bongso, A. (2000). Embryonic stem cell lines from human blastocysts: somatic differentiation *in vitro*. *Nat. Biotechnol.* **18**, 399–404. [Erratum in *Nat. Biotechnol* **18**(5), 559].

Robertson, E. J. (ed.) (1987). *Teratocarcinomas and embryonic stem cells: a practical approach.* IRL Press, Oxford.

Solter, D., and Knowles, B. B. (1975). Immunosurgery of mouse blastocyst. *PNAS* **72**, 5099–5102.

Thomson, J. A., Itskovitz-Eldor, J., Shapiro, S. S., Waknitz, M. A., Swiergiel, J. J., Marshall, V. S., and Jones, J. M. (1998). Embryonic stem cell lines derived from human blastocysts. *Science* **282**, 1145–1147. [Erratum in *Science* (1998). **282**(5395), 1827].

Isolation and Maintenance of Murine Embryonic Germ Cell Lines

Gabriela Durcova-Hills and Anne McLaren

Historical Introduction

Primordial germ cells (PGCs) are embryonic precursors of gametes (sperm in males, eggs in females) (see Chapter 13 of this volume). Early attempts to culture isolated mouse PGCs were performed under simple conditions on a plastic or glass substrate in standard culture media. Female PGCs that had already entered meiosis survived for a week or more and made some progress through meiotic prophase, but earlier stages failed to survive (De Felici & McLaren, 1983). When mouse fibroblast feeder cells were used as substrate, migratory and early postmigratory PGCs survived and proliferated for a few days. Many factors have been shown to increase the proliferation of PGCs *in vitro*, such as leukaemia inhibitory factor (LIF), oncostatin M, interleukin-6, and ciliary neurotrophic factor (reviewed by Buehr, 1997).

Stem cell factor (SCF) is important for the survival of PGCs (Dolci *et al.*, 1991). PGCs co-cultured with mouse feeder cells that express SCF (Sl^4-m220 or STO) in medium supplemented with LIF and fibroblast growth factor-2 (FGF-2) proliferated indefinitely and generated pluripotent cell lines (Matsui *et al.*, 1992; Resnick *et al.*, 1992) that resembled embryonic stem (ES) cell lines (see Chapter 38 of this volume). These "immortalized" pluripotent cells, termed *embryonic germ (EG) cells*, contributed to all cell lineages in chimeras, including the germ line. EG cell lines have been derived from PGCs isolated from embryos at 8.0 and 8.5 days postcoitum (dpc), shortly after establishment of the germ line at 9.5 dpc during migration in the wall of the hindgut and after entering the genital ridges between 11.5 and 12.5 dpc. PGCs isolated from older embryos failed to generate EG cell lines when cultured under similar conditions. EG cell lines have been derived from different mouse strains, including129Sv, C57BL/6, and strains of mixed background.

The first marker used for PGC identification *in vivo* and *in vitro* was tissue-nonspecific alkaline phosphatase (AP). Later cell-surface markers such as SSEA1, 4C9, EMA1, and c-kit were used to identify PGCs in mouse embryos as well as intracellular proteins including mouse *vasa* homolog (Fujiwara *et al.*, 1994), germ cell nuclear antigen 1 (Enders & May, 1994), and Oct 3/4. PGCs have been partially purified from the surrounding somatic cells by pricking the genital ridges

(De Felici & McLaren, 1982), by using antibodies for immunoaffinity purification, or by magnetic or FACS cell sorting.

The methods described in this chapter have been used successfully in our laboratory for several years. Other laboratories use slightly different procedures that also work well.

Derivation of EG Cell Lines from PGCs

The culture conditions for deriving EG cell lines are the same for PGCs from embryos of different ages. However, the efficiency of derivation of new EG cell lines is considerably lower with 12.5- than with 8.5- or 11.5-dpc PGCs. Derivation of EG cells involves the following steps:

1. Isolation of tissue containing PGCs. Purification of PGCs from somatic cells by magnetic cell sorting (optional).
2. Culture of PGCs leading to the derivation of new EG cell lines.
3. Maintenance of derived EG cell lines.
4. Characterization of derived EG cell lines.

ISOLATION OF TISSUES CONTAINING PGCS

Reagents — Equipment

- *Plastics:* 10-cm tissue culture dishes (Nunc) used for dissection of fetuses, and 1.5-ml Eppendorf tubes used for collecting tissue containing PGCs. Sterile tips.
- *Dissecting medium:* Ca^{2+}/Mg^{2+}-free phosphate-buffered saline (PBS, Invitrogen Corporation, Cat. No. 20012-019) containing 10% fetal calf serum (FCS, Sigma).
- *Trypsin–EDTA:* 0.25% trypsin, 1-mM EDTA·4Na (Invitrogen Corporation, Cat. No. 25200-056).
- *Equipment:* Stereomicroscope (Zeiss), forceps with sharp tips for dissection (sterilized prior to use by autoclaving or wiping with 70% alcohol), and a bench-top centrifuge.

Isolation of Tissue Containing PGCs from 8.5 dpc Fetuses

1. Using forceps, dissect the embryo from decidua and extraembryonic membranes in the dissecting medium.
2. Remove the posterior part of the embryo, starting from the end of the primitive streak and continuing to the base of the allantois, using sharp forceps. PGCs are located at the base of the allantois.
3. Collect PGC-containing fragments with a blue tip into a 1.5-ml Eppendorf tube containing dissecting medium. The whole dissection procedure should not exceed one hour. If

longer times are required, the Eppendorf tube containing samples should be kept on ice.

Isolation of Genital Ridges from 11.5 or 12.5 dpc Fetuses

1. Dissect the embryos from their extraembryonic membranes. Cut the head off each embryo in the dissecting medium by using sharp forceps. Make a cut along the ventral midline of embryos. Remove the internal organs.

2. The urogenital ridges (genital ridges with mesonephros) lie on the dorsal wall of the embryo fragment by the dorsal aorta. Remove them, for example, by holding the dorsal aorta at the anterior part of the embryo and peeling off both the aorta and the urogenital ridges. Collect the urogenital ridges in a 5-cm tissue culture dish filled with dissecting medium.

3. Separate the genital ridges from the mesonephros by using a needle or sharp forceps.

4. Collect the genital ridges in a 1.5-ml Eppendorf tube containing dissecting medium. By 12.5 dpc, female and male genital ridges can be distinguished by their morphology. Male 12.5 dpc genital ridges have "stripes" (i.e., testicular cords), whereas female genital ridges are "spotted."

Magnetic Cell Sorting of PGCs

REAGENTS–EQUIPMENT

- *MiniMACS Starting Kit:* The kit (Miltenyi Biotec, Cat. No.130-042-101) contains 1 MiniMACS separating unit, 1 MACS multistand, 25 MS columns, one 1-ml unit of MACS microbeads (rat–antimouse IgM). We use as primary antibody TG1 (given to us by Dr. Peter Beverly), which recognizes cell-surface antigen SSEA1. Antibody against SSEA1 is available commercially from the Developmental Studies Hybridoma Bank, University of Iowa.
- *Sorting Medium:* Cell sorting is done in cold Dulbecco's Modified Eagle Medium (DMEM, Invitrogen Corporation, Cat. No. 41965-039) supplemented with 2% FCS, Penicillin G/Streptomycin (Invitrogen Corporation, Cat. No. 15140-122), L-glutamine (200 mM, Invitrogen Corporation, Cat. no. 25030-024).
- *Equipment:* Laminar flow hood, pipette aid or equivalent, and a cold room.

This method is adapted from that published by Pesce and De Felici (1995).

1. Collect 11.5 dpc genital ridges as described previously. Wash them twice with PBS to remove the traces of FCS. (FCS inhibits the activity of trypsin.)

2. Incubate the genital ridges in trypsin–EDTA for 5 to 10 minutes at 37°C in a water bath and then pipette with a yellow tip to obtain a single-cell suspension.

3. Add 500-µl cold sorting medium. Centrifuge at 1000 rpm for 5 minutes.

4. Resuspend the pellet in 500-µl cold sorting medium to obtain a single-cell suspension. Remove small clumps if present in the cell suspension.

5. Add 50 µl of TG 1 antibody (the amount of primary antibody must be checked with each new lot of antibody) and incubate on a plate shaker for 45 minutes at 4°C.

6. Centrifuge at 700 rpm for three minutes. Resuspend the pellet in 500-µl cold sorting medium.

7. Add 20 µl of magnetically coated secondary antibody (microbeads conjugated with rat–antimouse IgM) and incubate on a plate shaker for 25 minutes at 4°C.

8. Under sterile conditions (e.g., in a laminar flow hood), pipette the magnetically coated cell suspension on top of the pre-washed column. The column is washed with 1000 µl of the cold sorting medium. Do not allow the column to dry off.

9. Collect the negative fraction (i.e., cells that were not magnetically stained).

10. Wash the column with 500 to 1000 µl of the cold sorting medium. Then remove the column from the magnet and flush magnetically stained cells into a 1.5-ml Eppendorf tube with 1 ml of the cold sorting medium by using the plunger. Collect as a positive fraction.

11. Determine the purity of the positive fraction. Stain 50 µl from the positive fraction with AP solution (See the section AP Staining and Figure 39-1). Count the number of AP+ cells with a haemocytometer. Seed cells from the positive fraction onto Sl^4-m220 or STO cells in medium used for derivation of EG cell lines.

CULTURE OF PGCS LEADING TO THE DERIVATION OF NEW EG CELL LINES

Reagents — Equipment

- *PGC growth medium:* This culture medium is supplemented with LIF and FGF-2 (also termed bFGF) (see Table 39-1).
- *EG cell growth medium:* This culture medium is supplemented only with LIF.
- *FCS:* FCS is tested to support the growth of normal ES cells and EG cells.
- *Growth factors:* LIF (Chemicon International) used at concentration of 1000–1200 units in 1 ml of culture medium.

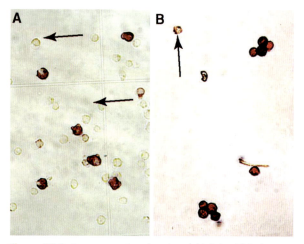

Figure 39-1. Immunomagnetic cell sorting of 11.5-dpc PGCs from surrounding somatic cells. (A) Cell suspension of PGCs and somatic cells before cell sorting. PGCs are identified by AP activity (in red). Somatic cells (arrows) are negative for AP activity. (B) PGCs were sorted by magnetic beads. The purity of sorted PGCs is checked by AP activity. One somatic cell (arrow) was observed.

TABLE 39-1

EG Cell Growth Medium

Component	Volume added for 40 ml
DMEM	32 ml
FCS	6 ml
L-glutamine 200 mM	0.4 ml
NEAA 100 mM	0.4 ml
Penicillin G + Streptomycin	0.4 ml
Sodium pyruvate 100 mM	0.4 ml
β-Mercaptoethanol	0.4 ml
LIF	1200 units/ml
FGF-2[a]	25 ng/ml

[a]FGF-2 is added to this medium during the primary culture of PGCs.

FGF-2 (Invitrogen) is dissolved in PBS at the concentration of 10 μg in 1 ml of PBS. Aliquots are stored at −20°C.

- *Mitomycin C:* Make a stock solution of mitomycin C (Sigma, Cat. No. M 4287) by dissolving 2 mg in 1 ml of tissue culture water by vortexing for 30 minutes. Aliquots are stored at −20°C.
- *Gelatin:* Make 0.1% gelatin (Swine Skin type II, Sigma, G 2500) solution in water.
- *Plastics:* Four-well culture dishes (Nunc) are used for primary culture of PGCs, and 10-cm culture dishes are used for culturing feeder cells. Also used are 15-ml Falcon tubes.
- *Feeder cells:* Two cell lines used as feeder cells support the EG cell-like colonies arising during primary culture. Sl⁴-m220 is a bone marrow stromal cell line that stably expresses only the membrane-bound form of stem cell factor. STO cells express both the membrane and the soluble form of stem cell factor. Primary mouse embryonic fibroblasts (MEFs) derived from 12.5 dpc fetuses are also used as feeder cells (see Chapter 38) for secondary and subsequent subcultures.
- *Equipment:* Standard humidified tissue culture incubator (37°C, 5% CO₂ in air), laminar flow, liquid nitrogen storage, finely drawn Pasteur pipettes, and forceps.

Primary Culture

1. Seventy-two hours before starting the culture of PGCs, Sl⁴-m220 cells are plated into five pregelatinised 10-cm culture dishes.
2. Treat subconfluent Sl⁴-m220 cells with 12.5 μl of mitomycin C from the stock solution in 5 ml of feeder culture medium for 2 hours in the incubator.
3. Wash the feeder cells three times with PBS to remove the traces of mitomycin C.
4. Incubate cells with trypsin for 5 minutes at room temperature and pipette with blue tip to harvest a single-cell suspension
5. Collect trypsinized cells in a 15-ml conical tube containing 9 ml of the feeder growth medium to inactivate and dilute the trypsin solution.
6. Centrifuge at 1000 rpm for 5 minutes, then remove solution.

7. Resuspend Sl⁴-m220 cells in fresh feeder culture medium and plate at the density of 5× in pregelatinized wells of four-well culture dishes.
8. The next day, first wash Sl⁴-m220 cells with PBS, then replace with the PGC growth medium supplemented with both LIF and FGF-2.
9. Collect tissues containing PGCs in a sterile 1.5-ml Eppendorf tube containing PBS supplemented with 10% FCS. When the isolation is completed, wash tissues twice with PBS.
10. Add 100–200 μl of trypsin (the volume depends on the amount of tissue) and incubate in a water bath at 37°C for 5 to 10 minutes. Cell suspension is obtained by pipetting tissues with a blue and then with a yellow tip. Add 800 μl of culture medium to trypsinized cell suspension.
11. Centrifuge as described previously. Resuspend the pellet in 500–1000 μl of the PGC growth medium. Plate cell suspension into Sl⁴-m220-containing wells of four-well culture dishes. Plate approximately 0.5 of 8.5-dpc embryo or 0.2–0.25 of one 11.5-dpc genital ridge into one well of a four-well dish. Two genital ridges from 12.5 dpc may be cultured in a 35-mm culture dish.
12. Culture at 37°C, 5% CO₂ in air. After two days, change the PGC growth medium for medium freshly supplemented with both LIF and FGF-2. Then change the culture medium every day.
13. By 8 to 10 days of primary culture, small colonies with EG cell-like morphology are observed (Figure 39-2). Mitotically inactive primary mouse feeder cells are prepared. After two more days of culture, colonies with EG-like morphology are ready to be transferred into a new well of a four-well dish.

Secondary and Following Subcultures

14. Pick EG cell-like colonies from culture dishes with a pulled glass mouth pipette and wash twice in PBS.
15. Incubate the colony in trypsin for 5 to 10 minutes at 37°C. Transfer the colony into a well of a four-well dish containing both the mitotically inactive MEF and the EG cell

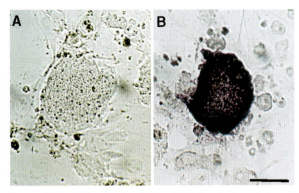

Figure 39-2. PGCs cultured with LIF and FGF-2 on feeder cells formed multicellular colonies. (A) A colony resembling an EG cell colony in appearance. (B) An EG cell-like colony expressing high AP activity. Scale bar 60 μm.

growth medium supplemented only with LIF. The colony is gently disaggregated into small clumps by pulling and pushing the colony up and down in the glass mouth pipette.

16. Culture in an incubator at 37°C, 5% CO_2 in air for a few days. After 1 to 2 days, the colonies with EG cell-like morphology will start growing on MEF feeders.

17. After two additional days, transfer colonies by glass mouth pipette onto fresh mitotically inactive MEF feeders in the EG cell growth medium in one well of a four-well dish.

18. After 3 to 4 days of culture, trypsinize the well and transfer the cell suspension into a 35-mm dish on mitotically inactive MEF feeders. If only a few EG cell-like colonies are observed, transfer them into a new well in a four-well dish. After 3 to 4 days, many EG cell-like colonies will be growing in the culture dish. Subculture them into a 35-mm culture dish as described previously. The transfer of EG cells into a 35-mm culture dish is counted as the first passage. Passage cells one or two more times to get enough EG cells to freeze in a few cryotubes in a liquid nitrogen storage tank (See the section Freezing of EG Cells.) Record it in cell line recording book.

MAINTENANCE OF DERIVED EG CELL LINE

This comprises maintenance, subculture, freezing, and thawing EG cell lines.

Equipment — Reagents

- *Medium:* Culture EG cells in the same medium used for derivation of EG cell-like colonies, but omit FGF2.
- *Plastics:* 35-mm culture dishes for growing EG cells on feeder cells prepared from primary MEFs. For some assays (i.e., Southern or an immunoblotting analysis), EG cells are grown on 10-cm gelatin-coated tissue culture dishes with a double amount of LIF (2400 U/ml).
- *FCS:* Use batch-tested FCS, tested both for supporting the growth of derived pluripotent cell lines (ES cells and EG cells) and for the ability to support derivation of EG cell-like colonies from cultured PGCs.
- *Trypsin — EDTA:* Use 0.25% solution (Invitrogen).
- *Freezing solution:* Add 4-ml DMSO to 6-ml DMEM and mix. Then transfer into 30-ml FCS and mix well. Keep the freezing solution in a −20°C freezer.

Maintenance of Newly Derived EG Lines

1. EG cells are cultured on mitotically inactive MEF cells in the EG cell growth medium (see Table 39-1) supplemented only with LIF. The medium is changed every day. After 2 to 3 days, cells are ready to be subcultured or frozen.

2. Do not allow the culture medium during the culture of EG cells to turn yellowish because it might induce differentiation of EG cells. EG cell colonies during this period enlarge but keep the undifferentiated morphology. The undifferentiated EG cell colony usually has a round or oval shape with a distinct border around the colony in which individual cells are tightly compacted (Figure 39-1). Signs of differentiation can be observed by the changing morphology of colonies and markers down-regulated during

the differentiation (AP, Oct3/4, and SSEA1). Signs of differentiation observed by changed morphology of EG colonies are as follows: (1) Colonies are surrounded by differentiated cells with flattened morphology. (2) Individual cells within the colony are recognized rather than a single clump. (3) Colonies adopt a more rounded shape, and some free-floating embryoid bodies are formed. When a culture starts spontaneously differentiating, it should be thrown away. If the culture is very important, subculture using a mouth pipette only those colonies that have "proper" EG cell phenotype. After 3 to 4 days, only EG cell-like colonies should be observed in the culture. Expand the cells and freeze.

Subculture of EG Cells

1. EG cell colonies grown on 35-mm culture dishes are usually ready to be subcultured 2 to 3 days after plating.

2. Wash cells with 1 ml of PBS. Harvest a single-cell suspension by treatment with trypsin (3 to 5 minutes at room temperature) and then pipetting with a blue tip.

3. Neutralize trypsin by adding 1 ml of EG cell growth medium. Transfer the cell suspension into a 15-ml tube containing 8 ml of culture medium.

4. Centrifuge at 1000 rpm for 5 minutes.

5. Resuspend the pellet in 6–8 ml of EG cell growth medium.

6. Plate 2 ml of EG cell suspension into one 35-mm culture dish. Culture at 37°C in an incubator.

Freezing of EG Cells

1. Remove the EG cell growth medium and wash cultures with PBS when EG cells are in a lag phase (growing phase)

2. Make a single-cell suspension by trypsin treatment, then pellet the cells as described previously.

3. Resuspend the pellet in a cold freezing solution. Use 1 ml of freezing solution for EG cells from one 35-mm culture dish. Transfer cells in freezing solution to cryotubes and label with cell name, passage number, and date.

4. Place cryotubes overnight into a −80°C freezer. The next day, transfer the cryotubes into LN_2 and record their position in cell database file.

Thawing of EG Cells

1. Thaw frozen EG cells quickly by placing the cryotube into a water bath at 37°C.

2. Transfer the cell suspension into a 15-ml tube containing 9 ml of DMEM.

3. Centrifuge at 1000 rpm for 5 minutes.

4. Resuspend the cell pellet in the EG cell growth medium and seed onto mitotically inactive MEF cells in a 35-mm culture dish.

5. Place the culture into the CO_2 incubator. The next day, small colonies of growing EG cells are observed. Change the EG cell growth medium to remove dead, floating cells.

CHARACTERIZATION OF DERIVED EG CELL LINES

Sexing by PCR

This technique is adopted from that previously published by Chuma and Nakatsuji (2001).

Equipment — Reagents

- Thermal cycler (e.g., PTC-100, MJ Research), electrophoresis equipment, tips, and tubes
- *PCR reagents:* dNTPs (10 mM stocks), reaction buffer (10 × PCR buffer), Taq polymerase, and Autoclaved distilled water
- *Primers:* Ube1XA: TGGTCTGGACCCAAACGCTGTC-CACA and Ube1XB: GGCAGCAGCCATCA-CATAATCCAGATG
- *DNA size ladder:* 1-kb DNA ladder (Invitrogen)
- Agarose, TBA buffer (1X)
  1. Isolate genomic DNA from derived EG cells (the EG cells were grown without feeder cells).
  2. Set up a PCR reaction in 0.2-ml thin wall tubes kept on ice:
- 15.5-µl distilled water
- 2.5-µl 10 × PCR buffer (15-mM MgCl$_2$)
- 1-µl 10-mM dNTPs
- 2.5-µl 10 pmol/µl Ube1XA primer
- 2.5-µl 10 pmol/µl Ube1XB primer
- 0.5-µl template DNA (genomic DNA from EG cells)
- 0.25-µl Taq polymerase (5 U/µl)
- Total volume is 25 µl
  3. Transfer the tubes to a PCR block and run the PCR (Table 39-2).
  4. After the amplification, add a loading buffer to 10 µl of the PCR products and run out on 2% agarose gel along with the 1-kb DNA marker. The sizes of PCR products are as follows: The male has two bands, 217 and 198 basepair (bp). The female has just the 217-bp band.

AP Staining
Equipment — Reagents

- *Staining solution:* A kit is available from Sigma (Cat. No. 86-R). Add 50 µl of sodium nitrite solution to 2.25 ml of autoclaved water and mix. Then add 50 µl of FRV-alkaline solution and 50 µl of naphthol AS-BI alkaline solution. The staining solution must be used immediately. PGCs (from 8.5 to 12.5 dpc) and EG cells express a high level of tissue-nonspecific AP.
  1. PGCs or EG cells are grown on feeder cells in a four-well culture dish.
  2. Aspirate the medium and rinse wells twice with 0.5 ml of PBS.

TABLE 39-2

Steps for Running the PCR

PCR step	Temperature (°C)	Time
1. Denaturing	94	1 min.
2. Denaturing	94	30 sec.
3. Annealing	66	30 sec.
4. Extension	72	1 min.
5. Cycle 29 times to step 2		
6. Extension	72	5 min.
7. Hold	4	

3. Fix cells by air drying for 15 to 20 minutes at room temperature.
4. Make the AP staining solution.
5. Add 200 µl of the staining solution to each well of a four-well dish. Keep in the dark for 10 to 15 minutes.
6. Identify PGCs or EG cells by their dark-red staining.

Immunofluorescence Staining for SSEA1 and Oct 3/4
Equipment — Reagents

- *Plastic — glass:* Use Lab-tek chambers (Nunc) for staining of cultured EG cells or PGCs. For freshly prepared cell suspensions, use multiwell microscope glass slides (C. A. Hendley, Cat. No. PH-136) precoated with poly-L-lysine to enhance the adherence of cells. Coverslips.
- *Antibodies:* SSEA1 antibody is available from the Developmental Studies Hybridoma Bank, University of Iowa. TG 1 (1:2) also recognizes SSEA1 antigen. Anti-Oct3/4 (1:200) from Chemicon Int. Antimouse IgM-FITC (1:50, Sigma) and antimouse IgG-Texas Red (1:100, Santa Cruz). Toto3 is a fluorescent DNA stain, detecting the nucleus in interphase cells and the chromosomes in mitotic cells (1:500, Molecular Probe).
- *Fixative solution:* Paraformaldehyde (PFA) fixation preserves structure as well as retaining antigen recognition. Use 4% PFA in PBS, 2-g PFA to 35-ml water, and warm to 60°C in the oven. Then add 20 µl of 10% NaOH (to dissolve PFA) and put back into the oven. After 10 minutes, when PFA is dissolved, fill with water to 40 ml, cool to room temperature, and add 5 ml of 10 × PBS. Titrate to pH 7.4 with 1N HCl. Fill to 50 ml with water. Filter using a 0.22-µm filter.
- *PBS-TX:* This solution is used for both blocking and permeabilization. Dissolve BSA (10 mg/1 ml) in PBS. Add Triton X-100 so that the final concentration is 0.1%. BSA is a blocking agent that reduces nonspecific antibody binding. Triton X-100 is a detergent used to permeabilize plasma and nuclear membranes.
- *Equipment:* Fluorescence microscope and a humidified dark chamber.
  1. Make a cell suspension of PGCs or EG cells by trypsin treatment.
  2. Wash cells twice with PBS. Dilute the pellet in PBS to obtain appropriate cell density.
  3. Apply cell suspension to poly-L lysine-treated wells of a multiwell slide. Wait 10 to 15 minutes until cells stick to wells at room temperature. Keep samples in a humidified chamber. If cells have been grown in Lab-tek chambers, wash twice with PBS. In either case, remove the excess PBS and add the fixative solution (4% PFA) for 15 minutes at room temperature.
  4. Remove the fixative solution and wash wells twice with PBS for 5 minutes each.
  5. Permeabilize and block cells with PBS-TX for 20 minutes at room temperature.
  6. Add TG-1 or Oct3/4 antibody diluted in PBS-TX. Incubate overnight at 4°C in a humidified chamber.

7. The next day, wash samples twice with PBS for 5 minutes each. Add secondary antibody coupled with fluorescence diluted in PBS-TX. Incubate in a humidified dark chamber for 60 minutes at room temperature.

8. Wash twice with PBS for 5 minutes each.

9. Counterstain nuclei with Toto3 for 15 minutes at room temperature. Place a drop of appropriate commercial fluorescent mounting medium on a slide and place a coverslip on top.

10. Examine samples under a fluorescence microscope (Figure 39-3) with the appropriate filters as soon as possible because the signal diminishes over time. Store slides at 4°C (short-storage) or in a freezer at −20°C (for a few days) for subsequent observation.

Questions for Future Study

Highly purified PGCs, with negligible contamination from neighboring somatic cells, can generate EG cell colonies and eventually EG cell lines, suggesting that somatic cells are not required for derivation of EG lines. However, PGCs convert into EG cells only when co-cultured with feeder cells (Sl⁴-m220 or STO). When PGCs are cultured without feeder cells — even in conditioned medium from feeder cells or supplemented with LIF, SCF, and FGF-2 — EG cell-like colonies are not observed. This implies that feeder cells must provide other unidentified factors important for the conversion of PGCs into EG cells. For future studies, factors and signaling pathways involved in the process of conversion of PGCs into the new cell status of EG cells should be investigated. Such studies could throw light on the molecular basis of pluripotency.

Other questions arise from a comparison of EG cells and ES cells. PGCs at any particular embryonic stage may be considered a more defined cell type than inner cell mass or epiblast cells. Is this related to the finding that the efficiency of derivation (the number of colonies containing pluripotent cells, as a proportion of number of starting cells) is higher for ES cells than for EG cells? And higher for 8.5 dpc than for 11.5 dpc PGCs, lower still at 12.5 dpc, and zero at 13.5 dpc and later? Is the efficiency the same for PGCs from male and female embryos? And why does EG cell derivation require the presence of factors such as SCF and FGF-2, but ES cell derivation does not? Why has it proved possible to make pluripotent porcine EG cell lines (Shim *et al.*, 1997), when attempts to derive ES cells from other farm animal species appear to have been tantalizingly unsuccessful?

ACKNOWLEDGMENT

We thank the Leverhulme Trust for financial support.

KEY WORDS

Bisulfite sequencing   A technique whereby the methylation of each CpG site in a defined stretch of DNA is determined.

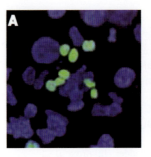
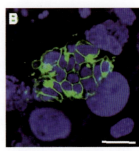

Figure 39-3. Immunofluorescence staining of EG cells. (A) An EG cell colony expresses Oct3/4 antigen localized in the nucleus of EG cells. (B) SSEA1 is expressed on the cell surface of EG cells. Nuclei are counterstained with Toto3 (blue). Oct3/4 and SSEA1 are identified by FITC. Scale bar 50 μm.

Epigenetic   Factors involved in influencing gene expression during development, without affecting DNA base sequence.

Gastrulation   The period of embryonic development during which the definitive body plan is laid down. Cells from the primitive ectoderm pass through the "primitive streak" region to form a third layer, the mesoderm, between ectoderm and endoderm.

Meiosis   The process by which germ cells (i.e., those in the ovaries or testes) divide to produce gametes. In meiosis I, homologous chromosomes exchange genetic material. In meiosis II, the two resulting diploid cells (i.e., which contain two sets of chromosomes) with their recombined chromosomes divide further to form two haploid gametes (i.e., which contain only one set of chromosomes).

Pluripotent   Capable of giving rise to all the cell types in the fetus but not able on its own to form a fetus.

FURTHER READING

Buehr, M. (1997). The primordial germ cells of mammals: some current perspectives. *Exp. Cell Res.* **232**, 194–207.

Chuma, S., and Nakatsuji, N. (2001). Autonomous transition into meiosis of mouse fetal germ cells *in vitro* and its inhibition by gp130-mediated signalling. *Dev. Biol.* **229**, 468–479.

De Felici, M., and McLaren, A. (1982). Isolation of mouse primordial germ cells. *Exp. Cell Res.* **142**, 476–482.

De Felici, M., and McLaren, A. (1983). In vitro culture of mouse primordial germ cells. *Exp. Cell Res.* **144**, 417–427.

Dolci, S., Williams, D. E., Ernst, M. K., Resnick, J. L., Brannan, C. I., Lock, L. F., Lyman, S. D., Boswell, H. S., and Donovan, P. J. (1991). Requirement for mast cell growth factor for primordial germ cells survival in culture. *Nature* **352**, 809–811.

Enders, G. C., and May, J. J., II. (1994). Developmentally regulated expression of a mouse germ cell nuclear antigen examined from embryonic day 11 to adult in male and female mice. *Dev. Biol.* **163**, 331–340.

Fujiwara, Y., Komiya, T., Kawabata, H., Sato, M., Fujimoto, H., Fususawa, M., and Noce, T. (1994). Isolation of a DEAD-family protein gene that encodes a murine homolog of *Drosophila vasa* and its specific expression in germ cell lineage. *Proc. Natl. Acad. Sci. USA* **91**, 331–340.

Matsui, Y., Zsebo, K., and Hogan, B. L. M. (1992). Derivation of pluripotential embryonic stem cells from murine primordial germ cells in culture. *Cell* **70**, 841–847.

Pesce, M., and De Felici, M. (1995). Purification of mouse primordial germ cells by MiniMACS magnetic separation system. *Dev. Biol.* **170**, 722–725.

Resnick, J. L., Bixler, L. S., Cheng, L., and Donovan, P. J. (1992). Long-term proliferation of mouse primordial germ cells in culture. *Nature* **359**, 550–551.

Shim, H., Gutiérrez-Adán, A., Chen, L. R., BonDurant, R. H., Behboodi, E., and Anderson, G. B. (1997). Isolation of pluripotent stem cells from cultured porcine primordial germ cells. *Biol. Reprod.* **57**, 1089–1095.

40

Derivation and Differentiation of Human Embryonic Germ Cells

Michael J. Shamblott, Candace L. Kerr, Joyce Axelman, John W. Littlefield, Gregory O. Clark, Ethan S. Patterson, Russell C. Addis, Jennifer N. Kraszewski, Kathleen C. Kent, and John D. Gearhart

Embryonic germ (EG) cells are pluripotent stem cells derived from primordial germ cells (PGCs) that arise in the late embryonic and early fetal period of development. EG cells have been derived from several species, including mouse, pig, chicken, and human. Mouse, pig, and chicken EG cells have been demonstrated to contribute to experimentally produced chimeric animals, including germ-line transmission in the latter two species. Furthermore, germ-line transmission of mouse and chicken EG cell derivatives have been demonstrated. Mouse and human EG cells can be differentiated *in vitro* to form embryoid bodies (EBs). Like EBs generated from embryonic stem (ES) cells, EG-derived EBs contain differentiated cells representing all three germ layers as well as mixed-cell populations of less differentiated progenitors and precursors. These human EB-derived (EBD) cells are capable of considerable cell proliferation and express a variety of lineage-specific markers. Human EBD cell cultures have a normal and stable karyotype and normal patterns of genomic imprinting, including X-inactivation. Transplantation studies have demonstrated that human EBD cells can engraft into a variety of rodent tissues and can participate in the recovery of rats following motor neuron injury.

Introduction

Pluripotent stem cells have been derived from two embryonic sources. ES cells were first derived from the inner cell mass of mouse preimplantation embryos, and EG cells were initially derived from mouse PGCs. Subsequently, EG cells have been derived from chicken, pig, and human PGCs. Pig, chicken, and mouse EG cells have been demonstrated to contribute to experimentally produced chimeric animals, including germ-line transmission in the latter two species.

PRIMORDIAL GERM CELLS

PGCs are the sole means of genetic transmission between parent and offspring, as they generate eggs and sperm. In many species, such as *C. elegans*, germ cells are segregated very early in development, during the first embryonic cleavages, and are marked by deposition of ribonucleoprotein

P-granules. In mammals, the process occurs later in development and seems to be directed more by extrinsic factors than by preprogrammed intrinsic differences. For example, in mice, cells that generate PGCs are located close to extraembryonic ectoderm during gastrulation. Rather than having a previously determined fate, cells in this location receive external signals to further differentiate into PGCs, as demonstrated by the observation that transplantation of cells from other parts of the epiblast to this region can take on a PGC fate. Several components of this signaling process have been identified. Initially, bone morphogenetic protein 4 (BMP4) and BMP8b are produced by extraembryonic ectoderm program cells from the epiblast to become extraembryonic mesoderm precursors, or PGCs. Cells destined to become PGCs express higher levels of membrane protein fragilis than nuclear protein stella.

In the mouse, PGCs are visible as alkaline phosphatase (AP) positive cells at the base of the allantois at 7.5 to 8.0 days postcoitus (dpc). They begin to associate with the endoderm that is invaginating to form the hindgut at 8.5 dpc. By 10.5 dpc, PGCs are associated with dorsal mesenteries and are translocated to the genital ridges. The migration of PGCs is caused by both cellular migration and association with moving tissues. Throughout this migration, PGCs expand from approximately 130 cells at 8.5 dpc to more than 25,000 at 13.5 dpc. Once they arrive at the genital ridge, PGCs continue proliferating until they enter prophase of the first meiotic division. In males, entry into meiosis is inhibited by signals from the developing testis, blocking PGCs at G_0 until after birth. In the absence of inhibitory signals, female PGCs undergo oogenesis.

Although not as thoroughly studied, much is known regarding the migratory path of human PGCs, including their association with gut endoderm and migration into developing genital ridges.

PGCs do not survive well under standard tissue culture conditions and are not pluripotent stem cells *in vivo* or *in vitro*. Early attempts to use various growth factors and feeder layers succeeded in prolonging their survival, but proliferation was limited. The combination of leukemia inhibitory factor (LIF), basic fibroblast growth factor (bFGF), and c-kit ligand (KL, also known as stem cell factor, mast cell factor, or steel factor) proved to result in an immortal cell population, especially if the KL was presented in the transmembrane form by a layer of "feeder" cells (see the section Feeder Layer). Instead of simply encouraging PGC proliferation, these factors cause the normally solitary PGCs to congregate and proliferate as multicellular colonies known as EG cells and to gain pluripotency. Mouse EG cell lines have been derived from PGCs prior to migration around 8.0–8.5 dpc, during migration at 9.5 dpc, and after entry into the genital ridges between 11.5 and 12.5 dpc.

The roles played by KL and the tyrosine kinase receptor for KL, c-Kit, in the *in vitro* derivation of EG cells from PGCs have parallels *in vivo*. c-Kit is expressed in PGCs, and KL is expressed along the PGC migratory pathway and in the genital ridges. The roles of KL and c-Kit in PGC survival were originally characterized through several mutations at their respective loci, *Sl* and *W*, which resulted in subfertile or sterile mice. PGCs are formed in homozygous mutant embryos of *W* and *Sl*, but mitosis is severely impaired, and the few PGCs that reach the gonad do not survive. KL is produced as a membrane-bound growth factor that can undergo proteolytic cleavage to generate a soluble form. Mice lacking the membrane-bound KL but not the soluble form maintain low PGC numbers and are sterile, suggesting that the membrane-bound form but not the soluble form is essential for PGC survival. The mechanism involved in KL-induced PGC survival has been shown to involve suppression of apoptosis. The c-kit receptor has also been shown to be involved in the adhesion of mouse PGCs to somatic cells *in vitro*. Other recent studies attempting to identify signaling pathways activated by KL binding to its receptor in mouse PGCs have shown activation of AKT kinase and telomerase.

In contrast to the embryologically early and relatively undifferentiated epiblast, PGCs arise late and have a specialized role during normal development. In this regard, it is somewhat surprising that exposure to three cytokines can convert PGCs to pluripotent stem cells *in vitro*. It is possible that the flexibility provided by extrinsic signaling during PGC specification, rather than intrinsic preprogramming, allows for this conversion.

COMPARISON TO ES CELLS

Both mouse ES and EG cells are pluripotent and demonstrate germ-line transmission in experimentally produced chimeras. Mouse ES and EG cells share several morphological characteristics such as high levels of intracellular AP and presentation of specific cell-surface glycolipids and glycoproteins. These properties are characteristic of, but not specific for, pluripotent stem cells. Other important characteristics include growth as multicellular colonies, normal and stable karyotypes, the ability to be continuously passaged, and the capability to differentiate into cells derived from all three EG layers: endoderm, ectoderm, and mesoderm.

Human EG Cell Derivation

Although many combinations of cytokines and feeder layers have been evaluated, the standard practice for derivation of human EG cells remains similar to techniques developed for the mouse. As of 2003, approximately 140 human EG cultures

have been derived in our laboratory using this general technique.

INITIAL DISAGGREGATION AND PLATING

Gonadal ridges and mesenteries of week 5–9 postfertilization human embryos (obtained as a result of therapeutic termination of pregnancies) are collected in 1-ml ice-cold growth media and rapidly transported to a sterile work space. The tissues are then soaked in calcium–magnesium-free Dulbecco's phosphate-buffered saline (DPBS) for 5 minutes and then transferred to 0.1-ml trypsin–EDTA solution. The concentration of trypsin and EDTA is varied such that at the earliest developmental stages, a gentler 0.05% trypsin–0.5-mM EDTA is used, and at later developmental stages, a stronger 0.25% trypsin–0.5-mM EDTA solution is used. The tissue is mechanically disaggregated thoroughly using a fine forceps and iris scissors. This process is carried out for 5 to 10 minutes at room temperature and then incubated at 37°C for 5 to 10 minutes. This disaggregation process often results in a single-cell suspension and large pieces of undigested tissue. To stop the digestion, serum containing growth media is added. The digested tissue is transferred to wells of a 96-well tissue culture plate that has been previously prepared with feeder layer (see the sections Plating an STO Feeder Layer Prior to Inactivation and Plating an STO Feeder Later after Inactivation). Usually, the initial plating occupies 4 to 10 wells of the 96-well plate. The plate is incubated at 37°C in 5% CO_2 with 95% humidity for seven days. Approximately 90% of the growth media is removed each day, and the plate is replenished with fresh growth media.

SUBSEQUENT PASSAGE OF EG CULTURES

In the first seven days of derivation (passage 0), most human EG cultures do not produce visible EG colonies. Staining for AP activity demonstrates the presence of solitary PGCs with either stationary or migratory morphology (Figure 40-1A and 40-1B). Often, colonies of cells that do not stain AP+ are seen (Figure 40-1D and 40-1E), as are small clumps of tissue remaining from the initial disaggregation (Figure 40-1C). After seven days, the media is removed, and the wells are rinsed twice with calcium–magnesium-free DPBS for a total of five minutes. Then, 40 μl of freshly thawed trypsin solution is added to each well, and the plate is incubated on a heated platform or in a tissue culture incubator for five minutes at 37°C. As previously described, the trypsin solution is 0.05% trypsin–0.5-mM EDTA, 0.25% trypsin–0.5-mM EDTA, or a mixture of these two solutions. The important point at this stage is to facilitate the complete disaggregation of the STO cell feeder layer (see the section Feeder Layer), which can be a significant challenge. After the incubation in trypsin, the edge of the 96-well culture plate is hit firmly against a solid surface until the STO cells have completely lifted off the growth surface. This process can be aided by scraping the well and gently triturated. After the STO cells have been loosened, fresh growth media is added to each well, and the contents are triturated. This phase is critical to successful disaggregation of STO feeder layer and EG cells.

All subsequent passages are done as described. After 14 to 21 days (during passage 1 or 2), large and recognizable EG colonies will arise in some of the wells (Figure 40-2). At this point, wells that do not have EG colonies are discarded.

Several common problems occur during the passage of human EG cells. One observation is that the STO feeder layer will sometimes not fully disaggregate. This can be observed by the presence of large cell aggregates immediately follow- ing disaggregation. If this occurs with regularity, it is a sign that the trypsinization method is insufficient or that the STO cells have become less contact inhibited and have overgrown during the seven days of culture. Another common occurrence is that the EG colonies do not fully disaggregate. The conse- quences of poor disaggregation are that the large pieces dif- ferentiate or die and fewer EG cells are available for continued culture expansion. Although much effort has been expended to find a solution to this problem, it remains the most difficult aspect and challenging hurdle to human EG cell biology.

To gain some insight into this problem, a series of electron microscopic images were taken to compare the cell–cell inter- actions found in mouse ES, mouse EG, and human EG cell colonies. It is evident from these images that cells within the human EG colonies adhere more completely to each other than cells within mouse ES and EG colonies (Figure 40-3). It is possible that this tight association within the colony limits the access of disaggregation reagents. At this time, neither the nature of the cell–cell interactions nor an effective solution to this problem is evident.

Because of incomplete disaggregation and other intrinsic or extrinsic signals, many human EG colonies (10 to 30% per passage) differentiate to form three-dimensional structures termed EBs (see the section EB Formation and Analysis) or flatten into structures that are AP− and do no continue to pro- liferate (Figure 40-4). EG colonies that are more fully disag- gregated go on to produce new EG colonies, and under the best circumstances, EG cultures can be expanded continu- ously for many months and have routinely exceeded 20 pas- sages. Inevitably, as large EG colonies are removed from the culture as a result of EB formation, the cultures become sparse and are discontinued for practical considerations. Efforts employ- ing standard dimethyl sulfoxide (DMSO) cryopreservation techniques have so far been unsuccessful.

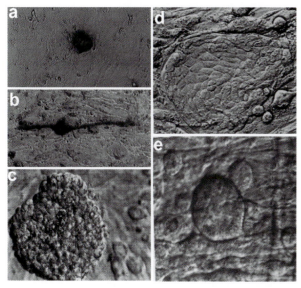

Figure 40-1. Cell morphologies seen in early passage human embryonic germ culture. (A and B) Alkaline phosphatase-positive (AP⁺) stationary and migrating primordial germ cells. (C) Multicellular piece of undisaggregated gonadal tissue. (D and E) Flat and round cell colonies that do not lead to human EG cells.

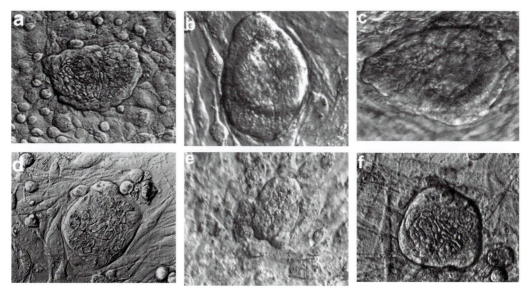

Figure 40-2. Human EG colonies growing on an STO cell feeder layer.

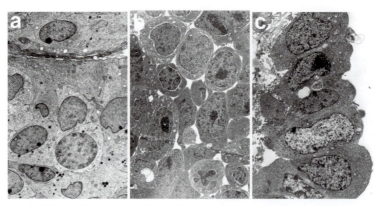

Figure 40-3. Electron micrograph of EG and ES colonies. (A) Human EG colony, (B) mouse EG colony, and (C) mouse ES colony.

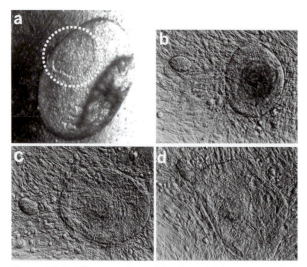

Figure 40-4. Differentiating human EG colonies. (A) Stereomicrograph of an EG colony with attached embryoid body (EB). The EG colony is circled. The diameter of the EB is approximately 0.5 mm. (B–D) Time-lapse study of EG cell flattening: (B) Four days after disaggregation, the small EG colony on the left results from more complete disaggregation. The large colony on the right is composed of a dark residual colony surrounded by new EG cell growth. (C) Five days after disaggregation, the large EG colony has begun to flatten. (D) Seven days later, the large EG colony has completely flattened and did not survive as a recognizable EG cell colony after disaggregation and replating. In a separate experiment, EBs and these flattened structures are shown to be largely AP⁻.

FEEDER LAYER

Unlike ES cells, EG cell derivation is highly dependent on a specific type of feeder layer. Mouse STO fibroblasts are a spontaneously transformed cell line from the Sandoz inbred mouse. They are Thioguanine- and Ouabain-resistant, features of historical interest but not used in this context. STO cells have been used to derive mouse embryonal carcinoma (EC), ES, and EG cells, and to date, they are the only cell type capable of generating human EG cells. The factor or factors provided exclusively by this cell type are not fully understood.

However, the transmembrane form of KL is present on STO fibroblasts but not on most other cell types evaluated in our lab.

Although STO is a clonal cell line, individual isolates vary greatly in their ability to support human EG derivation. This is further complicated by the known phenotypic variation of STO cells in continuous culture. Given the very limited supply of human tissue, it is prudent to screen STO cells for suitability prior to use. The most reliable screening method is to produce a number of clonal STO lines (by limiting dilution or cloning cylinder) and to evaluate them for their ability to support the *derivation* of mouse EG cells. The growth of existing mouse EG lines is not a sufficient method, as most mouse EG lines become feeder layer independent after derivation. Derivation of mouse EG cells is not a simple undertaking. In an effort to more rapidly screen STO cells and to investigate the role of transmembrane KL, we have begun to screen STO cell lines by using an immunocytochemical method to detect KL. Once a supportive STO fibroblast line is identified, it should be immediately cryopreserved in several low-passage aliquots. One of these aliquots can then be expanded to provide many medium passage aliquots, which are thawed and used with limited further expansion. Continuous passage of STO fibroblasts without frequent screening should be avoided.

Plating an STO Feeder Layer Prior to Inactivation

Two different methods can be used to prepare an STO feeder layer: plate-then-irradiate and irradiate-then-plate. Most human EG cultures are derived using the former method. This is practical when a large γ-irradiator is available. STO cells are passaged for short periods (not continuously) in PGC growth media without LIF, bFGF, or forskolin and are disaggregated using 0.05% trypsin–EDTA solution. One day prior to use, a 96-well culture dish is coated with 0.1% gelatin for 30 minutes. The gelatin is withdrawn, and 5×10^4 STO cells are plated per well in PGC growth media without the LIF, bFGF, or forskolin. Similar cell densities (~1.5×10^5 cells/cm²) can be achieved in other well configurations. The cells are grown overnight and then exposed to 5000 rads (1 rad = 0.01 Gy) γ-radiation or X ray. The cells are then returned to the tissue culture incubator until required. Prior to use, the growth media is removed, 0.1-ml PGC growth medium with added factors is added to each well (or half of the required well volume), and the dish is returned to the tissue culture incubator.

Plating an STO Feeder Layer after Inactivation

This method of STO cell preparation is used when a large γ-radiation unit is not available, when large amounts of cells are required, or if better control of STO cell density is required. STO cells are grown as described previously, trypsinized, counted, and resuspended in PGC growth media without added factors. The cells are placed into one or more 50-ml conical tubes and placed into the γ-irradiator or X ray device. Cells are exposed to 5000 rads γ-radiation or X ray. Following exposure, cells are adjusted to a convenient concentration with PGC growth media without added factors, counted, and plated into tissue culture dishes that have been previously coated with 0.1% gelatin for 30 minutes. Cells are allowed to adhere overnight; then the media is replaced with half the final volume of PGC growth media, including factors prior to use.

PGC GROWTH MEDIA COMPONENTS

Human EG cells are derived and maintained in Dulbecco's modified Eagle's medium (Gibco BRL) supplemented with 15% fetal bovine serum (FBS, Hyclone), 0.1-mM nonessential amino acids (Gibco BRL), 0.1-mM 2-mercaptoethanol (Sigma), 2-mM glutamine (Gibco BRL), 1-mM sodium pyruvate (Gibco BRL), 100 U/ml penicillin (Gibco BRL), 100 µg/ml streptomycin (Gibco BRL), 1000 U/ml human recombinant LIF (hrLIF, Chemicon), 1–2 ng/ml human recombinant bFGF (hrbFGF, R&D systems), and 10-µM forskolin (Sigma) prepared in DMSO.

EVALUATION OF EG CULTURES

Of 150 human PGC cultures initiated, 142 (~95%) demonstrated morphological, biochemical, and/or immunocytochemical characteristics consistent with previously characterized pluripotent stem cell lines. The easiest and most consistent method for evaluating EG cultures is to note the presence of tightly compacted multicellular colonies resembling early-passage mouse ES and EG cell colonies (Figure 40-2) rather than the flattened and more loosely associated colonies seen in human EC and rhesus ES cells. Under the best circumstances, the number of colonies should double or triple over a one-week passage. The trend is usually to start with small colonies and to end with larger colonies that result from incomplete colony disaggregation (see the earlier section Subsequent Passage of EG Cultures).

High levels of AP activity are associated with human EG cells. Under standard culture conditions, human EG colonies are >70–90% AP$^+$. As colonies differentiate, one can observe a lower staining percentage and weaker staining, sometimes restricted to the periphery of the colony.

Human EG cells have been further characterized by a bank of five monoclonal antibodies: SSEA-1, SSEA-3, SSEA-4, TRA-1-60, and TRA-1-81. Colonies stain strongly for four of the five antibodies. The antibody recognizing SSEA-3 antigen stains the cells inconsistently and weakly. As with the results of AP staining, the percentage of cells within a colony that stain positive is variable.

The histological profile of human EG cells (AP$^+$, SSEA-1$^+$, SSEA-3$^+$, SSEA-4$^+$, TRA-1-60$^+$, and TRA-1-81$^+$) differs from undifferentiated human EC and rhesus ES cells, which are SSEA-1–. The fact that differentiation of the human EC line NTERA2 leads to increased expression of SSEA-1 may suggest that this is indicative of differentiation in the human EG cultures. However, NTERA2 differentiation is accompanied by the loss of the other markers, which are not observed in these cultures.

Karyotypic analyses carried out at passage 8–10 (60–70 days in culture) indicated apparently normal human chromosomes at the 300-band level of resolution. Both XX and XY cultures have been derived.

Other markers of pluripotency, such as mRNA expression of the human ortholog of mouse Oct3/4 and telomerase enzyme activity, have been examined. Results differ greatly depending on the status of the EG colony, culture, or both. In general, relatively undifferentiated EG colonies are OCT4 mRNA$^+$ as detected by RTPCR and have detectable levels of telomerase.

AP and Immunocytochemical Staining

Cells are fixed for detection of AP activity in 66% acetone–3% formaldehyde and then stained with naphthol/FRV-alkaline AP substrate (Sigma). For immunocytochemistry, cells are fixed in 3% paraformaldehyde in DPBS. Cell-surface glycolipid- and glycoprotein-specific monoclonal antibodies are used at 1:15 to 1:50 dilution. MC480 (SSEA-1), MC631 (SSEA-3), and MC813-70 (SSEA-4) antibodies were supplied by the Developmental Studies Hybridoma Bank (University of Iowa). TRA-1-60 and TRA-1-81 were a gift from Dr. Peter Andrews (University of Sheffield, U.K.). Antibodies are detected by using biotinylated antimouse secondary antibody, streptavidin-conjugated horseradish peroxidase, and AEC chromogen (BioGenex).

EB Formation and Analysis

EBs form spontaneously in human EG cultures. Although this represents a loss of pluripotent EG cells from the culture, EBs provide evidence for the pluripotent status of the culture and provide cellular material for subsequent culture and experimentation (see the section EBD Cells). Initially, EBs provided the only direct evidence that human EG cultures were pluripotent, as all attempts to form teratomas in mice from human EG cells failed. To this day, there is no evidence of teratoma formation from human EG cells or their derivatives.

EB Embedding and Immunohistochemistry. The constituent cells of EBs can be identified most reliably by embedding them in paraffin and staining sections with a bank of well-characterized antibodies. This process avoids the significant problem of antibody trapping that complicates analyses of large three-dimensional structures when direct staining is attempted. EBs are collected from cultures and placed into a small drop of molten 2% low melting point agarose (FMC), prepared in DPBS, and cooled to 42°C. Solidified agarose-containing EBs are then fixed in 3% paraformaldehyde in DPBS and embedded in paraffin. Individual 6-µm sections are placed on microscope slides (ProbeOn Plus, Fisher Scientific). Routinely, immunohistochemical analysis is carried out by using a BioTek-Tech Mate 1000 automated stainer (Ventana-BioTek Solutions). Manual staining is also carried out using standard immunohistochemical methods; however, antigen retrieval is required for some of the antibodies employed. Cryosections of EBs generally produce less satisfactory cell morphology and have not been used extensively.

Antibodies used on paraffin sections include HHF35 (muscle specific actin, Dako), M 760 (desmin, Dako), CD34 (Immunotech), Z311, (S-100, Dako), sm311 (panneurofila-ment, Sternberger Monoclonals), A008 (α-1-fetoprotein), CKERAE1/AE3 (pancytokeratin, Boehringer Mannheim), OV-TL 12/30 (cytokeratin 7, Dako), and K_s20.8 (cyto-keratin 20, Dako). Primary antibodies are detected by using biotinylated antirabbit or antimouse secondary antibody, streptavidin-conjugated horseradish peroxidase, and DAB chromogen (Ventana-BioTek Solutions). Slides are counter-stained with hematoxylin. Using these and other antibodies, it can be demonstrated that when human EG cells differentiate, they form EBs comprised of endodermal, ectodermal, and mesodermal derivatives.

EBD Cells

Although a compelling demonstration of the potential of human EG cells, the limited growth characteristics of differ-entiated cells within EBs and difficulties associated with their isolation make extensive experimental manipulation difficult and limit their use in future cellular transplantation therapies. At least two possibilities exist to explain the presence of the observed differentiated cell types. These cells could be gen-erated directly from the pluripotent EG cell, or they could proceed through a series of precursor or progenitor cell types prior to the acquisition of a mature phenotype. It seems unlikely that cells within EBs would bypass normal pathways of differentiation, so efforts have been made to isolate and expand these populations. The hypothesis was that progeni-tor–precursor cells would have desirable proliferation charac-teristics and could be recognized by the expression of molecules known to mark progenitor–precursor populations, as well as by the simultaneous expression of markers normally considered part of a mature cellular expression repertoire. Simultaneous expression of neuronal and glial markers by neural progenitors[52,53] and the expression of a variety of lineage-affiliated transcription factors and cytokine recep-tors[54] by multipotent hematopoietic progenitors provide some basis for this hypothesis. In this model of differentiation, multilineage gene expression by precursor or progenitor cells defines a ground state from which cell-extrinsic and -intrinsic signals work to continuously define a differentiated expres-sion pattern and phenotype, resulting in the developmental

plasticity observed after the differentiation of bone marrow and central nervous system stem cells.

The method used to isolate cell populations from EBs is conceptually similar to microbiological selective media experiments. EBs are disaggregated and plated into several different cellular growth environments. These environments consist of combinations of a growth media and a matrix. Although many combinations were evaluated, most EBD cell cultures have been derived from one of six environments formed by combinations of two growth media and three plating surfaces. The growth media are an RPMI 1640 media supplemented with 15% FBS and a low (5%) FBS media supplemented with bFGF, epidermal growth factor (EGF), insulin-like growth factor 1 (IGF1), and vascular endothelial growth factor (VEGF). The plating surfaces are bovine type I collagen, human extracellular matrix extract, and tissue culture-treated plastic. These are not intended to be highly selective environments. Instead, they favor several basic themes: cells thriving in high serum and elevated glucose (10 mM) conditions versus cells proliferating in low glucose (5 mM) under the control of four mitogens. Surfaces included binding to type I collagen, a biomatrix often thought to favor undifferentiated proliferation versus human extracellular matrix, a more complex mixture of laminin, collagen, and fibronectin. The initial assay is to determine conditions that favor extensive cell proliferation, with the hypothesis that this condition will favor undifferentiated cell populations or at least provide numerical disadvantage to the terminally differ-entiated constituents of the EB.

EBD GROWTH AND EXPRESSION CHARACTERISTICS

Cell populations capable of long-term and robust proliferation can be isolated in this way from human EG-derived EBs. Embryoid body-derived is the generic term used to describe cells derived in this way *and* capable of extensive further pro-liferation. In general, the type I collagen and human extracel-lular matrices combined with the low serum media provide the most rapid and extensive cell proliferation. EBD cell lines and cultures are routinely maintained in the environment in which they were derived. The EBD naming convention aids this process. The first two letters of the name refer to the EG culture from which it was derived. The second letter indicates the growth media (E for EGM2MV and R for RPMI1640), and the last letter indicates the matrix (C for collagen, E for human extracellular matrix extract, and P for plastic). For example, the EBD culture SDEC was derived from EG culture SD in EGM2MV medium on type I collagen.

To distinguish EBD cells from a simple population of cells rapidly proliferating but presumably uninteresting in terms of molecules expressed, it is important to establish a robust expression profile assay. The assay should use redundant measures when possible and must combine breadth, sensitiv-ity, specificity, and speed. A series of 24 RTPCRs detecting products from five cell lineages (neuronal, glial, muscle, hematopoietic–vascular endothelia, and endoderm) combined with immunocytochemical staining provides a rapid measure of cell expression. Inevitably, molecular markers are not as definitively specific as desired, so multiple markers for each lineage are advisable.

Using mRNA and antibody expression profiling, we can demonstrate that most rapidly proliferating EBD cell cultures simultaneously express a wide array of mRNA and protein markers normally associated with distinct developmental lineages. This is not a surprising property considering that EBD cells are, at least during the derivation stage, a mixed-cell population. More remarkable is the finding that most (11 of 13) EBD cell *lines* isolated by dilution cloning also exhibit a broad multilineage gene expression profile. It can also be demonstrated that the expression profile for a given EBD culture remains stable throughout the lifespan of the culture. This normally exceeds 70 population doublings but is not unlimited since EBD cells are not immortal.

More than 100 EBD cell cultures and clonally isolated cell lines have been derived and characterized as described here. Most of these cultures share the properties of rapid and robust proliferation and broad multilineage gene expression. Less than 10% of the EBD cultures derived have a narrow expression profile, with one extreme case expressing only nestin, vimentin, and α-1-fetoprotein mRNA. Other general trends in EBD expression are that many cultures appear to be neurally biased, with strong expression of neuronal, glial, and neural progenitor markers, and relatively weak in expression of muscle markers.

Other general characteristics of EBD cells are the relative ease with which they can be genetically manipulated using lipofection and electroporation as well as retroviral, adenoviral, and lentiviral vectors. Adenoviral and lentiviral vectors are capable of nearly 100% transduction efficiency. These techniques have been used to generate EBD lines that constitutively and tissue-specifically express enhanced GFP and contain many different genetic selection vectors. EBD cells can be immortalized by retroviral-based expression of the telomerase RNA subunit (pBABE). Interestingly, after several hundred population doublings, these lines often become genetically unstable, generating at least two rearrangements: [47,XX,−1,$^+$ del (1)(q12), $^+$i (1)(q10)] and [46,XX, del(4)(p14)]. In addition, these EBD lines tend to have narrow expression profiles.

The imprinting pattern of several EBD cultures has been examined. In one study, expression levels of four imprinted genes (TSSC5, H19, SNRPN, and IGF2) were determined in five EBD cultures. Three of these genes (TSSC5, H19, and SNRPN) had normal monoallelic expression levels, and IGF2 had a partially relaxed imprinting pattern comparable to levels found in normal somatic cells. This study also determined that the imprinting control region that regulates H19 and IGF2 imprinting had a normal pattern of DNA methylation. A second study determined that two XX EBD cultures had a normal pattern of X-inactivation.

The proliferation and expression characteristics of EBD cells suggest that they may be useful in the study of human cell differentiation and as a resource for cellular transplantation therapies. One important property in this regard is that no tumor of human origin has arisen in any animal receiving EBD cells, although hundreds of mice, rats, and African green monkeys have received EBD transplants in a variety of anatomical locations, often consisting of more than 1 million cells injected. This is in contrast to the infrequent, yet significant, number of teratocarcinomas that have arisen following transplantation of cells produced through neural and hematopoietic differentiation of mouse ES cells.

EBD TRANSPLANTATION

Transplantation of EBD cells into animal models of human disease constitutes an active and promising research avenue. Studies with EBD cells and many other cell types have suggested that tissue injury can be highly instructive to transplanted cells. This provides a powerful method to test the potential of cells to differentiate without an initial understanding of the underlying mechanisms. These studies also suggest possibilities for the eventual treatment of patients suffering from these diseases.

One example of EBD cell transplantation is the use of an EBD culture named SDEC. This culture was initially selected for further study because of its strong neural expression bias. SDEC cells were introduced into the cerebrospinal fluid of normal rats and rats exposed to the neuroadapted Sindbis virus. This virus specifically targets spinal cord motor neurons, and infection results in permanent hind limb paralysis. SDEC cells transplanted into virally injured rats engrafted extensively the length of spinal cord and migrated into the cord parenchyma. Substantial engraftment was not observed in uninjured animals receiving SDEC cells. Engrafted SDEC cells took on expression characteristics of mature neurons and astrocytes. Remarkably, albeit at low frequency, engrafted SDEC cells became immunoreactive to choline acetyltransferase and sent axons into the sciatic nerve. Even more remarkably, after 12 and 24 weeks, paralyzed animals receiving SDEC cells partially recovered hind limb function. In this experiment, the frequency and total number of neurons generated from SDEC cells was not sufficient to easily explain the significant recovery of function, which is not surprising since the transplanted SDEC cells are a mixed-cell population rather than a line or culture grown or differentiated to promote neural outcome. The mechanism proposed to explain the functional recovery involves EBD cells protecting host motor neurons from death and facilitating host motor neuron reafferentation, possibly through the secretion of transforming growth factor-α and brain-derived neurotrophic factor.

This example illustrates several important points. The engraftment (and possibly one or more steps in cellular differentiation) of SDEC cells was promoted by an injury signal following viral infection. Once engrafted, cells within the SDEC population were capable of differentiation *in vivo* into mature astrocytes and neurons, some of which sent processes along the correct pathway to the sciatic nerve and were then capable of retrograde transport. It is difficult to see how this elaborate and spatially precise differentiation could be carried out *ex vivo* and then introduced into an animal or patient. The use of a mixed-cell population may have allowed the variety of cellular responses that ultimately resulted in functional recovery. Some cells within the population were capable of forming new neural cells, and others took on supportive and protective roles. Although in this experimental model the multiple roles carried out by use of a mixed-cell population may have resulted in functional recovery, future experiments will need to focus on isolation of subpopulations to increase the efficiency of differentiation into required cell types and to address issues of safety. Lastly, the rats in all treatment groups received immunosuppressive drugs to discourage rejection of the human EBD cells. In the near term, this will likely be a feature of all EBD-based cellular transplantation experiments and therapies.

EBD DERIVATION, GROWTH, AND CRYOPRESERVATION METHODS

EBs formed in the presence of PGC growth media are harvested in groups of 10 or more and are dissociated by digestion in 1-mg/ml Collagenase/Dispase (Roche) for 30 minutes to 1 hour at 37°C. Cells are then spun at 1000 rpm for 5 minutes and resuspended in various growth media–matrix environments. These include RPMI growth media [RPMI 1640 (LTI), 15% FCS, 0.1-mM nonessential amino acids, 2-mM L-glutamine, 100 U/ml penicillin, and 100 μg/ml streptomycin] and EGM2MV media (Clonetics) [5% FCS, hydrocortisone, hbFGF, hVEGF, R^3-IGF1, ascorbic acid, hEGF, heparin, gentamycin, and amphotericin B]. Matrices are bovine collagen I (Collaborative Biomedical, 10 μg/cm^2), human extracellular matrix (Collaborative Biomedical, 5 μg/cm^2), and tissue culture plastic. EBD cells are cultured at 37°C, 5% CO_2, 95% humidity and are routinely passaged 1 : 10 to 1 : 40 using 0.025% trypsin–0.01% EDTA (Clonetics) for 5 minutes at 37°C. Low serum cultures are treated with trypsin inhibitor (Clonetics) and then spun down and resuspended in growth media. EBD cells are cryopreserved in the presence of 50% FCS, 10% DMSO, in a controlled rate freezing vessel and stored in liquid nitrogen.

Summary

Human EG cells can be derived from PGCs by using methods similar to those used to derive mouse EG cultures. Like mouse ES and EG cells, human EG cells require LIF for proliferation as undifferentiated stem cells. Unlike mouse EG cells, however, human EG cells do not readily lose their dependence on exogenous cytokines and factors supplied by the feeder layer, and they have a higher frequency of spontaneous differentiation into EBs. Although EBs are a loss to the pluripotent stem cell population, they are a source of cells expressing markers of mature cellular phenotypes as well as their presumed progenitors and precursors. Cells that retain a high capacity for cell proliferation and express makers of multiple lineages can be isolated from EBs and can be used in a variety of *in vitro* and *in vivo* differentiation paradigms. The current challenges are to match individual EBD cultures to desired endpoints and to enrich or purify populations of cells within EBD cultures to more specifically address biological requirements.

KEY WORDS

Embryoid body   A multicellular structure comprised of multiple-cell types formed during the *in vitro* and *in vivo* differentiation of embryonic stem cells.

Embryonic germ cell   A pluripotent stem cell derived from primordial germ cells.

Primordial germ cell   A cell type that colonizes the developing gonad and turns into egg or sperm.

FURTHER READING

Kerr, D. A., Llado, J., Shamblott, M. J., Maragakis, N. J., Irani, D. N., Crawford, T. O., Krishan, C., Dike, S., Gearhart, J. D., and Rothstein, J. D. (2003). Human embryonic germ cell derivatives facilitate motor recovery of rats with diffuse motor neuron injury. *J. Neurosci.* **23**, 5131–5140.

Matsui, Y., Toksoz, D., Nishikawa, S., Nishikawa, S., Williams, D., Zsebo, K., and Hogan, B. L. (1991). Effect of steel factor and leukemia inhibitory factor on murine primordial germ cells in culture. *Nature* **353**, 750–752.

Matsui, Y., Zsebo, K., and Hogan, B. L. (1992). Derivation of pluripotential embryonic stem cells from murine primordial germ cells in culture. *Cell* **70**, 841–847.

McLaren, A. (2000). Establishment of the germ cell lineage in mammals. *J. Cell Physiol.* **182**, 141–143.

Resnick, J. L., Bixler, L. S., Cheng, L., and Donovan, P. J. (1992). Long-term proliferation of mouse primordial germ cells in culture. *Nature* **359**, 550–551.

Shamblott, M., Axelman, J., Littlefield, J., Blumenthal, P., Huggins, G., Cui, Y., Cheng, L., and Gearhart, J. (2001). Human embryonic germ cell derivatives express a broad range of developmentally distinct markers and proliferate extensively *in vitro. Proc. Nat. Acad. Sci. USA* **98**, 113–118.

Shamblott, M. J., Axelman, J., Wang, S., Bugg, E. M., Littlefield, J. W., Donovan, P. J., Blumenthal, P. D., Huggins, G. R., and Gearhart, J. D. (1998). Derivation of pluripotent stem cells from cultured human primordial germ cells. *Proc. Nat. Acad. Sci. USA* **95**, 13,726–13,731.

Tam, P. P., and Snow, M. H. (1981). Proliferation and migration of primordial germ cells during compensatory growth in mouse embryos. *J. Embryol. Exp. Morphol.* **64**, 133–147.

41

Growth Factors and the Serum-free Culture of Human Pluripotent Stem Cells

Alice Pébay and Martin F. Pera

Introduction

The development of human embryonic stem (ES) cells (HESC) in 1998, presaged by work on stem cell lines from human embryonal carcinoma (EC) and monkey ES cells, led to a great surge of interest in the biology and potential therapeutic applications of pluripotent stem cells. The early reports of derivation of both monkey and HESC used serum-containing medium in combination with mouse embryo feeder cell support to maintain stem cell renewal. This methodology did not differ very much from that used previously to establish and propagate mouse ES cells or human EC cells. However, several critical differences in growth properties of mouse and HESC soon emerged. The human cells display very low cloning efficiency when dissociated to single cells, and the addition of leukemia inhibitory factor (LIF) is without effect on the growth of human stem cells in the absence of a feeder cell layer. Both of these properties of human pluripotent stem cells were first noted in early efforts to characterize human EC cells. The findings meant that techniques that had been developed for efficient propagation and manipulation of mouse ES cells would find limited application in the human system.

The original methods for cultivation of HESC had several drawbacks. The use of fetal calf serum (FCS) and mouse feeder cells potentially exposes the human cells to unknown pathogens, in particular viruses or prions, which might be transmitted between species. Both serum and feeder cell layers also contain many undefined components that may have profound and undesirable effects on HESC differentiation. The poor growth of HESC following dissociation to single cells means that cloning at low densities, which is key to many experimental manipulations, is problematic. Moreover, because single cells survive poorly, and large aggregates of cells usually differentiate, the resulting requirement to subculture the cells as clusters of an appropriate size can limit the ability to scale up the culture while maintaining cells in an undifferentiated state.

Thus, the major challenges to the development of HESC culture systems are as follows: to eliminate the use of serum in the medium, to eliminate the use of feeder cells, to improve the cloning efficiency, and to enable scale-up of the cultures.

Since the first reports of the development of HESC, there have been several advances in our ability to grow these cells *in vitro*, but not all of these goals have been achieved to date. This chapter reviews the current status of development of culture systems for HESC.

Mouse Embryonic Stem Cells

The growth requirements of pluripotent stem cells have been studied in most detail in the mouse. The original reports of mouse ES cell derivation used embryonic fibroblast feeder cells and serum-containing medium. A key advance was the discovery of LIF as a critical component for maintenance of the pluripotent state. LIF and other members of this cytokine family that act through engagement of their particular receptors and the gp130 receptor maintain mouse ES cells in the pluripotent state through activation of JAK/STAT signaling. Production of LIF is one factor that accounts for the requirement of feeder cells to support mouse ES cell growth. Although many workers maintain mouse ES cells on a feeder layer in the presence of LIF, there are certainly a number of examples of germ-line competent mouse ES cell lines that can be maintained without a feeder cell layer in the presence of LIF. Notwithstanding the evidence that LIF is important for mouse ES cell renewal, it is important to recognize that mice deficient in the gp130 receptor can still undergo normal development through midgestation, that mouse ES cell lines lacking the LIF receptor can still form colonies on a feeder cell layer, and that there is evidence that factors other than LIF can maintain mouse and human EC cells in the pluripotent state. These factors remain to be identified, but the important message for HESC work is that extrinsic pathways independent of gp130 signaling may have a role in pluripotent stem cell maintenance.

Human Embryonal Carcinoma Cells

Human EC cell lines have been isolated by a number of investigators from different pathologic subclasses of these human tumors using different methodologies. Thus, although the stem cells of these cultures have certain common characteristics including marker expression, the lines do vary in their ability to differentiate and in their growth requirements *in vitro*. One general observation is that many human EC cells, like HESC, are difficult to passage as single cells at low density. Although some human EC cell lines do not require feeder cell support, other pluripotent cell lines show a strong feeder cell dependence that is not met by addition of LIF to the culture medium. There is a spectrum of growth requirements among human EC cell lines that ranges from cell lines displaying a feeder cell requirement similar to that of HESC to those EC lines that may be grown at clonal density at high efficiency in the absence of a feeder cell layer. In general, the more robust the human EC cell line in minimalist culture conditions, the more limited its capacity for differentiation.

Early efforts to analyze the growth requirements of human EC cells led to identification of several factors critical for their growth. One important attachment factor identified was the serum protein vitronectin, which on its own was able to support adhesion and subsequent growth of feeder independent EC cell lines in the presence of insulin, transferrin, and albumin. Studies on feeder cell-independent clones of the cell line Tera-2 revealed a key role for insulin-like growth factors in maintaining survival of pluripotent stem cells in the absence of serum. In addition to insulin-like growth factors, which are antiapoptotic for human EC cells, several factors active through tyrosine kinase surface receptors were identified with effect on human EC cells. Several groups reported a mitogenic effect of fibroblast growth factor-2 (FGF-2) on clones of the Tera-2 cell line grown under serum-free conditions; FGF-1 and FGF-4 were also active. At high doses, FGF-2 stimulated cell motility. Further studies pointed to a potential role for platelet-derived growth factor (PDGF) signaling in promoting growth of pluripotent human EC cells. The presence of receptors for this molecule was surprising because, in the mouse, expression of PDGF receptors occurs only after commitment of cells to mesoderm differentiation. There was indirect evidence for a potential autocrine role of this molecule in driving EC cell proliferation.

These observations on feeder independent human EC lines can be contrasted to work with pluripotent, feeder-dependent human EC cell lines. These cells cannot be maintained without a feeder cell layer even in the presence of serum. Insulin, PDGF, FGF-2, and LIF, along with many other known factors, fail to replace the requirement of a feeder layer for stem cell maintenance, at least in the presence of serum. Supernatants produced by a feeder-independent yolk sac carcinoma cell line capable of growth in the absence of serum could substitute for a feeder cell layer, but the active principle in these supernatants remains unknown.

Several candidate autocrine growth factor regulators were identified through studies of human EC cells. TDGF-1, also known as cripto, is an epidermal growth factor Cripto/FRL-1/Cryptic (EGF-CFC) superfamily member that is expressed in human EC cells and a few other cell types. Transcripts for a smaller, truncated form of this growth factor are more widely expressed in transformed cells and in normal human tissues. In a clone of the cell line Tera-2, recombinant TDGF-1 was found to be mitogenic. The transforming growth factor beta (TGFβ) superfamily member GDF-3 (also known as Vgr-1) is expressed in mouse and human EC cells and mouse ES cells, but there is no evidence for its expression in the peri-

implantation phase of mouse embryonic development, and distribution of its transcripts in adult tissues is fairly limited. GDF-3 is, therefore, a gene whose expression might represent adaptation of stem cells to a condition of permanent self-renewal in vitro. However, this member of the TGFβ superfamily is unusual because it lacks a key cysteine residue involved in dimerization of most superfamily members and because no biologic activity of GDF-3 either in vitro or in vivo has been described thus far. Finally, human EC cell lines and ES cell lines express receptors for the Wnt family of secreted regulators, and recent evidence implicates these molecules in the control of differentiation of mouse ES cells.

Human Embryonic Stem Cells

CRITICAL COMPONENTS OF THE ORIGINAL CULTURE SYSTEMS

Unlike mouse ES cells, HESCs do not respond to LIF. Nonetheless, the first successful derivations of HESC used LIF in combination with a feeder layer of early passage mouse embryonic fibroblasts in a medium optimized for the propagation of mouse ES cells, supplemented with serum (80% DMEM [no pyruvate, high-glucose formulation], 20% fetal calf serum, 1 mM glutamine, 0.1 mM β-mercaptoethanol and 1% nonessential amino acids). There have been few reports of the use of alternative basal media, and no groups have reported systematic studies of the role of low molecular weight components of the basal media in the support of HESC growth. Optimization of the basal medium can markedly affect the growth factor requirements of particular cell types in vitro, and more work in this area is certainly merited. In the original cell culture system used to establish HESC, omission of either feeder cells or serum leads to rapid differentiation of the HESC. Individual lots of serum or feeder cells can vary significantly in their ability to promote HESC growth. Moreover, the extent and range of HESC differentiation in vitro is also strongly influenced by the batch of serum or feeder cells used for growth.

Several groups have now described the growth of HESC on various types of human feeder cells, including fibroblasts, marrow stromal cells, and other cell types, and a recent report described derivation of HESC lines on STO cells, a permanent line of mouse embryo fibroblasts. Both human feeder cell layers and STO cells are much easier to maintain than primary mouse embryo fibroblasts, and the properties of the HESC grown on these cells appear similar to those grown on early-passage embryo fibroblasts.

Elimination of the use of a feeder cell layer will require some understanding of the molecular basis of the support of stem cell renewal. In an effort to identify proteins produced by feeder cells that might be responsible for stem cell maintenance, proteomic analysis of secreted products found in conditioned medium of STO feeder cells have been carried out. One secreted protein identified of interest was the insulin-like growth factor binding protein (IGFBP)-4, which may play a role in presenting insulin-like growth factors to the cell. Other factors found in the STO-conditioned medium included pigment epithelium-derived factor; a range of extracellular matrix components including SPARC, nidogen, and collagen alpha 1 and 2 chains; and several antioxidant proteins. This proteomic analysis probably was most likely to detect relatively abundant components of conditioned medium and may well have missed critical membrane-bound regulators of stem cell growth and differentiation. One example of this latter class of protein is the bone morphogenetic protein (BMP) antagonist gremlin, a protein produced by fibroblasts that is mainly cell associated; gremlin transcripts are found in mouse embryonic fibroblast feeder cells used to support HESC growth, and BMPs, which are produced by HESC, have a profound differentiation effect on HESC. It is likely, though unproven, that feeder cell layers have multiple roles in the support of HESC growth and that multiple proteins (cytokines, survival factors, antidifferentiative proteins, and extracellular matrix components) in either secreted or membrane-associated forms are involved in the interactions between feeder cells and stem cells.

Another approach to definition of ES cell growth requirements is to ask what receptors the stem cells themselves express. Microarray analysis showed that HESC express transcripts of all four FGF receptors, BMPR1A, and Frzd5, indicating possible roles for FGF, BMP, and Wnt family members in stem cell regulation.

FGF-2 AND KNOCKOUT SERUM REPLACER

Amit et al. (2000) examined the use of different media to increase the cloning efficiency of HESC. This study describes conditions allowing the use of a serum-free medium for the propagation of HESC. Basal medium (80% DMEM [no pyruvate, high glucose formulation], 1 mM glutamine, 0.1 mM β-mercaptoethanol, and 1% nonessential amino acids) was supplemented with either 20% fetal calf serum or 20% Knockout Serum Replacer, with or without 4 ng/ml FGF-2. Knockout Serum Replacer is a proprietary serum replacement consisting of amino acids, vitamins, transferrin or substitutes, insulin or insulin substitutes, trace elements, collagen precursors, and albumin preloaded with lipids. The composition of this serum replacement is not precisely specified by the manufacturer, and the description provides for the use of albumin substitutes, including other sources of animal protein such as embryo extracts.

In this study, the cells were passaged every week. The authors showed that FGF-2 was without effect in the presence of serum but increased the cloning efficiency of HESC when added to basal medium supplemented with serum replacer in the presence of a feeder cell layer. Under conditions of routine maintenance, FGF-2 blocked stem cell differentiation, as judged by cell morphology; in the presence of serum replacer and FGF-2, HESC appeared smaller and colonies were more compacted. The use of this system enabled long-term maintenance of diploid pluripotent cells expressing markers characteristic of HESC. The system has since been modified to incorporate DMEM:Nutrient Mixture F12 as the basal media and a reduced feeder cell layer density. As noted previously, FGF-2 and other members of this growth factor family were

shown to be mitogenic for human EC cells under serum-free conditions, but in the EC cell studies there was no indication that these factors inhibited differentiation.

The system described previously is now widely used, but the cloning efficiency of HESC in the system remains low, and feeder cells are still required. Xu et al. (2001) described a further modification of this system that replaced the feeder cell layer with a combination of conditioned medium from mouse embryonic fibroblasts and Matrigel, a commercially available basement membrane extract. This system also enabled long-term maintenance of HESC. However, both this system and the feeder cell-based culture method using FGF-2 supplemented media still incorporate an undefined mixture of animal products.

SPHINGOSINE-1-PHOSPHATE (S1P) AND PDGF

One goal for the development of the HESC culture system for use in potential clinical applications is the elimination of serum and feeder layers and their replacement with defined components consisting of no animal products. Our laboratory has, therefore, studied the components of serum likely to be critical to HESC growth. Thus, another approach to the propagation of HESC in the absence of serum is the use of S1P and PDGF, two bioactive molecules present in serum. When added to serum-free medium in the presence of mouse embryonic fibroblasts, S1P and PDGF are able to maintain the HESC in the undifferentiated state. The serum-free culture medium used in this technique consists of DMEM (without sodium pyruvate, high glucose formulation) supplemented with 1% insulin/transferrin/selenium (ITS), 0.1 mM, β-mercaptoethanol, 1% nonessential amino acids, 2 mM glutamine, 25 mM Hepes, penicillin (50 U/ml), streptomycin (50 mg/ml), 10 μM S1P, and 20 ng/ml PDGF-AB. Cells were passaged every seven days. The cells remain diploid, retain markers of HESC, and are pluripotent as evidenced by differentiation in vitro and by formation of teratomas when grafted into immunodeprived animals. Although this technique presents the advantage of using a totally defined serum-free medium, a feeder cell layer is still required and clonal growth is poor.

Future Prospects

To date, two different protocols are available to cultivate HESC in serum-free conditions. Although these developments represent important steps toward the development of robust culture methodologies for propagation of HESC and eventual therapeutic use, some challenges remain. Both techniques require the use of feeder cells. Neither technique allows for clonal growth at high efficiency, and neither has been shown to be adaptable to scale-up methodologies such as bioreactor cell production. Many experimental laboratory protocols often require large numbers of homogeneous populations of stem cells. Thus, even before clinical application of this technology, there is an urgent need to optimize the culture conditions of HESC.

Such a challenge might be achieved by identifying the factors synthesized and released by the feeder cells and by the stem cells themselves and by identifying molecules produced by cells at early stages of differentiation that might influence stem cell fate. Also critical to our understanding will be dissecting the signaling pathways involved in the maintenance and self-renewal of HESC. However, there is much to be learned about how extrinsic factors regulate networks of transcriptional factors that are thought to be critical to the maintenance of the pluripotent state, such as Oct-4, nanog, and FoxD3.

A further challenge will be to understand how modifications in the culture system might affect the properties of different ES cell isolates over the long term. Although we know that the systems developed to date can support long-term cultivation of pluripotent cells with normal karyotypes, no studies have examined the potential effects of various culture conditions on the long-term genetic stability of ES cells at a higher resolution level of analysis (microdeletion, amplification, or point mutations). We also know nothing of how culture conditions affect the epigenetic stability of ES cells, and how epigenetic changes might alter the potential for a given cell line to respond to extracellular or intracellular triggers for differentiation into particular lineages. The answers to these important questions will require systematic comparison of different HESC lines propagated under defined conditions over long periods, using appropriate assays for genetic and epigenetic stability and differentiation capacity.

KEY WORDS

Embryonal carcinoma stem cell   Pluripotent stem cell of teratocarcinoma (an embryonal tumor containing multiple types of differentiated cell), similar in phenotype to embryonic stem cell, but often more limited in differentiation capacity and of abnormal karyotype.

Knockout serum replacer   Proprietary serum replacement that consists of amino acids, vitamins, transferrin or substitutes, insulin or insulin substitutes, trace elements, collagen precursors, and albumin preloaded with lipids.

Matrigel   Commercially available basement membrane extract.

Mouse embryo fibroblast   Diploid, heterogeneous adherent cells of limited lifespan derived from primary cultures of midgestation mouse fetuses (most often eviscerated and decapitated, then dissociated mechanically or enzymatically), grown in standard tissue culture medium supplemented with serum.

FURTHER READING

Amit, M., Carpenter, M. K., Inokuma, M. S., Chiu, C. P., Harris, C. P., Waknitz, M. A., Itskovitz-Eldor, J., and Thomson, J. A. (2000). Clonally derived human embryonic stem cell lines maintain pluripotency and proliferative potential for prolonged periods of culture. *Dev. Biol.* **227**, 271–278.

Lim, J. W., and Bodnar, A. (2002). Proteome analysis of conditioned medium from mouse embryonic fibroblast feeder layers which support the growth of human embryonic stem cells. *Proteomics* **2**, 1187–1203.

Pébay, A., and Pera, M. F. (2004). Growth factors and the serum-free culture of human pluripotent stem cells. *In Handbook of stem cells*, Volume 1, Chapter 51 529–534 (Robert Lanza, ed.) Elsevier, Academic Press, New York.

Pébay, A., Wong, R. C. B., Pitson, S. M., Peh, G. S. L., Koh, K. L. L. Tellis, I., Nguyen, L. T. V., and Pera, M. F. Sphingosine-1-phosphate and platelet-derived growth factor inhibit the spontaneous differentiation of human embryonic stem cells. ASCB 43rd Annual Meeting, San Francisco. Abstract 2925.

Reubinoff, B. E., Pera, M. F., Fong, C. Y., Trounson, A., and Bongso, A. (2000). Embryonic stem cell lines from human blastocysts: Somatic differentiation *in vitro*. *Nat. Biotechnol.* **18**, 399–404.

Richards, M., Fong, C. Y., Chan, W. K., Wong, P. C., and Bongso, A. (2002). Human feeders support prolonged undifferentiated growth of human inner cell masses and embryonic stem cells. *Nat. Biotechnol.* **20**, 933–936.

Thomson, J. A., Itskovitz-Eldor, J., Shapiro, S. S., Waknitz, M. A., Swiergiel, J. J., Marshall, V. S., and Jones, J. M. (1998). Embryonic stem cell lines derived from human blastocysts. *Science* **282**, 1145–1147.

Xu, C., Inokuma, M. S., Denham, J., Golds, K., Kundu, P., Gold, J. D., and Carpenter, M. K. (2001). Feeder-free growth of undifferentiated human embryonic stem cells. *Nat. Biotechnol.* **19**, 971–974.

42

Feeder-free Culture

Holly Young, Chunhui Xu, and Melissa K. Carpenter

Introduction

Because of their remarkable proliferative capacity and differentiation potential, human embryonic stem (hES) cells may provide a source of cells for cell therapies, drug screening, and functional genomics applications. Derivation of hES cell lines has been accomplished by culturing cells from the inner cell mass of preimplantation embryos on mouse or human embryonic feeder cells. They are maintained in media containing serum or serum replacement supplemented with basic fibroblast growth factor (bFGF). In these conditions, the cells retain an undifferentiated state and show stability in long-term culture. In addition, hES cells maintained on or off feeders express integrins α6 and β1, which may form a laminin-specific receptor, suggesting that the cells may interact with matrix components. These findings indicate that hES cells require both soluble factors and matrix proteins.

In our laboratory, we have developed methods for maintaining the hES cells in feeder-free conditions. In this culture system, hES cells can be maintained on Matrigel or laminin-coated plates in serum-free medium conditioned by mouse embryonic fibroblast (MEF) feeders, as shown in Figure 42-1. Cells maintained using these feeder-free conditions retain a normal karyotype and a stable proliferation rate; express SSEA-4, TRA-1-60, TRA-1-81, alkaline phosphatase, hTERT, and OCT-4; and have the capacity to differentiate into cell types representing the three germ layers both *in vitro* and *in vivo*. The cells maintain these features even after long-term culture for up to 700 population doublings. The cells maintained in feeder-free conditions can also be induced to differentiate using specific culture conditions, into neural progenitors and neurons, endothelial cells, cardiomyocytes, trophoblast, hepatocyte-like cells, oligodendrocytes, and hematopoietic progenitors. In this chapter, we present detailed protocols for the maintenance of hES cells in feeder-free conditions and briefly describe methods for characterizing hES cells.

Materials for Feeder-free hES Cell Culture

Filtering is done through a 0.22 μM, Corning, cellulose acetate, low-protein-binding filter. Most solutions are stored

as aliquots at −20°C unless otherwise noted. Growth factors can be held at −80°C for long-term storage.

STOCK SOLUTIONS

1. Collagenase IV solution (200 units/ml). Dissolve 20,000 units of collagenase IV in 100 ml knockout Dulbecco's modified Eagle's medium (DMEM) and filter.
2. DMEM, high-glucose, without glutamine.
3. Dimethyl sulfoxide (DMSO).
4. Fetal bovine serum (FBS).
5. Gelatin (0.5%). Add 100 ml of 2% gelatin to 300 ml of water for embryo transfer and filter. Store at 4°C.
6. L-glutamine solution (200 mM).
7. Human basic fibroblast growth factor, recombinant (hbFGF, 10 μg/ml). Dissolve 10 μg hbFGF in 1 ml phosphate buffered saline (PBS) with 0.2% BSA and filter. Thawed aliquots can be kept at 4°C for up to one month.
8. Knockout DMEM.
9. Knockout serum replacement.
10. Laminin.
11. Matrigel. Growth factor-reduced or regular. To prepare Matrigel aliquots, slowly thaw at 4°C overnight to avoid the formation of a gel. Add 10 ml of cold knockout DMEM to the bottle containing 10 ml of Matrigel. Keep the mixture on ice and mix well with a pipette, aliquot 1–2 ml into prechilled tubes for storage.
12. β-mercaptoethanol (1.43-M). 14.3-M β-mercaptoethanol is diluted 1:10 in PBS for use. Note; do not use after thawing once.
13. Nonessential amino acids (10 mM).
14. Sterile PBS without $Ca^{2+} Mg^{2+}$.
15. Trypsin–EDTA (0.05% trypsin–0.53-mM EDTA).

MEDIA

1. *MEF medium.* Combine and filter all medium components listed: 450 ml DMEM, 50 ml FBS (final concentration 10%, heat inactivated), and 5 ml 200 mM L-glutamine (final concentration 2 mM). Store at 4°C and use within one month.
2. *hES serum-free medium.* Combine and filter all medium components listed: 400 ml knockout DMEM, 100 ml knockout serum replacement, 5 ml nonessential amino acids (final concentration 1%), 2.5 ml 200 mM L-glutamine (final concentration of 1 mM), 35 μl 0.14 M β-mercaptoethanol (final concentration of 0.1 mM). Store at 4°C for no longer than two weeks.

Feeder-free Human ES Cell Culture

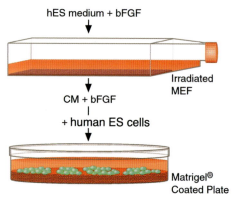

Characterization of Human ES Cells

Morphology

Markers
- OCT4 and hTERT
- Surface markers

Telomerase activity

Karyotyping

Proliferation

Differentiation
- *In vitro* differentiation
- Teratoma formation

Figure 42-1. Feeder-free hES culture. MEF cells are irradiated to 40 Gy and seeded into a flask for preparation of conditioned medium (CM). After at least 5 hours, the medium is exchanged with hES medium. CM is collected daily and supplemented with an additional 4–8 ng/ml of hbFGF before addition to the hES cells on Matrigel or Laminin. These cultures can then be assessed for the assays suggested in the table.

3. *Differentiation medium.* The differentiation medium is made by replacing knockout serum replacement with 20% FBS (not heat inactivated) in the hES serum-free medium described previously. Store at 4°C for no longer than two weeks.

4. *Cryopreservation medium.* Combine and filter 40 ml knockout DMEM (for ES cells) or 60 ml DMEM (for MEFs) and 40 ml knockout serum replacement (for ES cells) or 20 ml FBS (for MEFs) and then add 20 ml of DMSO.

Preparation of Conditioned Medium (CM) from MEFs

HARVESTING MEFs

(Modified from "Teratocarcinomas and Embryonic Stem Cells: A Practical Approach" by E. J. Robertson.) In our la-

boratory, we do not add antibiotics to any of our culture media. Briefly, the uterine horns are dissected out from 13-day pregnant mice (we use CF-1 mice), and the embryos are released into a Petri dish for processing. The placenta and embryonic membranes are removed, and the embryos are eviscerated before finely mincing the tissue. At each step, gentle and thorough washing in sterile PBS is essential. Further dissociation is attained by incubating the minced tissue in trypsin-EDTA. Finally, the dissociated cells can be plated into T150 flasks. Cells are maintained in a humidified incubator in a 5% CO_2–95% air atmosphere at 37°C. Cells are expanded when 80 to 85% confluent (usually the next day) and are ready to be banked (see Cryopreserving MEFs) or used as feeders (see Preparing and Storing CM) after the next split. As with any new cell in the laboratory, a representative sample should be tested for mycoplasma and sterility.

The following sections are detailed protocols for the maintenance and use of MEFs for preparing CM used in the feeder-free culture of hESCs.

CRYOPRESERVING MEFs

1. After removing the media and washing the cells once with PBS, detach the cells from the plate by adding trypsin–EDTA (1.0 ml/T75 and 1.5–2.0 ml/T150). Incubate for approximately 5 minutes at 37°C. Pipetting or tapping the flask against the heel of your hand will help release the cells.

2. Neutralize the trypsin by adding MEF medium (5 ml/T75 and 10 ml/T150). Serum in the MEF medium will expend the enzyme, preventing further dissociation.

3. Gently pipette the cell mixture to break up clumps of cells. If clumps remain, transfer the suspension to a 50 ml tube and allow the chunks to settle out before performing a cell count. We usually freeze 10 million to 20 million cells per milliliter per vial.

4. Pellet the cells by centrifugation for 5 minutes at $300 \times g$ and resuspend in 0.5 ml of MEF medium containing 20% FBS per vial — that is, one-half the final volume required for freezing.

5. Dropwise, add an equivalent volume (0.5 ml per vial) of cryopreservation medium and mix. The final DMSO concentration is 10%.

6. Place 1 ml of cell mixture into prelabeled cryovials and immediately transfer the cells to a Nalgene freezing container to be held at −80°C overnight. Transfer the cells to liquid nitrogen the next day for long-term storage. Alternatively, cells can be frozen using a controlled-rate freezer and transferred to liquid nitrogen at the completion of the freeze cycle.

THAWING AND MAINTAINING MEFs

1. Thaw a cryovial by gently swirling it in a 37°C water bath until only a small ice-crystal pellet remains in the tube, being careful not to submerge the cryovial. Sterilize the outside of the vial with 70% ethanol.

2. Remove the contents of the vial by gently pipetting the cell suspension up and down once, and transfer to a 15 ml conical tube.

3. Slowly add 10 ml warm MEF medium to the tube drop-wise, to reduce osmotic shock, and centrifuge at $300 \times g$ for 5 minutes. Remove the supernatant.

4. Resuspend the cell pellet in 10 ml (T75) or 20 ml (T150) of culture medium, and transfer to a flask to be held in a 37°C incubator. The plating density will need to be determined for each lot of MEF. Ideally, $\sim 5 \times 10^4$ to 1.5×10^5 cells/cm^2.

5. Replace the medium the day after thawing. The MEFs are expanded by splitting 1:2 when they become 80 to 85% confluent, which is approximately every other day. It is important to keep cells subconfluent so that they will not "contact inhibit." In our laboratory, MEFs are only used through passage 5.

IRRADIATING AND PLATING MEFs

1. Coat flasks or plates with 0.5% gelatin (15 ml/T225, 10 ml/T150, 5 ml/T75, and 1 ml/well of six-well plates) and incubate at 37°C. Use at least one hour to one day after coating, removing the gelatin solution immediately before use.

2. Prepare MEFs for irradiation by aspirating the medium and washing once in PBS without Ca^{2+}/Mg^{2+} (2–3 ml/T75 and 5–10 ml/T150).

3. Add trypsin–EDTA (1.0 ml/T75 and 1.5 ml/T150), and incubate at 37°C until cells are rounded up (usually 3 to 10 minutes). To loosen cells, either tap the flask against the heel of your hand or pipette them off.

4. Add 9 ml of MEF medium for one T150, transfer the cell suspension into a 15 ml conical tube, and gently triturate several times to dissociate the cells. Perform a cell count using a hemocytometer to prepare for plating.

5. Irradiate the cell suspension at 40–80 Gy. This number is variable among fibroblast sources. The goal is to stop proliferation without damaging the cells.

6. Spin down cells at $300 \times g$ for 5 minutes and discard the supernatant.

7. Resuspend the cells in 10 to 30 ml of MEF medium. Plate at 56,000 cells/cm^2 for cultures to be used for CM. The final volumes should be 3 ml per well for a six-well plate, 10 ml/T75, 20 ml/T150, and 50 ml/T225.

8. When placing the plate or flask in the incubator, gently shake it left to right and back to front to obtain an even distribution of cells. Do not swirl as all cells will accumulate in the center of the dish.

9. Irradiated MEFs can be used for CM production five hours to seven days after plating.

PREPARING AND STORING CM

For convenience, CM can be generated and stored prior to use.

1. Plate irradiated MEFs at 56,000 cells/cm^2 in MEF medium as described previously.

2. To generate CM, replace the MEF medium with hES medium (0.5 ml/cm^2) supplemented with 4 ng/ml hbFGF (0.4 ml/cm^2).

3. Collect medium after overnight incubation and add an additional 4 ng/mL hbFGF.

4. Add fresh hES medium containing 4 ng/mL hbFGF (0.4 mL/cm2) to the feeders. The MEFs can be used for one week, with CM collection once every day.

5. Pool and filter CM. Store at −20°C for up to five months.

6. Thaw CM at 37°C before using. Store CM at 4°C and use within one week after thaw.

Culture of hESCs on Matrigel or Laminin in CM

Derivation of human embryonic stem cell lines has been accomplished in several laboratories by culturing cells from the inner cell mass of preimplantation embryos on mouse or human embryonic feeder cells (see Chapter 38). Human embryos for this process are generally discarded embryos from *in vitro* fertilization (IVF) procedures. The initiation of the IVF procedure involves a series of cyclic cell divisions following successful fertilization of an oocyte by a sperm cell. The resulting cells, termed *blastomeres* or *cleavage cells*, are thought to be totipotent and equivalent. When 8 to 16 cells are generated, they go through compaction, a process through which the collection of individual blastomeres changes into a solid mass. Here, cell boundaries become tightly opposed and the individual cells are no longer equivalent. The resulting mass of cells is defined by an outer layer forming the trophectoderm and an inner spherical shell enclosing a fluid-filled cavity (the blastocoel). During hESC derivation, the ICM is dissected out and placed onto a layer of mitotically inactivated feeders. After the ICM-derived mass has proliferated sufficiently, it is expanded by serially plating the cells onto feeders. HESCs are maintained using media containing serum or serum replacement and supplemented with basic fibroblast growth factor. Cells can also be cultured using feeder-free conditions. The following protocols detail our preferred methods for maintenance of hESCs.

PREPARATION OF MATRIGEL OR LAMININ-COATED PLATES

1. Slowly thaw Matrigel or Laminin aliquots at 4°C to avoid gelling.

2. Dilute the Matrigel aliquots 1:15 in cold knockout DMEM (for a final dilution of 1:30) or the Laminin aliquots in PBS to a final concentration of 20 μg/ml.

3. Add 1 ml of Matrigel or Laminin solution per one well to be used.

4. Incubate the plates for at least one hour at room temperature, or overnight at 4°C. The plates can be stored at 4°C for one week.

5. Remove coating solution immediately before use. Laminin-coated plates should be washed once with knockout DMEM before introducing cells.

PASSAGE OF hES CELLS ON MATRIGEL OR LAMININ

1. Aspirate the medium from hES cells, and add 1 ml of 200 units/ml collagenase IV per well of a six-well plate.

2. Incubate cells for 5 to 10 minutes at 37°C. Incubation time will vary among batches of collagenase; therefore,

determine the appropriate incubation time by examining the colonies. Stop incubation when the edges of the colonies start to pull from the plate.

3. Aspirate the collagenase and add 2 ml of CM supplemented with hbFGF at a final concentration of 4 ng/ml into each well.

4. Gently scrape the cells with a cell scraper or a 5 ml pipette to collect most of the cells from the well.

5. For better cell adhesion and colony formation upon passage, dissociate the cells into small clusters (50–500 cells) by gently pipetting. Do not triturate the cells to a single-cell suspension.

6. Seed the cells into each well of the Matrigel or Laminin-coated plates. The final volume of medium should be 3 ml per well. In this system, the hES cells are maintained at high density. At confluence (usually one week in culture) the cells will be 300,000 to 500,000 cells/cm². We find that the optimal split ratio is 1:3 to 1:6. Using these ratios, we find that the seeding density is 50,000 to 150,000 cells/cm².

7. Maintain cells in a humidified incubator in a 5% CO2–95% air atmosphere at 37°C. Be sure to gently shake the plate left to right and back to front to obtain even distribution of cells. Do not swirl the dish, as the cells will collect in the middle.

8. Note that the day after seeding, undifferentiated cells are visible as small colonies. Single cells between the colonies will begin to differentiate. As the cells proliferate, the colonies will become large and compact, taking most of the surface area of the culture dish (Figure 42-2).

9. For maintenance of feeder-free cultures, feed the hES cells with 3–4 ml CM supplemented with hbFGF per well of a six-well plate daily. We find that one feeding can be skipped at day 2 or 3 after splitting. *Note:* prewarm media

in a 37°C water bath. Do not heat the entire bottle of medium, as the knockout DMEM and knockout serum replacement do not tolerate repeated warming and cooling.

10. Passage when cells are 100% confluent. At this time, the undifferentiated cells should cover ~80% of the surface area. The cells between the colonies of undifferentiated cells appear to be stroma-like cells (Figure 42-2). The colonies (but not the stroma-like cells) show positive immunoreactivity for SSEA-4, Tra-1–60, Tra-1–81, and alkaline phosphatase.

CRYOPRESERVING hES CELLS

1. Treat the cells with 200 U/ml collagenase IV for 5 to 10 minutes at 37°C (until the edges of the colonies curl). Remove the collagenase and add 2 ml hES medium per well.

2. With a 5 ml pipette, gently scrape colonies from the plate. Add the cell suspension to a 15 ml centrifuge tube and *gently* break up the colonies. It is important to be gentle in this step because larger clusters seem to result in a better yield upon thaw. Ideally, colonies meant for freezing are left slightly larger than they would be for splitting.

3. Centrifuge for 5 minutes at 300 × g and resuspend gently at 1 × 10⁶ in 0.5 ml hES serum-free medium containing 20% knockout serum replacement per vial — that is, one-half the final volume required for freezing.

4. Dropwise, add an equivalent volume (0.5 ml per vial) of cryopreservation medium and mix. The final DMSO concentration is 10%.

5. Place 1 ml of cell mixture into prelabeled cryovials and immediately transfer the cells to a Nalgene freezing container, place directly at −80°C overnight. Transfer cells to liquid nitrogen the next day for long-term storage. Alter-

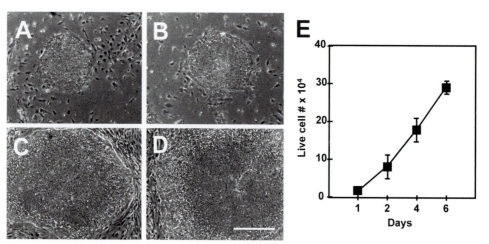

Figure 42-2. Growth of hES cells in feeder-free conditions. Phase images of H7 (passage 31) undifferentiated hES cell colonies maintained in feeder-free conditions were taken (A and B) 2 days and (C and D) 5 days after seeding. Bar = 200 μm. (E) Proliferation of H7 cells (passage 24) was measured by counting live cells at various days after seeding. H7 hES cells were dissociated with collagenase IV, resuspended in CM, and plated into Matrigel-coated 24-well plates. Cells were harvested with trypsin and counted at days 2, 4, and 6 after seeding. Each point on the graph represents the mean ± standard deviation of cell counts for three separate wells.

natively, cells can be frozen using a controlled-rate freezer and transferred to liquid nitrogen at the end of the freeze cycle.

THAWING hES CELLS

1. Thaw a cryovial by gently swirling it in a 37°C water bath until only a small ice-crystal pellet remains in the tube, being careful not to submerge the cryovial. Sterilize the outside of the cryovial in 70% ethanol.
2. Remove the contents of the vial very gently and transfer the cells into a 15 ml conical centrifuge tube.
3. Slowly add 10 ml of warm CM or hES cell medium dropwise (to reduce osmotic shock). While adding the medium, mix the cells in the tube by gently tapping the tube with a finger.
4. Centrifuge at $150 \times g$ for 5 minutes and resuspend in 0.5 to 1.0 ml of CM.
5. Seed 0.5 ml per well of a four-well plate.
6. Change the medium daily with CM. It may take two weeks before the cells are ready to be expanded.

CHARACTERIZATION OF FEEDER-FREE hES CELL CULTURES

A collection of markers, including glycoproteins, glycolipids, and transcription factors, has been compiled to aid in the characterization of hES cells. This panel consists largely of markers previously identified on human embryonic carcinoma (EC) cells, hematopoietic stem cells, and mouse ES cells. hES cells maintained in feeder-free cultures have similar expression patterns to hES cells maintained on feeders. Typical evaluation includes the assessment of (1) the expression of undifferentiated cell markers such as OCT-4, hTERT, SSEA-4, TRA-1–60, TRA-1–81, and alkaline phosphatase; (2) telomerase activity; (3) karyotype; (4) proliferation rate; (5) *in vitro* differentiation capacity; and (6) teratoma formation.

Quantitative RT-PCR Analysis of OCT-4 and hTERT

1. Lyse cells by adding 600 to 750 µl of lysis buffer per each well of a six-well plate and isolate RNA using the Qiagen RNeasy (Qiagen, Valencia, California) kit, following the animal cells isolation procedure with the QiaShredder (Qiagen, Valencia, California). Further removal of contaminating DNA can be attained by including the steps for on-column DNase digestion with the RNase-Free DNase set.
2. Set up TaqMan one-step RT-PCR using a master mix (Applied Biosystems, Foster City, California) and specific primers and probes for OCT-4 (forward GAAACCCA-CACTGCAGCAGA, reverse CACATCCTTCTCGAGC-CCA, and probe FAM-CAGCCACATCGCCCAGCAGC-TAM) and hTERT (purchase from Applied Biosystems). The reaction mixture contains 1X RT-Master Mix, 300 nM for each primer, 80 nM of probe, and approximately 50 ng of total RNA in nuclease-free water.

3. Perform real-time RT-PCR on the ABI Prism 7700 Sequence Detection System (Applied Biosystems) under the following conditions: reverse transcription at 48°C for 30 minutes, denaturation and AmpliTaq gold activation at 95°C for 10 minutes, and amplification for 40 cycles at 95°C for 15 seconds and 60°C for 1 minute.
4. Set up and run RT-PCR for the 18S ribosomal RNA as a control using a kit for TaqMan ribosomal RNA control reagents (Applied Biosystems) according to the manufacturer's instructions.
5. Analyze reactions using the ABI Prism 7700 Sequence Detection System. Relative quantitation of gene expression among multiple samples can be achieved by normalization against endogenous 18S ribosomal RNA using the $\Delta\Delta C_T$ method described in ABI User Bulletin #2, Relative Quantitation of Gene Expression, 1997.

Flow Cytometry Analysis of Surface Markers

1. When hES cell cultures reach confluence, remove the culture medium and wash with PBS.
2. Dissociate the cells by incubating with 200 units/ml of collagenase IV solution at 37°C for 5 to 10 minutes.
3. Further dissociate the cells with Versene at 37°C for 20 minutes to obtain a single-cell suspension.
4. Collect cells through a cell strainer and resuspend to about 5×10^5 cells in 100 µl blocking buffer (20% Goat serum in PBS). Incubate at 4°C for 15 minutes.
5. Centrifuge samples and add 100 µl per sample of primary antibodies — IgG isotype control (1 µl/test) for SSEA-1 and SSEA-4; IgM isotype control (1:20) for TRA-1–60 and TRA-1–81; and SSEA-1 (1:40), SSEA-4 (1:10), TRA-1–60 (1:40), and TRA-1–81 (1:80) in the diluent (2% heat inactivated FBS in PBS) — at 4°C for 30 minutes. In our laboratory, the antibodies against SSEA-1 and SSEA-4 were obtained from the Developmental Studies Hybridoma Bank (University of Iowa, Iowa City). TRA-1–60 and TRA-1–81 were a gift from Dr. Peter Andrews (University of Sheffield, U.K.). All of these antibodies are now available from Chemicon.
6. After washing with the diluent, incubate the cells with goat anti-mouse IgG$_3$γ-chain or goat anti-mouse IgM µ-chain antibodies conjugated with phycoerythrin or fluorescein (Southern Biotech) at 4°C for 30 minutes.
7. Wash the cells and analyze on FACScalibur Flow Cytometer (Becton Dickinson, San Jose, California) using CellQuest or FlowJo software.

Immunocytochemistry of Surface Markers

1. Passage the hES cells into Matrigel or Laminin-coated chamber slides or coverslips and maintain the cells in CM for 2 to 7 days.
2. Remove the medium and incubate the cells with primary antibodies — SSEA-1 (1:10), SSEA-4 (1:20), TRA-1–60

(1:40), and TRA-1–81 (1:80) — diluted in warm knock-out DMEM at 37°C for 30 minutes.

3. Wash the cells with warm knockout DMEM.
4. Fix the cells in 2% paraformaldehyde in PBS for 15 minutes.
5. Wash the cells with PBS (twice, 5 minutes each).
6. Block the cells by incubation in 5% normal goat serum (NGS) in PBS at RT for 30 minutes.
7. Incubate with the fluorescein isothiocyanate (FITC)-conjugated goat anti-mouse IgG (Sigma) diluted 1:125 in PBS containing 1% NGS at RT for 30 minutes.
8. Wash the cells with PBS (three times, 5 minutes each), Mount slides with Vectashield mounting media (Vector Laboratories, Inc.) for fluorescence with 4′,6-diamidino-2-phenylindole (DAPI).

Detection of Alkaline Phosphatase

1. Passage the hES cells into Matrigel or Laminin-coated chamber slides and maintain the cells in CM for 2 to 7 days.
2. Remove the culture medium, wash with PBS, fix the cells with 4% paraformaldehyde for 15 minutes, and wash with PBS.
3. Incubate the cells with an alkaline phosphatase substrate (Vector Laboratories, Inc., Burlingame, California) at RT in the dark for one hour.
4. Rinse the slides gently once for 2 to 5 minutes in 100% ethanol before mounting.

Formation of Teratomas

1. When cells reach confluence, harvest the cells by incubating them in 200 units/ml of collagenase IV at 37°C for 10 minutes. Be careful not to dissociate to single cells.
2. Wash the cells in PBS, resuspend them at 1×10^8/ml, and store the cells in PBS.
3. Inject the cells intramuscularly into severe-combined immunodeficiency, beige mice (~5×10^6 cells in 50 µl per site).
4. Monitor teratoma formation. After teratomas become visible (usually 70 to 90 days in our hands), they are excised and processed for histological analysis. Teratomas are benign tumors consisting of cell types representing all three germ layers.

Formation of EBs

In vitro differentiation can be induced by culturing the hES cells in suspension to form EBs. We have evaluated the differentiation of hES cells into neurons, cardiomyocytes, and hepatocyte-like cells by using EBs generated with the following procedure:

1. Aspirate the medium from the hES cells, add 1 ml/well of collagenase IV (200 units/ml) into six-well plates, and incubate at 37°C for 5 minutes.
2. Aspirate the collagenase IV and wash once with 2 ml PBS.
3. Add 2 ml of differentiation medium into each well.
4. Scrape the cells with a cell scraper or pipette and transfer them to one well of a low-attachment plate (1:1 split). Cells should be collected in clumps. Add 2 ml of differentiation medium to each well for total volume of 4 ml per well. Depending on the density of the hES cells, the split ratio for this procedure can vary.
5. After overnight culture in suspension, ES cells form floating aggregates known as EBs.
6. To change the medium, transfer the EBs into a 15 ml tube and let the aggregates settle for 5 minutes. Aspirate the supernatant, replace with fresh differentiation medium (4 ml/well), and transfer the EBs to low-attachment six-well plates for further culture.
7. Change the medium every 2 to 3 days. During the first few days, the EBs are small with irregular outlines; they increase in size by day 4 in suspension. The EBs can be maintained in suspension for more than 10 days. Alternatively, EBs at different stages can be transferred to adherent tissue culture plates for further induction of differentiation.

Immunocytochemical Analysis of Differentiated Cultures

To identify specific cell types in the differentiated culture, immunocytochemical or RT-PCR analysis can be used to detect cell type-specific markers. Here are a few examples of immunocytochemical analysis of β-tubulin III for neurons, alpha fetoprotein (AFP) for endoderm cells, and cardiac troponin I (cTnI) for cardiomyocytes:

1. For β-tubulin III or AFP analysis, fix EB outgrowth cultures in 4% paraformaldehyde at RT for 20 minutes, followed by permeabilization for 2 minutes in 100% ethanol. For cTnI staining, fix cells in methanol:acetone (3:1) for 20 minutes at −20°C.
2. After fixation, wash the cells twice with PBS (5 minutes each).
3. Block with 10% NGS in PBS at RT for 2 hours or at 4°C overnight.
4. Incubate the cells at RT for 2 hours with a monoclonal antibody against β-tubulin III, AFP (Sigma) diluted 1:500, or cTnI (Spectral Diagnostic INC, Toronto, Ontario, Canada) diluted 1:450 in 1% NGS in PBS.
5. Wash the cells three times with PBS on a shaker, 5 to 10 minutes/washing.
6. Incubate the cells with FITC-conjugated goat anti-mouse IgG (Sigma) diluted 1:128 in PBS containing 1% NGS at RT for 30 minutes.
7. Wash the cells three times with PBS on a shaker, 5 to 10 minutes/washing.
8. Mount slides with Vectashield mounting media (Vector Labortories, Inc.) for fluorescence with DAPI.

KEY WORDS

Cryopreservation Freezing of a cell population or cell line in specialized media that allows storage of the cells for prolonged periods of time. Frozen cells can then be thawed and propagated at a later date.

Feeder-free Cells that are cultured without direct contact with feeder layers.

Feeder layer A layer of cells that are growth-arrested, providing nutrients, growth factors, and matrix components to a second cell population.

Inner cell mass (ICM) The cluster of pluripotent cells located at one pole of the inner surface of the blastocyst; these cells will form the body of the embryo after implantation.

Split/Passage Transfer or subculture of cells from one culture vessel to another, resulting in the expansion of the cell population.

FURTHER READING

Carpenter, M. K., Rosler, E., and Rao, M. S. (2003). Characterization and differentiation of human embryonic stem cells. *Clon. Stem Cells* **5**, 79–88.

Chiu, A. Y., and Rao, M. S. (eds.) (2003). *Human embryonic stem cells.* Humana Press Totowa, NJ.

Gearhart, J., and P. J. Donovan. (2001). The end of the beginning for pluripotent stem cells. *Nature* **414**, 92–97.

Robertson, E. J. (ed.) (1987). *Teratocarcinomas and embryonic stem cells: a practical approach.* IRL Press, Washington, DC.

Smith, A. G. (2001). Embryo-derived stem cells: of mice and men. *Annu. Rev. Cell. Dev. Biol.* **17**, 435–462.

Xu, C., Inokuma, M. S., Denham, J., Golds, K., Kundu, P., Gold, J. D., and Carpenter, M. K. (2001). Feeder-free growth of undifferentiated human embryonic stem cells. *Nat. Biotechnol.* **19**, 971–974.

43

Genetic Manipulation of Human Embryonic Stem Cells

Yoav Mayshar and Nissim Benvenisty

Introduction

The new technology of establishing human embryonic stem (ES) cell lines has raised great hopes for breaking new ground in basic and clinical research. Genetic manipulation will play a pivotal role in applying this technology to biological research, as well as for the specific needs of transplantation medicine.

Genetic manipulation has proven to be a key experimental procedure in the field of mouse ES cell research since their initial isolation over 20 years ago. Many effective techniques have since been developed for bringing about specific genetic modifications. Some of these methodologies have recently been adapted for the manipulation of human ES cells. These include both transfection and infection protocols, as well as overexpression and homologous recombination procedures. The newly acquired availability of these techniques should make genetic engineering of human ES cells routine and enable great advances in the field of human ES cell research.

By introduction of genetic modifications to mouse ES cells, a great number of genetically modified animals have been produced, which have proven to be an extremely valuable tool for research. Obviously, no equivalent research can be conducted in humans. Thus, despite the great potential attributed to human ES cells, much of the basic developmental research using these cells may be limited to *in vitro* studies. Researchers of human ES cells are therefore forced to adopt developmental models such as embryoid bodies (EBs) for studying early embryogenesis. This chapter presents the principal aspects of *in vitro* genetic manipulation of human ES cells and the possible uses of the variety of methods currently available.

Methods of Genetic Manipulation

A variety of methods are currently used for manipulating mammalian gene expression. Many of these methods have been used in mouse ES cell research. Of these, several gene delivery methods have been shown to be effective in human ES cells. These include transfection by various chemical reagents, electroporation, and viral infection. Transfection is probably the most commonly used method for genetic manip-

ulation. The system is straightforward, and relatively easy to calibrate. It provides sufficient numbers of cells for clonal expansion and allows for insertion of constructs of virtually unlimited size. Recently, long-term transgene expression in ES cells was demonstrated using infection by lentiviral vectors. This technique has been shown to be highly effective and could therefore emerge as a popular tool in human ES cell genetic engineering.

TRANSFECTION

The method used for gene delivery is one of the major variables that directly affects the efficiency of genetic manipulation. The choice of the gene delivery method should be based first and foremost on the nature of the experiment to be performed. The duration of the genetic modification is one factor that will dictate the method of choice. For short-term induction, transient expression can be achieved through the introduction of supercoiled plasmid DNA, from which transcription has been shown to be more efficient than from the linear form. Transcription from transiently introduced plasmids usually peaks at around 48 hours post-transfection. For the most part, positive cells demonstrate relatively high expression levels, as a result of a high-copy number of plasmids being introduced into transfected cells. Stable transfection is accomplished by selecting for cells where the vector DNA has integrated into the genome. In this type of experiment, it is important to linearize the vector, as this significantly raises integration and targeting efficiencies. A positive selection marker under regulation of a strong constitutive promoter is frequently included in the insert when the gene target is nonselectable. Selection should not be carried out immediately following transfection. Delaying the selection gives the cells time to recuperate and express the resistance marker. Given that the various selection reagents are toxic to all mammalian cells, the feeder layer supporting the cells must also be resistant to the selection drug. To this end, lines of transgenic mice, carrying various genes conferring resistance to selection drugs, are currently available for the derivation of resistant feeder cells.

Transfection Reagents

In the case of chemical-based transfection, carrier molecules bind the foreign nucleic acids and introduce them into the cells through the plasma membrane. In general, the cellular uptake of the exogenous nucleic acids is thought to occur either through endocytosis, or in the case of lipid-based reagents, through fusion of a lipidic vesicle to the plasma membrane. This method can routinely produce transient transfection rates of approximately 10% and stable transfection efficiency of 10^{-5} to 10^{-6}. The main advantages of chemical transfection are, notably, the relative simplicity of the procedure and the fact that it can be performed on an adherent cell culture.

Electroporation

Transfection by electroporation was adopted early on as the method of choice for gene targeting in mouse ES cells. This process involves applying a brief high-voltage shock to cells in suspension, in the presence of DNA molecules. The electric shock is thought to cause transient pores to open up in the cell membrane enabling DNA to enter. A recent study performed by Zwaka and Thomson (2003) demonstrated the efficiency of electroporation for the production of transgenic HESC clones. The mouse electroporation protocol that proved inadequate for HESCs was customized for these cells. Through use of electroporation, a stable integration rate of $\sim 10^{-5}$ was achieved, and homologous recombination efficiencies were reported to be between 2 and 40% — subject to the properties of the vector employed. A substantial number of homologous recombination clones have thus been generated for a number of different vector constructs, demonstrating the feasibility of achieving efficient gene targeting in human ES cells.

Infection

Unlike transfection methods, viral vectors can produce a very high percentage of infected cells. Viral vectors also allow easy determination of the vector copy number, for both transient and stable expression. Several groups have recently reported transgene expression in human ES cells using lentiviral vectors. Lentiviral infection provides a very high proportion of stable integrants, although integration cannot be targeted, and vector size may be an additional limiting factor. Because of its efficiency, this method could prove useful for bypassing the need for selection and time-consuming clonal expansion, as well as for experiments intended to induce random insertional mutagenesis or gene trap.

Genetic Modification Approaches

One of the major roles of genetic engineering will be research into the early stages of human development and the genes involved in this process. The roles of many of these genes have been elucidated using genetic manipulation techniques, mostly with the aid of genetically modified animals. Several studies using EB models have shown that specific early embryonic developmental processes can to some extent be recapitulated *in vitro*. Human ES cells thus provide researchers with a unique system to investigate the early stages of embryonic development through the implementation of genetic manipulation techniques. These techniques can also be used to determine gene function at the cellular level.

Several approaches can be used to discover the activity of specific genes in ES cells through genetic manipulation. These include overexpression of cellular or foreign genes, use of reporter genes, and gene silencing. In this section, we attempt to summarize the major techniques currently used in human ES cells together with a number of other techniques that are likely to be implemented in human ES cell research in the near future (see Figure 43-1).

EXOGENOUS GENE EXPRESSION

The concept of directed differentiation is an example of the possible utility of introducing exogenous properties into human ES cells. Extensive studies are currently being carried out on directed differentiation in human ES cells. The addition of specific growth factors to EBs during the differentiation process has been shown to modify their differentiation potential. Use of various enrichment protocols has made it possible to obtain several different cell types such as neurons, hematopoietic cells, and cardiomyocytes. Although creation of substantial amounts of certain cell types has been reported, the outcomes of some differentiation protocols are often inconsistent and are currently unable to yield the pure cell populations required for transplantation. Genetic manipulation could complement these methods and facilitate the practical application of human ES cells in transplantation medicine.

Genetic Labeling

Introduction of constitutively expressed reporter genes, such as the gene encoding for green fluorescent protein (GFP), into the cells allows their detection in a mixed population. This method was applied in a transplantation experiment where undifferentiated human ES cells were introduced into an early stage chick embryo. By using the same reporter gene regulated by a promoter of an ES cell-enriched gene (*Rex1*), it was possible to follow undifferentiated cell populations during differentiation *in vitro*, at the same time screening out differentiated cells with a fluorescence activated cell sorter (FACS). Accordingly, Zwaka and Thomson created knockin cell lines using GFP or neomycin resistance promoterless constructs. Through this method, in the homologous recombination events that were selected, GFP expression was regulated by the endogenous OCT4 promoter. By placing the reporter under the regulation of the native gene, possible position effects that could confound the interpretation of the results are avoided, and the entire set of *cis*-acting elements is utilized.

Alternatively, the introduction of a conditional suicide gene could solve one of the principal concerns facing cellular transplantation. Adverse effects from transplants may occur as a result of malfunction or uncontrolled proliferation of grafted cells. Selective ablation of human ES cells expressing herpes simplex virus thymidine kinase (HSV-*tk*) has been demonstrated in an animal model. Growth of teratomas formed by transplanting PGK-HSV-*tk* transfected cells was successfully arrested by oral administration of the pro drug gancyclovir. Using this technique, we will be able to transplant "fail-safe"

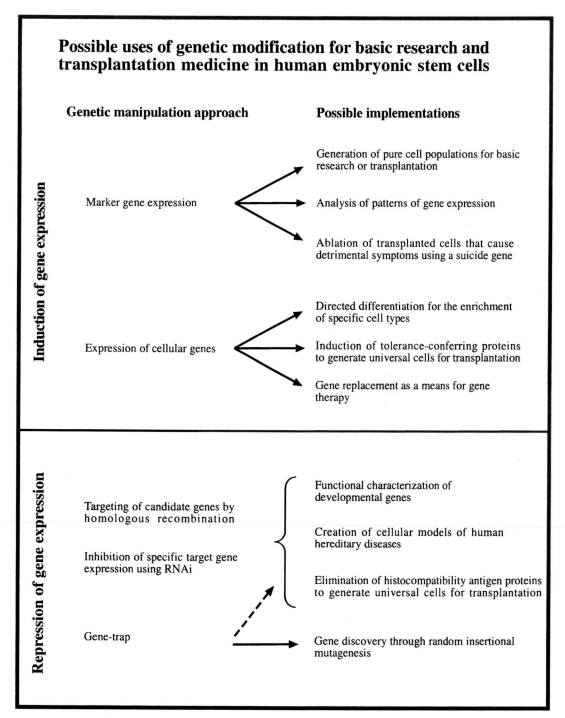

Figure 43-1. Possible uses of genetic modification for basic research and transplantation medicine in human embryonic stem cells.

cells into patients. These cells could then be eliminated upon the onset of harmful symptoms.

Probably the most direct way to achieve enrichment of a specific cell type using genetic manipulation is to isolate it using a tissue-specific promoter driving a reporter or resistance gene. This method has been demonstrated for the enrichment of undifferentiated human ES cells, but it can also be applied to isolate specific populations of differentiated cells. In mouse ES cells, through expression of a neomycin resistance gene regulated by the endogenous promoter of *sox2*, a neuronal progenitor specific gene, Li *et al.* (1998) produced highly enriched neural precursors from a heterogeneous population following selection. Similarly, mouse cardiac cells were isolated by using a reporter gene driven by the cardiac alpha actin promoter. Of course, the efficiency of such direct selection methods could be improved under culture conditions that allow greater quantities of the desired cell type to be produced.

Overexpression of Cellular Genes as Dominant Effectors of Cell Fate

The overexpression of developmentally important genes is currently being used as a means of achieving directed differentiation. Several regulatory genes have been introduced in various mouse experiments. Hepatocyte nuclear factor 3 (HNF3) transcription factors $-\alpha$ and $-\beta$ induced differentiation of mouse ES cells toward the early endodermal lineage. Ectopic expression of *HoxB4*, a homeotic selector gene, has enabled the use of mouse ES cells for the repopulation of myeloid and lymphoid lineages in immune-deficient mice. Similarly, the expression of the nuclear receptor related-1 (Nurr1) transcription factor generated a highly enriched population of midbrain neural stem cells that have been shown to ameliorate the phenotype of a mouse model of Parkinson's disease.

As yet, no single key factor or gene has been found that could cause undifferentiated ES cells to differentiate *en masse* to a specific lineage or cell type. Therefore, a combination of genetic manipulation and selective culture conditions would probably be necessary to generate differentiated transplant ready cells from undifferentiated ES cells.

GENE SILENCING

As opposed to the exogenous expression of various genes, gene silencing assesses the functionality of a particular gene of interest through its elimination. Currently, the most common method for generating silencing is homologous recombination, in which an exogenous sequence is inserted within the gene's coding region, thus disrupting it. Gene trap is an additional method through which a wide range of genomic sequences can be randomly mutated. Various other targeting techniques used to induce more subtle or controlled effects are available, and recently an RNA-based silencing approach has been gaining recognition.

Gene Targeting

Genetic modification has for many years been considered a cornerstone in mammalian biology research. Two major

advances in genetic manipulation of the mammalian genome were the isolation of mouse ES cells and the development of targeting vectors. Together these two methods have enabled the relatively straightforward production of genetically modified mice, carrying predesigned alterations to their genome. The subsequent proliferation of data has shed light on the functional characteristics of a very large number of genes and has led to the creation of various models for human disease. Although a number of mouse models have been created, some have proved inadequate in recapitulating some or all aspects of the human phenotype. Research is often limited by lack of an adequate mouse model, and working with primary cultures can often be problematic. In light of these difficulties, the creation of a cellular model of human disease, using human ES cells, will be very valuable.

Mutation of the hypoxanthine guanine phosphorybosyltransferase gene (HPRT) in humans is the basis for Lesch-Nyhan syndrome, a severe genetic disease. This syndrome is manifested in a variety of problems related to high uric acid levels. These include gout-like arthritis and renal failure. Neurological disorders are also associated with this disease, but the exact causes are still unknown. Mouse models for HPRT deficiency failed to fully recapitulate the human disease, possibly because uric acid metabolism is different in humans. Homologous recombination was recently used to disrupt HPRT using positive negative selection. Analysis of mutant human ES cell metabolism showed that, unlike their mouse counterparts, human cells mimic the Lesch-Nyhan phenotype.

HPRT is ideal for analyzing homologous recombination, as it is X linked. Thus, mutation of the single allele in male cells may produce a phenotype. Furthermore, cells lacking HPRT activity can be selected with 6-thioguanine (6TG). By culturing these cells in HAT medium they can be selectively eliminated. In many cases, the rate-limiting step toward obtaining targeted recombinants is the high number of nonhomologous integrations, making screening for homologous recombinants very time consuming. The additional negative selection is useful in order to confirm successful homologous recombination events and eliminate false positives. Thus, by using an HPRT⁻ cell line as a base for further genetic manipulation, an HPRT minigene can be used as an additional positive and negative selection, by correcting the HPRT deficiency. These advantages have made HPRT very useful in the initial study of homologous recombination. In cases other than HPRT targeting, a vector is usually constructed that contains a positive selection cassette (usually the neomycin-resistance gene) flanked by sequences homologous to the endogenous target gene. A negative selection marker, usually HSV-*tk*, is often placed in such a way that following linearization of the vector, it is located outside the region of homology. Thus, in those cases in which homologous recombination has successfully occurred, the transfected cells will not contain the HSV-*tk* gene and will be resistant to the drug gancyclovir.

When targeting expressed genes, promoterless constructs can be employed in order to increase the efficiency of homologous recombination by reducing the selectability of most random integrations. This method requires careful calibration of the concentration of the selection reagent subject to the strength of endogenous promoter. Using this approach, Zwaka and Thomson targeted the gene encoding OCT4 (*POU5F1*) and achieved homologous recombination rates of up to 40%.

As opposed to gene targeting, random gene disruption on a wide scale is currently being performed by gene trap. In this method, a reporter gene lacking an essential regulatory element is randomly inserted into the genome. Thus, in cases in which the reporter gene is expressed, often this is due to the use of an endogenous regulatory sequence. A large-scale disruption of genes is therefore made possible, allowing for the discovery of new genes and the creation of banks for a wide variety of mutations.

Induction of Homozygosity in Mutant Alleles

Several methods have been developed for the homozygotization of mutated alleles in mouse ES cells. When mutating the mouse genome, there is often an option of generating homozygosity through breeding two heterozygouts. However, research with human models will require *in vitro* manipulation. Two strategies can be employed to this end: (1) targeting the remaining allele by a second round of homologous recombination using distinct selection markers, and (2) growing transfected cells in high concentrations of the selection reagent, thus selecting for spontaneous homozygotization of the mutated allele containing the resistance gene.

The Effect of Extent and Nature of Homology on Targeting Efficiency

The length of the homologous regions has been shown to be a critical factor in determining the effectiveness of targeted recombination. By increasing the length of homology from 2 kb to 10 kb, an exponential increase in targeting efficiency is achieved. A less dramatic rate of increase is observed with up to ~14 kb of homology. The length of the short arm of homology should not be less than approximately 0.5 kb. Use of isogenic rather than nonisogenic DNA has also been shown to increase the rate of targeted recombination dramatically, although the need for isogenic DNA for human somatic cell lines and ES cells has recently been disputed.

ADVANCED METHODS OF GENE TARGETING

It has been shown that the strategy of inserting a selection cassette to generate a knockout phenotype can confound the interpretation of these experiments, as the exogenous promoter may disrupt endogenous gene expression at distances of over 100 kb from the point of insertion. This disruption of the local regulatory elements can produce an artifact phenotype where none exists. This is demonstrated by distinct phenotypic effects that may be caused by different constructs used to target the same gene. Removal of the selection cassette is therefore recommended. Clean mutations can be achieved either by the "hit and run" or "tag and exchange" methods, or by application of an exogenous recombinase.

"Hit and Run" and "Tag and Exchange" Methods

Several strategies have been demonstrated that can create specific substitutions. These strategies create subtle mutations in the gene as opposed to disrupting it with a large insert. The hit and run method is based on selection of intrachromosomal recombination events that excise the vector sequence. In another strategy, termed tag and exchange (double replacement), a series of targeting procedures is performed. In the first stage, the gene is targeted with a vector that contains both positive and negative selections in the insert. In the second stage, a vector containing the mutation of choice is inserted by homologous recombination, selected by the loss of the negative selection marker.

Exogenous Recombinases

By introducing highly efficient and specific exogenous recombination systems, both efficiency and tight control over the recombination event are achieved. The system most widely used is the Cre recombinase from phage P1. Cre identifies a highly specific 34 bp target sequence (*lox P*) and promotes reciprocal recombination between two such sites. The versatility of the system is such that, depending on the location of the *lox P* sequences, it can promote either reversible integration or excision while retaining the recombination sites' integrity. Thus, the Cre-*lox* system has been used for a variety of ends. These include the induction of chromosomal rearrangements, spatial or temporal inactivation of cellular genes by deletion, or activation of reporter genes. They also include multiple insertions into the same genomic site and the creation of clean mutations devoid of exogenous regulatory sequences. This is performed by flanking the sequence to be deleted with two *lox P* sites. Following the establishment of targeted clones, a second round of transfection with a plasmid containing Cre recombinase will remove the flanked region from the genome. Depending on the promoter under which Cre is expressed, this deletion can be constitutive, tissue specific or inducible.

RNA INTERFERENCE

While gene targeting remains the most powerful method for the silencing of endogenous gene expression, post-transcriptional gene silencing has lately been characterized as a potent method of down-regulating specific genes. Double-stranded RNA (dsRNA) molecules have been shown to initiate RNA interference (RNAi), causing sequence-specific mRNA degradation. The main advantage of RNAi is that it acts in *trans*. This means that RNAi inducing molecules can be directly introduced by transfection or synthesized within the cells using expression vectors, either transiently or by stable integration. Because RNAi does not entail any modification of the targeted gene itself, it should be relatively simple to generate transient or conditional gene silencing using this method.

The evolutionarily conserved machinery that underlies the phenomenon of homology-dependent gene silencing involves endonuclease (DICER) processing of dsRNA into small interfering RNA (siRNA). siRNAs are 21- to 23-nt long fragments that in turn are incorporated into RISC, a multicomponent nuclease that uses the siRNA as a guide for the degradation of mRNA substrate. Although the use of long dsRNA has been demonstrated to be effective, it is well established that dsRNA within mammalian cells may induce a cellular response by interferon, thus generating nonspecific effects. Direct use of siRNA has been shown to circumvent this problem. The use of RNAi technology has been demonstrated for both mouse ES and EC lines, and most recently, RNAi has been shown to faithfully recapitulate the knockout phenotype in ES cell-derived mouse embryos. In light of the aforementioned recent improvements in the field of RNAi, it seems that RNAi may quickly become a widely used method for rapid gene silencing in human ES cells.

Conclusions

One advantage of ES cells over other stem cell types currently under research is their accessibility for genetic manipulation. Progress so far indicates that genetic manipulation in human ES cells will closely resemble that of their veteran mouse counterparts. Thus, in addition to the methods used so far, the implementation of other advanced strategies for genetic manipulation can probably be expected in human ES cells. Possible uses for genetically modified human ES cells could be the creation of cellular models for disease, studying various aspects of early human development, transplantation medicine, and many more. Thus, in the near future a large number of studies using genetic manipulation techniques in human ES cells is foreseeable.

KEY WORDS

DNA transfection The introduction of foreign DNA molecules into cultured cells.

Gene targeting The modification of a gene in a specific and directed manner through *homologous recombination* with a DNA segment. Possible applications include elimination of gene expression or correction of a mutation in a gene.

Homologous recombination The substitution of one DNA segment by another based on exact or near identity between them.

Negative selection The application of growth conditions eliminating those cells expressing the selection marker.

Positive selection The application of specific growth conditions causing selective elimination of cells that do not contain an exogenous genetic element (selection marker) conferring resistance.

Stable transfection *Transfection* in which foreign DNA is incorporated into the host genome, generating a long-term genetic modification.

Transient transfection *Transfection* in which plasmid DNA is not physically incorporated into the genome and is retained as an episome, generating short-term gene expression of the foreign DNA.

FURTHER READING

Eiges, R., Schuldiner, M., Drukker, M., Yanuka, O., Itskovitz-Eldor, J., and Benvenisty, N. (2001). Establishment of human embryonic stem cell-transfected clones carrying a marker for undifferentiated cells. *Curr. Biol.* **11**, 514–518.

Gu, H., Marth, J. D., Orban, P. C., Mossmann, H., and Rajewsky, K. (1994). Deletion of a DNA polymerase beta gene segment in T cells using cell type-specific gene targeting. *Science* **265**, 103–106.

Hannon, G. J. (2002). RNA interference. *Nature* **418**, 244–251.

Li, M., Pevny, L., Lovell-Badge, R., and Smith, A. (1998). Generation of purified neural precursors from embryonic stem cells by lineage selection. *Curr. Biol.* **8**, 971–974.

Pfeifer, A., Ikawa, M., Dayn, Y., and Verma, I. M. (2002). Transgenesis by lentiviral vectors: lack of gene silencing in mammalian embryonic stem cells and preimplantation embryos. *Proc. Natl. Acad. Sci. USA* **99**, 2140–2145.

Rideout, W. M., III, Hochedlinger, K., Kyba, M., Daley, G. Q., and Jaenisch, R. (2002). Correction of a genetic defect by nuclear transplantation and combined cell and gene therapy. *Cell* **109**, 17–27.

Smithies, O., Gregg, R. G., Boggs, S. S., Koralewski, M. A., and Kucherlapati, R. S. (1985). Insertion of DNA sequences into the human chromosomal beta-globin locus by homologous recombination. *Nature* **317**, 230–234.

Stanford, W. L., Cohn, J. B., and Cordes, S. P. (2001). Gene-trap mutagenesis: past, present and beyond. *Nat. Rev. Genet.* **2**, 756–768.

Thomas, K. R., and Capecchi, M. R. (1987). Site-directed mutagenesis by gene targeting in mouse embryo-derived stem cells. *Cell* **51**, 503–512.

Urbach, A., Schuldiner, M., and Benvenisty, N. (2004). Modeling for Lesch-Nyhan disease by gene targeting in human embryonic stem cells. *Stem Cells* **22**, 635–641.

Zwaka, T. P., and Thomson, J. A. (2003). Homologous recombination in human embryonic stem cells. *Nat. Biotechnol.* **21**, 319–321.

Homologous Recombination in Human Embryonic Stem Cells

Thomas P. Zwaka and James A. Thomson

Biologists have long attempted by chemical means to induce in higher organisms predictable and specific changes as hereditary characters

Avery, 1944

Introduction

Homologous recombination provides a precise mechanism for defined modifications of genomes in living cells and has been used extensively with mouse ES cells to investigate gene function and to create mouse models of human diseases. The ability to modify the mouse genome in the intact mouse in a predetermined manner has revolutionized biomedical research involving mice. Two breakthroughs have made this possible: the derivation of embryonic stem (ES) cells and the identification of conditions for recombination between incoming DNA and the homologous sequences in the mammalian chromosome.

Recently, human ES cell lines have been derived from preimplantation embryos. Human ES cells can be maintained for a prolonged period of culture while retaining their ability to form extraembryonic tissues and advanced derivatives of all three embryonic germ layers. Human ES cells provide a new model for understanding human development, a renewal source of human cells either for *in vitro* studies or for transplantation therapies, and new *in vitro* models for understanding human disease.

Differences between mouse and human ES cells delayed the development of homologous recombination techniques in human ES cells. First, high, stable transfection efficiencies in human ES cells have been difficult to achieve, and in particular, electroporation protocols established for mouse ES cells do not work in human ES cells. Second, in contrast to mouse ES cells, human ES cells clone inefficiently from single cells, making screening procedures to identify rare homologous recombination events difficult. We have recently developed protocols to overcome these problems. Here we describe how these protocols have allowed us to successfully target an ubiquitously expressed gene (*HPRT1*), an ES cell-specific gene (*POU5F1*), and a tissue-specific gene (*tyrosine hydroxylase, TH*).

Targeted Ablation of the HPRT1 Gene as a Tool to Optimize Homologous Recombination Efficiency

The *HPRT1* gene is located on the X-chromosome, so a single homologous recombination event can lead to complete loss of function in XY cells. *HPRT1*-deficient cells can be selected based on their resistance to 2-amino-6-captopurine (6-TG), and thus, the frequency of homologous recombination events is easy to estimate. It was because of these properties that the *HPRT1* gene played such an important role in the initial development of homologous recombination in mouse ES cells.

Therefore, we designed a gene-targeting vector that was able to ablate parts of the human *HPRT1* gene, and we used this system to optimize transfection protocols in human ES cells. The gene-targeting vector was constructed by substituting the last three exons (exons 7, 8, and 9) of the *HPRT1* gene by a *neo* resistance cassette under the thymidine-kinase (tk) promoter control (Figure 44-1A). This cassette was flanked in the 5′ region by a 10-kb homologous arm and by a 1.9-kb homologous arm in the 3′ region. At the end of the 3′ homologous arm, the *tk* gene was added for negative selection.

Previously described problems with nonisogenic genomic DNA in mouse experiments encouraged us to use homologous DNA that was isogenic to the particular human ES cell line that we wanted to target (H1 subclone 1). This DNA was obtained by long-distance, high-fidelity, genomic polymerase chain reaction (PCR) and subcloned into a PCR-cloning vector.

For human ES cells, the best chemical reagents yield stable (drug-selectable) transfection rates of about 10^{-5}, and mouse ES cell electroporation procedures yield rates that are significantly lower. Therefore, we started to test our *HPRT1* gene-targeting vector system using FuGene-6 and ExGen 500, two chemical transfection reagents commonly used in human ES cell work.

Two days before transfection, human ES cells (H1 subclone 1) were trypsinized and plated on Matrigel at a density of 200,000 cells/6-well and cultured with fibroblast-conditioned medium in six-well plates. Transfection, using FuGene-6 or ExGen 500, was performed according to manufacturers' recommendations. Briefly, the vector was co-incubated with the plasmid (2 μg DNA and 6 μl FuGene-6 or 3 μg DNA and 10 μl ExGen 500) and added to the supernatant

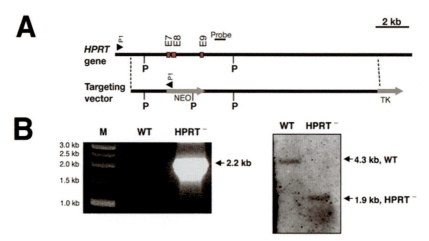

Figure 44-1. Knockout of the *HPRT1* gene. (A) Genomic organization of the human *HPRT1* gene and the gene-targeting vector. 3' probe for Southern blot analysis is shown, E = exon, P = PstI, P1 = primer pair, TK = thymidine-kinase. (B) PCR analysis of ES cell lines (HPRT = knockout; WT = wild-type cells, M = marker) with P1 (left) and Southern blot analysis with dedicated probe and PstI digest (right).

of each well of the six-well plate. In experiments using ExGen 500, we gently centrifuged the six-well plate after adding the DNA/ExGen mixture (300 g for 5 minutes). The medium was changed 4 hours after transfection, G418 selection was started two days after transfection, and two days later, selection with 6-TG (1 mM) and/or GANC (1 μM) was begun.

Although G418 (130 versus 261) and gancyclovir-resistant (35 versus 61) clones were obtained for both transfection reagents, none of these was G418- and 6-TG resistant (HPRT⁻), indicating that none was the result of homologous recombination (see Table 44-1). These results are consistent with the observation that transfection using lipid (FuGene-6) and cationic (ExGen 500) reagents results in inefficient homologous recombination in other mammalian cell types and that physical methods of introducing DNA, in general, are more effective. A possible explanation is that physical transfection induces a cellular stress reaction and activates the recombination machinery to a greater extent than do chemical transfection reagents.

Many parameters influence the transfection rate for electroporation, and usually, the procedure has to be optimized for each cell line. This is critical for human ES cells, for they are very sensitive and difficult to grow from individual cells. Indeed, the difference between the diameter of human and mouse ES cells suggests that successful protocols will be significantly different (Figure 44-2). This was confirmed by an experiment, using a typical mouse ES cell protocol, that clearly showed the overall transfection rate to be too low in human ES cells to expect recovery of homologous-recombinant clones (Table 44-1). In our hands, electroporation using a typical mouse ES cell protocol (220 V, 960 μF electroporation in PBS) yielded a stable transfection rate in human ES cells that was lower than ~10^{-7}.

We adapted electroporation parameters for human ES cells. To increase plating efficiency, we removed ES cell colonies from the culture dish as intact, small clumps and plated them out at a high density following electroporation.

TABLE 44-1

Numbers of Colonies Obtained by Positive and Negative Selection and Targeted Events in the *HPRT* Gene Locus (from 1.5×10^7 Electroporated Human ES Cells)

Selection Procedure	ExGen 500	Fugene 500	Electroporation
G418	130	261	350
G418 and gancyclovir	35	61	50
G418 and 6-TG	0	0	7

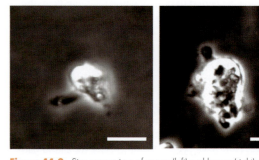

Figure 44-2. Size comparison of mouse (left) and human (right) embryonic stem cells (bar = 10 μm).

To increase survival, we used culture medium instead of PBS as an electroporation buffer and performed electroporation at room temperature. And finally, we changed the physical parameters of the electric field during electroporation. Applying this modified protocol, we were able to achieve stable, G418-resistant transfection rates that were 100-fold (or more) higher than standard mouse ES cell electroporation procedures.

In order to remove ES cells as clumps, we treated cultures with collagenase IV (1 mg/ml, Invitrogen) for 7 minutes, washed with media, and resuspended the cells in 0.5 ml of culture media ($1.5–3.0 \times 10^7$ cells). Just prior to electroporation, 0.3 ml of phosphate-buffered saline (PBS, Invitrogen) containing 40 µg linearized targeting vector DNA was added. Cells were then exposed to a single, 320-V, 200-µF pulse at room temperature, using the BioRad Gene Pulser II (0.4 cm gap cuvette). Cells were incubated for 10 minutes at room temperature and were plated at high density (3.0×10^7 cells per 10-cm culture dish) on Matrigel. G418 selection (50 µg/ml, Invitrogen) was started 48 hours after electroporation. After one week, G418 concentration was doubled, and 6-TG selection (1 mM, Sigma) was started. After three weeks, surviving colonies were picked, transferred into 48-well plates, and analyzed individually by PCR, using primers specific for the *neo* cassette and for the *HPRT1* gene just upstream of the 5′ region of homology, respectively. PCR-positive clones were re-screened by Southern blot analysis using PstI-digested DNA and a probe 3′ of the *neo* cassette. After transfection of 1.5×10^7 human ES cells, 350 G418-resistant clones were obtained. Among them, 50 were gancyclovir-resistant, and of those, 7 were also 6-TG-resistant, suggesting successful homologous recombination (Table 44-1). PCR and Southern blotting (Figure 44-2B) confirmed that homologous recombination had occurred in all of the 6-TG-resistant clones. Electroporation of human ES cells by a DNA construct containing a *neo* resistance cassette, under the control of the *tk* promoter, yielded a stable transfection rate of 5.6×10^{-5}.

Oct4 EGFP/*neo* Knockin

A useful application for "knockins" in human ES cells will be to generate cell lines with a selectable marker introduced into a locus with a tissue-specific expression pattern. Such knockins can be used, for example, to purify a specific ES cell-derived cell type from a mixed population. To test this approach, we introduced two reporter genes into the Oct4 coding gene *POU5F1* by homologous recombination. Oct4 belongs to the POU (Pit, Oct, Unc) transcription factor family, is expressed exclusively in the pluripotent cells of the embryo, and is a central regulator of pluripotency. We introduced two promoterless reporter/selection cassettes into the 3′ untranslated region (UTR) of the human Oct4 gene (Figure 44-3A). The first cassette contained an internal ribosomal entry site (IRES) sequence of the encephalomyocarditis virus and a gene encoding the enhanced green fluorescence protein (EGFP). The second cassette included the same IRES sequence and a gene-encoding neomycin resistance (*neo*). The cassettes were flanked by two homologous arms. (6.3 kb in the 5′ direction and 1.6 kb in the 3′ direction). After electroporation of 1.5×10^7 human ES cells with the linearized targeting vector, we obtained 103 G418-resistant clones. PCR and DNA Southern blotting demonstrated that 28 of these clones (27%) were positive for homologous recombination (Figure 44-3B). Using a second targeting vector with a longer 3′ homologous arm, the rate of homologous recombination increased to almost 40% (22 homologous clones out of 56 G418-resistant clones). Similar transfection experiments, using FuGene-6

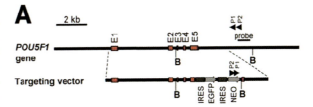

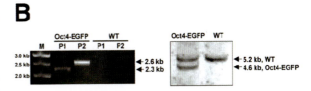

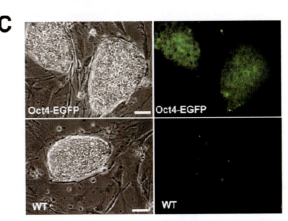

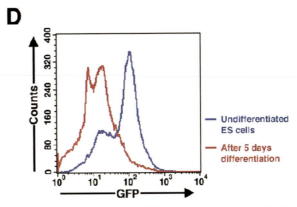

Figure 44-3. Knockin of an IRES-EGFP-IRES-*neo* cassette into the 3′ UTR of the Oct4-coding gene *POU5F1*. (A) Genomic structure of the human *POU5F1* gene and the gene-targeting vector. 3′ probe for Southern blot analysis is shown, E = exon, B = BamHI. P1 and P2 = primer pairs 1 and 2. (B) PCR analysis of ES cell lines (Oct4-EGFP = heterozygous knockin; WT = wild-type cells, M = marker) with P1 and P2 (left) and Southern blot analysis with dedicated probe and BamHI digest (right). (C) Fluorescence microscopy (right) and phase contrast (left) of Oct4 knockin and wild-type colonies, bar = 25 µm. (D) Flow cytometry of Oct4 knockin, undifferentiated (EGFP-positive) ES cells (blue), and their differentiated derivates after five days of differentiation (red).

with the same Oct4 gene-targeting vector, produced 11 G418-resistant clones, none of which resulted from homologous recombination.

Human ES cells with the Oct4 knockin expressed EGFP. After induction of differentiation with various stimuli, EGFP expression was turned off, indicating a down-regulation of Oct4. Both drug selection and flow cytometry for EGFP expression allowed purification of undifferentiated ES cells from a mixed, partially differentiated cell population. These properties will make the knockin cell lines useful for studying Oct4 gene expression during differentiation, *in vitro*, and optimizing culture conditions for human ES cells.

TH EGFP Knockin

Because Oct4 and HPRT are both expressed in human ES cells, we next asked if it is possible to target a gene that is not expressed in ES cells. TH is the rate-limiting enzyme in the synthesis of dopamine, and it is one of the most common markers used for dopaminergic neurons. In order to achieve bicistronic expression of both TH and EGFP, we constructed a gene-targeting vector that introduces an IRES-EGFP reporter gene cassette into the 3′ UTR region in the last exon of the *TH* gene (Figure 44-4A). The cassette is flanked by a short homologous arm 5′ of this exon (1.4 kb) and a long homologous arm in the 3′ region of the last exon (8 kb). The

long arm follows the gene for *tk* for negative selection of random integrated, stable transfected clones. Between the long homologous arm and the IRES-EGFP cassette, we cloned a PGK-driven *neo* resistance cassette flanked by two *loxP* sites. After transfection of 6.0×10^7 human ES cells (cell line H9), we obtained five PCR-confirmed, homologous-recombinant clones after double selection for the positive selection marker *neo* with G418 and the negative selection marker TK with gancyclovir (Figure 44-4B).

The positive selection marker in this experiment was a *neo* cassette under the PGK promoter. As this cassette is still present in the knockin cells line, it could interfere with the normal expression of the *TH* gene itself and the IRES-EGFP reporter gene. To delete the *neo* cassette, we transiently transfected two of the TH-EGFP knockin cell lines with a plasmid containing the phage recombinase Cre under the control of the EF1α promoter. We transfected with FuGene-6 and used the protocols described above. The cDNA of Cre was followed by an IRES-EGFP cassette. After transient transfection with this plasmid, Cre overexpressing cells could be easily identified by EGFP expression. Those EGFP-positive cells were purified by fluorescence-activated cell sorting (FACS). Individual clones were analyzed for successful recombination of the two *loxP* sites, and two clones were identified that have the *neo* cassette excised by PCR (Figure 44-5B). The clones were confirmed to have a normal karyotype.

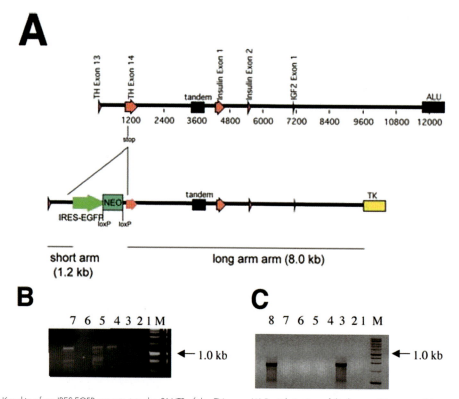

Figure 44-4. Knockin of an IRES-EGFP cassette into the 3′ UTR of the *TH* gene. (A) Partial structure of the human *TH* gene and the gene-targeting vector. (B) Some of the G418/GANC resistant clones (4, 5, and 7) are positive in the screening PCR, indicating successful homologous recombination. (C) After Cre transfection, two of the eight screened clones (3 and 8) show the PCR product that indicates a positive recombination event.

TABLE 44-2

Comparison of Gene-Targeting Frequencies in Different Experimental Setting in Human Embryonic Stem Cells (Data Normalized to 1.5 × 10^8 Electroporated Human ES Cells). Homologous DNA was Isogenic to H1 Genomic DNA (ND = Not Determined)

Selection Procedure	Old Mouse ES Cell Protocol	HPRT1 in H1.1	Oct4 Short Vector in H1.1	Oct4 Long Vector in H1.1	Oct4 Short Vector in H9	TH in H9
G418	1.0	230	69	37	48	ND
G418 and gancyclovir	ND	3.3	ND	ND	ND	3.8
Homologous recombinants	ND	5	19	15	2	1.3

We performed the *TH* gene targeting on cell line H9 using a targeting construct that was homologous to genomic DNA from the cell line H1 (see Table 44-2), suggesting that isogenic DNA is not absolutely required to obtain homologous recombination in human ES cells.

Conclusions

The overall targeting frequencies for the three genes discussed here suggest that homologous recombination is a broadly applicable technique in human ES cells (Table 44-2). The homologous recombination frequencies are roughly comparable to those observed for mouse ES cells and suggest that, although successful transfection strategies differ between human and mouse ES cells, homologous recombination itself may be similar. We also demonstrated that the Cre/lox system can be used to excise specific sequences from human ES cells.

Homologous recombination in human ES cells will be important for studying gene function *in vitro* and for lineage selection. It is a powerful approach for understanding the function of any human gene. For therapeutic applications in transplantation medicine, controlled modification of specific genes should be useful for purifying specific ES cell-derived, differentiated cell types from a mixed population and for altering the antigenicity of ES cell derivatives. In addition, it should be possible to give cells new properties (such as viral resistance) to combat specific diseases. Homologous recombination in human ES cells might also be used for recently described approaches combining therapeutic cloning with gene therapy.

In vitro studies will be particularly useful for learning more about the pathogenesis of diseases where mouse models have proven inadequate. For example, HPRT-deficient mice fail to demonstrate a Lesch-Nyhan-like phenotype. *In vitro* neural differentiation of HPRT⁻ human ES cells or transplantation of ES cell-derived neural tissue to an animal model could help in understanding the pathogenesis of Lesch-Nyhan syndrome. Another good example of where human ES cells and homologous recombination will be useful is in understanding the function of the human heart. Specific mutations and polymorphisms that predispose people to an increased risk for sudden death syndrome or severe arrhythmia at a young age have been identified. Mice generated from ES cells in which ion channel genes have been modified by homologous recom-

bination often have a normal heart, again reflecting basic, clinically significant species differences. With homologous recombination, one could generate human ES cell lines bearing mutations or polymorphisms in specific ion channels, and use ES cell-derived cardiomyocytes to better understand the effects of the mutations on the physiology of the heart. A panel of ion channel polymorphisms should be extremely useful for screening drugs for toxic side effects on the heart prior to clinical trials. Thus, homologous recombination in human ES cells would be useful for numerous *in vitro* models of human disease.

ACKNOWLEDGMENTS

We thank Henry Yuen and the Oscar Rennebohm Foundation for their gifts to the Wisconsin Foundation that supports this work. We thank S. Witowski, J. Antosiewicz, and K. Murphy for technical assistance. The double IRES cassette and the Cre-IRES-EGFP cassette were kindly provided by O. Weber and H. J. Fehling (University of Ulm, Germany).

KEY WORDS

Embryonic stem cells Pluripotent cell lines derived from mammalian blastocysts.

Homologous recombination The reciprocal exchange of sequences between two homologous DNA strands. It is widely used to introduce targeted and specific modifications in prokaryotic and eukaryotic genomes.

FURTHER READING

Doetschman, T., Gregg, R. G., Maeda, N., Hooper, M. L., Melton, D. W., Thompson, S., and Smithies, O. (1987). Targeted correction of a mutant HPRT gene in mouse embryonic stem cells. *Nature* **330**, 576–578.

Evans, M. J., and Kaufman, M. H. (1981). Establishment in culture of pluripotential cells from mouse embryos. *Nature* **292**, 154–156.

Martin, G. R. (1981). Isolation of a pluripotent cell line from early mouse embryos cultured in medium conditioned by teratocarcinoma stem cells. *Proc. Natl. Acad. Sci. USA* **78**, 7634–7638.

Nichols, J., Zevnik, B., Anastassiadis, K., Niwa, H., Klewe-Nebenius, D., Chambers, I., Scholer, H., and Smith, A. (1998). Formation of pluripotent stem cells in the mammalian embryo depends on the POU transcription factor Oct4. *Cell* **95**, 379–391.

Odorico, J. S., Kaufman, D. S., and Thomson, J. A. (2001). Multilineage differentiation from human embryonic stem cell lines. *Stem Cells* **19**, 193–204.

Smithies, O., Gregg, R. G., Boggs, S. S., Koralewski, M. A., and Kucherlapati, R. S. (1985). Insertion of DNA sequences into the human chromosomal beta-globin locus by homologous recombination. *Nature* **317**, 230–234.

Thomas, K. R., Folger, K. R., and Capecchi, M. R. (1986). High frequency targeting of genes to specific sites in the mammalian genome. *Cell* **44**, 419–428.

Thomson, J. A., Itskovitz-Eldor, J., Shapiro, S. S., Waknitz, M. A., Swiergiel, J. J., Marshall, V. S., and Jones, J. M. (1998). Embryonic stem cell lines derived from human blastocysts. *Science* **282**, 1145–1147.

Vasquez, K. M., Marburger, K., Intody, Z., and Wilson, J. H. (2001). Manipulating the mammalian genome by homologous recombination. *Proc. Natl. Acad. Sci. USA* **98**, 8403–8410.

Zwaka, T. P., and Thomson, J. A. (2003). Homologous recombination in human embryonic stem cells. *Nat. Biotechnol.* **21**, 319–321.

45

Surface Antigen Markers

Jonathan S. Draper and Peter W. Andrews

Introduction

The accessibility of molecules on the surface of cells makes them exceptionally convenient markers for characterizing cell types. Often recognized as antigens by specific antibodies, some of these surface molecules are common to many cells, but others show highly restricted patterns of expression characteristic of particular cell types. Boyse and Old (1996) called such antigens with restricted expression *differentiation antigens*, and the patterns of expression of these markers, or *surface antigen phenotypes*, have been widely used for analyzing cell relationships in the hematopoietic and lymphoid systems. Initially, polyclonal sera, produced by cross-immunizing mice of different strains bearing alternate alleles of specific surface antigens, were used to define subsets of lymphocytes with different functional characteristics. With the subsequent development of monoclonal antibodies, a wide array of so-called cluster determinant markers were characterized for defining cells of different lineages within the hematopoietic system of both mice and humans and for identifying the pluripotent hematopoietic stem cells from bone marrow.

Although many surface antigens are informative in certain defined contexts, many are frequently reused at different temporal and special points during the life of an organism. An apposite example is the expression of stage-specific embryonic antigen-3 and -4 (SSEA3 and SSEA4) on cells of cleavage and blastocyst stage human embryos and separately as part of the P-blood group system on adult red blood cells. Indeed, it can be difficult to establish whether expression of an antigen is restricted to a specific cell type. Thus, caution is required when interpreting the presence or absence of a particular antigen and extrapolating to the identification of the cell type or to the definition of lineage relationships between cell types. That two cell types express a common antigen is not evidence that they are related by a common lineage, although this erroneous argument is sometimes encountered. Nevertheless, surface antigens can provide powerful tools for analyzing and sorting cells that have particular characteristics within specific contexts. Added flexibility is provided by the ability to use antibodies directed to surface antigens for labeling and sorting live cells into discrete populations that retain viability.

Cell-Surface Embryonic Antigens of the Laboratory Mouse

Building on the work of Boyse and Old with differentiation antigens of the adult hematopoietic system, Artzt and her colleagues in the Institut Pasteur proposed that immunization of inbred adult mice with syngeneic embryonic cells should yield an antiserum that would recognize those antigens that are expressed only in the embryo and not in the adult. They immunized 129 mice with F9 embryonal carcinoma (EC) cells derived from a teratocarcinoma of a 129 male mouse. The resulting antiserum did react specifically with EC cells and some early embryonic cells, notably those of cleavage-stage embryos and the inner cell mass (ICM) of the blastocyst, but not with postblastocyst-stage embryos, differentiated derivatives of EC cells, or various cells of adult origins. This pattern of reactivity supported the notion that embryo-specific surface antigens exist and contributed to the argument that EC cells from teratocarcinomas are malignant counterparts of ICM cells, the "stem cells" of the early embryo.

Although some progress was made in characterizing the F9 antigen as a high-molecular-weight glycoprotein, the complexity and variability of such sera impeded detailed study. The development of monoclonal antibodies provided a route to much better defined reagents. Of these, the best known is the monoclonal antibody MC480, produced from a BALB/c mouse immunized with EC cells from a 129 teratocarcinoma. This antibody defines SSEA1, which shows a similar pattern of expression to the F9 antigen, and most likely, the epitope recognized by MC480 is a significant component of the F9 antigen.

SSEA1 was identified as a glycosphingolipid, and the epitope was shown to be the Lewis-X (Le^x) haptene. However, the epitope can also be carried by high-molecular-weight glycoproteins. Like the F9 antigen, SSEA1 was found to be expressed by blastomeres of late-cleavage- and morula-stage embryos, by cells of the ICM, and by trophectoderm. It is also characteristically expressed by mouse EC and embryonic stem (ES) cells and has become one of the hallmarks for identifying these undifferentiated stems cells; their differentiation is typically characterized by disappearance of this antigen. Nevertheless, variant EC lines lacking expression of SSEA1 have been isolated, so SSEA1 expression is not, in itself, a requirement for a pluripotent EC–ES phenotype.

Another embryonic antigen, SSEA3, was defined by a monoclonal antibody, MC631, raised by immunizing rats with four-cell stage mouse embryos. This antigen was found to be expressed by cleavage embryos of most strains of mice, but then it disappears from the ICM of the blastocyst to reappear on the primitive endoderm. Its reactivity was thus the reciprocal of SSEA1 expression. It also proved to be a glycosphingolipid but with a globoseries core structure rather than with the lactoseries core of SSEA1 (Table 45-1). A further antigen in the series, SSEA4, defined by monoclonal antibody MC813-70, shows an expression pattern similar to that of SSEA3 and is a globoseries glycosphingolipid similar to the SSEA3-reactive lipid but with a terminal sialic acid.

Although SSEA3 and SSEA4 are not expressed by murine EC cells and embryonic cells of the ICM, the Forssman antigen is expressed by these cells. The Forssman antigen is also a globoseries glycolipid, but it has a terminal galactosamine residue that can be added to globoside in rodent but not in human cells.

Capitalizing on these studies of teratocarcinoma-derived EC cells, in 1981 Evans and Kaufman and, independently, Martin demonstrated that it is possible to derive cell lines by explanting the ICM of a blastocysts onto feeder cell layers *in vitro* under conditions used in some of the early derivations of EC cell lines. Subsequently, embryonic germ cell lines closely resembling EC and ES cells were established. These embryo-derived cells closely resembled EC cells morphologically and in developmental potential. They also expressed similar markers, including SSEA1.

Human EC and ES Cells

Initially, it was assumed that human EC cells would closely resemble mouse EC cells; in the first studies of cell lines derived from human testicular cancers, it was reported that human EC cells, like murine EC cells, were F9 antigen-positive. However, upon more detailed examination, it was established that, although human EC cells closely resemble mouse EC cells in their general morphology and growth patterns, they do not express some of the markers characteristic of mouse EC cells; in particular, they lacked SSEA1. By contrast, human EC cells *in vitro* and in tumors express SSEA3 and SSEA4.

The functional significance of these glycolipid antigens, which provide excellent markers of the undifferentiated EC and ES cells in both mice and humans, remains obscure. They appear to be strongly developmentally regulated: SSEA1 expression is rapidly down-regulated on differentiation of mouse EC cells. In human EC and ES cells, not only is differentiation marked by down-regulation of the globoseries antigens SSEA3, but at least on some derivative cells, these antigens are replaced by SSEA1 or by ganglioseries antigens, notably ganglioside GT3 recognized by antibody A2B5 (Table 45-1). This change in expression seems to relate to switching expression of the glycosyl transferases responsible for synthesis of the globo-, lacto-, and ganglioseries core oligosaccharide structures. It might be that the difference between mouse and human EC–ES cells is smaller than at first sight,

TABLE 45-1
Glycolipid Antigens of Mouse and Human EC and ES Cells

Antigen	Oligosaccharide structure
Globoseries glycolipids	
SSEA3	Galβ1 → 3GalNAcβ1 → 3Galα1 → 4**Galβ1** → 4**Glcβ1** → **Cer**
SSEA3, SSEA4	NeuNAcα → 3Galβ1 → 3GalNAcβ1 → 3Galα1 → 4**Galβ1** → 4**Glcβ1** → **Cer**
Forssman	GalNAcα1 → 3GalNAcβ1 → 3Galα1 → 4**Galβ1** → 4**Glcβ1** → **Cer**
Lactoseries glycolipids	
SSEA1	Galβ1 → 4GlcNAcβ1 → 4**Galβ1** → 4**Glcβ1** → **Cer**
	$\qquad\qquad\qquad$ 3
	$\qquad\qquad\qquad \uparrow$
	$\qquad\qquad\qquad$ Fucα1
Ganglioseries glycolipids	
9-O-acetyl GD3 (ME311)	(9-O-acetyl) NeuNAcα2 → 8NeuNAcα2 → 3**Galβ1** → 4**Glcβ1** → **Cer**
GT3 (A2B5)	NeuNAcα2 → 8NeuNAcα2 → 8NeuNAcα2 → 3**Galβ1** → 4**Glcβ1** → **Cer**

All of these glycosphingolipid antigens are synthesized by extension from a common precursor, lactosylceramide. The likely parts of these structures comprising the epitopes recognized by the different antibodies are highlighted. The enzymes responsible for the addition of the third sugar residues (galactosyl, glucosaminyl, and sialyltransferase) appear to regulate the rate-limiting step that controls which synthetic pathway predominates. Human EC and ES cells are characterized by expression of the glycolipid antigens SSEA3 and SSEA4, and not the lactoseries antigen SSEA1 (Leˣ). By contrast, mouse EC and ES cells are typically characterized by their expression of SSEA1. Mouse cells also express the globoseries Forssman antigen, which is not synthesized by human cells. Thus, despite the apparent distinction between mouse and human EC and ES cells in terms of their terminal sugars that comprise the different epitopes, the undifferentiated cells of both species do seem to share in common the expression of globoseries core structures.

since globoseries glycolipids are prominent in both species. The differences are in terminal modifications to yield Forssman in the mouse and SSEA3 and SSEA4 in humans (Table 45-1). SSEA1 mouse EC cells and SSEA3/4 human EC cells have been reported, although at least in the case of the human cells, this is a consequence of changes to terminal modification rather than of expression of core glycolipid structures. However, a small number of humans, those with the p^k and pp blood groups, are unable to synthesize the elongated globoseries structures and do not express SSEA3 or SSEA4. Thus, lack of these structures is evidently not detrimental to embryonic development. More tellingly, it is possible to eliminate all glycolipid synthesis with the inhibitor of glucosylceramide synthetase, PDMP: NTERA2 human EC cells still survive and differentiate in the presence of this inhibitor, and embryos of medaka fish develop normally despite near-elimination of glycolipid synthesis. Interestingly, however, p^k and pp women frequently have high rates of early spontaneous abortions, which might be caused by an immune response to embryonic antigens, perhaps SSEA3, SSEA4, or both.

A different series of antigens of human EC and ES cells comprises the keratan sulfates, different epitopes of which are recognized by several antibodies, most notably TRA-1-60, TRA-1-81, GCTM2, K4, and K21. Of these antigens, TRA-1-60 is dependent upon a terminal sialic acid, and GCTM2 has been reported to recognize the internal protein structure. Unlike SSEA3 and SSEA4, expression of these antigens is unaffected by PDMP.

Human ES cells closely resemble human EC cells in expression of these antigens; they are generally SSEA3+, SSEA4+, TRA-1-60+, TRA-1-81+, GCTM2+, and SSEA1-. Similarly, the ICM cells of human blastocyst stage embryos are SSEA3+, SSEA4+, TRA-1-60+, and SSSEA1-, so human EC and ES cells resemble human ICM cells as mouse EC and ES cells resemble mouse ICM cells, even though these cells differ between the two species.

Other surface antigens that appear to be characteristic of human EC and ES cells include Thy1 and the tissue-nonspe-

cific form of alkaline phosphatase, which can be readily recognized by the monoclonal antibodies TRA-2-49 and TRA-2-54.

The differentiation of human EC and ES cells is characterized by the loss of expression of all of these various markers of undifferentiated cells. At the same time, various antigens appear. SSEA1, which seems to be absent from undifferentiated human EC and ES cells, frequently appears on differentiation. At least some of the cells that express this antigen may belong to the trophoblastic lineage. Ganglioside antigens, notably GT3, recognized by antibody A2B5, and GD2 and GD3, recognized by antibodies VIN2PB22 and VINIS56, all appear on subsets of cells upon human EC and ES differentiation and most likely identify cells of the neural lineages.

One set of antigens that can cause confusion in comparisons of mouse and human EC and ES cells is the class 1 major histocompatibility antigens, H-2 in mice and HLA in humans. H-2 antigens are not expressed by murine EC cells, although they do appear on differentiation. HLA is generally expressed by human EC and ES cells and may show a transient down-regulation upon differentiation. HLA class 1 antigen expression is also strongly induced in human EC and ES lines by interferon-γ, although an antiviral response does not occur.

Summary

Several surface marker antigens that show developmental regulation are found on both mouse and human EC and ES cells, although there are significant species differences in the particular antigens expressed (Table 45-2) and a greater array of such markers is available for the human cells. These antigens can provide important tools both for identifying undifferentiated stem cells and for monitoring their differentiation. A workshop comparison of the antigens expressed by a large panel of human EC cells suggested that SSEA3, an antigen rapidly lost as human EC cells differentiate, may be one of the more sensitive indicators of the undifferentiated state. However, a similar comparison of antigen expression by a panel of human ES lines remains to be undertaken. Further-

TABLE 45-2

Differential Expression of Characteristic Surface Marker Antigens of Mouse and Human EC and ES Cells

Antigen	Mouse	Human	Comments
Forssman	Yes	No	Globoseries glycolipid
SSEA1 (Le^x)	Yes	No	Lactoseries glycolipid
SSEA3	No	Yes	Globoseries glycolipid
SSEA4	No	Yes	Globoseries glycolipid
TRA-1-60	No	Yes	Keratan sulfate
TRA-1-81	No	Yes	Keratan sulfate
GCTM2	No	Yes	Keratan sulfate
Liver-alkaline phosphatase (L-ALP)	Yes	Yes	No antibody is available for murine ALP, but human L-ALP is recognized by antibodies TRA-2-54 and TRA-2-49.
MHC Class 1	No	Yes	H-2 in the mouse; HLA in humans

more, caution is warranted; some human EC cell sublines may lack SSEA3 expression but retain the expression of other markers of the undifferentiated state and display pluripotency as evidenced by an ability to differentiate *in vivo* and *in vitro*. Also, many of the key markers so far defined have oligosaccharide epitopes, the expression of which depend on a variety of complex interacting factors, not merely the expression or the lack of expression of specific glycosyl transferases. There remains, therefore, considerable scope for the identification of additional surface antigens for identifying and monitoring undifferentiated human ES cells.

Appendix: Methods

Indirect Immunofluorescence and Flow Cytometry

1. Harvest cells to obtain a single-cell suspension. Use any standard protocol; typically, 0.25% trypsin in 1-mM EDTA, in Ca^{++}/Mg^{++}-free Dulbecco's phosphate-buffered saline (PBS), is used. After pelleting, resuspend the cells in either HEPES-buffered medium or wash buffer (5% fetal calf serum in PBS plus 0.1% sodium azide) to 10^7 cells per milliliter. (If there are fewer cells, then 2×10^6/ml can be used conveniently.)

2. Choose antibodies and dilute as appropriate in wash buffer.

3. Distribute the primary antibodies (Table 45-3), diluted appropriately after a preliminary titration, at 50 μl per well of a round-bottom 96-well plate, 1 well for each assay point. To prevent carryover from one well to another, it is good practice to use every other well of the plate. As a negative control, we use the antibody from the original parent myeloma of most hybridomas, namely, P3X63Ag8. However, others may prefer a class-matched nonreactive antibody if one is available or no first antibody.

4. Add 50 μl of cell suspension (i.e., $10^5 - 5 \times 10^5$ cells) to each 50 μl of antibody.

5. Seal the plate by covering it with a sticky plastic cover, ensuring that each well is sealed, and incubate at 4°C, with gentle shaking, for 30 to 60 minutes.

6. Spin the plate at $280 \times g$ for 3 minutes using microtiter plate carriers in a tissue culture centrifuge. Check that the cells are pelleted, and remove the plastic seal using a sharp motion but holding the plate firmly to avoid disturbing the cell pellet. Dump the supernatant by inverting the plate with a rapid downward movement; blot the surface and turn the plate over. Provided that this is done in a single movement without hesitation, the cells remain as pellets at the bottom of the wells. If there are any concerns about pathogens, or any other contaminant of a culture, supernatants can be removed by aspiration rather than dumping.

7. Wash the cells by adding 100 μl of wash buffer to each well, seal, and agitate to resuspend the cells. Spin down as described previously. After removing the supernatant, repeat with two more washes.

8. After the third wash, remove the supernatant, and add 50 μl of fluorescent-tagged antibody, previously titered and diluted in wash buffer to each well. We routinely use FITC-tagged goat–antimouse IgM or antimouse IgG as appropriate to the first antibody. Antimouse IgM, but not antimouse IgG, usually works satisfactorily with MC631 (a rat IgM). Affinity-purified and/or F(ab')₂ second antibodies may be used if required to eliminate background.

9. Seal the plate as described previously, and repeat the incubation and washings as before.

10. Resuspend the cells, about 5×10^5 per milliliter, in wash buffer and analyze in the flow cytometer. The precise final cell concentration will depend upon local operating conditions and protocols.

TABLE 46–3

Common Antibodies Used to Detect Antigens Expressed by Human EC and ES Cells

Antibody	Antigen	Antibody species and isotype	Remarks
MC631	SSEA3	Rat IgM	Undifferentiated hES–EC
MC480	SSEA1 (Le^x)	Mouse IgM	Differentiated hES–EC; including trophoblast
A2B5	GT3	Mouse IgM	Neural
ME311	9-0-acetylGD3	Mouse IgG	Neural
VINI556	GD3	Mouse IgM	Neural
VIN2PB22	GD2	Mouse IgM	Neural
MC813-70	SSEA4	Mouse IgG	Undifferentiated hES–EC
TRA-1-60	TRA-1-60	Mouse IgM	Undifferentiated hES–EC
TRA-1-81	TRA-1-81	Mouse IgM	Undifferentiated hES–EC
TRA-2-54	L-ALP	Mouse IgG	Undifferentiated hES–EC
TRA-2-49	L-ALP	Mouse IgG	Undifferentiated hES–EC
TRA-1-85	Ok[a]	Mouse IgG	Panhuman
BBM1	β2-microglobulin	Mouse IgG	HLA
W6/32	HLA-A, -B, -C	Mouse IgG	HLA

NOTES

All antibodies, both primary and secondary, should be pretitered on a standard cell line to determine optimal concentrations. For monoclonal antibodies, we typically find that the dilution for ascites is between 1:100 and 1:1000, and for hybridoma supernatants, it is between 2X and 1:10.

A useful antibody is TRA-1-85 (Table 45-3). This antibody recognizes an antigen, Ok[a], that appears to be expressed by all human cells but not by mouse cells. It is therefore valuable as a tool for distinguishing human ES cells from the mouse feeder cells on which they are commonly grown.

Facs

1. Harvest the hES cells using trypsin–EDTA as before for analysis. Note, however, that lower concentrations of trypsin (0.05%) greatly improve cell viability. The cells should be pelleted in aliquots — 10^7 cells per aliquot is convenient.
2. Resuspend the cells in primary antibody, diluted in medium without added azide, as determined by prior titration (100 µl per 10^7 cells). The primary and secondary antibodies should be sterilized using a 0.2-micron cellulose acetate filter.
3. Incubate the cells, with occasional shaking, at 4°C for 20 to 30 minutes.
4. Wash the cells by adding 10 ml medium and pellet by centrifugation at 200 g for 5 minutes; repeat this wash step once.
5. Remove supernatant and gently flick to disperse the pellet. Add 100 µl of diluted secondary antibody per 10^7 cells and incubate, with occasional shaking, at 4°C for 20 minutes.
6. Wash the cells twice as described previously. After the final wash, resuspend the cells in medium at 10^7 cells per milliliter. Sort the cells using the flow cytometer according to local protocols. If the cells are to be cultured after sorting, sort into hES medium supplemented with antibiotics.

NOTES

Our standard is 0.25% trypsin–1-mM EDTA in PBS. However, for human ES cells, it is necessary to use lower trypsin concentrations (0.05% trypsin–1-mM EDTA) if cell viability is to be maintained for culturing after sorting.

In situ Immunofluorescence for Surface Antigens

1. Remove the medium from the cells and replace with primary antibody diluted as appropriate in hES medium. To maintain the cultures after staining, the antibody should be sterilized by filtration using a 0.2-micron cellulose acetate filter.
2. Place the cells in the incubator for 30 minutes at 37°C.
3. Remove the antibody–medium solution and wash three times with fresh medium.
4. Add the secondary antibody, diluted as appropriate in hES medium.
5. Place the cells in the incubator for 30 minutes at 37°C.
6. Remove the antibody–medium solution and wash three times with PBS (with Ca^{2+} and Mg^{2+}).
7. Visualize under PBS (with Mg^{2+} and Ca^{2+}). The cells are still alive at this point, so they can be recultured if necessary. Alternatively, the cells can be fixed by treatment with a 4% PFA solution for 20 minutes and then stored under a 50:50 solution of PBS:glycerol.

SOLUTIONS AND NOTES

The use of PBS, with Ca^{2+} and Mg^{2+}, or medium stops the cells from detaching from the tissue culture plastic.

4% PFA is 4% paraformaldehyde in PBS (without Ca^{2+} and Mg^{2+}). Heat to 65°C to dissolve, and then filter to remove particulates. This can be frozen at −20°C but should be used within two weeks. PFA is toxic, so take appropriate measures to protect yourself during preparation and usage (e.g., weigh out and make up in a fume cupboard).

ACKNOWLEDGMENTS

This work was supported partly by grants from the Wellcome Trust, Yorkshire Cancer Research, and the BBSRC.

KEY WORDS

Antibody An immunoglobulin protein that specifically binds to an antigen.

Antigen A molecule that is bound in a specific manner by an antibody.

Differentiation antigen (Typically used with respect to cell surface antigens) An antigen that is expressed by a restricted range of cells and shows changes in expression during cell differentiation.

Epitope That part of a macromolecule that interacts specifically with a given monoclonal antibody. Typically, it comprises 2 or 3 amino acid or monosaccharide moieties but may also depend on some aspect of the teriary structure of the antigen.

Hybridoma Typically used to describe a hybrid cell line produced by fusing an established myeloma cell line and primary lymphocyte.

Monoclonal Antibody An antibody produced by a cloned hybridoma cell line: typically, it will be a single immunoglobulin species with a defined antigen specificity.

FURTHER READING

Andrews, P. W. (2002). From teratocarcinomas to embryonic stem cells. *Phil. Trans. R. Soc. Lond. B* **357**, 405–417.

Andrews, P. W., Casper, J., Damjanov, I., Duggan-Keen, M., Giwercman, A., Hata, J. I., von Keitz, A., Looijenga, L. H. J., Millán, J. L., Oosterhuis, J. W., Pera, M., Sawada, M., Schmoll, H. J., Skakkebaek, N. E., van Putten, W., and Stern, P. (1996). Comparative analysis of cell surface antigens expressed by cell lines derived from human germ cell tumors. *Int. J. Cancer* **66**, 806–816.

Artzt, K., Dubois, P., Bennett, D., Condamine, H., Babinet, C., and Jacob, F. (1973). Surface antigens common to mouse cleavage embryos and primitive teratocarcinoma cells in culture. *Proc. Natl. Acad. Sci. USA* **70**, 2988–2992.

Boyse, E. A., and Old, L. J. (1969). Some aspects of normal and abnormal cell surface genetics. *Ann. Rev. Genet.* **3**, 269–290.

Draper, J. S., Pigott, C., Thomson, J. A., and Andrews, P. W. (2002). Surface antigens of human embryonic stem cells: Changes upon differentiation in culture. *J. Anat.* **200**, 249–258.

Erlebacher, A., Price, K. A., and Glimcher, L. H. (2004). Maintenance of mouse trophoblast stem cell proliferation by TGF-beta/activin. *Dev. Biol.* **275**, 158–169.

Evans, M. J., and Kaufman, M. H. (1981). Establishment in culture of pluripotential cells from mouse embryos. *Nature* **292**, 154–156.

Fenderson, B. A., Andrews, P. W., Nudelman, E., Clausen, H., and Hakomori, S. (1987). Glycolipid core structure switching from globo- to lacto- and ganglioseries during retinoic acid-induced differentiation of TERA-2-derived human embryonal carcinoma cells. *Dev. Biol.* **122**, 21–34.

Fenderson, B. A., Radin, N., and Andrews, P. W. (1993). Differentiation antigens of human germ cell tumors: Distribution of carbohydrate epitopes on glycolipids and glycoproteins analyzed using PDMP, an inhibitor of glycolipid synthesis. *Eur. Urol.* **23**, 30–37.

Henderson, J. K., Draper, J. S., Baillie, H. S., Fishel, S., Thomson, J. A., Moore, H., and Andrews, P. W. (2002). Preimplantation human embryos and embryonic stem cells show comparable expression of stage-specific embryonic antigens. *Stem Cells* **20**, 329–337.

Jacob, F. (1978). Mouse teratocarcinoma and mouse embryo. *Proc. Roy. Soc. Lond. B.* **201**, 249–270.

Race, R. R., and Sanger, R. (1975). *Blood Groups in Man*, 6th ed., pp. 169–171. Blackwell Scientific Publications, Oxford.

46

Lineage Marking

Andras Nagy

Definitions

Mammalian development starts from a singe cell, the zygote, which is the ancestor to all the somatic and germ cells of the entire later organism. This type of developmental capacity is called *totipotency*. In the first few divisions of preimplantation development, the cells (blastomeres) retain this totipotency, but shortly before implantation, at the blastocyst stage, cells become committed to certain tissues of the later conceptus. Three cell types are clearly distinguishable at this stage: the *trophectoderm*, committed to the trophoblast cells of the placenta; the *primitive endoderm*, which derivatives remain extraembryonic and will form the parietal and visceral endoderm; and the *primitive ectoderm* responsible for all cells in the embryo proper and for some internal extraembryonic membranes, such as allantois, amnion, and yolk sac mesoderm (the inner layer of this membrane). A given cell of the blastocyst is only capable of differentiating into a subset of cells of the conceptus; therefore, these cells are no longer totipotent. Instead they are referred to as *pluripotent*. After implantation, further subdivisions of developmental commitments occur in the embryo proper, as the embryonic ectoderm, mesoderm, and definitive endoderm form. The following organogenesis produces the final cellular diversity of an individual by differentiating highly specialized cell types (*differentiated* cells) for organ functions. The cellular diversity consists of approximately 260 distinguishable cell types in mammals, which make up an estimated total cell number of 1 trillion (10^{12}) in an adult human or 10 billion (10^{10}) in a mouse.

The progress of cell commitment to an ever-decreasing range of cell types can be viewed from the opposite direction: Every cell among the 260 cell types has its own "history" that can be traced. This trace is often referred to as the *lineage* (Figure 46-1). If the traces are looked at from a given *progenitor* or a *stem cell* (the origin) in the forward direction, the set of traces is the *fate* of the cell. If the fate remains within a single type of differentiated cells or results in two or more kinds, the origin is called *uni-*, *bi-*, or *multipotential*, respectively. With these definitions, cell lineage and fate are practically the same phenomenon, a set of developmental traces viewed either from the differentiated product or the origin. In Figure 46-1, therefore, the fate of p1 is d1 and d2, and the fate of p2 is d1. The progenitor cell, p1, is bipotential, and p2 is unipotential. For the differentiated cell, the lineage of d2 is

simple: It goes back to p1 only. The d1 lineage could be traced back to either the p1 or the p2 progenitor.

Questions to Ask

There are two basic questions to be asked about cell lineages and cell fates:

- The lineage question: *What are the traces and progenitors that lead to a certain differentiated cell type, for example, d1?*
- The fate question: *What are the possible differentiated products of a certain progenitor cell, such as p1?*

Markers and Lineage Marking

As cells go through the diversification process, they correspondingly change their *gene expression profile*. This profile determines the properties of the cells and therefore their *identities*. Cells of the same type express similar genes. Within this pool are some genes specific to one particular cell type. These genes are referred to as cell type-specific *markers*. For example, hepatocytes, astrocytes, neurons, and cardiac myocytes can be identified by the expression of albumin, GFAP, Neu-N, and α-myosin heavy chain, respectively.

Identifying traces during development or adult regeneration is no simple task. There is often no specific marker identifying a lineage along its complete trace. Instead, a set of genes is expressed at certain segments of the lineage. For example, oligodendrocytes go through several developmental stages, such as proliferative and differentiative phases. The immature oligodendrocytes can first be identified by a surface antigen expression recognized by an antibody, A2B5. As the cells mature, they acquire another specific surface marker recognized by the antibody called O4. Then, as they differentiate further, they start expressing myelin-specific genes such as MBP and PLP. Typically, a temporal series of specific markers, or the lack of certain gene expression, represents a particular lineage. If these markers are known and easy to detect, they can be used to characterize the fate of a progenitor or a stem cell. However, to follow all the descendants of a given progenitor, there is a need for a unique identifier characteristic of the progenitor under study and retained in the derivatives. Providing this identifier is *lineage marking*, the subject of this chapter.

MARKING WITH ENDOGENOUS IDENTIFIERS

It was pointed out earlier that existence of a unique marker gene or genes covering an entire lineage is not common. For-

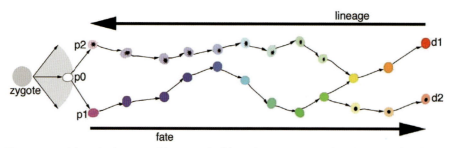

Figure 46-1. Relationship between cell lineage and cell fate. (Please see CD-ROM for color version of this figure.)

tunately, however, the useful genetic differences among individuals within or among species can be used as identifiers in cell–tissue transplantation or chimera studies. These types of lineage marking were important in the early studies of fate mapping. A great deal of knowledge was acquired using embryo transplantation chimeras between quail and chicken (see Teillet, Ziller, and Douarin, 1999). The nuclei of quail cells contain heterochromatin condensed into one (or sometimes two) large masses. This easily recognizable phenomenon was used to distinguish quail from chicken cells. In the mouse, *in situ* hybridization for a specific genomic sequence of *Mus caroli* can also discriminate between these and *Mus musculus* cells in chimeric tissues. Haplotype cell-surface markers have been frequently used to study hematopoietic lineages. *In situ* hybridization for the Y-chromosome can be used as an identifier when grafting male cells into female recipients. The visualization is relatively straightforward using Y-chromosome-specific probes on histological sections.

The unique advantage of the endogenous identifiers is that they are "built-in"; therefore, they do not require genetic or other modifications of cells. Grafting, chimera-making, and frequent requirements for bridging between genders or even species, however, limit the use of such endogenous markers. Modern transgenic methods have in recent years replaced these traditional approaches. Before detailing these methods, we will mention another historically important approach that provides a short-term, transient marking of a cell and its progenitors.

MARKING WITH DYES OR ENZYMES

Vital fluorescence dyes or an enzyme with a long half-life can mark cells in a temporary manner. Among these dyes, the most popular are DiI (a carbocyanine) and its derivatives because of their high photostability and low toxicity. These dyes label cell membranes by inserting two long hydrocarbon chains into the lipid bilayers, resulting in an orange-red (565 nm) fluorescence emission. DiI can be used with DiO, which gives a green fluorescent color. These dyes are usually applied to cells either from an ethanol solution or directly from the dye crystal. This simple labeling technique has frequently been used in nonmammalian model systems to successfully address lineage, potentiality, and fate questions.

Horseradish peroxidase and Rhodamine-conjugated dextran have also been used extensively for cell marking; how-

ever, these techniques are more invasive because they have to be microinjected into the cells.

The obvious limitations to all these methods are the temporary nature of marking and the need for free access to cells for labeling. In the mouse, the usefulness of these approaches is further limited by the requirement for *in vitro* culturing of postimplantation-stage embryos.

MARKING WITH EXOGENOUS GENETIC IDENTIFIERS

Rapidly evolving transgene-based approaches are currently the most versatile methods of lineage marking. This fact does not mean, however, that these approaches have replaced more traditional techniques. They all have their optimal applications, and they are still valid members of the arsenal of options. There are two categories of exogenous genetic identifiers: passive and active markers.

"Passive" Exogenous Genetic Markers

Passive exogenous markers are identified on the basis of genomic insertion of a known DNA sequence into the genome. Their detection requires techniques such as DNA *in situ* hybridization or Southern blotting. A typical example of this category is retroviral insertion. Probing with a retroviral sequence in a genomic Southern blot detects an integration site-specific band pattern. In the case of multiple integrations, the Southern blotting can serve as a fingerprint for a group of cells derived from a single precursor or stem cell. Such an identifier can be used as a tool to measure the turnover of active stem cells that generate an entire hematopoietic compartment. Unfortunately, such an approach is not practical at a single-cell level, which restricts its application in lineage studies. Large copy number transgene insertion can also be used as a passive genetic identifier. Such a transgene creates a long enough unique and known sequence to visualize with DNA *in situ* hybridization on histological sections (see Figure 46-2A).

"Active" Exogenous Genetic Markers

Active exogenous genetic markers are based on reporter gene expression from a transgene. Several reporters are available and have proven to be useful in lineage studies and fate mapping. Depending on the application, one may be better than the other. The most common reporters fall into two groups: enzymes and fluorescent proteins.

Enzyme Activity-Based Reporters. The β-*galactosidase* (*lacZ* gene of the *E. coli*) has had a very successful "career"

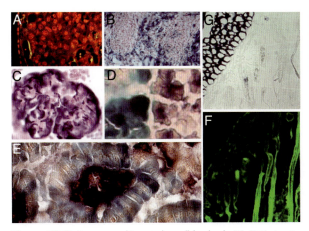

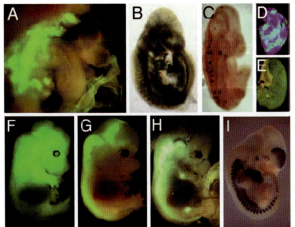

Figure 46-2. Lineage marking at the cellular level. (A) DNA *in situ* hybridization in trophoblast cells transgenic to a multicopy globin gene insertion. (B) Chimeric tissue where one of the components is N-myc oncogene deficient and LacZ tagged with the Rosa-26 gene trap. (C) Immunohistochemical staining for podocyte-specific expression of green fluorescent protein (GFP). (D and E) LacZ and human alkaline phosphatase double staining on mosaic intestine and pancreas. (G and F) LacZ staining and GFP visualization of an adjacent section of skin of a Z/EG, K14-Cre recombinase double transgenic animal.

Figure 46-3. Whole-mount embryos and organs with lineage-marked cells. (A) ES cell-derived embryo in the uterus. The trophoblast of the placenta and the yolk sac endoderm is GFP transgenic tetraploid embryo-derived. (B) Whole-mount LacZ stained embryo in which the *lacZ* gene is inserted into the flk-1, endothelial cell specifically expressed receptor kinase. (C) Z/AP and Cre recombinase double transgenic embryo with sporadic human placental alkaline phosphatase activation. (D) Heart of the chimeric embryo between cyan fluorescent protein expressor ES cells and GFP transgenic embryo. Z/EG and Cre recombinase double transgenic embryos with (E) GFP podocyte-specific, (F) complete, (G) differentiated neuron-specific, and (H) chondrocyte-specific activation. (I) LacZ-stained embryo derived from a Mef-2c gene trap ES cell line.

as a reporter for all sorts of eukaryotic cells. The expression of the enzyme by a heterologous promoter seems to be neutral for the functioning of cells. A variety of substrates can be used for detecting β-galactosidase. The most common is an indole derivative, 5-bromo–4-chloro–3-indolyl–β-D–galactoside (X-gal), which produces a blue color as it breaks down. This simple histochemical staining works on small tissues as whole mount and on histological cryostat sections. Many useful transgenic mouse lines have been made with different specificity for LacZ expression. Of those, the most well known is a gene trap insertion of *lacZ* into an endogenous locus named Rosa-26. This locus provides expression throughout the entire embryo and in most adult tissues. Figure 46-2B shows an example for use of Rosa-26 transgenic and N-myc-deficient embryonic stem (ES) cells in a chimera study. The derivatives of these cells (LacZ stained blue) are not able to contribute to the chondrocyte lineage of the embryo. Since the aim of most gene trap programs is to generate *lacZ* insertions randomly in all the genes of the mouse, cell type-or lineage-specific LacZ tagging is now created in huge numbers (see Stanford, Cohn, and Cordes, 2001). Figure 46-3I shows an example of a LacZ-stained embryo in which the trap vector "landed" in the *Mef-2c* gene. In addition to gene trap lines, many targeted alleles contain a *lacZ* gene placed into the target vector in such a way that the enzyme is expressed under the regulation of the endogenous gene. Figure 46-3B presents an example in which the *lacZ* gene was knocked in to the endothelial-specific *flk-1* locus. The LacZ protein is rather tolerant to N- and C-terminal modifications. Heterologous functional domains may be added, which changes the property of the enzyme, such as the intracellular localization.

Alkaline phosphatase is an essential, constitutively expressed enzyme in cells. Similar to LacZ, a simple histo-

chemical or whole-mount staining procedure can visualize enzyme activity in cells. Endogenous alkaline phosphatases can easily be inactivated by a brief heat treatment (at 70°C for 15 minutes). However, the alkaline phosphatase expressed by the human placenta (hPLAP) is unique in that it is not affected by such a heat treatment. This unique property has been utilized for detection of hPLAP activity when expressed from a transgene. Since heat treatment can be applied to both whole-mount and cryostat sections, hPLAP has become a convenient histological marker, which allows simple detection–recognition of enzyme-containing cells. Since hPLAP is efficiently transported into axons, it is an ideal reporter if nerve visualization is required. Figure 46-3C shows sporadic hPLAP+ cells in a 10.5-day postcoitus (dpc) embryo, and Figures 46-2D and 46-2E show double staining for LacZ and hPLAP on a cryostat section of intestine and pancreas of a mosaic newborn (for more details, see the next sections of this chapter). However, compared to LacZ, hPLAP stain penetration to whole-mount tissue is slightly more limited, and the quality of histological sections falls short of excellence because of the heat treatment of the tissue.

Fluorescent Proteins. The *green fluorescent protein* (GFP) isolated from the jellyfish (*Aequorea victoria*) has recently joined the group of reporters used to trace lineages and determine cell fate in mouse and other model organisms. Several mutant derivatives have been developed providing different levels of stability and intensity and a large spectrum of light emission. The gene that encodes for GFP can be applied in heterologous-transient or -stable integrant transgenic expression settings or its mRNA for transient production of the protein. This reporter is definitely unique because its visualization does not require cell fixation or staining. Instead, live specimens can be observed, and cells expressing GFP (or its derivatives) can be followed in their dynamic behavior. However, not even GFP could overcome the main obstacle in studying mammalian development — that the embryo needs an intra-*utero* environment. *Ex vivo* organ culture systems are gaining importance to overcome this limitation; however, it is not yet possible to follow the very early postimplantation time (4.5–6.5 dpc) *in vitro*.

Establishing Overall Expressor Transgenic Mouse Lines. When cell grafting is the approach to studying cell fate, it is vital that all derivatives of the graft be identifiable. For this purpose, the reporter has to be expressed in all the cells of the graft. For over 10 years now, the Rosa-26 gene trap line has been the best LacZ expressor reporter in the mouse. Second on the "popularity list" is a LacZ line that has the transgene integration on the X-chromosome. This line has been used successfully in several studies to address cell-fate and lineage determination questions during gastrulation.

Establishing a new overall expressor transgenic mouse line is a challenging task. Even the most promising promoter gives unstable, mosaic, or restricted expression from most insertion sites in the genome. Random insertion transgenic approach based on pronuclear DNA injection requires a very large number of founder animals to be established and tested for expression. For this reason, ES cell-mediated transgenesis may be a better alternative. If the same constructs are introduced into ES cells, rather than directly in the embryos, hundreds of transgenic clones can be isolated in a short time. These clones, which all represent a unique genomic integration site of the transgene, can be screened for the expression level and pattern of the reporter, both in undifferentiated ES cells and in *in vitro* differentiation assays. If only ES cell lines with overall expression are used to generate mice, the chance of obtaining satisfying reporter expression is high.

ES cells may also provide control of the transgene insertion site. The most common method is targeted integration of a transgene (reporter) into a well-characterized loci, for example, Rosa-26 or the X-chromosome-linked HPRT. Recombinase-mediated cassette exchange is an alternative that could be used to eliminate the uncertainty of transgene expression caused by random genomic insertion. Three recombinase–integrase systems are available for such an approach; the Cre, Flip (Flp), and PhiC31. By using any of these, a "docking" site can be prepared and characterized for expression permissiveness. The transgene is then equipped with a recognition site or sites of the recombinase and introduced into the cells together with the enzyme. Site-specific genomic integration of the transgene can thus be achieved at a high efficiency. Further details about this approach can be found in Chapter 9 in Nagy *et al.*, 2003.

The Chimera — A Tool for Fate Mapping

The possibility of combining two or more embryos in one chimera contributed tremendously to the understanding of lineage determination and cell fate in mouse development (for a review, see Nagy *et al.*, 2001). Identifiers described previously are essential to detecting and characterizing the contribution of a chimeric component to a lineage, cell type, or organ. Injection of cells isolated from the preimplantation embryo into a blastocyst and characterization of the allocation of derivatives of each compartment in later stages revealed the first specification events. Chimeras (Figure 46-3D) have also been very informative in addressing basic organization questions — for example, whether complex structures are derived from single progenitors (clonality in development).

The use of ES cells and tetraploid embryos has expanded the horizon of questions that could be addressed. ES cells are not able to contribute to the trophectoderm and primitive endoderm lineages of the developing chimeras. On the other hand, cells derived from tetraploid embryos are excluded from the primitive ectoderm lineage when they have to compete with diploid embryo or ES cell derivatives. For chimera production, one can choose any two of the following three sources as components: diploid embryo, tetraploid embryo, and ES cells. In addition, two diploid embryos can be chosen, increasing the number of possible combinations to four. Depending on the actual combination, different lineage allocation can be obtained for the selected components (see Nagy and Rossant, 2001) (see also Figure 46-4). The most extreme separation occurs when tetraploid embryos are combined with ES cells; practically no chimeric lineages are generated. The cells in the resulting embryo are either tetraploid derived (trophoblast lineage of the placenta, visceral, and parietal endoderm) or ES cell-derived (amnion, allantois, embryonic mesoderm component of the placenta and yolk sac, umbilical cord, and embryo proper) (Figure 46-3A). In a broader sense, the lineage restrictions in chimeras could be considered lineage marking.

Cell and Tissue Grafting

The classical chimera production restricts the components to the sources discussed in the previous paragraph and the timing to the preimplantation stages. Later stage cell mixing is also possible but demanding, and the possible applications are more limited. Small-tissue transplantation can be performed between two postimplantation stage embryos. If the graft is tagged with a unique identifier, the fate of these cells can be followed during development of the recipient. This technique has been in practice for almost 20 years. Developmental trajectories, as they are depicted in Figure 46-1, are dynamic processes not only at the level of gene expression but also that of cell allocations in the developing embryo. These two levels interact: allocation can induce gene expression changes, and gene expression changes can influence cell allocations. Knowing the complexity of mammalian gastrulation, one can easily imagine the heroic effort of creating a map of cell movements in the early postimplantation embryo using mostly LacZ-marked tissue grafting and still pictures showing the allocation of the graft derivatives.

Further limitations associated with this approach derive from our inability to successfully place the postimplantation-

A   B   C   D

Diploid<->Diploid   ES cell<->Diploid   Diploid<->Tetraploid   ES cell<->Tetraploid

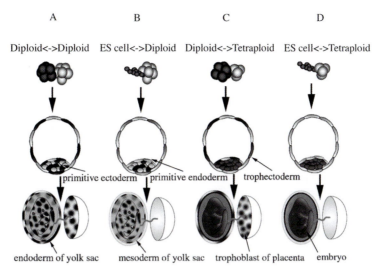

primitive ectoderm   primitive endoderm   trophectoderm

endoderm of yolk sac   mesoderm of yolk sac   trophoblast of placenta   embryo

Figure 46-4. Diagram of lineage contributions in different kinds of aggregation chimeras. Solid colors indicate nonchimeric tissues, and stripes indicate chimeric tissues.

stage embryo back into the uterine environment where it could develop further. *In vitro* postimplantation embryo culture sets a limitation on the "developmental stage window" of isolation from 6.5 dpc to 10.5 dpc and on survival for a period of only 2 to 3 days.

Using Genetic Switches

Grafting identifier-tagged tissues from one embryo to another creates a small population of progenitor cells whose derivatives can easily be recognized at later stages. The experimental achievement of such a scenario is rather invasive and is hampered by the limitations pointed out in the previous paragraph. State-of-art tools, however, can be used to achieve a similar situation with much less invasiveness. The key element is the Cre site-specific recombinase, which recombines DNA between two consensus loxP sites. If two loxP sites are similarly oriented in the same DNA strand, the result of such a recombination is the excision of the intervening sequence. This property allows the design of the Cre excision-conditional reporter *Rosa26R*, a knockin transgenic mouse line that uses the widespread expression from the Rosa26 locus, where now the LacZ expression is conditional to Cre excision.

A similar reporter line has been developed for the Flp recombinase system. The *Z/EG* transgene is a more sophisticated "switch" reporter with the following features: A strong overall expressor promoter is driving a *lacZ-neomycin* resistance fusion gene (βgeo) followed by three polyadenilation (pA) sites. The βgeo and the 3xpAs are flanked by loxP sites, and a coding region of an enhanced green fluorescent protein (EGFP) with its own pA follows this flanked region. This transgene was inserted as a single copy into a single site of the mouse genome leading to the expression of the βgeo only. However, EGFP expression can be activated by Cre recombinase; Cre activity removes βgeo+3xpA and moves the EGFP coding region under the transcriptional control of the pro-

moter (Figure 46-3E-H). The EGFP activation is, therefore, dependent on the presence and action of the Cre recombinase. Z/AP, another reporter line, can be used in the same way as Z/EG, except that the second reporter is hPLAP instead of EGFP. More recently, a similar system (Z/RED) has been established for red fluorescent protein (RFP or more specifically dsRed-T3) expression activation as well.

Cre recombinase has had a great career in mouse genetics (see Nagy, 2000). Dozens of transgenic mouse lines have been produced, expressing this phage enzyme with different spatial and temporal control. (http://www.mshri.on.ca/nagy/Cre-pub.html). Double transgenic combination of a Cre transgene and the Z/EG (or Z/AP, Z/RED) reporter turns on EGFP (hPLAP, RFP) expression not only in the actual Cre expressor cells but also in any cell that had an ancestor expressing the recombinase during the course of development. Therefore, the EGFP+ cells in these double transgenic embryos or animals are the union of all the cells expressing the recombinase and those that had a Cre recombinase expressor ancestor (Figure 46-3E through 3H). It is easy to recognize that an efficient Cre recombinase excision is not always advantageous in this situation, since the resulting set of EGFP+ (hPLAP+, RFP) cells could be too large, obscuring the recognition of fates derived from a single Cre+ progenitor. A less efficient Cre recombinase is preferable here. If the excision (activation of the second reporter) only occurs sporadically, the allocation of the descendent cells may still reflect the clonality (Figure 46-3C) and reveal the potential of the progenitor. Low-frequency activation of the Cre recombinase can be achieved by the use of either the tamoxifen- or the tetracycline-inducible Cre recombinase systems. Proper titering of the inducer can create informative excision–activation frequency in progenitors. In addition, withdrawal of the inducer may be used to create an upper developmental limit of the excision–activation of the reporter.

The induction of low-frequency or controlled activation of the conditional reporter is also possible by direct injection of the recombinase or its mRNA into cells of developing embryos.

Future Directions

It is difficult to predict what the distant future will bring into this exciting area. In the near future, however, efforts will certainly lead to novel features that will provide the means to follow the behavior of cells. New imaging techniques will allow the recording of the dynamic property of cells, such as movements, speed, direction, interactions with other cells in *in vitro* cultures, or formation of specific embryonic structures in *ex vivo* cultures of early postimplantation embryos. The study of later development may require slice cultures for recording cellular events, such as the birth of neurons from radial glia cells. Another technical development, ultrasound-guided embryonic transplantation, is expected to have an effect on lineage studies and fate mapping. It is becoming possible to graft reporter-tagged cells into early postimplantation-stage embryos while *in utero* and to recover the embryos at a later stage for studying the derivatives of the graft.

Remembering the enormous effort behind generating the complete map of lineage development in *C. elegans* — which consists of less than 1000 cells — the task of untangling the same system in the mammalian embryo may seem impossible. This conclusion comes not only from the dramatic increase in cell number but also from the increased plasticity of higher organisms. Plasticity creates a stochastic component in differentiation and cell determination processes. Therefore, the lineage studies and fate mapping have to take this uncertainty component into account.

Nevertheless, the increasing understanding of lineage determination and differentiation will place us in a good position to control these processes both *in vivo* and *in vitro*. ES cells have been playing an important role in this process. The acquired knowledge, in return, promotes the development of technologies aiming controlled *in vitro* differentiation of ES cells into therapeutically useful cell types.

ACKNOWLEDGMENTS

I am grateful to Patrick Tam for useful discussion, to Kristina Vintersten for very valuable comments and help in finalizing this chapter, and to Jody Haigh, Sue Quaggin, Bill Stanford Corrine Lobe, and Kristina Vintersten for providing images.

KEY WORDS

Cell fate Future possible course of differentiation of a given cell.

Cell lineage Set of courses of differentiation that leads to a given cell type.

Differentiation Changes in biological, physical, and/or chemical cell properties associated with the development of final cell types of an organism.

Proliferation Cell division without a change in cell properties.

Pluripotent Capability of a cell to develop into many but not all the cell types of an organism.

Totipotency Capability of a cell to develop all cell types of an organism.

FURTHER READING

Branda, C. S., and Dymecki, S. M. (2004, January). Talking about a revolution: the impact of site-specific recombinases on genetic analyses in mice. *Dev. Cell.* **6**(1), 7–28.

Hadjantonakis, A. K., Dickinson, M. E., Fraser, S. E., and Papaioannou, V. E. (2003, August). Technicolour transgenics: imaging tools for functional genomics in the mouse. *Nat. Rev. Genet.* **4**(8), 613–625.

Kinder, S. J., Tan, S. S., and Tam, P. P. (2002). Cell grafting and fate mapping of the early-somite-stage mouse embryo. *Methods Mol. Biol.* **135**:425–437.

Lewandoski, M. (2001). Conditional control of gene expression in the mouse. *Nat. Rev. Genet.* **2**, 743–755.

Nagy, A. (2000). Cre recombinase: the universal reagent for genome tailoring. *Genesis* **26**, 99–109.

Nagy, A., Gertsenstein, M., Vintersten, K., and Behringer, R. (2003). *Manipulating the mouse embryo, a laboratory manual*, 3rd ed. Cold Spring Harbor Press, Cold Spring Harbor, NY.

Nagy, A., and Rossant, J. (2001). Chimaeras and mosaics for dissecting complex mutant phenotypes. *Int. J. Dev. Biol.* **45**, 577–582.

Selleck, M. A., and Stern, C. D. (1991). Fate mapping and cell lineage analysis of Hensen's node in the chick embryo. *Development* **112**, 615–626.

Stanford, W. L., Cohn, J. B., and Cordes, S. P. (2001). Gene-trap mutagenesis: past, present and beyond. *Nat. Rev. Genet.* **2**, 756–768.

Tam, P. P., and Rossant, J. (2003, December). Mouse embryonic chimeras: tools for studying mammalian development. *Development* **130**(25), 6155–6163.

Teillet, M. A., Ziller, C., and Le Douarin, N. M. (1999). Quail-chick chimeras. *Methods Mol. Biol.* **97**, 305–318.

Genomic Reprogramming

M. A. Surani

Introduction

Most cells contain the same set of genes, and yet they are extremely diverse in appearance. They are all derived from a totipotent zygote generated after fertilization of an oocyte. William Harvey in 1651 was the first to propose that "everything comes from an egg," which also encapsulated the concept of *epigenesis*, or the gradual emergence of the embryo and fetus from an egg. Nearly 300 years later, conrad Waddington elaborated on epigenesis in his famous sketch depicting the "epigenetic landscape." Developmental biologists, beginning with Jacques Loeb and Hans Spemann, also began to consider whether all nuclei could be reprogrammed and become totipotent in nuclear transfer experiments. From many subsequent experiments it became clear that the maternally inherited factors contained within the oocyte also have the extraordinary property to restore totipotency to a differentiated somatic nucleus when transplanted into it. The components within the oocyte must have the property to alter the somatic nucleus so that it can recapitulate the entire developmental program and thus give rise to an exact genetic copy or clone of the individual who donated the transplanted nucleus. This transformation of differentiated cell to a totipotent state is probably the most widely understood meaning of genomic reprogramming. However, it is important to note that extensive epigenetic reprogramming of the genome also occurs in the germ line and during early development, which is essential for generating the totipotent zygote and for creating the pluripotent epiblast cells from which both germ cells and somatic cells are subsequently derived.

Specification of diverse cell types from pluripotent cells is determined by the expression of a precise set of genes while the rest are repressed. These newly acquired cell fates are propagated by heritable epigenetic mechanisms through modifications of chromatin and by DNA methylation. These epigenetic modifications, though they are heritable, are also reversible and can be erased, which is why it is possible to change the phenotypic characteristics of cells and restore totipotency to somatic nuclei under specific conditions (see Figure 47-1). To understand the mechanisms of reprogramming, it is important to know the nature of chromatin modifications and the mechanisms that can reverse or erase the existing modifications, and also how new modifications are imposed.

Because these reprogramming factors normally play a significant role during early development, it is important to determine their role in this context, and how these factors act on somatic nuclei during restoration of totipotency or pluripotency.

Genomic Reprogramming in Germ Cells

Germ cells provide the enduring link between generations, and for this reason, this lineage exhibits many unique properties, including the extensive epigenetic reprogramming of the genome prior to gametogenesis. This reprogramming is crucial for generating viable and functional gametes, which in turn generate a totipotent zygote. Primordial germ cells (PGCs) are among the first cells to undergo specification from pluripotent epiblast cells, when the distinction between germ cells and soma is established. PGCs as precursors of sperm and oocyte are highly specialized cells and the only cells that can undergo meiosis. However, PGCs retain expression of some markers of pluripotency such as *Oct4*. It is also possible to derive pluripotent embryonic germ cells (EG) from PGCs. In this context, it is interesting to determine both how PGC specification occurs and how these cells undergo dedifferentiation to pluripotent EG cells, which may provide some insights into genomic reprogramming.

STEM CELL MODEL FOR THE SPECIFICATION OF GERM CELLS IN MAMMALS

There are two key mechanisms for the specification of germ cells. The first involves inheritance of preformed germ plasm, which is found in *Drosophila* and *C. elegans*. In mammals, germ cell specification occurs according to the stem cell model where germ cells are derived from pluripotent epiblast cells in response to signaling molecules from the extraembryonic ectoderm. BMP4 and BMP8b are among the key signaling molecules in conferring germ cell competence on pluripotent epiblast cells in mice starting at E6.5 (Figure 47-2), an event that is detected by the expression of *fragilis*, a transmembrane protein. These germ cell competent cells are initially destined for a mesoderm somatic cell fate as they show expression of *Brachyury* and some region-specific *Hox* genes as they migrate toward the posterior proximal region. However, at around E7.25, cells that ultimately acquire a germ cell fate switch off the somatic program through repression of a number of genes, which continue to be expressed in the neighboring somatic cells. Cells that acquire germ cell fate continue to show expression of markers of pluripotency, including *Oct4*. A unique marker of germ cells at this time is

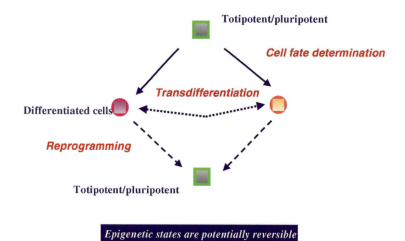

Figure 47-1. Genomic reprogramming involves heritable but reversible epigenetic modifications.

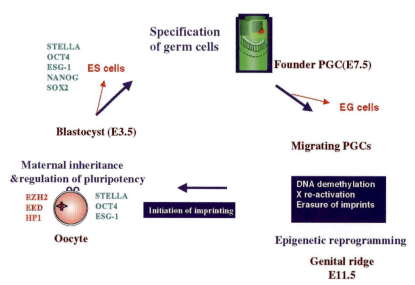

Figure 47-2. Genomic reprogramming in the oocyte, early embryos, and the germ line. The origins of pluripotent embryonic stem cells (ES) and embryonic germ cells (EG) are indicated.

stella, which is first detected in the 45–50 founder germ cells at E7.5. Thus, diversification between germ cell and somatic cell fate among neighboring cells occurs over approximately 6 to 10 hours between E7.25–E7.5, with the repression of somatic cell program being one of the major events during germ cell specification.

From Stem Cells to Germ Cells

Recent studies show that it is also possible to derive PGCs and gametes from pluripotent stem cells. In one study, ES cells with *gcOct4-GFP* reporter that drives expression specifically in germ cells were generated. These ES cells with the reporter were allowed to undergo differentiation when cells with GFP were detected. These cells showed expression of a variety of germ cell-specific markers. Further cultures following cell

sorting eventually produced oocyte-like cells that underwent development to form blastocyst-like structures. The latter show that mouse ES cells can under these conditions differentiate into oocyte and, subsequently, blastocysts. With further detailed characterization of germ cells and gametes, such an *in vitro* system may be useful for investigations concerning specific aspects of germ cell development.

Another study has similarly been carried out to generate spermatogenic cells from pluripotent ES cells. In this case, the endogenous mouse homolog of *Vasa*, *Mvh* was used to knock in the reporters, *LacZ* and *GFP*. In this study, germ cells were generated in embryoid bodies, which were detected through the expression of MVH-GFP. This process was greatly enhanced by the exposure of ES cells to BMP4. These MVH-GFP cells were aggregated with E12.5–13.5 male gonadal cells when germ cells within these aggregates developed into elongated spermatids.

These studies shows that it may be possible to generate an efficient *in vitro* system to derive germ cells from pluripotent ES cells. Such a system would be useful to study the mechanism of PGC specification *in vitro*, as well as other aspects, including the formation of gametes, and aspects of epigenetic reprogramming of the genome. Derivation of germ cells from human ES cells would be particularly useful for studies on this lineage. Furthermore, derivation of human oocytes from ES cells would greatly add to this scarce resource, thus providing opportunities for fundamental studies on somatic cell reprogramming and for the subsequent derivation of stem cell lines from human somatic cells for investigations of specific mutations and diseases.

From Germ Cells to Stem Cells

Embryonal carcinoma cells (EC) that are derived from PGCs *in vivo* were the first pluripotent stem cells to be identified, and several loci have been identified that have a critical role in this process. At the same time, derivation of pluripotent cells from PGCs has also been achieved *in vitro*. This conversion of germ cells into EG cells occurs in the presence of leukemia inhibitory factor (LIF), basic fibroblast growth factor (FGF2), and the Kit ligand (KL). The precise mechanism for conversion of the highly specialized germ cells into pluripotent stem cells is largely unknown. Further investigations would provide insights into de-differentiation of cells and on the mechanism of genomic reprogramming.

EPIGENETIC REPROGRAMMING IN GERM CELLS

One of the properties of germ cells of particular interest is the epigenetic reprogramming of the genome. This event occurs when PGCs enter into the developing gonads (Figure 47-2), when there is extensive erasure of epigenetic modifications including erasure of genomic imprints and reactivation of the inactive X-chromosome. New parental imprints are initiated later during gametogenesis, particularly oogenesis, and these modifications that are heritable after fertilization dictate parent of origin dependent gene expression.

As germ cells proliferate after the formation of the founder population at E7.5, they start to migrate to the developing gonads. At this stage, germ cells as well as somatic cells contain epigenetic marks associated with imprinted genes. During their migration, female germ cells also show inactivation of one X-chromosome. Upon the entry of germ cells into developing gonads at E10.5–E11.5, a major epigenetic reprogramming event occurs, which includes reactivation of the inactive X-chromosome and the erasure of epigenetic marks associated with imprinted genes. Indeed, there appears to be genomewide DNA demethylation of the genome at this time. This genomic reprogramming event occurs relatively rapidly, and it is completed by E12.5.

The mechanism involved in the erasure of epigenetic modifications in the germ line may provide insights on the erasure of epigenetic modifications from somatic nuclei that occurs after transplantation into oocytes, which restores totipotency.

If this is the case, similar factors may be transcribed during oocyte maturation at the germinal vesicle stage and translated, to be stored in oocytes as maternally inherited factors. The onset of genomic reprogramming in PGCs in the gonad may be triggered by a signal from somatic cells, or in response to a developmental timer such as the number of cell divisions in PGCs since the establishment of the founder population of PGCs. Because the gonads at E11.5 are bipotential, it is possible that the signal from somatic cells, if it exists, should be the same in both male and female embryos. In this case, it would be of interest to discover the nature of the signal and determine how such an external cue can trigger extensive genomic reprogramming. However, there is also support for the alternative developmental timer model since EG cells show erasure of imprints even when they are derived from PGCs where this process has not yet commenced. It is possible that erasure is initiated when germ cells complete a critical number of cell cycles, although as these cells are cultured in a complex medium, a role for an environmental cue cannot be entirely discounted. Whatever the case may be, it is important to note that EG cells themselves have the property to induce erasure of epigenetic modifications from somatic nuclei (see below).

Genomic Reprogramming in Oocytes.

Resumption of oocyte growth is accompanied by further epigenetic reprogramming events, particularly the initiation of genomic imprints. The majority of the epigenetic marks associated with imprinting are introduced during oocyte growth, although some genes acquire paternal-specific imprints in the male germ line. These epigenetic marks are eventually detected as DNA methylation of specific *cis*-control elements. Some of the marks, for example, in the *Igf2r* locus, ensure that the gene will be active only when maternally inherited while others such as *Peg3* will be silent in the female genome. *Dnmt3l* is a key gene involved in the initiation of parental imprints, which acts together with the *de novo* DNA methylase enzyme, Dnmt3a. A mutation in the *Dnmt3l* gene does not disrupt development or maturation of the oocyte except that these oocytes do not carry appropriate maternal imprints or epigenetic marks. Following their fertilization, the resulting embryos are unable to develop normally, and they die shortly after implantation. Other genes such as *H19* undergo DNA methylation in the paternal germ line, and this gene is repressed in the paternal genome. The topic is discussed comprehensively elsewhere.

MATERNAL INHERITANCE AND REPROGRAMMING OF PARENTAL GENOMES

As in other organisms, mouse oocytes contain a number of maternally inherited proteins and messages (Figure 47-2). In mammals, maternally inherited factors are essential for totipotency and pluripotency such as Oct4, Esg1, and Stella, although there is no Nanog. Maternal inheritance of Stella is apparently necessary for normal preimplantation development. The oocytes also contain epigenetic modifiers, including the *Polycomb* group proteins, Ezh2 and eed, as well as the heterochromatin factor, HP1. These factors are essential for regulating early development and for generating the pluripotent epiblast and trophectoderm cells of blastocysts. The oocyte is also likely to inherit some key chromatin remodeling factors.

In mammals, the parental genomes exhibit epigenetic asymmetry in the zygote as a result of imprinting, which confers functional differences between parental genomes. At fertilization, the maternal genome apparently has high methylated lysine 9 histone H3 (H3meK9). Immediately after fertilization, the hetrochromatin protein HP1β binds preferentially to the maternal genome. The Polycomb proteins, Ezh2 and eed, also bind preferentially to the maternal genome. While this takes place, the paternal genome that has relatively low levels of H3meK9 shows genomewide DNA demethyaltion, thus enhancing the epigenetic differences between the parental genomes. Ezh2 has the conserved suvar E2 trithorax (SET) domain with histone methylase activity for methylation of histone H3-lysine 27/lysine 9 (H3meK27/9). The maternal inheritance of Ezh2 per se is apparently important since depletion of this factor from oocytes results in development of very small neonates presumably because of an effect on placental development. This seems likely because the neonates eventually grow and acquire normal size, indicating a placental functional deficiency during development. Whether this is due to an effect on imprinted genes remains to be determined. These experiments show that factors present within the oocyte have the potential to exert a variety of epigenetic effects on development. Somatic nuclei transplanted into oocytes would be affected by the activities of these factors during reprogramming, but the variable expression of genes associated with totipotency and of imprinted genes argues that appropriate epigenetic reprogramming of the genome may not be accomplished in every case (see below).

Factors involved in chromatin remodeling are also likely to be important for early development and genomic reprogramming as they regulate accessibility to DNA (Tada and Tada, 2001). The SWI/SNF-like complexes consist of at least two ATPase subunits, BRG1 and BRM. *Brg1* is important during preimplantation development as loss of function is lethal during preimplantation development. It is also known that mutation in *ATRX*, a member of the SNF2 helicase/ATPase family has an effect on DNA methylation of highly repeated sequences. Mutation in *Lsh* similarly results in substantial demethylation of the genome. Lsh is related to the SNF2 subfamily; most members of the SNF2 family of proteins appear to have the capacity to alter chromatin structure. The activity of the nucleosome-dependent ATPase, ISWI, may be used in chromatin remodeling in nuclear reprogramming of somatic nuclei; if so, it is likely to be present in the oocyte and would have a role in the zygote. One of the earliest changes observed following fertilization (or indeed after transplantation of the somatic nucleus) is the apparent increase in the size of the nucleus. This morphological change may be in response to chromatin remodeling factors belonging to the ISWI complexes. This activity may be necessary for the initial unwinding of the chromosomes to facilitate epigenetic modifications of the chromatin.

REPROGRAMMING DURING EARLY DEVELOPMENT

Epigenetic reprogramming of the embryonic genome continues throughout preimplantation development as judged by the continual changes in histone modifications and a decline in the genomic levels of DNA methylation. During preimplantation development, both pluripotent epiblast and differentiated trophectoderm cells are formed. There are differences in the epigenetic reprogramming in these two tissues. For example, there is preferential paternal X inactivation in the trophectoderm, a process in which the *Polycomb* group proteins, Ezh2/eed complex, has a significant role. The cells of the late morula that are positive for the expression of *Nanog* and destined to form the inner cell mass cease to show Ezh2/eed accumulation at Xi, as seen with the paternal X-chromosome in the trophectoderm. Ezh2 is also detected in the inner cell mass, which may account for the presence of the overall H3meK27 staining of epiblast cells, which are positive for Oct4 expression.

It appears that histone modifications such as H3meK27 may have a role in the maintenance of epigenetic plasticity of the pluripotent epiblast cells since the loss of function of Ezh2 is early embryonic lethal, and it is not possible to derive pluripotent ES cells from blastocyst that are null for Ezh2. These experiments show the importance of appropriate epigenetic reprogramming of the genome for early development and for generating pluripotent epiblast cells that are the precursors of both somatic and germ cell lineages.

As we learn more about nuclear reprogramming events that occur normally in germ cells, oocytes, and early development, these studies are likely to be used to identify key candidates for genomic reprogramming.

Reprogramming Somatic Nuclei

NUCLEAR TRANSPLANTATION

Epigenetic reprogramming of somatic nuclei transplanted into oocytes must require erasure and initiation of appropriate epigenetic modifications compatible for development. This subject has already been reviewed extensively elsewhere. At least some of the key reprogramming events may be faulty to account for the very low success rate since somatic nuclei undergo variable reprogramming resulting in a wide variety of phenotypes. The effects of aberrant reprogramming are apparent, particularly soon after implantation and during postimplantation development. Both the embryo and extraembryonic tissue seem to be affected. Some epigenetic marks associated with imprinted genes are erased, resulting in aberrant expression of these genes. It seems likely therefore that a large number of genes fail to show appropriate temporal and spatial patterns of expression. Further studies on the mechanisms of genomic reprogramming during normal development and following nuclear transplantation are necessary to assess the reasons for faulty reprogramming of somatic nuclei.

REPROGRAMMING IN ES/EG-SOMATIC CELL HYBRIDS

Somatic nuclear reprogramming has also been demonstrated in hybrid cells between pluripotent ES/EG and somatic cells, which also restores pluripotency in somatic nuclei. These studies indicate that not only the oocytes but also pluripotent ES/EG cells must contain appropriate factors to reprogram the somatic nucleus. Reprogramming of somatic nuclei in ES/G-somatic cell hybrids is, however, relatively less complex compared to its transplantation into oocytes. This is because the somatic nucleus in the oocyte has to be reprogrammed to recapitulate the entire program of early development to the blastocyst stage. It is important to note that this donor somatic nucleus has to be reprogramd to generate both pluripotent epiblast cells as well as the highly differentiated trophectoderm cells. The latter should be viewed as a trans-differentiation event because somatic nuclei of diverse origin must direct differentiation of highly specialized trophectoderm cells after only a few cleavage divisions. Indeed, in some respects, this transdifferentiation event is more striking as a reprogramming event. By comparison, reprogramming of somatic nuclei in ES/EG–somatic cell hybrids is less complex as there is restoration of pluripotency without the necessity to recapitulate early events of development.

Although EG and ES cells on the whole have similar effects on somatic nuclei, there is at least one critical difference between them. Using the EG–thymocyte hybrid cells, it was shown that the somatic nucleus underwent extensive reprogramming, resulting in the erasure of DNA methylation associated with imprinted genes, and the inactive X-chromosome was reactivated. The somatic nucleus also acquired pluripotency as judged by the activation of the *Oct4* gene, and the hybrid cells could differentiate into all three germ layers in chimeras. This study shows that EG cells, apart from conferring pluripotency to somatic nucleus, retained a key property found only in germ cells, which is the ability to erase parental imprints, and indeed, induce genomewide DNA demethylation. Experiments using ES–thymocyte hybrid cells gave similar results, including the restoration of pluripotency to somatic nuclei as shown by the activation of the *Oct4–GFP* reporter gene and for the ability of these cells to differentiate into a variety of cell types. Unlike EG cells, however, ES cells do not cause erasure of imprints from somatic nuclei. Furthermore, in ES–EG hybrids, EG cells can induce erasure of imprints from ES cells, which shows that EG cells have dominant activity for the erasure of imprints and DNA demethylation. However, from these studies, it is clear that DNA demethylation activity at least for the erasure of imprints present in EG cells is not essential for restoring pluripotency to somatic cells. It is possible to use this system to design cell-based assays in search of key reprogramming factors.

The ability of ES/EG cells to restore pluripotency in somatic nuclei is significant because it also opens up possibilities to identify the molecules involved in reprogramming somatic nuclei. Such studies are difficult with mammalian oocytes partly because they are small compared to amphibian oocytes, and it is difficult to collect large numbers of

them. More importantly, as discussed earlier, oocytes are complex cells containing factors essential for pluripotency as well as for the early development and differentiation of trophectoderm cells. By contrast, pluripotent ES/EG cells are relatively less complex, and, more importantly, they can be grown indefinitely *in vitro*, Thus, they can provide a considerable source of material for analysis. For example, it is possible to use nuclear extracts from ES/EG cells to examine reprogramming of somatic nuclei as described in one experimental approach. The availability of relatively large amounts of nuclear extracts from ES/EG cells also makes it possible to undertake biochemical studies to identify the key reprogramming factors.

Conclusions

The evidence from studies on early mammalian development shows that there is dynamic and extensive reprogramming of the genome in the oocyte, zygote, and germ cells. Pluripotent stem cells also appear to show considerable potential for genomic reprogramming, and while there are differences between ES and EG cells, both of them can restore pluripotency to somatic nuclei. There is maternal inheritance of factors for pluripotency, epigenetic modifications, and chromatin remodeling in the oocyte. Reprogramming in the oocyte is relatively complex since the parental pronuclei exhibit epigenetic asymmetry in the zygote. The paternal genome becomes rapidly demethylated after fertilization, but the maternal genome does not, at least in part, because of the differential histone modifications such as the preferential H3meK9 of the maternal genome but not of the paternal genome and the preferential binding of HP1β and Ezh2/eed proteins to it. It is possible that epigenetic reprogramming of somatic nuclei may be affected by their original epigenetic state.

The phenomenon of reprogramming somatic nuclei is now well established in mammals both by nuclear transplantation studies and in heterokaryons, but the mechanisms and the key molecules involved are yet unknown. It is reasonable to assume that some of the factors involved in reprogramming the genome in the germ line are also present in the oocyte. It is also possible that some of the basic reprogramming factors present in pluripotent stem cells are also present in the oocyte.

A likely sequence of events for converting a somatic nucleus to a pluripotent nucleus may first require chromatin-remodeling activity. Many of these complexes are known to exist in mammals, but at this stage it is not known precisely which are important for reprogramming of somatic nuclei. This may be followed by changes in histone modifications compatible with pluripotency. What these changes are has yet to be fully determined, together with the identity of the histone modifiers.

As ES/EG cells apparently have the capacity for reprogramming somatic nuclei to pluripotency, they may be used for identifying key molecules necessary for genomic reprogramming through cell-based assays combined with appropriate biochemical and cellular analyses.

KEY WORDS

Chromatin DNA and associated proteins that create appropriate conditions for the expression or repression of genes during development and differentiation of cells.

Epigenesis The formation of entirely new structures during development of the embryo.

Epigenetic The study of mitotically and/or meiotically heritable changes in gene function that cannot be explained by changes in DNA sequences.

Genomic reprogramming The process that alters functions and properties of a cell through reversible modifications of DNA or associated proteins, without affecting the DNA sequence itself.

FURTHER READING

Collas, P. (2003). Nuclear reprogramming in cell-free extracts. *Philos. Trans. R. Soc. Lond. B. Biol. Sci.* **358**, 1389–1395.

Donovan, P. J., and de Miguel, M. P. (2003). Turning germ cells into stem cells. *Curr. Opin. Genet. Dev.* **13**, 463–471.

Li, E. (2002). Chromatin modification and epigenetic reprogramming in mammalian development. *Nat. Rev. Genet.* **3**, 662–673.

McLaren, A. (2003). Primordial germ cells in the mouse. *Dev. Biol.* **262**, 1–15.

Reik, W., and Walter, J. (2001). Genomic imprinting: parental influence on the genome. *Nat. Rev. Genet.* **2**, 21–32.

Saitou, M., Payer, B., Lange, U. C., Erhardt, S., Barton, S. C., and Surani, M. A. (2003). Specification of germ cell fate in mice. *Philos. Trans. R. Soc. Lond. B. Biol. Sci.* **358**, 1363–1370.

Surani, M. A. (2001). Reprogramming of genome function through epigenetic inheritance. *Nature* **414**, 122–128.

Surani, M. A. (2004). Stem cells: how to make eggs and sperm. *Nature* **427**, 106–107.

Tada, T., and Tada, M. (2001). Toti-/pluripotential stem cells and epigenetic modifications. *Cell Struct. Funct.* **26**, 149–160.

Identification and Maintenance of Cell Lineage Progenitors Derived from Human ES Cells

Susan Hawes and Martin F. Pera

Introduction

Human embryonic stem (ES) cells, derived from the inner cell mass (ICM) of human blastocysts are karyotypically normal and can be maintained in culture while sustaining their ability to differentiate into somatic cell types indicative of all three embryonal germ layers. Thus, human ES cells may be a renewable source of a wide range of embryonic or adult cells for the study of early human development, screening of pharmaceuticals, or cellular replacement of damaged and diseased tissue. Exploitation of human ES cells for research and therapy requires understanding of molecular signals that induce human embryonic cell fates, knowledge of cell-specific markers to identify progenitor cells, and development of methods to maintain these early cells in culture.

An ES cell might differentiate directly into a mature end cell with little proliferative potential. Alternatively, ES cells might give rise to cells committed to a restricted fate with some capacity for self-renewal. These progenitors might resemble either lineage-restricted cells in the embryo or stem cells in adult tissues, but in either case they represent intermediates between the ES cell and a terminally differentiated mature cell.

Differentiation of ES cells toward progenitor or somatic-cell types is assessed by morphology, by marker expression, or by their developmental potential following transplantation into animal models, but factors that control lineage-specific differentiation are often unknown. Identification of markers for stem or progenitor cells from differentiating human ES cells is in its infancy. This is in part due to the limitations in the current methodology for maintenance of human ES cells *in vitro* but also to our lack of understanding of these early cell populations within the developing mouse embryo and even more so during human embryogenesis. Thus, although lineage-specific markers, known to be involved in mouse embryo development or representative of particular cell lineages, have been defined, whether similar genes or markers identify similar cell populations during human ES cell differentiation remains to be seen. Moreover, whether a human ES cell-derived cell expressing characteristic cell-specific markers *in vitro* behaves functionally as its *in vivo* counter-part must be determined. Expansion and long-term cultivation of progenitor cells may result in aberrant expression of cell lineage-specific markers, making repeated testing of the multidevelopmental potential of progenitor cells essential.

Another limitation of our current methodologies is that the signals required to maintain and to expand progenitor cells in culture are not known. Limited expansion of hematopoietic stem cells (HSCs), one of the most studied adult stem cell populations, has been achieved using a number of cytokines. Long-term proliferation and maintenance of HSCs has not been reported, although recent studies by Bhardwaj *et al.* and Austin *et al.* have shown that sonic hedgehog and Wnt gene family members, respectively, have been implicated as factors that may achieve this goal. Two studies by Reya *et al.* and Willert *et al.* show that purified Wnt-3a may enable long-term expansion of HSCs *in vitro*. Thus, understanding of maintenance and expansion of stem, progenitor, and intermediate cell types from ES cells is limited.

Transplantation of neural precursors, isolated from differentiating human ES cells, into the brains of newborn mice verified their potential to integrate and differentiate into adult neural cell types. Such transplantation studies using normal or diseased animals will be important to assess the functional capability of human ES-derived cells.

Characteristics of Human Embryonic Stem Cells

Similar to ICM cells from which they are derived, human ES cells are pluripotent and are able, theoretically, to develop into all adult cell types. Following transplantation into immuno-compromised mice, human ES cells form teratomas containing differentiated cell types such as cartilage, glandular and squamous epithelia, as well as neural and muscle cells. Human ES cells seem to have the ability to self-renew. To date, the HES2 cell line, derived in 1998 by Reubinoff *et al.*, has been maintained in culture for up to 168 passages. Whether long-term cultivation of human ES cells under present culture conditions may result in subtle genetic or epigenetic changes affecting their underlying characteristics remains to be seen.

As the blastocyst implants and a proportion of ICM cells form the epiblast, pluripotent cells respond to extracellular signals and differentiate into committed cell types, losing their ability for self-renewal. Human ICM cells removed before exposure to differentiative signals can be maintained and expanded in culture as ES cells. However, it is not known what molecular signals mediate their self-renewal. Unlike mouse ES cells, human ES cells are not apparently responsive to leukemia inhibitory factor (LIF). Human ES cell maintenance requires co-culture with mouse embryonic fibroblast (MEF) cells. Routine passage of human ES cell colonies involves discarding cells with morphologic changes indicative of early differentiation or excluding larger differentiated cellular clumps by filtration following dissociation. Using these methods, we can avoid progressive differentiation of the culture, as assessed by morphology (Figure 48-1A and B) and immunofluorescent staining using antibody markers specific for undifferentiated embryonic cells. When human ES cell colonies are grown on a low density of MEF cells without serum, addition of basic fibroblast growth factor (bFGF) enables maintenance. However, if human ES cells are not passaged as they reach confluency, they spontaneously differentiate *in vitro*, with or without the presence of bFGF, generating many morphologically distinct cell types (Figure 48-1C). Unlike the coordinated establishment of germ cell lineages during embryogenesis, differentiation of ES cells in culture is apparently disorganized, resulting in colonies made up of progenitor cells of various cell types.

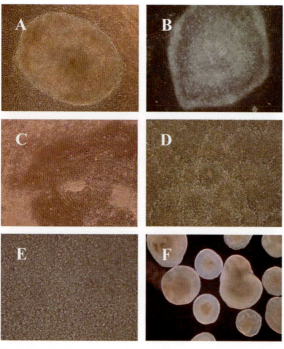

Figure 48-1. Human embryonic stem (ES) cell colonies of approximately 50,000 cells grown on mouse embryonic fibroblast (MEF) cells (A) and imaged by phase (B). Areas of morphologically distinct cells types within spontaneously differentiating human ES cells, 2-weeks following plating (C). Rosette-like cell clusters within 2-week spontaneously differentiating human ES cells (D). Human ES cells following 10 days of treatment with soluble noggin (E). Neurospheres formed following transfer of human ES cell-derived rosette cells into neural stem cell culture conditions (F).

DIFFERENTIATION OF HUMAN EMBRYONIC STEM CELLS

Strategies to promote differentiation of human ES cells into particular cell types have been adapted from studies undertaken using mouse ES cells. In many studies, mouse ES cells are transferred as small clumps into a suspension culture where they develop into aggregates called embryoid bodies (EBs). EB formation is assumed to recapitulate the three-dimensional complexity of the embryo, resulting in formation of progenitor cells representative of three embryonal germ layers. Following disaggregation of EBs, their constituent cells can be further cultivated to yield a variety of differentiated somatic cell types. The second method allows mouse ES cells to differentiate *in vitro* as an adherent cell layer without EB formation and has been utilized for derivation of neural and hematopoietic progenitors. Both prior aggregation of human ES cells into EBs and spontaneous differentiation of human ES cells as an adherent cell layer result in cultures of mixed populations of cells as evidenced by morphology and the expression of lineage-specific markers. This is assessed by studying gene (reverse transcriptase polymerase chain reaction [RT-PCR]) and protein expression (immunostaining).

To study the biology of human ES cell derivatives and to generate cells for cellular based transplantation therapies, pure populations of progenitor or somatic cell types need to be derived. Different strategies can be used to achieve this objective. First, the addition of soluble inducing factors to human ES cells may direct their differentiation and/or facilitate growth of particular lineages. Second, using markers expressed by lineage-committed progenitors during embryogenesis, isolation or depletion of particular cells from mixed cell populations can be achieved with antibodies raised against cell-surface markers and subsequent fluorescence-activated cell sorting (FACS) or immunomagnetic bead separation. Third, through genetic selection using selectable markers under the control of lineage-specific promoters, specific cell types can be isolated, generating pure cell populations. Thus, for example, using homologous recombination, Li and colleagues (1998) genetically modified mouse ES cells to express antibiotic resistance under the control of the promoter for Sox2. Subsequent culture of the mouse ES cells in the appropriate antibiotic resulted in higher proportions of Sox2-positive cells with a neural identity.

Presence of Early Precursors within Human Embryonic Stem Cell Colonies

Heterogeneity within human ES cell colonies may indicate early progenitor cells with potential to develop into differentiated cell types. Because present conditions to maintain human ES cells are suboptimal for stem cell maintenance, cells will differentiate under basal culture conditions. That human ES cell colonies may contain trophoblast cells is suggested by their secretion of human chorionic gonadotrophin (hCG) into the medium and enhancement of trophoblast differentiation following treatment with BMP-4. A subpopulation of human ES cells isolated using the mouse ES cell surface marker SSEA1 expressed hCG, suggesting enrichment for trophoblast cells. Subsequent analysis confirmed that in contrast to the mouse embryo, SSEA1 is localized to trophectoderm but not ICM cells of human blastocysts.

Whether all or only subpopulations of the cells within human ES cell colonies contribute to teratoma formation following transplantation into immunocompromised mice is uncertain. Clearly, clonal derivation of human ES cell colonies will be crucial for truly ascertaining a human ES cell ability for pluripotency. Clonal derivation of a human ES cell line by dissociation into single cells resulted in the establishment of two cell lines, propagated for eight months and capable of teratoma formation *in vivo*. However, in general, survival of human ES cells following dissociation into single cells is poor, making derivation of clonal cell lines difficult.

Mouse ES cells, like their embryonic precursors, differentiate into primitive endoderm. Following implantation, mouse ICM cells differentiate into progenitors of the epiblast and of extraembryonic tissues. In the mouse embryo, establishment of the extraembryonic lineages including the visceral endoderm (VE) is critical for subsequent specification of the three germ layers because the extraembryonic tissues produce spatially and temporally regulated signals that direct commitment of epiblast cells into specific fates. Thus, primitive endoderm cells within human ES cell colonies may contribute to early differentiation events by producing signals that direct the formation of early progenitor cell types.

Published studies of human ES cell phenotype have observed not only expression of genes known to be expressed by pluripotent cells but also markers most commonly associated with specification of early cell lineages (Houssami and Pera, personal communication). One study reported that human ES cells were immunoreactive to neuronal marker βIII-tubulin. Other workers have reported that human ES cells express hematopoietic markers, AC133, vascular endothelial growth factor receptor-1 (VEGFR-2, Flk-1), AC133, and c-kit, as well as CD34 at apparently low frequency. However, there are a number of explanations for the observed expression of differentiation lineage-specific markers in ES cell cultures. First, some markers of committed progenitors are also found on pluripotent stem cells. AC133 and c-kit are also expressed by ES cells and germ cells, respectively, and c-kit has recently been used as a marker to isolate putative oocyte-like cells from mouse ES cells. Alternatively, expression of these markers at a low level may reflect promiscuity in transcription of differentiation markers in ES cells or low-frequency spontaneous differentiation into particular lineages.

LINEAGE-SPECIFIC PROGENITOR CELLS DERIVED FROM HUMAN EMBRYONIC STEM CELLS

Blood

The most extensively characterized adult stem cell population with respect to knowledge of multipotentiality, differentiation stage-specific marker expression, and factors that govern growth and differentiation is the HSC. HSCs are rare cell populations residing within bone marrow or fetal liver and form all blood cell lineages. Two types of HSCs have been described, characterized by their long-term (lifetime) or short-term (8–10 weeks) differentiative and/or reconstitutive abilities and are identified by antibodies recognizing cell-surface proteins. Two studies have developed methods to derive human hematopoietic progenitor cells from human ES cells. Critical to both of these studies is the use of cell-surface markers previously shown to be expressed by subpopulations of HSCs. One study, undertaken by Kaufman and colleagues, derived early hematopoietic precursors that stained positively for the marker, CD34+ but not immunoreactive to CD45. CD45 is a leukocytic marker that, along with CD34, identifies definitive hematopoietic cells within 3- to 5-week human embryos. Using a CD34 antibody, these workers enriched for a hematopoietic precursor population.

The CD34 antigen is expressed by most human hematopoietic stem and progenitor cell populations. Following co-culture of human ES cells on mouse bone marrow or yolk sac stromal lines, hematopoietic stem and precursor cells were isolated by immunobeads conjugated with the CD34 antibody. This resulted in differentiation of human ES cell colonies into many morphologically distinct cell types after 3 to 5 days, 1–2% of which were immunoreactive for the CD34 antibody (CD34+) but not an antibody raised against CD38 (a marker for myeloid cells). Further analysis of the CD34+ cells revealed that 50% were immunoreactive to a CD31 antibody (CD31+), suggesting that they may be endothelial cell progenitors. Evidence that the isolated CD34+ were hematopoietic progenitors was obtained following transfer of cells into agar or methyl-cellulose containing cytokines. Colony-forming units (CFU) were generated from the CD34+ cells at an average of 270 colonies per 10^5 cells in comparison to 10 colonies derived from CD34− cells. The CD34+ cells developed into colonies containing mature erythroid (red blood), myeloid (white blood), and megakaryocyte (platelet progenitor) cells upon subsequent differentiation. Interestingly, RT-PCR analysis of the differentiated CFU colonies for erythroid gene expression showed adult globin expression.

In the second study, Chadwick *et al.* (2003) formed EBs from human ES cells before addition of growth factors and cytokines, known to enhance blood progenitor development. These authors used antibodies against CD34 and CD38, markers used previously to isolate human HSCs from umbilical cord blood, fetal liver, and fetal and adult bone marrow. Use of CD45 expression verified the presence of mature hematopoietic cells following 10 days of directed differentiation. Thus, by use of CD45 expression to monitor the effects of altering differentiation protocols, the authors showed that, by addition of BMP-4 to human ES-derived EB cultures, they could increase the proportion of cells immunoreactive to CD45 (CD45+). Addition of cytokines with BMP-4 increases the proportion of CD45+ cells that co-express CD34. Of the CD45+ cells derived from the differentiated human ES cells all (100%) formed CFU colonies, in comparison to none of the CD45− cells. These CFU colonies formed secondary CFU, following disaggregation and replating.

Identification and isolation of endothelial cells from differentiating human ES cells was achieved using the cell-surface marker platelet endothelial cell adhesion molecule (PECAM), or CD31. Four days following formation of EBs from human ES cells, CD31$^+$ cells appeared and increased in proportion by day 13. Subsequently, EBs, were dissociated, and cells FACS sorted. When these CD31$^+$ cells were cultured on Matrigel, cord-like structures formed. Other vascular endothelial markers were also expressed by EBs following 13 days in culture, including AC133 and vascular endothelial-cadherin (VE-cadherin). Verification of the identity of the CD31$^+$ cells was confirmed by transplantation of cells into subcutaneous tissue of immunodeficient mice. Immunoreactivity to human-specific antibodies against CD31 and CD34 confirmed that the human ES cell-derived CD31$^+$ cell population formed microvessels, some of which contained mouse blood cells. The authors did not use other human-specific markers to show whether these cells were committed within the endothelial cell lineage or are able to produce other cell types *in vivo*.

Neural Cells

MOUSE

There are several well-established protocols for deriving neural cells from mouse ES cells; most, but not all, of them involve EB formation. Protocols that first differentiate mouse ES cells into EBs allow for formation of progenitor cells of the three embryonic germ layers., Subsequently, selective enrichment for neural cells is achieved by removal of serum. Following this, addition of bFGF and/or epidermal growth factor (EGF) results in proliferation of neural stem and progenitor cells. Subsequent addition of FGF8, sonic hedgehog (SHH), and ascorbic acid results in derivation of midbrain neurons. Neural progenitors and their derivatives have also been generated from mouse ES cells without prior formation of EB.

HUMAN

A number of cell-specific markers have been used to identify neural stem and progenitor cell populations derived from human ES cells, including the intermediate filament protein nestin, musashi-1, and the neural-cell adhesion molecule (N-CAM). None of these markers is specific for neural stem or progenitor cells. Nestin, a marker characteristic for neural stem cells is localized to the neural tube and cerebral cortex of rat embryos, becoming restricted to a subpopulation of ependymal cells in the adult. However, nestin is likely present in all dividing neural populations, being temporally expressed by muscle and pancreas. Musashi-1, more indicative of neural progenitors rather than stem cells, has also been shown to be present in stem cells of nonneural cell lineages.

Putative neural stem or progenitor cells have been identified morphologically within differentiating human ES cell colonies by virtue of their formation into distinct rosette-like clusters, similar to structures within mouse neural tubes (Figure 48-1D). These cells appear within spontaneously differentiating human ES cell colonies, about 2 weeks after

plating or following culture of EBs with FGF2. These rosette cells stain positively for nestin, musashi, and N-CAM.

Further characterization of the human ES cell-derived rosette cells was undertaken by their propagation as neurospheres. Neural stem cells derived from fetal or adult brain are propagated following transfer into nonattachment conditions using defined medium with the addition of FGF and/or EGF. Most neurospheres in these cultures comprise a heterogeneous population of cells, containing few (<5%) with the ability for self-renewal.

Within the human ES cell-derived neurospheres, a high proportion of cells express neural markers nestin (97%), N-CAM (99%), and the neuronal and glial progenitor marker A2B5 (90%). The developmental potential of these neural stem and progenitor cells was revealed following their dissociation and attachment onto a adhesive substrate in the presence of serum. Further differentiation into the neuronal lineage was demonstrated by cellular staining with antibodies against βIII-tubulin, MAP2ab, neurofilament light chain protein (NF-L), glutamate, synaptophysin, and glutamic acid decarboxylase (GABA). The appearance of cells that stain positively for A2B5 and glial fibrillary acidic protein (GFAP) was indicative of astrocytes and a small proportion of O1, O4, and myelin binding protein (MBP) positive cells suggestive of oligodendrocyte formation. Reubinoff and colleagues observed that 57% of cells were immunoreactive to βIII-tubulin and 26% to GFAP. A low proportion of oligodendrocytes within these cultures may reflect limitations in the induction protocols toward this lineage.

Maintenance of a progenitor cell population within neurospheres was evidenced by serial cultivation of the spheres through approximately 20 to 25 population doublings. To assess differentiation *in vivo*, two studies have transplanted these human ES cell-derived neurospheres into lateral ventricles of newborn mice. The human neural stem and progenitor cells successfully engrafted into the host brain and differentiated into neuronal and glial cells, thus showing that cell derivatives from human ES cells can migrate, integrate, and differentiate *in vivo* in response to regionally specific environmental cues.

Other studies added soluble factors to human ES cell cultures to direct differentiation toward neural lineages. Retinoic acid (RA), a critical regulator of mouse embryogenesis, directs neural differentiation of mouse ES cells and human ES cells. Addition of RA to mouse ES cell-derived EB cultures increased the proportion of neural cell types from 15–30% to 100%. The mechanism by which RA potently induces neural cell populations from ES cells is not fully understood, but it involves up-regulation of genes involved in embryonic neural pathways, such as Shh, Pax6, mash-1, and neuroD. Down-regulation of the expression of brachyury, cardiac actin, and ξ globin suggests that RA may inhibit mesoderm induction while promoting neural differentiation.

Addition of RA and nerve growth-β (NGF) increases the proportion of neuronal cells that form within human ES cell-derived EBs as assessed by detection of NF-L (neural progenitors) by *in situ* hybridization. When either RA or NGF was added to EB cultures the number of NF-L positive cells within EBs increased from 54 to 74%. Carpenter and colleagues reported an increased proportion of A2B5 and polysialylated neural cell adhesion molecule (PS-NCAM) positive cells following addition of RA to EBs and subsequent disaggregation onto adhesive substrates. Subpopulations of nestin positive cells expressed PS-NCAM and A2B5. To a lesser proportion, astrocyte cells were detected within the cultures by expression of GFAP and GalC. Further differentiation of these progenitors resulted in mature neuronal cell types that stained with antibodies against MAP-2, β-tubulin III, and GABA as well as being capable of responding to neurotransmitters, GABA, acetylcholine, and ATP. About 3% of the neurons expressed tyrosine hydroxylase (TH) indicative of dopaminergic (DA) neurons.

The presence of cells expressing muscle actin and alphafetoprotein (AFP) showed that the induction protocol results in a heterogeneous cell population. Purification of the mixed cell population using magnetic bead sorting with antibodies against cell surface proteins A2B5 and PS-NCAM resulted in enriched neural progenitor cell populations, with 86% of cells being immunoreactive to PS-NCAM and 91% to A2B5.

Based on the effect of the BMP family and their antagonists on embryonic neuralization, Pera and Trounson (2004) antagonized BMP-2 by addition of soluble noggin to human ES cell cultures. Between 5 and 10 days after noggin treatment of human ES cells, an apparently homogeneous, morphologically distinct cell population formed. These noggin-treated cells form neurospheres at a tenfold greater frequency to rosette-like cells isolated from spontaneously differentiating human ES cells, and with prolonged treatment, the entire culture could be converted into neurosphere precursors. Expression of Sox2, musashi1, nestin, and Pax6, but not brachyury or endoderm markers, GATA-4, and AFP, suggests that these distinct cells are neural stem or progenitor cells. Their ability to differentiate into both neurons and astroglial cells confirmed this hypothesis.

Endodermal Cell Types

Compared with our understanding of neural and hematopoietic development, less is known about the factors that establish endodermal development in the mouse embryo and the signals that direct endoderm into different cellular fates such as hepatic and pancreatic differentiation. Use of specific markers for identifying and isolating endodermal cell types from human ES cells, with the ability to form hepatocytes, pancreatic β-islet cells, or lung, all indicative of definitive endoderm, is confounded by the similarity in molecular profiles between extraembryonic cells, VE, and embryonic or fetal liver. The lack of temporal and spatial organization of ES cell differentiation makes identification of definitive endoderm (as opposed to extraembryonic endodermal cells) problematic. Detailed characterization of endoderm-specific gene expression during EB formation from mouse ES cells showed that cells immunoreactive to AFP and TTR are present within yolk sac-like structures. These cells are restricted to the outer

layer of the EB and appear subsequent to expression of HNF1 and HNF3β; thus, they appear to be visceral and not embryonic endoderm.

To distinguish hepatocytes derived from ES cells from extraembryonic cells, Jones and colleagues used an ES cell line in which a gene-trapping event resulted in expression of β-galactosidase by embryonic liver cells. Using this ES cell line to generate a transgenic mouse, the authors established that β-gal expression was restricted to early liver with no apparent detection in the yolk sac endoderm. Thus, β-galactosidase expression was used to identify cells in vitro, expressing the target of the gene trap (a liver-specific ankyrin repeat containing protein) and that had, therefore, differentiated into hepatocytes. EBs were disaggregated after 5 days and plated onto gelatin for further differentiation. Following 9 days, up to 90% of disaggregated EBs were positive for β-galactosidase, compared with 27% of EBs in suspension. Immunocytochemistry showed that the β-galactosidase positive cells were positive for liver markers TTR, AFP, and albumin.

Liver

Hepatocytes have been derived following EB formation from mouse ES cells. After 5 to 9 days, after EB formation, mouse ES cells differentiate into cells expressing genes characteristic of early hepatic development, including HNF4-α, TTR, AFP, and albumin but no markers of mature hepatocytes. However, following attachment of EBs to an adhesive substrate and sequential addition of dexamethasone, oncostatin M, acidic FGF, and hepatocyte growth factor (HGF), expression of glucose-6-phosphate (G-6-P) and tyrosine aminotransferase (TAT) was observed, indicating differentiation toward mature hepatocytes.

Rambhatla and colleagues added sodium butyrate to either human ES cells or to day 4 EBs produces an apparent homogeneous population of cells with a distinctive epithelial morphology. This directed differentiation protocol results in about 60 to 90% of cells being immunoreactive to hepatocyte markers, albumin, AAT, glycogen, cytokeratin (CK)-8, CK-18, CK-19, but not to AFP. Cells staining positively for CK-19 may suggest the presence of biliary epithelial cells. The functional capacity of the differentiated hepatocyte-like cells was assessed by measuring cytochrome P450 enzymes, characteristic of hepatocytes. The mechanism by which sodium butyrate induced human ES cells to develop into hepatocytes is not understood. Sodium butyrate inhibits histone deacetylase activity regulating chromatin and is known to induce the differentiation of many different cell types. Whether the appearance of hepatocyte-like cells was due to directed differentiation by sodium butyrate or to the high proportion of cell death (about 80%) and selective survival of hepatocyte progenitor cells remains to be determined.

Pancreas

Successful derivation of pancreatic β-islet cells from human ES cells has not yet been reported. To date, Assady et al. have shown that differentiating human ES cells express genes characteristic of the establishment of the mouse embryonic pancreas following EB formation, but cell populations that produce insulin in culture have not been isolated. Expression of genes indicative of pancreas development, such as Pdx1, neurogenin3, glucose transporter (GLUT)-1, GLUT2, and glucokinase, was shown by RT-PCR analysis following 15 days in culture after EB formation. Immunoreactivity to insulin was detected within EBs after 19 days in culture, with secretion of insulin into the medium increasing from 21 to 31 days. However, only 2% of the total cells were immunoreactive to an antibody against insulin.

More controversial have been attempts to obtain pancreatic islet cells from mouse ES-derived neural progenitors. Using a protocol for generation of neural precursors, Lumelsky and colleagues expanded nestin-positive cells. Following further differentiation via pancreatic cell culture conditions, results in cellular clusters that secrete insulin *in vitro*, albeit at low levels, although other workers found that insulin uptake from the culture medium accounted for this phenomenon. Although other pancreatic markers were detected (glucagon and somatostatin) within these mouse ES cell-derived nestin-positive cells, Pdx1, critical for pancreatic formation, was not expressed. The cells contain mixed-cell populations of neuronal cells, as evidenced by their immunoreactivity to β_{III}-tubulin. The relationship between nestin-positive cells and islet cells has been the subject of a number of experimental studies and considerable discussion. Nestin expression has been detected in both embryonic and adult pancreas. Nestin-positive cells were isolated from rat and human pancreatic islets and expanded with the addition of bFGF and EGF to the cultures, conditions adapted from proliferation of nestin-positive neural cell types.

Further differentiation resulted in cells that expressed markers suggestive of hepatic lineages and pancreatic exocrine and endocrine cells, including insulin-secreting cells, suggesting a multipotent progenitor cell. However, examination of the human embryonic pancreas revealed that, in contrast to the adult pancreas, embryonic pancreatic epithelial cells do not stain positively for a nestin antibody, and culture of these cells in conditions used to differentiate pancreatic cells failed to direct them into insulin-producing cells. In the developing mouse embryo, mesenchymal cells surrounding the pancreatic bud are immunoreactive to nestin. These studies suggest that, during embryogenesis, islet precursor cells are not demarcated by nestin expression. Whether the developmental potential of ES-derived cells that are immunoreactive to nestin is different from the potential of those cells from the embryo must be clarified.

Another strategy, used by Blyszczuk *et al.* to direct mouse ES cells toward a pancreatic fate, involves forced expression of *Pax4* and *Pdx1*, genes that regulate endocrine pancreas development. Detection of genes expressed during pancreatic formation using RT-PCR analysis, namely, islet-1, neurogenin-3, insulin, islet amyloid polypeptide (IAPP), and GLUT-2, confirmed direction toward pancreatic cell types. In these experiments, the resulting cell populations were able to

normalize blood glucose levels after transplantation into streptozotocin-diabetic mice. Finally, by genetic selection for mouse ES cells expressing a selectable marker driven by the insulin promoter, Soria *et al.* isolated a purified population of insulin-producing cells that temporarily restores blood glucose homeostasis following transplantation into a streptozotocin-diabetic mouse.

Conclusions

Establishment of clinical therapies for replacement of damaged or diseased cells involves derivation of pure populations of progenitor or somatic cell types from human ES cells. This objective is greatly aided by development of protocols to expand and maintain these early human cell populations in culture for long periods. Critical to this work is the use of markers to identify and characterize the cells, as well as to select and isolate subpopulations. Application of markers and genes whose expression is characteristic of progenitor cells in the early mouse embryo or adult stem cells can facilitate this approach.

The appearance of ES cell-derived lineage-specific progenitor cells *in vitro* may reflect the regulation and timing of their development in the embryo *in vivo*. This has been suggested by observations for mouse neural cell derivation. Differences between mouse and human embryo development may become apparent as we accrue more knowledge about the differentiation pathways for human ES cells. Observed differences between the effects of Shh and BMP-4 on mouse and nonhuman primate ES cells in their acquisition of dorsal or ventral fates have been reported. Use of similar co-culture conditions to those applied for directing mouse ES cells into both dorsal and ventral fates results in primate ES-cell-derived ventral but not dorsal neural cells. This may reflect species differences in embryonic regulation or in expression of cell-specific markers.

As noted by Smith (2001), mouse and human embryogenesis develop on a different time scale. Establishment of the mouse embryonic germ layers, gastrulation, and organogenesis occur over 8–9 days. Similar developmental processes within the human embryo occurs between 2 and 5 weeks after fertilization. This difference in timing may influence how human ES cells differentiate *in vitro*. The timing of mouse ES cell neural induction *in vitro* corresponds to neurulation of the mouse embryo (3.5 days). Differentiation of mouse ES-cell-derived oligodendrocyte progenitor cells into oligodendrocytes was similar to differentiation of relative cells within the mouse spinal cord. Interestingly, the appearance of equivalent neural cell types from nonhuman primate ES cells was later, in comparison to the mouse (10 days) although significantly faster than in the embryo (5 weeks).

Use of animal models to fully characterize human ES cell-derived progenitor cells will be critical before their adoption for clinical use (Figure 48-2). Transplantation of human ES cell-derived progenitor cells into different sites of the developing chick embryo or recombination with mouse embryonic explants will be a means to ascertain their developmental

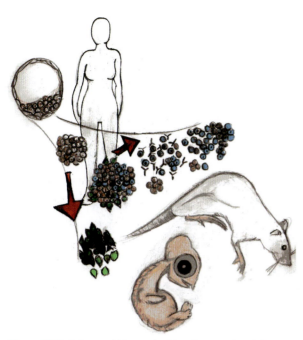

Figure 48-2. Before establishment of cell lines for use in clinical therapies, pure populations of progenitor cells from human embryonic stem (ES) cells derived from human blastocysts will be generated following directed differentiation or by isolation using cell-surface-specific markers. Before clinical use, their full biologic potential can be assessed by transplantation into embryonic and diseased or damaged animal models.

potential. This will verify and complement the use of cell-specific markers as *in vitro* functional assays, such as insulin secretion to identify β-cells of the pancreas and urea uptake by putative liver cells. Finally, transplantation into either diseased or damaged animal models will shed light on the functional ability of progenitor cells.

KEY WORDS

Commitment An epigenetic, stable, and heritable change in the developmental capacity of a cell, as defined operationally under a particular set of physiological or experimental conditions.

Differentiation Process of specialization wherein a cell, through changes in gene expression, acquires the correct shape, polarity, and orientation with respect to its neighbors and the extracellular matrix, and an appropriate set of organelles and proteins to enable it to carry out metabolic, signaling, transport, contractile, structural, or other specific functions in a particular tissue.

Progenitor cell Immature cell that is committed to a particular differentiation lineage and has some limited proliferative capacity.

FURTHER READING

Chadwick, K., Wang, L., Li, L., Menendez, P., Murdoch, B., Rouleau, A., and Bhatia, M. (2003). Cytokines and BMP-4 promote hematopoietic differentiation of human embryonic stem cells. *Blood* **102**, 906–915.

Levenberg, S., Golub, J. S., Amit, M., Itskovitz-Eldor, J., and Langer, R. (2002). Endothelial cells derived from human embryonic stem cells. *Proc. Natl. Acad. Sci. USA* **99**, 4391–4396.

Li, M., Pevny, R., Lovell-Badge, R., and Smith, A. (1998). Generation of purified neural precursors from embryonic stem cells by lineage selection. *Curr. Biol.* **8**, 971–974.

Pera, M. F., and Trounson A. O. (2004). Human embryonic stem cells: prospects for development. *Development* **131**(22), 5515–5525.

Pevny, L., and Rao, M. S. (2003). The stem-cell menagerie. *Trends Neurosci.* **26**, 351–359.

Reubinoff, B. E., Itsykson, P., Turetsky, T., Pera, M. F., Reinhartz, E., Itzik, A., and Ben-Hur, T. (2001). Neural progenitors from human embryonic stem cells. *Nat. Biotechnol.* **19**, 1134–1140.

Reubinoff, B. E., Pera, M. F., Fong, C. Y., Trounson, A., and Bongso, A. (2000). Embryonic stem cell lines from human blastocysts: Somatic differentiation in vitro. *Nat. Biotechnol.* **18**, 399–404.

Smith, A. G. (2001). Embryo-derived stem cells: Of mice and men. *Annu. Rev. Cell Dev. Biol.* **17**, 435–462.

Thomson, J. A., Itskovitz-Eldor, J., Shapiro, S., Waknitz, M. A., and Swiergiel, J. J., Marshall, V. S., and Jones, J. M. (1998). Embryonic stem cell lines derived from human blastocysts. *Science* **282**, 1145–1147.

Zhang, S. C., Wernig, M., Duncan, I. D., Brustle, O., and Thomson, J. A. (2001). In vitro differentiation of transplantable neural precursors from human embryonic stem cells. *Nat. Biotechnol.* **19**, 1129–1133.

49

Isolation and Characterization of Hematopoietic Stem Cells

Gerald J. Spangrude and William B. Slayton

Introdution

Central to the development of methods for stem cell isolation is the availability of quantitative techniques for assessment of stem cell function. In this regard, any approach to identification and isolation of adult stem cells can be no better than the assay used to detect function. This is a critical issue with blood stem cells because the robust nature of hematopoiesis requires massive expansion of very few stem cells to provide a continual source of replacements for mature cells that die every day. The degree of expansion that occurs from stem cell to progenitor cell to mature cell is vast enough to prevent absolute distinction between the most primitive stem cells and the differentiating progeny of these cells. Methods for hematopoietic stem cell (HSC) isolation described here include approaches that minimize co-isolation of non-stem cells, while also providing techniques to isolate populations of progenitor cells possessing remarkable proliferative potential in the absence of stem cell activity. Comparison of primitive HSCs with early progenitor cells provides interesting insights into the early stages of hematopoietic development.

Isolation of Hematopoietic Stem Cells from Mice

HISTORICAL PERSPECTIVE

Although transplantation experiments conducted in the 1950s established the ability of cells derived from bone marrow and spleen to reverse hematopoietic failure after radiation exposure, HSCs were studied mainly on a morphologic basis before the 1960s. A major breakthrough came in 1961, when Till and McCulloch published an article describing the spleen colony-forming assay. This marked the first attempt to describe and quantitate stem cell activity *in vivo*. Shortly after this, tissue culture systems were developed that allowed the assay of hematopoietic progenitor cells *in vitro*. These assays provided quantitative results because of their clonal nature, whereby single cells proliferate to form a colony of differentiating progeny. The mouse became the animal of choice for the early studies of HSCs, because of the ability to obtain bone marrow easily and inexpensively and to perform transplant

studies in which the behavior of primitive cells could be studied *in vivo*. In addition, the development of inbred strains of mice had already allowed the definition of the role of the major histocompatibility complex in determining graft acceptance or rejection. This animal model, in combination with the ability of clonal assays to determine enrichment factors for the colony-forming cells, provided the basis for approaches to the characterization and isolation of HSCs.

Till and McCulloch demonstrated that the subset of cells able to form spleen colonies was heterogeneous. When spleen colonies were harvested from animals and transplanted into a second set of irradiated recipients, only a subset of colonies could give rise to secondary colonies. This group of cells was also capable of reconstituting lethally irradiated animals. For approximately 20 years, the spleen colony-forming unit was equated with HSC activity. However, in the early 1980s it became clear that this assay measured the behavior of a number of cell types that were not by definition HSCs. Conversely, populations of cells that fit the definition of HSCs but did not make colonies in the colony-forming unit–spleen (CFU-S) assay were also identified. Prospective isolation of defined cell populations from the mixture of cell types found in normal bone marrow was required to resolve the differences between true HSCs and other types of blood progenitors.

Some of the earliest attempts to isolate stem and progenitor cell subsets from bone marrow involved separating cells based on size, density, and cell cycle characteristics. These experiments demonstrated that the cells that produced splenic colonies were largely separable from those that made colonies *in vitro* and that many spleen colony-forming cells were sensitive to killing by cycle-active drugs, whereas HSCs were largely spared. It soon became clear that the only definitive means by which HSCs could be distinguished from later stages of development was to evaluate the ability of a transplanted population of cells to maintain production of multiple types of blood cells over many months *in vivo*. Even by this assessment, false indications of stem cell function are possible. Because of the extensive proliferative potential of true stem cells, a single cell remaining within the bone marrow after radiation-induced ablation will proliferate and self-renew sufficiently to dominate long-term hematopoiesis. Although transplanted progenitor cells may provide sufficient hematopoietic function to rescue animals from hematopoietic failure after radiation conditioning, it is necessary to use markers to distinguish blood cells derived from the donor graft from those of recipient origin to definitively demonstrate HSC function. Examples of such markers include chromosomal abnormalities induced by radiation; retroviral insertion sites; and allelic forms of enzymes, intracellular molecules, and cell-surface antigens. Using these tools, researchers have been able to demonstrate that progenitor cells can be radioprotective without providing the long-term chimerism that is characteristic of HSCs. Furthermore, researchers using these tools have shown that a single transplanted or endogenous HSC is sufficient to maintain life-long hematopoiesis.

The development of fluorescence-activated cell sorting (FACS) allowed isolation of distinct cell populations that were too similar in physical characteristics to be separated by size or density. Flow cytometry uses fluorescent tags to detect surface proteins that are differentially expressed at particular stages of development, allowing fine distinctions to be made among physically homogeneous populations of cells. Although a number of surface proteins are expressed by HSCs, there is no single marker that absolutely defines HSCs in any species. This makes a multiparameter approach, using positive and negative selection for expression of a variety of markers, a necessary component of any enrichment scheme.

METHODS FOR ENRICHMENT OF MOUSE HEMATOPOIETIC STEM CELLS

A method using a combination of negative selection for proteins not expressed by HSCs with positive selection for proteins that are expressed has proven to be extremely useful for definition and isolation of HSCs from both mouse and human tissue. In the first applications of this technology, both positive and negative selections were applied using FACS. In more recent years, both types of selection have been adapted to magnetic technology. This advance in the field makes crude methods of HSC enrichment accessible to investigators who lack access to sophisticated FACS instruments. However, isolation of true HSCs from primitive progenitor cells requires FACS technology because of the application of intracellular fluorescent probes that indicate cellular quiescence or efflux pump activity. In addition, although magnetic isolation usually maximizes yields at the expense of purity, FACS isolations provide high purity with acceptable yields. A combination of magnetic and FACS selection allows isolation of HSCs in a routine manner.

Magnetic Selection Techniques

Regardless of the specific technology used for HSC enrichments, the most critical parameter to be considered is the specific marker to be used in negative and positive selections. For negative selections, a conventional approach has evolved that uses panels of antibodies specific for proteins expressed by most bone marrow cells. Caution must be exercised to select antibodies that do not react with HSCs because this would result in the loss of the HSCs during negative selection. It is particularly important to note that developmental stages and activation-specific responses can change the antigenic profile of HSCs. For example, the CD34 antigen fluctuates on and off in HSCs during development, as well as during a period of

time following the administration of chemotherapy. Other antigens are also known to fluctuate in a similar manner. Therefore, negative selection protocols must be tailored to match the known phenotype of HSCs in the particular tissue of interest. If the investigator wishes to work with tissues for which the antigenic phenotype of HSCs has not been described, appropriate assays should be used to evaluate the HSC content of populations fractionated into positive and negative cells to identify antibodies that will be useful in depleting significant numbers of cells without affecting the recovery of HSCs. As discussed previously, the most appropriate assay for this analysis would be a long-term engraftment study rather than one of the more rapid techniques such as spleen colony formation or *in vitro* culture systems.

Magnetic selections can be accomplished by two general approaches. The first approach, which is best suited for negative selections, involves mixing a cell population marked by the antibodies of interest with a suspension of paramagnetic particles that bind to the antibody tag. Application of a magnetic field will then separate bead–cell aggregates from the unmarked cells, which can be collected by simple aspiration. This technique is rapid and has the advantage of not perturbing the cells of interest through association with antibody molecules or magnetic particles. Because of this, the magnetic particles of choice will be relatively large (1–5 μm) to facilitate the rapid migration of the labeled cells in the magnetic field.

The second approach to magnetic selection is positive selection of the cells of choice. In this case, particles of a much smaller size (50 nm) are used to minimize potential effects of the antibody–particle complexes on the biology of the selected cells. The small size of the magnetic particles used in positive selection decreases the effectiveness of the magnetic field in attracting the labeled cells. To overcome this limitation, commercial positive selection systems use a flow column packed with a fibrous metal, into which the magnetic field is introduced by induction. This creates a high-flux magnetic field with very short distances between the labeled cells and the magnetized column matrix and results in the retention of labeled cells in the column. Once the unlabeled cells are passed through the column and washed out, the column is removed from the magnetic field and the selected cells can be collected.

The combination of negative with positive magnetic selection is a powerful technique for crude enrichment of a mixture of hematopoietic stem and progenitor cells. These cells may be used directly in studies evaluating hematopoietic function *in vivo* or *in vitro* or may be further processed to achieve higher purity of functionally distinct subsets of HSCs or progenitor cells (see later discussion). A specific method, originally reported using negative and positive selection by FACS, can be easily modified to substitute magnetic selections using the same combinations of antibodies as originally described (Table 49-1). Unconjugated rat antibodies are used in combination with magnetic beads conjugated to anti-rat immunoglobulin for negative selection, whereas avidin–biotin selection systems work very well for positive selection using

TABLE 49-1

Antibody Specificities for Use in a Negative–Positive Magnetic
Enrichment Protocol

Antigen	Antibody	Lineage
Negative selection		
CD11b	M1/70	Myeloid[a]
Ly6-G	RB6-8C5	Myeloid
CD45(B220 isoform)	RA3-6B2	Lymphoid
CD2	RM2.2	Lymphoid
CD3	KT3	Lymphoid
CD5	53-7.3	Lymphoid
CD8	53-6.7	Lymphoid
TER-119	TER-119	Erythroid
Positive selection		
Ly6A/E	E13 161-7	Various[b]

[a]CD11b is expressed by mouse fetal liver stem cells and, thus, should
not be included in negative selections from this tissue.

[b]Ly6A/E is expressed by lymphoid lineage cells and by hematopoietic
stem and progenitor cells. Depletion of the lymphoid lineage cells during
negative selection allows positive selection based on Ly6A/E to yield a
highly enriched population of stem and progenitor cells.

biotinylated antibodies followed by avidin-conjugated
microbeads. HSCs and progenitor cells can be enriched routinely by a factor of 1000 using this technique.

Fluorescence-activated Cell Sorting Selection

Although magnetic enrichment strategies are sufficient when
mixed populations of cells are an adequate product, the more
refined approaches to HSC and progenitor cell characterization require FACS selection. Modern commercial FACS
instruments can separate populations of cells based on 15
parameters, allowing the simultaneous application of positive
and negative selection for a variety of surface markers. High-
speed sorting improves the yield of cells obtained from this
approach, and purity is very high. Workers in the field often
use a negative selection before FACS sorting, because a good
negative selection reduces the cellularity of the sample and
thus the time needed for sorting by a factor of 50 to 100
without adversely affecting the yield of HSCs. Samples subjected to positive magnetic selection can subsequently be
processed to isolate specific cellular subsets by FACS. Pre-
enrichment of target populations before FACS can have a
significant impact on the final purity of the isolated cell
populations.

Selection Based on Cell-Surface Antigen Expression.
Recent studies have described antigenic profiles of a variety
of progenitor cell subsets in the mouse. Progenitors restricted
to lymphoid, myeloid, and erythroid lineages have been characterized for prospective isolation using FACS technology. In
addition, the hierarchy of HSCs, which includes cell populations possessing more or less potential for self-renewal, has
been fractionated based on surface antigen expression.

Selection Based on Supravital Stains. The combination
of multiple cell-surface markers with supravital fluorescent

dyes that indicate cell function has become a method of
choice for isolating HSCs. Baines and Visser first used the
fluorescent DNA probe Hoescht-33342 to separate quiescent
cells from the bone marrow. More recently, red and blue
emissions from Hoescht-33342 have been used to define a
small subset of bone marrow cells, called the side population,
which comprise approximately 0.07% of the whole bone
marrow, and are highly enriched for long-term reconstituting
HSCs. Recently, the molecular basis for the ability of
Hoescht-33342 to identify HSCs has been suggested to be
due to its role as a substrate for the multidrug resistance pump
ABCG2. Interestingly, expression of ABCG2 is a characteristic also observed in embryonic stem cells, suggesting that
expression of this protein may represent a general marker of
stem cells.

A second supravital fluorescent probe, rhodamine-123
(Rh-123), has also proven useful for distinguishing subsets of
primitive stem cells. Although Rh-123 is most widely used as
a substrate to evaluate multidrug resistance pump activity, a
second characteristic of the probe is its affinity for the inner
mitochondrial membrane. Hence, the intensity of Rh-123
staining reflects a balance between efflux by specific molecular pumps and accumulation in mitochondrial membranes.
Because Rh-123 is a charge-sensitive probe for mitochondria,
accumulation of the probe is directly proportional to cellular
activation state and inversely proportional to cellular efflux
activity. In combination, these two selective criteria achieve a
remarkable segregation of stem and progenitor cells previously isolated by negative/positive selection based on antigen
expression. Primitive HSCs, which express high levels of
efflux pump activity and contain few activated mitochondria,
accumulate very little Rh-123 during a 30-minute labeling
period. In contrast, progenitor cells that have become activated by the differentiation process contain many activated
mitochondria and have down-regulated efflux activity.

The combination of cell-surface and metabolic markers
provides a means to enrich HSCs to the point that fewer than
10 cells are required to reconstitute hematopoiesis in irradiated mice. In addition, enrichment based on antigenic expression combined with Rh-123 allows simultaneous isolation of
Rh-123[low] HSCs and Rh-123[high] primitive progenitor cells, permitting direct comparisons of the functions and molecular signatures of these two closely related populations of cells.

Isolation of Hematopoietic Stem Cells from Humans

HISTORICAL PERSPECTIVE

In contrast to the mouse, in which HSC behavior can be
defined in transplant experiments, defining and quantifying
human HSCs have been difficult because transplantation
experiments using limiting dilution or competitive repopulation approaches are impossible to perform in humans. *In vitro*
colony-forming assays were once used routinely to estimate
the number of HSCs in a bone marrow or peripheral blood
stem cell harvest. However, with the recognition that these

assays fail to distinguish different classes of primitive stem cells that have distinct functions on transplantation, other assays were developed. Several *in vitro* assays, including the long-term culture-initiating cell assay and the cobblestone area-forming cell assay, have been used to measure human stem cell number and behavior.

Researchers have addressed the inability to study HSC engraftment behavior in humans by developing xenogeneic transplant models. Several models that use immunodeficient mice as transplant recipients have been developed to measure engraftment and self-renewal of human HSCs *in vivo*. The nonobese diabetic/severe combined immunodeficiency (NOD/SCID) mouse model has emerged as a standard assay that allows engraftment of human cells in mouse bone marrow. Ultimately, however, studies that depend on xenogeneic transplantation rely on inference, and stem cells may behave differently in systems that are more genetically compatible.

The first attempts to separate human HSCs from bone marrow mirrored earlier studies in mouse. The recognition that primate hematopoietic progenitor cells expressed CD34 provided a way for scientists to study the HSC compartment in humans. Using techniques pioneered in the mouse, researchers used negative selection against lineage antigens and positive selection for CD34. The CD34$^+$ subset was found to be heterogeneous, with most of the cells being progenitors for the erythroid, lymphoid, or myeloid lineages. Further enrichment of HSCs was obtained by positive selection for additional surface antigens in combination with metabolic markers.

CD34 AS THE PRIMARY MARKER OF HUMAN HEMATOPOIETIC STEM CELLS

Recently, a controversy has evolved regarding the utility of CD34 as a general marker of HSCs in human tissue. The controversy arose as a result of studies in the mouse, in which various groups published conflicting results regarding the isolation of mouse HSCs based on CD34 expression. A large number of monoclonal antibodies have been developed to different epitopes on the human CD34 antigen. In contrast, the CD34 molecule in the mouse has been characterized only by one rabbit polyclonal antiserum that is monospecific and one monoclonal antibody directed against a CD34 epitope. These reagents provide contrasting results: the CD34$^+$ but not the CD34$^-$ cells in mouse bone marrow contained HSCs based on the polyclonal serum, whereas the mouse-specific monoclonal antibody RAM34 showed variable expression on HSCs defined by other markers. This issue was resolved in a series of experiments by Ogawa and colleagues, in which the expression of mouse CD34 was shown to be modulated from negative to positive depending on the proliferative state of the HSC population under analysis. Thus, although most HSCs found under normal steady-state conditions in adult mouse bone marrow lack CD34 expression, stimulation *in vivo* by chemotherapy or *in vitro* by cytokine stimulation resulted in recovery of most HSC activity among CD34$^+$ cells. The expression of CD34 was again lost as the cells engrafted in marrow and returned to steady-state conditions. Similar observations were reported in an analysis of mouse ontogeny. Further experiments showed that CD38, which is absent from human HSCs but present on mouse HSCs under steady-state conditions, also modulates during recovery from chemotherapy or during mobilization induced by granulocyte colony-stimulating factor (G-CSF). Thus, in the mouse the expression of CD34 and CD38 reflect the activation state of the HSCs. Interestingly, an analysis of the expression of human CD34 in a transgenic mouse model demonstrated that the human and mouse genes are differentially regulated, suggesting that the current practice of human bone marrow transplantation using only CD34$^+$ cells is justified despite the mouse data. Because the initial data from the mouse model could be interpreted to indicate that human patients receiving CD34$^+$ transplants are missing an important subset of HSCs, it is important to learn that CD34 expression is a characteristic of most HSCs in humans.

The question of lineage-marker fidelity between mouse and human is biologically interesting because it might imply important functions for these molecules in the behavior of HSC subsets. The functions of these molecules in mice and humans must be identified so that the gene program that specifies HSCs and their functions can be better understood. Enriched populations of CD34$^+$ cells have been used clinically for transplantation in humans following chemotherapy. Enriching marrow progenitors and purging T cells have been used to prevent graft versus host disease in allogeneic transplantation and as a means to enrich human HSCs for gene therapy applications.

Hematopoietic Stem Cells as a Paradigm for Stem Cell Biology

Robust production of blood is essential for survival because of the critical roles blood plays in oxygenation of tissues, control of vascular integrity, and distribution of cells responsible for immunity to infection. Furthermore, the large total number of blood cells coupled with relatively short cellular lifespans results in a need for ongoing replacement of senescent cells. The sustained daily rate of blood cell production reflects the potential of the stem cells that drive the process of hematopoiesis. These extreme demands of proliferation and differentiation, coupled with a long history of successful transplantation in humans, make the HSC a gold standard for all forms of stem cell therapy. The development of methods to enrich this class of stem cells serves as a robust paradigm by which other applications of stem cell therapy might be measured. It is entirely possible that organ systems such as the central nervous system, while harboring cells with stem cell-like characteristics, have not been subjected to evolutionary pressure favoring rapid replacement of dying cells. It will, therefore, be critical to understand more of the regulatory control over self-renewal, proliferation, and differentiation of HSCs to facilitate the application of stem cell therapies for tissues lacking the demands of blood production.

366

ACKNOWLEDGMENT

This work was supported by grants DK57899 (G.J.S.) and HL03962 (W.B.S.) from the National Institutes of Health.

KEY WORDS

Antigen A substance, usually a protein, that can be identified by an antibody molecule. Stem cells express a pattern of antigens that can be detected by fluorescent-labeled antibodies using a FACS instrument, or by unlabeled antibodies using magnetic selection methods.

Fluorescence-activated cell sorting (FACS) A technology that evolved from instruments designed to count blood cells, FACS combines detection of fluorescent probes with the isolation of single cells that are labeled with any combination of probes. This technology allows rapid separation of stem cells from a mixed population of other types of cells, provided that markers to distinguish the stem cells are available.

Hematopoietic progenitor cell A type of cell found in bone marrow and other tissues that has lost the potential for self-renewal but is still multipotent (can produce some or all types of blood cells).

Hematopoietic stem cell A type of cell found in bone marrow and other tissues that possesses the defining characteristics of self-renewal (produces more stem cells) and multipotency (produces all types of blood cells).

Spleen colony-forming assay The first biological assay used to detect hematopoietic stem cells. The assay detects macroscopic splenic nodules formed by proliferating stem and progenitor cells that lodge in the spleen of an irradiated mouse. The cells that initiate these nodules are termed colony-forming units-spleen (CFU-S) because it was initially not certain that each nodule represented the outgrowth of a single cell versus that of a cluster (unit) of cells.

Supravital fluorescent probe A fluorescent molecule that can enter a living cell and bind to specific subcellular components, such as DNA, RNA, or organelle membranes. These probes can be used in combination with labeled antibodies directed against cell-surface antigens in order to identify stem cells.

FURTHER READING

Baines, P., and Visser, J. W. M. (1983). Analysis and separation of murine bone marrow stem cells by H33342 fluorescence-activated cell sorting. *Exp. Hematol* **11**, 701–708.

Herzenberg, L. A., and Sweet, R. G. (1976). Fluorescence-activated cell sorting. *Sci. Am.* **234**, 108–117.

Nakano, T. (2003). Hematopoietic stem cells: generation and manipulation. *Trends Immunol.* **24**, 589–594.

Spangrude, G. J. (1989). Enrichment of murine haemopoietic stem cells: diverging roads. *Immunol. Today* **10**, 344–350.

Tajima, F., Deguchi, T., Laver, J. H., Zeng, H., and Ogawa, M. (2001). Reciprocal expression of CD38 and CD34 by adult murine hematopoietic stem cells. *Blood* **97**, 2618–2624.

Till, J. E., and McCulloch, E. A. (1961). A direct measurement of the radiation sensitivity of normal mouse bone marrow cells. *Radiat. Res.* **14**, 213–222.

50

Microarray Analysis of Stem Cells and Differentiation

Howard Y. Chang, James A. Thomson, and Xin Chen

Introduction

For the past several decades, biologists have only been able to tackle the analysis of one or a few genes at a time. However, the advent of complete genomic sequences of over 800 organisms (including the human and mouse genomes) and the development of microarray technology have revolutionized molecular biology. Microarrays enable biologists to perform global analysis on the expression of tens of thousand of genes simultaneously, and they have been widely used in gene discovery, biomarker determination, disease classification, and studies of gene regulation. Expression profiling using microarrays is generally considered "discovery research," although it can also be a powerful approach to test defined hypotheses. One advantage of microarray experiments is that, at the outset, microarray experiments need not be hypothesis driven. Instead, it allows biologists a means to gather gene expression data on an unbiased basis, and it can help identify genes that may be further tested as the targets in hypothesis-driven studies.

Overview of Microarray Technology

Two major microarray platforms in wide use are cDNA microarrays and oligonucleotide microarrays.

cDNA MICROARRAYS

The principle of cDNA microarray is illustrated in Figure 50-1. In brief, cDNA clones, which generally range from several hundred base pairs to several kilobases, are printed on a glass surface, either by mechanical or ink jet microspotting. Sample RNA, as well as a reference RNA, are differentially labeled with fluorescent Cy5 or Cy3 dyes, respectively, using reverse transcriptase. The subsequent cDNA are hybridized to the arrays overnight. The slides are washed and scanned with a fluorescence laser scanner. The relative abundance of the transcripts in the samples can be determined by the red/green ratio on each spotted array element.

One limitation of the cDNA microarray has been that it required a relatively large amount of total RNA ($\geq 10\,u.g$) for hybridization. However, significant progress has been made

recently for linear amplification of RNA, generally based on Eberwine's protocol. In this case, RNA is converted into cDNA with oligo dT primers, which contain T7 RNA polymerase promoter sequence at its 5' end. The cDNA can be subsequently used as the template for T7 RNA polymerase to transcribe into antisense RNA. The linear amplification protocol can produce 10^6-fold of amplification. Therefore, only a very small amount of samples are required in modern microarray experiments.

cDNA microarrays have several advantages. The two color competitive hybridization can reliably measure the difference between two samples since variations in spot size or amount of cDNA probe on the array will not affect the signal ratio. cDNA microarrays are relatively easy to produce. In fact, the arrayer can be easily built and allow the microarrays to be manufactured in university research labs. Also, cDNA microarrays are in general much cheaper compared with oligonucleotide arrays and are quite affordable to most research biologists.

This system also has some disadvantages. One is that the production of cDNA microarray requires the collection of a large set of sequenced clones. The clones, however, may be misidentified or contaminated. Second, genes with high-sequence similarity may hybridize to the same clone and generate cross hybridization. To avoid this problem, clones with 3' end untranslated regions, which, in general, are much more divergent compared with the coding sequences, should be used in producing the microarrays.

OLIGONUCLEOTIDE ARRAYS

The most widely used oligonucleotide arrays are "GeneChips" produced by Affymetrix, which uses photolithography directed synthesis of oligonucleotides on glass slides. Affymetrix GeneChip measures the absolute levels for each transcript in the sample. The principle of Affymetrix GeneChip is shown in Figure 50-2. In general, for each transcript, approximately 20 distinct and minimal overlapped 25-mer oligonucleotides are selected and synthesized on the array. For each oligonucleotide, there is also a paired mismatch control oligonucleotide, which differs from the perfect match probe by one nucleotide in the central position. Comparison of the hybridization signals from perfect match oligonucleotide with the paired mismatch oligonucleotide will allow automatic subtraction of background.

Oligonucleotide arrays have several advantages. The sequences of the oligonucleotide arrays are determined by

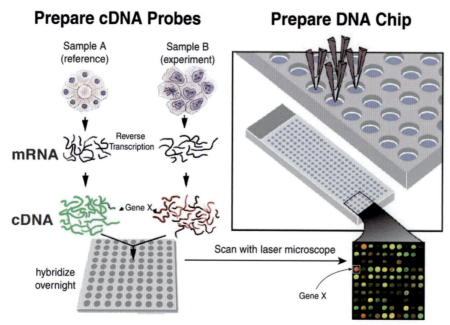

Figure 50-1. Principle of cDNA microarrays. PCR products are printed onto glass slides to produce high-density cDNA microarrays. RNA is extracted from experimental samples and reference samples, and differentially labeled with Cy5 and Cy3, respectively, by reverse transcriptase. The subsequent cDNA probes are mixed and hybridized to cDNA microarray overnight. The slides are washed and scanned with fluorescence laser scanner. The relative red/green ratio of gene X indicates the relative abundance of gene X in experimental samples versus reference.

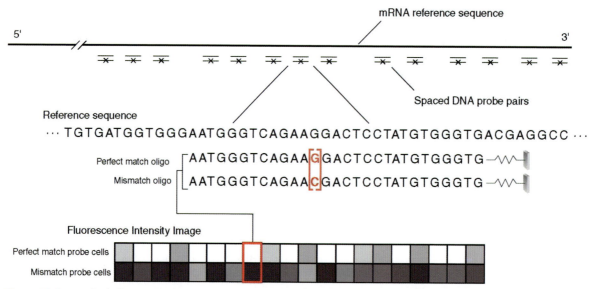

Figure 50-2. Principle of Affymetrix GeneChip. For each gene, approximately 20 distinct and minimally overlapped 25-mer oligonucleotides are selected and synthesized on the array. For each oligonucleotide, there is also a paired mismatch control oligonucleotide probe, which differs from the perfect match probe by one nucleotide in the central position. Comparison of the hybridization signals from perfect match oligonucleotide with the paired mismatch oligonucleotide allows automatic subtraction of background. The figure was kindly provided by Affymetrix, Santa Clara, California.

sequencing information from the database and synthesized *de novo* for the arrays. Therefore, there is no need to validate cDNA clones or PCR products that are used to print cDNA microarray. The use of multiple oligonucleotides for each transcript also allows for the detection of splice variants and helps to distinguish genes with high-sequence similarities.

The major disadvantage for oligonucleotide arrays is that they can only be produced by commercial manufacturers, and these prefabricated GeneChips are still very expensive. Currently, there is little work to systematically to compare the two microarray platforms. Both cDNA microarrays and oligonucleotide arrays have been widely used in biological research and produce data.

Next, we will discuss some basic issues of microarray experiments, including experimental design, data analysis, and gene validation. We will focus on our discussion for cDNA microarray, but most of the general principles can be applied to both cDNA microarrays and oligonucleotide arrays.

Experimental Design

cDNA microarrays use two-color competitive hybridization; therefore, the design of experiments (i.e., which samples are labeled for each color), is very important in the subsequent data interpretation. There are two commonly used experimental designs, and in this chapter, we refer them as type I or type II designs.

In the type I experimental design, the sample RNA (e.g., drug-treated cells, diseased tissues) is labeled with one dye, and another sample RNA (e.g., mock treated cells, normal tissues) is labeled with another dye. The two sample cDNA probes are mixed and co-hybridize to microarray. The data analysis for type I design can be quite straightforward. The relative red/green ratio represents the relative up-regulation or down-regulation of each gene. This design is most suited for experiments with a few perturbations — for example, to identify genes response to certain stimulus, genes affected by a genetic mutation. However, it is difficult to apply this pairwise "type I" design to a complicated system. For example, to identify genes differentially expressed in a disease and normal tissues, it may not be appropriate to compare diseased tissues versus corresponding normal tissues from the sample patients. Gene expression patterns in normal and diseased tissues are affected by many factors, including patient variation (ethnicity, sex, age, genetic background), sample variation (proximity to disease, anatomic location and developmental range), as well as heterogeneity of cell populations within the samples. In the simple pairwise type I design, it is impossible to distinguish all these variations within both normal and diseased tissues. Another drawback of type I experiment is that it cannot accurately measure the relative abundance of a transcript which is not expressed or expressed at very low levels in Cy3 labeling, as this would produce a green channel signal very close to the background signal.

Type II experimental design avoids most of the problems with the simple pairwise type I design. In the type II design, a common reference is used in Cy3 channel labeling, and each sample is compared with the same common reference. The two most important criteria for selection of common reference are gene coverage ("light up" most of the spots on the array) and reproducibility (can be relatively easily reproduced with minimum batch to batch variation). In most cases, the reference pool RNA is derived from mixture of tumor cell lines, and they are commercially available from Stragene and Clontech. Cell lines from different histological origins ensure the complex of the reference. This provides relatively good coverage of the spots on the array. Cell lines can be recultured to produce more reference RNA; however, growth conditions need to be tightly controlled to reduce batch-to-batch variation. The greatest advantage of the type II experimental design is that it allows the cross comparison of many samples collected over long period of time, by different persons, as well as samples from different sources.

Another situation in which type II designs are widely used is in the time series experiments — that is, to study gene expression variations in response to a stimulus (e.g., drug treatment and growth factor stimulation). A simple pairwise type I design using treated cells against untreated cells may be quite straightforward. However, some of the genes that are induced by the stimulus may not be expressed in the untreated cells. It is difficult to obtain an accurate description of the expression profile changes over the time for these genes. A better design is a type II design, in which the cells from different time points are mixed and used as reference. All the samples, including untreated cells ($t = 0$), can then be compared with this reference. This average $t = 0$ measurement can then be subtracted from each subsequent time-point measurement in order to depict the temporal response patterns of expression relative to $t = 0$ as the baseline. This design ensures that all the transcripts that present in different time points are represented in the reference RNA, and therefore accurately describes the temporal changes in response to the stimulus.

Although the type II design has many advantages, one has to realize that the design may make the experiments more complicated and sometimes inefficient. In all the analysis, the baseline reference signals have to be subtracted in order to extract biologically meaningful data. This undoubtfully will add more data variations. So the proper selection of experimental design is one of the most important steps toward the success of experiments.

Data Analysis

Microarray experiments produce large amounts of data; for example, 30,000 genes × 100 samples will generate 3 million data points. Data analysis may be one of the most challenging issues facing biologists in microarray experiments. Microarray data are often noisy and not normally distributed, and usually with missing values in its matrix. During the past several years, with the combined efforts of biologists, statisticians, and computer scientists, there has been great progress in the bioinformatic techniques that can be used for the analysis of genome-wide expression data. However, there is no standard or one-size-fits-all solution for interpretation of microarray data.

Current methods for microarray data analysis can be divided into two major categories: supervised and unsupervised methods. Supervised approaches try to identify gene expression patterns that fit a predetermined pattern. Unsupervised approaches characterize expression components without prior input or knowledge of predetermined pattern.

SUPERVISED ANALYSIS

The purpose of supervised analysis is to identify genes that are differentially expressed between groups of samples, as well as to find genes that can be used to accurately predict the characteristics of groups.

Many statistical tests have been developed for identification of differential expression genes in microarray data, for example, *t*-test for detecting significant changes between repeated measurements of a variable in two groups; and ANOVA (Analysis of Variance) F statistic for detecting significant changes in multiple groups. All these tests involve two parts: calculating a test statistic and determining the false discover rate (FDR), or the significance of the test statistic. Here are some commonly used statistical methods for two-group comparisons.

Nonparametric t-test: In the nonparametric *t*-test, the statistical significance of each gene is calculated by computing a *p* value for it without assuming the specific parametric form for the distribution of the test statistics. To determine the *p* value, a permutation procedure is used in which the class labels of the samples are permuted (10,000 to 500,000 times), and for each permutation, *t* statistics are computed for each gene. The permutation *p* value for a particular gene is the proportion of the permutations in which the permuted test statistic exceeds the observed test statistic. A *p* value cut off can be chosen for the dataset. And the FDR = number of genes test X cutoff *p* value/gene declared to be significant.

Wilcoxon (or Mann-Whitney) Rank Sum test: This test rank transforms the data and looks for genes with a skewed distribution of ranks. The rank transform smoothes the data by reducing the effect of outliers. This method is proven to be superior for decidedly nonnormal data. In general, the Wilcoxon Rank Sum test has been shown to be most conservative, with the lower FDR.

Ideal discriminator method: This method is based on the similarity of gene expression patterns on the array to a theoretical pattern that clearly discriminates between two groups. It potentially allows more flexibility in defining more complex theoretical pattern behavior.

Significant Analysis of Microarray (SAM): SAM uses a statistic that is similar to a *t*-statistic. However, it introduces a fudge factor at the denominator when calculating the *t* statistic; therefore, it underweights those genes that have a relatively small magnitude of differences and small variation within groups. It also permutes the whole dataset and sets a threshold for a FDR, instead of assigning an individual *p* value to each gene.

Overall, each statistical method has its own advantages. Each biologist may need to choose proper tests for her own dataset and, in some cases, try different statistical tests. In a situation where the most reliable list of genes is desirable, the best approach may be to examine the intersection of genes identified by different statistical test, or by the more conservative rank sum test and nonparametric *t*-test. However, if a more inclusive list of genes is desired, a higher *p* value cutoff or SAM may be more appropriate.

UNSUPERVISED ANALYSIS

Users of unsupervised analysis try to find internal structure in the dataset without any prior input knowledge. Here are some widely used analytical methods.

Hierarchical clustering: Hierarchical clustering is a simple but proven method for analyzing gene expression data by building clusters of genes with similar patterns of expression. This is done by iteratively grouping together genes that are highly correlated in their expression matrix. As a result, a dendrogram is generated. Branches in the dendrogram represent the similarities among the genes; the shorter the branch, the greater similarity of the gene expression pattern. Hierarchical clustering is also popular because it helps to visualize the overall similarities of expression profiles. The number and size of expression patterns within a dataset can be quickly estimated by biologists.

Self-organizing maps (SOM): In SOM, each biological sample is considered as a separate partition of the space, and after partitions are defined, genes are plotted using expression matrix as coordinate. To initiate SOM, the number of partitions to use must first be defined by the users as an input parameter. A map is set with the centers of each partition to be (known as centroids) arranged in an initially arbitrary way. As the algorithm iterates, the centroids move toward randomly chosen genes at a decreasing rate. The method continues until there is no further significant movement of these centroids. The advantage of SOM is that it can be used to partition the data with easy two-dimensional visualization of expression patterns. It also reduces computational requirements compared with other methods. The drawbacks of the method are that the number of partitions has to be user-defined, and genes can only belong to a single partition at a time.

Singular value deposition (SVD): SVD is also known as principal-components analysis by statisticians. SVD transforms the genomewide expression data into diagonalized "eigengenes" and "eigenarrays" space, where the eigengenes (or eigenarrays) are the unique orthogonal superpositions of the genes (or arrays). Sorting the gene expression data according to the eigengenes and eigenarrays will reduce the features of the data to their principal components and may help to identify the main patterns within the data. SVD is also a powerful technique to capture any patterns within the data that may be due to experimental artifacts. For example, array hybridization on different days or hybridization on different batches of arrays occasionally gives slight differences in gene expression patterns. SVD can also remove the artifacts without removing any genes or arrays from the dataset.

Post-data Analysis

GENE ONTOLOGY

Microarray experiments often generate large amounts of data. Even after statistical analysis, one may identify gene clusters that consist of hundreds of genes that are differentially expressed. One of the most commonly used methods to annotate the gene function is through Gene Ontology (GO, http://www.geneontology.org/). GO classifies gene function according to three organizing principles: molecular function, biological process, and cellular component. There are thousands of GO terms under each principle. Currently, there are over 9000 GO terms listed. Clusters of genes can therefore be classified by GO terms. When certain GO terms are statistically enriched in a cluster, it may suggest the possible functional significance of the cluster of the genes. However, one has to realize the limitation of GO annotation, as our understanding of the functions of the genes is still incomplete, and many genes may have multiple functions.

CONFIRMATION STUDIES

It is very important to verify array observations independently. There are two approaches: *in silico* analysis and experimental validation.

The *in silico* method compares array results with information available in the literature or other independent array expression database. Agreement between array results from other groups, especially using different array platforms, will validate the general performance of the system and provide confidence in the overall data.

Experimental validation uses independent experimental methods to assay the expression levels of genes of interest, preferably on another sample set other than the samples that have been used in the microarray analysis. The methods that have been used widely depend on the specific scientific questions. Commonly used techniques include at the mRNA level: semiquantitative RT-PCR, real-time RT-PCR (Taqman, Applied Biosystems), Northern blot, *in situ* hybridization (ISH); at the protein level: Western blot, fluorescence-activated cell sorting, enzyme-linked immunoabsorbent assays, immunofluorescence, and immunohistochemistry (IHC). Both ISH and IHC can be performed in a high throughput manner via tissue arrays. In addition, both methods provide additional information on the anatomic relationship and cellular origin of gene expression programs.

Examples of Microarray Experiments

Classically, differentiation is defined by the expression of lineage-specific markers, appearance of unique cell morphologies, or acquisition of specific biologic functions (e.g., hemoglobin synthesis in red blood cells). From a genomic perspective, differentiation can be considered as sets of gene expression programs. These gene expression programs may be self-reinforcing, sequential, or mutually exclusive depending on the specific biologic context. Thus, the biologic state of stem cells and their subsequent differentiated states are highly amenable to microarray analysis. In exploring stem cell biology, expression profiling offers a decided advantage as an experimental approach because no specific assumptions are necessary at the outset. By observing the activity of the entire genome, one can ask what major features define stem cells and characterize their differentiation programs toward specific lineages. Here we highlight several examples in the literature that apply microarray technology to tackle several of the main questions in stem cell biology.

Identification of "Stemness"

Stem cells share certain biologic properties — the capacity for self-renewal and multipotency. What are the molecular programs that underlie these properties? Several investigators have approached this problem by comparing the gene expression profiles of several embryonic and adult stem cells. By comparing the intersection of the relatively enriched genes in stem cells, a set of genes that are shared by several stem cells has been identified. It is reassuring that traditional markers of stem cells (e.g., CD34 for marrow-derived hematopoietic stem cells) were also found to be enriched in the stem cell transcriptome by microarray analysis. In addition, these results provide several candidate pathways that may be involved in regulating maintenance of stem cell fate. Intriguingly, it was observed that a large fraction of stem cell-enriched genes are genes or expressed sequence tags with no previously ascribed function, and that in the mouse an unexpected larger fraction of stem cell-enriched genes reside on chromosome 17. Similar chromosomal domains of muscle-specific transcription have been observed in *C. elegans*, and transcriptional profiling of *Drosophila* in a variety of conditions revealed significant correlation of the expression patterns for chromosomally adjacent genes. The large number and precise quantitative nature of gene expression measurements make microarray analysis particularly attractive to analyze the relationship between chromosomal clustering of genes and differentiation. Future studies comparing the transcriptional profiles of many stem cell types, particularly during a dynamic process such as the cell division cycle, will likely yield additional insights and lead to functional classification of stem cell-enriched genes.

Differentiation

The excitement about stem cells arises from their ability to differentiate into many lineages and cell types, but on a practical level, the pluripotency of stem cells present experimental difficulties in guiding and assessing their development into particular lineages *in vitro* and *in vivo*. Traditionally, one may rely on certain well-established markers of the cell types in question, and in some instances, one may verify the ability of one or more stem cells to repopulate a compartment by reconstitution experiments (e.g., reconstitution of peripheral blood cells by transplantation of hematopoietic stem cells, HSC). However, these approaches rely on a relatively small number of protein markers or are laborious and time consuming. The specificity of lineage markers has been explored to a limited extent in many cases, and many cell types of biologic and medical interest do not have well-defined markers. For instance, CD34, the classical marker for HSC, is also present in a number of other cell types and neoplasms (e.g., endothelial cells and the fibrohistiocytic tumor dermatofibrosarcoma protuberans). Expression profiling of stem cells and their differentiated derivatives can help to identify the direction and progress of the differentiation program as well as the interrelationships of the possible differentiation states to one another.

In a prime example of this strategy, Xu *et al.* observed that exposure of human embryonic stem cells to bone morphogenetic protein 4 (BMP4) induced a substantial number of trophoblast markers, including the placental hormone human chorionic gonadotropin. Thus, BMP4 is probably a key molecular switch that guides the first differentiation event of embryonic stem cells toward this extraembryonic lineage, and BMP4-treated ES cells provide a simple system to derive human trophoblastic cells for studying maternal-fetal interactions. In another powerful use of microarray technology, cell-surface markers can also be identified in a high-throughput fashion by isolating messenger RNA associated with membrane-bound polysomes. Hybridization of such selected RNAs to microarrays allows rapid identification of membrane proteins that are likely to be useful lineage markers and receptors that respond to environmental stimuli.

Stem Cell Niches

In many of the medical applications based on stem cells, introduction of stem cells into the diseased tissue somehow allows the engraftment and differentiation of the stem cells to replace the deficient cell type. Underlying these startling observations is the concept that stem cell behavior is intimately related to and controlled by their microenvironment. Specifically, the microenvironment that maintains the biologic properties of the stem cell is termed its niche, and depletion of endogenous stem cell populations in disease processes empties specific niches that guide new, exogenous stem cells to differentiate into the missing cell types.

Thus, one of the central questions in stem cell biology concerns the molecular features of the niches that govern the

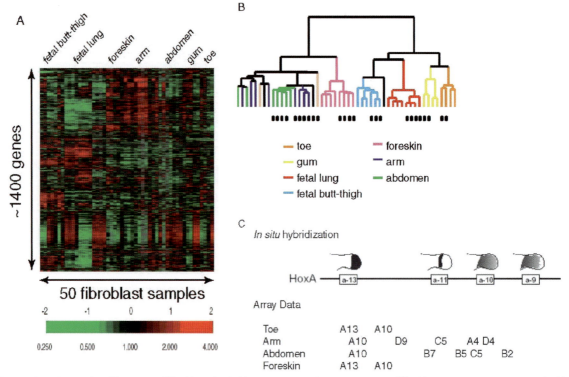

Figure 50-3. Topographic differentiation of fibroblasts identified by microarray analysis. (A) Heat map of fibroblast gene expression patterns. Fibroblasts from several anatomic sites were cultured, and their mRNAs were analyzed by cDNA microarray hybridization. Approximately 1400 genes varied by at least threefold in two samples. The fibroblast samples were predominantly grouped together based on site of origin. (B) Supervised hierarchical clustering revealed the relationship of fibroblast cultures to one another. Site of origin is indicated by the color code, and high- or low-serum culture condition is indicated by the absence (high) or presence (low) of the black square below each branch. Because fibroblasts from the same site were grouped together regardless of donor, passage number, or serum condition, topographic differentiation appeared to be the predominant source of gene expression variation among these cells. (C) *HOX* expression in adult fibroblasts recapitulates the embryonic Hox code. In a comparison of the *HOX* expression pattern in secondary axes, the schematic of expression domains of 5′ *HoxA* genes in the mouse limb bud at approximately 11.5 days postcoitum is shown on top. The *HOX* genes up-regulated in fibroblasts from the indicated sites are shown below. *HoxC5* is expressed in embryonic chick forelimbs, and *HoxD9* functions in proximal forelimb morphogenesis. (Discussed in detail in Chang *et al.*, 2002.)

behavior of fetal and adult tissue-specific stem cells. Although some of these key molecules have been identified genetically in amenable organisms, many of the molecular details that define stem cell niches, especially in mammalian systems, are incompletely understood and may be approached by microarray analysis. One approach to understanding stem niches is to explore the diversity of the normal tissue compartments, especially that provided by resident stromal cells.

An illustrative demonstration of this concept is a recent study of global gene expression patterns of fibroblasts derived from different anatomic sites of skin. Chang *et al.* cultured fibroblasts from multiple sites of human skin. Microarray analysis of their global gene expression patterns revealed that fibroblasts from different sites have distinct gene expression programs that have the stability and diversity characteristic of differentiated cell types (Figure 50-3). Some of the site-specific gene expression programs in fibroblasts included components of extracellular matrix and many cell fate signaling pathways, including members of transforming growth factor β, Wnt, receptor tyrosine kinase, and G-protein-coupled receptor signaling pathways. An intriguing hint to the specification program in fibroblasts is the maintenance of key features of the embryonic Hox code (which specifies the anterior-posterior body plan) in adult fibroblasts. Thus, stromal cells such as fibroblasts are likely to encode position-specific information in a niche that specifies the developmental potential of interacting stem cell populations. In the case of fibroblasts, because their positional identities are maintained *in vitro* (as evidenced by the fidelity of the Hox expression patterns), it is likely that efforts using stem cells to develop artificial tissue and organs will benefit from incorporation of site-specific fibroblasts or their molecular signatures that recreate the stem cell niche. Identification of the specific stem cell niches — and specific cell fate determining pathways — can thus be accelerated by a comprehensive description and understanding of the stromal cell diversity using microarray analysis.

Future Directions

The rapid evolution of microarray technology, bioinformatics techniques, and availability of new genome sequences in model organisms will present many opportunities for harnessing genomic information in stem cell and development research. Fuzzy clustering, a method that gives proportional weight to class assignment, is a valuable technique that may help to reveal more subtle and intricate relationships among various stem cells and their differentiated progenies. Additional methods of selecting mRNA or DNA fragments coupled to microarray analysis provide rapid and powerful techniques to elucidate protein subcellular localization or the interaction with DNA binding proteins. Because of its versatility and ability for revealing unexpected features of biology, microarray analysis is likely to become one of the main workhorses of the stem cell biologist.

KEY WORDS

DNA Microarray   A collection of genes spotted on a solid surface (often a glass slide), arranged in rows and columns, so that the origin of each spot is known. Depending on the type of microarray, the spots consist of cDNA sequences amplified by •• (PCR) or synthetic oligo nucleotides. Each DNA microarray may contain tens of thousands of gene spots and is therefore a powerful tool for the study of gene expression patterns on a genomic scale.

Hierarchical clustering   A method of data mining in which members of a data set are statistically grouped or clustered together according to a certain parameter (e.g., gene expression profile) in a way that maximizes intracluster similarity and minimizes intercluster similarity. In the clustering process, two clusters are merged only if the interconnectivity and closeness (proximity) between two clusters are high relative to the internal interconnectivity of the clusters and closeness of items within the clusters.

Supervised analysis   Statistical methods to identify genes that are differentially expressed between groups of samples, as well as to find genes that can be used to accurately predict the characteristics of groups.

FURTHER READING

Brown, P. O., and Botstein, D. (1999). Exploring the new world of the genome with DNA microarrays. *Nat. Genet.*, **21**(1 Suppl.), 33–37.

Butte, A. (2002). The use and analysis of microarray data. *Nat. Rev. Drug Discov.*, **1**(12), 951–960.

Chang, H. Y., Chi, J. T., Dudoit, S., *et al.* (2002). Diversity, topographic differentiation, and positional memory in human fibroblasts. *Proc. Natl. Acad. Sci. USA*, **99**(20), 12877–12882.

Chung, C. H., Bernard, P. S., and Perou, C. M. (2002). Molecular portraits and the family tree of cancer. *Nat. Genet.*, **32**(Suppl.), 533–540.

Gerhold, D. L., Jensen, R. V., and Gullans, S. R. (2002). Better therapeutics through microarrays. *Nat. Genet.*, **32**(Suppl.), 547–551.

Holloway, A. J., van Laar, R. K., Tothill, R. W., and Bowtell, D. D. (2002). Options available — from start to finish — for obtaining data from DNA microarrays II. *Nat. Genet.*, **32**(Suppl.), 481–489.

Slonim, D. K. (2002). From patterns to pathways: gene expression data analysis comes of age. *Nat. Genet.*, **32**(Suppl.), 502–528.

Xu, R. H., Chen, X., Li, D. S., *et al.* (2002). BMP4 initiates human embryonic stem cell differentiation to trophoblast. *Nat. Biotechnol*, **20**(12), 1261–1264.

51

Zebra Fish and Stem Cell Research

K. Rose Finley and Leonard I. Zon

Introduction

The utility of the zebra fish system in vertebrate biology has sparked the burgeoning development of a new and noteworthy field. Biologists across the world are taking advantage of *Danio rerio* for investigations into organogenesis, disease, and development. The small size, ease of care, and rapid generation time are but a few of the many aspects that make the zebra fish a particularly useful model organism. In addition, the optical transparency of zebra fish embryos is especially suited for investigations into hematopoiesis and stem cell biology.

Perhaps one of the most significant features of the zebra fish system is the ability to conduct large-scale forward genetic screens in an organism with high homology to the human genome. Screening imparts the ability to address specific developmental processes without any prior knowledge of what genes may be involved. The Zebra Fish Genome project performed by the Wellcome Trust Sanger Institute facilitates the positional cloning of genes uncovered via genetic screens in the zebra fish (www.sanger.ac.uk). With regard to stem cells, the zebra fish has been utilized largely to understand hematopoietic stem cells (HSCs). The success of the system makes it likely that stem cells from other organs will be explored in the zebra fish. This chapter will discuss attributes of zebra fish for investigations into hematopoiesis, strategies for conducting genetic screens in the fish, current research in the field, and finally, the future of the zebra fish system for investigations into stem cell biology.

The Zebra Fish System

MORPHOLOGICAL AND EMBRYONIC ATTRIBUTES

Zebra fish are relatively small (3 to 4 cm) and reach sexual maturity at 2 to 3 months. The fish mate year round, with females mating weekly. Because of the small size of the fish, facilities are able to maintain thousands of individuals in a compact laboratory environment. Few, if any, vertebrate model organisms allow for such population size or ease of care. Furthermore, females lay between 100 and 200 eggs per mating, permitting large-scale genetic screens as well as Mendelian approaches and analysis.

Of particular importance to the study of early blood development and hematopoietic stem cells is the unique embryonic

morphology of the zebra fish. Zebra fish embryos develop externally from the one-cell stage and are transparent, permitting embryonic development to be readily viewed under a dissecting microscope. Circulation begins by 24 hours postfertilization (hpf), and the number and morphology of blood cells may be identified under a microscope.

BLOOD FORMATION IN THE ZEBRA FISH

The hematopoietic program in vertebrates begins with mesodermal patterning along the dorsal-ventral axis. During the process of mesoderm patterning, a subset of cells gives rise to a putative population termed *hemangioblasts*. These cells are hypothesized to be a common precursor of blood and vascular lineages due to their close physical association and common gene expression during early development. Unlike many other vertebrates that form primitive HSCs in extraembryonic yolk sac blood islands, the initial site of blood formation in zebra fish is the inner cell mass (ICM), an intraembryonic tissue. The zebra fish ICM tissue is analogous to the extraembryonic mammalian blood islands in its role as the primary site of primitive erythropoiesis.

As in all vertebrates, a shift in hematopoietic sites occurs in the zebra fish. Cells within the ICM are hypothesized to populate the dorsal mesentery as well as the ventral wall of the dorsal aorta. This region is thought to be the zebra fish equivalent of the aorta–gonad–mesenephros region (AGM), an intraembryonic site believed to specify definitive HSCs in vertebrate organisms. These HSCs are then thought to colonize the developing kidney where blood formation will continue throughout the juvenile and adult life of the fish. In contrast, the primary sites of definitive hematopoiesis in mammals are the fetal liver and the adult bone marrow. Despite some differences (for example, fish erythrocytes remain nucleated), zebra fish blood is remarkably similar to mammalian blood, with all lineages—erythroid, myeloid, lymphoid, and thrombocytic—represented. Thus, the fish is a capable and applicable model organism for the study of hematopoiesis.

The identification and isolation of zebra fish homologs of factors deemed critical for normal hematopoiesis and vasculogenesis, such as the transcription factors *scl*, *lmo2*, *GATA-1*, *GATA-2*, and *c-myb*, have permitted investigations into many aspects of blood and vascular tissue formation. RNA *in situ* hybridization studies using whole zebra fish embryos provide gene expression patterns that may be used to investigate the timing and anatomy of hematopoiesis and vasculogenesis (Figure 51-1). The first appearance of vascular and blood markers is seen at the three-somite stage (11 hpf) with

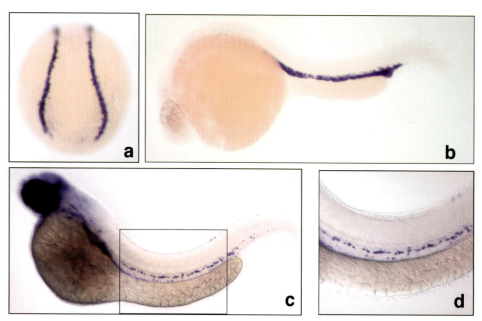

Figure 51-1. Expression patterns of specific hematopoietic factors in zebra fish embryos. (A) Whole embryo *in situ* hybridization for *scl* at 12 hpf shows the stripes of ventral mesoderm and marks putative hemangioblasts. (B) *GATA-1* expression in an embryo 24 hpf marks the ICM prior to circulation of erythroblasts. (C) *c-myb* expression at 36 hpf marks the wall of the dorsal aorta, and the zebra fish equivalent of mammalian AGM, shown enlarged in (D). (Courtesy of C. Erter.)

the expression of *GATA-2* and *scl* in two stripes of lateral plate mesoderm. Some cells within these stripes will become blood in the fish. By the 18-somite stage (18 hpf), these expression stripes have fused to form the tubular ICM structure. The ICM is fully formed by 23 hpf, just before the commencement of circulation. The presence of *c-myb* expressing cells located in the ventral wall of the dorsal aorta at 36 hpf suggests the presence of definitive HSCs in the zebra fish. Finally, by 4 days postfertilization (dpf), HSCs have migrated to the developing kidney.

Although the discrete steps necessary for the induction of hematopoietic precursors into differentiated blood lineages, as well as the program directing the different hematopoietic waves, are still unclear, the zebra fish is a powerful model organism with which to begin to unravel the story. It is evident that the process of differentiation of hematopoietic stem cells into the erythroid, myeloid, and lymphoid lineages is the result of a highly conserved gene program. It is also clear that the zebra fish, like other vertebrates, possesses discrete waves of hematopoietic events characterized by expression of specific and recognized transcription factors. As will be discussed below, zebra fish have played an important role in uncovering novel genes involved in these programs via large-scale genetic screens.

Genetic Screens in Zebra Fish

The first large-scale genetic screens in a vertebrate organism were conducted in zebra fish in Boston, Massachusetts, and Tübingen, Germany, in the mid-1990s. The ability to conduct

screens in a vertebrate organism, in addition to those conducted in such nonvertebrates as *C. elegans* and *Drosophila*, is clearly an important achievement. Such screens have already yielded scores of mutants with defects in varying stages of blood and vasculature formation, a few of which will be discussed in detail in this chapter. The power of screens in zebra fish is the ability to uncover novel genes involved in specific developmental processes in an unbiased manner.

The Boston and Tübingen ventures utilized F_2 screens in order to uncover developmental mutants (Figure 51-2). The screens began with the mating of a male fish that had been treated with the chemical ethylnitrosourea (ENU) to a wild-type female. Mutations present in the germ line of the ENU-treated male were passed to heterozygous individuals in the F_1 generation. The F_1 and F_2 generations were successively incrossed to obtain an F_3 generation. In the presence of a recessive mutation, half of an F2 family should be heterozygous for that mutation. When two F_2 fish are mated together and both are heterozygotes for the mutation, their progeny will display 25% wild-type, 50% heterozygous, and 25% homozygous mutant embryos. The Boston and Tübingen screens began with approximately 300 ENU-treated males and uncovered more than 2000 new developmental mutants.

An important aspect of the zebra fish is the ease with which mutations may be generated in the fish. The chemical ethylnitrosourea (ENU), which causes point mutations throughout much of the genome, is a popular and effective mutagen for zebra fish. However, many other chemicals and avenues are available in order to achieve mutations in the fish, all achiev-

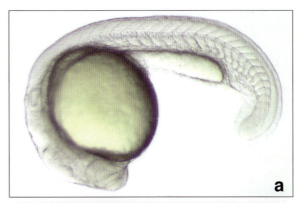

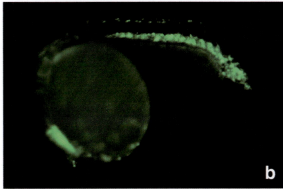

Figure 51-2. GATA-1 promoter driving GFP in transgenic zebra fish embryo at 24 hpf. (A) The transgenic embryo as seen under dissecting scope. (B) Under fluorescence the GATA-1 driving GFP expression localizes to the ICM. (GATA-1 GFP line courtesy of S. Lin; Photos courtesy of R. Wingert.)

ing slightly different results. Gamma and X rays may be utilized to create large genomic deletions and translocation events, as well as point mutations. In a newer approach, insertional mutagenesis has been employed in zebra fish. Mutations are achieved by injecting retroviruses into the 1000–2000 cell stage embryo. Although the degree of mutations achieved by this approach is significantly lower than ENU mutagenesis, the ease of cloning the mutated gene makes it an exciting resource for the field.

In addition to the F_2 diploid screens conducted in Tübingen and Boston, haploid and gynogenetic diploid screens have also been designed and carried out in zebra fish. Zebra fish embryos are able to survive three days postfertilization as haploid organisms. This allows a shortcut to uncovering mutations, as recessive alleles in F_1 fish may be uncovered in a single generation as opposed to the two generations needed in diploid screens. To create viable haploid embryos, a female fish is gently squeezed to release her eggs. Sperm from male fish is harvested from testes and UV-inactivated before adding to the unfertilized eggs. Because the UV irradiated sperm is unable to contribute any genetic material to the eggs, the embryos develop with only the genetic material of the mother. Haploid embryos have the same body plan as diploids; however, they are shorter in length and possess some devel-

opmental problems as well as a lifespan of only 3 to 5 days. Nonetheless, haploid screens have successfully uncovered many types of developmental mutations and have proven to be an extremely significant resource.

In addition to haploid embryos, gynogenetic embryos may also be utilized in zebra fish screens. By applying early hydrostatic pressure or with heat shock, unfertilized eggs from female fish can be forced to become diploid, comprised solely of the mother's genetic material. Like haploid embryos, gynogenetic diploid embryos represent the genome of one female fish, and the large size of the female's clutch allows the implementation of Mendelian genetics. Gynogenetic diploid embryos will survive to adulthood, unlike haploid embryos, and will allow screening for later developmental abnormalities in the fish. This method bypasses the lethal haploinsufficiency of haploid embryos while maintaining identical genomic information. With the use of haploid and gynogentic embryos, the genome of a female fish may be screened for recessive mutations displaying a phenotype in her progeny. These approaches provide a less burdensome method for uncovering mutations involved in specific developmental pathways.

Zebra Fish Blood Mutants

Many blood mutants have been uncovered in large-scale screens, and several others have occurred spontaneously or been discovered while screening for mutations in other developmental areas. Characterization and attempts to clone these genes have provided valuable information into hematopoiesis and its genetic program. Some mutants have also proved useful as new models for human blood diseases. The blood mutants have been divided into discrete categories based on their stage of disruption in the hematopoietic scheme. Categories of blood mutants include mesoderm patterning mutants, hematopoietic stem cell mutants, committed progenitor mutants, proliferation and/or maintenance mutants, hypochromic mutants, and photosensitive mutants. Mutants involved in the first three of these categories will be discussed in this section.

In the category of mesoderm patterning is the zebra fish mutant dubbed *spadetail* (*spt*). The *spt* gene was shown to be a T-box transcription factor necessary for correct release from mesoderm-inducing signals as well as proper formation of endoderm. The phenotype of *spt* is one of incorrect trunk formation, poorly formed dorsal somites, undifferentiated blood, disorganized dorsal vessel formation, and other characteristics indicating severe mesoderm and endoderm tissue deficiencies.

Other mesoderm patterning mutants have been identified in zebra fish, including *swirl* (*swr*) and *snailhouse* (*snh*), which have been shown to have increased dorsal tissue leading to scarcity of ventral blood. Both *swr* and *snh* phenotypes are caused by mutations in members of the bone morphogenetic protein (BMP) subgroup belonging to the TGF-β superfamily of growth factors. BMP signaling is critical for correct mesodermal patterning along the ventral axis and plays a key role in the antagonistic relationship between dorsalizing and ventralizing signals during mesoderm patterning and ultimately the correct formation of blood. Whereas these mutants are not blood mutants directly, *spadetail, swirl,* and *snailhouse* are prime examples of the importance of correct mesoderm patterning for both proper blood and vessel formation.

Another interesting blood mutant is *cloche* (*clo*), which has been shown in studies to possess severe depletion of blood and vasculature, as well as reduced expression of transcription factors necessary for both blood and vessel formation. The abnormally low expression of *scl, lmo2, GATA-2,* and *GATA-1* suggest that *clo* has a loss of blood progenitors. In addition, *clo* embryos have severely reduced expression of *flk1* or *tie1,* both shown to be expressed by blood and endothelial progenitors.

The almost complete lack of blood and vasculature in *cloche* suggests that the *clo* gene product may be responsible for maintenance or differentiation of the putative hemangioblast. Interestingly, overexpression of *scl* has been shown to partially rescue *GATA-1, flk1,* and *tie1* expression in *clo,* but has been ruled out as a candidate due to its linkage group position. The *cloche* gene has been shown to act upstream of *flk1* in the program of endothelial cell dedifferentiation. The *clo* mutant demonstrates the ability of genetic screens to uncover novel genes associated with stem cell biology.

The *bloodless* (*bls*) mutant, like *clo,* may be placed in the category of hematopoietic stem cell mutants. Embryos carrying the *bls* mutation have been shown to possess almost no erythroid cells at the onset of circulation but contain normal macrophages and carry out normal, though delayed, lymphopoiesis. Although dominant, the *bls* mutant is not totally penetrant in its display of an almost complete lack of primitive erythrocytes. Corresponding with the lack of embryonic erythroid cells are decreased levels of *scl* and *GATA-l* expression in *bls.* Levels of *GATA-2* expression are also reduced but present in *bls* mutants, likely due to the presence of *GATA-2* expressing endothelial precursors in the ICM. The *bls* mutant does in fact possess normal vessel formation and has normal *flk1* expression during embryogenesis. Interestingly, *bls* embryos produce normal red blood cells at approximately 5 dpf, after the initiation of the definitive wave of hematopoiesis. These results suggest that the *bls* gene product is responsible for maintenance or production of hematopoietic precursors during primitive hematopoiesis but does not affect production of HSCs. This mutant should provide much insight into the method of differentiation of stem cells from precursors and for the connection between primitive and definitive waves of embryonic hematopoiesis.

The *moonshine* (*mon*) mutant not only displays virtually no circulating red blood cells in early embryogenesis, but, unlike *bls,* was shown by Ransom *et al.* (1996) to fail to develop erythrocytes throughout the lifespan of the fish. GATA-2 expression in *mon* mutants is normal in the early ICM by 18 hpf while GATA-1 expression is reduced or absent corresponding to the strength of the allele. These results suggest that *mon* mutants have an early primitive erythroid lineage defect that plays a role in maintaining the lineage.

Cell Sorting and Transplantation

New techniques have recently been carried out in zebra fish by David Traver and Leonard Zon, allowing differentiated blood cell populations to be sorted via flow cytometry. Samples have been collected from the kidney, spleen, and blood and analyzed by their light scatter characteristics. Kidney tissue in the zebra fish, the site of adult definitive hematopoiesis, reveals distinct scatter populations when examined via this method.

Scatter populations of kidney tissue correspond to the major blood lineages and can be used to isolate pure populations of mature erythroid cells, myelomonocytic cells, lymphoid cells, and immature precursors. The percentage of cells in each category obtained by morphological cell counts in normal adults corresponds to the percentage of each cell type found via flow cytomoetry. Thus, the use of scatter profiles to evaluate and quantitate discrete populations of differentiated cells is an accurate and important tool in both adult wild-type and mutant zebra fish.

Cell sorting may be used to characterize the blood mutants obtained through genetic screens in the fish. The light scatter populations of mutant fish may be compared to those of wild-type fish, enabling the quantification of population deficits or increases in specific blood lineages in the mutant fish. Many primitive blood mutants exhibit a homozygous lethal phenotype making the analysis of adult homozygous mutants impossible. The analysis of blood lineages of adult heterozygous fish carrying only one copy of the mutant allele, however, can be revealed using this technique. This technique has exposed several cases of haploinsufficiency, suggesting that many genes required for embryonic and primitive hematopoiesis are additionally important in the adult fish.

The technique of flow cytometry also enables the collection and purification of GFP-tagged cells present in zebra fish transgenic lines. The creation of transgenic lines has become common in zebra fish studies as it allows visualization of specific tissues in the transparent embryo (Figure 51-2). GFP lines have been created with the GATA-1 promoter to produce fluorescent erythroid cells and the *lmo2* promoter creating zebra fish with fluorescent green blood and vasculature. GFP tagged cells in these transgenic fish can be isolated from the rest of the population of cells in the fish.

Once isolated, purified blood lineages may be collected and transplanted into recipient fish. Transplantation assays may be used to measure many different aspects of hematopoiesis in mutant and nonmutant fish alike. In one assay, hematopoietic stem cell activity in mutants is characterized by injecting kidney cells harvested from lines of fish in which the GATA-1 promoter drives GFP. The donor transgenic HSCs will produce GFP$^+$ erythrocytes, which, due to their short half-lives, are continuously made and replenished from upstream stem and progenitor cells. Many transplant recipients possess GFP+ blood cells up to six months post-transplantation, indicating the existence of a long-term repopulating stem cell in the zebra fish kidney.

Disease Models in Zebra Fish

Zebra fish mutants have been shown to resemble human diseases including hematopoietic, cardiovascular, and kidney disorders. Although other organisms such as *Caenorhabditis elegans* and *Drosophila* have been extensively studied with respect to developmental processes and may be used in screens, they do not address many of the morphological aspects unique to vertebrates. The zebra fish, however, is a vertebrate system and thus may be used to address developmental processes in the kidney, in multilineage hematopoiesis, in notochord as well as neural crest cells, among others. Zebra fish mutants have played an important role not only in clarifying vertebrate-specific developmental processes, but also as models for specific human disorders.

The hematopoietic mutant known as *sauternes* (*sau*) exhibits delayed maturation of erythroid cells and abnormal globin expression. The *sau* mutant fish develops a microcytic hypochromic anemia. Cloning of the *sau* gene revealed the gene product to be aminoleulinate synthase (ALAS2), an enzyme required for the first step of heme biosynthesis. Mirroring the fish genetic disease, humans with a mutation in ALAS2 gene have been found in a study by Edgar *et al.* to have congenital sideroblastic anemias. *Sau* is the first animal model of this disease.

The zebra fish mutant known as *retsina* (*ret*) has recently been shown by Paw *et al.* to exhibit an erythroid-specific mitotic defect with dyserythropoiesis comparable to human congenital dyserythropoietic anemia. This work identifies a gene (*slc4a1*) encoding an erythroid-specific cytoskeletal protein necessary for correct mitotic divisions of erythroid cells. In addition to modeling a human disease, this study demonstrates the notion of cell-specific mitotic adaptation.

Positional cloning of the zebra fish mutant *wiessherbst* identified the gene *ferroportin1*, a transmembrane protein responsible for the transport of iron from maternal yolk stores into circulation, solving the puzzle of the elusive iron exporter. This gene was first identified in zebra fish. Later research conducted by Njajou *et al.* determined that humans with mutations in *ferroportin1* develop the human disease type IV hemachromatosis.

Further studies are being conducted in zebra fish to tie mutant phenotypes to known as well as undocumented human disorders. The use of the zebra fish system to gain understanding of human disease states has been proven and is evolving as technologies advance. Screens continue to uncover mutant phenotypes with specific developmental abnormalities, many of which will play a role in deciphering human disease states.

The Future of the System

In a recent study conducted by Langenau *et al.* transgenic zebra fish lines were created which resulted in acute T cell leukemia in the fish. In this investigation, transgenic fish were created with the lymphoid-specific RAG2 promoter driving mouse *c-myc*, a gene known to function in the pathogenesis of leukemias and lymphomas. In addition, a chimeric transgene consisting of myc fused to GFP allowed the visualization of the spread of tumors from the thymus of affected fish. Tumors in these transgenic fish spread from origins in the thymus to skeletal muscle and abdominal organs. The leukemic transgenic fish from this study may now be used for suppressor/enhancer genetic screens in order to identify components either lessening or worsening the leukemic progress.

Another recent body of work has focused on zebra fish regeneration. The zebra fish is able to regenerate fins, a unique characteristic of some lower vertebrates. Little is known about the molecular and cellular processes involved in regeneration, and genetic screens have recently been employed to identify essential genes. The processes of de-differentiation, patterning, and proliferation, as well as the existence of pluripotent cells in adult tissues, are being examined in zebra fish and remain extremely compelling for the field of stem cell biology. Although at a very early stage, regeneration studies may also address human tissue repair, stem cell transplantation, and even tissue engineering.

Pioneering studies have been and are currently being conducted that reveal an exciting future for the zebra fish system. Large-scale genetic screens, cell-sorting techniques, and transgenic technologies have been utilized in elegant investigations into hematopoiesis and stem cell biology. Further dissection of developmental processes and their connections to human disease will surely benefit from the utility of this model organism.

ACKNOWLEDGMENTS

We thank Caroline Erter for providing (Figure 51-1) and Rebecca Winger and Shou Lin for providing (Figure 51-2). We thank David Traver, James Amatruda, and Elizabeth Patton for critical review of this manuscript, and Trista North, Jenna Galloway, Alan Davidson, and Noelle Paffet-Lugasy for beneficial discussions. K.R.F. is supported by grants from the National Institutes of Heath and the Howard Hughes Medical Institute. L.I.Z. is an associate investigator of the Howard Hughes Medical Institute.

KEY WORDS

Gynogenesis   Manipulation of ploidy by suppression of the polar body in an egg lacking paternal genetic contribution.

Hemangioblast   The postulated bipotential cell that would give rise to hematopoietic and endothelial cell lineages in vertebrates.

Hematopoiesis   The formation and development of blood cells.

FURTHER READING

Amatruda, J. F., and Zon, L. I. (1999). Dissecting hematopoiesis and disease using the zebrafish. *Dev. Bio.* **216**, 1–15.

Bahary N., and Zon, L. I. (1998). Use of the zebrafish (*Danio rerio*) to define hematopoiesis. *Stem Cells* **16**, 89–98.

Davidson, A. J., and Zon, L. I. (2000). Turning mesoderm into blood: the formation of hematopoietic stem cells during embryogenesis. *Curr. Top. Dev. Biol.* **50**, 45–60.

Dooley, K., and Zon, L. I. (2000). Zebrafish: a model system for the study of human disease. *Curr. Opin. Genet. Dev.* **10**, 252–256.

Gaiano, N., Amsterdam, A., Kawakami, K., Allende, M., Becker, T., and Hopkins, N. (1996). Insertional mutagenesis and rapid cloning of essential genes in zebrafish. *Nature* **383**, 829–832.

Patton, E. E., and Zon, L. I. (2001). The art and design of genetic screens: zebrafish. *Nat. Gen. Rev.* **2**, 956–966.

Poss, K., Keating, M. T., and Nechiporuk, A. (2003). Tales of regeneration in zebrafish. *Dev. Dyn. Rev.* **226**, 202–210.

Ransom, D. G., Haffter, P., Odenthal, J., Brownlie, A., Vogelsang, E., Kelsh, R. N., Brand, M., van Eeden, F. J. M., Furutani-Seiki, M., Granato, M., Hammerschmidt, M., Heisenberg, C–P., Jiang, Y–J., Kane, D. A., Mullins, M. C., and Nusslein-Volhard, C. (1996). Characterization of zebrafish mutants with defects in embryonic hematopoiesis. *Development* **123**, 311–319.

Thisse, C., and Zon, L. I. (2002). Organogenesis—heart and blood formation from the zebrafish point of view. *Science* **295**, 393–572.

Traver, D., and Zon, L. I. (2002). Walking the walk: migration and other common themes in blood and vascular development. *Cell* **108**, 731–734.

52

Neural Stem Cells: Therapeutic Applications in Neurodegenerative Diseases

Rodolfo Gonzalez, Yang D. Teng, Kook I. Park, Jean Pyo Lee, Jitka Ourednik, Vaclav Ourednik, Jaimie Imitola, Franz-Josef Mueller, Richard L. Sidman, and Evan Y. Snyder

Introduction

Despite the presence of endogenous neural stem cells (NSCs) in the mammalian brain, it is recognized that intrinsic "self-repair" activity for the most devastating of injuries is inadequate or ineffective. This poor "regenerative" ability, particularly in the adult central nervous system (CNS), may because of the limited number and restricted location of native NSCs and/or limitations imposed by the surrounding microenvironment, which may not be supportive or instructive for neuronal differentiation. NSCs expanded *ex vivo* in culture and then implanted into regions needing repair may overcome those limitations related simply to inadequate numbers of NSCs near the defective region. Whether the environment may also inhibit exogenous NSCs from surviving or differentiating toward replacement cells is a possibility. However, several transplantation experiments have suggested that neurogenic cues are transiently elaborated during degenerative processes (perhaps recapitulating developmental cues) and that exogenous NSCs are able to sense, home in, and respond appropriately to those cues. In other words, NSCs appear to respond *in vivo* to neurogenic signals not only when they occur appropriately during development but even when induced at later stages by certain neurodegenerative processes.

In this chapter, we review some of the work that has been performed in animal models of CNS diseases, where transplanted NSCs have mediated a therapeutic effect. These disorders include rodent models of genetic and acquired (e.g., traumatic and ischemic) neurodegeneration, inheritable metabolic disorders, age-related degeneration, and neoplasms. These conditions often have widespread neural cell loss, dysfunction, or both. The disseminated nature of the pathology in these diseases is not readily treated by conventional transplantation approaches in which a solid tissue graft or a limited number of nonmotile cells are delivered to a restricted area. Similarly, most gene therapy approaches tend to fall short because of their limited "sphere of influence" following injection of the vector into the CNS parenchyma. The use of bone marrow transplantation, even with hematopoietic stem cells, is typically inadequate because the cells do not efficiently broach the blood–brain barrier. Cells from nonneural organs, if they transdifferentiate into neural cells at all (a controversial prospect), do so far too inefficiently to be therapeutically reliable for cell replacement in these conditions.

NSCs, on the other hand, circumvent many of these obstacles. Because they differentiate robustly into neural cells, integrate seamlessly into neural parenchyma as multiple neural cell types (both neuronal and glial), respond to normal developmental and regeneration cues, and migrate (even long distances) to multiple, disseminated areas of neuropathology, NSCs appear to be ideally suited for the molecular and cellular therapies required by extensive, diffuse (even "global") degenerative processes. Examples of such widespread neurodegenerative conditions include myelin disorders, storage diseases, motor neuron degeneration, dementing conditions such as Alzheimer's disease, and ischemic and traumatic pathologies such as stroke. Some diseases appear to be restricted in their involvement — Parkinson's disease is localized to the mesostriatum, Huntington's to the caudate, spinal cord contusion to a few spinal segments, and cerebellar degeneration to the hindbrain. However, even these disorders require that cell replacement be distributed evenly over relatively large terrain, that multiple neural cell types be replaced even in a given region, or both. Again, these needs are best accomplished by a migratory, responsive, multipotent neural progenitor even if transplantation is directed toward a more circumscribed CNS region.

Therapeutic Potential of Neural Stem Cells

The clinical potential of the NSC is rooted in its inherent biologic properties. The ability of NSCs and their progeny to develop into integral cytoarchitectural components of many regions throughout the host brain as neurons, oligodendrocytes, astrocytes, or even immature neural progenitors makes them capable of replacing a range of missing or dysfunctional neural cells. Although the field of neural repair has tended to place emphasis on the replacement of missing neurons in acute and chronic neurodegenerative diseases or oligodendrocytes in demyelinating disorders, it is becoming increasingly evident that simultaneously replacing a diseased cell's "neighboring cells" — typically astrocytes — may be of equal importance because of the indispensable trophic, guidance, and detoxification role such "chaperone" cells may play. Fortuitously, part of the biological repertoire of an NSC is to generate the variety of cells that constitute the "fabric" of a given neural region. NSCs in particular differentiation states will spontaneously produce a variety of neurotrophic factors that may serve trophic, protective, or both functions — for example, glial cell line-derived neurotrophic factor (GDNF), brain-derived neurotrophic factor (BDNF), nerve growth factor (NGF), and neurotrophin-3 (NT-3). NSCs also inherently express most of the "housekeeping" enzymes and factors necessary for any cell to maintain normal metabolism.

NSCs may also be readily engineered *ex vivo* to express a variety of molecules that either are not produced by NSCs or are not produced in therapeutically adequate quantities. The NSCs are amenable to various types of viral vector transduction as well as to other gene transfer strategies, such as lipofection, electroporation, and calcium-phosphate precipitation. Following transplantation, such engrafted NSCs may then be used as cellular vectors for the *in vivo* expression of exogenous genes of developmental and/or therapeutic relevance. Such gene products can be delivered either to circumscribed regions or, if necessary, to more widely disseminated areas throughout the host CNS. Because they display significant migratory capacity as well as an ability to integrate widely throughout the brain when implanted into germinal zones, NSCs may help to reconstitute enzyme and cellular deficiencies in a global fashion (Figure 52-1).

The ability of true NSCs to adjust to their regions of engraftment probably obviates the need for abstracting stem cells from specific CNS regions. Furthermore, NSCs appear to possess a tropism for degenerating CNS regions (Figure 52-2). NSCs may be attracted — across long distances — to

regions of neurodegeneration in brains of all ages, including old age. Under some circumstances, an environment in which a particular type of neural cell has degenerated creates a milieu that directs the differentiation of an NSC toward maintaining homeostasis, including replenishment of a specific deficient neural cell type. Through mechanisms yet to be determined, it appears that some neurodegenerative processes (e.g., those associated with apoptosis) elaborate neurogenic signals that recapitulate developmental cues to which NSCs can respond (Figure 52-3).

It would seem that NSC-mediated cell replacement would only be feasible in "cell autonomous" disease states — that is, diseases in which the pathology is restricted to a particular cell whose lifespan is short-circuited but within an extracel-

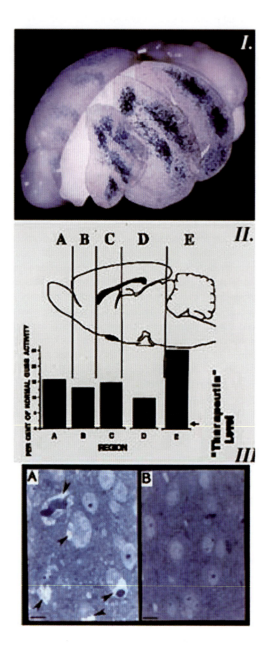

Figure 52-1. Widespread engraftment of NSCs expressing GUSB throughout the brain of the MPS VII mouse. (A) Brain of a mature MPS VII mouse after receiving a neonatal intraventricular transplant of murine NSCs expressing GUSB. Donor NSC-derived cells, identified by their X-gal histochemical reaction for expression of the *LacZ* marker gene, have engrafted throughout the recipient mutant brain. Coronal sections — placed at their appropriate level by computer — show these cells to span the rostral-caudal expanse of the brain. (B) Distribution of GUSB enzymatic activity throughout brains of MPS VII NSC transplant recipients. Serial sections were collected from throughout the brains of transplant recipients and assayed for GUSB activity. Sections were pooled to reflect the activity within the demarcated regions. The regions were defined by anatomical landmarks in the anterior-to-posterior plane to permit comparison among animals. The mean levels of GUSB activity for each region (n = 17) are presented as the percentage of average normal levels for each region. Untreated MPS VII mice show no GUSB activity biochemically or histochemically. Enzyme activity of 2% of normal is corrective based on data from liver and spleen. (C) Decreased lysosomal storage in a treated MPS VII mouse brain at eight months. (1) Extensive vacuolation representing distended lysosomes (arrowheads) in both neurons and glia in the neocortex of an 8-month-old, untransplanted control MPS VII mouse. (2) Decrease in lysosomal storage in the cortex of an MPS VII mouse treated at birth from a region analogous to the untreated control section in panel 1. The other regions of this animal's brain showed a similar decrease in storage compared with untreated, age-matched mutants in regions where GUSB was expressed. Scale bars = 21 μm.

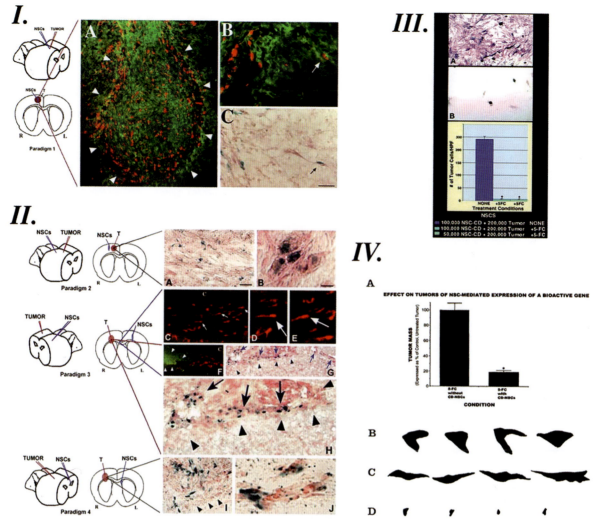

Figure 52-2. Neural stem cells display extensive tropism for pathology in the adult brain and can express bioactive genes within such pathological situations: Evidence from intracranial gliomas. (A) NSCs migrate extensively throughout a brain tumor mass *in vivo* and "trail" advancing tumor cells. Paradigm (*1*), in which NSCs are implanted directly into an established experimental intracranial glioblastoma, is illustrated schematically. (*1*) The virulent and aggressively invasive CNS-1 glioblastoma cell line, used to create the tumor, has been labeled *ex vivo* by transduction with GFP cDNA. The NSCs stably express *lacZ* and produce βgal. This panel, processed for double immunofluorescence using an anti-βgal antibody (NSCs) and an anti-GFP antibody (glioblastoma cells), shows a section of brain (under low power) from an adult nude mouse sacrificed 10 days after NSC injection into the glioblastoma; arrowheads mark where the tumor mass interfaces with normal tissue. Donor βgal + NSCs can be seen extensively distributed throughout the mass, interspersed among the tumor cells. This degree of interspersion by NSCs after injection occurs within 48 hours. Interestingly, although NSCs have extensively migrated and distributed themselves within the mass, they largely stop at the junction between tumor and normal tissue except where a tumor cell is infiltrating normal tissue; then, NSCs appear to follow the invading tumor cell into surrounding tissue. (*1 and 2*) Detail of the trailing of individual glioblastoma cells migrating from the main tumor bed (*2*). High-power view, under fluorescence microscopy, of single migrating infiltrating GFP+ tumor cells in apposition to βgal+ NSCs (white arrow). (*3*) Costaining with Xgal (the arrow points to the *lacZ*-expressing NSCs) and with neutral red (for the elongated glioblastoma cells). The NSC is in direct juxtaposition to a single migrating, invading the neutral red+, spindle-shaped tumor cell (arrow), with the NSC "piggy backing" the glioma cell. Scale bar = 60 μm. (B) NSCs implanted at various intracranial sites far from the main tumor bed migrate through normal adult tissue toward glioblastoma cells. (*1 and 2*) Same hemisphere: A section through the tumor from an adult nude mouse six days following NSC implantation caudal to tumor. Panel *1* shows a tumor populated as pictured under low power in Figure 52-1A. Note Xgal+ NSCs interspersed among neutral red+ tumor cells. (*2*) High-power view of NSCs in juxtaposition to islands of tumor cells. (*3–8*) Contralateral hemisphere. (*2–5*) Views through the corpus callosum (c), where βgal+ immunopositive NSCs (arrows) are migrating from their site of implantation on one side of the brain toward the tumor on the other. Two NSCs indicated by arrows in panel *3* are viewed at higher magnification in panels *4* and *5* to show the classic elongated morphology and leading process of a migrating neural progenitor oriented toward its target. (*6*) βgal+ NSCs are "homing in" on the GFP+ tumor, having migrated from the other hemisphere. In panel *7*, and magnified further in panel *8*, the Xgal+ NSCs (arrows) have entered the neutral red+ tumor (arrowheads) from the opposite hemisphere. (*9 and 10*) Intraventricular: A section through the brain tumor of an adult nude mouse six days after NSC injection into the contralateral cerebral ventricle. (*9*) Xgal+ NSCs are distributed within the neutral red+ main tumor bed (edge delineated by arrowheads). (*10*) At higher power, the NSCs are in juxtaposition to migrating islands of glioblastoma cells. Fibroblast control cells never migrated from their injection site in any paradigm. All Xgal positivity was corroborated by anti-βgal immunoreactivity. Scale bars = 20 μm (*1*; also applies to *3*) 8 μm (*2*), 14 μm (*4 and 5*), 30 μm (*6 and 7*), 15 μm (*8*), 20 μm (*9*), and 15 μm (*10*). (C) Bioactive transgene (cytosine deaminase, *CD*) remains functional (as assayed by *in vitro* oncolysis) when expressed within NSCs. CNS-1 glioblastoma cells were cocultured with murine CD-NSCs. (*1*) Co-cultures unexposed to 5-FC grew healthily and confluent, (*2*) whereas plates exposed to 5-FC showed dramatic loss of tumor cells represented quantitatively by the histograms (* = p < 0.001). The oncolytic effect was identical whether 1×10^5 CD-NSCs or half that number were co-cultured with a constant number of tumor cells. (Subconfluent NSCs were still mitotic at the time of 5-FC exposure and thus subject to self-elimination by the generated 5-FU and its toxic metabolites.) (D) Expression of *CD* delivered by NSCs *in vivo* as assayed by reduction in tumor mass. The size of an intracranial glioblastoma populated with CD-NSCs in an adult nude mouse treated with 5-FC was compared with that of tumor treated with 5-FC but lacking CD-NSCs. These data, standardized against and expressed as a percentage of a control tumor populated with CD-NSCs receiving no treatment, are in the histograms in panel *1*. These measurements were derived from measuring the surface area of tumors; camera lucidas of them are in panels *2–4*. Note in the large areas of panel *2* a control non-5-FC-treated, tumor-containing CD-NSCs and in panel *3* a control 5-FC-treated, tumor-lacking CD-NSCs as compared to panel *4* the dramatically smaller tumor areas of the 5-FC-treated animal, which also received CD-NSCs (~80% reduction, * = p < 0.001), suggesting both activity and specificity of the transgene. (*3*) The lack of effect of 5-FC on tumor mass when no CD-bearing NSCs were within the tumor was identical to panel *4* the effect of CD-NSCs in the tumor without the gene being employed.

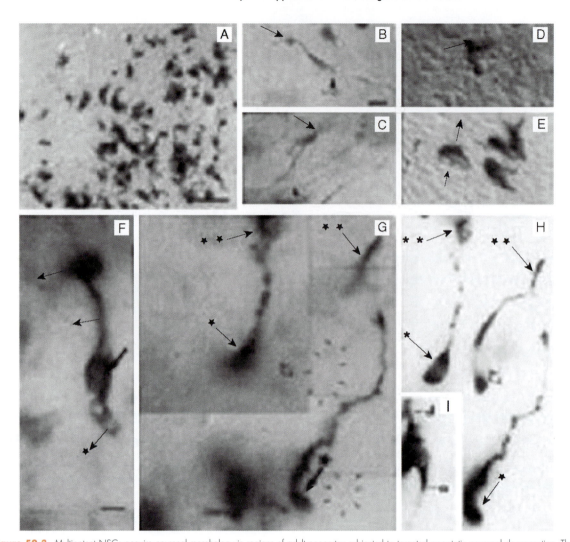

Figure 52-3. Multipotent NSCs acquire neuronal morphology in regions of adult neocortex subjected to targeted apoptotic neuronal degeneration. They differentiate into only glia or remain undifferentiated in intact control cortex. (A) Engrafted X-Gal⁺ glia at low magnification 6 weeks following transplantation at 12 weeks; (B–E) higher magnification. (B and C) Donor-derived cells with astroglial features: Small, ovoid cell bodies (arrows) with few, short processes often extending as perivascular end-feet (arrowheads). (D) Small soma of a donor-derived presumptive glial cell (arrow) compared with a much larger (≈30 μm), unlabeled, host pyramidal neuron (small arrows). (E) Donor-derived cells with oligodendroglial features (arrow): Multiprocessed, ensheathing neuronal processes (short arrows). (F–I) A total of 15 ± 7% of engrafted cells in regions of neurodegeneration developed neuronal morphology, resembling pyramidal neurons within layer II/III 6 weeks following transplantation at 12 weeks. (F and G) Donor-derived cells with neuronal morphology (large arrows): Large somata (20–30 μm diameter) comparable to residual host pyramidal neurons (small arrows in panel g outline two host neurons, visualized under DIC, each with the characteristic large nucleus and prominent nucleolus of a pyramidal neuron), 300–600 μm presumptive apical dendrites positioned between host (small arrows) and donor (large arrows) neurons of similar morphology and size, and presumptive axons. (F) The dark object at the upper end of the presumptive dendrite is another X-Gal⁺ cell out of the plane of focus. (H) DCM image of panel g, collapsing multiple planes of focus. The cell has the characteristic large nucleus of a pyramidal neuron. All but the terminal dendrite of the soma of the cell is out of the plane of focus in panel G. Cells and neurons are indicated to cross-reference views of the same field in panels G and H. (I) Two overlapping donor-derived pyramidal neurons in different focal planes show a characteristic large nucleus, prominent nucleolus, and axon of the overlying cell and a prominent dendrite of the underlying cell, imaged through multiple focal planes of this thick section under DCM. The identification of the previously mentioned cells was supported by immunocytochemical and ultrastructural analysis under electron microscopy. In addition, one could determine that donor-derived neurons received synaptic input from the host and were myelinated by host oligodendrocytes, further supporting their incorporation into the host cytoarchitecture (a: axon; d: dendrite; cells: *, ★★, **, and open arrow; and ★: neurons). Bars = 25 μm.

lular milieu is otherwise normal. Conversely, a "cell nonautonomous" condition, where normal cells die because of an inhospitable microenvironment, would appear not to be amenable to NSC therapy given that "replacement" cells would presumably meet a similar fate. Surprisingly, however,

NSCs may nevertheless be useful in such cell-extrinsic pathological conditions. It appears that NSCs, particularly in their immature, undifferentiated state, may be more resistant to certain stressors — various toxic metabolites, oxidizing agents — than more mature cells. Furthermore, if the NSCs are not inherently more resistant, they can readily be genetically engineered *ex vivo* to become more resistant.

Despite their extensive plasticity, NSCs never produce cell types inappropriate to the brain (e.g., muscle, bone, or teeth) or yield neoplasms. The use of NSCs as graft material in the CNS may be considered almost analogous to hematopoietic stem cell-mediated reconstitution of the bone marrow.

Therefore, the biological repertoire of the NSC, if harnessed, may provide multiple strategies for addressing CNS dysfunction. Some of these approaches have already shown promise experimentally in animal models of neurodegeneration. Some illustrative examples are briefly described in the next sections of this chapter.

Gene Therapy Using Neural Stem Cells

As stated previously, the ability of NSCs to deliver therapeutic gene products in an immediate, direct, sustained, and perhaps regulated fashion as normal cytoarchitectural components throughout the CNS may overcome some of the limitations of standard viral and cellular vectors (reviewed by Park *et al.*, 2002). The feasibility of this strategy was first demonstrated in a mouse mutant characterized by a single gene defect in all cells, including those in the CNS, leading to their death. The particular mouse modeled the lysosomal storage disease mucopolysaccharidosis type VII (MPS VII) caused by a deletion mutation of the β-glucuronidase (*GUSB*) gene. This incurable inheritable condition is characterized by neurodegeneration in mice and by progressive mental retardation in humans. Although the particular disease was rare, it served as a model for neurological diseases whose etiology stemmed from a genetically based loss of function. GUSB-secreting NSCs were implanted into the cerebral ventricles of newborn MPS VII mice, allowing the cells access to the subventricular zone (SVZ), a germinal zone from which the cells were disseminated throughout the brain (creating "chimeric forebrains"). These enzyme-producing cells, now in residence as normal cerebral constituents, not only metabolized lysosomal storage normally for themselves but also cross-corrected mutant cells throughout the brains of recipient mice (Figure 52-1). Employing a similar strategy, retrovirally transduced NSCs implanted into the brains of fetal and neonatal mice (particularly into periventricular regions) have successfully mediated brainwide expression of other enzymes — for example, β-hexosaminidase, a deficiency of which leads to the pathological accumulation of GM_2 ganglioside.

Findings such as these have helped to establish the paradigm of using NSCs for the transfer of other factors of therapeutic or developmental interest into and throughout the CNS. Although NSCs express baseline amounts of particular enzymes and neuroprotective factors, NSCs can be genetically modified to enhance their production of these molecules or to produce additional molecules that might enhance their therapeutic potential. For example, NSCs have been used to deliver NT-3 within the hemisectioned rat spinal cord, to express NGF and BDNF within the septum and basal ganglia, to provide tyrosine hydroxylase to the parkinsonian striatum,

and to express myelin basic protein (MBP) throughout the dysmyelinated cerebrum. In one illustrative set of experiments, NSCs that overexpressed NGF were transplanted into the striatum of a rat lesioned with quinolinic acid, a toxin used to emulate some of the neuronal loss see in Huntington's disease. Delivery of NGF by the engrafted NSCs appeared to reduce the size of the lesion and to promote sparing of host striatal neurons. When implanted into the septum of aging rats, NGF-overexpressing NSCs appeared to blunt the typical cognitive decline by preventing age-related atrophy of forebrain cholinergic neurons. Well integrated into the host tissue, the engrafted NSCs continued to produce NGF until nine months after grafting.

That NSCs appear to have a strong affinity for sites of pathology and will migrate extensive distances (even from the opposite cerebral hemisphere) to home in on them makes the NSC unique and valuable as a gene delivery vehicle. For example, a type of pathology particularly elusive to gene therapeutic interventions has been the brain tumor, especially the glioma. Gliomas are so exceptionally migratory and infiltrative that they elude even the most effective surgical, radiation, or gene therapeutic strategy. However, the ability of transgene-expressing NSCs to "track down" and deliver therapeutic gene products directly to these widely dispersed and invasive neoplastic cells makes them a potentially valuable adjunct in the treatment of these aggressive tumors.

Although this therapeutic approach — exploiting the normal biological behavior of NSCs for transplantation-based gene therapy — is being extended to animal models of many neurological disorders, it is important to recognize that each pathophysiological process, each animal model, and each therapeutic molecule must be assessed and optimized individually.

Cell Replacement Using Neural Stem Cells

It is postulated that NSCs persist within the CNS long beyond cerebrogenesis to maintain homeostasis following perturbations in the CNS. Presumably the mechanisms that allow the endogenous NSC to perform this function are preserved when the cells are isolated from the CNS, expanded in culture, and reimplanted into the damaged CNS. Indeed, it appears that when NSCs are so implanted, they respond by shifting their pattern of differentiation toward replenishing the missing cell type.

The study that first demonstrated this phenomenon was performed on a model of experimentally induced apoptosis of selectively targeted pyramidal neurons in the adult mammalian neocortex. When transplanted into this circumscribed, neuron-depleted region, 15% of the NSCs altered the differentiation path they would otherwise have pursued in the intact adult, nonneurogenic mammalian neocortex (i.e., to become glia). Instead, they differentiated specifically into pyramidal neurons, partially replacing that lost neuronal population (Figure 52-3). Of these, a subpopulation spontaneously sent axonal projections to their proper targets in the contralateral cortex. Outside the borders of this small region of selective neuronal death, the NSCs yielded only glia. Thus, neurodegeneration appeared to create a milieu that recapitulated embryonic development cues (e.g., for cortical neurogenesis), and the NSCs were sufficiently sensitive to detect and respond to this micromolecular alteration — possibly to therapeutic advantage.

Apoptosis is implicated in a growing number of both neurodegenerative and normal developmental processes. Whether this differentiation shift occurs only in response to signals associated with apoptosis or in response to other types of cell death as well remains to be determined. Most neurological diseases are characterized by a mixture of neuropathological processes, making dissection of the key stimuli complex. For example, we have observed that following an ischemic insult, NSCs will robustly repopulate infarcted regions of the postdevelopmental cortex and differentiate into cortical neurons in these regions. However, even hypoxic-ischemic cerebral injury — the quintessential example of necrotic and excitotoxic injury — is characterized by an apoptotic phase. An added level of complexity may be the tempo at which apoptotic signals are elaborated — an acute burst may be more instructive than a more languid burst.

Studies are ongoing to identify the key molecules that direct NSC fate during and following various kinds of neural injury. Intriguingly, we are starting to learn that cytokines released during an inflammatory reaction — those emanating from microphages and microglia as well as from parenchymal damage (e.g., SDF1-α) — play a pivotal role in "beckoning" NSCs. Coincidentally, we are starting to recognize that many processes regarded as neurodegenerative — including amyotrophic lateral sclerosis (ALS), Alzheimer's disease, tumors, and stroke — are characterized by a prominent inflammatory signature. Although each is different in its etiology, region of involvement, and course, inflammation may be their common denominator from the "viewpoint" of the NSC.

"Global" Cell Replacement Using Neural Stem Cells

The targeted-apoptosis model described in the previous section exemplifies a circumscribed type of neural cell loss. However, the pathologic lesions of many neurological disorders are often widely — even globally — dispersed throughout the brain. Such conditions include the neurodegenerative disorders of childhood (e.g., inborn errors of metabolism, storage diseases, leukodystrophies, and neuronal ceroid lipofuscinoses) and hypoxic-ischemic encephalopathy as well as some adult CNS diseases (e.g., multiple sclerosis, Alzheimer's disease, and ALS). Treatment for these disorders requires widespread replacement of genes, cells, or both as well as the regeneration, protection, or both of broad networks of neural circuitry. The ability of inherently migratory NSCs to integrate into germinal zones from which they can be "launched" and the inclination of NSCs to travel long distances to home in on pathologic regions make these cells ideally — and perhaps uniquely — suited for this task.

Mouse mutants characterized by CNS-wide white matter disease provided the first models for testing the hypothesis that NSCs might be useful against neuropathologies requiring widespread neural cell replacement. The oligodendrocytes — the myelin-producing cells — of the dysmyelinated *shiverer* (*shi*) mouse are dysfunctional because they lack MBP, a molecule essential for effective myelination. Therapy, therefore, requires widespread replacement with MBP-expressing oligodendrocytes. NSCs transplanted at birth (employing the intracerebroventricular implantation technique described previously for the diffuse engraftment of enzyme-expressing NSCs to treat global metabolic lesions) resulted in engraftment throughout the *shi* brain with repletion of significant amounts of MBP. Of the many donor cells that differentiated into oligodendrocytes, a subgroup myelinated ~40% of host neuronal processes. In some recipient animals, the symptomatic tremor decreased (Figure 52-4). Therefore, "global" cell replacement was shown to be feasible for some pathological conditions if cells with stem-like features are employed. This approach has been extended to other myelin-impaired animal models, such as the experimental allergic encephalomyelitis mouse model of multiple sclerosis as well as rodent mutant models of Palezeus-Merzbacher, Krabbe, and Canavan leukodystrophies. Recently, it was demonstrated that neural progenitor-stem cells of human origin have a similar remyelinating capacity in the *shi* mouse brain. Viewed more parochially, the ability of NSCs to remyelinate is of significant importance because impaired myelination plays such an important role in many genetic (e.g., leukodystrophies and inborn metabolic errors) and acquired (e.g., traumatic, infectious, asphyxial, ischemic, and inflammatory) neurodegenerative processes. However, viewed more broadly, the ability of the NSC to replace this particular type of neural cell throughout the brain bodes well for its ability to replace other classes of neural cell across the broad terrains often demanded by other categories of complex neurodegenerative diseases.

Neural Stem Cells Display an Inherent Mechanism for Rescuing Dysfunctional Neurons

The examples cited previously highlight the idea that an abnormal environment can direct the behavior of a grafted NSC. However, they leave the impression that the exogenous NSC alone fills gaps. The situation is more complex and richer. We are beginning to learn that the NSC and the injured host engage in a dynamic series of ongoing reciprocal interactions, each instructing the other. Under instruction from exogenous NSCs, the injured host nervous system also contributes to its own repair. These important stem cell phenomena were first illustrated in a few examples.

The effect of NSCs in directly rescuing endangered host neurons was first evinced in a series of experiments in aged rodents in which the nigrostriatal system was impaired (Figure 52-5). Parkinson's disease is a degenerative disorder charac-

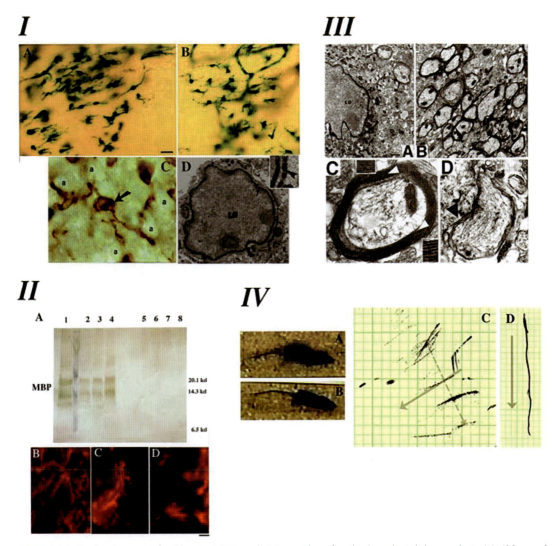

Figure 52-4. "Global" cell replacement is feasible using NSC transplantation: evidence from the dysmyelinated *shi* mouse brain. (A) NSCs engraft extensively throughout the *shi* dysmyelinated brain, including within white tracts, and differentiate into oligodendrocytes. *LacZ*-expressing, βgal-producing NSCs were transplanted into newborn *shi* mutants and analyzed systematically at intervals between two and eight weeks following engraftment. Coronal sections through the *shi* brain at adulthood demonstrated widely disseminated integration of Xgal⁺ donor-derived cells throughout the neuraxis, similar to the pattern seen in Figure 52-1A in the MPS VII mutant mouse. (*1 and 2*) Donor-derived Xgal⁺ cells (higher magnification) in sections through the corpus callosum possessed characteristic oligodendroglial features (small, round, or polygonal cell bodies with multiple fine processes oriented toward the neural fiber tracts). (*3*) Close-up of a donor-derived anti-β immunoreactive oligodendrocyte (arrow) extending multiple processes toward and beginning to enwrap large, adjacent axonal bundles (a) viewed on end in a section through the corpus callosum. That cells such as those in panel A *1–3* and in panel B *2–4* were oligodendroglia was confirmed by the electron micrograph in panel A *4* and in panel C, demonstrating their defining ultrastructural features. A donor-derived Xgal⁺ oligodendrocyte (LO) can be distinguished by the electron dense Xgal precipitate typically localized to the nuclear membrane (arrow), ER (arrowhead), and other cytoplasmic organelles. The ER is magnified in the inset to demonstrate the unique crystalline nature of individual precipitate particles. (B) MBP expression in mature transplanted and control brains. (*1*) Western analysis for MBP in whole brain lysates. The brains of three transplanted *shi* mutants (lanes 2–4) expressed MBP at levels close to those of an age-matched unaffected mouse (lane 1, positive control) and significantly greater than the amounts seen in untransplanted (lanes 7–8, negative control) or unengrafted (lanes 5–6, negative control) age-matched *shi* mutants. (Identical total protein amounts were loaded in each lane.) (*2–4*) Immunocytochemical analysis for MBP. (*2*) The brain of a mature unaffected mouse was immunoreactive to an antibody to MBP (revealed with a Texas red-conjugated secondary antibody). (*3 and 4*) Age-matched engrafted brains from *shi* mice similarly showed immunoreactivity. Because untransplanted *shi* brains lack MBP, MBP immunoreactivity has classically been a marker for normal donor-derived oligodendrocytes in transplant paradigms. (C) NSC-derived "replacement" oligodendrocytes are capable of myelination of *shi* axons. In regions of MBP-expressing NSC engraftment, *shi* neuronal processes became enwrapped by thick, better-compacted myelin. (*1*) Two weeks after transplant, a donor-derived, labeled oligodendrocyte (LO), recognized by extensive Xgal precipitate (p) in the nuclear membrane, cytoplasmic organelles, and processes, was extending processes (arrowheads) to host neurites and was beginning to ensheathe them with myelin (m). (*2*) If engrafted *shi* regions (*1*) were followed to four weeks, the myelin began to appear healthier, thicker, and better-compacted (arrows) than that in age-matched untransplanted control mutants. (*3*) By six weeks after transplant, these matured into even thicker wraps; ~40% of host axons were ensheathed by myelin (white arrowheads; MDLs are evident) that was dramatically thicker and better-compacted than that of *shi* myelin (*4*, black arrowhead) from an unengrafted region of an otherwise successfully engrafted *shi* brain). (D) Functional and behavioral assessment of transplanted *shi* mutants and controls. The *shi* mutation is characterized by the onset of tremor and a "shivering gait" by postnatal week 2 or 3. The degree of motor dysfunction in animals was gauged in two ways: by blindly scoring periods of standardized, videotaped cage behavior of experimental and control animals and by measuring the amplitude of tail displacement from the body's rostral-caudal axis (an objective, quantifiable index of tremor). Video freeze-frames of (4.4a) unengrafted and (4.4b) successfully engrafted *shi* mice. (4.4a) The whole body tremor and ataxic movement observed in the unengrafted symptomatic animal causes the frame to blur, a contrast with (*1*) the well-focused frame of the asymptomatic transplanted *shi* mouse. Of transplanted mutants, 60% evinced nearly normal-appearing behavior (*2*) and attained scores similar to normal controls. (*3 and 4*) Whole body tremor was mirrored by the amplitude of tail displacement (dotted gray arrow), measured perpendicularly from a line drawn in the direction of the animal's movement. Measurements were made by permitting a mouse, whose tail had been dipped in India ink, to move freely in a straight line on a sheet of graph paper. (*3*) Large degrees of tremor cause the tail to make widely divergent ink marks from the midline, representing the body's axis (solid gray arrow). (*4*) Absence of tremor allows the tail to make long, straight, uninterrupted ink lines on the paper congruent with the body's axis. The distance between points of maximal tail displacement from the axis was measured and averaged for transplanted and untransplanted *shi* mutants and for unaffected controls (dotted gray arrow). Panel *3* shows data from a poorly engrafted mutant that did not improve with respect to tremor, whereas panel *4* reveals lack of tail displacement in a successfully engrafted asymptomatic mutant. Overall, 64% of transplanted *shi* mice examined displayed at least a 50% decrement in the degree of tremor or "shiver." Several showed no displacement.

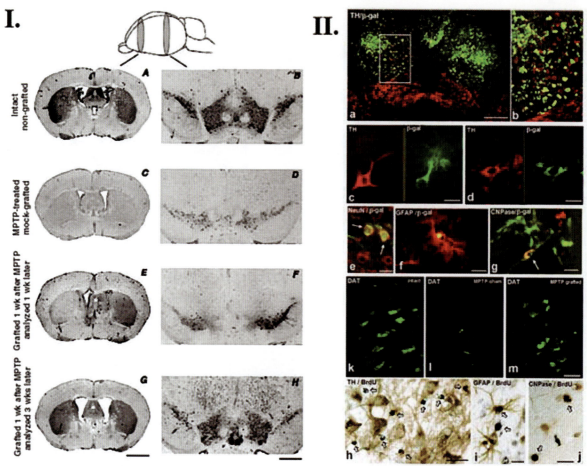

Figure 52-5. NSCs possess an inherent mechanism for rescuing dysfunctional neurons: evidence from the effects of NSCs in the restoration of mesencephalic dopaminergic function. (A) TH expression in mesencephalon and striatum of aged mice following MPTP lesioning and unilateral NSC engraftment into the substantia nigra-ventral tegmental area (SN–VTA). A model that emulates the slow dysfunction of aging dopaminergic neurons in SN was generated by giving aged mice repeated high doses of MPTP. Schematic indicates the levels of the analyzed transverse sections along the rostrocaudal axis of the mouse brain. Coronal sections are through the striatum in the left column and through the SN–VTA area in the right column. Immunodetection of TH (black cells) shows the normal distribution of DA-producing TH+ neurons in (2) coronal sections in the intact SN–VTA and (1) their projections to the striatum. Within one week, MPTP treatment caused extensive and permanent bilateral loss of TH immunoreactivity in both (3) the mesostriatal nuclei and (4) the striatum. Shown in this example, and matching the time point in 7 and 8, is the situation in a mock-grafted animal 4 weeks after MPTP treatment. Unilateral (right side) stereotactic injection of NSCs into the nigra is associated, within one week after grafting, with substantial recovery of TH synthesis within (6) the ipsilateral DA nuclei and (5) their ipsilateral striatal projections. By three weeks after transplant, however, the asymmetric distribution of TH expression disappeared, producing TH immunoreactivity in (8) the midbrain and (7) striatum of both hemispheres that approached the immunoreactivity of (1 and 2) the intact controls and gave the appearance of mesostriatal restoration. Similar observations were made when NSCs were injected four weeks after MPTP treatment (not shown). Bars: 2 mm (left), 1 mm (right). Note the ectopically placed TH+ cells in panel 8. These are analyzed in greater detail with the entire SN in B. (B) Immunohistochemical analyses of TH, DAT, and BrdU+ cells in MPTP-treated and grafted mouse brains. The initial presumption was that the NSCs had replaced the dysfunctional TH neurons. However, examination of the reconstituted SN with dual βgal and TH ICC showed that (1 and 3) 90% of the TH+ cells in the SN were rescued, host-derived cells, and (4) only 10% were donor-derived cells. Most NSC-derived TH+ cells were just above the SN ectopically (blocked area in panel 1; enlarged in panel 2). These photomicrographs were taken from immunostained brain sections from aged mice exposed to MPTP, transplanted one week later with NSCs and sacrificed after three weeks. The following combinations of markers were evaluated: (1–4) TH with βgal, (5) NeuN with βgal, (6) GFAP with βgal, (7) CNPase with βgal, (8) TH with BrdU, (9) GFAP with BrdU, and (10) CNPase with BrdU. Anti-DAT-stained areas are revealed in the SN of (11) intact, (12) mock-grafted, and (13) NSC-grafted brains. Fluorescence filters for Alexa Fluor 488 and Texas red and a double-filter for both types of fluorochromes were used to show antibody binding. (3, 4, and 8–10) Single-filter exposures; (1, 2, and 5–7) double-filter exposures. (1) Low-power overview of the SN–VTA of both hemispheres. Most TH+ cells within the nigra are of host origin (~90%), a much smaller proportion (~10%) are donor derived (close-up in panel 4). Although a significant proportion of NSCs differentiated into TH+ neurons, many of these resided ectopically, dorsal to the SN (boxed area in panel 1 enlarged in panel 2; and a high-power view of the donor-derived cell that was also TH+ in panel 3), where the ratio of donor-to-host cells was inverted: ~90% donor derived compared with ~10% host derived. Note the near absence of a βgal-specific signal in the SN–VTA, whereas ectopically, many of the TH+ cells were double labeled and thus NSC derived. (5) NSC-derived non-TH neurons (NeuN+, arrow), (6) astrocytes (GFAP+), and (7) oligodendrocytes (CNPase+, arrow) were also seen, both within the mesencephalic nuclei and dorsal to them. (10) The DAT-specific signal suggests that the reconstituted mesencephalic nuclei in the NSC-grafted mice were functional DA neurons comparable to those seen (8) in intact nuclei but not (9) in MPTP-lesioned, sham-engrafted controls. This further suggests that the TH+ mesostriatal DA neurons affected by MPTP are functionally impaired. Note that (9) sham-grafted animals contain only punctate residual DAT staining within their dysfunctional fibers, whereas DAT staining (8) in normal and (10) in engrafted animals was normally and robustly distributed both within processes and throughout their cell bodies. (11–12) Any proliferative BrdU+ cells after MPTP insult, graft, or both were confined to glial cells, whereas (11) the TH+ neurons were BrdU−. This finding suggested that the reappearance of TH+ host cells was not the result of neurogenesis but rather the recovery of extant host TH+ neurons. Bars = 90 μm (1), 20 μm (3–5), 30 μm (6), 10 μm (7), 20 μm (8–10), 25 μm (11), 10 μm (11), and 20 μm (12).

terized by a loss of midbrain dopamine (DA) neurons with a subsequent reduction in striatal DA. The disease, in addition to incapacitating thousands of patients, has long served as a model for testing neural cell replacement strategies. Transplantation therapy for this CNS disorder has a long history (for a review, see Isacson *et al.*). It was the neural disease first treated clinically by neural transplantation, using primary tissue from human fetal ventral mesencephalon to replace DA-expressing cells. Indeed, in this disease, the limitations of fetal tissue grafts in not only rodent and primate models of Parkinson's disease but also in clinical trials were first recognized. These limitations include, on one hand, short graft survival and limited integration of the grafts and, on the other hand, the possibility of unregulated DA production in improper regions leading adversely to dyskinesias. Given the storied role of Parkinson's disease in the development of cellular therapies, it is appropriate that a model of this disease should have also played a pivotal role in revealing a little suspected but powerful therapeutic action that NSCs may play in preserving degenerating host cells by some heretofore unheralded mechanisms that are nevertheless inherent to stem cell biology.

In the hope that NSCs might spontaneously differentiate into DA neurons when implanted into a DA-depleted region of the CNS, unmanipulated murine NSCs were implanted unilaterally into the substantia nigra of aged mice exposed systemically to high-dose 1-methyl-4-phenyl-1,2,3,6-tetrahydropyridine (MPTP), a neurotoxin that produces a persistent impairment of mesencephalic DA neurons and their striatal projections. The NSCs not only migrated from their point of implantation and integrated extensively within both hemispheres but also were associated with a dramatic reconstitution of DA function throughout the mesostriatal system. Although there was spontaneous conversion of a subpopulation of donor NSCs into dopaminergic neurons in DA-depleted areas, contributing to nigral reconstitution, most (80–90%) dopaminergic neurons in the "reconstituted midbrain" were actually host cells that had been "rescued" by factors produced constitutively by the NSCs with which they were juxtaposed and that had not become neurons. These chaperone cells constitutively produce substantial amounts of neurosupportive agents. One such prominent molecule was GDNF, a factor known to be neuroprotective of ventrally located neurons (including DA neurons and spinal-ventral horn cells). A similar observation is beginning to emerge from the implantation of human NSCs into the MPTP-lesioned, subhuman primate model of Parkinson's disease.

A sense for the extent of the cross-talk also became evident when examining rodent models of hypoxia–ischemia, a common cause of neurological disability in adults and children. Hypoxia–ischemia causes much of its damage from extensive loss of cerebral parenchyma and the cells and connections that reside there. When NSCs are implanted into these regions of extensive degeneration (particularly when transiently supported by biodegradable scaffolds), robust reciprocal interactions ensue spontaneously between the exogenous implant and the injured host brain, substantially reconstituting parenchyma and anatomical connections as well as reducing parenchymal loss, secondary cell loss, inflammation, and scarring. Similar results are observed in the hemi-resected adult rodent spinal cord in which evidence of an up-regulated host neuronal regenerative response is noted, resulting in significant functional improvement.

Indeed, the ability of engrafted NSCs to exert a protective and regenerative influence on degenerating host neural systems because of their intrinsic expression of trophic factors is being observed in an increasing number of conditions. For example, the implantation of murine and human NSCs into the spinal cords of the SOD1 transgenic mouse model of ALS, a disease characterized by virulent motor neuron degeneration, has been pivotal in protecting these ventral horn cells from death, preserving motor and respiratory function, blunting disease progression, and extending life. NSCs can similarly protect other neuron pools, promote motor axonal outgrowth following traumatic spinal cord injury, preserve infarcted regions of cerebrum, induce vascularization of reconstituted regions of cortical parenchyma, and inhibit inflammation and scarring following traumatic or ischemic insult.

Together, these observations suggest that exogenous NSCs may not only replenish inadequate pools of endogenous NSCs to compensate for missing neural cells but also reactivate or enhance endogenous regenerative and protective capacities. It also highlights a heretofore unanticipated mechanism by which NSCs may exert a therapeutic effect — to be added to their more traditional roles in direct cell replacement and gene therapy.

Neural Stem Cells as the Glue That Holds Multiple Therapies Together

Experimentally, one may eliminate a particular cell type, lesion a particular region of CNS, knock out a particular gene, or choose a mouse strain in which, by chance, certain mutations have spontaneously occurred. However, most human neurodegenerative diseases are not as "clean." They are quite complex. And complex diseases, such as those affecting the nervous system, will require complex and multifaceted solutions — including pharmacological, gene and molecular, cell replacement, tissue engineering, angiogenic, anti-inflammatory, antiapoptotic, pro-regenerative, pro-neurite outgrowth-promoting therapies. The NSC, as a key player in a set of fundamental developmental mechanisms, may serve as the glue that holds many of these strategies together. How they may be intelligently, effectively, safely, and practically orchestrated in actual patients will require a good deal of careful investigation.

For example, as our sophistication about disease processes grow, we are beginning to learn that more than one neural cell type probably needs to be replaced in a disease. For example, in a disease like ALS, a disorder characterized by progressive motor neuron degeneration, we are beginning to learn that astrocyte replacement may be just as critical as motor neuron replacement. Conversely, in multiple sclerosis, a white matter disease characterized by oligodendrocyte degeneration, replacing neurons and their axonal connections may be critical for the restoration of function. These same caveats may apply to many diseases in which replacement of multiple cell types may be the key to neurological reconstitution and to damaged milieu reconstruction. Because of their ability to develop into multiple integral, cytoarchitectural components of many regions throughout the host brain, NSCs may be able to replace a range of missing or dysfunctional neural cell types within the same region. This is important in the likely situation in which return of function may require the reconstitution of the milieu of a given region (e.g., not just the neurons but also the chaperone cells — the glia and support cells) to nurture, detoxify, guide, and/or myelinate the neurons. As noted previously, the NSCs, especially in particular differentiation states, express certain genes of intrinsic interest (many neurotrophic factors, lysosomal enzymes, angiogenic factors, anti-inflammatory molecules, antioxidants, etc.), or they can be engineered *ex vivo* to do so.

The challenge remains, however, to coordinate these multifaceted therapies so that they work in concert synergistically and not at cross purposes.

Summary

The ability of NSCs to migrate and integrate throughout the brain as well as to disseminate a foreign gene product is of great significance for the development of new therapies for neurodegenerative diseases in humans. Lethal, hereditary neurodegenerative diseases of childhood, such as the gangliosidoses, leukodystrophies, neuronal ceroid lipofuscinoses, and other storage diseases, result in lesions throughout the CNS. Diseases of adult onset (e.g., Alzheimer's disease) are also diffuse in their pathology. Even acquired diseases such as spinal cord injury, head trauma, and stroke are broader in their involvement than is typically assumed. Such abnormalities may benefit from the multifaceted approach NSCs may enable. First, they may replace a range of cell types — not only neurons to reconstitute neural connections but also oligodendrocytes to elaborate myelin and astrocytes to serve trophic, guidance, protective, and detoxification functions. Second, NSCs may deliver exogenous genes that might restore normal metabolism, complement a factor deficiency, support the survival of a damaged host neuron, neutralize a hostile milieu, counteract a growth-inhibitory environment, promote neurite regrowth, and reform stable and functional contacts. As noted previously, many molecules are produced inherently by NSCs when they are in particular states of differentiation. The production of other molecules might require *ex vivo* genetic engineering. NSCs appear to have their greatest therapeutic effect when used in the early stages of a degenerative process or in the subacute phase following injury.

A major requirement for the better use of NSCs will be a better understanding of the pathophysiology of the diseases to be targeted — that is, knowing what aspects require repair and which cell type or types require replacement or rescue.

Another challenge will be determining how to exploit stem cell biology for chronic conditions — in other words, how the acute milieu can be recreated in the chronic environment such that NSCs behave therapeutically there as well.

Among the methodological hurdles will be to devise how, when, where, and with what frequency to deliver NSCs to adults with disseminated diseases in regions not fed by readily accessible germinal zones, for example, the spinal cord in which the central canal is no longer patent. Linked to this is the need for a better understanding of the methods for augmenting yet controlling the propagation and the phenotype specification of NSCs *in vitro* and *in vivo*.

Another question has come to dominate the stem cell field: What is the most effective way to obtain NSCs for therapy? Should they be obtained directly from regions of neuroectodermal origin, and if so, should that be from a fetal source? Can an adult source be equally effective? Should stable lines of NSCs be established that can be used for all patients, or should NSCs be abstracted from each patient to be used as an autograft on a case-by-case basis? The degree to which the immune system presents a barrier to stem cell transplantation will likely determine the answers to this question and its subquestions. If ES cells (from the inner cell mass of blastocysts) are directed to become NSCs *in vitro*, will such cells be equally as safe and effective?

The success in isolating stem-like cells from the CNS (the first "solid organ stem cells" discovered) and their therapeutic potential also gave rise to the search for and successful isolation of stem-like cells from other nonneural organ systems, including bone marrow mesenchyme, muscle, skin, retina, and liver, for the purposes of repairing those tissues. Whether such *nonneural* stem cells can yield *neural* stem cells — through metaplasia or transdifferentiation — remains unresolved and exceptionally controversial. Can NSCs be derived efficiently and effectively from nonneural organs with outcomes as good as from NSCs themselves?

A better understanding of fundamental NSC biology will be required before human NSCs can be transplanted efficaciously in true clinical settings. Nevertheless, progress in this regard is being made. Several NSC lines have been established from the human fetal telencephalon and spinal cord that seem to emulate many of the appealing properties of their rodent counterparts: They differentiate *in vitro* and *in vivo* into neurons, astrocytes, and oligodendrocytes; they follow appropriate developmental programs and migrational pathways similar to endogenous precursors following engraftment into developing rodent and subhuman primate brain (Figure 52-6); they express foreign genes *in vivo* in a widely disseminated manner; and they can replace missing neural cell types when grafted into various mutant mice or rodent models of injury. The principal differences between human NSCs and mouse NSCs so far seem to be the length of the cell cycle (up to four days in the human) and the predilection of human NSCs to senesce (after ~100 cell divisions), obstacles that are being actively addressed. If human NSCs behave in lesioned subhuman primate brains with safe and effective engraftment and foreign gene expression as they seem to do in rodents, then

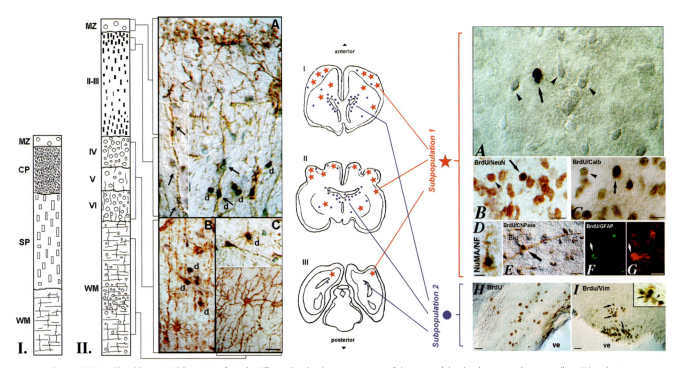

Figure 52-6. Clonal human NSCs migrate from the VZ into the developing neocortex. Schematics of the developing monkey neopallium (A) at the time of transplantation. (*1*) 12–13 wpc and (*2*) at the time of death (16–17 wpc). (*3–5*) Photomicrographs from selected locations spanning the neopallium. (Their location relative to the schematic is indicated by brackets.) (*3*) Injected into the left lateral ventricle and having integrated throughout the VZ, the human NSC-derived cells (d), identified by their BrdU immunoreactivity (black nuclei), migrated along the monkey's radial glial processes through the neopallial wall to reach their temporally appropriate destination in the nascent superficial layers II and III (*3*), where they detached from the radial glia and took up residence as neurons. Arrows indicate climbing (donor- and host-derived) cells positioned along the processes of the vimentin-positive host radial glia. Some cells (inset) are still attached to these fibers and in the process of migration. (*4 and 5*) Immature, donor, human NSC-derived astrocytes intermixed with host-derived astrocytes in deeper cortical lamina, having differentiated as expected for that site and time. (B) Segregation of the fates of human NSCs and their progeny into two subpopulations in the brains of developing Old World monkeys. Schematics (left) and photomicrographs (right) illustrate the distribution and properties of clonal human NSC-derived cells. Human NSCs (labeled with BrdU) dispersed throughout and integrated into the VZ. From there, clonally related human NSC-derived cells pursued one of two fates. Those donor cells that migrated outward from the VZ along radial glial fibers into the developing neocortex constituted one pool or subpopulation. (*4–12*) The differentiated phenotypes of cells in this subpopulation (*1*) (stars in the schematic), particularly in layers II and III. (*4*) An human NSC-derived BrdU+ cell (black nucleus, arrow) — likely a neuron according to its size, morphology, large nucleus, and location — is shown (under Nomarski optics) intermingled with the monkey's similar neurons (arrowheads) in neocortical layers II and III. The neuronal identity of such donor-derived cells is confirmed by immunocytochemical analysis. High-power photomicrographs of human donor-derived cells integrated into the monkey cortex double-stained with antibodies against BrdU and cell type-specific markers: (*5*) NeuN and (*6*) calbindin for neurons (arrows, donor-derived cells; arrowheads, host-derived cells). (*8*) CNPase for oligodendroglia (arrow). (BrdU+ is the black nucleus in the CNPase+ cell; the arrowhead indicates the long process emanating from the soma.) (*9, 10*) GFAP for astroglia antibody to BrdU revealed through fluorescein; (*9*) antibody to GFAP revealed through Texas red. The human origin of the cortical neurons is independently confirmed in panel *7*, where the human-specific nuclear marker NuMA (black nucleus) is co-localized in the same cell with neurofilament immunoreactivity. Progeny from this same human NSC clone were allocated to a second cellular pool — subpopulation 2 (dots in the schematic; arrows in panels *9* and *10*) — that remained mainly confined to the SVZ and stained only for an immature neural marker. (Vimentin co-localized with BrdU is easily seen in inset (arrows); the arrowhead indicates the host vimentin-positive cell.) Some members of subpopulation 2 were identified within the developing neocortex (dots) intermixed with differentiated cells. Panels *9* and *10* use immunofluorescence; the other immunostains use a DAB-based color reaction. Scale bars = 30 µm (*4–6*) and 20 µm (*7–12*). (d: human NSC-derived cells, MZ: marginal zone, CP: cortical plate, SP: subplate, WM: white matter, and II–VI: cortical layers).

human trials might be warranted for testing their value against certain genuine clinical neurodegenerative diseases. Through a careful and circumspect series of experiments and trials, we may learn whether we, indeed, have found within Nature's own toolbox a powerful and versatile therapeutic tool — one of the goals of experiments that started more than 15 years ago.

KEY WORDS

Acute degeneration Damage restricted to a circumscribed region (both in space and time); multiple neural cells types must be replaced (for example, stroke, CNS injury, and spinal injury).

Chronic degeneration Progressive neurodegenerative diseases requiring the replacement of all neural cell types in large brain regions, in the face of ongoing degeneration and possibly the presence of toxic molecules.

Demyelination disorders Mainly of two classes; disorders in which endogenous myelinating cells are genetically impaired and fail to myelinate during development, and disorders that result from loss/failure of remyelination in the adult (for example, multiple sclerosis, allergic encephalomyelitis, and injury).

Neural stem cells (NSCs) The most primordial cells of the nervous system. They generate the array of specialized cells throughout the CNS (and probably the peripheral, autonomic, and enteric nervous systems as well). NSCs are operationally defined by their ability: (1) to differentiate into cells of all neural lineages (i.e., neurons, ideally of multiple subtypes; oligodendroglia; and astroglia) in multiple regional and developmental contexts; (2) to self-renew; and (3) to populate developing and/or degenerating CNS regions.

FURTHER READING

Lindvall, O., Kokaia, Z., and Martinez-Serrano, A. (2004). Stem cell therapy for human neurodegenerative disorders — how to make it work. *Nat. Med.* **10**(Suppl.), S42–50.

Park, K. I., Ourednik, V., Taylor, R. M., Aboody, K. S., Auguste, K. I., Lachyankar, M. B., Redmond, D. E., and Snyder, E. Y. (2002). Global gene and cell replacement strategies via stem cells. *Gene Ther.* **9**, 613–624.

Snyder, E. Y., Daley, G. Q., and Goodell, M. (2004). Taking stock and planning for the next decade: realistic prospects for stem cell therapies for the nervous system. *J. Neurosci.* **76**, 257–168.

Spinal Cord Injury

John W. McDonald, Daniel Becker, and James Huettner

Spinal cord injury (SCI) is a major medical problem because there currently is no way to repair the central nervous system (CNS) and restore function. In this chapter, we focus on embryonic stem (ES) cells as an important research tool and potential therapy. We review studies that have used ES cells in spinal cord repair. We conclude that progress has been good, that knowledge is still too limited, and that harnessing the potential of ES cells will be important for solving the problem of SCI.

Problem

Nearly 12,000 people suffer a traumatic SCI every year, and about a quarter of a million Americans are living with this devastating condition. There are also four to five times as many spinal cord injuries caused by medical conditions such as multiple sclerosis, ALS (Lou Gehrig's disease), polio, HTLV-1, metabolic deficits, stenosis, and disc herniation.

The consequences of SCI depend on the level at which the cord is damaged. In its severest form, SCI causes paralysis and loss of sensation throughout the body, inability to control bowel and bladder function, trouble controlling autonomic functions such as regulation of blood pressure, and inability to breathe or cough. The long-term disability from SCI results not only from the initial loss of function but also from the complications that accumulate. Up to 30% of individuals with SCI are hospitalized every year for medical complications. Patients who maintain their body in the best condition for nervous system repair will benefit most from future therapeutic strategies.

Spinal Cord Organization

Unlike the brain, the spinal cord has its white matter (nerve axons) on the outside and gray matter (cell bodies) on the inside (Figure 53-1). The gray matter houses neurons that project to the periphery at that level to control movement and receive sensory signals. The white matter carries axonal connections to and from the brain, and half of all those axons are myelinated. Most traumatic SCIs also injure the incoming and outgoing (or afferent and efferent) peripheral nerves at the injury level. In most SCIs, however, the caudal cord remains intact beginning several segments below the injury level. Cir-

cuitry in those segments can produce reflexes by activating a ventral motor neuron (MN) to produce muscle contraction when it receives a sensory stimulus from the periphery of the body (Figure 53-1).

The distal spinal cord also contains groups of nerve cells that generate the patterns of activity needed for walking and running. Because the upper spinal cord plays only a limited role in controlling these pattern generators, people with injury to the cervical spine can walk or ride a stationary bicycle if their muscles are stimulated appropriately.

Injury

Traumatic injury occurs when broken fragments of bone and ligament impinge on the soft spinal cord, which is no wider than the thumb. The cord responds by swelling and the swelling compounds the initial injury by reducing venous blood flow, causing a secondary venous infarct in the central part of the cord. This causes the initial injury site to rapidly enlarge into a hole in the middle of the spinal cord (Figure 53-2). Because a donut-like rim of viable tissue usually remains at the level of injury, SCI affects preferentially gray matter.

A second wave of delayed cell death occurs during the weeks after a SCI. It removes mostly oligodendrocytes from adjacent white matter tracts. Since each oligodendrocyte myelinates 10 to 40 different axons, loss of one oligodendrocyte leads to exponential loss of myelin and function.

The problem does not stop with the secondary wave of cell death, however. An injured, underactive nervous system may be unable to adequately replace cells, particularly glial cells that normally turn over. Therefore, individuals with SCI may experience a slow, progressive loss of neurological function over long periods in addition to complications from their initial injury.

Spontaneous Regeneration

A growing body of evidence indicates that the capacity for spontaneous regeneration may be much greater than previously perceived. Data obtained since the 1960s has demonstrated that new cells are added to the nervous system continually; these cells include neurons in at least two brain regions, the hippocampus and olfactory bulb. Moreover, glial cells are frequently born and are capable of regeneration.

Nonetheless, our ability to maximize spontaneous regeneration is limited. The recent detection of endogenous stem cells within the spinal cord raises hope that such cells can be

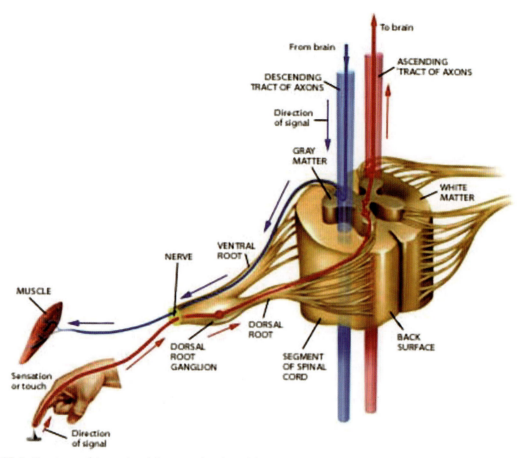

Figure 53-1. Organization of the spinal cord. A segment of cord reveals the butterfly-shaped gray matter at the core and a ring of white matter. The main components of the gray matter are neuronal cell bodies and glial cells and blood vessels. The outer white matter also contains astrocytes and blood vessels, but it consists mostly of axons and oligodendrocytes glial cells that wrap axons in white, insulated myelin. Axonal tracts that ascend in the cord convey sensory messages received from the body; the descending tracts carry motor commands to muscles. Reproduced with permission of Alexander and Turner Studio, FL, © 2002 Edmond Alexander.

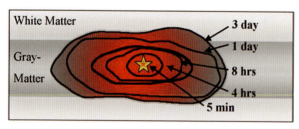

Figure 53-2. Evolution of primary to secondary injury in the spinal cord. Following the initial trauma, the injury rapidly enlarges as a consequence of secondary injury, particularly over the first 24 hours.

harnessed to repair the damaged spinal system. Although the birth of new neurons from these progenitors has never been demonstrated within the spinal cord, new glial cells, including oligodendrocytes and astrocytes, are continually being added. In fact, injury stimulates cell birth. Our laboratory recently showed that nearly 2 million cells are born in the spinal cord each day, though most eventually die. Collectively, these data suggest that cell turnover does occur in the nervous system, albeit much more slowly than in other organs of the body. Moreover, we recently showed that patterned neural activity can stimulate cell birth, which suggests that behavioral modification could be important for maximizing cord regeneration and functional recovery.

Limitations and Approaches to Repair and Redefining Goals

Although some limited spontaneous regeneration is known to occur, dramatic self-repair of the nervous system does not take place. A growing body of evidence suggests that factors in the nervous system actively inhibit regeneration. Such factors include inhibitory proteins in the cord, which guide regrowing connections, and scar tissue, which contains chondroitin sulfate and proteoglycans.

Figure 53-3 and Figure 52-15 outline the general strategies for repairing the damaged spinal cord. Complete cure of nervous system injury is not practical or required. Partial

396

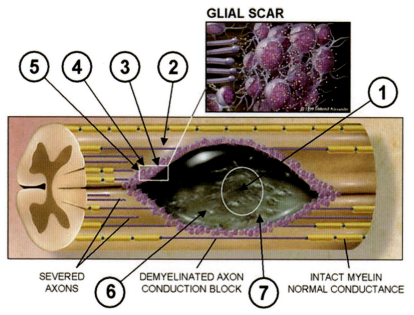

GLIAL SCAR

SEVERED AXONS

DEMYELINATED AXON CONDUCTION BLOCK

INTACT MYELIN NORMAL CONDUCTANCE

Figure 53-3. Common strategies toward regeneration of the damaged spinal cord. (A) Prevention of progression of secondary injury: Necrotic and apoptotic cell death would be prevented by antiexcitotoxic drugs and antiapoptotic treatments. (B) Compensation for demyelination: Chemicals that prevent conduction block in demyelinated areas, and agents that encourage surviving oligodendrocytes to remyelinate axons, would be provided. Lost oligodendrocytes would be replenished. (C) Removal of inhibition: Agents that block the actions of natural inhibitors of regeneration or drugs that down-regulate expression of inhibitory proteins would be provided. (D) Promotion of axonal regeneration: Growth factors that promote regeneration (sprouting) of new axons would be provided. (E) Direction of axons to proper targets: Guidance molecules would be provided, or their expression would be increased in host cells. (F) Creation of bridges: Bridges would be implanted into the cyst, which would provide directional scaffolding that encourages axon growth. (G) Replacement of lost cells: Cells capable of generation of all cell types (progenitor cells or ES cells) would be implanted. Substances that induce undifferentiated cells to replace dead cells would be provided. Also, transplanted cells to deliver regenerative molecules would be used. Reproduced with permission of Alexander and Turner Studio, FL, © 2002 Edmond Alexander.

repair translates into proportionally greater recovery of function. For example, only about 10% of the functional connections are required to support locomotion in cats, and humans missing more than half the gray matter in the cervical spinal cord can walk fairly normally.

One strategy for spinal cord repair is to transplant stem cells or other biomaterials to fill the cyst that forms at the eye of the injury. This cyst acts as a physical barrier to the growth of anatomically intact axons in the surrounding donut of white matter.

As well as filling in the cyst, it may be necessary to repair axons that no longer function properly because they have lost their myelin or have been inappropriately myelinated. Several approaches have been used to overcome this problem, which can manifest itself as a segmental conduction block. For example, potassium channels on dysfunctional axons can be blocked pharmacologically, and preexisting or transplanted oligodendrocytes and their progenitors can be encouraged to make new myelin.

Another strategy is to remove or mask the effects of proteins in the glial scar around the cyst that actively inhibit the regrowth of new connections. Antibodies that block the inhibitory effects of these proteins promote sprouting. Moreover, certain growth factors promote the self-repair of physically broken connections.

It is important to understand the feasibility of various repair strategies and to redefine appropriate goals of repair. If we rank the preceding strategies in terms of likely success, it becomes clear that it will be difficult to persuade axons to regrow across a lesion, extend down the spinal cord, and make connections with the appropriate target cells. Remyelination seems more feasible because it occurs continually at a steady rate in the damaged spinal cord. It is important to investigate all the possible strategies, however, because some will materialize sooner and some later. Most importantly, we must understand that multiple strategies delivered over time will be more important than the elusive magic bullet.

When defining appropriate goals for therapy, improving patients' quality of life must rank first. To this end, strategies that limit complications are important. Moreover, individuals with SCI often value small gains in function more highly than larger gains, such as walking. Thus, the top goal for most patients is recovery of bowel, bladder, or sexual function. Distant seconds include recovery of respiratory function and use of a single hand.

Spinal Cord Development

To understand spinal cord regeneration, it is necessary to understand spinal cord development. In humans, oligodendrocytes are found in cultures of fetal spinal cord at 7 and 12 weeks of gestation. Myelination begins at 10 to 11 weeks of gestation and continues throughout the second year of life. Signals derived from axons regulate the growth of progenitor cells and the survival of oligodendrocytes. Other locally synthesized growth factors appear to control the balance between OP proliferation and OP differentiation.

Many neuronal phenotypes arise from various progenitor pools during CNS development, but most of the pathways are poorly understood. One of the best described is the generation of spinal MNs, which involves several steps. Through bone morphogenetic protein, fibroblast growth factor (FGF), and Wnt signaling, ectodermal cells obtain a rostral neural character. In response to caudalizing signals, such as retinoic acid (RA), these progenitors acquire a spinal positional identity. Through the ventralizing action of Sonic hedgehog (Shh), spinal progenitor cells gain their MN phenotype.

These findings raise the question of whether ES cells can be shunted along specific pathways to produce specific neural cell populations for CNS repair. Wichterle and colleagues (2002) demonstrated that signaling factors operating along the rostrocaudal and dorsoventral axes of the neural tube to specify MN fate *in vivo* could be harnessed *in vitro* to direct the differentiation of mouse ES cells into functional MNs.

ES Cells

ES cells have unique features that are important for spinal cord repair. They represent every cell type in the body, are the earliest stem cells capable of replicating indefinitely without aging, and their DNA can be modified easily even in single cells. They also fulfill two criteria essential for nervous system repair: Transplantable cells should be cells that normally belong in the spinal cord, and transplantable cells should contain the recipient's own DNA to avoid the need for immunosuppression, which cannot be used in spinal cord patients because of the increased incidence of infections.

Several methods are available for obtaining ES cells. The most common is *in vitro* fertilization, which requires a fertilized egg and therefore produces cells that differ genetically from the host. A second strategy involves somatic nuclear transfer, which takes a nucleus from a somatic cell such as skin and transfers it to an enucleated fertilized egg (Figure 53-4A). The subsequent ES cells therefore contain only the genetic material of the recipient. Another possibility for women is parthenogenesis, which tricks an egg into thinking it is fertilized, allowing it to begin duplicating its DNA (Figure 53-4B). Although this process is unable to create a viable embryo (because factors derived from sperm are necessary for preimplantation), it can produce normal cells.

Many studies of spinal cord repair have involved differentiated cells. The earliest studies used peripheral nerve grafts to demonstrate that the nervous system has the capacity to regenerate but that the environment in the CNS is not permissive. Substantial progress has been made with the transplantation of peripheral nervous system cells and nonnervous system cells. In general, fetal sources of cells have been the most successful because derivatives of postnatal and adult cells are less able to withstand neural transplantation. One recent exception has been neural stem cells derived from the adult CNS.

ES Cells and the Neural Lineage

Several protocols are available for converting ES cells into neural lineage cells. However, protocols for mouse and human cells differ (Figure 53-5). Differentiating murine ES cells traditionally begins with floating spheres called embry-

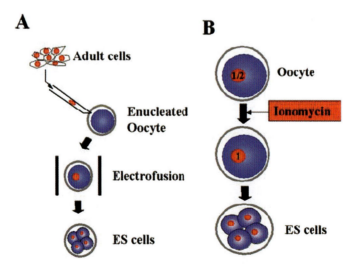

Figure 53-4. ES cells can be made by *in vitro* fertilization and two additional methods. (A) Somatic nuclear transfer: Donor cells are placed under the zona pellucida into the perivitelline space of enucleated oocytes. The cell nucleus is introduced into the cytoplast by electrofusion, which also activates the oocyte. It can then be grown to the blastocyst stage to harvest the ES cells. (B) Parthenogenic activation: For activation, oocytes are briefly exposed to a calcium ionophore such as ionomycin. The resulting cell can mature for ES cell harvest.

oid bodies (EBs). These bodies are akin to the neural spheres of stem cells obtained from the adult CNS. RA is a key induction agent for producing neural progenitors from mouse ES cells. Neural cells resembling anatomically normal neurons, astrocytes, and oligodendrocytes from the CNS can be easily derived from mouse ES cells using these protocols (Figure 53-6).

ES Cell Transplantation

Use of ES cells for neural transplantation is in its infancy, and only a limited amount of work has been completed with the spinal cord. Overall, they demonstrate that ES cells have a remarkable ability to integrate into the injured region of the cord and differentiate appropriately.

ES cells can participate in the normal development of spinal cord cells, including MNs. It demonstrated that developmentally relevant signaling factors can induce mouse ES cells to differentiate into spinal progenitor cells and then MNs through the normal developmental pathway (Figure 53-7). Thus, the signals that promote the differentiation of neural stem cells *in vivo* are effective when applied to ES cells. MNs derived from ES cells can populate the embryonic spinal cord, extend axons, and form synapses with target muscles (Figure 53-8). Therefore, they not only participate in normal development but also grow appropriately when transplanted into the embryonic spinal cord, targeting muscle. Thus, inductive signals in normal pathways of neurogenesis can direct ES cells to form specific classes of CNS neurons.

Brustle, Duncan, and McKay demonstrated that ES cells transplanted into the brain and the spinal cord of normal adult animals can differentiate into oligodendrocytes that can myelinate axons. They generated OPs efficiently by supplementing cultures of ES cells with FGF and epidermal growth

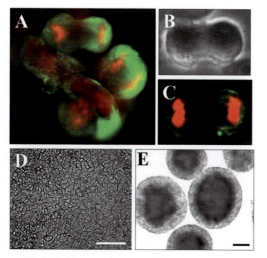

Figure 53-5. Undifferentiated mouse and human es cells dividing in a culture dish. (A and C) Immunofluorescence images of mouse ES cells demonstrate anti-β-tubulin and anti-DNA (Hoechst). The phase image of (C) the identical field is shown in panel B. Panel D illustrates undifferentiated human ES cells. EBs derived from human ES cells are shown in panel E. Scale bar for panels D–E = 100 μm.

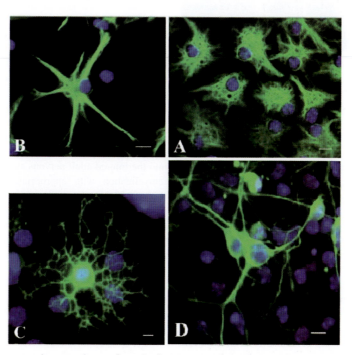

Figure 53-6. ES cells differentiate into the principal types of neural cells. (A) Type I and (B) type II astrocytes (anti-GFAP), (C) oligodendrocytes (anti-O1), and (D) neurons (anti-β-tubulin). Scale bar = 10 μm. Reproduced with the permission of Becker *et al.*, 2003.

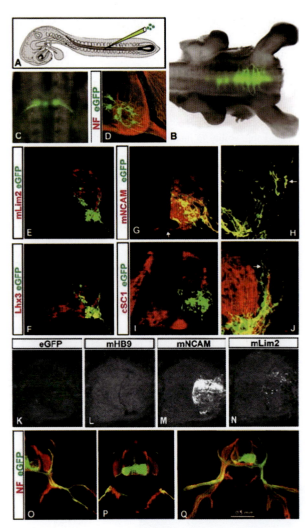

Figure 53-7. Embryonic transplantation of MNs derived from mouse ES cells. Integration of transplanted ES cell-derived MNs into the spinal cord *in vivo*. (A) Implantation of HBG3 ES cell-derived MN-enriched EBs into stage 15–17 chick spinal cord. (B) Bright-field–fluorescence image showing eGFP+ MNs in thoracic and lumbar spinal cord, assayed at stage 27 (ventral view). (C and D) Location of FACS, ES cell-derived, eGFP+ MNs in thoracic spinal cord, assayed at stage 27. (D) eGFP+ MNs are clustered in the ventral spinal cord. (E–J) Transverse sections through stage 27 chick spinal cord at rostral cervical levels after transplantation of MN-enriched EBs. (E) MNs are concentrated in the ventral spinal cord and are segregated from transplanted interneurons, labeled by a mouse-specific Lim2 antibody. (F) Many ES cell-derived MNs coexpress eGFP and Lhx3. (G) ES cell-derived MNs and (H) axons (arrow) are labeled by (I and J) rodent-specific anti-NCAM antibody, but do not express the chick MN marker protein SC1. (G and H) eGFP–, NCAM+ axons cross the floor plate but do not project out of the spinal cord (arrows). (K–N) Transverse sections of thoracic spinal cord at stage 27, after grafting EBs grown with RA (2 µM) and anti-Hh antibody (5E1, 30 µg/ml). No mouse-derived MNs were detected either (K) by eGFP or (L) by a mouse-specific anti-HB9 antibody. In contrast, many mouse-derived (M) NCAM+ and (N) Lim2+ interneurons are present. (O–Q) Transverse sections through stage 27 spinal cord at (O and P) thoracic and (Q) lumbar levels after grafting MN-enriched EBs. eGFP+ MNs are concentrated in the ventral spinal cord. Ectopic eGFP+ MNs are located within the lumen of the spinal cord. eGFP+ axons exit the spinal cord primarily through the ventral root and project along nerve branches that supply (O–Q) axial, (O and P) body wall, and (Q) dorsal and ventral limb muscles. The pathway of axons is detected by neurofilament (NF) expression. eGFP+ axons are not detected in motor nerves that project to sympathetic neuronal targets. Reproduced with the permission of Wichterle *et al.*, 2002.

factor and later including PDGF. To investigate whether these oligodendrocytes could myelinate *in vivo*, cells grown in the presence of bFGF and PDGF were injected into the spinal cord of 1-week-old myelin-deficient rats. Two weeks after transplantation, numerous myelin sheaths were detected in six of the nine affected rats. The original 100,000 cells had migrated widely and made myelin with appropriate ultrastructure.

This group also transplanted ES cell-derived precursors into the cerebral ventricles of developing rodents (embryonic day 17). Three weeks later, proteolipid protein-positive myelin sheaths were evident in a variety of brain regions. The cells' exogenous origin was confirmed by *in situ* hybridization with a probe to mouse satellite DNA. There was no observable evidence of abnormal cellular differentiation or tumor formation.

In the same year, McDonald, Gottlieb, and Choi demonstrated for the first time that ES cells that had been induced to become neural cell precursors could be successfully transplanted into the injured spinal cord. Examination of the spinal

cord nine days after 1 million precursor cells were transplanted into a cyst caused by contusion injury showed that the cells had survived, grafted, migrated long distances, and differentiated into the three principal neural cell types: neurons, astrocytes, and oligodendrocytes (Figures 53-9 and 53-10). Transplantation was associated with a significant and sizable improvement in function. Their study was the first demonstration that transplanted embryonic precursors can successfully repair the damaged adult nervous system. This is an important finding given that conditions in the damaged adult nervous system are much less favorable than those in the neonatal spinal cord. Subsequently, Liu and colleagues (2000) demonstrated that stem cell-derived precursors transplanted into the injured adult nervous system can achieve substantial remyelination with appropriate anatomical characteristics (Figure 53-11). Furthermore, they showed that ES cell-derived oligodendrocytes functioned normally, myelinating multiple axons in culture, just as they do in the normal nervous system (Figure 53-12). Using patch clamp analysis, Huettner and McDonald later demonstrated that the physiologic characteristics of ES cell-derived oligodendrocytes are similar to those of oligodendrocytes taken from the adult spinal cord (Figures 53-12C and 53-13). ES cell-derived oligodendrocytes represent the entire oligodendrocyte lineage, from early OPs to mature, myelinating oligodendrocytes.

In culture, ES cell-derived neurons rapidly differentiate and spontaneously create neural circuits with anatomical (Figure 53-14) and physiological evidence of excitatory and

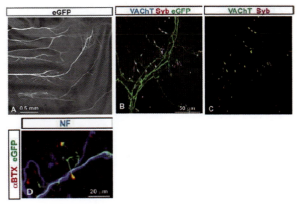

Figure 53-8. Anatomic integration of MNs derived from transplanted mouse ES cells. Synaptic differentiation of ES cell-derived MNs *in vivo*. (A) Whole-mount preparation of stage 35 chick embryonic rib cage. ES cell-derived eGFP+ motor axons contact intercostal muscles. (B and C) Coexpression of synaptobrevin (Syb) and vesicular ACh transporter (VAChT) in the terminals of eGFP+ motor axons at sites of nerve contact with muscle. The anti-Syb and VAChT antibodies recognize mouse but not chick proteins. (D) NF and eGFP expression in motor axons that supply intercostal muscles. eGFP+ axons lack NF expression. The terminals of eGFP+ axons coincide with clusters of ACh receptors, defined by α-bungarotoxin (αBTX) labeling. (E) Coincidence of synaptotagmin (Syn) expression in eGFP+ motor axon terminals and αBTX labeling. (F) Coincidence of Syb expression in eGFP+ motor axon terminals and αBTX labeling. Reproduced with the permission of Wichterle *et al.*, 2002.

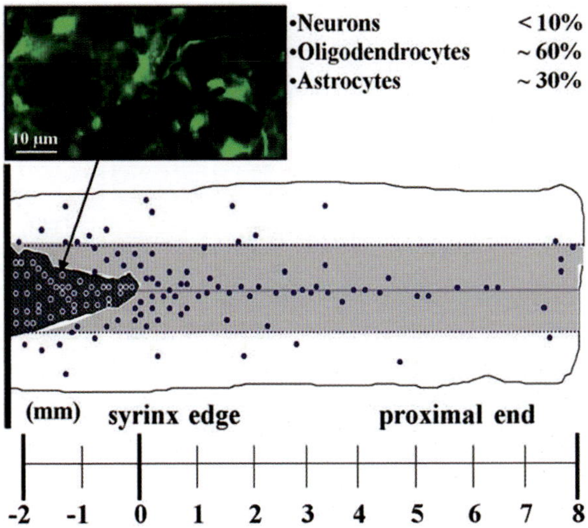

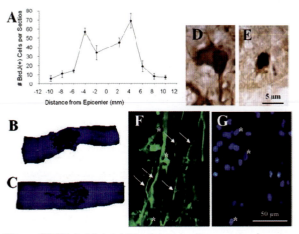

Figure 53-9. In the contusion-injured spinal cord of the rat, ES cell-derived neural precursors survive, migrate, and differentiate following transplantation. Schematic demonstrates the relative distribution of ES cell-derived cells two weeks after transplantation into the forming central cavity, which was done nine days after spinal contusion injury. The cavity is partially filled, and cells migrate long distances. Most distant cells are identified as oligodendrocytes, and astrocytes and neurons remain restricted to the site of transplantation. The inset shows GFP-expressing ES cell-derived neural cells.

inhibitory synapses. Substantial neural differentiation is also evident *in vivo* in the model of contusion injury, where neurons show extensive axonal outgrowth with presumptive morphological evidence of synapse formation. Moreover, immunological evidence of cholinergic, serotonergic, GABAergic, glycinergic, and glutamatergic neurons has been obtained *in vitro* and *in vivo* (data not shown).

Implantation, survival, and migration of transplanted ES cell-derived precursors in the damaged spinal cord have been verified by using magnetic resonance imaging to track oligodendrocytes prelabeled with paramagnetic agents as well as real-time polymerase chain reaction. In both cases, migration up to 1 cm from the transplantation site was evident.

Novel Approaches to CNS Repair

Most neural transplantation studies, including those using ES cell precursors, have used cells to replace cells lost after injury. Given that we do not know which types of differentiated cells are required for repair, neural precursors may be ideal for this purpose because environmental clues can decide their developmental fate. Moreover, ES cell-derived precursors can serve as bridges to support axonal regrowth.

In an attempt to overcome the constraints of the injured nervous system, ES cells are often genetically altered so that they will deliver growth molecules, such as NT3, after transplantation, and they are particularly well suited to this task. McDonald and Silver recently adopted a novel approach to overcoming restraints by showing that early ES cell-derived

Figure 53-10. BrdU-labeled ES cell-derived cells two weeks after transplantation. Mean ± SEM BrdU-labeled nuclei per 1 mm segment in longitudinal sections. (A) Hoechst 33342-labeled sections 42 days after injury, transplanted with (B) vehicle or (C) ES cells 9 days after injury. (D) BrdU+ cell co-labeled with GFAP. (E) BrdU-labeled cell co-labeled with APC CC-1. (F) The mouse-specific marker EMA indicates processes (arrows) emanating from ES cells. (G) Corresponding nuclei are marked by asterisks. Modified from McDonald *et al.* with permission.

progenitors can phagocytize key inhibitory molecules in glial scar tissue. These cells therefore created an inhibitor-free bridge over which axons could rapidly sprout from the transplant into normal cord. By nine days after transplantation, axons from the graft had grown up to 1 cm — a rate of more

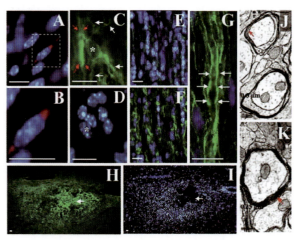

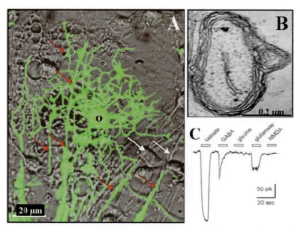

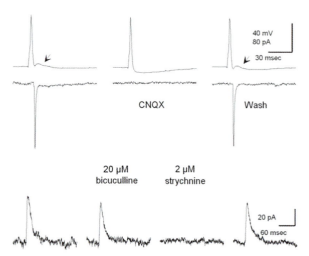

Figure 53-11. Cells derived from ES oligospheres can migrate and myelinate axons when transplanted into dysmyelinated spinal cords of adult *Shiverer* mice. Such mice lack the gene for myelin basic protein (MBP). Transplanted cells were identified by Cell Tracker Orange (CTO), epifluorescence, or immunoreactivity for MBP. (A and B) CTO-labeled cells aligned with native intrafascicular oligodendrocytes in white matter. (C and D) An ES cell-derived (MBP+) oligodendrocyte (asterisk) with longitudinally oriented processes (white arrows) is shown. (C) Arrows mark probable myelination around an adjacent axon. (E) Little MBP immunoreactivity is seen in white matter in a longitudinal section of spinal cord from a mouse that received a sham transplantation. (F) A gradient of MBP immunoreactivity centers on the site of ES cell transplantation. (G) High magnification of intrafascicular oligodendrocyte nuclei and MBP immunoreactivity, two indications of axonal myelination (white arrows), in white matter from a mouse transplanted with ES cells. (H) The spatial distribution of MBP immunoreactivity 1 month after ES cell transplantation is shown at low magnification, with (I) corresponding Hoechst 33342 counterstaining. White arrows indicate the center of the transplant site. (J) Transmission EM shows four loose wraps of myelin, the maximal number of layers typically seen around axons in control animals (arrow), and (K) nine or more compact wraps around axons from the transplanted area (arrow). *Shiverer* mutant mice lack a functional *MBP* gene required to form mature compact myelin; therefore, the presence of mature compact myelin is a standard for transplant oligodendrocyte-associated myelin. Scale bars = 10 μm (A–I) and 0.3 μm (J and K). Reproduced with the permission of McDonald *et al.*

Figure 53-12. ES cells produced mature oligodendrocytes with normal anatomical features of myelination and physiological response to neurotransmitters. Reproduced in adapted form with permission from Liu *et al.*, 2000.

Figure 53-13. Excitatory and inhibitory synaptic transmission among neurons derived *in vitro* from mouse ES cells. Action potentials were evoked in the presynaptic neuron by current injection. Excitatory (top) postsynaptic currents, blocked by superfusion with the selective glutamate receptor antagonist 6-cyano-7-nitroquinoxaline-2,3-dione (CNQX; 10 μM), were recorded under a voltage clamp. Arrows point to an autaptic excitatory synaptic potential also blocked by CNQX. Inhibitory (bottom) synaptic currents were evoked in a different ES cell pair and tested sequentially with the antagonists bicuculline and strychnine. For this cell pair, only the postsynaptic currents are shown. Transmission was blocked by strychnine, indicating glycine was the transmitter. Other presynaptic cells evoked bicuculline-sensitive synaptic currents, indicating transmission mediated by GABA (not shown).

than 1 mm per day! This rate is similar to that seen in the normal embryo.

Another novel approach is to use stem cell transplantation to limit the secondary injury that occurs after nervous system injury. For example, we recently demonstrated that transplanting ES cells can limit the delayed death of neurons and oligodendrocytes. Since most transplantation studies are completed at the time of injury, it is possible that many of their results may be attributable to this neuroprotective role. It seems clear that even genetically unmodified ES cells release large quantities of growth factors.

Although replacing lost neurons is difficult, it might be possible to use neurons to create bridging circuits across the injury site. Descending axons could synapse onto ES cell-derived neurons that subsequently synapse onto key pattern generators in the lower spinal cord. A more global delivery of neurotransmitters using synaptic or nonsynaptic mechanisms might enhance functions such as locomotion. Previous work

by others has demonstrated that release of noradrenergic and serotonergic neurotransmitters can stimulate and enhance the central pattern generator for locomotion. More recent work indicates that release of the neurotrophins BDNF and NT3 can perform this function.

Because ES cells are embryonic in nature, their progenitors may also be able to reprogram the adult CNS so that

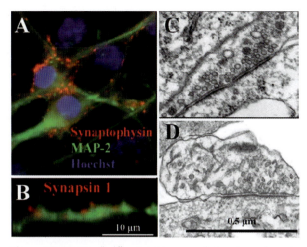

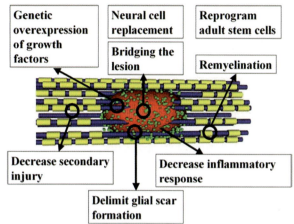

Figure 53-14. ES cells differentiate into neurons that spontaneously create neural circuits. (A and B) Presumptive presynaptic sites oppose dendrites. (C and D) Ultrastructural characteristics of synaptic profiles *in vitro*.

Figure 53-15. Novel approaches to spinal cord repair using ES cells. Remyelination is one of the most pragmatic approaches to restoring function to the damaged spinal cord. ES cell-derived precursors can serve as bridges to support axonal regrowth. They can phagocytize key inhibitory molecules in glial scar tissue. These cells therefore created an inhibitor-free bridge over which axons could rapidly sprout from the transplant into normal cord. Newly generated neurons can be used to create bridging circuits across the injury site. ES cell transplantation can limit the secondary injury. To try to overcome the constraints of the injured nervous system, ES cells are often genetically altered so they will deliver growth molecules. Because ES cells are embryonic in nature, their progenitors may also be able to reprogram the adult CNS to optimize spontaneous host regeneration. Replacement of nonneural cells as well as neural cells will be important to regain normal CNS function, such as neovascularization.

damage can be repaired. ES cells are also unique in that they have the potential to replace cells derived from multiple embryonic germ layers. In the CNS, it will be important to replace lost nonneural cells as well as neural cells to regain normal CNS function, such as neovascularization. The growing link between neovascularization and neurogenesis strengthens this approach. We recently found that ES cells in transplanted 4–/4+ EBs differentiate into both neural cells and vascular endothelial cells.

Finally, remyelination (Figure 53-15) is one of the most pragmatic approaches to restoring function to the damaged spinal cord. Because many potentially functional connections remain in the outer donut of surviving tissue, appropriate remyelination could substantially improve function. One must consider, however, that dysmyelination rather than demyelination is the biggest problem in the damaged nervous system.

Toward Human Trials

Some of the animal studies described previously have prompted early human safety trials. Most have focused on Parkinson's disease. Phase I human trials for repairing the spinal cord are also under way; they include transplantation of porcine-derived stem cells for the purpose of remyelination (see GenVec at http://www.genvec.com/) as well as allogeneic transplantation of olfactory ensheathing glia (Carlos Lima, Lisbon, Portugal). Although safety data are not yet available, it is important to note that such trials are paving the way for further uses of ES cells. With the recent advent of human ES cells, human transplantation appears more feasible, though somatic transplantation will be required to ensure a genetic match. Such manipulation of cells raises many regulatory concerns that must be addressed before clinical trials can move forward. The recent demonstration of *in vitro* oocyte genera-

tion from mouse ES cells raises the possibility that ethically acceptable procedures for human nuclear transfer may soon be available.

Summary

Studies of neurotransplantation for repairing the damaged nervous system are making good progress. Early animal studies showed that mouse ES cells can replace neurons, astrocytes, and oligodendrocytes; instigate appropriate remyelination; and even improve locomotion. As is the case for all transplantation studies, however, the mechanisms underlying functional improvement remain unclear. Nevertheless, ES cells offer a novel approach to deciphering these mechanisms.

Although it is impossible to predict the future of the field, it is clear that murine and human ES cells are tools that will revolutionize neurobiology and neural transplantation by providing the unprecedented ability to selectively deliver key regulatory factors. Also, ES cells promise to be one of the greatest therapies for chronic nervous system disorders. Only the future will reveal their full potential for treating human disease and disability, but it is possible to say with confidence that ES cells will have a major effect on repairing the human CNS within our lifetimes.

KEY WORDS

Embryoid body   A collection of hundreds of cells that resemble the early embryo and that is produced by growing ES cells in the presence of retinoic acid. These bodies contain neural tube-like structures and have been successfully used to repair the damaged spinal cords of rats.

Parthenogenesis   The process of tricking a nonfertilized egg into duplication of DNA to produce an all-female-derived embryonic stem cell.

Remyelination   The process of repair that is predicted to most likely lead to return of limited function without requiring new circuits to form. Oligodendrocytes are the glial cells that wrap (mylinate) neuronal axons. New data suggests that remyelination is one of the processes that can occur spontaneously after injury.

Somatic nuclear transfer   A method that involves transferring a nucleus from a somatic (e.g., skin cell) to an enucleated fertilized egg. This produces an embryonic stem cell clone of the somatic cell DNA.

FURTHER READING

Aguayo, A. J., David, S., and Bray, G. M. (1981). Influences of the glial environment on the elongation of axons after injury: transplantation studies in adult rodents. *J. Exp. Biol.* **95**, 231–240.

Becker, D., Sadowsky, C. L., and McDonald, J. W. (2003). Restoring function after spinal cord injury. *Neurologist* **9**(1), 1–15.

Corbetta, M., Burton, H., Sinclair, R. J., Conturo, T. E., Akbudak, E., and McDonald, J. W. (2002). Functional reorganization and stability of somatosensory-motor cortical topography in a tetraplegic subject with late recovery. *Proc. Natl. Acad. Sci.* **99**, 17066–17071.

Horner, P. J., Power, A. E., Kempermann, G., Kuhn, H. G., Palmer, T. D., Winkler, J., Thal, L. J., and Gage, F. H. (2000). Proliferation and differentiation of progenitor cells throughout the intact adult rat spinal cord. *J. Neurosci.* **20**, 2218–2228.

Liu, S., Qu, Y., Stewart, T. J., Howard, M. J., Chakrabortty, S., Holekamp, T. F., and McDonald, J. W. (2000). Embryonic stem cells differentiate into oligodendrocytes and myelinate in culture and after spinal cord transplantation. *Proc. Nat. Acad. Sci. USA* **97**, 6126–6131.

McDonald, J. W., and the Research Consortium of the Christopher Reeve Paralysis Foundation. (1999). Repairing the damaged spinal cord. *Scientific American* **281**, 64–73.

McDonald, J. W., Liu, X. Z., Qu, Y., Liu, S., Mickey, S. K., Turetsky, D., Gottlieb, D. I., and Choi, D. W. (1999). Transplanted embryonic stem cells survive, differentiate, and promote recovery in injured rat spinal cord. *Nat. Med.* **5**, 1410–1412.

McDonald, J. W., Becker, D., Sadowsky, C. L., Jane, J. A. Sr., Conturo, T. E., and Schultz, L. M. (2002). Late recovery following spinal cord injury. Case report and review of the literature. *J. Neurosurg.* **97**, 252–265.

McDonald, J. W., and Sadowsky, C. (2002). Spinal-cord injury. *Lancet* **359**, 417–425.

Wichterle, H., Lieberam, I., Porter, J. A., and Jessell, T. M. (2002). Directed differentiation of embryonic stem cells into motor neurons. *Cell* **110**, 385–397.

Use of Embryonic Stem Cells to Treat Heart Disease

Joshua D. Dowell, Robert Zweigerdt, Michael Rubart, and Loren J. Field

Introduction

It is now well established that cardiomyocytes can be stably transplanted into normal or injured adult hearts. Recent studies have demonstrated that transplanted donor cells can form a functional syncytium with the host myocardium. It is also well established that embryonic stem (ES) cells can differentiate into functional cardiomyocytes *in vitro* and that these ES cell-derived cardiomyocytes form stable intracardiac grafts when transplanted into the myocardium. ES cells might thus be a suitable source of donor cardiomyocytes for cell transplantation therapies aimed at restoring lost myocardial mass in diseased hearts. In this chapter, proof of concept studies demonstrating that donor cardiomyocyte can successfully engraft and participate in a functional syncytium with the host heart following transplantation are described. In addition, approaches to generate pure cultures of ES cell-derived cardiomyocytes suitable for transplantation are discussed. Finally, studies demonstrating the large-scale *in vitro* production of ES cell-derived cardiomyocytes suitable for cellular transplant are examined.

Cardiomyocyte Transplantation as a Paradigm for Treating Diseased Hearts

Cell transplantation has emerged as a potential therapeutic intervention to enhance angiogenesis, provide structural support, and perhaps even restore lost myocardial mass in diseased hearts. Several donor cell types have been transplanted in efforts to attain these goals, including angioblasts and vascular precursors, bone marrow and mesenchymal stem cells, skeletal myoblasts, and fetal or ES cell-derived cardiomyocytes. Clinical studies using skeletal myoblasts or crude mononuclear hematopoietic stem cell preparations from either the bone marrow or peripheral circulation have been initiated. These studies collectively established that cell transplantation in injured hearts is relatively safe, although the presence of arrhythmias in patients receiving skeletal myoblast transplants has necessitated defibrillator implantation. Some degree of improvement in cardiac function has been observed in these initial clinical studies, but the underlying mechanism remains unclear.

Essentials of Stem Cell Biology
Copyright © 2006, Elsevier, Inc.
All rights reserved.

Cardiomyocyte transplantation as a potential intervention to restore cardiac mass in diseased hearts was initially explored in the early 1990s. In these studies, single-cell suspensions generated by enzymatic digestion of fetal hearts were transplanted directly into the ventricle of histocompatibility-matched or immune-suppressed recipient animals. The transplanted cells were identified by virtue of reporter transgene activity or, alternatively, by the expression of an endogenous gene product absent in the recipient myocardium. The donor cells were stably engrafted into the recipient hearts, where they terminally differentiated and expressed many of the molecular and morphological attributes typical of normal adult cardiomyocytes. Ultrastructural analyses indicated that the transplanted donor cells formed intercalated disks with host cardiomyocytes comprised of fascia adherens, desmosomes, and gap junctions. Subsequent studies have demonstrated that donor cardiomyocytes could also stably engraft into injured hearts and that in some cases functional improvement could be obtained.

Recent studies by Rubart and colleagues (2003) have demonstrated that transplanted cardiomyocytes formed a functional syncytium with the host myocardium. In these experiments, fetal cardiomyocytes from transgenic mice expressing enhanced green fluorescent protein (eGFP) were transplanted into the ventricular septum of histocompatibility-matched nontransgenic recipient animals. A typical graft is shown in Figure 54-1A. To demonstrate the formation of a functional syncytium, transplanted hearts were harvested and perfused on a Langendorff apparatus with rhod-2 (a calcium indicator dye that exhibits increased fluorescence with increased intracellular calcium content) and cytochalasin D (an excitation-contraction uncoupler). The perfused hearts were then subjected to two-photon molecular excitation laser scanning microscopy under conditions that permitted simultaneous imaging of eGFP status (to distinguish the donor and host cardiomyocytes) and rhod-2 fluorescence (to monitor intracellular calcium transients). Examination of intracellular calcium transients in neighboring donor and host cardiomyocytes revealed that they occurred simultaneously and were indistinguishable (Figure 54-1B and 54-1C). These data indicated that donor cells were able to directly participate in a functional syncytium with the host heart. Cardiomyocyte transplantation was not associated with any anomalies in intracellular calcium handling in either the donor or the host cardiomyocytes.

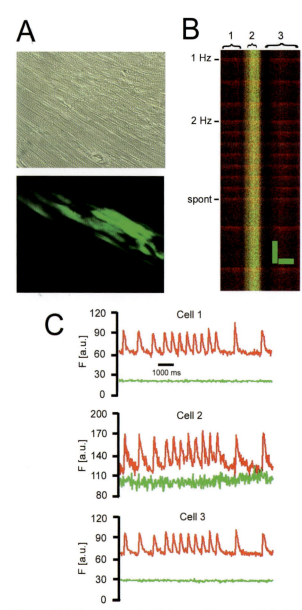

Figure 54-1. Functional coupling of donor cardiomyocytes with the host myocardium. (A) Bright-field (top) and epifluorescence (bottom) image of an area of a mouse left ventricle containing eGFP-expressing donor cardiomyocytes and nonexpressing host cardiomyocytes. The heart was harvested 27 days after engraftment. (B) Simultaneous imaging of rhod-2 and eGFP fluorescence in a nontransgenic heart carrying an MHC–eGFP fetal cardiomyocyte graft. The image shows vertically stacked line scans that traverse three juxtaposed cardiomyocytes. Cells 1 and 3 are host cardiomyocytes, whereas cell 2 is an eGFP-expressing donor cardiomyocyte. The heart was paced using point stimulation at a remote site at the rates indicated. Spon. indicates spontaneous $[Ca^{2+}]_i$ transient. Scale bars = 20 microns horizontally, 1000 ms vertically. Time runs from top to the bottom. (C) Spatially integrated traces of the changes in rhod-2 (upper trace) and eGFP (lower trace) fluorescence for the three cardiomyocytes in panel B. The signal across the entire cell was averaged and plotted as a function of time. Note that both stimulated and spontaneous $[Ca^{2+}]_i$ transients occur simultaneously in donor and host cardiomyocytes and exhibit similar kinetics.

Cardiomyogenic Differentiation of ES Cells *in Vitro*

Based on the above discussion, cardiomyocyte transplantation might be a useful mechanism to restore lost function in diseased hearts. Since clinical use of fetal donor cells is currently impractical, a surrogate source of donor cardiomyocytes would be required for clinical application. ES cells are multipotent cells derived from preimplantation embryos. ES cells can be amplified and maintained indefinitely in an undifferentiated state under appropriate conditions. For mouse ES cells, this entails culturing in the presence of leukemia inhibitory factor (LIF). Molecular analysis indicated that LIF enhanced the expression of Oct-4, a key transcription factor associated with maintenance of the multipotent state. Growing mouse ES cells in suspension culture and in the absence of LIF resulted in the formation of multicellular aggregates known as embryoid bodies (EBs). Stochastic cell–cell interactions during EB formation mimicked normal developmental induction cues and resulted in the random differentiation of ecto-, endo-, and mesodermal cell lineages. Cardiomyogenic differentiation was frequently observed in differentiating EBs and was easily recognized because of spontaneous contractile activity. ES cells with cardiomyogenic potential were also generated from rat, rabbit, monkey, and human.

Cardiomyocytes derived from ES cells have been fairly well characterized *in vitro* by many groups. Molecular and electrophysiological analyses indicated that cells with characteristics of atrial, ventricular, and perhaps conduction system cardiomyocytes can be generated *in vitro*. The temporal pattern of terminal differentiation *in vitro* for ES cell-derived cardiomyocytes mirrored that of embryonic development. Moreover, most gene expression qualitatively paralleled that seen during bona fide cardiac development. Although the preponderance of these studies were performed with mouse ES cells, preliminary assessment of cardiomyogenic induction in human ES cells revealed essentially similar results. When maintained under typical culture conditions, ES cell-derived cardiomyocytes did not acquire the gross morphologic attributes, or the quantitative levels of gene expression, characteristic of neonatal or adult cardiomyocytes. However, since the introduction of ES cells into tetraploid blastocysts generated entirely ES cell-derived mice with normal hearts, murine ES cells are clearly capable of bona fide cardiomyogenic differentiation when placed in the appropriate environment. Thus, the observed differentiated status of ES cell-derived cardiomyocytes *in vitro* likely reflects limitations in the cell culture conditions. It remains to be seen if the use of tissue bioengineering approaches that enhance fetal cardiomyocyte differentiation *in vitro* will have a similar effect on ES cell-derived cardiomyocytes.

Cardiomyocytes have been reported to comprise from 0.5 to 5% of the cells in randomly differentiating ES cultures. Accordingly, considerable effort has been invested to enhance cardiomyogenic yield during *in vitro* differentiation. Toward that end, several growth factor and cell culture conditions have been identified that can be used to enhance cardiomyogenic induction in differentiating mouse and human ES cultures. An alternative strategy to enhance cardiomyogenic yield in differentiating ES cultures would be to exploit genetic pathways that potentiate cardiomyocyte cell cycle activity. Many genes have been identified that either positively or negatively influenced cardiomyocyte cell cycle activity. Genetic modification of the ES cells, or development of small molecule agonists, antagonists, or both of the genetic pathways in question, could be exploited to enhance cardiomyogenic yield in differentiating ES cultures.

Transplantation of ES Cell-Derived Cardiomyocytes

Although the local tissue environment may have a minor influence on cell fate during differentiation, the ability to form multiple-tissue lineages following transplantation is a hallmark of ES cells. Accordingly, the use of ES cells to treat heart disease would require *in vitro* differentiation and purification of donor cardiomyocytes prior to transplantation. Although cytokine- and growth factor-based approaches can enhance cardiomyogenic yield, more rigorous methodologies would be required to generate cultures of sufficient purity for therapeutic transplantation. Elimination of multipotent cells would be absolutely required for clinical applications.

One approach to generate pure ES cell-derived cardiomyocyte cultures relied on genetic modification of the progenitor cells such that the resulting cardiomyocytes could be separated from nonmyocytes (Figure 54-2). For example, undifferentiated ES cells carrying a transgene comprising the cardiac-restricted α-myosin heavy-chain promoter driving expression of a cDNA encoding aminoglycoside phosphotransferase were allowed to differentiate *in vitro*. After cardiomyogenic differentiation was apparent (by the appearance of spontaneous contractile activity), the cultures were treated with G418. Since the myosin heavy-chain promoter was only active in cardiomyocytes, only those cells expressed aminoglycoside phosphotransferase and survived G418 treatment. Therefore, virtually all of the nonmyocytes were eliminated from the culture. A similar approach was used to isolate cardiomyocytes by targeted expression of eGFP followed by fluorescence-activated cell sorting (FACS). Klug and colleagues (1996) transplanted cardiomyocytes purified by the G418 purification method described above into the hearts of adult mice. The ES cell-derived cardiomyocytes formed stable grafts in the host heart and were well aligned with host cardiomyocytes. Noncardiomyocyte outgrowths were not observed in this study, indicating that the purification method efficiently removed any residual multipotent cells from the differentiating ES cultures.

Large-Scale Generation of ES Cell-Derived Cardiomyocytes

Successful clinical implementation of stem cell-derived cardiomyocyte transplantation would face several technical hurdles, of which the purity and safety of the stem cell-derived

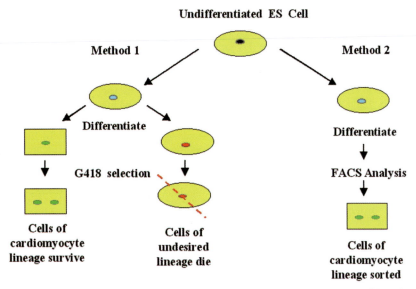

Figure 54-2. Selection method to produce pure ES cell-derived cardiomyocyte cultures. Two methods to separate ES cell-derived cardiomyocytes from nonmyocytes by cardiac-specific expression of a transgene. In Method 1, undifferentiated ES cells carrying a transgene comprising the cardiac-restricted α-myosin heavy-chain promoter driving expression of aminoglycoside phosphotransferase were allowed to differentiate *in vitro*. Cultures were treated with G418 following cardiomyogenic differentiation to select for cardiomyocytes, since the transgene promoter was only active in cardiomyocytes. In Method 2, undifferentiated ES cells were transfected with a transgene comprising the ventricular-specific myosin light-chain-2v promoter driving expression of eGFP and allowed to differentiate *in vitro*. Following cardiomyogenic differentiation, FACS was used to isolate a cardiomyocyte-pure population.

cardiomyocytes are among the most important. In addition, from a commercial point of view, the ability to generate clinically relevant cell numbers through an economically viable bioprocess would be another fundamental prerequisite. In this context, it is interesting to make some projections of cell number requirements. It has been estimated that the left ventricle of the human heart contains about 6×10^9 cardiomyocytes. Consequently, cell transplantation therapies aimed at replacing tissues lost to myocardial infarction would require the successful seeding of as many as $5–10 \times 10^8$ donor cardiomyocytes per patient (assuming limited cardiomyocyte proliferation after transplantation). The robust generation of such large cardiomyocyte numbers would be only feasible in controlled bioreactors capable of high-density differentiation of ES cells in clinical scale.

Published studies have indicated that EB formation is an efficient strategy to induce cardiomyogenic differentiation in both mouse and human ES cells. The main challenge to achieving large-scale generation of ES cell-derived cells is to establish conditions that would initially support the formation of numerous small EBs from undifferentiated ES cells and that upon subsequent culture would support the further growth of the resulting EBs in the absence of shear-induced damage. This goal is particularly challenging because EB characteristics change continuously during the differentiation process: size, density, and robustness all change from the initial formation of EBs to the ultimate appearance of differentiated cells.

The hanging drop technique permitted the formation of EBs with highly uniform size and cell content. Similarly, bulk EB induction on Petri dishes with or without semisolid media, as well as culturing of ES cell suspensions on rotation devices, have been used to affect controlled EB formation. Although these approaches were useful tools for bioprocess development (for example, to test the effect of compounds on cell proliferation and differentiation and to optimize ES cell inoculation density), they were unsuitable for the large-scale production of differentiated cells. Spinner flask cultures have been used in other systems to effect hydrodynamic control of cell aggregate formation. Large-scale cultures of neonatal hamster kidney and neuronal stem cell aggregates were previously produced with this approach. However, these studies focused on the expansion of cell aggregates and were not concerned with establishing conditions amenable to differentiation.

It has been noted that spinner cultures with a bulb-shaped impeller and inoculated with relatively high ES cell numbers can successfully generate EBs. Although the diameter of the resulting EBs was two to three times larger than that of EBs generated by traditional protocols, efficient induction of microvessel-like structures and endothelial cells was observed. Spontaneous contractile activity indicative of cardiomyogenic induction was also noted in this experiment; however, cell proliferation kinetics were not presented. Despite these encouraging results, studies from our laboratory using spinner flasks with a paddle-type impeller resulted in the formation of large aggregates instead of homogeneous

EBs, even though numerous conditions including various inoculation densities and various stirrer speeds were tested. The aggregates were essentially devoid of active cell proliferation and differentiation. It has been reported that fusion of EBs in the early stage of culture in spinner flasks was dependent at least in part on high levels of E-cadherin expression in undifferentiated ES cells and that E-cadherin was downregulated upon further differentiation. Thus, by simply transferring EBs that were preformed on Petri dishes for three days into spinner flasks, EB aggregation was avoided. This technique has already been employed to efficiently form cardiomyocytes in "floating EBs."

Although spinner flasks constituted an essential step for the transition from lab- to clinical-scale applications, the approach has numerous deficiencies, including stringent limits on maximal cell density as well as the inability to adequately control oxygen concentration and pH levels. A single-step culture protocol to generate ES cell-derived cardiomyocytes in an actively aerated, fully controlled tank reactor was therefore developed. Zweigerdt and colleagues (2003) used a common two-liter tank reactor equipped with a blade impeller creating axial flow to determine the effects of agitation rates on resulting EB formation, cell yield, and differentiation toward cardiomyocytes. The cellular content of EBs was highly dependent upon the agitation rate (Figure 54-3A). The average EB size for cultures maintained at 65 rpm was in agreement with that obtained with traditional approaches. In terms of total yield, the maximal cell density was also obtained in cultures maintained at 65 rpm.

Using these culture parameters, researchers performed additional experiments to generate pure cardiomyocyte cultures. Bioreactors were seeded with ES cells that carried the cardiac myosin heavy-chain aminoglycoside phosphotransferase transgene described above. After nine days of culture under conditions suitable for EB formation and differentiation and for cell expansion and differentiation, G418 was added to eliminate the noncardiomyocytes. Cell density prior to the addition of G418 was approximately 7×10^6 cells/ml. The expected cell density dropped dramatically upon antibiotic addition and stabilized around 6×10^5 cells/ml. Immune cytochemical analysis revealed that more than 99% of the cells obtained from the antibiotic-treated bioreactor cultures expressed sarcomeric myosin (Figure 54-3B and 54-3C), indicating that essentially pure cardiomyocyte cultures were generated. Overall, this correlated to a yield of more than 10^9 cardiomyocytes from a two-liter reaction.

Summary

The data summarized in this chapter indicated that donor cardiomyocytes were stable following transplantation into normal or injured recipient hearts; furthermore, they were capable of forming a functional syncytium with the host myocardium. In addition, the data demonstrated that highly differentiated cardiomyocytes could be derived from ES cells and that these cells could be stably transplanted into recipient hearts. Growth factors and enrichment methods were

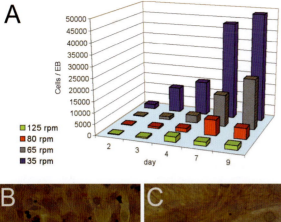

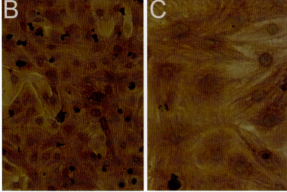

Figure 54-3. Generation of EBs and ES cell-derived cardiomyocytes in bioreactors. (A) Relationship between agitation rate, time in culture, and EB cellular content in a single-step culture protocol. (B and C) Low- and high-power images, respectively, of ES cell-derived cardiomyocytes generated in a bioreactor maintained at an agitation rate of 65 rpm. Cells from the bioreactor were enzymatically dispersed, plated onto chamber slides, and then processed for sarcomeric myosin immune reactivity (HPR-conjugated secondary antibody, signal indicates the presence of sarcomeric myosin). Virtually all cells present expressed sarcomeric myosin, although the level of expression per cell was variable under the culture conditions employed.

achieve high levels of *de novo* cardiomyocyte seeding. Toward that end, numerous cardiomyocyte prosurvival pathways have been identified. Moreover, genes that affect cardiomyocyte proliferation during development, and in some cases result in cell cycle reentry in adult cardiomyocytes, have been identified. Manipulation of these pathways in donor cells should greatly enhance seeding efficiencies. The genetic trackability of ES cells *in vitro* will likely facilitate this process.

Issues associated with the use of allogeneic donor cells constitute an additional potential hurdle for the widespread application of ES cell-derived cardiomyocyte transplantation. Although some degree of intervention is likely to be required, the evolution of comparably mild immune suppression protocols used for type 1 diabetic patients following cadaveric β-cell transplantation are quite encouraging. Hope in this regard is bolstered by the observation that mouse ES cell-derived cardiomyocytes did not appear to undergo acute rejection when transplanted into rat hearts with experimental infarctions. Moreover, a novel system based on ablation of IL-15 responsive T-cells might ultimately be useful to induce tolerance to allogeneic ES cell-derived donor cells. Finally, the potential use of nuclear transfer to generate autologous ES cells for the generation of donor cells could circumvent the need for chronic immune suppression, although this approach is not without ethical issues.

In summary, the potential use of cellular transplantation for the treatment of heart disease is gaining greater acceptance. The notion of using stem cell-derived, and in particular ES cell-derived, donor cardiomyocytes for partial restoration of lost cardiac mass in diseased hearts has been validated in experimental animals. Additional research is required to determine whether it is possible to seed sufficient numbers of donor cells to directly affect cardiac function and to develop viable strategies to prevent donor cell rejection. If these latter hurdles are overcome, cellular transplantation with ES cell-derived cardiomyocytes will emerge as a promising intervention for the treatment of heart disease.

ACKNOWLEDGMENTS

Loren J. Field thanks the NHLBI and Joshua D. Dowell thanks the AHA for support. We thank our many colleagues working in the field. We apologize in advance for any relevant views or studies that were inadvertently not included.

KEY WORDS

Cardiomyocytes Muscle cells in the heart which are interconnected via gap junctions and which participate in a functional syncytium.

Embryoid bodies Three-dimensional structures obtained when ES cells are allowed to differentiate while grown in suspension culture.

Embryonic stem (ES) cells Multipotent cells derived from the inner cell mass of blastocysts and able to differentiate under appropriate conditions into cells of ecto-, endo- and mesodermal origin.

described that enhanced cardiomyogenic differentiation and facilitated the generation of essentially pure cardiomyocyte cultures, respectively. Finally, conditions were established that permitted the generation of clinically relevant numbers of cardiomyocytes using bioreactor systems. Thus, the available data support the notion that transplantation of ES cell-derived cardiomyocytes for the treatment of heart disease might be both therapeutically useful and technically feasible. Nonetheless, several obstacles remain that must be overcome for clinical implementation.

A major limitation of current cardiomyocyte transplantation protocols is the relatively low level of donor cell seeding in the host myocardium. Using fetal cardiomyocytes as donor cells, several groups have demonstrated that the preponderance of donor cardiomyocytes die following transplantation into the heart, with typically less than 5% of the cells successfully seeding the myocardium. It is likely that interventions aimed at enhancing donor cell survival, post-transplantation proliferation, or both may be required to

FURTHER READING

Dowell, J. D., Rubart, M., Pasumarthi, K. B., Soonpaa, M. H., and Field, L. J. (2003). Myocyte and myogenic stem cell transplantation in the heart. *Cardiovasc. Res.* **58**, 336–350.

Kehat, I., Kenyagin-Karsenti, D., Snir, M., Segev, H., Amit, M., Gepstein, A., Livne, E., Binah, O., Itskovitz-Eldor, J., and Gepstein, L. (2001). Human embryonic stem cells can differentiate into myocytes with structural and functional properties of cardiomyocytes. *J. Clin. Invest.* **108**, 407–414.

Klug, M. G., Soonpaa, M. H., Koh, G. Y., and Field, L. J. (1996). Genetically selected cardiomyocytes from differentiating embryonic stem cells form stable intracardiac grafts. *J. Clin. Invest.* **98**, 216–224.

Koh, G. Y., Soonpaa, M. H., Klug, M. G., Pride, H. P., Cooper, B. J., Zipes, D. P., and Field, L. J. (1995). Stable fetal cardiomyocyte grafts in the hearts of dystrophic mice and dogs. *J. Clin. Invest.* **96**, 2034–2042.

Mummery, C., Ward-van Oostwaard, D., Doevendans, P., Spijker, R., van den Brink, S., Hassink, R., van der Heyden, M., Opthof, T., Pera, M.,

de la Riviere, A. B., Passier, R., and Tertoolen, L. (2003). Differentiation of human embryonic stem cells to cardiomyocytes: role of coculture with visceral endoderm-like cells. *Circulation* **107**, 2733–2740.

Rubart, M., Pasumarthi, K. B., Nakajima, H., Soonpaa, M. H., Nakajima, H. O., and Field, L. J. (2003). Physiological coupling of donor and host cardiomyocytes after cellular transplantation. *Circ. Res.* **92**, 1217–1224.

Soonpaa, M. H., Koh, G. Y., Klug, M. G., and Field, L. J. (1994). Formation of nascent intercalated disks between grafted fetal cardiomyocytes and host myocardium. *Science* **264**, 98–101.

Thomson, J. A., Itskovitz-Eldor, J., Shapiro, S. S., Waknitz, M. A., Swiergiel, J. J., Marshall, V. S., and Jones, J. M. (1998). Embryonic stem cell lines derived from human blastocysts. *Science* **282**, 1145–1147.

Zandstra, P. W., Bauwens, C., Yin, T., Liu, Q., Schiller, H., Zweigerdt, R., Pasumarthi, K. B., and Field, L. J. (2003). Scalable production of embryonic stem cell-derived cardiomyocytes. *Tiss. Eng.* **9**, 767–778.

Zweigerdt, R., Burg, M., Willbold, E., Abts, H., and Ruediger, M. (2003). Generation of confluent cardiomyocyte monolayers derived from embryonic stem cells in suspension. *Cytotherapy* **5**, 399–413.

Insulin-Producing Cells Derived from Stem Cells: A Potential Treatment for Diabetes

Susan Bonner-Weir, Gordon C. Weir, and Alexandra Haagensen

The Need for Insulin-Producing Cells

THE IMPORTANCE OF β-CELL REPLACEMENT THERAPY

Although insulin was discovered over 75 years ago, the complications of diabetes still produce devastating consequences. The link between high blood glucose levels and the complications of retinopathy, nephropathy, and neuropathy is now established beyond doubt. An obvious path to prevention of complications is some form of β-cell replacement therapy in the form of transplantation. Since 1978, over 12,000 pancreas transplants have been performed, and now over 1000 are done yearly, but these require major surgery. It would be preferable to transplant only the pancreatic islet cells which comprise only about 1% of the pancreas, but islet transplantation, lagging behind whole organ transplants, only began to obtain encouraging results in the late 1980s. Generally, poor outcomes were obtained throughout the 1990s, but the introduction of the Edmonton protocol in 2000 provided far better results, the improvement being due to better islet preparations, transplantation of more islets, and improved immunosuppression. Islets are introduced into the liver through the portal vein via transhepatic angiography. For most patients, more than one cadaver donor is usually required. By 2004, over 150 patients worldwide have received transplants using Edmonton-like approaches, with similar results. However, over a period of about two years, most of the patients slip back to mild diabetes, although their control is much easier due to continuing insulin production, and they are largely free of severe hypoglycemia.

LIMITED SUPPLIES OF INSULIN-PRODUCING CELLS

In spite of this success, the supply of insulin-producing cells, currently only from heart-beating cadaver donors is insufficient and limits the extension of this therapy. In the United States, it would be a major challenge to obtain 3000 usable cadaver pancreases per year. Much attention is focused on finding new sources of insulin-producing cells that can be used for transplantation. The quest includes exploring the potential of embryonic and adult stem cells, transdifferentiation such as directing acinar cells or hepatocytes to make insulin, altering cells with bioengineering, developing human cell lines, and using β-cells from other species as xenografts.

Defining β-Cells, Stem Cells, and Progenitor Cells

For the purposes of this chapter, a β-cell is defined as a cell with the phenotype of a mature insulin-producing cell found in pancreatic islets. It is possible to have insulin-producing cells that are immature and lack the full phenotype of a true β-cell. Some of these are β-cell precursors, which at some point can be called young β-cells. However, there are cells containing insulin that will never become β-cells, such as those identified in thymus, brain, retina, liver, and yolk sac.

It is also necessary to distinguish between the words "stem," "progenitor," and "precursor" cells. All new insulin-producing cells originate from precursor cells, which are not necessarily true stem cells. Stem cells can be defined as precursor cells capable of indefinite self-renewal. An embryonic stem cell is pluripotent, being capable of differentiating into the three embryonic germ layers, ectoderm, endoderm, and mesoderm, and then any cell type of the body even oocytes. Typical adult stem cells, such as are found in bone marrow, intestine, and skin, are multipotent, being able to form a variety of cell types restricted to a defined lineage. For example, cells originating from bone marrow stem cells are of mesodermal origin and form blood cell elements and endothelial cells. Then there are precursor cells with more restricted products, such as hepatic oval cells, which can form either new hepatocytes or bile duct epithelia. Facultative progenitor cells may also include pancreatic duct cells that can form pancreatic acini and probably islets.

The Potential of Embryonic Stem Cells as a Source of Insulin-Producing Cells

As a guide for exploiting the potential of human embryonic stem cells (ESC) to make pancreatic β-cells, mouse ESC should become an important model. With their potential to develop into virtually any cell type, they can be kept in an undifferentiated state being grown on feeder layers of irradiated mouse embryonic fibroblasts with leukemia inhibitory factor (LIF). To use human ESC for clinical purposes, it will probably be necessary to grow them on something other than mouse cells, which might include gelatin or human fibroblasts. Once removed from feeder layers, ESCs develop embryoid bodies, expressing their pluripotency to develop the three germ lines, ectoderm, endoderm, and mesoderm. Another potentially useful source of pluripotent cells is embryonic germ cells derived from primordial gonads. There is also a report of pluripotent cells being derived from adult bone marrow cells.

DIRECTING DIFFERENTIATION

ESC can be directed toward different progeny by a variety of factors, including soluble growth and differentiation mediators, matrix materials and cell–cell contacts. A good example of such directed differentiation by changing the environment is provided by studies showing the production of postmitotic neurons from mouse ESC. These results were made possible by the five-stage differentiation process of the McKay group, which has been successful in making neurons. The elements of this process include: Stage 1: expansion of ES cells in the presence of LIF; Stage 2: generation of embryoid bodies in suspension; Stage 3: Selection of nestin-positive cells using serum-free medium, ITSFn; Stage 4: expansion of cells with growth factors such as basic fibroblast growth factors (bFGF); Stage 5: induction of differentiation. A notable finding from Kubo et al. is that mouse ESC can be pushed to express endoderm markers by culture in the presence of activin A using serum-free conditions.

Genetic manipulations can also be used. Recently, Kim et al. stably transfected mouse ESC with a plasmid expressing the transcription factor nuclear receptor related −1 (Nurr1) driven by CMV. This intervention led to the generation of postmitotic neurons producing dopamine that produced symptomatic improvement in a mouse model of Parkinson's disease.

Another approach is called "trapping" with antibiotic selection. A cell-specific promoter linked with a gene for antibiotic resistance, aminoglycoside phosphotransferase, has been used to select remarkably pure cultures of cardiomyocytes. The same approach was used by the group of Soria with the reported generation of insulin-producing cells.

KEY FACTORS INVOLVED IN PANCREATIC AND ISLET DEVELOPMENT

Knowledge of pancreatic and islet differentiation provides important clues for work with ESC. The finding that Pdx-1 knockouts have virtually no pancreatic development identified a key early role for this transcription factor, which is likely to be different from its role at a later stage in maintaining β-cell differentiation. Another key point is that Ngn3 is expressed in precursor cells committed to differentiating into islets. Transient Ngn3 expression is thought to be induced by a combination of inhibition of notch signaling and stimulation of signaling by molecules of the TGF-β family. Pax4 appears to play a key role in post-Ngn3 β-cell differentiation. Knockout of Pax4 leads to increased α-cell number but no β- or delta-cells. Particularly noteworthy were data at the German Endocrine Society Meeting in Feb 2002 of Drs. Peter Gruss and Xunlei Zhou in which mice were generated with Pax4 being driven by Pax6, which is known to be important at an earlier stage. At age 6 weeks these mice had fourfold the number of β-cell containing islets, with extensive neogenesis also being seen. Other transcription factors important for β-cell development are B2/NeuroD, Nkx2.2, Nkx6.1, and the recently discovered MafA.

GENERATION OF INSULIN-PRODUCING CELLS FROM MOUSE ESC

In the past few years, a number of papers have appeared describing the generation of insulin-producing cells from mouse ESC. These papers have received a great deal of attention, with concerns being raised about the identity and potential of these cells.

The first notable paper, from Soria et al., employed a trapping approach to obtain cells expressing the insulin gene. The insulin content was low, but the cells released insulin in response to glucose and, when transplanted, normalized glucose levels in diabetic mice. It has been very difficult, even for the original authors, to generate additional clones of cells to allow confirmation of the data.

This was followed by a paper by Lumelsky et al. reporting that functional insulin-producing cells could be derived from mouse ESC using a variation of the McKay nestin selection protocol. This study has been criticized with the demonstration that the insulin staining can be an artifact of dying cells taking up insulin present in high concentrations in the media. In spite of this artifact, the Lumelsky paper did show the appearance of islet cell markers with RT-PCR, including glucagon, GLUT2, and IAPP, as well as immunostaining for glucagon. Moreover, enhanced insulin secretion in response to glucose, tolbutamide, IBMX, and carbachol was found.

A different strategy was reported by Hori et al. in which the phosphoinositide 3-kinase (PI3 kinase) inhibitor (LY294002) was used to induce differentiation by inhibiting cell growth, a strategy previously used by others. Using this inhibitor along with nicotinamide in Stage 5 of the McKay method, researchers generated insulin-producing cell clusters (IPCCs) with 95–97% of the cells immunostained for insulin while only 2–3% were stained for glucagon; no staining for somatostatin or pancreatic polypeptide (PP) was found. Unless something was, wrong with the specificity of the staining as suggested by Rajagopal et al., it appeared that insulin-containing cells result from a remarkably efficient default or survival pathway. When LY294002 was not used, cells from Stage 5 when transplanted resulted in formation of tumors that resembled teratomas, suggesting that undifferentiated cells capable of forming all three germ-line derivatives were present.

These IPCCs contained a notable quantity of insulin, 5 ng/IPCC, but they are larger than typical rodent islets, being 300 to 400 um in diameter compared to 150 um for a typical rodent islet. It was suggested that the insulin content was about 10% of islets, but it is probably much less when the volume of these large IPCCs is taken into account, perhaps only 1 to 2% of normal. When transplanted, these IPCCs improved the survival of diabetic mice, had some impact upon blood glucose levels, and were able to increase plasma insulin levels. Other indications of success were that these IPCCs expressed a variety of β-cell markers by PCR and could be stained red with dithizone, which reacts with the zinc of β-cells.

Important questions remain about this study, in particular: (1) What exactly happened to cells during the treatment with LY294002? Is there enrichment of islet cells by loss of other cells or by directed differentiation? The lack of acinar cells is not surprising because they are known to die rapidly in conventional tissue media. (2) What will a more detailed *in vitro* evaluation of the IPCCs reveal? (3) What are the characteristics of these cells after transplantation? The paper does not provide enough information to know if these are real β-cells. We know that immature β-cells in rodents have undeveloped secretory machinery, but maturation occurs in only a few days. One might expect β-cells derived from mouse ES cells to develop with the same timing.

In another variation, Blyszczuk *et al.* transduced mouse ESC with Pax4 or Pdx1 expressed constitutively using a CMV promoter. Other important aspects in this study were the use of serum (20% FCS), selection for nestin, and generation of "spheroids" (aggregates of cells). Both transcription factors appeared to lead to the development of insulin-containing cells, with Pax4 working somewhat better than Pdx1. Maximum insulin content with Pax4 was 455 ng/mg of protein, which is much less (0.5%) than normal β-cells (about 100 ug/mg of protein), and somewhat less than that reported by Hori *et al.* The transduced cells and wild-type cells subjected to the nestin selection protocol expressed insulin, GLUT2 and IAPP, by RT-PCR. Insulin secretion was stimulated by a glucose concentration of 27.7 mM and by tolbutamide. Finally, when cells were transplanted under the kidney capsule and into the spleen, streptozocin-induced diabetes was corrected. Insulin staining of cells was shown in the grafts, but insulin content of the graft was not determined and surgical removal of the grafts to test the source of insulin was not done.

Since these early papers, others have employed a variety of strategies, but none has been able to convincingly generate insulin-producing cells that closely resemble true β-cells.

SUMMARY OF PROGRESS-TO-DATE WITH MOUSE ES CELLS

Attempts to make β-cells from mouse ESC have been frustrating, in part because too much has been expected too soon. In spite of the concerns that insulin staining may be artifactual, it appears that mouse ESC can be directed to make some insulin-containing cells that express other gene characteristics of β cells such as GLUT2 and IAPP. The problem remains that no one has shown that these cells are truly similar to normal β-cells. The differentiation seen in embryoid bodies and in later stages may only progress far enough to produce some kind of primitive cells that makes a small amount of insulin. Small amounts of insulin mRNA have been found in yolk sac, brain, retina, fetal liver, and thymus but do not become β cells, so it would be interesting if these mouse ESC derivatives had a similar primitive phenotype, as has been suggested by Edlund.

NEED FOR RIGOROUS ASSESSMENT OF INSULIN-PRODUCING CELLS DERIVED FROM ESC

It is essential that future studies rigorously compare ESC-derived cells with normal β cells both in culture and in a transplant graft. The cells should be assessed for insulin content, insulin secretory response to see if a normal glucose dose response curve is obtained, EM characterization of granules to determine whether dense core granules are present, characteristic staining for IAPP, C-peptide, GLUT2 and insulin, and, finally, gene expression profiles with a variety of techniques including gene arrays.

WORK WITH HUMAN ESC

Several laboratories are intently working with human ESC, and there is a published demonstration of immunostained insulin-containing cells in embryoid bodies. This study also showed expression of various β cell genes in the embryoid bodies using RT-PCR, including glucokinase, GLUT2, ngn-3, and PDX-1. Unfortunately, the insulin-staining cells were few in number and not further characterized. More recent studies using basic fibroblast growth factor (bFGF) followed by nicotinamide led to increased numbers of insulin-expressing cells.

WILL TRANSPLANTED INSULIN-PRODUCING CELLS DERIVED FROM ESC FUNCTION NORMALLY WITHOUT ISLET NON-β-CELLS?

When developing strategies to transplant β-cells derived from ESC, it is important to consider the complexities of transplanting even normal islets. The pancreatic islet is a highly organized micro-organ in which β-cells are typically found in the core with non-β-cells forming a surrounding mantle. The blood supply comes from arterioles that penetrate gaps in the islet mantle and then form a glomerular-like network of capillaries in the islet core. The capillaries leave an islet through the mantle to enter the acinar tissue to form the islet-acinar portal system. The peptides produced by the different islet cells have the capacity to influence other islet cells. For example, glucagon can stimulate the secretion of insulin and somatostatin; somatostatin can inhibit glucagon and insulin secretion; and insulin can inhibit glucagon secretion. In spite of this potential, some of these interactions probably never take place *in vivo*. As fits with the known vascular pattern of islets, the upstream β-cells of the central core are probably protected from the potentially powerful effects of glucagon and somatostatin because blood and interstitial fluid flow from core to mantle. In contrast, the mantle cells, being downstream, are bathed in high concentrations of insulin, such that glucagon suppression by glucose appears to some extent indirectly mediated by local insulin secretion. Ingenious physiological experiments by Stagner and Samols indicate that these intra-islet interactions are unidirectional, B to A to D (B–A–D), which refer to β, α, and δ cells, respectively. Additional data supporting this concept have recently been generated by Moens *et al.* using a glucagon antagonist.

Little is known about the intra-islet interrelationships of transplanted islets. It is clear that substantial remodeling of transplanted islets occurs and that the vascular pattern of the new microvascuture does not faithfully recapitulate the pre-isolation pattern. A key point is that the *in vivo* B–A–D (β-α-δ) relationship is completely disrupted when islets are isolated and may not be reestablished after transplantation, which could allow β-cells to be locally influenced by α-cells.

If α-cells were upstream in a graft, the local glucagon secretion could potentiate insulin secretion with a beneficial effect. On the other hand, it could lower the set point for glucose stimulated insulin secretion (GSIS) by potentiating the stimulatory effect of normal concentrations of glucose and thus possibly cause hypoglycemia. We have found that rats transplanted with islets in the liver become hypoglycemic with exercise. Hypoglycemia has not been described with human transplants but may become an issue when it is possible to transplant a larger β-cell mass. There are, however, potential benefits from antiapoptotic effects of local glucagon upon β-cells. Two papers show that GLP-1 signaling, which is similar to that of glucagon, has an antiapoptotic influence on β-cells, and this effect seems likely to be exerted through cAMP.

It is important to know if pure insulin-producing cells derived from ESC will function appropriately when transplanted without non-β-cells. Based upon the B–A–D concept, our bias is that β-cells of the pancreas normally function well without ever seeing the secretory products from the mantle. The most valuable studies to date are those from Pipeleers and co-workers who used flow cytometry to sort highly enriched populations of β-cells on the basis of endogenous fluorescence. When transplanted into the liver, they did almost as well as intact islets. These carefully done studies provide some assurance that ESC-derived β-cells will function well when transplanted without islet non-β-cells, but many questions remain. What will be the ideal site for the transplantation of such cells? An obvious site would be for the cells to be aggregated *in vitro* and then placed into the liver via the portal vein. Other candidate sites might include the peritoneal cavity, a subrenal capsular site, a subcutaneous pocket, or even the pancreas.

WHAT IMMUNOLOGICAL BARRIERS WILL BE FACED BY TRANSPLANTED INSULIN-PRODUCING CELLS DERIVED FROM ESC?

Cells transplanted into subjects with type 1 diabetes will face both allograft rejection and autoimmunity. Both could be controlled with present-day immunosuppression, but the side effects of these drugs would limit the number of potential recipients. Improvements in the tolerability of these and other drugs should occur with time, and eventually the goals of transplantation tolerance and safe control of autoimmunity should be realized. Many subjects with type 2 diabetes should also benefit from β-cell replacement therapy. They might be an easier population to target because autoimmunity would not be an issue. Selection of ESC with a particular MHC match to make the insulin-producing cells may make it pos-

sible to limit or avoid allorejection. Another possibility would be to employ therapeutic cloning to obtain a perfect match. Other possibilities for the future include genetic manipulation of the cells to allow them to elude both allorejection and autoimmunity.

CONCLUSIONS ABOUT ESC

For the purposes of making insulin-producing cells that are true β-cells, the potential of ESC is tantalizing, especially considering the clinical need for cells that could solve the problem of diabetes. ESC are capable of generating any specialized cell in the body, so they must have the capability to generate cells with a β-cell phenotype. There must be some combination of environment, coupled with growth and differentiation factors, that could make this happen. At present, no simple default pathway has emerged to provide an easy route to this goal. Nonetheless, more work with ESC aided by better insights into the mechanisms that govern the development of the endocrine pancreas should eventually point investigators in the right direction.

The Potential of Adult Stem/Progenitor Cells as a Source of Insulin-Producing Cells

NEW β-CELLS ARE FORMED THROUGHOUT ADULT LIFE

Maintenance of β-cell mass is a dynamic process of continuing apoptosis that is balanced by the formation of new β-cells by replication of existing β-cells and probably neogenesis, which is the formation of new islets from precursor cells. The neonatal period in rodents is characterized by very active β-cell replication and neogenesis, and near the time of weaning there is an increase in β-cell apoptosis that leads to a remodeling of the endocrine pancreas. Through measurement of rates of β-cells replication, cell size, and β-cell mass, and making assumptions about neogenesis and apoptosis, it has been possible to estimate the lifetime of a rat β-cell as approximately 58 days. The endocrine pancreas of rats has considerable capacity for regeneration as shown by the 90% pancreatectomy. Even by 4 weeks following this surgical reduction in β-cell mass, the β-cell mass has increased from 10% to 42% of controls through the combined contributions of β-cell replication and neogenesis. There seem to be important species differences in how β-cell mass expands. Mice with genetically induced peripheral insulin resistance develop gigantic islets mainly through β-cell replication. Humans with the insulin resistance of obesity have increased β-cell mass, but islets are not particularly enlarged. Moreover, the frequent finding of insulin-positive clusters of cells appearing to emerge from pancreatic ducts in adult pancreases and the very low rate of β-cell replication as judged by staining with Ki67 suggest that neogenesis is an important contributor to the increase in mass in adult humans. Another mechanism for increasing β-cell mass, β-cell hypertrophy, has be found in rats in a variety of situations, including glucose infusions, pregnancy, and partial pancreatectomy.

WHAT IS THE CELLULAR ORIGIN OF ADULT ISLET NEOGENESIS?

Although it is clear that new islets are formed in adult pancreas, the source of these new islets continues to be a matter of debate. One hypothesis is that islets are derived from differentiated duct cells in the pancreas, with these cells serving as facultative progenitor cells. Others argue that new islets come from some other, as yet to be identified precursor cells, which may or may not be true stem cells. A recent study by Dor *et al.* used lineage tracing with the insulin promoter in mice to conclude that no new islets were formed after birth or after 70% pancreatectomy, but that new β-cells could be formed by replication of preexisting β-cells. The study is likely correct in the observation that β-cell replication is the dominant mechanism for β-cell expansion in adult mice but does not convincingly show that new islets are not formed during the neonatal life or after partial pancreatectomy. Two recent reports have found that single cells from adult mouse pancreas can expand to produce clusters of cells that contain insulin and express a variety of β-cell markers. It has not yet been shown, however, that these progeny are true β-cells.

ARGUMENTS FAVORING THE DUCTAL ORIGIN OF NEW ISLETS

Morphologically, new islets bud from the ducts, breaking through the epithelial basement lamina and migrating away from the duct. The process in the adult pancreas closely resembles embryonic new islet formation, which also is seen as budding off from ducts. The retention of the ability to differentiate into islet cells in adult duct tissue was shown experimentally by Dudek *et al.* showing islet hormone cells after transplantation of adult pancreatic ducts and fetal mesenchyme. In the RIP interferon gamma transgenic, newly formed islets were often seen within the lumen of the ducts, again suggesting their origin from ducts.

Other evidence supporting the concept is that after partial pancreatectomy in rats, ductal cells rapidly replicate and de-differentiate with a marked increase in PDX-1 protein, a factor known to be important for pancreas development and β-cell function. This return to more embryonic phenotype precedes the differentiation of whole new lobes of pancreas. Yet another finding strengthening this hypothesis is the marked change in composition of porcine neonatal pancreatic cell clusters transplanted under the kidney capsule of immunodeficient mice. When transplanted, the majority of cells can be stained with the duct marker cytokeratin 7, but after several months 94% of the cells were β-cells. During this engraftment, as well as in neonatal pancreas, cells co-stained for both insulin and cytokeratin 7 are found, suggesting residual duct markers as duct cells become β-cells. This finding is extended by gene profiling with microarrays of new versus mature islets 7 days after partial pancreatectomy in rats. These new islets, which can be identified by markers of new lobe formation, are about 3 days of age. Their β-cells contain a variety of duct markers, supporting the duct precursor hypothesis.

Other supporting data are that human duct-cell rich fractions remaining after islet isolation and purification can be cultured and form new islets after exposure to various growth and environmental factors. It must be said, however, that although these new islets bud from duct cells with a cystic structure, it has not been proven that the precursor cells are duct cells. However, PANC1, a human pancreatic duct cell line, has been manipulated in culture to give islet hormone-expressing cells. Similarly, studies from the group of Peck have reported on cells derived from mouse pancreatic ducts that can be markedly expanded and still contain insulin, albeit in small amounts. Although results have been presented that these cells could reverse diabetes in mice, concerns have been raised about whether these results are compatible with the very low insulin content of the cells.

A hypothetical argument supporting the concept of duct cells serving as facultative precursor cells comes from estimating how many precursor cells must be needed to form a new islet. After partial pancreatectomy in rats, fully formed new islets can be seen by 72 h after surgery, with these new islets consisting of 1000 to 1500 cells. No increased replication in the pancreas is seen until 24 h after surgery, so they must form within 48 h. With cell cycle times in rodents being 10 to 20 h, 16 starting cells with a doubling time of 10 h (5 doublings over 48 h) would be required to make 1000 cells. If the doubling time were 12 h, then 64 cells would be required. Because so few candidate nonduct precursor cells can be found morphologically, these calculations fit with the local abundance of duct precursor cells. Moreover, this concept fits with the demonstration that β-cells in islets are of polyclonal origin.

ARGUMENTS FOR NONDUCT CELLS BEING ISLET PRECURSOR CELLS

Because of the intense interest in identifying islet precursor cells, a variety of studies have made a case for the presence of precursor cells that originate from islets. One postulate for a β-cell precursor in islets comes from streptozocin-treated rats in which cells stained for both insulin and somatostatin are thought to differentiate into β-cells. Although all these studies are provocative and potentially important, more work is required to understand whether these are true pathways for β-cell regeneration.

There has been considerable controversy about whether the intermediate filament protein, nestin, which is known to be present in neural stem cells, is also found in islet precursor cells. Work with mouse embryonic stem cells suggested that insulin-producing cells could be generated through nestin-containing precursors. Another study reported on isolated cells that are stained for nestin derived from human islets. These cells also have capacity for expansion and express a variety of islet markers, although they too have very little insulin. Additional data have shown nestin-positive cells in the pancreas co-localize with mesenchymal, endothelial, and stellate cell markers, not with endocrine nor epithelial markers. While considerable controversy exists about nestin being a marker of β-cell development, the hypothesis remains viable. A related concept is that adult β-cells can serve as precursor cells by undergoing altered differentiation through an epithelial-mesenchymal transition, whereby they can be greatly expanded and then redifferentiate to express β-cell markers.

Only one lineage-tracing study has addressed the ductal origin of new islet cells. This study, which used two sets of cell specific Cre transgenic mice crossed with floxed reporter gene mice, found no evidence of a ductal origin for islets after birth. With crosses with inducible PDX-1: cre mice, no duct cells were labeled after E12.5, but all islets were labeled, leading to the conclusion that the adult duct population was segregated at an early age and did not contribute to islet formation. However, the expression level of Pdx-1 in ducts after E13 is known to be markedly reduced from that in βcells or earlier embryonic ducts, so the expression level of the Cre may not have been adequate to mark the ductal cells. The second set of experiments using postnatal inducible NGN3: cre mice did not find any labeled cells within the ducts when studied 1, 4, or 7 days after induction. Here the main criticism is that the labeled cells may have moved rapidly from the duct and by the time of examination were no longer within the ductal compartment. Further lineage-tracing studies are needed to resolve the issue of ductal origin of the islets after E12.5.

Transdifferentiation of Nonislet Cells to Islet Cells

A variety of studies support the concept that other cells can be directed to become insulin-producing cells. Strictly speaking, the concept that differentiated duct cells can form new islets, as was discussed above, could be considered a form of transdifferentiation. Current claims about transdifferentiation now include pancreatic acinar cells, hepatocytes, and bone marrow. Recently, splenocytes have been implicated, although co-staining of splenocyte and β-cell markers was not demonstrated. The possibility that transdifferentiation within pancreatic tissues occurs is not surprising in view of the ease with which the phenotype of islet and pancreatic cell lines can shift from one predominant gene product to another. Examples include clonal RIN cells varying their expression of insulin, glucagon, and somatostatin over time in culture; a pancreatic adenocarcinoma cell line (AR42J) that can be pushed to produce insulin by either hepatocyte growth factor or a combination of betacellulin and activin A; growing evidence that progenitor cells in adult pancreas can form hepatocytes and that liver cells can form pancreatic cells.

PANCREATIC ACINAR CELL TRANSDIFFERENTIATION

It has been difficult to prove that differentiated adult pancreatic cells are converted to insulin-producing cells. Suggestive evidence is provided by studies in a rat duct ligation model of regeneration in which gastrin infusions produce changes consistent with transdifferentiation of acinar cells to duct cells

that can then serve as precursors for β-cell neogenesis. The same pathway for neogenesis has been postulated to occur in rats receiving glucose infusions.

BONE MARROW CELLS AS A SOURCE OF INSULIN-PRODUCING CELLS

There has been much interest in the possibility that circulating bone marrow cells could serve as precursors for a wide variety of cells scattered throughout the body. There are several fundamental issues. One is whether hematopoietic stem cells (HSC) can transdifferentiate into nonhematopoietic cells as has been suggested by various studies. However, this appears not to be the case as shown in rigorous studies by Wagers et al. There remains the possibility that cells other than true HSC derived from bone marrow can serve as circulating stem/precursor cells and whether there is true transdifferentiation or cell fusion. It has recently been shown that impressive repopulation of destroyed liver tissue can be generated from bone marrow cells, but this process was more recently shown to be due to cell fusion. Nonetheless, a provocative report has appeared suggesting that cells from bone marrow can become glucose-responsive insulin-secreting cells in islets with this not being due to fusion. The question of the potential of bone marrow cells remains open, especially considering the finding by Verfaillie and coworkers that adult bone marrow cells can be pluripotent, capable of progressing to ectoderm, mesoderm, and endoderm fates. Another way in which bone marrow cells could contribute to new β-cell formation is as a facilitator rather than a precursor. In a study by Hess et al., when c-kit positive adult bone marrow cells were given to mice made diabetic with streptozocin there was evidence of regeneration of the endocrine pancreas and return to normoglycemia.

LIVER AS A SOURCE OF INSULIN-PRODUCING CELLS

As suggested above, liver, being of endodermal origin, might be expected to be a prime candidate for transdifferentiation. One of the first suggestions of this promise was the finding of the group of Ferber that in vivo transduction of hepatocytes in mice with an adenovirus containing pdx-1 could induce endogenous pdx-1 and associated expression of other β-cell markers and substantial amounts of insulin. A similar approach in mice employing helper-dependent adenoviruses expressing betacellulin and neuro-D appeared to result in the production of enough insulin in the liver to reverse streptozocin diabetes. These results were not interpreted as being transdifferentiation of hepatocytes but the promotion of islet neogenesis in the liver. The use of adenoviruses expressing pdx-1 in these experiments resulted in hepatitis, which contradicts the Ferber work. Yang et al. reported that long-term exposure to high glucose drove hepatic oval cells to an islet phenotype. Another provocative finding was that human fetal liver cells transduced with human telomerase, and then pdx-1, produced a remarkable amount of stored and secreted insulin. These cells not only secreted insulin in a regulated manner but also reversed diabetes when transplanted into immunodifficient diabetic mice.

ENGINEERING OTHER NON-β-CELLS TO PRODUCE INSULIN

The possibility that non-β cells might be engineered to make insulin and even secrete insulin in response to glucose was pioneered by the group of Newgard using cultured cells. This approach employs the introduction of genes expressing insulin and proteins involved in glucose recognition, but there is concern about what level of engineering will be required to obtain the complex and sophisticated phenotype of β-cells. It may be that the introduction of transcription factors serving as master genes will be required to obtain β-cells with adequate function for therapy. Another interesting example of this approach, reported by the group of Lipes, had insulin expressed in the cells of the intermediate lobe of the pituitary in transgenic mice. Remarkably, not only were these cells able to secrete insulin, but they were resistant to autoimmune destruction by autoimmunity. The engineering of glucose-dependent insulinotropic polypeptide (GIP)-secreting K cells in the intestine to produce insulin is another novel approach. These cells can secrete insulin along with GIP into the bloodstream after the oral administration of glucose. Concern must be raised about how this means could produce insulin in a manner that will be precise enough to be useful clinically.

One should be open-minded to the possibility that various other stem or precursor cells might be converted to useful insulin-producing cells. Neural or skin stem cells would seem to be a challenge because of their ectodermal origin, but we do not understand the restrictions of germline fates well enough to rule this out. Intestinal stem cells are of endodermal origin, which makes them attractive targets. Hopes for this pathway are raised by the demonstration that intestinal epithelioid IEC-6 cells can produce and secrete insulin after transfection with pdx-1 followed either by transplantation or by treatment with betacellulin.

ATTEMPTS TO DELIVER INSULIN VIA CONSTITUTIVE RATHER THAN REGULATED SECRETION

The essence of this concept is that glucose could be used to induce the promoter of a gene that could drive the synthesis of insulin followed by constitutive release by a cell such as the hepatocyte. An example of this approach employed use of adeno-associated virus to transduce liver with the L-type pyruvate kinase promoter that responds to glucose to induce the production of single-chain insulin. The advantage of single-chain insulin is that it does not need to be cleaved by the convertases normally needed for proinsulin to become biologically active insulin. It was shown that high plasma glucose levels could stimulate increases of plasma insulin that were sufficient to bring glucose levels of diabetic mice into the normal range. Important questions have been raised, however, about whether this approach could ever be clinically useful. A recent perspective article has pointed out that the timing of insulin release by a gene-dependent constitutive secretory mechanism seems highly unlikely to be useful for type 1, or even type 2 diabetes, in which more rapid insulin delivery or suppression will be needed to cope with the normal dynamics of metabolism found with eating, fasting, and exercise.

Conclusion on Adult Sources of New Islet Cells

The quest to find insulin-producing cells that might be used for transplantation is intense. Rapid improvements in our understanding of the mechanisms of cellular development and a wide array of potential stem or precursor cell candidates provide fuel for optimism that adult cells could solve.

ACKNOWLEDGMENTS

This research that provides the background for this review has been supported by the National Institutes of Health, the Juvenile Diabetes Foundation Research Foundation, the Diabetes Research and Wellness Foundation and an important group of private donors.

KEY WORDS

Beta cell Insulin-producing cell contained within pancreatic islets.

Islet neogenesis Generation of new pancreatic islets from precursor cells.

Islet transplantation Transplantation of isolated pancreatic islets with the intent of curing diabetes.

FURTHER READING

Antinozzi, P. A., Berman, H. K., O'Doherty, R. M., and Newgard, C. B. (1999). Metabolic engineering with recombinant adenoviruses. *Annu. Rev. Nutr.* **19**, 511–544.

Blyszczuk, P., Czyz, J., Kania, G., Wagner, M., Roll, U., St-Onge, L., and Wobus, A. M. (2003). Expression of Pax4 in embryonic stem cells promotes differentiation of nestin-positive progenitor and insulin-producing cells. *Proc. Natl. Acad. Sci. USA* **100**, 998–1003.

Bonner-Weir, S., Toschi, E., Inada, A., Reitz, P., Fonseca, S. Y., Aye, T., and Sharma, A. (2004). The pancreatic ductal epithelium serves as a potential pool of progenitor cells. *Pediatr. Diabetes* **5** (Suppl. 2), 16–22.

Colman, A. (2004). Making new beta cells from stem cells. *Semin. Cell. Dev. Biol.* **15**, 337–345.

Dor, Y., Brown, J., Martinez, O. I., and Melton, D. A. (2004). Adult pancreatic beta-cells are formed by self-duplication rather than stem-cell differentiation. *Nature* **429**, 41–46.

Dudek, R. W., Lawrence, I. E. Jr., Hill, R. S., and Johnson, R. C. (1991). Induction of islet cytodifferentiation by fetal mesenchyme in adult pancreatic ductal epithelium. *Diabetes* **40**, 1041–1048.

Edlund, H. (1998). Transcribing pancreas. *Diabetes* **47**, 1817–1823.

Ferber, S., Halkin, A., Cohen, H., Ber, I., Einav, Y., Goldberg, I., Barshack, I., Seijffers, R., Kopolovic, J., Kaiser, N., and Karasik, A. (2000). Pancreatic and duodenal homeobox gene 1 induces expression of insulin genes in liver and ameliorates streptozotocin-induced hyperglycemia. *Nat. Med.* **6**, 505–506.

Habener, J. F. (2004). A perspective on pancreatic stem/progenitor cells. *Pediatr. Diabetes* **5** (Suppl. 2), 29–37.

Hess, D., Li, L., Martin, M., Sakano, S., Hill, D., Strutt, B., Thyssen, S., Gray, D. A., and Bhatia, M. (2003). Bone marrow-derived stem cells initiate pancreatic regeneration. *Nat. Biotechnol* **21**, 763–770.

Hori, Y., Rulifson, I. C., Tsai, B. C., Heit, J. J., Cahoy, J. D., and Kim, S. K. (2002). Growth inhibitors promote differentiation of insulin-producing tissue from embryonic stem cells. *Proc. Natl. Acad. Sci. USA* **99**, 16105–16110.

Kim, J. H., Auerbach, J. M., Rodriguez-Gomez, J. A., Velasco, I., Gavin, D., Lumelsky, N., Lee, S. H., Nguyen, J., Sanchez-Pernaute, R., Bankiewicz, K., and McKay, R. (2002). Dopamine neurons derived from embryonic stem cells function in an animal model of Partkinson's disease. *Nature* **418**, 50–56.

Kubo, A., Shinozaki, K., Shannon, J. M., Kouskoff, V., Kennedy, M., Woo, S., Fehling, H. J., and Keller, G. (2004). Development of definitive endoderm from embryonic stem cells in culture. *Development* **131**, 1651–1662.

Lipes, M. A., Cooper, E. M., Skelly, R., Rhodes, C. J., Boschetti, E., Weir, G. C., and Davalli, A. M. (1996). Insulin-secreting non-islet cells are resistant to autoimmune destruction. *Proc. Natl. Acad. Sci. USA* **93**, 8596–8600.

Lumelsky, N., Blondel, O., Laeng, P., Velasco, I., Ravin, R., and Mckay, R. (2001). Differentiation of embryonic stem cells to insulin-secreting structures similar to pancreatic islets. *Science* **292**, 1389–1393.

Moens, K., Berger, V., Ahn, J. M., Van Schravendijk, C., Hruby, V. J., Pipeleers, D., and Schuit, F. (2002). Assessment of the role of interstitial glucagon in the acute glucose secretory responsiveness of in situ pancreatic beta-cells. *Diabetes* **51**, 669–675.

Rajagopal, J., Anderson, W. J., Kume, S., Martinez, O. I, and Melton, D. A. (2003). Insulin staining of ES cell progeny from insulin uptake. *Science* **299**, 363.

Shapiro, A. M., Nanji, S. A., and Lakey, J. R. (2003). Clinical islet transplant: current and future directions towards tolerance. *Immunol. Rev.* **196**, 219–236.

Soria, B., Roche, E., Berna, G., Leon-Quinto, T., Reig, J. A., and Martin, F. (2000). Insulin-secreting cells derived from embryonic stem cells normalize glycemia in strepotozotocin-induced diabetic mice. *Diabetes* **49**, 157–162.

Stagner, J. I., and Samols, E. (1992). The vascular order of islet cellular perfusion in the human pancreas. *Diabetes* **41**, 93–97.

Stoffel, M., Vallier, L., and Pedersen, R. A. (2004). Navigating the pathway from embryonic stem cells to beta cells. *Semin. Cell. Dev. Biol.* **15**, 327–336.

Street, C. N., Sipione, S., Helms, L., Binette, T., Rajotte, R. V., Bleackley, R. C., and Korbutt, G. S. (2004). Stem cell-based approaches to solving the problem of tissue supply for islet transplantation in type 1 diabetes. *Int. J. Biochem. Cell. Biol.* **36**, 667–683.

Wagers, A. J., Sherwood, R. I., Christensen, J. L., and Weissman, I. L. (2002). Little evidence for developmental plasticity of adult hematopoietic stem cells. *Science* **297**, 2256–2259.

Weir, G. C. (2004). Can we make surrogate beta-cells better than the original? *Semin. Cell. Dev. Biol.* **15**, 347–357.

Yang, L., Li, S., Hatch, H., Ahrens, K., Cornelius, J. G., Petersen, B. E., and Peck, A. B. (2002). In vitro trans-differentiation of adult hepatic stem cells into pancreatic endocrine hormone-producing cells. *Proc. Natl. Acad. Sci. USA* **99**, 8078–8083.

Burns and Skin Ulcers

Edward Upjohn, George Varigos, and Pritinder Kaur

Introduction

Burns and skin ulcers are major causes of morbidity and, in the case of burns, mortality in both the developed and developing world. Epidermal cells have been used in the therapy of these conditions for decades in the form of autologous skin grafts. For the last 20 years, advanced cell culture techniques have permitted the development of methods to identify and even isolate viable skin stem cells that could potentially be used in the treatment of these two conditions. These advances provide a strong foundation for building more advanced and exciting therapies based on stem cells to treat these challenging conditions.

The epidermis of the skin is a constantly renewing stratified squamous epithelium. It consists mostly of keratinocytes but also of Langerhans cells, melanocytes, and Merkel cells resting on a supporting dermis that contains the nerve and vascular networks, which nourish the epidermis. The dermis is also the location of epidermal appendages, fibroblasts, mast cells, macrophages, and lymphocytes. Epidermal stem cells are responsible for the ability of the epidermis to replace itself both in normal circumstances and in traumatic skin loss such as from burns and skin ulceration.

Burns and Skin Ulcers: The Problem

As with many medical conditions, it is difficult to quantify the burden and impact of these two conditions in a meaningful and tangible way. A study estimating the lifetime costs for injuries in the United States in 1985 rated fire and burns as the fourth largest cause of lifetime economic loss at $3.8 billion. Only motor vehicles, firearms, and poisonings were responsible for greater losses. Burns are also an important cause of injury in the developing world, where more traditional methods of lighting and cooking, such as oil lamps and open fires, are still commonplace. Skin ulcers may be caused by several pathological processes, including infection, trauma, diabetes, and venous ulcer disease. Chronic venous ulceration is a common cause of skin ulcers, with an estimated prevalence of 1 to 1.3%. Skin ulcers are also difficult and expensive to manage because of their slow rates of healing and the requirement for expensive and labor-intensive dressing regimes. A cost of care analysis by the Visiting Nurse Associa-

tion in Boston found the average cost per month in 1992 of an unhealed ulcer to be $1927.89. Diabetic foot ulcers, like venous ulcers, often have a chronic course. In one study, the cost estimates for foot ulcer care over a two-year period in a population of type 1 and type 2 diabetics who developed an ulcer was $27,987. These figures, especially when extrapolated to include lost productivity, are almost beyond comprehension. What is easily comprehended, however, is the personal cost and burden borne by patients suffering from burns or skin ulcers.

Epidermal Stem Cells

The epidermis of the skin is a multilayered, continuously self-renewing tissue replaced every 30 to 60 days in human skin. *In vivo* cell turnover studies in mice have shown that all proliferative activity is restricted to the basal layer, which generates the mature functional suprabasal keratinocytes. Thus, epidermal stem cells reside within the basal layer and are characterized by their capacity for self-maintenance and self-renewal. In addition to producing more stem cells, the stem cells generate transient amplifying cells that divide three to five times before producing terminally differentiated keratinocytes *in vivo*. The ability to identify slowly cycling stem cells *in situ*, visualized as ^{3}H-Tdr or bromodeoxyuridine (BrdU) label-retaining cells, has permitted their localization at specific niches including the deep rete ridges of the interfollicular epidermis and the bulge region of hair follicles (Figure 56-1). The identification of epidermal stem cells *ex vivo* is a more controversial subject because there is no unequivocal assay for these cells *in vitro*.

Heterogeneity in the growth capacity of keratinocytes was first reported by Barrandon and Green in 1987; they used clonal analysis to retrospectively identify cells capable of generating large colonies of cells exhibiting limited differentiation in culture, termed *holoclones*. These investigators proposed that holoclones were derived from stem cells because of their greater proliferative capacity. This work did not provide a means to prospectively isolate keratinocyte stem cells (KSCs). Subsequent work from Watt and colleagues established that cell-surface markers, specifically β_1 integrin, could be used to distinguish basal keratinocytes, with high (β_1 bright) or low (β_1 dim) proliferative capacity measured in terms of short-term colony-forming efficiency. Despite the initial conclusion that β_1 integrin was a marker of KSCs, subsequent work from many laboratories has shown that, although most basal cells are integrin bright, only a small sub-

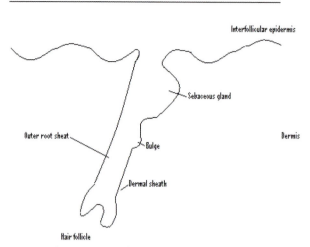

Figure 56-1. Epidermis, dermis, and other structures.

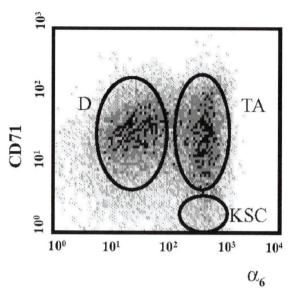

Figure 56-2. This FACS plot of primary human, neonatal, foreskin keratinocytes — labeled with antibodies to the cell-surface marker α_6 integrin (FITC) and the transferrin receptor (CD71-PE) — reveals phenotypically discrete subpopulations of cells corresponding to stem cells (KSC) with the phenotype $\alpha_6^{bri}CD71^{dim}$, transit-amplifying (TA) cells with $\alpha_6^{bri}CD71^{bri}$, and early differentiating (ED) cells with α_6^{dim}.

population of these exhibit quiescence as defined by their ability to retain a ^{3}H-Tdr or BrdU label.

Work from our laboratory has further established that ^{3}H-Tdr label-retaining cells can be purified from integrin-bright keratinocytes on the basis of a second cell-surface marker — that is, CD71, or the transferrin receptor. Thus, epidermal cells exhibiting the phenotype $\alpha_6^{bri}CD71^{dim}$ represent the stem cell population of both human neonatal and adult murine epidermis. This fraction is enriched for label-retaining cells that are small (~9 μm) with a blast-like morphology, display a high nuclear-to-cytoplasmic ratio, and exhibit the greatest long-term proliferative capacity to regenerate keratinocytes *in vitro* and comprise about 5% of total basal cells (Figure 56-2). Moreover, we have demonstrated that the progeny of KSCs can be distinguished by their cell-surface phenotype: Transit-amplifying (TA) cells exhibit high levels of CD71 ($\alpha_6^{bri}CD71^{bri}$), are enriched for actively cycling cells defined as pulse-labeled cells in murine epidermis, and exhibit intermediate keratinocyte cell regeneration capacity; early differentiating (ED) cells are identifiable as α_6^{dim} cells exhibiting the poorest long-term proliferative capacity and expressing keratin 10 and involucrin — both markers of keratinocyte differentiation (Figure 56-2). This work permits the prospective isolation of KSCs and their immediate progeny by fluorescence-activated cell sorting (FACS), so that the contribution of distinct classes of epidermal progenitors to tissue regeneration during homeostasis or in wound healing can be directly assessed. This work is an important prerequisite to the development of therapeutic strategies using KSCs for skin conditions, including gene therapy. The identification of growth factors that recruit KSCs to proliferate and regenerate tissue will be important to the development of techniques for rapid *ex vivo* expansion that facilitate earlier transplantation for burn victims.

Stem Cells in Burns and Skin Ulcers: Current Use

BURNS

Autografts

Epidermal cells have been used in the treatment of burns since the introduction of the split skin graft by Karl Thiersh in the late 1800s. Skin grafting to cover defects caused by burns or skin ulceration is limited by the area of skin that may be harvested on any one occasion. Full-thickness grafts (including all of the epidermis and dermis) provide good cosmetic results but require a primary closure of the donor site, limiting the area that may be grafted. To overcome this problem, the use of split skin grafts was developed, whereby epidermis and underlying dermis is shaved from the donor site to provide a graft. The donor site then reepithelializes from hair follicles, a process that is easily seen with the unaided eye and that takes two to three weeks, after which the donor site can be reharvested. It is thought that stem cells of the hair follicle are responsible for healing both the donor site and the grafted area. Experimental evidence in favor of this notion was recently provided by elegant experiments performed by Barrandon and colleagues (2001), who demonstrated that microdissected hair follicle bulges (enriched for stem cells) from transgenic mice expressing the β-galactosidase gene could regenerate interfollicular epidermal tissue, as well as entire hair follicles, when transplanted.

The limitations of skin grafting techniques are the area that can be covered by them and the many weeks it may take to

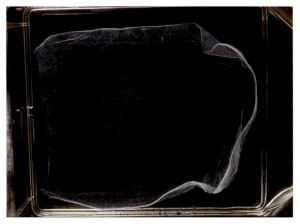

Figure 56-3. Detached epithelial graft in a tissue culture flask. Reproduced with the permission of Joanne Paddle-Ledinek.

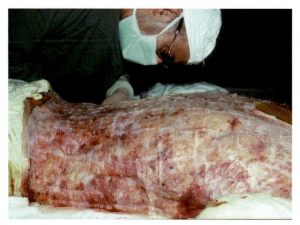

Figure 56-5. Application of cultured autograft epidermis to a burns patient. Reproduced with the permission of Joanne Paddle-Ledinek.

Figure 56-4. Secondary cultures of the epithelium in an incubator. Reproduced with the permission of Joanne Paddle-Ledinek.

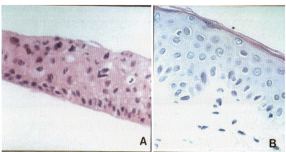

Figure 56-6. (A) Side-by-side histological comparison of cultured epidermis and (B) epidermis *in vivo*. Reproduced with the permission of Joanne Paddle-Ledinek.

cover a large area of burn with autologous split skin. Burns of 80 to 90% are survivable in the short term with resuscitation, but if coverage of the wounds is delayed because of a lack of grafts, then a high morbidity and mortality is the result.

In the mid-1970s, Rheinwald and Green developed a technique for serial cultivation of epidermal cells, producing a 1000- to 10,000-fold area of graftable epidermis than the initial biopsy (Figure 56-3 and Figure 56-4). These epidermal sheets can then be grafted onto clean wound beds, but they are sensitive to loss by bacterial infection and blistering. In full-thickness burns where the dermis has been lost, the cultured epidermal autograft may be placed directly onto muscle or fascia (Figure 56-5). These cultured epidermal autografts form a permanent covering, suggesting that the stem cells initially cultured and then transplanted have been maintained as stem cells and therefore retain their crucial role in epidermal maintenance. Histological examination of cultured epithelium

reveals the structural similarity to normal epithelium *in vivo* (Figure 56-6). Culturing epidermis is a time- and labor-intensive exercise estimated to cost between $600 and $13,000 per 1% of the body surface area covered (depending on the proportion of the grafts that successfully take).

BURNS

Allografts

The development of allografts has been driven by the lack of available donor sites for split skin grafting in patients with massive burns and by the time taken to grow cultured autologous skin from these patients. Burn therapy requires coverage of the burnt areas to prevent secondary sepsis and other complications. An alternative to split skin grafts is needed that is immediately available, plentiful, effective, and affordable. Cadaveric skin is such an alternative; it is a true allograft and is always eventually rejected by the recipient. As mentioned earlier, skin that lacks a dermis is less able to resist trauma and is prone to contraction, resulting in a poor functional and cosmetic outcome. Alloderm (Lifecell, Branchburg, New Jersey) is a processed human dermis from which the epider-

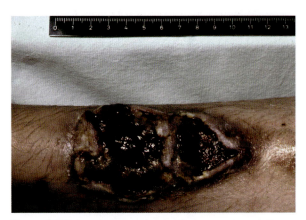

Figure 56-7. Leg ulcer. Reproduced with the permission of George Varigos.

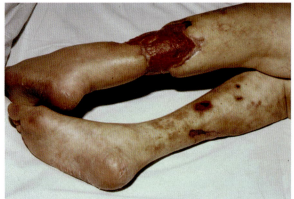

Figure 56-8. Circumferential leg ulcer healing by granulation. Reproduced with the permission of George Varigos.

mal and dermal cells have been removed, leaving only the connective tissue matrix. Alloderm can then be applied to burns, and cultured autograft may be placed on top of it. Integra (Integra LifeSciences, Plainsboro, New Jersey) is another dermal substitute developed through the coprecipitation and lyophilization of bovine collagen, chondroitin-6-sulfate, and an artificial epidermal layer of synthetic, polysiloxane polymer. One to two weeks after application, the artificial epidermis is removed, and an ultrathin epidermal autograft (0.003–0.005 inches thick) is placed on the burn area.

All currently available examples of artificial dermis lack a vascular plexus for the nourishment of the epidermis and require host vasculogenesis into the dermis graft to supply nourishment to the grafted epidermis. Efforts have therefore been focused on encouraging the process of vasculogenesis by genetic engineering of grafts to produce growth factors and cytokines vital to this process (see later sections of this chapter).

SKIN ULCERS

Therapy for skin ulcers is based on treating the precipitating and perpetuating factors. This includes antibiotic use for infective ulcers, rigorous pressure care for decubitus ulcers, and compression stockings for venous ulcers (for examples of skin ulcers, see Figures 56-7 and 56-8). The use of occlusive dressing techniques has greatly expanded in recent decades, and these form the foundation for the treatment of many ulcers. In spite of advances like occlusive dressings, ulcer healing often takes months, if not years, to achieve. In burns therapy, the impetus behind the use of stem cells or cultured epidermal autografts is the need to cover large areas quickly. In ulcer therapy, the time constraints are not as severe, and for definitive closure, split skin grafts remain the gold standard. Cultured skin has been used in the treatment of skin ulcers — in particular, in the use of cultured allografts as a "living dressing." Cultured, autologous outer root sheath cells used in the treatment of chronic decubitus ulcers have been found to

produce an *edging effect* — the contraction of the chronic wounds edges in response to the graft, believed to be caused by a release of growth factors, cytokines, and hormones from the outer root sheath cells. Apligraf (Organogenesis, Canton, Massachusetts) is a cultured, bilayered living skin equivalent derived from neonatal foreskin keratinocytes, fibroblasts, and bovine type I collagen. It is indicated for the treatment of venous ulcers and neuropathic diabetic foot ulcers. Chronic wounds (e.g., those with dormant edges) reepithelialize when exposed to living allograft material. This edge effect, like that seen with outer root sheath cells, is most probably caused by the presence of stimulatory factors. Chronic wounds heal better after repeated application of skin grafts, suggesting that growth factors in the grafts are responsible. Bioengineered tissues such as Apligraf probably act as biologic systems for delivering growth factors to wounds.

Future Developments

GENE THERAPY APPROACHES IN WOUND HEALING

Epidermal cells can be genetically modified both *in vivo* and *ex vivo* by both viral and nonviral methods (e.g., recombinant retro and adenovirus infection, liposomes, plasmid injection, and particle bombardment). Initially, gene therapy of epidermal cells was pursued to correct inherited genetic defects, but it is now used to treat wound healing by using genetically engineered keratinocytes as a source of cytokines and growth factors. Fluid from chronic wounds has been found to be inhibitory to cell proliferation and to contain degradation products that inhibit keratinocyte migration. The beneficial effect that allografts have on wound healing is thought to be at least partly because of the production of cytokines and growth factors, although the host eventually rejects the allografts. Cultured autografts engineered to produce these cytokines and growth factors in supranormal quantities may expedite wound healing. Epidermal cells can be engineered to express a gene either permanently or transiently. These genes could encode growth factors or cytokines, as mentioned earlier, or they may antagonize some of the inhibitory factors found in chronic wounds.

The vascularization of cultured skin autografts is often delayed when compared to split skin grafting and may be a contributing factor to the failure of the cultured autograft to take. Cultured skin has been genetically modified *in vitro* to produce vascular endothelial growth factor (VEGF). These modified, cultured skin substitutes have been shown to secrete elevated levels of VEGF and to have decreased time to vascularization when grafted onto athymic mice.

TISSUE ENGINEERING

Tissue engineering of skin is an active area of research and development. Efforts to develop temporary skin substitutes began in the early 1960s. The skin is one of the first organs to have been successfully generated *ex vivo*, and there are several tissue-engineered skin equivalents available today. In bioengineered skin, the epidermal component is either cultured allograft or cultured autograft. The keratinocyte sheets are then combined with the dermal component. The dermal component may be acellular, or it may contain allogeneic or autogenic fibroblasts or other cells. The optimum skin equivalent would be readily available, could be stored or frozen and ready for use, would be inexpensive to make, would have excellent take, and would give good cosmetic results. Such an ideal skin equivalent does not exist. Split skin grafts fulfill many of these requirements, but unfortunately, they are often not available in a sufficient amount. Currently, the culturing of autografts requires at least three weeks, is labor intensive, and is not available in many developing parts of the world. It would be a great advance if the technique could be made faster and easier to perform. Given our recent ability to use FACS methodology to separate KSC versus TA populations from the epidermis, we have been testing the hypothesis that the stem cell-enriched keratinocytes may provide faster and reliable skin regeneration in *in vitro* and *in vivo* transplantation model systems. Our recent studies indicate that both epidermal stem cells and TA cells can regenerate epithelial tissue in the short term fairly rapidly and that dermal cells, and specific components of the basement membrane (i.e., Laminins), are critical regulators of this tissue-regenerative capacity. Current studies in our laboratory are aimed at understanding how to recruit greater numbers of epidermal stem cells to proliferate *in vitro* as well as identifying factors capable of promoting stem cell expansion and renewal *ex vivo*. These studies will be important to the development of cellular therapies using epidermal stem cells as vehicles.

Summary

Investigators of epidermal stem cells have the advantage of easy access to their cells of interest. Keratinocytes are also relatively easy to culture and engineer into their normal "organ form" (i.e., sheets of epidermis). As stated earlier, there is an established role for epidermal stem cells in the therapy of human diseases and injuries. The concept of epidermal stem cells and their application to health care could form a conceptual framework for the education of the general public about stem cells and could help to demystify what is commonly perceived to be a complicated science. Epidermal stem cells are excellent targets for gene therapy seeking to correct genetic deficiencies permanently. This could form the basis of new therapies for previously untreatable genodermatoses (such as epidermolysis bullosa) as well as for burns and skin ulcers.

Epidermal appendages like hair, sweat glands, and sebaceous glands are often destroyed by burns, and their replacement with bioengineered equivalents is still to be achieved. The ability to develop a hair-bearing skin replacement, for example, would address many of the cosmetic problems caused by burns to the face and scalp.

The stem cells of the dermis have yet to be definitively identified and localized within the skin; this can be attributed to the complex cellular heterogeneity of this tissue. Tissue engineering of a replacement dermis to graft onto burns or other defects lacking a dermis would clearly benefit from the incorporation of dermal stem cells. Given that the dermis provides growth factors regulating epidermal growth and morphogenesis, as well as "hair inductive" capacity, further elucidation of its molecular, cellular, and functional components is essential to the development of cellular therapies.

Risk of viral transmission remains a concern with allografts, but they can be frozen, thawed, and used when needed. This is very convenient and to contributes to their lifesaving potential. Tissue engineering with bovine or other sources of collagen must be carefully managed to reduce the risk of transmission of prion and other diseases to humans. For these reasons, research into cultured autografts will continue, with the goal being a cultured skin equivalent made from epidermal and dermal stem cells, tissue engineered to provide rapid wound coverage with excellent take. The great advances made in the last quarter of a century have saved many lives, in the case of burns, and made lives worth living again, in the case of skin ulcers. Our expanding understanding of stem cells and their manipulation will build on these advances to enable even better treatment of these conditions in the future.

ACKNOWLEDGMENTS

We wish to thank Joanne Paddle-Ledinek and Heather Cleland.

KEY WORDS

Allograft Cultured allogeneic keratinocyte graft grown from a donor's skin cells, which is eventually rejected by the recipient's immune system (may also have a dermal component).

Autograft Cultured autologous keratinocyte graft grown from the recipient's own skin cells, not rejected by the recipient's immune system.

Holoclone A colony of cultured epithelial cells thought to be produced by keratinocyte stem cells.

Epidermal stem cells Slow-cycling and self-renewing cells found in the basal layer of the epidermis responsible for its constant repopulation.

FURTHER READING

Barrandon, Y., and Green, H. (1987). Three clonal types of keratinocyte with different capacities for multiplication. *Proc. Nat. Acad. Sci. USA* **84**, 2302–2306.

Eaglstein, W., and Falanga, V. (1998). Tissue engineering and the development of Apligraf, a human skin equivalent. *Cutis* **62**, 1–8.

Eming, S., Davidson, J., and Krieg, T. (2001). Gene transfer strategies in tissue repair. *In The skin and gene therapy* (U. Hengge *et al.*, eds.), pp. 117–137. Springer, Berlin.

Jeschke, M. G., Richter, W., and Ruf, S. G. (2001). Cultured autologous outer root sheath cells: a new therapeutic alternative for chronic decubitus ulcers. *Plastic Reconst. Surg.* **107**, 1803–1806.

Li, A., Pouliot, N., Redvers, R., and Kaur, P. (2004). Extensive tissue regenerative capacity of neonatal human keratinocyte stem cells and their progeny. *J. Clin. Invest.* **113**, 390–400.

Li, A., Simmons, P. J., and Kaur, P. (1998). Identification and isolation of candidate human keratinocyte stem cells based on cell surface phenotype. *Proc. Nat. Acad. Sci. USA* **95**, 3902–3907.

Oshima, H., Rochat, A., Kedzia, C., Kobayashi, K., and Barrandon, Y. (2001). Morphogenesis and renewal of hair follicles from adult multipotent stem cells. *Cell* **104**, 233–245.

Pellegrini, G., Bondanza, S., Guerra, L., and De Luca, M. (1998). Cultivation of human keratinocyte stem cells: current and future clinical implications. *Med. Biol. Eng. Comp.* **36**, 778–790.

Rheinwald, J. G., and Green, H. (1975). Serial cultivation of strains of human epidermal keratinocytes: The formation of keratinizing colonies from single cells. *Cell* **6**, 331–344.

57

Stem Cells and Heart Disease

Piero Anversa, Annarosa Leri, and Jan Kajstura

The Heart Is a Self-Renewing Organ

A fundamental issue concerning the ability of the heart to sustain cardiac diseases of ischemic and nonischemic origin is whether myocardial regeneration can occur in the adult organ, or whether this growth adaptation is restricted to prenatal life severely limiting the response of the heart to pathologic states. The concept of the heart as a terminally differentiated organ unable to replace working myocytes has been at the center of cardiovascular research and therapeutic developments for the last 50 years. The accepted view has been and remains that the heart reacts to an increase in workload only by hypertrophy of the existing myocytes during postnatal maturation, in adulthood and senility. When myocardial hypertrophy is exhausted, cell death and heart failure supervene. In contrast to the widespread acceptance of this dogma, there are numerous findings indicating that the mammalian heart has replication potential and that its response to abnormal elevations in load can be accompanied by myocyte proliferation. Despite the increasing consistency of accumulating data in favor of the formation of new myocytes in a variety of physiological and pathological conditions, the notion of myocyte proliferation in the adult heart continues to be challenged. It is rather remarkable that cardiologists and cardiovascular scientists are willing to believe that ventricular myocytes in human beings can work and live over 100 years until death of the organism. The possibility of a low turnover of cells as part of cardiac homeostasis is constantly being rejected without any scientific basis. Moreover, myocyte death occurs as a function of age, and the progressive loss of cells in the absence of myocyte multiplication should result in the disappearance of the entire organ over a period of a few decades.

Recent findings demonstrate that the heart belongs to the group of self-renewing organs such as the hematopoietic system, the intestine, the skin and the brain. We have documented that the adult rat heart contains a population of undifferentiated cells (Figure 57-1A–1C), which express surface antigens typically found in hematopoietic stem cells (HSCs): c-kit, MDR1, and Sca-1. Similar results have been obtained in the canine, pig, and, most importantly, the human heart.

As it occurred for the central nervous system (CNS), cardiac stem cells (CSCs) can be isolated from the ventricular myocardium, and long-term cultures can be developed. So far,

only Lin⁻c-kit^POS cells have been shown to have the components of stemness: self-renewal, clonogenicity, and multipotentiality. Properties of MDR1 and Sca-1-like cells remain to be determined. *In vitro*, Lin⁻c-kit^POS cells grow as a monolayer when seeded in substrate-coated dishes or form spheroids when in suspension, mimicking the biology of neural stem cells. Single clonogenic cells are able to differentiate into myocytes, endothelial cells, smooth muscle cells, and fibroblasts (Figure 57-2A–2C). Therefore, primitive cells with properties of stem cells are present in the myocardium, either as a resident population of embryonic origin or as a blood-borne population that continuously seeds the tissue. This arrangement points to a mechanism for the continuous renewal of myocytes and coronary vessels throughout life of an individual.

Distribution of CSCs in the Heart

Differences in the organization of myocyte bundles and in the levels of mechanical forces are present in the anatomical parts of the heart: atria, and base, midregion, and apex of the ventricle. Wall stress is higher at the base and midregion of the ventricle and lower at the apex. Stress is further reduced in the atria. These variables may play a critical role in dictating the function and fate of CSCs. Such information is crucial for understanding the effects of structural and physical factors on the behavior of adult CSCs. Although stem cells possess strong cellular defense mechanisms, they are sheltered in specialized structures called *niches*. Stem cell niches provide a microenvironment designed to preserve the survival and replication potential of stem cells. Primitive cells rarely divide, and those endowed in a niche have a much higher probability of self-renewal. For example, the slow-cycling cells of the hair follicle are confined to the bulge, where niches are located, and give rise to the epithelia of the follicle and contribute to the renewal of the epidermis. This region ensures good physical protection and is rich in melanin to prevent DNA damage induced by ultraviolet light. Niches are present in all self-renewing organs. Whether CSCs are dispersed in the myocardium or are nested in pockets with a structural organization typical of niches is an important question, which has recently been answered. Qualitative and quantitative results indicate that CSCs are stored in niches, which are preferentially located in the atria and apex but are also detectable in the ventricle (Figure 57-3A–3C). The recognition that stem cells are clustered in specific regions of the heart exposed to moderate and minimal mechanical forces and that they are stored in niches favors the notion that these cells are organ

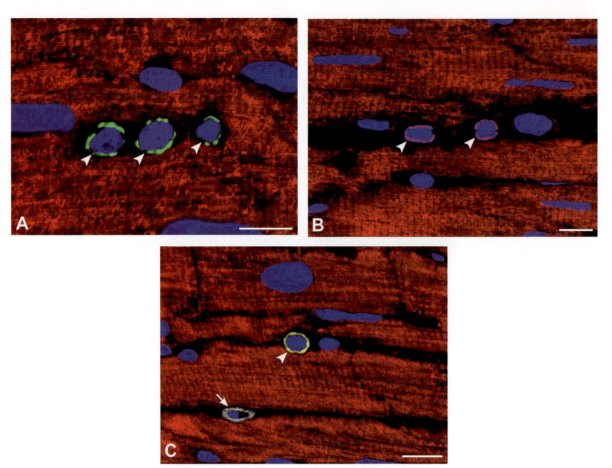

Figure 57-1. Sections of adult rat ventricular myocardium showing three c-kit (A, green fluorescence, arrowheads), two MDR-1 (B, magenta fluorescence, arrowheads) and one Sca-1-like (C, yellow fluorescence, arrowhead) positive cells. White fluorescence in panel C corresponds to von Willebrand factor labeling of an endothelial cell (arrow). Myocytes are identified by the red fluorescence of α-sarcomeric actin antibody staining. Nuclei are labeled by the blue fluorescence of propidium iodide. Confocal microscopy, scale bar = 10 μm.

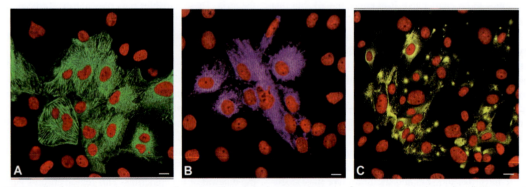

Figure 57-2. Differentiation of cardiac primitive cells *in vitro*. Clone-derived myocytes (A, green fluorescence), smooth muscle cells (B, magenta fluorescence), and endothelial cells (C, yellow fluorescence); the cytoplasm is recognized by α-sarcomeric actin, α-smooth muscle actin, and von Willebrand factor antibody staining, respectively. Nuclei are labeled by the red fluorescence of propidium iodide. Confocal microscopy, scale bar = 10 μm.

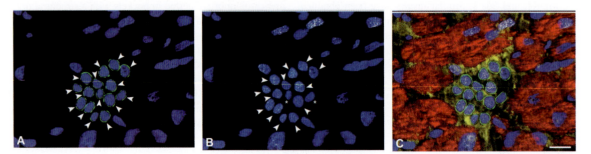

Figure 57-3. Atrial niche in the adult rat heart. A cluster of 15 c-kit positive cells (A, green fluorescence, arrowheads), 13 of which express GATA-4 (B, white fluorescence). Asterisks indicate two GATA-4 negative cells. These cells are nested in fibronectin (C, yellow fluorescence). Myocytes are identified by the red fluorescence of α-sarcomeric actin antibody staining. Nuclei are labeled by the blue fluorescence of propidium iodide. Confocal microscopy, scale bar = 10 μm.

specific. This possibility would speak against the view that stem cells in the myocardium are replenished by cells of bone marrow origin, which migrate chronically to the heart. In summary, the localization of primitive cells in the myocardium is inversely related to the distribution of stress in the anatomical components of the heart.

Repair of Myocardial Damage by Nonresident Primitive Cells

Major discoveries have been made concerning the biology of adult stem cells: (1) They can differentiate into cell lineages distinct from the organ in which they reside and into cells derived from a different germ layer. These properties were considered to be restricted to embryonic stem cells. (2) Adult stem cells can migrate to sites of injury, repairing damage in various organs. Neural stem cells have been identified in selective regions of the brain that, like the heart, was considered a postmitotic organ. Thus, adult stem cells may exist in other unexpected organs, and stem cell behavior is not dictated by the source. HSCs can replace bone marrow and lymphoid organs by migrating to these sites across the vascular endothelium and along stromal pathways. Moreover, HSCs regenerate skeletal muscle, while skeletal muscle stem cells can repopulate the bone marrow. However, these particular skeletal muscle stem cells are of bone marrow origin. HSCs can also differentiate in functioning hepatocytes. Hematopoietic precursors, injected intravenously, reach the brain where they divide, infiltrate the entire organ and give rise to the cell types of the CNS. Conversely, CNS stem cells, which derive from the ectoderm, can transdifferentiate into blood cells, which generate from the mesoderm, whereas HSCs can assume the characteristics of CNS cells. Together, this information supports the notion that injury to a target organ promotes alternate stem cell differentiation emphasizing the plasticity of these cells. This very controversial issue of transdifferentiation will be addressed later in this section. This is because the early belief in stem cell transdifferentiation has prompted the utilization of bone marrow cells for the reconstitution of the infarcted myocardium. In addition, this approach was considered superior to other forms of cellular therapy of the damaged heart.

In the last few years, effort has been made to restore function in the infarcted myocardium by transplanting cultured fetal myocytes or tissue, neonatal and adult myocytes, skeletal myoblasts and bone marrow-derived immature cardiomyocytes. When incorporation of the engrafted cells or tissue was successful, some improvement in ventricular performance occurred. However, these interventions failed to reconstitute healthy myocardium, integrated structurally and functionally with the spared portion of the wall. This defect was particularly evident with skeletal myoblasts, which did not express connexin 43, a surface protein responsible for the formation of ion channels and electrical coupling between cells. Moreover, the vascularization of the implants remained an unresolved issue. These problems pointed to the identification of new therapeutic strategies for the regeneration of

dead myocardium. The generalized growth potential and differentiation of adult HSCs injected in the circulation or locally delivered in areas of injury suggested that these primitive cells sense signals from the lesion foci, migrating to these sites of damage. Subsequently, homed HSCs proliferate and differentiate, initiating growth processes and resulting in the formation of all cellular components of the originally destroyed tissue.

On this basis, a population of bone marrow cells enriched with Lin⁻c-kit^{POS} cells was implanted into the viable myocardium in the proximity of an acute infarct. This was done to facilitate the translocation of these cells to the necrotic region of the left ventricular wall, to reconstitute myocardium, and to interfere with healing and cardiac decompensation. In nine days, numerous small cardiomyocytes and vascular structures developed within the infarcted zone and partially replaced the dead tissue (Figure 57-4). The newly formed young myocardium expressed transcription factors required for myogenic differentiation, cardiac myofibrillar proteins, and connexin 43. Coronary arterioles and capillaries were distributed within the regenerated portion of the left ventricle; these vessels were functionally connected with the primary coronary circulation. For the first time, bone marrow cells were shown to develop myocardium in vivo reducing infarct size and ameliorating cardiac performance.

We will now discuss the role of cytokines and growth factors in the mobilization of stem cells and their translocation to damaged organs and to the heart in particular. This issue is highly relevant clinically because its understanding may permit the application of strategies, which do not require local implantation of exogenous stem cells or their preventive storage from the recipient patient. Two critical determinants seem to be necessary to obtain the maximal activation

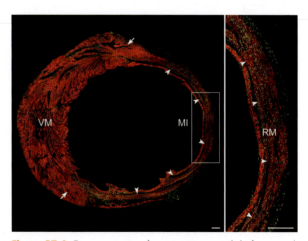

Figure 57-4. Transverse section of an extensive myocardial infarct treated with the injection of bone marrow cells in the border zone (arrows). Nine days later, a band of regenerating myocardium was identified (arrowheads). A portion of this band, included in the rectangle, is illustrated at higher magnification in the adjacent panel. Red fluorescence corresponds to cardiac myosin heavy-chain labeling of myocyte cytoplasm, and green fluorescence reflects propidium iodide staining of nuclei. VM = viable myocardium, MI = myocardial infarct, RM = regenerating myocardium. Confocal microscopy, scale bar = 300 μm. From Orlic *et al.* Nature 410, 701–705 (2001).

of the therapeutic potential of undifferentiated cells in organs from which they are not derived: organ damage and high levels of primitive cells in the circulation. It has been shown that these two conditions are not always essential. Cell fate transition has also been observed in uninjured organs and following utilization of a single bone marrow cell. However, the degree of engraftment of individual cells is much lower and decreases further in unaffected tissues. It is well established that stem cell factor (SCF) and granulocyte colony-stimulating factor (G-CSF) are powerful cytokines, which markedly increase the number of circulating HSCs. These mobilized cells can reconstitute the lymphohematopoietic system of lethally irradiated recipient mice, raising the likely possibility that, in the presence of myocardial infarction, they may home to the heart and follow a differentiation pathway, which includes myocytes and coronary vessels. As anticipated, this protocol was very successful; tissue reconstitution included parenchymal cells and vascular profiles. Again, *de novo* myocardium was obtained, and the repaired area of the wall was functionally competent, improving cardiac pump function and significantly attenuating the negative remodeling of the postinfarcted heart.

Our results on myocardial regeneration after infarction utilizing bone marrow cells implantation or cytokine administration do not address the critical issues of HSCs plasticity and transdifferentiation. The hypothesis has been advanced that the donor stem cells might fuse with the parenchymal cells of the host tissue, giving the wrong appearance of transdifferentiation. So far, the accumulated data indicate that the bone marrow contains cells capable of regenerating dead tissue, but which cell(s) in the bone marrow or which cytokine-mobilized cell(s) has the potential of triggering a reparative response within the infarct has not been determined. In addition, the contribution of cell fusion cannot be completely excluded. However, the size of newly formed myocytes is only one-fifteenth to one-twentieth of the surrounding spared myocytes, and this factor argues against cell fusion. Similarly, the donor-derived cells divide rapidly while tetraploid cells divide slowly and might not divide at all if the partner cell is a terminally differentiated cardiomyocyte.

Repair of Myocardial Damage by Resident Primitive Cells

An important question to be raised concerns whether HSCs have to be considered the stem cells of choice for cardiac repair or whether resident CSCs could be selectively mobilized and, ultimately, employed to replace damaged myocardium. CSCs can be expected to be more effective than HSCs in rebuilding dead ventricular tissue. This is because HSCs have to reprogram themselves to give rise to progeny differentiating into cardiac cell lineages. Such an intermediate phase is completely avoided by the direct activation and migration of CSCs to the site of injury. Moreover, CSCs may be capable of reaching in a short time functional competence and structural characteristics typical of mature myocytes and coronary vessels. It is intuitively apparent that the attraction

of this concept and approach is its simplicity. Cardiac repair might be accomplished by merely enhancing the normal turnover of myocardial cells. Although this is a gray area, results in our laboratory have demonstrated that cell regeneration occurs throughout the lifespan of the heart and the organism. This process continuously replaces old dying cells with new, younger, better functioning units. Cell renewal is not restricted to parenchymal cells but involves all the cell populations of the heart. For example, the presence of CSCs has provided a logical explanation for the wide heterogeneity among myocytes in the adult ventricular myocardium. Old hypertrophied myocytes are mixed together with smaller fully differentiated cells and cycling amplifying myocytes. The latter group of cells can create significant amounts of new myocardium by dividing rapidly and differentiating simultaneously, until the adult phenotype has been reached.

Stem cells divide rarely while committed transient amplifying cells are the actual group of replicating cells in self-renewing organs. The less primitive amplifying cells possess a unique property; they undergo rounds of doublings and simultaneously differentiate. Stem cells can divide symmetrically and asymmetrically. When stem cells divide symmetrically, two self-renewing daughter cells are formed. The purpose of this division is cell proliferation, that is, expansion of the stem cell pool. When stem cells engage themselves in asymmetric division, one daughter-stem cell and one daughter-amplifying cell are obtained. The objective of this division is cell differentiation — that is, production of committed progeny. Stem cells can also divide symmetrically into two committed amplifying cells, decreasing the number of primitive cells.

CSCs undergo lineage commitment, and myocytes, smooth muscle cells, and endothelial cells are generated. CSCs express c-Met and insulin-like growth factor-1 receptors (IGF-1R) and thereby can be activated and mobilized by hepatocyte growth factor (HGF) and insulin-like growth factor-1 (IGF-1) (Figure 57-5A–5C). *In vitro* mobilization and invasion assays have documented that the c-Met-HGF system is responsible for most of the locomotion of these primitive cells. However, the IGF-1-IGF-1R system is implicated in cell replication, differentiation, and survival. These observations have promoted a series of experiments in which primitive and progenitor cells have been mobilized from the site of storage in the atria to the infarcted ventricular myocardium in rodents. The reconstituted infarcted ventricular wall is composed of contracting cells and blood-supplying coronary vessels, resembling the composition and characteristics of early postnatal myocardial tissue. Since the period for regeneration was very short, it is reasonable to assume that with time the formed parenchymal cells would develop into mature myocytes.

As discussed in the following section, the transition from putative CSCs to cardiac progenitors, myocyte progenitors, and precursors and, ultimately, to amplifying myocytes has been seen, in combination with aspects resembling clonogenic growth, in the hypertrophied heart of patients with chronic aortic stenosis. The identification of these forms of growth in humans was critical for the significance of the observations

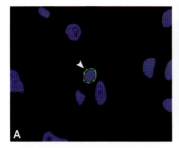 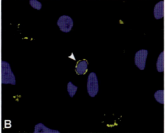 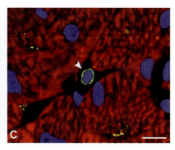

Figure 57-5. A c-kit positive cell (A, green fluorescence, arrowhead) expresses IGF-1 receptors (B, yellow fluorescence, arrowhead). Panel C shows the colocalization of c-kit and IGF-1 receptors on the same cell (yellow-green fluorescence, arrowhead). Myocytes are identified by the red fluorescence of α-sarcomeric actin antibody staining. Nuclei are labeled by the blue fluorescence of propidium iodide. Confocal microscopy, scale bar = 10 μm.

made in animal models. The magnitude of growth in the pressure-overloaded left ventricle exceeds the extent of cardiac regeneration detected in rodents and provides us with the first demonstration that the human heart can repair itself. Thus, the heart is a self-renewing organ in which the replenishment of its parenchymal and nonparenchymal cells has to be regulated by a stem cell compartment and by the ability of these primitive cells to self-renew and differentiate. This is because regeneration conforms to a hierarchical archetype in which slowly dividing stem cells give rise to highly proliferating, lineage-restricted progenitor cells, which then become committed precursors that eventually reach growth arrest and terminal differentiation.

In summary, primitive cells resident in the heart can be activated and translocated to damaged portions of the ventricle where, after homing, initiate an extensive reparative process leading to the reconstitution of functioning myocardium supplied by newly formed vessels connected with the primary coronary circulation. This approach is superior to that obtained by the utilization of bone marrow cells for its simplicity and immediate accessibility to the sites of stem cell storage in the heart. In addition, the differentiation of cardiac primitive cells is very rapid and does not require the reprogramming necessary for cells used to make blood into a state that would allow them to give rise to cardiac cell lineages.

Myocardial Regeneration in Humans

According to the dogma, ventricular myocytes in human beings are terminally differentiated cells, and their lifespan corresponds to that of the individual. The number of myocytes attains an adult value a few months after birth, and the same myocytes are believed to contract 70 times per minute throughout life. Because a certain fraction of the population reaches 100 years of age or more, an inevitable consequence of the dogma is that cardiac myocytes are immortal functionally and structurally. This assumption contradicts the concept of cellular aging and programmed cell death, and the logic of a slow turnover of cells with the progression of life in the mammalian heart. Conversely, several reports have provided unequivocal evidence that myocytes die and that new ones are

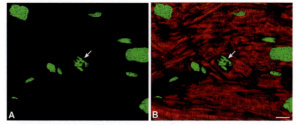

Figure 57-6. Dividing myocyte in the left ventricle of a human heart affected by idiopathic dilated cardiomyopathy. Metaphase chromosomes are shown by green fluorescence of propidium iodide (A, arrow), and the myocyte cytoplasm is stained by the red fluorescence of α-sarcomeric actin antibody. Confocal microscopy, scale bar = 10 μm.

constantly formed in the normal human heart at all ages. Both processes are markedly enhanced in pathologic states, and the imbalance between cell growth (Figure 57-6A, 6B) and cell death may be an important determinant of the onset of ventricular dysfunction and its evolution to terminal failure and death of the organism. Recent observations have indicated that the human heart contains a population of primitive cells prevalently located in the atria, mimicking the results in rodents. Through growth and differentiation, primitive cells contribute to remodeling the stressed heart by generating myocytes, coronary arterioles, and capillary profiles. These newly formed structures acquire the adult phenotype, are well integrated in the existing myocardium, and become indistinguishable from the preexisting tissue components. Such a phenomenon has been carefully documented in cardiac chimerism caused by the migration of primitive cells from the recipient to the grafted heart.

A relevant issue is that the growth potential of the diseased human heart decreases in relation to the duration of the overload, suggesting that primitive cells undergo lineage commitment, and this process reduces the stem cell pool size. For example, a myocyte mitotic index of nearly 0.08% is detected in the surviving myocardium of the border zone acutely after infarction. However, in postinfarction end-stage failure this parameter becomes 0.015%. Concurrently, myocyte death, apoptotic and necrotic in nature, markedly increases outsetting cell multiplication. Therefore, the question concerns whether interventions can be applied locally to expand the stem cell compartment or whether this amplification has to be performed outside of the organ in culture systems. The latter is a less favorable strategy because it requires time and interferes with the urgency of therapy in most patients with advanced cardiac decompensation.

KEY WORDS

Myocardial repair Regeneration of dead myocardium.
Niche Structural and biological entity where stem cells are located, proliferate, and differentiate.
Self-renewing organ Organs that possess stem cells.

FURTHER READING

Anversa, P., and Kajstura, J. (1998). Ventricular myocytes are not terminally differentiated in the adult mammalian heart. *Circ. Res.* **183**, 1–14.

Beltrami, A. P., Barlucchi, L., Torella, D., Baker, M., Limana, F., Chimenti, S., Kasahara, H., Rota, M., Musso, E., Urbanek, K., Leri, A., Kajstura, J., Nadal-Ginard, B., and Anversa, P. (2003). Adult cardiac stem cells are multipotent and support myocardial regeneration. *Cell* **114**, 763–776.

Blau, H. M., Brazelton, T. R., and Weimann, J. M. (2001). The evolving concept of a stem cell: entity or function? *Cell* **105**, 829–841.

Chien, K. R., and Olson, E. N. (2002). Converging pathways and principles in heart development and disease. *Cell* **110**, 153–162.

Kondo, M., Wagers, A. J., Manz, M. G., Prohaska, S. S., Scherer, D. C., Beilhack, G. F., Shizuru, J. A., and Weissman, I. L. (2003). Biology of hematopoietic stem cells and progenitors: Implications for clinical application. *Annu. Rev. Immunol.* **21**, 759–806.

Korbling, M., and Estrov, Z. (2003). Adult stem cells for tissue repair: a new therapeutic concept? *N. Engl. J. Med.* **349**, 570–582.

Orlic, D., Kajstura, J., Chimenti, S., Jakoniuk, I., Anderson, S. M., Li, B., Pickel, J., McKay, R., Nadal-Ginard, B., Bodine, D. M., Leri, A., and Anversa, P. (2001). Bone marrow cells regenerate infarcted myocardium. *Nature* **410**, 701–705.

Quaini, F., Urbanek, K., Beltrami, A. P., Finato, N., Beltrami, C. A., Nadal-Ginard, B., Kajstura, J., Leri, A., and Anversa, P. (2002). Chimerism of the transplanted heart. *N. Engl. J. Med.* **346**, 5–15.

Spradling, A., Drummond-Barbosa, D., and Kai, T. (2001). Stem cells find their niche. *Nature* **414**, 98–104.

van Praag, H., Schinder, A. F., Christie, B. R., Toni, N., Palmer, T. D., and Gage, F. H. (2002). Functional neurogenesis in the adult hippocampus. *Nature* **415**, 1030–1034.

Stem Cells for the Treatment of Muscular Dystrophy: More Than Wishful Thinking?

Maurilio Sampaolesi, M. Gabriella Cusella De Angelis, and Giulio Cossu

This chapter deals with the use of stem cells to treat muscular dystrophy. It reviews previous attempts at cell therapy in animal models and patients through the use of donor myoblasts, explaining the likely reasons for their failure. It then reports the identification of myogenic progenitors in the bone marrow and the related trials, this time in mice only, to treat dystrophy by bone marrow transplantation. It proceeds to describe recently identified stem cells in relationship to their ability to home within a dystrophic muscle and to differentiate into skeletal muscle cells. Different known features of various stem cells are compared in this perspective, and the few available examples of their use in murine models of dystrophy are reported. On the basis of current knowledge and waiting for a rapid advance in stem cell biology, an audacious prediction of clinical translation for these cell therapy protocols is outlined.

Introduction

Muscular dystrophies are caused by progressive degeneration of skeletal muscle fibers. Lack of one of several proteins either at the plasma membrane or, less frequently, within internal membranes increases the probability of damage during contraction and eventually leads to fiber degeneration, although the molecular mechanisms are not yet understood in detail. Fiber degeneration is counterbalanced by the regeneration of new fibers at the expense of resident myogenic cells, located underneath the basal lamina and termed *satellite cells*. Although satellite cells were considered the only cell endowed with myogenic potential, evidence has accumulated showing that other progenitor cells may participate to muscle regeneration. These latter cells are probably derived from distinct anatomic sites, such as the microvascular niche of the bone marrow, and they reach skeletal muscle through the circulation and possibly with the incoming vessels. Balance between fiber degeneration and fiber regeneration dictates the cellular and the clinical outcome. In the most severe forms, such as Duchenne's muscular dystrophy, regeneration is exhausted and skeletal muscle is progressively replaced by fat and fibrous tissue. This condition leads the patient to progressive weakness and eventual death by respiratory failure, cardiac failure, or both.

Current therapeutic approaches involve steroids and result in modest beneficial effects. Novel experimental approaches can be schematically grouped in three major areas. The first is gene therapy aiming at the production of new viral vectors (mainly Adeno, Adeno-associated, lenti, and herpes vectors) that are less antigenic and more efficient in transducing adult muscle fibers. Second, novel pharmacologic approaches focus on high-throughput screens for molecules that may interfere with pathogenetic pathways. The search should identify molecules that cause the skipping of the mutated exon or up-regulate utrophin synthesis, a cognate protein that compensates for dystrophin absence when overexpressed in dystrophic mice. The final group is cell therapy, based initially on myoblast transplantation and more recently on the transplantation of stem–progenitor cells. The recent identification of novel types of stem cells opens new perspectives for cell therapy, which is the topic of this chapter. However, the limited knowledge of stem cell biology is an obstacle that must be overcome to devise protocols with a significant prospect of clinical improvement in patients affected by severe forms of muscular dystrophy.

Myoblast Transplantation: Reasons for Failure

This field was opened by a pioneer study by Partridge and his collaborators (Partridge *et al.* 1989), who showed that intramuscular injection of C2C12 cells, an immortal myogenic cell line derived from adult satellite cells, would reconstitute with high-efficiency dystrophin-positive, apparently normal fibers in dystrophic *mdx* mice. This result caused immediate hopes for therapy that within a few months led to several clinical trials in the early 1990s. Myogenic cells were isolated from immune-compatible donors, expanded *in vitro*, and injected in a specific muscle of the patient. These clinical trials failed for several reasons, some of which may have been predicted. (At variance with C2, human myogenic cells do not have an unlimited lifespan and are not syngeneic between donor and host.) Others became apparent long after the start of the trials. For example, most injected cells (up to 99%) succumb first to an inflammatory response and then to an immune reaction against donor cell antigens. Cells that survive do not migrate more than a few millimeters from the injection site, indicating that innumerable injections should be performed to provide a homogeneous distribution of donor cells. Even though donor myogenic cells can survive for a decade in the injected muscle, the overall efficiency of the process soon made it clear that, though biologically interesting, this approach would have been clinically hopeless.

In the following years, the laboratory of Tremblay focused on these problems and in a stepwise manner produced a progressive increase in the survival success of injected myoblasts and in their colonization efficiency. Immune suppression and injection of neutralizing antibodies (directed against surface molecules of infiltrating cells such as LFA-1), pretreatment of myoblasts *in vitro* with growth factors, and modification of the muscle-connective tissue all contributed to this improvement. Extension of this protocol to primates has led to a new clinical trial whose initial results have just been published. On a parallel route, several laboratories, including our own, developed strategies to expand *in vitro* myoblasts from the patients, transduce them with viral vectors encoding the therapeutic gene, and inject them back in at least a few life-essential muscles of the patients from which they were initially derived. This approach would solve the problem of donor cell rejection but not that of an immune reaction against the vector and the therapeutic gene, a new antigen in genetic diseases. In this case, however, it soon became clear that it was difficult to produce an integrating vector that would accommodate the enormous cDNA of dystrophin. The generation of a microdystrophin (containing both protein ends but missing most of the internal exons) appears to have recently solved at least part of this problem. However, the limited lifespan of myogenic cells isolated from dystrophic patients appeared as an even worse problem, and all the attempts to solve it, ranging from immortalization with oncogenes or telomerase to myogenic conversion of nonmyogenic cells, again produced biologically interesting but clinically inadequate results.

Myogenic Stem Cells in the Bone Marrow and Bone Marrow Transplantation

The search for cells that may be converted to a myogenic phenotype led us to identify a cryptic myogenic potential in a large number of cells within the mesoderm, including the bone marrow. For these studies, we took advantage of a transgenic mouse expressing a nuclear *lacZ* under the control of muscle-specific regulatory elements (MLC3F-n*lacZ*) only in striated muscle. Since direct injection into the muscle tissue is an inefficient and impractical cell delivery route, bone marrow-derived progenitors immediately appeared as a potential alternative to myoblasts and satellite cells, since they could be systemically delivered to a dystrophic muscle. To test this possibility, we transplanted MLC3F-n*lacZ* bone marrow into lethally irradiated *scid/bg* mice and, when reconstitution by donor bone marrow had occurred, induced muscle regeneration by cardiotoxin injection into a leg muscle *(tibialis anterior)*. Histochemical analysis unequivocally showed the presence of β-gal-staining nuclei at the center and periph-

ery of regenerated fibers, demonstrating for the first time that murine bone marrow contains transplantable progenitors that can be recruited to an injured muscle through the peripheral circulation and can participate in muscle repair by undergoing differentiation into mature muscle fibers. The publication of this report raised new interest in myogenic progenitors and in their possible clinical use. It was reasoned that, although the frequency of the phenomenon was low, in a chronically regenerating, dystrophic muscle, myogenic progenitors would have found a favorable environment and consequently would have contributed significantly to regeneration of dystrophin-positive normal fibers. This, however, did not turn out to be the case. In the following year, several groups showed that *mdx* mice transplanted with the bone marrow side population, or SP (a fraction of the total cells separated by die exclusion and containing stem–progenitor cells able to repopulate the hematopoietic system upon transplantation), of syngeneic C57BL/10 mice develop, within several weeks, a small number of dystrophin-positive fibers containing genetically marked (Y-chromosome) donor nuclei. Even many months after the transplantation, the number of fibers carrying dystrophin and the Y-chromosome never exceeded 1% of the total fibers in the average muscle, thus precluding a direct clinical translation for this protocol. Recently, retrospective analysis in a Duchenne patient who had undergone bone marrow transplantation confirmed the persistence of donor-derived skeletal muscle cells over many years, again at a low frequency.

Reasons for this low efficiency may be: (1) the paucity of myogenic progenitors in the bone marrow; (2) inadequate transplantation, a procedure optimized for hematopoietic reconstitution; (3) insufficient signals to recruit myogenic progenitors from the bone marrow; (4) inadequate environment to promote survival, proliferation, differentiation, or a combination of these for these progenitors because of competition by resident satellite cells (that sustain regeneration for most of the *mdx* mouse life span) or because of an unfavorable environment created by inflammatory cells; and (5) difficulties in reaching regenerating fibers because of the increased deposition of fibrous tissue and the reduced vascular bed of the dystrophic muscle. Although in Duchenne's muscular dystrophy patients regeneration by endogenous satellite cells is exhausted much earlier in life than in the mouse, and therefore muscle colonization by blood-borne progenitors may be different, the data by Gussoni *et al.* (2002) suggests that in any case the process does not occur frequently.

Multipotent Stem Cells in the Bone Marrow: What Can Be Expected from Them?

The bone marrow hosts several multipotent cells, which include hematopoietic stem cells (HSCs), mesenchymal stem cells (MSCs) and, recently added to the list, multipotent adult progenitor cells (MAPCs), endothelial progenitor cells (EPCs) and mesoangioblasts.

These cells are described in detail in different chapters of this book; here, we briefly review their features concerning their myogenic potential and consequently their possible use in preclinical models of muscular dystrophy.

HEMATOPOIETIC STEM CELLS

Until very recently, no data were available on the ability of purified HSC to differentiate into skeletal muscle. When bone marrow was fractionated into CD45 positive and negative fractions, the muscle-forming activity after bone marrow transplantation was associated with the CD45+ fraction, suggesting that a myogenic potential is present in the hematopoietic stem cell itself or in a yet to be identified cell that expresses several markers in common with true HSC. In the following two years, several reports have convincingly demonstrated that bone marrow SP cells can be recruited to dystrophic or regenerating muscle and differentiated into skeletal muscle cells upon exposure to differentiating muscle cells or in response to Wnt molecules secreted by recruiting cells. Moreover, a fraction of SP cells localizes in a position (between the basal lamina and the sarcolemma) typical of satellite cells and indeed expresses markers of satellite cells. Moreover, circulating human AC133 positive cells differentiate into skeletal muscles *in vitro* and *in vivo* when injected into dystrophic immunodeficient mice.

Two papers recently provided final evidence that a single hematopoietic stem cell is able to reconstitute the hematopoietic system of a recipient mouse upon BMT and at the same time gives rise to a progeny that differentiates into skeletal muscle *in vivo*. The two papers, however, disagree on the mechanism of such a phenomenon: one, in agreement with previous data, identifies donor cells also as bona fide "satellite cells"; the other fails to identify donor-derived satellite cells and claims that fusion of a myeloid intermediate progenitor is the cause of differentiation. While further work will be required to solve this issue, the presence of a cryptic myogenic potential in true hematopoietic stem cells suggests that it should be possible to import a huge amount of knowledge from both experimental and clinical hematology toward the design of clinical protocols for muscular dystrophy. Still, additional steps would have to be devised since we already know that bone marrow transplantation is likely to be insufficient to ameliorate the dystrophic phenotype. Specifically, homing of progenitor cells to muscle would have to be stimulated since HSC naturally home to the bone marrow; equally important, HSC would have to be directed to a myogenic fate with much higher efficiency than what is currently observed. Transient expression of leukocyte adhesion proteins and inducible expression of MyoD may be steps in this direction.

MESENCHYMAL STEM CELLS

MSCs, originating mainly from pericytes, are located in the perivascular district of the bone marrow stroma and are the natural precursors of bone, cartilage, and fat, the constituent tissues of the bone.

Although MSCs were reported to produce myotubes in culture upon induction with 5-azacytidine, they do not differentiate into muscle under normal conditions. When transplanted in sheep fetus *in utero*, human MSCs colonized most tissues, including skeletal muscle, although their effective muscle differentiation was not demonstrated. Even though MSCs can be easily expanded *in vitro*, they are not currently considered among the best candidates in protocols aimed at reconstituting skeletal muscle in primary myopathies.

ENDOTHELIAL PROGENITOR CELLS

Initially identified as CD34+, Flk-1+ circulating cells, endothelial progenitor cells (EPCs) were shown to be transplantable and to participate actively in angiogenesis in a variety of physiologic and pathologic conditions. *In vitro* expansion of EPCs is still problematic, and few laboratories have succeeded in optimizing this process. The clear advantage of EPCs would be their natural homing to the site of angiogenesis that would target them to the site of muscle regeneration. However, their ability to differentiate into skeletal muscle has not been tested.

MULTIPOTENT ADULT PROGENITOR CELLS

The group of Verfaillie recently identified a rare cell within MSC cultures from human or rodent bone marrow: the multipotent adult progenitor cell (MAPC). This cell can be expanded for more than 150 population doublings and differentiates not only into mesenchymal lineage cells but also endothelium, neuroectoderm, and endoderm. Similar cells can be selected from mouse muscle and brain, suggesting that they may be associated with the microvascular niche of probably many, if not all, tissues of the mammalian body. Furthermore, when injected into a blastocyst, MAPCs colonize all the tissues of the embryo, with a frequency comparable to that of ES cells. Because of their apparently unlimited lifespan and multipotency, MAPCs appear to be obvious candidates for many cell replacement therapies, although complete differentiation into the desired cell type still needs to be optimized. As concerns skeletal muscle, neither the frequency at which MAPCs differentiate into skeletal muscle cells after azacytidine treatment nor attempts to optimize this process have been reported. In addition, the ability of MAPCs to travel through the body using the circulatory route has not been formally demonstrated, although the general features of these cells strongly suggest this to be the case. It will be interesting to see whether MAPCs may restore a normal phenotype in mouse models of muscular dystrophy.

MESOANGIOBLASTS

Searching for the origin of the bone marrow cells that contribute to muscle regeneration, we identified, by clonal analysis, a progenitor cell derived from the embryonic aorta that shows similar morphology to adult satellite cells and expresses several myogenic and endothelial markers expressed by satellite cells. To test the role of these progenitors *in vitro*, we grafted quail or mouse embryonic aorta into host chick embryos. Donor cells, initially incorporated into the host vessels, were later integrated into mesoderm tissues, including bone marrow, cartilage, bone, smooth, skeletal, and cardiac muscle. When expanded on a feeder layer of embryonic fibroblasts, the clonal progeny of a single cell from the mouse dorsal aorta acquires unlimited lifespan, expresses hemangioblastic markers (CD34, Sca-1, and Thy-1), and maintains multipotency in culture or when transplanted into a chick embryo. We concluded that these newly identified, vessel-associated stem cells, the mesoangioblasts, participate in postembryonic development of the mesoderm, and we speculated that postnatal mesodermal stem cells may be rooted in a vascular developmental origin.

Inasmuch as mesoangioblasts can be expanded indefinitely, are able to circulate, and are easily transduced with lentiviral vectors, they appeared as a potential novel strategy for the cell therapy of genetic diseases. In principle, mesoangioblasts can be derived from the patient, expanded and transduced *in vitro*, and then reintroduced into the arterial circulation leading to colonization of the downstream tissues. In the past few months, we succeeded in isolating mesoangioblasts from juvenile tissues of the mouse and from human fetal vessels. Attempts to isolate the same cells from postnatal human vessels are in progress.

When injected into the blood circulation, mesoangioblasts accumulate in the first capillary filter they encounter and are able to migrate outside of the vessel, but only in the presence of inflammation, as in the case of dystrophic muscle. We thus reasoned that if these cells were injected into an artery, they would accumulate in the capillary filter and from there in the interstitial tissue of downstream muscles. Indeed, intra-arterial delivery of wild-type mesoangioblasts in the α-sarcoglycan knockout mouse, a model for limb-girdle muscular dystrophy, corrects morphologically and functionally the dystrophic phenotype of all the muscle downstream of the injected vessel (Figure 58-1).

Furthermore, mesoangioblasts, isolated from small vessels of juvenile (P15) α-sarcoglycan null mice, were transduced with a lentiviral vector expressing the α-sarcoglycan–enhanced green fluorescent protein fusion protein, injected into the femoral artery of null mice, and reconstituted skeletal muscle similarly to wild-type cells. These data represent the first successful attempt to treat a murine model of limb-girdle myopathy with a novel class of autologous stem cells. Widespread distribution through the capillary network is the distinct advantage of this strategy over alternative approaches of cell or gene therapy.

Stem Cells with a Myogenic Potential from Tissues Other Than Bone Marrow

In the last few years, many reports have described the presence of multipotent stem cells in a variety of tissues of the mammalian body. Besides bone marrow, described previously, and skeletal muscle, described in the next section, stem cells able to form skeletal muscle have been described in the nervous system, adipose tissue, and the synovia. Although these reports provided evidence for muscle repair *in vivo*, no data were reported of efficacious amelioration of dystrophic phenotype.

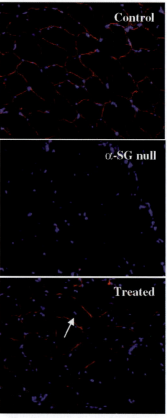

Figure 58-1. Expression of α-Sarcoglycan in α-Sarcoglycan null mice after intra-arterial delivery of wild-type mesoangioblasts. (A) A control quadriceps (CTR) from a wild-type mouse, (B) a dystrophic quadriceps, and (C) a dystrophic quadriceps injected with wild-type mesoangioblasts two months before sacrifice were stained with anti-α-sarcoglycan polyclonal antibody (white lines) and Dapi (white blotches). Large areas of the treated muscle expressed α-sarcoglycan at the fiber plasma membrane; the arrow shows a centrally located nucleus of a regenerating fiber.

In conclusion, of the many types of stem cells with a myogenic potential (Table 58-1), only mesoangioblasts have been shown so far to be effective in restoring to a significant extent the dystrophic phenotype in a murine model of dystrophy.

Unfortunately, human mesoangioblasts from patients have not yet been isolated (although preliminary work suggests this to be possible). Several other types of stem cells, already isolated from adult human tissues, show the potential of achieving similar results, since they can be expanded *in vitro*, delivered systemically, and induced to differentiate into skeletal muscle. It will be interesting in the future to see which of these cell types maintains this promise.

Cellular Environment of a Dystrophic Muscle: Which Cells Are There and Which Cells Get There

Before the appearance of stem cells, the histologic situation of a dystrophic muscle could be described with relatively few

434

TABLE 58-1

Potential of Types of Mesodermal Stem Cells to Contribute to Muscle Regeneration in a Dystrophic Muscle[a]

Stem Cell	Source	Growth in vitro	Crossing Muscle Endothelium	Myogenic Differentiation	Effect in Primary Myopathy
HSC	Bone marrow	Yes	Upon induction	Upon induction	Unknown
MSC	Bone marrow	Yes	Unknown	Poor	Unknown
EPC	Microvessels	No	Unknown	Unknown	Unknown
MAPC	Bone marrow	Yes	Unknown	Upon induction	Unknown
MAB	Microvessels	Yes	Upon induction	Upon induction	Yes
NSC	Subventricular	Yes	Unknown	Upon induction	Unknown

[a]A schematic and simplified overview of the features known, presumed, or under scrutiny of different adult stem cells in the perspective of use for cell therapy (HSC: hematopoietic stem cell, MSC: mesenchymal stem cell, EPC: endothelial progenitor cell, MAPC: multipotent adult progenitor cell, MAB: mesoangioblast, and NSC: neural stem cell). Source is generic, since the exact location within bone marrow is uncertain for some stem cells.

players. Muscle fibers undergo degeneration for the lack of one important protein, as described previously. This leads to activation of resident satellite cells that proliferate and produce myogenic cells that repair damaged fibers or replace dead fibers with new ones. Satellite cells should also generate new satellite cells (though this has not yet been formally demonstrated) to ensure a reserve population for further cycles of regeneration; failure of this process, especially in the most severe forms of dystrophy, is one major cause of anatomic loss of muscle tissue and functional impairment. However, recent data suggest that even bona fide satellite cells (i.e., residing underneath the basal lamina) may be heterogeneous and contain a subpopulation of cells that do not express typical markers such as CD34, Myf5, or M-cadherin and thus represent higher progenitors.

Dystrophic muscle is also infiltrated by inflammatory cells, mainly lymphocytes and macrophages, followed by fibroblasts that deposit large amounts of collagen contributing to progressive sclerosis of the muscle and adipocytes that replace muscle with fat tissue. The local microcirculation is progressively lost in this process, leading to a hypoxic condition for surviving or regenerated fibers that activates a vicious circle, increasing the chance of further degeneration.

A series of studies, which followed the original observation of muscle regeneration by bone marrow-derived cells, demonstrated that additional types of progenitor cells are present in muscle. These studies suggested, though not conclusively, a lineage relationship with bone marrow and resident satellite cells. Work by the Huard group led to the identification of a separate population of late-adhering myogenic cells with several features of self-renewing stem cells. These cells are particularly efficient in restoring numerous dystrophin positive fibers when injected into mdx muscles. Recent work from Rudnicki group has shown that in skeletal muscle the SP is Sca-1- and CD45-positive and can generate hematopoietic colonies but not differentiated muscle unless co-cultured with myoblasts or injected into regenerating muscle. This population is different from resident satellite cells that are fractionated as main population, do not express Sca-1 or CD45, and do not form hematopoietic colonies but that express myogenic markers and spontaneously differentiate into skeletal muscle. Interestingly, several genetically labeled SP cells could also be found as satellite cells (i.e., underneath the basal lamina and positive for c-Met and M-cadherin) after injection into regenerating muscle, strongly suggesting that at least a fraction of resident satellite cells may have derived from another cell type with SP features. Further work by La Barge and Blau (2002), supported the notion that bone marrow stem cells are at least one source of muscle stem cells and produce satellite cells besides undergoing immediate muscle differentiation. In the course of this progression, blood-borne, bone marrow-derived cells may respond to local and perhaps systemic signals and cross the vessel layers, using a mechanism analogous to leukocytes, to localize initially in a perivascular interstitial district. From there, they may be directly recruited to a muscle-forming population or may replenish the pool of satellite cells (Figure 58-2).

Acquiring a myogenic identity should imply simultaneous loss of hematopoietic potency, as suggested by the work discussed previously. Like hematopoietic cells, other incoming cells, possibly EPCs, may be incorporated into the endothelium of growing vessels that accompany and eventually contribute to muscle regeneration. The existence of endothelial-myogenic progenitors in adult muscle supports this possibility. Since this process probably occurs continuously during life, it is possible that any resident satellite cell may derive from the somite-derived myogenic population during embryogenesis or, as we hypothesized previously, from incoming cells that may have followed a circulatory route or may have reached developing muscle with the incoming vasculature during fetal angiogenesis. If the latter is the case and these progenitors have a major role in tissue histogenesis, they may play only a minor role in adult tissue homeostasis, being just a remnant of fetal life that is progressively lost after birth. This would explain the low numbers of differentiated muscle cells derived from bone marrow or other

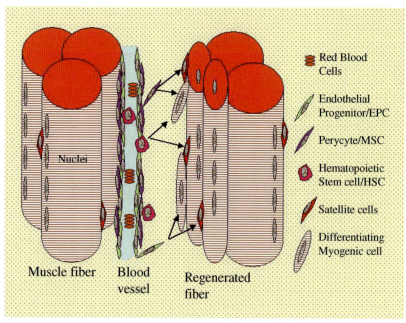

Figure 58-2. Representation of the possible origin of myogenic cells in regenerating skeletal muscle. Regenerated fibers (characterized by centrally located nuclei) are formed by fusion of resident satellite cells that may derive, at least partly, from cells derived from the microvasculature as circulating cells (HSCs), endothelial cells (EPCs), or pericytes (MSCs) of the vessel wall.

tissue and the previous failure of bone marrow transplantation in the mouse model. On the other hand, the case of "mesoangioblasts," which after *in vitro* expansion and genetic correction can be delivered in large numbers to the microvasculature of dystrophic muscle, strongly suggests that this route may lead to a successful cell therapy by exploiting different features of different stem cells.

Future Perspectives

To be successful, a future cell therapy for muscular dystrophy should probably use stem cells derived from either the patient (autologous) or a healthy, immunocompatible donor. In the first case, cells should meet as many of the following criteria as possible: (1) an accessible source (e.g., blood, bone marrow, fat aspirate, muscle, or skin biopsy), (2) the ability to be separated from an heterogeneous population on the basis of antigen expression, (3) the potential to grow *in vitro* for extended periods without loss of differentiation potency (since it appears unlikely that cells may be acutely isolated in numbers sufficient for therapeutic purposes), (4) a susceptibility to *in vitro* transduction with vectors encoding therapeutic genes (these vectors should themselves meet criteria of efficiency, safety, and long-term expression), (5) the ability to reach the sites of muscle degeneration–regeneration through a systemic route and in response to cytokines released by dystrophic muscle, (6) the ability to differentiate *in situ* into new muscle fibers with high efficiency and to produce physiologically normal muscle cells, and (7) the ability to be ignored by the immune system despite the presence of a new protein

(the product of the therapeutic gene) and possibly of some residual antigen from the viral vector.

In the case of donor stem cells from an immunocompatible donor, some of the criteria would be less stringent and some more rigorous. For example, the source of the cells would not need to be easily accessible since cells from aborted fetal material also may be obtained from sites, such as the central nervous system, not easily accessible in a patient. Gene transduction would not be necessary since donor stem cells would possess a normal copy of the mutated gene. On the other hand, immune suppression would become a major issue to prevent the rejection of a classic allograft. Although the vast experience accumulated in bone marrow transplantation may be essential to devise appropriate protocols, it should be remembered that some stem cells, such as mesoangioblasts, do not express class II MHC, making control of the immune response less complex than in the case of HSC transplantation.

Either way, we are not yet close to meeting all of the required criteria. Nevertheless, in the last few years, considerable progress has been made in the identification and initial characterization of novel classes of stem cells, some of which have been shown to be able to form skeletal muscle. Furthermore, a replenishment of skeletal muscle myogenic cells by nonresident stem cells has been convincingly demonstrated. Finally, in a mouse model of muscular dystrophy, significant morphologic, biochemical, and functional rescue has been achieved by intra-arterial delivery of stem cells.

Functional correction of a large muscle may be a different problem, however, probably calling for experimentation in a large animal model such as the dystrophic dog. Despite the elevated cost, the need to produce several species-specific reagents, the high variability among different dystrophic pups, and the ethical concerns related to the use of dogs for research, it is likely that cell therapy protocols in this animal will bring further information about the feasibility of similar protocols in dystrophic patients. Hopefully, in a few years, phase I clinical trials with stem cells may start and may set the stage for another, and at least partly successful, attack to defeat these genetic diseases.

ACKNOWLEDGMENTS

We thank G. Ferrari, M. Goodell, J. Huard, F. Mavilio, and M. Rudnicki for critical reading of this chapter and helpful suggestions. We are especially grateful to T. Partridge for thoughtful and provocative insights.

KEY WORDS

C2C12 Immortal myogenic cell line derived from adult murine satellite cells.
MAPCs Multipotent adult progenitor cells.
MCS Mesenchymal stem cell.
Mesoangioblasts Vessel-associated stem cells.
Satellite cells Resident myogenic progenitors of postnatal muscle.

FURTHER READING

Blake, D. J., Weir, A., Newey, S. E., and Davies, K. E. (2002). Function and genetics of dystrophin and dystrophin-related proteins in muscle. *Physiol. Rev.* **82**, 291–329.

Burton, E. A., and Davies, K. E. (2002). Muscular dystrophy — Reason for optimism? *Cell* **108**, 5–8.

Cossu, G., and Mavilio, F. (2000). Myogenic stem cells for the therapy of primary myopathies: wishful thinking or therapeutic perspective? *J. Clin. Invest.* **105**, 1669–1674.

Emery, A. E. (2002). The muscular dystrophies. *Lancet.* **317**, 991–995.

Ferrari, G., Cusella De Angelis, M. G., Coletta, M., Stornaioulo, A., Paolucci, E., Cossu, G., and Mavilio, F. (1998). Skeletal muscle regeneration by bone marrow-derived myogenic progenitors. *Science* **279**, 1528–1530.

Gussoni, E., Bennett, R. R., Muskiewicz, K. R., Meyerrose, T., Nolta, J. A., Gilgoff, I., Stein, J., Chan, Y. M., Lidov, H. G., Bonnemann, C. G., Von Moers, A., Morris, G. E., Den Dunnen, J. T., Chamberlain, J. S., Kunkel, L. M., and Weinberg, K. (2002). Long-term persistence of donor nuclei in a Duchenne muscular dystrophy patient receiving bone marrow transplantation. *J. Clin. Invest.* **110**, 807–814.

LaBarge, M. A., and Blau, H. M. (2002). Biological progression from adult bone marrow to mononucleate muscle stem cell to multinucleate muscle fiber in response to injury. *Cell* **111**, 589–601.

Partridge, T. A., Morgan, J. E., Coulton, G. R., Hoffman, E. P., and Kunkel, L. M. (1989). Conversion of mdx myofibres from dystrophin-negative to -positive by injection of normal myoblasts. *Nature* **337**, 176–179.

Sampaolesi, M., Torrente, Y., Innocenzi, A., Tonlorenzi, R., D'Antona, G., Pellegrino, M. A., Barresi, R., Bresolin, N., Cusella De Angelis, M. G., Campbell, K. P., Bottinelli, R., and Cossu, G. (2003). Cell therapy of alpha sarcoglycan null dystrophic mice through intra-arterial delivery of mesoangioblasts. *Science* **301**, 487–492.

Seale, P., Asakura, A., and Rudnicki, M. A. (2001). The potential of muscle stem cells. *Dev. Cell* **1**, 333–342.

Skuka, D., Vilquin, J. T., and Tremblay, J. P. (2002). Experimental and therapeutic approaches to muscular dystrophies. *Curr. Opin. Neurol.* **15**, 563–569.

59

Regeneration of Epidermis from Adult Keratinocyte Stem Cells

Ariane Rochat and Yann Barrandon

Keratinocyte stem cells, along with hematopoietic stem cells, have the longest record in cell therapy. Following the seminal work by Green and colleagues in the early 1980s, hundreds of burned patients worldwide had their lives saved by the transplantation of cultured autologous keratinocyte stem cells. Reconstitution of a functional epidermal barrier, self-renewal of the epidermis, and the unexpected regeneration of a papillary dermis are major accomplishments. Gene therapy of hereditary, disabling skin diseases using genetically modified keratinocyte stem cells is down the road. Nevertheless, many challenges lie ahead. The functionality, the mechanical properties, and the aesthetic of the regenerated skin must improve. Sweat glands and hair follicles must be reconstructed, and the pigmentation of the skin better controlled. It is then necessary to thoroughly comprehend the cellular and molecular mechanisms involved in skin morphogenesis, epidermal renewal, stem cell interactions, dermal remodeling, and fetal wound healing. Only then will stem cell therapy become a major therapeutic option in plastic and reconstructive surgery.

Introduction

The skin of vertebrates is covered with scales, feathers, or hairs, whose main function is to protect the body from environmental hazards; the skin also contributes to protection, tactile sensation, thermoregulation, and camouflage. Furthermore, it participates in body homeostasis – for example, through water balance in the frog or vitamin D synthesis in the human. Most importantly, the appearance of the skin shapes emotional and social attitudes, which is best illustrated in the human by the incessant advertising campaigns promoting a young and smooth skin and by the distress of patients with disabling skin diseases or scars.

The human skin represents about 16% of body weight and an area of $1.7\,m^2$ in the adult. It is composed of three layers that are, from the outer to the inner surface, the epidermis, the dermis, and the subcutis. The epidermis is a stratified, keratinized epithelium in continuity with the epithelial part of the skin appendages (i.e., the hair follicles, the sebaceous glands, and the sweat glands) (Figure 59-1). The epidermis contains several specialized cell types that contribute to body protec-

tion. The keratinocytes, which represent 90% of the epidermal cells, are of ectodermal origin and are responsible for epidermal renewal, cohesion, and barrier function. The melanocytes are pigment-forming cells of neural crest origin and are responsible for the color of the skin and UV protection, whereas the Langerhans and epidermal T-cells are of marrow origin and are involved in immunologic defense. The nonvascularized epidermis and epidermal appendages are separated from the highly vascularized dermis by a complex basement membrane, which the nutrients originating from the dermal vessels cross. The dermis is a connective tissue that contains fibroblasts producing proteoglycans, collagen, and elastic fibers, and it is responsible for the mechanical properties of the skin. The subcutis that mainly contains adipocytes (fat cells) is involved in thermoregulation and lipid storage. The skin is also highly innervated, with numerous free nerve endings in the epidermis and specialized nerves endings in the dermis. The function of the neuroepithelial Merkel cells, which rest in the basal layer of the epidermis, is largely unknown. Altogether, the skin contains more than 25 specialized cell types originating from at least six distinct stem cells. Most importantly, the skin is constantly renewed and remodeled. In the human, the epithelial part of the epidermis is renewed once every three weeks, and hair follicles undergo regular cycles of growth, regression, and rest about 25 times in a lifetime. Similarly, the dermis is constantly remodeled with new collagen fibers being produced as old ones are degraded.

Keratinocyte Stem Cell

The keratinocyte stem cell is one of the adult stem cells that inhabits the skin and contributes to skin function and renewal. Adult stem cells are best defined by their capacity to self-renew and to maintain tissue function for a long period. Keratinocyte stem cells, like hematopoietic stem cells, match this functional and rigorous definition. Moreover, unipotent (single-lineage) and multipotent (multiple-lineage) keratinocyte stem cells coexist in adult skin. Compiling evidence demonstrates that the basal cell layer of the epidermis contains unipotent keratinocyte stem cells that are each responsible for the renewal of a tiny portion of the epidermis. Multipotent keratinocyte stem cells are located in the upper region of the hair follicle below the sebaceous glands, but sweat glands and sweat ducts may also contain multipotent keratinocytes. Unipotent and multipotent stem cells are

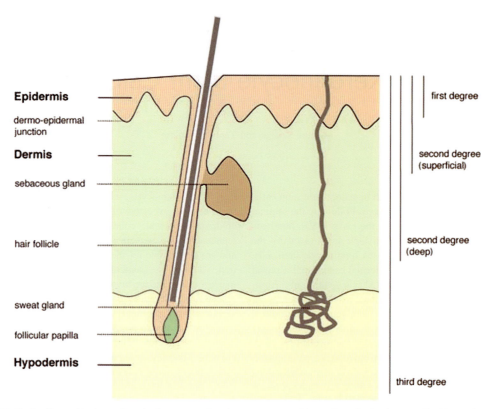

Figure 59-1. Depth of burns. First-degree burns heal spontaneously without a scar; second-degree burns heal with a scar. Third-degree burns do not heal and may necessitate the transplantation of autologous split-thickness skin grafts and autologous cultured keratinocyte stem cells depending on the extent of the burns.

clearly related to each other. Indeed, keratinocytes can migrate out of a hair follicle or a sweat gland to adopt an epidermal fate and to regenerate a wounded epidermis. Most importantly, keratinocyte stem cells, which can divide every 16 to 18 hours, have a remarkable growth capacity in culture, far better than that of dermal fibroblasts. This remarkable growth capacity is at the origin of cutaneous cell therapy.

Keratinocyte stem cells have other remarkable, though circumstantial, properties. They divide infrequently *in vivo* in contrast to their transient amplifying progeny, which multiplies rapidly, and they express high levels of β1 and α6 integrin and low levels of CD71 (transferrin receptor). Theoretically, a keratinocyte stem cell can generate a progeny with variable fate depending on circumstances, the most probable fates being those of a stem cell (symmetric divisions), of a transient amplifying cell, or both (asymmetric division). However, more research is needed to fully comprehend the mechanisms involved in stem cell decision making.

CLONAL ANALYSIS

The growth potential of individual multiplying keratinocytes is best evaluated by clonal analysis that can be performed either with cells freshly isolated from the skin or with cultured cells. A clonal analysis experiment typically implies the iso-

lation and the cultivation of 100 to 200 individual cells. Briefly, each single cell is isolated with an elongated Pasteur pipet under direct vision using an inverted microscope before it is carefully deposited at the center of a 35-mm Petri dish containing lethally irradiated 3T3 cells. After 7 days of cultivation, each dish is carefully searched for the presence of a clone under an inverted microscope. Each clone is then dissociated with trypsin to obtain a single-cell suspension, which is inoculated into several indicator dishes (usually 2 × 100-mm Petri dishes) containing lethally irradiated 3T3 cells. Cultures are fixed after 12 days of cultivation, stained with Rhodamine B, and examined under a dissecting microscope (Figure 59-2). Three types of colonies are easily distinguished: (1) Growing colonies are large and round, have a smooth perimeter, and mostly contain small cells. These colonies can be passaged many times and are formed by keratinocytes with significant growth potential. (2) Aborted colonies are small, have an irregular shape, and contain large, flattened differentiated squame-like cells. These colonies cannot be passaged and are formed by cells with restricted growth potential. (3) Mixed colonies usually have a jagged perimeter and contain intermingled areas of small and large differentiated cells. These colonies can only be passaged a limited number of times and are formed by cells with limited growth potential.

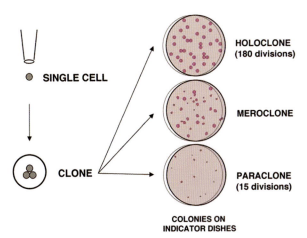

SINGLE CELL

CLONE

HOLOCLONE
(180 divisions)

MEROCLONE

PARACLONE
(15 divisions)

**COLONIES ON
INDICATOR DISHES**

Figure 59-2. Clonal types. A single cell is isolated with a Pasteur pipet and cultivated onto a feeder layer of irradiated 3T3 cells. If a clone forms, it is passaged while still growing exponentially. The number and shape of colonies on the indicator dishes determine the clonal type of the founding cell. The progeny of a holoclone generates large progressively growing colonies, the progeny of a meroclone generates a mixture of large progressively growing colonies and aborted colonies, and that of a paraclone generates only aborted colonies or no colony. The proportion of those clonal types varies with donor age. Fetal skin mostly contains holoclones, young skin mostly contains holoclones and meroclones, and old skin mainly contains meroclones and paraclones.

CLONAL TYPES

The growth potential of a clonogenic keratinocyte is best evaluated by the frequency of aborted colonies formed by its progeny in indicator dishes (Figure 59-2). A cell whose progeny forms almost exclusively growing colonies is termed *holoclone*. A cell whose progeny forms growing, mixed colonies and aborted colonies is termed *meroclone*, and a cell whose progeny forms exclusively aborted colonies or no colony is termed *paraclone*. Serial cultivation experiments have demonstrated the tremendous growth capacity of holoclones (up to 180 divisions), whereas transplantation experiments have demonstrated the ability of holoclones to form an epidermis. Indeed, a single holoclone can generate a large enough progeny to entirely restore the epidermis of an adult human. Moreover, clonal analysis experiments have demonstrated the self-renewal capacity of holoclones in culture and *in vivo*. Hence, holoclones fulfill the fundamental definition of adult stem cells. The epithelial cells of stratified squamous epithelia are all termed keratinocytes, even if the latter differ in their differentiation capacities. Therefore, all stratified epithelia contain holoclones that theoretically should be referred to as epidermal holoclones, vaginal holoclones, or limbal holoclones. Most importantly, the term *holoclone* only refers to the growth capacity of a keratinocyte – not to the lineage capability of the cell. Therefore, some holoclones are multipotent (i.e., follicular holoclones), whereas others are unipotent (i.e., epidermal holoclones).

Meroclones, whose growth potential is always lower than that of holoclones, are best characterized by an increased capacity to generate paraclones, possibly by asymmetric division. Most importantly, the growth potential of meroclones directly correlates with the frequency of paraclone formation. Meroclones with a significant growth potential generate few paraclones and can form an epidermis when transplanted. Meroclones that generate a significant number of paraclones have a limited growth potential and are unlikely to contribute to long-term renewal of the epidermis. Thus, meroclones are not strictly self-renewing and can be viewed as the skin equivalent of short-term hematopoietic stem cells.

The growth potential of paraclones is restricted between 1 and 15 divisions, meaning that the size of the progeny of a paraclone ranges between 2 and 32,000 cells. Consequently, paraclones are transient amplifying cells. However, paraclones are unlikely to be the transient amplifying cells identified *in vivo* by kinetic experiments. Indeed, the number of paraclones in a suspension of freshly isolated keratinocytes, whatever the age of the donor, is far below the expected number of transient amplifying cells, which supposedly includes most multiplying cells in the epidermis. Paraclones may then be a population of multiplying cells generated in response to special circumstances such as wound healing and cultivation.

CLONAL CONVERSION

A culture of human keratinocytes is thus composed of cells with differing capacity for growth that are termed *paraclone*, *meroclone*, and *holoclone* and are best identified by clonal analysis. Clonal analysis experiments demonstrated that holoclones self-renew for few passages in culture, but they stop self-renewing for unknown reasons at some point and entirely convert to meroclones, which in turn generate paraclones. Ultimately, the culture ends without multiplying cells. This phenomenon is termed *clonal conversion* (Figure 59-3). The conversion of holoclone to meroclone to paraclone is irreversible under normal conditions and is greatly influenced by the conditions of cultivation. For instance, it takes more than 23 weeks to exhaust the growth potential of a holoclone cultivated onto a feeder layer of irradiated 3T3 cells and in presence of fetal calf serum, an otherwise impossible achievement with suboptimal culture conditions or in defined medium. This is why culture conditions of keratinocytes must be constantly monitored to maintain the highest level of cultivation, especially when the cells are to be transplanted. Clonal conversion is also accelerated by many parameters, including age. In our hands, the number of paraclones in a suspension of freshly isolated keratinocytes is independent of donor age, but it becomes age related after the primary keratinocytes are passaged. Indeed, secondary cultures initiated from aged donors contain many more paraclones than cultures initiated from young donors, and they cannot be passaged as many times. Hence, the formation of paraclones reveals the effect on cell growth of age-related genetic modification or that of an adverse epigenetic phenomenon. This is of paramount importance when cells are cultivated for therapy, as clonal conversion can adversely affect the engraftment (take) of cultured epithelium grafts and the long-term renewal of the regenerated epidermis.

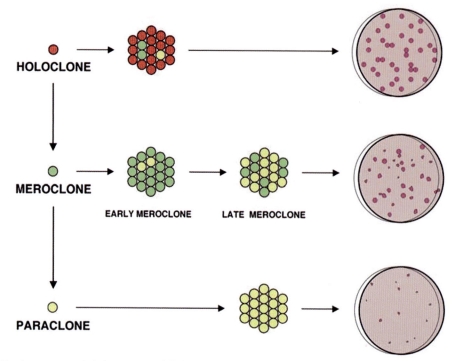

Figure 59-3. Clonal conversion. A holoclone generates holoclones, a few meroclones, and almost no paraclones. A meroclone generates more paraclones with time. A paraclone generates only aborted colonies. Clonal conversion increases with passages and is drastically enhanced by suboptimal culture conditions.

Keratinocyte Cultivation

GENERALITIES

In 1975, Rheinwald and Green reported in a seminal paper the cultivation of human keratinocytes using a feeder layer of irradiated 3T3 cells and envisioned its potential application for the treatment of extensive skin defects. It took Green and his colleagues eight more years of intensive work before the first transplantation of cultured autologous keratinocytes onto a human was accomplished. This period was paved with several landmarks, including the use of the enzyme dispase to detach the cultured epithelium as a transplantable homogeneous sheet and the demonstration that cultured human keratinocytes can generate an epidermis once transplanted onto athymic mice. The technology has been successfully used worldwide to treat hundreds of extensively burned patients. Green's method to cultivate transplantable keratinocytes is extremely efficient, and we have always been able to initiate cultures, even from tiny biopsies. The cultivation of epidermal stem cells for therapy requires the use of a feeder layer of lethally irradiated mouse 3T3 cells and serum. To our knowledge, no clinical success has been obtained with cells grown in defined medium or in other conditions. The use of cells of animal origin (xenotransplantation) can theoretically induce adverse reactions and transmit pathogens. However, the risk is significantly minimized by using 3T3 cells originating from a master bank that tested negative for a battery of viruses according to guidelines of regulatory agencies. Animal serum, widely used in the biotechnology industry, is also a potential life hazard, both through the induction of adverse reactions to foreign proteins and the transmission of pathogens. Again, the risk is minimal if lots of serum with traceable origin and for which the manufacturer maintains a master file in accordance with the guidelines of regulatory agencies are purchased.

Most importantly, no adverse reaction to 3T3 cells and serum has been reported in the hundreds of patients treated with epidermal stem cells cultivated in the presence of serum and 3T3 cells. The monitoring of culture conditions includes the testing of serum lots, media, and supplements, and is best accomplished by using a reference strain of keratinocytes, the growth characteristics of which are well known. Indeed, any modification in the method of cultivation (e.g., a new lot of serum or supplements) should be systematically tested in a colony-forming assay using a reference strain to determine the extent to which the modification affects the ratio of growing colonies (stem cells and close progeny) to aborted colonies (paraclones). It also necessitates strict adherence to instructions and protocols. For instance, 3T3 cells are best grown in the presence of bovine serum, and switching to fetal bovine serum is a common mistake. In our experience, fiddling with protocols in an uncontrolled manner rarely results in better culture performances and often has deleterious consequences, such as enhanced clonal conversion resulting in rapid exhaustion of the growth capacity of the stem cells and, consequently, in failure of engraftment.

A cutaneous wound ruptures skin integrity and can have serious consequences, such as infection, loss of proteins, and even death. Moreover, chronic wounds favor improper scarring and the development of skin cancers. Burn wounds are a perfect illustration (Figure 59-1). A first-degree burn wound, which only affects the epidermis, heals spontaneously and without a scar. Reconstitution of the epidermal integrity then results from the migration of keratinocytes originating from the nearby epidermis or appendages. A second-degree burn wound, defined as superficial or deep upon the depth of the dermal burn, destroys the epidermis and part of the dermis. It heals spontaneously by the migration of cells originating from the epidermal appendages. However, a permanent scar will form. A third-degree burn destroys the entire thickness of the skin (full-thickness burn). Although a tiny third-degree burn wound can possibly heal with a scar by the migration of cells from the surrounding unwounded skin, a large one will not. The transplantation of split-thickness autografts obtained from unburned donor areas is then the unique therapeutic option, with the added risk of improper healing of donor sites, including infection, hypertrophic scars, and cheloids. Moreover, the recurrent cropping of donor sites can become a serious problem, especially for patients burned over 60% of the total body surface area (TBSA). Progress in medical resuscitation and intensive care allows patients, whose skin has almost completely burned, to remain alive. The transplantation of autologous epidermal stem cells then becomes lifesaving for those patients who do not have enough donor skin for conventional split-thickness autografts.

METHODOLOGY

The methodology has been reported in great detail.[16,22,35] Briefly, a full-thickness skin biopsy is obtained aseptically from normal skin to isolate stem cells from both the epidermis and the epidermal appendages. The biopsy is then stored in culture medium and transferred to the laboratory, where it is minced with scissors and treated with trypsin to obtain a single-cell suspension. The cells are seeded at a seeding density of 2.4×10^4 cells/cm^2 onto a feeder layer of irradiated 3T3 cells in medium containing fetal bovine serum and supplemented with cholera toxin, hydrocortisone, insulin, and triodothyronine according to published protocols. Cultures are supplemented with epidermal growth factor (10 ng/ml) at the first medium change. Under these conditions, clonogenic keratinocytes divide every 16 to 18 hours, and colonies quickly expand and push away the irradiated 3T3 cells, which then detach from the culture vessel. If the culture lasts long enough, the colonies of keratinocytes will fuse to form a stratified epithelium. The method of cultivation selects for clonogenic keratinocytes, but some melanocytes and dermal fibroblasts can also adhere and multiply, although at a much slower rate than the keratinocytes. Primary cultures are usually passaged while still in exponential growth – that is, before confluence or after nine days of cultivation, whichever comes first. If

primary cultures yield enough cells, cultures for transplantation are then initiated at a seeding density of 1.2×10^4 cells/cm^2 and the leftover cells are frozen for future use. If the primary cultures do not yield enough cells, it is then necessary to further expand the cells by setting up secondary cultures at a seeding density of 1.2×10^3 cells/cm^2. Cells are then cultivated for no more than a week before they are used to set up cultures for transplantation. Under optimal conditions, a 10,000-fold expansion is obtained within two to three weeks. Confluent cultured epithelia then need to be detached enzymatically from the bottom of the culture vessels using dispase or thermolysine. However, the enzymatic treatment significantly shrinks the cultured epithelium. Once rinsed, each epithelial graft is mounted onto petroleum gauze backing, with special care taken to maintain the correct orientation of the epithelium (i.e., the basal layer facing away from the backing). This step of the procedure is cumbersome, tedious, and manpower demanding, especially when hundreds of cultured epidermal autografts (CEAs) need to be prepared in a reasonable amount of time. This is why many groups have explored alternative methods of cultivating keratinocytes.

TRANSPLANTABLE MATRICES

Clonogenic human keratinocytes can be cultivated with variable success on various matrices, including human fibrin, bovine, or human collagen; hyaluronic acid; shark proteoglycans; and polymers. Therefore, it is crucial to thoroughly evaluate the effects of each of these matrices on the growth of the keratinocyte stem cells, since suboptimal culture conditions may result in enhanced clonal conversion, a reduced keratinocyte lifespan, and an irreversible degradation of the quality of the graft. This is best accomplished by clonal analysis and by the transplantation of the cultured cells onto athymic or severe-combined immunodeficiency mice using reference strains of keratinocytes, the performances of which are well known. Using this approach, we have demonstrated that fibrin matrices alleviate the need for the detachment of the grafts, eliminate shrinkage, and hence have considerable advantages over plastic for the culture of epidermal stem cells for grafting. Use of a fibrin matrix considerably shortens the duration of cultivation required to generate many CEAs, making it possible to transplant the entire body area of an adult human with CEAs as soon as 15 days after injury. Most importantly, the long-term regeneration of human epidermis on third-degree burns transplanted with autologous, cultured epithelium grown on a fibrin matrix has been demonstrated. Fibrin matrices are also extremely useful in the transplantation of autologous limbal stem cells to restore a corneal epithelium.

The transplantation of small colonies of keratinocytes (preconfluent cultures) grown on fibrin matrices or of individual keratinocytes suspended in fibrin gels has been proposed. This may be useful to quickly transplant small full-thickness burns or to enhance the healing of superficial burn wounds. However, it remains to be demonstrated that individual cells or small colonies generate an epidermal barrier faster than confluent and organized cultures when directly exposed to the hostile environment of the grafting bed in large full-thickness burns. Composite grafts combining autologous keratinocytes and fibroblasts embedded in a collagen gel, a fibrin matrix, or polymers may also be useful to generate closer-to-normal skin by enhancing the formation of dermis. However, with today's technology, the population of fibroblasts in a small skin biopsy cannot be sufficiently expanded in a time frame compatible with the transplantation of acute burn wounds. This is why composite grafts associating autologous fibroblasts and keratinocytes are reserved for chronic wounds. Many groups have proposed using allogeneic fibroblasts, large stocks of which can be prepared in advance. However, allogeneic fibroblasts do not permanently engraft, despite contradictory claims in the literature. To our knowledge, composite grafts may be efficient to treat chronic skin defects such as leg ulcers, but they have not proved useful for the successful treatment of extensive burn wounds. Hopefully, progress in the cultivation of human fibroblasts will be made; then, composite grafting will have a major effect.

Transplantation of Keratinocyte Stem Cells

Various surgical procedures have been developed over the years to transplant cultured keratinocytes. CEAs were initially transplanted directly on the granulation tissue that spontaneously forms after full-thickness burns were excised to fascia. Success was then variable, ranging from excellent to poor, and surgeons quickly stressed that the quality of the wound bed was crucial. In that regard, one of the most appreciated procedures is known as Cuono's, in which full-thickness burns are excised to fascia and temporarily covered with meshed, expanded fresh or cryopreserved split-thickness human allografts obtained from organ donors. At grafting time, which usually occurs two to three weeks after the initial biopsy was taken for expansion of stem cells, allografts are dermabraded or tangentially excised and a careful hemostasis is performed. Cultured epithelium grafts, with the basal cells oriented toward the grafting bed, are then gently applied and covered with petroleum gauze dressings to prevent desiccation (Figure 59-4). Sliding and displacement of the graft should be strictly avoided. CEAs are then overlaid with bridal veil gauze, sterile dry compresses, and elastic bandage. Skin substitutes like Integra® or Biobrane® are useful alternatives to cadaver skin when the latter is not available or cannot be used for religious reasons.

Cuono's procedure has greatly increased the rate of CEA engraftment, which is best appreciated clinically. At the time of the first dressing ("takedown"), the engrafted epithelium appears shiny and translucid. With time, the regenerated epithelium matures to resemble a normal epidermis (Figures 59-4, 59-5). CEAs perform best in areas in which shear forces or frictions are minimal and in non-weight-bearing surfaces. Successful engraftment now averages 70% of the transplanted area, but it is not unusual to obtain a 100% take. In a recent study, younger age was significantly associated with better CEAs take, possibly reflecting clonal type differences between young and old people. Most importantly, the transplantation of CEAs necessitates a highly trained surgical and nursing staff. For instance, four to six nurses require from 60 to 90 minutes to perform each dressing change following CEA application at the Percy Burn Center in France, which has one

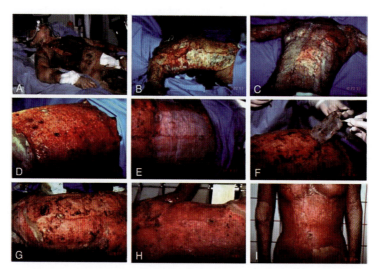

Figure 59-4. Transplantation of human cultured epithelium autografts. A 35-year-old woman was burned by flames; over 50% of her body has third-degree burns. (A) Admission. (B) Excision to fascia of full-thickness thoracic and abdominal burn wounds the day after injury. (C) Temporary coverage of the excised area with split-thickness meshed human cadaver allografts. (D) Appearance of the dermabraded grafting bed. (E) Transplantation of control-cultured epithelia (top) and cultured epithelia grown on fibrin matrices (bottom). (F) Appearance of the transplanted area 8 days later (takedown). (G) Appearance of the same area 21 days later. Note the excellent take of both control (bottom) and experimental (top) cultured epithelia (the photograph was taken from the opposite side to that shown in panels E, H, and I). Appearance of the same transplanted area (H) 57 days and (I) 4 months later. Reproduced with permission from Ronfard et al. (2000).

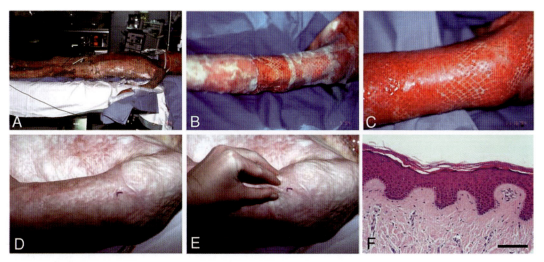

Figure 59-5. Long-term followup of cultured epithelia transplanted on a fibrin matrix. A 9-year-old boy was burned by flames; over 95% of his body has third-degree burns. (A) Admission at Percy Burn Centre a month after injury. (B) Transplantation of cultured epithelia grown on a fibrin matrix on the left arm. (C) Appearance of the transplanted area at takedown. (D and E) Clinical appearance of the skin 3.5 years after the transplantation. The skin is elastic when pinched and has a smoother appearance than the neighboring split-thickness skin autografts. (F) Histologic appearance of the skin 3.5 years after transplantation. The epidermis is histologically normal. Note the presence of rete ridges and a superficial neodermis with vascular arcades. Similar results were obtained with cultured epithelia grown in absence of fibrin. Bar = 100 μm. Reproduced with permission from Ronfard et al. (2000).

of the largest and most successful records in CEA transplantation. Infection of the grafting bed is the enemy of CEAs and provokes a quick disappearance of the grafts. The fight of infection relies on systemic antibiotics and local antiseptics, which are clinically efficient and usually not harmful to the CEAs, despite an alleged toxicity *in vitro*. Fragility and blistering of CEAs are also commonly observed during the weeks following transplantation, most likely as a consequence of a delayed maturation of the anchoring fibrils. In some cases, lysis of CEAs, which had otherwise nicely engrafted, occurs after a few weeks with no rational explanation. Bleeding of the grafting bed can also result in the loss of CEAs, further emphasizing the importance of thorough hemostasis at grafting time. Another complication of CEAs is epidermal hyperkeratosis, which may need repeated treatment with moisturizing and keratolytic ointments. Wound contracture is also a problem, as in all patients with large burns, but there seems to be less hypertrophic scarring with CEAs than with conventional split-thickness autografts.

Regeneration of Epidermis

Cultured epithelia prepared according to the methods of Green and colleagues are stratified but lack the uppermost-differentiated epidermal layers necessary for barrier function. However, the horny layers mature rapidly once the CEAs are transplanted, indicating that the cells adequately express their program of differentiation. This epidermis always contains suprabasal Langerhans cells and may contain melanocytes, although at a variable density in the basal layer. The presence of Merkel cells has been reported. The most surprising observation is the presence of rete ridges, vascular arcades, and elastic fibers, which are not observed during normal wound healing. However, the formation of a neodermis that closely resembles the papillary dermis is a slow process that takes years (see Figure 59-5). This extremely intriguing observation suggests that mesenchymal stem cells may have been transplanted with the epidermal stem cells and that epidermal-mesenchymal interactions reminiscent of fetal wound healing have occurred. This awaits confirmation. Most importantly, hair follicles and sweat glands are not regenerated. There are several explanations for this: Multipotent epithelial stem cells are not present in CEAs, possibly because current culture conditions favor the epidermal lineage or because multipotent stem cells are absent in adult skin, and the inductive signals necessary for appendage morphogenesis are missing. However, recent results from our laboratory indicate that the skin of adult mammals contains multipotent epithelial cells with the capacity to generate epidermis, sebaceous glands, and all follicular lineages. Other groups have demonstrated that dermal papilla cells obtained from adult hair follicles have inductive capacities. These are important findings that open the path to the reconstruction of hair follicles and other epidermal appendages.

The long-term self-renewal of epidermis generated from CEAs is a concern in view of recent findings demonstrating a shortening of telomeres in keratinocytes obtained from transplanted patients. This further illustrates that cells should be cultivated according to state-of-the-art technology and that cell culture, as good as it can be, is a stress to cells. It is impossible as yet to distinguish between the consequences of clonal conversion, which can occur quickly in relation to poor cultivation and independently of the mitotic clock, and the shortening of telomeres, a difference expressed in a recent review as mitotic clock or culture shock. The expansion of the epidermal stem cells in culture represents, at the most, 22 cell doublings, a small number when holoclones can undergo at least 180 doublings, meaning that holoclones and early meroclones have plenty of growth potential left to renew their share of epidermis for a long time. The number of stem cells (holoclones) or cells with significant growth potential (meroclones) in the CEAs that can engraft and self-renew the epidermis becomes important. Surprisingly, there are no data in the literature on this subject even though hundreds of patients have been transplanted. Nevertheless, clinical followup indicates that epidermis generated from CEAs self-renews for years, suggesting that holoclones and meroclones are present in sufficient number. However, this possibility should be thoroughly investigated.

Indications of CEAs

Indications of CEAs have been extensively reviewed. The main indication of CEAs is full-thickness burns over 60% of TBSA in which CEAs are highly beneficial and lifesaving. CEAs can also be useful in reconstructive surgery to enhance the healing of donor sites, to prevent recurrence of cheloids, and to cover wounds made by the excision of giant congenital nevi and of tattoos. CEAs may be useful in urology and oral surgery. But one of the most impressive uses of CEAs is in ophthalmology. Pellegrini *et al.* (1998) have beautifully demonstrated that limbal stem cells (limbal holoclones) can be cultivated on a fibrin matrix and transplanted to restore the integrity of the corneal epithelium. This approach, coupled with elective keratoplasty, has restored vision to patients whose sight was impaired because of corneal burn wounds. Gene therapy of skin-disabling diseases or of hereditary diseases of the corneal epithelium will also certainly benefit from the experience gained with the treatment of large burns. Indeed, epidermal stem cells or limbal stem cells can be successfully transduced by defective retroviral or lentiviral vectors and engineered to express a protein of medical interest. For instance, transduction of a cDNA encoding laminin 5 or collagen VII in epidermal stem cells of patients with junctional epidermolysis bullosa or dystrophic epidermolysis bullosa, respectively, should permit the treatment of elective areas to enhance the quality of life of those patients.

Fragility, lack of dermal components, and high costs have significantly limited the use of CEAs. As an example, a commercial CEA (Epicel®) manufactured according to good manufacturing practices costs $14 per cm^2 (including biopsy kit, culture, and shipping), which prices the transplantation of a square meter of CEA at $140,000. However, the transplantation of keratinocyte stem cells is lifesaving, and its cost should be compared to that of other transplantation procedures. Furthermore, progress has been made. For instance, use of a fibrin matrix diminishes the fragility of the CEA and is user friendly. Moreover, it should reduce costs as the number of CEAs necessary to cover an area is two times less, the duration of cultivation is significantly shortened, and there are no large personnel requirements to prepare the grafts. Composite grafts including autologous mesenchymal and keratinocyte stem cells are also foreseen, but they are more complex to manufacture. Their cost will certainly be higher than that of CEAs.

Summary

A challenge in *ex vivo* skin cell therapy is to obtain a massive expansion of the stem cells from a small skin biopsy in the shortest time as possible while preserving stemness. The method of cultivation described by Rheinwald and Green in 1975 achieves this goal and has not been matched. Cell therapy using autologous keratinocyte stem cells cultivated according to the preceding method has saved the lives of hundreds of burned patients and has restored sight to vision-impaired patients. But progress can be made, such as improving the cultivation technology both for keratinocytes stem cells and for mesenchymal stem cells, reconstructing hair follicles and sweat glands, and improving the mechanical properties and the aesthetic of the regenerated skin. The answer to those challenges lies in a better understanding of the cellular and molecular events involved in skin renewal and morphogenesis as well as in a better understanding of fetal and adult wound healing. In that regard, the use of embryonic stem cell lines will be extremely helpful. But the skin experience clearly indicates that scientific knowledge is not enough and that it is necessary to adapt the requirements of the stem cells to the preoccupations of the regulatory agencies. Moreover, high manufacturing costs are a major preoccupation of biotechnology and pharmaceutical companies as well as of health insurance providers. This economical aspect should not be underestimated; it can adversely affect stem cell therapy, as illustrated by the recent difficulties of several biotechnology companies specialized in skin cell therapy.

ACKNOWLEDGMENTS

We are grateful to Daniel Littman for proofreading the manuscript. This work was supported by funds from the Swiss Federal Institute Lausanne (EPFL) and from the Lausanne University Hospital (CHUV) to Y. Barrandon and by a grant from the Swiss National Foundation to A. Rochat.

KEY WORDS

Clonal conversion Conversion of a holoclone to a meroclone or to a paraclone irreversible under normal conditions; affected by culture conditions and age.

Holoclone (Gk *holos* whole + Gk *klôn* growing) Colony-forming cell with tremendous growth potential (up to 180 divisions), whose progeny forms almost exclusively growing colonies.

Meroclone (Gk *mero* part + Gk *klôn* growing) Colony-forming cell whose growth potential is lower than the growth potential of an holoclone but superior to 15 divisions; forms growing and aborted colonies.

Paraclone (Gk para *akin* + Gk *klôn* growing)   Colony-forming cell committed to a maximum of 15 divisions; forms only aborted colonies.

FURTHER READING

Barrandon, Y., and Green, H. (1987). Three clonal types of keratinocyte with different capacities for multiplication. *Proc. Natl. Acad. Sci. USA.* **84**, 2302–2306.

Carsin, H., Ainaud, P., Le Bever, H., Rives, J., Lakhel, A., Stephanazzi, J., Lambert, F., and Perrot, J. (2000). Cultured epithelial autografts in extensive burn coverage of severely traumatized patients: a five year single-center experience with 30 patients. *Burns* **26**, 379–387.

Gallico, G. G., 3rd, O'Connor, N. E., Compton, C. C., Kehinde, O., and Green, H. (1984). Permanent coverage of large burn wounds with autologous cultured human epithelium. *N. Engl. J. Med.* **311**, 448–451.

Gambardella, L., and Barrandon, Y. (2003). The multifaceted adult epidermal stem cell. *Curr. Opin. Cell Biol.* **15**, 771–777.

Green, H. (1991). Cultured cells for the treatment of disease. *Sci. Am.* **265**, 96–102.

Oshima, H., Rochat, A., Kedzia, C., Kobayashi, K., and Barrandon, Y. (2001). Morphogenesis and renewal of hair follicles from adult multipotent stem cells. *Cell* **104**, 233–245.

Pellegrini, G., Bondanza, S., Guerra, L., and De Luca, M. (1998). Cultivation of human keratinocyte stem cells: current and future clinical applications. *Med. Biol. Eng. Comput.* **36**, 778–790.

Rheinwald, J. G., and Green, H. (1975). Serial cultivation of strains of human epidermal keratinocytes: The formation of keratinizing colonies from single cells. *Cell* **6**, 331–343.

Ronfard, V., Rives, J. M., Neveux, Y., Carsin, H., and Barrandon, Y. (2000). Long-term regeneration of human epidermis on third degree burns transplanted with autologous cultured epithelium grown on a fibrin matrix. *Transplantation* **70**, 1588–1598.

60

Orthopedic Applications of Stem Cells

Jerry I. Huang, Jung U. Yoo, and Victor M. Goldberg

Introduction

More than 33 million musculoskeletal injuries occur in the United States each year. Bone is capable of regeneration, and defects often heal spontaneously. However, cartilage, tendon, and ligament injuries usually result in replacement of the site by organized scar tissue, which is inferior to the native tissue. An increased understanding of cell biology and various tissue types may lead to the future possibility of using tissue engineering techniques to recapitulate the embryonic events that result in the development of native tissue. The goal of tissue engineering is to generate biologic substitutes for repair or replacement of injured tissue. Three basic elements are required. First, appropriate cells must be present to give rise to the structural tissue. Second, appropriate growth factors and differentiation stimuli must exist for the cells to proceed down the proper lineage. Third, a scaffolding matrix must act as a building block for cellular attachment, differentiation, and maturation into the desired tissue. The generated construct must be site-specific and integrate well at the host–graft interface.

Technologic advances have resulted in the advent of a wide array of biomaterials and synthetic growth factors. Clinicians and scientists face the enormous task of generating biologic constructs that emulate the interaction between cells and the complex extracellular matrix they secrete and encapsulate them. A thorough understanding of the biology of each tissue type (material properties, ratio of extracellular matrix components, and cellular profile) is essential to construction of functional tissue. The first element of tissue engineering is the use of cells. Different approaches include the use of differentiated lineage-specific cells (osteoblasts, chondrocytes, tenocytes, meniscal fibrochondrocytes, etc.) or the use of progenitor cells.

Pluripotent mesenchymal stem cells (MSCs) with the ability to differentiate into multiple mesodermal lineages have been isolated from bone marrow, adipose tissue, and synovium. Stem cells have the advantage that they are unlimited in supply, easily harvested, and can be expanded in tissue culture to large numbers. Moreover, it is postulated that induction of stem cells down various mesodermal lineages will result in tissue that more closely resembles native tissue and recapitulates embryonic development. MSCs isolated from human bone marrow aspirates have been shown to retain

pluripotentiality and proliferative ability through long-term passaging. Another advantage of using MSCs for tissue engineering strategies is that they can be used in allogeneic transplantation. Human MSCs cells do not induce a mixed lymphocyte reaction when incubated with allogeneic donor lymphocytes and have the ability to suppress an ongoing mixed lymphocyte reaction. Allogeneic MSC-based tissue constructs would be limitless in supply and have enormous economic advantages.

A number of growth factors have been described that enhance angiogenesis, promote cell proliferation, and induce differentiation of cells down various mesodermal lineages. The delivery of single doses of recombinant growth factors has been demonstrated to be effective in the healing of segmental bone defects and articular cartilage defects. Similarly, the structural properties of ligaments during the repair process can be improved by the application of growth factors. However, in many clinical situations a sustained exposure to growth factors is necessary. Moreover, leakage of supraphysiologic doses to adjacent sites can pose huge problems such as heterotopic ossification from osteoinductive factors.

In addition to being a powerful tool in cell-based therapy, MSCs may also be useful as a delivery vehicle for growth factors through genetic manipulation. Cells can be transduced with a retrovirus so that the target gene can be integrated into the cell's DNA. This allows for propagation of gene expression when the cell replicates so that long-term expression of the target gene would occur. Alternatively, gene therapy strategies using adenoviral vectors would be useful in clinical situations where only transient gene expression is desired. *In vivo* gene therapy in which the viral vector is introduced to a tissue by direct injection has many disadvantages. A host response to the virus may result in immunologic rejection. Moreover, transduction efficiencies are often low, especially if the target tissue contains slowly growing cells. Diffusion of the vector to adjacent areas may also lead to potential complications. *Ex vivo* gene therapy using MSCs overcomes some of these potential problems. Use of autologous cells as delivery vehicles would prevent the host immune response. In addition, cells can be seeded onto matrix scaffolds customized to the shapes and sizes required at the target tissue.

This chapter provides a brief summary of the biology and properties of various musculoskeletal tissue types, highlights the present tissue engineering strategies, and draws speculations as to the future direction of the application of MSCs in tissue engineering as it applies to each tissue. MSCs can serve as the building blocks for tissue regeneration by their ability to differentiate down various lineages. They also hold great promise as delivery vehicles for sustained release of growth factors to injured tissue sites. The growth factors may enhance tissue repair by promoting migration of MSCs, inducing differentiation of progenitor cells, and enhancing vascularization of the newly formed tissue.

Bone

Bone has the ability to regenerate itself with functional tissue with properties similar to the original tissue. However, bone tissue engineering still has many clinical applications, including fracture nonunion, congenital malformations requiring bone-lengthening, tumors, and bone loss secondary to trauma or osseous infections. Moreover, bone regeneration in the setting of bony ingrowth is important in joint arthrodesis for osteoarthritis, spinal fusion, and more rigid fixation of prosthetic implants.

The extracellular matrix of bone is primarily type I collagen and calcium phosphate. Osteoblasts line the periphery and are active in matrix deposition. In the clinical setting, autogenous bone grafts harvested from the iliac crest and fibular grafts are often used in fracture repair. However, some of the disadvantages of harvesting autogenous bone graft are donor-site morbidity and their limited supply. Clearly, there is a need for alternative methods of harvesting autologous bone substitutes. In their study, Connolly et al. harvested autologous marrow aspirates from the posterior iliac crests as a graft substitute for patients with tibial nonunions and injected them into fracture sites. A considerable amount of new bone was evident, and the patient was able to fully bear weight at 5 months. Percutaneous harvest of bone marrow aspirates is less invasive, and bone marrow suspensions may serve as useful bone graft substitutes.

MSCs isolated from bone marrow aspirates and processed lipoaspirate (PLA) cells from adipose tissue differentiate into osteoblasts when cultured in the presence of dexamethasone, ascorbic acid, and beta-glycerophosphate. Bone marrow-derived MSCs have also shown the ability to form heterotopic bone in animal models. MSCs seeded onto hydroxyapatite (HA) and tricalcium phosphate (TCP) ceramic discs implanted into subcutaneous pockets show evidence of bone formation within the pores of the ceramic scaffolds. Yoshikawa et al. demonstrated that preincubation of human MSCs in a porous ceramics in the presence of osteoinductive medium, followed by intraperitoneal implantation into athymic nude mice, resulted in the formation of thick layers of lamellar bone and active osteoblasts that line the ceramic surface.

In another preclinical animal study, Bruder et al. showed the ability of autologous MSCs loaded into HA/TCP carriers to heal a critical-sized segmental defect in a canine model. Significant bone formation occurred at the host–implant interface, and a continuous span of bone was seen across the defect. Evidence of both woven and lamellar bone was seen. Periosteal calluses formed around the implant. During the 16-week period of the study, the callus remodeled and resulted in healing of the defect with bone that was similar in shape and size to the original segment of bone that was resected. Human MSCs loaded onto HA/TCP carriers were implanted into femoral defects in athymic rats in a study by the same group. Radiographic and histologic evidence of new bone formation was seen at 8 weeks. Biomechanical testing showed that cell-loaded ceramic implants had more than twice the stiffness and torque to failure as ceramic implants that had no cells.

Growth factors in the bone morphogenetic protein (BMP) family, including BMP-2 and BMP-7, also induce progenitor cells to differentiate down the osteogenic lineage. Current strategies for bone tissue engineering include the use of osteoconductive matrix devices that promote bony ingrowth and the delivery of osteoinductive growth factors to bony defect sites. Matrix materials include calcium HA, type I collagen gel, polylactic acid polymers, and demineralized bone matrix. Yasko et al. demonstrated histologic and radiographic evidence of healing in segmental bone defects using demineralized bone matrices implanted with recombinant human bone morphogenetic protein-2 (rhBMP-2). In a study involving a larger animal model, a sheep femoral defect model was used by Gerhart et al. to show the efficacy of BMP-2 in healing critical-sized long bone defects in a large animal model. A prospective, randomized trial of 450 patients looking at the safety and efficacy of rhBMP-2 in improving the outcome of open tibial fractures showed that patients treated with intramedullary nailing and rhBMP-2 had significantly lower risks of delayed union and need for more invasive intervention such as bone grafting and nail exchange. Moreover, they tend to have significantly faster fracture healing and lower risks of infection and need for hardware removal. Gene therapy using genetically manipulated MSCs is an attractive alternative method for delivery of growth factors. Not only can they safely deliver sustained release of growth factors to anatomic sites but their osteogenic potential allows them to serve as substrates for osteoinductive factors and building blocks for newly formed bone.

Regional gene therapy using MSCs as a vehicle for localized expression of osteoinductive proteins has shown promising results in animal models. Lieberman et al. showed that human bone marrow stromal cells transfected with adenoviral-BMP-2 led to formation of more robust, trabecular bone in an athymic rat femoral defect model compared with the thin, lace-like bone that formed in defects filled with matrix carrying localized recombinant human BMP-2 (rhBMP-2). Moreover, the femurs from the adenoviral BMP-2 showed no statistically significant difference in biomechanical strength with respect to ultimate torque to failure and energy to failure compared with control femurs. The difference in characteristics of the healing response may be related to the increased efficacy from sustained release of BMP-2 from the adenovirally transfected bone marrow cells compared with the single-dose response seen in the rhBMP-2 group. Moreover, the implanted bone marrow stromal cells may themselves contribute to osteogenesis in the defect site. Dragoo et al. successfully transfected stem cells isolated from adipose tissue with the BMP-2 gene and showed rapid induction of the cells into the osteoblast phenotype in *in vitro* cell cultures. Collagen matrices seeded with the transduced cells were able to produce heterotopic bone in the hind limbs of SCID mice. Adipose tissue is plentiful and easily accessible and may well serve as another alternative bone graft substitute in an orthopedic surgeon's armamentarium.

Vascular invasion is a vital step in endochondral ossification. Vascular endothelial growth factor (VEGF) is one of the best-characterized angiogenic factors. Studies have also shown that it is essential during embryogenesis, skeletal development, and endothelial function. In their study, Peng *et al.* demonstrated that muscle-derived MSCs transduced with retroviral VEGF work synergistically with cells transfected with BMP-4 to enhance bone healing in critical-sized calvarial defects. VEGF was found to be important for endochondral bone formation through enhancement of angiogenesis, cell recruitment, improved cartilage formation, and accelerated cartilage resorption. This study demonstrated another strategy for use of stem cells in gene therapy-based treatment of skeletal defects.

Several applications of MSCs in the repair of local bone defects have been described previously. Bone marrow-derived MSCs may also hold great potential for treatment of diffuse musculoskeletal disease. Osteoporosis and osteogenesis imperfecta (OI) are two of the more attractive candidates. OI is a genetic disorder of MSCs characterized by a defect in the type I collagen gene that results in children with growth retardation, short stature, and numerous fractures secondary to fragile bone. Six children have already been enrolled in a clinical trial at St. Jude Children's Research Hospital that involve the intravenous administration of unmanipulated bone marrow from human leukocyte antigen (HLA)-identical or single-antigen-mismatched siblings. Five of the six patients showed engraftment in one or more sites, including bone, skin, and stroma. More importantly, these five patients all showed acceleration of growth velocity during the six-month postinfusion. The authors attributed the increase in growth to the generation of normal osteoblasts from the MSCs that engrafted in skeletal sites.

Cartilage

Osteoarthritis is one of the most prevalent chronic conditions in the United States, accounting for as many as 39 million physician visits a year. According to a National Health Interview Survey, it is estimated that approximately 70% of the population older than 65 years will have activity limitation or require medical attention because of osteoarthritis. Moreover, a significant number of adolescents and young adults suffer from chondral defects secondary to trauma, sports-related injuries, and osteochondritis dissecans. Of 31,516 knee arthroscopies in a survey conducted by Curl *et al.*, 63 % of the knees had chondral lesions, with an average of 2.7 hyaline cartilage lesions per knee. Cartilage has poor intrinsic healing ability, and superficial defects typically do not heal spontaneously. This is related to both the lack of vascular supply and the poor proliferative ability of chondrocytes. When the lesion extends into the subchondral bone, mesenchymal cell recruitment occurs from the synovium and subchondral marrow and a healing response ensues in which the defect is filled with repair tissue resembling fibrocartilage. Unfortunately, this tissue is structurally inferior to the native cartilage.

Articular cartilage is highly acellular with cell volume averaging only approximately 2% of the total cartilage volume in adults. The extracellular matrix is composed of a highly complex network of collagen fibrils and proteoglycans. Type II collagen is the dominant collagen subtype found in cartilage. The molecule is a triple helix composed of three $\alpha 1$ chains and with multiple cross-links. The collagen fibrils contribute to the tensile strength of cartilage. Load-transmission capacity and compressive strength arise mainly from the proteoglycans, the other main constituent of the extracellular matrix of articular cartilage. The core protein of aggrecan contains a large number of chondroitin sulfate and keratin sulfate side chains that become highly hydrated. Aggrecan molecules also contain a hyaluronic acid-binding region.

Clinically, current treatment options for cartilage defects can be categorized into cartilage stimulation and cartilage replacement strategies. Cartilage stimulation techniques include abrasion arthroplasty, subchondral drilling, and the microfracture technique. However, the repair tissue never achieves the hyaline architecture of the native tissue. Osteochondral autografts and allografts represent the other spectrum where the defect is filled with plugs taken from normal regions of articular cartilage. The main disadvantages associated with use of autografts are paucity of tissue and donor-site morbidity and the possible long-term complications. Allografts, on the other hand, face the risk of donor rejection from immunogenicity and disease transmission. Joint prosthesis remains the mainstay for symptomatic relief of pain and improvement of daily function of patients suffering from severe articular cartilage defects. In 1999 alone, more than 244,000 total knee arthroplasties were performed.

An exciting development has been the introduction of tissue engineering strategies for cartilage defects. Cell-based therapy for treatment of cartilage defects is already in use clinically with the use of autologous chondrocytes. Chondrocytes are isolated and culture expanded from arthroscopically harvested cartilage and reimplanted into deep articular cartilage defects. Since 1987, more than 950 patients have been treated with this technique. Long-term (mean 7.5 years) followup of patients showed good to excellent rating with the Cincinnati rating score. Biopsy specimen resembled hyaline cartilage with extracellular matrices that consisted primarily of type II collagen and aggrecan. Some of the disadvantages of this technology include a limited supply and donor site morbidity associated with the initial cartilage harvest.

MSCs capable of chondrogenesis are present in bone marrow, periosteum, synovium, and adipose tissue. Human MSCs can be culture expanded more than 1 billion-fold and retain their multilineage potential. Moreover, they proliferate rapidly and have the advantage that they can recapitulate the embryonic events present in chondrogenesis. The importance of MSCs in cartilage repair is illustrated by the lack of a repair response in defects that do not penetrate subchondral bone. On penetration, recruitment of pluripotential marrow mesenchymal cells ensues. The influence of local cytokines from inflammatory cells and synovial fluid results in differentiation of the progenitor cells into chondrocytes and filling of the space with fibrocartilage.

The ability of MSCs to heal full-thickness articular cartilage defects was shown to be possible in a rabbit femoral condyle model. Wakitani et al. isolated bone marrow-derived and periosteum-derived MSCs from New Zealand White rabbits and embedded them into a collagen-based scaffold. The cellular constructs were implanted into 6 mm long × 3 mm wide × 3 mm deep full-thickness cartilage defects in the weight-bearing portion of the medial femoral condyle. Healing of the defect by hyaline cartilage was evident as early as 2 weeks with filling of the subchondral space with dense, highly vascularized new bone. After 4 weeks, the neocartilage was thicker, and excellent integration was noted at the interface between the new subchondral bone and the host tissue. Moreover, further analysis of the histologic specimen at the various time points showed that the formation of subchondral bone resulted from a recapitulation of the embryonic process of enchondral ossification with progression of chondrocytes to a hypertrophic state, followed by vascular invasion and ossification. No evidence of osteoarthrosis was noted in any of the knees. At later time points, however, significant remodeling of the cartilage occurred with loss of metachromatic staining. Genetic modification of the MSCs may enhance the healing response and allow for better long-term repair tissue.

In a study by Grande et al. (1999), periosteal cells transfected with a retroviral vector containing BMP-7 showed in vitro and in vivo gene expression for at least 8 weeks after seeding onto polymer grafts. When the cell-based constructs were placed into 3-mm circular osteochondral defects in the intertrochlear grooves of rabbit knees, the defects were completely filled with predominantly hyaline-like tissue that persisted as along as 12 weeks after implantation. The subchondral portion of the defects showed rapid reconstitution of bone. In a separate study by Sellers et al., restoration of the tidemark and formation of subchondral bone occurred in articular cartilage defects when they were treated with collagen scaffolds containing another member of the transforming growth factor (TGF) superfamily BMP-2.

As we reach a better understanding of the biology and biomechanical properties of articular cartilage and its response to different growth factors, gene-modified tissue engineering can become a more powerful tool for cartilage resurfacing. A number of growth factors have been described that enhance chondrocyte proliferation and chondrogenic differentiation, including fibroblast growth factor-2 (FGF-2), insulin-like growth factor-I (IGF-I), TGFβ-1, growth hormone (GH), BMP-7 (also known as osteogenic protein-1), and BMP-2. Induction of chondrogenesis using growth factors can be accomplished by direct injection into the defect site and application of gene therapy techniques.

With direct injections, single supraphysiologic doses of growth factors may often be insufficient for proper healing.

Degradation of the proteins may occur as a result of an inflammatory reaction and the recruitment of macrophages and neutrophils. Moreover, dilution and clearance of growth factors by the surrounding synovial fluid occurs over time. Prolonged exposure to growth factors may be necessary for chondrogenic differentiation of resident progenitor cells. In addition, by using chondrocytes or MSCs as a delivery vehicle for growth factors via gene therapy, the cells can in turn be the building blocks for the regenerated cartilage and respond to autocrine and paracrine growth factors. The newly formed cartilage can be a chimera of the original cell construct and new mesenchymal progenitor cells from the recruitment and migration that occurs from the actions of the cytokines secreted by the transduced cells.

Mandel et al. demonstrated the feasibility of successful transduction of bone marrow-derived mesenchymal progenitor cells with both retroviral and adenoviral TGFβ-1 vectors. These genetically modified progenitor cells were able to undergo chondrogenesis in vitro with and without exogenous TGFβ-1 addition. Nixon et al. constructed a replication-incompetent adenovirus vector expressing IGF-I and successfully transfected chondrocytes, MSCs, and synoviocytes. Significant increases in proteoglycan production in the extracellular matrix secondary to IGF-I induction were noted. Moreover, cells expressed high levels of the IGF-I for up to 28 days in culture. Both of these studies verify the possibility of genetically modifying MSCs so that they have the ability to secrete chondroinductive growth factors over sustained time periods.

Animal models for ex vivo gene therapy of articular cartilage defects using MSCs have already been carried out. Gelse et al. investigated the ability of adenoviral-mediated expression of perichondrial MSCs to heal an articular cartilage defect in a rat model. The effects of AdBMP-2 and AdIGF-1 transfected cells suspended in fibrin glue were evaluated with respect to their ability to heal the defects and integrate with the host tissue. The partial thickness lesions healed in both the AdBMP-2 and AdIGF-1 groups with repair cartilage exhibiting hyaline morphology that were composed of type II collagen in the extracellular matrix but no type I collagen. The defects in the nontransfected cell group primarily filled with fibrous tissue rich in type I collagen. Again, this is consistent with the observations seen in the natural repair process in which MSCs from marrow cavity progress to repair subchondral cartilage defects with fibrocartilage-like tissue. A complication of the AdBMP-2 group that was not seen in the AdIGF-1 group was osteophyte formation secondary to leakage of cells outside the construct.

Chondroprotective approaches to preservative of articular cartilage have also been explored by several groups. Instead of trying to stimulate cartilage repair and replace cartilage defects, an alternative is prophylactic treatment and cessation of disease progression in its earlier stages. Interleukin-1 receptor antagonists (IL-1Ra), tumor necrosis factor (TNF) blockers, and interleukin-4 (IL-4) have all been reported to protect against degenerative changes in articular cartilage. The anti-inflammatory drugs infliximab (Remicade) and etanercept (Enbrel), both inhibitors of TNF, have been approved by the Food and Drug Administration for patients suffering from rheumatoid arthritis. Animal models have already shown the suppression of osteoarthritic changes using gene therapy to effect delivery of IL-1Ra. Transduced MSCs expressing IL-1Ra or antagonists of TNF may both enhance the reparative process through their inherent chondrogenic potential and retard the degradative process in cartilage lesions.

Meniscus

The annual incidence of meniscal tears is approximately 60 to 70 per 100,000. The menisci of the knee are semilunar fibrocartilaginous structures that are integral to the normal function of the knee. The extracellular matrix consists of collagenous fibers that are mostly oriented circumferentially, with interspersing of radial fibers that contribute to its structural integrity. The circumferential fibers help disperse compressive forces while the radial fibers protect against tearing from tensile forces. The matrix contains mainly type I, II, and III collagens, with type I collagen being the most prevalent. The cells of the meniscus are fibrochondrocytes because of their chondrocyte-like appearance and synthesis of fibrocartilaginous matrices. The main function of the meniscus is load transmission during weight-bearing. In extension, approximately 50% of the load in the knee is transmitted to the menisci. At 90 degrees of flexion, this increases to almost 90% of the total load. The medial meniscus also acts as an important secondary restraint to anteroposterior translation, especially in the anterior cruciate ligament (ACL) deficient knee. Finally, the meniscus is important in shock absorption and joint lubrication.

Early in prenatal development, the menisci are very cellular and highly vascularized. After skeletal maturity, the vascular zone is generally confined to the peripheral one-third of the meniscus. Longitudinal tears in the peripheral zone can generally be repaired, whereas radial tears are usually not amenable to repair. Defects that cannot be repaired are usually debrided to avoid the irritation from loose meniscal flaps. Total meniscectomy is usually contraindicated because it leads to a number of osteoarthritic changes originally described by Fairbank. Long-term followup studies showed that a high percentage of patients with a history of meniscectomy go on to develop knee instability and considerable amounts of radiographic degenerative changes in the knee.

Clearly, there is a need for new treatment options. Currently, meniscal allograft transplantation is one of the few alternatives for replacement of large meniscal defects. Unfortunately, there is a risk of rejection and disease transmission with allograft tissue. Moreover, the graft size of the donor must match the recipient, which can sometimes be difficult. Long-term problems with meniscal allografts include graft shrinkage, decreased cellularity, and loss of normal biologic activity. Other experimental methods for meniscal repair include the use of fibrin clots, collagen-based polymers, and a number of polyurethane-based meniscal prostheses.

MSCs may provide novel treatment strategies in both growth-factor-based and cell-based treatment options for meniscal tears. A number of cytokines including platelet-derived growth factor (PDGF), BMP-2, hepatocyte growth factor (HGF), epidermal growth factor (EGF), IGF-I, and endothelial cell growth factor have been implicated in the proliferation and migration of meniscal cells within the different zones. Single doses of growth factors may not provide adequate stimulus in the repair tissue. Successful adenoviral-mediated expression of *BMP-2*, *PDGF*, *EGF*, and *IGF-1* have already been demonstrated in culture systems of MSCs and fibroblasts. MSCs can serve as delivery vehicles for sustained release of growth factors that contribute to cell proliferation, recruitment, and differentiation and an increase in localized vascularization. Martinek *et al.* demonstrated the feasibility of gene therapy in tissue-engineered meniscal tissue by showing evidence of transduced gene expression in meniscal allografts at 4 weeks. In another study by Hidaka *et al.*, bovine meniscal cells were transfected with an adenovirus vector encoding *HGF*, seeded onto polyglycolic acid scaffolds, and placed into subcutaneous pouches of athymic mice. Gene expression of *HGF* was associated with a significant increase in neovascularization.

The repair of meniscal defects by fibrocartilage has been advocated by some authors. Lesions in the avascular zone of canine menisci repaired by implantation of polyurethane polymers showed evidence of fibrocartilaginous tissue compared with only fibrous tissue in the control group. In a small clinical series, Rodkey *et al.* implanted resorbable collagen scaffolds into medial meniscus lesions and showed preservation of the joint surfaces and evidence of regeneration of tissue regeneration two years postoperatively. The chondrogenic potential of bone marrow-derived MSCs has been well-characterized. An alternative strategy in meniscal repair would be a prefabrication of a cartilage-like construct using stem cells and a biodegradable matrix. Data from a rabbit partial meniscectomy model showed that MSCs embedded in a collagen sponge can enhance the formation of fibrocartilage tissue at the defect site. Although the tissue would not have the identical properties of native menisci, it would serve a similar function in joint surface preservation and load transmission across the knee joint.

Ligaments and Tendons

Ligaments and tendons are bands of dense connective tissue that lend stability and provide movement of joints. Injury that leads to inflammation or tear of these structures can result in significant functional deficits and development of degenerative joint disease. During the normal healing process, various growth factors such as PDGF, FGF, and TGFβ are released by macrophages and platelets to stimulate fibroblast proliferation and tissue remodeling. These growth factors contribute to proliferation, extracellular matrix secretion, and recruitment of cells. Tissue-engineered cellular constructs and delivery of growth factors via gene therapy offers great potential in augmenting the healing process.

In general, most of the dry weight of skeletal ligaments is made up of collagen. Greater than 90% of this is type I with a small percentage of type III. Glycosaminoglycans and elastin also make up a small proportion of the biochemical makeup. Ligaments are oriented to resist tensile forces along their long axes. The diameter of collagen fibril diameters and the number of collagen pyridinoline cross-links correlate with the tensile strength of the healed ligament. Ligaments work in conjunction with muscle-tendon forces, bony intra-articular constraints, and other soft tissues to help stabilize joints and prevent nonphysiologic movements. There are differences in the intrinsic healing capacity of various ligaments. After injury, ligaments heal in a series of stages: hemorrhagic, inflammatory, proliferation, and remodeling. Clinically, medial collateral ligament (MCL) injuries heal reliably without surgical intervention, whereas ACL tears usually do not heal spontaneously. Histologic studies of human ACL tears show that, unlike extrasynovial ligaments, no evidence of bridging occurs between remnants of the ACL from the femoral and tibial sides. Typically, ACL ruptures are treated with a variety of different tendon autografts and allografts. Unfortunately, autograft harvest involves damage to previously healthy tissue, and allografts carry the risk of disease transmission. Moreover, although isolated MCL tears do well with nonoperative treatment, biochemical analysis shows that untreated MCL scar has a higher than normal proportion of type III collagen, higher collagen turnover rates, and increases in total glycosaminoglycan. Histologically, the extracellular matrix never completely approaches the highly organized appearance of normal ligament substance. Studies have also shown that MCL fibroblasts migrate more rapidly and repopulate cell-free areas more rapidly than those from ACLs. Structural differences exist between cells of the MCL and the ACL.

Growth factors play a large role in healing and remodeling of musculoskeletal tissue. Schmidt et al. showed exposure to EGF, and basic FGF leads to a significantly higher rate of proliferation in fibroblasts from the MCL and ACL. Fibroblast proliferation is a major component of the normal ligament healing process. Similarly, Marui et al. showed that FGF and EGF treatment of ligament fibroblast cultures leads to increased collagen synthesis. IGF-I also stimulates type I collagen synthesis in fibroblasts. However, complex interactions exist between different growth factors because some act synergistically, whereas others antagonize one another. The structural and functional properties of tissues are largely dependent on the composition of the extracellular matrix.

The use of MSCs in gene therapy strategies for ligamentous and tendon injuries has many potential clinical uses. Growth factors can be delivered locally to injury sites that can promote each of the four stages of normal healing. Increased fibroblast proliferation effected by FGF and EGF can be promoted through adenoviral-mediated gene expression. Short bursts of growth factors will enable the healing process to occur more rapidly in the initial stages. TGFβ-1 promotes wound healing in many animal models. Its ability to augment cellular proliferation and increase secretion of collagen in the extracellular matrix hold great promises as a useful target gene for regional growth factor delivery. Animal studies by Hildebrand et al. and Batten et al. have already shown that MCLs exposed to PDGF leads to stiffness and breaking energy similar to their respective controls. Moreover, Letson and Dahners showed that combination treatment with IGF-I and FGF further enhances these structural properties. In a canine model by Kobayashi et al., basic FGF leads to increased neovascularization and better orientation of collagen fibers in a partially ruptured ACL. The feasibility of using cells as a delivery vehicle of growth factors into the knee joint was confirmed by a study by Bellicampi et al. showing cell-seeded collagen scaffolds that remained viable for up to 4 weeks when implanted in the knee joint.

In addition to the use of autografts and allografts, current treatment options that have been proposed for tendon defects include use of synthetic polymers and acellular biodegradable scaffolds. Scaffolds serve as a conduit for recruitment of cells during the initial healing process. As they degrade over time, the mechanical loading forces are transferred to the new repair tissue. Despite the promising data regarding repair strength and biomechanical properties, the immunogenicity and biocompatibility of the materials over time are not well understood. It would be ideal to use autologous sources of cells in combination with biodegradable matrices as a tissue-engineered construct to reconstruct ligaments or tendons.

Fibroblasts generate tension and change their orientation along tensile forces when cultured in three-dimensional collagen gels. MSCs display a similar fibroblastic property. The combination of MSCs with biodegradable scaffolds could be useful for bridging large tendon defects. Autologous cell-based tendon repair was performed in a hen flexor tendon model by Cao et al. in which tenocytes seeded onto polyglycolic acid scaffolds were able to heal a 3- to 4-cm defect. The healed tissue resembled native tendon grossly and histologically. Biomechanically, the experimental group had 83% of the breaking strength of the normal tendon. Similarly, MSC-based repairs of Achilles tendon in animal models showed improvement of the biomechanics and function of the tendon. However, the histologic appearance and biomechanical strength are inferior to normal tendon controls.

In one study by Awad et al., collagen gels were seeded with MSCs in a full-thickness patellar tendon defect model to compare its effectiveness compared with the natural repair process. At 26 weeks after the original implantation, composite constructs showed significantly higher moduli and maximum stresses compared with the natural repair group. When compared with native tissue, however, the maximum stress was only one-fourth of the control. Moreover, 28% of the cell–matrix constructs formed bone in the tendon repair site

Young et al. seeded rabbit MSCs onto a pretensioned polyglyconate suture to create a contracted construct. At 40 hours, cell nuclei were spindle shaped, and a cell viability of approximately 75% was noted in the constructs. Repair tissue treated with MSCs had a significantly larger cross-sectional area at the repair site than the contralateral control (suture only) and untreated native tissue. Cell-seeded repairs showed superior biomechanical properties with a twofold increase in load-related properties. Compared with native tissue, the treated tendons had almost two-thirds of the structural properties at 12 weeks. A rapid rate of increase in load-related material properties may reflect the remodeling stage of healing mediated by the MSCs. Histologically, increases in tenocytes and collagen crimping pattern also occurred over time.

The two studies reviewed previously demonstrate the potential of MSCs to help augment the tendon repair process. Tissue-engineered constructs combining stem cells with biodegradable scaffolds enhance the biomechanical properties of the repair tissue and remodels over time to more closely resemble the complex organization of normal tendon. Preincubation of the constructs *in vitro* with growth factors may produce better biologic substitutes. Moreover, transfection of MSCs with genes such as *TGFβ-1*, *EGF*, and *FGF* may lead to a synergistic effect and to stronger structural repair tissue. Successful transfection of myoblasts and ACL fibroblasts with adenovirus and subsequent introduction into the rabbit ACL have been demonstrated. Expression of the *lacZ* reporter gene was noted at 7 days and persisted as long as 6 weeks. *BMP-12* gene transfer into lacerated chicken tendon resulted in augmentation of ultimate force and stiffness of the repaired tissue. Lou *et al.* previously demonstrated that mesenchymal progenitor cells form tendon-like tissue ectopically when transfected with the *BMP-12* gene.

Spine

Posterolateral lumbar intertransverse process fusions are commonly performed for spinal disorders secondary to degenerative changes and trauma. However, the nonunion rates have been reported to be as high as 40% with single-level fusions and even higher in multiple-level procedures. Currently, nonunions are prevented by application of instrumentation such as pedicle screws, rods or plates, and various interbody fusion devices to achieve better correction of deformities and biomechanical stability during the healing period. However, a significant number of nonunions and pseudoarthrosis still exists in fusions with instrumentation, and some have shown no statistically significant decrease in nonunions with use of screw fixation.

More recently, a large number of clinical studies have compared the efficacy of osteoinductive proteins with autogenous iliac crest bone graft in lumbar fusions. Johnsson *et al.* showed in a small randomized clinical series that osteogenic protein-1 (OP-1), also known as BMP-7, was as effective as autogenous bone graft in achieving single-level lumbar fusions. In a rabbit model by Patel *et al.*, *BMP-7* was found to overcome the inhibitory effects of nicotine on spinal fusion. In a prospective, randomized, clinical trial, Boden *et al.* compared Texas Scottish Rite Hospital (TSRH) pedicle screw instrumentation with rhBMP-2 and with TSRH pedicle screw instrumentation and rhBMP-2 without instrumentation and showed that the groups with rhBMP-2 had 100% fusion

rates compared with only 40% (2/5) fusion in the TSRH pedicle screw instrumentation-only group. Improvement of the Oswetry score at followup was highest in the rhBMP-2 only group. The use of *rhBMP-2* as an adjunct in fusion using lumbar interbody devices was demonstrated by Sandhu *et al.* in a sheep model. Recombinant BMP-2 was shown to be superior to autogenous bone graft. Minimally invasive spinal fusion and relief of discogenic back pain was possible in a series of 22 patients undergoing laparoscopic placement of *rhBMP-2*.

Direct introduction of growth factors into the intervertebral space for spinal fusion, however, does have its risks. Heterotopic bone formation could occur from leakage of osteoinductive proteins outside its carrier into the surrounding tissue. This could be devastating, especially in the setting of dural tears. Moreover, most osteoinductive growth factors are rapidly metabolized so that single-bolus injections may not achieve ideal efficacy in clinical settings. Gene therapy using MSCs offers the ability to deliver sustained release of growth factors to a local site. Successful spinal fusion has been shown by Boden *et al.* in a rat study using bone marrow cells transfected with a novel osteoinductive protein, LIM mineralization protein-1 (*LMP-1*). LMP-1 is thought to be a soluble osteoinductive factor that induces expression of other BMPs and their receptors. Wang *et al.*, in an animal model of spinal fusion using adenovirus-mediated gene therapy, showed that radiographic evidence of fusion was present by 4 weeks postoperatively in rats treated with MSCs expressing BMP-2. Moreover, the group receiving *AdBMP-2*-transfected cells showed repair with coarse trabecular bone compared with thin, lace-like bone in the rhBMP-2 group. The long-term sustained release of osteoinductive proteins was perhaps more effective in producing a stronger biologic response than a single supraphysiologic dose. Another possibility could be that the MSCs differentiated down the osteogenic lineage and are now contributing to the new bone in the fusion mass.

Similar to cartilage repair strategies, another potential strategy in the treatment of spinal disorders would be the suppression of disc degeneration. Fraying, splitting, and loss of collagen fibers in the intervertebral discs and calcification of the cartilage in the endplates occur with age. Matrix metalloproteinases and aggrecanases are also thought to be important in extracellular matrix degradation in the disc. Proteoglycans are important for maintenance of disc height and its compressive ability, and their preservation is essential for the load-bearing capacity of intervertebral discs. Thompson *et al.* studied the effect of various growth factors on proliferation and proteoglycan synthesis of annular pulposus and nuclear pulposus cells isolated from canine intervertebral discs and found that TGFβ-1 and EGF induced a fivefold increase in proteoglycan production. Retardation of programmed cell death in annulus cells and a dose-dependent increase in proteoglycan synthesis have been shown in discs exposed to IGF-1. Delivery of *TGFβ-1* and/or *IGF-I* to intervertebral discs using MSCs as a vehicle for gene therapy, therefore, represent two possible strategies in the treatment of disc disorders.

Successful retroviral-mediated gene transfer of *lacZ* and *IL-1Ra* of cultured chondrocytes has been performed by Wehling *et al*. The authors suggested the harvest of endplate tissue cartilage from patients with degenerative discs and the subsequent reintroduction of genetically modified cells into the disc space. This same strategy can be applied using MSCs that can be easily isolated from the iliac crest. As mentioned previously, adenoviral-mediated expression of *IGF-1* by MSCs has already been successfully shown. Gruber *et al*. (2003) showed the efficacy of *in vivo TGFβ-1* gene transfer to intervertebral discs and its ability to increase the level of proteoglycan synthesis.

Summary

MSCs hold great potential for the development of new treatment strategies for a host of orthopedic conditions. Animal models have demonstrated the wide spectrum of clinical situations in which MSCs could have therapeutic effects. Clinical application of the principles of cell-based tissue engineering is already seen with the Carticel (Genzyme, Cambridge, Massachusetts) program that uses culture-expanded autologous chondrocytes in the repair of articular cartilage defect. The multilineage potential and plasticity of MSCs allow them to be building blocks for a host of non-hematopoietic tissues including bone, cartilage, tendon, and ligament. Advances in fabrication of biodegradable scaffolds that serve as beds for MSC implantation will hopefully lead to better biocompatibility and host tissue integration. Minimal toxicity has been observed in animal models involving genetically manipulated stem cells transduced with retroviral and adenoviral vectors. Gene therapy using stem cells as delivery vehicles is a powerful weapon that can be used in a plethora of clinical situations that would benefit from the osteoinductive, chondroinductive, proliferative, and angiogenic effects of growth factors.

KEY WORDS

Gene therapy   Novel strategy in treating diseases by modifying the expression of specific genes toward a therapeutic goal.

Osteoconductive   Scaffolds or materials that are permissive to bone formation but do not attract osteoprogenitors that initiate bone formation.

Osteoinductive   Containing growth factors that signal attraction of osteoprogenitors that lead to osteoblastic differentiation and bone formation.

Pluripotent stem cells   Cells not capable of growing into a whole organism but able to differentiate into any cell type in the body.

Tissue engineering   The development of biologic substitutes for regeneration and/or repair of defective tissue by combining cells with appropriate bioactive scaffolds in combination with differentiation stimuli.

FURTHER READING

Caplan, A. I., and Mosca, J. D. (2000). Orthopaedic gene therapy. Stem cells for gene delivery. *Clin. Orthop.* **379**(Suppl.), S98–100.

Caplan, A. I., Elyaderani, M., Mochizuki, Y., Wakitani, S., and Goldberg, V. M. (1997). Principles of cartilage repair and regeneration. *Clin. Orthop.* **342**, 254–269.

Frank, C., Amiel, D., Woo, S. L., and Akeson, W. (1985). Normal ligament properties and ligament healing. *Clin. Orthop.* **196**, 15–25.

Grande, D. A., Breitbart, A. S., Mason, J., Paulino, C., Laser, J., and Schwartz, R. E. (1999). Cartilage tissue engineering: current limitations and solutions. *Clin. Orthop.* **369**(Suppl.), S176–185.

Gruber, H. E., Hanley, E. N., Jr. (2003). Recent advances in disc cell biology. *Spine* **28**, 186–193.

Jackson, D. W., and Simon, T. M. (1999). Tissue engineering principles in orthopaedic surgery. *Clin. Orthop.* **367**(Suppl.), S31–45.

McCarty, E. C., Marx, R. G., and DeHaven, K. E. (2002). Meniscus repair: Considerations in treatment and update of clinical results. *Clin. Orthop.* **402**, 122–134.

Praemer, A. F. S., and Rice, D. P. (1992). *Musculoskeletal conditions in the United States*, pp. 85–113. American Academy of Orthopedic Surgeons, Park Ridge, IL.

Woo, S. L., Hildebrand, K., Watanabe, N., Fenwick, J. A., Papageorgiou, C. D., and Wang, J. H. (1999). Tissue engineering of ligament and tendon healing. *Clin. Orthop.* **367**(Suppl.), S312–323.

61

Embryonic Stem Cells in Tissue Engineering

Shulamit Levenberg, Ali Khademhosseini, and Robert Langer

Introduction

Traditionally, approaches to restore tissue function have involved organ donation. However, despite attempts to encourage organ donations, there is a shortage of transplantable human tissues such as bone marrow, hearts, kidneys, livers and pancreases. Currently, more than 74,000 patients in the United States are awaiting organ transplantation, and only 21,000 people receive transplants annually.

Tissue engineering-based therapies may provide a possible solution to alleviate the current shortage of organ donors. In tissue engineering, biological and engineering principles are combined to produce cell-based substitutes with or without the use of materials. One of the major obstacles in engineering tissue constructs for clinical use is the limit in available human cells. Stem cells isolated from adults or developing embryos are a current source for cells for tissue engineering. The derivation of human embryonic stem (hES) cells in 1998 generated great interest in their potential application in tissue engineering. This is because of the ES cells' ability to grow in culture and to give rise to differentiated cells of all adult tissues. However, despite their therapeutic potential, both adult and ES cells present a number of challenges associated with their clinical application. For example, although adult stem cells can be directly isolated from the patient and are therefore immunologically compatible with the patient, they are typically hard to isolate and grow in culture. In contrast, ES cells can be easily grown in culture and differentiated to a variety of cell types, but ES cells derived cells may be rejected by the patient and undifferentiated ES cells may form tumors.

This chapter analyzes the potential of ES cells in tissue engineering and discusses the importance of ES cells as a source of cells for tissue engineering by using examples from the current research in the field. Also discussed are some of the fundamental principles and seminal work in tissue engineering.

Tissue Engineering Principles and Perspectives

Tissue engineering is an interdisciplinary field that applies the principles of engineering and life sciences to develop biological substitutes, typically composed of biological and syn-

thetic components, that restore, maintain or improve tissue function. Tissue-engineered products would provide a lifelong therapy and would greatly reduce the hospitalization and health-care costs associated with drug therapy while simultaneously enhancing the patients' quality of life.

In general, there are three main approaches to tissue engineering: (1) to use isolated cells or cell substitutes as cellular replacement parts; (2) to use acellular materials capable of inducing tissue regeneration; and (3) to use a combination of cells and materials (typically in the form of scaffolds). Although host stem cells could be a part of all these approaches, ES cells can be directly involved in the first and the third approach.

ISOLATED CELLS OR CELL SUBSTITUTES AS CELLULAR REPLACEMENT PARTS

Isolated cells have been used as a substitute for cell replacement parts for many years. In fact, the first application of stem cells as a cellular replacement therapy is associated with bone marrow transplantation or blood transfusion studies in which donor hematopoietic stem cells repopulated the host's blood cells. Other stem cells have demonstrated their potential in various diseases. For example, bone marrow-derived cells have been shown to (1) give rise to endothelial progenitor cells that were used to induce neovascularization of ischemic tissues, (2) to regenerate myocardium, (3) to give rise to bone, cartilage, and muscle cells, and (4) to migrate into the brain to give rise to neurons. In addition, myoblasts isolated from skeletal muscle that upon injection into the heart restored heart muscle function, and neural stem cells that resulted in the treatment of Parkinson disease are some examples of other potential adult stem cell-based therapies. Tissue engineering products based on cells have been developed in the form of skin substitutes through the use of allogeneic cells (by the Organogenesis and Advanced Tissue Sciences companies). In addition, the injection of mesenchymal stem cells is under way for cartilage and bone regeneration.

ES provides an alternative source of cells for cellular substitutes. *In vitro* ES cells have been shown to give rise to cells of hematopoietic, endothelial, cardiac, neural, osteogenic, hepatic, and pancreatic tissues. Although ES cells provide a versatile source of cells for generations of many cell types, so far only a few experiments have demonstrated the use of ES cells to replace functional loss of particular tissues. One such example is the creation of dopamine-producing cells in animal models of Parkinson's disease. These ES cell-derived, highly enriched populations of midbrain neural stem cells generated neurons that showed electrophysiological and behavioral properties similar to neurons. Although functional properties of neurons derived from human ES cells still need to be investigated, it has been shown that human ES cell-derived neural precursors can be incorporated into various regions of the mouse brain and differentiate into neurons and astrocytes. Also, human ES cells that were differentiated to neural precursors were shown to migrate within the host brain and differentiate in a region-specific manner. ES cells were also tested for future use in the heart. It was shown that mouse ES-derived cardiomyocytes were morphologically similar to neighboring host cardiomyocytes. In addition, mouse ES cells that were transfected with an insulin promoter (driving expression of the neo gene, a marker for antibiotic resistance) have been shown to give rise to insulin-producing cells that can restore glucose levels in animals. Although these functional data were obtained by using genetically modified ES cells, insulin production from ES cells suggests that these cells may potentially be used for the treatment of diabetes.

ES cells have also been shown to give rise to functional vascular tissue. Early endothelial progenitor cells isolated from differentiating mouse ES cells were shown to give rise to three blood vessel cell components: hematopoietic, endothelial, and smooth muscle cells. Once injected into chick embryos these endothelial progenitors differentiated into endothelial and mural cells and contributed to the vascular development. We have shown that human ES cells can differentiate into endothelial cells, and we have isolated these cells using platelet endothelial cell adhesion molecule-1 antibodies. *In vivo*, when transplanted into immunodeficient mice, the cells appeared to form microvessels.

USING COMBINATIONS OF CELLS AND MATERIALS

Tissue engineering approaches that use cells and scaffolds can be categorized into two categories: open and closed systems. These systems are distinguished based on the exposure of the cells to the immune system upon implantation.

Open Systems

In open tissue engineering systems, cells are immobilized within a highly porous, three-dimensional scaffold. The scaffold could be comprised of either synthetic or natural materials or composites of both. Ideally, this scaffold provides the cells with suitable growth environment, optimum oxygen and nutrient transport properties, good mechanical integrity, and suitable degradation rate. The use of scaffolds provides three-dimensional environments and brings the cells in close proximity so that the cells have sufficient time to enable self-assembly and formation of various components associated with the tissue microenvironment. Ideally, the material is degraded as cells deposits their extracellular matrix molecules. The materials used for tissue engineering are either synthetic biodegradable materials (such as poly(lactic acid) (PLA), poly(glycolic acid) (PGA), poly lactic-glycolic acid (PLGA), poly(propylene fumarate), poly ethylene glycol (PEG) and polyarylates, or natural materials such as collagen, hydroxyapatite, calcium carbonate, and alginate. Natural materials are typically more favorable to cell adherence, whereas the properties of synthetic materials such as degradation rate, mechanical properties, structure, and porosity can be better controlled.

Open tissue engineering systems have been successfully used to create a number of biological substitutes such as bone, cartilage, blood vessels, cardiac, smooth muscle, pancreatic, liver, tooth, retina, and skin tissues. Several tissue-engineered products are under clinical trials for FDA approval. Engineered skin or wound dressing and cartilage are two of the most advanced areas with regards to clinical potential. For example, a skin substitute that consists of living human dermis cells in a natural scaffold consisting of type I collagen already received FDA approval to be used for a diabetic foot ulcer. In addition, various cartilage and bone are also currently in clinical stages, and bladder and urologic tissue are being tested in various stages of research.

Despite the ability of stem cells to differentiate to cells with phenotypic and morphological structure of desired cell types, very few scaffold-based tissue engineering studies use ES cells. For the case of adult stem cells, scaffolds have been utilized in conjunction with mesenchymal stem cells, neural stem cells, and oval cells. One such example is the transplantation of neural stem cells onto a polymer scaffold that was subsequently implanted into the infarction cavities of mouse brains injured by hypoxia–ischemia. These stem cells generated an intricate meshwork of many neurites and integrated with the host. We have seeded neural stem cells onto specialized scaffolds and have demonstrated spinal cord regeneration and improved hind-leg function of adult rats from a hemisection injury model. Also, MSCs have been differentiated on polyethylene glycol (PEG) or PLGA scaffolds and have been shown to give rise to cartilage or bone depending on the medium conditions.

ES cells may be differentiated in culture, with desired cell types selected and subsequently seeded onto scaffolds. We have used this technique to study the behavior of ES cell-derived endothelial cells in tissue engineering and constructs. Human ES cell-derived endothelial progenitors that were seeded onto highly porous PLLA/PLGA biodegradable polymer scaffolds formed blood vessels that appeared to merge with the host vasculature when implanted into immunodeficient mice (Figure 61-1).

There may be other approaches to using ES cells or their progeny with scaffold-based tissue engineering systems. For example, it may be possible to directly differentiate ES cells on scaffolds in culture. Finally, it may be possible to differentiate genetically engineered ES cells seeded onto scaffolds *in vivo* (see Figure 61-2).

Coercing cells to form tissues while differentiating is an important issue that has not been explored greatly. This may be achieved by seeding ES cells directly onto the scaffolds, followed by inducing their differentiation *in situ*. Porous biodegradable polymer scaffolds can be used to support the ES cells, and they represent a promising system for allowing formation of complex 3D tissues during differentiation. The scaffold provides physical cues for cell orientation and

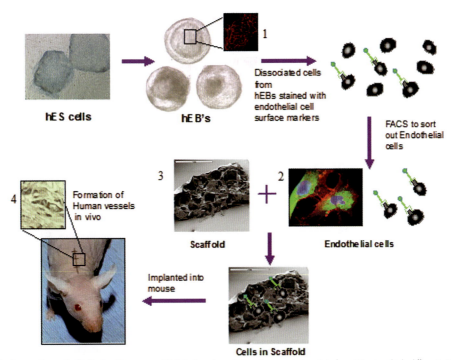

Figure 61-1. Embryonic endothelial cells on scaffolds *in vivo*. hES cells were induced to form EBs, in which differentiation into endothelial cells and formation of vessel-like network was observed. (1) Embryonic endothelial cells were isolated from hEBs by staining dissociated EB cells with endothelial surface marker and sorting out positive cells using flow cytometry (FACS). Isolated endothelial cells (2) were seeded on polymer scaffolds (3) and implanted into immunodeficient mice. The embryonic endothelial cells appeared to form vessels *in vivo* (4).

1. Confocal image of vessel network formation within 13-day-old hEB, stained with PECAM-1 antibodies.
2. Isolated embryonic endothelial cells grown in culture stained with PECAM-1 (red) and VWF (green) antibodies.
3. Scanning electron microscopy (SEM) of PLLA/PLGA scaffolds.
4. Immunoperoxidase (brown) staining of 7-day implants with antihuman PECAM1 antibodies showing vessels lined with human endothelial cells.

Source: Levenberg *et al.*

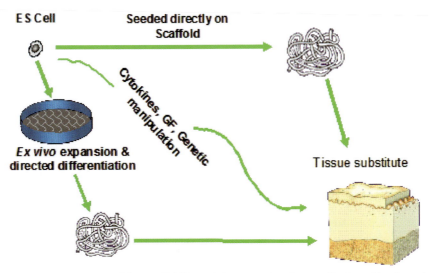

Figure 61-2. Approaches for using ES cell for scaffold-based tissue engineering applications. ES cells can be used in tissue engineering constructs in a variety of methods. ES cells can be expanded in culture and then seeded directly onto scaffold where they are allowed to differentiate. Alternatively, stem cells can be directed to differentiate into various tissues and enriched for desired cells prior to seeding the cells onto scaffolds.
Source: Levenberg *et al.*

spreading, and pores provide space for remodeling of tissue structures. These scaffolds should ideally provide the cells with cues to guide their differentiation into desired cell types. The possible advantages of this system could involve the assembly of the cells as they differentiate. This differentiation pattern may mimic the developmental differentiation of the cells much more closely and therefore may induce differentiation into desired tissue. Ultimately, *in vitro* differentiated constructs can potentially be used directly for transplantation.

An approach that has not been considered as an alternative to *in vitro* differentiation of ES cells is to use the adult body's microenvironment to induce the differentiation of ES cells. *In vivo* differentiation of ES cells is not yet a feasible option due to the tumourgenic nature of ES cells as well as the heterogeneous cell population that results from their nondirected differentiation. However, it may be possible to use a cell's apoptotic response mechanism to induce selective pressure for the desired cells *in vivo*. Thus, genetically modified ES cells that undergo apoptosis upon differentiation into undesirable cell types could be used to direct the differentiation of these cells, while similar approaches could be adopted to control the proliferative behavior of these cells.

Closed Systems

One of the main difficulties associated with open tissue engineering systems is the potential immunological issues associated with the implanted cells. Closed systems aim to overcome this difficulty by immobilizing cells within polymeric matrices that provide a barrier for the immunological components of the host. For example, cells can be immobilized within semipermeable membranes that are permeable to nutrients and oxygen while providing a barrier to immune cells, antibodies, and other components of the immune system. Furthermore, the implants can either be implanted into the patient or be used as extracorpical devices. Closed tissue engineering systems have been used particularly for the treatment of diabetes, liver failure, and Parkinson's disease. This system may prove to be especially useful in conjunction with ES cells since the immobilization of ES cells within closed systems may overcome the immunological barrier that faces ES cell-based therapies. For example, ES cell-derived β-cells that can respond to insulin, or domaine-producing neurons can be used in clinics without the fear of rejection. In addition, closed systems protect the host against potentially tumourgenic cells as it limits the cells within the polymeric barrier. Currently, engineering and biological limitations such as material biocompatibility, molecular weight cutoff, and the immune system's reaction to shed antigens by the transplanted cells are some of the challenges that prevent these systems from widespread clinical applications.

Limitations and Hurdles of Using ES Cells in Tissue Engineering

Despite significant progress in the field of tissue engineering, a number of challenges are limiting the use of ES cells in tissue engineering. These challenges range from understanding stem cell biology questions to how to control stem cell fate, to engineering challenges on scale-up, and to business questions of feasibility and pricing.

DIRECTING THE DIFFERENTIATION OF ES CELLS

Perhaps the biggest challenge in using ES cells in clinical applications is the lack of knowledge in directing their differentiation ability. All studies that have shown the generation of specific cell types have not shown a uniform differentiation into a particular cell type. This may be attributed to the intrinsic property of ES cells to differentiate stochastically in the absence of proper temporal and spatial signals from the surrounding microenvironment.

Techniques can be used to increase the ratio of cells that give rise to the desired lineages that include genetic and microenvironmental manipulations. Genetic techniques can be categorized into positive or negative regulators. The positive regulators include the constitutive or controlled expression of transcription factors that have been shown to derive the differentiation into particular tissues. For example, the overexpression of Nurr transcription factor has been shown to increase the frequency of ES cells that differentiate into functional neural cells. Alternatively, the negative regulators could be incorporated to induce the apoptosis of cells that differentiate to varying pathways. For example, neomycin selection and suicide genes that are activated by certain transcription factors can be used. Clearly, these techniques will benefit from further understanding of the inner workings of transient cells and knowledge of the differentiation pathways and lineages. Further analysis into stem cell and progenitor hierarchy through high-throughput analysis of microarray or proteomics data should accelerate this process.

Another important criterion is the functionality of ES cell-derived cells as a source of tissues. The importance of rigorous testing has become clear in studies in which nestin positive putative pancreatic cells stained positive for insulin using antibodies due to cellular uptake from the surrounding medium. Thus, the incorporation of protein and functional tests should accompany the morphological and phenotypic analysis that is often used in ES cell literature to characterize differentiated cells.

ISOLATING THE DESIRED CELL TYPES FOR THERAPY

One of the main problems with ES cell-based therapies is finding suitable techniques to isolate desired cells from a heterogeneous population of cells. One approach is to allow for the random differentiation of the ES cells followed by isolation using cell-surface markers. We have used this method for isolating ES cell-derived endothelial cells using PECAM-1 receptor. Also, ES cell-derived hematopoietic progenitors have been isolated in a similar manner using the CD34 marker. Another potential method is through reporter gene knockin modifications. These modifications have already been used on ES cells to permit labeling of cells at various stages of differentiation. The utility of other techniques such as magnetic separation, or the use of neomycin selection, must be further examined for selecting various ES cell-derived progeny.

SCALE-UP OF ES CELLS IN TISSUE ENGINEERING

Although laboratory scale ES cell cultures have been shown to produce differentiated progeny for both rodent and human ES cells, it is generally acceptable that these culturing methods are not feasible for large-scale production of ES cells for therapeutic applications. Production of a sufficient quantity of differentiated cells from ES cells is an important challenge in realizing the clinical potential of ES cells. The large-scale production of ES cells will likely be specific to the type of tissue being generated and must remain reproducible, sterile, and economically feasible. Furthermore, this scale-up process must maintain the appropriate control over bioprocess conditions such as mechanical stimuli, medium conditions, physichochemical parameters (such as temperature, oxygen, pH, and carbon dioxide levels), as well as growth factor and cytokine concentrations.

ES cell differentiation protocols have generally used two-dimensional cultures and/or embryoid bodies. Although each technique provides specific advantages, differentiation of ES cells in embryoid bodies produces a wider spectrum of cell types. This effect has been attributed to the embryoid bodies' ability to better mimic the temporal pattern of cell differentiation as seen in the embryo. However, in some applications the combined use of EBs and adherent cultures has resulted in better cell yields. For example, to induce ES cells to differentiate to cardiomyocytes, a EB formation in suspension cultures followed by a differentiation in adhesion cultures has been shown to optimize the percentage of cells that give rise to cardiomyocytes. Similarly, the production of hepatocytes has been shown to be induced by first culturing the cell in EBs followed by culturing on 2D cultures.

The formation of EBs in labs has been generally performed using techniques that have not been ideal for large-scale production. For example, many studies have used the *hanging drop protocol*, in which ES cells are placed within a "hanging drop" and are allowed to form an aggregate that can then be differentiated. Other techniques have formed EBs by placing the cells on nonadherent tissue culture dishes that once more provide a limitation on the quantity of cells produced. A technique that may allow for the scale-up of the embryoid body cultures is the use of suspension cultures using spinner flasks. Such cultures have been shown to enhance the supply of oxygen and nutrients to the cells within the embryoid body by exposing the surface of the cell aggregate to a continuous supply of fresh medium.

To prevent the difficulties associated with EB heterogeneity, EBs have been immobilized in alginate microbeads. The microencapsulation of cells within these microbeads resulted in differentiation of cells into cardiomyocytes and smooth muscle cells. In addition, ES cells may be adhered to beads with desired extracellular matrices and differentiated. This approach also enhances the transport of medium and oxygen to the cells in comparison to two-dimensional cultures and provides additional mechanical stimuli that

may be an improved alternative to two-dimensional culture systems.

To enhance the supply of medium to tissue-engineered scaffolds or embryoid bodies, methods other than passive diffusion may be required. In perfusion systems the medium is flown through the scaffold. Perfusion bioreactors have already been developed for a variety of tissue engineering applications such as cartilage and cardiac. For example, perfusion through scaffolds has been generated in rotating wall vessels or through pumping medium directly through the scaffolds to grow chondrocytes for cartilage generation.

Mechanical forces affect the differentiation and functional properties of many cell types; thus, ES cell-based cultures that aim to direct the differentiation of ES cells require proper mechanical stimuli for the tissue. Our understanding of the effects of mechanical stimuli on ES cell differentiation is still primitive, but tissue engineering systems have been developed that incorporate the effects of mechanical forces. For example, functional autologous arteries have been cultured using pulsatile perfusion bioreactors. Thus, the use of mechanical stimuli may further enhance the ability of these cells to respond to exogenous signals. Other environmental factors that may be required is the use of electrical signals and spatially regulated signals to induce the differentiation and allow for maturation of the desired tissues. Hopefully, with time such techniques will become particularly important in allowing for scaled-up ES cell-based tissue engineering applications. The development of bioreactors that control the spatial and temporal signaling that induces ES cell differentiation requires collaborative effort between engineers and biologists and is currently in the early stages of its development.

TISSUE ENGINEERING LIMITATIONS

Synthetic scaffolds that support tissue growth by serving as the extracellular matrix for the cells do not represent the natural ECM associated with each cell type and tissue. ES cells and their progeny during development reside in a dynamic environment; thus, synthetic or natural substrates that aim to mimic the developing embryo must present a similar signaling and structural elements. A number of approaches are currently under development which may prove useful for ES cells. For example, the use of "smart" scaffolds that release particular factors and/or control the temporal expression of various molecules released from the polymer could be used to induce differentiation of ES cell within the scaffolds. For example, by dual delivery of vascular endothelial growth factor (VEGF)-165 and platelet-derived growth factor (PDGF)-BB each with distinct kinetics, from a single, structural polymer scaffold, it has been shown that a mature vascular network can be formed. An alternative approach to modify the surface that is exposed to the cells is to immobilize desired ligands onto the scaffold. For example, RGD peptides, the adherent domain of fibronectin, can be incorporated into polymers to provide anchorage for adherent cells.

Another difficulty associated with the current materials is their lack of control over the spatial organization within the scaffold. To create tissues that resemble the natural structure of biological tissues, the spatial patterning of cells must be recapitulated. For ES cells differentiated in scaffolds, this modeling and structure may be directly obtained as cells differentiate. The spatial arrangement of cells grown in EBs is typically organized together with cells of particular tissues appearing in clusters. For example, blood precursors occur in the form of blood islands similar to their normal appearance in embryonic development. In the system in which ES cell-derived cells are plated onto scaffolds, spatial rearrangement can occur via direct patterning or cell "reorganization." In the direct cell-patterning system, cells can be seeded into the scaffold at particular regions within the cells. For example, the direct attachment of two different cell types on the different sides of the scaffold has been used to generate cells of the bladder. Cell-patterning techniques have been developed for soft lithography for controlled co-culture of hepatocytes, and fibroblasts could be scaled up to tissue engineering scaffolds to permit more controlled and complex direct patterning.

Conclusions and Future Perspectives

ES cells have generated a great deal of interest as a source of cells for tissue engineering. However, a number of challenges exist in making ES cell-based therapy a reality. These include directing the differentiation of ES cells (using controlled microenvironments or genetic engineering) to ensure their safety and efficacy *in vivo*, to ensure that the cells are immunologically compatible with the patient and will not form tumors, to improve protocols for isolating desired cell types from heterogeneous populations, and to enhance current tissue engineering methods. Further research is required to control and direct the differentiation of ES cells, in parallel with developing methods to generate tissues of various organs, which may lead to realizing the ultimate goal of tissue engineering. We are getting close to the day when ES cells can be manipulated in culture to produce fully differentiated cells that can be used to create and repair specific organs. Clearly, significant challenges remain, and the ability to overcome these difficulties does not lie within any scientific discipline but rather involves an interdisciplinary approach. Innovative approaches to solve these challenges could lead to improved quality of life for a variety of patients who could benefit from tissue engineering approaches.

KEY WORDS

Biomaterials   Any natural or synthetic materials that interface with living tissue and/or biological fluids.

Cell therapy   The transplantation of cells from various sources to replace or repair damaged tissue and/or cells.

Differentiation   The developmental process in which cells change their genetic programs to become more restricted in their potential and to mature into functional cells of various tissues and organs.

Embryonic stem cells   Cells derived from the embryo that can self-renew and give rise to all the cells of the adult organism.

Scaffold   A porous structure typically made out of degradable polymers within which cells are seeded. Scaffolds provide geometrical structure for cells to reorganize and form 3D multicellular tissues.

Tissue engineering   The interdisciplinary field that applies the principle of engineering and the life science toward the development of biological substitutes that restore, maintain, or improve tissue function.

FURTHER READING

Griffith, L. G., and Naughton, G. (2002). Tissue engineering — current challenges and expanding opportunities. *Science* **295**(5557), 1009–1014.

Langer, R., and Vacantio, J. P. (1993). Tissue engineering. *Science* **260**(5110), 920–926.

Langer, R. S., and Vacantio, J. P. (1999). Tissue engineering: the challenges ahead. *Sci. Am.* **280**(4), 86–89.

Lanza, R., Moore, M. A. *et al.* (2004). Regeneration of the infarcted heart with stem cells derived by nuclear transplantation. *Circ Res.* ••.

Lanza, R. P., Langer, R. S. *et al.* (1997). *Principles of tissues engineering.* Academic Press Austin M TX.

Levenberg, S., Golub, J. S. *et al.* (2002). Endothelial cells derived from human embryonic stem cells. *Proc. Natl. Acad. Sci. USA* **99**(7), 4391–4396.

Levenberg, S., Huang, N. F. *et al.* (2003). Differentiation of human embryonic stem cells on three-dimensional polymer scaffolds. *Proc. Natl. Acad. Sci. USA* **100**(22), 12741–12746.

McKay, R. (2002). Building animals from stem cells. *Ann. NY Acad. Sci.* **961**, 44.

Petit-Zeman, S. (2001). Regenerative medicine. *Nat. Biotechnol.* **19**(3), 201–206.

Shin, H., Jo, S. *et al.* (2003). Biomimetic materials for tissue engineering. *Biomaterials* **24**(24), 4353–4364.

Strauer, B. E., and Kornowski, R. (2003). Stem cell therapy in perspective. *Circulation* **107**(7), 929–934.

62

Stem Cells in Tissue Engineering

Pamela Gehron Robey and Paolo Bianco

Introduction

At its first inception, tissue engineering was based on the use of natural or synthetic scaffolds seeded with organ-specific cells *ex vivo*. This approach was somewhat distinct from guided tissue regeneration, which utilized scaffolds and/or bioactive factors to encourage local cells to repair a defect *in situ*. These two approaches are now merged in the current field of tissue engineering that encompasses multiple and diverse disciplines to use cells, materials, and bioactive factors in various combinations to restore and even improve tissue structure and function. Stem cell-based tissue engineering represents a major turn in the conceptual approach to reconstruction of tissues. By expanding the repertoire of available cells, of envisioned targets, and of the technological means of generating functional tissues *ex vivo*, and above all by making it a possibility to engineer tissues otherwise not amenable to reconstruction, advances in stem cell biology have a profound impact on tissue engineering at large.

In many cases, tissue engineering seems to have more use for stem cells than nature itself. This is especially apparent in the case of stem cells derived from tissues with low turnover, or no apparent turnover at all. If neural stem cells were able to repair the loss of dopaminergic neurons in the intact brain *in vivo*, there would be no Parkinson's disease for which to envision stem cell-based therapies. Dental pulp stem cells do not regenerate primary dentin *in vivo*, but enough dentin can be made from the pulp of a single extracted tooth to generate *ex vivo* the amount of dentin required to fabricate dentures for a marine platoon. Current definitions of stem cells, in fact, include, in many cases, a technological dimension — that is, a cell that can be expanded and encouraged to generate differentiating cells *ex vivo*. The two neighboring fields of cell therapy (reconstruction of functional tissues *in vivo* using cells) and tissue engineering proper (reconstruction of functional tissues using cells and something else) merge significantly once a stem cell angle is adopted for either one, not only with respect to the ultimate goals, but also to several biotechnological aspects.

The Reservoirs of Postnatal Stem Cells

Both for ethical constraints and for ease of harvest and control, current approaches to tissue engineering using stem cells rely largely on the use of postnatal stem cells. Once restricted to a handful of constantly (and rapidly) self-renewing tissues, the repertoire of stem cells has expanded to include perhaps every single tissue in the body, regardless of the rate of tissue turnover. Not all of these tissue-specific stem cells, however, are equally accessible for safe harvest, or available in sufficient quantity (or amenable for *ex vivo* expansion) to generate the number of cells needed for tissue regeneration. However, lessons on the dynamics of tissue homeostasis (growth, turnover) that can be learned from these cells have an obvious impact in the design of future tissue engineering strategies nonetheless. In addition, mechanisms whereby postnatal cells maintain differentiated functions in tissues and organs are relevant even to future embryonic stem cell-based approaches and can only be learned from postnatal cells.

DON'T BE RIGID ABOUT PLASTICITY

The recent clamor over the potential "plasticity" of certain classes of postnatal stem cells unquestionably adds further questions, once the perspective of usage for tissue reconstruction is directly addressed. It would be impractical to say the least, to use neural stem cells for bone marrow transplantation, and a tissue engineer's enthusiasm for this unexpected finding may remain lukewarm. In contrast, deciding whether liver regeneration is successfully accomplished using previously unknown "hepatogenic" stem cells, or by reprogramming of a donor lymphocyte nucleus following cell fusion, may not disrupt a tissue engineer's sleep, as long as the goal is met. Unquestionably, some examples of effective tissue reconstruction *in vivo* based on "unorthodox" differentiation of postnatal stem cells have been noted. However, further investigation is needed to confirm not only the concept of postnatal stem cell plasticity, but also specifically the technological aspects of its effective translation into clinical application, once proof of principle has been demonstrated.

"BREAK UP THE BONES AND SUCK THE SUBSTANTIVE MARROW" — RABELAIS, *GARGANTUA AND PANTAGRUEL*

It has long been known that bone marrow is the home of at least two different types of stem cells, the hematopoietic stem cell (HSC) and the bone marrow stromal stem cell (BMSSC), also known as the mesenchymal stem cell, or, more accurately, the skeletal stem cell (SSC), each being able to reconstitute the hematopoietic and skeletal system, respectively. Both systems are thought to be able to contribute differentiated cell types outside of their physiological progeny. Highly purified HSCs are believed to give rise to cardiomyocytes, as well as hepatocytes, and a host of epithelial tissues. BMSSCs have been reported to generate functional cardiomyocytes *in vitro* and to be capable of neural differentiation. A rare subset of murine BMSSCs (multipotent adult progenitor cells, MAPCs) may be as multipotent as ES cells. AC133 (CD133) positive endothelial progenitors are found in the marrow, and endothelial cells themselves may generate cardiomyocytes *in vitro*. Circulating, marrow-derived cells appear to contribute to regeneration of skeletal muscle in response to injury, and in mouse models of muscular dystrophy. Following bone marrow or mobilized peripheral blood transplantation, donor-derived cells have also been detected in neuronal tissue, newly formed vasculature, in the kidney, and even in the oral cavity. Ideally, the identification of a single accessible site that would contain a multitude of cells with pluri- and multipotentiality that are easily harvested in large quantities would mark a major advantage in tissue engineering. If substantiated, all of these observations would make the bone marrow the central organ of tissue engineering and stem cell therapy (Figure 62-1).

Current Approaches to Tissue Engineering

Approaches to the regeneration of functional tissue using postnatal stem cells can be envisioned by three different scenarios: (A) expansion of a population *ex vivo* prior to transplantation into the host, (B) *ex vivo* recreation of a tissue or organ for transplantation, and (C) design of substances and/or devices for *in vivo* activation of stem cells, either local or distant, to induce appropriate tissue repair (Figure 62-2). In all of these cases, considerable knowledge of the stem cell population's dynamics is required in order to predict and control their activity under a variety of different circumstances.

EX VIVO CULTURE OF POSTNATAL STEM CELLS

Ex vivo expansion of tissue- or organ-specific cells, used either alone or added to carriers or scaffoldings at the time of transplantation, has been the primary approach in tissue engineering to date. However, *ex vivo* expansion of postnatal stem cells in a fashion that maintains an appropriate proportion of stem cells within the population is a significant hurdle that must be overcome. For example, in spite of enormous effort, the culture conditions for maintaining HSCs (let alone expanding their number) are as yet undefined. It is perhaps for this very reason that currently in only a handful of examples have *ex vivo* expanded postnatal stem cells been used successfully to restore structure and function. The key to successful expansion will lie in understanding cell proliferation kinetics (asymmetric versus symmetric division). The efficacy achieved by the use of *ex vivo*-expanded populations, whether stem or more committed in character, may also depend, at least in part, on the nature of the tissue under reconstruction.

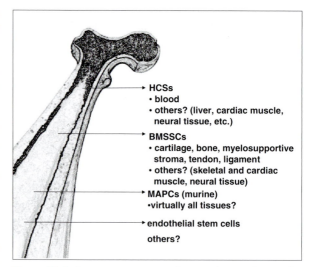

HCSs
• blood
• others? (liver, cardiac muscle, neural tissue, etc.)

BMSSCs
• cartilage, bone, myelosupportive stroma, tendon, ligament
• others? (skeletal and cardiac muscle, neural tissue)

MAPCs (murine)
• virtually all tissues?

endothelial stem cells

others?

Figure 62-1. Bone marrow as a central source of postnatal stem cells. Bone marrow consists of at least two well-defined populations of postnatal stem cells, the hematopoietic stem cell (HSC) and the bone marrow stromal stem cell (BMSSC). Both of these cells form numerous phenotypes within their cellular system but may also form cells outside of them. Multipotential Adult Progenitor Cells (MAPCs) are a subset of BMSSCs and have been reported to form virtually all cell types in mouse. Endothelial precursors (AC133+) have recently been identified, and other types, such as a hepatocyte-like stem cell may also exist. These remarkable findings, if verified, place bone marrow high on the list of tissues that are easily accessible in sufficient quantity for use in tissue engineering.

Within this context, the rate of tissue turnover most likely defines the rate of success. In tissues with a high rate of turnover, such as blood and skin, it is clear that long-term success depends on the persistence of a stem cell within the transplanted population. More committed progenitors may provide some short-term advantage, but without a self-renewing population, failure is ultimate. However, in tissues that turn over more slowly (e.g., bone), long-lasting benefit may be achieved with more committed populations of cells.

Although optimizing culture conditions represents one hurdle, the issues of time and quantities represent others. In most cases, the amount of time required to generate the number of autologous cells sufficient to repair defects induced by trauma or disease is in the order of weeks. In cases of trauma, this poses a large problem if the use of autologous stem cells is considered. For that reason, the use of allogenic populations that could be used "off the shelf" from unrelated donors would be preferable, but raises the issue of rejection, as in organ transplantation. Although it is suggested that many postnatal stem cells appear to escape from immune surveillance in allogenic settings, definitive proof is lacking. Furthermore, differentiation of stem cells would necessarily imply the expression of a mature tissue-specific phenotype, including a complete histocompatibility profile. Use of allogenic cells would most likely require concomitant immunosuppressive therapy, which has its own list of side effects. Co-transplantation of allogeneic bone marrow has been proposed. Recent studies in organ transplantation in conjunction

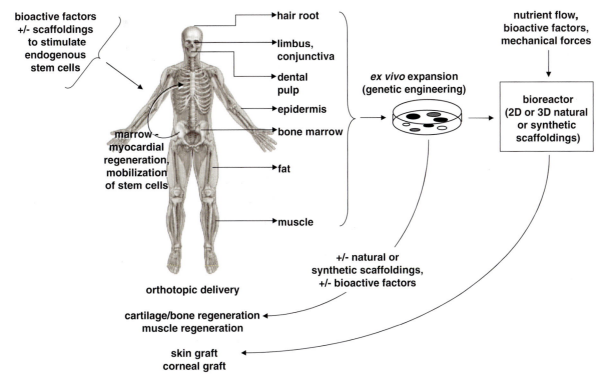

Figure 62-2. Current applications of postnatal stem cells in tissue engineering.
While virtually all tissues in the body have cells with some regenerative capabilities, current postnatal stem cells-based strategies, or those in the foreseeable future, rely on a relatively limited number of source tissues. Autologous bone marrow is currently in trial for myocardial regeneration. Most approaches utilize *ex vivo*-expanded cell populations that are then delivered orthotopically in various combinations with bioactive factors and scaffolds, with skeletal and muscle regeneration. Generation of 2D and 3D structures *ex vivo* requires the use of bioreactors in which cells are seeded onto scaffoldings and subjected to nutrient flow, bioactive factors, and mechanical forces to induce formation of functional tissue for transplantation.

with bone marrow or mobilized blood transplantation after immune ablation indicate a substantial improvement in long-term survival. In this type of approach, allogenic cells could be better envisioned, provided that a single donor would be the source of cells for the tissue to be restored, along with reconstitution of the immune system. But perhaps a better approach would entail molecular engineering strategies to delete the offending histocompatability-associated genes to render the cells nonimmunogenic, such that "one size" fits all.

DELIVERY OF STEM CELLS

Delivery of bone marrow or mobilized blood by systemic infusion for restoration of the hematopoietic system by HSCs is an example of the efficacy of this method of delivery, but it is perhaps the only example. The success of systemic infusion is based primarily on the fact that in terms of dimensions, blood is a simple, fluid structure. Reconstruction of two- or three-dimensional structures requires different approaches. Currently, there are a number of clinical applications, or applications soon to emerge, for the delivery of cell populations directly into a site for tissue regeneration. In some cases, specific populations of cells are expanded *ex vivo* and delivered either alone, or in conjunction with a natural or synthetic material. As important as it is to control the *ex vivo* behavior

of the cell population, it is also essential to take into consideration what their response will be to the host environment. It is the host that must provide the signals that will dictate the differentiation if uncommitted populations and/or noninstructive scaffolds are utilized. Furthermore, the recipient host tissue must also support at least the maintenance of the stem cell's niche, if not the creation of it de novo.

RECONSTRUCTION OF THE SKELETON — BONE, CARTILAGE, AND EVEN TEETH

Preclinical studies in a variety of animal models have demonstrated that bone marrow stromal cells, when used in conjunction with appropriate carriers, can regenerate new bone in critical-sized defects that would never heal without intervention. A wide variety of carriers have been tested, ranging from collagen sponges, synthetic biodegradable polymers, to synthetic hydroxyapatite derivatives, although to date none appears to be optimal. A carrier that provides immediate stability, especially in the case of weight-bearing bones, and yet can be completely resorbed as new bone is formed and remodeled, has not yet been fabricated. Nonetheless, a few patients have been treated with *ex vivo* expanded bone marrow stromal cells in conjunction with hydroxyapatite/tricalcium phosphate ceramic particles with good outcome, and other clinical studies are in progress. In addition to healing segmental defects, these cells can also be used to construct vascularized bone flaps. In cases where morbidity of the recipient site is an issue, it can be envisioned that a new bone rudiment could be grown elsewhere in the body and then transferred, with vasculature intact. More recently, adipose tissue has been identified as a source of stem cells that also have the potential to differentiate into bone and cartilage at least and may be as multipotent as BMSSCs. Naturally, if these studies were to be further substantiated, fat would become another prime source of easily accessible stem cells.

Current cell-based therapy for cartilage defects relies on growth of chondrocytes from a biopsy followed by transplantation onto articular surfaces in conjunction with a periosteal flap. Although this procedure is practiced worldwide, long-term efficacy is questionable, and there is a clear need for better procedures. Although BMSSCs and stem cells from adipose tissue are capable of forming cartilage in a test tube, their use in reconstruction of cartilage on articular surfaces will rely on inhibiting their further differentiation into hypertrophic chondrocytes. In addition, further development of suitable carriers is also required. Natural polymeric gels such as hyaluronic acid, collagen, alginate, and chitosan provide an adequate three-dimensional structure to maintain the chondrocyte phenotype, but are not well modeled into specific shapes and have very poor biomechanical properties. For these reasons, synthetic biodegradable polymers such as polylactic acid and polyglycolic acid, and mixtures of both, that can be cross-linked and molded to form porous three-dimensional structures are thought to provide more adequate scaffolding, and have been used to construct cartilage in the shapes of noses and ears. In designing constructs of stem cells and scaffolds, the differences between the elastic cartilages of the ear and nose, and articular cartilage, must also be considered to ensure that tissues form with the appropriate biomechanical characteristics.

Using techniques developed for isolation and characterization of BMSSCs, researchers have found that dental pulp from the permanent dentition contains a population of stem cells (dental pulp stem cells, DPSCs) that have the ability to form copious amounts of primary dentin and a pulp-like complex upon *in vivo* transplantation with hydroxyapatite/tricalcium phosphate ceramic particles. Stem cells are also present in the pulp of deciduous teeth (SHED, Stem cells from Human, Exfoliated, Deciduous teeth) that have a high proliferative capacity, and in addition to forming dentin, may be able to form neuronal cells (perhaps reflective of their neuroectodermal origin) *in vitro* and *in vivo*. Further characterization of their differentiation potential is required, but SHED, like cord blood, may represent an extremely useful source of stem cells.

SKELETAL AND CARDIAC MUSCLE REGENERATION

Muscle contains a population of muscle-specific stem cells, the satellite cell, and perhaps a more multipotential stem cell (the so-called side population cell). Early attempts to correct muscular dystrophies by using *ex vivo*-expanded populations of allogeneic myoblastic cells derived from satellite cells were all uniformly unsuccessful due to immunological responses of the host to allogenic cells, the expression of a foreign protein by the donor cells, and the inability of donor cells to repopulate extensively. In other studies, bone marrow transplantation provided a cell that could participate in formation of new myotubes in a muscle injury model, and in a model of muscular dystrophy. However, the level of incorporation of donor cells was very low. Recently, "mesenchymal" stem cells isolated from the synovial membrane, and sharing resemblances with bone marrow-derived stromal cells, have been shown to repair dystrophic muscle in the *mdx* mouse. More dramatic results were obtained when mesoangioblasts, which are blood vessel-associated stem cells, were isolated from young mice with a form of muscular dystrophy (alpha-sarcoglycan deficiency). Once the gene defect was corrected *ex vivo*, cells injected into the femoral artery of affected mice restored skeletal muscle, suggesting that mesoangioblasts may be useful in skeletal muscle regeneration.

Although skeletal muscle cells are functionally quite distinct from cardiomyocytes, autologous myoblastic cells have been viewed as a potential therapy for the treatment of myocardial infarct. Initial results suggest that there is functional improvement, albeit the mechanisms are unclear due to recent evidence that engrafted cells are not electromechanically linked with recipient cells. The utility of autologous bone marrow for myocardial regeneration is also currently receiving a great deal of attention based on studies that attribute the ability of HSCs and BMSSCs to form cardiomyocytes, and the ability of marrow-derived endothelial precursors to participate in development of new blood vessels in a number of animal models. Because of these encouraging findings, transendocardial injection for the treatment of ischemic heart failure using marrow-derived cells is being tested in a number of small clinical trials. Patients receiving an autologous bone marrow transplant in conjunction with cardiac bypass surgery, and those receiving AC133+ endothelial precursor cells alone, are reported to have increased cardiac perfusion and enhanced ventricular function. While substantial improvement using these different populations of cells is evident, whether bone marrow-derived cells engraft at a substantial level, and in fact differentiate into cardiomyocytes, remains to be proven. It should be considered that within this context, clinical benefit may be conveyed by a "helper" effect on putative local progenitors, or by improvement of perfusion following angiogenesis induced by transplanted endothelial progenitors. Considering the high prevalence of ischemic heart disease, further testing of this application is warranted.

Ex Vivo Reconstructions — Cells, Scaffolds, and Bioreactors

The creation of tissues and even organs prior to transplantation is a rapidly expanding area of research. This approach relies on expanding cell populations on, or in, either natural or synthetic scaffolds in a bioreactor. Whether they are natural or synthetic, scaffolds must be biocompatible, bioresorbable, and nonimmunogenic. Furthermore, they must also be instructive, and dictate appropriate cell growth and differentiation, and support the recapitulation of a stem cell niche, which is essential for tissue renewal upon transplantation. The most commonly used natural scaffolds include either individual purified extracellular matrix (ECM) proteins such as collagen, fibronectin, and laminin (or peptides derived from them), or devitalized ECM from skin, submucosa of the small intestine, urinary bladder and others, and can be autologous, allogeneic, or even xenogeneic. Such devitalized ECMs contain not only the structural proteins (which also have important biological activities) that define the three-dimensional organization of the tissue, but also the local repertoire of growth factors that are stored within them.

Various chemical and nonchemical treatments have been applied to ECMs and their components in attempts to modify their biomechanical and immunological properties upon transplantation, but less than desirable effects result due to inactivation of their conducive and inductive properties, and inhibition of their resorption and replacement. If left unmodified, ECMs generally promote infiltration and proliferation of cells and differentiation, *in vitro* and *in vivo*, and are ultimately turned over upon transplantation. Synthetic scaffolds are designed not only to mimic the biological properties of ECMs, but also to have enhanced material properties appropriate for a particular tissue. Polyglycolic acids, polyhyroxy-alkonates, and hydrogels are the most common examples, and all can be manipulated to form a broad range of structures with varying degrees of rigidity. Scaffolds can also be formulated to include morphogenetic and growth factors, or even naked plasmid DNA to transduce in-growing cells to produce appropriate factors.

In addition to having appropriate scaffolds, the ability to construct tissues and organs *ex vivo* also depends on the design of appropriate bioreactors. The initial designs primarily provided nutrient flow through the developing tissue. However, many tissues grown under these circumstances fail to achieve the biomechanical properties required for tissue function *in vivo*. For example, the construction of blood vessels using a variety of biomaterials and cell populations using culture perfusion yields structures that histologically resemble native blood vessels. Yet when transplanted, these constructs fail due to their inability to withstand changes in pressure. A substantial improvement in such constructs has been achieved by subjecting the developing tissues to pulsatile flow conditions. Perfusion-type bioreactors support the growth of tissue with only a nominal number of cell layers and are amenable for construction of relatively simple structures. Construction of larger, more complex three-dimensional structures will require the development of bioreactors that will support cell survival to achieve a significant cell mass. Another critical aspect of using large constructs is the development of supporting vasculature. While constructs can be placed in vascular beds in some cases, it will be essential to design constructs that rapidly induce vascularization, perhaps by including the angiogenic factor, VEGF, or cells that

produce it, or even by including endothelial cells and their precursors, which tend to assemble themselves into primitive tubular structures that may allow for more rapid establishment of organ perfusion.

Skin grafts perhaps provide the best example of the impact of postnatal stem cell biology on the success of cell-based tissue reconstruction. Although the use of *ex vivo*-expanded keratinocytes for the generation of skin grafts was first brought to a clinical setting by the pioneering work of Howard Green and co-workers, persistent engraftment has not been routine. It is now recognized that both the expansion culture conditions and also the scaffolding upon which the cells are grown and placed must support self-renewal of epidermal stem cells (cells that give rise to holoclones as opposed to transiently amplifying meroclones or more differentiated paraclones). Expansion of keratinocyte populations on fibrin substrates appears to maintain the stem cell population and to improve long-term maintenance of skin grafts. Currently, there are several commercially available products that incorporate dermal fibroblasts in a collagen gel, which is then layered with *ex vivo*-expanded epidermal cells. Although these constructs are allogenic, they provide coverage of severe wounds and are maintained while the recipient generates new skin. While the standard starting material for establishing epidermal cultures is skin, recent evidence indicates that a stem cell located in the bulge region of the root sheath of hair follicles has the ability to form interfollicular epidermis. These cells, which can be obtained noninvasively, are highly proliferative and have been used successfully to treat leg ulcers. Based on lessons learned from skin regeneration, stem cells in the limbus of the sclera and conjunctiva have been expanded in culture on fibrin substrates to generate sheets of epithelial cells that are able to reconstitute a damaged corneal surface, with dramatic beneficial clinical results.

ACTIVATION OF LOCAL AND DISTANT ENDOGENOUS STEM CELLS

While the *ex vivo* reconstruction of entire organs that are functional is the goal the bioengineering world, what is perhaps even more challenging is the goal of inducing endogenous stem cells to become activated to reconstruct a tissue. Most tissues in the body display at least some sort of regenerative capability, which, however, in many cases remains insufficient to mount a spontaneous repair response *in vivo*. In this scenario, the application or induction of morphogens and growth factors, perhaps in combination with appropriate scaffolds, might be envisioned to "encourage" local or distant progenitors to regenerate a functional tissue. This *in situ* process could occur through several different pathways, including transdifferentiation or reprogramming, or activation of stem cells to generate adequate numbers of committed precursors. However, the mechanisms of these pathways are as yet poorly defined. In all cases, a regulated morphogenetic process is needed to establish normal structure and function. Definition of the intrinsic properties of regenerative cells and extrinsic signals that may trigger a recapitulation of developmental processes is critical.

Local Cells

Based on reasoning that following an injury, signals that normally activate a local cell for tissue turnover would be either obliterated or not of a high enough magnitude, numerous studies have attempted to provide appropriate morphogenetic and growth factors. This approach has proved uniformly disappointing, most likely due to the short half-life of such soluble factors and the lack of appropriate scaffolding. Scaffolding provides not only a template for the organized outgrowth of local cells, but also a substrate on which to stabilize and/or orient factors (or parts thereof) for appropriate presentation to a cell. However, in using "smart" constructs, tissue regeneration of large defects may not be complete due to exhaustion of the local stem cell population, hence the need to maintain a stem cell niche. Furthermore, some clues as to the negative impact of local constraints upon the native, albeit incomplete, repair capacity of cells and tissues may be found in studies of spinal cord regeneration. Following injury, there is a phase characterized by neurite outgrowth from the severed end. However, it is short-lived due to the production of inhibitory factors by cells within the myelin sheath, and the ensuing gliosis impedes neurite outgrowth. Thus, regeneration by endogenous cells must not only take into consideration bioactive factors and scaffolding that are necessary to maintain the in-growing population with the appropriate balance of stem cells, transiently amplifying cells and differentiating cells, but also the inactivation of local inhibitory factors that work against the process.

Recruitment and Mobilization of Distant Cells

In some instances, once introduced into the circulation, many types of cells have the propensity to end up at a diseased or injured site, and participate in local regeneration. Ongoing tissue damage and repair, at the time when systemically infused stem cells reach a target organ, may be the critical determinant for triggering their homing, engraftment, and differentiation to local cell types. Cytokine administration has become a well-established procedure for mobilization of HSCs into peripheral blood, but is not clear that other stem cell populations present in marrow would be equally as amenable to liberation by current procedures. However, mobilization of all types of marrow stem cells might provide a mechanism to enhance a local population and improve tissue regeneration, a hypothesis that must be tested in the context of extant local, tissue-specific damage and repair.

Current Challenges

Recent advances in embryonic and postnatal stem cell biology have captured the public's attention and have raised expectations for miraculous cures in the near future. Although certain applications that utilize postnatal stem cells are in practice or will be shortly, the majority are in their infancy and will take much more effort. However, the current sense of urgency in translating recent findings into clinical applications should not lead this field to repeat the mistakes experienced in the field of gene therapy. The development of tissue engineering as a viable medical practice must proceed in an evidence-based fashion in all of the associated disciplines that are involved, and several hurdles are yet to be overcome. First, our current understanding of postnatal stem cell biology is rudimentary, at best. The manner in which we isolate stem cells and manipulate them and their progeny appropriately relies heavily on a clear understanding of cell population kinetics. This also requires a complete understanding of their response not only to bioactive factors, scaffoldings, and delivery systems, but also to the host microenvironment in which they must survive and function. Second, more efficient bioreactors that adequately model the microenvironment and also allow for scale-up of the tissue engineering process must be developed. Third, in order to translate what we learn into clinical application, development of appropriate preclinical models to prove the principle that stem cells do indeed have a positive biological impact is absolutely essential. In analyzing these models, stringent criteria must be defined to determine efficacy in order to bring what started off as a scientific curiosity into medical reality.

ACKNOWLEDGMENTS

The support of Telethon Fondazione Onlus Grant E1029 (to P.B.) is gratefully acknowledged.

KEY WORDS

Bioreactor An apparatus designed for the *ex vivo* construction of tissue or organ-like structures that provides appropriate structure and nutrition to cells that are seeded within it. Current designs may also include the application of appropriate physiological stresses to "condition" the construct for better performance upon *in vivo* transplantation.

Morphogen A substance produced primarily during embryogenesis that influences the pattern of cellular differentiation during development, but may also be elicited during the process of tissue turnover and repair.

Scaffold A temporary or permanent framework used to support cell growth and differentiation that may be natural or synthetic in nature and may also include morphogens and/or growth factors that influence the pattern of proliferation and differentiation (biomaterial).

Skeletal stem cell A cell derived from adult bone marrow that has the ability to recapitulate the formation of a complete bone/marrow organ, also known as a mesenchymal stem cell, or a bone marrow stromal stem cell.

Tissue engineering A currently emerging field that encompasses multiple and diverse disciplines to use cells, materials, and bioactive factors in various combinations to restore and even improve tissue structure and function.

FURTHER READING

Asakura, A. (2003). Stem cells in adult skeletal muscle. *Trends Cardiovasc. Med.* **13**, 123–128.

Bianco, P., and Robey, P. G. (2001). Stem cells in tissue engineering. *Nature* **414**, 118–121.

Cossu, G., and Mavilio, F. (2000). Myogenic stem cells for the therapy of primary myopathies: wishful thinking or therapeutic perspective? *J. Clin. Invest.* **105**, 1669–1674.

Griffith, L. G., and Naughton, G. (2002). Tissue engineering–current challenges and expanding opportunities. *Science* **295**, 1009–1014.

Halvorsen, Y. D., Franklin, D., Bond, A. L., Hitt, D. C., Auchter, C., Boskey, A. L., Paschalis, E. P., Wilkison, W. O., and Gimble, J. M. (2001). Extracellular matrix mineralization and osteoblast gene expression by human adipose tissue-derived stromal cells. *Tissue Eng.* **7**, 729–741.

Jiang, Y., Jahagirdar, B. N., Reinhardt, R. L., Schwartz, R. E., Keene, C. D., Ortiz-Gonzalez, X. R., Reyes, M., Lenvik, T., Lund, T., Blackstad, M., Du, J., Aldrich, S., Lisberg, A., Low, W. C., Largaespada, D. A., and Verfaillie, C. M. (2002). Pluripotency of mesenchymal stem cells derived from adult marrow. *Nature* **418**, 41–49.

Pellegrini, G., Ranno, R., Stracuzzi, G., Bondanza, S., Guerra, L., Zambruno, G., Micali, G., and De Luca, M. (1999). The control of epidermal stem cells (holoclones) in the treatment of massive full-thickness burns with autologous keratinocytes cultured on fibrin. *Transplantation* **68**, 868–879.

Preston, S. L., Alison, M. R., Forbes, S. J., Direkze, N. C., Poulsom, R., and Wright, N. A. (2003). The new stem cell biology: something for everyone. *Mol. Pathol.* **56**, 86–96.

Ringe, J., Kaps, C., Burmester, G. R., and Sittinger, M. (2002). Stem cells for regenerative medicine: advances in the engineering of tissues and organs. *Naturwissenschaften* **89**, 338–351.

Stem Cell Gene Therapy

Brian R. Davis and Nicole L. Prokopishyn

In this chapter, we focus specifically on *ex vivo* gene therapeutic modification of autologous stem cells from patients with genetic disease and discuss the elements crucial for successful clinical application of these therapies. In particular, we contrast gene addition and genome editing approaches for stem cell gene therapy, highlighting the particular challenges that each approach faces to achieve therapeutic benefit.

Introduction

Primitive stem cells capable of self-renewing proliferation and single- or multiple-cell lineage progeny generation have been identified in several human and mouse tissues. For example, various stem cells individually capable of producing hematopoietic, mesenchymal, endothelial, or liver cells have been identified in adult bone marrow. Although the biologic characterization of various nonhematopoietic stem cells is still in its early stages, laboratory and therapeutic clinical experience with hematopoietic stem cells (HSCs) suggests that other stem cell types will likely have successful clinical application.

The HSC has exhibited the ability to establish a normal, healthy blood system in patients following transplantation of normal blood stem cells from closely matched individuals — treating disorders of the blood system including immune deficiency, thalassemia, and leukemia. HSC gene therapy offers significant promise for treating various hematopoietic diseases because genetic correction of autologous HSCs could result in long-term correction of blood system cells while avoiding the immune system complications resulting from nonidentical transplantation. For example, thalassemia and sickle cell anemia, the most common genetic diseases of blood, could potentially be treated either by delivery of the globin transgene to HSCs or by direct repair of a specific globin gene mutation in the HSCs.

Significant attention has been devoted to the isolation, culture, and genetic modification of human bone marrow-derived mesenchymal stem cells (MSCs) because they generate cells of cartilage, bone, adipose, marrow stroma, and possibly muscle. Provided that these genetically modified cells or their differentiated progeny can be efficiently delivered to the required tissues *in vivo*, genetic modification of these cells is a potential treatment for various genetic diseases affecting mesenchymal cells such as osteogenesis imperfecta,

Marfan's syndrome, and muscular dystrophy. Identification of somatic stem cells giving rise to liver, pancreas, and brain raises the possibilities of future application of *ex vivo* stem cell gene therapy to treatment of diseases affecting these organ systems. Furthermore, recent identification and isolation of multipotential adult progenitor cells (MAPCs) capable of significant *ex vivo* expansion and differentiation to multiple lineages, including neurons, hepatocytes, and endothelial cells, potentially makes these cells ideal targets for *ex vivo* stem cell gene therapy.

Although one day it may be possible to specifically target various stem cells *in vivo* (e.g., based on a stem cell–specific surface phenotype), this possibility is not available today. Instead, it is more likely that the relevant autologous stem cells will first be isolated from tissues of affected individuals and genetically modified *ex vivo*. The ability to isolate stem cells (hematopoietic and nonhematopoietic) from patients with genetic disease, genetically correct the stem cells, possibly expand them *ex vivo*, and transplant them back into patients with the goal of producing genetically corrected cells *in vivo* offers significant potential for the genetic treatment of human disease.

Genetic modification in stem cells will be designed to either completely correct the genetic defect or at least compensate for the genetic defect. Gene defects occur in a variety of forms, ranging from a simple base pair mutation to complete absence of a gene. Two general approaches can be used to correct defective genes within cells: gene addition and genome editing (outlined in Figure 63-1).

Gene Addition

Gene addition involves the delivery of corrective DNA (usually composed of the entire coding region of a gene and appropriate regulatory sequences) that compensates for or overrides the defective gene (Figure 63-1). The defective gene, unless it is completely absent, remains in the affected cells. This approach is particularly applicable when the endogenous gene product is not expressed (e.g., because of extensive deletion of the gene).

Successful application of stem cell gene therapy requires that therapeutic genes delivered to stem cells persist in the self-renewing stem cells and in their mature and differentiated progeny. This requirement for transgene persistence is critical because (1) transgenes present only transiently in stem cells undergoing self-renewing proliferation will only be of short-term therapeutic benefit and (2) expression of the corrective

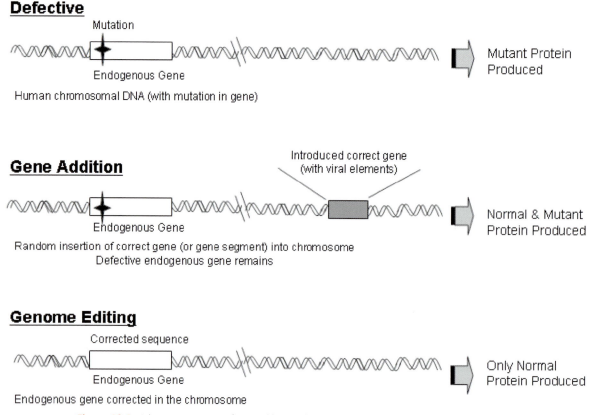

Defective

Mutation

Endogenous Gene

Human chromosomal DNA (with mutation in gene)

Mutant Protein
Produced

Gene Addition

Introduced correct gene
(with viral elements)

Endogenous Gene

Random insertion of correct gene (or gene segment) into chromosome
Defective endogenous gene remains

Normal & Mutant
Protein Produced

Genome Editing

Corrected sequence

Endogenous Gene

Endogenous gene corrected in the chromosome

Only Normal
Protein Produced

Figure 63-1. Schematic comparison of gene addition and genome editing approaches to stem cell gene therapy.

gene likely exerts its effect in the differentiated progeny, which may be numerous cell divisions downstream of the genetically modified stem cell. Transgene persistence can be accomplished, in principle, either by integration into one of the existing chromosomes or by incorporation of the transgene in a synthetic human microchromosome.

VIRAL VECTORS

Over the past two decades, significant attention has been devoted to the development of viral vectors and transduction protocols capable of stable introduction of genetic information into stem cells. Retrovirus, lentivirus, and adeno-associated virus (AAV) vectors are the most common vectors used in transduction of HSCs. After several years of disappointing results, recent reports in humans and other primates, most particularly the French report of successful treatment of X-linked severe combined immune deficiency (SCID), indicate that viral approaches can be successful in treating specific HSC-based diseases; however, a number of potential difficulties must be overcome to make this a safe and effective approach for stem cell gene therapy (discussed later).

NONVIRAL INTEGRATION STRATEGIES

Other approaches have recently been developed that allow for integration into the genome without use of viral vectors.

Stable integration of large DNA sequences (including necessary regulatory elements) into specific "attP sites" in the genome can be accomplished using integrase from bacteriophage phiC31. This method may allow genetic correction of inherited diseases caused by mutations in large genes. In addition, transposon-based systems (e.g., Sleeping Beauty) have been designed to allow for the insertion of transgenes into the human genome without the use of integrating viral vectors. However, optimal transposon size may prevent the inclusion of larger genes or necessary regulatory elements.

SYNTHETIC MICROCHROMOSOMES

Synthetic microchromosomes (SMCs), containing centromeric and telomeric sequences derived from functional human chromosomes, can accommodate insertion of genetic regions that are sufficient to house therapeutic transgenes, together with crucial intron/exon structure and regulatory sequences conferring appropriate transgene regulation. Whether transgene sequences maintained on SMCs are subject to the dysregulated expression and silencing that affect integrated sequences remains to be determined. Crucial issues with application of these SMCs to stem cell gene therapy will be the efficient delivery of single-copy SMCs to target cells and persistence of these SMCs during cell division and differentiation.

Genome Editing

The second approach, genome editing, uses DNA repair and/or homologous recombination processes to correct an existing defective gene sequence so that the defective or mutated area of the gene is restored to a corrected normal state. Genome editing (as shown in Figure 63-1) involves the delivery of small DNA fragments, hybrid DNA/RNA molecules, and/or modified DNA polymers that are homologous to the target gene sequence, with the exception of the base or bases intended for alteration. The genome editing process is directed by endogenous cellular machinery (potentially including mismatch repair and homologous recombination) acting at these target bases and the sequence mismatches created. The target bases are exchanged for the bases present on the introduced DNA fragment — correcting or repairing the gene. The genetic alterations exacted are specific, targeted, and permanent. Repairing the defective sequence itself maintains the corrected genetic material within its normal chromatin environment, ensuring appropriate genetic regulation and expression in the cell. Genetic diseases resulting from well-defined, limited alterations in the DNA sequence, such as sickle cell disease, are ideal candidates for gene therapy strategies based on genome editing. As well, genome editing may be the only suitable strategy in situations in which mutant gene product exercises a dominant negative influence over the normal gene product. For example, overexpression of normal collagen cannot surmount the harmful effects of mutant collagen chains produced in the disease osteogenesis imperfecta.

Several classes of genome editing molecules display conversion frequencies that are significantly higher than traditional gene-targeting methodologies in mammalian cells and have potential for clinical benefit. In addition, recent advances in AAV vector development suggest these constructs may be capable of gene repair at frequencies sufficient for therapy.

RNA–DNA HYBRIDS

These molecules, containing a central stretch of DNA bases with the "correcting" sequence flanked by short stretches of RNA that form hairpin loops at the molecule ends, have demonstrated repair of both single base-pair substitutions and deletions. The gene-correction mechanism is believed to involve mismatch repair. RNA–DNA hybrid molecules have been shown able to mediate correction of the sickle cell mutation in lymphoblastoid cell lines and to introduce the sickle cell mutation into CD34[+] cells from normal individuals.

SINGLE-STRANDED DNA OLIGONUCLEOTIDES

Modified synthetic single-stranded DNA oligonucleotides (25 to 74 bases) have also been found capable of single-nucleotide exchange in a variety of organisms. Studies in yeast suggest that these molecules act by creating a single mismatched base pair on hybridization with the complementary sequence in the chromosome, which is then recognized and corrected by endogenous DNA repair machinery. Successful application of these molecules to genome editing in mammalian cells has recently been reported.

SMALL FRAGMENT HOMOLOGOUS REPLACEMENT (SFHR)

SFHR uses short (typically 50 to 800 nucleotides) single- or double-stranded DNA fragments (SDFs) to alter one or more nucleotides of a specific sequence in the chromosome of a living cell. The SDFs typically span the exon targeted for correction, with the terminal DNA sequences extending into nonexon regions. The precise cellular mechanism(s) responsible for SFHR-mediated genome editing remains to be elucidated. SDFs have been shown to efficiently (~0.1 to 10%) correct or modify specific sequences in genes known to be responsible for disease in several human and mouse transformed and primary cell lines. Recent reports describing the targeting of the *CFTR* gene in mouse lung epithelial cells and the β-globin gene in human hematopoietic stem/progenitor cells (HSPCs) suggest therapeutic potential for this technology. Furthermore, studies have demonstrated that SFHR-mediated genome editing can exact the modification of up to five bases within a give region.

TRIPLEX FORMING OLIGONUCLEOTIDES (TFOs)

Triplex DNA can be used to introduce mutations or genetic modifications in certain gene sequences. Triple helices, formed when the TFOs bind in the major groove of duplex DNA at polypurine–polypyrimidine sequences, have the ability to induce mutations in mammalian cells. A major limitation of this technology is the requirement for specific GC-rich repeats for sequence recognition.

ADENO-ASSOCIATED VIRUS VECTORS

Although initial attempts to use appropriately constructed linear single-stranded AAV vectors in gene repair resulted in low levels of targeted gene repair, recent improvements have provided for targeted replacement of up to 1 kb into human chromosomes without additional mutation to the genome. AAV vectors have also demonstrated efficient correction of single-base mutations in marker genes. The random integration of AAV vectors into chromosomal DNA represents a potential drawback to this approach.

Addition of accessory molecules may also provide for increased genome editing rates. For example, up to 100-fold increases in gene targeting mediated by AAV have been observed following the introduction of DNA double-strand breaks with sequence specific nuclease. Indeed, induced DNA double-strand breaks using sequence specific nucleases have also been shown effective in increasing the efficiency of non-AAV targeted gene repair in human cells.

Much debate surrounds the reported efficiencies of genome editing by many of the previously listed molecules. For example, it is well recognized that the residual presence of significant quantities of genome editing molecules for several days post-transfection can result in polymerase chain reaction (PCR) artifacts, giving false evidence for conversion. Indeed, application of these technologies to stem cell gene therapy will require definitive proof of their merit in genome editing of chromosomal targets. Required demonstration of efficacy includes molecular evidence for genome editing at later times post-transfection, phenotypic demonstration of genome editing, Southern blot confirmation of the genome editing event, and demonstration of the absence of nonspecific gene conversion and absence of random integration of genome editing molecules in target cells. Such evidence will allow for further application of these powerful tools.

Requirements for Successful Stem Cell Gene Therapy

GENETIC MODIFICATION DIRECTLY IN STEM CELLS

For long-term therapeutic benefit, it will typically be insufficient to modify only progenitor or differentiated cells. Instead, direct genetic modification in stem cells will be required for sustained availability and production of corrected progeny cells. It is also essential that this genetic modification be accomplished without loss of stem cell activity.

Gene Addition

Since the mid-1980s, retrovirus vectors have been the vehicles of choice for delivering transgenes (typically cDNAs together with limited transcriptional regulatory sequence) to cells, because they facilitate efficient transgene delivery and integration into the chromosomal DNA of proliferating target cells. The particular challenge of transducing quiescent HSCs was a primary impetus for significant improvements in transduction technology, optimization of *in vitro* transduction conditions including choice of cytokines and use of retronectin, use of various viral envelopes, and development of lentivirus vectors and AAV vectors. The limited packaging size of retroviral, lentiviral, and AAV vectors precludes packaging of certain cDNAs (e.g., the complete dystrophin cDNA is too large to be incorporated into the previously mentioned vectors). The packaging of cDNAs means that only one gene product will typically be expressed, as opposed to the possible expression of alternatively spliced forms from the normal genomic configuration.

These transduction methods allow for delivery without loss of stem cell activity. However, questions still remain as to the ability of these constructs to maintain long-term expression in the stem cells and their differentiated progeny, especially because the limited packing size of the vectors often precludes inclusion of key regulatory elements, exon/intron structures, and necessary insulator sequences.

Genome Editing

Stem cell genome editing requires delivery of genome editing molecules to stem cells without loss of stem cell function and successful editing of genomic sequences directly in the stem cells. Ideally, one would want efficient, quantitative delivery of genomic editing molecules to the nuclei of stem cells. Typical nonviral macromolecule delivery methodologies face inherent and/or potential limitations in human stem cells. For example, electroporation or liposome-mediated transfection conditions have not yet been reported that allow for the efficient delivery of macromolecules to HSCs without significant

loss of viability or stem cell function. In addition, these methods typically give rise to cells having significant cell-to-cell variation in the number of DNA molecules transfected. The microinjection technology described in the section on genome editing of human HSPCs was developed for delivery of macromolecules to both human HSCs and MSCs in an effort to alleviate the problems inherent to traditional nonviral methods.

It is not presently known whether the efficiency of editing chromosomal sequences is affected by the cycling status of stem cells. Although not an issue for those stem cells that can apparently be maintained in a proliferative state *in vitro* without loss of stem cell function (e.g., MAPCs and MSCs), the question of whether replication of the target gene chromosomal DNA is required for successful genome editing is important for quiescent HSCs. It is perhaps important to note that the various genome editing molecules summarized previously may use different mechanisms for genome editing and may, therefore, exhibit different requirements for cycling. It also remains to be determined whether efficient genome editing requires active transcription and/or an open chromatin conformation of the target gene. Although some genes requiring modification will already be expressed in stem cells (e.g., housekeeping genes), there are other target genes that would normally not be expressed until a certain state of differentiation or activation has been reached. Although it is assumed that stem cells in a quiescent state normally perform ongoing surveillance and repair of chromosomal mutations in both transcriptionally active and inactive genes, to our knowledge this has not been directly examined.

GENETICALLY CORRECTED STEM CELLS AND THEIR RELEVANT DIFFERENTIATED PROGENY CONSISTENTLY PRESENT AT SUFFICIENT FREQUENCY *IN VIVO*

A critical issue in stem cell gene therapy is achieving and maintaining a level of corrected stem cells and/or their progeny *in vivo* sufficient to achieve the desired therapeutic effect. Transplanted genetically modified stem cells and their progeny will exist *in vivo* in a background of transplanted unmodified cells and endogenous, unmodified stem cells and their progeny. The percentage of "corrected" cells required for therapeutic value will differ for various diseases. Factors that will influence the percentage of genetically modified cells *in vivo* are (1) the percentage and number of genetically modified stem cells *ex vivo* before transplantation; (2) the ability to expand the stem cells *ex vivo;* (3) the ability to selectively expand or select for the genetically modified stem cells *ex vivo* or *in vivo* (at the expense of the endogenous, defective stem cells); and (4) whether the genetically modified cells (either stem cells or their differentiated progeny) have a proliferative, survival, or functional advantage *in vivo*.

If there was a capability for significant *ex vivo* expansion of stem cells, without loss of stem cell biologic activity, the number of genetically modified stem cells delivered to the patient could be significantly increased, potentially increasing the frequency of genetically corrected versus endogenous defective stem cells *in vivo* following transplant. In addition, if there existed the ability to obtain, *ex vivo*, expanded populations of cells all derived from a single stem cell clone, cells for transplantation could be prepared that were precharacterized for their site of retroviral integration (to eliminate clones with problematic insertional mutagenesis sites or to identify integrants likely to demonstrate appropriately regulated transgene expression) or for successful genome editing (i.e., only delivering appropriately corrected stem cells). Studies performed with human MSCs or human MAPCs suggest that these classes of stem cells may be capable of significant *ex vivo* expansion.

Gene Addition

Inclusion of a selectable marker gene (e.g., chemotherapeutic drug resistance gene), in addition to the therapeutic transgene, in the same viral vector will, in principle, permit *ex vivo* or *in vivo* selection of genetically modified stem cells and their progeny. Although studies have demonstrated excellent *in vivo* selection for the genetically modified cell, a concern remains with regard to potential *in vivo* toxicity at time of chemotherapeutic treatment or incidence of cancer in later years because of the chemotherapeutic treatment(s).

Genome Editing

Genome editing will most likely first find application when there is a natural *in vivo* selective advantage for genetically corrected stem cells and/or their progeny (e.g., Wiskott Aldrich syndrome [WAS], Fanconi's anemia, or X-SCID). There should be a strong selective advantage *in vivo* for the corrected HSCs and their T-lymphocyte progeny in WAS as evidenced by studies of X-chromosome inactivation patterns in female WAS carriers and the selective reversion to a corrected WASP allele in a patient with WAS. These data strongly suggest that even the correction of a small number of HSCs with the capability of differentiating into gene-corrected T-cell progenitors and mature T lymphocytes may lead to significant clinical benefit in WAS. This suggestion is supported by recent efforts to treat X-SCID via stem cell gene therapy. It is believed that strong *in vivo* selective pressure for T and B cells expressing the gamma chain of the interleukin (IL)-2 receptor contributed to the success of the X-SCID trial in France.

For those diseases in which the genome edited stem cells and/or their progeny do not have a selective advantage *in vivo*, successfully genome edited stem cells could be subsequently transduced with constructs having selectable genes. These selectable genes could be excised by systems such as Cre-Loc following *in vivo* selection.

PHYSIOLOGICALLY APPROPRIATE EXPRESSION LEVELS OF THE CORRECTED GENE PRODUCT IN THE RELEVANT CELLS

Correction of genetic deficiencies will typically require that the therapeutic gene be expressed at the appropriate level in the relevant cells for the life of the patient. Genetic treatment of some diseases will have less stringent requirements for either the level of required expression (e.g., chronic granulomatous disease) or the ability to tolerate indiscriminate expression (e.g., adenosine deaminase deficiency, Gaucher disease). In contrast, successful treatment of other diseases (e.g., hemoglobinopathies, X-linked agammaglobulinemia) will likely require expression that is both of sufficient level and cell-type specific. For example, it is expected that clinical benefit in hemoglobinopathies will require that nonerythroid cells remain absent of globin expression and that most erythrocytes will express the inserted gene at greater than 10 to 20% of normal globin levels for thalassemia and 20 to 40% of HbF levels for sickle cell disease. Finally, ectopic constitutive expression of therapeutic transgenes may be harmful in cases in which expression of the endogenous gene is tightly regulated (e.g., with respect to the cell's activation state).

Gene Addition

Significant problems have been encountered in satisfying the requirement for long-term, cell type-specific transgene expression. The expression of retrovirus-transduced transgenes is frequently silenced or exhibits significant variability in the level of expression from cell to cell in the progeny of transduced cells. The difficulty in achieving appropriate expression (proper level, regulation, and cell-type specificity) is due to integration position effects, uneven distribution of the number of integrated gene copies per cell, and the inability to package sufficient regulatory sequences within the viral vector. As a consequence of this dysregulated expression, cells, though genetically modified with corrective genes, may not efficiently display the corrected phenotype. Several recent studies have reported that multiple retroviral or lentiviral integration events per cell were required for adequate levels of transgene expression. The requirement for multiple integration events creates at least two problems: (1) increased likelihood of insertion mutagenesis (see later) and (2) potential for inappropriately high expression of transgene in the cells (e.g., an overexpression of β globin may manifest as α-thalassemia).

Interestingly, these same features of dysregulated transgene expression were also observed in the early transgenic mouse expression studies. Subsequent studies demonstrated that long-term, position-independent, copy number-dependent, cell-type specific expression required strong promoter/enhancer elements, sufficient genomic sequences to dominantly confer the appropriate chromatin configuration and sufficient intron/exon structure and sequences for high-level expression. The strict packaging requirements for retrovirus, lentivirus, and AAV vectors may preclude inclusion of sufficient regulatory sequences and/or intron/exon structure for therapeutic applications requiring highly regulated gene expression.

The complex and coordinate regulation of globin gene expression *in vivo* raises significant challenges for treating sickle cell anemia or β-thalassemia by the addition of a normal or specially designed globin gene using retroviral, lentiviral, or AAV vectors. Much effort has been dedicated to identifying and including in viral vectors the regulatory sequences that are necessary and sufficient for appropriately regulated transgene expression. Recently, use of optimized vector constructs have yielded significant improvements in long-term, erythroid-specific globin expression in mouse models. Whether these very promising results will be confirmed in human clinical trials remains to be determined.

A significant challenge for achieving appropriately regulated transgene expression via stem cell gene therapy is that the transgene, together with regulatory sequences, is delivered into a chromatin environment (specifically, stem cell) that may be significantly different from the chromatin environment in which the transgene expression is ultimately required (e.g., erythroid for globin expression). For example, a transgene may integrate into an active chromatin locus in stem cells that may subsequently become an inactive chromatin locus in the differentiated cell type(s). Furthermore, it is not known whether transcriptional regulatory sequences directly incorporated into stem cell chromatin will show the same loading of chromatin remodeling and transcription factors as when the transcriptional regulatory sequences are present in the chromatin from the stage of embryonic stem cell onward. In other words, is chromatin structure (including histones, transcription factors, etc.) a consequence of sequential steps of factor addition and removal, originating in the fertilized egg or embryonal stem cell? Or can it be created *de novo* from factors present in the HSC?

Genome Editing

Repairing the mutation itself within the defective gene (e.g., β-globin gene) would maintain the correct genetic material within its normal chromatin environment and in principle ensure appropriate genetic regulation and expression in the progeny-differentiated cells (e.g., erythroid cells). This is one of the primary advantages of the genome editing approach for genes in which the transcriptional regulatory sequences are intact.

ABSENCE OF INTERFERENCE FROM ENDOGENOUS DEFECTIVE GENE PRODUCT ON THE ACTIVITY OF THE CORRECTED GENE PRODUCT

For those situations in which there is ongoing expression of the endogenous defective gene, a critical issue is whether this mutant gene product will interfere with the function of the introduced correct transgene. For example, it is possible that endogenous expression of a truncated or mutant protein could interfere with the functioning of an introduced normal protein in a dominant negative manner (e.g., collagen or WASP). Persistence of the mutant protein may be a particular problem in cases in which the gene product normally forms homodimers or heterodimers or trimers.

Gene Addition

Addition of a normal gene into a stem cell means that the defective gene is still present within that cell and capable of action. If ongoing expression of the endogenous defective gene is detrimental to the cell, it may be necessary either to express a specific transgene product that is designed to specifically counteract the activity of the mutant gene (e.g., expression of an antisickling gene) or to overexpress the normal

transgene product at a level sufficient to dilute out the effect of the mutant gene product. These approaches may not be possible in all cases because overexpression of the normal gene product may itself generate negative side effects or unexpected results.

Genome Editing

Genome editing may be the only suitable genetic modification in situations in which the mutant gene product exercises a dominant negative influence over the normal gene product — because this approach would convert the endogenous gene to a normal form — ablating the detrimental mutant gene.

ABSENCE OF ADVERSE EFFECTS

Of critical importance in any therapy is minimization of side effects or adverse effects caused by the treatment itself. A partial listing of potential adverse effects from stem cell gene therapy would include insertional mutagenesis from transgene integration, nonspecific gene editing introducing untoward mutations, and inadvertent carcinogenesis from the *in vivo* selection method for corrected cells. Although a complete absence of any adverse effects is the goal of any genetic therapy, the potential risks of any treatment must be assessed in concert with the potential benefits. For some of the diseases potentially amenable to stem cell gene therapy, the current prognosis is very poor, with no other therapy available. It should also be remembered that the introduction of other groundbreaking therapies (e.g., heart transplantation, marrow transplantation) resulted in numerous initial failures, including death of patients. In addition, even today, therapies such as allogenic marrow transplantation for primary immunodeficiency disease are only 90% successful.

Gene Addition

A difficulty of the gene addition approach is the potential leukemogenicity resulting from random integration of the introduced viral vector into chromosomal DNA (e.g., either activating oncogenes or disabling tumor suppressor genes). This possibility has been reported in the otherwise successful X-SCID human trial in France.

Genome Editing

One of the crucial criteria for genome editing is the specificity of the editing event. A balance between specific and nonspecific genome editing may occur, and as such it is imperative to develop methods that favor specific actions. In addition, use of methods that introduce double-strand DNA breaks (believed to increase the rates of repair) requires strict regulation to ensure that these breaks are limited and specific to the target sequence (e.g., through engineering of site-specific nucleases). Some of the molecules used in genome editing also have the potential for random integration within the genome and, therefore, possess the potential to disrupt genes and or cause mutagenesis. Assessment of this risk or application of molecules that do not randomly integrate into the genome will be necessary for therapeutic application.

Genome Editing of Human Hematopoietic Stem/Progenitor Cells

One of the current limitations to development of genome editing strategies in human adult stem cells, and particularly HSPCs, has been the difficulty in achieving efficient delivery of genome editing molecules to the nucleus of stem cells without loss of stem cell activity. This section summarizes our development of a novel method for delivering macromolecules to the nuclei of HSPCs and MSCs and our experience, in particular, with genome editing of HSPCs.

MICROINJECTION-MEDIATED DELIVERY OF MACROMOLECULES TO ADULT STEM CELLS

Although therapeutic application will likely require the correction of a large number of HSPCs (given current identification procedures for engrafting cells) and, therefore, a robust method for bulk delivery of macromolecules to the patient's cells, microinjection provides an ideal experimental tool for quantitative delivery of macromolecules to the nuclei of HSPCs. With the development of strategies for attachment of HSPCs to extracellular matrix-coated dishes without affecting cell function and the development of injection needles with very small outer tip diameters (OTDs) (~0.2 μ) that do not damage these relatively small cells (~6 μ diameter for stem cells), glass needle-mediated microinjection technology has been successfully applied to HSPCs. Macromolecule delivery is accomplished with high postinjection cell viability (up to 87% postinjection viabilities in CD34$^+$/CD38$^-$ cells), no discernible impact on stem/progenitor cell proliferation or biologic activity, and a high frequency of cells expressing injected transgenes.

Microinjection can deliver macromolecules into cells regardless of whether the cells are in a quiescent or cycling state, and by regulating concentration of DNA injected, flow rate, and injection time per cell, the number of molecules delivered to a cell can be approximately controlled. Microinjection has been demonstrated to be a well-tolerated method for the delivery of various genome editing molecules to HSPCs. These microinjection technologies have been expanded to MSCs.

GENOME EDITING OF THE β-GLOBIN GENE IN HEMATOPOIETIC STEM/PROGENITOR CELLS

Genome editing strategies have been used to introduce the sickle cell disease lesion, a single base-pair transversion (A → T), in codon 6 of the β-globin gene in normal HSPCs. Targeted modification of the β-globin gene has been assessed following delivery via microinjection of both RNA–DNA hybrids and SFHR molecules.

RNA–DNA Hybrids

Liu *et al.* recently described the application of RNA–DNA hybrids to site-specific nucleotide exchange in the human globin gene in primitive human blood cells. RNA–DNA hybrids delivered via microinjection to HSPCs resulted in the

A to T nucleotide exchange in 23% of experiments analyzed. Furthermore, conversion of the β-globin gene was detected in the erythroid progeny of HSPCs at the mRNA level. Interestingly, conversion rates as high as 10 to 15% were seen in some experimental samples, suggesting that levels of conversion may be sufficient to achieve therapeutic benefit in patients with sickle cell disease.

Small Fragment Homologous Replacement (SFHR)

The feasibility of using SFHR in the modification of normal human β (β^A) -globin sequences was assessed in engrafting normal HSPCs. Short DNA fragments (β^S SDF, 559 bp) made up of sickle β (β^S)–globin sequence were microinjected into the nuclei of HSPCs. Site-specific conversion (β^A → β^S-globin) was observed in 42% of the experiments (70 experiments total) as determined by DNA and RNA analysis at 2 to 7 weeks postinjection. The percent fraction of β-globin alleles that were converted in each experiment ranged between 1 and 13%.

Lin$^-$/CD38$^-$ cells, microinjected with β^S SDF, were also transplanted into irradiated NOD/SCID/β$_2$ microglobulin knockout mice to assess whether β^A to β^S conversion had actually occurred in cells capable of engrafting the bone marrow of these mice. Successful engraftment of genetically modified primitive human blood cells in immune-deficient mice was observed with significant conversion of the β-globin gene (β^A to β^S). Evidence for the presence of sickle globin protein in modified cells and/or demonstration of genome editing of HSPCs isolated from sickle cell patients will provide further support for application of genome editing technologies in *ex vivo* stem cell gene therapy.

Conclusion

Gene addition and genome editing approaches as applied to human stem cells offer significant therapeutic potential. It is most likely that clinical benefit will be initially demonstrated in diseases in which there is a clear selective advantage (e.g., proliferative, functional, or survival) for the corrected stem cells or their progeny and/or in which the requirements for regulation of gene expression are less stringent. Significant improvements in both technologies will be required to bring about therapeutic benefit to a wider range of inherited genetic diseases.

KEY WORDS

Ex vivo stem cell gene therapy   The process of isolating stem cells (hematopoietic and nonhematopoietic) from patients with genetic disease, genetically correcting the stem cells, possibly expanding them *ex vivo*, and transplanting them back into patients with the goal of producing genetically corrected cells *in vivo*.

Gene addition   Delivery of corrective DNA (usually composed of the entire coding region of a gene and appropriate regulatory sequences) to cells containing the defective gene.

Genome editing   Specific, targeted, and permanent correction of an existing defective gene sequence so that the defective or mutated area of the gene is restored to a corrected normal state by delivery of molecules that are homologous to the target gene sequence with the exception of the base or bases intended for alteration. The genome editing process is directed by endogenous cellular machinery (potentially including mismatch repair and homologous recombination) acting at these target bases and the sequence mismatches created.

FURTHER READING

Cavazzana-Calvo, M., Hacein-Bey, S., de Saint, B. G., Gross, F., Yvon, E., Nusbaum, P., Selz, F., Hue, C., Certain, S., Casanova, J. L., Bousso, P., Deist, F. L., and Fischer, A. (2000). Gene therapy of human severe combined immunodeficiency (SCID)-X1 disease. *Science* **288**, 669–672.

Davis, B. R., Brown, D. B., Prokopishyn, N. L., and Yannariello-Brown, J. (2000). Micro-injection-mediated hematopoietic stem cell gene therapy. *Curr. Opin. Mol. Ther.* **2**, 412–419.

Davis, B. R. and Prokopishyn, N. L. (2004). Stem cell gene therapy. *In Handbook of Stem Cells* Volume 2 (R. Lanza, ed.), pp. 793–804. Elsevier Academic Press, New York.

Hawley, R. G. (2001). Progress toward vector design for hematopoietic stem cell gene therapy. *Curr. Gene Ther.* **1**, 1–17.

McInerney, J. M., Nemeth, M. J., and Lowrey, C. H. (2000). Erythropoiesis: review article. Slow and steady wins the race? Progress in the development of vectors for gene therapy of beta-thalassemia and sickle cell disease. *Hematology* **4**, 437–455.

Persons, D. A., and Nienhuis, A. W. (2000). Gene therapy for the hemoglobin disorders: past, present, and future. *Proc. Natl. Acad. Sci. USA* **97**, 5022–5024.

The Ethics of Human Stem Cell Research: Immortal Cells, Moral Selves

Laurie Zoloth

Introduction

In any discussion of the tantalizing promise of stem cell research, in which a new field is in the process of inventing itself, and in so doing may reinvent the way illness and injury are treated, why is it that ethical issues are often the first issues to be raised? Why is it that stem cell research was debated on the floor of most houses of government, a subject for the United Nation General Assembly, and a factor in the U.S. presidential elections? Stem cell researchers are confronting the most serious religious and ethical challenge to science since the debate arose over nuclear fission. In the largely secular pluralistic world of the academic science community, one is led to ask: Why do religious and philosophical arguments so dominate the debate on stem cells? What is the warrant for listening to such ethical arguments in biology? Should science be a matter of politics or ethics at all? In this chapter, I give an account of how this came to be the case, summarize the leading arguments held by different sides in the debate, and describe the policy options suggested by these sides.

In part, stem cells have reawakened two types of contentious debates that have long captivated the popular imagination. The first debate concerns genetic research. In its determination to seek the fundamental answers to phenomenological puzzles by seeking biologically mediated answers, genetic research has raised questions about what we mean by freedom, moral choice, and justice for the last 50 years, ever since Watson, Crick, Wilkins, and Franklin first caught glimpses of nuclear DNA (Watson, 2003). Human embryonic stem cell research — research that fully uses the insights and essential framework of genomics but seeks to understand far more about how cells signal, repress, and express the proteins that shape them — is linked to this larger series of intellectual and social debates (Shamblott *et al.*, 1998; Thomson *et al.*, 1998). The debate about the meaning, telos, and nature of the work is a challenging one, for fears about genetic interventions have signified and symbolized the way we think about our human future, the nature of life itself, and our place in it as humans. As each piece of the puzzle of humanness, being, and behavior is set in place by biology, however, it threatens the power of the classic disciplines of faith and reason. The

struggle over the authority to name the nature, means, and ends of scientific knowledge is an old one in which, until the late twentieth century, the hard sciences steadily gained ground. Many core issues and definitions were first contested and then largely ceded to biology by the late nineteenth century. The revolution in molecular biology accelerated the process as new frontiers were continually explored. However, the current ethical and moral debates on human stem cells mark both a new centrality of ethical reflection and a collective sense of caution about how we cross these new frontiers. In this debate — about genetics, genomics and natural boundaries — both human embryonic stem cells (hES) and "adult" or somatic cells committed to a lineage (precourser) cells raised ethical questions.

The second important moral debate is about the use of the early human embryo, a debate that has focused on the use of hES cells. For those who consider the human embryo to be a protected, fully ensouled human person from the moment of conception and the first recombined DNA, all such research is akin to murder and is prohibited.

When the terms of the debate are so passionately felt, it can be difficult to find common ground. In stem cell research, in a global scientific world, and in the twenty-first century, much of what had been previously agreed upon, even the very ability to find facts, or how to obtain gametes for donation, or whether private or public resources can be fairly used, are subjects of debate (Callahan, 1995; Brock, 1987). We share a deeply divided pluralistic society, a new sort of science, and a true moral uncertainty about both the limits of nature and the nature of limits. Religions have become central to the debate because they offer one consistent set of moral rules — yet religions differ sharply on this issue, with some permitting, some prohibiting, and some encouraging the research.

We are confronted with uncertainty in our national — and any international — science policy. Who should make and then monitor such a policy? What is the scope of its power, and what is the right approach for a normative policy?

Policy statements conflict with the deliberations of each new commission, national organization, and science board. The National Academy of Sciences (2001), the United States' most authorative scientific deliberative body, has issued a series of reports on stem cells, all recommending a cautious use, with national regulation and oversight — including permission for nuclear transfer ("cloning") for research purposes. When U.S. federal policy bodies debated the issue, individual states deliberated and then pursued their own policies:

Wisconsin, New Jersey, and California (after a series of reports from the California Cloning Commission (2002) in 2001 and a ballot initiative passed by the voters) are funding stem cell research.

Two presidential committees have offered conflicting advice to respective American administrations. In 1999, the National Bioethics Advisory Committee, appointed by President Bill Clinton to advise him on bioethical issues, supported the use of human embryonic stem cells in research and the limited use of embryos to produce them. Then, in late 2000, the National Institutes of Health (NIH) crafted a compromise to allow the use of stem cells but not their derivation, a policy taken "under advisement" by President Clinton. After the election of George W. Bush, new policy was proposed. President Bush's attention was turned fully to the debate in his first months in office, and it was the subject of his first public national address on August 9, 2001.

Such uncertainty led to a congressional debate, after which the House of Representatives passed a bill in August 2001 (265 to 162) that banned and criminalized research on the making of tissue or products from human embryonic stem cells as well as the use and import of any products from that tissue. The Senate considered a complementary bill, the Brownback Bill, in June 2002 but did not come to a final resolution. In August 2003, the Senate faced widespread opposition to such a bill from 70 Nobel laureates, the American Association for the Advancement of Science (AAAS), scholars in religion and ethics, and more than 100 deans of medical schools. The AAAS and the Hastings Center had offered reports in 1999.

Internationally, the debate has been profound, with policies of research support, together with ethical and legal regulatory oversight, coming from Britain, Israel, Australia, Korea, China, Singapore, France, Switzerland, Sweden, Holland, and India. More limited support has come from Italy, Germany, and several countries in Latin America and the Middle East.

The most profound debate has been the American one: an American polity has arisen expressing disagreement about the role and reach of the state, and concurrent divisions about family and reproduction, illness, aging, and death; reflecting competing moral understandings of the nature of a good act; and debating life's beginnings. This debate was already strained by the explosion in advanced reproduction technology.

Furthermore, health care since the AIDS epidemic has been shaped largely by patient advocacy and consumer groups, who have emerged as a new force in the debate. Should patient advocacy matter? What role should activists play in promoting research, and should such advocacy groups have, as many argue, a larger role in the debate because they bear the costs of disease most directly and vividly?[1]

The polity erupted in controversy that became politicized — yet in unusual ways, with opponents and supporters from all points of the traditional political spectrum. The American Heart Association (AHA) rescinded its initial support of stem cells after Catholic donors expressed concern and withdrew support for AHA funding in this research. Prominent leaders on the left, such as ecologist William McKibben and William Kuchinic, opposed the research; former First Lady Nancy Reagan, Senator Orrin Hatch (R-UT), Senator Arlen Specter (R-PA), and others traditionally of conservative leanings came out in public support. Like political leaders, religious groups differed in their approach. For Islamic, Jewish, and liberal Protestant scholars, the research was either permissible or mandated as a positive act of healing, and the embryo was understood to lack full moral status as a person. For many Roman Catholic and evangelical Protestants, the moral status of the embryo as fully ensouled at the moment of conception rendered acts of research on it sinful violations, and its destruction in research was regarded as the murder of a human being. Genetic research had already queried matters about the boundaries of life and our warrant and providence, and stem cell manipulation was seen merely as an extension of this hubris.

"There may be valuable scientific knowledge which it is morally impossible to obtain . . ."

Paul Ramsey, 1970

Question of Limit

When the Christian moral theologian Paul Ramsey made the arguments for limits on our knowledge, he was partly concerned about genetic research. He asked that scientific researchers first address certain ethical questions: What is the proper object of our desire? What can one trust? In this way, he was assuming that science was embedded in communities of responsibility. Science's implications bear upon us all; hence, we all participate in these implications and have a right and responsibility to comment on them. When the mapping of the human genome was proposed, funding was set aside to allow ethics discourse and research to continue mapping the ethical, social, and legal issues raised by science (ELSI Projects). Ethics and the attendant question of limits have been a potent force in the debate since Ramsey first raised the question.

Forbidden Knowledge

In several critical ways, research on human embryonic stem cells recapitulates old arguments about our faith in science, progress, and technology. Science in general, and genetics in particular, lays claim to topics that are controversial by their very nature — not only because they are new to our established normative narratives, causing scholars such as Francis Fukayama, or Leon Kass to claim that we are "moving rapidly toward a posthuman future," but also because they deconstruct the narrative of creation. In some ways, this research raises fears about forbidden and new knowledge; in other ways, it potentiates fears about violations of "mother nature," an argument engaged in by both fundamentalists and envi-

---

[1]The case that advocacy groups have a larger role in the debate was made by an activist from the Christopher Reeves Foundation at the first meeting, to be held annually, in June 2003 of the International Society of Stem Cell Research in Washington, DC.

ronmentalists. The forbidden nature or the speed of research is part of a larger debate about modernity, its pace, and its uses. This debate is being waged around many other new areas of knowledge: anesthesia, vaccination, and electricity, among others.

Furthermore, new knowledge is threatening to an established order and to the nature of order (and the order of nature) envisioned as reflected in the nuclear family. Many have been adamant in their opposition to research on embryos or research in molecular genetics because they have felt it destabilizes families and natural reproduction. Hence, Kass has argued against stem cells because it may cause "moral harms" to the family unit.

This idea of a lost, beneficent, and essential primitive order, in which nature is seen as morally normative, is a familiar one in philosophy and in political life. In many religions, the same period that has witnessed fierce opposition to science has seen the rise of fundamentalism, the questioning of evolution, and the rejection of facticity. In politics, a fear of the future has often replaced an earlier optimism about the future. And in all of these arguments, a return to the past, or an imagined purer past, is seen as a logical response to fears about technology.

Four Sets of Questions about the Ethics of Research

In thinking about how to respond to these questions, I suggest an ordering along these lines:

1. Ethical questions about the moral status of the embryo.
2. Ethical questions about how women are convinced to donate eggs for research.
3. Ethical questions about the long-term consequences of regenerative medicine on our ideas about illness or aging.
4. Ethical questions about how to share any therapies that might develop fairly and without unfair business practices (a feature of many current debates about biotechnology).

What Is the Moral Status of the Blastocyst?

This question has dominated the controversy about any research on embryos and, in particular, about stem cell research. First, what is the essential nature of these cells? And second, what are our duties toward the blastocyst? If the cells are fully ensouled humans (like newborns), one needs to regard them as such, and one's duties to the cells are morally equivalent to duties toward any other dependent, vulnerable human. If they are regarded as tissue of worth (like hearts one can transplant), one has duties such as respect, care, and prudence to consider. If they are regarded as tissue like any other body tissue or like tissue to be discarded (like placentas), then one's duties are largely those of attention to the symbolic dignity of anything human.

When Does Life Begin?

Human eggs are alive (in that they are not inanimate objects and in that they are cells with the ability to divide). All eggs are potentially fertilizable. Brigid Hogan (2001) has noted that when "life begins" is a complex question — think of a blastocyst as origami paper, she argues, that needs a genetic signal to be folded correctly. This signal is one in a cascade of biological events and could be one mark of human life. Conversely, one could point to other moments within the activity called fertilization in nineteenth-century terms — that is, the many moments in which the one-cell being differentiates and individuates. Moral status is contended ground and can be defined relative to many factors, including when biological individuality is established, when a certain level of organization is achieved in the blastocyst, and standard temporality (ranging from 1 second to 40 days of maturity). One could define moral status by analyzing intent — by the reason for the creation and the existence of the embryo (intended to become a baby or intended to be used for research), by the physical location of the embryo (in the womb or in a test tube), by the potentiality of the embryo, by the likelihood that it is destined for destruction, or as some have suggested, by determining the rates of loss in the human reproduction of all embryos (close to 90%). Since moral status has been important in political, religious, and legal systems for centuries (Feldman, 1968),[2] and since a pregnancy is not perceptible to the external world of the polity for months, most textual traditions rooted in antiquity assume an ancient tradition of an "unformed fetus." Such a term is found in the writings of Aristotle, in both the Hebrew Scripture and the subsequent Talmudic discourse, in the Sharia, in Islamic legal commentaries that interpret the Koran, in Augustine, and in Thomas Aquinas. The Vatican held the interpretation that an embryo had a null to limited moral status until 40 days into the pregnancy; for Muslims, it was until the bones had "knit," and for the Aristotelian tradition, it was until the menstrual blood had "congealed." In canon law (the Catholic legal authority), this idea held until 1917, when, following the new science that could observe eggs and sperm, the idea of a homunculus at the head of a sperm being implanted like a seed into a woman was dispelled. Until the late nineteenth century, most theologians had an idea that pregnancy was not an established and protected fact until this time; they ruled on cases and wrote as if this were indeed the case. Recent debates may tend to obscure this history, but the idea of personhood beginning at "the moment of conception" is a relatively new idea that grounds the primacy of fetal rights over maternal ones, It is an idea that has changed dramatically from earlier perceptions (Feldman, 1968).

Is This Research Like Abortion?

The central concern about moral status, however, has led many to think of the ethical issue of stem cells as one synonymous with abortion. This change has led to the use of similar language in both debates: women's rights, babies, fetuses, reproduction, and choice versus life. It is not the only linguistic uncertainty. Several people have raised the issues of whether an artificially created or a very early blastocyst is

[2]For matters of compensation in the loss of a pregnancy, for issues of when and under what circumstances a community mourns for the loss of a pregnancy, etc., see Feldman, (1968).

properly called an embryo and whether this term, when improperly used, merely confuses the public. Use of the term *cloning* presents further confusion, since the goal of human or animal cloning is to produce offspring, but the goal of cloning for research is to create genetically identical duplicates (of cells or regions of DNA). In this case, the goal of cloning is to make many exact copies of cells to study how very early development occurs. Narrative uncertainty has also been introduced into the discourse. By this term I would argue that there has been a break in the outstanding cultural story of the nuclear family, the miracle of birth through loving sexual intercourse. Significant changes in the essential and primal narrative of human reproduction have raised the following questions: What is a family now? What does it mean to make embryos with a series of unions of parts from variable sources? What if the narrative has alternative possible endings? The traditional narrative of human reproduction — one man, one woman, a meaningful Adamic cleaving leading to progeny that carry the story forward — is at the heart of many faith traditions. Indeed, it is through this human story that Western traditions and several of the traditions of Eastern and indigenous religions[3] create a core narrative about the meaning, nature, and goal of being human.

Our understanding of ourselves as a part of this narrative, as children and then parents, strengthens the meaning of many theological constructs: natural law theory, the begotten creation of children, the pronatalist imperative, and the obligations and relationships in families and communities. However, since the early 1970s, the idea of the natural process of sexual reproduction has been disrupted by emerging scientific technology, which has created many possible origins for any human embryo: it may be fabricated by mixing eggs and sperm, or by injecting an egg with a selected sperm. The course of development may be altered as well: sperm may be "spun" and separated by weight to select for gender; the egg may be altered with extra mitochondrial DNA; embryos may be deselected by genetic trait; or the embryo may be implanted in a surrogate, the egg obtained from another woman, and the resulting child given to a third family, which may itself be constituted in a variety of genders and permutations. All of these disruptions in the core narrative elicited considerable alarm initially, and at each stage, social discourse has emerged as new possibilities are discovered. In many societies, the narrative has been reimagined and retold to account for these new variants. Regenerative medicine offers not only another set of beginnings for the narrative of reproduction but also other possible telos for the embryo. Now, a blastocyst fabricated in an *in vitro* fertilization (IVF) clinic faces at least four fates: (1) it might be transferred to a human womb, where it might implant successfully, grow, and be born into childhood; (2) it might be transferred but not develop; (3) it might be frozen indefinitely; or (4) it might be discarded. Alternately, the embryo might be destroyed in a lab in the process of being used to make stem cells — and some (such as Lebacqz and

Peters) argue that this allows the DNA of that cell immortal replication.

Once our society allowed the first four outcomes, the last, in the lab, can be understood as an alternate ending or an alternative goal. For many, such a deconstructed narrative, with the possibility of origins other than monogamous union and ends other than reproduction, elicits a sense of moral repugnance, the ultimate horror of a scientific, desacralized world. But for others, this revised narrative elicits a curiosity and awe at the new possibilities for human understanding and at the possibility of altering other key aspects of what had been understood as moral fixities — the nature and scope of human suffering, the "natural" span of a human life, and the capacity for human reach (Shattuk, 1996).[4]

How Can Eggs for Research Be Obtained from Women Justly?

Ethical questions have emerged not only about the moral status and origins of the tissue but also about the "harvesting" of the gametes needed to fabricate the blastocyst. All eggs come from a particular woman, all sperm from a particular man. How are these obtained justly and safely from human subjects — and does this change if the women and men whose gametes are at stake are voluntarily (even desperately) trying to achieve a pregnancy and in so doing create "excess" embryos they do not choose to use? Dresser and others (Dresser, 2002) have raised concerns that women might be exploited or manipulated into using their bodies to make money and be placed at undue risk if they are hormonally stimulated to produce eggs. Suzanne Holland (1999) raises concerns about the reduction of women to the value of their reproductive capacity.

Others have raised yet more issues in the process. What does it mean to "make" embryos with a series of unions of parts from variable sources? Will such disaggregating of the pieces of the person lead us toward a world of commodified, exchangeable selves — a sort of warehouse supply store that would cheapen unique human lives? Would disabled persons be seen as poor products and be discarded as some have claimed? What are we to make of the practice — already in place — of advertising for gametes from women of privileged social or intellectual status and competing for the "best eggs?" Since some marketplace relations have, in the past, understood human bodies as at least potential commodities, what protections might be instituted to protect human subjects from the pressures of the market?

"No Truth But the Thing Itself"

A second problem associated with the process of the design of all biological research is that one cannot make an abstract model of the problem, as one can in other sciences. Unlike physics or chemistry, the model is the actual human event. Hence, even making models creates the problems one needs to tentatively explore. Even a proof-of-principle experiment requires a blastocyst.

---

[3]Variants include heroic or divine-human conceptions, but all are based on sexual union, gestation, and birth.

[4]I wrote the first drafts of this work and delivered the speech on which it is based prior to reading the seminal work by Shuttuk on this central idea.

A third problem is the slippery-slope issue, or the "trigger" problem. It is argued that making and perfecting a small part of the technology that can be used for cloning or genetic engineering, which may in itself be only an incremental shift in knowledge, can allow a desensitized and facile acceptance to the next (troubling) step of the science. Here, the concern is that setting up a project that allows the "harvesting" of gametes, cloning, and so on, would set the stage for cloning for reproductive purposes, genetic engineering for "designer" babies, or other such scenarios.

A final problem is a structural one. In the past, the public understood research on embryos as being instrumental toward the goal of reproduction (hence the support of IVF research). Here, the process is geared toward a more abstract telos; hence, the charge arises that embryos are only being made to be destroyed. If the embryos used are created for reproductive purposes, then for many, their destruction in research is an event that occurs along the inevitable trajectory toward destruction and is a different ethical question than that raised by embryos newly created for research. Yet it is precisely this sort of experimental use that promises to yield an important understanding of early stages of cell signaling, cell programming, and genetic control mechanizes in both normal and disease states.

Does This Research Aim Toward a World that Devalues Aging or the Disabled?

Thinking about the ends of research on human stem cells has initiated a discussion on the nature of the ends and goals of health care itself and has led to a critical split in how we consider aging, human frailty, and illness. Kass, Gil Meilander, Fukayama, and others have raised serious concerns that if the goal of this research is to eliminate illness or human suffering, it is a flawed goal and it is a tragic error to pursue it. Kass has spoken of the character-building engagement of a life lived well as one ages, of lessons learned through the suffering and subjection of the creaturely body, and of the virtues enhanced with the acceptance of even serious disability. What will happen, ask these critics, to our sense of compassion if its objects — vulnerable, frail, and elderly — are enhanced to robust, cheerful perfection? Yet others, such as Stock, Silver, and Caplan, disagree, arguing for a world progressively liberated from such limitations. Others have raised issues of unintended consequences — unknown and unknowable chaos that may result if this research is pursued. Clearly, since we are witnessing only the earliest stages of research that, although interesting, is still largely theoretical, the civic discourse will have to welcome such concerns and attend to the immediate issues of how investigators need to act now to structure such attention should such choices ever confront us.

CONTEXTUAL FRAMING OF THE ISSUES

Can Just Research Be Conducted in a World of Injustice?

The context of all research is that health care is an unfinished project of social justice. In America, the uninsured with minimal access to basic health care continue to vex political policy. International issues of distributive justice render the problem of access to new research and the therapies that will emerge from such research as a central ethical concern. Moreover, as noted previously, stem cell research has been placed in the context of the abortion debate, and so the unsettled and volatile nature of the discourse about embryos is based in the unfinished debate about abortion. Like slavery, such a debate is about religion, moral status, civil rights, and civil duties, but it is also about health-care funding and services. The debate about abortion has defined and has been thematic of American politics since 1973; hence, there was no *public* funding in the first debates about fetal tissue, and most of the first research projects were privatized, funded by independent capital. This created labs that by federal mandate could not be located in any building or institution that used federal funds, leading to new concerns about secrecy, profits, and the like. The need to separate controversial research from research that could be supported by a polity and their tax revenues led to a further separation than some critics are now comfortable with — hence the call for more federal oversight.

The second contextual problem is that stem cell research is taking place against a background of four decades of unease about all things genetic. From genetic manipulation and the creation of genetically modified foods to issues of genetic testing and privacy, Americans, and to a larger extent Europeans, have been vocally mistrustful of the motives and aims of research genetics. This has risen to a level of concern that has been taken, literally, to the public square and linked with globalization and colonization. Protesters of international banking policy routinely appear in butterfly costumes, alluding to a report (never replicated) that genetically modified corn negatively affects the reproduction of the butterfly.

Furthermore, there is unease over human-subject research because research errors have occurred at major medical centers such as the University of Pennsylvania, Johns Hopkins University, and Duke University. Further mistrust has been engendered by the failures of the marketplace to self-regulate, as in the case of Enron. That several of these scandals in research (gene therapy) and in the market (Martha Stewart) are linked to genetics heightens the context of anxiety.

Finally, many have raised the issue of the context of the marketplace and the increasing involvement — through patenting, technology transfer, and licensing agreements — of biotechnology in the life of the academic science lab, an issue that Derek Bok and others have actively queried.

Normative Issues: Three "Bright Lines" Have Long Limited Research

Social concerns have thus driven the ethical debate, and ethicists have responded with recourse to the traditional sanctions suggested by bioethics' first principles: autonomy, beneficence or nonmalfeasance, and justice, with bioethics' deep affection for autonomy as a premier principle. Hence, policies have been developed with strong privacy and informed consent requirements, and reproductive medicine has long operated with private, parental desire as both the main driver and the main funding source. Ethical boundaries were established in the 1970s to limit technologies seen then as remote. These three "bright lines" were a reluctance to sanction possible intervention in human-inheritable genetic material, a ban on the fabrication of human embryos for research alone, and a ban on cloning (scnt) for any purpose. Human embryonic stem cell inquiry challenges each of these norms, and even a close examination of several IVF methodologies reveals that here, too, such "bright lines" have long been crossed. Normative oversight (civil committees, state, federal, or scientific) has been called for by nearly every deliberative body that has considered the issues of the regulation of stem cell research. But in so doing, six types of policy problems will have to be decided: How will differences in strongly held religious and moral stances be expressed and defended? How will the freedom of the scientific pursuit be limited? What of the power of the ends expressed by patient advocacy groups? What will happen to violators? Who will fund such oversight? And who will be chosen to be on such committees?

In Summary

ARGUMENTS FOR PROCEEDING: ETHICAL RESEARCH ON HUMAN STEM CELLS CAN BE DONE

Let me summarize the central arguments for actively supporting, funding, and pursuing research in stem cells.

Teleological, Consequential, and Largely Utilitarian Arguments

Research on stem cells has a nearly unlimited potential for good ends. Various diseases that affect millions worldwide have as their case the disadvantages of cell growth or cell death. Thus, understanding how cells grow, how they are genetically regulated, and how they develop both normally and abnormally will be key to therapy. It is this vision of future therapies and this attention to human suffering that ought to lie at the core of the medical endeavor. A correlative research end will be met by research on embryonic development. Stem cell cultures will allow an ability to test toxicity — pharmaceuticals in early embryos and in human tissues, a task that is dubious in animal models and ethically unacceptable in human pregnancies. A final, related, telos-based argument is that such research is of itself a good end, for it allows study of the process of genetic diseases at the cellular level, using the full power of recent genomic advances in understanding causality. Others note that the funding of open research is a valid goal and that they are eager to embrace a fully open telos as a principle of science policy.

Many diseases that affect millions worldwide would be cured — not merely treated — by the use of tissue transplants. Cardiac diseases, cardiovascular diseases, degeneration, or trauma to the spinal or central nervous system are obvious first candidates. That such tissue transplants have shown promise in early testing in animal models drives this argument into a central location in the debate.

Equivalency Arguments

Stem cell research is very much like other research on embryos that is already being done in universities and medical centers all over the country — IVF research in which many eggs are tested, injected with sperm, given growth factors to stimulate growth, and used as tools in teaching physicians their craft as infertility specialists. All such embryo experiments are approved by an institutional research board if the work, and the embryos created therein, are destroyed at 14 days of life, just prior to the development of an individuated primitive streak. Linked to this argument is the larger one that much of early IVF research (some would say all) is a vast experiment, and that many embryos are created with the clear understanding that few would survive. The protocols for IVF originally called for the implantation of up to eight embryos in the womb in the hopes that not all would die, thus building embryonic waste directly into the research and currant clinical practice. If more than three embryos do implant, the couple is routinely offered "embryo reduction," meaning targeted and selective abortion of the "excess" embryos, in the name of saving or enhancing the lives of the remaining sibling twins.

A variant of this is based on a naturalistic premise, which permits research on blastocysts since so many are simply nonviable in the natural course of things. Thus, embryonic loss is like the loss that occurs in nature, and many of the blastocysts would be lost in any case.

Deontological (Duty-Based) Arguments

In many religions, and in secular medicine's premise, there is a duty to heal. Obligations are correlative with rights. In this argument, the limited moral status of the *in vitro* blastocyst determines duties to it, and the relatively larger (some say unlimited) duties to the ill and vulnerable may be primary ones. We have a duty to heal, and this is expressed in legal and social policy. To turn from the possibility of healing would be an abrogation of an essential duty. Furthermore, justice concerns may mandate this research because, unlike whole organ transplants, tissue transplants and pharmaceuticalized stem cell tissues may be made scaleable, universal, and affordable, thus allowing a widely applicable use for these transplants. Serious issues of histocompatibility may in theory block this path for now, but the duty to justice would mandate a fully expressed research effort in this direction.

Making the claim for duty can be religiously motivated or can come from sources such as the determinates of biology (that we need to protect kin, that we are dependent as neonates and need protection, and that primates have a long period of parenting until adulthood). Other sources include our shared aspirational duties to improve our situation of suffering, as is argued in Christianity; a divine command, as is posited by Judaism; our "experiences," as argued in American pragmatism; and our ability to be social beings making social contracts, as Locke and Jefferson suggest.

What are such duties? In other work, I have suggested six (Zoloth, 2003):

1. *Duties to make justice:* Judged by social contracts that attend to healing the most vulnerable in our society and to making therapies accessible to all.
2. *Duties to discern and judge:* Assessed by our ability to be coherent moral actors, to set limits, and to see differences in moral status and ability.
3. *Duties to heal the ill, save lives if we can, and care for the dying if we cannot:* Enacted by the inherent duty of medicine that we must extrapolate to societies, in which no self can be exempted.
4. *Duties to guardianship:* Enacted by attention to a world unfinished and in need of protease inhibitors, vaccines, yeasted bread, eyeglasses, and so on. This duty of rational discourse grounds a thoughtful civil debate.
5. *Duties to be readers of text:* Meaning that interpretation and analysis of the phenomenological world is suggested by the very way knowledge is structure — imperfect, mutable, and unrevealed.
6. *Duties toward solidarity:* Taken from European debates on genetic issues, this term means that we have a duty to social cohesion. Activities that merely instrumentally use one another (exploitative relationships with gamete donors, etc.) are a violation of this duty.

Arguments from Legal and Historical Precedents

Here, one can turn to the example of times when a severely divided country moved ahead on an issue of policy despite the deeply held moral opposition of many — Mennonites who opposed World War II and Quakers who opposed the war in Vietnam offer examples of how democracies must act for the majority and how the minority view must continue to be expressed, even if such dissent carries the risk of civil disobedience. America, from the time of Thoreau, has understood democracy as a serious matter of dissent as well as assent.

Arguments That Are Political in Nature

Here, the arguments are as follows: If research is not funded publicly, it could be driven into private and unconsidered spheres or could limit the goods of the research to particular sectors, specific groups, or the needs of the market (one thinks of Viagra instead of pediatric diseases, for example.)

ARGUMENTS FOR STOPPING: STEM CELL RESEARCH IS IMMORAL AND ILL CONSIDERED

The arguments against stem cell research, whether they favor banning it for a time or banning it permanently, are summarized in the following sections.

Deontological Objections

First, stem cell research is murder of nascent humans and is deontologically forbidden. In the report from the President's Council on Bioethics, the majority argued for a moratorium on such research, with strong opposition from a significant portion of the commission. The members opposed to stem cell research argued largely deontologically, stating that the moral status of a cloned embryo is nascent human life and is thus a member of our shared humanity. As such, they argued, Americans had a special obligation to protect vulnerable members of our social contract, the most vulnerable being entities such as a blastocyst. Furthermore, to use an embryo would be an exploitative use of human life as a tool, constituting a serious moral wrong in addition to the moral wrong of killing. Such violations of essential duties to care thus create serious moral harm to society — coarsening our ideas of family union and exposing our culture to the uncertainties of asexual reproduction. They also argued that to think that suffering can be cured or alleviated, especially with the sacrifice of life, is a misunderstanding of our duty to heal. Here is employed the caution that there is no moral obligation to treat all disease — and it is a moral error to think we can do so — and the complementary idea that our ability to suffer and to feel compassion for the suffering stranger is at the base of our shared humanity. The fulcrum of this sort of deontological argument rests on the view of suffering, frailty, and limitation as central to our human creatureliness and our human nature.

Teleological Vein

It is argued that such research will engender a terrifying, "posthuman" set of consequences. Since we face a lack of moral consensus about the family and reproduction, allowing research on this volatile and contentious issue will create political chaos. Others fear that it *will not work* and that hopes for cures will be cruelly dashed; that it *will work* and be unsafe and dangerous; or that it *will work* and give parents powerful, morally repugnant choices such as the elimination of all imperfect children, creating "designer babies" that may be, in this argument, very skillful and very beautiful but cruel and soulless. Such choices are disturbing and, some argue, "inherently, essentially morally repugnant." ("The yuck factor" is the term Callahan (1995) and others use to describe this phenomena.)

Slippery-Slope Arguments

Slippery-slope arguments are key to the opposition to stem cell research — a series of classic arguments that maintains that, although the particular act may be marginally permissible, the road to which it leads will be a dark and downward descent. Powerful historical precedents in the form of American and German eugenics, as have been soundly exposed, document a slope of precisely this sort in which technology was used to marginalized and eliminate the ill, the disabled, and the socially different in the years prior to the elimination of the Jews of Europe. Manipulation of embryos or cloning could lead down the slope to the possibility that governments will determine which sort of life is a good one, that cloning people will lead to two classes of human, or that human–animal chimeric monsters will be created.

Concerns of Justice, Especially Feminist and Environmentalist Ones

It is feared that stem cell research may exploit women for their eggs, that women may be coerced, that huge "embryo farms" will be needed to make enough stem cell cultures, or that human tissue will be merely another scarce commodity to which the poor contribute but do not have access. Some raise the fear that Americans already spend too much on research such as this, especially on research for the privileged elderly, and not enough on preventative health-care clinics for the poor; others fear that profit-driven private pharmaceutical companies or illegal offshore labs will have too much control over the processes and the products of the research. Some raise the fear that such a violation of natural limits and borders is too closely akin to the errors made in the use of nature in the nineteenth century, and that human ecology, or a human "gene pool," may be disrupted by stem cell research. In this argument (partly deontological and partly teleological), nature is seen as normative, morally stable, and instructive.

Regulatory Concerns

Some fear that scientists cannot be trusted to self-regulate, since a proportion of the research community believes that nature is flawed and in need of their ministrations. This can too easily segue into the research community "playing God with creation," a fear raised about all genetic research. The fear that the technology will be impossible to regulate is behind the policy of absolute bans.

FROM ETHICS TO POLICY IN HUMAN STEM CELL RESEARCH: EIGHT POLICY OPTIONS

Stem cell research is global in character,[e] with 8 of 12 sources and most lines named in the Bush administration's August 9, 2003, compromise plan for the use of stem cells in research outside of the United States. Yet, as the 2004 election that swept stem cells, and their funding and regulation, to the forefront of California politics demonstrated, stem cell research policy is also profoundly local — and personal in nature. Wherever it is debated, core ideas about informed consent vary, core cultural and social meanings of IVF differ, and core notions of the polity and process of oversight vary. Therefore, how can one speak of coherent, reflective public policy to adjudicate between the powerful arguments noted previously in this chapter?

Even as ethicists debate a clear moral path for policy, the science and availability of the cells themselves change daily, as research advance and political options shift quickly across the global. What are the possible directions for stem cell ethics and policy?

Leroy Walters (2004) has suggested six options; I would argue that eight choices of policy have already been employed globally.

Ban Outright All Research Involving Stem Cells — an extreme measure but under consideration in some states in the United States.

Permit Use Only of Cells from "Excess" Embryos Derived by Others from IVF Clinics.

This was the option suggested by the National Institutes of Health under President Clinton in 1999 and 2000, by

Senator Bill Frist in July 2001 (with a limit on number), and by President Bush in 2003 (with a limit on time of derivation).

Permit Derivation from "Excess" IVF Embryos by Stem Cell Researchers.

This was suggested by President Clinton's National Bioethics Advisory Board in September 1999; by the European Union Group on Ethics on Science and the New Technology in November 2000; by the Advisory Group to Canadian Institutes of Health Research in March 2001; by Deutsche Forschungsgemeinschaft in May 2001; by two national advisory groups in France in January and June 2001; by Japan's Expert Panel on Bioethics in August 2001; by the Australian House of Representatives' Committee on Constitutional and Legal Affairs in September 2001; by the Bioethics Advisory Board for the Howard Hughes Medical Institute; by Canada, Italy, Spain, and the Netherlands; and by 2 states in Australia and 40 states in the United States.

Permit Derivation and Use from Embryos Created Just for This Research.

This is the policy of the United Kingdom, China, Sweden, Belgium, Israel, Korea, and California and is under consideration by a growing set of countries.

Permit Nonreproductive Cloning to Create Embryos for Research and Use.

This is an option allowed by the United Kingdom, China, Belgium, Saudi Arabia, and Israel; by California; by the Bioethics Advisory Board of the National Academy of Science and Humanities (September 9, 2001); and by the U.S. National Academy of Sciences Task Force Report (September. 11, 2001). It is under consideration by the NAS, and several countries as well as several states as this chapter is written (Illinois, Connecticut).

Permit a Separation Compromise.

This tack would allow different populations or jurisdictions to do different things. In the United States, this is understood as a possible model for many controversial policies, often as a transitional policy ("the laboratory of the states") until consensus can be held federally (as in civil rights laws). This idea has gained ground in countries such as the United States that find themselves at odd regionally over the best sort of policy to develop.

Allow All Ideas Uncovered in Research to Be Fully Explored.

This option is no longer being seriously debated, for all countries, including China, now suggest some sort of regulation and oversight on stem cell research.

Create a Limited Year, or a Limited Technique, Moratorium.

The President's Council on Bioethics has recommended such a policy, and the term varies. The point of a moratorium on various parts or all of the processes from use to application would be to have a more open debate in the political arena so that all views could be fully aired. It should be noted that all of the other seven options also call for a robust debate.

POINTS OF CONVERGENCE: WHAT CAN BE AGREED ON?

Clearly, then, we have deeply held beliefs and widely divergent policies. Can we agree on any point so convergent? I will argue that we can, and I will present a few beginning candidates.

First, science is a kind of free speech, but free science is for the public good; hence, it must be honest, freely open, and regulated in some way by the very public sphere in which it aspires to be considered. Second, science must be just, with its social goods available to all without discrimination. It must never coerce or exploit human subjects. Third, science must be prudent and safe, taking care to protect the environment even as it alters it. Fourth, *medical* research must aim at beneficence toward patients, whose futures and interests must be protected. Fifth, disability, aging, and illness must not be dishonored. Finally, although each human person has core human rights, such rights suggest correlative duties that must be fulfilled.

POINTS OF DIVERGENCE: WHAT CAN NOT BE AGREED ON?

I would argue that there are four matters we will not come to agreement on, and we must find ways of negotiating our serious differences, which are ultimately serious religious matters. We must come to understand that we will likely not agree on the moral status of fetuses and embryos. Nor will we agree on the definition of a family. We will not agree on the meaning and content of what is "repugnant" in science. We will not agree on the place of suffering in our theo–social world view.

Conclusion and Recommendations: Creating a Civic Witness

Bioethics can be faulted if, after raising a chapter of questions, concerns, and inquiries, it does not offer a thoughtful recommendation of a way forward. How can we now apply ethics? How do we go beyond a call for justice or a call for deepening the public debate? Here are some specific recommendations. First, I would argue for the development of a range of civic responses to science research beyond the "red light–green light" approach. Research can be (rarely, I think) prohibited when it is abusive, deadly, or coercive (as has been done in certain human-subject research); permitted and regulated closely by citizenship oversight; or permitted with institutional oversight. Finally, research should be encouraged, funded, and socially rewarded. Each project needs our assessment, rather than just the projects that are given special scrutiny in the press. Many, such as Alta Charo, have noted that this largely is already our practice through the review board at every university (the IRB system) and the NIH review process, especially in genetic research, but the mechanisms need to be more fully explained to the American public so that they can be assured of research transparency. This will mean that the public will have to come to understand, without panic or hype, that all great research is inherently risky, given

to failure and error, and may not yield success suddenly or ever. (That is why it is called "research.") Patience will have to be taught as a duty if public oversight is to be wise. Public accountability is a model for the Recombinant DNA Advisory Committee — a process begun with researchers at Asilomar, querying their own direction, and used to regulate genetic intervention. That such a limited regulatory model is in place, as opposed to the broader model used in the United Kingdom, is a result of different regulatory etiologies. In the United States, regulation emerged after more than 15 years of debate after *Roe v. Wade,* the unregulated growth of IVF industry (1979–1994), the commission of the Human Embryo Research Report, the rejection of findings, and the move to regulation at the state level.

But in the United Kingdom, IVF was debated in the government-sponsored Warnock Report, which recommended the Human Fertilization and Embryology Authority (HFEA), which has been in place since 1991 (issuing both public and private licenses and providing oversight to all uses of all embryos, from IVF clinics to stem cells).

A public oversight committee built along these lines would make a strong contribution to the resolution of the controversy. It will require public members and full, open, public debate; it will need to publish reports on the construction of standards, the conduct of trials, and the setting up of local oversight committees. It will have to foster a wide educational campaign, as does the HFEA, and provide oversight of all IVF procedures, use of eggs, and research protocols. It will have to decide whether ongoing research will be supported and which will need more careful review. It should enforce the global ban on all cloning for reproduction.

DEVELOPMENT OF A THEORY OF VIRTUE FOR RESEARCH

Beyond such a committee, as in Britain, I would argue for making the question of moral agency in research a central one. That a serious debate about research asks a great deal from us is a good thing, and it is a development that researchers should not fear but should welcome. The question in the foreground is the moral status of the persons involved and of the witnessing public: what does this research make of us? Thinking about the moral relationships between researchers and donors is central. How does this work shape us as a society? Since many of the objections warn of the problem of complicity, one must ask: How does one avoid such evil? How does the gravitas of research itself suggest answers? When is civil disobedience and moral dissent rightly used?

NOTES TOWARD A RECOMMENDATION

If the British oversight model is employed, and if we make serious efforts to create a virtue ethics for basic research and to articulate it clearly, I would state that is reasonable to support stem cell research. I would state that this is a morally defensible gesture, one that may hold extraordinary promise for a shared human future. I would further argue that we may use early human embryos in research before 14 days — not frivolously, but where important new scientific knowledge can be gained; where new therapies may be able to be developed; and where, if used judiciously, well-designed research has the informed consent of the genetic providers and full and transparent public oversight.

The creativity and the tenacity of the biologist community have created an explosion in science knowledge. This calls us to respond with creative and tenacious ethical discourse.

ACKNOWLEDGMENTS

I wish to thank Leroy Walters for work on policy options and international research. I thank Brigid Hogan, Leonard Zon, Doug Melton, Tom Okarma, John Gearheart, Ted Friedman, Larry Goldstein, Ron McKey, Jamie Thompson, David Anderson, and Irv Weissman for explanations of the scientific basis of the problems.

KEY TERMS

Deontological Duty based reasons to perform an action. This may be a sense that rights and duties are intertwined (as in the idea that persons have a right to life and each citizen has a duty to protect that right), or a duty may arise from a sense of religious command or may be role-specific ("doctors must heal the sick.")

IRB, ACUC, DSMB Respectively, the Institutional Review Board, Animal Care and Use Committees, and Data Safety and Monitoring Boards. These committees of lay and professional scientists, academicians, doctors, veterinarians, and statisticians provide publicly accountable oversight and regulation to all human and animal experimentation in the United States. The *RAC* is the Recombinant DNA Advisory Committee, which provides an additional level of review for all research involving these uses of synthetically created DNA.

Moral status A description of how we understand our relationship toward the being in question, what we owe the being, and what the being may rightfully expect of us in our society. Often, this is expressed in relative terms and is linked to our understanding that many living beings (frogs, mice, dogs) are sentient, but that their moral status relative to human primates allows us to do things to them (own them, for example) that we cannot do to a person of moral status equivalent to our own.

Potentiality The concept that if a being possesses the elements that would allow it to potentially develop into another, more complex thing (acorn to oak tree, embryo to baby), it is morally equivalent to that thing.

Primary texts of religion and religious law–Hebrew Scripture, the New Testament, the Koran (Hadith) Classic texts that form the basis of, respectively, Jewish, Christian, and Muslim thought. For all three of these *Abrahamic* traditions, a monotheistic and commanding God in history leaves a direct covenant that forms the basis for how duties are enacted. The *Talmud and Responsa, Canon Law* and the *Sharia* are later codes of law in each tradition that explain how judges and scholars have interpreted these commands. The use of these texts and others in the Eastern traditions form the basis for how many religions respond to the questions of when life begins, what a family is, what the meaning of suffering is, and how to address suffering.

Roe v. Wade The United States Supreme Court decision in 1973 that allows termination of a pregancy and hence the destruction of an embryo and fetus in the first 12 weeks of life at the request of the mother, without permission from the state, hence, understanding that neither the embryo nor the fetus is a person with the requisite full moral status.

Teleological A reason to perform an action based on the end or outcome one desires. One such formulation is "the greatest good for the greatest number," if one assumes that great good for the majority is a good end.

FURTHER READING

Brock, D. (1987). Truth or consequences: the role of philosophers in policy making. *Ethics* **97**, 786–791.

California Cloning Commission. (2002). Report on human embryonic stem cells.

Callahan, D. (1995, January–February). The puzzle of profound respect. *Hasting Center Report* **25,** 39–40.

Dresser, R. (2002). *In* "Report of the President's Council on Bioethics."

Feldman, David M. (1968). *Birth control in jewish law: marital relations, contraception, and abortion as set forth in the classic texts of jewish law*. New York University Press, New York.

Green, Ron. (2001). *The human embryo research debate: bioethics in the vortex of controversy*. Oxford University Press, New York.

Hogan, Brigid. (2001, October). Talks at Vanderbilt University.

Holland, Suzanne, Lebacqz, Karen, and Zoloth, Laurie. (1999). *The human embryonics stem cell debate: science, ethics, and public policy*. MIT Press, Cambridge, MA.

The impact of california's stem cell policy on the biomedical industry. (2002). Hearing of the Senate Health & Human Services Committee, May 10.

National Academy of Sciences. (2001). Special Committee on Establishing Standards for Stem Cell Research.

National Bioethics Advisory Committee Report on Ethical Issues in Human Stem Cell Research. (1999, September). Vol. III: *Religious Perspectives*. www.georgetown.edu/research/nrcbl/nbac/pubs.html.

President's Council on Bioethics. (2004, January). Monitoring stem cell research. www.bioethics.gov.

Shamblott, G., *et al.* (1998). Derivation of pluripotent stem cells from cultured human primordial germ cells. *Proceedings of the National Academy of Science of the United States of America* **95(23)**: 13,726–13,731.

Shattuck, R. (1996). *Forbidden knowledge: from prometheus to pornography*. St. Martin's Press, New York.

The Raelians (2002). *NY Times* **Dec. 23**, A.

Thomson, J., *et al.* (1998, November). Embryonic stem cell lines derived from human blastocysts. *Science* **282(5391)**: 1145–1147.

Walters, LeRoy. (2004, March). Human embryonic stem cell research: an intercultural perspective. *Kennedy Institute of Ethics Journal,* **14(1)**, 3–38.

Watson, J. (2003). "DNA."—Press.

Zoloth, L. (2003). Freedoms, duties, and limits: the Ethics of Stem Cell Research. In *God and the embryo: religious voices on stem cells and cloning*, Brent Waters and Ronald Cole-Turner, eds., pp. 141–151. Center of Bioethics, Northwestern University, Chicago, IL.

65

Ethical Considerations

Ronald M. Green

Introduction

There is consensus in the scientific community that human embryonic stem cell (hES) research holds great promise for developing new treatments for a variety of serious and currently untreatable disease conditions, ranging from juvenile diabetes to Parkinson's disease. However, because hES cell research requires the manipulation and destruction of human embryos, it has also been a focus of ethical controversy and opposition. In the course of these debates, several challenging ethical questions have been raised. Scientists, clinicians, or patients involved in hES cells research or therapies must formulate their answers to these questions. Society, too, must address them to determine the extent to which hES cell research may require oversight and regulation. This chapter presents these questions and examines some of the leading answers that have been given to them.

Is it Morally Permissible to Destroy a Human Embryo?

Human ES cell lines are made by chemically and physically disaggregating an early, blastocyst-stage embryo and removing its inner cell mass. At this stage the embryo is composed of approximately 200 cells, including an outer layer of differentiated placental material and the undifferentiated (totipotent or pluripotent) cells of the inner cell mass. The embryo inevitably dies as a result of this procedure. Proposals have been made for approaches that would eliminate the need to produce and destroy a viable blastocyst. Some anticipate the direct conversion of a fertilized egg into a population of hES cells; others look to modifying the genes of a very early embryo so that it could never develop into a fetus or baby. But these approaches either have not been scientifically demonstrated or pose ethical problems of their own. Hence the question: Can we intentionally kill a developing human being at this stage to expand scientific knowledge and potentially provide medical benefit to others?

At one end of the spectrum are those who believe that, in moral terms, human life begins at conception. For those holding this view, the early embryo is morally no different from a child or adult human being. It cannot be used in

research that is not to its benefit and without its consent. Furthermore, proxy consent by parents in such cases is inadmissible since it is an accepted rule of pediatric research that parents may not volunteer a child for studies that are not potentially beneficial and that risk the child's life. Many Roman Catholics, evangelical Protestants, and some Orthodox Jews take the position that life (morally) begins at conception and they oppose hES cell research.

At the other end of the spectrum are those who believe that the embryo is not yet fully a human being in a moral sense. They hold a *developmental* or *gradualist* view of life's beginning. They do not deny that the early embryo is alive and has the biological potential to become a person. Nevertheless, they believe that other features are needed for the full and equal protections we normally accord children and adults and that these features only develop gradually across the full course of a pregnancy. These features include such things as bodily form and the ability to feel or think. The early embryo, they maintain, does not have these features or abilities. They note, as well, that the very early embryo lacks human individuality, since it can still undergo twinning at this early stage and two separate embryos with distinct genomes can still fuse to become a single individual. The very high mortality rate of such embryos (the majority never implant) also reduces the force of the argument from potentiality. Those who hold this view do not agree on the classes of research that warrant the destruction of embryos, but most support some form of hES cell research. Their reasoning is that although the early embryo may merit some respect as a nascent form of human life, the lives and health of children and adults outweigh whatever claim it possesses.

Each individual faced with involvement in hES cell research must arrive at his or her own answer to this first question. Legislators and others must also wrestle with these issues. Because American law (and the laws of most other nations) does not regard the early embryo as a person meriting the legal protections afforded to children and adults, it is hard to see how one can justify legal or regulatory prohibitions on privately financed hES cell research or clinical applications. Such prohibitions would interfere with individual liberty on grounds that are inconsistent with the lesser view of the embryo shown elsewhere in the law. However, because publicly funded research rests on more narrowly political considerations, including the way that a majority chooses to spend public funds, it may be expected that public support for hES cell research will depend on how a majority of citizens answers this first question.

Should We Postpone hES Cell Research?

Some who oppose the destruction of human embryos maintain that hES cell research should be deferred at least until science provides a better view of the likely benefits of adult stem cell research. They maintain that such research is as promising as hES cell research, and they argue that the moral acceptability of this alternative justifies any delay in the availability of therapies. Others add that scientific uncertainty and the ethically controversial nature of this research warrant a moratorium on hES cell studies in both the private and public sectors. Scientific studies of adult stem cells continue to be equivocal, sometimes supporting and sometimes contradicting claims to these cells' plasticity, ability to proliferate without senescence, and usefulness. This raises the question of whether it is justified to delay the development of therapies and cures for children and adults to protect embryos and to respect the sensitivities of those opposed to embryo research. Many people feel that such a delay is not warranted and that it is scientifically and morally preferable to keep open multiple pathways to stem cell therapies. In the words of the National Research Council, "The application of stem cell research to therapies for human disease will require much more knowledge about the biological properties of all types of stem cells."

Can We Benefit from Others' Destruction of Embryos?

It might seem that a negative answer to the first question ends discussion. If the embryo is morally as human as you and me, what could justify the use of cells derived from its deliberate destruction for other people's benefit? However, hES cell research or therapy has many steps, not all of which involve the destruction of embryos. This raises the question of whether downstream researchers, clinicians, or patients may *use* the stem cell lines that others have derived. Ethically, this is the question of whether we can ever benefit from deeds with which we disagree morally or regard as morally wrong.

This question arises partly because most embryos used to produce stem cell lines are left over from infertility procedures. Couples using *in vitro* fertilization (IVF) routinely create more embryos than can safely be implanted. There are hundreds of thousands of these embryos in cryogenic freezers around the country and around the world. Since very few frozen embryos are made available for adoption, most of these supernumerary embryos will be destroyed. In 1996, British law mandated the destruction of 3600 such embryos. This destruction continues regardless of whether some embryos are diverted to hES cell research. Some ethicists have appealed to a "nothing is lost" principle to justify the use in stem cell research of embryos that are otherwise destined for destruction, but critics of this approach see it as nevertheless implicating stem cell researchers or patients in morally offensive acts.

This raises the question of when, if ever, it is morally wrong to benefit from others' wrongdoing. One answer is that, it is wrong when, by doing so, we encourage similar deeds in the future. This explains why we are morally and legally prohibited from receiving stolen goods or why it may be wrong to benefit from research produced by scientists who choose to ignore human subject constraints. However, it seems less objectionable to benefit from others' wrongdoing when their deeds are independently undertaken, when our choices are not connected to theirs, and when these choices do not encourage the wrongful deeds. For example, few people would object to using the organs from a teenage victim of a gang killing to save the life of another dying child. The use of such organs benefits one young person and does not encourage teen violence. Similar logic might apply to stem cell research using spare embryos remaining from infertility procedures. A downstream researcher, clinician, or patient may abhor the deeds that led to the existence of an hES cell line, including the creation and destruction of excess human embryos in infertility medicine. But nothing that a recipient of an hES cell line chooses to do is likely to alter, prevent, or discourage this continuing creation or destruction of human embryos or to make the existing lines go away. Those who use such embryos also know that if they refuse to use an hES cell line they forego great therapeutic benefit. A researcher will fail to develop a lifesaving or health-restoring therapy. A clinician's decision may threaten a patient's life. People in this position will struggle with the question of whether it is worthwhile to uphold a moral ideal when doing so has no practical effect and when it risks injury to others.

Religious views on the question of whether one may ever benefit from others' wrongdoing are diverse. The Roman Catholic moral tradition, with its staunch opposition to complicity with wrongdoing, presents different answers, including some that permit one to derive benefit in particular cases. This suggests that some researchers, clinicians, or patients who morally oppose the destruction of human embryos may nevertheless conscientiously conclude that they can *use* hES cell lines derived from embryos otherwise slated for destruction.

It is noteworthy that in his August 2001 address to the nation President George W. Bush adopted a conservative version of the position that allows one to benefit from acts one morally opposes. Stating his belief that it is morally wrong to kill a human embryo, the president nevertheless permitted the use of existing stem cell lines on the grounds that the deaths of these embryos had already occurred. Presumably, the president believed that because these acts were in the past and would not further be permitted under federal auspices, his permission to use the existing lines would not encourage further destruction of embryos. The president did not go so far as to permit the use of lines derived from embryos slated for destruction in the future. However, it is possible that if existing lines prove inadequate, as some scientists fear, many people would support a slightly more expansive version of this willingness to benefit from what one regards as objectionable deeds.

Can We Create an Embryo to Destroy it?

A fourth question takes us into even more controversial territory. Is it ever morally permissible to deliberately create an embryo to produce a stem cell line? This was done in the summer of 2002 at the Jones Institute in Norfolk, Virginia. Those in favor of this research defend it on several grounds. First, they say that in the future, if we seek to develop stem cell lines with special properties and perhaps closer genetic matches to tissue recipients, it will be necessary to produce stem cell lines to order using donor sperm and eggs. Second, they argue that it is *ethically* better to use an hES cell line created from embryos that have been produced just for this purpose, with the full and informed consent of their donor progenitors, than to use cell lines from embryos originally created for a different, reproductive purpose.

Those who believe the early embryo is our moral equal oppose the deliberate creation of embryos for research or clinical use. They are joined by some that do not share this view of the embryo's status but who believe that it is morally repellant to deliberately create a potential human being only to destroy it. Such people argue that this research opens the way to the "instrumentalization" of all human life and the use of children or adult human beings as commodities. Some ask whether such research does not violate the Kantian principle that we should never use others as "a means only."

On the other side of this debate are those who believe that the reduced moral status of the early embryo permits its creation and destruction for lifesaving research and therapies. Proponents of this research direction ask why it is morally permissible to create supernumerary embryos in IVF procedures to help couples have children, but morally wrong to do the same thing to save a child's life. They are not persuaded by the reply that the status of the embryo is affected by its progenitors' intent, such that it is permissible to create excess embryos for a "good" (reproductive) purpose but not for a "bad" (research) purpose. They point out that the embryo is the same entity. We do not ordinarily believe that a child's rights are dependent on its parents' intent in conceiving it. They conclude that it is not parental intent that warrants the creation of excess of embryos in such cases, but the embryos' lesser moral status and the likelihood of significant human benefit from their use. These same considerations, they argue, justify deliberately creating embryos for stem cell research.

Should We Clone Human Embryos?

A fifth question is whether we are willing to support human cloning for stem cell research. This question arises in connection with a specific stem cell technology known as *human therapeutic cloning*. It involves the deliberate creation of an embryo by somatic cell nuclear transfer technology (cloning) to produce an immunologically compatible (isogenic) hES cell line.

Immune rejection may occur if the embryo used to prepare a line of hES cells for transplant does not share the same genome as the recipient. This will be the case whether the cell line was created from a spare embryo or from one made to order. Therapeutic cloning offers a way around this problem. In the case of a diabetic child, the mother could donate an egg whose nucleus would then be removed. A cell would be taken from the child's body and its nucleus inserted into the egg cytoplasm. With stimulation, the reconstructed cell would divide, just like a fertilized egg. If the resulting embryo were transferred back to a womb, it could go on to birth and become a new individual — a clone of the child. But in therapeutic cloning, the blastocyst would be dissected and an hES cell line prepared. Growth factors could be administered to induce the cells to become replacement pancreatic cells for the child. Because these cells contain the child's own DNA and even the same maternal mitochondrial DNA, they would not be subject to rejection. Recent research has shown that the small amount of alien RNA from the mitochondria of an egg other than the mother's might not provoke an immune rejection.

Although this is a very promising technology, it raises a host of novel questions. One is whether the embryonic organism produced in this way should be regarded as a "human embryo" in the accepted sense of that term. Those who believe that "life begins at conception" tend to answer this question affirmatively, even though cloned "embryos" are not the result of sexual fertilization. They base their view on the biological similarities between cloned and sexually produced embryos and on the argument that both have the *potential* to become a human being. Nevertheless, the very high mortality rate of cloned embryos suggests significant biological differences from sexually produced embryos. Furthermore, if their status rests on their potential, this potential is greatly reduced. In an era of cloning, some degree of potentiality attaches to all bodily cells.

The promise of this technology rests on the ability to make stem cell lines "to order" for a specific patient. If you hold the view that the cloned organism is morally equivalent to a human embryo, therapeutic cloning research and therapies again raise the question of whether it is morally permissible to deliberately create an embryo in order to destroy it.

Another question often raised in this context is whether therapeutic cloning will create an enormous demand for human eggs. If it does, some believe this may create substantial problems of social justice, since collecting these eggs may involve hundreds or thousands of women, many of whom are likely to be poor women of color attracted by the financial rewards of egg donation. Those who discount this problem offer several arguments. They point out that even if many eggs are needed for therapeutic cloning procedures, they are likely to be provided by relatives of patients. This reduces the likely size of a market in oocytes. Others observe that therapeutic cloning may well be "transitional research"; it may lead the way to direct somatic cell reprogramming, eliminating the need for eggs altogether.

Finally, there is a moral question specific to cloning itself. The more scientists are able to perfect therapeutic cloning, the more likely it is that they will sharpen the skills needed to accomplish reproductive cloning, which aims at the birth of a cloned child. There is a broad consensus in the scientific and bioethics communities that, at this time, the state of cloning technology poses serious health risks to any child born as a result of it. There are also unresolved questions about the psychological welfare of such children. Finally, there is the possibility that embryos created for therapeutic cloning research might be diverted to reproductive cloning attempts. All these concerns raise the question: Do we really want to develop cloning technology for the production of isogenic stem cells if doing so hastens the advent of reproductive cloning?

In 2001 and again in 2003, the U.S. House of Representatives answered "no" to this question and passed a bill introduced by Representative James Weldon that banned *both* reproductive and therapeutic cloning. Similar bills have been introduced in the Senate. Although the Senate initiatives have been stalled for some time, this may soon change. If a Senate bill passes, therapeutic cloning research and therapies will be outlawed in the United States. Similar prohibitions are either in effect or being considered for passage in continental Europe. This would leave only a relatively small number of countries, including Great Britain, Israel, South Korea and China, in which such research would be allowed.

Those who oppose these prohibitions believe that therapeutic and reproductive cloning research can be decoupled. They observe that strict regulations and governmental oversight, of the sort provided in Great Britain by the Human Fertilisation and Embryology Authority, make it unlikely that embryos produced for therapeutic cloning will be diverted to reproductive purposes. They also point out that several researchers or groups with minimal qualifications in cloning research have announced their intention to clone a child or have even tried to do so. Such attempts are likely to continue whether or not therapeutic cloning research is banned. As a result, a ban on therapeutic cloning will not protect children and will only have the negative effect of interrupting beneficial stem cell research.

What Ethical Guidelines Should Govern hES Cell and Therapeutic Cloning Research?

Mention of the need to prevent the diversion of cloned embryos to reproductive purposes raises the larger question of what guidelines should apply to the conduct of hES cell research and therapeutic cloning research. In August 2000, the U.S. National Institutes of Health released a series of guidelines for hES cell research that never went into effect because they were largely preempted by President Bush's decision to limit hES cell research to existing cell lines. Guidelines have also been developed by the Chief Medical Officer's Expert Group in Great Britain and by private Ethics Boards at the Geron Corporation and Advanced Cell technology in the United States. The recommendations share several features.

Donor Issues. Because hES cell and therapeutic cloning research require a supply of human gametes or embryos, steps must be taken to elicit the informed consent of donors, to protect their privacy, and to minimize any risks to which they might be subject. Informed consent requires that donors fully understand the nature of the research being undertaken and that they explicitly consent to that research. For example, it is morally impermissible to elicit sperm, eggs, or embryos for the production of hES cell lines without informing donors that an immortalized, pluripotent cell line might result that could be widely used in research or therapeutic applications. If there are likely to be commercial benefits flowing from the research, donors must also be informed of this and their rights (if any) in such benefits should be clearly specified. If the research involves therapeutic cloning, both egg and somatic cell donors must be informed that a cloned embryo and a cell line with the somatic cell donor's nuclear DNA and egg's donor's mitochondrial DNA will result.

In conducting research, efforts should be made to preserve donor privacy by removing identifying information from gametes, embryos, and cell lines and keeping this information apart in a secure location. In view of the controversy surrounding much of this research, donors can be subjected to harassment or embarrassment if their association with the research is revealed without their consent.

Ovulation induction is an invasive medical procedure with known and undetermined risks. Not only must egg donors be informed of these risks, but steps must also be taken to preserve the voluntary nature of their consent. This includes preventing them from being pressured into producing excess eggs or embryos for research in return for discounts on infertility services. It also includes avoiding undue financial incentives. Although it is unreasonable to expect even altruistically motivated donors to undergo the inconvenience and risks of these procedures without some form of compensation, payment should not be so high as to lead a donor to ignore the risks involved. For Advanced Cell Technology's therapeutic cloning egg donor program, its Ethics Advisory Board set a payment level similar to that established for reproductive egg donors in the New England area. Payments were also prorated for the degree of participation in the program to allow donors to drop out of the program at any time. Levels of stimulatory medications were maintained on the low side of current regimens, and payments were never attached to the number of eggs harvested. Additional protections assured that donors had the educational level and backgrounds needed to appreciate the risks involved. A study monitor was employed to ensure that donors' consent was free and informed.

Research Conduct. Guidelines also apply to the actual conduct of research. These include the requirement that no embryo used in hES cell or therapeutic cloning research be allowed to develop beyond 14 days *in vitro*. This limit is based on the substantial changes that occur at gastrulation, which marks the beginning of individuation and organogenesis. Also required are supervision and accountability of all staff and scientists involved in this research to prevent any diversion of gametes or embryos to reproductive purposes.

Transplantation Research. If and when hES cell lines become available for research in transplant therapies, researchers and institutional review boards will have to address ethical questions raised by their use. For example, if such lines are cultured on mouse or other feeder layers, as all current federally approved lines are, there will be a safety question regarding the possible introduction of retroviruses or other pathogens into the human population. Basing their judgments on adequate preliminary animal studies, researchers and oversight bodies will have to assess the risks of rejection, including graft-versus-host disease. Issues surrounding the tumorigenicity of hES cells will also have to be resolved. Experience with fetal cell transplants for Parkinson's disease shows that cells can behave in unexpected ways when transplanted into the human body. None of these problems is insurmountable, however, and their existence itself constitutes an argument for continued research in this area under careful ethical oversight.

Conclusion

Answering all of the questions identified in this chapter would require an ethical treatise. However, by noting that I currently serve in a *pro bono* capacity as chairman of the Ethics Advisory Board of Advanced Cell Technology, I indicate my own answers to the most controversial of these questions. I would not have accepted this position if I did not believe in the importance and ethical acceptability of hES cell and therapeutic cloning research. In my view, the moral claims of the very early embryo do not outweigh those of children and adults that can be helped by hES cell and therapeutic cloning technologies. I also do not believe that therapeutic cloning research by itself will lead to reproductive cloning, which we can take firm steps to forbid. I recognize that others may disagree with these conclusions. As our debates move forward, continuing dialogue about these questions and clearer scientific research results will bring us closer to a national consensus on these issues.

KEY WORDS

Bioethics   The study of the norms of conduct that should govern research and clinical applications of biomedical technologies.

Stem cells   Self-renewing (immortalized), unspecialized cells that can give rise to multiple types of all specialized cells.

Therapeutic cloning   The deliberate creation of a human organism by nuclear transfer (cloning) technology in order to produce an embryo that can be used to create a stem cell line that will not provoke an immune response and rejection.

FURTHER READING

Chief Medical Officer's Expert Group. (2000). *Stem cell research: medical progress with responsibility.* Available at http://www.doh.gov.uk/cegc/stemcellreport.htm.

Committee on Guidelines.

Doerflinger, R. (1999). The ethics of funding embryonic stem cell research: a Catholic viewpoint. *Kennedy Inst. of Ethics J.* **9**, 137–150.

Geron Advisory Board. (1999). Research with human embryonic stem cells: ethical considerations. *Hastings Cent. Rep.* **29**(2), 31–36.

Green, R. M. (2001). *The human embryo research debates.* Oxford University Press, New York.

Green, R. M. (2002). Benefiting from "evil": an incipient moral problem in human stem cell research. *Bioethics* **16**(6), 544–556.

Holland, S., and Zoloth, L. (2001). *The human embryonic stem cell debate.* MIT Press, Cambridge, MA.

Landry, D. W., and Zucker, H. A. (2004). Embryonic death and the creation of human embryonic stem cells. *Jour. of Clin. Invest.* **114**, 1184–1186.

Lanza, R. P., *et al.* (2000). The ethical validity of using nuclear transfer in human transplantation. *JAMA* **284**(24), 3175–3179.

National Research Council. (2001). *Stem cells and the future of regenerative medicine.* National Academy Press, Washington, DC.

Outka, G. The ethics of human stem cell research. *Kennedy Inst. of Ethics J.* **12**(2), 175–213.

The President's Council on Bioethics. (2002). *Human cloning and human dignity: an ethical inquiry.* Washington, D.C. Available at www.bioethics.gov.

66

Stem Cell Research: Religious Considerations

Margaret A. Farley

Introduction

Religious traditions have endured in large part because they help people to make sense of their lives. Major religious traditions become major because they offer some response to the large human questions of suffering and death, hope and transcendence, history and community, as well as the everyday issues of how we are to live together with some modicum of harmony and peace. Adherents to traditions of faith generally experience their shared beliefs and practices not as irrational but as part of the effort of reason to understand the actual and the possible in human life and all that is around it. Although religious sources of insight may reach beyond empirical data and logical reasoning, they need not ultimately do violence to either. When we ask profound questions such as how humans ought to reproduce themselves, or how extreme suffering can be alleviated, or whether the meaning of the human body changes when we exchange its parts, we can be helped by looking to the source and the substance of the attitudes, beliefs, and practices that animate people's lives.

Of course, when we look for guidance in the faith of believers, we do not find univocal voices — not from believers in general and not from believers within particular traditions. Roman Catholics will often not agree with spokespersons from the United Church of Christ; many Jews will not agree with many Catholics about abortion; and so on. But the division within faith traditions today is often wider than between or among them. So, there is more than one view regarding stem cell research among Roman Catholics; more than one among the different strands of Judaism; more than one among Muslims. Such disagreements do not make religious voices useless for societal and religious discernment, no more than disagreements among scientists render scientific research and the voices of scientists useless.

This chapter aims to review religious perspectives on stem cell research. It is not possible here to survey every religious or theological argument for and against stem cell research. Something of an overview is presented, and then particular issues of concern articulated by representatives of some religious traditions are examined.

Mapping the Terrain

SOURCES FOR STEM CELLS

The major disputed questions surrounding stem cell research are focused not on stem cells as such but on the sources from which stem cells are derived. Hence, debates rage regarding the moral status of human embryos and the permissibility of taking stem cells from already dead fetuses. Religious thinkers who consider the human embryo sacrosanct from its inception will consistently oppose the extraction of embryonic stem cells as a form of killing (of embryos). Those who do not appraise the value of an embryo as being on a par with that of a human person are much more willing to favor embryo research in general, including the taking of stem cells for research. Still others who emphasize the ambiguity of the moral status of embryos are likely to weigh the advantages of embryo cell research over the possible violations of the embryo as an entity in itself. All religious traditions recognize an aborted fetus as a cadaver; hence, the derivation of stem cells from the gonadal ridge of dead fetuses may be as permissible as the harvesting of tissues or organs from any human cadavers. Nonetheless, representatives of traditions that prohibit abortion worry about complicity with and support of the moral evil of abortion when stem cells are derived from electively aborted fetuses.

Alternative sources for stem cells are championed by those who are opposed to embryo stem cell and fetal stem cell research. The least objectionable source often cited is stem cells taken from adult tissue (given that this does not ultimately harm the donors of the tissue). A compromise source (though still controversial) is embryonic stem cells that have already been harvested (that is, the life of the original embryo has already been taken, and no new moral agency is involved in the killing of more embryos). Problems with both of these sources continue, however. Research on adult stem cells is less advanced than research on embryonic cells (and in any case, scientists argue that it is better to proceed on all fronts rather than only one). Moreover, the still undetermined (undifferentiated) nature of early embryonic cells is believed to be more readily suitable for the goals of research. Not surprisingly, the proposed use of already existing embryonic cell lines encounters the same objections of complicity with evil that is involved in the use of fetal cells.

Other issues complicate moral assessments of the sources of stem cells and the aims of stem cell research. For example, almost all religious thinkers argue that some form of respect is due to embryos, even if it is not the same form of respect required for human persons. Yet there is lack of clarity about what respect can mean when it is aligned with the killing of the respected object. It is not difficult to get minimal consensus on prohibitions against buying and selling human embryos or on safeguards of informed consent from the donors of embryos for research. For many religious believers, it makes a difference that these embryos are "spare" embryos, left over from *in vitro* fertilization procedures and destined to be discarded if they are not used in research. From other religious perspectives, however, this fact in itself does not lift the prohibition against killing them.

QUESTIONS OF JUSTICE

Quite another kind of ethical issue is raised by people from almost all religious traditions. This is the issue of justice. Who will be expected to be the primary donors of embryos or aborted fetuses? Will gender, race, and class discrimination characterize the whole process of research on stem cells? And will the primary goals of research be skewed toward profit rather than toward healing? Who will gain from the predicted marvelous therapeutic advances achieved through stem cell research — the wealthy but not the poor? the powerful but not the marginalized? What will be the overall results for respect for human persons if human embryos and stem cells are commodified, wholly instrumentalized, and in some ways trivialized?

All of these questions are questions for secular philosophical, scientific, and medical ethics as well as for religious ethics. When they are identified by religious and theological thinkers, however, they are always lodged in the larger questions with which religious traditions are concerned, and they almost always appeal to sacred texts and faith community traditions, teachings, and practices, as well as to secular sources and forms of moral reasoning.

RELIGIOUS BELIEFS AND ETHICAL QUESTIONS

Anchoring religious concerns for the moral status of the human embryo, the nature of moral evil and immoral complicity, and questions of distributive justice are beliefs about the nature and destiny of human life, the meaning of bodily existence, the interaction between divine and human in the progress and healing of creation; the importance of consequences in moral reasoning along with the importance of religious laws and ontologically based norms, the basic equality of all human people before God, and so forth. Even when there is no direct line between the most profound beliefs of a person and the answer to a particular moral question, such a belief makes a difference for moral discernment and conviction. There are no world religions today that oppose all human intervention in "nature," yet all religions recognize some limitations on what humans may do — either to themselves or to other created beings. All consider humility before concrete reality an important antidote to arrogance or pride (and in the wake of ecological disasters, all now take account of the risks of some interventions made only for the sake of the flourishing of some human generations). No religions favor illness over health, or death over life, yet each has a perspective in which health may not be an absolute value

and even death may be welcomed. In the present context of stem cell research, there is no religious tradition that does not take seriously the beneficial results promised by such research.

Hence, positions of religious believers regarding stem cell research tend to be highly complex. Each tradition needs to be understood in its complexity and its diversity. This chapter cannot track all of this, but it can try to show how some beliefs are coherent, whether or not others can agree on them in the same tradition or across faiths and cultures.

Particular Traditions: An Overview

A beginning understanding of positions taken by religious scholars and spokespersons in mainline traditions is, fortunately, available to some extent in the testimony and writings that have appeared in relation to the public debate on stem cell research. We have here, of course, just the proverbial tip of the iceberg. Nonetheless, these documents and essays point us to both the concrete moral positions being debated and the larger rationales behind them. In brief summaries, then, I survey some of these positions and rationales. Following this, I focus on specific arguments taken by representatives of the Roman Catholic tradition. I do this not to privilege this tradition, but because its articulation of arguments is more prolific than most; it claims to be trying to be persuasive in a secular public sphere; its central positions are representative of strands of other religious as well as philosophical traditions; and it offers an ongoing critical conversation on both sides of the stem cell debate.

HINDUISM AND BUDDHISM

Representatives of the Hindu and Buddhist traditions have not been major players in the contemporary debate about stem cell research, at least not in the United States. There are historical reasons for this, no doubt, including the fact that neither is as yet a majority religion in the West. Yet there is much in both of these traditions that may, for their adherents, be relevant to issues of stem cell research. Given the complexity of the traditions, it remains for their scholars and spokespersons to render accessible to outsiders the foundational beliefs and moral concerns that may shed light on such issues. Briefly, however, some elements in these faith traditions can be noted.

Both Hinduism and Buddhism incorporate beliefs in reincarnation. This complicates issues of what can be done with human embryos. According to some Hindu traditions, human life begins prior to conception; the "soul" may be present even in sperm, or it may be in some other life form before and after its human existence. Human incarnation, however, offers a unique opportunity to influence the future of an individual. Great caution must be taken whenever actions are considered that may destroy this opportunity. On the other hand, for Hindus, a belief in reincarnation is also tied to motivation to be compassionate toward others. Good actions that will change the course of suffering for oneself and others are exhorted not only so that one can advance morally (and thereby improve one's karma) but because other people are worthy of beneficent deeds. Although the forms of compassion include nonviolence, there is in Hinduism also a broad tradition of sacrifice, wherein one human life can be taken for the sake of a higher cause. It is not impossible that these convictions provide a rationale for embryonic stem cell research.

For Buddhists, a goal of self-transcendence involves a process of self-forgetting. This undergirds a requirement for compassion, one of the Four-fold Holy Truths of Buddhism. In Mahayana Buddhism the ideal is the Bodhisattva, one who achieves self-emptying but then returns to help those in need. Again, here is a possible rationale for embryonic stem cell research. Moreover, some Buddhists believe that a soul has come choice as to where it will be incarnated. It has even been suggested that souls will not elect to embed in "spare" embryos that will only be destroyed. Hence, early embryos may not ever have a potential to become a human person and can therefore be used for some other purpose.

Whether or not such considerations are of overall significance in decisions to pursue embryonic or fetal stem cell research remains to be seen. At the very least, it can be said that they represent large concerns for the value of human life — whether at its inception or in response to later injury and illness.

ISLAM

Like other world religions, there are many schools of thought among Muslims. There are also multiple sources for moral guidance, including the Qur'an and its commentaries, the *hadiths* (a second source of moral indicators supplementary to the Qur'an), Muslim philosophies that range from a form of ontological realism to a version of divine command theory, Sharia (formulations of Islamic law), and centuries of juridical literature. Testifying before the U.S. National Bioethics Advisory Commission in 1999, the Islamic scholar, Abdulaziz Sachedina, attempted to provide some general insights from the tradition, taking into account the diverse interpretations of major Sunni and Shi'i schools of legal thought. His analysis of various texts led him to infer guidelines for stem cell research from the rulings of the Sharia on fetal viability and the moral status of the embryo.

Islamic traditions have given serious attention to the moral status of the fetus and its development to a particular point when it achieves human personhood with full moral and legal status. The early embryo has been variously valued in different eras, depending on the information available from science. Throughout the history of legal rulings in this regard, however, a developmental view of the fetus — according to a divine plan — has been sustained. This suggests that the moral status of personhood is not achieved at the earliest stages of embryonic life but only after sufficient biological development has taken place — the kind of development that includes a recognizable human anatomy and the possibility of "voluntary" movement. Most Sunni and some Shi'ite scholars therefore distinguish two stages in pregnancy. The first stage, pre-ensoulment, is human biological life but not yet human personal life. It is only after the fourth month (120 days, or the time of quickening) that the "biological human" becomes a "moral person." This is to say that at this stage of development the fetus achieves the status of a moral and legal person. In the first stage, the biological entity is to be respected, yet abortion is allowed for grave reasons. In the second stage, killing the fetus is homicide. The conclusion to be drawn from this is that there may be room for early embryonic stem cell extraction without violating divinely given laws or the embryo itself. In Islam, therefore, there are elements of the tradition that affirm embryonic stem cell research, though reasons can also be adduced against it.

JUDAISM

In the twenty-first century, Orthodox, Conservative, and Reform Jews take different positions on many issues of applied ethics. These views are not easily specified when it comes to stem cell research. What is possible here is to present the opinions of some Jewish scholars, with the caution that not all views are represented. In the context of debates about stem cells, Reform Judaism may be the clearest supporter of this research even when it entails the derivation of stem cells from human embryos or from aborted fetuses. Jewish law does not give legal status to the fertilized ovum outside a mother's womb. When the embryo is inside a uterus, it has legal status but not that of a human person. In general, however, this strand of Judaism favors the use only of so-called spare embryos, those that will be otherwise discarded.

Rabbi Elliott Dorff has identified theological assumptions useful for understanding a Jewish response to stem cell research. These include the following: moral discernment (of what God wants of God's people) must be based on both Jewish theology and Jewish law; all human beings are created in the image of God and are to be valued as such; human bodies belong to God and are only on loan to the individuals who have them; human agency is important in responses to human illness, so that both "natural" and "artificial" means are acceptable in overcoming disease. Indeed, there is a duty to develop and use any therapies that can help in the care of the human body. Yet, because humans are not, like God, omniscient, humility and caution are essential, especially when human science and technology press to the edges of human knowledge.

There are grounds in Jewish law and theology for permitting the derivation of stem cells both from aborted fetuses and from human embryos. Just as abortion (though generally forbidden) can be justified for serious reasons, so also can the use of fetuses for important research. A fetus *in utero* is considered not as a human person but as the "thigh of its mother." One is not allowed to amputate a part of one's body (in this case, one's thigh) except for good reasons (for example, to remove a gangrenous limb, save the life of a mother or remove other serious risk, and perhaps remove a fetus that is genetically malformed or diseased). When an abortion is thus justified, the abortus may be used as a source of stem cells.

Embryos can also be legitimate sources for stem cells extracted for purposes of research. During the first 40 days of gestation, *in utero* embryos (and fetuses) are only like "water." They are "non-souled," with only a liminal status. Although abortion during this time is permitted only for good reasons, the situation is different when embryos are not in the womb. Here there is no potential for embryos to develop into human persons; hence, they may be discarded, frozen, or used for promising research. Both Rabbi Dorff and Jewish scholar Laurie Zoloth argue that the duty to care for human bodies, the duty to heal, may undergird not just permission to pursue research on stem cells but an ethical duty to do so.

The Jewish tradition places all of these considerations in a wider context of responsibilities to community, fairness in distribution of benefits and burdens, and caution about the connections of stem cell research to invidious programs of eugenics. Ethical norms that safeguard the common good as well as the good of individuals provide boundaries for the development and future uses of stem cell research; they focus it, perhaps obligate it, but do not forbid it.

CHRISTIANITY

The diversity and pluralism to be expected in Christian responses to stem cell research mirrors to some extent the historical institutional divisions within Christianity. From the early schisms between East and West to the great Reformations in the West and the subsequent proliferations of Protestant denominations, to the Anglican and Roman Catholic church traditions, to the rise of new forms of Pentecostalism in the world's South, Christians have diversified in their beliefs and in their moral convictions and practices. There remains some important commonality in basic doctrines, largely expressed in creeds, and these are not irrelevant to issues of stem cell research. Examples include belief in divine revelation, particularly through sacred scripture; reconciliation made possible between humans and God through Jesus Christ; affirmation of the goodness of creation despite the damage sustained as the result of moral evil or sin; acceptance of the importance of human agency (in practice if not always in theology); a call to unconditional love of God and love and service of neighbor; a basic view of human equality; and the obligation to promote justice. Theological anthropologies vary importantly in terms of understandings of freedom and grace, virtue and sin. Strands of Christian moral theologies diverge when it comes to the basis and force of moral norms — depending primarily either on God's command or God's will manifest in creation itself and holding to absolute moral norms or only *prima facie* norms relativized according to circumstances, consequences, or priorities among aims. Law and gospel are important to each Christian tradition, though the emphasis on one or the other may vary.

Articulating an Eastern Orthodox view of embryonic stem cell research, Demetrios Demopulos begins with concern for the alleviation of suffering. "Medical arts" are to be encouraged, with the proviso that they be practiced as gifts from God ordered to divine purposes. Created in the image and likeness of God, people are destined for participation in the life of God — hence the telos of the human person is referred to as *theosis* or "deification." People are authentic insofar as they struggle to grow into this life; they remain "potential" persons until this is achieved. But even zygotes are potential persons in this sense; that is, they, too, are in a developmental process that ends in deification. Hence, human life is sacred from the beginning and at every step along the way, entitled to protec-tion and the opportunity to seek its destiny. Even though it is not yet a person, an embryo should not be used for or sacrificed in research. Correlatively, practices of *in vitro* fertilization that yield "surplus" embryos that will be discarded cannot be condoned.

Although the development of alternative techniques and sources is preferable, Demopulos allows the use of already harvested embryonic stem cells for research. Wishing that something had not been done will not undo it, he maintains. The compromise position, then, of continuing research on available cell lines is accepted as long as the research does not violate other norms of justice (by maximizing profits rather than the health of people, fostering trivial medical procedures for cosmetic purposes alone, or contributing to a questionable agenda for eugenics). The use of cadaver fetal cells is also acceptable if the abortion that provides them is spontaneous. But human life — potential human personal life — from zygote to a life beyond this world, remains sacred; what ultimately ends in God ought to be inviolable among humans.

Among the Protestant denominations in the United States, some oppose and some support embryonic stem cell research. Many churches do not yet have an official policy on this issue, though their representatives indicate a kind of majority view. Southern Baptists are as a group generally opposed to this research, though there is some openness to the compromise position of using already harvested cells. Some leaders among United Methodists, without an official church position, nonetheless asked President George W. Bush to oppose federal funding for embryonic stem cell research. Presbyterians in General Assembly articulated their support of research on cells extracted from embryos that would otherwise be discarded and affirmed the use of federal funds to make this possible. Acknowledging the significance of concepts like "potential personhood" and the need to "respect" embryos, they nonetheless determined that whatever form this respect takes, it ought not to have priority over alleviating the suffering of actual people. Members of the United Church of Christ tend to approve embryonic stem cell research as long as it is motivated by a clear and attainable healing benefit.

Ethicists from a variety of Protestant backgrounds (though seldom speaking for their denominations) also vary in their approaches to stem cell research. Ted Peters of the Pacific Lutheran School of Theology comes down on the side of supporting embryonic cell research because he believes that, on balance, concern for the human dignity not of embryos but of future real persons is more compelling. Ronald Cole-Turner agrees with others in the United Church of Christ by approving the extraction of stem cells from embryos, but he sets conditions for research in terms of justice issues. Gene Outka and Gilbert Meilaender, both Lutherans, share certain theological and philosophical convictions about the irreducible value of the early embryo. Both want to affirm the continuity of human life between early embryo and an actualized full person. This means that both will consider the taking of the life of an embryo as a violation of the prohibition against killing. Both also want to distinguish the question of embryo stem cell extraction from the question of abortion (in a variety of ways,

498

but primarily because in the former there is no direct conflict between a woman and a fetus).

Outka, however, finally approves of the use of some embryos for the derivation of stem cells. He does so by introducing two conditions that can exempt one from the absolute force of the prohibition. The first and most distinctive is the *nothing is lost* condition. That is, it is possible to relativize the prohibition against killing when those who are to be killed will die anyway. This is precisely the condition that characterizes the situation of spare (or excess) embryos (from *in vitro* fertilization procedures) destined to be discarded. The second condition is that *other lives will be saved* through this act of killing. Given the ultimate therapeutic aims of stem cell research, this condition is also present. It is the *nothing is lost* condition (or principle) that in the end allows Outka both to continue to affirm the status of the embryo as an end and not a means only and yet allow the taking of its life for purposes of research. It becomes, then, the basis of ethical acceptance of the derivation of stem cells from some human embryos and therefore a carefully circumscribed but nonetheless positive approval of embryonic stem cell research.

Meilaender, on the other hand, remains firm (though cautious and reluctant) in the conviction that embryonic stem cell research ought not to go forward, or better, that it is not morally justifiable to use human embryos as the source for stem cells. Referencing three other Protestant theologians (Karl Barth, John Howard Yoder, and Stanley Hauerwas), Meilaender poses three arguments: (1) People should be considered persons not on the basis of potentiality or actuality, not on the basis of capacities, but simply as members of the human community. To think inclusively about the human species should lead us to honor the dignity of even the weakest of living human beings — the embryo — and thus be able to appreciate the mystery of humanity and the mystery of our own individuality. (2) Immediately to opt for embryonic stem cell research is to prevent us from finding better solutions to the medical problems this research is designed to address. Given the ethical compromises involved in following this "handy" route, it would be better to "deny ourselves" this remedy and look to ones that do not denigrate the dignity of people however and wherever they live. (3) The church should bear witness to its beliefs by refusing to make sophistic distinctions such as those between funding embryonic stem cell research and funding the "procurement" of these cells. Meilaender is skeptical that only excess embryos or available cell lines will be used.

Along with evangelical Protestants, Roman Catholic voices have been prominent in the debates surrounding embryonic stem cell research. The position most frequently articulated in the public forum is that of opposition to the derivation of stem cells from embryos. This is the position of the Catholic hierarchy, promulgated through both Vatican and national episcopal channels. It is a position ably supported theologically by spokespersons such as Richard Doerflinger. Yet it is not the only position espoused by Roman Catholic ethicists and moral theologians or by all Roman Catholics. The alternative position is basically the "14-day" position, or

the theory that an embryo in the first 14 days, prior to implantation in the uterus, is not yet even a "potential person," and hence need not be protected in the same way as human persons (or even the same as fetuses, considered potential persons after implantation). Neither of these positions goes the route of declaring absolutely that the early embryo is a person, or the route of approving the exploitation of fetuses *in utero* in later stages of development. Since the debate between these two positions reflects similar philosophical and theological debates in other religious traditions and in the public forum, it is worth looking at it in closer detail.

Roman Catholic Contributions and the 14-Day Theory

Before turning to the technicalities of the 14-day theory, it should be noted that Roman Catholics (including the official leaders of the church) tend to worry, like people in other traditions, about issues of justice, ecology, and the well-being of the whole Earth. Apart from the moral status of the embryo, Catholic concerns are focused on questions of equity in the shared lives of persons across the world. For all the pressures to go forward with embryonic and fetal tissue stem cell research, there is little assurance anywhere that the benefits of this research will be shared in the human community among the poor as well as the wealthy, across racial and gender lines, and in countries of the world's South as well as North. Moreover, as the research goes forward (for it surely will), some safeguards will need to be in place to keep the goals of the research focused on the healing of human persons rather than on the commodification of human bodies, their tissues and cells. And the specter of human genetic engineering for enhancement purposes (and not only the treatment of diseases and injuries) is never far from the horizon. The other side of tremendous positive advances of modern technology is not to be ignored, not even if our present anxieties lead us desperately to seek means to lift the burdens of one generation or one group of people.

It is important to take preliminary note, also, of the shared community of discourse among Roman Catholic scholars, church leaders, and people. No matter how divided they may be among themselves — regarding the moral status of the early embryo or other particular moral questions — common moral convictions are nonetheless expressed in common language. The Catholic tradition is undivided in its affirmation of the goodness of creation, the role of human persons as agents in cooperation with ongoing divine creative activity, the importance of not only the individual but the community, the responsibilities of human persons to promote the health and well-being of one another. At the same time, Catholics can disagree profoundly; a key example of disagreement is the debate about the status of the embryo and the moral evil or goodness of extracting stem cells from it in a way that leads to its demise.

The debate within the Catholic community regarding stem cell research incorporates opposition over something like the 14-day theory because the Roman Catholic moral tradition has consistently embraced a form of moral realism. At the heart of this tradition is the conviction that creation is itself revelatory. Embedded in an ultimately intelligible (though only partially so to humans) created reality is the possibility of perceiving and discerning moral claims of respect in response to every entity, especially human persons. This is what is at stake in the Catholic tradition of natural law. Although there are historical aberrations and contradictions in this long tradition, for the most part, natural law has not meant that morality can simply and fully be "read off" of nature, not even with the help of the special revelation of the Bible (though this is a help). What natural law theory does is tell persons where to look — that is, to the concrete reality of the world around us, to the basic needs and possibilities of human persons and to the world as a whole. "Looking" involves discerning, deliberating, structuring insights, interpreting meanings, taking account not only of what is similar among entities but also what is particular in their histories, contexts, relationships.

Roman Catholic thinkers engaged in discernment about human embryonic stem cell research tend, therefore, to "look" to the reality of stem cells and their sources — that is, human embryos and aborted fetuses, as well as adult tissues. All answers are not in the Bible, official church teachings, or individual experiences of the sacred. Discernment incorporates all of these sources of insight, but it also requires the knowledge available from the sciences, philosophy, and whatever other secular disciplines provide some access to the reality that is being studied. The ongoing intensity of the debate about the status of embryos bears witness to the notion that not all inquiries can be settled by one discipline or one interpreter or within one epistemological perspective.

The argument against procuring stem cells from human embryos is primarily that it entails the death of the embryos. No one disputes this fact, but disagreement rages as to whether this death can be justified. Those who oppose its justification argue that for each human person there is a biological and ontological continuum from the single-cell stage to birth, to whatever one has of childhood and adult life, and then to death. Since a new and complete genetic code is present after fertilization, there is already a unique individual human person, potential in an important sense but concretely and really (inherently) already directed to full actualization. The zygote itself, as a living organism, is self-organizing. The moral status of the early embryo is, therefore, that of a human person; it does not achieve this status by degrees or at some arbitrary point in development; nor is this status simply bestowed upon it by social recognition or convention.

But if the embryo has the status of a human person, then killing it even in order to treat the illnesses of other people cannot — from this Catholic point of view — be justified. No human person, not even an inchoate one like an embryo, can be reduced to a pure means in relation to other ends. Creating or using embryos for research purposes is wrong because it treats this distinct human being as a disposable instrument for the benefit of someone else. Those who oppose the extraction of stem cells from embryos on the grounds that it involves

the destruction of a human person are not unsympathetic to the suffering that may be alleviated ultimately through research on stem cells. Part of the argument against the use of embryo sources for stem cells rests on the identification of alternative sources (particularly adult cells).

On the other side of this debate are Catholic moral theologians who do not consider the human embryo in its earliest stages (prior to the primitive streak or to implantation) to be a human person, potential or actual. They hold the same meaning for "potency" as do their opponents — that is, the Aristotelian and Thomistic meaning, signifying not an extrinsic or "sheer possibility" but rather an intrinsic principle already as such actual within a being, directing it to individualized species-specific development (even though the being may not yet be developmentally fully actualized). These moral theologians argue, then, precisely *against* the view that the early embryo has this inherent potency to become a person. Their argument harks back to a centuries-old Catholic position that a certain degree of biological development is necessary in order for the human spirit (or soul) to be embodied. The conceptus, in other words, must have a baseline organization before it can embody a human person; it must embody the potentiality as well as the actuality that will lead to the formation of a human person. Although previous theories of embryological development were grounded in minimal (and to a great extent, erroneous) scientific evidence, proponents of embryonic stem cell research find in contemporary embryology insights that tend to support this developmental theory rather than defeat it.

Australian Catholic moral theologian Norman M. Ford has presented perhaps the most detailed argument in support of the view that an early embryo is not a potential human person. The plausibility of "delayed personhood" is based on scientific evidence that suggests that the embryo prior to implantation is not yet a self-organizing organism. Fertilization itself does not take place in a "moment," as is sometimes rhetorically claimed; it comes about through a process that takes approximately 24 hours, finally issuing in the one-celled zygote. When the zygote then divides, there comes to be not a definitive, singular organism, but a collection of cells, each with a complete genetic code. In other words, the first two cells, even if they interact, appear to be distinct cells, not a two-cell ontological individual. The same is true as division continues to four, eight, sixteen cells, each with its own life cycle and nutrients for sustaining its life. Cells multiply and differentiate as they are gradually specified in their potential. Some cells form membranes that will finally enclose an organized individual entity approximately 14 days after fertilization. It is only then that a unified being can be said to be "self-organizing." Until this time, cells are "totipotential" in that they can become any part of what may ultimately be a human being or even a human being as a whole.

The genotype of the zygote does not have a built-in blueprint for development, as if it were a miniature person. Only with the formation of a single primitive streak (after implantation) does the totipotency of the embryo actually become a potency to develop into one human person. As Ford puts it, fertilization is not the beginning of the development *of* the human individual but the beginning of the formative process and development *into* one (or more) human individuals. Many who find this argument plausible tend to express it more simply: Since at its early stages an embryo can twin and recombine, there is not here an individualized human entity with the inherent settled potential to become a human person. Critics of the position respond that *if* an embryo does *not* twin or recombine, then it may be supposed that an individual is already existing. Ford's analysis goes further than this, however, suggesting that the very undifferentiation of cells prior to implantation disallows the identification of individuality, whether or not there is twinning and recombining.

In addition to the argument from lack of individuation, some theologians have noted that the high rate of spontaneous early embryo loss undermines the credibility of the claim that human individuals begin at fertilization. In unassisted human reproduction, the development and loss of early embryos is estimated to be as high as 50 to 80%. From a Catholic theological point of view, it seems unbelievable that more than half of the individual human persons created by God should populate heaven without ever having seen the light of day in this world. This is a speculative observation, however, and it is generally not used today by those who favor embryonic stem cell research.

There are, then, two opposing cases articulated within the Roman Catholic tradition. This need not leave Catholics (or others) with a kind of "draw." Rather, it opens and sustains a significant conversation that takes account of scientific discoveries as well as of a larger set of theological and ethical insights. Moral theologians and ethicists attempt to provide reasons for their views that will be open to the scrutiny of all. Both sides claim a certain amount of epistemic humility, since there are only degrees of certainty available regarding the ontological status of the early embryo. Indeed, even official church documents acknowledge that there is no definitive answer to the question of when human individual life begins. In 1974, the "Declaration on Procured Abortion" stated in a footnote that the question of the moment when the spiritual soul is infused is an open question. It acknowledged that there is not a unanimous tradition on this point, and authors are as yet in disagreement. In 1987, an instruction on "The Gift of Life" admitted that its authors were "aware of" current debates concerning the beginning of human life, the individuality of the human being, and the identity of the human person. And Pope John Paul II made no decision on the question of delayed ensoulment (or "hominization") in his 1995 encyclical letter, "Evangelium Vitae."

Uncertainty is dealt with in different ways by those who oppose and those who support embryonic stem cell research. Those who oppose it argue that even if the presence of the "personal soul" in an embryo is only "probable," this suffices to prohibit *risking* the killing of a human person. In other words, probability (if not certitude) warrants the safer course. Hence, the embryo from fertilization on is to be *treated as* a human person, with the kind of unconditional respect due to all members of the human community.

For those who conclude that embryonic stem cell research can be permitted, uncertainty works the other way. First, the probability of an early embryo's actually representing an individualized human being is much lower than is argued by opponents of embryonic stem cell research. Although there is not absolute certitude to be gained at this point in time from scientific evidence, the weight of the evidence *against* there being an individualized human being (incarnated in an embryo prior to implantation) is greater and more persuasive than the evidence for it. Moreover, the low level of uncertainty (for those who hold this position), when placed in relation to the prospect of great benefits for human healing (and some assurance that these benefits are not promised unrealistically) makes it possible — without yielding to a full-blown utilitarianism — to justify using early embryos as sources for stem cells for research. To *prohibit* all such research on the basis of a low probability of fact, or on a theological stipulation of moral status in the face of questionable accuracy of appraisal, seems itself not ethically justifiable.

Neither side in this debate wants to sacrifice the tradition's commitments to respect the dignity inherent in human life, promote human well-being, and honor the sacred in created realities. When a move forward is advocated for embryonic stem cell research, it need not soften the tradition's concerns to oppose the commercialization of human life and to promote distributive justice in the provision of medical care. The ongoing Roman Catholic conversation on all of these matters can be of assistance to others in a pluralistic society as long as it remains open to wider dialogue and respectful of all dialogue partners, while retaining its own integrity. This is probably a lesson for all religious and secular traditions.

KEY WORDS

Moral realism   Theory of knowledge that affirms the possibility of at least partial correspondence between what is in reality and what is in the mind.

Ontological   Having to do with the structure of a being.

Primitive streak   The initial band of cells from which the embryo begins to develop; it is present at approximately 14 days after fertilization.

FURTHER READING

Ford, Norman M. (2002). *The prenatal person: ethics from conception to birth.* Blackwell Publishing, Oxford.

Holland, S., Lebacqz, K., and Zoloth, L., eds. (2001). *The human embryonic stem cell: science, ethics, and public policy.* MIT Press, Cambridge, MA.

National Bioethics Advisory Commission. (2000). *Ethical issues in stem cell research.* Vol. III. National Bioethics Advisory Commission, Rockville, MD.

Outka, Gene. (2002). The ethics of human stem cell research. *Kennedy Inst. of Ethics Jour.* **12**, 175–213.

Shannon, Thomas, and Walter, James. (1990). Reflections on the moral status of the pre-embryo. *Theological Studies* **51**, 603–626.

Warnock, Mary. (1984). *A question of life: the Warnock Report on Human Fertilization and Embryology.* Basil Blackwell, Oxford.

Stem Cell-based Therapies: FDA Product and Preclinical Regulatory Considerations

Donald W. Fink and Steven R. Bauer

Introduction

Stem cell-based therapies are cellular constructs consisting of, or derived from, populations of stem cell progenitors. They represent examples of complex and dynamic biological entities. Stem cells are endowed with the capacity for expansion, through self-renewal, and with the capacity for differentiation into a variety of cell types that is limited primarily through constraints imposed by the lineage-forming potential of the stem cell progenitors. Once transplanted into patients, stem cell-based preparations are expected to interact intimately with and be influenced by the physiologic milieu of the recipient. While elucidation of the molecular mechanisms that control stem cell self-renewal and commitment to any of a variety of cell fates poses vexing challenges to the field of developmental biology, it is, nonetheless, the responsibility of the Food and Drug Administration (FDA) to confront these complexities as the agency conducts oversight of stem cell-based therapies. Numerous fundamental scientific questions pertaining to regulation of gene expression, cell–cell interaction, and control of cellular differentiation remain unresolved. Accordingly, the science-based regulatory approach adopted by the Center for Biologics Evaluation and Research/Office of Cellular, Tissue, and Gene Therapies (CBER/OCTGT) is consistent with practices in evidence throughout the FDA. This regulatory approach is designed to facilitate progress in the development of new biological treatments ensuring to the greatest extent possible the safety of patients enrolled in investigative clinical studies conducted to support potential future licensure.

Developing Recommendations for the Manufacture and Characterization of Stem Cell-Based Products

An Investigational New Drug application (IND) must be compiled and submitted to CBER/FDA for evaluation prior to the initiation of a clinical trial in humans designed to generate information pertaining to the safety and efficacy of a previously untested stem cell-based therapy. An IND is a request for authorization from the FDA to administer an investigational drug or biological stem cell-based product to patients. Permission to administer an investigational stem cell-based therapy may be granted only upon thorough assessment of an IND application. Among the essential items included in the IND submission document is detailed information that describes the manufacture and characterization of the investigational stem cell-based therapy along with a compilation of results obtained from preclinical studies conducted in animals. Due to the inherent complexity of stem cell-based therapies coupled with the continuous accumulation of science-generated information about their biological properties, individual investigators and corporate sponsors alike are encouraged to request a meeting with CBER/FDA before submitting a completed IND application for evaluation. Within the context of a pre-IND meeting, sponsors are able to receive input from CBER/FDA staff that represents current thinking with respect to regulatory expectations that are to be addressed by the information provided in an IND application. Recommendations and requirements put forth by the FDA for characterization of stem cell-based therapies are derived from a variety of sources, including product-specific regulations stipulated in the Code of Federal Regulations (CFR), relevant guidance documents, and direct interactions with the scientific community. The legal authority for FDA to regulate stem cell-based biological therapies is derived from two congressional acts, the Federal Food, Drug and Cosmetic Act (FD&C Act) and the Public Health Services Act (PHS Act).

CODE OF FEDERAL REGULATIONS (CFRS)

The CFRs represent the structural backbone of FDA's regulatory requirements. Among the most appropriate regulations applicable to stem cells are those that outline requirements for submission of an investigational new drug application (21 CFR 312) and adherence to current good manufacturing practices (21 CFR 210 & 211). During new product development, the objective intent of this body of regulations is to ensure, to the greatest extent possible, that patients enrolled in clinical trials receive investigational stem cell products that primarily have an acceptable safety profile and, in addition, possess evidence of a relevant biological activity (potency) that could possibly convey a therapeutic effect. Ultimately, the goal of performing clinical trials is to collect sufficient safety and efficacy data in accordance with regulatory requirements to permit licensure of a biologic therapy, thus allowing marketing and/or widespread use of a product in patients.

RELEVANT GUIDANCE DOCUMENTS

Regulations specified in the CFRs are intended to be applicable to a diversity of therapeutic products. In contrast, guidance documents are often developed for the purpose of reflecting FDA's current thinking and expectations concerning development of specific product classes. In addition to its own efforts, FDA interacts with outside regulatory organizations and participates in more global initiatives such as the International Conference on Harmonization (ICH), resulting in development of instructive documents that pertain to the manufacture and clinical testing of most types of biological products. The principles defined in a number of these collaboratively developed guidance documents are applicable to the evaluation of stem cell-based therapies.

SCIENTIFIC INTERACTIONS

To facilitate the development of guidance documents and regulatory review approaches that are pertinent to the development of stem cell-based biological therapies, CBER/OCTGT review staff participates in regularly scheduled internal meetings for the purpose of discussing relevant scientific issues pertaining to specific biologic products and to promote consistency in their regulation. In addition, the OCTGT staff is involved in various outreach activities, including regulatory and scientific society meetings where interested parties are afforded the opportunity to interact directly with individuals involved in the regulation of stem cell-based therapies. FDA staff and members of the regulated community benefit mutually from interactions that occur outside of the formal regulatory process. These exchanges allow for an enhanced understanding of and greater appreciation for each other's perspectives. A number of CBER/OCTGT review scientists also maintain active laboratory-based research programs in cellular or developmental biology as well as in other areas relevant to the regulation of stem cell-based therapies. This staffing approach, designated the research–reviewer model, is critical to FDA's effort to sustain state-of-the-art scientific expertise and credibility as it conducts the complex task of regulating stem cells and developing regulatory policy that affects the stem cell therapy field.

The experiences of academic clinical institutions and corporate entities involved in the development of stem cell therapies play a key role in shaping FDA recommendations pertaining to the regulation of stem cell-based products. Experience communicated through meetings with CBER/OCTGT staff and presentations made at scientific society meetings or in front of CBER's Cellular, Tissue and Gene Therapy Advisory Committee (CTGTAC), formerly known as the Biological Response Modifiers Advisory Committee (BRMAC), are vital to informing the regulatory process. Within the context of a CTGTAC meeting the FDA is able to obtain advice in a public forum on scientific issues that may impact its approach to the regulation of stem cell-based therapies.

Another important scientific resource used to facilitate development of recommendations pertaining to the regulation of stem cells involves interagency interactions between FDA and the National Institutes of Health (NIH) through formalized Memoranda of Understanding agreements (MOUs) that allow for information exchange on the subject of translational research. The FDA benefits from the leveraging of scientific expertise provided by NIH extramural program managers and intramural research scientists, while NIH derives benefit from FDA's advice concerning the development of important NIH grant funding initiatives designed to expedite translation of research from bench to bedside. These MOU agreements serve the FDA-regulated community, the NIH-funded scientific research community, and the public by permitting the focusing of critical resources on development of novel technologies such as stem cell therapies to address existent challenges to the public health as well as unmet medical needs.

In summary, FDA is the recipient of input from a variety of sources facilitating formulation of recommendations that pertain to the evaluation of stem cell-based therapies for evidence of safety and effectiveness. Guidance documents that contain recommendations encompassing the agency's current thinking with respect to the regulation of cellular therapies, including those composed of stem cells, are constructed so that they might evolve with continuing advances made in technology and the accumulation of regulatory experience. In making specific decisions pertaining to investigational new drug applications for stem cell-based therapies, the FDA takes into consideration the potential risks versus any benefit that may be realized for each stem cell-based therapy and proposed clinical trial.

Critical Elements for Developing a Safe Stem Cell-Based Therapy

The current, comprehensive approach CBER uses to evaluate investigational stem cell-based therapies is depicted schematically in Figure 67-1. Captured thematically are the essential elements that must receive consideration in order for

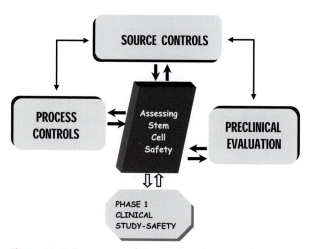

Figure 67-1. Representation of elements deemed necessary for assessing the safety of novel human embryonic stem cell-derived cellular therapies.

CBER/OCTGT review staff to perform an adequate safety assessment of a stem cell-based therapy prior to authorizing permission for initiation of a Phase 1 clinical safety trial. The multifaceted review process encompasses the primary objective of assuring the safety of stem cell-based cellular therapies. Beginning with "Source Controls," every step involved in producing a stem cell-based cellular therapy intended for clinical testing is subjected to careful scrutiny. At its core, assuring safety of a stem cell-based biologic cellular therapy is all about establishing and maintaining control over every facet involved in the process of producing a stem cell-based preparation intended for clinical study. Only in this manner is the degree of safety maximized and consistency from one cell preparation to the next assured.

Elements of this comprehensive approach are not to be viewed as isolated components but rather as interrelated with each of the other elements that constitute the review framework. Acquisition of the starting biological materials obtained from qualified consented donors, derivation, expansion, manipulation, and characterization of established stem cell lines to generation of differentiated cellular preparations and preclinical proof-of-concept/toxicological testing conducted in appropriate animal models are among the items considered in a contiguous fashion. An example of the interrelated nature of this review approach is the principle of traceability, or tracking. For those situations when an unanticipated adverse event occurs during the course of an investigational clinical trial, tracking allows investigators and study reviewers alike the opportunity to trace back from the patient through the stem cell preparation process all the way to the initial acquisition of biologic materials used to generate the founder stem cell

cultures. It is through tracking that an unexpected adverse event may be evaluated for its relationship to clinical, manufacturing, and preclinical testing issues allowing informed decisions to be made that assist in ensuring continued patient safety.

MANUFACTURING AND CHARACTERIZATION ISSUES PERTAINING TO STEM CELL PRODUCTS

One primary objective of FDA's oversight of stem cell-based therapies is to ensure the safety of patients who will receive an investigational product not previously tested in humans. Strict attention to the details of product manufacturing and characterization is crucial to safety assurance. The information provided in this section describes various recommendations currently applied by CBER/OCTGT reviewers to ensure that investigational stem cell products administered to patients during the course of a clinical trial meet agency expectations in terms of safety, purity, identity, and potency. Figure 67-2 illustrates a number of the complexities inherent in stem cell biology and summarizes a sampling of the challenges associated with development of stem cell-based therapies.

Control of Source Materials Is the First Step in Safety Assurance

Demonstrating an acceptable safety profile for an investigational stem cell-based therapy prior to commencing an investigative clinical study begins at the stage of tissue donor qualification. In the case of products composed of cells derived from human embryonic stem (hES) cells, this encompasses evaluation of donors providing egg and sperm used to generate embryos from which embryonic stem cells are to be

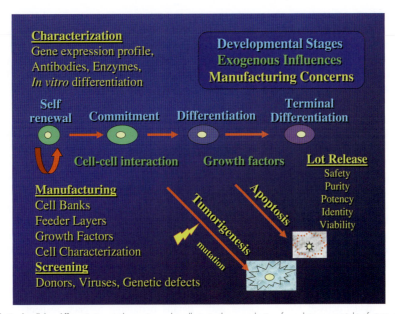

Figure 67-2. A hypothetical cellular differentiation pathway is used to illustrate the complexity of regulatory oversight of stem cell products. Mechanisms that control biologic processes such as self-renewal or lineage commitment include changes in gene and protein expression that lead to differentiation (steps in development, blue text) and interactions with the host through microenvironmental cell–cell interactions or growth factors (influences on development, green text). Regulatory concerns associated with these processes during *in vitro* stem cell manufacture are shown (yellow text).

isolated. Description of the specific sources from which stem cells are acquired should include the tissue of origin and cell type, such as hematopoietic, neuronal, or embryonic stem cells, along with details indicating whether or not donor cells are obtained following mobilization or *in vivo* activation.

Determining Donor Eligibility. Preventing the introduction, transmission, and spread of communicable diseases is one of the FDA's principal public health obligations. Biologic therapies comprised of stem cells pose a potential risk for transmitting communicable disease from donor to recipient due to their nature as derivatives of the human body. As described in the FDA's recently published final rule on the eligibility determination for donors of human cells, tissues, and cellular and tissue-based products, acceptance of tissues obtained from allogeneic (non-self) donors as suitable source material for producing stem cell-based products depends principally on screening and appropriate testing of donors for human infectious diseases including, at a minimum, human immunodeficiency virus types 1 and 2, hepatitis B virus (surface and core antigen), hepatitis C virus, treponema pallidum (syphilis), human T-lymphotropic virus types I and II, and cytomegalovirus (CMV). FDA-licensed or FDA-approved test kits are to be used to perform this testing, when available. Within the autologous setting, where patients serve as their own stem cell donors, testing for infectious disease or adventitious agents is relevant if *ex vivo* culture methods are used during processing that could promote propagation of an adventitious agent within a given stem cell product or increase the risk for cross-contamination of other biological products manufactured within the same facility.

In addition to testing for evidence of infectious diseases, IND applications to the agency should include a description of other serologic, diagnostic, and clinical history information obtained for the donor. Furthermore, it may be important to conduct other donor characterizations such as typing for genetic polymorphisms and major histocompatibility complex (MHC) loci that may play a role in determining whether or not transplanted stem cell products engraft successfully or could fail due to donor-host immunologic incompatibility.

Determining the eligibility for raw biologic materials, namely, unfertilized eggs and sperm, used to generate human embryonic stem cells (hES) derived from the inner cell mass of 5- to 7-day-old blastocysts can prove more challenging. In circumstances when freshly obtained blastocysts are designated to serve as the starting source material, volunteer donors themselves may be screened for high-risk behaviors and tested serologically for evidence of infectious disease using FDA licensed test kits. When cryopreserved fertilized embryos serve as source material and the original tissue donors are no longer available for communicable disease testing, cultures of undifferentiated hES cells themselves may be assessed directly for evidence of infectious disease once cell populations have expanded to numbers that are sufficient to support sampling. Archiving of donor-derived blood samples will permit screening for additional disease markers as new information and diagnostic techniques become available, thus providing an added measure of safety assurance.

In addition to screening for infectious diseases, a role for molecular genetic testing might one day be integrated into an overall assessment program to establish the suitability of particular hES cell preparations for use in developing a stem cell-based therapy. Information obtained from this type of evaluation could serve to determine whether a preparation of hES cells represents the best-qualified source material for generating a cellular therapy in the context of a particular disease setting. As a hypothetical example, genetic testing could, in theory, detect mutations in specific target genes such as α-synuclein that have been linked to autosomal dominant familial Parkinson's disease. Detection of such genetic anomalies in a population of hES cells could negate their use in generating neuronal progenitor cells for treating a number of neurodegenerative conditions, including Parkinson's disease. Admittedly, at this point a relatively small number of genes are known to be directly responsible for causing disease in a monogenetic fashion or to operate as suspected contributors to aberrant physiologic function. Undoubtedly this number will grow, perhaps substantially, due to advances in techniques for identifying, isolating, and analyzing genes, coupled with an increasing abundance of information certain to become available as one outcome of the human genome sequencing projects. In addition, it is expected that a great deal more will be learned as to how multiple gene products, each contributing incrementally to an overall outcome, will predispose individuals to development of certain diseases. Barring significant improvements to existing technology, it is impractical to expect screening of every donor or hES cell preparation for the entire panoply of disease-associated genes. In this context, molecular genetic testing would be conducted only for targeted genes that are relevant to a specific clinical indication. It is important to note that, at present, molecular genetic testing is not expected by the FDA as part of the approach used to evaluate of the safety of an investigational stem cell-based therapy intended for use in a clinical trial.

Manufacturing Components and Procedures. The approach adopted by CBER/OCTGT for regulation of biologic products, including stem cell-based therapies, involves scrutiny of the manufacturing process used to produce the stem cell-based therapy and analysis of final product testing. Often the production of a stem cell-based product requires a multistep process that must be performed using aseptic techniques. An error in the execution of virtually any step could impact the safety, purity, and potency of the stem cell therapy along with allowing for the introduction of undesired adventitious agents. CBER/OCTGT review staff analyzes all procedures used during the production, purification, and harvest of a cellular therapy product. Details describing methods used for cell collection, processing, and culturing conditions, including mechanical or enzymatic dispersion, cell selection methods, and a description of any cell culture system, are important. In addition, CBER/OCTGT review staff examines details pertaining to each reagent used to manufacture a product. Examples include materials that are essential for promoting cellular growth and eliciting differentiation, and those used for selecting target cell populations, specifically, bovine serum, trypsin, peptide growth factors, cytokines, monoclonal antibodies, antibiotics, cell selection devices, and culture media.

An issue that receives considerable attention is the inclusion of bovine-sourced serum or other critical components as media supplements due to the risk for contamination by the causative agent of bovine spongiform encephalopathy (BSE). A hierarchy of options exists to safeguard against inadvertent transmission of BSE to patients through transplantation of contaminated stem cell-based therapies. First, cell cultures may be maintained in medium supplemented with bovine serum once it is demonstrated that the serum reagent is produced from donor cows reared for the entirety of their lives in herds maintained in countries certified to be free of BSE. A second alternative is to replace bovine serum with clinical grade serum sourced from humans. In this case, testing of the serum donors or final pooled serum product for infectious human diseases is critical. The third option is development of a chemically defined, serum-free medium, thus obviating altogether risks associated with serum supplementation. Use of FDA-approved reagents, when available, is encouraged, however, when this is not possible, additional testing of a reagent may be necessary to ensure its safety and purity. Add-on testing may include assessment of sterility, endotoxin levels, mycoplasma, and the presence of adventitious agents, along with functional analysis, evaluation of purity, and other analytical tests designed to demonstrate the absence of potentially harmful substances.

Derivation of Stem Cell Lines

General Principles. An IND application should contain information about the origin of any stem cells used to produce an investigational therapy and include the source tissue, manner of collection, and methods for cell expansion. Demonstrating rigorous control of standardized practices and procedures used during isolation, derivation, and maintenance of stem cell lines ensures the integrity, uniformity, and reliability of candidate stem cell-based therapies. When a stem cell therapy is derived from a nonhuman source, it meets the definition of a xenotransplantation product. As defined in the Public Health Service (PHS) guideline on this topic, xenotransplantation is any procedure that involves the transplantation, implantation, or infusion into human recipients of either (a) live cells, tissues, or organs from a nonhuman animal source, or (b) human body fluids, cells, tissues, or organs that have had *ex vivo* contact with live nonhuman animal cells, tissues, or body organs. Guidance documents provided by the agency describe additional testing requirements necessary to address potential risks posed to humans through the use of xenotransplantation products. Moreover, details that pertain to animal husbandry practices, herd health surveillance, adventitious agent testing of derivative stem cell products, and archiving of animal and patient specimens are all taken into consideration by the agency in association with the use of xenotransplantation products.

Finally, once a stem cell line is established, a description of prospective testing used to characterize that stem cell line should be provided in an IND submission to FDA/CBER. A panel of tests that includes, but is not limited to, the following parameters is recommended: (1) for hES cell lines, demonstration of pluripotency, (2) karyotypic and chromosomal analysis, (3) assessment of proliferative potential and growth characteristics, (4) evaluation for the expression of markers indicative of undifferentiated stem cells versus differentiated derivatives, and (5) analysis of the stability of cell lines over time in tissue culture and following prolonged periods of cryopreservation storage.

Human Embryonic Stem Cells. The derivation of an hES cell line requires that additional considerations be taken into account. Because hES cells represent a complex, dynamic biological entity, failure to standardize procedures for expansion and maintenance of cultured hES cells could result in unintended alterations of their intrinsic properties. For example, the density of the initial cell seeding, the frequency with which culture medium is replenished, the quality of the reagents used to supplement the culture media, and the density cell cultures are permitted to achieve prior to subdividing could all have an impact on the characteristics of cultured hES cells. Standardization of procedures used to acquire egg and sperm from eligible volunteer donors, perform *in vitro* fertilization, and isolate cleavage-stage embryos is important to the generation of quality, undifferentiated hES cell cultures. When cryopreserved embryos serve as the starting source material for deriving self-renewing hES cell cultures, it is essential that critical information is documented and that records are maintained regarding the reagents and procedures used to perform embryo cryopreservation. This is the case particularly when more than one infertility clinic is providing cryopreserved embryos as source material for hES cell derivation. Differences in the concentration and type of cryoprotectant used, rate of initial cell freeze down, and routine length of cryopreservation storage may result in variability that precludes consistency from preparation to preparation. Perhaps the most significant issue with respect to the use of frozen fertilized embryos is specification of prospective criteria used to ascertain the acceptability of a thawed, cryopreserved embryo as source material for generating hES cell cultures.

Historically, undifferentiated hES cells teased from the inner cell mass of cleavage-stage embryos by mechanical or enzymatic means have been propagated on layers of irradiated, nonreplicating feeder cells frequently comprised of fibroblasts obtained from fetal mice. This is because simply adding growth factors and cytokines to the culture media appears insufficient to maintain hES cells in an undifferentiated form, and thus the presence of the underlying feeder cells is required. Accordingly, as indicated earlier, stem cell therapies derived from hES cells cultured on layers of nonhuman cells meet the definition of xenotransplantation products and are subject to requirements specified in various guidance documents pertaining to the regulation of xenotransplantation products. This circumstance persists even if hES founder cell lines are migrated from nonhuman feeder layers to feeder-free conditions during the process of developing a stem cell-based product. Typically, undifferentiated hES cells are expanded and maintained on irradiated murine fetal fibroblasts. Regardless of the species of origin for hES cell culture feeder layers, detailed information should be supplied that indicates whether or not feeder cells are derived from primary cell cultures or established cell lines, describes maintenance of the feeder cells, provides an assessment of feeder line stability (if derived from established cell lines), and includes details regarding the establishment of feeder cell banks (the passage number for cells stored as a cell bank as well as the passage number for cells used during production of the stem cell product).

Histocompatibility constitutes an additional safety issue meriting consideration with respect to the characterization of cell lines established following differentiation of hES cells. Initial reports suggested that hES cells derived from the inner cell mass of blastocyst stage embryos do not express immune-recognition proteins, raising the possibility that hES cells might be immuno-privileged and therefore unrecognized by the recipient's immune system. This engendered the hope that therapies derived from hES cells would be immunologically silent and thus remain undetected by the recipient's immune surveillance system following transplantation. However, low-level expression of major histocompatibility complex (MHC) class I proteins on the surface of hES cells has been reported, thus providing evidence that it might not be the case that hES cells are or remain immunologically inert. MHC class I proteins are involved in tissue rejection mediated by cytotoxic T lymphocytes. In addition to the detection of low levels of MHC class I proteins on the surface of hES cells, moderate increases in their expression are noted in conjunction with *in vitro* or *in vivo* differentiation. At this point it is uncertain whether the level of MHC class I expressed in hES cells is sufficient to elicit a vigorous, let alone even a tepid, immunologic rejection response. However, it does appear that characterization of MHC profiles for stem cell therapies derived from hES cells is warranted.

Testing of Stem Cell Products: Safety, Identity, Purity, and Potency

A challenge to the evaluation of stem cell-based therapies is accurate analysis of product safety, identity, purity and potency. Testing of process intermediates and final stem cell products intended for patient administration is comprehensive and includes, but is not limited to, microbiological testing (sterility, mycoplasma, and adventitious viral agents), as well as assessments of various product parameters such as unique product identity, purity (freedom from contaminants introduced during the manufacturing process), viability, and potency (biological activity). Appropriate analytical tests selected from this testing battery should be performed at various points throughout the process of manufacturing a stem cell product and thus provide a check on the process itself in order to ensure the quality and consistency of stem cell products produced over time.

Identity and purity testing may involve biologic, biochemical, and/or functional characterization. Tests to demonstrate stem cell product identity and purity should be designed

to determine the degree to which the characteristics of a manufactured stem cell-based product conform to expected and desired properties. Potency, as defined in the Code of Federal Regulations, is the "specific ability of the product, as indicated by appropriate laboratory tests . . . to effect a given result." In many circumstances, the mechanism of action for a stem cell-based therapy is unknown. However, the outcome of testing for product potency is expected to correlate well with a desired clinical effect. Because stem cells are expected to undergo additional differentiation prior to mediating a measurable clinical effect, the problem of deciding what to evaluate as an index of potency and how to correlate particular measurements with a stem cell product's final outcome is a notable challenge. Since direct measurement of the potency of a stem cell-based product may prove daunting because of a lack of appropriate *in vitro* or *in vivo* assays, alternative approaches that assess surrogate biomarkers or activities providing an estimate of potency are considered adequate for initiation of early-phase clinical trials. To comply with regulations for FDA licensure, however, development of a valid potency assay for release of the final stem cell-based product will be required. It is expected generally that a suitable potency assay will be in place prior to commencing a pivotal Phase 3 clinical trial to confirm safety and determine efficacy.

GENERAL EXPECTATIONS FOR CONTROL OF THE MANUFACTURING PROCESS

Control of the Manufacturing Process

A principle that applies broadly to the entire range of complex biologic products, including stem cell-based therapies, is demonstrated control over the manufacturing process used to produce a biologic product. In terms of importance, this principle is equivalent to the objectives of thorough product characterization and adequate final product testing for ensuring the safety, identity, purity, and potency of stem cell products. Consequently, strict adherence to standard operating procedures and quality control of manufacturing product intermediates, in addition to final product testing, is essential to ensure that the final stem cell-based product will be acceptable for administration to human subjects. Exerting control over the manufacturing process is crucial to the consistent production of a stem cell-based therapy.

Current Good Manufacturing Practices (cGMPs)

Current Good Manufacturing Practices (cGMPs) as set forth in 21 CFR 210 and 211 of the Code of Federal Regulations convey principles that apply to stem cell products encompassing both the physical attributes of the facility where products are manufactured and the processes and procedures specific to the cellular products prepared within a particular manufacturing facility. The following are representative examples of expectations set forth in the cGMPs. Evidence of appropriate written protocols applicable to each stage of product manufacturing and characterization is expected. Standard operating procedures (SOPs) documenting all critical information pertaining to stem cell-based therapy production should be established and followed. Quality oversight exercised by the sponsor is important at each stage of product development and involves two elements, quality control (QC) and quality assurance (QA). Through these mechanisms, individuals assure that manufacturing and characterization testing have been properly performed and that the finished product meets specified release criteria and maintain independence from and are not directly subordinate to individuals responsible for conducting the specific tests and filing detailed reports. During the latter stages of biological product development, validation testing of the processes and assays used during product manufacturing and characterization should be performed.

PRECLINICAL EVALUATION SUPPORTS SAFETY ASSESSMENT: PROOF-OF-CONCEPT AND TOXICITY TESTING IN ANIMAL MODELS

In addition to careful management of the process used to manufacture a stem cell-based product along with intentionally designed analytical testing used to characterize a stem cell product and assure its quality, preclinical evaluation of cellular therapies consisting of, or derived from, stem cells in appropriate animal models is a key element in the overall safety assessment process. Preclinical testing focuses on two primary objectives: supportive scientific justification for the use of a particular cellular product for a specified disease indication (i.e., demonstrating "proof of concept") and assessment of the potential toxicity of the stem cell-based therapy. With respect to the core biology of stem cells, namely, a propensity for self-renewal and differentiation potential, it is crucial that experimental animals are inspected carefully following transplantation of stem cell therapies for evidence of unregulated growth and genesis of inappropriate cell types.

Demonstrating Proof-of-Concept

Investigations performed using animal transplant models of human disease constitute an essential component of the assessment paradigm used for judging the safety and rationale for administration of stem cell-based therapies in humans. Results from these proof-of-concept animal studies serve to substantiate a rationale for conducting a proposed clinical trial with an investigational stem cell therapy. Although the ideal animal model should reflect the pathophysiology of the targeted human disease as closely as possible, it is recognized that unless the disease occurs spontaneously in an animal species, such models will not replicate the clinical situation in an identical manner. Thus, chemical, surgical, and immunologic methods are employed to create these models of disease, resulting in damage to neural tissue; induction of diabetes; simulation of myocardial infarction, stroke, and hypertension; or compromised organ function. In selected situations when focal monogenetic lesions are known to cause disease, it is possible to create animal models in which the culpable gene is either eliminated or over-expressed, resulting in the generation of animal disease models that can recapitulate features of human disease-specific pathologies.

Demonstration of proof-of-concept for a stem cell-based therapy in an animal model of human disease is encouraged for the following reasons: (1) to provide information concerning feasibility and establish rationale that supports initiation of clinical trials, (2) to permit concurrent measurement of bioactivity and safety endpoints, (3) to explore the dose–response relationship between a stem cell-based product and an activity/safety outcome, and (4) to facilitate optimization of the route of administration. Demonstrating that a restorative effect occurs following implantation of an investigational stem cell-based therapy into an animal construct of human disease provides circumstantial evidence for anatomic/functional integration within the host physiology. If corroborated by histopathological data, this proof-of-concept information serves to provide support for hypotheses underlying the mechanism of action of a stem cell-based product.

Toxicological Assessment

Appropriate evaluation of the safety of an implanted stem cell-based therapy is an important aspect of the product development program. Toxicity endpoints are selected that focus on the potential for both local and systemic effects to occur. Depending on the nature of the stem cell-based product being tested, studies conducted in normal animals or animal models of disease should take into consideration assessment of the following in response to implantation of the stem cell-based product: (1) implant site reaction, (2) any inflammatory response observed in target or nontarget tissues, (3) host immune response, (4) differentiation/phenotype expression of the stem cell-based product postimplant, (5) capacity for cells to survive, migrate, and integrate into the host physiology post-implant, (6) observed behavioral changes, (7) morphological alterations in either target or nontarget tissues, and (8) evidence of tumorigenicity.

Completing a comprehensive microscopic evaluation following implantation of a stem cell-based therapy into animals provides a fundamental assessment of the overall safety of the product intended for administration into humans. The morphological data obtained from such an analysis provides potentially considerable insight into other, nonterminal evaluations of the animals (i.e., behavioral, neurologic, cardiac, and clinical laboratory changes). Histopathological evaluation can encompass assessment of acute and long-term cell survival, the extent and pattern of cell migration, evidence for differentiation and plasticity, indication of hyperplastic growth or tumorigenicity, and anatomical/functional integration. The extent of interpretable safety information extracted from investigations conducted in animals will be a direct function of the experimental study design as well as technical limitations associated with the specific test methodologies. For example, for methodologies used to identify and track implanted populations of stem cells, investigators have recently reported monitoring of stem cell migration *in vivo* following implantation using high-resolution *in vivo* magnetic resonance imaging or sensitive methods for detecting visible light emitted from marked stem cells.

Tumorigenicity or hyperplastic, unregulated cellular proliferation is an extremely important safety issue for stem cell-based therapies, particularly those derived from hES cells. Concern as to the ramifications of dysregulated cellular replication following implantation arises from one of the fundamental characteristics of true stem cells, namely, their capacity to effect expansion through self-renewing proliferation. For hES cells, concern is heightened due to their pluripotential character and proclivity for generating malignant teratomas when injected in sufficient numbers into immunodeficient strains of mice. This property alone provides a persuasive argument against the administration into humans of stem cell-based therapies consisting exclusively of unmodified, undifferentiated hES cells. The critical issue is to determine at what point during stem cell differentiation do risks attributable to tumorigenic potential and hyperplasia become insignificant, if ever. Successful identification of the precise stage in the differentiation process of these hES cells, when the risk for tumor formation is no longer a safety concern, will depend on whether or not cellular differentiation proceeds in a unidirectional manner or is reversible. Because of inherent inefficiencies in the biologic process of differentiation, it is unlikely that phenotypic maturation induced in cultures of undifferentiated stem cells will be complete. Accordingly, it is essential that analytical techniques be developed that permit careful evaluation of cell preparations derived from stem cells in order to determine the number of residual undifferentiated cells and partially differentiated intermediates that are present. Judicious preclinical studies of appropriate duration that involve implantation of undifferentiated or partially differentiated stem cells into immunocompromised animals should be carried out prior to initiating clinical trials in order to address the issue of unregulated growth potential of a candidate stem cell-based therapy and its relationship to the process of cell differentiation.

As previously stated, proof-of-concept studies are important for characterizing the dose-response relationship of the intended stem cell-based therapy as it pertains to both activity and safety outcomes. Unlike small-molecule pharmaceuticals or recombinant therapeutic proteins, standard kinetic profiles for product metabolism and excretion are not applicable to stem cell-based therapies. Moreover, once delivered, the stem cell-based therapy is not readily retrievable, and thus, dose-dependent adverse effects may be difficult to control. It is important to bear in mind that the fate of cells post-implant (i.e., survival, migration, structural and functional integration, and phenotypic plasticity) could potentially impact the safety and/or activity profile for a stem cell-based product. Thus, employment of a range of doses of the stem cell-based therapy in all animal studies is highly recommended in order to establish both a safety and activity profile. Similarly, developing strategies for effectively controlling cell dose levels after implantation of stem cell-based therapies could serve to improve the safety profile for this category of products. Potential solutions include ongoing efforts to introduce genetic switches into preparations of stem cells that are capable of regulating cellular proliferation and that may be turned on or off using noninvasive, extrinsic means after the cells have been implanted.

Summary

Stem cells are among the most complex biologic therapeutic entities proposed for clinical use to date. This chapter describes a regulatory review approach used by the FDA to assess the safety of novel stem cell-based therapies. Primary responsibility for evaluating the safety of investigational new treatments comprised of stem cells and their derivatives is assigned to FDA's Center for Biologics Evaluation and Research, Office of Cellular, Tissue, and Gene Therapies. Outfitted with a comprehensive approach developed for the regulation of tissue and cellular therapies, staffed by scientific reviewers with expertise in cell and developmental biology, and endowed with a wealth of experience in the evaluation of submissions involving administration of stem cell preparations to patients, the FDA is well positioned to successfully meet its obligation for assuring that investigative human clinical trials are conducted in as safe a manner as is reasonably achievable. With respect to experimental treatments involving stem cell-based therapies, safety assurance is fundamentally about demonstrating control over every facet of the cell therapy preparation process beginning with the acquisition of source donor materials and carried forward through derivation of an hES cell line, its characterization, and ultimately the implementation of adequate and appropriate preclinical testing. In counseling academic clinical investigators and biotechnology sponsors about particulars concerning individual stem cell development programs, the FDA recognizes the need to maintain flexibility in its recommendations, taking into consideration a variety of impact factors including the intended target patient population and the severity of the specific disease under study, along with the potential risks and benefits attributable to the investigational stem cell-based product. The agency regularly updates and reevaluates recommendations applicable to production and testing of stem cell products based on an accumulating scientific and clinical experience, along with feedback from a variety of valued sources. In order to be credible and effective, any approach taken for assessing the safety of stem cell-based therapies must consider the dynamic complexity of their biology. Operating within the jurisdiction of its mandated legal authority, the FDA will meet its obligation to ensure the safety of nascent biological therapies consisting of or derived from stem cells by relying on an approach that encourages iterative and collaborative interactions with the innovators and which is updated based on the best available science.

ACKNOWLEDGMENTS

The authors wish to acknowledge the thoughtful contributions of numerous colleagues in the Office of Cellular, Tissue, and Gene Therapies. We are especially grateful to Drs. Kimberly Benton, Keith

Wonnacott, Richard McFarland, and Mercedes Serabian, M. S. for their critical review of the manuscript during its preparation.

KEY WORDS

cGMPs (Current Good Manufacturing Practices)   The most up-to-date scientifically sound methods, practices, or principles that are implemented and documented during product development and production to ensure consistent manufacture of safe, pure, and potent products.

Manufacture   All steps involved in the propagation or processing and preparation of therapeutic products including, but not limited to filling, testing, labeling, packaging, and storage as performed by the manufacturer [21 CFR 600.3(t)].

Potency   The specific ability or capacity of a therapeutic product to elicit an expected result, as indicated by appropriate laboratory tests or by adequately controlled clinical data obtained through the administration of the product in the manner intended [21 CFR 600.3(s)].

Purity   Relative freedom from extraneous matter in the finished product introduced during the manufacturing process, whether or not harmful to the recipient or deleterious to the product [21 CFR 600.3(r)].

Stability   Ability of a product to retain its sterility, identity, purity, and potency over time when stored and shipped in designated packaging and under specified conditions.

FURTHER READING

Draft Guidance for Reviewers: Instructions and Template for Chemistry, Manufacturing, and Control (CMC) Reviewers of Human Somatic Cell Therapy Investigational New Drug Applications (INDs). August 15, 2003. Available at http://www.fda.gov/cber/gdlns/cmcsomcell.pdf (Accessed December 30, 2004).

Drukker, M., Katz, G., Mandelboim, O. and Benvenisty, N. (2004). Immunogenicity of human embryonic stem cells. *In Handbook of stem cells — volume 1, Embryonic stem cells* (R. Lanza, J. Gearhart, B. Hogan, D. Melton, R. Pedersen, J. Thomson, and M. West, eds.), pp. 663–674. Elsevier Academic Press, New York.

Eligibility determination for donors of human cells, tissues, and cellular and tissue-based products — Final Rule. (2004). NARA, *Federal Register* **69**, 101, 29786–29833.

Gardner, R. L. (2004). Pluripotential stem cells from vertebrate embryos: present perspective and future challenges. *In Handbook of stem cells — volume 1, Embryonic stem cells*, (R. Lanza, J. Gearhart, B. Hogan, D. Melton, R. Pedersen, J. Thomson, and M. West, eds.), pp. 15–26. Elsevier Academic Press, New York.

Guidance for Industry: Guidance for Human Somatic Cell Therapy and Gene Therapy. (1998, March). Available at http://www.fda.gov/cber/gdlns/somgene.pdf (Accessed December 30, 2004).

PHS Guidelines on Infectious Disease Issues in Xenotransplantation. (2001 January 19). Available at http://www.fda.gov/cber/gdlns/xenophs0101.pdf (Accessed December 30, 2004).

Points to Consider in the Characterization of Cell Lines Used to Produce Biologicals. (1993, July 12). Available at http://www.fda.gov/cber/gdlns/ptccell.pdf (Accessed December 30, 2004).

Weber, D. J. (2004). Biosafety considerations for cell-based therapies. Available at http://www.biopharminternational.com/biopharm/article/articleDetail.jsp?id=104336 (Accessed December 30, 2004).

It's Not About Curiosity, It's About Cures
Stem Cell Research: People Help Drive Progress

Mary Tyler Moore with S. Robert Levine

Choosing Life

Ah, but a man's reach should exceed his grasp. Or what's a heaven for?

Robert Browning, "Adrea del Sarto"

Many of you know that I have had type 1 (juvenile) diabetes for more than 35 years. As a consequence, I struggle every day, like millions of others, to do what happens naturally for people who do not have diabetes: achieve a balance between what I eat, the energy I expend, and the amount of insulin I inject. Although to most, metabolic balance is as automatic as breathing, to people with type 1 diabetes, like me, it requires constant vigilance, constant factoring and adjusting, frequent finger sticks to check blood sugars, and multiple daily insulin injections just to stay alive. Even with the greatest of care and closest of personal scrutiny, I find that often I am unable to achieve good balance — my sugars are dangerously low or frighteningly high. Yes, dangerous and frightening — because, frankly, serious lows can lead to seizures, coma, and death, and highs, over time, result in life-limiting and life-shortening complications like blindness, amputation, kidney failure, heart disease, and stroke. Diabetes is an all too personal time bomb that can go off today, tomorrow, next year, or 10 years from now — a time bomb that affects millions and must be defused.

This reality is made all too clear by the recent sudden death of a young friend, Danielle Alberti. Danielle was 31. She was an aspiring artist. Although rapidly losing her vision because of diabetic retinopathy, Danielle stuck to her dream of being a painter and was pursuing her career when she, like too many young adults with type 1 diabetes, developed kidney failure. People with diabetes-related kidney failure do not do well on dialysis, so kidney transplant was her only real option. With her doctor's guidance, she and her mother decided to return home together to Australia where her chances for a transplant were greater. However, Danielle did not survive the flight. She died at 30,000 feet, seeking comfort in her mother's arms — her last words, "Mum, hold me."

Most of us share at least a piece of this experience — our loved ones, in times of pain or need, reaching out, looking to us for comfort, for a way to stop their suffering. At that moment, we would do anything in our power to change their reality, to take their pain from them. When given the choice or the power to effect change, we would choose to protect the lives of those we love. This is the quest we join, together, when we contemplate the promise of stem cell research, debate its proper methods, and work toward making our hopes a reality — curing disease and disability through stem cell–derived therapies. We choose the idea of a better life and reach beyond our grasp to achieve it.

Size of the Promise

It is not unrealistic to say that [stem cell] research has the potential to revolutionize the practice of medicine and improve the quality and length of life.

Dr. Harold Varmus, Nobel laureate and former Director of the U.S. National Institutes of Health (NIH)

Judgment does not only require choosing between the good and the bad. It often challenges us to balance more than one good or choose between the bad and the worse. Good judgment, therefore, demands that we make efforts to understand the relative impact of our choices, to come to an understanding of the greater good.

However, measuring the promise of stem cell research, for me, does not start with the recitation of the literally hundreds of millions of people who could benefit from the insights gained and therapies derived. Instead, it starts with understanding its potential for you and me, our parents and children, our friends and families, neighbors, and co-workers.

For people with type 1 diabetes, we look first to stem cell research as a means to help us replace the insulin-producing cells of the pancreas that are destroyed by our disease. However, it also may provide insights into the genetic basis of diabetes, including the differences between type 1 diabetes, which is an autoimmune disease like lupus and multiple sclerosis, and the more common, obesity-related, type 2 diabetes. It may also provide solutions to the devastating complications of diabetes: blindness, kidney failure, amputation, and cardiovascular disease. For people with Parkinson's disease, stem cell research holds the promise of replacing destroyed specialized brain cells and thereby freeing patients from the prison of disease-induced rigidity. For spinal injury patients, it offers the potential for regeneration of neural tissue, which would reconnect the pathways of sensation and motor control and allow them to walk again, or talk again, or hug their child again. For people with heart failure, stem cell research may mean sleeping through the night without struggling for breath, dancing with a spouse, working in the garden, or sustaining one's job and independence. For the person with macular degeneration it might offer sight. Stem cell research offers hope for people with a great diversity of illnesses, for people of all ages and genders and all backgrounds. It offers hope for each of us, and that hope is not measured by numbers. It is ultimately personal.

Personal Promises Fuel Progress

Never, never, never give up.

Sir Winston Churchill

I have had the privilege of serving as the International Chairman of the Juvenile Diabetes Research Foundation (JDRF) (www.jdrf.org) since the mid 1980s (Fig. 68-1). JDRF was founded in 1970 by the parents of children with type 1 diabetes. They were not satisfied with the only option offered to their children by health professionals — a lifetime of insulin injections just to stay alive, and the constant fear of the life-stealing complications they would face in the future. Insulin was not a cure, they knew it, and they wanted someone to do something about it.

So they challenged the established professional associations to do something, to think anew, and to invest in more research (especially type 1 diabetes research). They were brushed aside but not bowed. They may have been "just moms

Figure 68–1. Juvenile Diabetes Research Foundation (JDRF) founding moms, Lee Ducat and Carol Lurie.

and dads," but they had a purpose that was highly personal. Each of them had promised their child, their loved one, that they would do all they could to find a cure. They intended to keep that promise, and they have. In the years since its founding, JDRF has grown to be the largest charitable contributor to diabetes research in the world, providing over $800 million in direct funding since 1975, including well over $400 million in the last 5 years alone. However, the impact of this "people-driven" effort to find a cure has been far greater than just the dollars raised for research. JDRF families (supported by an extraordinary professional staff, many of whom have a direct connection to diabetes) have been key leaders in public advocacy that has resulted in the following:

1. The Diabetes Research and Education Act (mid-1970s) that established The National Institute of Diabetes Digestive and Kidney Diseases at the National Institutes of Health (NIH) and called for substantive increases in funding of diabetes research. At the time the NIH was only investing $18 million per year in diabetes research. It now spends more than $800 million per year.
2. Congressional earmarks for research into the genetics of diabetes and diabetes-related kidney disease (1980s).
3. Lifting the ban on fetal tissue research in 1993, which President Clinton did "for Sam," a young man with type 1 diabetes.
4. Doubling the NIH budget in 5 years. Between 1998 and 2003 the budget increased from approximately $13 billion per year to more than $26 billion per year.
5. Establishment of the congressionally mandated Diabetes Research Working Group that reports to Congress periodically on progress in diabetes research, on research needs and opportunities for the future, and on adequacy of funding (1998).
6. The Special Diabetes Initiative (1998–2008) that will provide by fiscal year 2008, more than $1 billion in supplemental funding for special initiatives in type 1 diabetes research (on top of usual NIH appropriations) and an equal dollar amount (more than $1 billion) to fund special initiatives in diabetes care and education for Native Americans with diabetes.

JDRF was also a founder of the Coalition for the Advancement of Medical Research (CAMR), a diverse coalition of health-related organizations committed to sustaining federal funding of stem cell research. Working through CAMR and independently, JDRF volunteers and staff played a crucial role in 2001 in convincing the George W. Bush Administration and congressional leadership not to support a total ban on embryonic stem cell research in the U.S. (Fig. 68-2) and more recently leading public advocacy in building bipartisan support for passage of H.R. 810–the Castle-DeGette Stem Cell Research Enhancement Act of 2005. JDRFers were also lead proponents of California's newly enacted $3 billion stem-cell research/regenerative medicine initiative.

The experience of the JDRF along with the HIV and AIDS community, the women's health movement, the Parkinson's Action Network, and other grass roots organizations proves

Figure 68–2. Mary Tyler Moore (with Michael J. Fox) testifying before Congress in support of federal funding of stem cell research. (Photo courtesy of Larry Lettera/Camera 1.)

that in their quests to find a cure for their children and loved ones for whatever pains them, in their personal wars against disease, moms and dads, partners and spouses, and people who care will never give up. In fact, people personally affected by illness are the natural and necessary leaders of any global cure movement. They understand the urgency and are uniquely willing do anything required to ensure that their loved ones are freed of the burden of disease as soon as possible. For them, "failure is not an option," because their very survival is at stake.

Hope Versus Hype

I am not discouraged, because every wrong attempt discarded is another step forward.

Thomas A. Edison

What we know about any area of health or science is as much the "wisdom" accumulated through countless errors as it is the outcome of our research "successes." Furthermore, progress is often gained as much through the accidental collision of an unexpected finding and a willing mind as it is through the careful and detailed application of all that is "known."

What, then, drives the hope versus hype discussions regarding stem cell research? Do we know enough to project possibilities? Yes. Do we know enough to make assertions of a particular outcome by a date certain? No. Should this diminish our commitment to pushing the field forward to prove its potential? Certainly not. Too much is at stake for us to delay or to apply unreasonable constraints because we are worried we might be wrong or might be overestimating the potential. What if we are underestimating it?

Just as with any endeavor at the frontier of new worlds, there is risk in taking the next steps. However, we cannot shrink from these risks. Rather, we must — with proper deliberation and due consideration of the risks — chart our course, prepare ourselves for dealing with the unexpected, and move forward.

Giving Life

If you save one life, you save the world.

The Talmud

I understand that embryonic stem cell research raises concerns among people of good will, each trying to do what is right based on their very personal religious and moral beliefs. I have not shied from that personal soul searching, nor has JDRF in its policy making, nor should anyone. I have found comfort in my heartfelt view that human stem cell research is truly life affirming. It is a direct outcome of a young family making a choice, without coercion or compensation, to donate a fertilized egg not used for *in vitro* fertilization, for research. An egg that otherwise would have been discarded or frozen forever. Because of the great potential of stem cell research, donating unused fertilized eggs is much like the life-giving choice a mother, whose child has died tragically in an automobile accident, makes when donating her child's organs to save another mother's child. It is the true pinnacle of charity to give so totally, so freely of oneself, to give life to another. Public support for stem cell research is an extension of this affirmation of life and is the best way to ensure that it is undertaken with the highest ethical standards.

515

FRIENDS IN NEED AND DEED

LISA, 12 YEARS OLD
El Paso, Texas

Lisa has had type 1 diabetes since she was 5 years old. "It makes me sad to watch my parents cry when they tell people about my diabetes," she says. "I feel really sad when I think about the stuff that can happen if a cure isn't found soon. I don't want to go blind and never again see the faces of the people I love or the really pretty sunsets in El Paso. I don't want my legs amputated because I love to dance. I don't want to die early and have heart and kidney disease. I want a cure before any of these terrible complications can happen."

NICHOLAS, 4 YEARS OLD
Boca Raton, Florida

Nicholas was diagnosed with type 1 diabetes when he was 20 months old. Most nights, his mother, Rose Marie, must intervene in some way to keep Nicholas' blood sugar in the normal range. "As I hold him in the middle of the night, trying to coax him awake so I can feed him, I realize just how delicately his life hangs in the balance," Rose Marie says. "I rededicate myself to doing everything within my power to find a cure for this disease, which robs my son, and millions of others like him, of a healthy, carefree childhood, and which carries the constant threat of danger — like a thief in the night lurking to strike."

KYLIE, 12 YEARS OLD
Ogden, Utah

Kylie was diagnosed at age 8. "Over Christmas vacation last year, I had a friend in my fifth grade class who died," she says. "He had diabetes just like me. During the middle of the night he went into insulin shock and never woke up." Kylie fears the same thing could happen to her. "I would love to have a guarantee that the rest of my life will be long and normal, but I know that there is a lot of work still to be done before they can find a cure."

BRENNAN, 6 YEARS OLD, AND TANNER, 8 YEARS OLD
Round Rock, Texas

Brennan, 6, was diagnosed with type 1 diabetes at the age of 2, just 7 months before his older brother Tanner was diagnosed. "Each day is filled with adversity and challenge," mom Amy says. "Will the disease win today? Will Brennan succumb to too-low blood sugar levels and lose consciousness? Will Tanner have his first seizure?"

COREY, 11 YEARS OLD
Secaucus, New Jersey

Corey has had type 1 diabetes since the age of 5. Sitting at home and complaining about diabetes, he says, won't change things or get your voice heard or get us closer to a cure. Since the age of 7, he has spoken at schools and fundraisers to tell his story of what it is like to be very young and have to live with diabetes. "I've found that while millions have this disease, many people don't have a clue about what it really means to live with it each day," he says. He talks about the adjustments and fears that diabetes has brought to him and his family, adding that "it has not made me afraid to do what I want to do in life, but I known it *will* make things harder." At the moment of his sixth grade graduation, Corey says, "I want to leave a legacy to help . . . other children as they strive to live a normal life and hope they'll never forget that we will find a cure."

EMILY, 15 YEARS OLD
Houston, Texas

Diagnosed with type 1 diabetes 9 years ago, Emily says each day is a battle. At first the finger pricks and shots were the most difficult thing that Emily had to handle, but now she has bigger fears. "Now the hardest part is facing the reality that there is no cure, and that my life could be determined by this awful disease." Emily is a straight A student and plays on her school's field hockey team. Her goal is to one day be an orthopedic surgeon — and to help find a cure for diabetes. "I and many others will not be able to rest until we find a cure," she says. "A cure for diabetes will not just happen; it must be pursued, researched, and fought for."

Each of these courageous children, supported by their families, were delegates to the JDRF's 2003 Children's Congress where they met their elected leaders and advocated for a cure for themselves and all children (and adults) like them with type 1 diabetes (Fig. 68-3).

Excerpted, with permission, from the JDRF 2003 Children's Congress Yearbook.

Figure 68–3. Mary Tyler Moore at the Juvenile Diabetes Research Foundation's (JDRF's) Children's Congress of 2001. Joined on stage by Larry King, Tony Bennett, John McDonough, George Nethercutt, R-WA, Alan Silvestri, and child delegates from 50 states. (Photo courtesy of Larry Lettera/Camera 1.)

People Drive Progress

I know of no safe depository of the ultimate powers of the society but the people themselves; and if we think them not enlightened enough to exercise their control with a wholesome discretion, the remedy is not to take it from them, but to inform their discretion by education.

Thomas Jefferson

In science, politics, even religion, the public often cedes decisions, important decisions, to an "expertocracy" or to dogma. Perhaps this is out of respect, humility, fear, or unfamiliarity — even out of a presumption of incapacity. It is, however, a well-informed public that is most capable of making decisions in its own interest and that, uniquely, has the power to effect change. It is, therefore, a true test of leadership to cede discretion back to people and bring comfort to public decision-making through careful and objective expert counsel, access to a broad base of information, and support for taking specific actions to achieve goals.

This approach has defined the JDRF success. From its inception, the JDRF has been unique in the way that it conducts its review of research grants and how it decides what research is funded. Scientific experts (peer reviewers) are joined by people personally affected by diabetes (lay reviewers) in the discussion of all proposed research. Scientific merit is established in these peer/lay collaborative review sessions. Then the lay reviewers meet separately to discuss which of the meritorious grants are most responsive to the needs of people with diabetes — that is, which are most likely to have the greatest impact on finding a cure or reducing the burden of diabetes and its complications. It is the people personally affected by diabetes who make the final decisions of what to fund and what not to fund. The decisions of this lay review group are made in the context of cure goals and research priorities established by the JDRF Board, which is itself predominantly made up of people personally affected by diabetes. These goals and priorities were derived by the Board via a process of knowledge mapping, which was conducted by JDRF volunteers, staff, and expert advisors. Knowledge mapping identifies the current state of science along the many potential paths to a cure, where obstacles remain, and where there is the greatest opportunity for JDRF investment to make a difference.

The experience at JDRF infers that decisions regarding embryonic stem cell research are best made in the open, with the full engagement of the public and with particular attention to presentation of the broadest breadth of available information and opinion. We can be confident that the powers of society (in overseeing the conduct of science) can be safely ceded to the discretion of a well-informed populace.

517

Better Health For All

William Bradford, speaking in 1630 of the founding of the Plymouth Bay Colony, said that all great and honorable actions are accompanied with great difficulties, and both must be enterprised and overcome with answerable courage. If . . . our progress teaches us anything, it is that man, in his quest for knowledge and progress, is determined and cannot be deterred.

President John F. Kennedy, speech at Rice University
September 12, 1962

People like me who struggle daily with disease or disability, and the people who love us, recognize the difficulties of scientific advancement, accept the challenges, and every day answer with courage — if not for our own sake, for our children and our children's children. We are motivated not by curiosity, but by our dedication to finding cures. New therapies derived from embryonic stem cell research, conducted with public support by scientists from all areas of the globe, and made available to all who might benefit, are part of our broader vision of better health for all.

Patient Advocacy

Christopher Reeve*

When I became a patient advocate in 1995, I thought that the major obstacles to achieving a cure for spinal cord injury would be a lack of funding and a shortage of scientists willing to dedicate their careers to an orphan condition. Those turned out not to be the problems. The budget of the National Institutes of Health (NIH) actually doubled. In 1998 the NIH research budget was $12 billion, and by fiscal year 2003 it had grown to just over $27.2 billion. Today researchers all over the world believe that effective therapies for Parkinson's disease, Alzheimer's disease, diabetes, heart disease, spinal cord injuries, and a wide variety of other afflictions can be achieved. Instead, the main obstacle is the controversy over embryonic stem (ES) cells and therapeutic cloning. The NIH has not been allowed to fund ES cell research using excess embryos from *in vitro* fertilization (IVF) clinics. The House banned therapeutic cloning for the second time in as many years, and the Senate remains gridlocked on the issue. The frustration of investigators all over the country became front-page news in the April 22, 2003 issue of *The Washington Post*:

> A series of important advances have boosted the potential of human embryonic stem cells to treat heart disease, spinal cord injuries and other ailments, but researchers say they are unable to take advantage of the new techniques under a two-year-old administration policy that requires federally supported scientists to use older colonies of stem cells. Now pressure is building from scientists, patient advocates and members of Congress to loosen the embryo-protecting restrictions imposed by President Bush.

The article goes on to say that the White House "has no intention of changing its position."

It appears that elected officials and average citizens have learned to fear "human cloning," and are failing or unwilling to appreciate the distinction between reproductive and therapeutic cloning. It is likely that somatic cell nuclear transplantation, the proper terminology for therapeutic cloning, will become standard medical procedure in the future. Perhaps there will be more progress when the public becomes aware of the advances at home and major breakthroughs overseas. However, while we wait, hundreds of thousands will have to endure prolonged suffering. An untold number will die.

Having lived with a spinal cord injury for nearly 9 years, I still have to emerge every morning from dreams in which I am completely healthy and adjust to the reality of paralysis. In the weeks and months after my injury that transition was often very difficult. After a few years it became less so, because I believed that the scientists were progressing well, that more funding would become available, and that the light at the end of the tunnel would continue to shine brighter every day. I never imagined that a heated political debate over the insertion of a patient's DNA into an unfertilized egg to derive genetically matched stem cells would have such an effect on me. Now instead of waking up just to rediscover that I am paralyzed, I wake up shocked by the realization that I may remain paralyzed for a very long time, if not forever.

Once that moment passes, I begin my day. Rationality and hope return. I am able to focus on what might be accomplished. Perhaps through education we can change people's minds, or even reverse the positions of powerful opponents.

The first task is to dispel misinformation. For example, the idea has been put forward that adult stem cells are better than ES cells because they have the same therapeutic potential without the controversy. However, adult tissue stem cells appear to have a much more restricted path for development, limiting their usefulness in therapies for disease. Eighty Nobel laureates sent a letter to President Bush stating that "it is premature to conclude that adult stem cells have the same potential as embryonic stem cells." The Department of Health and Human Services released a report, "Stem Cells: Scientific Progress and Future Research Directions" in June 2001. The report confirms the incredible potential of ES cells. It also stresses that there is limited evidence that adult stem cells can generate mature, fully functional cells or that the cells have restored lost function *in vivo*. As stated in Chapter 5 of the Committee on the Biological and Biomedical Applications of Stem Cell Research, Board on Life Sciences, National Research Council, Board on Neuroscience and Behavioral Health, Institute of Medicine of *Stem Cells and the Future of Regenerative Medicine* published by the National Academy of Sciences in 2002:

> A substantial obstacle to the success of transplantation of any cells, including stem cells and their derivatives, is the immune-mediated rejection of foreign tissue by the recipient's body. In current stem cell transplantation procedures with bone marrow and blood, success hinges on obtaining a close match between donor and recipient tissues and on the use of immunosuppressive drugs, which often have severe and potentially life-threatening side effects. To ensure that stem cell-based therapies can be broadly applicable for many conditions and people, new means of overcoming the problem of tissue rejection must be found.

*Deceased

Although ethically controversial, the somatic cell nuclear transfer technique promises to have that advantage.

It is extremely disturbing that the George W. Bush Administration prefers the opinions of social and religious conservatives over those of scientists on this issue. Those of us who hoped for a fair debate leading to governmental approval of therapeutic cloning were especially disheartened by a media opportunity that took place at the White House on April 10, 2002. The President urged the Senate to pass the Brownback Bill, S. 1899, which would ban all forms of cloning. He stated the following:

> I believe all human cloning is wrong, and both forms of cloning ought to be banned, for the following reasons. First, anything other than a total ban on human cloning would be unethical. Research cloning would contradict the most fundamental principle of medical ethics, that no human life should be exploited or extinguished for the benefit of another.

At the President's side, in full dress uniform, was former New York City police officer Stephen McDonald, still confined to a wheelchair 16 years after suffering a gunshot wound that left him paralyzed from the shoulders down. Officer McDonald is a devout Catholic. After the press conference he told the media that his accident was "God's will" and echoed the Pope's position on the sanctity of human life.

Stephen McDonald does not represent most Americans living with paralysis. He was there purely as a prop for the Administration. In light of the fact that the Council on Bioethics that the President had appointed to advise him had not yet issued its opinion, the President's press conference was an inappropriate attempt to tip the balance of the debate.

I was one of many patient advocates who were greatly relieved to find an op-ed piece in the *New York Times* on April 25 by Michael Gazzaniga, Ph.D., Director of the Center for Cognitive Neuroscience at Dartmouth College. A fellow of the American Association for the Advancement of Science and the American Neurological Association, he is also one of the distinguished members of the President's advisory panel. He wrote:

> It was a surprise when, on April 10, the President announced his decision to ban cloning of all kinds. His opinions appeared fully formed even though our panel has yet to prepare a final report. . . . Some religious groups and ethicists argue that the moment of transfer of cellular material is an initiation of life and establishes a moral equivalency between a developing group of cells and a human being. This point of view is problematic when viewed with modern biological knowledge. We wouldn't consider this clump of cells even equivalent to an embryo formed in normal human reproduction. And we now know that in normal reproduction as many as 50 percent to 80 percent of all fertilized eggs spontaneously abort and are simply expelled from the woman's body. It is hard to believe that under any religious belief system people would grieve and hold funerals for these natural events. . . . The biological clump of cells produced in biomedical cloning is the size of the dot on this i. It has no nervous system and is not sentient in any way. It has no trajectory to becoming a human being; it will never be implanted in a women's uterus. What it probably does have is the potential for the cure of diseases affecting millions of people. When I joined the panel, officially named the President's Council on Bioethics, I was confident that a sensible and a sensitive policy might evolve from what was sure to be a cacophony of voices of scientists and philosophers representing a spectrum of opinions, beliefs and intellectual backgrounds. I only hope that in the end the President hears his council's full debate.

Moral and ethical questions have always attended the birth of new ideas and new technologies. However, the questions and concerns surrounding the use of stem cells have remained unresolved in this country for too long. It is painful to contemplate what might have been achieved if the nation had rallied behind the scientists who isolated human ES cells for the first time in 1998. Now we can only hope that the federal government will not impose a ban or that other states will follow California and New Jersey and pass enabling legislation on their own. We can hope that Sweden, the United Kingdom, Israel, China, Singapore, and other countries that have decided in favor of government funded research using stem cells derived from any source will succeed in filling the void created by American public policy.

Will the United States surrender its preeminence in science by effectively killing stem cell research, the future of medicine, in its infancy? Before we make that unthinkable mistake we should remember Robert F. Kennedy, who said:

> The future does not belong to those who are content with today, apathetic toward common problems and their fellow man alike, timid and fearful in the face of bold projects and new ideas. Rather, it will belong to those who can blend passion, reason and courage in a personal commitment to the great enterprises and ideals of American society.

Glossary

Acute degeneration Damage restricted to a circumscribed region (both in space and time); multiple neural cell types must be replaced (for example, stroke, central nervous system injury, and spinal injury).

Adult neurogenesis The addition of new neurons to the adult brain, a process consisting of the proliferation, differentiation, and survival of neural stem cells.

Adult stem cell Cell derived from postembryonic stage tissue with self-renewal ability and ability to generate all cell types of the tissue from which it was derived but not other tissues.

AGM region Aorto-gonad mesonephros region. The first site of intraembryonic hematopoiesis and the site of origin of hematopoietic stem cells that can function in the adult.

Allograft A graft transplanted from one individual to another genetically different member of the same species.

Aneuploidy Any deviation from the normal haploid number of chromosomes.

Angiogenesis Maturation and remodeling of the primitive vascular plexus into a complex network of large and small vessels.

Antibody An immunoglobulin protein that specifically binds to an antigen.

Antigen A substance, usually a protein, that can be identified by an antibody molecule. Stem cells express a pattern of antigens that can be detected by fluorescent-labeled antibodies using a fluoresence-activated cell sorting (FACS) instrument, or by unlabeled antibodies using magnetic selection methods.

APC The human *adenomatous polyposis coli* gene, a negative regulator of Wnt signaling. Also, a structural protein associated with the cytoplasmic face of adherens junctions. Mutations in the *APC* gene are linked to the development of colorectal cancer.

Autograft A graft obtained from an individual's own cells or body.

Beta cell Insulin-producing cell contained within pancreatic islets.

Bioethics The study of the norms of conduct that should govern research and clinical applications of biomedical technologies.

Biomaterials Any natural or synthetic materials that interface with living tissue and/or biological fluids.

Bioreactor An apparatus designed for the *ex vivo* construction of tissue or organ-like structures providing appropriate structure and nutrition to cells that are seeded within it. Current designs may also include the application of appropriate physiological stresses to "condition" the construct for better performance upon *in vivo* transplantation.

Bisulfite sequencing A technique whereby the methylation of each CpG site in a defined stretch of DNA is determined.

521

Blastocyst Very early animal embryo consisting of a spherical outer epithelial layer of cells known as the trophectoderm (that forms the placenta), a clump of cells attached to the trophectoderm known as the inner-cell mass (from which stem cells are derived), and a fluid-filled cavity, the blastocoel. When fully expanded, the blastocyst "hatches" from the zona pellucida in which it has developed and implants in the uterus of the mother.

C2C12 Immortal myogenic cell line derived from adult murine satellite cells.

Cardiac hypertrophy Cellular response of cardiac cells to an increase in biomechanical stimuli, characterized by increased cell size, enhanced protein synthesis, and sarcomere organization. Prolonged maladaptive hypertrophy is associated with heart failure.

Cardiomyocytes Muscle cells in the heart which are interconnected via gap junctions and which participate in a functional syncytium.

CDK inhibitors (CKIs) Intracellular molecules with low molecular weight specifically inhibiting the activities of cyclin-dependent kinases during cell cycle progression. There are two subfamilies of CKIs, including Cip/Kip (p21$^{Cip1/Waf1}$, p27^{Kip1}, and p57^{Kip2}) and INK4 (p16^{INK4A}, p15^{INK4B}, p18^{INK4C}, and p19^{INK4D}).

Cell fate Future possible course of differentiation of a given cell.

Cell fusion The process whereby the cell membrane between two juxtaposed cells breaks down and re-forms to incorporate the cytoplasm and nuclei of both cells into a single, viable cell.

Cell lineage Set of courses of differentiation that leads to a given cell type.

Cell therapy The transplantation of cells from various sources to replace or repair damaged tissue and/or cells.

cGMPs (Current Good Manufacturing Practices) The most up-to-date scientifically sound methods, practices, or principles that are implemented and documented during product development and production to ensure consistent manufacture of safe, pure, and potent products.

Chemokine A family of cytokines originally defined by their chemoattractant properties but having additional functional activity on cells.

Chimera Organism made up of cells from two or more different genetic donors. In general, two genetically different, very early embryos (at the morula stage) are aggregated to form one single embryo, or, alternatively, embryonic stem cells are allowed to attach to a morula or are injected inside a blastocyst. Embryonic stem cells integrate preferentially in embryonic tissues, whereas the extraembryonic tissues are derived from the recipient embryo.

Chondrocyte Cell type of mesenchymal origin present in cartilage; can be nonhypertrophic or hypertrophic.

Chromatin DNA and associated proteins that create appropriate conditions for the expression or repression of genes during development and differentiation of cells.

Chronic degeneration Progressive neurodegenerative disease requiring the replacement of all neural cell types in large brain regions, in the face of ongoing degeneration and possibly the presence of toxic molecules.

Ciliary margin The peripheral region of the eye where retinal stem cells reside. In fish and amphibians, retinal cells are generated *in vivo* from these stem cells throughout life. It is also the area from which retinal stem cells have been isolated in mammals.

Clonal conversion Conversion of a holoclone to a meroclone or to a paraclone irreversible under normal conditions; affected by culture conditions and age.

Commitment An epigenetic, stable, and heritable change in the developmental capacity of a cell, as defined operationally under a particular set of physiological or experimental conditions.

Conditoned medium Medium left in contact with cultured cells, usually for a prolonged period of time.

Cord blood transplantation Infusion of umbilical cord blood containing hematopoietic stem and progenitor cells into recipients to replace the blood cell system.

Cryopreservation Freezing of a cell population or cell line in specialized media that allows storage of the cells for prolonged periods of time. Frozen cells can then be thawed and propagated at a later date.

Crypt clonality Intestinal crypts, wherein all cell lineages are originally derived from a single cell during embryonic development.

Cyst progenitor cell (CPS) Stem cell for the somatic cyst cells in the *Drosophila* testis.

Cytokine Biologically active molecule that occurs naturally and influences the survival, proliferation, migration, or other aspects of immature or mature cells.

De-differentiation and transdifferentiation The loss of genetic information that provides cell identity with acquisition of a more primitive (de-differentiation, as in "Dolly") or different (as in apparent stem cell plasticity) cell identity. See also stem cell plasticity.

Demyelination disorders Mainly of two classes; disorders in which endogenous myelinating cells are genetically impaired and fail to myelinate during development, and disorders that result from loss/failure of remyelination in the adult (for example, multiple sclerosis, allergic ancephalomyelitis, and injury).

Dentate gyrus   Subregion of the hippocampus in the medial temporal lobe, which contains the subgranular zone, the neurogenic site of the dentate gyrus.

Deontological   Duty-based reasons to perform an action. This may be a sense that rights and duties are intertwined (as in the idea that persons have a right to life and each citizen has a duty to protect that right), or a duty may arise from a sense of religious command, or a duty may be role-specific ("doctors must heal the sick").

Differentiation   Process of specialization wherein a cell, through changes in gene expression, acquires the correct shape, polarity, and orientation with respect to its neighbors and the extracellular matrix, and an appropriate set of organelles and proteins to enable it to carry out metabolic, signaling, transport, contractile, structural, or other specific functions in a particular tissue.

Differentiation antigen   Typically used with respect to cell surface antigens; an antigen that is expressed by a restricted range of cells and shows changes in expression during cell differentiation.

Dividing transit cells   The amplifying cells derived from stem cells that continue to divide several times before undergoing terminal differentiation (maturation) into the functional cells of the tissue.

DNA methylation   A biological process of modifying DNA molecules by covalently adding a methyl group to position 5 of the pyrimidine ring of cytosines in DNA. DNA methylation plays important roles in controlling genome stability, regulating gene expression, and silencing viral genomes.

DNA microarray   A collection of genes spotted on a solid surface (often a glass slide), arranged in rows and columns, so that the origin of each spot is known. Depending on the type of microarray, the spots consist of cDNA sequences amplified by (PCR) or synthetic oligo nucleotides. Each DNA microarray may contain tens of thousands of gene spots and is therefore a powerful tool for the study of gene expression patterns on a genomic scale.

DNA transfection   The introduction of foreign DNA molecules into cultured cells.

Embryoid body   A collection of hundreds of cells that resemble the early embryo and that is produced by growing ES cells in the presence of retinoic acid. These bodies contain neural tube-like structures and have been successfully used to repair the damaged spinal cord of rats.

Embryonal carcinoma stem cell   Pluripotent stem cell of teratocarcinoma (an embryonal tumor containing multiple types of differentiated cells), similar in phenotype to embryonic stem cell, but often more limited in differentiation capacity and of abnormal karyotype.

Embryonic germ cell   A pluripotent stem cell derived from primordial germ cells.

Embryonic stem cells   Stem cells derived from the preimplantation embryo with an unrestricted capacity for self-renewal that have the potential to form all types of adult cells including germ cells. Thus far, such cells have only been obtained from certain strains of mice.

Embryonic stem cell-like cells   Stem cells that resemble embryonic stem cells but whose potential to form all types of adults cells including germ cells has yet to be established.

Epidermal proliferative unit   The functional group of proliferative basal cells derived from a single stem cell, together with the distally arranged functional differentiated cells.

Epidermal stem cells   Slow-cycling and self-renewing cells found in the basal layer of the epidermis responsible for its constant repopulation.

Epigenesis   The formation of entirely new structures during development of the embryo.

Epigenetic   Factors involved in influencing gene expression during development, without affecting DNA base sequence.

Epigenetic reprogramming   A process by which the epigenetic state of the genome of a cell is partially or completely converted to that of another type of cell. In nuclear transfer, it generally refers to the process of reversing the epigenetic state of a differentiated donor genome back to a totipotent embryonic state.

Epigenetics   A term describing the phenomenon that certain inheritable, yet reversible, traits of a cell are not determined by the primary DNA sequence but by specific higher-order chromatin organizations of its genome.

Epitope   That part of a macromolecule that interacts specifically with a given monoclonal antibody. Typically, it comprises 2 or 3 amino acid or monosaccharide moieties but may also depend on some aspect of the teriary structure of the antigen.

***Ex vivo* stem cell gene therapy**   The process of isolating stem cells (hematopoietic and nonhematopoietic) from patients with genetic disease, genetically correcting the stem cells, possibly expanding them *ex vivo*, and transplanting them back into patients with the goal of producing genetically corrected cells *in vivo*.

Feeder-free   Cells that are cultured without direct contact with feeder layers.

Feeder layer   A layer of cells that are growth-arrested, providing nutrients, growth factors, and matrix components to a second cell population.

Fibrosis   Formation of fibrous scar tissue after damage of an organ with complete substitution of the cells constituting the organ with collagenous material. Extended fibrosis leads to organ impairment and loss of functionality.

Fluorescence-activated cell sorting (FACS) A technology that evolved from instruments designed to count blood cells, FACS combines detection of fluorescent probes with the isolation of single cells that are labeled with any combination of probes. This technology allows rapid separation of stem cells from a mixed population of other types of cells, provided that markers to distinguish the stem cells are available.

Fluorescent activated cell sorter A technology extremely useful for purifying intermediates during ES cell differentiation. Gene expression profile is distinct among distinct cell types. As some of such genes characterizing a particular cell type are expressed on the cell surface, fluorescence dye-conjugated antibodies to those molecules can be used for distinguishing and purifying the living cells in terms of expression of a set of molecules on the cell surface.

Gastrointestinal stem cells Founder cells capable of limitless self-renewal throughout the lifetime of their host. Gastrointestinal stem cells are also responsible for the maintenance of tissue homeostasis, by producing daughter cells committed to multipotent differentiation to form all adult lineages within the gastrointestinal tract.

Gastrointestinal stem cell signaling Elucidation of the components involved in molecular pathways, such as Wnt/b-catenin, which regulate cell proliferation and differentiation in the gastrointestinal tract, will provide an insight into the mechanisms of normal epithelial cell turnover and neoplastic development.

Gastrointestinal tumor morphogenesis A topic of heated debate, with two main conflicting theories. The first "bottom-up" theory proposes that neoplastic lesions arise in the base of the crypt, the "stem cell zone", and proliferate upwards. The alternative "top-down" theory states that lesions arise within an "intercryptal" zone, and spread laterally and down into adjacent crypts.

Gastrulation The period of embryonic development during which the definitive body plan is laid down. Cells from the primitive ectoderm pass through the "primitive streak" region to form a third layer, the mesoderm, between ectoderm and endoderm.

Gene addition Delivery of corrective DNA (usually composed of the entire coding region of a gene and appropriate regulatory sequences) to cells containing the defective gene.

Gene targeting The modification of a gene in a specific and directed manner through *homologous recombination* with a DNA segment. Possible applications include elimination of gene expression and correction of a mutation in a gene.

Gene therapy Novel strategy in treating diseases by the modifying the expression of specific genes toward a therapeutic goal.

Genetic lineage tracing The use of permanent labeling, usually DNA recombination, to mark a defined cellular population and trace its progeny.

Genome editing Specific, targeted, and permanent correction of an existing defective gene sequence so that the defective or mutated area of the gene is restored to a corrected normal state by delivery of molecules that are homologous to the target gene sequence with the exception of the base or bases intended for alteration. The genome editing process is directed by endogenous cellular machinery (potentially including mismatch repair and homologous recombination) acting at these target bases and the sequence mismatches created.

Genomic reprogramming The process that alters functions and properties of a cell through reversible modifications of DNA or associated proteins, without affecting the DNA sequence itself.

Gynogenesis Manipulation of ploidy by suppression of the polar body in an egg lacking paternal genetic contribution.

Hair cycle Succession of growth, regression, and rest that the hair follicle undergoes during the life of an animal.

Hemangioblast Common progenitor of both hematopoietic and endothelial lineage. Present in the yolk sac blood islands and in the floor of the dorsal aorta in early development. Presence in the adult marrow is controversial.

Hemangioblast (blast colony forming cell — BL-CFC) A long-theorized but only recently identified embryonic cell representing a common progenitor of both the endothelial and hematopoietic lineages. Can be readily detected in differentiating cultures of embryonic stem cells as the first mesodermal element committed to the hematopoietic lineage. Assayed by colony formation in methycellulose supplemented with vascular endothelial growth factor and stem cell factor (the BL-CFC).

Hematopoiesis The formation and development of blood cells.

Hematopoietic progenitor cell The progeny of a stem cell that has little or no self-renewal capacity but is committed to produce mature blood cell types of one or more lineages (e.g., granulocytes, macrophages, erythrocytes, megakaryocytes, lymphocytes, etc.). It is found in bone marrow and other tissues that have lost the potential for self-renewal but are still multipotent.

Hematopoietic stem cell (HSC) A type of cell found in bone marrow and other tissues that possesses the defining characteristics of self-renewal (produces more stem cells) and multipotency (produces all types of blood cells).

Heterokaryon The immediate result of fusing two different cells which yields a single cell with two or more identifiable intact nuclei, one from each parental cell.

Hierarchical clustering A method of data mining in which members of a data set are statistically grouped or clustered together according to a certain parameter (e.g., gene expression profile) in a way that maximizes intracluster similarity and minimizes intercluster similarity. In the clustering process, two clusters are merged only if the interconnectivity and closeness (proximity) between two clusters are high relative to the internal interconnectivity of the clusters and closeness of items within the clusters.

Holoclone (Gk *holos* whole + Gk *klôn* growing) Colony-forming cell with tremendous growth potential (up to 180 divisions), whose progeny forms almost exclusively growing colonies.

Homologous recombination The reciprocal exchange of sequences between two homologous DNA strands. It is widely used to introduce targeted and specific modifications in prokaryotic and eukaryotic genomes.

Hybrid cell The cell that arises when a heterokaryon resulting from cell fusion enters mitosis, with breakdown of the nuclear membranes of the individual nuclei and formation of a single nucleus in each daughter cell containing the genetic material from the original parental cells. Note, however, that there is often significant genome loss during this and subsequent divisions, so the hybrid cell may not contain all of the genetic material from each parental cell.

Hybridoma Typically used to describe a hybrid cell line produced by fusing an established myeloma cell line and primary lymphocyte.

Immunosurgery A method of removing the trophectoderm of a blastocyst using antibodies bound to the surface antigens of the trophectoderm and complement.

Implantation Process by which the mammalian blastocyst physically connects with the uterus of the mother. It involves the displacement of the uterine epithelium cells and extracellular matrix degradation by proteolytic enzymes secreted by the embryo. Implantation occurs only in mammalian development and is necessary to provide the growing embryo with sufficient nutrition (and protection).

Imprinting Differential expression of an allele depending upon maternal versus paternal inheritance.

Inner-cell mass (ICM) The cluster of pluripotent cells located at one pole of the inner surface of the blastocyst; these cells will form the body of the embryo after implantation.

Intermediate mesoderm A region of mesoderm extending anterior-posterior between the axial, or somitic, mesoderm and the lateral plate mesoderm, marked by expression of the *Pax2* and *Pax8* genes.

IRB, ACUC, DSMB Institutional Review Board, Animal Care and Use Committees, and Data Safety and Monitoring Boards, respectively. These are committees of lay and professional scientists, academicians, doctors, veterinarians, and statisticians that provide publicly accountable oversight and regulation to all human and animal experimentation in the United States.

Islet neogenesis Generation of new pancreatic islets from precursor cells.

Islet transplantation Transplantation of isolated pancreatic islets with the intent of curing diabetes.

Karyotype A standard arrangement of the chromosome complement. A normal human female karyotype would have each of the 22 pairs of autosomes (nonsex chromosomes arranged in numerical order together with the two X-chromosomes.

Kidney induction The activation of the epithelial specific program in the metanephric mesenchyme in response to signals emanating from the invading ureteric bud.

Knockout serum replacer Proprietary serum replacement that consists of amino acids, vitamins, transferrin or substitutes, insulin or insulin substitutes, trace elements, collagen precursors, and albumin preloaded with lipids.

Label retaining cells Cells that incorporate labeled nucleotides during the S phase and retain it preferentially over extended periods of time due to slow or infrequent divisions.

Liver stem cell A poorly characterized cell proposed as the precursor of oval cells, or to comprise the most primitive subset of oval cells; the term is also used to refer broadly to all liver progenitor cells.

Master switch gene A transcription factor that is sufficient to induce the transdifferentiation of one cell type to another; is associated with the loss of one set of differentiated properties (e.g., pancreatic) and the gain of another phenotype (e.g., hepatic). Master Switch Genes are normally involved in coordinating the differentiation of a particular cell fate through the activation of various downstream targets.

Matrigel Commercially available basement membrane extract.

Meiosis The process by which germ cells (i.e., those in ovaries or testes) divide to produce gametes. In meiosis I, homologous chromosomes exchange genetic material. In meiosis II, the two resulting diploid cells (i.e., which contain two sets of chromosomes) with their recombined chromosomes divide further to form tow haploid gametes (i.e., which contain only one set of chromosomes).

Meroclone (GK *mero* part + GK *klôn* growing) Colony-forming cell whose growth potential is lower than the growth potential of an holoclone but superior to 15 divisions; forms growing and aborted colonies.

Mesenchymal condensation Aggregation of undifferentiated mesenchymal cells prefiguring a future skeletal element.

Mesenchymal stem cells (MSC) Cells that have the capacity to form different mesenchymal tissues given exposure to different inductive agents.

Mesenchyme The middle layer of an embryo; the mesoderm and the tissues that form (bone, cartilage, muscle, fat, etc.).

Mesoangioblasts   Vessel-associated stem cells.

Mesoderm   The middle embryonic germ layer, between the ectoderm and endoderm, from which connective tissue, muscle, bone, cartilage, and blood vessels develop.

Metanephric mesenchyme   The region of posterior inter-mediate mesoderm adjacent to the nephric ducts that is able to generate epithelial cells upon induction.

Metaplasia   The conversion of one cell type to another; can include conversions between tissue-specific stem cells.

Monoclonal antibody   An antibody produced by a cloned hybridoma cell line; typically, it will be a single immunoglob-ulin species with a defined antigen specificity.

Moral realism   Theory of knowledge that affirms the possi-bility of at least partial correspondence between what is in reality and what is in the mind.

Moral status   A description of how we understand our rela-tionship toward the being in question, what we one the being, and what the being may rightfully expect of us in our society. Often, this is expressed in relative terms and is linked to our understanding that many living beings (frogs, mice, dogs) are sentient, but that their moral status relative to human primates allows us to do things to them (own them, for example) that we cannot do to a person of moral status equivalent to our own.

Morphogen   A substance produced primarily during embryogenesis that influences the pattern of cellular differen-tiation during development, but may also be elicited during the process of tissue turnover and repair.

Mouse embryo fibroblast   Diploid, heterogeneous adherent cells of limited lifespan derived from primary cultures of midgestation mouse fetuses (most often eviscerated and decapitated, then dissociated mechanically or enzymatically), grown in standard tissue culture medium supplemented with serum.

Multipotent adult progenitor cell (MAPC)   Cell cultured from postembryonic stage bone marrow, and possibly other tissues, with extensive self-renewal ability and ability to gen-erate most, if not all, cell types from the three germ layers of the embryo (i.e., mesoderm, endoderm, and ectoderm).

Muscle stem cells (MuSC)   Monomucleate cells that func-tion to regenerate damaged myofibers through a process of proliferation, followed by subsequent fusion with existing or nascent myofibers.

Myocardial repair   Regeneration of dead myocardium.

Myofiber   Large multinucleate cells that contain a contrac-tile apparatus, composed primarily of actin and myosin family proteins, to facilitate movement. Muscle groups are composed bundles of myofibers.

Negative selection   The application of growth conditions eliminating those cells expressing the selection marker.

Neural induction   Early events in the vertebrate embryo that lend to the formation of the neuroepithelium of neural tube, thus creating the progenitor cells that give rise to the neurons and glia comprising the central nervous system.

Neural patterning   Developmental processes in which neural precursors are endowed with a positional identity, thus enabling them to give rise to subtypes of neurons that comprise distinct regions of neural tissue along the anterior-posterior and doral-ventral axis of the nervous system.

Neural rosettes   Term used to describe the tube-like colum-nar epithelial structures appearing during early stages of neural differentiation from ES cells. Particularly prominent are these structures during neural differentiation of primate ES cells. Rosettes correspond developmentally to the neural plate and early neural tube stage.

Neural stem cells (NSCs)   The most primordial cells of the nervous system. They generate the array of specialized cells throughout the central nervous system (and probably the peripheral, autonomic, and enteric nervous systems as well). NSCs are operationally defined by their ability (1) to differ-entiate into cells of all neural lineages (i.e., neurons, ideally of multiple subtypes; oligodendroglia; and astroglia) in mul-tiple regional and developmental contexts; (2) to self-renew; and (3) to populate developing and/or degenerating CNS regions.

Neurogenesis   The events that occur when neural precursors leave the cell cycle and activate a program of terminal neu-ronal differentiation.

Neurogenin3   A transcription factor expressed in endocrine progenitor cells of the embryonic pancreas.

Neurospheres   Nonadherent clusters of neural progenitor cells grown *in vitro*; culture system for the propagation of neural stem cells.

Niche   Structural and biological entity where stem cells are located, proliferate, and differentiate.

Ontological   Having to do with the structure of a being.

Osteoblasts   Cell type of mesenchymal origin present in osteoblast, formed by a multistep differentiation sequence, and responsible for bone formation.

Osteoconductive   Scaffolds or materials that are permissive to bone formation but do not attract osteoprogenitors that ini-tiate bone formation.

Osteoinductive   Containing growth factors that signal attraction of osteoprogenitors that lead to osteoblastic differ-entiation and bone formation.

Otosphere   Spherical colonies generated by dissociation and culturing of inner ear sensory cells from the newborn rat organ of Corti.

Oval cell   A small cell with an oval-shaped nucleus and a high nuclear/cytoplasmic size ratio that emerges in the portal zone under conditions of chronic liver damage; can differentiate into the hepatocyte or bile duct epithelial lineages and thus is a bipotential liver progenitor.

Pancreatic liver stem cell   A cell located in the pancreas that can generate hepatocytes to support liver repair upon transplantation; may be a bipotential hepato-pancreatic precursor or an extrahepatic progenitor committed to hepatic differentiation.

Paraclone (Gk para *akin* + Gk *klôn* growing)   Colony-forming cell committed to a maximum of 15 divisions; forms only aborted colonies.

Parthenogenesis   The process of tricking a nonfertilized egg into duplication of DNA to produce an all-female-derived embryonic stem cell.

pdx1   Pancreatic and duodenal homeobox gene-1; a transcription factor essential for pancreas development. During embryogenesis, pdx1 is expressed in stem/progenitor cells, giving rise to all cell types in the adult pancreas. During postnatal life, its expression becomes restricted to beta cells.

Pluripotency   A differentiation ability to differentiate varieties of cells that belong to all three germ layers, at least one cell type for each. In contrast, totipotency is defined as an ability to generate a whole animal autonomously, whereas multipotency is defined as an ability to give multiple cell types that belong to particular germ layers, not all three. Unipotency means an ability to give a single differentiation cell type that shows the most limited ability of stem cells such as germ-line stem cells.

Pluripotent   Able to give rise to differentiated cells of all three germ layers. Cells of the inner-cell mass and ES cells are pluripotent.

Pluripotent stem cells   Cells not capable of growing into a whole organism but able to differentiate into any cell type in the body.

Polycomb group silencing   Chroatin-based gene silencing mechanism, originally identified for its role in *Hox* gene repression during developing.

Position effect variegation (PEV)   Mosaic silencing of a gene due to a chromosomal rearrangement.

Positive selective   The application of specific growth conditions causing selective elimination of cells that do not contain an exogenous genetic element (selection marker) conferring resistance.

Potency   The specific ability or capacity of a therapeutic product to elicit an expected result, as indicated by appropriate laboratory tests or by adequately controlled clinical data obtained through the administration of the product in the manner intended [21 CFR 600.3(s)].

Potentiality   The concept that states that if a being has the elements that would allow it to potentially develop into another, more complex thing (acorn to oak tree, embryo to baby), then it is morally equivalent to that thing.

Potential stem cell   Cell that retains the capacity to function fully as a stem cell if needed. Normally, these cells are displaced with time into the dividing transit population, but they retain the undifferentiated status of the ultimate stem cell until such time that they are displaced into the dividing transit populations.

Primary mouse embryonic fibroblasts (PMEF)   A mixed population of cells derived from dispersed mouse embryos and cultured for a limited number of passages under conditions favoring the growth of fibroblasts.

Primary texts of religion and religious law   Hebrew Scripture, the New Testament, and the Koran (Hadith) are classic texts that form the basis of, respectively, Jewish, Christian, and Muslim thought. For all three of these Abrahamic traditions, a monotheist and commanding God in history leaves a direct covenant that forms the basis for how duties are enacted. The Talmud and Responsa, canon law, and the Sharia are later codes of law in each tradition that explain how judges and scholars have interpreted these commands. Use of these texts, together with others in the Eastern traditions, form the basis for how many religions respond to the questions of when life begins, what the meaning of suffering is, and how to address suffering.

Primitive endoderm   An extraembryonic lineage that forms part of the yolk sac. The primitive endoderm does not contribute to endodermal tissues, such as the gut, liver, or pancreas. These are derived from definitive endoderm, a distinct lineage from primitive endoderm.

Primitive streak   The initial band of cells from which the embryo begins to develop; it is present at approximately 14 days after fertilization.

Primitive versus definitive hematopoiesis   The first wave of embryonic blood development. It occurs in the yolk sac and consists principally of nucleated erythrocytes that express embryonic globins. The primitive wave is believed to supply the needs of the embryo and to be later supplanted by a definitive class of hematopoietic stem cells that arise in the para-aortic region of the developing embryo proper. Definitive hematopoiesis generates mature myeloid and lymphoid lineages for the life of the animal.

Primordial germ cell   A cell type that colonizes the developing gonad and turns into egg or sperm.

Progenitor cells   In development, a parent cell that gives rise to a distinct cell lineage by a series of cell divisions.

Proliferation   Cell division without a change in cell properties.

Proneural proteins   A calls of basic–helix–turn–helix transcription factors that are known to play critical roles in promoting neurogenesis within progenitor cells.

Prospective isolation   A strategy to isolate defined cell types based on a set of predetermined cell-specific markers using fluorescence activated cell sorting (FACS) or other techniques that allow physical separation of cells.

Purity Relative freedom from extraneous matter in the finished product introduced during the manufacturing process, whether or not harmful to the recipient or deleterious to the product [21 CFR 600.3(r)].

The RAC The Recombinant DNA Advisory Committee provides an additional level of review for all research that involves the uses of synthetically created DNA.

Regeneration Capacity of certain adult vertebrates to repair damaged organs in a well-defined spatial and temporal plan that reconstitutes the original organ.

Remyelination The process of repair that is predicted to most likely lead to return of limited function without requiring new circuits to form. Oligodendrocytes are the glial cells that wrap (myelinate) neuronal axons. New data suggests that remyelination is one of the processes that can occur spontaneously after injury.

Renal stem cell A hypothetical metanephric mesenchymal cell that could potentially generate all the derivatives found in the nephron, including tubular epithelia, mesangia, stroma, and endothelia.

Renal vesicle The first polarized epithelial derivative of the induced metanephric mesenchyme that is found abutting the branching tips of the ureteric bud.

Roe v. Wade The U.S. Supreme Court decision in 1973 that allows termination of a pregnancy and hence the destruction of an embryo and fetus in the first 12 weeks of life at the request of the mother, without permission from the state. Hence, the embryo and fetus are not regarded as a person with full moral status.

Satellite cell An anatomical designation of a mononuclear cell that sits upon a myofiber juxtaposed to the plasma membrane of the myofiber, yet is ensheathed by the basal laminal membrane, which also surrounds the myofiber.

Scaffold A porous structure typically made out of degradable polymers within which cells are seeded. This framework is used to support cell growth and differentiation that may be natural or synthetic in nature and may also include morphogens and/or growth factors that influence the pattern of proliferation and differentitation (biomaterial).

Self-maintenance probability The probability that stem cells make other stem cells on division. It applies to populations of stem cells rather than individual cells. In steady state it is 0.5, but during situations where stem cell populations expand it can be between 0.5 and 1.0.

Self-renewal A style of cell division characteristic of stem cells that give at least one daughter cell with the same differentiation ability as the parental stem cells. Symmetric self-renewal generates two stem cells, whereas the asymmetric one gives one-stem cell and one differentiated progeny. On the molecular level, this event can be divided into two mechanisms for preventing differentiation and promoting cell division.

Self-renewing organ An organ that possess stem cells.

Skeletal stem cell A cell derived from adult bone marrow that has the ability to recapitulate the formation of a complete bone/marrow organ, also known as mesenchymal stem cells, or bone marrow stromal stem cells.

Somatic nuclear transfer A method that involves transferring a nucleus from a somatic (e.g., skin cell) to an enucleated fertilized egg. This produces an embryonic stem cell clone of the somatic cell DNA.

Spleen colony-forming assay The first biological assay used to detect hematopoietic stem cells. The assay detects macroscopic splenic nodules formed by proliferating stem and progenitor cells that lodge in the spleen of an irradiated mouse. The cells that initiate these nodules are termed colony-forming units-spleen (CFU-S) because it was initially not certain that each nodule represented the outgrowth of a single cell versus that of a cluster (unit) of cells.

Split/Passage Transfer or subculture of cells from one culture vessel to another, resulting in the expansion of the cell population.

Stability Ability of a product to retain its sterility, identity, purity, and potency over time when stored and shipped in designated packaging and under specified conditions.

Stable transfection Transfection in which foreign DNA is incorporated into the host genome, generating a long-term genetic modification.

Stem cell activation Process that stimulates stem cells to exit their dormant, quiescent, undifferentiated state, and begin to produce more committed cells to participate in tissue growth or repair.

Stem cell plasticity Apparent ability of a stem cell/progenitor cell fated to a given tissue to acquire a differentiated phenotype of a different tissue. In the gastrointestinal tract, evidence suggests that a population of stem cells in the adult bone marrow can engraft within the gastrointestinal tract and form both epithelial and mesenchymal lineages, with an increased propensity in damaged tissue. However, the mechanisms by which bone marrow cells form gastrointestinal lineages are unresolved.

Stem cell quiescence Mitotic quiescence or slow cycling of adult stem cells in comparison with the lineage-committed progenitor cells. It is generally associated with G_0 or prolonged G_1 phase in cell cycle.

Stem cells Self-renewing, undifferentiated cells that can give rise to multiple types of specialized cells.

Stem cell self-renewal The process by which a stem cell replicates itself. It is not a synonym of stem cell proliferation since the proliferation may be also accompanied by cell differentiation.

Subventricular zone Regions of the forebrain bordering the cerebral ventricles and identified as a neurogenic site.

Supervised analysis Statistical methods to identify genes that are differentially expressed between groups of samples, as well as to find genes that can be used to accurately predict the characteristics of groups.

Supravital fluorescent probe A fluorescent molecule that can enter a living cell and bind to specific subcellular components, such as DNA, RNA, or organelle membranes. These probes can be used in combination with labeled antibodies directed against cell surface antigens in order to identify stem cells.

Teleological A reason to perform an action based on the ends or outcome you desire. One such formulation is "the greatest good for the greatest number," if one assumes that great good for the majority is a good end.

Telomere A specialized nucleic acid structure found at the ends of linear eukaryotic chromosomes.

Teratoma Malignant tumor thought to originate from primordial germ cells or misplaced blastomeres that contains tissues derived from all three embryonic layers, such as bone, muscle, cartilage, nerve, tooth buds, and various glands.

Therapeutic cloning The use of somatic cell nuclear transfer (cloning) to generate stem cells that genetically match a patient.

Tissue engineering The interdisciplinary field that applies the principle of engineering and the life science toward the development of biological substitutes that restore, maintain, or improve tissue function.

Tongue proliferative unit A modified version of the epidermal proliferative unit identified in the filiform papillae on the dorsal surface of the tongue.

Totipotency Capability of a cell to develop all cell types of an organism.

Totipotent Able to give rise to all cell types, including cells of the trophectoderm lineage. In mammals only the fertilized egg and early cleavage stage blastomeres are totipotent.

Transdifferentiation The conversion of one differentiated cell type to another; a subset of metaplasia.

Transient transfection *Transfection* in which plasmid DNA is not physically incorporated into the genome and is retained as an episome, generating short-term gene expression of the foreign DNA.

Trithorax group activation Chromatin-based mechanism required to counteract Polycomb group silencing and originally identified for its role in *Hox* gene expression during development.

Trophectoderm The outer layer of cells of the blastocyst. Trophectoderm cells are the sole precursors to the trophoblast lineage of the placenta, and they do not contribute cells to the embryo proper.

Trophic factor Growth substance produced by target cells, which promotes the proliferation of progenitor cells and/or the survival of their neuronal progeny by preventing programmed cell death (i.e., apoptosis).

Trophoblast A general term used to describe all the cell types of the developing and mature placenta derived from the trophectoderm. In the mouse, the trophoblast lineage would include extraembryonic ectoderm, ectoplacental cone, trophoblast giant cells, chorionic ectoderm, labyrinthine trophoblast, and spongiotrophoblast.

Vascular endothelial growth factor (VEGF) An essential molecule for proliferation and differentiation of endothelial cells. VEGF binds to both VEGF-receptor (VEGFR)1 and VEGFR2. For inducing endothelial cells in ES cell differentiation cultures, VEGFs have to be added when VEGFR2$^+$ cells appear, which is 3.5 to 4 days after induction of ES cell differentiation.

Vasculogenesis Generation of new blood vessels in which endothelial cell precursors undergo differentiation, expansion, and coalescence to form a network of primitive tubules.

Wnt A secreted growth factor that binds to members of the Frizzled (Fz) family of cell surface receptors.

X-Chromosome inactivation Mammalian dosage compensation mechanism involving inactivation of one X-chromosome in females.

Yolk sac blood islands First site of both primitive and definitive hematopoietic development in mammals.

Index